50 Years of Integer Programming 1958–2008

Michael Jünger · Thomas Liebling ·
Denis Naddef · George Nemhauser ·
William Pulleyblank · Gerhard Reinelt ·
Giovanni Rinaldi · Laurence Wolsey
Editors

50 Years of Integer Programming 1958–2008

From the Early Years to the State-of-the-Art

Springer

Editors

Michael Jünger
Universität zu Köln
Institut für Informatik
Pohligstraße 1
50969 Köln
Germany
mjuenger@informatik.uni-koeln.de

Denis Naddef
Grenoble Institute of Technology - Ensimag
Laboratoire G-SCOP
46 avenue Félix Viallet
38031 Grenoble Cedex 1
France
denis.naddef@grenoble-inp.fr

William R. Pulleyblank
IBM Corporation
294 Route 100
Somers NY 10589
USA
wp@us.ibm.com

Giovanni Rinaldi
CNR - Istituto di Analisi dei Sistemi
ed Informatica "Antonio Ruberti"
Viale Manzoni, 30
00185 Roma
Italy
rinaldi@iasi.cnr.it

Thomas M. Liebling
Ecole Polytechnique Fédérale de Lausanne
Faculté des Sciences de Base
Institut de Mathématiques
Station 8
1015 Lausanne
Switzerland
thomas.liebling@epfl.ch

George L. Nemhauser
Industrial and Systems Engineering
Georgia Institute of Technology
Atlanta, GA 30332-0205
USA
george.nemhauser@isye.gatech.edu

Gerhard Reinelt
Universität Heidelberg
Institut für Informatik
Im Neuenheimer Feld 368
69120 Heidelberg
Germany
gerhard.reinelt@informatik.uni-heidelberg.de

Laurence A. Wolsey
Université Catholique de Louvain
Center for Operations Research and
Econometrics (CORE)
voie du Roman Pays 34
1348 Louvain-la-Neuve
Belgium
laurence.wolsey@uclouvain.be

Additional material to this book can be downloaded from http://extras.springer.com.

ISBN 978-3-662-50181-8 ISBN 978-3-540-68279-0 (eBook)
DOI 10.1007/978-3-540-68279-0
Springer Heidelberg Dordrecht London New York

Mathematics Subject Classification (2000): 01-02, 01-06, 01-08, 65K05, 65K10, 90-01, 90-02, 90-03, 90-06, 90-08, 90C05, 90C06, 90C08, 90C09, 90C10, 90C11, 90C20, 90C22, 90C27, 90C30, 90C35, 90C46, 90C47, 90C57, 90C59, 90C60, 90C90

Cover design: WMXDesign, Heidelberg

Printed on acid-free paper

Springer is part of Springer Science+Business Media (www.springer.com)

We dedicate this book to the pioneers of Integer Programming.

Preface

The name integer programming refers to the class of constrained optimization problems in which some or all of the variables are required to be integers. In the most widely studied and used integer programs, the objective function is linear and the constraints are linear inequalities. The field of integer programming has achieved great success in the academic and business worlds. Hundreds of papers are published every year in a variety of journals, several international conferences are held annually and software for solving integer programs, both commercial and open source, is widely available and used by thousands of organizations. The application areas include logistics and supply chains, telecommunications, finance, manufacturing and many others.

This book is dedicated to the theoretical, algorithmic and computational aspects of integer programming. While it is not a textbook, it can be read as an introduction to the field and provides a historical perspective. Graduate students, academics and practitioners, even those who have spent most of their careers in discrete optimization, will all find something useful to learn from the material in this book. Given the amount that has been accomplished, it is remarkable that the field of integer programming began only fifty years ago.

The 12th Combinatorial Optimization Workshop AUSSOIS 2008 took place in Aussois, France, 7–11 January 2008. The workshop, entitled *Fifty Years of Integer Programming*, and this book, which resulted from the workshop, were created to celebrate the 50th anniversary of integer programming. The workshop had a total of 136 participants from 14 countries ranging in experience from pioneers who founded the field to current graduate students. In addition to the formal program, the workshop provided many opportunities for informal discussions among participants as well as a chance to enjoy the spectacular Alpine setting provided by Aussois.

The book is organized into four parts. The first day of the workshop honored some of the pioneers of the field. Ralph Gomory's path-breaking paper, showing how the simplex algorithm could be generalized to provide a finite algorithm for integer programming and published in 1958, provided the justification of the aniversary celebration. The activities of the first day, led by George Nemhauser and Bill Pulleyblank, included a panel discussion with the pioneers who attended the workshop (Egon Balas, Michel Balinski, Jack Edmonds, Arthur Geoffrion, Ralph Gomory and Richard Karp) as well as three invited talks by Bill Cook, Gérard Cornuéjols and Laurence Wolsey on integer programming and combinatorial optimization from the beginnings to the state-of-the-art. The whole day is captured in two Video DVDs which come with the book (Part IV). Parts I, II, and III contain 20 papers of historical and current interest.

Part I of the book, entitled *The Early Years*, presents, in order of publication date, reprints of eleven fundamental papers published between 1954 and 1979. Ten of these papers were selected by one or more of the authors of the paper, who also wrote new introductions to the papers that explain their motivations for working on the problems addressed and their reason for selecting the paper for inclusion in this volume. The authors are Egon Balas, Michel Balinski, Alison Doig, Jack Edmonds, Arthur Geoffrion, Ralph Gomory, Alan Hoffman, Richard Karp, Joseph Kruskal, Harold Kuhn, and Ailsa Land. Each of these heavily cited papers has had a major influence on the development of the field and lasting value. The eleventh selection, which starts this section, is a groundbreaking paper by George Dantzig, Ray Fulkerson, and Selmer Johnson, with an introduction by Vašek Chvátal and William Cook. The introduction to Part I closes with a list, in chronological order, of our selection of some of the most influential papers appearing between 1954 and 1973 pertaining to the many facets of integer programming.

Part II contains papers based on the talks given by Cornuéjols, Cook, and Wolsey. The paper *Polyhedral Approaches to Mixed Integer Programming* by Michele Conforti, Gérard Cornuéjols, and Giacomo Zambelli presents tools from polyhedral theory that are used in integer programming. It applies them to the study of valid inequalities for mixed integer linear sets, such as Gomory's mixed integer cuts. The study of combinatorial optimization problems such as the traveling salesman problem has had a significant influence on integer programming. *Fifty-plus Years of Combinatorial Integer Programming* by Bill Cook discusses these connections. In solving integer programming problems by branch-and-bound methods, it is important to use relaxations that provide tight bounds. In the third paper entitled *Reformulation and Decomposition of Integer Programs*, François Vanderbeck and Laurence Wolsey survey ways to reformulate integer and mixed integer programs to obtain stronger linear programming relaxations. Together, these three papers give a remarkably broad and comprehensive survey of developments in the last fifty-plus years and their impacts on state-of-the-art theory and methodology.

Six survey talks on current hot topics in integer programming were given at the workshop by Fritz Eisenbrand, Andrea Lodi, François Margot, Franz Rendl, Jean-Philippe P. Richard, and Robert Weismantel. These talks covered topics that are actively being researched now and likely to have substantial influence in the coming decade and beyond.

Part III contains the six papers that are based on these talks. *Integer Programming and Algorithmic Geometry of Numbers* by Fritz Eisenbrand surveys some of the most important results from the interplay of integer programming and the geometry of numbers. *Nonlinear Integer Programming* by Raymond Hemmecke, Matthias Köppe, Jon Lee, and Robert Weismantel generalizes the usual integer programming model by studying integer programs with nonlinear objective functions. *Mixed Integer Programming Computation* by Andrea Lodi discusses the important ingredients involved in building a successful mixed integer solver as well as the problems that need to be solved in building the next generation of faster and more stable solvers. Symmetry is a huge obstacle encountered in solving mixed integer programs efficiently. In *Symmetry in Integer Programming*, François Margot presents several techniques that have been used successfully to overcome this difficulty. Semidefinite programming is a generalization of linear programming that provides a tighter relaxation to integer programs than linear programs. In *Semidefinite Relaxations for Integer Programming*, Franz Rendl surveys how semidefinite models and algorithms can be used effectively in solving certain combinatorial optimization problems. In the 1960s Ralph Gomory created a new tight relaxation for integer programs based on group theory. Recently the group theoretic model has been revived in the study of two-row integer programs. In *The Group-Theoretic Approach in Mixed Integer Programming*, Jean-Philippe P. Richard and Santanu S. Dey provide an overview of the mathematical foundations and recent theoretical and computational advances in the study of the group-theoretic approach.

We close with the hope that the next fifty years will be as rich as the last fifty have been in theoretical and practical accomplishments in integer programming.

November 2009

Cologne, Germany — *Michael Jünger*
Lausanne, Switzerland — *Thomas Liebling*
Grenoble, France — *Denis Naddef*
Atlanta, USA — *George Nemhauser*
New York, USA — *William Pulleyblank*
Heidelberg, Germany — *Gerhard Reinelt*
Rome, Italy — *Giovanni Rinaldi*
Louvain-la-Neuve, Belgium — *Laurence Wolsey*

About the Cover Illustration

The four figures on the cover illustrate adding Gomory mixed integer cuts to a polyhedron of dimension 3. The x-axis is horizontal, the y-axis is vertical and the z-axis is orthogonal to the cover. The starting polyhedron P shown in Fig. 1(a) is a cone with a square base and a peak having $y = 4.25$. P contains twelve integer lattice points. Suppose we solve the linear program: maximize y, subject to $y \in P$. The unique optimum will have $y = 4.25$. However, if we add the constraint that y be integral, then there are four optima, the lattice points illustrated on the edges of P having $y = 2$.

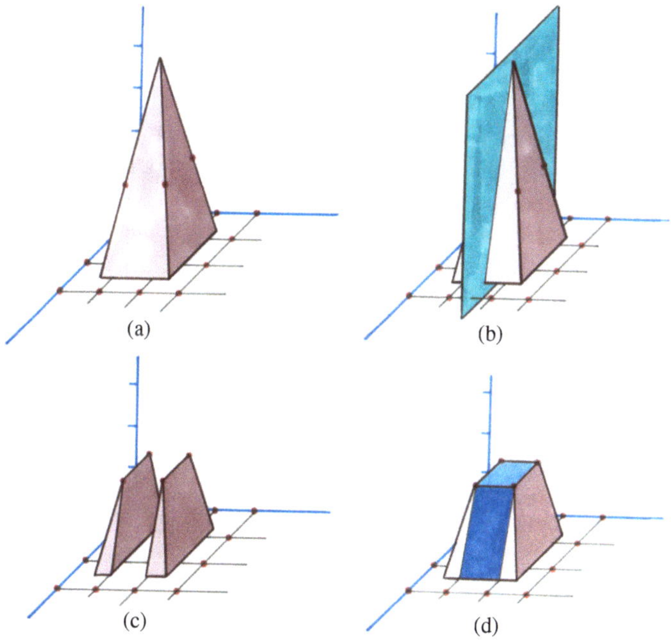

Fig. 1 The Cover Illustration.

This example is a 3-D version of a 2-D example, first shown to us by Vašek Chvátal, which Bill Cook told us that Vašek attributes to Adrian Bondy. A "standard" Chvátal-Gomory cut (CG cut) is obtained by taking a hyperplane that supports a polyhedron and which contains no lattice points in space, then moving in a direction orthogonal to the hyperplane into the polyhedron until it hits a lattice point somewhere in space (not necessarily in the polyhedron). This gives a new valid inequality for all lattice points in the polyhedron, and which cuts off part of the original polyhedron. Gomory's fundamental result described a finite algorithm that, given any integer program, would automatically generate a finite sequence of CG cuts such that when they were added, the resulting linear program would have an integer optimum.

What cuts must be added to P to remove all points having $y > 2$? How do we generate the inequality $y \leq 2$ which must be added if the resulting linear program is going to have an integral optimum? The Bondy-Chvátal example showed that, even for dimension 2, the number of CG cuts that would have to added was unbounded, depending only on the height of the peak of the pyramid (provided that we adjust the base so that the lattice points in P having $y = 2$ continue to lie on the edges). In particular, the number of CG cuts that need to be added to solve an integer program is independent of the dimension of the polyhedron, and is not polynomial in the size of a linear system necessary to define the original polyhedron.

In 1960, Gomory described a method to generate so-called mixed integer cuts. These cuts have turned out to be very powerful in practice, both for integer and mixed integer programs. They work as follows: Take a hyperplane that intersects the polyhedron and passes through no lattice points in space. In Fig. 1(b), we chose the hyperplane $x = 1.5$. Note that it passes right through P. Consider the inequalities $x \leq 1$ and $x \geq 2$ which are obtained by shifting the hyperplane left and right respectively, until it hits a lattice point in space. We construct two new polyhedra P_1 and P_2 from P, one by adding the inequality $x \leq 1$ and one by adding $x \geq 2$. Then every lattice point in P will belong to one of P_1 and P_2.

These two polyhedra are the two wedges shown in Fig. 1(c). Note that every lattice point contained in P is in one of the two wedges.

The final step is to take the convex hull of the union of P_1 and P_2. This is the polyhedron shown in Fig. 1(d). Note that one hyperplane was used to create two subproblems. Then by maximizing y over these two subproblems, we get the solution we are seeking. Balas, Ceria and Cornuéjols describe a method called *lift-and-project* for generating a cut after a polyhedron has been split into two subpolyhedra. This is discussed in Balas' introduction to Chapter 10.

Also, everything we have done remains valid if x and z are allowed to be continuous variables and only y is required to be integral. For this reason, these types of cuts are usually called "mixed integer cuts".

Acknowledgements

We gratefully acknowledge the sponsors of the 12th Combinatorial Optimization Workshop, Aussois, France, 7–11 January 2008, on which this book is based:

In addition, we would like to express our gratitude to

- Martin Peters of Springer Verlag who enthusiastically supported this book project from the first phone call in which we roughly sketched the idea,
- Ruth Allewelt of Springer Verlag who accompanied us all the way from the early stages to the final book,
- Marc Egger of the University of Cologne for his technical support during the organization of the 12th Combinatorial Optimization Workshop AUSSOIS 2008 on which this book is based,
- Michael Belling of the University of Cologne for tracking down printed originals of the pioneering articles reprinted in Part I at various libraries and scanning them at high resolution,
- Mark Sprenger of the University of Cologne who spent many hours adjusting angles and removing specks and handwritten remarks and corrections in the scans of the pioneering articles,
- Thomas Lange of the University of Cologne for technical advice and various "quick hacks" whenever needed,
- Mauro Pioli and his team at PGM Video in Turin for the care they put in producing and editing the material on the two DVDs,
- Manfred Bender of WMXDesign in Heidelberg for turning our rough sketches of "artwork" into a beautiful cover design,
- the authors of Part II and Part III, who patiently dealt with our many requests and strict deadlines, and
- the pioneers, without whom …

Contents

Part II From the Beginnings to the State-of-the-Art

Part III Current Topics

Part I
The Early Years

In 1947, George Dantzig created the simplex algorithm, which was subsequently published in 1951. This landmark paper described a finite method for optimizing a linear objective function subject to a finite set of linear constraints. It was already recognized that this type of problem, called a linear programming problem, occurred in a great many situations. Moreover, Dantzig's simplex method was proving to be very effective in practice.

It was recognized that adding integrality constraints on some or all of the variables significantly increased the applicability of these models. A great many problems, including combinatorial optimization problems, could be modeled using linear functions and integer variables, but no method was known for modeling these problems using linear functions and continuous variables. (It is noteworthy that now, more than fifty years later, it is still not known whether integer programming is more powerful than linear programming.) Moreover, no general method was known for solving this type of problem, called mixed-integer linear programming problem.

In 1958, Ralph Gomory published a short paper which described how, with relatively straightforward modifications, Dantzig's simplex algorithm could be adapted to provide a finite algorithm for finding an optimal integral solution to a linear program. He showed how the simplex tableau could be used to generate new inequalities which were valid for all solutions satisfying the integrality constraints, but which were violated by the current linear program's optimum solution. The study of these inequalities, called cuts, quickly became a major area of activity both for theoretical reasons and because of the promise they showed as a computational tool. Recall that at the end of the decade of the 50s, digital computers were emerging as a force in the way that business was conducted with the potential to actually optimize business processes.

The year 2008 marked the 50th anniversary of the appearance of Gomory's groundbreaking paper. There is an annual meeting on combinatorial optimization and integer programming held each year at the French ski resort of Aussois. The editors of this volume proposed dedicating the January 2008 meeting to a celebration of the development of the field of integer programming together with an overview of the field today, including state-of-the-art surveys and recent results on selected hot topics. Our plan was to invite a number of pioneers of the field of integer programming who had been active in the fifties and sixties to provide a historical perspective and to participate in the scientific agenda. The first person we contacted was, of course, Ralph Gomory who enthusiastically accepted. (This may have been influenced by the fact that Ralph is an avid skier.) Each of these pioneers agreed to select one of their papers for inclusion in this volume, and to write a new introduction for it that would provide a historical and mathematical perspective.

We include two of Gomory's foundational papers on the cutting plane method for integer programming. The second dealt with the mixed integer problem and introduced a method of cut generation that has proved to be very effective in practice. Previously, this paper was only available as a Rand report.

The earliest paper we reprint here contains the solution to a 49 city traveling salesman problem using linear programming and cuts by George Dantzig, Ray Fulkerson, and Selmer Johnson. In addition to showing how a small set of cuts could be sufficient to prove optimality of a solution to an integer programming problem, this paper laid a foundation for much of the subsequent work on computational polyhedral combinatorics. Because the authors are deceased, Vašek Chvátal and William Cook, two of the coauthors of a recent book on the traveling salesman problem, volunteered to write the introduction.

The 1955 paper by Harold Kuhn describes a combinatorial algorithm for a specially structured integer program, the assignment problem. This work provided an early example of a specialized method for solving a structured problem and was one of the first uses of a primal-dual linear programming algorithm.

Alan Hoffman and Joseph Kruskal's 1956 paper showed the importance of the notion of total unimodularity to finding integer solutions to linear programs. They showed that this property characterized when this would happen automatically for all linear objective functions and choices of integral right-hand sides.

The 1960 paper by Ailsa Land and Alison Doig introduced the other method that has been so important in obtaining solutions to integer programming problems, branch-and-bound. In fact, most successful modern computer codes integrate cuts with branch-and-bound.

Michel Balinski's 1965 paper described the power of integer programming models to a range of real world problems. It provided the first comprehensive survey and introduced integer programming to a much broader audience.

Jack Edmonds' 1968 paper on matroid partition is one of a remarkable series of papers that he wrote showing a number of cases for which a combinatorially described set of cuts added to a linear program would yield the integer hull and would provide the basis for a polynomial run-time algorithm to solve the integer problem.

The importance of polynomial algorithms for combinatorial algorithms reached a broader audience in the early 1970s with the introduction of the classes P (polynomial) and NP (nondeterministic polynomial) in the theoretical computer science community. Steven Cook's fundamental result showed that there was a set of so-called NP-complete problems with the property that if any were solvable in polynomial time, then so too were all problems in the class NP. Richard Karp's 1972 paper highlighted the importance of these results to the mathematical programming community and showed that a long list of specially structured integer programs, for which no polynomially bounded algorithm was known, belonged to the class of NP-complete programs.

Art Geoffrion's 1974 paper showed how Lagrangean methods provided an alternative method for solving integer programming problems by incorporating certain constraints into the objective function and then alternating between solving primal and dual problems. He also established connections between the Lagrangean approach and Dantzig-Wolfe decomposition.

Egon Balas' 1979 paper showed that the class of integer programming problems could be extended to a much broader class defined by considering disjunctions of polyhedra, and that methods for this broader framework had specializations to integer programming that have turned out to have computational as well as theoretical importance.

We conclude with a list, in chronological order, of our selections of some of the most influential papers pertaining to the many facets of integer programming appearing between 1954 and 1973.

20 YEARS OF MIXED-INTEGER PROGRAMMING: MILESTONES (1954–1973)

G.B. Dantzig, D.R. Fulkerson, and S.M. Johnson, *Solution of a large scale traveling salesman problem*, Operations Research 2 (1954) 393–410.

H.W. Kuhn, *The Hungarian method for the assignment problem*, Naval Research Logistics Quarterly 2 (1955) 83–97.

A.J. Hoffman and J.B. Kruskal, *Integral boundary points of convex polyhedra*, Linear Inequalities and Related Systems (H.W. Kuhn and A.J. Tucker eds.), Princeton University Press, 1956, pp. 223–246.

G.B. Dantzig, *Discrete variable extremum problems*, Operations Research 5 (1957) 266–277.

R.E. Gomory, *Outline of an algorithm for integer solutions to linear programs*, Bulletin of the American Mathematical Society 64 (1958) 275–278.

G.B. Dantzig, *On the significance of solving linear programs with some integer variables*, Econometrica 28 (1960) 30–34.

A.H. Land and A.G. Doig, *An automatic method for solving discrete programming problems*, Econometrica 28 (1960) 497–520.

R.E. Gomory, *An algorithm for the mixed integer problem*, Research Memorandum RM-2597, The Rand Corporation, 1960.

J.F. Benders, *Partitioning procedures for solving mixed variables programming problems*, Numerische Mathematik 4 (1962) 238–252.

H. Everett III, *Generalized Lagrange multiplier method for solving problems of optimal allocation of resources*, Operations Research 11 (1963) 399–417.

R.E. Gomory, *An algorithm for integer solutions to linear programs*, Recent Advances in Mathematical Programming (R.L. Graves and P. Wolfe, eds.), McGraw-Hill, 1963, pp. 269–302.

J.D.C. Little, K.G. Murty, D.W. Sweeney, and C. Karel, *An algorithm for the traveling salesman problem*, Operations Research 11 (1963) 972–989.

M. Balinski and R. Quandt, *On an integer program for a delivery problem*, Operations Research 12 (1964) 300–304.

E. Balas, *An additive algorithm for solving linear programs with zero-one variables*, Operations Research 13 (1965) 517–546.

M. Balinski, *Integer programming: methods, uses, computation*, Management Science 12 (1965) 253–313.

R.J. Dakin, *A tree-search algorithm for mixed integer programming problems*, The Computer Journal 8 (1965) 250–254.

J. Edmonds Paths, trees, and flowers, Canadian journal of Mathematics 17 (1965) 449–467.

J. Edmonds, *Maximum matching and a polyhedron with 0,1- vertices*, Journal of Research of the National Bureau of Standards, Section B 69 (1965) 125–130.

R.E. Gomory, *On the relation between integer and noninteger solutions to linear programs*, Proceedings of the National Academy of Sciences of the United States of America 53 (1965) 260–263.

R.D. Young, *A primal (all integer) integer programming algorithm*, Journal of Research of the National Bureau of Standards, Section B 69 (1965) 213–250.

R. Brooks and A.M. Geoffrion, *Finding Everett's Lagrange multipliers by linear programming*, Operations Research 14 (1966) 1149–1153.

P.C. Gilmore and R.E. Gomory, *The theory and computation of knapsack functions*, Operations Research 14 (1966) 1045–1074.

J. Edmonds, *Optimum branchings*, Journal of Research of the National Bureau of Standards, Section B 71 (1967) 233–240.

J. Edmonds, *Matroid partition*, Mathematics of the Decision Sciences: Part 1 (G.B. Dantzig and A.F. Veinott, eds.), American Mathematical Society, 1968, pp. 335–345.

A.M. Geoffrion, *An improved implicit enumeration approach for integer programming*, Operations Research 17 (1969) 437–454.

R.E. Gomory, *Some polyhedra related to combinatorial problems*, Linear Algebra and its Applications 2 (1969) 451–558.

E.M.L. Beale and J. Tomlin, *Special facilities for nonconvex problems using ordered sets of variables*, Proceedings of the 5th International Conference on Operational Research (J. Lawrence, ed.), Tavistock Publications, 1970, pp. 447–454.

D.R. Fulkerson, *The perfect graph conjecture and pluperfect graph theorem*, Proceedings of the Second Chapel Hill Conference on Combinatorial Mathematics and its Applications (R.C. Bose, ed.), University of North Carolina Press, 1970, pp. 171–175.

M. Held and R.M. Karp, *The traveling salesman problem and minimum spanning trees*, Operations Research 18 (1970) 1138–1162.

E. Balas, *Intersection cuts – A new type of cutting plane for integer programming*, Operations Research 19 (1971) 19–39.

J. Edmonds, *Matroids and the greedy algorithm*, Mathematical Programming 1 (1971) 125–136.

D.R. Fulkerson, *Blocking and antiblocking pairs of polyhedra*, Mathematical Programming 1 (1971) 168–194.

M. Held and R.M. Karp, *The traveling salesman problem and minimum spanning trees: Part II*, Mathematical Programming 1 (1971) 6–25.

R.S. Garfinkel and G.L. Nemhauser, *Integer Programming*, Wiley, 1972.

A.M. Geoffrion, *Generalized Benders decomposition*, Journal of Optimization Theory and Applications 10 (1972) 237–260.

R.M. Karp, *Reducibility among combinatorial problems*, Complexity of Computer Computations (R.E. Miller and J.W. Thatcher, eds.), Plenum Press, 1972, pp. 85–103.

L. Lovász, *Normal hypergraphs and the perfect graph conjecture*, Discrete Mathematics 2 (1972) 253–267.

V. Chvátal, *Edmonds polytopes and a hierarchy of combinatorial problems*, Discrete Mathematics 4 (1973) 305–337.

J. Edmonds and E.L. Johnson, *Matching, Euler tours and the Chinese postman*, Mathematical Programming 5 (1973) 88–124.

S. Lin and B.W. Kernighan, *An effective heuristic algorithm for the traveling salesman problem*, Operations Research 21 (1973) 498–516.

M.W. Padberg, *On the facial structure of set packing polyhedra*, Mathematical Programming 5 (1973) 199–215.

References to the history of integer programming

J.K. Lenstra, A.H.G. Rinnooy Kan, and A. Schrijver, eds., *History of Mathematical Programming: A Collection of Personal Reminiscences*, North-Holland, 1991.

K. Spielberg and M. Guignard-Spielberg, eds., *History of Integer Programming: Distinguished Personal Notes and Reminiscences*, Annals of Operations Research 149, 2007.

Chapter 1
Solution of a Large-Scale Traveling-Salesman Problem

George B. Dantzig, Delbert R. Fulkerson, and Selmer M. Johnson

Introduction by *Vašek Chvátal* and *William Cook*

The birth of the cutting-plane method

The RAND Corporation in the early 1950s contained "what may have been the most remarkable group of mathematicians working on optimization ever assembled" [6]: Arrow, Bellman, Dantzig, Flood, Ford, Fulkerson, Gale, Johnson, Nash, Orchard-Hays, Robinson, Shapley, Simon, Wagner, and other household names. Groups like this need their challenges. One of them appears to have been the traveling salesman problem (TSP) and particularly its instance of finding a shortest route through Washington, DC, and the 48 states [4, 7].

Dantzig's work on the assignment problem [1] revealed a paradigm for minimizing a linear function $f : \mathbb{R}^n \to \mathbb{R}$ over a finite subset $\mathscr{S}$ of $\mathbb{R}^n$: first describe the convex hull of $\mathscr{S}$ by a system $Ax \leq b$ of linear constraints and then solve the linear programming problem

$$\text{minimize } f(x) \text{ subject to } Ax \leq b$$

by the simplex method. Attempts by Heller and by Kuhn to apply this paradigm to the TSP indicated that sets of linear constraints describing the convex hull of all tours are far too large to be handled directly. Undeterred, Dantzig, Fulkerson, and Johnson bashed on. The preliminary version of their paper [2] includes a discussion of the convex hull of all tours, nowadays called "the TSP polytope". The version submitted for publication four months later (and eventually published and

Vašek Chvátal
Canada Research Chair in Combinatorial Optimization
Department of Computer Science and Software Engineering, Concordia University, Canada
e-mail: chvatal@cse.concordia.ca

William Cook
School of Industrial and Systems Engineering, Georgia Institute of Technology, Atlanta, USA
e-mail: bico@isye.gatech.edu

M. Jünger et al. (eds.), *50 Years of Integer Programming 1958-2008*,
DOI 10.1007/978-3-540-68279-0_1, © Springer-Verlag Berlin Heidelberg 2010

reproduced here) breaks free of the dogma: without letting the TSP polytope obscure their exposition, the authors just go ahead and solve the 49-city instance. (Regarding this change, Fulkerson writes in a September 2, 1954, letter to *Operations Research* editor George Shortly "In an effort to keep the version submitted for publication elementary, we avoid going into these matters in any detail.")

This case study ushered in the *cutting-plane method*. To solve a problem

$$\text{minimize } f(x) \text{ subject to } x \in \mathscr{S} \tag{1.1}$$

where $f : \mathbb{R}^n \to \mathbb{R}$ is a linear function and $\mathscr{S}$ is a finite subset of $\mathbb{R}^n$, choose a system $Ax \le b$ of linear inequalities satisfied by all points of $\mathscr{S}$ and use the simplex method to find an optimal solution x^* of the linear programming problem

$$\text{minimize } f(x) \text{ subject to } Ax \le b, \tag{1.2}$$

called the *linear programming relaxation* of (1.1). If x^* belongs to $\mathscr{S}$, then it is an optimal solution of (1.1); else there are linear inequalities satisfied by all points of $\mathscr{S}$ and violated by x^*, called *cutting planes*. Find one or more such inequalities, add them to $Ax \le b$, and iterate. (The method actually used by Dantzig, Fulkerson, and Johnson—described also in [2, 3]—is a slight variation on this theme: rather than introducing cutting planes only when an optimal solution x^* of (1.2) lies outside $\mathscr{S}$, they introduce them after each simplex pivot leading to a basic feasible solution x^* of (1.2) that lies outside $\mathscr{S}$.)

The role played by the convex hull of $\mathscr{S}$ in this new paradigm is only implicit: we have to be able to find a cutting plane whenever one exists, which is the case if and only if x^* lies outside the convex hull of $\mathscr{S}$. In particular, the number of linear constraints in a description of the convex hull of $\mathscr{S}$ is irrelevant here. Another important difference between the two paradigms is that the cutting-plane method is an engineering rather than mathematical method: unlike the simplex method, it carries no guarantee that the sequence of its iterations will terminate. (But then again, a guarantee of termination after finitely many iterations is a far cry from a guarantee of termination before the end of our solar system.) Our three authors write "... what we shall do is outline a way of approaching the problem that sometimes, at least, enables one to find an optimal path and prove it so."

Until 1954, no one had an inkling of a way to solve large instances of the TSP. The lament about the number of tours through n cities being too large to allow their listing one by one marked the vanguard of scientific progress on this front. Then Dantzig, Fulkerson, and Johnson let the light in and inaugurated a new era. All successful TSP solvers echo their breakthrough. This was the Big Bang.

This Big Bang reverberates far beyond the narrow confines of the TSP. It provides a tempting template for coping with any NP-complete problem of minimizing a linear function over a finite set $\mathscr{S}$. For each problem of this kind, the challenge lies in finding cutting planes quickly. In the special case of integer linear programming, where $\mathscr{S}$ consists all integer solutions of a prescribed set of linear constraints, this challenge was met with remarkable elegance (and termination after finitely many iterations guaranteed) by Gomory in a series of papers beginning with [5].

Great new ideas may transform the discipline they came from so profoundly that they become hard to discern against the changed background. When terms such as "defense mechanism" and "libido" are in the common vocabulary, it is easy to forget that they came from Sigmund Freud. The cutting-plane method of George Dantzig, Ray Fulkerson, and Selmer Johnson had the same kind of impact on the discipline of mathematical programming.

References

1. G.B. Dantzig, *Application of the simplex method to a transportation problem*, Activity Analysis of Production and Allocation (T.C. Koopmans, ed.), Cowles Commission Monograph No. 13. John Wiley & Sons, Inc., New York, N. Y.; Chapman & Hall, Ltd., London, 1951, pp. 359–373.
2. G.B. Dantzig, D.R. Fulkerson, and S.M. Johnson, *Solution of a large scale traveling salesman problem*, Technical Report P-510, RAND Corporation, Santa Monica, California, USA, 1954.
3. G.B. Dantzig, D.R. Fulkerson, and S.M. Johnson, *On a linear-programming, combinatorial approach to the traveling-salesman problem*, Operations Research 7 (1959) 58–66.
4. M.M. Flood, *Merrill Flood (with Albert Tucker)*, Interview of Merrill Flood in San Francisco on 14 May 1984, The Princeton Mathematics Community in the 1930s, Transcript Number 11 (PMC11), Princeton University, 1984.
5. R.E. Gomory, *Outline of an algorithm for integer solutions to linear programs*, Bulletin of the American Mathematical Society 64 (1958) 275–278.
6. M. Grötschel and G.L. Nemhauser, *George Dantzig's contributions to integer programming*, Discrete Optimization 5 (2008) 168–173.
7. J. Robinson, *On the Hamiltonian game (a traveling salesman problem)*, RAND Research Memorandum RM-303, RAND Corporation, Santa Monica, California, USA, 1949.

The following article originally appeared as:

G.B. Dantzig, D.R. Fulkerson, and S.M. Johnson, *Solution of a Large-Scale Traveling-Salesman Problem*, Operations Research 2 (1954) 393–410.

SOLUTION OF A LARGE-SCALE TRAVELING-SALESMAN PROBLEM*

G. DANTZIG, R. FULKERSON, AND S. JOHNSON
The Rand Corporation, Santa Monica, California

(Received August 9, 1954)

It is shown that a certain tour of 49 cities, one in each of the 48 states and Washington, D. C., has the shortest road distance.

THE TRAVELING-SALESMAN PROBLEM might be described as follows: Find the shortest route (tour) for a salesman starting from a given city, visiting each of a specified group of cities, and then returning to the original point of departure. More generally, given an n by n symmetric matrix $D=(d_{IJ})$, where d_{IJ} represents the 'distance' from I to J, arrange the points in a cyclic order in such a way that the sum of the d_{IJ} between consecutive points is minimal. Since there are only a finite number of possibilities (at most $\frac{1}{2}(n-1)!$) to consider, the problem is to devise a method of picking out the optimal arrangement which is reasonably efficient for fairly large values of n. Although algorithms have been devised for problems of similar nature, e.g., the optimal assignment problem,[3,7,8] little is known about the traveling-salesman problem. We do not claim that this note alters the situation very much; what we shall do is outline a way of approaching the problem that sometimes, at least, enables one to find an optimal path and prove it so. In particular, it will be shown that a certain arrangement of 49 cities, one in each of the 48 states and Washington, D. C., is best, the d_{IJ} used representing road distances as taken from an atlas.

* HISTORICAL NOTE: The origin of this problem is somewhat obscure. It appears to have been discussed informally among mathematicians at mathematics meetings for many years. Surprisingly little in the way of results has appeared in the mathematical literature.[10] It may be that the minimal-distance tour problem was stimulated by the so-called Hamiltonian game[1] which is concerned with finding the number of different tours possible over a specified network. The latter problem is cited by some as the origin of group theory and has some connections with the famous Four-Color Conjecture.[9] Merrill Flood (Columbia University) should be credited with stimulating interest in the traveling-salesman problem in many quarters. As early as 1937, he tried to obtain near optimal solutions in reference to routing of school buses. Both Flood and A. W. Tucker (Princeton University) recall that they heard about the problem first in a seminar talk by Hassler Whitney at Princeton in 1934 (although Whitney, recently queried, does not seem to recall the problem). The relations between the traveling-salesman problem and the transportation problem of linear programming appear to have been first explored by M. Flood, J. Robinson, T. C. Koopmans, M. Beckmann, and later by I. Heller and H. Kuhn.[4,5,6]

In order to try the method on a large problem, the following set of 49 cities, one in each state and the District of Columbia, was selected:

1. Manchester, N. H.
2. Montpelier, Vt.
3. Detroit, Mich.
4. Cleveland, Ohio
5. Charleston, W. Va.
6. Louisville, Ky.
7. Indianapolis, Ind.
8. Chicago, Ill.
9. Milwaukee, Wis.
10. Minneapolis, Minn.
11. Pierre, S. D.
12. Bismarck, N. D.
13. Helena, Mont.
14. Seattle, Wash.
15. Portland, Ore.
16. Boise, Idaho
17. Salt Lake City, Utah
18. Carson City, Nev.
19. Los Angeles, Calif.
20. Phoenix, Ariz.
21. Santa Fe, N. M.
22. Denver, Colo.
23. Cheyenne, Wyo.
24. Omaha, Neb.
25. Des Moines, Iowa
26. Kansas City, Mo.
27. Topeka, Kans.
28. Oklahoma City, Okla.
29. Dallas, Tex.
30. Little Rock, Ark.
31. Memphis, Tenn.
32. Jackson, Miss.
33. New Orleans, La.
34. Birmingham, Ala.
35. Atlanta, Ga.
36. Jacksonville, Fla.
37. Columbia, S. C.
38. Raleigh, N. C.
39. Richmond, Va.
40. Washington, D. C.
41. Boston, Mass.
42. Portland, Me.

A. Baltimore, Md.
B. Wilmington, Del.
C. Philadelphia, Penn.
D. Newark, N. J.
E. New York, N. Y.
F. Hartford, Conn.
G. Providence, R. I.

The reason for picking this particular set was that most of the road distances between them were easy to get from an atlas. The triangular table of distances between these cities (Table I) is part of the original one prepared by Bernice Brown of The Rand Corporation. It gives $d_{IJ} = \frac{1}{17}(d'_{IJ} - 11)$,* $(I, J = 1, 2, \cdots, 42)$, where d'_{IJ} is the road distance in miles between I and J. The d_{IJ} have been rounded to the nearest integer. Certainly such a linear transformation does not alter the ordering of the tour lengths, although, of course, rounding could cause a tour that was not optimal in terms of the original mileage to become optimal in terms of the adjusted units used in this paper.

We will show that the tour (see Fig. 16) through the cities $1, 2, \cdots, 42$ in this order is minimal for this subset of 42 cities. Moreover, since in driving from city 40 (Washington, D. C.) to city 41 (Boston, Massachusetts) by the shortest road distance one goes through A, B, $\cdots$, G, successively, it follows that the tour through 49 cities $1, 2, \cdots, 40$, A, B, $\cdots$, G, 41, 42 in that order is also optimal.

PRELIMINARY NOTIONS

Whenever the road from I to J (in that order) is traveled, the value $x'_{IJ} = 1$ is entered into the I, J element of a matrix; otherwise $x'_{IJ} = 0$ is entered. A (directed) tour through n cities can now be thought of as a permutation matrix of order n which represents an n-cycle (we assume

* This particular transformation was chosen to make the d_{IJ} of the original table less than 256 which would permit compact storage of the distance table in binary representation; however, no use was made of this.

TABLE I

Road Distances between Cities in Adjusted Units

The figures in the table are mileages between the two specified numbered cities, less 11, divided by 17, and rounded to the nearest integer.

	1	2	3	4	5	6	7	8	9	10	11	12	13	14	15	16	17	18	19	20
2	8																			
3	39	45																		
4	37	47	9																	
5	50	49	21	15																
6	61	62	21	20	17															
7	58	60	16	17	18	6														
8	59	60	15	20	26	17	10													
9	62	66	20	25	31	22	15	5												
10	81	81	40	44	50	41	35	24	20											
11	103	107	62	67	72	63	57	46	41	23										
12	108	117	66	71	77	68	61	51	46	26	11									
13	145	149	104	108	114	106	99	88	84	63	49	40								
14	181	185	140	144	150	142	135	124	120	99	85	76	35							
15	187	191	146	150	156	142	137	130	125	105	90	81	41	10						
16	161	170	120	124	130	115	110	104	105	90	72	64	34	31	27					
17	142	146	101	104	111	97	91	85	86	75	51	59	29	53	48	21				
18	174	178	133	138	143	129	123	117	118	107	83	84	54	46	35	26	31			
19	185	186	142	143	140	130	126	124	128	118	93	101	72	69	58	58	43	26		
20	164	165	120	123	124	106	106	105	110	104	86	97	71	93	82	62	42	45	22	
21	137	139	94	96	94	80	78	77	84	77	56	64	65	90	87	58	36	68	50	30
22	117	122	77	80	83	68	62	60	61	50	34	42	49	82	77	60	30	62	70	49
23	114	118	73	78	84	69	63	57	59	48	28	36	43	77	72	45	27	59	69	55
24	85	89	44	48	53	41	34	28	29	22	23	35	69	105	102	74	56	88	99	81
25	77	80	36	40	46	34	27	19	21	14	29	40	77	114	111	84	64	96	107	87
26	87	89	44	46	46	30	28	29	32	27	36	47	78	116	112	84	66	98	95	75
27	91	93	48	50	48	34	32	33	36	30	34	45	77	115	110	83	63	97	91	72
28	105	106	62	63	64	47	46	49	54	48	46	59	85	119	115	88	66	98	79	59
29	111	113	69	71	66	51	53	56	61	57	59	71	96	130	126	98	75	98	85	62
30	91	92	50	51	46	30	34	38	43	49	60	71	103	141	136	109	90	115	99	81
31	83	85	42	43	38	22	26	32	36	51	63	75	106	142	140	112	93	126	108	88
32	89	91	55	55	50	34	39	44	49	63	76	87	120	155	150	123	100	123	109	86
33	95	97	64	63	56	42	49	56	60	75	86	97	126	160	155	128	104	128	113	90
34	74	81	44	43	35	23	30	39	44	62	78	89	121	159	155	127	108	136	124	101
35	67	69	42	41	31	25	32	41	46	64	83	90	130	164	160	133	114	146	134	111
36	74	76	61	60	42	44	51	60	66	83	102	110	147	185	179	155	133	159	146	122
37	57	59	46	41	25	30	36	47	52	71	93	98	136	172	172	148	126	158	147	124
38	45	46	41	34	20	34	38	48	53	73	96	99	137	176	178	151	131	163	159	135
39	35	37	35	26	18	34	36	46	51	70	93	97	134	171	176	151	129	161	163	139
40	29	33	30	21	18	35	33	40	45	65	87	91	117	166	171	144	125	157	156	139
41	3	11	41	37	47	57	55	58	63	83	105	109	147	186	188	164	144	176	182	161
42	5	12	55	41	53	64	61	61	66	84	111	113	150	186	192	166	147	180	188	167

	21	22	23	24	25	26	27	28	29	30	31	32	33	34	35	36	37	38	39	40	41
22	21																				
23	27	5																			
24	54	32	29																		
25	60	40	37	8																	
26	47	36	39	12	11																
27	44	32	36	9	15	3															
28	31	36	42	28	33	21	20														
29	38	47	53	39	42	29	30	12													
30	53	61	62	36	34	24	28	20	20												
31	60	64	66	39	36	27	31	28	28	8											
32	62	71	78	52	49	39	44	35	24	15	12										
33	67	76	82	62	59	49	53	40	29	25	23	11									
34	75	79	81	54	50	42	46	43	39	23	14	14	21								
35	85	84	86	59	52	47	51	53	49	32	24	24	30	9							
36	98	105	107	79	71	66	70	70	60	48	40	36	33	25	18						
37	121	97	99	71	65	59	63	67	62	46	38	37	43	23	13	17					
38	108	102	103	73	67	64	69	75	72	54	46	49	54	34	24	29	12				
39	118	102	101	71	65	65	70	84	78	58	50	56	62	41	32	38	21	9			
40	113	95	97	67	60	62	67	79	82	62	53	59	66	45	38	45	27	15	6		
41	134	119	116	86	78	84	88	101	108	88	80	86	92	71	64	71	54	41	32	25	
42	140	124	119	90	87	90	94	107	114	77	86	92	98	80	74	77	60	48	38	32	6

$n>2$ throughout). For example, for $n=5$, the first matrix displayed below

$$\|x'_{IJ}\| = \left\| \begin{array}{ccccc} 0 & 1 & 0 & 0 & 0 \\ 0 & 0 & 0 & 1 & 0 \\ 0 & 0 & 0 & 0 & 1 \\ 0 & 0 & 1 & 0 & 0 \\ 1 & 0 & 0 & 0 & 0 \end{array} \right\|, \qquad \|x'_{IJ}\| = \left\| \begin{array}{ccccc} 0 & 1 & 0 & 0 & 0 \\ 1 & 0 & 0 & 0 & 0 \\ 0 & 0 & 0 & 0 & 1 \\ 0 & 0 & 1 & 0 & 0 \\ 0 & 0 & 0 & 1 & 0 \end{array} \right\|.$$

is a tour since it represents visiting the cities in the 5-cycle (1 2 4 3 5), while the other matrix is not a tour since it represents visiting the cities by means of two sub-cycles (1 2) and (3 5 4).

It is clear that all representations for directed tours satisfy the relations

$$\sum_I x'_{IJ} = \sum_J x'_{IJ} = 1, \qquad x'_{II} = 0, \qquad x'_{IJ} \geq 0.$$

The matrix may be made into a triangular array by reflecting the numbers above the diagonal in the diagonal. The sum of corresponding elements is denoted by $x_{IJ} = x'_{IJ} + x'_{JI}$. Then the matrices above become

$$\|x_{IJ}\| = \left\| \begin{array}{ccccc} \cdot & & & & \\ 1 & \cdot & & & \\ 0 & 0 & \cdot & & \\ 0 & 1 & 1 & \cdot & \\ 1 & 0 & 1 & 0 & \cdot \end{array} \right\|, \qquad \|x_{IJ}\| = \left\| \begin{array}{ccccc} \cdot & & & & \\ 2 & \cdot & & & \\ 0 & 0 & \cdot & & \\ 0 & 0 & 1 & \cdot & \\ 0 & 0 & 1 & 1 & \cdot \end{array} \right\|.$$

Consequently, the sum along the Kth row plus the sum along the Kth column must now be 2. This may be written

$$\sum_{J<I=K} x_{IJ} + \sum_{I>J=K} x_{IJ} = 2, \qquad (K=1, \cdots, n;\ x_{IJ} \geq 0) \quad (1)$$

This device yields a representation for undirected tours and is the one used throughout this paper. It will be noted that the second array above does not represent a tour but nevertheless satisfies the relation (1).

For undirected tours, the symbol x_{IJ} will be treated identically with x_{JI} so that we may rewrite (1) as

$$\sum_{J=1}^{n} x_{IJ} = 2. \qquad (x_{IJ} \geq 0;\ I=1,2,\cdots,n;\ I \neq J;\ x_{IJ} \equiv x_{JI}) \quad (2)$$

The problem is to find the minimum of the linear form

$$D(x) = \sum_{I>J} d_{IJ}\, x_{IJ}, \qquad (3)$$

where the $x_{IJ}=0$ or 1 and the $x_{IJ}=1$ form a tour, and where the summation in (3) extends over all indices (I,J) such that $I>J$.

To make a linear programming problem out of this (see ref. 2) one needs, as we have observed, a way to describe tours by more linear restraints than that given by (2). This is extremely difficult to do as illustrated by work of I. Heller[4] and H. Kuhn.[6] They point out that such relations always exist. However, there seems to be no simple way to characterize them and for moderate size n the number of such restraints appears to be astronomical. In spite of these difficulties, this paper will describe the techniques we have developed which have been successful in solving all the problems we have tried by this approach. A surprising empirical observation is the use of only a trivial number of the many possible restraints to solve any particular problem. To demonstrate the procedure, we shall attempt to use direct elementary proofs even though they were originally motivated in many places by linear programming procedures.

There are possibly four devices we have used which have greatly reduced the effort in obtaining solutions of the problems we have attempted.

First of all, we use undirected tours. This seems to simplify the characterization of the tours when n is small and certainly cuts down the amount of computation, even for large n. Secondly, and this is decisive, we do not try to characterize the tours by the complete set of linear restraints, but rather impose, in addition to (2), just enough linear conditions on the x_{IJ} to assure that the minimum of the linear form (3) is assumed by some tour. For the 49-city problem and also for all the smaller problems we have considered, such a procedure has been relatively easy to carry through by hand computation. This may be due in part to the fact that we use a simple symbolism which permits direct representation of the algebraic relationships and manipulations on a map of the cities. This third device speeds up the entire iterative process, makes it easy to follow, and sometimes suggests new linear restraints that are not likely to be obtained by less visual methods. Finally, once a tour has been obtained which is nearly optimal, a combinatorial approach, using the map and listing possible tours which have not yet been eliminated by the conditions imposed on the problem, may be advantageous. This list can be very much shorter than one would expect, due to the complex interlocking of the restraints. However, except for short discussion in the section below, "An Estimation Procedure," this method will not be described in detail although it has worked out well for all examples we have studied.

An important class of conditions that tours satisfy, which excludes many non-tour cases satisfying (2), are the 'loop conditions.' These are linear inequality restraints that exclude subcycles or loops. Consider a non-tour solution to (2) which has a subtour of $n_1 < n$ cities; we note that the sum of the x_{IJ} for those links (I,J) in the subtour is n_1. Hence we can eliminate this type of solution by imposing the condition that the sum of x_{IJ} over all links (I,J) connecting cities in the subset S of n_1 cities be less than n_1, i.e.,

$$\sum_{S} x_{IJ} \leq n_1 - 1 \tag{4}$$

where the summation extends over all (I,J) with I and J in the n_1 cities S. From (2) we note that two other conditions, each equivalent to (4), are

$$\sum_{\bar{S}} x_{IJ} \leq n - n_1 - 1, \tag{5}$$

where $\bar{S}$ means the summation extends over all (I,J) such that neither I nor J is in S, and

$$\sum_{S\bar{S}} x_{IJ} \geq 2, \tag{6}$$

where $S\bar{S}$ means that the summation extends over all (I,J) such that I is in S and J not in S.

There are, however, other more complicated types of restraints which sometimes must be added to (2) in addition to an assortment of loop conditions in order to exclude solutions involving fractional weights x_{IJ}. In the 49-city case we needed two such conditions. However, later when we tried the combinatorial approach, after imposing a few of the loop conditions, we found we could handle the 49-city problem without the use of the special restraints and this would have led to a shorter proof of optimality. In fact, we have yet to find an example which could not be handled by using only loop conditions and combinatorial arguments.

THE METHOD

The technique will be illustrated by a series of simple examples.

Example 1

First consider a five-city map forming a regular pentagon of unit length per side and with length $\frac{1}{2}(\sqrt{5}+1) \doteq 1.7$ on a diagonal (Fig. 1). Sup-

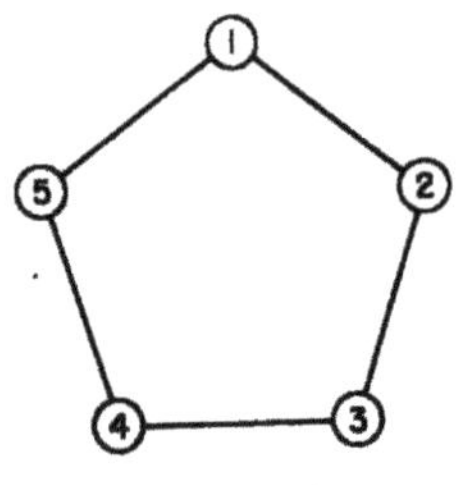

FIGURE 1

pose that the problem is to minimize (3) subject only to (2). Start with a tour which is conjectured to be optimal, obviously (1 2 3 4 5). In this case the values of x_{IJ}, denoted by $\bar{x}_{IJ}$, are $\bar{x}_{12}=\bar{x}_{23}=\bar{x}_{34}=\bar{x}_{45}=\bar{x}_{51}=1$ and all other $\bar{x}_{IJ}=0$. The variables x_{IJ} corresponding to links on the tour are called 'basic variables.' The length of the tour given by the linear form (3) for $x=\bar{x}$ is $D(\bar{x})=5$. There are five equations in (2). Multiply each by a parameter π_I to be determined, and then subtract the sum from (3). Thus, we are led to

$$D(x)=\sum_{I>J} d_{IJ}x_{IJ}-\sum_{I=1}^{n}\pi_I\left(\sum_{J=1}^{n}x_{IJ}-2\right)\qquad (x_{IJ}\equiv x_{JI};\ I\neq J)$$

$$=-\sum_{I>J}(\pi_I+\pi_J-d_{IJ})x_{IJ}+2\sum_{1}^{n}\pi_I.$$

Denote the coefficients of x_{IJ} by δ_{IJ} so that

$$D(x)=-\sum_{I>J}\delta_{IJ}x_{IJ}+2\sum_{1}^{n}\pi_I.\qquad (\delta_{IJ}=\pi_I+\pi_J-d_{IJ})\quad (7)$$

Now determine the five π_I values so that δ_{IJ} corresponding to basic variables vanish: .

$$\delta_{IJ}=0,\qquad (\text{for } \bar{x}_{IJ}=1)\quad (8)$$

i.e., if the link (I,J) is on the tour in question. Note that to solve for the π_I we have five linear equations in five unknowns.

If now we set $x_{IJ}=\bar{x}_{IJ}$ in (7), then $\bar{x}_{IJ}\delta_{IJ}=0$ for *all* (I,J) and

$$D(\bar{x})=2\sum_{1}^{n}\pi_I=5.\qquad (9)$$

Subtracting (9) from (7) we have finally

$$D(x)-D(\bar{x})=-\sum_{I>J}\delta_{IJ}x_{IJ}.\qquad (10)$$

For the regular pentagon $\pi_I=\frac{1}{2}$ for $I=1, 2, 3, 4, 5$ solves (8), and so $\delta_{IJ}=\frac{1}{2}(1-\sqrt{5})<0$ on a diagonal, i.e., $\delta_{IJ}\leq 0$ for every (I,J). Thus, the right side of (10) is always nonnegative or $D(x)\geq D(\bar{x})$ for all x satisfying (2), and in particular all other tours are longer than the tour represented by $\bar{x}$.

Example 2

Next, take another five-city problem whose map is not a regular pentagon (Fig. 2). We start with the tour (1 2 3 4 5) of length $D(\bar{x})=32$ where the basic variables take on the values $\bar{x}_{12}=\bar{x}_{23}=\bar{x}_{34}=\bar{x}_{45}=\bar{x}_{51}=1$ and all other $\bar{x}_{IJ}=0$. Repeat the steps in the previous problem leading to (10)

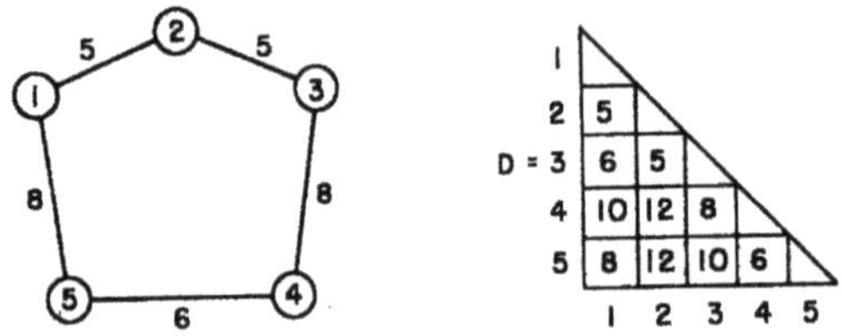

FIGURE 2

where, as before, calculate the π_I by setting $\delta_{IJ}=0$ for δ_{IJ} corresponding to basic variables x_{IJ}. The five equations that the π_I must satisfy are

$$\pi_1+\pi_2=5, \quad \pi_2+\pi_3=5, \quad \pi_3+\pi_4=8, \quad \pi_4+\pi_5=6, \quad \pi_5+\pi_1=8$$

By alternately subtracting and adding these equations one obtains.

$$2\pi_1=d_{12}-d_{23}+d_{34}-d_{45}+d_{51}=5-5+8-6+8=10,$$

or
$$\pi_1=5, \quad \pi_2=0, \quad \pi_3=5, \quad \pi_4=3, \quad \pi_5=3.$$

The factors π_I which multiply equations (2) to form (10) are called 'potentials.'* There is one such potential associated with each city I, and these are readily computed by working directly on the map of the cities (see Fig. 3).

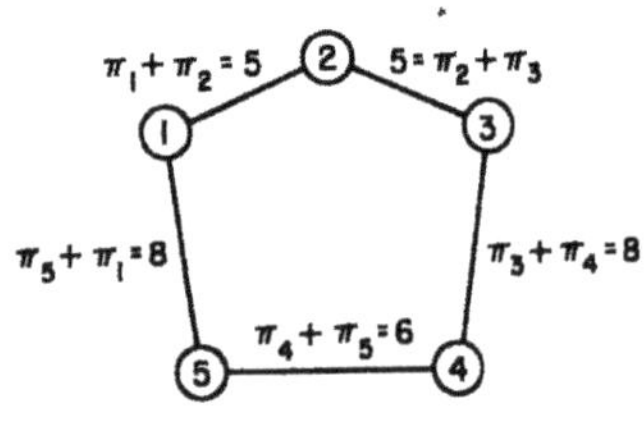

FIGURE 3

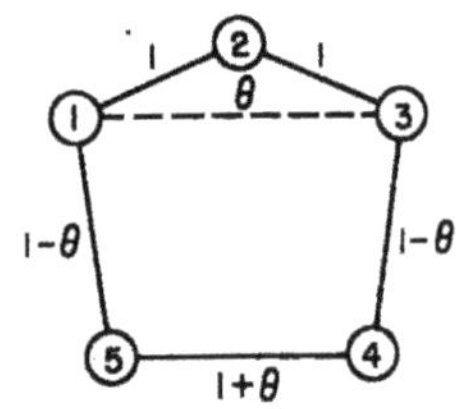

FIGURE 4

To form other δ_{IJ}, add the π_I and π_J of city I and city J and subtract off the distance d_{IJ} between them. In this case we note that except for $\delta_{31}=5+5-6=+4$, all the other δ_{IJ} are ≤ 0.

We see from (10) that if x_{31} were to take on a positive value, $x_{31}=\theta$, the other nonbasic variables remaining at zero, this may lead to a better solution. We let θ be the largest value consistent with (2). Thus, the weights x_{IJ} must add up to 2 on links from each city and no weight is negative. However, in setting $x_{31}=\theta$ we adjust only the basic set of variables, leaving all other nonbasic variables at zero value. This is

* The term potential is used by T. C. Koopmans in an analogous connection for the transportation problem.[5]

worked out on the map shown in Fig. 4. Here the maximum value of θ is 1, and this leads to a 3-cycle (1 2 3) and a 2-cycle (4 5) (Fig. 5).

This is not a tour, so we add a loop condition which excludes this solution but which is satisfied by all tours. In this case $x_{45} \leq 1$ or

$$x_{45} + y_6 - 1 = 0, \qquad (y_6 \geq 0) \quad (11)$$

is such a condition. Accordingly, we start over again using the five equations (2) and the sixth equation (11). This time we will need six basic variables and it will be convenient to have x_{13} (the one we set equal to θ previously) included with those associated with the tour. Thus, the

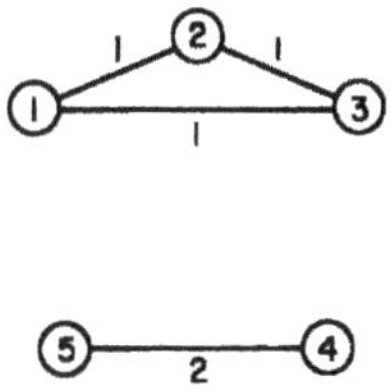

FIGURE 5

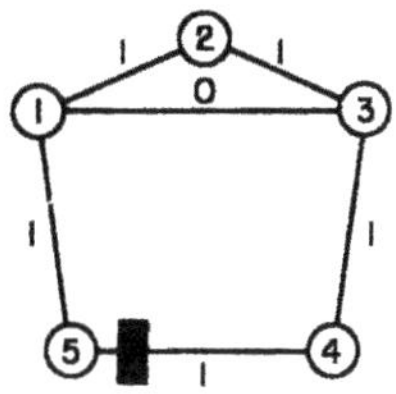

FIGURE 6

starting solution is as follows: The basic variables have values $\bar{x}_{12} = \bar{x}_{23} = \bar{x}_{34} = \bar{x}_{45} = \bar{x}_{51} = 1$, $\bar{x}_{13} = 0$. All other $\bar{x}_{IJ} = 0$. This solution is shown in Fig. 6. The presence of an upper bound on x_{45} or relation (11) is depicted in Fig. 6 by a block symbol on (4, 5). Now we multiply equation (11) by π_6, add it to $\sum_{I=1}^{5} \pi_I \left(\sum_{J=1}^{5} x_{IJ} - 2 \right)$, subtract the sum from $\sum d_{IJ} x_{IJ}$ and collect terms in x_{IJ} as before. The result is

$$\sum_{I>J} d_{IJ} x_{IJ} = - \sum_{I>J} \delta_{IJ} x_{IJ} + 2 \sum_{I=1}^{5} \pi_I + \pi_6 (1 - y_6) \qquad (12)$$

where $\delta_{IJ} = \pi_I + \pi_J - d_{IJ}$ except $\delta_{45} = \pi_4 + \pi_5 - (d_{45} - \pi_6)$.

Now determine the six values of π_I by setting $\delta_{IJ} = 0$ corresponding to basic variables x_{IJ}:

$$\delta_{12} = \delta_{23} = \delta_{34} = \delta_{45} = \delta_{51} = \delta_{13} = 0, \qquad (13)$$

from which it follows that

$$D(x) - D(\bar{x}) = - \sum \delta_{IJ} x_{IJ} - \pi_6 y_6. \qquad (14)$$

To evaluate π_I we note that there are six equations in six unknowns. These are shown on the map below (Fig. 7). The three conditions about the triangular loop (1, 2, 3) permit us to solve for π_1, π_2, π_3. Branching out from the triangle we get next π_4 and π_5 and finally π_6. Thus, we determine first that $2\pi_1 = d_{12} - d_{23} + d_{31} = 5 - 5 + 6$ so that $\pi_1 = 3$, $\pi_2 = 2$, $\pi_3 = 3$.

Working down, $\pi_4=5$, $\pi_5=5$. Thus, $\pi_4+\pi_5=-\pi_6+6$, so $\pi_6=-4$. These values are shown adjacent to each city in Fig. 7.

With these values of π_I all remaining $\delta_{IJ}=(\pi_I+\pi_J-d_{IJ})\leq 0$; hence, with $\pi_6<0$ we have the right side of (14) always positive, so the tour (1 2 3 4 5) is minimal. This illustrates the use of the simplest of the loop conditions, namely, an upper bound on the variable x_{45}.

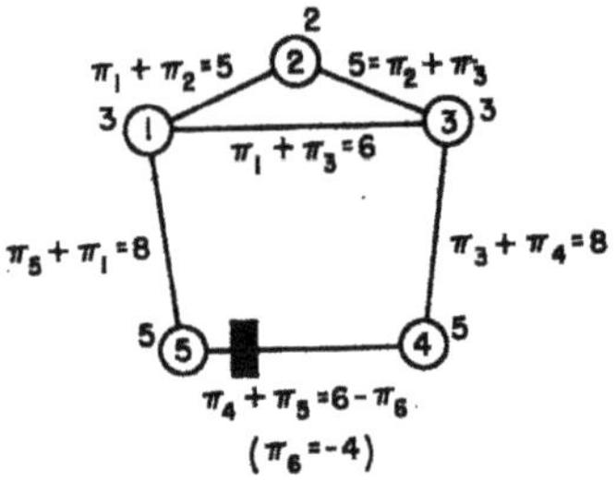

FIGURE 7

Example 3

Here we consider a six-city case (Fig. 8) where the optimal tour is not our initial choice. Let the starting tour be (1 2 3 4 5 6) of length $D(\bar{x})=23$. If we proceed as before, relation (8) implies that the π_I satisfy the relations shown in Fig. 8. In this case (and this is generally true for

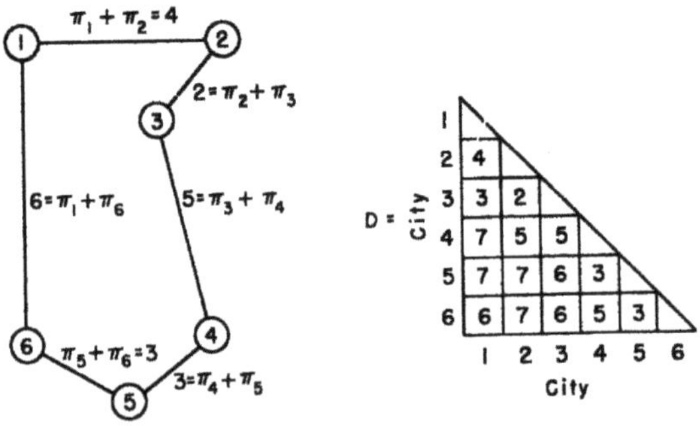

FIGURE 8

loops with an even number of links) the sum of equations on links (1, 2), (3, 4), (5, 6) is identical with the sum for (2, 3), (4, 5), (6, 1) except for different constant terms, so that the system of equations in π_I is inconsistent.

This difficulty can be avoided if the following general rule is followed: The set of basic variables must be so selected that when the remaining

x_{IJ} are fixed, the values of the basic variables are uniquely determined. This means the matrix of coefficients of the basic variables is nonsingular (i.e., their determinant is nonvanishing). Since the π_I satisfy a system of equations whose coefficient matrix is the transpose of this matrix, the π_I will be uniquely determined also. In the six-city case, one may augment system (2) with the additional upper-bound condition

$$x_{45}+y_7=1 \qquad (y_7 \geq 0) \quad (15)$$

and select x_{13} as a basic variable in addition to the basic variables x_{IJ} corresponding to (I, J) on the tour. Then, letting π_7 be the weight associated with restriction (15), the π_I satisfy relations in Fig. 9.

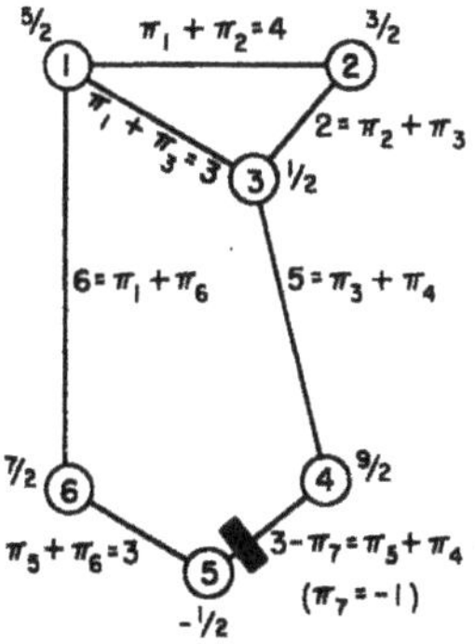

FIGURE 9

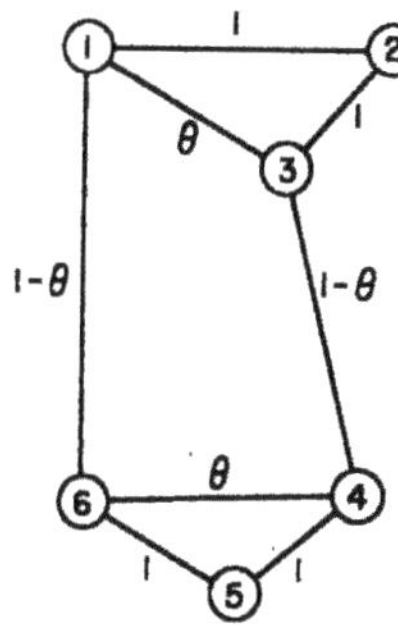

FIGURE 10

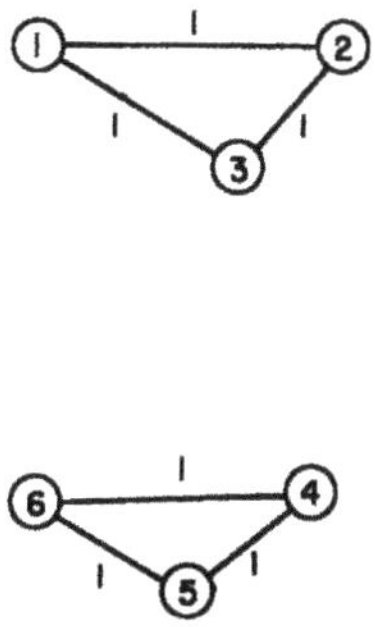

FIGURE 11

The value of $\pi_1 = 5/2$ can be determined from the odd loop (1 2 3) by alternately adding and subtracting the equations around the loop. The others can then be evaluated immediately. In this case, we have, analogous to (14),

$$D(x)-D(\bar{x}) = -\sum \delta_{IJ}\, x_{IJ} - \pi_7\, y_7, \qquad (16)$$

where $\delta_{IJ}=0$ if x_{IJ} is a basic variable and $\delta_{IJ}=\pi_I+\pi_J-d_{IJ}$ otherwise. Since $\delta_{46}=3$, increasing the value of x_{46} to θ (while all other nonbasic variables remain zero), with corresponding adjustments in the basic variables, will yield $D(x)-D(\bar{x})=-3\theta<0$. In Fig. 10 it is seen that the largest value of $\theta=1$ and the resulting solution is Fig. 11, which is not a new tour, but two loops. However, we can exclude this solution by imposing the additional restriction satisfied by all tour solutions

$$x_{12}+x_{23}+x_{31} \leq 2, \quad \text{or} \quad x_{12}+x_{23}+x_{31}+y_8=2, \qquad (y_8 \geq 0) \quad (17)$$

since in Fig. 11 the inadmissible solution has $x_{12}+x_{23}+x_{31}=3$. We now start all over again augmenting relations (2) by (15) and (17). Let the basic variables be the same as before but include x_{46} (i.e., the one we set equal to θ in Fig. 10). Let π_I for 1, 2, $\cdots$, 8 be the weights assigned to these relations respectively in forming $D(x)-D(\bar{x})$; then the π_I satisfy the relations shown in Fig. 12, where the loop condition (17) is symbolized by the dotted loop in the figure.

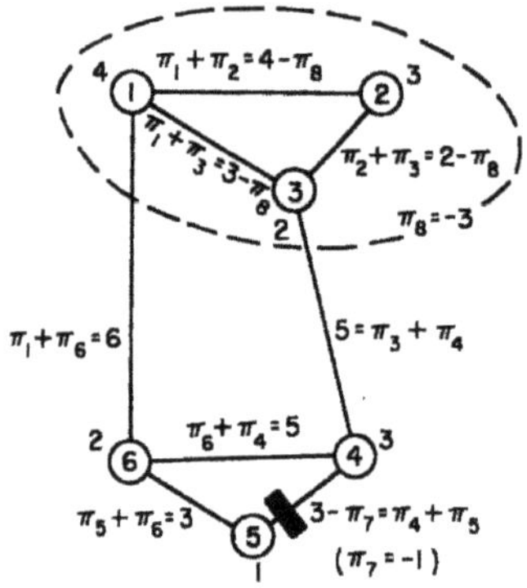

FIGURE 12

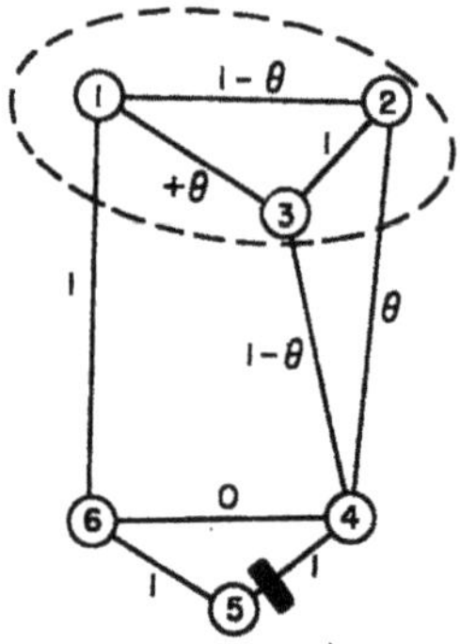

FIGURE 13

The value of $\pi_6=2$ may be evaluated from the odd loop (6 4 3 2 1) by alternately adding and subtracting the equations in π_I shown on this loop. The other π_I can then be immediately determined. This time

$$D(x)-D(\bar{x})=-\sum \delta_{IJ}x_{IJ}-\pi_7 y_7-\pi_8 y_8 \tag{18}$$

where $\delta_{IJ}=0$ for x_{IJ} a basic variable and $\delta_{IJ}=\pi_I+\pi_J-d_{IJ}$ otherwise. Since $\delta_{24}=1$ while all other $\delta_{IJ}\leq 0$, we set $x_{24}=\theta$; then the adjustments in the values of the basic variables necessary to satisfy (2), (15), (17) are

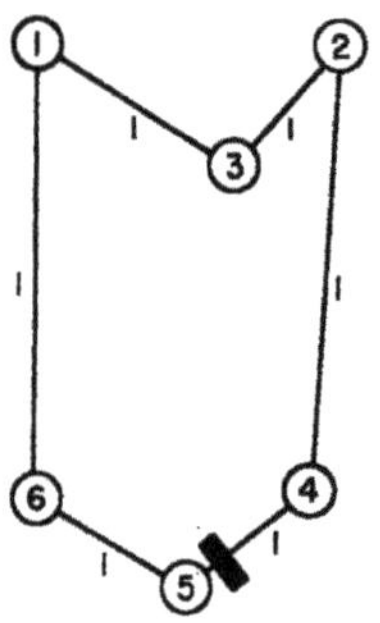

FIGURE 14

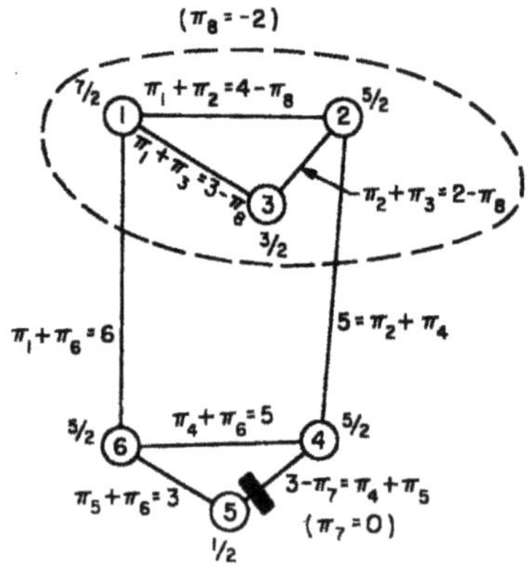

FIGURE 15

shown in Fig. 13 and the new solution for $\theta=1$ is a new tour $\bar{\bar{x}}$ with length $D(\bar{\bar{x}})=D(\bar{x})-1=22$, Fig. 14. We may now drop $x_{34}=0$ from the basic set of variables (or alternatively x_{12}) and replace it by x_{24} as a new basic variable. This yields the relations for π_I of Fig. 15. The expression for $D(x)-D(\bar{\bar{x}})$ is similar to (18). It can now be tested that all $\delta_{IJ}\leq 0$ corresponding to non-basic x_{IJ}, and the coefficients of y_7 and y_8 are $\pi_7\leq 0$, $\pi_8\leq 0$, so that the new tour is established as optimal.

AN ESTIMATION PROCEDURE

In any linear programming problem with bounded variables, an estimate is available of how much a basic solution differs from an optimal solution. Let $D(x)$ represent a linear form to be minimized and $D(\bar{x})$ be the value for some basic solution $\bar{x}$ where variables $(x_1, x_2, \cdots, x_{n'})$, represented by the symbol x, satisfy a system of equations as well as bounds $0\leq x_J\leq r_J$. If the equations are multiplied by weights π_I and substracted from $D(x)$, then (as we have noted earlier)

$$D(x)-D(\bar{x})=-\sum_{J=1}^{n'}\delta_J x_J \qquad (x_J\geq 0) \quad (19)$$

where π_I are chosen such that $\delta_J=0$ if the corresponding x_J is a basic variable. We may now split the right side of (19) into positive and negative parts and obtain a lower bound for the difference by dropping the positive part, i.e.,

$$D(x)-D(\bar{x})=-\sum_{\delta_J>0}\delta_J x_J-\sum_{\delta_J\leq 0}\delta_J x_J, \qquad (x_J\geq 0) \quad (20)$$

$$D(x)-D(\bar{x})\geq -\sum_{\delta_J>0}\delta_J x_J\geq -E, \qquad (E\geq 0) \quad (21)$$

where $-E$ is some estimate for the negative part. By setting $x_J=r_J$, we obtain in particular

$$D(x)-D(\bar{x})\geq -\sum_{\delta_J>0}\delta_J r_J. \quad (22)$$

For the traveling-salesman problem the variables x_{IJ} must be either 0 or 1 if x represents a tour. From (20), no link (I, J) can occur in an optimal tour if

$$\delta_{IJ}<-E, \quad (23)$$

hence all corresponding variables x_{IJ} can be dropped from further consideration.

During the early stages of the computation, E may be quite large and very few links can be dropped by this rule; however, in the latter stages often so many links are eliminated that one can list all possible tours that

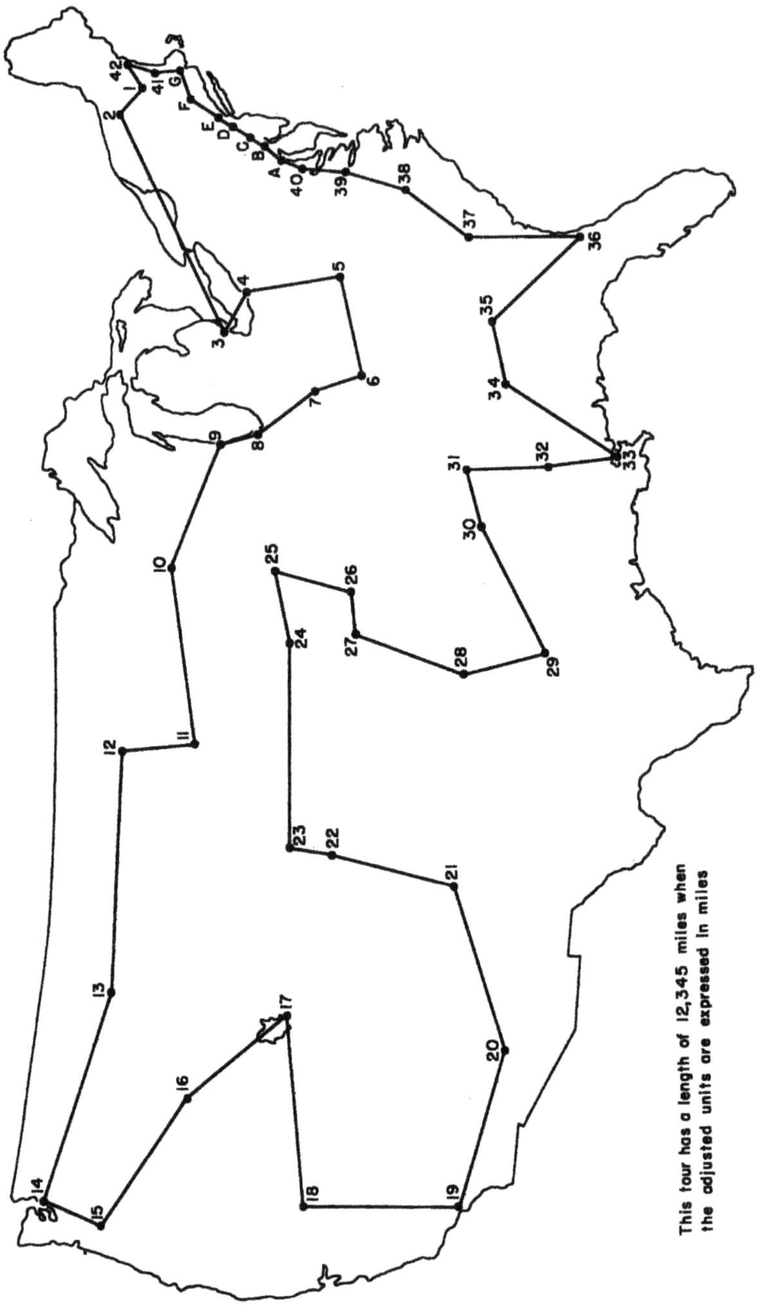

FIG. 16. The optimal tour of 49 cities.

406

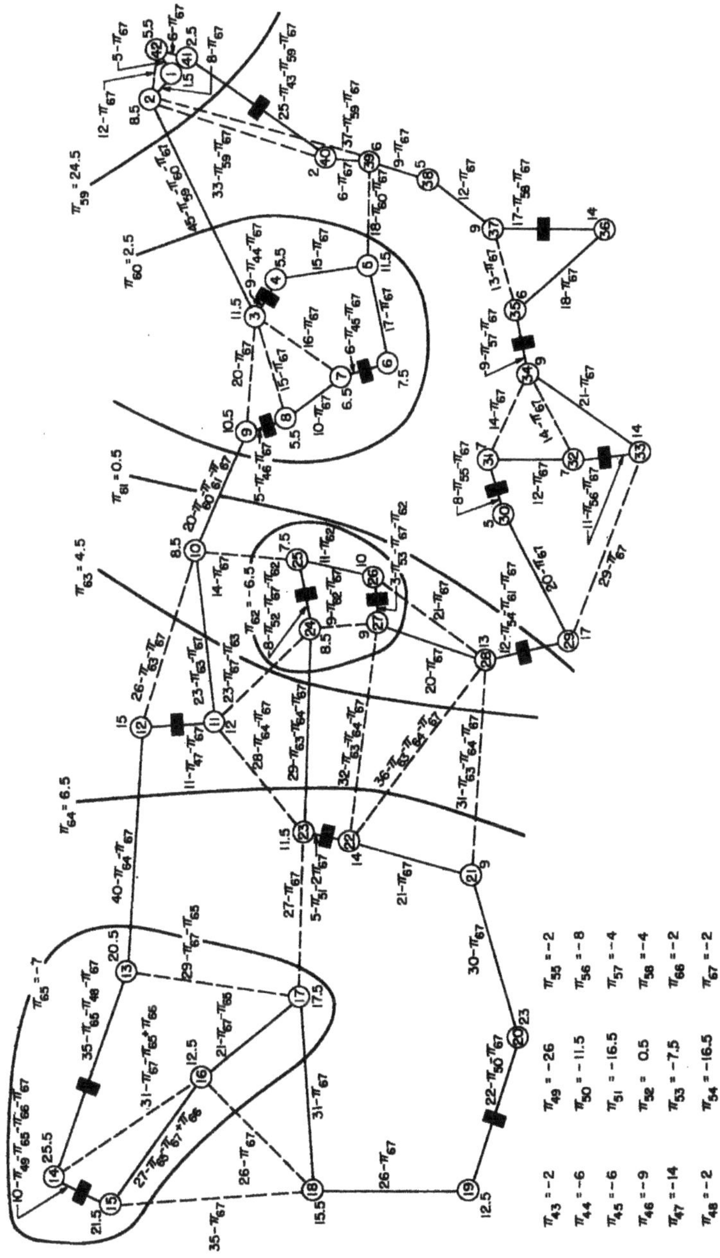

FIG. 17. Only the right-hand side of the equations satisfied by π_I are shown on the map; the left-hand side on line (I,J) is $\pi_I+\pi_J$. Dotted links (I,J) correspond to additional basic variables x_{IJ}.

407

use the remaining admissible links. By extending this type of combinatorial argument to the range of values of the 'slack' variables y_K, it is often possible at an earlier stage of the iterative algorithm to rule out so many of the tours that direct examination of the remaining tours for minimum length is a feasible approach.

THE 49-CITY PROBLEM*

The optimal tour $\bar{x}$ is shown in Fig. 16. The proof that it is optimal is given in Fig. 17. To make the correspondence between the latter and its programming problem clear, we will write down in addition to 42 relations in non-negative variables (2), a set of 25 relations which suffice to prove that $D(x)$ is a minimum for $\bar{x}$. We distinguish the following subsets of the 42 cities:

$$S_1 = \{1, 2, 41, 42\} \qquad S_5 = \{13, 14, \cdots, 23\}$$
$$S_2 = \{3, 4, \cdots, 9\} \qquad S_6 = \{13, 14, 15, 16, 17\}$$
$$S_3 = \{1, 2, \cdots, 9, 29, 30, \cdots, 42\} \qquad S_7 = \{24, 25, 26, 27\}.$$
$$S_4 = \{11, 12, \cdots, 23\}$$

Except for two inequalities which we will discuss in a moment, the programming problem may now be written as the following 65 relations:†

$$\sum_J x_{IJ} = 2 \ (I = 1, \cdots, 42), \qquad x_{41,1} \leq 1, \qquad x_{4,3} \leq 1, \qquad x_{7,6} \leq 1,$$
$$x_{9,8} \leq 1, \qquad x_{12,11} \leq 1, \qquad x_{14,13} \leq 1, \qquad x_{15,14} \leq 1, \qquad x_{20,19} \leq 1,$$
$$x_{23,22} \leq 1, \qquad x_{25,24} \leq 1, \qquad x_{27,26} \leq 1, \qquad x_{29,28} \leq 1, \qquad x_{31,30} \leq 1,$$
$$x_{33,32} \leq 1, \qquad x_{35,34} \leq 1, \qquad x_{37,36} \leq 1, \qquad \sum_{S_1,\bar{S}_1} x_{IJ} \geq 2, \qquad \sum_{S_2,\bar{S}_2} x_{IJ} \geq 2,$$
$$\sum_{S_3,\bar{S}_3} x_{IJ} \geq 2, \qquad \sum_{S_4,\bar{S}_4} x_{IJ} \geq 2, \qquad \sum_{S_5,\bar{S}_5} x_{IJ} \geq 2, \qquad \sum_{S_6} x_{IJ} \leq 4, \qquad \sum_{S_7} x_{IJ} \leq 3.$$

The remaining two relations (*66* and *67*) are perhaps most easily described verbally.

66: $x_{14,15}$ minus the sum of all other x_{IJ} on links out of 15, 16, 19, except for $x_{18,15}$, $x_{18,16}$, $x_{17,16}$, $x_{19,18}$, and $x_{20,19}$, is not positive.

67: $\Sigma a_{IJ} x_{IJ} \leq 42$, where $a_{23,22} = 2$, $a_{26,25} = 0$, all other $a_{IJ} = 1$ except $a_{IJ} = 0$ if x_{IJ} is a non-basic variable and either (a) I is in S_3, J not in S_3, or (b) I or J is 10, 21, 25, 26, 27, or 28.‡

These two inequalities are satisfied by all tours. For example, if a tour were to violate the first one, it must have successively $x_{15,14} = 1$,

* As indicated earlier, it was possible to treat this as a 42-city problem.

† $\Sigma_{S,\bar{S}}\, x_{IJ}$ means the sum of all variables where only one of the subscripts I or J is in S. $\Sigma_S\, x_{IJ}$ means the sum of all variables such that I and J are in S—see relations (4), (5), (6).

‡ We are indebted to I. Glicksberg of Rand for pointing out relations of this kind to us.

$x_{18,15}=1$, $x_{18,17}=1$, but also $x_{19,18}=1$, a contradiction. The argument that each tour satisfies the second inequality is similar. If a tour x exists with $\Sigma\, a_{IJ}x_{IJ}>42$, then clearly $x_{23,22}=1$, and also $x_{10,9}=x_{29,28}=1$, since by (a) these are the only links connecting S_3 and $\bar{S}_3$ having non-zero a_{IJ}. (See Fig. 17 to distinguish between basic and non-basic variables.) Moreover, since $a_{26,25}=0$, it follows from (b) that $x_{25,10}=x_{25,24}=x_{27,26}=x_{28,27}=1$. Again, (b) and the fact that $x_{28,21}=0$ imply $x_{21,20}=x_{22,21}=1$. Now look at city 27. There are three possibilities: $x_{27,24}=1$, $x_{27,22}=1$, or $x_{28,27}=1$. But each of these contradicts the assumption that x is a tour.

These relations were imposed to cut out fractional solutions which satisfy all the conditions (2) and (4). A picture of such a fractional solution, which gives a smaller value for the minimizing form than does any tour, is shown in Fig. 18. Notice that it does not satisfy relation *67*.

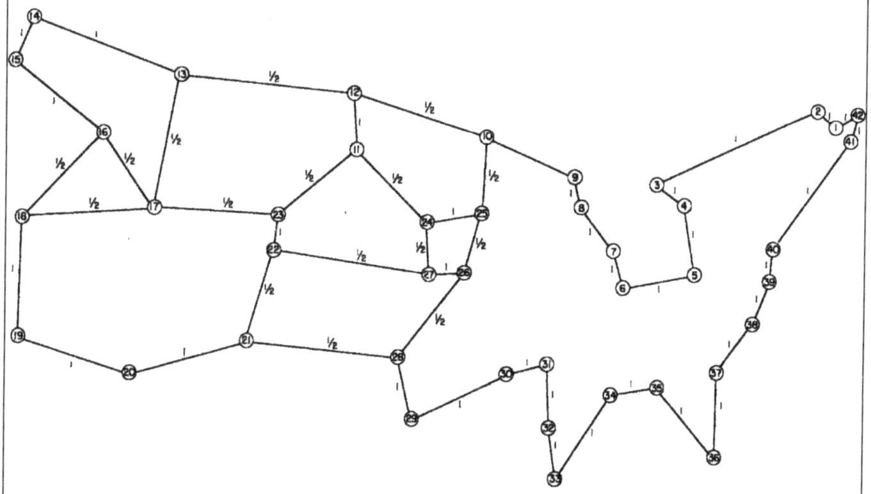

FIG. 18. A fractional solution x satisfying all loop conditions with $\Sigma\, d_{IJ}\, x_{IJ}=698$.

We assert that if the weights π_I are assigned to these restraints in the order presented above, then the values as given in Fig. 17 satisfy $\delta_{IJ}=0$ for all variables x_{IJ} in the basis. With these values of π_I in the expression for $D(x)-D(\bar{x})$, all $\delta_{IJ}\leq 0$ corresponding to variables x_{IJ} and π_{43}, π_{44}, $\cdots$, π_{67} corresponding to variables y_{43}, $\cdots$, y_{67} are appropriately positive or negative (positive if its y occurs with a minus sign in the relation, negative otherwise) with the exception of $\pi_{52}=\frac{1}{2}$ where $x_{25,24}+y_{52}=1$. This proves, since $E=\frac{1}{2}$ and all the d_{IJ} are integers, that $\bar{x}$ is minimal. The length $D(\bar{x})$ is 699 units, or 12,345 miles except for rounding errors.

It can be shown by introducing all links for which $\delta_{IJ} \geq -\frac{1}{2}$ that $\bar{x}$ is the unique minimum. There are only 7 such links in addition to those shown in Fig. 17, and consequently all possible tying tours were enumerated without too much trouble. None of them proved to be as good as $\bar{x}$.

CONCLUDING REMARK

It is clear that we have left unanswered practically any question one might pose of a theoretical nature concerning the traveling-salesman problem; however, we hope that the feasibility of attacking problems involving a moderate number of points has been successfully demonstrated, and that perhaps some of the ideas can be used in problems of similar nature.

REFERENCES

1. W. W. R. BALL, *Mathematical Recreations and Essays*, as rev. by H. S. M. Coxeter, 11th ed., Macmillan, New York, 1939.
2. G. B. DANTZIG, A. ORDEN, AND P. WOLFE, *The Generalized Simplex Method for Minimizing a Linear Form under Linear Inequality Restraints*, Rand Research Memorandum RM-1264 (April 5, 1954).
3. G. B. DANTZIG, "Application of the Simplex Method to a Transportation Problem," *Activity Analysis of Production and Allocation*, T. C. Koopmans, Ed., Wiley, New York, 1951.
4. I. HELLER, "On the Problem of Shortest Path Between Points," I and II (abstract), *Bull. Am. Math. Soc.* **59**, 6 (November, 1953).
5. T. C. KOOPMANS, "A Model of Transportation," *Activity Analysis of Production and Allocation*, T. C. Koopmans, Ed., Wiley, New York, 1951.
6. H. W. KUHN, "The Traveling-Salesman Problem," to appear in the *Proc. Sixth Symposium in Applied Mathematics* of the American Mathematical Society, McGraw-Hill, New York.
7. D. F. VOTAW AND A. ORDEN, "Personnel Assignment Problem," *Symposium on Linear Inequalities and Programming*, Comptroller, Headquarters U. S. Air Force (June 14–16, 1951).
8. J. VON NEUMANN, "A Certain Zero-sum Two-person Game Equivalent to the Optimal Assignment Problem," *Contributions to the Theory of Games II*, Princeton University Press, 1953.
9. W. T. TUTTE, "On Hamiltonian Circuits," *London Mathematical Society Journal* **XXI**, Part 2, No. 82, 98–101 (April, 1946).
10. S. VERBLUNSKY, "On the Shortest Path Through a Number of Points," *Proc. Am. Math. Soc.* **II**, 6 (December, 1951).

Chapter 2
The Hungarian Method for the Assignment Problem

Harold W. Kuhn

Introduction by *Harold W. Kuhn*

This paper has always been one of my favorite "children," combining as it does elements of the duality of linear programming and combinatorial tools from graph theory. It may be of some interest to tell the story of its origin.

I spent the summer of 1953 at the Institute for Numerical Analysis which was housed on the U.C.L.A. campus. I was supported by the National Bureau of Standards and shared an office with Ted Motzkin, a pioneer in the theory of inequalities and one of the most scholarly mathematicians I have ever known. I had no fixed duties and spent the summer working on subjects that were of interest to me at the time, such as the traveling salesman problem and the assignment problem.

The Institute for Numerical Analysis was the home of the SWAC (Standards Western Automatic Computer), which had been designed by Harry Huskey and had a memory of 256 words of 40 bits each on 40 Williamson tubes. The formulation of the assignment problem as a linear program was well known, but a 10 by 10 assignment problem has 100 variables in its primal statement and 100 constraints in the dual and so was too large for the SWAC to solve as a linear program. The SEAC (Standard Eastern Automatic Computer), housed in the National Bureau of Standards in Washington, could solve linear programs with about 25 variables and 25 constraints. The SEAC had a liquid mercury memory system which was extremely limiting.

During that summer, I was reading König's book on graph theory. I recognized the following theorem of König to be a pre-linear programming example of duality:

If the numbers of a matrix are 0's and 1's, then the minimum number of rows and columns that will contain all of the 1's is equal to the maximum number of 1's that can be chosen, with no two in the same row or column.

Harold W. Kuhn
Princeton University, USA
e-mail: kuhn@math.princeton.edu

M. Jünger et al. (eds.), *50 Years of Integer Programming 1958-2008*,
DOI 10.1007/978-3-540-68279-0_2, © Springer-Verlag Berlin Heidelberg 2010

Indeed, the primal problem is the special case of an assignment problem in which the ratings of the individuals in the jobs are only 0's and 1's. In a footnote, König refers to a paper of E. Egerváry (in Hungarian), which seemed to contain the treatment of a more general case. When I returned to Bryn Mawr, where I was on the faculty in 1953, I took out a Hungarian grammar and a large Hungarian-English dictionary and taught myself enough Hungarian to translate Egerváry's paper. I then realized that Egerváry's paper gave a computationally trivial method for reducing the general assignment problem to a 0-1 problem. Thus, by putting the two ideas together, the Hungarian Method was born. I tested the algorithm by solving 12 by 12 problems with random 3-digit ratings by hand. I could do any such problem, with pencil and paper, in no more than 2 hours. This seemed to be much better than any other method known at the time.

The paper was published in Naval Research Logistics Quarterly. This was a natural choice since the project in Game Theory, Linear and Nonlinear Programming, and Combinatorics at Princeton, with which Al Tucker and I were associated from 1948 to 1972, was supported by the Office of Naval Research Logistics Branch. Many mathematicians were benificiaries of the wise stewardship of Mina Rees as head of the ONR and Fred Rigby as chief of the Logistics branch. We were also fortunate to have Jack Laderman, the first editor of the journal, as our project supervisor.

I have told much of the same story in my paper [1]. Large sections of this account are reproduced in the book by Alexander Schrijver [2]. Schrijver's account places the Hungarian Method in the mathematical context of combinatorial optimization and rephrases the concepts in graph-theoretical language.

References

1. H.W. Kuhn, *On the origin of the Hungarian Method*, History of mathematical programming; a collection of personal reminiscences (J.K. Lenstra, A.H.G. Rinnooy Kan, and A. Schrijver, eds.), North Holland, Amsterdam, 1991, pp. 77–81.
2. A. Schrijver, *Combinatorial optimization: polyhedra and efficiency*, Vol. A. Paths, Flows, Matchings, Springer, Berlin, 2003.

The following article originally appeared as:

H.W. Kuhn, *The Hungarian Method for the Assignment Problem*, Naval Research Logistics Quarterly 2 (1955) 83–97.

THE HUNGARIAN METHOD FOR THE ASSIGNMENT PROBLEM[1]

H. W. Kuhn
Bryn Mawr College

Assuming that numerical scores are available for the performance of each of n persons on each of n jobs, the "assignment problem" is the quest for an assignment of persons to jobs so that the sum of the n scores so obtained is as large as possible. It is shown that ideas latent in the work of two Hungarian mathematicians may be exploited to yield a new method of solving this problem.

1. INTRODUCTION

Stated informally, the problem of personnel-assignment asks for the best assignment of a set of persons to a set of jobs, where the possible assignments are ranked by the total scores or ratings of the workers in the jobs to which they are assigned. Variations of this problem, both mathematical and non-mathematical, have a long history (see the Bibliography appended). However, recent interest in the question, when posed in the terms of linear programming, seems to stem from the independent work of Flood, J. Robinson, Votaw, and Orden. Flood's work [12], begun in 1949, regards the problem as the most "degenerate" case of the transportation problem. Robinson regarded it as a relative of the travelling salesman problem; her work is available only in the form of RAND Corporation memoranda. The problem was discussed from various points of view in the work of Votaw and Orden (see [9]) presented to the SCOOP Symposium on Linear Inequalities and Programming, June 14-16, 1951. The computational advantages to be gained by considering the problem in combination with the dual linear program have been stressed by Dantzig, von Neumann and others (see [8], [10], and [12]). The purpose of this paper is to develop a computational method that uses this duality in a particularly effective manner. One interesting aspect of the algorithm is the fact that it is latent in work of D. Kőnig and E. Egerváry that predates the birth of linear programming by more than 15 years (hence the name, the "Hungarian Method").

The theoretical basis of the algorithm is laid in Sections 2 and 3. Section 2 (which is derived from the proof of Kőnig in "Theorie der Graphen" (1936) Chelsea, 1950, pp. 232-233) treats the problem of assignment when there are but two ratings, 1 and 0, indicating that a worker is qualified or not. Section 3 (which is derived from the work of Egerváry in [3]) shows that the general problem of assignment can be reduced to this special case by a procedure that is computationally trivial.

The algorithm is given an independent (and self-contained) statement in Section 4 and Section 5 is devoted to a detailed example to illustrate its application.

2. THE SIMPLE ASSIGNMENT PROBLEM

The problem of Simple Assignment is illustrated by the following miniature example:

Four <u>individuals</u> (denoted by $i = 1, 2, 3, 4$) are available for four <u>jobs</u> (denoted by $j = 1, 2, 3, 4$). They qualify as follows:

[1]The preparation of this report was supported, in part, by the ONR Logistics Project, Department of Mathematics, Princeton University.

$$\text{Individual}\begin{cases}1\\2\\3\\4\end{cases}\text{qualifies for job(s)}\begin{cases}1,\ 2,\ \text{and}\ 3\\3\ \text{and}\ 4\\4\\4\end{cases}$$

This information can be presented effectively by a qualification matrix

$$Q = \begin{bmatrix}1&1&1&0\\0&0&1&1\\0&0&0&1\\0&0&0&1\end{bmatrix}$$

in which horizontal rows stand for individuals and vertical columns for jobs; a qualified individual is marked by a 1 and an unqualified individual by a 0. Then the Simple Assignment Problem asks:

> What is the largest number of jobs that can be assigned to qualified individuals (with not more than one job assigned to each individual)?

This may be stated abstractly in terms of the matrix Q:

> What is the largest number of 1's that can be chosen from Q with no two chosen from the same row or column?

It is clear that we can start an assignment by placing unassigned individuals in any unassigned jobs for which they qualify. Thus, we might assign individuals 1 and 2 to jobs 3 and 4, respectively; this information is entered in the matrix below by asterisks.

$$\begin{bmatrix}1&1&1^*&0\\0&0&1&1^*\\0&0&0&1\\0&0&0&1\end{bmatrix}$$

Since it is impossible to improve this assignment by placing an unassigned individual in an unassigned job for which he qualifies, this assignment is said to be complete. If an assignment is complete, it is natural to attempt an improvement by means of a transfer. For instance, the transfer:

Move individual 1 from job 3 to job 1
" 2 " " 4 " " 3,

results in the following incomplete assignment:

$$\begin{bmatrix}1^*&1&1&0\\0&0&1^*&1\\0&0&0&1\\0&0&0&1\end{bmatrix}.$$

Here we may assign either individual 3 or 4 to job 4 to complete the assignment. Either result, say

$$\begin{bmatrix}1^*&1&1&0\\0&0&1^*&1\\0&0&0&1^*\\0&0&0&1\end{bmatrix},$$

is optimal, since there all qualified pairs involve either individual 1 or job 3 or job 4, and hence four assignments would involve one of these twice. Thus, although there is a transfer possible in this optimal assignment (move 1 from job 1 to job 2), it leads to a complete assignment. The discussion to follow establishes that this situation holds in general, namely, that one can always construct an optimal assignment by a successsion of transfers followed by additional assignments until this is no longer possible.

Suppose n individuals ($i = 1, \ldots, n$) are available for n jobs ($j = 1, \ldots, n$) and that a qualification matrix $Q = (q_{ij})$ is given, where $q_{ij} = 1$ if individual i qualifies for job j and $q_{ij} = 0$ otherwise. If an assignment (not necessarily optimal) of certain qualified individuals to jobs is given, then the easiest way to improve it is to assign any unassigned individual to an unassigned job for which he qualifies. If this is possible, the given assignment is said to be incomplete; otherwise, it is complete. If the assignment is complete, then it is reasonable to attempt an improvement by means of a transfer. A transfer changes the assignment of r distinct individuals $i_1, \ldots, i_r$ employed in jobs $j_1, \ldots, j_r$. It moves i_1 into an unassigned job j_o and i_k into job j_{k-1} for $k = 2, \ldots, r$. All of the new assignments (i_k to j_{k-1}) are assumed to be qualified for $k = 1, \ldots, r$. It is convenient to call the result of leaving all assignments unchanged a transfer also. A useful notation for transfers that change some assignment is

$$\begin{matrix} & & i_1 & & i_2 & \cdots & i_{r-1} & & i_r \\ & \swarrow & & \swarrow & & & & \swarrow & \\ j_o & & j_1 & & j_2 & \cdots & j_{r-1} & & j_r \end{matrix} \ .$$

We shall call every (assigned) individual involved in such a transfer an essential individual and every job assigned to an inessential individual an essential job. Thus:

LEMMA 1. For a given assignment, if an individual is assigned to a job, then either the individual or the job is essential, and not both.

COROLLARY 1. For all assignments, the number of individuals assigned to jobs equals the number of essential individuals and jobs.

The motivation of the definition of essentiality is partially explained by the next two lemmas.

LEMMA 2. For a given assignment, if an individual is assigned to a job and qualifies for another, unassigned, job then the individual is essential.

PROOF: The transfer of the individual to the unassigned job establishes him as essential.

LEMMA 3. For a given assignment, if every transfer leaves a job assigned then the job is essential.

PROOF. Assume the job j to be inessential. Then some individual i_k is assigned to it and involved in a transfer that moves $i_1, i_2, \ldots, i_k$ in order. Symbolically,

$$\begin{matrix} & & i_1 & & i_2 & \cdots & i_{k-1} & & i_k \\ & \swarrow & & \swarrow & & & & \swarrow & \\ j_o & & j_1 & & j_2 & \cdots & j_{k-1} & & j \end{matrix} \ ,$$

and j is unassigned. This proves the lemma.

These lemmas, in combination, establish the key result:

THEOREM 1. For a given assignment, if every transfer leads to a complete assignment then, for every individual qualified for a job, either the individual or the job is essential, and possibly both.

PROOF. Let individual i be qualified for job j. If i is assigned to j then Lemma 1 asserts that one or the other is essential. If i is assigned to another job then j is unassigned and Lemma 2 asserts that the individual i is essential. If i is unassigned then every transfer leaves j assigned (otherwise the assignment is incomplete) and Lemma 3 asserts that j is essential. This proves the theorem.

Starting with any assignment (say, of one individual to a job for which he is qualified), either every transfer leads to a complete assignment or at least one more individual can be assigned after some transfer. Since at most n individuals can be assigned, this proves:

THEOREM 2. There is an assignment which is complete after every possible transfer.

The problem will now be viewed from another, dual, aspect. Consider a possible <u>budget</u> to account for the value of an individual assigned to a job for which he is qualifed. Such a budget will allot either one unit or nothing to each individual and to each job. A budget is said to be <u>adequate</u> if, for every individual qualified for a job, either the individual or the job is allotted one unit, and possibly both.

THEOREM 3. The total allotment of any adequate budget is not less than the largest number of jobs that can be assigned to qualified individuals.

PROOF. If the part of the adequate budget allotted to jobs assigned in an optimal assignment is counted, it is seen to be not less than the number of jobs assigned because these jobs are all assigned to qualified individuals. Since the total budget is not less than this amount, this proves the theorem.

Consider any assignment that is complete after every possible transfer (by Theorem 2, there are such) and consider the budget that allots one unit to each essential individual or job and zero otherwise. Theorem 1 asserts that this budget is adequate. Taking account of Corollary 1, we have proved:

THEOREM 4. There is an adequate budget and an assignment such that the total allotment of the budget <u>equals</u> the number of jobs assigned to qualified individuals.

Since Theorem 3 implies that the assignment of Theorem 4 is optimal, we have provided the following answer to the Simple Assignment Problem:

The largest number of jobs that can be assigned to qualified individuals is equal to the smallest total allotment of any adequate budget. Any assignment is optimal if and only if it is complete after every possible transfer.

3. THE GENERAL ASSIGNMENT PROBLEM

Suppose n individuals ($i = 1, \ldots, n$) are available for n jobs ($j = 1, \ldots, n$) and that a rating matrix $R = (r_{ij})$ is given, where the r_{ij} are positive integers, for all i and j. An assignment consists of the choice of one job j_i for each individual i such that no job is assigned to two different men. Thus, all of the jobs are assigned and an assignment is a permutation

$$\begin{pmatrix} 1 & 2 & \ldots & n \\ j_1 & j_2 & \ldots & j_n \end{pmatrix}$$

of the integers 1, 2, ..., n. The General Assignment Problem asks:

For which assignments is the sum

$$r_{1j_1} + r_{2j_2} + \ldots + r_{nj_n}$$

of the ratings largest?

The dual problem considers adequate budgets, that is, allotments of non-negative integral amounts of u_i to each individual and v_j to each job in such a manner that the sum of the allotments to the ith individual and the jth job is not less than his rating in that job. In symbols,

$$u_i + v_j \geqq r_{ij} \qquad (i, j = 1, \ldots, n). \tag{1}$$

The problem dual to the General Assignment Problem is then:

What is the smallest total allotment

$$u_1 + \ldots + u_n + v_1 + \ldots + v_n$$

possible for an adequate budget?

The following analogue of Theorem 3 is immediate.

THEOREM 5. The total allotment of any adequate budget is not less than the rating sum of any assignment.

PROOF. Since each individual and job occurs exactly once in an assignment the sum of the allotments to individuals and jobs in an assignment is exactly the total allotment. However, the budget is adequate and therefore this is not less than the sum of the ratings of the individuals in their assigned jobs. In symbols,

$$u_1 + v_{j_1} \geqq r_{1j_1}, \ldots, u_n + v_{j_n} \geqq r_{nj_n}$$

by the condition that the budget is adequate. Adding these inequalities, we have

$$u_1 + \dots + u_n + v_{j_1} + \dots + v_{j_n} \geq r_{1j_1} + \dots + r_{nj_n} .$$

However, the integers $j_1, \dots, j_n$ appearing in the assignment

$$\begin{pmatrix} 1 & 2 & \dots & n \\ j_1 & j_2 & \dots & j_n \end{pmatrix}$$

are merely an arrangement of $1, \dots, n$ and the theorem is proved.

It is an immediate consequence of this theory that, if an adequate budget and an assignment can be exhibited such that the total allotment equals the rating sum, then they must be simultaneously a solution of the assignment problem and its dual. We shall now show that this is always possible and can be achieved by solving certain, related, Simple Assignment Problems.

Associate with each adequate budget for the rating matrix $R = (r_{ij})$ a Simple Assignment Problem by the following rule:

The individual i is qualified for the job j if $u_i + v_j = r_{ij}$; otherwise, he is not qualified.

We see immediately that:

THEOREM 6. If all n individuals can be assigned to jobs for which they are qualified in the Simple Assignment Problem associated with an adequate budget, then the assignment and the budget solve the given General Assignment Problem and the rating sum equals the total allotment.

PROOF. For the given budget and assignment, we have

$$u_1 + v_{j_1} = r_{1j_1} \quad , \quad \dots \quad , \quad u_n + v_{j_n} = r_{nj_n}$$

Adding these equations,

$$u_1 + \dots + u_n + v_1 + \dots + v_n = r_{1j_1} + \dots + r_{nj_n}$$

and this proves the theorem.

If not all individuals can be assigned to jobs for which they are qualified in the Simple Assignment Problem associated with an adequate budget, then the budget can be improved by a simple procedure. Before this procedure can be described, it must be noted that an adequate budget must allot either a positive amount to every individual or a positive amount to every job since otherwise it would not be enough for the positive rating of some individual in some job. We shall assume, without loss of generality since rows and columns enter symmetrically, that every individual is allotted a positive amount; in symbols

$$u_i > 0 \qquad (i = 1, \dots, n) .$$

Assume that the largest number of individuals that can be assigned to jobs for which they are qualified is $m < n$. Choose an optimal assignment and let the essential individuals be $i = 1, \dots, r$

and the essential jobs be $j = 1, \ldots, s$ (possibly renumbering individuals and jobs). Corollary 1 asserts that

$$r + s = m .$$

Then the rule for changing the budget is:

$$u_1' = u_1, \ldots, u_r' = u_r, u_{r+1}' = u_{r+1} - 1, \ldots, u_n' = u_n - 1$$

$$v_1' = v_1 + 1, \ldots, v_s' = v_s' + 1, v_{s+1}' = v_{s+1}, \ldots, v_n' = v_n .$$

(The u_i' are still non-negative because the u_i were positive integers.) We must check that

(a) the new budget is adequate, and

(b) the total allotment has been decreased.

The adequacy is checked by inequalities (1) which can only fail where u_i has been decreased and v_j has been left unchanged. But this means that both the individual i and the job j are inessential. Theorem 1 then asserts that individual i is not qualified for job j and hence

$$u_i + v_j > r_{ij}$$

by the rule for constructing the associated Simple Assignment Problem. Since all the numbers involved are integers,

$$u_i' + v_j' = (u_i - 1) + v_j = (u_i + v_j) - 1 \geqq r_{ij}$$

and the new budget is adequate.

The total allotment has been decreased by $n - r$ and increased by s, thus has been decreased by $n - (r + s) = n - m > 0$. Summarizing:

THEOREM 7. If at most $m < n$ individuals can be assigned to jobs for which they are qualified in the Simple Assignment Problem associated with an adequate budget, then the total allotment of the budget can be decreased by a positive integral amount.

Starting with any adequate budget (say, that which allots to every individual his highest rating and nothing to the jobs), either it is optimal, and Theorem 6 applies, or it can be decreased by Theorem 7. Since it can be improved at most a finite number of times, we have provided the following answer to the General Assignment Problem:

The largest possible rating sum for any assignment is equal to the smallest total allotment of any adequate budget. It can be found by solving a finite sequence of associated Simple Assignment Problems.

4. THE HUNGARIAN METHOD

In this section we shall assemble the results of the two preceding sections, abstracted from the context of actual assignments, and state explicitly the algorithm implicit in the arguments of those sections. In certain cases where it seems advisable to use a different terminology, the discrepancy will be noted parenthetically.

As considered in this paper, the General Assignment Problems asks: Given an n by n matrix $R = (r_{ij})$ of positive integers, find the permutation $j_1, \ldots, j_n$ of the integers $1, \ldots, n$ that maximizes the sum $r_{1j_1} + \ldots + r_{nj_n}$. It is well known (see references [3] and [10] in the Bibliography) that the linear program dual to this problem can be stated: Find non-negative integers $u_1, \ldots, u_n$ and $v_1, \ldots, v_n$ subject to

$$u_i + v_j \geq r_{ij} \qquad (i, j = 1, \ldots, n) \tag{1}$$

that minimize the sum $u_1 + \ldots + u_n + v_1 + \ldots + v_n$. A set of non-negative integers satisfying (1) will be called a <u>cover</u> (or an <u>adequate budget</u>) and the positions (i, j) in the matrix for which equality holds are said to be <u>marked</u> (or <u>qualified</u> in the associated Simple Assignment Problem); otherwise (i, j) is said to be <u>blank</u>. A set of marks is called <u>independent</u> if no two marks from the set lie in the same line (the term "line" is used here to denote either a row or column). Then a fundamental result of König says: If the largest number of independent marks that can be chosen is m then m lines can be chosen that contain all of the marked positions. (This is precisely the conclusion of Section 1 with "jobs assigned to qualified individuals" playing the rôle of "independent marks.")

The algorithm to be described in this report is based on these remarks in the following manner. If a cover for R is given, a largest set of independent marks is found; if this set contains n marks then obviously the marked (i, j) constitute the desired assignment (Theorem 6). If the set contains less than n marks then a set of less than n lines containing all of the marked (i, j) is used to improve the cover (Theorem 7).

The construction of an initial cover and an initial set of independent marks can be made quite conveniently as follows:

Let $a_i = \max_j r_{ij}$ for $i = 1, \ldots, n$ and $b_j = \max_i r_{ij}$ for $j = 1, \ldots, n$. Further let $a = \Sigma_i a_i$ and $b = \Sigma_j b_j$.

$$\text{If } a \leqq b \text{ define } \begin{cases} u_i = a_i & \text{for } i = 1, \ldots, n \\ v_j = 0 & \text{for } j = 1, \ldots, n. \end{cases}$$

$$\text{If } a > b \text{ define } \begin{cases} u_i = 0 & \text{for } i = 1, \ldots, n \\ v_j = b_j & \text{for } j = 1, \ldots, n. \end{cases}$$

At this stage, as at all subsequent stages, there is associated with the matrix R and the cover $\{u_i, v_j\}$ a matrix $Q = (q_{ij})$ where

$$q_{ij} = \begin{cases} 1 & \text{if } u_i + v_j = r_{ij} \\ 0 & \text{otherwise.} \end{cases}$$

At each stage we shall also need a set of independent 1's from Q which will be distinguished by asterisks. To provide such a set at the first stage, in the first case ($a \leq b$) the rows are examined in order and the first 1 in each row without a 1* in its column is changed to a 1*. In the second case (a > b), the same instructions are followed with rows and columns exchanging rôles.

The two basic routines of the algorithm will be called Routine I and Routine II. A schematic description of the order of their repetition is given in Figure 1.

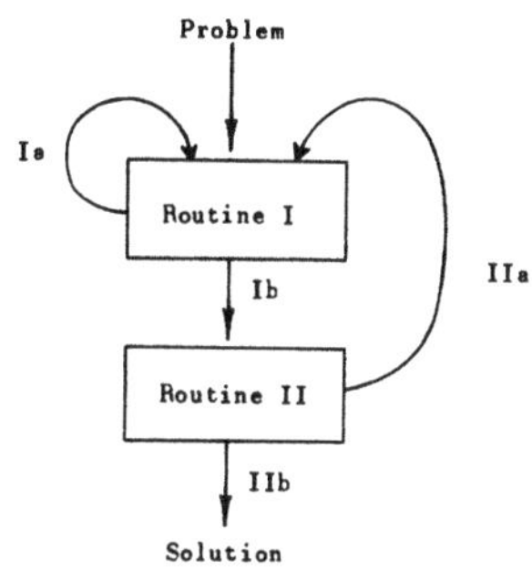

Figure 1

Every occurrence of Ia will increase the number of assignments (i.e., of asterisks in Q) by one and every occurrence of IIa will decrease the current covering sum $(\sum_i u_i + \sum_j v_j)$ by at least one. Since the number of assignments is bounded from above by n and the covering sums are bounded from below by zero, this insures the termination of the combined algorithm.

Routine I

Routine I works with a fixed matrix Q associated with a fixed cover $\{u_i, v_j\}$. The input also includes a certain set of asterisks marking 1's in Q.

The computation begins with the search of each column of Q in turn for a 1*. If a 1* is found, we proceed to the next column (no columns left = Alternative Ib). If a 1* is not found in the column, then the column is called eligible and is searched for a 1. If a 1 is not found, we proceed to the next column (no columns left = Alternative Ib). If a 1 is found in (i_1, j_o), we record i_1 and j_o and start a process that constructs a sequence of the following form:

$$
\begin{array}{ll}
1 & \text{in } (i_1, j_o) \\
1^* & \text{in } (i_1, j_1) \\
1 & \text{in } (i_2, j_1) \\
\ldots & \ldots \quad \ldots .
\end{array}
$$

The routine then divides into two cases according to the parity of the number of terms currently in the sequence. In Case 1, we have just found a 1 in (i_k, j_{k-1}) and have recorded i_k and j_{k-1}. We then search the row i_k for a 1*. If a 1* is not found then we change each 1 in the sequence to 1* and each 1* in the sequence (if any) to a 1. This is Alterhative Ia and means that we start

Routine I again. In Case 2, we have just found a 1^* in (i_k, j_k). We then search column j_k for a 1. If a 1 is not found, then row i_k is recorded as essential, i_k and j_{k-1} are deleted from the record and we go back to Case 2 with the last two terms of the sequence deleted and searching for a 1 in column j_{k-1} from row $i_k + 1$ on. Note that, if $k = 1$, then we go back to our preliminary search for a 1 in the eligible column j_o from row $i_1 + 1$ on. Completing Case 2, if a 1 is found in (i_{k+1}, j_k) we test whether i_{k+1} is distinct from $i_1, \ldots, i_k$. If it is distinct then we record i_{k+1} and j_k and are back in Case 1. If it is not distinct, we go on searching for a 1 in column j_k from row $i_{k+1} + 1$ on.

(This routine is connected with Section 2 in the following way. Given an assignment, we enumerate all possible transfers. Such a transfer starts at an eligible column. If there are no eligible columns, there are no transfers and the given assignment is complete. The occurrence of Alternative Ia means that we have found a transfer that frees a column that contains a 1 that is unassigned. In this event, we carry out the transfer:

$$\begin{array}{ccccccc} & i_1 & i_2 & \cdots & i_{k-2} & i_{k-1} \\ \swarrow & \swarrow & & & \swarrow & \\ j_o & j_1 & j_2 & \cdots & j_{k-2} & j_{k-1} \end{array}$$

and assign (i_k, j_{k-1}). If a transfer is developed that cannot be continued and which yields a complete assignment, the last row involved is recorded as essential, following which the enumeration of the transfers is continued. If the enumeration of the transfers is completed without the occurrence of Alternative Ia, this is Alternative Ib and we have an assignment in which all transfers yield complete assignments.)

The output of Routine I in Alternative Ib is an optimal assignment for Q and a set of essential rows. Every 1 lies either in an essential row or in the column of a 1^* in an essential row (Theorem 1).

A tentative flow diagram for Routine I is given in Figure 2. For this arrangement of the routine, we use the following notation:

Symbol	Use in Routine
i	Index of rows of Q.
j	Index of columns of Q.
k	Tally of length of sequence of 1's and 1^*'s. Tally to clear essential rows in Alternative Ia.
ℓ	Tally to test distinctness of i_{k+1} from $i_1, \ldots, i_k$.
$i_1, i_2, \ldots, i_n$	Record of rows in sequence of 1's and 1^*'s.
$j_o, j_1, \ldots, j_{n-1}$	Record of columns in sequence of 1's and 1^*'s.
$\epsilon_1, \epsilon_2, \ldots, \epsilon_n$	Record of essential rows.

The values of these quantities for the input of Routine I are:

$$i = j = k = \ell = 1, \quad i_\nu = \epsilon_\nu = 0 \text{ for } \nu = 1, \ldots, n.$$

The values of these quantities for the output of Alternative Ib are:

$$i = j = k = \ell = 1, \quad i_\nu = j_{\nu-1} = 0 \text{ for } \nu = 1, \ldots, n.$$

and

$$\epsilon_i = \begin{cases} 1 \\ 0 \end{cases} \text{ if row } i \text{ is } \begin{cases} \text{essential} \\ \text{inessential.} \end{cases}$$

The symbol "A ⟶ B" is to be read "replace the value of A by the value of B".

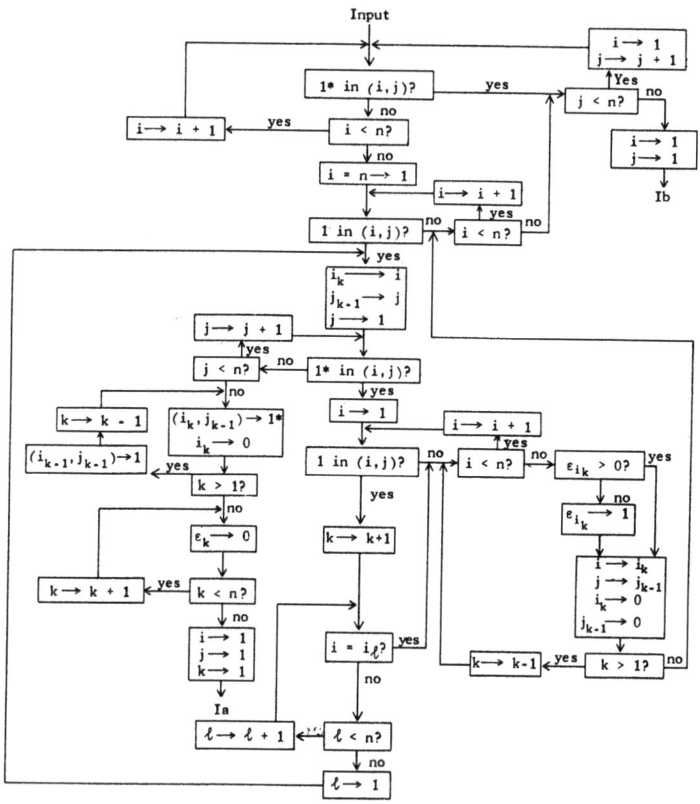

Figure 2

Routine II

The input of Routine II consists of a cover $\{u_i, v_j\}$ and a set of essential rows and columns (a column is essential if it contains a 1^* in an inessential row). We first compute d, the minimum of $u_i + v_j - r_{ij}$ taken over all inessential rows i and columns j. If there are no such (i, j) then the set of 1^* in Q constitutes a solution to the General Assignment Problem (Theorem 6). Otherwise, $d > 0$ and there are two mutually exclusive cases to be considered.

Case 1. For all inessential rows i, $u_i > 0$. Compute m, the minimum of d and u_i taken over all inessential i. Then

$$u_i \longrightarrow u_i - m \text{ for all inessential rows i, and}$$

$$v_j \longrightarrow v_j + m \text{ for all essential columns j.}$$

Case 2. For some inessential row i, $u_i = 0$. Compute m, the minimum of d and v_j taken over all inessential j. Then

$$u_i \longrightarrow u_i + m \text{ for all essential rows i, and}$$

$$v_j \longrightarrow v_j - m \text{ for all inessential columns j.}$$

After these changes have been made in the cover, we are in Alternative IIa and should return to Routine I.

5. AN EXAMPLE

The following example, although small in size, illustrates all of the possibilities of the routines (except Case 2 of Routine II):

$$R = \begin{bmatrix} 8 & 7 & 9 & 9 \\ 5 & 2 & 7 & 8 \\ 6 & 1 & 4 & 9 \\ 2 & 3 & 2 & 6 \end{bmatrix}$$

Sum of row maxima $= 9 + 8 + 9 + 6 = 32$.
Sum of column maxima $= 8 + 7 + 9 + 9 = 33$.
Hence, the initial cover is provided by the row maxima. The next table shows the successive covers obtained from the algorithm (reading out from the matrix):

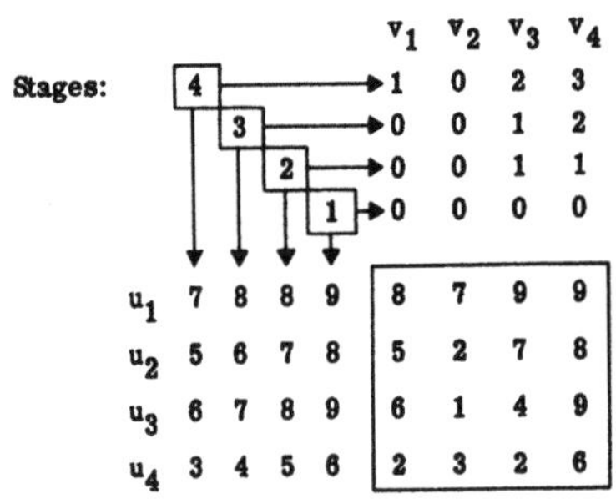

The following tables explain the construction of the successive covers and of the corresponding assignments:

Stage 1. Remarks

		1	1
			1
			1
			1

This matrix marks (with 1) those positions for which $u_i + v_j = r_{ij}$ in the first cover.

		1*	1
			1*
			1
			1

Assign in each row the first 1, if any, not in the column of a previous assignment. Assignments are marked by asterisks. No transfers are possible and hence all assigned columns and no assigned rows are essential.

	0	0		
9	8	7		
8	5	2		
9	6	1		
6	2	3		

Thus, the algorithm decreases <u>all</u> u_i and increases v_3 and v_4 by the minimum of $u_i + v_j - r_{ij}$ on the part of the matrix shown at left. The second cover is:

$$u_1 = 8,\ u_2 = 7,\ u_3 = 8,\ u_4 = 5 \text{ and } v_1 = v_2 = 0,\ v_3 = v_4 = 1.$$

Stage 2.

1 ←	←	1*	1
			1*
			1
			1

The change in the cover has introduced a new 1 at (1,1) and there is one possible transfer, indicated by an arrow. Thus, row 1 and column 4 are essential.

	0	0	1	
7	5	2	7	
8	6	1	4	
5	2	3	2	

Thus, the algorithm decreases u_2, u_3, and u_4 and increases v_4 by the minimum of $u_i + v_j - r_{ij}$ on the part of the matrix shown at left. The third cover is:

$$u_1 = 8,\ u_2 = 6,\ u_3 = 7,\ u_4 = 4 \text{ and } v_1 = v_2 = 0,\ v_3 = 1,\ v_4 = 2.$$

Stage 3.

1 ←	←	1*	
		1 ←	1*
			1
			1

The change in the cover has introduced a new 1 at (2,3) and eliminated the 1 at (1,4). The possible transfers are indicated by arrows.

1*		1	
		1*	1
			1*
			1

The transfer $1 \overset{1}{\swarrow} 3 \overset{2}{\swarrow} 4$ leads to an incomplete assignment (column 4 is unassigned and (3,4) is qualified). The matrix at left completes it. All assigned columns and no assigned rows are essential because there are no transfers.

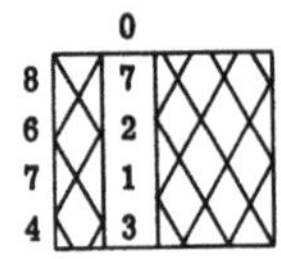

Thus, the algorithm decreases all u_i and increases v_1, v_3, and v_4 by the minimum of $u_i + v_j - r_{ij}$ on the part of the matrix shown at left. The fourth cover is:

$$u_1 = 7,\ u_2 = 5,\ u_3 = 6,\ u_4 = 3 \text{ and } v_1 = 1,\ v_2 = 0,\ v_3 = 2,\ v_4 = 3.$$

Stage 4.

1*	1	1	
		1*	1
			1*
	1		1

The change in the cover has introduced new 1's at (1,2) and (4,2). Thus the assignment is incomplete and is completed by assigning (4,2).

1*	1	1	
		1*	1
			1*
	1*		1

The assignment shown is optimal.

Check: $u_i + v_j \geq r_{ij}$ for all i, j.

$$r_{11} + r_{23} + r_{34} + r_{42} = 8 + 7 + 9 + 3 = 27.$$

$$u_1 + \ldots + u_4 + v_1 + \ldots + v_4 = 7 + 5 + 6 + 3 + 1 + 0 + 2 + 3 = 27.$$

BIBLIOGRAPHY

[1] König, D., "Über Graphen und ihre Anwendung auf Determinantentheorie und Mengenlehre." Math. Ann. 77 (1916) 453-465.

[2] Frobenius, G., "Über zerlegbare Determinanten," Sitzungsber., Preuss. Akad. Wiss. (1917) 274-277.

[3] Egerváry, J., "Matrixok kombinatorius tulajdonságairól." Mat. Fiz. Lapok (1931) 16-28 (translated as "Combinatorial Properties of Matrices" by H. W. Kuhn, ONR Logistics Project, Princeton (1953), mimeographed).

[4] Hall, P., "On Representatives of Subsets," J. London Math. Soc. 10 (1935) 26-30.

[5] Easterfield, T. E., "A Combinatorial Algorithm," J. London Math. Soc. 21 (1946) 219-226.

[6] Birkhoff, Garrett, "Tres Observaciones sobre el Algebra Lineal," Univ. Nac. Tucumán. Revista A5 (1946) 147-151.

[7] Thorndike, R. L., "The Problem of Classification of Personnel," Psychometrika 15 (1950) 215-235.

[8] Dantzig, G. B., "Application of the Simplex Method to a Transportation Problem," Chapter XXIII in Activity Analysis of Production and Allocation, Cowles Commission Monograph No. 13, ed. T. C. Koopmans, New York, 1951.

[9] Votaw, D. F. and Orden, A., "The Personnel Assignment Problem," Symposium on Linear Inequalities and Programming, SCOOP 10, USAF (1952) 155-163.

[10] Neumann, J. von., "A Certain Zero-sum Two-Person Game Equivalent to the Optimal Assignment Problem," Contributions to the Theory of Games II, Ann. Math. Study 28 (1953) 5-12.

[11] Dwyer, P. S., "Solution of the Personnel Classification Problem with the Method of Optimal Regions," Psychometrika 19 (1954) 11-26.

[12] Flood, M. M., "On the Hitchcock Distribution Problem," Pacific J. Math. 3 (1953) 369-386.

[13] Motzkin, T. S., "The Assignment Problem" in Proc. of Sixth Symposium on Applied Mathematics, to appear.

* * *

Chapter 3
Integral Boundary Points of Convex Polyhedra

Alan J. Hoffman and Joseph B. Kruskal

Introduction by *Alan J. Hoffman* and *Joseph B. Kruskal*

Here is the story of how this paper was written.

(a) Independently, Alan and Joe discovered this easy theorem: if the "right hand side" consists of integers, and if the matrix is "totally unimodular", then the vertices of the polyhedron defined by the linear inequalities will all be integral. This is easy to prove and useful. As far as we know, this is the only part of our theorem that anyone has ever used.

(b) But this was so easy, we each wanted to generalize it. Independently we worked hard to understand the cases where there are no vertices, i.e., the lowest dimensional faces of the polyhedron are 1-dimensional or higher. This was hard to write and hard to read.

(c) At this point, Alan benefitted greatly from simplifications suggested by David Gale and anonymous referees, but it was still not so simple.

(d) Independently, we both wondered: If the vertices were integral for every integral right hand side, did this mean the matrix was totally unimodular? This is discussed in our paper, and also in References [1] and [2], especially the latter.

Harold Kuhn and Al Tucker saw drafts from both of us and realized that we were working on the same problem, so they suggested that we start working together, and that Alan should send his latest draft to Joe to take the next step. For Joe, this turned out to be the most exciting collaboration he had ever experienced; and he still feels that way today.

Alan J. Hoffman
IBM Research, Yorktown Heights, New York, USA
e-mail: ajh@us.ibm.com

Joseph B. Kruskal
Bell Laboratories, Murray Hill, New Jersey, USA
e-mail: jkruskal@comcast.net

M. Jünger et al. (eds.), *50 Years of Integer Programming 1958-2008*,
DOI 10.1007/978-3-540-68279-0_3, © Springer-Verlag Berlin Heidelberg 2010

We must have met casually before the collaboration, but we never saw each other during it nor for a long time afterwards. Joe knew nothing about Alan's work. However Alan, who was working for the Navy, knew something about Joe's work through a Navy report on a real operations research project on which Joe and Bob Aumann had gotten impressive results. (Many years later, in 2005, Bob won the Nobel Prize for Economics.)

As it turned out, Joe merged our two papers—but did much more; Alan's ideas were very stimulating. When Alan got that version, he was also stimulated and made substantial improvements. Then Joe made further improvements, and finally Alan did the same. We could probably have made much more progress, but the deadline for publication cut off further work.

(e) One of our discoveries when collaborating was a new general class of totally unimodular matrices ... but several years later we were chagrined to learn from Jack Edmonds that in the 1800's Gustav Kirchoff (who was the inventor of Kirchoff's Laws) had constructed a class of totally unimodular matrices of which ours was only a special case.

(f) The term "totally unimodular" is due to Claude Berge, and far superior to our wishywashy phrase "matrices with the unimodular property". Claude had a flair for language.

(g) We had no thought about computational questions, practical or theoretical, that could be influenced by our work. We also did not imagine the host of interesting concepts, like total dual integrality, lattice polyhedral, etc. that would emerge, extending our idea. And we never dreamed that totally unimodular matrices could be completely described, see [3], because we didn't anticipate that a mathematician with the great talent of Paul Seymour would get interested in these concepts.

After we wrote the paper, we met once in a while (a theater in London, a meeting in Washington), but our interests diverged and we never got together again professionally. The last time we met was almost 15 years ago, when Vašek Chvátal organized at Rutgers a surprise 70th birthday party cum symposium for Alan. Joe spoke about this paper and read some of the letters we wrote each other, including reciprocal requests that each of us made begging the partner to forgive his stupidity.

References

1. A.J. Hoffman, *Some recent applications of the theory of linear inequalities to extremal combinatorial analysis*, Proceedings Symposium Applied Mathematics 10, American Mathematical Society, 1960, pp. 113–128.
2. A.J. Hoffman, *Total unimodularity and combinatorial theorems*, Linear Algebra and its Applications 13 (1976) 103–108.
3. P.D. Seymour, *Decomposition of regular matroids*, Journal of Combinatorial Theory B 29 (1980) 305–359.

The following article originally appeared as:

A.J. Hoffman and J.B. Kruskal, *Integral Boundary Points of Convex Polyhedra*, Linear Inequalities and Related Systems (H.W. Kuhn and A.J. Tucker, eds.), Princeton University Press, 1956, pp. 223–246.

INTEGRAL BOUNDARY POINTS OF CONVEX POLYHEDRA

A. J. Hoffman and J. B. Kruskal

INTRODUCTION

Suppose every vertex of a (convex) polyhedron in n-space has (all) integral coordinates. Then this polyhedron has the integral property (i.p.). Sections 1, 2, and 3 of this paper are concerned with such polyhedra.

Define two polyhedra[1]:

$$P(b) = \{x \mid Ax \geqq b\} ,$$

$$Q(b, c) = \{x \mid Ax \geqq b, x \geqq c\} ,$$

where A, b, and c are integral and A is fixed. Theorem 1 states that $P(b)$ has the i.p. for every (integral) b if and only if the minors of A satisfy certain conditions. Theorem 2 states that $Q(b, c)$ has the i.p. for every (integral) b and c if and only if every minor of A equals 0, $+1$, or -1. Section 1 contains the exact statement of Theorems 1 and 2, and Sections 2 and 3 contain proofs.

A matrix A is said to have the unimodular property (u.p.) if it satisfies the condition of Theorem 2, namely if every minor determinant equals 0, $+1$, or -1. In Section 4 we give Theorem 3, a simple sufficient condition for a matrix to have the u.p. which is interesting in itself and necessary to the proof of Theorem 4. In Section 5 we state and prove — at length — Theorem 4, a very general sufficient condition for a matrix to have the u.p. Finally, in Section 6 we discuss how to recognize the unimodular property, and give two theorems, based on Theorem 4, for this purpose.

Our results include all situations known to the authors in which the polyhedron has the integral property independently of the "right-hand sides" of the inequalities (given that the "right-hand sides" are integral of course). In particular, the well-known "integrality" of transportation

[1] Unless otherwise stated, we assume throughout this paper that the inequalities defining polyhedra are consistent.

type linear programs and their duals follows immediately from Theorems 2 and 4 as a special case.

1. DEFINITIONS AND THEOREMS

A point of n-space is an <u>integral point</u> if every coordinate is an integer. A (convex) polyhedron in n-space is said to have the <u>integral property</u> (i.p.) if <u>every</u> face (of every dimension) contains an integral point. Of course, this is true if and only if every <u>minimal</u> face contains an integral point. If the minimal faces happen to be vertices[2] (that is, of dimension 0), then the integral property simply means that the vertices of P are themselves all integral points.

Let A be an m by n matrix of integers; let b and b' be m-tuples (vectors), and c and c' be n-tuples (vectors), whose components are integers or $\pm\infty$. We will let $\infty(-\infty)$ also represent a vector all of whose components are $\infty(-\infty)$; this should cause no confusion. The vector inequality $b < b'$ means that strict inequality holds at every component. Let $P(b; b')$ and $Q(b; b'; c; c')$ be the polyhedra in n-space defined by

$$P(b; b') = \{x \mid b \leqq Ax \leqq b'\} ,$$

$$Q(b; b'; c; c') = \{x \mid b \leqq Ax \leqq b' \quad \text{and} \quad c \leqq x \leqq c'\} .$$

Of course $Q(b; b'; -\infty, +\infty) = P(b; b')$. If S is any set of rows of A, then define

$$\gcd(S) = \begin{cases} 0, & \text{if each minor determinant in } S \text{ which has as many rows as } S \text{ equals } 0, \\ \text{greatest common divisor (g.c.d.) of all those minor determinants in } S \text{ which have as many rows as } S, & \text{otherwise.} \end{cases}$$

THEOREM 1. The following conditions are equivalent:

(1.1) $P(b; b')$ has the i.p. for every b, b';

(1.2) $P(b; \infty)$ has the i.p. for every b;

(1.2') $P(-\infty; b')$ has the i.p. for every b';

(1.3) if r is the rank of A, then for every set S of r linearly independent rows of A, $\gcd(S) = 1$;

[2] It is well known (and, incidentally, is a by-product of our Lemma 1) that all minimal faces of a convex polyhedron have the same dimension.

(1.4) for every set S of rows of A, $\gcd(S) = 1$ or 0.

The main value of this theorem lies in the fact that condition (1.3) implies condition (1.1). However the converse implication is of esthetic interest. If it is believed that (1.3) does not hold, (1.4) often offers the easiest way to verify this, for it may suffice to examine small sets of rows.

A matrix (of integers) is said to have the unimodular property (u.p.) if every minor determinant equals 0, $+1$, or -1. We see immediately that the entries in a matrix with the u.p. can only be 0, $+1$, or -1.

THEOREM 2. The following conditions are equivalent:

(1.5) $Q(b; b'; c; c')$ has the i.p. for every b, b', c, c';

(1.6) for some fixed c such that $-\infty < c < +\infty$, $Q(b, \infty; c; \infty)$ has the i.p. for every b;

(1.6') for some fixed c such that $-\infty < c < \infty$, $Q(-\infty; b'; c; \infty)$ has the i.p. for every b';

(1.6'') for some fixed c' such that $-\infty < c' < \infty$, $Q(b; \infty; -\infty; c')$ has the i.p. for every b;

(1.6''') for some fixed c' such that $-\infty < c' < \infty$, $Q(-\infty; b'; -\infty; c')$ has the i.p. for every b';

(1.7) the matrix A has the unimodular property (u.p.).

The main value of this theorem for applications lies in the fact that condition (1.7) implies condition (1.5), a fact which can be proved directly (with the aid of Cramer's rule) without difficulty. However the converse implication is also of esthetic interest. The relationship between Theorems 1 and 2 is that Theorem 2 asserts the equivalence of stronger properties while Theorem 1 asserts the equivalence of weaker ones. Condition (1.5) is clearly stronger than condition (1.1), and condition (1.7) is clearly stronger than condition (1.3).

For A to have the unimodular property is the same thing as for A transpose to have the unimodular property. Therefore if a linear program has the matrix A with the u.p., both the "primal" and the dual programs lead to polyhedra with the i.p. This can be very valuable when applying the duality theorem to combinatorial problems (for examples, see several other papers in this volume).

2. PROOF OF THEOREM 1

We note that (1.1) $\Longrightarrow$ (1.2) and (1.2') trivially. Likewise (1.4) $\Longrightarrow$ (1.3) trivially. To see that (1.3) $\Longrightarrow$ (1.4), let S and S' be sets of rows of A. If $S \subset S'$, then the relevant determinants of S' are integral combinations of the relevant determinants of S. Hence gcd(S) divides gcd(S'). From this we easily see that (1.3) $\Longrightarrow$ (1.4).

As (1.2) and (1.2') are completely parallel, we shall only treat the former in our proofs.

Let the rows of A be $A_1, \dots, A_m$ and the components of b and b' be $b_1, \dots, b_m$ and $b_1', \dots, b_m'$. Suppose that we know that (1.3) for any matrix A_* implies (1.2) for the corresponding polyhedra $P_*(b; \infty)$. Also, suppose that (1.3) holds for the particular matrix A. Then setting

$$A_* = \begin{bmatrix} A \\ \\ -A \end{bmatrix},$$

we see immediately that (1.3) holds for A_*. Consequently

$$P_*(b_1, \dots, b_m, -b_1', \dots, -b_m'; \infty)$$

has the i.p. But it is easy to see that this polyhedron is identical with P(b; b'); hence the latter also has the i.p. Therefore if for every matrix (1.3) implies (1.2), then (1.3) implies (1.1) for every matrix.

Let $P(b) = P(b; \infty)$ for convenience.

It only remains to prove that (1.2) is equivalent to (1.3)[3]. If S is any set of rows A_i of A, we define

$$F_S = F_S(b) = \{x \mid Ax \geqq b \text{ and } A_i x = b_i \text{ if } A_i \text{ in } S\},$$

$$G_S = \text{the subspace of n-space spanned by the rows } A_i \text{ in } S.$$

If $F_S(b)$ is not empty, it is the face of P(b) corresponding to S. (We do not consider the empty set to be a face of a polyhedron.) We easily see that $F_S(b)$, if non-empty, corresponds to the usual notion of a face. Of course $F_\emptyset(b) = P(b)$, where $\emptyset$ is the empty set. We shall use the letter A to stand for the set of all rows of the matrix A. In general we will use the same letter to denote a set of rows and to denote the matrix formed by these rows. (This double meaning should cause no confusion.)

[3] The authors are indebted to Professor David Gale for this proof, which is much simpler than the original proof.

LEMMA 1. If $S \subset S'$, and if $F_S(b)$ and $F_{S'}(b)$ are faces (that is, not empty), then $F_{S'}(b)$ is a subface of $F_S(b)$. If $F_S(b)$ is a face, then it is a minimal face if and only if $G_S = G_A$, that is, if and only if S has rank r, where r is the rank of A.

PROOF. The first sentence of the lemma follows directly from the definitions. To prove the rest of the lemma, let S' be all rows of A which are in G_S. Then $G_S = G_{S'}$, and A_j is a linear combination of the A_i in S if and only if A_j is in S'. Clearly $G_S = G_A$ if and only if $S' = A$.

If $S' \neq A$, there is at least one row A_k in $A - S'$. Then there is a vector y such that $A_i y = 0$ for A_i in S, $A_k y < 0$. Let x be in F_S. As $A_k x \geqq b_k$, there is a number $\lambda_k \geqq 0$ for which $A_k(x + \lambda_k y) = b_k$. For every A_j in $A - S'$ such that $A_j y < 0$, the equation $A_j(x + \lambda y) = b_j$ has a non-negative solution. Let λ_j be that solution. Define $\lambda = \text{minimum } \lambda_j$, and let j' be a value such that $\lambda = \lambda_{j'}$. As λ_k exists, there is at least one λ_j, so λ exists. By the definition of λ,

$$A(x + \lambda y) \geqq b \;,$$
$$A_i(x + \lambda y) = b_i \qquad \text{for } A_i \text{ in } S,$$
$$A_{j'}(x + \lambda y) = b_{j'} \;.$$

Thus $F_{S \cup A_{j'}}$ is not empty, and is therefore a subface of F_S. Furthermore as $A_{j'}$ is not a linear combination of the A_i in S, $F_{S \cup A_{j'}}$ is a proper subface of F_S. Therefore F_S is not minimal.

On the other hand, if F_S is not minimal it has some proper subface $F_{S \cup A_k}$. Then there must be x_1 and x_2 in F_S such that $A_k x_1 = b_k$ and $A_k x_2 > b_k$. Therefore $A_k x$ varies as x ranges over F_S. But for A_i in S, $A_i x = b_i$ is constant as x varies over F_S. Hence A_k cannot be a linear combination of the A_i in S, so A_k is in $A - S'$. Hence $S' \neq A$. This proves the lemma.

If b, as usual, is an m-tuple and S is a set of r rows of A, then b_S is the "sub-vector" consisting of the r components of b which correspond to the rows of S. Let $\tilde{b}$ always represent an (integral) r-tuple. The components of $\tilde{b}$ and b_S will be indexed by the indices used for the rows of S, not by the integers from 1 to r. Let

$$L_S(\tilde{b}) = \{x \mid Sx = \tilde{b}\} \;.$$

LEMMA 2. Suppose S is a set of r linearly independent rows of A. Then for any $\tilde{b}$ there is a b such that

(2.1) $$b_S = \tilde{b} ;$$

(2.2) $F_S(b)$ is a minimal face of $P(b)$.

PROOF. As S is a set of linearly independent rows, the equation $Sx = \tilde{b}$ has at least one solution: call it y. Define b as follows:

$$b_i = \begin{cases} \tilde{b}_i & \text{if } A_i \text{ in } S, \\ [A_i y] & \text{if } A_i \text{ not in } S . \end{cases}$$

Clearly $b_S = \tilde{b}$, so (2.1) is satisfied. Obviously b is integral. Furthermore y is seen to be in $F_S(b)$, so $F_S(b)$ is not empty, and hence is a face of $P(b)$. By Lemma 1, $F_S(b)$ is a minimal face, so (2.2) is satisfied.

LEMMA 3. Suppose S' is a set of rows of A of rank r, and $S \subset S'$ is a set of r linearly independent rows. For any b such that $F_{S'}(b)$ is a face (that is, not empty),

$$F_{S'}(b) = L_S(b_S).$$

PROOF. Let y be a fixed element in $F_{S'}(b)$, and let x be any element of $L_S(b)$. As $F_{S'}(b) \subset L_S(b_S)$ is trivial, we only need show the reverse inclusion. Thus it suffices to prove that x is in $F_{S'}(b)$.

As S has rank r, any row A_k in A can be expressed as a linear combination of the rows A_i in S:

$$A_k = \Sigma \alpha_{ki} A_i .$$

Then

$$A_i x = b_i = A_i y$$

for A_i in S, so

$$A_k x = \Sigma \alpha_{ki} A_i x = \Sigma \alpha_{ki} A_i y = A_k y .$$

Then as y is in $F_{S'}(b)$, x must be also. This completes the proof of the lemma.

LEMMA 4. Any minimal face of $P(b)$ can be expressed in the form $F_S(b)$ where S is a set of r linearly independent rows of A.

PROOF. Suppose the face is $F_{S'}(b)$. By Lemma 1, S' must have rank r. Let S be a set of r linearly independent rows of S'. Then by applying Lemma 3 to both $F_{S'}(b)$ and $F_S(b)$, we see that

$$F_{S'}(b) = L_S(b_S) = F_S(b).$$

This proves the lemma.

LEMMA 5. If S is a set of r linearly independent rows of A, then the following two conditions are equivalent:

(2.3) $L_S(\tilde{b})$ contains an integral point for every (integral) $\tilde{b}$;

(2.4) $$\gcd(S) = 1.$$

PROOF. We use a basic theorem of linear algebra, namely that any integral matrix S which is r by n can be put into the form

$$S = UDV$$

where D is a (non-negative integral) diagonal matrix, and U and V are (integral) unimodular matrices. (Of course U is r by r, V is n by n, and D is r by n.) As U and V are unimodular, they have integral inverses. Furthermore $\gcd(S) = \gcd(D)$. (For proofs of these facts, see for example [3].)

Let the diagonal elements of D be d_{ii}. Clearly $\gcd(D) = d_{11}d_{22} \cdots d_{rr}$. Therefore condition (2.4) is equivalent to the condition that every $d_{ii} = 1$. Now we show that (2.3) is also equivalent to this same condition.

Suppose that some diagonal element of D is greater than 1. For convenience we may suppose that this element is $d_{11} = k > 1$. Let $\tilde{e}$ be the r-tuple $(1, 0, \dots, 0)$, and let $\tilde{b} = U\tilde{e}$. Then $L_S(\tilde{b})$ contains no integral point. To see this, let x be in $L_S(\tilde{b})$. Then

$$Sx = UDVx = \tilde{b} = U\tilde{e} ,$$

so $DVx = \tilde{e}$. Clearly the first component of $y = Vx$ is $1/k$, so y is not integral. Hence x cannot be integral. This shows that (2.3) cannot hold if some d_{ii} is greater than 1.

Suppose every $d_{ii} = 1$. Let x be in $L_S(\tilde{b})$ and set

$$Vx = (y_1, \ldots, y_r, y_{r+1}, \ldots, y_n).$$

Then

$$U^{-1}\tilde{b} = DVx = (y_1, \ldots, y_r),$$

and so $y_1, \ldots, y_r$ are integral. Let $y = (y_1, \ldots, y_r, 0, \ldots, 0)$. Then $V^{-1}y$ is integral, and since $Dy = DVx$,

$$S(V^{-1}y) = UDV(V^{-1}y) = UDy = UDVx = \tilde{b} .$$

Thus $V^{-1}y$ is in $L_S(\tilde{b})$. This shows that (2.3) does hold if every $d_{ii} = 1$, and completes the proof of the lemma.

Now it is easy to prove that (1.2) $\Longleftrightarrow$ (1.3). First we prove $\Longrightarrow$. Let S be any set of r linearly independent rows of A. Let $\tilde{b}$ be any (integral) r-tuple. Choose a b which satisfies (2.1) and (2.2). By (1.2), $F_S(b)$ must contain an integral point x. By Lemma 3 and (2.1),

$$F_S(b) = L_S(b_S) = L_S(\tilde{b}) .$$

Hence $L_S(\tilde{b})$ contains x. Therefore (2.3) is satisfied, so by Lemma 5, $\gcd(S) = 1$. This proves $\Longrightarrow$.

To prove $\Longleftarrow$, let $F_{S'}(b)$ be some minimal face of $P(b)$. By Lemma 4 this face can be expressed as $F_S(b)$ where S consists of r linearly independent rows of A. By Lemma 3, $F_S(b) = L_S(b_S)$. By (1.3), $\gcd(S) = 1$, and by Lemma 5 $L_S(b_S)$ must contain an integral point x. Hence $F_{S'}(b)$ contains the integral point x. Therefore every minimal face of $P(b)$ contains an integral point, and hence also every face. This proves $\Longleftarrow$, and completes the proof of Theorem 1.

3. PROOF OF THEOREM 2

The role of (1.6) and its primed analogues are exactly similar, so we treat only the former in our proofs. For convenience we let

$$Q(b; c) = Q(b; \infty; c; \infty).$$

It is not hard to see that (1.7) $\Longrightarrow$ (1.5). For suppose that A has the u.p. (that is, satisfies (1.7)). Then

$$A_* = \begin{bmatrix} A \\ -A \\ I \\ -I \end{bmatrix}$$

satisfies (1.3). By Theorem 1, the associated polyhedron

$$P_*(b_1, \ldots, b_m, -b'_1, \ldots, -b'_m, c_1, \ldots, c_n, -c'_1, \ldots, -c'_n)$$

has the i.p. But it is easy to see that this polyhedron is identical with $Q(b; b'; c; c')$. Therefore the latter has the i.p., so (1.7) $\Longrightarrow$ (1.5). (An alternate proof of this can easily be constructed using Cramer's Rule.)

Clearly (1.5) $\Longrightarrow$ (1.6). Hence it only remains to prove that (1.6) $\Longrightarrow$ (1.7). We shall prove[4] this by applying Theorem 1 to the matrix

$$A^* = \begin{bmatrix} I \\ A \end{bmatrix}.$$

Let d be any (integral) (n+m)-tuple, and let

$$c \cup b = (c_1, \ldots, c_n, b_1, \ldots, b_m) .$$

Then $P^*(c \cup b) = Q(b, c)$.

To verify condition (1.2) for A^*, we need to show that $P^*(d)$ has the i.p. for every d. Condition (1.6) yields only the fact that $P^*(d)$ has the i.p. for every d such that $d_I = c$. To fill this gap, note that A^* has rank n as it contains the n by n identity matrix, and let $F^*_{S'}(d)$ be any face of $P^*(d)$. This face contains some minimal face, which by Lemma 4 can be expressed as $F^*_S(d)$ where S consists of n linearly independent rows of A^*. By Lemma 3,

$$F^*_S(d) = L^*_S(d_S) \equiv \{x \mid Sx = d_S\} .$$

As S is an n by n matrix of rank n, $F^*_S(d)$ consists only of a single point. Call this point x. We shall show that x is integral.

Let I_1 be the rows of I in S, I_2 the rows of I not in S, A_1 the rows of A in S, and A_2 the rows of A not in S. We wish to pick an integral vector q such that

[4] The authors are indebted to Professor David Gale for this proof, which is much simpler than the original proof.

(3.1) $$x + q \geqq c,$$

(3.2) $$(x + q)_{I_1} = c_{I_1} .$$

Let $q = c - d_I$. Then q satisfies these requirements, for

$$x_i + q_i \left\{ \begin{matrix} = \\ \geqq \end{matrix} \right\} d_i + (c_i - d_i) = c_i \begin{cases} \text{if the i-th row of } I \text{ is in } I_1, \\ \text{otherwise.} \end{cases}$$

Define $d' = c \cup (d_A + Aq)$. Then d' is integral, and $d'_I = c$, so by (1.6) the polyhedron $P^*(d')$ has the i.p.

Now $F^*_S(d')$ is not empty because it contains $(x + q)$, as we may easily verify:

$$A^*(x + q) = I(x + q) \cup A(x + q) \geqq c \cup (d_A + Aq) = d' ,$$

$$S(x + q) = I_1(x + q) \cup A_1(x + q) = c_{I_1} \cup (d_{A_1} + A_1 q) = d'_S.$$

Therefore $F^*_S(d')$ must contain an integral point. However $F^*_S(d')$ can contain only a single point for the same reasons that applied to $F^*_S(d)$. Hence $x + q$ must be that single point, so $x + q$ must itself be integral. As q is integral, x must be integral also. Thus $F^*_S(d)$, and a fortiori $F^*_{S'}(d)$, contains the integral point x. This verifies condition (1.2) for A^*.

By Theorem 1, (1.3) holds for A^*. As the rank of A^* is n, $\gcd(S) = |S| = 1$ for every set S of n linearly independent rows of A^*. From this we wish to show that A has the u.p. Suppose E is any non-singular square submatrix of A. Let the order of E be s. By choosing S to consist of the rows of A^* which contain E together with the proper set of $(n - s)$ rows of I, and by rearranging columns, we can easily insure that

$$S = \begin{bmatrix} I & 0 \\ F & E \end{bmatrix}$$

where I is the identity matrix of order $(n - s)$, F is some s by $(n - s)$ matrix, and 0 is the $(n - s)$ by s matrix of zeros. Then $|S| = |E| \neq 0$, so S is non-singular. Therefore S consists of n linearly independent rows, so

$$|E| = |S| = \gcd(S) = 1.$$

This completes the proof of Theorem 2.

4. A THEOREM BY HELLER AND TOMPKINS

In this and the remaining sections we give various sufficient conditions for a matrix to have the unimodular property.

THEOREM 3. (Heller and Tompkins). Let A be an m by n matrix whose rows can be partitioned into two disjoint sets, T_1 and T_2, such that A, T_1, and T_2 have the following properties:

(4.1) every entry in A is 0, + 1, or - 1;

(4.2) every column contains at most two non-zero entries;

(4.3) if a column of A contains two non-zero entries, and both have the same sign, then one is in T_1 and one is in T_2;

(4.4) if a column of A contains two non-zero entries, and they are of opposite sign, then both are in T_1 or both in T_2.

Then A has the unimodular property.

This theorem is closely related to the central result of the paper by Heller and Tompkins in this Study. The theorem, as stated above, is given an independent proof in an appendix to their paper.

COROLLARY.[5] If A is the incidence matrix of the vertices versus the edges of an ordinary linear graph G, then in order that A have the unimodular property it is necessary and sufficient that G have no loops with an odd number of vertices.

PROOF. To prove the sufficiency, recall the following. The condition that G have no odd loops is well-known to be equivalent to the property that the vertices of G can be partitioned into two classes so that each edge of G has one vertex in each class. If we partition the rows of A correspondingly, it is easy to verify the conditions (4.1)-(4.4). Therefore A has the u.p.

If A has an odd loop, let A' be the submatrix contained in the rows and columns corresponding to the vertices and edges of the loop. Then

[5] The authors are indebted to the referee for this result.

it is not hard to see that $|A'| = \pm 2$. This proves the necessity.

5. A SUFFICIENT CONDITION FOR THE UNIMODULAR PROPERTY

We shall consider oriented graphs. For our purposes an oriented graph G is a graph (a) which has no circular edges, (b) which has at most one edge between any two given vertices, and (c) in which each edge has an orientation. Let V denote the set of vertices of G, and E the set of edges. If (r, s) is in E (that is, if (r, s) is an edge of G), then we shall call (s, r) an inverse edge. (Note that by (b), and inverse edge cannot be in E; thus an inverse edge cannot be an edge. This slight ambiguity in terminology should cause no confusion.) We shall often use the phrase direct edge to denote an ordinary edge.

Diagram 1

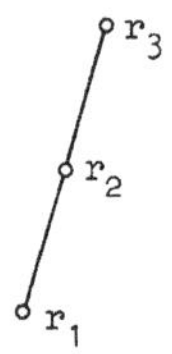

A direct path

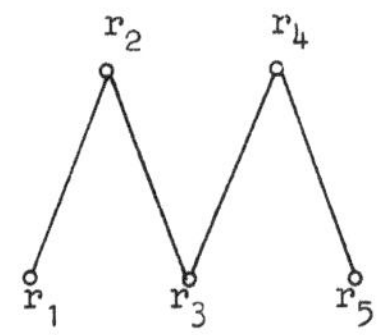

A(+ -) alternating path

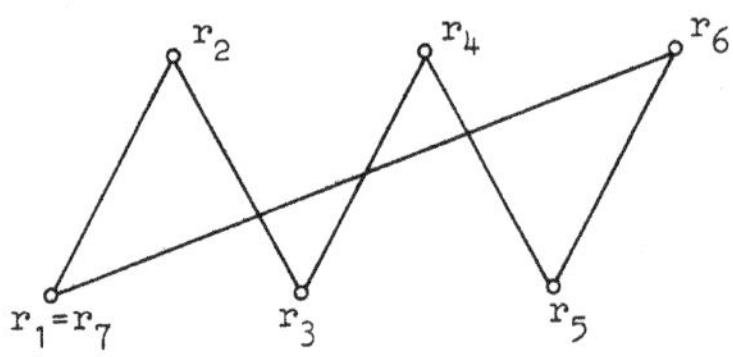

An alternating loop

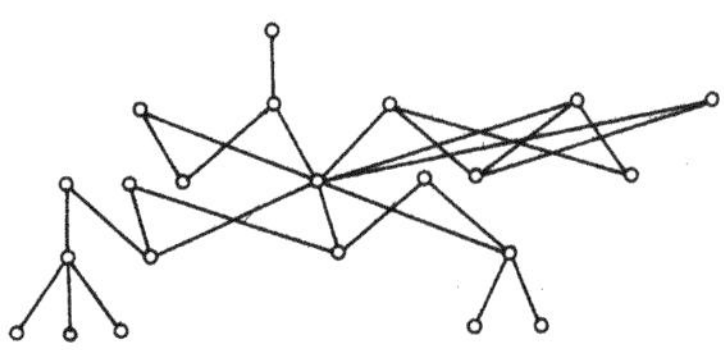

An alternating graph (arrows omitted - all should be upward)

A path is a sequence of distinct vertices $r_1, \ldots, r_k$, such that for each i, from 1 to $k - 1$, (r_i, r_{i+1}) is either a direct or an inverse edge. A path is directed if every edge is oriented forward, that is, if every edge (r_i, r_{i+1}) in the path is a direct edge. A path is alternating if successive edges are oppositely oriented. More precisely, a path is alternating if its edges are alternately direct and inverse. An alternating path may be described as being (++), (+-), (-+), or (--). The first sign indicates the orientation of the first edge of the path, the second sign the orientation of the last edge of the path. A + indicates a direct edge; a - indicates an inverse edge. A loop is a path which closes back on itself. More precisely, a loop is a sequence of vertices $r_1, \ldots, r_k$ in which $r_1 = r_k$ but which are otherwise distinct, and such that for each i (r_i, r_{i+1}) is either a direct or an inverse edge. A loop is alternating if successive edges are oppositely oriented and if the first and last edges are oppositely oriented. An alternating loop must obviously contain an even number of edges.

A graph is <u>alternating</u> if every loop in it is alternating. Let $V = \{v_1, \ldots, v_m\}$ be the vertices of G, and let $P = \{p_1, \ldots, p_n\}$ be some set of directed paths in G. Then the incidence matrix $A = \|a_{ij}\|$ of G versus P is defined by

$$a_{ij} = \begin{cases} 1 & \text{if } v_i \text{ is in } p_j, \\ 0 & \text{if } v_i \text{ is not in } p_j. \end{cases}$$

We let A_v represent the row of A corresponding to the vertex v and A^p represent the column of A corresponding to the path p. We often write a_{vp} instead of a_{ij} for the entry common to A_v and A^p.

THEOREM 4. Suppose G is an oriented graph, P is some set of directed paths in G, and A is the incidence matrix of G versus P. Then for A to have the unimodular property it is sufficient that G be alternating. If P consists of the set of <u>all</u> directed paths of G, then for A to have the unimodular property it is necessary and sufficient that G be alternating.

This theorem does not state that every matrix of zeros and ones with the u.p. can be obtained as the incidence matrix of an alternating graph versus a set of directed paths. Nor does it give necessary and sufficient conditions for a matrix of zeros and ones to have the unimodular property. (Such conditions would be very interesting.) However it does provide a very general sufficient condition. For example, the coefficient matrix of the i by j transportation problem (or its transpose, depending on which way you write the matrix) is the incidence matrix of the alternating graph versus the set of all directed paths. Hence this matrix has the u.p., from which by Theorem 2 follows the well-known i.p. of transportation problems and their duals. The extent to which alternating graphs can be more general than the graph shown to the left is a measure of how general Theorem 4 is.

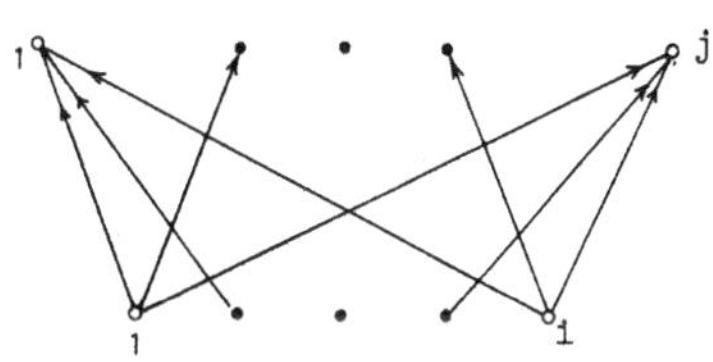

Diagram 2

So that the reader may follow our arguments more easily, we describe here what alternating graphs look like.

(As logically we do not need these facts and as the proofs are tedious, we omit them.) An integral height function $h(v)$ may be defined in such a way that (r, s) is a direct edge only when (but not necessarily when) $h(r) + 1 = h(s)$. If we define $r \leqq s$ to mean that there is a directed path from r to s, then $\leqq$ is a partial order. Then (r, s) is a direct edge if and only if both $r < s$ and there is no element t such that $r < t < s$.

PROOF OF NECESSITY. We consider here the case in which P is the set of all directed paths in G, and we prove that for A to have the u.p. it is necessary that G be alternating. It is easy to verify that the matrix (shown below) of odd order which has ones down the main diagonal and sub-diagonal and in the upper right-hand corner, and zeros elsewhere, has determinant $+2$.

$$\begin{bmatrix} 1 & & & & 1 \\ 1 & 1 & & & \\ & 1 & \ddots & & \\ & & \ddots & 1 & \\ & & & 1 & 1 \end{bmatrix}$$

We shall show that if G is not alternating then it contains this matrix, perhaps with rows and columns permuted, as a submatrix.

Let ℓ be a non-alternating loop in G. If ℓ has an odd number of distinct vertices, consider the rows in A which correspond to these vertices, and consider the columns in A which correspond to the one-edge directed paths which correspond to the edges in ℓ. The submatrix contained in these rows and columns is clearly the matrix shown above, up to row and column permutations. Hence in this case A does not have the u.p. If ℓ has an even number of distinct vertices, then find in it three successive vertices r, s, t such that (r, s) and (s, t) are both direct (or both inverse) edges. (To find r, s, t it may be necessary to let s be the initial-terminal vertex of ℓ, in which case r, s, t are successive only in a cyclic sense.) Consider the rows of A which correspond to all the vertices of ℓ except s. Consider the columns of A which correspond to the following directed paths: the two-edge path r, s, t (or t, s, r) and the one-edge paths using the other edges in ℓ. The submatrix contained in these rows and columns is the square matrix of odd order shown above, up to row and column permutations. Hence in this case also A does not have the u.p. This completes the proof of necessity.

The proof of the sufficiency condition, when P may be any set of directed paths in G, occupies the rest of this section. As this proof is long and complicated, it has been broken up into lemmas.

If $r_1, \ldots, r_k$ is a loop, then $r_1, \ldots, r_k, r_2, \ldots, r_1$ is called a cyclic permutation of the loop. Clearly a loop is alternating if and only if any cyclic permutation is alternating.

LEMMA 6. Suppose A is the incidence matrix of an alternating graph G versus some set of directed paths P in G. For any submatrix A' of A, there is an alternating graph G' and a set of directed paths P' in G' such that A' is the incidence matrix of G' versus P'.

PROOF. Any submatrix can be obtained by a sequence of row and column deletions. Hence it suffices to consider the two cases in which A' is formed from A by deleting a single column or a single row. If A' is formed from A by deleting the column A^p, let G' = G, and P' = P - {p}. Then A' is clearly the incidence matrix of G' versus P', and G' is indeed an alternating graph.

Suppose now that A' is formed from A by deleting row A_t. Define

$$V' = V - \{t\},$$

$$E' = \{(v, w) \mid v, w \text{ in } V' \text{ and either } (v, w) \text{ in } E \text{ or } (v, t) \text{ and } (t, w) \text{ in } E\},$$

G' = the graph with vertices V' and edges E',

$$P' = \{p - \{t\} \mid p \text{ in } P\}.$$

Clearly A' is the incidence matrix of G' versus P'. We shall prove (a) that P' is a collection of directed paths and (b) that G' is alternating.

The proof of (a) is quite simple. Suppose v, w are successive vertices of p' = p - {t} in P'. It may or may not happen that p contains t. In either case, however, if v, w are successive vertices in p, then (v, w) is a

Diagram 3

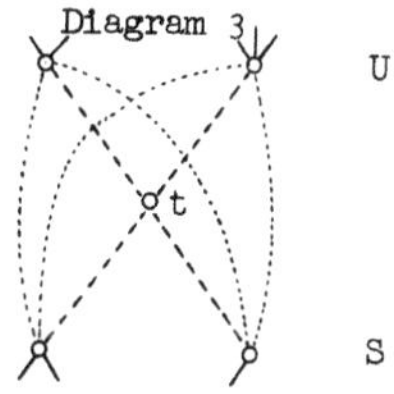

Solid edges - G and G'
Dashed edges - G only
Dotted edges - G' only

direct edge in G, so (v, w) is a direct edge in G'. If v, w are not successive vertices in p, then necessarily v, t, w are successive vertices in p. In this case (v, t) and (t, w) are direct edges in G, so (v, w) is a direct edge in G'.

The proof of (b) is more extended. Define

$$S = \{s \mid (s, t) \text{ in } E\}$$
$$U = \{u \mid (t, u) \text{ in } E\} .$$

Then each "new" edge in E', that is, each edge of E' - E, is of the form (s, u) with s in S and u in U. Let ℓ be any loop in G'. If ℓ contains no new edge, then ℓ is also a loop in G and hence alternating. If ℓ contains a new edge, it contains at least two vertices of $S \cup U$. Hence the vertices of $S \cup U$ break ℓ up into pieces which are paths of the form

$$p = v, r_1, \ldots, r_k, v'$$

where v and v' are in $S \cup U$ and the r's are not.

CASE (U, U): both v and v' belong to U. In this case

$$t, v, r_1, \ldots, r_k, v', t$$

is a loop in G, hence alternating. Therefore p is an alternating path. As (t, v) is a direct edge and (v', t) is an inverse edge in G, p must be a (- +) alternating path in G'.

CASE (S, S): both v and v' belong to S. In this case dual argument to the above proves that p must be a (+ -) alternating path in G'.

CASE (U, S): v belongs to U and v' belongs to S. In this case p must be exactly the one-edge path v, v'. For if not, p consists solely of edges in E, so the loop which we may represent symbolically v', t, p is a loop in G. But as (v', t) and (t, v) are both direct edges in G this loop is not alternating, which is impossible. As (v, v') is an inverse edge in G', p is a (- -) alternating path in G'.

CASE (S, U): v belongs to S and v' belongs to U. In this case dual argument to the above proves that p must be exactly the one-edge path v, v' and hence a (+ +) alternating path in G'.

Using these four cases, we easily see that the pieces of ℓ are alternating and fit together in such a way that ℓ itself is alternating - except for one technical difficulty, namely the requirement that the initial and terminal edges of ℓ must have opposite orientations. However, if we form a cyclic permutation of ℓ and apply the reasoning above to this new loop, we obtain the necessary information to complete our proof that ℓ is alternating. This completes the proof of (b) and Lemma 6.

In view of Lemma 6, the sufficiency condition of Theorem 4 will be proved if we prove that every square incidence matrix of an alternating graph versus a set of directed paths has determinant, 0, + 1, or - 1. We prove this by a kind of induction on two new variables, c(G) and d(G), which we shall now define:

c(G) = the number of unordered pairs {st} of distinct vertices of G which satisfy

(5.1) there is a vertex u such that (s, u) and (t, u) are direct edges of G;

d(G) = the number of unordered pairs {st} of distinct vertices of G which satisfy

(5.2) there is no directed path from s to t nor any directed path from t to s.

Though not logically necessary the following information may help orient the reader to the significance of these two variables. Assume G is alternating. Then using the partial-order $\leqq$ introduced informally earlier, d(G) is the number of pairs of vertices which are incomparable under $\leqq$. Any pair {st} which satisfies (5.1) also satisfies (5.2), so $c(G) \leqq d(G)$. If c(G) = 0, then each vertex of G has at most one "predecessor", and G consists of a set of trees, each springing from a single vertex and oriented outward from that vertex. If d(G) = 0, then G is even more special: it consists of a single directed path.

LEMMA 7. If G is alternating, and {st} satisfies (5.1), then it also satisfies (5.2). Hence $c(G) \leqq d(G)$.

PROOF. Let u be a vertex such that (s, u) and (t, u) are direct edges of G. Suppose there is a directed path

$$s,\ r_1,\ \ldots,\ r_k,\ t\ .$$

If none of the r's is u, then

$$s,\ r_1,\ \ldots,\ r_k,\ t,\ u,\ s$$

is a loop, hence alternating. As (t, u) is a direct edge, (r_k, t) is an inverse edge, so the path is not directed, a contradiction. If one of the r's is u, take the piece from u to t. By renaming, we may call this directed path

$$u, r_1, \ldots, r_k, t .$$

Then

$$t, u, r_1, \ldots, r_k, t$$

is a loop, hence alternating. As (t, u) is a direct edge, (u, r_1) must be an inverse edge, so the path is not directed, a contradiction. Therefore, there can be no directed path from s to t. By symmetrical argument, there can be no directed path from t to s. Therefore $\{st\}$ satisfies (5.2). It follows trivially that $c(G) \leqq d(G)$. This completes the proof of Lemma 7.

The induction proceeds in a slightly unusual manner. The "initial case" consists of all graphs G for which $c(G) = 0$. The inductive step consists of showing that the truth of the assertion for a graph G such that $c(G) > 0$ follows from the truth of the assertion for a graph G' for which $d(G') < d(G)$. It is easy to see that by using the inductive step repeatedly, we may reduce to a graph G^* for which either $c(G^*)$ or $d(G^*)$ is 0. But as $d(G^*) = 0$ implies $c(G^*) = 0$ by the inequality between c and d, we are down to the initial case either way.

We now treat the initial case.

LEMMA 8. Let A be the incidence matrix of an alternating graph G versus some set of directed paths P. Suppose that P contains as many directed paths as G contains vertices, so A is square. Suppose that $c(G) = 0$. Then $|A| = 0, +1$, or -1.

PROOF. If (r, s) is a direct edge of G, we call r a <u>predecessor</u> of s and s a <u>successor</u> of r. The fact that $c(G) = 0$ means that each vertex of G has at most one predecessor. If V' is a subset of V, and r is in V' but has no predecessor in V', then r is called an <u>initial vertex</u> of V'.

Every non-empty subset V' of V has at least one initial vertex. For if V' has none, then we can form in V' a sequence $r_1, r_2, \ldots$ of vertices such that for every i, r_{i+1} is a predecessor of r_i. Let r_j be the first term in the sequence which is the same as a vertex picked earlier, and let r_i be the earlier name for this vertex. Then r_i, $r_{i+1}, \ldots, r_j$ is a loop all of whose edges are inverse. As G is alternating, this is impossible.

Let $U(r) = \{s \mid s$ is a successor of $r\}$. Let r_1 be an initial vertex in V. Recursively, let r_i be an initial vertex of $V - \{r_1, r_2, \ldots, r_{i-1}\}$. Then define matrices $B(i)$ recursively:

$$B(0) = A,$$

$$B(i) = B(i - 1)$$

with the row $B_{r_i}(i - 1)$ replaced by

$$B_{r_i}(i - 1) - \sum_{s \text{ in } U(r_i)} B_s(i - 1) .$$

Let B be the final $B(i)$. We see immediately that $|A| = |B(1)| = \ldots = |B|$. Thus we only need show that $|B| = 0, + 1,$ or $- 1$.

We claim that each column B^p of B consists of zeros with either one or two exceptions: if w is the final vertex of the directed path p, then $b_{wp} = 1$, and if v is the unique predecessor to the initial vertex of p, then $b_{vp} = - 1$. As the initial vertex of p may have no predecessor at all, the $- 1$ may not occur.

We shall not prove in detail the assertions of the preceding paragraph. We content ourselves with considering the column corresponding to a fixed path p during the transition from $B(i - 1)$ to $B(i)$. Only one entry is altered, namely $b_{r_ip}(i - 1)$. There are four possible cases.

CASE (i): neither r_i nor any of its successors is in p.

CASE (ii): r_i is not in p but one of its successors is in p.

CASE (iii): both r_i and one of its successors is in p.

CASE (iv): r_i is in p but none of its successors is in p.

At most one successor of a vertex can be in a directed path because G is alternating, so these cases cover every possibility. In case (i), the entry we are considering starts as 0 and ends as 0. In case (ii), it starts as 0 and ends as $- 1$. In case (iii), it starts as 1 and ends as 0. In case (iv), it starts as 1 and ends as 1. From these facts, it is not hard to see that B satisfies our assertions.

From our assertions about B it is trivial to check that B satisfies the hypotheses of Theorem 3. It is only necessary to partition the rows of B into two classes, one class being empty and the other class containing every row. Then by Theorem 3, B has the u.p. Therefore, $|B| = 0, +1,$ or -1. As $|A| = |B|$, this completes the proof of Lemma 8.

We now prove the inductive step.

LEMMA 9. Suppose that A is the square incidence matrix of an alternating graph G versus a set of directed paths P. Suppose that $c(G) > 0$. Then there is a square matrix A' such that $|A'| = |A|$ and such that A' is the square incidence matrix of an alternating graph G' versus a set of directed paths P', where $d(G') < d(G)$.

PROOF. As $c(G) > 0$, G contains a vertex u which has at least two distinct predecessors, s and t. Define

$$A' = A \text{ with row } A_t \text{ replaced by } A_s + A_t.$$

Clearly $|A'| = |A|$. Define

$$V' = V,$$

$$E_s = \{(s, w) \mid (s, w) \text{ in } E\},$$

$$E_t = \{(t, w) \mid (s, w) \text{ in } E\},$$

$$E' = E \cup E_t \cup \{(s, t)\} - E_s,$$

$$G' = \text{the graph with vertices } V' \text{ and edges } E',$$

$$p' = \begin{cases} p \text{ if } p \text{ does not contain } s, \\ p \text{ with } t \text{ inserted after } s \text{ if } p \text{ does contain } s, \end{cases}$$

$$P' = \{p' \mid p \text{ in } P\}.$$

We shall prove (a) that G' is alternating, (b) that P' is a set of directed paths of G', (c) that $d(G') < d(G)$, and (d) that A' is the incidence matrix of G' versus P'.

Diagram 4

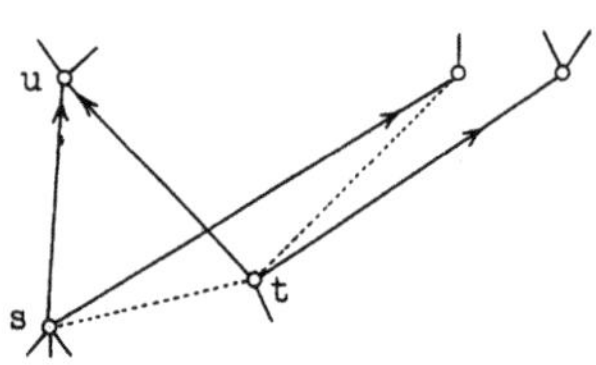

Graph G

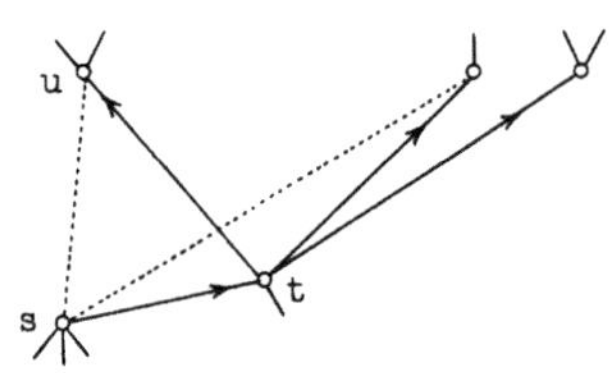

Graph G'

The proof of (b) is simple. If p does not contain s, then every edge in p' = p is in E', so p' is a directed path in G'. If p does contain s, write p thus:

$$r_1, \ldots, r_i, s, r_{i+1}, \ldots, r_j .$$

Then p' is

$$r_1, \ldots, r_i, s, t, r_{i+1}, \ldots, r_j .$$

Each edge of p except (s, r_{i+1}) is also in E'. Hence to show that p' is a directed path in G', we only need show that (s, t) and (t, r_{i+1}) are in E'. The former is in E' by definition, and the latter is in E_t because (s, r_{i+1}) must be in E. This proves (b).

To prove (c), let r_1 and r_2 be any pair of vertices such that there is a directed path p from one to the other in G. Then p' is a directed path from one to the other in G'. Hence every pair of vertices which satisfies (5.2) in G' also satisfies (5.2) in G. Furthermore, {st} does not satisfy (5.2) in G' because (s, t) is in E', while {st} does satisfy (5.2) in G by Lemma 7. This proves that $d(G') < d(G)$.

To prove (d), we first show that A' consists entirely of zeros and ones. The only way in which this could fail to happen is if A_s and A_t both contained ones in the same column. But if this were the case, then the directed path corresponding to this column would contain both s and t, which cannot happen by Lemma 7. To see that A' is the desired incidence matrix, consider how P' differs from P. Each directed path which did not contain s remains unchanged; each directed path which did contain s has t inserted in it. Thus the change from A to A' should be the following. Each column which has a zero in row A_s should remain unchanged; each column which has a one in row A_s should have the zero in row A_t changed to a one. But adding row A_s to A_t accomplishes exactly this. Therefore (d) is true.

The proof of (a) is more complicated.[6] Define S' to be the set of successors of s in G which are not also successors of t. Note that every edge in G' which is not in G terminates either in t, or in a vertex of S'. Let ℓ be any loop of G'. If ℓ is already a loop of G, then it is alternating. If not, it must contain either the edge (s, t) or an edge (t, s') with s' in S'. (Of course, ℓ might contain the inverse of one of these edges instead. If so, reversing the order of ℓ brings us to the situation above.) Ignoring trivial loops, that is,

[6] We are indebted to the referee for this proof, which replaces a considerably more complicated one.

loops of the form aba (which are alternating trivially), ℓ must have the form

$$str_1 \cdots r_k s$$

or

$$ts'r_1 \cdots r_k t, \quad \text{with} \quad s' \text{ in } S'.$$

The first form is impossible. To prove this, first suppose that no r_i is in $S' \cup \{u\}$. Then $sutr_1 \cdots r_k s$ is a loop of G, hence alternating. Thus (r_k, s) is an inverse edge and belongs to both G and G', which is impossible. Now suppose that some r_i is in $S' \cup \{u\}$, and let r_j be the last such r_i. Then $sr_j \cdots r_k s$ is a loop of G, hence alternating. Hence (r_k, s) is inverse, which is impossible as before.

We may now assume that ℓ is $ts'r_1 \cdots r_k t$. No r_i can be s. Clearly r_1 cannot be s, and if $r_j = s$, $j > 1$, then $ss'r_1 \cdots r_{j-1}s$ is a loop of G, hence alternating, so (r_{j-1}, s) is inverse and belongs to both G and G', which is impossible. Thus $r_1, \ldots, r_k$ are distinct from s. Suppose that r_k is in S'. Then $ss'r_1 \cdots r_k s$ is a loop of G, hence alternating. Consequently, so is ℓ. Suppose that r_k is not in S' and that no r_i is u. Then $ss'r_1 \cdots r_k tus$ is a loop of G, hence alternating. Thus $s'r_1 \cdots r_k t$ is a (- -) alternating path in G and also in G'. Hence ℓ is alternating. Finally, suppose that r_k is not in S' and that r_j is u. Then $ss'r_1 \cdots r_{j-1}us$ and $tur_{j+1} \cdots r_k t$ are loops of G, hence alternating. Thus $s'r_1 \cdots r_{j-1}u$ is a (- +) alternating path, and $ur_{j+1} \cdots r_k t$ is a (- -) alternating path. Fitting these paths together and adjoining t at the beginning, we see that ℓ is alternating. This completes the proof of (a), of Lemma 9, and of the sufficiency condition of Theorem 4.

6. HOW TO RECOGNIZE THE UNIMODULAR PROPERTY

To apply Theorem 3 is easy, although even there one point is important. To say that A has the unimodular property is the same thing as to say that A^T, the transpose of A, has the unimodular property. However the hypotheses of Theorem 3 or 4 may quite easily be satisfied for A^T but not for A. Consequently it is desirable to examine both A and A^T when using these theorems.

To apply Theorem 4 is not so easy: how shall we recognize whether matrix A (or matrix A^T) is the incidence matrix of an alternating graph versus some set of directed paths? We point out that in actual applications the graph G generally lies close at hand. For example, it was pointed out in Section 5 that the coefficient matrix A of the i by j transportation problem is the incidence matrix of the alternating graph shown in Diagram 2 (at the beginning of Section 5) versus all its directed paths. This graph is no strange object - it portrays the i "producing points", the j "consuming points", and the transportation routes between them.

In a given linear programming problem there will often be one (or several) graphs which are naturally associated with the coefficient matrix. Whenever the problem can be stated in terms of producers, consumers, and intermediate handlers, this is the case. It may well be possible in this situation to identify the matrix as a suitable incidence matrix.

However it is still useful to have criteria available which can be applied directly to the matrix A and which guarantee that A can be obtained as a suitable incidence matrix. The two following theorems give such conditions. Each corresponds to a very special case of Theorem 4. Theorem 5, historically, derives from the integrality of transportation-type problems, and finds application in [2]; Theorem 6 from the integrality of certain caterer-type problems (see [1]).

We shall write $A_i \geqq A_j$ to indicate that row A_i is componentwise $\geqq$ row A_j.

THEOREM 5. Suppose A is a matrix of 0's and 1's, and suppose that the rows of A can be partitioned into two disjoint classes V_1 and V_2 with this property: if A_i and A_j are both in V_1 or both in V_2, and if there is a column A^k in which both A_i and A_j have a 1, then either $A_i \leqq A_j$ or $A_i \geqq A_j$. Then A has the unimodular property.

This theorem corresponds to a generalized transportation situation, in which each upper vertex of the transportation graph has attached an outward flowing tree and each lower vertex has attached an inward flowing tree. Only directed paths which have at least one vertex in the original transportation graph can be represented as columns of the matrix A.

PROOF. Briefly the proof is this: Let vertices v_i in V correspond to the rows A_i of A. Define a partial-order $\leqq$ on the vertices:

$$\begin{aligned} v_i \leqq v_j \quad & \text{if } A_i \text{ in } V_1 \text{ and } A_j \text{ in } V_2 \\ & \text{or } A_i, A_j \text{ in } V_1 \text{ and } A_i \leqq A_j \\ & \text{or } A_i, A_j \text{ in } V_2 \text{ and } A_i \geqq A_j . \end{aligned}$$

Let G be the graph naturally associated with this partially-ordered set. We leave to the reader verification of the fact that G is alternating, and that the columns of A represent directed paths in P.

Say that two column vectors of the same size consisting of 0's and 1's are in accord if the portions of them between (in the inclusive sense) their lowest common 1 and the lower of their highest separate 1's are identical.

THEOREM 6. Suppose A is a matrix of 0's and 1's, and suppose that the rows of A can be rearranged in such a way that every pair of columns is in accord. Then A has the unimodular property.

This theorem corresponds to a situation in which $c(G) = 0$, that is, every vertex has at most one predecessor (or to the dual situation in which every vertex has at most one successor). The columns of A may represent any directed paths in the graph.

PROOF. Let vertices v_i in V correspond to the rows A_i in A. Assume that the rows are already arranged as described above. Define E as follows:

(v_i, v_j) is in E if $i > j$ and if there is a column A^k of A such that a_{ik} and a_{jk} are both 1 while all intervening entries are 0's.

Let G be the graph with vertices V and edges E. We leave to the reader verification of the fact that G is an alternating graph in which every vertex has at most one successor, and that the columns of A represent directed paths in G.

BIBLIOGRAPHY

[1] GADDUM, J. W., HOFFMAN, A. J., and SOKOLOWSKY, D., "On the solution of the caterer problem," Naval Research Logistics Quarterly, Vol. 1, 1954, pp. 223-227. See also JACOBS, W., "The caterer problem," ibid., pp. 154-165.

[2] HOFFMAN, A. J., and KUHN, H. W., "On systems of distinct representatives," this Study.

[3] JACOBSON, NATHAN, Lectures in Abstract Algebra, Vol. II, (1953), D. Van Nostrand Co., pp. 88-92.

National Bureau of Standards — A. J. Hoffman

Princeton University — J. B. Kruskal

Chapter 4
Outline of an Algorithm for Integer Solutions to Linear Programs *and* An Algorithm for the Mixed Integer Problem

Ralph E. Gomory

Introduction by *Ralph E. Gomory*

Later in 1957, as the end of my three-year tour of duty in the Navy was approaching, Princeton invited me to return as Higgins Lecturer in Mathematics. I had been a Williams undergraduate and a then a graduate student at Cambridge and Princeton while getting my Ph.D. I had published 4 papers in non-linear differential equations, a subject to which I had been introduced by two wonderful people whose support and encouragement made an unforgettable and wonderful difference in my life: Professor Donald Richmond of Williams College and Professor Solomon Lefschetz of Princeton.

Because of my interest in applied work I had planned to look for an industrial position rather than an academic one on leaving the Navy, but I decided instead to accept this attractive offer and spend a year or two at Princeton before going on. When I returned to Princeton late in the fall of 1957, I got to know Professor A. W. Tucker, then the department head, who was the organizer and prime mover of a group interested in game theory and related topics. This group included Harold Kuhn and Martin (E. M. L.) Beale.

As the Navy had kept me on as a consultant I continued to work on Navy problems through monthly trips to Washington. On one of these trips a group presented a linear programming model of a Navy Task Force. One of the presenters remarked that it would be nice to have whole number answers as 1.3 aircraft carriers, for example, was not directly usable.

I thought about his remark and determined to try inventing a method that would produce integer results. I saw the problem as clearly important, indivisibilities are everywhere, but I also thought it should be possible. My view of linear programming was that it was the study of systems of linear inequalities and that it was closely anal-

Ralph E. Gomory
Alfred P. Sloan Foundation, New York, USA
e-mail: gomory@sloan.org

M. Jünger et al. (eds.), *50 Years of Integer Programming 1958-2008*,
DOI 10.1007/978-3-540-68279-0_4, © Springer-Verlag Berlin Heidelberg 2010

ogous to studying systems of linear equations. Systems of linear equations could be solved in integers (Diophantine equations), so why not systems of linear inequalities? Returning to the office I shared with Bob Gunning (later Dean of the Faculty at Princeton), I set to work and spent about a week of continuous thought trying to combine methods for linear Diophantine equations with linear programming. This produced nothing but a large number of partly worked out numerical examples and a huge amount of waste paper.

Late in the afternoon of the eighth day of this I had run out of ideas. Yet I still believed that, if I had to, in one-way or another, I would always be able to get at an integer answer to any particular numerical example. At that point I said to myself, suppose you really had to solve some particular problem and get the answer by any means, what would be the first thing that you would do? The immediate answer was that as a first step I would solve the linear programming (maximization) problem and, if the answer turned out to be 7.14, then I would at least know that the integer maximum could not be more than 7. No sooner had I made this obvious remark to myself than I felt a sudden tingling in two of my left toes, and realized that I had just done something different, and something that certainly was not a part of classical Diophantine analysis.

How exactly had I managed to conclude, almost without thought, that, if the LP answer was 7.14, the integer answer was at most 7? As I was working with equations having integer coefficients and only integer variables, it did not take long to conclude that the reasoning consisted of two simple steps. First that the objective function was maximal on the linear programming problem and therefore as large or larger than it could ever be on the integer problem. Second that the objective function was an integer linear form and therefore had to produce integer results for any integer values of the variables, including the unknown integer answer. Therefore the objective function had to be an integer less than 7.14. Clearly then it was legitimate to add an additional constraint that confined the objective function to be less than or equal to 7. I thought of this as "pushing in" the objective function. It was also immediately clear to me that there would always be many other integer forms maximal at that vertex in addition to the given objective function and that they could be "pushed in" too.

Greatly excited I set to work and within a few days had discovered how to generate maximal integer forms easily from the rows of the transformed simplex matrix. It became clear rapidly that any entry in a given row of the tableau could be changed by an integer amount while remaining an integer form, that these changes could be used to create a form that was maximal, as that simply meant that all the row entries had to become negative (in the sign convention I was then using). It also was clear that, once an entry became negative, it strengthened the new inequality if the entry was as small as possible in absolute value; so all coefficients were best reduced to their negative fractional parts. This was the origin of the "fractional cut."

Within a very few days, I had worked out a complete method using the fractional cuts. I thought of this method as the "The Method of Integer Forms." With it I was steadily solving by hand one small numerical example after another and getting the right answer. However, I had no proof of finiteness. I also observed that the fractional rows I was creating seemed to have a lot of special properties, all of which were explained later in terms of the factor group.

Just at this time I ran into Martin Beale in the hall. He was looking for a speaker for the seminar we had on game theory and linear programming. I said I would be glad to give a talk on solving linear programs in integers. Martin said "but that's impossible." That was my first indication that others had thought about the problem. During the exciting weeks that followed, I finally worked out a finiteness proof and then programmed the algorithm on the E101, a pin board computer that was busy during the day but that I could use after midnight. The E101 had only about 100 characters of memory and the board held only 120 instructions at one time, so that I had to change boards after each simplex maximization cycle and put in a new board that generated the cut, and then put the old board back to re-maximize. It was also hard work to get the simplex method down to 120 E101 instructions. But the results were better and more reliable than my hand calculations, and I was able to steadily and rapidly produce solutions to four- and five-variable problems.

During these weeks I learned that others had thought about the problem and that George Dantzig had worked on the traveling salesman problem and had applied special handmade cuts to that problem. Professor Tucker, who was enormously helpful to me during my entire stay at Princeton, gave me the time he had for himself on the program of a mathematical society meeting. There early in 1958 I made the first public presentation of the cutting plane algorithm. This produced a great deal of reaction, many people wrote to me, and Rand Corporation invited me to come out to California for the summer.

In the summer of 1958 I flew west to Los Angeles, where Rand was located, carrying the first edition of the manual for Fortran, then a brand new language. I spent one month at Rand and succeeded in producing a working Fortran version of the algorithm for the IBM 704. During my stay at Rand, I renewed my acquaintance of graduate student days with Lloyd Shapley and with Herb Scarf and met for the first time George Dantzig, Dick Bellman, and Phil Wolfe. Phil, already well known for his work on quadratic programming, generously took on the assignment of orienting me during my visit at Rand. He helped me in every conceivable way.

The Fortran program seemed to be debugged about two days before I left Rand so I was able to do larger examples. Larger meant something like ten to fifteen variables. Most of these problems ran quickly, but one went on and on and producing reams of printout but never reaching a final answer. I thought at the time that perhaps there were still bugs left in the program, but in fact it was the first hint of the computational problems that lay ahead.

It seems likely that it was during that summer that I worked out the mixed integer method, which I never sent in to a journal but appeared later as a Rand report. At the time I regarded it as a pretty straightforward extension of the original cutting plane method. Having done so many hand problems I was aware that, despite its obvious strengths in some of its computational detail it lacked some attractive properties of the all integer calculation. However at this late date I am quite reconciled to the mixed integer cut by (1) its computational success in a world of larges scale computing and (2) a rather recent result in which I have shown that it provides the only facet for the one dimensional corner polyhedron problem that is a facet both for the continuous and for the integer variables case. This finally locates the mixed cutting plane in its proper theoretical setting.

The following article originally appeared as:

R.E. Gomory, *Outline of an Algorithm for Integer Solutions to Linear Programs*, Bulletin Of the American Mathematical Society 64 (1958) 275–278.

RESEARCH ANNOUNCEMENTS

The purpose of this department is to provide early announcement of significant new results, with some indications of proof. Although ordinarily a research announcement should be a brief summary of a paper to be published in full elsewhere, papers giving complete proofs of results of exceptional interest are also solicited.

OUTLINE OF AN ALGORITHM FOR INTEGER SOLUTIONS TO LINEAR PROGRAMS

BY RALPH E. GOMORY[1]

Communicated by A. W. Tucker, May 3, 1958

The problem of obtaining the best integer solution to a linear program comes up in several contexts. The connection with combinatorial problems is given by Dantzig in [1], the connection with problems involving economies of scale is given by Markowitz and Manne [3] in a paper which also contains an interesting example of the effect of discrete variables on a scheduling problem. Also Dreyfus [4] has discussed the role played by the requirement of discreteness of variables in limiting the range of problems amenable to linear programming techniques.

It is the purpose of this note to outline a finite algorithm for obtaining integer solutions to linear programs. The algorithm has been programmed successfully on an E101 computer and used to run off the integer solution to small (seven or less variables) linear programs completely automatically.

The algorithm closely resembles the procedures already used by Dantzig, Fulkerson and Johnson [2], and Markowitz and Manne [3] to obtain solutions to discrete variable programming problems. Their procedure is essentially this. Given the linear program, first maximize the objective function using the simplex method, then examine the solution. If the solution is not in integers, ingenuity is used to formulate a new constraint that can be shown to be satisfied by the still unknown integer solution but not by the noninteger solution already attained. This additional constraint is added to the original ones, the solution already attained becomes nonfeasible, and a new maximum satisfying the new constraint is sought. This process is repeated until an integer maximum is obtained, or until some argument shows that a nearby integer point is optimal. What has been needed to transform this procedure into an algorithm is a systematic method for generating

[1] This work has been supported in part by the Princeton-IBM Mathematics Research Project.

the new constraints. A proof that the method will give the integer solution in a finite number of steps is also important. This note will describe an automatic method of generating new constraints. The proof of the finiteness of the process will be given separately.

Let us suppose that the original inequalities of the linear program have been replaced by equalities in nonnegative variables, so that the problem is to find nonnegative integers, $w, x_1, \cdots, x_m, t_1, \cdots, t_n$, satisfying

$$
\begin{aligned}
w &= a_{0,0} + a_{0,1}(-t_1) \cdots a_{0,n}(-t_n),\\
x_1 &= a_{1,0} + a_{1,1}(-t_1) \cdots a_{1,n}(-t_n),\\
&\vdots\\
x_m &= a_{m,0} + a_{m,1}(-t_1) \cdots a_{m,n}(-t_n)
\end{aligned} \tag{1}
$$

such that w is maximal. Using the method of pivot choice given by the simplex (or dual simplex) method, successive pivots result in leading the above array into the standard simplex form,

$$
\begin{aligned}
w &= a'_{0,0} + a'_{0,1}(-t'_1) \cdots a'_{0,n}(-t'_n),\\
x'_1 &= a'_{1,0} + \quad \cdots \quad \cdots a'_{1,n}(-t'_n),\\
&\vdots\\
x'_m &= a'_{m,0} + \quad \cdots \quad \cdots a'_{m,n}(-t'_n)
\end{aligned} \tag{2}
$$

where the primed variables are a rearrangement of the original variables and the $a'_{0,j}$ and $a'_{i,0}$ are nonnegative. From this array the simplex solution $t'_j = 0$, $x'_i = a'_{i,0}$ is read out.

An additional constraint can now be formulated. The constraint which will be generated is not unique, but is one of a large class that can be produced by a more systematic version of the following procedure.

If the $a'_{i,0}$ are not all integers, select some i_0 with $a'_{i_0,0}$ noninteger, and introduce the new variable

$$
s_1 = -f'_{i_0,0} - \sum_{j=1}^{j=n} f'_{i_0,j}(-t'_j) \tag{3}
$$

where $f'_{i_0,j} = a'_{i_0,j} - n'_{i_0,j}$, with $n'_{i_0,j}$ the largest integer $\leqq a'_{i_0,j}$. This new equation is added to the Equations (2), obtaining a new set which will be referred to as (2*). A feasible solution to (2*) is a vector, $w', x'_1, \cdots, x'_m, t'_1, \cdots, t'_n, s_1$ of nonnegative components. The values of $x'_1, \cdots, x'_m, t'_1, \cdots, t'_n$ determine the s_1 value through (3), so there is a natural correspondence between a solution

$x_1', \cdots, x_m', t_1', \cdots, t_n'$ of (2) and the (not necessarily feasible) solution that these values determine for (2*). Clearly any feasible solution to (2*) determines a feasible solution to the equations (2) simply by dropping the s_1.

It should be noted that if $f_{i_0,0}$ is $\neq 0$, then there is at least one $f_{i_0,j} \neq 0$, with $j \neq 0$, otherwise the equation

$$x_{i_0}' = a_{i_0,0}' + \sum_{j=1}^{j=n} a_{i_0,j}'(-t_j')$$

can have no solution in integers, and the program has no integer solution.

Since the simplex solution to (2), $t_j' = 0$, $x_i' = a_{i,0}'$ determines, through Equation (3), a negative value, $-f_{i_0,0}$ for s_1, the corresponding solution to (2*) is not feasible, i.e. the new restraint cuts off the old maximum. However, any nonnegative *integer* solution to (2) does give rise to a nonnegative integer solution to the equations (2*).

To see this suppose $w'', x_1'', \cdots, x_m'', t_1'', \cdots, t_n''$ is any solution in nonnegative integers to (2). The s_1'' determined is

$$s_1'' = -f_{i_0,0}' - \sum_{j=1}^{j=n} f_{i_0,j}'(-t_j'')$$

$$= n_{i_0,0}' + \sum_{j=1}^{j=n} n_{i_0,j}'(-t_j'') - a_{i_0,0}' - \sum_{j=1}^{n} a_{i_0,j}'(-t_j'')$$

which using (2) becomes $s_1'' = n_{i_0,0}' + \sum_{j=1}^{n} n_{i_0,j}'(-t_j'') - x_{i_0}''$. Since the $n_{i_0,j}'$, the t_j'' and the x_{i_0}'' are all integers, the s_1'' determined is also an integer. Furthermore, since the $f_{i_0,j}'$ and the t_j'' are all nonnegative, (3) shows that s_1'' is $\geqq -f_{i_0,0}' > -1$. Since s_1'' is an integer, this shows it must be nonnegative.

This reasoning establishes a one-one correspondence between nonnegative integer solutions $w'', x_1'', \cdots, x_m'', t_1'', \cdots, t_n''$ to (2) and the corresponding nonnegative integer solutions $w'', x_1'', \cdots, x_m'', t_1'', \cdots, t_n'', s_1''$ to (2*). Since the w value is the same for both solutions, the problem of maximizing w over nonnegative integer solutions to (2) can be replaced by the problem of maximizing w over the nonnegative integer solutions to (2*). The solution to the original problem is obtained by dropping the s_1.

The procedure now is to maximize w over the solutions to (2*). This is done using the dual simplex method because all the $a_{0,j}'$ and $a_{i,0}'$ are already nonnegative, and $-f_{i_0,0}$ is the only negative entry in the zero column of the equations (2*). This fact usually makes remaximization quite rapid. The process is then repeated if the new simplex maximum is noninteger.

Of course the Equations (2*) involve one more equation than the Equations (2), and an equation is added after each remaximization. However, the total number need never exceed $m+n+2$. For if an s-variable, added earlier in the computation reappears among the variables on the left hand side of the equations after some remaximization, the equation involving it can simply be dropped, as the only equations that need be satisfied are the original ones. This limits the total number of s-variables to $n+1$ or less.

It should be noted that even the process just described involves an element of choice, any of the rows i of (2) with $a_{i,0}$ noninteger might be chosen to generate the new relation. Some choices are better than others. A good rule of thumb based on the idea of "cutting" as deeply as possible with the new relation, and borne out by limited computational experience, is to choose the row with the largest fractional part $f_{i,0}$ in the zero column.

The class of possible additional constraints is not limited to those produced by the method described here since it is easily seen that some simple operations on and between rows preserve the properties needed in the additional relations. These operations can be used to produce systematically a family of additional relations from which a particularly effective cut or cuts can be selected. A discussion of this class of possible additional constraints together with a rule of choice of row which can be shown to bring the process to an end in a finite number of steps—thus providing a finite algorithm—require some space and will be given as part of a more complete treatment in another place.

REFERENCES

1. George B. Dantzig, *Discrete-variable extremum problems*, J. Operations Res. Soc. Amer. vol. 5, no. 2 (1957).

2. G. Dantzig, R. Fulkerson, and S. Johnson, *Solution of a large-scale traveling-salesman problem*, J. Operations Res. Soc. Amer. vol. 2, no. 4 (1954).

3. Harry M. Markowitz and Alan S. Manne, *On the solution of discrete programming problems*, Econometrica vol. 25, no. 1 (1957).

4. Stuart E. Dreyfus, *A comparison of linear programming and dynamic programming*, Rand Report P-885, June, 1956.

PRINCETON UNIVERSITY

The following article originally appeared as:

R.E. Gomory, *An Algorithm for the Mixed Integer Problem*, Research Memorandum RM-2597, The Rand Corporation, 1960.

U. S. AIR FORCE

PROJECT RAND

RESEARCH MEMORANDUM

AN ALGORITHM FOR THE MIXED INTEGER PROBLEM

Notes on Linear Programming and Extensions—Part 54

Ralph Gomory

RM-2597

7 July 1960

Assigned to ______________________

This research is sponsored by the United States Air Force under contract No. AF 49(638)-700 monitored by the Directorate of Development Planning, Deputy Chief of Staff, Development, Hq USAF.

The RAND Corporation
1700 MAIN ST. • SANTA MONICA • CALIFORNIA

RM-2597
ii

SUMMARY

An algorithm is given for the numerical solution of the "mixed integer" linear programming problem, the problem of maximizing a linear form in finitely many variables constrained both by linear inequalities and the requirement that a proper subset of the variables assume only integral values. The algorithm is an extension of the cutting plane technique for the solution of the "pure integer" problem.

RM-2597
iii

CONTENTS

RM-2597
1

AN ALGORITHM FOR THE MIXED INTEGER PROBLEM

The problem discussed here is an integer programming problem, i.e., the problem of maximizing

$$z = a_{o,o} + \sum_{j=1}^{j=n} a_{o,j}(-t_j),$$

subject to the inequalities

$$(1) \qquad \sum_{j=1}^{j=n} a_{i,j}t_j \leq a_{i,o}, \qquad i = 1, \ldots, m$$

and subject to the additional condition that some specified subcollection of the variables appearing above should be integers.

If the inequalities above are changed into equations in nonnegative variables by the addition of m "slack" variables, and the whole set is enlarged to form a set in which <u>all</u> the variables are expressed in terms of the independent or "nonbasic" ones, we have

$$z = a_{o,o} + \sum_{j=1}^{j=n} a_{o,j}(-t_j)$$

$$s_i = a_{i,o} + \sum_{j=1}^{j=n} a_{i,j}(-t_j) \qquad i = 1, \ldots, m$$

$$t_j = -1(-t_j) \qquad j = 1, \ldots, n.$$

For the sake of a more uniform notation we will rewrite this as

$$(2) \qquad x_i = a_{i,o} + \sum_{j=1}^{j=n} a_{i,j}(-t_j) \qquad i = 0, \ldots, m+n,$$

where the x_i now are <u>all</u> the variables and the $a_{i,j}$ are <u>all</u> the coefficients.

The usual[1] linear programming problem is solved by applying G. B. Dantzig's simplex method. In this method a series of "pivot steps," "Gaussian eliminations," "changes of basis," or "changes to different sets of nonbasic variables" bring the equations (2) into a form in which, denoting the new coefficients in the equations by primes,

$$\text{(i)} \qquad a'_{i,o} \geq 0 \qquad i = 1, \ldots, m+n$$

and

$$\text{(ii)} \qquad a'_{o,j} \geq 0 \qquad j = 1, \ldots, n.$$

The first condition is the condition that in the "trial solution" obtained by putting all the nonbasic variables equal to zero, the values that result for all the variables are nonnegative. The second condition makes certain that the objective function is in fact maximal when the variables are given the values they attain in this trial solution. The solution obtained is of course

---

[1]The usual method terminates when conditions (ii) first hold. It is necessary here that the pivoting continue until all columns $j > 0$ become lexicographically positive. The procedure for doing this is described in [1].

$$x_i = a'_{i,o} \qquad i = 0, \ldots, m+n.$$

This solution may very well <u>not</u> satisfy the integer requirement; i.e., some x_i that is required to be an integer is assigned the noninteger value $a'_{i,o}$.

If this occurs we will be able to deduce a new inequality that will be satisfied by any integer solution, i.e., by any solution having integers where they are required, but will not be satisfied by the current trial solution.

Then, just as in [1] and [2], this new inequality will be added to the original set of inequalities, and the new set then remaximized by the simplex method. This remaximization is usually quite rapid, as adding the new inequality maintains dual feasibility and introduces just the one unsatisfied inequality.

If the new maximum solution still contains integer variables which are assigned noninteger values the process is repeated.

To deduce this new inequality we make use of the equation

$$x_i = a'_{i,o} + \Sigma\, a'_{i,j}(-t_j) \tag{3}$$

where the x_i is an integer variable, $a'_{i,o}$ is noninteger, and the t_j are the current set of nonbasic variables. Since $a'_{i,o}$ is noninteger it can be written uniquely as the sum of an integer $n'_{i,o}$ and a fractional part $f'_{i,o}$, $0 < f'_{i,o} < 1$.

RM-2597
4

We now imagine that we have an integer solution to the problem and use x_i', t_j' to denote the values given to the variables in (3) by this solution. Hence

$$x_i' = a_{i,o}' + \Sigma\, a_{i,j}'(-t_j')$$

and using $a \equiv b$ to mean a and b differ by an integer (equivalence modulo 1), we have, since $x_i' \equiv 0$ and $a_{i,o}' \equiv f_{i,o}'$,

$$(4) \qquad \Sigma\, a_{i,j}' t_j' \equiv f_{i,o}' .$$

We will group the constants on the left in (4) according to their sign. Let S^+ be the set of indices j for which $a_{i,j}' \geq 0$, and S^- the set for which $a_{i,j}' < 0$. Then

$$(5) \qquad \sum_{j \in S^+} a_{i,j}' t_j' + \sum_{j \in S^-} a_{i,j}' t_j' \equiv f_{i,o}'.$$

There are now two possibilities to consider. The expression on the left is either (i) nonnegative or (ii) negative.

Case (i). Since the left side is nonnegative and equivalent to f_o, its value can only be f_o', or $1 + f_o'$, or $2 + f_o'$, etc. Hence

$$f_{i,o}' \leq \sum_{j \in S^+} a_{i,j}' t_j' + \sum_{j \in S^-} a_{i,j}' t_j' \leq \sum_{j \in S^+} a_{i,j}' t_j'.$$

Case (ii). If the right-hand side is negative and equivalent to $f'_{i,o}$ it can only be $f_o - 1$, $f_o - 2$, etc. So in every case

$$f'_{i,o} - 1 \geq \sum_{j \in S^+} a'_{i,j} t'_j + \sum_{j \in S^-} a'_{i,j} t'_j \geq \sum_{j \in S^-} a'_{i,j} t'_j ,$$

or, multiplying by $- f'_{i,o}/1 - f'_{i,o}$,

$$f'_{i,o} \leq \sum_{j \in S^-} \frac{f'_{i,o}}{1 - f'_{i,o}} (-a'_{i,j}) t'_j .$$

Now either (i) holds or (ii) holds so always

$$(6) \qquad f'_{i,o} \leq \sum_{j \in S^+} a'_{i,j} t'_j + \sum_{j \in S^-} \frac{f'_{i,o}}{1 - f'_{i,o}} (-a'_{i,j}) t'_j ,$$

since the right side is the sum of two nonnegative numbers, one of which is $\geq f'_{i,o}$.

This inequality then is satisfied by any integer solution but not by the present trial solution, since substituting $t_j = 0$ for all j into (6) makes the right-hand side 0.

Of course the inequality (6) can be rewritten as an equation by introducing a nonnegative slack s. Then (6) becomes

$$s = -f'_{i,o} - \sum_{j \in S^+} a'_{i,j} (-t_j) - \sum_{j \in S^-} \frac{f'_{i,o}}{1 - f_{i,o}} (-a'_{i,o})(-t_j).$$

RM-2597
6

In obtaining (6) we have used only the fact that x_i was required to be an integer. If some of the nonbasic variables t_j are also integer variables, the inequality (6) can be improved in a manner entirely analogous to the reduction that is always possible in the strictly integer problem. The improvement will take the form of a decrease in the coefficients on the right in the resulting inequality (6). It is clear that for fixed $f'_{i,o}$ the smaller these coefficients, the stronger the inequality.

Let us suppose then that some t_{j_o} is required to be integer and hence is assigned an integer value t'_{j_o} in (5). Changing a'_{i,j_o} by an integer amount then changes the left side of (5) by an integer, and hence preserves the equivalence. Thus we may replace a'_{i,j_o} by any new value $a^* \equiv a_{i,j_o}$ and proceed just as before to deduce an inequality like (6).

If $a^* \geq 0$, the coefficient of t_{j_o} in the resulting inequality is simply a^*. If a^* is < 0, it is $-f'_{i,o}/(1 - f'_{i,o})a^*$. Among $a^* \geq 0$, $a^* = f'_{i,j_o}$ the fractional part[1] of a'_{i,j_o} clearly gives the smallest coefficient to t_{j_o} in the resulting inequality. (This may even be 0.) Among $a^* < 0$, the smallest coefficient is obtained from $a^* = f'_{i,j_o} - 1$, and is

$$\frac{f'_{i,o}}{1 - f'_{i,o}}\,(1 - f'_{i,j_o}). \tag{7}$$

[1]By the fractional part of both positive and negative numbers $a_{i,j}$ we will mean the nonnegative fraction $f_{i,j} < 1$ such that $a_{i,j} = n_{i,j} + f_{i,j}$ with $n_{i,j}$ integer.

To obtain the smallest possible coefficient we choose the smaller of f'_{i,j_o} and (7) which, because an expression of the form x/1 − x increases monotonically with x is seen to be

$$f'_{i,j_o} \text{ if } f'_{i,j_o} \leq f'_{i,o}$$

and

$$\frac{f'_{i,o}}{1 - f'_{i,o}} \left(1 - f'_{i,\, j_o}\right) \text{ if } f'_{i,j_o} > f'_{o,o}.$$

It follows that the strongest inequality is obtained by a simple two–stage process. (i) First replace coefficients of integer variables by their fractional parts if these are less than $f'_{i,o}$, or by the fractional parts less 1 if they are greater than $f'_{i,o}$. (ii) Then deduce the inequality (6) as before. The final result obtained from the equation

$$x_i = a'_{i,o} + \Sigma\, a'_{i,j}(-t_j)$$

by this procedure is the inequality represented by the equation

$$s = -f'_{i,o} - \Sigma\, f^*_{i,j}(-t_j) \tag{8}$$

where the $f^*_{i,j}$, all nonnegative, are given by the following formulae:

RM-2597 8

$$f^*_{i,j} = \begin{cases} a'_{i,j} & \text{if } a'_{i,j} \geq 0 \text{ and } t_j \text{ noninteger variable} \\ \dfrac{f'_{i,o}}{1 - f'_{i,o}} (-a'_{i,j}) & \text{if } a'_{i,j} < 0 \text{ and } t_j \text{ noninteger variable} \\ f'_{i,j} & \text{if } f'_{i,j} \leq f'_{i,o} \text{ and } t_j \text{ integer variable} \\ \dfrac{f'_{i,o}}{1 - f'_{i,o}} (f'_{i,j}-1) & \text{if } f'_{i,j} > f'_{i,o} \text{ and } t_j \text{ integer variable} \end{cases}$$

Equation (8) is then added and the problem is remaximized. It seems sensible to use the dual simplex method at this point as all the $a'_{o,j}$, $j \geq 1$, are nonnegative, and there is only one negative element, $-f'_{i,o}$, in the 0-column.

If the dual simplex method is applied, the 0-column is decreased lexicographically at the next step, and furthermore, denoting by double primes the coefficients after the next pivot step and by j_o the column in which the pivot step takes place, we have

$$(9) \qquad \begin{aligned} a''_{i,o} &\leq n'_{i,o} & &\text{if } a'_{i,j_o} > 0 \\ a''_{i,o} &\geq n'_{i,o} + 1 & &\text{if } a'_{i,j_o} < 0 \end{aligned}$$

where $n'_{i,o}$ is the integer part of $a'_{i,o}$, the index i in (9) is that of the row figuring in equations (3) through (8).

RM-2597
9

This means that after the next pivot step the value assigned to x_1 by the new trial solution is either $\geq$ the next highest integer, or $\leq$ the next lowest integer.

To see this we consider the mechanism of the dual simplex method. The dual simplex method will pick a pivot in the new row represented by (8). If the pivot element is chosen in this row and in the j_o column then the formula for the $a''_{1,o}$ that results after a pivot step is

$$a''_{1,o} = a'_{1,o} - \frac{f'_{1,o} a_{1,j_o}}{f^*_{1,j_o}} \, .$$

Now the formulas for $f'_{1,j}$ show that if a_{1,j_o} is positive and t_{j_o} noninteger we have

$$(10) \qquad a''_{1,o} = a'_{1,o} - \frac{f'_{1,o} a'_{1,j_o}}{a'_{1,j_o}} = n'_{1,o} \, .$$

If a'_{1,j_o} is negative and t_{j_o} noninteger we have

$$(11) \qquad a''_{1,o} = a'_{1,o} - \frac{f'_{1,o} a'_{1,j_o}}{\left(\dfrac{f'_{1,o}}{1 - f'_{1,o}}\right)\left(-a'_{1,j_o}\right)}$$

$$= a'_{1,o} - f'_{1,o} + 1 = n'_{1,o} + 1 \, .$$

RM-2597
10

To cover the cases when t_{j_o} is an integer variable we need only remember that in this case the f^*_{i,j_o} is deduced by a two-stage process, part (ii) of which is exactly the same as the process used to deduce the f^*_{i,j_o} when t_{j_o} is noninteger. Consequently if part (i) leaves $a'_{i,j}$ unchanged, either (10) or (11) holds just as above. Part (i) will have $a'_{i,j}$ unchanged only if either

$$a'_{i,j} = f'_{i,j}, \qquad \text{and } f'_{i,j} \le f'_{i,o}$$

or

$$a'_{i,j} < 0,\ a'_{i,j} = f'_{i,j} - 1, \qquad \text{and } f'_{i,j} > f'_{i,o}$$

Otherwise part (i) makes a change which results in a strictly smaller final $f^*_{i,j}$. So in these cases we have the strict inequalities

$$a''_{i,o} < n'_{i,o} \qquad \text{if } a'_{i,j_o} > 0$$

$$a''_{i,o} > n'_{i,o} + 1 \qquad \text{if } a'_{i,j_o} < 0.$$

The remaining possibility, $a'_{i,j_o} = 0$, can not occur because $a'_{i,j_o} = 0$ implies $f^*_{i,j_o} = 0$ and so f^*_{i,j_o} can not be the pivot element. Thus (9) holds in all cases.

Now (9) is exactly the property required for a finiteness proof—i.e., a proof that the solution is attained in a finite number of steps—<u>provided that the objective function</u> z <u>is one of the integer variables</u>. To see this we arrange the original equations so that the integer variables on the left in (2) are the first rows following the objective function z. (This means that they rank higher lexicographically in the dual simplex method.) Given property (9), the reasoning in the first finitness proof in [1] (pp. 33–35) now goes through unchanged. Of course one must stop now on attaining the required integer values in the 0-column, as an all-integer matrix is not generally obtained.

RM-2597
12

REFERENCES

1. Gomory, Ralph E., "An Algorithm for Integer Solutions to Linear Programs," Princeton-IBM Mathematics Research Project Technical Report No. 1, November 17, 1958.

2. Beale, E. M. L., "A Method of Solving Linear Programming Problems When Some But Not All of the Variables Must Take Integral Values," Statistical Techniques Research Group, Princeton University (manuscript).

Chapter 5
An Automatic Method for Solving Discrete Programming Problems

Ailsa H. Land and Alison G. Doig

Introduction by *Ailsa H. Land* and *Alison G. Doig*

In the late 1950s there was a group of teachers and research assistants at the London School of Economics interested in linear programming and its extensions, in particular Helen Makower, George Morton, Ailsa Land and Alison Doig. We had considered the 'Laundry Van Problem' until we discovered that it was known as the Traveling Salesman Problem, and had looked at aircraft timetabling, until quickly realizing that even the planning for the Scottish sector was beyond our capability! Alison Doig (now Harcourt) had studied the paper trim problem for her Masters project in Melbourne before coming to England.

At the same time, British Petroleum was developing linear programming models for their refinery operations. They had ambitions to extend the model to deal also with the planning of world movement of oil from source to refinery, but knew that the capacity restrictions on the ships and storage tanks introduced discrete variables into their models. BP contracted with LSE to pay the salaries of Alison Doig and Ailsa Land for one year to investigate the possibility of incorporating discrete variables into linear programming models.

We rapidly decided that the oil transport model was much too big to tackle until we had a working method to handle discrete variables. We easily found papers with smaller examples, ones with known optimal solutions, to use for testing purposes [1]. We are pretty sure that we got the approval of BP for this switch of attention. At the time, BP did not want their sponsorship acknowledged for any publication we might make on discrete variables. We suppose they did not want

Ailsa H. Land
London School of Economics and Political Science, United Kingdom
e-mail: a.land@lse.ac.uk

Alison G. Doig (now Harcourt)
Department of Mathematics and Statistics, University of Melbourne, Australia
e-mail: harc@ms.unimelb.edu.au

M. Jünger et al. (eds.), *50 Years of Integer Programming 1958-2008*,
DOI 10.1007/978-3-540-68279-0_5, © Springer-Verlag Berlin Heidelberg 2010

to alert competitors to their interest in the area, but we assume this prohibition no longer applies!

We were very well aware that the solution of this type of problem required electronic computation, but unfortunately LSE at that time did not have any access to such a facility. However, we had no doubt that using the same approach to computing could be achieved, if rather painfully, on desk computers, which were plentifully available. We became quite skilful at doing vector operations by multiplying with the left hand, and adding and subtracting with the right hand on another machine! Storage of bases and intermediate results did not present a limitation since it was all simply recorded on paper and kept in a folder. Hence we found it efficient to pursue each branch of our tree until its bound was no longer the best bound. To that extent our implementation was not exactly as we would later come to code it on a computer. It was efficient to make an estimate on the next bound in each direction of a branch before putting it aside for possible later development.

As well as solving the original zero-one model from the Markowitz and Manne paper, we felt we had also to publish one for which the solution was not already known. Hence we solved also an 'any integer' model using the same data set. We did not initially think of the method as 'branch and bound', but rather in the 'geometrical' interpretation of exploring the convex feasible region defined by the LP constraints. We are not sure if 'branch and bound' was already in the literature, but, if so, it had not occurred to us to use that name. We remember Steven Vajda telling us that he had met some French people solving ILPs by 'Lawndwa', and realising that they were applying a French pronunciation to 'Land-Doig', so we don't think they knew of it as branch and bound either. Much later someone wrote a paper about 'shoulder branch and bound' (no reference, we are afraid), which in fact was what we were doing by not leaving a node of the tree without bounding it on the upper and lower direction of the current integer variable. This, of course, isn't much help in the zero-one ILP, but we were very much thinking of the 'any-integer' discrete variable problem.

The paper we submitted to Econometrica described the thinking that led to our development of the algorithm, but much clearer expositions of Branch and Bound have been published since then and we wouldn't recommend the paper to students for learning the algorithm!

References

1. H.M. Markowitz and A.S. Manne, *On the solution of discrete programming problems*, Econometrica 25 (1957) 84–110.

The following article originally appeared as:

A.H. Land and A.G. Doig, *An Automatic Method for Solving Discrete Programming Problems*, Econometrica 28 (1960) 497–520.

ECONOMETRICA

VOLUME 28 July, 1960 NUMBER 3

AN AUTOMATIC METHOD OF SOLVING DISCRETE PROGRAMMING PROBLEMS

BY A. H. LAND AND A. G. DOIG

In the classical linear programming problem the behaviour of continuous, nonnegative variables subject to a system of linear inequalities is investigated. One possible generalization of this problem is to relax the continuity condition on the variables. This paper presents a simple numerical algorithm for the solution of programming problems in which some or all of the variables can take only discrete values. The algorithm requires no special techniques beyond those used in ordinary linear programming, and lends itself to automatic computing. Its use is illustrated on two numerical examples.

1. INTRODUCTION

THERE IS A growing literature [**1**, **3**, **5**, **6**] about optimization problems which could be formulated as linear programming problems with additional constraints that some or all of the variables may take only integral values. This form of linear programming arises whenever there are indivisibilities. It is not meaningful, for instance, to schedule 3 – 7/10 flights between two cities, or to undertake only 1/4 of the necessary setting up operation for running a job through a machine shop. Yet it is basic to linear programming that the variables are free to take on any positive value,[1] and this sort of answer is very likely to turn up.

In some cases, notably those which can be expressed as transport problems, the linear programming solution will itself yield discrete values of the variables. In other cases the percentage change in the maximand[2] from common sense rounding of the variables is sufficiently small to be neglected. But there remain many problems where the discrete variable constraints are significant and costly.

Until recently there was no general automatic routine for solving such problems, as opposed to procedures for proving the optimality of conjectured solutions, and the work reported here is intended to fill the gap. About the time of its completion an alternative method was proposed by Gomory [**5**] and subsequently extended by Beale [**1**]. Gomory's method

[1] Or more generally, any value within a bounded interval.

[2] We shall speak throughout of maximisation, but of course an exactly analogous argument applies to minimisation.

is based on the systematic addition of new constraints which are satisfied by a discrete variable solution but not by a continuous variable solution. At present, the published results apply only to problems in which all the variables are discrete, but a generalisation to the mixed case (i.e., in which not all the variables are required to be discrete) is known to exist. The mixed problem has been solved by Beale using a method in which the discrete variables appear as the parameters of a subsidiary linear programme which is expressed entirely in terms of continuous variables; the parameters of this continuous problem are themselves required to satisfy a pure discrete problem for which Gomory's technique may be employed. The method described here applies also to the mixed problem and although we have in fact worked only on a desk computer, we have borne in mind throughout that the algorithm should be susceptible to programming on an electronic computer. It is not suggested that this method should supersede successful ad hoc methods for particular problems. It may, in fact, be chiefly useful for testing the validity of proposed ad hoc methods for new problems.

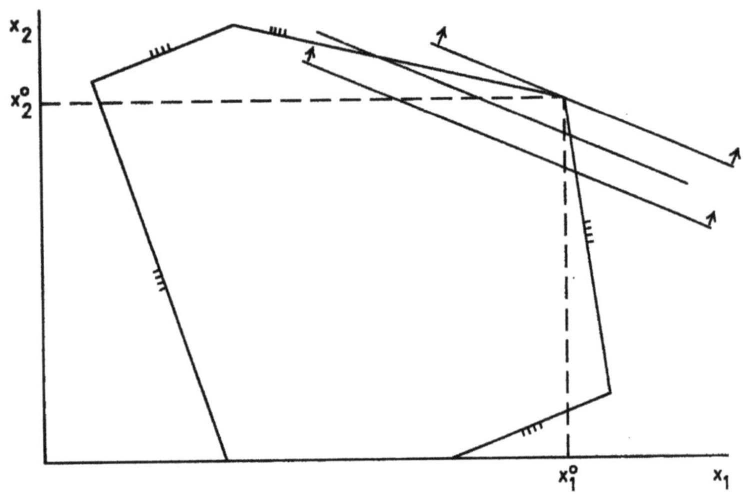

FIGURE 1

2. FORMULATION

The discrete programming problem can be expressed as follows:
Maximize

(1) $$c'x + \bar{c}'y = \gamma,$$

subject to the constraints

(2) $$Ax + \bar{A}y \leq b,$$

(3) x is a column vector with nonnegative integral components

(4) $$y \geq 0,$$

where y, the maximand, is a scalar, b is a column vector of m rows, c and x are column vectors of n_1 rows, $\bar{c}$ and y are column vectors of $(n-n_1)$ rows, A is a matrix of order $m \times n_1$, and $\bar{A}$ is a matrix of order $m \times (n-n_1)$. A feasible solution of the problem is one which satisfies (2), (3), and (4).

A two variable linear programming problem without discrete variable constraints is illustrated in Figure 1 where it can be seen that the functional (represented by the family of parallel lines) reaches its maximum at (x_1^0, x_2^0). The discrete variable constraints limit the set of feasible solutions to points within the original set for which both coordinates are integers; in Figure 2

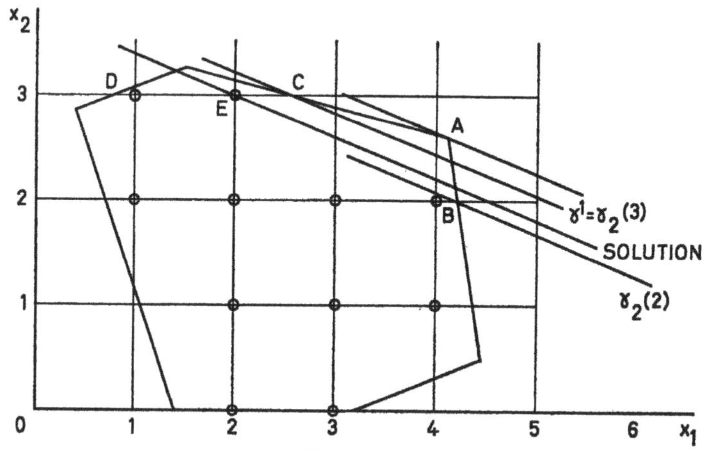

FIGURE 2

the complete set of feasible solutions to the discrete programming problem consists of the eleven points which have been circled. It is easy to see in this two dimensional case that the maximum solution is $x_1 = 2$, $x_2 = 3$. The procedure for arriving at this conclusion could be described as "pushing down the functional line until it meets an integral point." The algorithm to be described here is the generalisation of this procedure to many dimensions.

3. DESCRIPTION OF THE METHOD

As the set of feasible solutions to the discrete variable problem is not convex, any method which examines merely the immediate neighbourhood of a proposed solution can prove the existence only of a local optimum. The method used here makes systematic parallel shifts in the functional hyperplane in the direction of a reduction of the maximand, until a point within the ordinary linear programming set is found which has integral coordinates in the specified (x variable) dimensions. This is obvious in principle but cannot be so readily done in practice as one does not have the faculty of "seeing" a hyperplane in n-dimensional space in order to determine if it contains a point whose x coordinates are all integers. Numerical methods can only examine one point at a time. Rules must therefore be devised to switch attention from one region of the falling functional hyperplane to another to ensure that no integer point has been passed.

In other words an upper bound to the functional γ^0, is first obtained by solving the ordinary linear programming problem without the discrete variable constraints, and then successively more restrictive upper bounds, $\gamma^1, \gamma^2, \ldots, \gamma^k$, are found. If the upper bound of the functional at any stage is γ^k, then it has been proved that there is no discrete variable solution with a higher value of the maximand than γ^k.

Consider the convex set of solutions of an ordinary linear programming problem, such as might be represented for a two-dimensional case by Figure 2. Any feasible value of the maximand, say γ^k, is uniquely associated with a definite position of the functional hyperplane which in general cuts through the n-dimensional convex set, and in the special case of the optimum value just touches the convex set. The intersection of the hyperplane with the

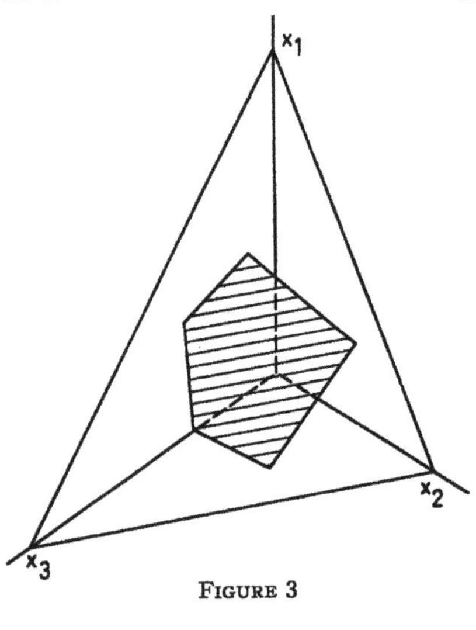

FIGURE 3

original convex set is itself a convex set of $n-1$ dimensions. For example, Figure 3 represents such an intersection with a three-dimensional set. The points x_1, x_2, and x_3 are the intercepts of the hyperplane on the 3 axes respectively and the shaded area represents the convex set which lies within the constraints of the ordinary linear programming problem.

In such an $(n-1)$-dimensional set there would be a minimum and a maximum value for each variable as shown in Figure 4.

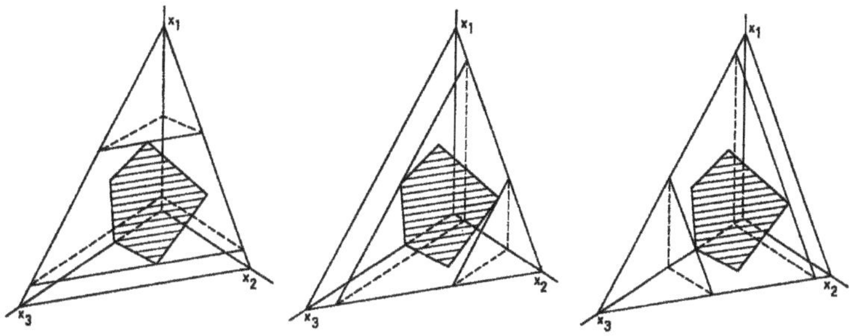

FIGURE 4

If $\gamma^k = \gamma^0$, i.e., if the functional hyperplane is that associated with the optimum solution of the ordinary linear programming problem, this convex set would normally consist of a single point (unless there were multiple optimum solutions). It is still true to say, however, that for every value of the maximand (position of the functional hyperplane) there is a minimum and a maximum value (which may coincide) for each x variable, consistent with the ordinary linear programming constraints.

Knowing for a particular value of γ the minimum and maximum value for each variable one knows also whether there is a possible integer value of each variable for that position of the hyperplane. If there is not at least one possible integer value for every x-variable, then one can say immediately that there is no solution to the problem at that value of the maximand. Unfortunately the converse is not true. The fact that each x variable takes an integer value at some point or points within the set is not a sufficient condition for there being an *intersection* of these integral coordinates within the set, and hence a feasible solution.

Since there is a unique minimum and maximum value for any variable x_k at a particular value of γ, it follows that one can define two functional relationships, min x_k and max x_k, between x_k and the falling functional hyperplane (value of γ). The connection between these two functions and the fundamental problem may be demonstrated by means of the following system of inequalities:

(1′) $$c'x + \bar{c}'y - \gamma = 0,$$

(2) $$Ax + \bar{A}y \leq b,$$

(3′) $$x \geq 0,$$

(4) $$y \geq 0,$$

(5) $$\gamma \gtreqless 0,$$

This system defines a convex polyhedral set in $(n+1)$-dimensional space and since the plane projection of a convex set is convex, the projection of this set onto the (x_k, γ) plane will yield a convex polygon whose upper (lower) boundary is a concave (convex) function of the abscissa. One such polygon is shown in Figure 5.

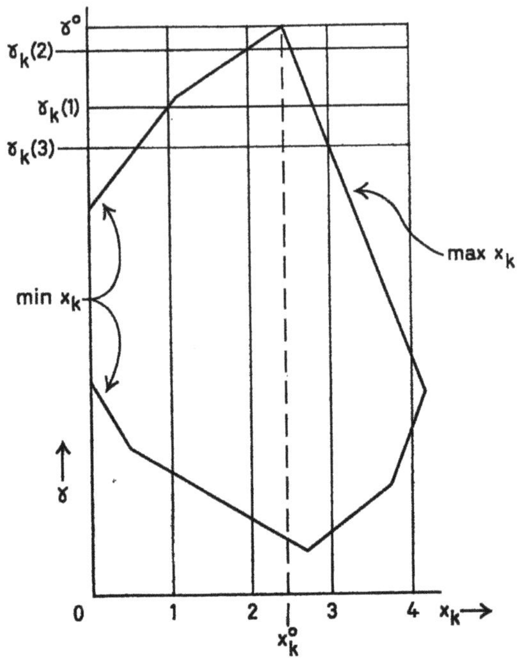

FIGURE 5

The value γ^0 which γ takes at the highest point of the polygon is the maximum value γ can take subject to the given inequality system. This is equivalent to saying that γ^0 is the optimal value of the functional γ in a linear programming problem with constraints (2), (3′), and (4). Further, x_k takes the value x_k^0 in this solution (if x_k were not in the optimum basis, the peak of the polygon would lie on the γ axis). The boundary of the polygon consists of the functions[3] min x_k and max x_k. These functions could be calculated by solving the two families of linear programming problems defined by (1′), (2), (3′), (4), and (5) with functionals "minimize x_k" and

[3] In a minimization problem, we are interested in the *lower* boundary of the polygon; if all coefficients of A, $\bar{A}$ and b are nonnegative, the point (x_k^0, γ^0) is at infinity and the polygon is unbounded; in this case, the functions min x_k and max x_k are monotonic.

"maximize x_k," for all possible values of γ using straightforward parametric linear programming [4, 7, 8]. A two-dimensional polygon of this type could be determined for every x variable of the original discrete problem.

In Figure 5, x_k^0 lies between the values 2 and 3. If min x_k and max x_k are traced out by reducing γ from γ^0 until an integer value of x_k is first encountered on each, the two values $\gamma_k(2)$ and $\gamma_k(3)$ on Figure 5 are obtained. This is equivalent to examining the path traced out by two specific corners of the intersection of the functional hyperplane with the convex set of feasible solutions when that hyperplane is systematically pushed back from its maximum position, γ^0. It may be said of $\gamma_k(2)$ that it is an upper bound on the maximand since at no higher value of γ can x_k take an integer value.

The solution of the discrete programming problem will be developed by the systematic use of this argument. For convenience of exposition, we define a set of subsidiary problems, $P(j)$, as follows:

$P(j)$: Maximize γ subject to constraints (2), (3′), and (4) above, and the additional constraints that j of the x variables be nonnegative integers ($j = 0,1,2,\ldots, n_1$).

Let S_j be the set of all feasible solutions to problems of type $P(j)$ and let $\bar{S}$ be the set of nonfeasible solutions to any problem of type $P(j)$. The particular problem in which, for example, x_2 and x_4 are constrained to be nonnegative integers, will be written $P(2; 2,4)$. and the set of its feasible solutions is written $S_2(2,4)$. In this notation, the required solution is the element of S_{n_1} for which γ is maximized. The γ value of this solution is bounded above by the maximum value of γ over the set S_{n_1-1} which is itself bounded by the maximum value of γ over S_{n_1-2}, etc. The ultimate upper bound reached in this manner is γ^0 ,the maximum value of γ over S_0.

The solution will be constructed in the form of a tree graph whose vertices are elements either of one of the sets S_j, $j = 0,1,2,\ldots, n_1$, or of the set $\bar{S}$. The steps of this construction are:

Step 0. The first vertex of the tree is the optimum solution of $P(0)$, which is now given the *label* γ^0. If this solution also satisfies $P(n_1)$, it is the required solution.

Step 1. If γ^0 is not the required solution then, according to the rules given below, an arc is drawn to each of two points in S_1 (or, possibly, in $\bar{S}$). γ is evaluated at these points if they are in S_1, but the γ value of a point in $\bar{S}$ need never be calculated.

Step 2. If vertices $\gamma^0, \gamma^1, \gamma^2, \ldots, \gamma^{k-1}$, have already been labelled, the highest (according to the value of γ) unlabelled vertex not in $\bar{S}$ is given the label γ^k.

Step 3. The final solution has been reached when for the first time a labelled vertex is an element of S_{n_1}; this occurs as soon as all the x variables of a solution are nonnegative integers. If γ^k is not such a solution, suppose it to be an element of S_j. A new arc is drawn originating at the (labelled) vertex immediately above γ^k (this is not necessarily γ^{k-1}) and terminating at a point in the same subset of solutions (S_j) as γ^k or in $\bar{S}$. If this new vertex is in S_j, its γ value is necessarily less than or equal to γ^k.

Step 4. Two arcs are drawn from γ^k to points in S_{j+1} or in $\bar{S}$. Again, if these points are in S_{j+1}, their γ values are less than or equal to γ^k. The last two steps add three new arcs and vertices to the tree. Return to Step 2.

The labelled vertices form a sequence of nonincreasing upper bounds on the γ value of the final solution. In addition, the γ value of an unlabelled vertex of a set S_j cannot exceed that of a labelled vertex and therefore, at the time that it is labelled, γ^k is the greatest current upper bound on the optimal value of γ. Thus, provided the method used to add arcs and vertices to the graph covers all possibilities, the algorithm must lead to the optimum solution of $P(n_1)$, provided such a solution exists.

Rules for adding arcs and vertices. At Step 0, the tree consists of the single labelled vertex γ^0 which is an element of S_0.

The decision is now made that x_r, which is in the basis at γ^0, is to be constrained to nonnegative integral values, i.e., attention is focused on the problem $P(1;r)$. Let $x_r = \overset{0}{x_r}$ at $\overset{0}{\gamma}$. The functions min x_r and max x_r are traced as far as the points $([\overset{0}{x_r}], \gamma_{rm})$ and $([\overset{0}{x_r}] + 1, \gamma_{rM})$ respectively, where:

$[\overset{0}{x_r}]$ is the greatest integer less than or equal to $\overset{0}{x_r}$,
γ_{rm} is the maximum value of γ subject to $P(0)$ and the condition $x_r = [\overset{0}{x_r}]$, and
γ_{rM} is the maximum value of γ subject to $P(0)$ and the condition $x_r = [\overset{0}{x_r}] + 1$.

The vertices γ_{rm} and γ_{rM}, if they exist, are elements of S_1. It is seen, however, that both of them may be evaluated by solving a standard linear programming problem in $(n - 1)$ variables. If one of these problems, say "maximize γ, subject to $P(0)$ and to $x_r = [\overset{0}{x_r}]$" is not feasible, γ_{rm} does not exist and a vertex is added to $\bar{S}$ instead. This implies that S_1 does not contain an element for which $x_r = [\overset{0}{x_r}]$, which in turn implies that only values of x_r which satisfy $x_r \geq [\overset{0}{x_r}] + 1$ need be considered in developing the tree. If neither γ_{rm} nor γ_{rM} exist, x_r cannot be constrained to an integral value and the problem possesses no feasible solution. We have now added the two arcs and vertices of Step 1.

From Step 2, $\gamma^1 = \max(\gamma_{rm}, \gamma_{rM})$. In calculating γ_{rm} and γ_{rM}, we have in effect examined every value of γ between γ^0 and γ^1, and have shown that γ^1 is the maximum value of γ which is compatible with an integral value of x_r. Suppose $x_r = v$ at γ^1. This equation is added to the problem as a new constraint which applies to any branch of the tree which can be traced back to γ^1. The variable x_r is dropped from the calculations, thus reducing the dimensions of the set of solutions being explored to $(n - 1)$, and each boundary value b_i is reduced by an amount $a_{ir}v$. This new constraint is certainly valid with respect to solutions of $P(1; r)$ over the range of values of γ for which the only permissible integer value of x_r is v. The upper end of this range is γ^1 and, since the upper boundary of the polygon of Figure 5 is a concave function of the abscissa, the lower end must be one of $\gamma_r(v - 1)$ and $\gamma_r(v + 1)$. The larger of these two is the second best solution of $P(1; r)$.[4] One of the neighbouring values is already known (it is min $(\gamma_{rm}, \gamma_{rM})$) and an upper bound to the other can be found by extrapolating the appropriate function of x_r.[5] If this upper bound exceeds the other neighbouring value, the corresponding γ value should be determined exactly. The arc and vertex of Step 3 have now been added to the tree.

To illustrate the argument, consider the two-dimensional problem shown in Figure 2.

Step 0. The solution at A provides the first vertex of the tree, γ^0.

Step 1. The variable x_2 is selected and the points B $(x_2 = 2, \gamma = \gamma_{2m} = \gamma_2(2))$ and C $(x_2 = 3, \gamma = \gamma_{2M} = \gamma_2(3))$ are determined.

Step 2. $\gamma^1 = \gamma_{2M}$ for which $v = 3$.

Step 3. γ^1 is not the final solution. Since $x_2 = 4$ lies entirely outside the convex set S_0, $\gamma_2(4)$ is in $\bar{S}$. The only possible integral value of x_2 for all values of γ satisfying $\gamma_2(2) < \gamma \leq \gamma_2(3)$ is $x_2 = 3$; this constraint is valid for the line segment CD which is a one-dimensional subset of the set of feasible solutions to $P(0)$.

[4] Should it happen that, for example, $\gamma_r(v - 1) = \gamma^1$, $\gamma_r(v - 2)$ must be determined, and this step down of the integral argument of γ_r is continued until a value of $\gamma_r(v - w)$, $(1 \leq w \leq v)$, is strictly less than γ^1, or until a minimum integral value of x_r consistent with the ordinary linear programming constraints is reached. A parallel calculation will be needed for each integer value of x_r for which $\gamma_r(x_r) = \gamma^1$.

[5] This procedure is used to calculate γ for $x_1 = 0$ in Appendix 1.

Step 4. x_1 is now constrained to integral values. One of the new vertices (γ_{1m}) is in S_2 and the other is in $\bar{S}$.

Step 2. $\gamma^2 = \gamma_{1m} > \gamma_{2M}$.

Step 3. γ^2 is an element of S_2; therefore the point E, corresponding to γ^2 is the required solution. At this point, $x_1 = 2$, $x_2 = 3$.

Clearly at any stage of the solution, the only x variables which may be violating the discrete constraints are those in the current basis. Therefore, by successively constraining each such variable in the manner described above, either a feasible solution of $P(n_1)$ will be found or else it will be shown that no such solution exists.

In the general case, the first upper bound on the functional, γ^0, is the optimum solution to $P(0)$. The second upper bound is $\gamma^1 \leq \gamma^0$; γ^1 is the optimum solution to $P(1; r)$. This second upper bound could have been sharpened by finding the optimum solution to all problems of type $P(1)$ and selecting the *least* of these as γ^1. Just as in the simplex method, however, it is not normally useful to determine which basis change yields the greatest increase in the functional, in this analysis it is usually more economical of computing effort to trace the min and max functions of one variable only at each stage. The criteria for selecting the appropriate variable are discussed in the appendices.

If γ^1 is not a solution to $P(n_1)$, a new x variable is chosen from those in the basis at γ^1. Step 4 consists of tracing the max and min functions of this variable to the nearest integer values respectively above and below its value at γ^1. This adds two new vertices to the tree, and, from Step 2, γ^2 is the maximum value of γ taken over these and the neighbouring values $\gamma_r(v-1)$ and $\gamma_r(v+1)$ of γ^1. The whole argument can now be repeated with γ^2 replacing γ^1 as the current upper bound on the optimal value of γ. By continuing this process, a tree is formed each of whose vertices represents a known set of integer constraints (γ^1, for example, represents $x_r = v$). A branch terminates if it reaches a vertex in $\bar{S}$. Ultimately, either all branches have terminated in $\bar{S}$ and the problem has no feasible solution, or else a vertex, γ^f say, is reached for which all x variables are nonnegative integers. In principle, appropriate constraints are now applied to any x variable which has not yet been *constrained* at an integral value, and a vertex γ^F in S_{n_1} is reached which has the same γ value as the labelled vertex γ^f. Since the labelled vertices are upper bounds on the functional, γ^F must be the required optimal solution.

In practical terms, one would not necessarily stop a block of computations in the middle once the upper bound of γ with the current set of integral constraints had been proved to be lower than some value of γ which is still possible with a different set of constraints, since it may later be necessary to return to that branch. So long as a discrete solution when found is checked to determine that the upper bound of γ on every branch of the computation is at least as low as the solution value, the algorithm must yield the maximum solution. In automatic computation, where there is the problem of storing the bases associated with each branch of the tree, the best procedure may be to carry each branch in turn down to some predetermined "cut off" value of γ. Since in many discrete programming problems, it is easy to find a "good" as opposed to an optimum solution, there should be no difficulty in selecting a cut off value sufficiently low to ensure that the tree extends as far as a solution, and hence that the optimum solution has not been excluded from the tree. The cut off value can be raised during the course of the calculation as soon as better discrete solutions are found. Alternatively, if a high cut off value were chosen, the output of the computer could include sufficient information on each branch to enable the calculation to be restarted if no discrete solution were reached above that value.

4. NUMERICAL EXAMPLE

To illustrate the preceding sections, consider the following problem:[6]
Maximize

(1) $$\gamma = c'x + \bar{c}'y$$

subject to

(2) $$b = Px + Qy\,,$$

(3) x to be a vector of nonnegative integers,

(3′) $$x \geq 0\,,$$

(4) $$y \geq 0\,,$$

where:

$$\begin{array}{rcl}
x',y' & = & [\;\; x_1 \quad x_2 \quad x_3 \;\;\vdots\;\; y_1 \quad y_2 \quad y_3 \quad y_4 \;\;] \\
P,Q & = & \left[\begin{array}{rrr:rrrr} 10.9 & 3.6 & -40.8 & 43.9 & 7.1 & 1 & 0 \\ -86.8 & 32.7 & 24.3 & 13.8 & -12.6 & 0 & 1 \\ 60.9 & 68.9 & 69.0 & -56.9 & 22.5 & 0 & 0 \end{array}\right] \quad b = \left[\begin{array}{r} 82.3 \\ 77.3 \\ 86.5 \end{array}\right] \\
c',\bar{c}' & = & [\;\; 77.9 \quad 76.8 \quad 89.6 \;\;\vdots\;\; 97.1 \quad 31.3 \quad 0 \quad 0 \;\;]
\end{array}$$

[6] Two possible computational routines, one based on the solution of many simple linear programmes and the other on parametric linear programming are given in Appendix 1.

The tree corresponding to the sequence of the calculations is given in Figure 6.

Step 0. The solution to the problem defined by (1), (2), (3′) and (4) is

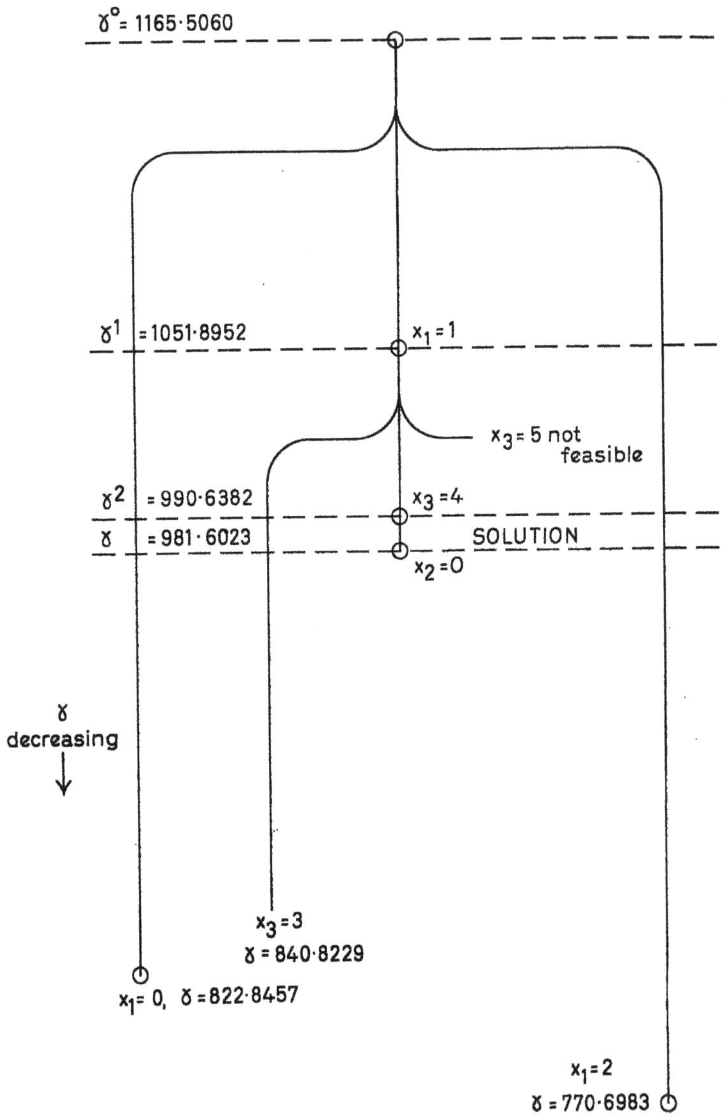

FIGURE 6

used as a starting point to the solution of the complete problem. In this solution

$$\gamma = \gamma^0 = 1165.5060 .$$[7]

Step 1. The upper bound of γ and the current solution at the start of this step are

$$\begin{aligned} \gamma^0 &= 1165.5060 \\ x_1 &= 1.4960 \\ x_3 &= 5.0210 \\ y_1 &= 6.1697 . \end{aligned}$$

As x_1 is furthest from an integer value, find the first integer value on the two functions min x_1 and max x_1.

1.1. Min x_1. As γ decreases the first integer value of x_1 which is encountered is $x_1 = 1$ at $\gamma = 1051.8952$.
1.2. Max x_1. The first integer encountered is $x_1 = 2$ at $\gamma = 770.6983$.

Step 2. The greater of these two is $\gamma = 1051.8952$ at $x_1 = 1$, and this is given the label γ^1. To complete this stage it is necessary to know the values of γ for the integers on either side of the one which is to be pursued.[8] Therefore:

Step 3. Min x_1 is continued as far as its second integer value, $x_1 = 0$, at $\gamma = 822.8457$.

These three values for γ are recorded in a list of values of γ which will be pursued in their order of magnitude from the highest down. Thus, at the beginning of each Step 2, one value will be taken from the list and at the end of Steps 3 and 4, three (lower) values will have been added to it. The computation is complete when all γ values remaining on the list are lower than a γ value associated with an integer solution. At this stage, the list is:

γ		*obtained in step*
1051.8952	γ^1	1
770.6983		1
822.8457		3

[7] For ease of presentation, the results have been rounded to four decimal places.
[8] In a larger problem, where several branches have to be pursued, it may be necessary to continue the min or max functions further. E.g., if $x_1 = 2$, $\gamma = 770.6983$ had to be picked up and explored (there being no integer solution for $\gamma > 770.6983$), it would be necessary to compute γ for $x_1 = 3$.

The highest value is 1051.8952 which has therefore been labelled γ^1 (Step 2). The solution and constraints at this point are:

Constraints:	$x_1 =$	1.
Solution:	$\gamma^1 =$	1051.8952 ,
	$x_1 =$	1 ,
	$x_3 =$	4.4089 ,
	$y_1 =$	5.4838 ,
	$y_2 =$	1.4851 .

Step 4. 4.1. Min x_3. At $x_3 = 4, \gamma = 990.6382$.

4.2. Max x_3. The first integer encountered is $x_3 = 4$ at $\gamma = 803.4034$.[9] The list of γ values is now:

γ		*obtained in step*
1051.8952	γ^1	1
770.6983		1
822.8457		3
990.6382		4

Step 2. The highest unlabelled value is 990.6382, which therefore becomes γ^2.

Step 3. Carry min x_3 one step further. At $x_3 = 3$, $\gamma \leq 840.8229$.[10] The solution and constraints are now:

Constraints:	$x_1 =$	$1, x_3 = 4$.
Solution:	$\gamma^2 =$	990.6382 ,
	$x_1 =$	1 ,
	$x_2 =$	0.2768 ,
	$x_3 =$	4 ,
	$y_1 =$	5.1513 ,
	$y_2 =$	1.0507 .

Step 4. 4.1. Min x_2. At $x_2 = 0, \gamma = 981.6023$.

4.2. Max x_2. This is a falling function. I.e., $x_2 = 1$ is not feasible, and another point is added to $\bar{S}$. Hence we have a solution at:

[9] This apparently anomalous result arises because max x_3 is a falling function of γ. It implies that the point for which $x_1 = 1$, $x_3 = 5$ lies in $\bar{S}$ since the problem with these constraints added is infeasible. It also implies that no feasible solution can be found in which $x_1 = 1$, $x_3 = 4$ and γ is less than 803.4034.

[10] This is an upper bound to the branch value of γ which has been obtained by ignoring the fact that y_2 would be negative at this point. The true branch value cannot exceed 840.8229, but it will be determined exactly only if the branch becomes active.

$$\begin{aligned} \gamma &= 981.6023 \\ x_1 &= 1 \\ x_2 &= 0 \\ x_3 &= 4 \\ y_1 &= 5.0702 \\ y_2 &= 1.6930\,. \end{aligned}$$

As this value of γ is higher than any remaining on the list, this is the optimum solution.

APPENDIX 1

COMPUTATIONAL METHOD

The starting point for both methods described in this appendix is the solution of the linear programming problem described by (1), (2), (3′) and (4) of Section 4, hereafter referred to as the initial problem. The integer restrictions (3) must now be considered. Since the two-dimensional set enclosed by min x_k and max x_k in the (x_k, γ)-plane is convex it follows that the first points for which γ must be determined are the integers on either side of the (nonintegral) value of one of the x variables. In principle, the routine will terminate irrespective of the x variable selected. But the limited experience afforded by the solution of two moderate size problems (Appendix 2) suggests that the criterion for selecting x_r has a considerable effect on the amount of computation needed. At first, x_r was chosen so as to maximize the decrease in γ at each step, which meant that the max and min functions had to be traced out for every x variable which was still nonintegral. As the calculations proceeded, it appeared that a more economical criterion (from the computational viewpoint) at each step would be to select the x variable which was furthest from an integer, and this rule was used in the remainder of the work. Another criterion which, however, has not been tested chooses the largest x variable with the object of making the sum of the constraints as large as possible. In some sense it must be true to say that the greater the sum of the constrained variables at any stage the more restricted is the remaining linear programming problem, and therefore the easier it is to solve.

In the example worked γ was determined for the points:

1. intersection of min x_1 with $x_1 = 1$; and
2. intersection of max x_1 with $x_1 = 2$.

Method 1. These two values of γ are found by solving the two linear programmes obtained by adding each of the above integral constraints in turn to the initial problem.

Additional constraint	γ
$x_1 = 1$	1051.8952
$x_1 = 2$	770.6983.

The remaining points on the tree of Figure 6 can be found similarly. In all, the nine linear programmes which have to be solved are:

1. Constraints (2), (3′) and (4) $\qquad \gamma_0$

2. Constraints (2), (3′) and (4) and $x_1 = 1$ γ_1
3. Constraints (2), (3′) and (4) and $x_1 = 2$
4. Constraints (2), (3′) and (4) and $x_1 = 0$
5. Constraints (2), (3′) and (4) and $x_1 = 1$, $x_3 = 4$ γ^2
6. Constraints (2), (3′) and (4) and $x_1 = 1$, $x_3 = 5$ (not feasible) $\bar{S}$
7. Constraints (2), (3′) and (4) and $x_1 = 1$, $x_3 = 3$
8. Constraints (2), (3′) and (4) and $x_1 = 1$, $x_3 = 4$, $x_2 = 0$ SOLUTION
9. Constraints (2), (3′) and (4) and $x_1 = 1$, $x_3 = 4$, $x_2 = 1$ (not feasible) $\bar{S}$

The result for programme 6 indicates that max x_3 (with $x_1 = 1$) is a falling function of γ. The intersection of max x_3 with $x_3 = 4$ lies on the lower boundary of the $(n-2)$-dimensional set of feasible solutions for which $x_1 = 1$, $x_3 = 4$. It is represented by the point $(4, \gamma_u)$ in Figure 7. Since γ cannot be lowered below this unknown value γ_u,

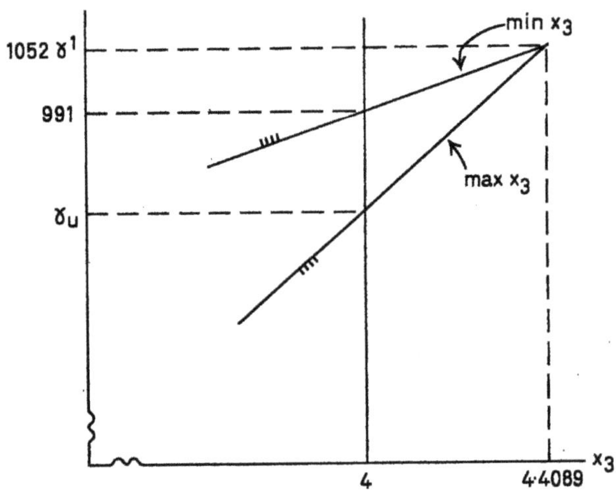

FIGURE 7

whilst still keeping x_3 fixed at 4, the point $x_1 = 1$, $x_3 = 4$, $\gamma = \gamma_u$ cannot be used as a starting point for exploration. Thus the fact that this method does not evaluate γ_u is irrelevant[11].

This method has the advantage that it uses only existing standard linear programming routines. If the time required to reach a solution is of paramount importance, e.g., for regular computer work or in desk computations, this method is slower than that based on parametric linear programming which is described in Method 2, below.

Method 2. The functional hyperplane is introduced as an additional (variable) constraint. This adds an extra row and column to the inverse basis, the entries in which can be easily deduced from the corresponding inverse basis in the initial problem. In the latter, let the first tableau be written in the form:

[11] If this value were desired, it could be obtained by ***minimizing*** γ subject to the the restrictions (2), (3′) and (4) and the imposed constraints $x_1 = 1$, $x_3 = 4$.

(A.1)
$$\begin{array}{r|c|c|c|}\hline & A & B & I_m \\ \hline \text{functional row} & c'_A & c'_B & 0' \\ \hline\end{array} \qquad \begin{array}{|c|}\hline b \\ \hline c_b = 0 \\ \hline\end{array}$$

where I_m is the $m \times m$ unit matrix, A is a possible basis, and, allowing for the appropriate reordering of the columns:

$$\begin{array}{|c|c|}\hline A & B \\ \hline c'_A & c'_B \\ \hline\end{array} = \begin{array}{|c|c|}\hline P & Q \\ \hline c' & \bar{c}' \\ \hline\end{array}$$

If the entire matrix (A.1) is premultiplied by

$$\begin{bmatrix} A & 0 \\ c'_A & -1 \end{bmatrix}^{-1} = \begin{bmatrix} A^{-1} & 0 \\ c'_A A^{-1} & -1 \end{bmatrix}$$

the result is:

(A.2)
$$\begin{array}{r|c|c|c|}\hline & I_m & A^{-1}B & A^{-1} \; (\text{inverse basis}) \\ \hline \text{functional row} & 0' & c'_A A^{-1}B - c'_B & c'_A A^{-1} \\ \hline\end{array} \qquad \begin{array}{|c|}\hline A^{-1}b \\ \hline c'_A A^{-1} b \\ \hline\end{array}.$$

When γ is added as a constraint, the basis corresponding to A for the enlarged problem is:

$$\left[\begin{array}{c|c} A & 0 \\ \hline c'_A & 1 \end{array}\right]$$

with inverse:

(A.3)
$$\left[\begin{array}{c|c} A^{-1} & 0 \\ \hline -c'_A A^{-1} & 1 \end{array}\right].$$

A unit column, with unit entry in the functional row is inserted in (A.1) between the inverse basis and the b column. The functional row now will have the form of the new constraint if the entry c_b is put equal to the current value of γ, i.e., if A is the current basis for the initial problem, $c_b = c'_A A^{-1} b$.

The convex set of feasible solutions to the initial problem is governed by the structure of the first m rows of (A.1). In the enlarged problem, the augmented matrix (A.1) plays the same role, and the result of multiplying this by (A.3) is:

(A.4)

		inverse basis			
I_m	$A^{-1}B$	A^{-1}	0		$A^{-1}b$
0	$c'_B - c'_A A^{-1}B$	$-c'_A A^{-1}$	1		0

Provided that A yielded a feasible solution to the initial problem, (A.4) represents a feasible solution to the enlarged problem for which as yet no functional has been specified. In particular, if A is an optimal basis for the initial problem, the entries in the bottom row of (A.4) which, with the exception of the elements in the last column of the inverse basis and in the b column, are obtained by multiplying the functional row of (A.2) by -1, are all nonpositive.

Throughout the calculation it is necessary to compute only the inverse basis and those rows of the rest of the tableau associated with the variable going out of the basis and with the x variable for which the functions min x and max x are currently being computed. In the larger problems summarised in Appendix 2, this represents a great economy over working with the whole tableau.

In the numerical example, the inverse basis and b column of (A.4) at the end of the solution of the initial problem are:

	I_1	I_2	I_3	I_4	b
x_1	0.0125	−0.0038	0.0088	0	1.4960
y_1	0.0401	0.0174	0.0176	0	6.1697
x_3	0.0220	0.0177	0.0213	0	5.0210
I_4	−6.8468	−2.9781	−4.2983	1	0

Columns I_1 and I_2 of the inverse basis correspond to the real variables y_3 and y_4, respectively, and at a later stage may enter the basis at a positive level. The remaining two columns correspond to disposal variables on equality constraints and therefore may appear in the basis *only* at zero level.

The next steps in the calculations consist of tracing the functions min x and max x by means of parametric programming.

Min x_1. In the enlarged problem, let the functional be x_1 and let the aim be to minimize this functional. The x_1 row becomes the functional row and the condition for a minimum will be satisfied if all the relevant elements in this row are negative. It can be seen that this requirement does not apply to columns I_3 and I_4 neither of which is a possible candidate for coming into the basis. The entire x_1 row[12] is calculated by premultiplying the matrix

$$\begin{bmatrix} P & Q \\ c' & \bar{c}' \end{bmatrix}$$

[12] Note that elements outside the inverse basis are not required to a very high degree of accuracy—only sufficient to enable a clear choice of a new basic variable to be made by the ratio rule given below. The only exception is the column corresponding to the new vector to be brought in when a change of basis has to be made, all the elements of which must be calculated to the same degree of accuracy as the inverse.

by the x_1 row vector of the inverse

$$[0.0125 \quad -0.0038 \quad 0.0088 \quad 0]$$

to give

x_1	x_2	x_3	y_1	y_2	$y_3(=I_1)$	$y_4(=I_2)$
[1	0.524	0	0	0.334	0.0125	−0.0038] .

Not all the entries are negative, so a basis change is required. The variable which is to go out of the basis is I_4, the elements of which are[13]

x_1	x_2	x_3	y_1	y_2	y_3	y_4
[0	−341.4	0	0	−76.5	−6.8468	−2.9781] .

The variable chosen to come in is y_2 since this ensures that all elements in the new x_1 row except the I_4 column are negative.[14] When this change has been made, the I_4 column and the b column are:

	I_4	b
x_1	0.0044	1.4960
y_1	0.0060	6.1697
x_3	0.0054	5.0210
y_2	−0.0131	0

If now y is decreased by d units, the elements b_i of the b column will alter according to the rule,

$$b_i^* = b_i - dI_{4i} \qquad (i = 1,2, \ldots, 4),$$

and, in particular

$$x_1^* = 1.4960 - d(0.0044)\,.$$

Therefore the value $y_1(1)$ of y for which the line $x_1 = 1$ intersects the function min x_1, cannot exceed

$$1165 - d_{\min} = 1165 - \frac{0.4960}{0.0044} = 1051.8952\,.$$

If any of the other variables should become negative for some value of d smaller than $d_{\min}$, a change of basis will be required. The variable which is brought in as the result of a change of this type is chosen so as to maintain the optimal character of the functional row by a rule of the same nature as that described in the footnote. From the convexity of the set in Figure 5, such a change reduces the gradient of min x_1 as a function of y and $y_1(1)$ would then be less than 1051.8952.

The remaining elements of the inverse basis and the b column need be calculated only if it becomes necessary actually to carry out the change of basis concerned.

If the entry in the I_4 column and x_1 row had been negative, this would have demonstrated the impossibility of reducing x_1 as a function of falling y and would mean that the constraint $x_1 = 1$ was not feasible.

[13] From the solution to the initial problem.

[14] The variable chosen is the one for which $-x_{1j}/I_{4j}$ reaches its maximum value. Similarly, if max x_1 were under consideration, the choice would be governed by the maximum value of x_{1j}/I_{4j}.

A similar method is used to determine the value of γ for the max x_1 function. The value of γ corresponding to $x_1 = 0$ has also to be determined. An upper bound to it can be obtained from the min x_1 calculation:

$$1165.5060 - \frac{1.4960}{0.0044} = 822.8457 .$$

Since neither y_1 nor x_3 has decreased to zero for this value of γ, this is exactly the value of γ for which $x_1 = 0$.

γ^1 is selected as the starting point for Step 4 and the complete inverse basis and b column for $\gamma = \gamma^1 = 1051.8952$ are worked out. An extra constraint, $x_1 - F = 1$, where F is a disposal variable which must always be zero, is now added to the programme. Theoretically this adds a row (F) and a column (I_5) to the inverse basis. The variable F (which is at present in the basis) plays the same role at this stage of the calculations as I_4 did in the previous stage. It can easily be shown that apart from the column I_5 the elements in the F row of the inverse basis are exactly the same as the elements of the x_1 row and therefore as soon as any basis change is made, the elements of the x_1 row in columns I_1 to I_4 vanish. It is now impossible for any basis change to affect the x_1 row and hence both it and the column I_5 may be dropped from the basis. The x_1 column in the main part of the tableau may also be deleted, since x_1 is now constrained to be equal to one.

In the succeeding steps of the calculation, the same method is employed to trace out the functions min x and max x, care being taken always to work down from the greatest value of γ which is still compatible with the constraints so far imposed. In the small numerical example, the solution was found without having to examine more than one branch of the tree, but in general this will not occur nor, of course, is there any guarantee that a solution exists.

Towards the end of a calculation following the second method, it may be helpful to use the first method to investigate some of the branches.The value of the first method is most apparent when the integral constraints have reduced the entries in the b column so much that a solution can almost be found by inspection.

Although the dual method for adding constraints used by Markowitz and Manne [**6**] can be adapted to the requirements of this algorithm, it appears to involve more work than the parametric method described above.

APPENDIX 2

The methods described in this paper have been tried out on the data published by Markowitz and Manne [**6**] as a hypothetical production problem. The problem they solved is:

$$\text{Maximize } c'x = \gamma ,$$
$$\text{subject to } Ax \leq b ,$$
$$x_j = 0 \text{ or } 1 ,$$

where A is a 6×21 matrix of technical coefficients, all of which are positive integers ranging from 3 to 99.

An alternative problem restricts the variables only to being nonnegative integers:

$$x_j = \text{any nonnegative integer.}$$

"ANY INTEGER" PROBLEM

We have solved both problems, but have chosen to present only the second in any detail since it is the more general problem. The main description of the calculations is the "tree" figure (Figure 8, p. 518), the text being in the nature of an annotation to the tree.

The solution of the primary problem is:

$$\begin{aligned} &\text{Maximize } c'x = \gamma\,, \\ &\text{subject to } Ax \leq b\,, \\ &\qquad\qquad\quad x \geq 0\,, \end{aligned}$$

$$\begin{aligned} \gamma^0 &= 643.99,^{15} \\ x_8 &= 1.77\,, \\ x_{11} &= 2.54\,, \\ x_{12} &= 1.04\,, \\ x_{13} &= 2.06\,, \\ x_{15} &= 0.97\,. \end{aligned}$$

It will be seen that x_{11} is the value furthest from an integer. In fact the min and max functions were computed for all 5 variables and x_{11} provided the least upper bound. Thus in this case the a priori selection of x_{11} as the variable to be considered would have given the same result as considering all variables. This is not always the case, however, and on the figure some γ's will be found marked with an asterisk showing that in the following step the x variable is not the one which would have been chosen following the "furthest from an integer" rule. During the later calculations this rule was followed, and the min and max functions were not obtained for every x variable. The calculations followed the procedure described as Method 2 in Appendix 1 except that where the b values had been reduced so much that only a few integer solutions were possible (i.e., when the sum of the constraints reached 5, 6 or 7) Method 1 was tried. Since, in general, Method 1 discovers an interior point of the convex set enclosed by the min and max functions, successively higher integer values of the x variable being examined must be explored until an integer is reached for which the constraints are no longer feasible. These nonfeasible branches are omitted from Figure 8. The other branches for which Method 1 was used are marked M.1. and for these the highest integer solution compatible with the constraints has been determined and is shown. The γ values on the other branches are upper bounds on the value of the functional, obtained by Method 2.[16]

Two optimum solutions were discovered at $\gamma = 594$. These are:

1st solution				*2nd solution*
1	=	x_5		
1	=	x_6		
		x_8	=	2
2	=	x_{11}	=	1
1	=	x_{12}	=	2
2	=	x_{13}	=	1
1	=	x_{15}	=	1
		x_{20}	=	1

[15] Results have been rounded to two decimal places.

[16] There may be other nonfeasible constraints for which an upper bound is shown. This merely means that the Method 2 calculation has not been taken far enough to show up the infeasibility.

It will be seen from the tree that the solution of the problem involved 37 steps (apart from the primary solution). This involved exploring 74 7 × 7 inverse bases, in

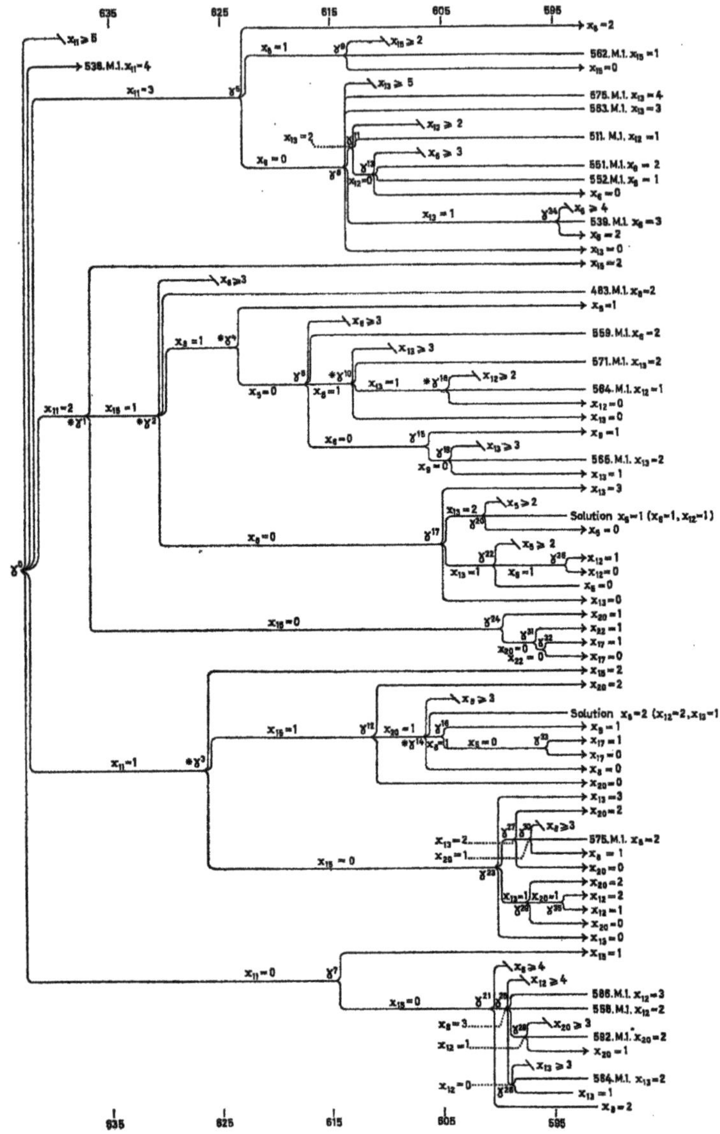

FIGURE 8

25 of which only the γ column was needed, and using Method 1 39 times, including 19 times in which the constraints were not feasible.

"0 OR 1" PROBLEM

The primary problem in this case was:

$$\begin{aligned} \operatorname{Max} c'x &= \gamma\,, \\ \text{subject to } Ax &\le b\,, \\ 0 \le x &\le 1\,. \end{aligned}$$

Dantzig's method for bounded variables [2] was used so that here again the inverse bases were of order 7 × 7. The primary solution is

$$\begin{aligned} \gamma^0 &= 594.40\,, \\ x_5 &= 0.69\,, \\ x_6 &= 1\,, \\ x_8 &= 1\,, \\ x_9 &= 0.29\,, \\ x_{11} &= 1\,, \\ x_{12} &= 1\,, \\ x_{13} &= 1\,, \\ x_{15} &= 0.94\,, \\ x_{20} &= 0.96\,. \end{aligned}$$

It will be noted that even in the primary solution there are 5 variables already at the level one, and it remains true throughout the calculation that at each γ value there are not only variables which are constrained to be equal to zero or one, but also others which are at the value one because of the constraints $x_j \le 1$. This has the unfortunate result that Method 1 cannot be used even when there are seven variables at the unit level, since they are not constrained to be equal to one, and therefore cannot be subtracted from the b column. Indeed at one point when approaching an integer solution at $\gamma = 538$ (the optimum being $\gamma = 540$) every unused x variable had to be successively constrained to be equal to zero before the upper bound of γ on that branch could be pushed below 540. This is not quite so disastrous as it sounds since in this situation one simply explored a series of bases with only two variables other than slack variables.

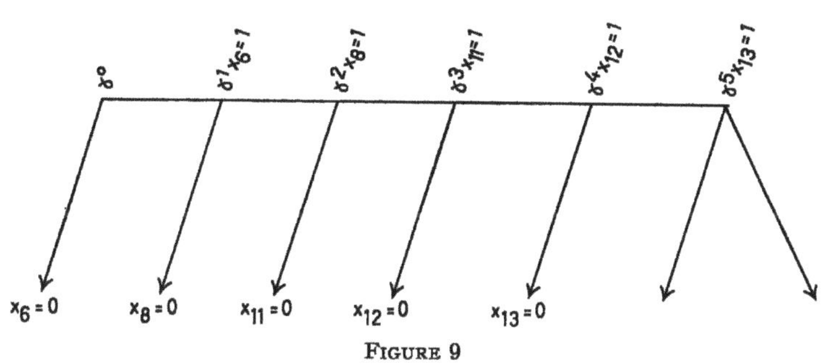

FIGURE 9

In summary there were 53 steps in this calculation. The total number of different bases was 165 in which 2 had 6 slack variables, 42 had 5, 63 had 4, 41 had 3, 15 had 2, and 2 had only 1. In 42 of the bases only the γ column was computed.

It appears to us almost certain that a different criterion for selecting the x variable at each step would have considerably reduced this work. In particular it appears that one should select not merely the greatest noninteger variable but the greatest unconstrained variable. In the problem starting with bounded variables, in other words, one would start by selecting (arbitrarily) x_6, x_8, x_{11}, x_{12}, or x_{13} and investigate the effect of reducing it to zero. The other side of the fork would be constraining it to be equal to unity, and hence would not reduce y at all. The start of the tree, therefore, would appear as:
The testing of this hypothesis, however, will be postponed until it is programmed for an electronic computer.

Acknowledgement. We wish to thank Mr. Neil Swan for his valuable assistance in solving the numerical examples.

London School of Economics and Political Science

REFERENCES

[1] BEALE, E. M. L.: "A Method of Solving Linear Programming Problems When Some But Not All of the Variables Must Take Integral Values," Statistical Techniques Research Group Technical Report No. 19, Princeton, N. J., July, 1958.

[2] DANTZIG, G. B.: "Recent Advances in Linear Programming," *Management Science*, 2, (1956), pp. 131–44.

[3] DANTZIG, G. B.: "Discrete-Variable Extremum Problems," *Operations Research*, 5, (1957), pp. 266–77.

[4] GASS, S. I. AND T. L. SAATY: "Parametric Objective Function, II, Generalisation," *Journal of the Operations Research Society of America*, 3, (1955), pp. 395–401.

[5] GOMORY, R. E.: "Outline of an Algorithm for Integer Solutions to Linear Programs," *Bulletin of the American Mathematical Society*, 64, (1958), pp. 275–8.

[6] MARKOWITZ, H. M. AND A. S. MANNE: "On the Solution of Discrete Programming Problems," *Econometrica*, 25, (1957), pp. 84–110.

[7] ORCHARD-HAYS, W.: "Revisions and Extensions to the Simplex Method," RAND Corporation, P–562, 2 September, 1954.

[8] SAATY, T. L. AND S. I. GASS: "Parametric Objective Function, I," *Journal of the Operations Research Society of America*, 2, (1954), pp. 316–9.

Chapter 6
Integer Programming: Methods, Uses, Computation

Michel Balinski

Introduction by *Michel Balinski*

This article exists because Robert Thrall, at the time Editor of Management Science, invited me to write a survey of the then new area of "integer programming." First printed in 1965, it was subsequently reprinted in 1968 in Mathematics of the Decision Sciences (edited by George B. Dantzig and Arthur F Veinott, Jr., a volume of the AMS's Lectures in Applied Mathematics) and in 1970 in Proceedings of the Princeton Symposium on Mathematical Programming (edited by Harold W. Kuhn, a Princeton University Press publication). In its last resurrection it was supplemented by 16 pages of "Recent Developments." By then the bibliography had grown from the original 105 items to 232 items: the field was burgeoning.

I now believe that this article was the first virtual text book on integer programming, for it seems to have been used as such in a rich variety of university courses. In fact, I was well on my way to integrating, supplementing and transforming the material of the article and other more recent developments into a bona fide textbook when in June 1967 a particularly violent summer storm decided otherwise: lightning caused a surge of current in the main electric line connected to our house, it skipped over the fuse, destroyed a part of the insulation in my study on the top floor, allowing a fire to ignite and warm up slowly undetected until a noise of slamming doors turned out to be the booms of falling beams. All of my work, books, notes, papers and articles were consumed in the flames. The burning desire to write a book extinguished: the effort to remember and redo was at once too daunting and too boring. I looked elsewhere, though not so far afield.

Why Bob Thrall asked me I really don't know! I had written several papers with Ralph Gomory, but not in integer programming. I belonged to the Princeton crowd that worked in linear programming and its extensions. My formal thesis adviser had

Michel Balinski
Département d'Économie, École Polytechnique, Palaiseau, France
e-mail: michel.balinski@polytechnique.edu

M. Jünger et al. (eds.), *50 Years of Integer Programming 1958-2008*,
DOI 10.1007/978-3-540-68279-0_6, © Springer-Verlag Berlin Heidelberg 2010

been Albert Tucker—though he was away on a sabbatical leave in 1958-59 while I wrote the thesis—so the person to whom I turned for advice and solace was Ralph Gomory, then a Lecturer in mathematics at Princeton, working on his first algorithm for general integer programs. Ralph tells me that—given this premise—I can claim to be his only student, a distinction that I accept with pleasure. The year that Jack Edmonds spent at Princeton we shared an office. I had also spent several summers at the RAND Corporation, so with Philip Wolfe, George Dantzig, Ray Fulkerson, Lloyd Shapley and others.

The decade of the 1960's was a very exciting one for a budding mathematician. It was a golden age: the idea pervaded that every problem could be solved with mathematics (or operations research, its applied arm)! There were not that many mathematicians: I remember noticing once that I was one of slightly more than two hundred persons in the entire USA who had earned a Ph.D. in the discipline in 1959. In addition to teaching first in mathematics at Princeton, then in economics at the University of Pennsylvania, finally in mathematics at the Graduate Center of C.U.N.Y., I also worked with Mathematica, a consulting company, located in Princeton. I began working there as a graduate student before it was named. I still have a brochure describing its services and personnel (which dates from later on in the 1960's). The people pictured include Oskar Morgenstern (Chairman of the Board), Tibor Fabian (President, trained in O.R. and a pioneer in the use of linear programming in the iron and steel industry), Harold Kuhn, Al Tucker and me. In earlier days Ralph Gomory was involved (indeed, to the best of my memory, he was the person who proposed the name of the company), in later days George Dantzig and Robert Aumann were involved in one or another contract. The brochure's list (complete and in the order given) of

"Characteristic Assignments in Mathematical Analysis and Programming:

- Basic research in algorithms for discrete optimization problems.
- Development of nonlinear programming algorithms for transportation and routing problems.
- Development of an algorithm and computer program for reliability analysis of electromechanical networks.
- Devising computer programs for the solution of decomposable linear programs to handle up to 30,000 inequalities.
- Establishing game theoretic models of bargaining processes.
- Game theoretic analyses of antisubmarine warfare."

This article was frequently used to claim up-to-date expertise in discrete optimization and, in particular, helped obtain important contracts to study "fixed-cost transportation problems" and to redistrict the state of New Jersey's legislative districts, among others.

Optimization and game theory and their applications were the big themes of those days. They have been the big themes for me too, though in research game theory followed optimization. More and more, as the years past, equity—fairness and how to achieve it—came to the fore as the focus of my work. I first worked on problems described in this article: apportioning the seats of the (or a) House of Representatives among the various states. Then, I worked on another problem described in this article, also in the political realm, the districting or redistricting problem: how to cut up a state into congressional districts. Later I extended the "vector" apportionment problem to the "matrix" apportionment problem (a method of voting which has recently been adopted in some Swiss elections). Early on I worked on the matching problem (the Edmonds type), later on—with equity on my mind—I worked on stable matching (the Gale-Shapley type). Now I have come full circle and work on the more fundamental problem of how to elect one person and how to rank several competitors when more than one judge or voter has opinions in the matter. Here optimization plays only a minor role, whereas game theory plays a crucial part.

In retrospect what seems to be the most likely explanation for Bob Thrall's invitation to write this article is the company I kept. Whether this be true or not, I am indebted to him: the problems I found and talked about in it have kept me busy for a lifetime—if I am allowed to construe that the earlier political questions engendered the later one and that game theory is but a multi-function extension of finding an optimum solution.

The following article originally appeared as:

M. Balinski, *Integer Programming: Methods, Uses, Computation*, Management Science 12 (1965) 253–313.

MANAGEMENT SCIENCE
Vol. 12, No. 3, November, 1965
Printed in U.S.A.

INTEGER PROGRAMMING: METHODS, USES, COMPUTATION*†

M. L. BALINSKI‡[1]

University of Pennsylvania and MATHEMATICA

This paper attempts to present the major methods, successful or interesting uses, and computational experience relating to integer or discrete programming problems. Included are descriptions of general algorithms for solving linear programs in integers, as well as some special purpose algorithms for use on highly structured problems. This reflects a belief, on the author's part, that various clever methods of enumeration and other specialized approaches are the most efficacious means existent by which to obtain solutions to practical problems. A serious try at gathering computational experience has been made—but facts are difficult to uncover.

The paper is written with intent to enable readers to read selected sections without having to read the whole.

A. Methods

1. *Introduction*

The integer programming problem is: find $x_0, x_1, \cdots, x_{n+m}$ with x_i integer valued for some specific set of indices $i \varepsilon J$ in order to

$$\text{(A.1)} \quad \begin{array}{ll} \text{maximize} & x_0 = a_{00} + \sum_1^n a_{0j}(-x_j) \\ \text{when} & x_{n+i} = a_{i0} + \sum_1^n a_{ij}(-x_j) \geqq 0, \qquad x_j \geqq 0 \end{array}$$

where, for simplicity we assume all a_{ij} are integer valued. Except for the requirement that a certain set of variables x_i must have integer values in a solution the problem is simply a linear program. In fact, for some problems (A.1) its solution as a linear program will already have the required integer property. This can happen fortuitously and might be expected to happen fairly often in certain classes of problems (see Section C.5).

In certain circumstances it is possible to predict that a linear program (A.1) will necessarily have an integer valued solution. The well known case of this occurrence is the class of mathematically equivalent problems which includes the assignment, transportation, and network flow problems. Dantzig [25] pointed out in 1949 that the transportation problem—and hence this class of problems—

* Received May 1965.

† This work was supported, in part, by MATHEMATICA under contract with the Army Research Office, United States Army, Contract No. DA-31-124-ARO(D)-31. Reproduction, translation, publication, use and disposal in whole or in part by or for the United States Government is permitted.

‡ Now at Graduate Mathematics Center, City University of New York.

[1] This is the third in a series of 12 expository papers commissioned jointly by the Office of Naval Research and the Army Research Office under contract numbers Nonr-4004(00) and DA-49-092-ARO-16, respectively.

always has integer solutions given integer valued "demands" and "supplies." In other terms, the "obvious" linear programming formulation of these integer programming problems leads to convex polyhedral constraint sets whose extreme points have integer components. Experience indicates that (so far) all real or practical problems which have been found to possess this property can—if properly interpreted—be displayed as a problem of the transportation or network type. However, there do exist (mathematical) problems which have this property but are not of transportation or network type.

If the solution of (A.1) as a linear program does not have the required integer property then techniques for achieving it must be invoked. The basic approach common to most methods is that of successively deducing supplementary linear constraints, from the linear constraints of (A.1) and the integer requirements, until a "new" linear program is obtained whose solution satisfies the integer requirements. The primitive idea of new constraint generation appears to have been first advanced by Dantzig, Fulkerson, and Johnson [31] in 1954 in their work on solving a travelling salesman problem, and, subsequently, by Markowitz and Manne [76] in 1957. In both cases the idea was used in an ad hoc manner on specific problems. In 1958 Gomory ([49] [53] [54]) developed a systematic method for new constraint generation, applicable to any "pure" problem (A.1) in which all variables are required to be integer valued, which guarantees that an integer solution is found (if any exists) in a finite number of steps. Then, Beale [9] and Gomory [50] developed algorithms for the "mixed" integer problem in which only a subset of the variables are required to be integer valued. In 1960 Gomory [48] generalized the new constraint generation idea to obtain a method for the "pure" problem which requires only addition and subtraction in computation (an "all-integer integer" method) provided the data (a_{ij}) of the problem is integer valued. Recently, papers on "nonlinear" integer problems have appeared. In [68] Kunzi and Oettli describe a method for minimizing a quadratic positive semidefinite function subject to linear and integer constraints which is based on solving a sequence of "mixed" integer problems; in [100] Witzgall generalizes the idea of Gomory's all-integer integer method to solve a problem in which the objective is linear, the variables required to be integer, and the constraints linear or "parabolic," i.e., of a form which may be transformed linearly into a constraint of type $t \geqq \sum_i t_i^2$.

In the above work the basic approach has been one of "convexification." Geometrically, the processes for solving problems (A.1) isolate a feasible point with the required integer property by making it an extreme point of a new polyhedral convex constraint set at which the linear form x_0 is maximized. The remarkable mathematical fact—and unquestionably the major technical accomplishment—is that it has been possible to devise new constraints in such a way that a finite number guarantees finding a linear program whose solution has the required integer property (if such a solution exists). However, these algorithms are "dual methods": no feasible solution to the problem of interest is obtained until an optimal solution is found . . . and this may require more iterations than can be afforded. There are, however, at least two proposals which offer primal integer methods. One, announced by Gomory in 1962 but never described in writing, has had poor computational success in some limited experimentation. Another is described by Ben-Israel and Charnes [15] but it requires, as a substep, the solution of a system of linear inequalities in integers. Finding a good primal method is most certainly an outstanding problem of great practical interest.[2] This situation is remedied, though in a perhaps artificial way, for the class of "mixed" problems through a partitioning approach due to Benders [13]. In this approach the problem is partitioned and every stage of the computation involves the solution of two subproblems, a pure integer problem and a linear programming problem.

Other approaches to solving "pure" and "mixed" problems (A.1) have been proposed. In 1960 Land and Doig [69] developed an enumerative type algorithm. It has not, as of February 1964, been tested computationally save by hand. Recently, Little and others [73] have used the same kind of idea in developing a very successful approach to solving the travelling salesman problem. Other proposals for special problems, e.g., problems (A.1) with variables x_i, $i \varepsilon J$, constrained to be 0 or 1 have been advanced by Bellman [10], by Benders and others [14], and, lately by Healy [60] (though no guarantee of finding an optimal solution is given in this last paper). Another special problem, the "knapsack problem" [26], has been solved by dynamic programming and by a special enumerative technique [43] which is closely allied to the ideas advanced in [14].

Some of the more interesting, important and useful of the above ideas are described in this part of the paper.

2. Cutting Methods

The basic technical vehicle for the "cutting methods," or the new constraint generation methods, is a dual simplex method. For the sake of being complete, we describe it briefly here.

The linear program (A.1) may be conveniently exhibited in tableau form as follows:

$$
\text{(A.2)} \qquad
\begin{array}{rc|c|ccc|}
 & & 1 & -x_1 & \cdots & -x_n \\
\hline
x_0 & = & a_{00} & a_{01} & \cdots & a_{0n} \\
\hline
x_{n+1} & = & a_{10} & a_{11} & \cdots & a_{1n} \\
\vdots & & \vdots & & \vdots & \\
x_{n+m} & = & a_{m0} & a_{m1} & \cdots & a_{mn} \\
\hline
x_1 & = & 0 & -1 & & \\
\vdots & & \vdots & & \ddots & \\
x_n & = & 0 & & & -1 \\
\hline
\end{array}
\qquad \text{all } x_j \geqq 0, j \neq 0.
$$

Here the "basic" variables are $x_0, x_{n+1}, \cdots, x_{n+m}$; the "nonbasic" variables are $x_1, \cdots, x_n$. A pivot step with pivot entry $a_{ij} \neq 0$, $i, j \neq 0$, is a complete

[2] Recently, another primal method has been announced [101].

elimination step which solves the equation of row i for the variable x_j in terms of x_{n+i} and the remaining nonbasic variables, and uses this equation to eliminate x_j from the remaining equations. x_j thereby becomes a basic and x_{n+i} a nonbasic variable of the new tableau. Thus, a pivot step is the simple algebraic operation so frequently encountered in elementary linear algebra. Its mechanics are most easily described in terms of an elementary column operation which takes one tableau of form (A.2) into another of the same form as indicated here:

$$\begin{array}{l|c|c|} & -x_j & \\ \hline x_0 = & & \\ x_{n+1} = & & \\ \vdots & \alpha_j & \alpha_k \\ x_{n+m} = & & \\ x_1 = & & \\ \vdots & & \\ x_n = & & \\ \hline \end{array} \quad \rightarrow \quad \begin{array}{l|c|c|} & -x_{n+i} & \\ \hline x_0 = & & \\ x_{n+1} = & & \\ \vdots & \bar{\alpha}_j & \bar{\alpha}_k \\ x_{n+m} = & & \\ x_1 = & & \\ \vdots & & \\ x_n = & & \\ \hline \end{array} \qquad \begin{array}{l} \bar{\alpha}_j = (-1/a_{ij})\alpha_j \\ \\ \bar{\alpha}_k = \alpha_k + a_{ik}\bar{\alpha}_j \\ \quad = \alpha_k \\ \qquad -(a_{ik}/a_{ij})\alpha_j \end{array}$$

where α_j is the column containing the pivot entry and α_k is any other column. The remaining "marginal" labels and variables stay in the same positions. For example, in Display I below, tableau (1) is transformed into tableau (2) with pivot entry -3; x_1 thereby becomes a basic and x_5 a nonbasic variable in tableau (2) as is indicated by the switch in these marginal labels. Successive tableaus, or sets of equations, obtained in this way are simply different representations of the same linear program but in terms of different partitions of the variables into basic and nonbasic sets.

A dual simplex method is initiated with a tableau of form (A.2) having $a_{01} \geqq 0, \cdots, a_{0n} \geqq 0$ or, more specifically, having lexicographically positive columns $\alpha_1 > 0, \cdots, \alpha_n > 0$, i.e., having the first nonzero entry of every column $\alpha_j (j \neq 0)$ positive. The aim of the computational process is to pivot on successive tableaus in order to obtain a tableau whose zero-column α_0 has all nonnegative components, except possibly the first which corresponds to the "current" value of the objective, while maintaining the lexicographic positive property of the remaining columns. If such a tableau is obtained then $(x_0, x_{n+1}, \cdots, x_{n+m}, x_1, \cdots, x_n)^T = \alpha_0$ is an optimal solution. Suppose, then, that in some tableau (A.2) there exists an entry $a_{i0} < 0$ for some i. Then either (a) every entry in the row of a_{i0} is nonnegative or (b) there exist negative entries. In case (a) that row of the tableau shows the noncompatibility of the constraints. In case (b) choose as the next pivot entry $a_{il} < 0$ where l is chosen by the usual ratio rule, $a_{0l}/a_{il} = \max_s \{a_{0s}/a_{is} \mid a_{is} < 0\}$; when ties occur resolve them by the corresponding ratio test between the next column entries a_{1s}/a_{is} of the tied columns, etc. This choice of pivot maintains lexicographically positive columns and assures termination of the algorithm in a finite number of pivot steps.

If an initial tableau (A.2) does not have $\alpha_1 > 0, \cdots, \alpha_n > 0$ then some special device is necessary. One particular such device is to append the extra constraint $x_{n+m+1} = a_{m+1,0} - x_1 - \cdots - x_n \geqq 0$ with $a_{m+1,0}$ very large; then pivot in this row and in the smallest lexicographic column. Clearly, the solution to this enlarged linear program is the solution to the original one if $x_{n+m+1} > 0$, i.e., if there exists a (bounded) solution to the original program.

We turn, now, to the cutting methods of Gomory, and first to the "pure" integer problem (A.1).

Suppose that at some stage of the computation a linear program or tableau is encountered all of whose variables are required to be integer and nonnegative, and all of whose variables are expressed as linear combinations of some nonbasic set, say $x_1, \cdots, x_n$. A typical equation of the tableau is (omitting row subscripts)

$$x = a_0 + \sum a_j(-x_j) \geqq 0. \tag{A.3}$$

From (A.3) we show how to derive a new linear constraint which must be satisfied by feasible integer solutions; or, equivalently, how to define a new nonnegative integer variable x' implied by (A.3).

Let $[t]$ denote the integer part of t, i.e., $t = [t] + r$, where $[t]$ is integer and $0 \leqq r < 1$. Define for $\lambda > 0$

$$a_j/\lambda = [a_j/\lambda] + r_j/\lambda, \quad 0 \leqq r_j < \lambda \quad (j = 0, 1, \cdots, n). \tag{A.4}$$

Then (A.3) may be rewritten, after dividing through by λ,

$$x/\lambda + (1/\lambda)\sum r_j x_j = \{[a_0/\lambda] + \sum [a_j/\lambda](-x_j)\} + r_0/\lambda. \tag{A.5}$$

Now, the left side of (A.5) is nonnegative, hence so is the right; but $r_0/\lambda < 1$ and the term in curly brackets is integer. Therefore, the term in curly brackets must be nonnegative as well as integer, i.e.,

$$x' = [a_0/\lambda] + \sum [a_j/\lambda](-x_j) \geqq 0 \tag{A.6}$$

is a new constraint or x' a new nonnegative integer variable implied by (A.3) and the fact that $x_1, \cdots, x_n$ are integer constrainted variables. This is the constraint developed in [48].

Suppose, now, that $\lambda = 1$. Rearrange (A.5) to obtain

$$\sum r_j x_j = \{[a_0] + \sum [a_j](-x_j) - x\} + r_0 . \tag{A.7}$$

The left side of (A.7) is nonnegative, hence so is the right; but $r_0 < 1$ and the term in curly brackets is integer. Therefore, the term in curly brackets must be nonnegative as well as integer, or, what is the same thing,

$$x' = -r_0 + \sum r_j x_j = -r_0 + \sum (-r_j)(-x_j) \geqq 0 \tag{A.8}$$

is a new constraint or x' a new nonnegative integer variable implied by (A.3) and the fact that $x_1, \cdots, x_n$ and x are integer constrained variables. This is the constraint developed in [49].

Suppose, now, that we are dealing with the "mixed" integer problem (A.1) in which only x_i for $i \varepsilon J$ are constrained to be integer valued. Consider a typical equation (A.3) of a tableau corresponding to a variable x which is integer constrained. Then, making the substitution (A.4) for $j = 0$ and $j \varepsilon J$, let $\lambda = 1$, and rewrite (A.3)

$$\sum_{j \varepsilon J} r_j x_j + \sum_{j \not\varepsilon J} a_j x_j = \{[a_0] + \sum_{j \varepsilon J} [a_j](-x_j) - x\} + r_0 . \tag{A.9}$$

The term in curly brackets is integer and either (a) greater or equal to 0 or (b) less than or equal to -1; i.e., in case (a)

$$\text{(A.10)} \quad \sum_{j\varepsilon J} r_j x_j + \sum_{j\notin J} a_j x_j \geqq r_0 \quad \text{hence} \quad \sum_{j\varepsilon J} r_j x_j + \sum_{\{j\notin J, a_j\geqq 0\}} a_j x_j \geqq r_0$$

and in case (b)

$$\text{(A.11)} \quad \begin{aligned} &\sum_{j\varepsilon J} r_j x_j + \sum_{j\notin J} a_j x_j \leqq -1 + r_0 \quad \text{hence} \quad \sum_{\{j\notin J, a_j<0\}} a_j x_j \leqq -1 + r_0 \\ &\qquad \text{or} \quad \sum_{\{j\notin J, a_j<0\}} (r_0/(1-r_0))(-a_j)x_j \geqq r_0 . \end{aligned}$$

Therefore, since the coefficients of all the variables in the last inequalities of (A.10) and (A.11) are nonnegative, we have

$$\text{(A.12)} \quad \begin{aligned} x' = -r_0 + \sum_{j\varepsilon J} (-r_j)(-x_j) + \sum_{\{j\notin J, a_j\geqq 0\}} (-a_j)(-x_j) \\ + \sum_{\{j\notin J, a_j<0\}} (r_0 a_j/(1-r_0))(-x_j) \geqq 0 \end{aligned}$$

is a new constraint or x' a new nonnegative (not integer) constrained variable implied by (A.3) and the fact that x and x_j for $j \varepsilon J$ are integer valued. This is the constraint developed in [50].

Using these new constraints Gomory proposed two algorithms for the pure problem, and one for the mixed problem.

Algorithm I ([49] 1958) for the "pure" problem. Obtain an optimal solution to the problem (A.1) considered as a linear program (assuming one exists, since otherwise no optimal integer solution exists). Then, if the solution is not in integers, generate from some row (A.3) with a_0 noninteger, a new row (A.8); adjoin it to the existing constraints to form a new linear program which is (lexicographically) dual feasible but not primal feasible; and use a dual simplex method to re-optimize. If x' of (A.8) becomes basic in any subsequent tableau, drop it and its defining equation or row. Continue to cycle through these steps until either an integer optimal solution is found or a constraint appears implying no feasible solution exists.

In geometric terms this algorithm considers, successively, optimal extreme point solutions to linear programs; if an extreme point X of some cycle has a noninteger component a closed half-space $x' \geqq 0$ defined by (A.8) is introduced which "cuts off" X, i.e., makes the last linear programming solution infeasible since at X the value of x' is negative. The constraint $x' \geqq 0$ can cut off no feasible integer solution since $x' \geqq 0$ and x' integer are implied by already imposed conditions. (See Display I for an example).

Algorithm II ([48] 1960) for the "pure" problem. In this algorithm a dual (lexicographic) simplex method is again used, but all computation is carried out through addition and subtraction of columns, i.e., the only pivot entries are -1. Given columns $\alpha_1, \cdots, \alpha_n$ in (A.1) or (A.2) lexicographically positive choose a row with $a_0 < 0$ for (A.3) and generate from it a new constraint (A.6). In it, take λ large enough, but as small as possible, to insure that the generated row or constraint (A.6) contains an entry -1 which is an eligible pivot entry for a dual simplex method. This is clearly possible unless $a_j \geqq 0$ for all $j \neq 0$

in (A.3); but, then, no solution exists. Rules for a choice of λ are: (a) the pivot entry $[a_i/\lambda] = -1$ must lie in the lexicographically smallest column having $\alpha_j < 0$; (b) for each column k with $\alpha_k < 0$ find the largest integer μ_k for which $(1/\mu_k)\alpha_k \geqq \alpha_j$ and let $\lambda_k = -a_k/\mu_k$ (the choice of $\lambda \geqq \lambda_k$ would assure that the new column $\bar{\alpha}_k$ in the subsequent tableau be lexicographically positive); and (c) take $\lambda = \max \{\lambda_k \mid \alpha_k < 0\}$ to assure preservation of dual feasibility (and if $\lambda = 1$ take $x' = x$). Pivot on the -1 entry of the new row and continue to repeat this step until either a primal solution is found or a constraint appears implying no feasible solution exists.

In geometric terms this algorithm generates, successively, new halfspace constraints $x' \geqq 0$ which are implied by already imposed conditions and takes as the next "trial" solution or point an integer point on the hyperplane $x' = 0$. (See Display II for an example.)

Algorithm III ([50] 1960) for the "mixed" problem. Obtain an optimal solution to the problem (A.1) considered as a linear program. If the solution does not satisfy the integer requirements then there must be a row (A.3) with x required to be integer and $a_0 > 0$ noninteger. Generate from it a new constraint (A.12); adjoin it to the existing constraints to form a new linear program which is (lexicographically) dual feasible but not primal feasible; and use a dual simplex method to reoptimize. If x' of (A.12) becomes basic in any subsequent tableau, drop it and its defining equation or row. Continue to cycle through these steps until either an optimal solution satisfying the integer requirements is found or a constraint appears implying no feasible solution exists.

The geometric interpretation is the same as for Algorithm I (see Display III for an example).

The remarkable mathematical fact is that these algorithms converge to an optimal solution with the required integer property in a finite number of steps if such a solution exists. Actually, "cuts" are rather subtle subjects despite their intuitively simple origins. For example, in Algorithm I, if instead of (A.8) the valid constraint $\sum_j x_j \geqq 1$ was used (i.e., the current basic solution is not all integer, hence the sum of the current nonbasic variables must be positive; but since these variables are integer constrained this means $\sum x_j \geqq 1$ [28]) or the new integer variable $x' = -1 + \sum_j (-1)(-x_j) \geqq 0$ were introduced, then it is known that convergence is not assured and will not occur for large classes of problems [56]. In the case of the "mixed" integer programming Algorithm III one extra stipulation must be made; namely, that $0 \in J$, i.e., that the value of the objective function x_0 must itself be required to be integer valued. It has been shown that if x_0 is not so constrained then it is possible for the process not to converge in a finite number of steps [99].

The proof of convergence for Algorithm I may be given as follows, where we assume that x_0 is bounded below and that the choice of row (A.3) from which to generate a new constraint is either always or periodically the topmost row with noninteger a_0. Let α_0^k be the column of $m + n + 1$ entries corresponding to the "current" values of the variables after concluding the kth reoptimization.

By the dual simplex method $\alpha_0^k > \alpha_0^{k+1}$ for all k. Assume the process does not converge in a finite number of steps, i.e., that no all integer α_0 is found. Since a_{00}^k is bounded below and $a_{00}^k \geqq a_{00}^{k+1}$ there must be a greatest integer $n_{00} \leqq a_{00}^k$ for all k. But, then, for some k, $a_{00}^k = n_{00} + r_{00}^k$ with $0 \leqq r_{00}^k < 1$; and if $r_{00}^k > 0$ the new constraint would be generated from

$$x_0 = n_{00} + r_{00}^k + \sum a_{0j}^k(-x_j) \quad \text{where} \quad a_{0j}^k \geqq 0$$

and hence would be

$$x' = -r_{00}^k + \sum (-r_{0j}^k)(-x_j) \geqq 0. \tag{A.13}$$

The next step of the dual simplex method would choose as pivot some entry of row (A.13), say $(-r_{0j}^k)$ which results in a new value for x_0

$$a_{00} = n_{00} + r_{00}^k - (a_{0j}^k r_{00}^k / r_{0j}^k) \leqq n_{00} \quad \text{since} \quad a_{0j}^k / r_{0j}^k \geqq 1. \tag{A.14}$$

Thus x_0 attains the integer value $a_{00} = n_{00}$ and can be no smaller. Repeat the same argument on each successive entry of α_0^k; each x_j, $j \neq 0$, is bounded below by the integer zero. If no constraint showing feasible solutions do not exist are encountered, every entry of α_0^k, k large enough, must become integer, thus contradicting the assumption of nontermination [49].

The proof of termination in a finite number of steps for Algorithms II and III may be given in much the same "lexicographic" terms.

The underlying approach of these "cutting methods" may be viewed in many ways. One such view—and perhaps the most "natural" or most immediate—is as a process of convexification. Another is as an instance of the general idea embodied in decomposition approaches [51]: this is the attempt to organize computation in large (and structured) linear programs in such a way that data is generated only when it is needed in the course of computation. In this view one imagines the integer program (A.1) as really being a linear program in which many (implied) linear constraints of rows of data have not yet been specified but are generated when needed by (A.6), (A.8) or (A.12). In the more familiar decomposition approaches columns of data have not yet been specified and are generated when needed (e.g., by a linear programming computation in [32], by a knapsack problem computation in [43], by a shortest path problem computation in [38]), and a primal simplex method is used instead of a dual simplex method.

In still another view, the integer programming algorithms for the "pure" problem may be interpreted as a generalization of the Euclidean algorithm for finding the greatest common divisor of a set of integers [3]; this is of pedagogical if not of practical interest. An all-integer algorithm of Glover [47]—which, in its very simplest form, is remarkably easy to explain but represents a considerable weakening of the constraints of Algorithm II—together with a recent explanation of the Euclidean Algorithm [17] makes clear this connection. The Glover all integer integer method is derived from two very simple observations. If a tableau is encountered in the course of computation which contains an equation (A.3) with

$$x = a_0 + \sum a_j(-x_j) \geqq 0, \quad \text{where} \quad a_j \geqq 0 \tag{A.15}$$
$$\text{all} \quad j \neq 0, k, \quad \text{and} \quad a_k < 0, a_0 < 0$$

then, clearly, $x_k \geqq a_0/a_k > 0$ and, since x_k must be integer valued, $x_k \geqq [a_0/a_k] + 1$, if that quotient is not integer. This lower bound on x_k is used as a new constraint on which to pivot instead of (A.6). If no such equation (A.15) is present in a tableau then one is created by picking some row (A.3) with $a_0 < 0$, finding among the entries $a_j < 0$ that one corresponding to the least lexicographic column α_k, subtracting the column α_k from other columns α_j with $a_j < 0$, and repeating until the row has form (A.15). In fact this sequence of operations can be summarized by a constraint of type (A.6) with constant term $[a_0/\lambda] = 0$. Finiteness of the procedure is easily proved by lexicography once again; however, it seems that the method is very much weaker than Algorithm II and would be considerably less successful in computation.

Display I

Consider the "pure" problem: find integers x_i ($i = 0, \cdots, 5$) maximizing x_0 and satisfying the constraints of tableau (1) below and $x_i \geqq 0$, $i \neq 0$. Pivot entries are designated by stars; rows (A.3) used for the generation of a new constraint (A.8) by arrows.

(1)	1	$-x_1$	$-x_2$
$x_0 =$	0	4	5
$x_3 =$	−2	−3	−1
$x_4 =$	−5	−1	−4
$x_5 =$	−7	−3*	−2
$x_1 =$	0	−1	0
$x_2 =$	0	0	−1

(2)	1	$-x_5$	$-x_2$
$x_0 =$	−28/3	4/3	7/3
$x_3 =$	5	−1	1
$x_4 =$	−8/3	−1/3	−10/3*
$x_5 =$	0	−1	0
$x_1 =$	7/3	−1/3	2/3
$x_2 =$	0	0	−1

(3)	1	$-x_5$	$-x_4$	
$x_0 =$	−112/10	11/10	7/10	
$x_3 =$	42/10	−11/10	3/10	
$x_4 =$	0	0	−1	
$x_5 =$	0	−1	0	
$x_1 =$	18/10	−4/10	2/10	←
$x_2 =$	8/10	1/10	−3/10	
$x'_1 =$	−8/10	−6/10*	−2/10	

(4)	1	$-x'_1$	$-x_4$	
$x_0 =$	−76/6	11/6	2/6	
$x_3 =$	34/6	−11/6	4/6	←
$x_4 =$	0	0	−1	
$x_5 =$	8/6	−10/6	2/6	
$x_1 =$	14/6	−4/6	2/6	
$x_2 =$	4/6	1/6	−2/6	
$x'_2 =$	−4/6	−1/6	−4/6*	

(5)	1	$-x'_1$	$-x'_2$
$x_0 =$	−13	7/4	2/4
$x_3 =$	5	−2	1
$x_4 =$	1	1/4	−6/4
$x_5 =$	1	−7/4	2/4
$x_1 =$	2	−3/4	2/4
$x_2 =$	1	1/4	−2/4

The new constraints, expressed in terms of x_1 and x_2 are $x'_1 = -6 + 2x_1 + 2x_2 \geqq 0$ and $x'_2 = -5 + x_1 + 3x_2 \geqq 0$. The progression of the "trial" solutions corresponding to the five tableaus may be graphed in (x_1, x_2)-space. The feasible region (ignoring integer constraints) lies to the northeast; the shaded region represents that part of the feasible region "cut off" by the new constraints. The numbered points show the correspondence with the tableaus.

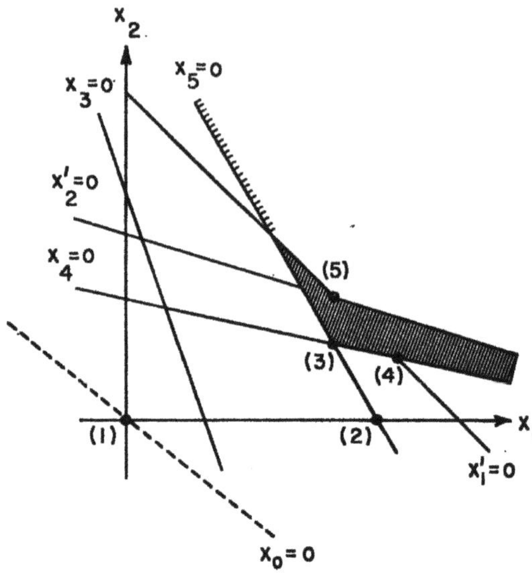

Display II

Consider the same "pure" problem as stated in "Display I." The same notation is used.

(1)	1	$-x_1$	$-x_2$	
$x_0 =$	0	4	5	(a) Pivot column 1
$x_3 =$	-2	-3	-1	(b) $(1/1)4 \geqq 4, \lambda_1 = 3$
$x_4 =$	-5	-1	-4	$(1/1)5 \geqq 4, \lambda_2 = 2$
$x_5 =$	-7	-3	-2	←
$x_1 =$	0	-1	0	(c) $\lambda = 3$
$x_2 =$	0	0	-1	
$x'_1 =$	-3	-1^*	-1	

(2)	1	$-x'_1$	$-x_2$	
$x_0 =$	-12	4	1	(a) Pivot column 2
$x_3 =$	7	-3	2	
$x_4 =$	-2	-1	-3	←
$x_5 =$	2	-3	1	(b) $(1/4)4 \geqq 1, \lambda_1 = 1/4$
$x_1 =$	3	-1	1	$(1/1)1 \geqq 1, \lambda_2 = 3$
$x_2 =$	0	0	-1	(c) $\lambda = 3$
$x'_2 =$	-1	-1	-1^*	

(3)	1	$-x'_1$	$-x'_2$
$x_0 =$	-13	3	1
$x_3 =$	5	-5	2
$x_4 =$	1	2	-3
$x_5 =$	1	-4	1
$x_1 =$	2	-2	1
$x_2 =$	1	1	-1

The new constraints, expressed in terms of x_1 and x_2 are $x'_1 = -3 + x_1 + x_2 \geqq 0$ and $x'_2 = -4 + x_1 + 2x_2 \geqq 0$. The progression of the three "trial" solutions corresponding to the tableaus is graphed below.

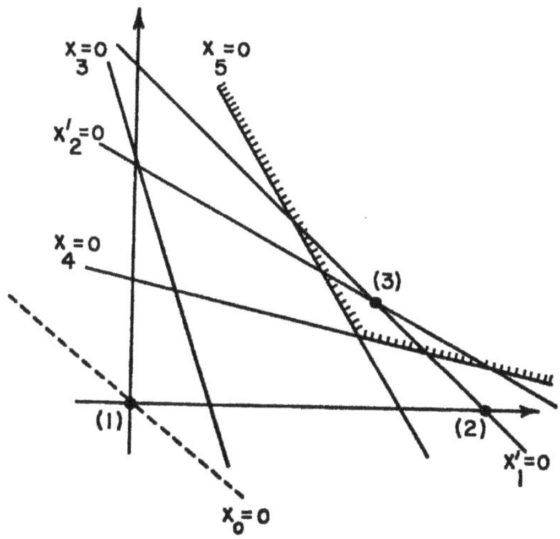

Display III

Consider the "mixed" problem which is identical with the problem of "Displays I and II" except that only x_0 and x_1 are constrained to be integer, i.e., $J = \{0, 1\}$. The same notation is used. Since in this algorithm the first step is to solve the linear program, tableaus (1), (2), (3) are identical with those of "Display I."

(3)	1	$-x_3$	$-x_4$	
$x_0 =$	−112/10	11/10	7/10	
$x_3 =$	42/10	−11/10	3/10	
$x_4 =$	0	0	−1	
$x_5 =$	0	−1	0	
$x_1 =$	18/10	−4/10	2/10	←
$x_2 =$	8/10	1/10	−3/10	
$x'_1 =$	−8/10	−16/10*	−2/10	

(4)	1	$-x'_1$	$-x_4$	
$x_0 =$	−188/16	11/16	9/16	←
$x_3 =$	76/16	−11/16	7/16	
$x_4 =$	0	0	−1	
$x_5 =$	1/2	−10/16	2/16	
$x_1 =$	2	−4/16	4/16	
$x_2 =$	12/16	1/16	−5/16	
$x'_2 =$	−4/16	−11/16*	−9/16	

(5)	1	$-x'_2$	$-x_4$	
$x_0 =$	−12	1	0	
$x_3 =$	5	−1	1	
$x_4 =$	0	0	−1	
$x_5 =$	8/11	−10/11	7/11	
$x_1 =$	23/11	−4/11	5/11	←
$x_2 =$	8/11	1/11	−4/11	
$x'_3 =$	−1/11	−4/110	−5/11*	

(6)	1	$-x'_2$	$-x'_3$
$x_0 =$	−12	1	0
$x_3 =$	24/5	−54/50	11/5
$x_4 =$	1/5	4/50	−11/5
$x_5 =$	3/5	−48/50	7/5
$x_1 =$	2	−2/5	1
$x_2 =$	4/5	6/50	−4/5

The progression of the six "trial" solutions corresponding to the tableaus is graphed below. Again, the shaded region has been "cut off" by the new constraints $x'_1 \geqq 0$, $x'_2 \geqq 0$, $x'_3 \geqq 0$. The new constraints, expressed in terms of x_1 and x_2 are $x'_1 = -13 + 5x_1 + 4x_2 \geqq 0$, $x'_2 = -12 + 4x_1 + 5x_2 \geqq 0$, $x'_3 = -14/5 + 3/5\, x_1 + 2x_2 \geqq 0$.

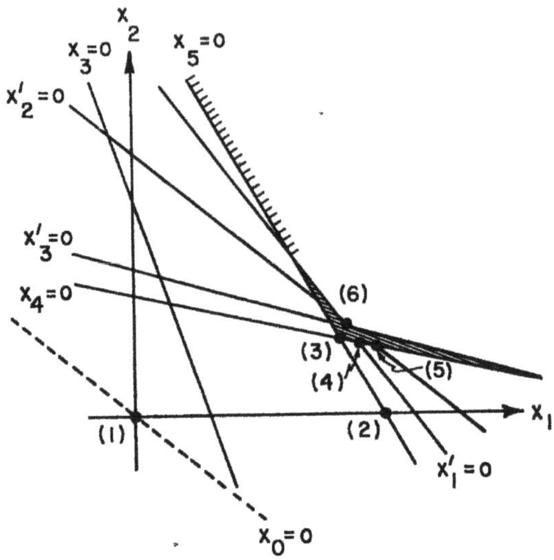

3. Branch and Bound Methods of Enumeration

It is natural enough to consider enumeration of feasible solutions to discrete or integer problems where there are only a finite number of candidates for an optimal solution. However, finite may be large and, hence, enumeration needs to be cleverly devised in order to avoid inspection of solutions which are dominated. "Branch and bound" is the very suggestive term given by Little et al [73] to their method for solving the travelling salesman problem; it is also an apt designation for the Land and Doig method [69] for solving pure and mixed integer programming problems (A.1). We describe the Land-Doig procedure.

The general approach is to start out as though to enumerate all feasible integer solutions according to the following scheme. First, solve (A.1) as a linear program. Either the solution, $X^0 = (x_0^0, \cdots, x_{m+n}^0)$, satisfies the integer requirements, or not. If it does, then X^0 is the required solution; if not, the value of the objective function, $\delta^0 = x_0^0$ provides an upper bound on the value of the optimal solution, $\max x_0 \leqq \delta^0$.

Consider some variable x_k, $k \varepsilon J$, which is required to be integer valued, but currently is not. It must be forced to take an integer value and hence decreased to at least $[x_k^0]$ or increased to at least $[x_k^0] + 1$. In fact, the entire range of possible integer values for x_k may be found by solving the two linear programs $\max x_k$ and $\min x_k$ subject to the constraints of (A.1). For each possible integer value in the range the $\max x_0$ subject to the constraints of (A.1) with x_k fixed can also be computed by linear programming. Thus, one can think of enumerating these possibilities in a directed tree, with rooted node 0 corresponding to X^0, and directed branches $(0, i)$ with node i corresponding to the solution X^i of the linear program max x_0 subject to the constraints of (A.1) and x_k set at some specific integer value, say x_k^i. From each new node the same procedure can be applied wherever integer restricted variables have noninteger values. Thus, if x_l^i, $l \varepsilon J$, is noninteger, its range of possible integer values can be computed for $x_k = x_k^i$, and for each possible integer value in the range the max x_0 subject to the constraints of (A.1) with $x_k = x_k^i$ and x_l fixed can also be computed, thereby developing new directed branches and nodes. In the tree a path from node 0 to any other node defines a sequence of integer valued choices made for certain integer restricted variables. Continuing until no further branching is possible, every terminal node will either correspond to a feasible integer solution or to some sequence of integer valued choices which do not admit a feasible solution to (A.1). The feasible node with max x_0 provides the optimal solution—for all feasible integer solutions have been enumerated.

The Land-Doig procedure involves precisely this enumeration—except that it attempts to search only a subset of the potentially immense (possibly infinite) tree described above. The idea is to develop only that part of the tree which contains the optimal solution and sufficient information to prove its optimality. Notice the following two facts. Any node in the above directed tree must have a value of x_0 no greater than its predecessor's, since branching means imposing an additional variable to some integer value, i.e., imposing an additional constraint. Second, consider any node X^i of the tree and suppose x_l is integer restricted and branching will be made by forcing x_l to be integer since x_l^i is nonintegral. Then, it is clear that the integer values for x_l which will result in the least decrease in x_0 after branching must be either $[x_l^i]$ or $[x_l^i] + 1$; in fact, max x_0 constrained by (A.1) and the past fixed integer values for certain variables is a concave function of x_l with maximum value at $x_l = x_l^i$. Thus, the further away the value of x_l from x_l^i the worse the resulting value of max x_0.

Algorithm IV ([69] 1960) for "pure" and "mixed" problems. *Step 0.* Solve (A.1) as a linear program to obtain a solution X^0 and bound $\delta^0 = x_0^0 \geqq \max x_0$. Construct node 0 with bound δ^0. *Step i.* In the tree select node l with bound $\delta^i = x_0^l \geqq \max x_0$. If all integer restricted variables have integer values in X^l then it constitutes an optimal solution. Otherwise, some integer restricted variable x_k has a noninteger value x_k^l.

(a) *Branch.* Solve the two linear programs, max x_0 subject to $x_k = [x_k^l]$ in one case and $x_k = [x_k^l] + 1$ in the other case, the constraints of (A.1) and the other fixed integer values for certain variables inherited in the tree. Call these solutions X^{l_1} and X^{l_2}; if a program has no solution assign it $(-\infty, \cdots)$. Adjoin nodes l_1 and l_2 and arcs (l, l_1) and (l, l_2).

(b) *Branch.* Let l_0 be the unique predecessor to l in the tree, where x_q was forced to the integer value x_q^l. l_0 has one or more immediate successors corresponding to other integer choices for x_q. Take x_q to be a new integer value $x_q^{l_3}$ which is either just smaller or larger than the integer choices already made, and leading to the larger x_0, say $x_0^{l_3}$ ($x_0^{l_3} \geqq \max x_0$ over integer choices for x_q not already made) by solving the two corresponding linear programs. Call the solution X^{l_3}; again, if there is no solution let $X^{l_3} = (-\infty, \cdots)$. Adjoin node l_3 and arc (l_0, l_3).

(c) *Bound.* Find the unbounded node having largest x_0 in the entire tree, say node r, and bound it with $\delta^{i+1} = x_0^r$.

It is only necessary to argue that $\max x_0 \leqq \delta^{i+1}$. This is true because every nonterminal node n in the tree so far constructed has one or two immediate successors which are terminal nodes, and which correspond to the best or/and next best choice of integer value for the variable in question at node n. Therefore *any* node of the entire tree which would be constructed if all feasible solutions were enumerated would have to have values of x_0 no greater than δ^{i+1}, i.e., $\max x_0 \leqq \delta^{i+1}$.

As mentioned in the introduction, this algorithm has not been programmed and hence not tried on large problems (recent computer programming of this approach is reported in [104]). On the face of it the approach seems unpromising for general or large integer programming problems. For problems where few variables are required to be integer or where some other special structure or conditions are present, it would seem that these ideas could be put to fruitful use. These comments seem only reasonable; they can, however, be supported by the recent success of Little et al [73] in solving the travelling salesman problem. Their algorithm is very similar to Algorithm IV: the same basic idea of branch and bound is used together with the obvious fact that a given route (i, j) from one city i to another j either belongs to the cycle of the travelling salesman ($x_{ij} = 1$) or not ($x_{ij} = 0$). This observation provides the "branching." "Bounding" is more intricate, but is based on the fact that changing the length of all departing routes (i, j) from a city i, or all arriving routes (j, i) to a city i, by the same amount produces a new problem which possesses the same optimal solutions as the original problem. Systematic use of this observation provides a method for obtaining bounds.

Another approach similar to these is contained in a paper of Healy [60]. This approach is concerned with a special type of mixed or pure problem (A.1) in which variables x_j, $j \varepsilon J$, are constrained to be 0 or 1 and $\sum_J x_j = 1$ (or $J = UJ_k$, i.e., the integer constrained variables belong to disjoint subsets J_k, $\sum_{J_k} x_j = 1$); hence the title of the paper, "Multiple Choice Programming."

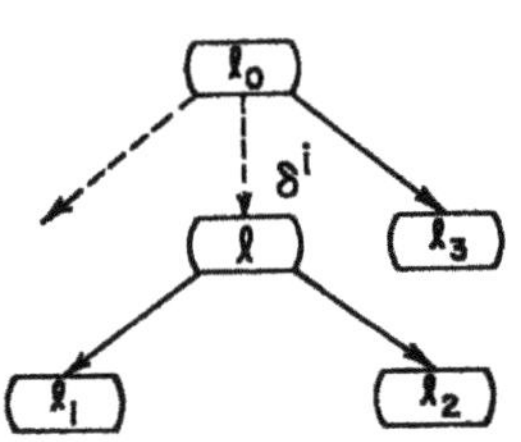

FIG. A.1. Step i (new arcs designated by solid arrows)

The central idea is this. Solve the problem as a linear program. If $x_j = 0, 1$ for all $j \varepsilon J$, then the optimal solution is found. Otherwise, some $x_j, j \varepsilon J$, must be made equal to 1. Consider each such variable $x_j, j \varepsilon J$, in turn, and compute a linear (over)estimate of the value of maximum x_0 given $x_j = 1$ from the optimal (noninteger) tableau. This is easily done: if x_j is nonbasic then an (under)estimate of the decrease in x_0 is a_{0j} (see tableau (A.2)), and if x_j is basic the corresponding estimate is $(1 - x_j) \min \{a_{0k}/-a_{jk} \mid x_k \text{ not basic}, a_{jk} < 0, k \not\varepsilon J\}$ (if no such minimum exists take a very large number as the estimate of decrease). Letting the respective estimates for x_0 be e_j^1 the constraint (max) $x_0 \leqq \sum_J e_j^1 x_j$ is valid. Repeat the procedure by solving the linear program with the new constraint. If the solution is not integer obtain new estimates e_j^2 and replace the last constraint by (max) $x_0 \leqq \sum_J e_j^2 x_j$. Convergence has not been proved; in fact, if in the course of computation a noninteger solution were found with the same value for x_0 as for the optimal integer solution, then the process would terminate since no improved constraints on max x_0 could be found. No significant computational success is cited for this method.

Display IV

Consider the pure problem given in Displays I and II. The branch and bound method of Land and Doig leads to the following tree.

Node 0: (−112/10, 18/10, 8/10, 42/10, 0, 0)
Node 1: (−14, 1, 2, 3, 4, 2)
Node 2: (−47/4, 2, 3/4, 19/4, 3/4, 1/2)
Node 3: (−∞, ···)
Node 4: (−13, 2, 1, 5, 1, 1)
Node 5: (−29/2, 3, 1/2, 15/2, 0, 3)

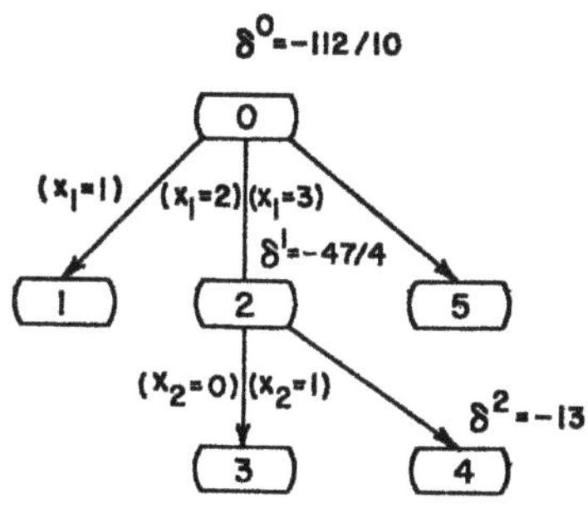

Step 0. Node 0 corresponds to the linear programming solution (see Display I, tableau (3)) and has bound $\delta^0 = -112/10$. *Step 1.* x_1 must be integer so branch with $x_1^1 = 1$ and $x_1^2 = 2$. $x_0^1 = -14$ and $x_0^2 = -47/4$, so bound node 2. *Step 2.* x_2 must be integer, so branch with $x_2^3 = 0$ and $x_2^4 = 1$, and branch from the predecessor to node 2 to inspect the next best choice for $x_1 = x_1^5 = 3$. $\max(x_0^3, x_0^4, x_0^5) = x_0^4 = -13 = \delta^2$, so bound node 4. *Step 3.* The bounded node 4 has X^4 all integer, stop.

4. Branch and Exclude Methods of Enumeration

The methods described here are for rather restricted types of integer programming problems. Their extensions and applications to the general problem (A.1) are subject to the same hesitations as made with regard to the Land and Doig method. The reason these methods are described here is this: clever enumerative schemes—despite their lack of glamour—are in certain situations, at least, the best known methods of solution.

Consider the special integer program: find $X = (x_1, \cdots, x_n)$ to

$$\text{(A.16)}\quad \begin{aligned} &\text{maximize} && CX \text{ or } \sum c_j x_j \\ &\text{when} && AX \leqq B, \text{ or } \sum a_{ij} x_j \leqq b_i \\ & && (i = 1, \cdots, m),\ x_j = 0 \text{ or } 1 \text{ all } j, \end{aligned}$$

where all a_{ij} and b_i are integers. Let α_j, $j = 1, \cdots, n$ be the columns of A and adjoin (for special purposes below) a column of zeros α_0 with $c_0 = 0$ and corresponding variable $x_0 = 0$ or 1. Further, let $A^k = (\alpha_0, \alpha_1, \cdots, \alpha_k)$, $C^k = (c_0, c_1, \cdots, c_k)$, and $X^k = (x_0, \cdots, x_k)$.

Bellman's [10] [11] functional equation approach, for the case $a_{ij} \geqq 0$, $b_i \geqq 0$, is to define for any $k = 0, 1, \cdots, n$ and m-vector Y,

$$\text{(A.17)}\quad \begin{aligned} F_k(Y) &= \max \{C^k X^k \mid A^k X^k \leqq Y, \text{ each } x_j = 0, 1\} && \text{if feasible,} \\ &= -\infty && \text{if nonfeasible.} \end{aligned}$$

Then, the recursion

$$\text{(A.18)}\quad F_k(Y) = \max_{x_k=0,1} \{F_{k-1}(Y - \alpha_k x_k) + c_k x_k\}$$

with "initial conditions" $F_0(Y) = 0$ if $Y \geqq 0$, $F_0(Y) = -\infty$ otherwise, provides the familiar computational scheme of dynamic programming. Here, $F_n(B)$ is the value of the optimal solution and $F_k(Y)$, $k \leqq n$, $0 \leqq Y \leqq B$, is the value of the optimal solution to subproblems of (A.16). The number of values computed and used in the course of computation is of the order $(n + 1)\prod_1^m (b_i + 1)$, for $F_k(Y)$ must be computed for all $0 \leqq k \leqq n$ and $0 \leqq Y \leqq B$. If the a_{ij} and b_i are not restricted to be nonnegative, a modification of the same approach may be used but the number of operations becomes very large. This is a branch and exclude method in the sense that for given Y and $k - 1$, i.e., for given Y and current choice of values of X^{k-1} only one choice of value for x_k needs to be considered; the other may be excluded immediately.

The idea of Benders et al [14] has certain similarities to that above. It is this. Begin as though to enumerate all 2^n possible (0, 1)-solutions X according to a branching scheme. The overall organization of the scheme is to start with the "trial" $X = 0$, which is defined to be the set S_0 of trials; then to enumerate the one solution of the set S_1 of trials having $x_1 = 1$ and $x_2 = \cdots = x_n = 0$; then to enumerate all 2^1 solutions of the set S_2 having $x_2 = 1$ and $x_3 = \cdots = x_n = 0$; and so forth, with the kth stage to enumerate all 2^{k-1} solutions of $S_k = \{X \mid x_k = 1 \text{ and } x_{k+1} = \cdots = x_n = 0\}$. Specifically, a tree of labelled nodes is developed. The nodes correspond to solutions X with associated columns $B - AX$ and values CX. Given that all nodes corresponding to solutions $X \varepsilon \bigcup_0^{k-1} S_j$ have been found, the nodes of S_k are found by adjoining a branch and node to each node of $\bigcup_0^{k-1} S_j$ (see Figure A.2). Namely, to any such node X with associated $B - AX$ and value CX adjoin a branch and a node corresponding to a trial solution X', where $x'_j = x_j$, $j \neq k$, and $x'_k = 1$, with associated $B - AX' = (B - AX) - \alpha_k$ and value $CX' = CX + c_k$. If all 2^n such nodes are enumerated then the node X with maximum value CX having $B - AX \geqq 0$ is the optimal solution. The number of values computed and used is of the order $(m + 2)2^n$, for the adjoining of a new node involves subtracting a column α_k, changing $x_k = 0$ to $x'_k = 1$, and adding c_k.

It is not necessary, however, to enumerate all nodes. Suppose that X^* with $B - AX^* \geqq 0$ has the maximum value CX^* of nodes so far enumerated. Let $m^k = \max_{j \geq k} \{c_j \mid \text{if } c_j \leqq 0 \text{ all } j \geqq k\}$ or $m^k = \{\sum c_j \mid \text{the sum extending over } c_j > 0, j \geqq k\}$. Then m^k is an upper bound on the increase that can be given to a node of $\bigcup_k^n S_j$ in excess of the value of a node of $\bigcup_0^{k-1} S_j$. Thus, if $X \varepsilon \bigcup_0^{k-1} S_j$ and $CX + m^k < CX^*$ then all successors of X in $\bigcup_k^n S_j$ may be excluded. This exclusion is possible because no such successor of X, feasible or not, can be a candidate for an optimal solution. Another rule for exclusion is as follows. Define a column vector α^k with components $\alpha_i^k = \min \{a_{ij} \mid \text{if } a_{ij} \geqq 0$ for all $j \geqq k\}$ or $\alpha_i^k = \{\sum a_{ij} \mid \text{the sum extending over } a_{ij} < 0, j \geqq k\}$. Then $-\alpha^k$ is a column of upper bounds on the increases that can be given to the values $B - AX$ of nodes $X \varepsilon \bigcup_k^n S_j$ in excess of those of $\bigcup_0^{k-1} S_j$. Thus, if $X \varepsilon \bigcup_0^{k-1} S_j$ and $B - AX - \alpha^k \ngeqq 0$ then all successors of X may be excluded. This is possible since no such successor could be feasible. Application of this method to general (0, 1)-problems is reported [13] to have been successful for problems involving up to 30 or 40 variables.

These two approaches have been used in solving the knapsack problem. Experiments [43] have shown that the adaptation of the Benders et al idea is five times as fast as the straightforward dynamic programming method. However, a more clever formulation of the problem as a dynamic program [45] should prove competitive since counts on the number of operations are about the same as for the Benders et al approach.

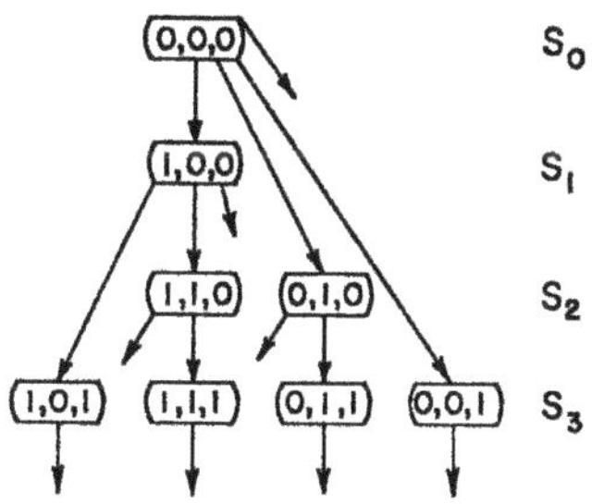

FIGURE A.2

The knapsack problem [26] is the integer program

$$\text{(A.19)}\qquad \begin{array}{ll} \text{maximize} & \sum c_j x_j \\ \text{when} & \sum a_j x_j \leqq b, \quad x_j \geqq 0, \quad x_j \text{ integer.} \end{array}$$

The reason it has received considerable attention is that its solution provides new columns in the decomposition algorithm for the cutting stock problem ([43] [45]; also see Section C.1). The name comes from the following interpretation of the problem: given n items of per item value $c_1, \cdots, c_n$ and size $a_1, \cdots, a_n$, and a knapsack of capacity b, pack the knapsack in order to maximize the value of the load.

The straightforward dynamic programming method of computation [11] is contained in the recursion

$$\text{(A.20)}\quad F_k(y) = \max_{x_k} \{F_{k-1}(y - a_k x_k) + c_k x_k \mid 0 \leqq x_k \leqq y/a_k, x_k \text{ integer}\}.$$

This requires the construction of an $(n + 1)$ by $(b + 1)$ table of values $F_k(y)$, $k = 0, 1, \cdots, n$ and $y = 0, 1, \cdots, b$; and, once $F_n(b)$ is computed, the optimal values for x_j are found by backtracking through the table. This can make serious demands on the memory capacity of a computer.

The clever dynamic programming computation [45] is contained in the recursion (where, actually, $F_k(y) = G_k(y)$, as shown below)

$$\text{(A.21)}\qquad G_k(y) = \max \{G_{k-1}(y), G_k(y - a_k) + c_k\}.$$

Here, each computation requires only a comparison of two quantities. Again $(n + 1)(b + 1)$ values need to be computed; however, the entire table need not be kept because backtracking to find the optimal values for x_j can be done on the $b + 1$ values $G_n(y)$, $y = 0, 1, \cdots, b$ alone. This is accomplished as follows. In performing (A.21) always store two values, $G_k(y)$ and $i(k, y)$, where

$$\text{(A.22)}\qquad i(k, y) = \begin{cases} i(k-1, y) & \text{if } G_{k-1}(y) \geqq G_k(y - a_k) + c_k \\ k & \text{otherwise} \end{cases}$$

Thus, $i(k, y)$ corresponds to the highest j with $x_j > 0$ for the problem with k items and capacity y. Then, to find the optimal values for the x_j, begin with $G_n(b)$ and $x_j = 0$ all j. At any step of the backtracking there will be a value $G_n(y)$ and current values for the x_j; maintain all current values except to increase $x_{i(n,y)}$ by 1 and to change $G_n(y)$ to $G_n(y - a_{i(n,y)})$. Stop when $G_n(0)$ is reached.

It should be clear that $F_k(y) = G_k(y)$ for all k, y. To see this directly, suppose it is true for $k < k^*$ all y and for $k = k^*$ and $y < y^*$. Then, using (A.20), (A.21), and the hypotheses,

$$\begin{aligned} G_k(y^*) &= \max \{G_{k-1}(y^*), G_k(y^* - a_k) + c_k\} = \max \{F_{k-1}(y^*), F_k(y^* - a_k) + c_k\} \\ &= \max \{F_{k-1}(y^*), \max_{x'_k} \{F_{k-1}(y^* - a_k - a_k x'_k) + c_k(x'_k + 1)\}\} \\ &= \max_{x_k} \{F_{k-1}(y^* - a_k x_k) + c_k x_k\} = F_k(y^*). \end{aligned}$$

By induction, this establishes the truth of the statement.

The modification of the Benders et al idea is given in [43]. It may be summarized in the following terms. Reorder the items according to their value per unit size, i.e., so that $c_1/a_1 \geqq c_2/a_2 \geqq \cdots \geqq c_m/a_m$. Then, begin as though to enumerate all possible solutions $X = (x_1, \cdots, x_m)$, starting with the lexicographically largest possible vector until $X = 0$ is reached. An ordered list of possible solutions is thus constructed; the X which maximizes $CX = \sum c_j x_j$ in the list is the optimal solution. The lexicographically largest possible solution is X^0 with components $x_1^0 = [b/a_1]$, $x_2^0 = [(b - a_1 x_1^0)/a_2], \cdots, x_n^0 = [(b - \sum_1^{n-1} a_j x_j^0)/a_n]$, i.e., as many units of item 1 as possible are loaded, then as many as possible of item 2 are loaded in the remaining space, and so forth. Given an X^i in the list the next solution of the list X^{i+1} is the lexicographically largest vector smaller than X^i; it is found by decreasing the last positive component of X^i by 1 and then increasing the next component as much as possible, and then the next, and so forth. But, again, it is not necessary to enumerate the entire list of such vectors. Suppose that X^* maximizes CX in the current list, and that X^i is the last solution of the list, with $x_s^i > 0$ and $x_{s+1}^i = \cdots = x_m^i = 0$. Define Y^i to be equal to X^i except that $y_s^i = x_s^i - 1 \geqq 0$. Then, if the integer requirements on the variables $x_{s+1}, \cdots, x_m$ were temporarily ignored, the maximum value of CX over all vectors X whose first s components were equal to Y^i, would be $\delta^i = CY^i + c_{s+1}(b - AY^i)/a_{s+1}$, since $(b - AY^i)$ is the remaining capacity and $c_{s+1}/a_{s+1} \geqq \cdots \geqq c_m/a_m$. If the integer requirements were imposed, then the maximum value could only be less than or equal to δ^i. So, if $\delta^i \leqq CX^*$ none of the vectors X which agree in the first s components with Y^i need to be considered and can thus be excluded from the list. In this case X^{i+1}, the vector following X^i in the list, is set equal to Y^i. Otherwise, X^{i+1} is set equal to the next lexicographic vectors. For many purposes it is desirable to solve (A.19) for many values of b; the method is easily and directly extendable to solve this problem in an efficient manner [43].

5. Partitioning in Mixed Problems

Consider, now, a mixed integer programming problem in which some variables are constrained to be integer valued and others not. Specifically, suppose that the problem is: find $X \geqq 0$ and Y integer valued to

$$\text{(A.23)} \quad \begin{array}{ll} \text{minimize} & y_0 = C_1X + C_2Y \\ \text{when} & A_1X + A_2Y \geqq B, \quad X, Y \geqq 0. \end{array}$$

Then a partitioning idea of Benders [13] seems to hold special promise. It may be explained in the following terms [6]. Denote the permissible values for Y by R. Then, (A.23) may be written

$$\text{(A.24)} \quad \min_{Y \in R} \{C_2Y + \min_X \{C_1X \mid A_1X \geqq B - A_2Y, X \geqq 0\}\}.$$

But, for given values of Y, the minimum problem "inside" the curly brackets is a linear program in X for which, by the duality theory of linear programming, the following statement holds:

$$(A.25)\quad \begin{aligned}&\min_X \{C_1X \mid A_1X \geqq B - A_2Y,\ X \geqq 0\}\\ &\qquad = \max_U \{U(B - A_2Y) \mid UA_1 \leqq C_1,\ U \geqq 0\},\end{aligned}$$

where the left(right) hand side has "value" $-\infty$ $(+\infty)$ if unbounded below(above), and "value" $+\infty$ $(-\infty)$ if undefined. Thus, (A.24) may be re-expressed as the problem

$$(A.26)\quad \min_Y \{C_2Y + \max_U \{U(B - A_2Y) \mid UA_1 \leqq C_1,\ U \geqq 0\}\}.$$

Consider, now, the convex polyhedral set $S = \{U \mid UA \leqq C_1,\ U \geqq 0\}$. It is independent of the values for Y. If it is empty then, by (A.25), no solution to the original problem (A.23) exists (in fact, this is the statement of Farkas' Theorem). Otherwise, the maximum value of $U(B - A_2Y)$, no matter what values of Y, attains its optimum at an extreme point of S or grows without bound along an extreme ray of S. But the extreme points or vertices of S are finite in number and can be enumerated, call them U^l, $l \varepsilon L$; and the extreme rays of S are finite in number and can also be enumerated by finding all extreme rays of $UA_1 \leqq 0$, $U \geqq 0$, call them U^k, $k \varepsilon K$. Now, if for some Y there exists a U^k, $k \varepsilon K$, with $U^k(B - A_2Y) > 0$ then the "value" of (A.25) is $+\infty$, showing the left side of (A.25) has no feasible solution; thus

$$(A.27)\quad U^k(B - A_2Y) \leqq 0, \qquad k \varepsilon K,$$

provide necessary and sufficient conditions on Y to admit a feasible X. Furthermore, this permits us to rewrite (A.26) as

$$(A.28)\quad \begin{aligned}&\min_{Y\varepsilon R} \{y_0 \mid y_0 \geqq C_2Y + \max_{l\varepsilon L} U^l(B - A_2Y),\\ &\qquad \text{and}\quad U^k(B - A_2Y) \leqq 0 \quad \text{all}\quad k \varepsilon K\}.\end{aligned}$$

Thus this partitioning has transformed the original mixed integer programming problem into a pure integer programming problem in variables Y containing potentially vast numbers of linear constraints on y_0 and Y. As in decomposition [32] (of which this is a dual form) the hope is that not all of these constraints need be enumerated, but rather that only some small subset need ever be considered. The computational procedure consists in solving (A.23) by iteration through a sequence of steps, each consisting in the solution of two subproblems.

Partition Algorithm V (1960, [14]). (i) Given a finite subset of U^j, $j \varepsilon Q$, where $Q \subset K \cup L$, solve the *restricted integer program* (A.28), i.e., solve (A.28) over the restricted set of constraints associated with U^j for $j \varepsilon Q$. If no feasible solution exists, then (A.23) has no feasible solution either; otherwise, let y_0^*, Y^* be an optimal solution—or y_0^* small and Y^* feasible if y_0 is unbounded below. (ii) Given the trial solution y_0^*, Y^* determine whether or not it is optimal by solving the *linear programs* (A.25) with $Y = Y^*$, and call their common value $f(Y^*)$. If $f(Y^*) = +\infty$ then the current Y^* does not admit a feasible X and a new U^k, $k \varepsilon K$, has been found as the "solution" to the right side of (A.25). If $f(Y^*) = -\infty$, then no feasible solution exists. If $f(Y^*)$ is finite, let X^* and U^* be the solutions to the left and right problem, respectively of (A.25). (X^*, Y^*) is a feasible solution to (A.23). If, in addition, $y_0^* \geqq C_2Y^* + f(Y^*) = C_2Y^* + C_1X^*$ then it is also an optimal solution; for y_0^*, Y^* is feasible for the entire integer program (A.28) while y_0 can be taken no smaller than its minimum value y_0^* over the restricted problem associated with U^j, $j \varepsilon Q$. Finally, if $y_0^* < C_2Y^* + f(Y^*)$ then some one or more linear constraints of (A.28) are not satisfied by y_0^*, Y^*; the most "violated" one corresponds to U^*, i.e., is $y_0 \geqq C_2Y + U^*(B - A_2Y)$, so U^* is adjoined to the set Q. This completes the description of a step of Algorithm V (see Display V in Section B.3).

The finiteness of the procedure is assured due to the finite number of extreme points and rays of S.

Consider, now, some of the properties of this algorithm. First, at every step one is provided with upper and lower bounds on the value of the minimum y_0. Substep (i) provides a lower bound y_0^* since minimization of y_0 is accomplished over a restricted set of constraints. Substep (ii) provides an upper bound since either $f(Y^*)$ is infinite or $f(Y^*)$ is finite implying (X^*, Y^*) is feasible with $C_1X^* + C_2Y^* = f(Y^*) + C_2Y^* \geqq \min y_0$. Thus, if computation becomes too expensive it can be stopped and the best feasible solutions (X^*, Y^*) so far found taken, together with an estimate of how far it is from optimum. In this regard Algorithm V may be said to have at least one of the advantageous attributes of a "primal" method. Second, if S is bounded, i.e., the set of rays U^j, $j \varepsilon K$ is void, then in fact every substep (ii) provides an (X^*, Y^*) which is feasible. Third, if S is not bounded, then it may be possible to enumerate its extreme rays initially and thus impose conditions on Y sufficient to assure that every Y^* of (i) admits a feasible X^* in (ii). Fourth, the fact of partitioning rather than applying some direct mixed integer algorithm preserves the structure of the matrix A_1 for the computation of substep (ii). Thus, if A_1 is of transportation type, then every occurrence of (ii) involves a problem of transportation type. All of these properties are desirable; nevertheless, the key question regards the number (or per cent) of extreme points and rays of S which enter the computational procedure. If this number is large, the approach is not successful.

It should, perhaps, be pointed out that this procedure, though algebraic in nature, has the "flavor" of a branch and bound method. In fact, every occurrence

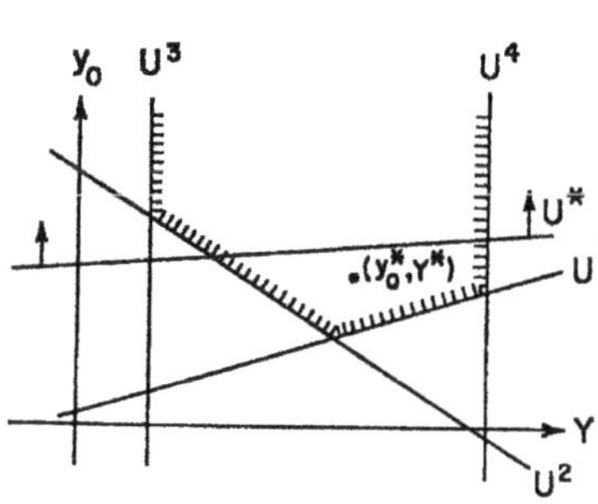

FIGURE A.3

of subset (ii) in which $f(Y^*)$ is finite provides a new lower bound on the value of the objective function y_0. Schematically, or geometrically, the process builds up a convex feasible region in (y_0, Y)-space, each new constraint "cutting off" the current integer solution. This interpretation is illustrated in Figure A.3. $U^1, \cdots, U^4$ correspond to the restricted set of constraints gotten from U^j, $j \varepsilon Q$; and U^3 and U^4 are rays of S. (y_0^*, Y^*) is the integer solution to a substep (i) and U^* the subsequent solution of substep (ii) which "cuts off" y_0^*, Y^* and gives a new bound on min y_0. U^* is then adjoined to Q. In Section B this method is applied to the plant location problem and given a direct interpretation.

B. Uses

1. *Introduction*

Many problems of practical and other interest can be formulated as linear programs by using one or more additional constraints which state that some variables are required to be integer valued in a solution. Thus, optimization problems over non-convex and/or disconnected polyhedral sets of points, the minimization of non-convex piecewise linear functions, the travelling salesman and salesmen problems [79], the $p - (p = ?)$ coloring map problem, the search for orthogonal latin squares and other statistical designs can all be formulated as pure or mixed integer programs. A survey of these and other formulations may be found in the paper of Dantzig [29].

There is the question as to whether or not these problems "should" be so formulated: although mathematically correct some seem awkward and unnatural posed as integer programs since they lead to very large problems (in numbers of variables and constraints). The same may be said about various industrial scheduling problems ([41] [74] [88] [96] [97], see especially the book [80]): very large integer programs are needed to pose rather "small" scheduling problems. Computational experience [41] appears to confirm this observation; in fact, for some small scheduling problems a version of Algorithm II is reported to have required more steps than would be needed for a solution by complete enumeration (though this may have been due, in part, to the particular code used, see Section C). This experience has discouraged many potential users of integer programming. In the author's opinion this experience should be heeded with respect to the application of integer programming to these scheduling problems only—and not to other classes of problems.

Other problems of practical interest which have been stated as integer programs are fixed cost problems (e.g., [2]); warehouse location problems ([6] [75] [89] [93]) and their uses in an apportionment problem [98]; the minimum disjunctive normal form problem or the network design problem ([20] [21] [22] [84]) and its generalization to a delivery problem [5]; problems in the construction of codes ([65] [78]); and of course any economic activity analysis or linear programming problem which concerns indivisible activities or commodities (e.g., [24] [34]). Since not discussed below—due to the relatively extensive mathematics needed to obtain a formulation—it should be said that a very successful application has been made to the construction of minimum-redundancy (minimum cost per bit) prefix codes for the discrete noiseless channel [65]. Computational experiments are reported on to show that the solution of the problem formulated as an integer program is a practical approach.

A host of combinatorial problems may be stated as integer programs: the marriage problem, the system of distinct representatives problem, the assignment problem and its generalization to the transportation, transshipment, and network flow problems, the caterer problem, and others (see [39] [61] for a survey of these and other applications). However, these problems are cases in which solutions to their "obvious" linear programming formulations are automatically integer valued: the extreme vertices of the convex polyhedral constraint sets are known to have integer components given integer data (e.g., given integer "supplies" and "demands" in the transportation problem). This useful property is due, of course, to the very special structure of these so-called "unimodular" constraint matrices: all subdeterminants have value 0, +1, or −1. This means that in the course of doing a simplex method the only pivot entries that can ever be encountered are ±1; hence, only addition and subtraction are needed in the course of computation. There has been considerable effort to characterize the set of unimodular matrices in a manner which would allow easy identification; but, in the large, this effort has failed. Classes of such matrices have been identified: the most inclusive of these has been, essentially, all matrices which are either of the transportation type or can be obtained from such a matrix through a finite number of pivot steps (see [3], and particularly [61] for more extensive discussions). On the other hand, it is of considerable practical interest to note that all unimodular matrices found (so far to the author's knowledge) "in nature," i.e., arising from some problem, have been of the transportation type. Nevertheless, there are examples of matrices which are unimodular but cannot be interpreted as being of the transportation type.

Other combinatorial problems which can be formulated as integer programs are not so well behaved. The travelling salesman problem, for example, requires additional cuts. An ad hoc method of generating additional cuts ([30] [31]) proved successful in solving some numerical problems: these cuts were derived in order to eliminate "loops" in the salesman's itinerary; i.e., to insure that the salesman makes a complete tour without returning home until each city is visited exactly once. It should also be said that in one special case of the problem an algorithm exists ([44] [46]) for obtaining the optimal tour in the order of n^2 steps, where n is the number of cities. The complexity of the general travelling salesman problem may be illustrated by the following fact [66]: a system of 390 irredundant inequalities describes the convex hull of the feasible tours (in 25-space) for the 5-city case (i.e., the extreme points of this convex polyhedron correspond to the 120 feasible tours). Other examples of problems which are not well behaved are the hierarchy of covering problems of graph theory—of which the delivery problem [5] cited above is a generalization—and their close relatives, the hierarchy of matching problems. These are discussed in the next section.

2. *Covering Problems*

Covering problems of various types represent a class of specialized integer programming problems which are of considerable importance in that many practical problems can be so formulated. In fact, the most successful computational experience to date with the integer programming algorithms of Gomory has been with problems of covering type. The term arises from the fact that these problems can be interpreted or explained in terms of graphs.

A graph $G = (Q, R)$ is a finite set of nodes Q together with a set of edges R joining certain pairs of nodes (or unordered pairs of nodes). In the *simple covering problem* a cover is defined to be a subset of edges such that each node of the graph is incident to some edge of the subset; the problem is to find a cover with the minimum number of edges. As an integer program, the problem may be stated

$$\text{(B.1)}\qquad \begin{array}{lll} \text{minimize} & \sum_{j=1}^{n} x_j & \\ \text{when} & \sum_{j=1}^{n} \alpha_j x_j \geqq \alpha_0\,, & x_j = 0 \text{ or } 1, \end{array}$$

where α_0 is a column of m 1's and each $\alpha_j = (a_{1j}, \cdots, a_{mj})^T$ has exactly two entries equal to 1 and the remainder equal to zero. α_j corresponds to the edge of G joining nodes i and k if $a_{ij} = a_{kj} = 1$. The matrix $A = (\alpha_1, \cdots, \alpha_n)$, with one column for each edge and one row for each node of G, is called the node-edge incidence matrix of the graph G.

Closely allied to the simple covering problem is the *simple matching problem*: Given G, a matching is defined to be a subset of edges such that no two edges of the subset are incident to the same node; the problem is to find a matching with the maximum number of edges. Clearly, the maximum matching contains no more edges than the minimum cover. Further, it is not difficult to see that to obtain a maximum matching from a minimum cover (or vice versa) it is only necessary to delete edges from the cover which "overcover" a node (or adjoin edges to obtain a cover) [81]. As an integer program, then, the simple matching problem is to

$$\text{(B.2)}\qquad \begin{array}{lll} \text{maximize} & \sum_{j=1}^{n} x_j & \\ \text{when} & \sum_{j=1}^{n} \alpha_j x_j \leqq \alpha_0\,, & x_j = 0 \text{ or } 1 \end{array}$$

where, again, $A = (\alpha_1, \cdots, \alpha_n)$ is the node-edge incidence matrix of G.

The complexity of these problems considered as integer programs can be illustrated in much the same terms as that of the travelling saleman problem. Since the basic integer programming algorithms are convexification processes it may be reasonable to measure the complexity of the problem by counting the number of linear inequalities which, when adjoined, would describe the convex hull of the integer solutions (if possible). Of course, such sets are not necessarily unique. Nevertheless, it is of interest to note [36] that the feasible integer solutions to (B.2), i.e., the matchings of G, are precisely the extreme points of the convex polyhedral set

$$\text{(B.3)}\qquad \begin{array}{l} \sum_j \alpha_j x_j \leqq \alpha_0\,, \quad x_j \geqq 0, \quad \text{and} \\ \sum_{j \in R_k} x_j \leqq r_k \quad \text{for all sets of edges} \quad R_k \quad \text{and} \quad r_k = 1, 2, \cdots \end{array}$$

where R_k is a set of edges of G whose nodes belong to a set S_k consisting of $2r_k + 1$ nodes of G. If the x_j have values 0 or 1 these new inequalities simply say that no more than r_k edges can be chosen from among the edges of $2r_k + 1$ nodes; for otherwise the edges would not form a matching since some node would necessarily be incident to more than one edge. These constraints, however, may be great in number.

Nevertheless—as might be expected—there are more direct algorithms for these special integer programs. The underlying idea which leads to these is that of the alternating path—an idea which goes back to 1891 [83]. A path of G is a sequence of pairwise distinct successively adjacent edges of G. Consider the simple covering problem. Suppose that a set of edges E is found which covers G. Then the following *theorem* [81] holds: E is a minimum cover of the nodes if and only if there exist no alternating paths (paths whose successive edges alternately belong to and do not belong to E) with the added property that the first and last edges, e_f and e_l, of the path belong to E and the first and last nodes of the path are also covered by edges of E other than e_f and e_l. Call such an alternating path a reducing path: if one does exist then it is easy to see that by reversing the assignment of edges in the alternating path to the covering set a new cover E' results which contains fewer edges. Thus, the "only if" part of the theorem is simple.

The "if" part is slightly more involved. Suppose E is a cover of G for which no reducing path exists and that C is a minimum cover of G. We show how to alter E and/or C, without changing the number of edges in either set, until they coincide; i.e., we show E must also constitute a minimum cover.

If $E = C$, the proof is complete. Otherwise there must exist an edge $e_1 \varepsilon E \cap C^{\sim}$ (an edge in E but not in C) for, if not, $E \subset C$ implying $E = C$. e_1 must have an adjacent edge $e_2 \varepsilon E^{\sim} \cap C$ because each node of e_1 must be covered by edges of C and if these edges also belonged to E, then there would be three adjacent edges of E with edge e_1 constituting a reducing path. Continue to construct a path according to the following rule. Given $e_i \varepsilon E \cap C^{\sim}$ and an adjacent edge $e_{i+1} \varepsilon E^{\sim} \cap C$ (or $e_i \varepsilon E^{\sim} \cap C$ and $e_{i+1} \varepsilon E \cap C^{\sim}$) adjoin to the path $e_1, \cdots, e_i, e_{i+1}$ an edge $e_{i+2} \varepsilon E \cap C^{\sim}$ (or $e_{i+2} \varepsilon E^{\sim} \cap C$) adjacent to e_{i+1}. Continue until either (a) no further edge can be adjoined or (b) an edge e_{i+2} is adjoined which is adjacent to some other edge e_k of the path, $k \leqq i$.

(a) The path is $e_1, e_2, \cdots, e_j$ $(j \geqq 2)$. Suppose $e_j \varepsilon E \cap C^{\sim}$ (or $e_j \varepsilon E^{\sim} \cap C$). There must exist an edge $e_{j+1} \varepsilon E \cap C$ adjacent to e_j since C and E are covers and there is no adjacent edge in $E^{\sim} \cap C$ (in $E \cap C^{\sim}$). But then change E (change C) by withdrawing e_j and adjoining e_{j-1} to obtain a new cover E (a new cover C) with the same number of edges (see Figure B.1). This results in covers E and C which coincide in one more edge than before.

(b) A cycle of adjacent edges $e_i, e_{i+1}, \cdots, e_j, e_i$ has been formed $(j \geqq i + 2 \geqq 3)$. The cycle contains (b_1) an even number of edges or (b_2) an odd number of edges. (b_1) Change E by withdrawing every edge in E of the cycle and adjoining every edge in C of the cycle to obtain a new cover E with the same number of edges which coincides with C in two or more edges than

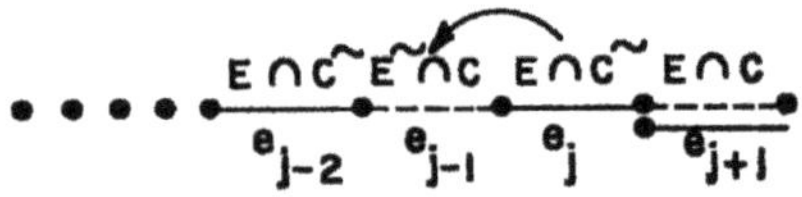

FIGURE B.1

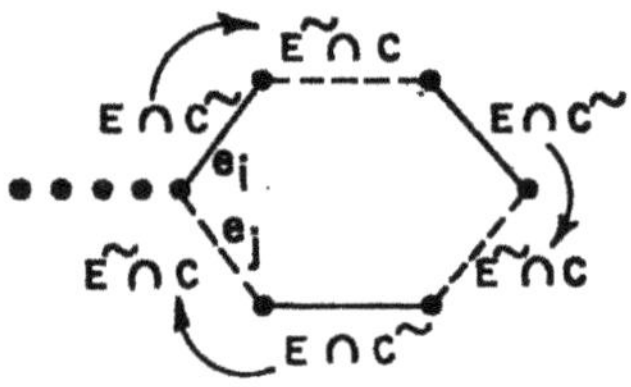

FIGURE B.2

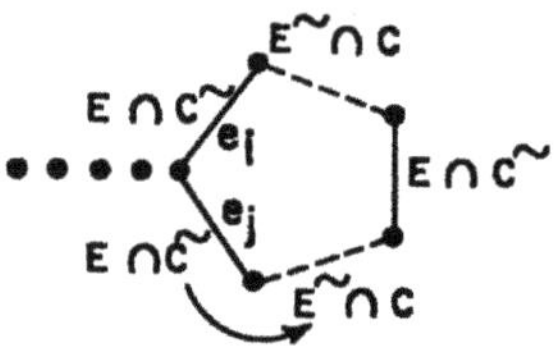

FIGURE B.3

before (see Figure B.2). (b_2) e_i and e_j both belong to $E \cap C^{\sim}$ (to $E^{\sim} \cap C$). Change E (change C) by withdrawing e_j and adjoining e_{j-1} to obtain a new cover E (a new cover C) with the same number of edges (see Figure B.3). This results in covers E and C which coincide in one more edge than before.

This completes the proof, for every construction results in covers E and C which coincide in at least one more edge but which each contain the same number of edges as before.

Algorithm VI ([81], [37] for the simple covering problem). This theorem provides the basis for an algorithm for finding a minimum cover. Namely, find a cover then search for reducing paths. If none exist, the cover is minimum; if one is found, inverting the assignment of edges to the covering set in the reducing path results in a new cover with one fewer edge. But how is the search for a reducing path to be made? This question was largely ignored until Edmunds [37] pointed out that on it hinges the efficiency of any resulting algorithm. In fact, Edmunds proposed a method which has, as an upper bound on the number of "steps" required for obtaining a solution, a function linear in the number of edges of the graph. His discussion is given in terms of the simple matching problem; however, the same ideas may be carried over or adopted for the simple covering problem.

A procedure for finding—if one exists—a reducing path given a cover E may be given in the following terms. If E contains less than two nodes which are more than once covered, then E must be a minimum cover. Otherwise, choose a node v_0 which is more than once covered. The following gives rules to construct alternating paths which branch out from v_0. At any stage of the construction let S be the set of edges which form these paths.

To initiate the procedure, let S consist of the edges (v_0, v_i) which belong to the cover E, $(v_0, v_i) \varepsilon E$, and label these nodes with $\{v_0, E\}$. Then, at any stage of the procedure apply the following rules. It should be kept in mind that if a node v_j has a label $\{v_i, E\}$ or $\{v_i, E^{\sim}\}$, then this means that there exists an alternating path of edges of S from v_j to v_0 beginning with the edge $(v_j, v_i) \varepsilon E$ or $(v_j, v_i) \varepsilon E^{\sim}$, respectively. If a node v_j also has a second label $\{v_i, E\}^{\alpha}$ then it means that there is a defined cycle α consisting of an odd number of edges in S which contains an alternating path from v_j to v_i and thence to v_0 beginning with an edge of E. Furthermore, the labels together with lists $\alpha, \beta, \cdots$ of cycles are sufficient to locate any such alternating paths.

(1) Suppose a node v_j has a label $\{v_i, E\}$ or second label $\{v_i, E\}^{\alpha}$. Then, either (1a) there exists an edge $(v_j, v_k) \varepsilon E$, $v_k \neq v_i$ or $(v_j, v_k) \not\varepsilon \alpha$; or (1b) there exists an edge $(v_j, v_k) \varepsilon E^{\sim} \cap S^{\sim}$ with v_k having a label $\{v_l, E\}$ or $\{v_l, E\}^{\beta}$ $(\beta \neq \alpha)$; or (1c) not. In case (1a) a reducing path exists beginning with edge $(v_j, v_i) \varepsilon E$ and may be located (see (4) below). In case (1b) adjoining (v_j, v_k) to S forms a new cycle consisting of an odd number of edges in S (see (3) below). In case (1c) adjoin to S an edge (v_j, v_k) if v_k has not been labelled, and label v_k with $\{v_j, E^{\sim}\}$. Then continue to label, if possible.

(2) Suppose a node v_j has one label $\{v_i, E^{\sim}\}$ (i.e., has no label $\{v_h, E\}^{\alpha}$). Then, since E is a cover there must exist some edge $(v_j, v_k) \varepsilon E$. Either (2a) v_k has a label $\{v_l, E\}$ or $(v_l, E\}^{\beta}$; or (2b) v_k has one label $\{v_l, E^{\sim}\}$; or (2c) v_k has no label. In case (2a) a reducing path exists beginning with (v_k, v_j) and may be located (see (4) below). In case (2b) adjoining (v_j, v_k) to S forms a new cycle consisting of an odd number of edges in S (see (3) below). In case (2c) adjoin to S the edge (v_j, v_k) and label v_k with $\{v_j, E\}$. Then, continue to label, if possible.

(3) Odd cycle. Adjoin the edge (v_j, v_k) to S. Then, an odd cycle of edges is formed in S (in cases (1b) and (2b)). This is because the labels on v_j and v_k show there exist alternating paths to v_0 beginning at v_j and v_k, respectively, whose first edges are either in E (case (1b)) or in $E^{\sim}$ (case (2b)). Therefore, these paths must first come together at a node v_p whose incident edges in the cycle are either both in E or both in $E^{\sim}$. This means that the sum of the edges in these two paths v_k to v_p and v_j to v_p is even, and hence the number of edges in the cycle odd since (v_j, v_k) was not counted. Either (3a) $v_p = v_0$ or (3b) the

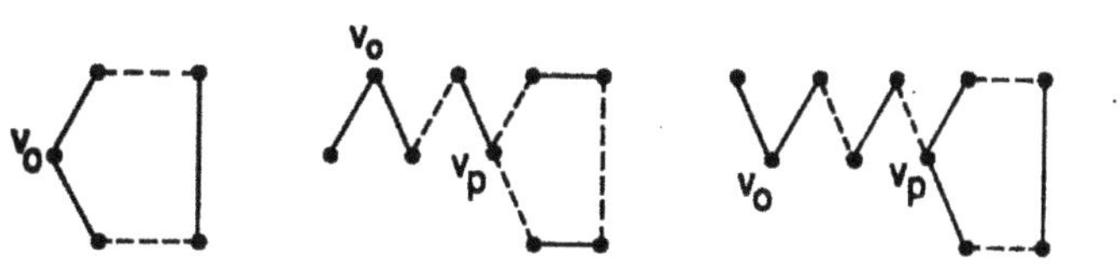

(3a) (3b) (3c)

edges in the cycle incident to v_p are both in $E^{\sim}$ or (3c) the edges in the cycle incident to v_p are both in E. In case (3a) or (3b) give as second label $\{v_0, E\}^{\nu}$ or $\{v_p, E\}^{\nu}$, respectively, to all nodes in the (ν^{th}) cycle which have labels containing $E^{\sim}$ but have no second labels. Record the nodes of the νth cycle in cyclic order starting with v_p. These labels reflect the following fact: starting with any node in the cycle it is possible to follow an alternating path back to v_0 beginning with an edge in E by going either in one or the other direction towards v_p. In case (3c) a reducing path exists and may be located (see (4)).

(4) Locating an alternating path. Given a labelled node v_f, it is always possible to follow an alternating path from v_f to v_0 in S. Namely, a label of v_f is either $\{v_g, E\}$, meaning that $(v_f, v_g) \varepsilon E$ is in a path and v_g has a label $\{v_h, E^{\sim}\}$; or $\{v_g, E^{\sim}\}$ meaning that $(v_f, v_g) \varepsilon E^{\sim}$ is in a path and v_g has a label $\{v_h, E\}$ or $\{v_h, E\}^{\alpha}$; or $\{v_g, E\}^{\alpha}$ meaning there are alternating paths beginning at v_f with an edge in E or in $E^{\sim}$ to v_g which can be found by looking at the record of the αth cycle, one of which can be continued from v_g to v_0. In case (1a) the path so located beginning with $(v_j, v_i) \varepsilon E$ and ending at v_0 is a reducing path since v_0 and v_j have incident edges in E and not in the path. In case (2a) the same is true with the path beginning with (v_k, v_j). In case (3c) the path beginning with an edge incident to v_p in the νth cycle, going around the cycle back to v_p and thence to v_0 is also a reducing path since the node v_p remains covered after reversing assignments (as well as the other nodes).

The rules above must terminate by finding a reducing path or by being unable to apply them further in a finite number of steps not exceeding the number of edges in the entire graph, since each step results in adjoining one new edge to S. In the latter case the cover E is a minimum cover. For suppose not, i.e., suppose there exists a reducing path from one overcovered node v_p to another overcovered v_q. Let $v_q \neq v_0$ (where v_0 is the initial node of the construction). Then, v_q must have a label (assuming the graph to be connected). Suppose it has a label $\{v_i, E\}$ or $\{v_i, E\}^{\alpha}$; then, since v_q is overcovered, step (1a) would have located a reducing path. Therefore, v_q must have one label $\{v_i, E^{\sim}\}$. But, then, either step (2a) occurred, meaning a reducing path was located; or step (2b) occurred, meaning a cycle was found and hence v_q must also have a label $\{v_q, E\}^{\alpha}$. Thus in all cases a contradiction is reached, showing the validity of the procedure as a means for obtaining a minimum cover.

The difficulty in the above procedure and in the simple covering problem arises from the presence of odd cycles of edges in the graph. In fact, if none existed then a linear programming procedure could be applied directly to (B.1) and an optimal solution in integers found. This is because of the following *theorem*: any square non-decomposable matrix of 0's and 1's with at most two 1's in any column has determinant 0, ± 1, or ± 2. This means that in a linear programming solution to (B.1) variables can have values 0, $\frac{1}{2}$, or 1. Thus, if only cycles of even number of edges are present and a solution has variables with values $\frac{1}{2}$, these must correspond to edges which form a cycle(s); increasing and decreasing these values by $\frac{1}{2}$ around the cycle(s) a cover is obtained with the same objective function value. The proof of the above statement is easily given by induction on the order of the matrix m. It is obvious for $m = 2$. If some row or column of an m by m matrix contains less than two 1's, induction establishes the fact. Otherwise, every row and every column contains exactly two 1's. But such a matrix has determinant 0 or ± 2, for evaluating it by computing all $m!$ products with appropriate sign and summing will result in only two nonzero terms which are each either $+1$ or -1. This is because the choice of any one of the two 1's in the first row of the matrix uniquely determines the choice of other 1's if one wants to obtain a nonzero product. In fact, if m is odd the value is ± 2, while if m is even the value is 0. It should be noted that the assumption of non-decomposability is the same as assuming the matrix is the node-edge incidence matrix of a connected graph.

Consider now the *set covering problem*: given a set of elements and a collection of subsets of these elements, find a minimum family of the subsets such that every element of the set is contained in some subset of the family. In graph language define a graph G with nodes v_i and F_j and edges (v_i, F_j): let nodes v_i represent elements of the set and nodes F_j subsets of the collection with an edge (v_i, F_j) in the graph meaning $v_i \varepsilon F_j$. Then the set covering problem is to find a minimum number of nodes F_j in G whose incident edges together meet or cover every v_i vertex. As an integer program the problem is

$$\text{(B.4)} \quad \begin{array}{lll} \text{Minimize} & \sum_1^n x_j & \\ \text{when} & \sum_1^n \alpha_j x_j \geqq \alpha_0, & x_j = 0, \text{ or } 1 \end{array}$$

where $\alpha_j = (a_{1j}, \cdots, a_{mj})^T$ has $a_{ij} = 1$ if $v_i \varepsilon F_j$ and $a_{ij} = 0$ otherwise. The matrix $A = (\alpha_1, \cdots, \alpha_n)$ with one column for each subset F_j and one row for each element v_i, is the (0, 1)-incidence matrix of elements versus subsets. It contains, in general, an arbitrary number of 1's in each column in contrast with the simple covering problem whose matrix has exactly two 1's in each column. A generalization of the alternating path ideas [35] has been given in the following terms. First, define a tree T to be a graph every pair of whose nodes are joined by exactly one path; or, equivalently, as a connected graph which contains one more node than edges. An alternating tree T in G with respect to a cover C (of nodes F_j) is a tree such that: (a) if T consists of exactly one edge (v_i, F_j) then $F_j \varepsilon C$; or (b) if T consists of more than one edge then $v_i \varepsilon T$ implies v_i has exactly two incident edges, (v_i, F_j) and (v_i, F_k) in T with $F_j \varepsilon C$ and $F_k \varepsilon C^\sim$, and $F_k \varepsilon T \cap C^\sim$ implies F_k has exactly two incident edges in T as well. This means that the tree T contains one more node of C than of $C^\sim$. *Theorem* [35]: C is a minimum cover if and only if there exist no alternating trees T with the added property that the set of F_j-nodes C' obtained by reversing the assignment of nodes F_j in T to C form a cover ($C' = (C - T \cap C) \cup (C^\sim \cap T)$). As before it is clear that the existence of an alternating tree T in G with the added property shows the cover is not a minimum. The other direction is more involved. But of particular interest here is the fact that the only known way of applying this theorem as a method of solution is to actually inspect every alternating tree and verify for each whether or not the added property obtains. Therefore, there is no known algorithm for searching for a tree with the required properties or showing none exist in a way which requires "looking at" each edge or each node of G once (as is the case in the simple covering problem).

This is a regrettable fact for the set covering problem, and its slight generalization, the ***weighted set covering problem***

$$\text{(B.5)}\quad \begin{array}{lll} \text{Minimize} & \sum_1^n c_j x_j & \\ \text{when} & \sum_1^n \alpha_j x_j \geqq \alpha_0, & x_j = 0 \text{ or } 1 \end{array}$$

where each $\alpha_j = (a_{1j}, \cdots, a_{mj})^T$ has $a_{ij} = 0$ or 1, are models for real problems of considerable interest and importance. These problems are faced by the engineer in designing switching circuits for digital computers or telephone centrals or control systems in automated manufacturing; by the logician in looking for a simplest disjunctive normal form equivalent to a given formula; by the businessman responsible for the delivery or distribution of goods from a warehouse to a host of clients.

These problems of the engineer and of the logician are exactly the same and have been extensively studied by both [62]. Consider the point of view of the engineer: a combinational circuit is a circuit having k binary (e.g., off-on) inputs and one binary output which is solely a function of the 2^k possible arrays of inputs. The designer specifies the switching function, i.e., the output he wishes for each of the 2^k inputs, and is then faced with the problem of constructing a network which realizes the desired function. There are, in general, many circuits which realize the function; the problem is to choose one which is "cheapest." The problem is often limited to that of constructing the circuit through interconnections of and-gates, circuits which are on only when all inputs or switches are on (are in series), and of or-gates, circuits which are on when at least one switch is on (are in parallel). In particular, circuits realizing any switching function can be made by building and-gates whose inputs are the circuit inputs, and connecting these into one or-gate. It is these special circuits with which we shall deal here.

Consider the point of view of the logician. A literal is a variable or the negation of a variable. A formula is a sequence of literals and/or formulae connected by ands (conjunctions) and ors (disjunctions), or the negation of such. If p and q are literals or formulae, then their disjunction ($p \vee q$) is true if either p or/and q are true but is false otherwise; their conjunction (pq) is true if both are true but is false otherwise; and if p is true then its negation ($p^\sim$) is false and vice versa. Thus, through the use of these laws the dependence of the truth value of a formula on the truth value of each of its k variables may be displayed in a truth table consisting of one line for each of the 2^k possible ways of assigning true and false to the variables. Formulae are said to be equivalent if their truth tables are the same. In general, there are many equivalent formulae; the problem is to choose a "simplest" one. Often, the problem is reduced to that of finding a simplest equivalent formula in disjunctive normal form, that is, a formula which is a disjunction of clauses, each clause being a conjunction of literals. It is clear that this problem is the same as the circuit problem: on corresponds to true, off to false, an and-gate to a conjunction, an or-gate to a disjunction, the special circuits to a disjunctive normal form. In fact, both of these are models for a Boolean algebra in which there are two elements 0 (off, false) and 1 (on, true); addition (or-gate, disjunction) is defined by $0 + 0 = 0, 0 + 1 = 1 + 0 = 1 + 1 = 1$; multiplication (and-gate, conjunction) is defined by $0 \cdot 0 = 0 \cdot 1 = 1 \cdot 0 = 0, 1 \cdot 1 = 1$.

The two problems can, therefore, be formulated in a unified fashion. Let $x_1, x_2, \cdots, x_k$ represent inputs or variables with values 0 or 1 in a Boolean algebra. Given a switching function or truth table or Boolean function F, that is, an output value of 0 or 1 corresponding to each of the 2^k possible assignments of 0, 1 to the variables, find a "cheapest" or "simplest" expression or function in the x_j which is a sum of products of terms, every term being either an x_j or an $(1 - x_j)$, whose values over the domain of definition agree with the output values. It is clear that there exist solutions: if $(x_1^0, \cdots, x_k^0)$ is a k-tuple of 0's and 1's with value 1, then include the product $\prod_1^k x_j^\sim$ in the sum, where $x_j^\sim = x_j$ if $x_j^0 = 1$ and $= 1 - x_j$ if $x_j^0 = 0$. Each such product is called a fundamental product or canonical term. To formulate the problem it is still necessary to specify what is meant by "cheapest" or "simplest." It suffices, in what follows, to impose any nonnegative integer costs on the products as long as any factor of the product has a cost no larger than the product (asking for rational costs is, of course, no more than asking for integer costs). For example, it may be of interest to find the solution containing the fewest number of products, or the solution which contains the fewest number of terms.

Given the output function F define a prime implicant of F to be a product P of terms over some subset S of indices $\{1, 2, \cdots, k\}$, $P = \prod_S x_j^\sim$ which has the property that $P = 1$ implies $F = 1$ but no proper factor of P has the same property. Thus, a prime implicant is a product containing a minimal number of terms which has value 1 only when F has value 1. It is easy to see [85], in view of the cost structure stated above, that it suffices to consider functions which are sums of prime implicants to find a cheapest. Thus, the problem may be reduced to finding all prime implicants and then to choosing a cheapest subset of them whose sum realizes F. The enumeration problem is quite simple and may be described as follows [85]. Begin with the sum of all fundamental products of F. Then, at any stage, apply rule (i) if possible; if not possible, apply rule (ii); if neither is possible, stop. Rule (i): if $P + P'$ is in the sum and P is a factor of P', drop P' from the current sum. This is clearly permissible since $P = 1$ if and only if $P + P' = 1$. Rule (ii): if $x_jP + (1 - x_j)Q$ is in the sum then add the product PQ to the sum, omitting any duplicate terms, (P or Q might be identically 1) provided PQ does not contain a factor $x_l(1 - x_l)$ (since then $PQ \equiv 0$) and does not contain a factor which is already a product in the sum (since then a cycle of rules (i) and (ii) could result). This rule is also permissible since $x_jP + (1 - x_j)Q = 1$ if and only if $x_jP + (1 - x_j)Q + PQ = 1$. Thus these rules represent a sequence of transformations each resulting in a new expression realizing F. It remains to show that the rules produce *all* prime implicants and *only* prime implicants. Suppose some prime implicant P of F is omitted in the final sum obtained. Let P^* be the product containing the largest number of terms such that P is a factor of P^* but no product in the sum is a factor of P^* (such a P^* must exist since P itself satisfies the constraints on P^*). P^* cannot contain all k variables for then, since no product in the sum is a factor of P^*, P^* would have a value different from every product in the sum. Suppose P^* does not contain the variable x_j. Consider x_jP^* and $(1 - x_j)P^*$. These products contain more terms than P^*, and P is a factor of them; thus some products of the sum must be factors of them. These products of the sum contain x_j and $1 - x_j$ since they were not factors of P^*. Thus, one of three cases obtains. First, the products are x_j and $1 - x_j$ alone. But, then, the problem is trivial since this implies that the Boolean function F is identically 1. Second, the products are x_j and $(1 - x_j)R$ (or $(1 - x_j)$ and x_jR) with R a factor of P^*. But, then, by rule (ii) the product R is part of the sum, contradicting the hypothesis on P^*. Third, the products are x_jR_1 and $(1 - x_j)R_2$, with R_1 and R_2 factors of P^*. Again, by rule (ii) the product R_1R_2 (with duplicates omitted) is part of the sum and a factor of P^*, contradicting the hypothesis on P^*. In either of these cases the product R or R_1R_2 might not have been added since some other product was already present in the sum which was a factor of R or R_1R_2 : the same contradiction derives. The *only* part of the argument is trivial: since all prime implicants are generated no other products would be present by rule (i).

To formulate the problem list the prime implicants of the sum, $P_1, P_2, \cdots, P_n$ and compute their costs $c_1, c_2, \cdots, c_n$. Then, list the fundamental products of F, $Q_1, Q_2, \cdots, Q_m$, or, what is the same thing, list the points of the domain of F where $F = 1$ (i.e., there are m such points out of the 2^k possible points or inputs). Define $a_{ij} = 1$ if $Q_i = 1$ implies $P_j = 1$, i.e., if P_j is a factor of Q_i and $a_{ij} = 0$ otherwise. Then, to any feasible solution to (B.4) there corresponds a subset of prime implicants which realize F; and an optimal solution selects the cheapest such subset.

The delivery problem [5] may be stated in the following terms. Commodities, located at a central warehouse, are to be shipped by common carrier to m clients at various destinations within some region. The shipper's objective—in any given time period—is to deliver the clients' orders at a least total transportation cost. Typically, the order of a given client cannot be shipped in separate deliveries; however, the shipper may specify the delivery schedule to be followed by the carrier in view of carrier constraints and rates. Carriers can be instructed to combine a number of orders (at most k) to be delivered together, i.e., by dropping-off one order and going on to drop-off another, etc., provided that these clients lie along one of a number of permissible geographical routes (at most r) or/and that certain weight or capacity limitations are not surpassed. Typically, a client may receive deliveries via a number of different routes. The carrier has a well-defined schedule of rates which is a function of the total weight of a combined shipment, the number of stop-offs required, and the routes or destinations involved. Usually, fixed costs are applied for each stop-off; rates decrease in jumps with larger loads (e.g., car-load, less than car-load); rates applied to particular combined shipments are based on rates applying to the furthest (i.e., most expensive) destination; and so forth. Thus, total cost is a complicated combination of rates which usually is a piecewise linear convex and concave in parts function.

To formulate the problem suppose that in a given period m orders of total weight w_i, $i = 1, \cdots, m$ are received and must be met. Define an activity to be a single feasible combination of orders with deliveries for some subset of clients (perhaps only one). If r, k, and m are small it is feasible to generate all possible activities. Suppose these are n in number. In some real problems it has been observed that $r \leqq 5$, $k = 4$, $m \leqq 15$ and $n \leqq 388$. Of course, in general, n is an involved function of r, k, and m and any supplementary constraints which may be in force for the particular problem being considered. If n_s $(s = 1, \cdots, r)$ is the number of destinations in each permissible route, then $n \leqq \sum_{s=1}^{r} \sum_{i=1}^{k} \binom{n_s}{i}$. In any case, list all activities $j = 1, \cdots, n$ and compute the cost c_j of each activity j. Define $a_{ij} = 1$ if activity j delivers order i and $a_{ij} = 0$ otherwise. Then, to any feasible solution to (B.5) there corresponds a subset of activities which together deliver all client orders; and an optimal solution selects the least cost delivery schedule. It is assumed implicitly, here, than an optimal solution will satisfy all inequalities as equations and not strict inequalities, since the latter occurrence would mean two identical orders were shipped to a same client. It is sufficient to assume for this to be true that if two activities j and k are exactly the same except that j delivers to i and k does not, then $c_j \leqq c_k$.

These three problems all lead to the weighted set covering problem (B.5). In any of these formulations m and/or n may be large, indeed. However, it is usually possible to simplify (B.5) considerably. There are four operations which can lead to simplifications. First, if any row i contains only one entry $a_{ij} = 1$, the other entries being zero, then, clearly $x_j = 1$ in a solution (if no entries in row i are 1, no feasible solution exists). Thus, row i and column j may be eliminated, and α_0 replaced by $\alpha_0 - \alpha_j$; if any entries of $\alpha_0 - \alpha_j$ are zero, their rows may be eliminated. Second, if there exist two rows, i_1 and i_2, so that $a_{i_1 j} = 1$ implies $a_{i_2 j} = 1$ for all j, then row i_2 may be eliminated since $\sum_j a_{i_1 j} x_j \geqq 1$ implies $\sum_j a_{i_2 j} x_j \geqq 1$, i.e., the row i_2 constraint is weaker than the row i_1 constraint. Third, if there exists a set of columns S and a column α_j so that $\sum_{i \in S} \alpha_i \geqq \alpha_j$ and $\sum_{i \in S} c_i \leqq c_j$, then activity j is "dominated" and it is clear that it is permissible to take $x_j = 0$ in an optimal solution. Thus, column j may be eliminated. Fourth, if the matrix $A = [\alpha_1, \cdots, \alpha_n]$ can be written, after rearrangements of rows and/or columns in the form

$$A = \begin{bmatrix} A' & 0 \\ 0 & A'' \end{bmatrix}$$

where the zeros represent blocks of zeros, then (B.5) may be written as two separate problems of the same type, but each of considerably smaller dimension. Clearly, these operations may be applied in any order successively. In practical problems related to Boolean functions [20] [21] and to delivery problems [5] it was found that these operations resulted in considerable simplifications.

A recent proposal ([70][71]) makes systematic use of these simplifications to specify an enumerative method for the set covering problem (B.4) (the method may be extended to the weighted case). Consider the first row of A and suppose $a_{1j} = 1$ for $j \leqq k$ and $a_{1j} = 0$ for $j > k$. The inequality $\sum a_{1j}x_j \geqq 1$ implies that $x_1 = 1$, or $x_1 = 0, x_2 = 1$, or, $\cdots$, or $x_1 = \cdots = x_{k-1} = 0, x_k = 1$. Consider the jth of these mutually exclusive possibilities. Discard rows i of A having $a_{ij} = 1$ and replace the first $j - 1$ columns by 0's; call the resulting matrix $A^{(j)}$. Clearly, a minimum cover for A may be found by computing a minimum cover $x^{(j)}$ for $A^{(j)}$ and adjoining $x_j = 1$, for each j, and then choosing a minimum of these k covers. The idea of this approach is to construct a matrix A' whose set of covers is the same as the set of the covers of $A^{(1)}, A^{(2)}, \cdots$, and of $A^{(k)}$. Let $a^{(j)}_{i(j)}$ be a row of $A^{(j)}$ and $a'_i = \sum_j a^{(j)}_{i(j)}$, addition being Boolean. Then $a'_i x \geqq 1$ if and only if $a^{(j)}_{i(j)}x \geqq 1$ for at least one j. Thus, the matrix A' formed by making all possible Boolean sums of k rows, one from each of the matrices $A^{(j)}$, enables the following statement: if x' is a minimum cover for A', then it is a minimum cover for some $A^{(j)}$, and x' together with $x_j = 1$ is a minimum cover of A. The covering problem A' may be considerably simplified by at least the second rule given above; also, at least one column of A' is composed entirely of 0's (in our discussion, the first column). Continue to apply the same ideas until $x = 0$ is a minimum cover for the matrix composed entirely of 0's. Then "backtrack".

3. Location Problems

Location problems of various types constitute another class of specialized integer programs having wide applicability. A typical problem is the following. A product is produced at one plant and is distributed to warehouses in various locations. Subsequently, the product is delivered to customers, each with known or estimated per period demands, from storage in the warehouses. The problem is to choose some set of warehouse locations from a finite set of possibilities in order to effect the distribution, storage and delivery from plant to customers at a least total cost. If costs are linear, or convex, the problem is straightforward; in fact, solutions in this case would usually tend to use all possible storage locations. More realistically, suppose that the costs of shipments from plant to warehouses and the costs of storage are concave, i.e., that economies of scale in shipments and storage are present (per unit costs are non-increasing), but that the subsequent delivery costs are linear.

A slight generalization of this problem with many plants rather than one was considered in 1958 [7] by Baumol and Wolfe and a method for determining a *local* minimum is given there. This model assumes that each plant's production is limited while warehouse capacities in the various locations are (for practical purposes) unlimited. Thus, the problem is one of location and flow of distribution. The idea behind the method for obtaining a local minimum is easily stated. Consider any feasible solution. Then, determine the marginal costs associated with every variable in the problem, and take these as the (temporary) per unit or average costs of their variables. This results in a (temporary) linear total cost function—the differential of the cost function—which approximates the change in total cost in a neighborhood of the last feasible solution. Solve the linear problem: it is, since in this case warehouse capacities are unlimited, a transportation problem. Repeat the procedure, beginning with the last transportation problem solution, until the same solution repeats itself. This last is a sought for local minimum. A difficulty here is that marginal costs are not always unambiguously defined. Also, no real *global* considerations have been made—and the local minimum found may be a very expensive one.

Consider, now, the one plant problem, thus eliminating the transportation part of the problem. Suppose there are m possible warehouse locations ($i = 1, \cdots, m$) and n customers with per period demands of d_j ($j = 1, \cdots, n$). Then the problem is to

$$\text{(B.6)}\quad \begin{array}{ll} \text{Minimize} & f(x) = \sum_i f_i(\sum_j x_{ij}) + \sum_{i,j} c_{ij}x_{ij} \\ \text{when} & \sum_i x_{ij} \geqq d_j, \qquad x_{ij} \geqq 0 \end{array}$$

where x_{ij} is the number of units of the product shipped from the plant via warehouse i to customer j, and $f_i(z_i)$ is a concave function of z_i with $f_i(0) = 0$. The first point to note is that since the $f_i(z_i)$ are concave functions, delivery costs are linear, and warehouses have unlimited capacities, every customer can receive his entire demand from exactly one warehouse in a least cost solution. In other terms, a concave function attains a minimum at an extreme point of the convex constraint region. The extreme points of the constraint set of (B.6) are easily described: since each of the equations $\sum_i x_{ij} = d_j$ are independent, an extreme point has, for each j, $x_{ij} = d_j$ for some one i and $x_{kj} = 0$, $k \neq i$. Thus, the extreme points of (B.6) are m^n in number and one possible method of solution is via enumeration: for each assignment of customers to warehouses compute the total cost and choose one with minimum cost. Though possible this approach is certainly not feasible for m or n moderately large.

Suppose, now, that the functions $f_i(z_i)$, representing cost of shipment to and storage at warehouse i of $z_i = \sum_j x_{ij}$ units, are continuous, piecewise linear functions of z_i, i.e., are of form

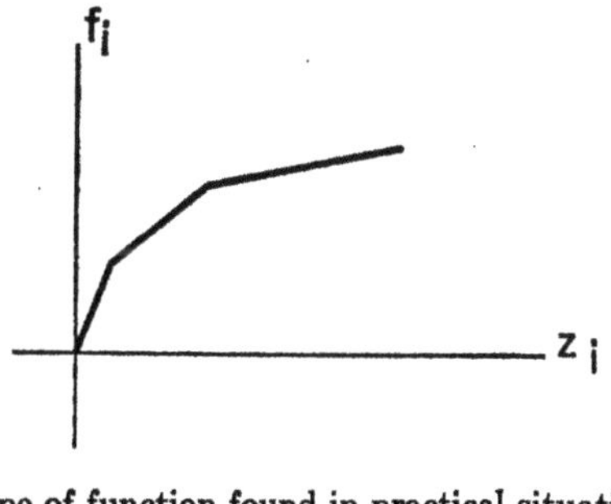

This is a common type of function found in practical situations, with the breaks in df_i/dz_i corresponding to rate breaks as larger quantities become involved. The local minimum approach could be easily applied, though some attention to

ambiguities at points corresponding to changes in rate would be necessary. However, R. L. Taylor [89] has suggested the following approach to the global problem. Suppose f_i has b_i different marginal rates (3 in the function drawn above). Then, (B.6) could be solved via another kind of enumeration. Namely, consider all possible marginal costs that could be assigned to variables x_{ij}. The local minima are found among the solutions to all linear distribution problems resulting from every possible assignment of marginal costs and a global minimum must, of course, be a local minimum. However, the number of possible rate assignments is $\prod b_i$, which is, for most practical problems, a number considerably smaller than the m^n alternatives of the previous enumeration. Taylor then proposed (and used) the following method of solution. Enumerate, in some systematic fashion, all possible rate assignments. Given any one, solve the resulting trivial distribution problem (for each j take $x_{ij} = d_j$ if its rate $r_{ij} = \min_k r_{kj}$, and $x_{kj} = 0, k \neq i$). Either the resulting solution x^0 is such that $\partial f(x^0)/\partial x_{ij} = r_{ij}$ or not, i.e., either the marginal costs of the solution match the rate assignment or not. If it does not then reject it. If it does then either it is a local minimum or some ambiguity is present due to some x_{ij}^0 corresponding to a point at which the rate changes. In the latter case iterate to find a local minimum. Continuing, all possible local minima are found. Taylor reports having programmed this approach and solving real problems, among them one with $m = 6$, $n = 40$, and $b_1 = 2$, $b_2 = 5$, $b_3 = 7$, $b_4 = 8$, $b_5 = 9$, $b_6 = 10$. This took approximately seven minutes of IBM 7090 time. Further, 12,754 vertices or solutions were found not to match the rate assignments; 25,389 vertices or solutions did but were not local minima; and 160 local minima were found. In this case solutions involving all warehouses were excluded. In another example with $m = 12$, $n = 40$, and $b_i = 3$ all i, forty minutes was required. It should be noted that the size of n is of no real significance. The lesson of this experience is that in many practical problems enumeration, intelligently done, is feasible and of great value.

Consider a still more special form of the location problem (B.6) in which $f_i(z_i) = f_i$ (a constant) if $z_i > 0$ and $f_i(0) = 0$ [75]. Then, by replacing x_{ij} by $d_j x_{ij}$ and appropriately redefining the c_{ij}, the problem may be stated as the integer program

$$\text{(B.7)} \quad \begin{array}{lll} \text{Minimize} & y_0 = \sum_i f_i y_i + \sum_{i,j} c_{ij} x_{ij} & \\ \text{when} & \sum_i x_{ij} \geqq 1, & 0 \leqq x_{ij} \leqq y_i,\ y_i = 0, 1. \end{array}$$

If $y_i = 1$, then warehouse i is used and a fixed cost of f_i incurred; otherwise, $y_i = 0$ and warehouse i is not used (or built). Manne [75] considers this problem and reports on some computational results of using a certain heuristic program for obtaining "near optimal" solutions. A number of problems are solved exactly by enumeration and the solutions compared to those obtained by heuristic means.

It might be conjectured that a solution to (B.7) with integer constraints ignored (i.e., $y_i = 0$ or 1 replaced by $y_i \geqq 0$) would automatically result in each y_i taking an integer value. The following example (of R. E. Gomory) shows this to be false. In this example $m = n = 3$, $f_i = 1$ all i, and $c_{ii} = 2$, $c_{ij} = 1$, $i \neq j$. Due to symmetry there are three possible solutions: use one warehouse (fixed cost of 1 plus linear cost of 4 for total cost of 5), use two warehouses (fixed cost of 2 plus linear cost of 3 for total cost of 5), or use three warehouses (fixed cost of 3 plus linear cost of 3 for total cost of 6). However, the following noninteger solution has a lower cost: use three $\frac{1}{2}$ plants (take $y_i = \frac{1}{2}$ all i, and $x_{ij} = \frac{1}{2}$, $i \neq j$, $x_{ii} = 0$ for a total cost of $\frac{9}{2}$).

The partition algorithm (see Section A.5) appears to be particularly well suited for application to (B.5) [6]. Partitioning into integer constrained variables Y and noninteger constrained variables X, the dual linear programs for given values $Y^* = (y_i^*)$ (see (A.23')) are

$$\text{(B.8)}\quad \begin{array}{lll} \text{Minimize} & \sum_{i,j} c_{ij}x_{ij} & \\ \text{when} & \sum_i x_{ij} \geqq 1, & 0 \leqq x_{ij} \leqq y_i^* \\ \text{Maximize} & \sum_j v_j - \sum_{i,j} y_i^* u_{ij} & \\ \text{when} & v_j - u_{ij} \leqq c_{ij}, & v_j \geqq 0, u_{ij} \geqq 0. \end{array}$$

Consider, first, the problem of finding all extreme rays which can lead to a nonfeasible Y^*, or, what is the same thing, of finding necessary and sufficient conditions that Y^* admit a feasible distribution X. The conditions are that for extreme rays of $v_j - u_{ij} \leqq 0$, v_j, $u_{ij} \geqq 0$ the inequality $\sum_j v_j - \sum_{i,j} y_i^* u_{ij} > 0$ is false. If $Y^* = 0$, then any $\{u_{ij}, v_j\}$ with some $v_j > 0$ makes the inequality true. Otherwise, if $Y^* \neq 0$ but has components 0 or 1, then $v_j \leqq \sum_i y_i^* u_{ij}$ all j, and the inequality is false. Thus, a necessary and sufficient condition on Y^* is that $Y^* \neq 0$; in words, at least one warehouse must be used.

Now, given some solution to the restricted integer program Y^* with value y_0^*, the *linear programs* which ask whether or not Y^* is optimal are the programs (B.8). These are easily solved explicitly. Namely optimal solutions are to take $v_j^* = \min_i \{c_{ij} \mid y_i^* = 1\}$, $u_{ij}^* = \max\{0, v_j^* - c_{ij}\}$; and $x_{ij}^* = 0$ if $y_i^* = 0$ while $x_{ij}^* = 1$ for one index i if $y_i^* = 1$ and $c_{ij} = \min_k c_{ik}$, each j. In words, for each sink j supply its unit by ordering from an open source with cheapest transportation cost. Let the common optimal value of (B.8) be $f(Y^*)$. Then, if $y_0^* \geqq \sum f_i y_i^* + f(Y^*)$ $\{x_{ij}^*, y_i^*\}$ constitutes an optimal solution. Otherwise, adjoin $\{u_{ij}^*, v_j^*\}$ to the set of extreme points and adjoin the corresponding inequality to the restricted integer program.

Consider the *restricted integer program*. Given $\{u_{ij}^k, v_j^k\}$, $k = 1, \cdots, K^*$, a subset of extreme points to the right hand problem of (B.8), it is: find $Y \neq 0$ in integers, and y_0 to

$$\text{(B.9)}\quad \begin{array}{ll} \text{Minimize} & y_0 \\ \text{when} & y_0 \geqq \sum_j v_j^k - \sum_{i,j} u_{ij}^k y_j + \sum_i f_i y_i, \quad (k = 1, \cdots, K^*). \end{array}$$

The inequalities of (B.9) have a direct interpretation. To give this interpretation, suppose some solution $Y^* \neq 0$ is given, and that the resulting solution to the right hand linear program of (B.8) is $\{u_{ij}^*, v_j^*\}$. Then the new inequality

$$\text{(B.10)}\qquad y_0 \geqq \sum_j v_j - \sum_{i,j} u_{ij}^* y_i + \sum_i f_i y_i$$

is adjoined to the restricted integer problem.

Now, as was seen above, $u_{ij}^* = 0$ if $y_i^* = 1$; and if $y_i^* = 0$, then u_{ij}^* is the difference between the present linear cost of supplying sink j, v_j^*, from the open sources and the linear cost of supplying j from a presently closed source i, if that cost is smaller. Thus, u_{ij}^* is the linear saving made possible in supplying j if the source i is opened, and $\sum_j u_{ij}^*$ is the total linear saving made possible if i is opened.

Let $I = \{i \mid y_i^* = 1\}$; it is the set of open sources. Then (B.10) may be written

$$\text{(B.11)} \quad y_0 \geqq \textstyle\sum_j v_j^* + \sum_{i \varepsilon I} f_i + \sum_{i \varepsilon I} f_i(y_i - 1) + \sum_{i \notin I} [f_i - \sum_j u_{ij}^*] y_i .$$

The sum of the first two terms is the total of the current solution $\{x_{ij}^*, y_i^*\}$: the linear cost is $\sum_j v_j^*$ and the fixed cost is $\sum_{i \varepsilon I} f_i$. The sum of the last two terms provides a lower bound on the change in total cost incurred in moving from Y^* to a new solution Y.

The fourth term of (B.11) is associated with the set of currently closed sources, $i \notin I$. If Y is chosen so that $y_i^* = 0$ implies $y_i = 0$, then this term is zero. If $y_i^* = 0$ implies $y_i = 0$ for all $i \notin I$ except for one source, say i_0, ($y_{i_0}^* = 0$ and $y_{i_0} = 1$), while $y_i^* = 1$ implies $y_i = 1$, then $f_{i_0} - \sum u_{i_0 j}^*$ is exactly the change in cost incurred in moving from Y^* to Y. Clearly, the total fixed cost is increased by f_{i_0}. But $y_{i_0} = 1$ and it is possible for some one or more sinks j to be supplied its units at a lower linear cost if $c_{i_0 j} < v_j^*$; the saving obtained at sink j is precisely $u_{i_0 j}^* = \max\{0, v_j^* - c_{i_0 j}\}$ since each sink has a demand of exactly one unit. However, if two or more sources are opened the contribution of linear savings in the fourth term is overly optimistic. The reason for this is that total linear savings obtained by opening two or more sources is necessarily less than (or equal to) the sum of the linear savings obtained by opening each of the sources separately. If some sources are closed and some opened then, again, the linear savings of the fourth term are overly optimistic.

The third term is associated with the set of currently open sources, $i \varepsilon I$. If Y is chosen so that $y_i^* = 1$ implies $y_i = 1$ then this term is zero. Otherwise some fixed costs f_i, $i \varepsilon I$, are saved (if $y_i^* = 1$ and $y_i = 0$). However, (B.11) does not take into account the fact that if a previously open source is closed then the linear cost may increase.

Therefore, the new inequalities generated by the algorithm may be described as "loose." In particular, it appears that the estimate of savings given in the third term may be "tightened"—although in a rather weak sort of way. Namely, let $J_k = \{j \mid x_{kj}^* = 1\}$. Then, if source k becomes closed, i.e., if $y_k = 0$ although $y_k^* = 1$, then all sinks $j \varepsilon J_k$ which received their supply from k must receive shipment from elsewhere. Since Y is unknown—the set of open sources is unpredictable—the strongest statement that can be made concerning going from source k open to source k closed is that the linear cost is increased by at least $\sum_{j \varepsilon J_k} \max\{\min_{l \neq k}(c_{lj} - v_j^*), 0\}$. Thus the third term of (B.11) may be replaced by

$$\textstyle\sum_{i \varepsilon I} [f_i - \sum_{j \varepsilon J_i} \max\{\min_{l \neq i}(c_{lj} - v_j^*), 0\}](y_i - 1)$$

which appears to be a weak improvement. However, if all sources are currently open, $y_i^* = 1$ all i, then this term accurately represents the change in cost if exactly one source is closed down (see Display V for an example).

Another form of location problem arises in the political sphere. This is the problem of drawing voting districts in a state for the purposes of providing citizens with equal representation in a state house of representatives. This problem is of considerable current interest due to recent Supreme Court rulings concerning citizens' rights for equal representation and the right of federal courts to intervene and review the constitutionality of state apportionments. The *districting problem* is the following. A house of representatives is to contain exactly r representatives: define r geographical districts in the state in order that each district contain nearly the same populations but in such a way that each district be a connected region with reasonable frontier (e.g., that familiar gerrymandered contours be eliminated). Weaver and Hess [98] have considered a formulation of this problem and, for a districting problem arising in the State of Delaware, have computed local optima via the spirit of the Baumol-Wolfe approach described above.

To formulate the problem define, first, some minimum geographic unit in which to make a population count and call this unit a precinct. For example, a precinct might be an "enumeration district," the unit used by the census bureau which is a well-defined area whose population is counted by one census enumerator. Suppose there are n such precincts, with p_j the population of district j. Further, for each, locate (by some means) a geographic center. A district is now defined as some collection of precincts, the center of the district being the center of some precinct in the collection. A feasible districting is some assignment of precincts to r disjoint districts. Ideally a district contains $1/r \sum p_j = p$ population; assume that a relative disparity of 2δ, $0 < \delta < \frac{1}{2}$, in population between districts is tolerable. Finally, to meet the need for "compactness"—a heretofore undefined geographical notion—define a measure of the compactness of a district as follows. Consider any district $D(l)$, with center the center of precinct l, and precincts $k \varepsilon D(l)$. Let r_{kl} be some geographical distance between centers of precincts k and l and define $c_{kl} = r_{kl}^2$. The compactness of district $D(l)$ is $\sum_{k\varepsilon D(l)} c_{kl}p_k$. The problem may now be formulated as that of defining a feasible districting to satisfy the population restrictions while minimizing the total compactness measure:

$$\begin{aligned}
&\text{Minimize} && \textstyle\sum_{i,j} c_{ij}p_i x_{ij} \\
\text{(B.12)}\quad &\text{when} && \textstyle\sum_j x_{ij} = 1, \qquad (1-\delta)py_j \leqq \sum_i p_i x_{ij} \leqq (1+\delta)py_j \\
&\text{and} && x_{ij} = 0, \text{ or } 1, \qquad y_j = 0, \text{ or } 1, \qquad \textstyle\sum_j y_j = r.
\end{aligned}$$

$y_j = 1$ means a district with center the center of precinct j is defined; $x_{ij} = 1$ means precinct i is assigned to the district with that center. The equation constraints ensure, respectively, that each precinct is assigned to some district, that the relative disparity of population between districts not exceed 2δ, and that exactly r districts are defined.

In [98] this problem is treated in an approximate fashion, and only a local minimum is found. The approach is this. Some collection of r y's is chosen and each of these y's is set to 1, i.e., some r district centers are chosen. Then, the x_{ij} are allowed to vary between 0, 1 and δ is set to zero, $\delta = 0$, and, thus, a transportation problem results and is solved. The x_{ij} are then rounded to 0 and 1, thus obtaining a set of districts. At this point, it is determined whether or not the district centers can be reassigned (the collection of r y's changed) in such a manner as to decrease the objective function: if so, the changes are made and the step repeated; if not, the solution found is kept.

The districting problem is distinct from an allied political problem, which is that of determining the size of the congressional delegation of each state in the U. S. House of Representatives. This problem, and schemes for its solution, has a long history and a very reasonable method of solution [63] [64]. The *apportionment problem* is exactly this: given n states, state j having a population of P_j, $j = 1, \cdots, n$, and a house of M members, determine the number of representatives or seats $x_j \geqq 1$ that each state j "should have." Controversy over this problem has centered on the meaning of "should have": what is a fair criterion on which to base the apportionment of seats to states? The criterion now in use has led to "the method of equal proportions." Define, for any apportionment, i.e., any set of $x_j \geqq 1$, integers and with $\sum x_j = M$, the congressional district of state j to be P_j/x_j; this is the number of citizens represented by one member of state j. The point of view adopted in the current method is that any solution $\{x_j\}$ should have the property that the percentage difference between the congressional districts of any two states cannot be reduced by a transfer of a seat from one state to the other. The percentage difference between congressional districts of states i and k is defined by $\{(P_i/x_i)/(P_k/x_k) - 1\}\ 100\%$, where $P_i/x_i \geqq P_k/x_k$. The method for constructing a solution is easily stated. First, assign one seat to each state. Then, at any stage, some $L(\leqq M)$ seats have been assigned which, in fact, constitute a solution satisfying the criterion to the problem were the house to contain only L members. Suppose this apportionment is given by $\{x_j\}$, $\sum x_j = L$. Then, if $L < M$, assign one more seat to state i if $P_i/(x_i(x_i + 1))^{1/2} = \max_j P_j/(x_j(x_j + 1))^{1/2}$. This priority scheme obtains a solution satisfying the criterion because every time a new seat is assigned the criterion is maintained. The reason is that the rule means $(P_j/x_j)/(P_i/x_i + 1) \leqq (P_i/x_i)/(P_j/x_j + 1)$ for all j. If $(P_j/x_j) \geqq (P_i/x_i + 1)$ then $(P_i/x_i) \geqq (P_j/x_j + 1)$, while, otherwise, $(P_i/x_i) > (P_i/x_i + 1) \geqq (P_j/x_j) > (P_j/x_j + 1)$ and the criterion is satisfied in either case for all j. Furthermore, the criterion leads to a unique solution except for possible "ties." To see this suppose $\{x_j\}$ and $\{y_j\}$ are different apportionments of the M seats which satisfy the criterion of "equal proportions": the priority scheme could lead to different ones if ties occurred in the max evaluation. There must be some pair of delegations satisfying $x_i \geqq y_i + 1$ and $x_k \leqq y_k - 1$. Then $P_i/x_i \geqq P_k/x_k$ (or $P_i/y_i \leqq P_k/y_k$, by symmetry) implies $1 \leqq (P_i/x_i)/(P_k/x_k) \leqq (P_i/y_i + 1)/(P_k/y_k - 1) < (P_i/y_i)/(P_k/y_k)$ thus contradicting the assumption that $\{y_j\}$ satisfies the criterion. Otherwise, $P_i/x_i < P_k/x_k$ and $P_i/y_i > P_k/y_k$. Clearly, $(P_i/y_i + 1)/(P_k/y_k - 1) \leqq (P_i/y_i)/(P_k/y_k)$, and since $\{y_j\}$ is assumed to satisfy the criterion this means $(P_i/y_i + 1)/(P_k/y_k - 1) < 1$. By the same argument, $(P_k/x_k + 1)/(P_i/x_i - 1) < 1$. Therefore,

$$1 \leqq \frac{P_i/y_i}{P_k/y_k} \leqq \frac{P_k/y_k - 1}{P_i/y_i + 1} \leqq \frac{P_k/x_k}{P_i/x_i} \leqq \frac{P_i/x_i - 1}{P_k/x_k + 1} \leqq \frac{P_i/y_i}{P_k/y_k}$$

a contradiction, unless all inequalities hold as equations. But, then, $x_i = y_i + 1$ and $x_k = y_k - 1$ and the percentage difference between congressional districts of states i and k is the same in either apportionment.

An important feature of the criterion and method of equal proportions—and one which seems to have been instrumental in leading to its adoption—is that if the size of the house M is increased while the total population of the country remains fixed then there is always a solution having the property that the delegation of no state is decreased. Some other methods which have been used or proposed do not have this feature and are, thus, liable to produce the "Alabama Paradox" (so named due to an occurrence of this phenomenon in an apportionment of the late 19th century, the loser being Alabama). For example, the "test of minimum range" proposes that the apportionment problem be formulated as

$$\text{(B.13)}\quad \begin{array}{lll} \text{Minimize} & \text{Max}_{i,j} \mid P_j/x_j - P_i/x_i \mid & \\ \text{when} & \sum_j x_j = M, & x_j \geqq 1,\ x_j \text{ integer,} \end{array}$$

that is, minimize the maximum disparity in congressional districts. This criterion has recently been re-proposed [42], but, as shown by numerical example [64], it admits the Alabama Paradox.

Display V.

Consider the problem (B.7) with $m = n = 4, f_i = 7$ all i, and

$$(c_{ij}) = \begin{bmatrix} 0 & 12 & 20 & 18 \\ 12 & 0 & 8 & 6 \\ 20 & 8 & 0 & 6 \\ 18 & 6 & 6 & 0 \end{bmatrix}$$

(an example occurring in [6] and taken from [75]).

The constraints of form (B.10) generated for the restricted integer program (B.9) are given in the following table. In a line of this table Y^* is an optimal solution to the restricted integer program defined by the constraints given in all the above lines and having the value Min y_0 given in the immediately preceding line (e.g. $Y^* = (1, 0, 1, 0)$ is an optimal solution with value Min $y_0 = 20$ to the program with three inequalities). $\sum_j v_j^* + \sum_i f_i y_i^*$ is the optimal value to (B.7) given that $Y = Y^*$ (e.g., $Y^* = (1, 0, 1, 0)$ has least cost 28). Thus, at any stage, Min y_0 provides a lower and Min $\sum_j v_j^* + \sum_i f_i y_i^*$ an upper bound on the true least cost solution.

$Y^* = (y_1^*, y_2^*, y_3^*, y_4^*)$	$\Sigma_j v_j^* + \Sigma_i f_i y_i^*$	$y_0 \geqq (\ ) + (\ )y_1 + (\ )y_2 + (\ )y_3 + (\ )y_4$					Min y_0
(0, 1, 0, 0)	33	26	−5	7	−1	−1	19
(1, 0, 1, 1)	27	6	7	1	7	7	20
(1, 0, 0, 1)	26	12	7	1	1	7	20
(1, 0, 1, 0)	28	14	7	−1	7	−1	21
(1, 0, 0, 0)	57	50	7	−29	−29	−31	24
(0, 0, 1, 1)	38	24	−11	−5	7	7	26
(1, 0, 0, 1)	26						

C. Computation

1. *Introduction*

Users of general integer programming algorithms I and II have met with erratic computational performance: some problems have been solved rapidly, others not for numbers of iterations surpassed limitations of time and/or expense. It is, thus, natural to ask why the experience has been erratic.

One reason resides in the fact that algorithms I and II allow a wide range of arbitrary choice which can seriously effect convergence. There is evidence to support this statement. Ouyahia [82] remarks that in a problem (A.1) with $m = 20, n = 29$ one version of a computer code for II found no solution in over 30,000 iterations; while in another version an optimal solution was found in 70 iterations. Computational experience in solving the "cutting stock" problem [43] is also germane to this statement. The algorithm used to solve that problem is a "primal" decomposition approach to solving a linear program with relatively few rows or constraints and potentially vast numbers of columns or variables (e.g., millions), just as an integer programming algorithm is (in one interpretation) a "dual" decomposition approach to solving a linear program with relatively few columns or variables and potentially vast numbers of rows or constraints (the "convexifying constraints"). In integer programming the basic step is the generation of a new row or constraint. In the cutting stock algorithm the basic step is the generation of a new column or new cutting pattern which will yield an improvement relative to the "current" feasible solution. This step is accomplished by solving a "knapsack problem." In doing this, however, it is possible to obtain (a) a column which leads to the greatest per unit improvement, or to (b) stop the knapsack computation earlier by accepting the first column found which yields some improvement. The effect of these alternate methods of choice of new column on a set of 21 test problems is given in [43]. It is conclusively demonstrated that (a) is superior to (b). For example, the total number of pivot steps necessary for solution in a set of 5 problems with these alternate methods of choice were:

Problems		#1	#2	#3	#4	#5
Column Choice	(a)	148	233	161	101	90
	(b)	541	1104	586	545	370

The inference to be drawn is that convergence in integer programming will be highly sensitive to the rule of choice of row (A.3) from which to generate a new constraint (A.6), (A.8) or (A.12).

Further evidence of sensitivity of convergence to choices in computation abounds. For example, in recent work [4] regarding algorithms I and II for solving generalized set covering problems (B.5), a problem having $m = 15$, $n = 50$ was not solved in 200 iterations using I, but after a shuffling of rows (i.e., a simple unpremeditated rearrangement or reordering of the constraints) the problem was solved in 34 pivots by the identical code. Such behavior—which has been encountered by others—suggests at least two lines of action. One is to develop choice mechanisms which are "row-independent" or which depend on a "more natural" kind of lexicography for convergence; another is to find ways for ranking rows or, more generally, for pre-conditioning problems for computation. The measure of iterations or number of pivot steps is not necessarily one which increases with total time of computation, or with any other measure, for some types of pivot steps require much time and others little in computation. Nonetheless, it is a measure which is not tied to type of hardware used and which does have direct mathematical interest and at least indirect computational interest.

In this section some further mathematical structure which underlies integer programming and, more particularly, the choice problem, is discussed, and a new "rounding method" based on the structure is described. Also, some possible rules of choice and their role in some of the current computer codes are discussed. Finally, an attempt is made at rallying the known experience in solving problems, and some reasons are tentatively advance for explaining the observed performance of the algorithms.

2. The Group of Alternate Cuts in Algorithm I

In the derivation of the new constraint or cut (A.8) from an equation (A.3) (we restore row subscripts)

$$x_{i+n} = a_{i0} + \textstyle\sum_j a_{ij}(-x_j) \tag{C.1}$$

appearing in some tableau equivalent to problem (A.1), the following facts were used. All x_j nonnegative and integer, and x_{i+n} integer but not constrained in sign. This permitted deduction of a new cut (A.8)

$$x'_{i+n} = -r_{i0} + \textstyle\sum_j (-r_{ij})(-x_j) \geqq 0 \tag{C.2}$$

with x'_{i+n} integer, where $r_{ij} \geqq 0$ is the fractional part of a_{ij}. But since x_{i+n} is only required to be integer in this argument, any equation which is an integer combination of equations of form (C.1) such as

$$x^* = \textstyle\sum_i u_i x_{i+n} = (\sum_i u_i a_{i0}) + \sum_j (\sum_i u_i a_{ij})(-x_j), \quad u_i \text{ integer} \tag{C.3}$$

could also be used to derive a cut (C.2) since x^* is integer constrained being by definition an integer combination of integer constrained variables. The cut (C.2) resulting therefrom has r_{ij} as the fractional part of $\sum_i u_i a_{ij}$, i.e., $r_{ij} = \sum u_i a_{ij} (\text{mod } 1) = \sum_i u_i r_{ij} (\text{mod } 1)$.

Suppose, now, that algorithm I is to be used to solve (A.1) and that the linear programming solution has been found. If the product of the pivot entries of the computation is D then it means that a matrix has been inverted in the course of the computation whose determinant is D; and every entry and every minor of the tableau can be written in the form h/D where h and D are integer, since the original data of the problem is integer. Let the rows $R_i = (r_{i0}, \cdots, r_{in})$ represent the cuts (C.2) derived directly from the equations (C.1) of the tableau as before. Then, all possible cuts resulting from forming equations (C.3) implied by the tableau are generated by choosing arbitrary integer values for u_i and forming rows $\sum_i u_i R_i \pmod 1) = (\sum_i u_i r_{i0} \pmod 1), \cdots, \sum_i u_i r_{in} \pmod 1))$. These rows form an additive group with respect to vector addition entries reduced to their fractional parts.

Does this group G of all cuts derived as above from the constraints in a tableau have any particular structure? Yes: by the "fundamental theorem of abelian groups" (see e.g., [92]) G is the direct sum of a finite number of cyclic subgroups. This means that G contains some set of rows $R^k = (r_0^k, \cdots, r_n^k)$, $k = 1, \cdots, l$, each R^k itself generating a subgroup G^k consisting of all rows $uR^k \pmod 1$, $u \geqq 0$ and integer, such that any row of G is uniquely expressible as a sum of rows (mod 1), one taken from each subgroup G_k. Notice that this immediately implies that G is finite, for each G_k can contain at most D rows (since any r_0^k has a denominator at most equal to $|D|$ and $|D|R^k = 0 \pmod 1$). Not so evident, but true, is the fact that G consists of exactly $|D|$ rows. Thus, if there is a row $R = (r_0, \cdots, r_n)$ containing an $r_i = h/D$, with h and D relatively prime (i.e., containing no common factors) then R itself generates the entire group of cuts since it must generate $|D|$ of them. In fact, if D has no repeated primes in its factorization then there must be one row R which generates all of G. In this case G is said to have rank 1; if l subgroups G_k generate G, then G is said to have rank l. Further results concerning bounds on the rank of G have also been derived [49].

In summary: investigation of the derivation of cuts led to the discovery that all possible cuts derived from an optimal linear programming tableau as above form a group G containing exactly $|D|$ rows and that in "many" cases G has rank 1, i.e., all rows can be generated by one row. The discussion may appear to be predicated on deriving cuts from an optimal linear programming tableau of (A.1). However, the same remarks hold for any stage of computation in algorithm I; for any cuts or rows that are adjoined at any stage can be re-expressed in all integer data in terms of the original nonbasic variables. In other terms, reverting the pivot process to return to the original tableau would mean reinverting a matrix of determinant D and thereby transforming any cut into an equivalent constraint having integer coefficients. Thus, in later stages of I the same statements are true if the "underlying" linear program is interpreted to contain among its original constraints those cuts previously adjoined.

Facts pertaining to the group structure of cuts have practical bearing. Given a choice of new cuts to introduce in a stage of I, how should one or more be picked? This is a question analogous to that of how to choose a new column to enter the basis in a simplex method and it appears that answers to both questions must be of a statistical rather than a mathematical nature. The general intent in deriving a new cut is to choose that which is strongest, i.e., for which the ratios r_0/r_j are large as possible for this means that the new constraint has intercepts with the x_j-axes (where the x_j are the current nonbasic variables) which are deep. Clearly, it is impossible, in general, to find cuts which are uniformly strong for all j, so other criteria for choice are necessary. Essentially, three intuitively derived criteria have been proposed. One is to choose a cut $R = (r_0, r_1, \cdots, r_n)$ having r_0 large; another having r_0/r_k large for some given k: the third, having $r_0/(\sum_{j\neq 0} r_j)$ large. If only directly derived cuts are admitted then each of these criteria involve inspecting each constraint in the current tableau from which a cut may be generated and choosing one as (C.1) which yields the cut with the desired property. However, it seems reasonable to believe that the group often has rank 1 but is always small. If the rank is actually 1 then the choice of a constraint in the tableau which generates the entire group is easy, and, moreover, finding that cut which maximizes r_0 or r_0/r_k is accomplished through an application of the Euclidean Algorithm [17]. Thus, if one believes the rank to be 1 often, then the choice of row (C.1) from which to generate cuts is relatively unimportant. Of course, one of the criteria assumes a choice of column k or x_k-axis along which the deepest cut is desired. It seems reasonable to choose that x_k which has the smallest a_{0k} or price attached to it for then a deepest cut is being made in a direction along which the objective decreases the least per unit. In this way the objective does not—it is felt—hinder the effectiveness of the deepness of the cut along the x_k-axis.

There are three codes[3] which are essentially variants of algorithm I. IPM3 [72] is an all in core Share code in which a number of intuitive rules are incorporated. For one, an attempt is made to pick cuts having $r_0/\sum r_j$ large; for another, given a cut with $r_0/\sum r_j$ large a further cut is derived from it using (A.6), that is a cut $([r_0/\lambda], \cdots, [r_n/\lambda])$, for some choice of λ, if the new cut enables making the ratio $[r_0/\lambda]/\sum_j [r_j/\lambda]$ larger; for still another, a number of cuts are adjoined at once (as many as storage allows) in order to allow the simplex method computations to "choose." The CEIR code [77] is a general code which is not all in-core and, hence, able to accept problems of greater dimension than any other code. It chooses a row (C.1) whose directly derived cut R has r_0 a maximum; then, it chooses k to be that column which would be the pivot column if R were the cut used; finally, it generates from R that row having r_0/r_k a maximum. Furthermore, after pivoting on such a cut, this code continues to derive new cuts, without obtaining a new linear programming solution, until an all integer tableau is obtained. Only at this point is a new linear programming solution found. LIPI [59] is an all in-core code to be released in Share which is, as is IPM3, an experimental code. In it the linear programming calculation is done by a primal, phase I-phase II, revised simplex method which causes, it would appear, computation to require rather more pivot steps than if a dual simples method were used (since when a cut is introduced the tableau is already dual feasible and "almost" optimal). The choice of cut is done lexicographically: after a linear programming solution is found the topmost row (including objective) of the tableau having a noninteger basic variable is chosen and from it the directly derived cut is found and used. However if the cut is not taken from the objective, then before generation of a cut pivoting is done to make all columns lexicographically positive (the first basic integer variable having nonpositive row entries is "suboptimized," values of variables higher in the tableau remaining unchanged, meaning such pivoting occurs in columns having zeros above the pivot entry). The authors of this code tried various row selection rules and concluded this was the best after

[3] Hards facts regarding codes and computational experience are difficult to obtain.

some experimentation. In particular, they rejected a choice of row leading to large r_0 and a choice of row leading to large $r_0/\sum r_j$.

3. The Group of Columns and a "Rounding" Algorithm

Given a linear program to be solved in integers it is tempting to solve the linear program and then—by some process—"round" the noninteger valued variables into an optimal integer solution. Recent particularly appealing results of Gomory [52] have led to a dynamic programming process for rounding a linear programming solution into an optimal integer solution. The inherent appeal of the results resides in their crucial dependence on the structure of a group of columns—a group which is essentially the same as (is isomorphic to) the group of cuts described above—which is just enough to enable specification of a finite recursive procedure.

It is convenient to alter notation. Consider the integer program: find integer X to

$$\text{(C.4)} \quad \begin{array}{lll} \text{Maximize} & CX & \\ \text{when} & AX = B, & X \geqq 0 \end{array}$$

where C is an $m + n$-vector, A an m by $m + n$ matrix of integers containing an m by m identity matrix (i.e., the constraints are immediately reducible to inequalities), and B an m-vector of integers. Let A_N be a basis, i.e., a set of m linearly independent columns of A, and partition the data of (C.4) by letting $A = (A_N, A_I)$, $C = (C_N, C_I)$ and $X = (X_N, X_I)$. Then (C.4) may be rewritten as

$$\begin{array}{lll} \text{Maximize} & C_N X_N + C_I X_I & \\ \text{when} & A_N X_N + A_I X_I = B, & X_N, X_I \geqq 0. \end{array}$$

Solving for X_N in terms of X_I (C.4) becomes the same problem as

$$\text{(C.5)} \quad \begin{array}{lll} \text{Maximize} & C_N A_N^{-1} B + (C_I - C_N A_N^{-1} A_I) X_I & \\ \text{when} & X_N + A_N^{-1} A_I X_I = A_N^{-1} B, & X_N, X_I \geqq 0. \end{array}$$

If $A_N^{-1}B \geqq 0$ and $C_I - C_N A_N^{-1} A_I \leqq 0$ then an optimal solution to (C.5), and hence (C.4), is $X_I = 0$ and $X_N = A_N^{-1}B$, for taking the independent variables X_I to be anything larger than zero can only decrease (or leave the same) the value of the objective. A simplex method locates precisely such a partition or basis A_N.

To simplify the notation, let $d = C_N A_N^{-1} B$, $A_N^{-1} A_I = (\alpha_1, \cdots, \alpha_n)$, $C_I - C_N A_N^{-1} A_I = (c_1, \cdots, c_n)$, $A_N^{-1} B = \alpha_0$, and rename variables so that $X_I = (x_1, \cdots, x_n)$ and $X_N = (x_{n+1}, \cdots, x_{n+m})$. Then, as we have seen, (C.4) is equivalent to: find integer X_I and X_D to

$$\text{(C.6)} \quad \begin{array}{lll} \text{Maximize} & d + \sum_j c_j x_j & \\ \text{when} & X_n = \alpha_0 - \sum_j \alpha_j x_j \geqq 0, & x_j \geqq 0, \end{array}$$

where we may assume $\alpha_0 \geqq 0$ and $c_j \leqq 0$ all j by a previous linear programming computation or basis A_N.

Consider the problem (C.6). If α_0 is all integer then $X_I = 0$, $X_N = \alpha_0$ is an optimal integer solution. Otherwise, the independent variables $X_I = (x_1, \cdots, x_n$ of (C.6) must take on some non-negative integer values ($X_I \neq 0$) in such a way that $\alpha_0 - \sum \alpha_j x_j$ is integer and non-negative. The proposal of Gomory may be said to be this: ignore, for the moment, the requirement that this difference be non-negative. Then, the problem of solving (C.6) reduces to: find x_j ($j = 1, \cdots, n$) integer to

$$\text{maximize} \quad \textstyle\sum_j c_j x_j$$
$$\text{when} \quad \textstyle\sum_j \alpha_j x_j = \alpha_0 \pmod 1, \qquad x_j \geqq 0$$

where the (mod 1) is applied row by row. But, clearly, any α_j ($j = 0, 1, \cdots, n$) may be replaced by a column $\bar{\alpha}_j$ so long as $\alpha_j = \bar{\alpha}_j \pmod 1$; in particular, $\bar{\alpha}_j$ can be taken to be the column of fractional parts of α_j and the problem defined as: find x_j integer to

$$\text{(C.7)} \quad \begin{array}{ll} \text{maximize} & \sum_j c_j x_j \\ \text{when} & \sum_j \bar{\alpha}_j x_j = \bar{\alpha}_0 \pmod 1, \qquad x_j \geqq 0. \end{array}$$

If there exists an integer feasible solution to (C.4) then there must exist a solution to (C.7) since every $c_j \leqq 0$. But, now, the group of cuts discussed above is essentially the same as the group of columns H which is generated by $\bar{\alpha}_j$, $j = 1, \cdots, r$ (a fact easily verified by considering the argument which led to the group theoretic results [49] for cuts and simply talking about columns instead of rows). Therefore, letting the determinant of A_N be the integer D, there are at most $|D|$ columns which can be generated by $\sum \bar{\alpha}_j x_j \pmod 1$ for all integer values x_j; and, moreover, all of these may be found by generating the cyclic subgroups of H and then taking all sums of the columns, one coming from each subgroup. Again, if some one column $\bar{\alpha}_j$ contains an entry h/D with h relatively prime to D then it generates the entire group.

Clearly it is possible to ask that $x_j \leqq |D| - 1$ in an optimal solution to (C.7). Slightly less clear is that there exist optimal solutions with $\sum_j x_j \leqq |D| - 1$. For suppose not. Form the sequence of vectors $\bar{\alpha}_j$ with which $\bar{\alpha}_j$ appearing x_j times. Then, the null column, the first column, the sum of the first two, of the first three, etc., form a sequence of more than D columns, implying that at least two of these sums must be the same and hence columns $\bar{\alpha}_j$ present in one sum but not the other may be deleted (since they sum to the zero column).

To compute solutions to (C.7) apply the idea of dynamic programming: let

$$\text{(C.8)} \qquad \varphi_s(\bar{\alpha}) = \{\max \textstyle\sum_1^s c_j x_j : \sum_1^s \bar{\alpha}_j x_j = \bar{\alpha}, \text{ for } \bar{\alpha} \in H\}.$$

Then

$$\text{(C.9)} \qquad \varphi_s(\bar{\alpha}) = \max\{\varphi_s(\bar{\alpha} - \bar{\alpha}_s) + c_s, \varphi_{s-1}(\bar{\alpha})\}$$

(by the same reasoning which led to the knapsack computation, see Section

A.4). Assuming $\varphi_s(0) = 0$ (i.e., if $\bar{\alpha}$ is the zero column 0 the solution must have value 0: take all $x_j = 0$) and that $\varphi_{s-1}(\bar{\alpha})$ is known for all $|D|$ columns $\bar{\alpha}$ of H, application of (C.9) gives

$$\text{(C.10)}\quad \begin{aligned} \varphi_s(\bar{\alpha}_s) &= \max\{\varphi_s(0) + c_s, \varphi_{s-1}(\bar{\alpha}_s)\} \\ \varphi_s(r\bar{\alpha}_s) &= \max\{\varphi_s(r\bar{\alpha}_s - \bar{\alpha}_s) + c_s, \varphi_{s-1}(r\bar{\alpha}_s)\} \quad r = 2, \cdots, D-1. \end{aligned}$$

If $\bar{\alpha}_s$ generates the entire group, this computation (C.10) finds $\varphi_s(\bar{\alpha})$ for all $\bar{\alpha} \varepsilon H$. Otherwise, $d\bar{\alpha}_s \equiv 0 \pmod 1$ for some d which is a factor of D. In this case the computation of the necessary values is slightly more involved.

Begin by choosing some $\bar{\alpha}$ of H which is not a column of the cyclic subgroup generated by $\bar{\alpha}_s$, $\bar{\alpha} \neq r\bar{\alpha}_s$ all integer r. The idea is to compute $\varphi_s(\bar{\alpha} + r\bar{\alpha}_s)$ for $r = 1, 2, \cdots, d$. The difficulty encountered is that it is not possible to compute $\varphi_s(\bar{\alpha})$ directly (by (C.9)) for $\varphi_s(\bar{\alpha} - \bar{\alpha}_s)$ is not known. Assume for the moment that $\varphi_s(\bar{\alpha})$ is $\varphi_{s-1}(\bar{\alpha})$ and (to indicate this special assumption which can also be stated as assuming $\varphi_s(\bar{\alpha} - \bar{\alpha}_s) + c_s < \varphi_{s-1}(\bar{\alpha})$) let $\varphi'_s(\bar{\alpha}) = \varphi_{s-1}(\bar{\alpha})$ and

$$\text{(C.11)}\quad \varphi'_s(\bar{\alpha} + r\bar{\alpha}_s) = \max\{\varphi'_s(\bar{\alpha} + r\bar{\alpha}_s - \bar{\alpha}_s) + c_s, \varphi_{s-1}(\bar{\alpha} + r\bar{\alpha}_s)\}, \quad r = 1, 2, \cdots, d.$$

If the assumption is correct then (C.11) provides the sought for values $\varphi_s(\bar{\alpha} + \bar{\alpha}_s)$. In fact, if $\varphi'_s(\bar{\alpha} + q\bar{\alpha}_s) = \varphi_s(\bar{\alpha} + q\bar{\alpha}_s)$ for some q then all subsequent values $\varphi'_s(\bar{\alpha} + r\bar{\alpha}_s) = \varphi_s(\bar{\alpha} + r\bar{\alpha}_s)$, $r > q$, as well; taking $r = d$, $\varphi'_s(\bar{\alpha} + d\bar{\alpha}_s) = \varphi'_s(\bar{\alpha}) = \varphi_s(\bar{\alpha})$ (which may be different than $\varphi_{s-1}(\bar{\alpha})$) and thus (C.11) may be used with primes dropped to compute $\varphi_s(\bar{\alpha} + r\bar{\alpha}_s)$ for $r < q$. Thus, in at most $2d$ uses of (C.11) all values $\varphi_s(\bar{\alpha} + r\bar{\alpha}_s)$ are found, if (C.11) once provides a correct value.

It only remains to show that $\varphi'_s(\bar{\alpha} + r\bar{\alpha}_s) = \varphi_s(\bar{\alpha} + r\bar{\alpha}_s)$ for some $1 \leqq r \leqq d$. Note that $\varphi'_s(\bar{\alpha} + r\bar{\alpha}_s) \leqq \varphi_s(\bar{\alpha} + r\bar{\alpha}_s)$ all r, i.e., if the original guess at $\varphi_s(\bar{\alpha})$ is an underestimate then (C.11) cannot overestimate any values φ_s. Suppose (a) $\varphi_s(\bar{\alpha} + r\bar{\alpha}_s) = \varphi_{s-1}(\bar{\alpha} + r\bar{\alpha}_s)$ for some r: then $\varphi'_s(\bar{\alpha} + r\bar{\alpha}_s) \geqq \varphi_{s-1}(\bar{\alpha} + r\bar{\alpha}_s) = \varphi_s(\bar{\alpha} + r\bar{\alpha}_s) \geqq \varphi'_s(\bar{\alpha} + r\bar{\alpha}_s)$ and equality must hold throughout. Otherwise, (b) $\varphi_s(\bar{\alpha} + r\bar{\alpha}_s) = \varphi_s(\bar{\alpha} + r\bar{\alpha}_s - \bar{\alpha}_s) + c_s$ for all r: then $\varphi_s(\bar{\alpha}) = \varphi_s(\bar{\alpha} + d\bar{\alpha}_s) = \varphi_s(\bar{\alpha} + d\bar{\alpha}_s - \bar{\alpha}_s) + c_s = \cdots = \varphi_s(\bar{\alpha}) + dc_s$, a contradiction if $c_s \neq 0$. If $c_s = 0$ then $\varphi_{s-1}(\bar{\alpha} + r\bar{\alpha}_s) = \varphi_s(\bar{\alpha})$ for some r, $1 \leqq r \leqq d$ (by (C.8)) and, hence, $\varphi_{s-1}(\bar{\alpha} + r\bar{\alpha}_s) = \varphi_s(\bar{\alpha} + r\bar{\alpha}_s)$, which is again case (a).

In summary: if $d\bar{\alpha}_s = 0$, $d < D$, choose some $\bar{\alpha} \neq r\bar{\alpha}_s$ any r, and apply (C.11) d or more times until $\varphi'_s(\bar{\alpha} + r\bar{\alpha}_s)$ agrees with its previously computed value. This means applying (C.11) at most $2d - 1$ times. The final values $\varphi'_s(\bar{\alpha} + r\bar{\alpha}_s) = \varphi_s(\bar{\alpha} + r\bar{\alpha}_s)$. Then, choose some new $\bar{\alpha}$ for which $\varphi_s(\bar{\alpha})$ has not been computed, and repeat. This procedure is applied for $|D|/d$ starting points $\bar{\alpha}$ to obtain all $|D|$ values $\varphi_s(\bar{\alpha})$, $\bar{\alpha} \varepsilon H$. All of this depends on the fundamental fact that the group of columns H is the direct sum of a finite number of cyclic subgroups.

Algorithm VII ([52] 1965). Let $\varphi_0(\bar{\alpha}) = m$, m very small (e.g., $m \leqq \min_i |D| c_i$). Apply (C.10) and/or (C.11) to compute $\varphi_n(\bar{\alpha})$, for all $\bar{\alpha} \varepsilon H$. At each stage it is only necessary to record and keep $\varphi_s(\bar{\alpha})$ together with $\sum_i x_i$ corresponding to each such maximizing solution. When $\varphi_n(\bar{\alpha})$ is found for all $\bar{\alpha}$, look up $\varphi_n(\bar{\alpha}_0)$ and backtrack (as described in the similar method for the knapsack problem) to obtain values $X_I^* = (x_1^*, \cdots, x_n^*)$ to (C.7). *If* these values are such as to make all values $X_N^* = \alpha_0 - \sum \alpha_j x_j^*$ nonnegative, $X_N^* \geqq 0$, then $X^* = (X_I^*, X_N^*)$ constitutes an optimal solution to (C.4).

The obvious question is: when does this algorithm provide a solution? I.e., under what circumstances is $X_N^* \geqq 0$? In geometric terms $X_N^* \geqq 0$ whenever the integer solution to (C.4) is determined by the half-space requirements alone where $X_I = 0$ determines an optimal linear programming solution; in other words, whenever the constraints $X_N \geqq 0$ are superfluous to both the linear and the integer problem. But this is not an answer.

In the absence of knowing an optimal basis A_N nothing can be said. However, once A_N is known sufficient conditions can be given. Notice, first, that A_N remains an optimal basis for all $B \varepsilon K = \{Z \mid A_N^{-1}Z \geqq 0\}$. The condition on B cannot be carried over to the integer case for given some B nonnegative integer multiples of columns α_j are subtracted from, rather than arbitrary nonnegative multiples, to obtain values X_N. What is the worst that could happen? Each $x_j \leqq D - 1$ is such a solution. Thus, if α is defined to be a vector whose i^{th} entry is the maximum of the ith entries of $\alpha_1, \cdots, \alpha_n$, then $X_N^* \geqq 0$ whenever $B \varepsilon K' = \{Z \mid A_N^{-1}Z \geqq (D-1)\alpha\}$, for $(D-1)\alpha$ is the most that could possibly be subtracted. Another sufficient condition is this [52]: if B belongs to K but is distant at least $(D-1) \max_j \| \alpha_j \|$ from the boundary of K then again $X_N^* \geqq 0$, for B cannot be displaced in K by more than a vector of that magnitude. Neither of these conditions are necessary.

In conclusion notice that if B is changed by adding any integer combination of columns of A_N to it, then the solution to (C.7) remains the same, since this addition only changes $A_N^{-1}B = \alpha_0$ by integer quantities. Essentially, therefore, only $D - 1$ different B's need or can ever be considered: namely those B which lead to each possible column $\bar{\alpha}$ of the group H. And, of course, these have all been considered in the dynamic programming computation since $\varphi_n(\bar{\alpha})$ is evaluated for every $\bar{\alpha} \varepsilon H$.

Display VII

Consider, again, the example of Display I. In the notation of this section it may be written as

x_1	x_2	x_3	x_4	x_5		
−4	−5	0	0	0	$= x_0$	
−3	−1	1			$= -2$	
−1	−4		1		$= -5$	$x_i \geqq 0, i = 1, \cdots, 5$
−3	−2			1	$= -7$	

and its linear programming solution displayed as

x_1	x_2	x_3	x_4	x_5	
0	0	0	−7/10	−11/10	$= x_0 + 112/10$
1			2/10	−4/10	$= 18/10$
	1		−3/10	1/10	$= 8/10$
		1	3/10	−11/10	$= 42/10$

Thus (C.7) becomes (where $y_1 = x_4$ and $y_2 = x_5$)

$$\text{Maximize} \quad -7/10\, y_1 - 11/10\, y_2$$

$$\text{when} \quad \begin{pmatrix} 2/10 \\ 7/10 \\ 3/10 \end{pmatrix} y_1 + \begin{pmatrix} 6/10 \\ 1/10 \\ 9/10 \end{pmatrix} y_2 = \begin{pmatrix} 8/10 \\ 8/10 \\ 2/10 \end{pmatrix}, \qquad y_1, y_2 \geqq 0, \text{ integer.}$$

The computation (C.10) is contained in the following table. In each entry the first number is the appropriate value $\varphi_s(\bar{\alpha})$, the second the sum $\sum y_i$ corresponding to that optimal solution, the third the index i of the last variable made equal to a 1.

	$\varphi_1(\,)$	y_1	i	$\varphi_2(\,)$	$y_1 + y_2$	index
$\bar{0}$	0	0	—	0	0	—
$\bar{\alpha}_1 = 7\bar{\alpha}_2$	−7/10	1	1	−7/10	1	1
$2\bar{\alpha}_1 = 4\bar{\alpha}_2$	−14/10	2	1	−14/10	2	1
$3\bar{\alpha}_1 = \bar{\alpha}_2$	−21/10	3	1	−11/10	1	2
$4\bar{\alpha}_1 = 8\bar{\alpha}_2$	−28/10	4	1	−18/10	2	2
$5\bar{\alpha}_1 = 5\bar{\alpha}_2$	−35/10	5	1	−25/10	3	2
$6\bar{\alpha}_1 = 2\bar{\alpha}_2$	−42/10	6	1	−22/10	2	2
$7\bar{\alpha}_1 = 9\bar{\alpha}_2$	−49/10	7	1	−29/10	3	2
$8\bar{\alpha}_1 = 6\bar{\alpha}_2$	−56/10	8	1	−36/10	4	2
$9\bar{\alpha}_1 = 3\bar{\alpha}_2$	−63/10	9	1	−33/10	3	2

$\bar{\alpha}_0 = 4\bar{\alpha}_1 = 8\bar{\alpha}_2$; therefore, backtracking, $y_2 = 1$, $y_1 = 1$ is an optimal solution. Hence, by substitution, an optimal solution is $x_0 = -13$, $x_1 = 2$, $x_2 = 1$, $x_3 = 5$, $x_4 = 1$, $x_5 = 1$. Notice that if, for example, $(-2, -5, -7)^T$ were altered to $(-5, -6, -10)^T$ by addition of the column of x_1 then the optimal linear programming tableau would be the same except that the right hand side would become (28/10, 8/10, 42/10). Thus problem (C.7) remains the same and an optimal solution is easily read off.

Finally, if the $(-2, -5, -7)^T$ were altered to, say, $(-6, -7, -8)^T$, then α_0^T becomes (18/10, 13/10, 7/10) and the solution to (C.7) is already contained above for $\bar{\alpha}_0 = 3\bar{\alpha}_2 = 9\bar{\alpha}_1$; $y_2 = 3$, $y_1 = 0$. Thus, $x_1 = 3$, $x_2 = 1$, $x_3 = 4$, $x_4 = 0$, $x_5 = 3$ is an optimal solution in this case.

4. *Choices in Algorithms II and III*

In the derivation of the new constraint (A.6) from an equation (A.3) (we restore row subscripts)

$$\text{(C.11)} \qquad x_{i+n} = a_{i0} + \textstyle\sum_j a_{ij}(-x_j)$$

appearing in some tableau equivalent to (A.1), the following facts were used. All x_j nonnegative and integer, and x_{i+n} nonnegative but not necessarily integer constrained. This permitted derivation of a new constraint

$$\text{(C.12)} \qquad x'_{i+n} = [a_{i0}/\lambda] + \textstyle\sum [a_{ij}/\lambda](-x) \geqq 0$$

with x'_{i+n} integer constrained. But since x_{i+n} is only required to be nonnegative in this argument, any equation which is a nonnegative combination of equations of form (C.1) such as

$$\text{(C.13)} \quad x^* = \sum_i u_i x_{i+n} = (\sum_i u_i a_{i0}) + \sum_j (\sum_i u_i a_{ij})(-x_j), \qquad u_i \geqq 0$$

could also be used to derive a constraint (C.12), since x^* is nonnegative being a nonnegative combination of nonnegative variables. The constraint (C.12) resulting therefrom is easily written down—but not in terms of the constraints resulting directly from equations (C.11) of the tableau.

In algorithm II a choice of row (C.11) with $a_{i0} < 0$ or (C.13) with $\sum_i u_i a_{i0} < 0$ from which to generate a constraint of type (C.12) uniquely specifies the smallest possible λ and the resultant -1 pivot entry if dual feasibility and integer data are to be maintained. The pivot entry must be in the smallest lexicographic column α_j which contains $a_{ij} < 0$ or $\sum_i u_i a_{ij} < 0$. Thus a reasonable approach to a choice of row is to find one which permits a choice of pivot in a large lexicographic column α_j, for this means that the pivot step leads to a change in the constant column at least as large as column α_j. Experiments have shown that concern with pivoting in large columns is considerably more important than concern with pivoting on rows with $a_{i0} < 0$ or $\sum u_i a_{i0} < 0$ as small as possible (contrary to the usual linear programming criteria). Of course, it is also desirable to pivot in a large lexicographic column α_j in such a way that $[a_{i0}/\lambda]$ or $[\sum u_i a_{i0}/\lambda]$ is as small as possible for this leads to a change in the constant column which is a large integer multiple of the column α_j.

What rules, then, should or can be adopted? Again, the answers (which are not known) must be of a statistical rather than a mathematical nature. Some suggestions have been made. Assume that (or rank) the columns of a given tableau are arranged lexicographically with $\alpha_1 \geqq \cdots \geqq \alpha_n$. Then the rows may be ranked by assigning to each row the index (or rank) of the column which would have to be its pivot column, i.e., row i has rank $r(i)$ where $\alpha_{r(i)} = \min_j \{\alpha_j \mid a_{ij} < 0\}$. Essentially the one intuitive criterion which has been used is to choose or construct a row having $r(i)$ small. A direct rule is this: choose row i in the current tableau having $a_{i0} < 0$ with minimum rank $r(i)$. More involved are iterative schemes for producing a row having small rank. For ex- example, choose row i, or x_{i+n}, in the tableau having $a_{i0} < 0$ with minimum $r(i)$ and generate its constraint x'_{i+n}. Then, for rows k having $r(k) < r(i)$ (and hence having $a_{k0} \geqq 0$) form a new row $x^* = x_{k+n} + ux'_{i+n}$, $u \geqq 0$, with negative constant term and zero entry corresponding to the pivot column $r(i)$ of x'_{i+n} (thus $u = a_{kr(i)}$). If it is possible to find such an x^* then it is an eligible row with lower rank, and the process may be repeated. A slight (untried) amendment to this approach would be to compare the total lexicographic change in α_0 resulting from use of x'_{i+n}, $[a_{i0}/\lambda]\alpha_{r(i)}$, with that resulting from use of x^*, $b\alpha_{r(k)}$, $b < 0$ and integer, and continue the process only when this change increases (the first scheme acts as though $[a_{i0}/\lambda] = b = -1$). A logical extension to this procedure is to assign weights $u_i \geqq 0$ to each row i of the tableau and form $x^* = \sum_i u_i x_{i+n}$ in such a way that $\sum_i u_i a_{i0} < 0$ and $\sum_i u_i a_{ij} \geqq 0$ for the largest possible set of smallest lexicographic columns α_n, α_{n-1}, $\cdots$. This can be accomplished through a computation analogous to obtaining a first feasible solution in linear programming, and represents an attempt at generating a row which leads to pivoting in the largest possible column. Of course, combinations and variations of these rules may be applied; in fact, it may be desirable to combine methods of algorithms I and II. For example, it may be efficient to first solve the integer program as a linear program, then use I to obtain an all integer dual feasible tableau, and then use II. Or, begin with a linear programming solution to obtain an upper bound on the value of the objective function to be introduced as a constraint in II.

Three codes of algorithm II are known. IPM 1 [94] is an all in core share code which uses a row choice rule which is derivative to the direct choice of row with smallest $r(i)$ interspersed with a choice rule which picks out that row with most negative a_0. This code seems to have had the least computational success of those mentioned here. IPM 2 [95] is an all in core share code which uses a variant of the second procedure described above for producing a row with small rank. Finally, IPLP 6 is an IBM experimental code which uses the last described linear programming type computation for producing a row with smallest rank. This code has led to convergence in remarkably fewer steps than IPM 2; however, the linear programming technique for generating a new row has not been perfected and has taken relatively long times.

In the derivation of (A.12), the constraint used in algorithm III, from (A.3) it can again be observed that only limited information was used; however, no ideas or theory concerning the development of strong constraints have been advanced. One slight improvement is immediate; namely, it may be worthwhile to replace r_j by $(r_j - 1)$ in (A.9) and then interpret $r_j - 1$ as belonging to the set $j \varepsilon J$, $a_j < 0$ in the arguments (A.10) and (A.11). The effect of this is that the coefficient of x_j in (A.12) becomes $(1 - r_j)r_0/1 - r_0$ instead of r_j; hence, this is worthwhile doing if $r_j > (1 - r_j)r_0/1 - r_0$. In considering the problem of how to choose a row (A.3) in the tableau from which to generate a constraint three ideas have been discussed [16]: choose that one which leads to the largest r_0, choose that one which leads to an r_0 closest to $\frac{1}{2}$, and choose the topmost row with noninteger a_0. The first of these has proven to be bad, perhaps because a large r_0 makes certain coefficients $-r_0a_j/(1 - r_0)$, where $j \varepsilon J$, $a_j < 0$ very large and hence the cut weak along the x_j-axis. This is the reason for choosing the second criterion. The third criterion is the lexicographic one and limited experimentation seems to indicate it is the best of the three [16].

5. *On Computational Experience*

Computational experience with integer programming algorithms must be a function of the code used and the type of problem solved. No distinct comparisons or conclusions are possible. The attempt here is to rally some principal and useful facts pertaining to experience with special codes and particular problems.

The first computational successes in the use of integer programming algorithms seems to have been with set covering problems. In [20][21][22] experiments in solving problems of type (B.4)—in which the "cost" of each column is 1—arising from randomly generated normal disjunctive form problems are discussed. The codes used are IPM 2 and IPLP 6 (versions of II) and IPM 3 (a version of I). Of the 999 problems generated all but 83 were solved by IPM 2 and/or IPM 3—most of the 83 exceeded the storage capacities of the codes. Generally speaking IPM 2 appeared better in number of iterations and time. For small and/or easy problems IPM 2 required perhaps slightly fewer pivots; but as problems became larger and/or are harder to solve the performance and reliability of IPM 3 came to the fore. The authors give $\frac{2}{3}n$ as a "reasonable" rule of thumb for number of pivots required for convergence in problems (B.4) having[4] $(m, n) \leqq (125, 125)$ using IPM 2. In terms of time on an IBM 7090 using the same code "reasonable" estimates are: .21 minutes for (60, 60); .38 minutes for (90, 90); .82 minutes for (150, 150). Among a group of problems requiring an unusually large number of pivots using IPM 2 were these: (104, 117), 391 pivots; (77, 61), 373 pivots in one run and 53 in another after a rearrangement of problem data; (53, 58), 852 pivots in one run and 979 in another after data rearrangement. Experiments with a very few problems showed IPLP 6 to be much superior to either of these other codes in numbers of pivots; however, the linear programming computation for generation of a new row required relatively long times (its pivots were not counted). All in all, these reports conclude that integer programming is an effective means of dealing with the minimization of Boolean function problems.

In an ongoing project, work is being done [4] in an attempt to establish some ranking of rules for selection or generation of a new constraint in algorithms I and II. This work is currently confined to generalized set covering problems (B.5). The approach has been to generate random (m, n) matrices A, (m, n) specified, with each column α_j containing a random number of 1's from given range placed randomly in the rows, and each cost c_j a random integer from some given range. Then, sets of identical problems are solved using various ad hoc rules of generating new constraints in algorithms I and II. Generally speaking the results showed little difference between the various rules or between I and II. In fact, in 25 problems (20, 50) with the density of 1's in A about 24% and $1 \leqq c_j \leqq 4$, a variant of II in which that direct row leading to a pivot in the largest lexicographic column is used, seemed slightly better than all other methods. On closer inspection the reason for this was easily found: most problems are "easy," i.e., in no case did a problem require more than four cuts with at least one variant of I. On the average, about 18 pivot steps was needed in the variant of II mentioned; while about 24 pivot steps was needed in a variant of I in which the row leading to a maximum $r_0/\sum r_j$ is chosen.

Why are such problems "easy"? And why has integer programming been successful in so many covering type problems? Subsequent work in which an attempt has been made to generate hard problems indicates a partial explanation: by following the same procedure as above but specifying the costs c_j to be precisely the number of 1's in its column α_j, all problems become relatively very difficult to solve for they become "purely" combinatorial problems. It seems,

[4] (m, n) refers to a problem having m constraints and n nonnegative variables.

thus, that past successes are at least partly due to the fact that in real or random problems columns α_j containing many 1's with small c_j or containing few 1's with large c_j arise, and "linearity forces" x_j to be 1 or 0, respectively. A problem (15, 50) was attempted four times, each representing a new arrangement of data, using the variant of I mentioned above with these results: (i) no solution in 200 pivots, (ii) optimal solution 34 pivots, 1 constraint adjoined, (ii) optimal solution 29 pivots, no constraint adjoined, and (iv) no solution in 400 pivots. The same problem required 39 pivots using the variant of II mentioned above. On the other hand in 10 problems (10, 50) with $2 \leqq c_j \leqq 8$, the variant of II solved all, while the variant of I failed in 150 pivots on four. For example, consider two of these problems, P_1 and P_2, each of which were solved three times under different arrangements of rows using the variants of I and II mentioned, with these results:

	P_1			P_2		
	(i)	(ii)	(iii)	(i)	(ii)	(iii)
I Pivots	52	150	26	15	14	19
I Cuts	5	no solution	2	0	0	1
II Pivots	24	24	15	52	36	19

In a word these experiments are still inconclusive except that they have led to some possible explanations for success in solving covering type problems. It may be worthwhile to mention that D, the product of the pivot entries leading to a linear programming solution, always seems to be small: in 25 problems (20, 50) the average $|D| = 8$ and $1 \leqq |D| \leqq 20$. $|D|$ is a crucial parameter of the amount of computational work necessary in rounding a noninteger solution to an integer solution in Gomory's new algorithm [52].

The CEIR Code [77] is the only known operating code which is not all in core and, hence, problems of large size have been solved with this code only. In nine problems of the generalized set covering type arising from practical delivery problems [5] solutions were successfully found where $(m, n) \leqq (15, 305)$, usually within about 40 pivot steps. For example,

Problem	(9, 102)	(11, 305)	(15, 270)
Pivots	142	36	23
Cuts	20	1	1

where the problem (9, 102) was the hardest of the nine. A tenth problem was not completed after more than 200 iterations. In other work Martin reports solving combinatorial problems arising from scheduling problems with $(70, 160) \leqq (m, n) \leqq (1600, 2200)$ on a fairly regular basis. These problems have constraint matrices whose entries are 85% 0's; and of the nonzeros, about 85%

are 1's. Most problems take in the order of 100 pivot steps and one hour of 7090 computation (in a set of 20 problems[5] the longest took 157 pivots). Lately,[6] in a revised version of the code, Martin reports solving a problem (80, 2400) of the same type in 24 minutes of 7094 computation—this being an example of a successful run. The largest single problem (215, 2600) was solved in about the same order of pivots and time as the bulk of the others; however, it required only one new constraint. In fact, most of the problems require few new cuts (of the set of 20 none required more than 10). Thus, it is easy to see why these problems are successfully solved: optimal solutions lie on a low dimensional face of the underlying linear constraint set.

LIP 1 has been tested on three small sets of problems [59], and the results compared with the performance of IPM 1, IPM 2 and IPM 3. One set of problems consists of six integer programs arising from 6-item, 3-machine scheduling problems [41]—problems which can be solved by enumerating all 6! possible sequences. Generally speaking, LIP 1 performed better than IPM 1, although the two most difficult problems required 798 and 3249 iterations, respectively (numbers larger than 6!), thus leading one to believe that the integer programming formulation of these scheduling problems is somehow not "right." Another set of problems consists of ten fixed charge problems [58] of very modest size, i.e., not larger than (10, 12). LIP 1 solves all ten, while IPM 2 and IPM 3 do not; LIP 1 converges faster than IPM 3 uniformly, and is slightly faster than IPM 2, but not on all problems. For example, consider three of the hardest problems for LIP 1:

Problem	(4, 5)	(6, 5)	(10, 12)
LIP 1—Pivots	164	126	110
LIP 1—Cuts	38	31	11
IPM 2—Pivots	731	20	4000, no solution
IPM 3—Pivots	3000, no solution	3000, no solution	3000, no solution

The last set of problems consists of nine "IBM Test Problems." Results are very mixed: each algorithm, LIP 1, IPM 2 and IPM 3, is better for some problems than the others (for the specific numerical problems see [58]).

It appears that almost no computational experience has been obtained with the mixed-integer algorithm III. The only known report describes very limited computation with a code written in 1961 for SILLIAC in Australia [16], [24]. The code is one which chooses a row leading to an r_0 closest to zero. In six problems P_i, arising from a scheduling problem in a power generation system, of form (A.1) with $m = 32$, $n = 39$ and 44 of the 71 variables and x_0 required to be integer the following results obtained:

[5] Privately communicated, February 1b.

[6] Privately communicated, March 1965.

	P_1	P_2	P_3	P_4	P_5	P_6
Pivots	43	49	70	101	228	53
Cuts	11	19	22	30	52	6
Pivots to L.P. Solution	?	?	44	54	52	46

In using the lexicographic row choice rule P_5 was solved in 75 pivots with 15 cuts being introduced; and in using a rule which introduces several cuts simultaneously P_5 was solved in 76 pivots with cuts introduced on 5 occasions, a total of 19 cuts being introduced.

Some recent work has been done in an attempt to determine the performance of algorithm V in solving a type of non-linear distribution problem which is slightly more general than the plant location problem (B.7) [18]. The problem is that of producing and distributing, at least cost, a homogeneous product from m plants to n customers given customer demands and maximum plant capacities, where transportation costs are linear but plant production costs are piecewise linear for some plants, i.e., economies of scale may occur. Given any such problem it is possible to obtain an optimal solution by enumeration by locating every local optimum: solve every possible "derived marginal problem" resulting from the choice of one marginal production cost from each plant taken as an average cost (see Section B.3). The approach taken was to generate problems at random and compare the number N_c of cycles of algorithm V, that is, the number of times an integer program is solved, with the number N_d of possible "derived marginal problems." Generally speaking the authors concluded that as N_d becomes large, i.e., as the problem becomes hard, the algorithm becomes efficient in the sense that $100\ N_c/N_d$ becomes small: $100\ N_c/N_d \leqq 5$ for $N_d > 500$ although usually for such problems $100\ N_c/N_d \approx 1$. 135 problems were solved having $7 \leqq m \leqq 19$, $7 \leqq n \leqq 30$, the number of plants m_s subject to economies of scale, $2 \leqq m_s \leqq 6$, the total number of marginal production rates r over all such plants, $5 \leqq r \leqq 24$ (e.g., if two such plants have 3 and 4 different marginal rates, respectively, $r = 3 + 4 = 7$ while $N_d = 3.4 = 12$). Other data was generated randomly from fixed ranges. Some examples of results for problems P_i are:

	P_1	P_2	P_3	P_4	P_5	P_6	P_7	P_8
m_s	3	4	4	5	5	6	6	6
r	7	19	20	17	22	16	22	23
m	10	13	13	16	16	19	19	19
n	14	18	19	24	26	22	28	30
N_c	12	14	21	3	43	38	52	8
N_d	12	450	405	432	1200	288	1944	2592
$100\ N_c/N_d$	100	3.1	5.2	.7	3.6	13	2.7	.3

Although N_c/N_d is a reasonable measure of the efficiency of the basic idea of algorithm V, the actual usefulness of this approach depends on the efficiency of the integer programming calculation that is a part of it. The number of pivots required for integer programming is not specified in the data given here. However, in related work [93] in which the problem is the plant location problem (B.7) a 5-plant problem had $N_e = 535$, with a total of 2133 pivots being made in integer programming calculations: this seems incredibly disheartening.

It is unfortunate that no more computational experience can be related. This paucity of information indicates potential for considerable improvement in integer programming codes and in discovering classes of problems which can be solved efficiently.

References

1. Alcaly, Roger E. and Klevorick, Alvin K., "A Note on the Dual Prices of Integer Programs," to appear in *Econometrica*.
2. Balinski, Michel L., "Fixed-Cost Transportation Problems," *Naval Research Logistics Quarterly*, Vol. 8 (1961), pp. 41–54.
3. ——, "On Finding Integer Solutions to Linear Programs," in *Proceedings of the IBM Scientific Symposium on Combinatorial Problems*, Herbert Koenig, Ed., to appear, 1965.
4. —— and Norman, Morris, "Experiments With Set Covering Problems," Mathematica Working Paper, January 1965.
5. —— and Quandt, R. E., "On an Integer Program for a Delivery Problem," *Operations Research*, Vol. 12 (1964), pp. 300–304.
6. —— and Wolfe, Philip, "On Benders Decomposition and a Plant Location Problem," Mathematica Working Paper, ARO-27, December 27, 1963.
7. Baumol, William J. and Wolfe, Philip, "A Warehouse Location Problem," *Operations Research*, Vol. 6 (1958), pp. 252–263.
8. Bauer, F. L., "Algorithm 153, Gomory," *Communications of the Association for Computing Machinery*, Vol. 6, No 2 (1963).
9. Beale, E. M. L., "A Method of Solving Linear Programming Problems when some but not all of the Variables must take Integral Values," Statistical Techniques Research Group, Technical Report No. 19, Princeton University, July 1958.
10. Bellman, Richard, "Maximization over Discrete Sets," *Naval Research Logistics Quarterly*, Vol. 3 (1956), pp. 67–70.
11. —— and Dreyfus, Stuart E., *Applied Dynamic Programming*, Princeton University Press, 1962.
12. —— and Hall, Marshall, Jr., Eds., *Combinatorial Analysis*, Proceedings of the Tenth Symposium in Applied Mathematics of the American Mathematical Society, 1960.
13. Benders, J. F., "Partitioning Procedures for Solving Mixed-Variables Programming Problems," *Numerische Mathematik*, Vol. 4 (1962), pp. 238–252.
14. ——, Catchpole, A. R. and Kuiken, C., "Discrete Variables Optimization Problems," *Rand Symposium on Mathematical Programming*, March 16–20, 1959.
15. Ben-Israel, A. and Charnes, A., "On Some Problems of Diophantine Programming," *Cahièrs du Centre d'Etudes de Rècherche Operationelle* (Bruxelles), Vol. 4 (1962), pp. 215–280.
16. Bennet, J. M. and Dakin, R. J., "Experience with Mixed Linear Programming Problems," University of Sydney, mimeographed report, October 26, 1961.
17. Blankinship, W. A., "A New Version of the Euclidean Algorithm," *American Mathematical Monthly*, Vol. 70 (1963), pp. 742–745.
18. Buzby, B. R., Stone, B. J. and Taylor, R. L. "Computational Experience with a Non-Linear Distribution Program," Union Carbide Corporation, privately communicated, January 5, 1965.
19. Camion, P., "Une Mèthode de Résolution par l'Algebre de Boole des Problèmes Com-

binatoires ou Interviennent des Entiers," *Cahièrs du Centre D'Etudes de Rècherche Operationelle*, Vol. 2 (1960), pp. 234–289.
20. COBHAM, A., FRIDSHAL, R. AND NORTH, J. H., "An Application of Linear Programming to the Minimization of Boolean Functions," Research Report RC-472, IBM Research Center, June 9, 1961.
21. ——, —— AND ——, "A Statistical Study of the Minimization of Boolean Functions Using Integer Programming," Research Report RC-756, IBM Research Center, June 21, 1962.
22. —— AND NORTH, J. H., "Extensions of the Integer Programming Approach to the Minimization of Boolean Functions," Research Report RC-915, IBM Research Center, April 2, 1963.
23. CORD, JOEL, "A Method of Balancing Assembly Lines Using Linear Programming," A. O. Smith Corporation, privately circulated paper, October 1963.
24. DAKIN, R. J., "Application of Mathematical Programming Techniques to Cost Optimization in Power Generating Systems," Technical Report No. 19, Basser Computing Department, School of Physics, The University of Sydney, December 1961.
25. DANTZIG, G. B., "Application of the Simplex Method to a Transportation Problem, pp. 359–373 in *Activity Analysis of Production and Allocation*, T. C. Koopmans, Ed., Cowles Commission Monograph No. 13, John Wiley and Sons, Inc., 1951.
26. ——, "Discrete-Variable Extremum Problems," *Operations Research*, Vol. 5 (1957), pp. 266–277.
27. ——, *Linear Programming and Extensions*, Princeton University Press, Princeton, N. J., 1963.
28. ——, "Note on Solving Linear Programs in Integers," *Naval Research Logistics Quarterly*, Vol. 6 (1959), pp. 75–76.
29. ——, "On the Significance of Solving Linear Programming Problems with some Integer Variables," *Econometrica*, Vol. 28 (1960), pp. 30–44.
30. ——, FULKERSON, D. R. AND JOHNSON, S. M., "On a Linear Programming, Combinatorial Approach to the Travelling Salesman Problem," *Operations Research*, Vol. 7 (1959), pp. 58–66.
31. ——, —— AND ——, "Solution of a Large Scale Travelling Salesman Problem," *Journal of the Operations Research Society of America*, Vol. 2 (1954), pp. 393–410.
32. —— AND WOLFE, PHILIP, "The Decomposition Algorithm for Linear Programs," *Econometrica*, Vol. 29 (1961), pp. 767–778.
33. D'ESOPO, D. A. AND LEFKOWITZ, B., "Certification of Algorithm 153," *Communications of the Association for Computing Machinery*, Vol. 6, No. 8 (1963).
34. —— AND ——, "Note on an Integer Linear Programming Model for Determining a Minimum Embarkation Fleet," *Naval Research Logistics Quarterly*, Vol. 11 (1964), pp. 79–82.
35. EDMONDS, JACK, "Covers and Packings in a Family of Sets," *Bulletin of the American Mathematical Society*, Vol. 68 (1962), pp. 494–499.
36. ——, "Maximum Matching and a Polyhedron with 0, 1-vertices," Princeton University and National Bureau of Standards, March 1963, mimeographed report.
37. ——, "Paths, Trees, and Flowers," National Bureau of Standards and Princeton University, February 1963, mimeographed report.
38. FORD, L. R., JR. AND FULKERSON, D. R., "A Suggested Computation for Maximal Multi-Commodity Network Flows," *Management Science*, Vol. 5 (1958), pp. 97–101.
39. —— AND ——, *Flows in Networks*, Princeton University Press, 1963.
40. FULKERSON, D. R. AND RYSER, H. J., "Widths and Heights of (0, 1)-Matrices," *Canadian Journal of Mathematics*, Vol. 13 (1961), pp. 239–255.
41. GIGLIO, RICHARD J., AND WAGNER, HARVEY M., "Approximate Solutions to the Three Machine Scheduling Problem," *Operations Research*, Vol. 12 (1964), pp. 305–324.
42. GILBERT, E. J. AND SCHATZ, J. A., "An Ill-Conceived Proposal for Apportionment of

the U. S. House of Representatives," *Operations Research*, Vol. 12 (1964), pp. 768–773.

43. Gilmore, P. C. and Gomory, R. E., "A Linear Programming Approach to the Cutting Stock Problem—Part II," *Operations Research*, Vol. 11 (1963), pp. 863–888.

44. —— and ——, "A Solvable Case of the Travelling Salesman Problem," *Proceedings of the National Academy of Sciences*, Vol. 51 (1964), pp. 178–181.

45. —— and ——, "Multi-Stage Cutting Stock Problems of Two and More Dimensions," *Operations Research*, Vol. 13 (1965), pp. 94–120.

46. —— and ——, "Sequencing a One State Variable Machine: A Solvable Case of the Travelling Salesman Problem," IBM Watson Research Center, Report RC-1103, January 24, 1964.

47. Glover, Fred, "A Bound Escalation Method for the Solution of Integer Linear Programs," Graduate School of Industrial Administration, Carnegie Institute of Technology, Pittsburgh, Pennsylvania, December 1963.

48. Gomory, R. E., "All-Integer Integer Programming Algorithm," pp. 193–206 in [80]. First issued as IBM Research Center, Research Report RC-189, January 29, 1960.

49. ——, "An Algorithm for Integer Solutions to Linear Programs," pp. 269–302 in [57]. First issued as Princeton-IBM Mathematics Research Project, Technical Report Number 1, November 17, 1958.

50. ——, "An Algorithm for the Mixed Integer Problem," RM-2597 Rand Corporation, July 7, 1960.

51. ——, "Large and Non-Convex Problems in Linear Programming," in *Experimental Arithmetic, High Speed Computing and Mathematics*, Proceedings of Symposium in Applied Mathematics, Volume XV, American Mathematical Society, 1963.

52. ——, "On the Relation Between Integer and Non-Integer Solutions to Linear Programs," *Proceedings of the National Academy of Sciences*, Vol. 53 (1965), pp. 260–265.

53. ——, "Outline of an Algorithm for Integer Solutions to Linear Programs," *Bulletin of the American Mathematical Society*, Vol. 64 (1958), pp. 275–278.

54. ——, "Solving Linear Programming Problems in Integers," pp. 211–216 in [12].

55. —— and Baumol, William J., "Integer Programming and Pricing," *Econometrica*, Vol. 28 (1960), pp. 521–550.

56. —— and Hoffman, A. J., "On the Convergence of an Integer-Programming Process," *Naval Research Logistics Quarterly*, Vol. 10 (1963), pp. 121–123.

57. Graves, Robert L. and Wolfe, Philip, Editors, *Recent Advances in Mathematical Programming*, McGraw-Hill Book Company, Inc., 1963.

58. Haldi, John, "25 Integer Programming Test Problems," Working Paper No. 43, Graduate School of Business, Stanford University, December 1964.

59. —— and Isaacson, Leonard M., "Linear Integer Programming," Working Paper No. 45, Graduate School of Business, Stanford University, December 1964.

60. Healy, W. C., Jr., "Multiple Choice Programming," *Operations Research*, Vol. 12 (1964), pp. 122–138.

61. Hoffman, A. J., "Some Recent Applications of the Theory of Linear Inequalities to Extremal Combinatorial Analysis," pp. 113–128 in [12].

62. Hohn, Franz, "Some Mathematical Aspects of Switching," *American Mathematical Monthly*, Vol. 62 (1955), pp. 75–90.

63. Huntington, Edward V., "The Apportionment of Representatives in Congress," *Transactions of the American Mathematical Society*, Vol. 30 (1928), pp. 85–110.

64. Huntington, Edward V., *A Survey of Methods of Apportionment in Congress*, 76th Congress, 3rd Session Document No. 304, U. S. Government Printing Office, Washington, D. C., 1940.

65. Karp, Richard M., "Minimum-Redundancy Coding for the Discrete Noiseless Channel," *IRE Transactions of the Professional Group on Information Theory*, Vol. IT-7, No. 1 (1961), pp. 27–38.

66. KUHN, HAROLD W., "On Certain Convex Polyhedra," *Bulletin of the American Mathematical Society*, Vol. 61 (1955), pp. 557–558.
67. —— AND TUCKER, A. W., Eds., *Linear Inequalities and Related Systems*, Annals of Mathematics Study No. 38, Princeton University Press, 1956.
68. KUNZI, HANS P. AND OETTLI, WERNER, "Integer Quadratic Programming," pp. 303–308 in [57].
69. LAND, A. H. AND DOIG, A. G., "An Automatic Method of Solving Discrete Programming Problems," *Econometrica*, Vol. 28 (1960), pp. 497–520.
70. LAWLER, E. L., "An Algorithm for Solving Covering Problems," University of Michigan, mimeographed report, December 11, 1964.
71. ——, "Duality of Covering Problems," University of Michigan, mimeographed report, December 7, 1964.
72. LEVITAN, R. E., *IPM 3*, Share Distribution Number 1190, September 1961.
73. LITTLE, JOHN D. C., MURTY, KATTA G., SWEENEY, DURA W. AND CAROLINE, KAREL, "An Algorithm for the Travelling Salesman Problem," *Operations Research*, Vol. 11 (1963), pp. 972–989.
74. MANNE, ALAN S., "On the Job-Shop Scheduling Problem," *Operations Research*, Vol. 8 (1960), pp. 219–223. Also, pp. 187–192 in [80].
75. ——, "Plant Location under Economics of Scale-Decentralization and Computation," *Management Science*, Vol. 11 (1964), pp. 213–235.
76. MARKOWITZ, HARRY M. AND MANNE, ALAN S., "On the Solution of Discrete Programming Problems," *Econometrica*, Vol. 25 (1957), pp. 84–110.
77. MARTIN, GLENN T., "An Accelerated Euclidean Algorithm for Integer Linear Programming," pp. 311–318 in [57].
78. MCCLUSKEY, E. J., "Error Correcting Codes—A Linear Programming Approach," *The Bell System Technical Journal*, Vol. XXXVIII (1959), pp. 1485–1512.
79. MILLER, C. E., TUCKER, A. W. AND ZEMLIN, R. A., "Integer Programming Formulation of Travelling Salesman Problems," *Journal of the Association for Computing Machinery*, Vol. 7 (1960), pp. 326–329.
80. MUTH, JOHN F. AND THOMPSON, GERALD L., Eds., *Industrial Scheduling*, Prentice-Hall, New York, 1963.
81. NORMAN, R. Z. AND RABIN, M. O., "An Algorithm for the Minimum Cover of a Graph," *Proceedings of the American Mathematical Society*, Vol. 10 (1959), pp. 315–319.
82. OUYAHIA, AIT, "Programmes Lineaires a Variables Discretes," *Revue Francaise de Récherche Operationelle*, Vol. 6 (1962), pp. 55–75.
83. PETERSEN, J., "Die Theorie der regulären Graphen," *Acta Mathematica*, Vol. 51 (1891), pp. 193–220.
84. PYNE, I. B. AND MCCLUSKEY, E. J., JR., "An Essay on Prime Implicant Tables," *Journal of the Society for Industrial and Applied Mathematics*, Vol. 9 (1961), pp. 604–631.
85. QUINE, W. V., "A Way to Simplify Truth Functions," *American Mathematical Monthly*, Vol. 62 (1955), pp. 627–631.
86. ROTH, J. P., "Algebraic Topological Methods for the Synthesis of Switching Systems, I," *Transactions of the American Mathematical Society*, Vol. 88 (1958), pp. 301–326.
87. SHANNON, C. E., "A Symbolic Analysis of Relay and Switching Circuits," *Transactions of the American Institute of Electrical Engineers*, Vol. 57 (1938), pp. 713–723.
88. STORY, A. E. AND WAGNER, H. M., "Computational Experience With Integer Programming for Job-Shop Scheduling," pp. 207–220 in [80].
89. TAYLOR, R. L., "Carbide Warehouse Selection Program," Union Carbide Corporation, privately communicated, April 1964.
90. THOMPSON, GERALD L., "The Stopped Simplex Method: Basic Theory for Mixed Integer Programming," *Revue Francaise de Récherche Operationelle*, Vol. 8 (1964), pp. 159–182.
91. TUCKER, A. W., "A Combinatorial Equivalence of Matrices," pp. 129–140 in [12].

92. van der Waerden, B. L., *Modern Algebra*, Volume II, Frederick Ungar Publishing Co., New York, 1950.
93. Vietorisz, Thomas, "Industrial Development Planning Models with Economics of Scale and Indivisibilities," IBM Watson Research Center, RC-1061, September 30, 1963.
94. Wade, C. S. and Gomory, R. E., *IPM 1*, Share Distribution Number 1192, September 1961.
95. —— and ——, *IPM 2*, Share Distribution Number 1191, September 1961.
96. Wagner, H. M., "An Integer Linear Programming Model for Machine Shop Scheduling," *Naval Research Logistics Quarterly*, Vol. 6 (1959), pp. 131–140.
97. ——, Giglio, Richard J. and Glaser, R. George, "Preventive Maintenance Scheduling by Mathematical Programming," *Management Science*, Vol. 10 (1964), pp. 316–334.
98. Weaver, James B. and Hess, Sydney W., "A Procedure for Non-Partisan Districting: Development of Computer Techniques," *Yale Law Journal*, Vol. 73 (1963), pp. 288–308.
99. White, William W., "On Gomory's Mixed Integer Algorithm," Senior Thesis, Princeton University, May 1, 1961.
100. Witzgall, Christoph, "An All-Integer Programming Algorithm with Parabolic Constraints," *Journal of the Society for Industrial and Applied Mathematics*, Vol. 11 (1963), pp. 855–871.
101. Young, Richard D., "A Primal (All Integer) Integer Programming Algorithm: Antecedents, Description, Proof of Finiteness, Exemplification," Working Paper No. 52, Graduate School of Business, Stanford University, December 1964.
102. Balas, Egon, "Un algorithme additif pour la resolution des programmes lineares en variables bivalentes", *Comptes Rendus, Academie des Sciences, Paris*, Vol. 258 (1964), pp. 3817–3820.
103. ——. "Extension de l'algorithme additif a la programmation en nombres entiers et a la programmation nonlineare," *Comptes Rendus, Academie des Sciences, Paris*, Vol. 258 (1964), pp. 5136–5139.
104. Beale, E. M. L. and Small, R. E., "A Branch and Bound Method for Mixed Integer Programming", *Proceedings of the Third IFIP Conference, Vol. 2, 1965*, to appear.
105. Harris, P. M. J., "An Algorithm for Solving Mixed Integer Linear Programs", *Operational Research Quarterly*, Vol. 15 (1964) pp. 117–132.

Chapter 7
Matroid Partition

Jack Edmonds

Introduction by *Jack Edmonds*

This article, "Matroid Partition", which first appeared in the book edited by George Dantzig and Pete Veinott, is important to me for many reasons: First for personal memories of my mentors, Alan J. Goldman, George Dantzig, and Al Tucker. Second, for memories of close friends, as well as mentors, Al Lehman, Ray Fulkerson, and Alan Hoffman. Third, for memories of Pete Veinott, who, many years after he invited and published the present paper, became a closest friend. And, finally, for memories of how my mixed-blessing obsession with good characterizations and good algorithms developed.

Alan Goldman was my boss at the National Bureau of Standards in Washington, D.C., now the National Institutes of Science and Technology, in the suburbs. He meticulously vetted all of my math including this paper, and I would not have been a math researcher at all if he had not encouraged it when I was a university drop-out trying to support a baby and stay-at-home teenage wife. His mentor at Princeton, Al Tucker, through him of course, invited me with my child and wife to be one of the three junior participants in a 1963 Summer of Combinatorics at the Rand Corporation in California, across the road from Muscle Beach. The Bureau chiefs would not approve this so I quit my job at the Bureau so that I could attend. At the end of the summer Alan hired me back with a big raise.

Dantzig was and still is the only historically towering person I have known. He cared about me from a few days before my preaching at Rand about blossoms and about good algorithms and good characterizations. There were some eminent combinatorial hecklers at my presentation but support from Dantzig, and Alan Hoffman, made me brave.

Jack Edmonds
Department of Combinatorics and Optimization, University of Waterloo, Canada
e-mail: jackedmonds@rogers.com

M. Jünger et al. (eds.), *50 Years of Integer Programming 1958-2008*,
DOI 10.1007/978-3-540-68279-0_7, © Springer-Verlag Berlin Heidelberg 2010

I think of Bertrand Russell, Alan Turing, and George Dantzig as the three most important philosophers of the last century. During an infrequent visit to California from Washington, D.C., sometime in the 60s, Dantzig took me, a wife, and three kids, to Marineland and also to see a new shopping mall in order to prove to us that having a ceiling of a certain height in his carefully planned Compact City is as good as a sky.

One time when I unexpectedly dropped in on Dantzig, the thrill of my life was him asking me to lecture to his linear programming class about how the number of pivots of a simplex method can grow exponentially for non-degenerate linear programming formulations of shortest path problems, and also asking me to vet contributions for a math programming symposium which he was organizing.

One of my great joys with George Dantzig was when a friend working at Hewlett-Packard asked me to come discuss the future of operations research with his artificial intelligence colleagues. I was discouraged when no one I knew in O.R. seemed interested in helping—that is, until I asked George. He told my second wife Kathie and me that he was a neighbor and had socialized with Mr. Hewlett, or was it Mr. Packard, for years, and had never been invited to HP, two blocks away. George took over the show and was wonderful. Kathie video-taped it. The next morning he asked if she had made him a copy yet.

Al Tucker made me a Research Associate and put me in charge of his Combinatorics Seminar at Princeton during 1963–64. Combinatorists whom I wanted to meet accepted paying their own way to speak at my 'Princeton Combinatorics and Games Seminar'. However, except for Ron Graham who came over from Bell, and Moses Richardson who came down from City University, they were unable to schedule their visits. So I hastily organized a Princeton Conference in the spring of 1964 where the eminent seminar invitees could lecture to each other.

At that conference I met Al Lehman who led me, by his matroidal treatment of what he called the Shannon switching game, to see that matroids are important for oracle-based good algorithms and characterizations. I persuaded Al, along with Chris Witzgall, to come work at the Bureau of Standards, and immediately we started looking for people to participate in a two-week Matroid Workshop at the Bureau of Standards in autumn 1964. We didn't find more than six who had even heard of the term 'matroid'. About twenty serious people came to it, including Ray Fulkerson, George Minty, Henry Crapo, Dan Younger, Neil Robertson, and Bill Tutte. Within a year it seemed the whole world was discovering matroids.

The Bureau was delighted at the prospect of hiring Al Lehman. However, an aftermath of McCartheism left the Bureau with the rule that new employees had to take an oath of loyalty. The early computer-guru, Ida Rhodes, actually tugged at Al's arm to try to get him to take the oath but he wouldn't. Instead he took a research job with a Johns Hopkins satellite of the U.S. Army which did not require such an oath. He literally picketed the Matroid Workshop, speaking to whomever would listen about the 'Bureau of Double Standards'. We stayed friends for the many years until his recent death in Toronto.

At the same workshop, Gian-Carlo Rota conceived of and started organizing the Journal of Combinatorial Theory. He also insisted that the 'ineffably cacophonic word matroid' be replaced by 'combinatorial geometry'.

George Minty was an especially sweet and brilliant participant. He wrote a paper which Bob Bland credits with being a precursor of oriented matroids. He spent years afterwards on successfully extending the good algorithm for optimum matchings in a graph to optimum independent sets in a clawfree graph. His work is still the most interesting aspect of matching theory.

During the year after the Matroid Workshop, Ray Fulkerson and I regularly spent hours talking math by government telephone between Santa Monica and Washington. Ray and I never did learn how to work computers, and though I think the prototype of email did exist back then in our government circles, he and I didn't know about it. One of the outcomes of our talk was combining a version of the matroid partitioning algorithm described in the paper here with Ray's interest in doing everything possible by using network flow methods.

My huff about him and Ellis Johnson calling the blossom method "a primal-dual method" led me to look for algorithms for network flow problems which were polytime relative to the number of bits in the capacities as well as in the costs. The reason I had presented the blossom method only for 1-matchings is that for b-matchings I could not call it a "good algorithm" until I had figured out how to do that for network flows. Once it's done for flows, it's easy to reduce optimum b-matchings to a flow problem and a b-matching problem where the b is ones and twos. Dick Karp was independently developing good algorithms for network flows and so much later I published with Dick instead of, as intended, with Ray and Ellis. I enjoyed working with Ray and I coined the terms "clutter" and "blocker". I can't remember who suggested the term "greedy" but it must have been Alan Goldman and probably Ray as well.

It was important to me to ask Ray to check with the subadditive set function expert he knew about submodular set functions. When the answer came back that they are probably the same as convex functions of additive set functions, I knew I had a new tiger by the tail.

Ray and I liked to show off to each other. I bragged to him about discovering the disjoint branchings theorem, mentioned later. Trouble is, I then became desperate to find quickly a correction of my faulty proof. I think I would have done a better job on the theorem if I had not been frantic to cover my hubris.

During a phone call, Ray mentioned that one day later, four months after the Matroid Workshop, there would be a combinatorics workshop in Waterloo. My boss Alan Goldman rescued me as usual and I quickly hopped a plane to Canada to sleep along with George Minty on sofas in Tutte's living room.

Neil Robertson, a meticulous note-taker, had reported to Crispin Nash-Williams on my Matroid Workshop lectures. Crispin, by his own description, was too enthusiastic about them. He was giving a keynote lecture about matroid partitioning on the first morning of this Waterloo workshop. I felt compelled immediately after his talk to speak for an impromptu hour on the following:

Theorem 1. A non-negative, monotone, submodular set function, $f(S)$, of the subsets S of a finite set E, is called a polymatroid function on E. For any integer-valued polymatroid function on E, let F be the family of subsets J of E such that for every non-empty subset S of J, the cardinality of S is at most $f(S)$. Then $M = (E,F)$ is a matroid. Its rank function is, for every subset A of E, $r(A)$, meaning max[cardinality of a subset of A which is in F] = min[$f(S)$ + cardinality of $(A \setminus S)$ for any subset S of A].

After this opening of the Waterloo meeting I urgently needed a mimeographed abstract handout and so I submitted Theorem 1.

The theorem is dramatic because people had only seen matroids as an axiomatic abstraction of algebraic independence, and not as something so concrete as a kind of linear programming construction quite different from algebraic independence.

I tried to explain on that snowy April Waterloo morning how the theorem is a corollary of a theory of a class of polyhedra, called polymatroids, given by non-negative vectors x satisfying inequality systems of the form:

For every subset S of E, the sum of the coordinates of x indexed by the j in S is at most $f(S)$.

However, even now, this is often outside the interest of graph theorists, or formal axiomatists. I am sorry when expositions of matroid theory still treat the subject only as axiomatic abstract algebra, citing the mimeographed abstract of that Waterloo meeting with no hint about the linear programming foundations of pure matroid theory.

What does Theorem 1 have to do with matroid partitioning? Well—the rank function of a matroid is a polymatroid function, and hence so is the sum of the rank functions of any family of matroids all on the same set E. Hence a special case of Theorem 1, applied to this sum, yields a matroid on E as the 'sum' of matroids on E. I had hoped to understand the prime matroids relative to this sum, but, so far, not much has come of that.

Suppose we have an oracle which for an integer polymatroid function $f(S)$ on E gives the value of $f(S)$ for any subset S of E. Then the theorem gives an easy way to recognize when a given subset J of E is not a member of F, in other words not independent in the matroid determined by Theorem 1. Simply observe some single subset S of J having cardinality greater than $f(S)$.

Does there exist an easy way to recognize when a set J is independent? The answer is yes. For a general integer polymatroid function f, this easy way needs some of the linear programming theory which led me to Theorem 1, which I will describe in a moment.

However for the special case of Theorem 1 where f is the sum of a given family, say H, of matroid rank functions, an easy way to recognize that a set J is independent, which even the most lp resistant combinatorist can appreciate, is given by the 'matroid partition theorem' of the present paper: a set J is independent if and only if it can be partitioned into a family of sets, which correspond to the members of H, and which are independent respectively in the matroids of H.

Thus, relative to oracles for the matroids of H, for the matroid M determined as in Theorem 1 by the f which is the sum of the rank functions of H, we have a 'good characterization' for whether or not a subset J of E is independent in M. To me this meant that there was an excellent chance of proving the matroid partition theorem by a good algorithm which, for a given J, decides whether or not J is independent in matroid M. That is what the present paper does.

Having an instance of a good characterization relative to an oracle, and having a good algorithm relative to the oracle which proves the good characterization, was the main point and motivation for the subject.

One reason I like the choice of "Matroid Partition" for the present volume is that, as far as I know, it is the first time that the idea of what is now called NP explicitly appears in mathematics. The idea of NP is what forced me to try to do some mathematics, and it has been my obsession since 1962.

I talked about it with Knuth at about that time and ten years later he asked me to vote on whether to call it NP. I regret that I did not respond. I did not see what non-deterministic had to do with it. NP is a very positive thing and it has saddened me for these many years that the justified success of the theory of NP-completeness has so often been interpreted as giving a bad rap to NP.

Let me turn my attention to linear programming which gave me Theorem 1, which led to the present paper.

Given the enormous success that the marriage problem and network flows had had with linear programming, I wanted to understand the goodness of optimum spanning trees in the context of linear programming. I wanted to find some combinatorial example of linear programming duality which was not an optimum network flow problem. Until optimum matchings, every min max theorem in combinatorics which was understood to be linear programming was in fact derivable from network flows—thanks in great measure to Alan Hoffman and Ray Fulkerson. Since that was (slightly) before my time, I took it for granted as ancient.

It seemed to be more or less presumed that the goodness of network flow came from the fact that an optimum flow problem could be written explicitly as a linear program. The Farkas lemma and the duality theorem of linear programming are good characterizations for explicitly written linear programs. It occurred to me, preceding any success with the idea, that if you know a polytope as the hull of a set of points with a good, i.e., easily recognizable, description, and you also know that polytope as the solution-set of a set of inequalities with a good description, then using lp duality you have a good characterization. And I hoped, and still hope, that if you have good characterization then there exists a good algorithm which proves it. This philosophy worked for optimum matchings. It eventually worked for explicitly written linear programs. I hoped in looking at spanning trees, and I still hope, that it works in many other contexts.

The main thing I learned about matroids from my forefathers, other than Lehman, is that the edge-sets of forests in a graph are the independent sets of a matroid, called the matroid of the graph. What is it about a matroid which could be relevant to a set of linear inequalities determining the polytope which is the hull of the 0-1 vectors of independent sets of the matroid? The rank function of course. Well what is it about the rank function of a matroid which makes that polytope extraordinarily nice for optimizing over? That it is a polymatroid function of course. So we're on our way to being pure matroid theorists.

A "polymatroid" is the polytope $P(f)$ of non-negative solutions to the system of inequalities where the vectors of coefficients of the vector of variables is the 0-1 vectors of subsets S of E and the r.h.s. constants are the values of the polymatroidal function $f(S)$. It turns out that it is as easy, relative to an oracle for f, to optimimize any linear function over $P(f)$, as it is to find a maximum weight forest in an edge-weighted graph. Hence it is easy to describe a set of points for which $P(f)$ is the convex hull. Where f is the rank function of a matroid, those points are the 0-1 vectors of the independent sets of the matroid, in particular of the edge-sets of the forests for the matroid of a graph.

A polymatroid has other nice properties. For example, one especially relevant here is that any polymatroid intersected with any box, $0 \leq x \leq a$, is a polymatroid. In particular, any integer-valued polymatroid function gives a polymatroid which intersected with a unit cube, $0 \leq x \leq 1$, is the polytope of a matroid. That is Theorem 1.

So what? Is this linear programming needed to understand Theorem 1? Not to prove it, though it helps. For Theorem 1, rather than for any box, the lp proof can be specialized, though not simplified, to being more elementary. However linear programming helps answer "yes" to the crucial question asked earlier: Does there exist an easy way to recognize when a set J is independent?

It is obvious that the 0-1 vector of the set J is in the unit box. Using the oracle for function f we can easily recognize if J is not independent by seeing just one of the inequalities defining $P(f)$ violated by the 0-1 vector of J. But if the vector of J satisfies all of those inequalities, and hence J is independent in the matroid M described by Theorem 1, how can we recognize that? Well using linear programming theory you can immediately answer. We have mentioned that we have a very easy algorithm for optimizing over polytope $P(f)$ and so, where n is the size of the ground set E which indexes the coordinates of the points of $P(f)$, we have an easy way to recognize any size $n+1$ subset of points each of which optimizes some linear objective over $P(f)$. Linear programming theory tells that the 0-1 vector of J is in $P(f)$, and hence J is independent, if and only if it is a convex combination of some $n+1$ points each of which optimizes some linear function over $P(f)$.

That's it. We have a good characterization of whether or not a set J is independent in the matroid described by Theorem 1. It takes a lot more work to say that directly without linear programming. We do that in the paper here with the matroid partition theorem for the case where f is the sum of some given matroid rank functions.

For concreteness assume that a is any vector of non-negative integers corresponding to the elements of finite set E. Of course Theorem 1 is the special case for a unit box of the theorem which says that box, $0 \leq x \leq a$, intersected with integer polymatroid, $P(f)$, is an integer polymatroid. Call it $P(f,a)$.

The rank $r(f,a)$ of $P(f,a)$, meaning the maximum sum of coordinates of an integer valued vector x in $P(f,a)$ is equal to the minimum of $f(S)$ + the sum of the coordinates of a which correspond to $E \setminus S$. If you know the meaning of a submodular set function, the proof of this is very easy. At the same time, you prove that the max sum of x is achieved by taking any integer valued x in $P(f,a)$, such as the zero vector, and pushing up the value of its coordinates in any way you can while staying in $P(f,a)$. (By analogy with matroid, having this property is in fact the way we define polymatroid.) The only difficulty with this otherwise easy algorithm is deciding how to be sure that the x stays in $P(f,a)$. Hence the crux of the problem algorithmically is getting an algorithm for deciding whether or not a given x is in $P(f,a)$. We do get a good characterization for recognizing whether or not an x is in $P(f,a)$ in the same way we suggested for characterizing whether or not a subset J of E is member of matroid M. Hence from this we have a good characterization of the rank $r(f,a)$ without necessarily having a good algorithm for determining the rank $r(f,a)$.

Any integer-valued submodular set function $g(S)$, not necessarily monotone or non-negative, can be easily represented in the form constant + $f(S)$ + the sum of the coordinates of a which correspond to $E \setminus S$, where f is an integer polymatroid function and a is a vector of non-negative integers. Hence, since the mid sixties, we have had a good characterization of the minimum of a general integer-valued submodular function, relative to an oracle for evaluating it. Lovász expressed to me a strong interest in finding a good algorithm for it in the early seventies. He, Grötschel, and Schrijver, showed in the late seventies that the ellipsoid method for linear programming officially provides such an algorithm. However it has taken many years, many papers, and the efforts of many people, to get satisfying direct algorithms, and this currently still has wide research interest. We have observed here how the matroid partitioning algorithm was a first step. The methods by which Dick Karp and I got algorithms for network flows was another first step.

There are other interesting things to say about matroid and submodular set-function optimization theory which I won't mention, but there is one I would like to mention. Gilberto Calvillo and I have developed good direct algorithms for the optimum branching system problem, which might have some down to earth interest. Given a directed graph G, a value $c(j)$ and a capacity $d(j)$ for each edge, find a family of k branchings which together do not exceed the capacity of any edge and which together maximize total value. A branching in G is a forest such that each node of G has at most one edge of the forest directed toward it. Of course there are a number of equivalent problems but this one is convenient to say and to treat. By looking at the study of branchings and the study of optimum network flow in chapters of combinatorial optimization you might agree that the optimum branching systems problem is a natural gap. The analogous problem for forest systems in an undirected graph is solved by the matroid partitioning algorithm here together with the matroid greedy algorithm. The optimum branching system problem is quite different. It is solved in principle by a stew of matroid ideas including the ones here, and was first done that way, but it is better treated directly.

The following article originally appeared as:

J. Edmonds, *Matroid Partition*, Mathematics of the Decision Sciences: Part 1 (G.B. Dantzig and A.F. Veinott, eds.), American Mathematical Society, 1968, pp. 335–345.

Jack Edmonds

Matroid Partition

1. Introduction. Matroids can be regarded as a certain abstraction of matrices. They represent properties of matrices which are invariant under elementary row operations, namely properties of dependence among the columns. For any matrix over any field, there is a matroid whose elements correspond to the columns of the matrix and whose independent sets of elements correspond to the linearly independent sets of columns. A matroid M is completely determined by its elements and its independent sets of elements.

There are matroids which do not arise from any matrix over any field, so matroid theory does truly generalize an aspect of matrices. However, matroid theory is justified by new problems in matrix theory itself, in fact by problems in the special matrix theory of graphs (networks). It happens that an axiomatic matroid setting is natural for viewing these problems and that matrix machinery is superfluous for viewing them.

Much of matroid theory has been motivated by graphs. A graph G may be regarded as a matrix $N(G)$ of zeroes and ones, mod 2, which has exactly two ones in each column. The columns are the edges of the graph and the rows are the nodes of the graph. An edge and a node are said to meet if there is a one located in that column and that row. Of course a graph can also be regarded visually as a geometric network. It is often helpful to visualize statements on matroids for the case of graphs, though it can be misleading. Matroids do not contain objects corresponding to nodes or rows.

Another motivation here will be another source of matroids which is an extensive theory in its own right. It is well known in various contexts, including systems of distinct representatives, (0, 1)-matrices, network flows, matchings in graphs, and marriages. We will refer to it here as transversal theory.

2. **Problem.** The following definition of matroid has certain intrinsic interest.

A *matroid*, $M = (E, F)$, is a finite set E of elements and a nonempty family F of subsets of E, called *independent sets*, such that (1) every subset of an independent set is independent; and (2) for every set $A \subset E$, all maximal independent subsets of A have the same cardinality, called the *rank* $r(A)$ of A.

Any finite collection of elements and nonempty family of so-called independent sets of these elements which satisfies axiom 1 we shall call an *independence system*. This also happens to be the definition of an abstract simplicial complex, though the topology of complexes will not concern us.

It is easy to describe implicitly large independence systems which are apparently very unwieldy to analyze. For example, given a graph G, define an independent set of nodes in G to be such that no edge of G meets two nodes of the set.

The minimum coloring problem for an independence system is to find a partition of its elements into as few independent sets as possible.

A problem closely related to minimum coloring is the "packing problem". That is to find a maximum cardinality independent set. More generally the "weighted packing problem" is, where each element of the system carries a real numerical weight, to find an independent set whose weight-sum is maximum.

For any independence system, any *subsystem* consisting of a subset A of the elements and all of the independent sets contained in A is an independence system. Thus, a matroid is an independence system where the packing problem is postulated to be trivial for the system and all of its subsystems. After having spent much labor on packing problems, it is pleasant to study such systems. Matroids have a surprising richness of structure, as even the special case of graphic matroids shows.

A main result of this paper is a solution of the minimum coloring problem for the independent sets of a matroid. Another paper will treat the weighted packing problem for matroids.

3. **Ground rules.** One is tempted to surmise that a minimum coloring can be effected for a system by some simple process like extracting a maximal independent set to take on the first color, then extracting a maximal independent set of what is left to take on the second color, and so on till all elements are colored. This is usually far from being successful even for matroids.

Consider the class of matroids implicit in the class Π of all matrices over fields of integers modulo primes. (For large enough prime, this class includes the matroid of any matrix over the rational field.) We seek a good algorithm for partitioning the columns (elements of the matroid) of any one of the matrices (matroids) into as few sets as possible so that each set is independent. Of course, by carrying out the monotonic coloring procedure described above in all possible ways for a given matrix, one can be assured of encountering such a partition for the matrix, but this would entail a horrendous amount of work. We seek an algorithm for which the work involved increases only algebraically with the size of the matrix to which it is applied, where we regard the size of a matrix as increasing only linearly with the number of columns, the number of rows, and the characteristic of the field. As in most combinatorial problems, finding a finite algorithm is trivial but finding an algorithm which meets this condition for practical feasibility is not trivial.

We seek a good characterization of the minimum number of independent sets into which the columns of a matrix of Π can be partitioned. As the criterion of "good" for the characterization we apply the "principle of the absolute supervisor." The good characterization will describe certain information about the matrix which the supervisor can require his assistant to search out along with a minimum partition and which the supervisor can then use "with ease" to verify with mathematical certainty that the partition is indeed minimum. Having a good characterization does not mean necessarily that there is a good algorithm. The assistant might have to kill himself with work to find the information and the partition.

Theorem 1 on partitioning matroids provides the good characterization in the case of matrices of Π. The proof of the theorem provides a good algorithm in the case of matrices of Π. (We will not elaborate on how.) The theorem and the algorithm apply as well to all matroids via the matroid axioms. However, the "goodness" depends on having a good algorithm for recognizing independence.

4. **Theorem.** Let $\{M_i\}$, $i = 1, \cdots, k$, be an indexed family of matroids, $M_i = (E, F_i)$, all defined on the same set E of elements. Let $r_i(A)$ denote the rank of $A \subset E$ relative to M_i. Let $|A|$ denote the cardinality of A.

THEOREM 1. *Set E can be partitioned into a family $\{I_i\}$, $i = 1, \cdots, k$, of sets $I_i \in F_i$, if and only if there is no $A \subset E$ such that*

$$|A| > {}_i\sum r_i(A).$$

In particular, where the M_i's are the same matroid M, we have that: *The elements of a matroid M can be partitioned into as few as k sets, each independent in M, if and only if there is no set A of elements such that*

$$|A| > k \cdot r(A).$$

Proof of the "only if" part is easy. Suppose that $\{I_i\}$, $i = 1, \cdots, k$, is a partition of E such that I_i is independent in M_i. Then for any $A \subset E$,

$$|A| = {}_i\sum |A \cap I_i| \leqq {}_i\sum r_i(A).$$

5. **Lemmas.** A set $A \subset E$ is called *dependent* relative to a matroid $M = (E, F)$ if it is not a member of F.

Let A be any subset of the elements of a matroid M. Let I be any independent subset of A (relative to M). The set $S \subset A$, consisting of I and all elements $e \in A$ such that $I \cup e$ is dependent, is called the *span of I in A (with respect to M)*.

LEMMA 1. *For any set A of the elements of a matroid M, and any independent set $I \subset A$, the span of I in A is the unique maximal set S such that $I \subset S \subset A$ and $r(S) = |I|$.*

PROOF. Let S be any maximal set such that $I \subset S \subset A$ and $r(S) = |I|$. Consider any $e \in A - I$. By the definition of rank, I is a maximal independent subset of S. Thus, if $e \cup I$ is independent, then $e \notin S$. And thus, on the other hand, if $e \cup I$ is dependent, then I is a maximal independent subset of $e \cup S$. Hence, in this latter case, by matroid-axiom 2, $r(e \cup S) = |I|$ and so $e \in S$. Thus, S is the span of I in A, and the lemma is proved.

Where I is any independent set of matroid M, not necessarily contained in set A, we will denote the span of $I \cap A$ in A, relative to matroid M, by $T(I, A, M)$.

A minimal dependent set of elements of a matroid M is called a *circuit* of M.

LEMMA 2. *The union of any independent set I and any element e of a matroid M contains at most one circuit of M.*

PROOF. Suppose $I \cup e$ contains two distinct circuits C_1 and C_2. Assume I is minimal for this possibility. We have $e \in C_1 \cap C_2$. There is an element $e_1 \in C_1 - C_2$ and an element $e_2 \in C_2 - C_1$. Set $(I \cup e) - (e_1 \cup e_2)$ is independent since otherwise $I - e_1$ is a smaller independent set than I for which $(I - e_1) \cup e$ contains more than one circuit. Set I and set $(I \cup e) - (e_1 \cup e_2)$ are maximal independent subsets of set $I \cup e$. This contradicts axiom 2.

6. **Algorithm.** Let $\{I_i\}$ $(i = 1, \cdots, k)$ be a family of mutually disjoint subsets of E such that I_i is independent in matroid $M_i = (E, F_i)$. Any number of these may be empty. Denote their union by $H = \cup(\{I_i\})$. Set H is said to be *partitionable* (relative to $\{M_i\}$).

Suppose there is an $e \in E - H$. We shall show how either to find an $A \subset H \cup e$ such that $|A| > {}_i\sum r_i(A)$, or else partition $H \cup e$, i.e., rearrange elements among the sets I_i to make room for e in one of them while preserving their mutual disjointness and their independence, respectively, in the matroids, M_i. This will prove the theorem.

This algorithm uses as a "primitive operation" the following: for any given index i, for any given set $I \subset E$ which is known to be independent in M_i, and for any element $e \in E - I$, determine that $I \cup e$ is independent in M_i or else find the $C \subset I \cup e$ such that C is a circuit of M_i. It is easy to see how this operation can be reduced to operations of the following type: determine whether or not $I \cup e$ is independent in M_i. In view of axiom 2, it is easy to see how, using either one of these types of operation, to determine $r_i(S)$ for any $S \subset E$.

PHASE 1 OF THE ALGORITHM. Let $S_0 = E$. For each $j - 1$, starting with $j - 1 = 0$, see if there is some i, call it $i(j)$, such that

$$|I_{i(j)} \cap S_{j-1}| < r_{i(j)}(S_{j-1}).$$

If so, let

$$S_j = T(I_{i(j)}; S_{j-1}; M_{i(j)})$$

be the span in S_{j-1}, with respect to matroid $M_{i(j)}$, of $I_{i(j)} \cap S_{j-1}$. Then repeat the above with $j - 1$ one greater. In this way, we construct a sequence $(I_{i(1)}, S_1), \cdots, (I_{i(n)}, S_n)$. The labels $i(j)$ are not necessarily distinct.

Set S_j is a proper subset of S_{j-1} since, by Lemma 1,

$$r_{i(j)}(S_j) = |I_{i(j)} \cap S_{j-1}| < r_{i(j)}(S_{j-1}).$$

Therefore, in order for the sequence to stop, we must reach an S_n such that for every i, $|I_i \cap S_n| = r_i(S_n)$. Because the I_i's are disjoint, this is equivalent to

$$|H \cap S_n| = {}_i\sum r_i(S_n).$$

Now, if $e \in S_n - H$, then where

$$A = (H \cap S_n) \cup e \subset S_n,$$

we have

$$|A| = |H \cap S_n| + 1 > {}_i\sum r_i(S_n) \geqq {}_i\sum r_i(A).$$

Therefore, according to the "only if" part of Theorem 1, since $A \subset H \cup e$ and $|A| > {}_i\sum r_i(A)$, set $H \cup e$ can not be partitioned.

On the other hand, where $e \in E - (H \cup S_n)$, we shall show how $H \cup e$ can be partitioned.

PHASE 2. Since $e \notin S_n$ and $e \in S_0$, and since the S_j's are nested, there is some S_h such that $e \notin S_h$ and such that $e \in S_j$ for $0 \leqq j < h$.

If $e \cup I_{i(h)}$ is independent, in $M_{i(h)}$, then adjoin e to $I_{i(h)}$ and we are done. Otherwise, let C be the circuit of matroid $M_{i(h)}$ which is contained in $e \cup I_{i(h)}$.

Set C is not contained in S_{h-1} because then, by the definition of the span function T and the construction of S_h, we would have $e \in S_h$. Let m be the smallest integer, $0 < m < h$, such that C is not contained in S_m.

Let e' be some member of $C - S_m$. By Lemma 2, $I'_{i(h)} = e \cup I_{i(h)} - e'$ is independent in $M_{i(h)}$. Replacing $I_{i(h)}$ by $I'_{i(h)}$, and letting $I'_i = I_i$ for $i \neq i(h)$, we now need to dispose of e' instead of e.

We know that $e' \notin S_m$ and that $e' \in S_j$ for $0 \leqq j < m$. We can also show that sequence, $(I'_{i(1)}, S_1), \cdots, (I'_{i(m)}, S_m)$, is of the same construction as sequence, $(I_{i(1)}, S_1), \cdots, (I_{i(h)}, S_h)$:

Consider the terms, $j = 1, \cdots, m$, in order. If $i(j) = i(h)$, then $I'_{i(j)} = e \cup I_{i(j)} - e'$. Since $C \subset S_{j-1}$, the set $D = (I'_{i(j)} \cup e') \cap S_{j-1} = (I_{i(j)} \cup e) \cap S_{j-1}$ is dependent in $M_{i(j)}$, and thus

$$r_{i(j)}(D) = |I'_{i(j)} \cap S_{j-1}| = |I_{i(j)} \cap S_{j-1}|.$$

Therefore by the uniqueness asserted in Lemma 1 we have that

$$T(I'_{i(j)}, S_{j-1}; M_{i(j)}) = T(I_{i(j)}; S_{j-1}; M_{i(j)}) = S_j.$$

This relation obviously also holds if $i(j) \neq i(h)$, since then $I'_{i(j)}$

$= I_{i(j)}$. Thus we can treat e' in the same manner that we treated e, and so on. Since m is a positive integer strictly less than h, we will be done with this application of Phase 2 in less than h iterations. That completes the description of the algorithm and the proof of Theorem 1.

7. **Some applications.** For any integer t, $0 \leqq t \leqq |E|$, let F^t_* be the family of subsets of E which have cardinality at most t. Clearly, $M^t_* = (E, F^t_*)$ is a matroid, and the rank function r^t_* of M^t_* is $r^t_*(A) = \min(t, |A|)$.

Application 1. For any indexed family $\{M_i\}$ of matroids, $M_i = (E, F_i)$, $i = 1, \cdots, k$, consider Theorem 1 applied to the indexed family consisting of $\{M_i\}$ together with matroid M^t_*. Clearly, set E can be partitioned into sets which are independent respectively in matroid M^t_* and matroids M_i if and only if at least some $|E| - t$ members of E can be partitioned into sets which are independent respectively in matroids M_i. Thus Theorem 1 says that the latter can be done if and only if there is no $A \subset E$ such that

$$|A| > r^t_*(A) + {}_i\sum r_i(A), \quad \text{i.e.,} \ |A| > t + {}_i\sum r_i(A).$$

Application 2. In particular where $|E| - t = {}_i\sum r_i(E)$, we have that: there exist mutually disjoint subsets of E which are bases (maximum cardinality independent sets) respectively of matroids M_i if and only if there is no $A \subset E$ such that $|A| > |E| - \sum r_i(E) + \sum r_i(A)$, i.e., such that $|E - A| < {}_i\sum (r_i(E) - r_i(A))$.

Where the M_i's are the matroids of graphs, this result is equivalent, using rank-properties of graphs, to a theorem of Tutte [11], and also of Nash-Williams [7] where the graphs are identical.

Similarly, Theorem 1 itself implies the following theorem of Nash-Williams [8].

The edges of a graph G can be partitioned into as few as k forests if and only if there is no subset U of nodes in G such that, where E_U is the set of edges in G which have both ends in U,

$$|E_U| > k(|U| - 1).$$

Nash-Williams' theorem follows (we omit the proof) from Theorem 1 by using the following characterization of the rank function of a graph due to Whitney:

The rank $r(A)$ of any subset A of edges in G, i.e., the rank of the matroid subset corresponding to A, equals the number of nodes minus the number of connected components in the subgraph consisting of the edges A and the nodes they meet, or equivalently in the subgraph consisting of the edges A and all the nodes in G.

For the case where the M_i's are identical sets of vectors in a vector space (with respect to linear independence), Theorem 1 is proved by Horn [5] and Rado [9]. In fact, Rado posed the problem of deciding whether or not the result extends to matroids. I did not know of their work until the present work was completed.

For any integer $t \geqq 0$, and any matroid $M = (E, F)$, let F^t consist of the members of F which have cardinality at most t. Clearly, $M^t = (E, F^t)$ is a matroid, called the *truncation* of M at t.

APPLICATION 3. For any matroid M and for any prescribed integers $t(i)$, $i = 1, \cdots, k$, such that $0 \leqq t(i) \leqq r(E)$, let $M_i = M^{t(i)}$. Applying Theorem 1 to this family of matroids gives n. and s. conditions for there to be a family of independent sets of M of specified sizes $t(i)$, $i = 1, \cdots, k$, whose union is E.

APPLICATION 4. Applying the result of Application 2, to this same family of matroids gives n. and s. conditions for there to be a family of mutually disjoint independent sets of M having specified sizes $t(i)$.

For any matroid $M = (E, F)$ and any prescribed independent set $J \in F$, let F^J consist of sets I such that $J \cap I = \emptyset$ and such that $(J \cup I) \in F$. Clearly, $M^J = (E, F^J)$ is a matroid.

APPLICATION 5. For any matroid $M = (E, F)$ and any prescribed family of mutually disjoint independent sets $J(i) \in F$, $i = 1, \cdots, k$, let $M_i = M^{J(i)}$. Applying Theorem 1 to this family of matroids gives n. and s. conditions for there to be a partition of E into independent sets $I_i \in F$ such that $J_i \subset I_i$.

APPLICATION 6. Applying the result of Application 2 to this same family of matroids gives n. and s. conditions for there to be a family of mutually disjoint bases I_i of M such that $J_i \subset I_i$.

For any matroid $M = (E, F)$ and any $A \subset E$, let F^A consist of the sets $I \in F$ such that $I \subset A$. Clearly $M^A = (E, F^A)$ is a matroid.

For any matroid $M = (E, F)$ and any prescribed family of sets $A(i) \subset E$, $i = 1, \cdots, k$, let $M_i = M^{A(i)}$ and so on.

One can combine these and later constructions in other ways. The algorithm of course applies as well as the theorem.

8. **A class of Abelian semigroups.** Relative to any indexed family $\{M_i\}, i = 1, \cdots, k$, of independence systems, $M_i = (E, F_i)$, on the same set E, a set $H \subset E$ is called *partitionable* if it can be expressed in the form $H = I_1 \cup \cdots \cup I_k$ where $I_i \in F_i$. We may, of course, without further restricting H, require that these I_i's be mutually

disjoint. Let F denote the family of subsets H of E which are partitionable relative to $\{M_i\}$. Clearly, $M = (E, F)$ is an independence system. It is denoted as $M = {}_i\sum M_i$ and called the *sum* of the systems M_i.

It is easy to show that this sum is associative, and commutative, and has a unique identity element. Thus, the independence systems on a given set E form an abelian semigroup, say G, under this operation. The theorem below shows that all the matroids on set E form a sub-semigroup of G.

THEOREM 2. *Where the M_i's are matroids, ${}_i\sum M_i$ is a matroid.*

We have only to prove that this system satisfies matroid-axiom 2.

Let H be any member of F. Letting this H be the H of the matroid-partition algorithm, we get a certain set $S_n \subset E$ such that

$$|H \cap S_n| = {}_i\sum r_i(S_n).$$

We showed during Phase 1 of the algorithm that, for any $e \in S_n - H$, the set $H \cup e$ is not partitionable. Thus H is a maximal partitionable subset of $H \cup S_n$. We must show the stronger fact that H is a maximum-cardinality partitionable subset of $H \cup S_n$. That is, for any partitionable subset H' of $H \cup S_n$, $|H'| \leqq |H|$.

Since $H' \subset H \cup S_n$, we have

$$|H' - S_n| \leqq |H - S_n|.$$

Since H' is partitionable, we have, using the "only if" part of Theorem 1,

$$|H' \cap S_n| \leqq {}_i\sum r_i(H' \cap S_n) \leqq {}_i\sum r_i(S_n) = |H \cap S_n|.$$

Adding these two inequalities together, we get $|H'| \leqq |H|$.

Let A be any subset of E. Suppose that the above H is any maximal partitionable subset of A. Phase 2 of the partitioning algorithm shows that, for any $e \in E - (H \cap S_n)$, the set $H \cup e$ is partitionable. Therefore, $A \subset H \cup S_n$. Therefore, H is a maximum-cardinality partitionable subset of A, as well as of $H \cup S_n$. Thus, system M satisfies matroid-axiom 2, and so Theorem 2 is proved.

(There are at least two other proofs of Theorem 2 which do not use the partition-algorithm. They appear respectively in the as yet unpublished papers *Matroids and the greedy algorithm* and *Submodular set functions*. The latter paper describes a more general construction, from which the present matroid-sums, as well as the "transversal matroids" which we are about to describe, were originally derived.)

"Matroid-sums" is a useful "nonmatric" way to construct matroids. Let M_1 and M_2 be matroids determined by matrices N_1 and N_2 where the columns of both N_1 and N_2 are indexed by the set E. It can be shown that if N_1 and N_2 are matrices over different fields, say the integers modulo different primes, then matroid $M_1 + M_2$ is not generally the matroid of any matrix over any field.

It can be shown using an extension of the technique in [**2**], that if N_1 and N_2 are matrices over the same commutative field, Ψ, then $M_1 + M_2$ is the matroid of a matrix, say N, over a field extension of Ψ by about $|E|$ indeterminates. However, as indicated in [**2**], even in this case one would not determine whether a set is indepe-pendent in $M_1 + M_2$ by pivoting in N, but rather by applying the matroid-partition algorithm to M_1 and M_2.

9. **Transversal matroids.** An interesting special case of matroid-sum, ${}_i\sum M_i$, is where all the summands are rank-one matroids. Clearly, $r_i(E) = 1$ if and only if every member of F_i, besides the empty set, consists of a single element of E.

Assume $r_i(E) = 1$, $i = 1, \cdots, k$. Let Q_i consist of the elements of E which are not themselves dependent in M_i. A set $H \subset E$ which is partitionable relative to $\{M_i\}$ is called a partial transversal, or a partial *SDR*, of the indexed family $\{Q_i\}$. Matroid $M = {}_i\sum M_i$, for this case, is called a *transversal matroid.*

Clearly, the partial transversals of any indexed family, $\{Q_i\}$, $i = 1, \cdots, k$, (of not-necessarily-distinct sets $Q_i \subset E$) are the sets $H \subset E$ which can be expressed in the form $H = I_1 \cup \cdots \cup I_k$ where I_i either consists of a single element of Q_i or else is empty. We may, of course, without further restricting H, require that these I_i's be mutually disjoint. Set H is called a *transversal* or an *SDR of* $\{Q_i\}$ if the I_i's are mutually disjoint and nonempty, i.e., if $|H| = k$. Thus a partial transversal of family $\{Q_i\}$ is a set which is a transversal of some subfamily of $\{Q_i\}$.

Theorem 2 shows that the partial transversals of a $\{Q_i\}$ are the independent sets of a matroid $M = (E, F)$, a transversal matroid. Given $\{Q_i\}$, the partition algorithm provides a good algorithm for deciding whether or not any given subset of E is independent in M, for computing the rank of M, and so on.

Matroid theory and transversal theory enhance each other via transversal matroids, as do matroid theory and graph theory via graphic matroids.

P. J. Higgins [**4**] gives *n.* and *s.* conditions for an indexed family

$\{Q_i\}$, $i = 1, \cdots, k_0$, of sets to have k_1 mutually disjoint partial transversals of prescribed sizes $t(1), t(2), \cdots, t(k_1)$. This is the subject of Application 4 in §7 by taking the M there to be the transversal matroid of $\{Q_i\}$.

Much of the present article consists of extractions from [**1**] and [**3**]. Paper [**3**] contains another view of transversal matroids, and also an alternative approach to the partition of transversal matroids using the max-flow min-cut theorem and the integrity theorem of Ford and Fulkerson.

For other related material see [**6**], [**10**], [**12**], the preceding article, and the next one.

References

1. J. Edmonds, *Minimum partition of a matroid into independent subsets*, J. Res. Nat. Bur. Standards, Sect. B. **69** (1965), 67-72.

2. ———, *Systems of distinct representatives and linear algebra*, J. Res. Nat. Bur. Standards, Sect. B. **71** (1967), 241-245.

3. J. Edmonds and D. R. Fulkerson, *Transversals and matroid partition*, J. Res. Nat. Bur. Standards, Sect. B. **69** (1965), 147-153.

4. P. J. Higgins, *Disjoint transversals of subsets*, Canad. J. Math. **11** (1959), 280-285.

5. A. Horn, *A characterization of unions of linearly independent sets*, J. London Math. Soc. **30** (1955), 494-496.

6. Alfred Lehman, *A solution of the Shannon switching game*, J. Soc. Indust. Appl. Math. **12** (1964), 687-725.

7. C. St. J. A. Nash-Williams, *Edge-disjoint spanning trees of finite graphs*, J. London Math. Soc. **36** (1961), 445-450.

8. ———, *Decomposition of finite graphs into forests*, J. London Math. Soc. **39** (1964), 12.

9. R. Rado, *A combinatorial theorem on vector spaces*, J. London Math. Soc. **37** (1962), 351-353.

10. ———, *A theorem on independence relations*, Quart. J. Math. **13** (1942), 83-89.

11. W. T. Tutte, *On the problem of decomposing a graph into n connected factors*, J. London Math. Soc. **36** (1961), 221-230.

12. ———, *Lectures on matroids*, J. Res. Nat. Bur. Standards, Sect. B. **69** (1965), 1-47.

National Bureau of Standards

Chapter 8
Reducibility Among Combinatorial Problems

Richard M. Karp

Introduction by *Richard M. Karp*

Throughout the 1960s I worked on combinatorial optimization problems including logic circuit design with Paul Roth and assembly line balancing and the traveling salesman problem with Mike Held. These experiences made me aware that seemingly simple discrete optimization problems could hold the seeds of combinatorial explosions. The work of Dantzig, Fulkerson, Hoffman, Edmonds, Lawler and other pioneers on network flows, matching and matroids acquainted me with the elegant and efficient algorithms that were sometimes possible. Jack Edmonds' papers and a few key discussions with him drew my attention to the crucial distinction between polynomial-time and superpolynomial-time solvability. I was also influenced by Jack's emphasis on min-max theorems as a tool for fast verification of optimal solutions, which foreshadowed Steve Cook's definition of the complexity class NP. Another influence was George Dantzig's suggestion that integer programming could serve as a universal format for combinatorial optimization problems.

Throughout the '60s I followed developments in computational complexity theory, pioneered by Rabin, Blum, Hartmanis, Stearns and others. In the late '60s I studied Hartley Rogers' beautiful book on recursive function theory while teaching a course on the subject at the Polytechnic Institute of Brooklyn. This experience brought home to me the key role of reducibilities in recursive function theory, and started me wondering whether subrecursive reducibilities could play a similar role in complexity theory, but I did not yet pursue the analogy.

Cook's 1971 paper [1], in which he defined the class NP and showed that propositional satisfiability was an NP-complete problem, brought together for me the two strands of complexity theory and combinatorial optimization. It was immediately apparent to me that many familiar combinatorial problems were likely to have the

---

Richard M. Karp
The University of California at Berkeley, USA
e-mail: karp@icsi.berkeley.edu

M. Jünger et al. (eds.), *50 Years of Integer Programming 1958-2008*,
DOI 10.1007/978-3-540-68279-0_8, © Springer-Verlag Berlin Heidelberg 2010

same universal role as satisfiability. I enjoyed constructing the polynomial-time reductions that verified this intuition. Most of them were easy to find, but I failed to prove the NP-completeness of the undirected hamiltonian circuit problem; that reduction was provided independently by Lawler and Tarjan. I was also frustrated by my inability to classify linear programming, graph isomorphism and primality.

As I recall I first presented my results at an informal seminar at Don Knuth's home, and a few months later, in April 1972, I exposed the work more broadly at an IBM symposium. In the next couple of years many results more refined than my own were added to the accumulation of NP-completeness proofs, and later the Garey-Johnson book [2] presented the concepts to a wide audience.

Heuristic algorithms often find near-optimal solutions to NP-hard optimization problems. For some time in the mid-1970s I tried to explain this phenomenon by departing from worst-case analysis, and instead analyzing the performance of simple heuristics on instances drawn from simple probability distributions. This work was technically successful but gained limited traction, because there was no way to show that the problem instances drawn from these probability distributions are representative of those arising in practice. The surprising success of many heuristics remains a mystery.

References

1. S.A. Cook, *The complexity of theorem proving procedures*, Proceedings of the 3rd Annual ACM Symposium on Theory of Computing, 1971, pp. 151–158.
2. M.R. Garey and D.S. Johnson, *Computers and Intractability: A Guide to the Theory of NP-Completeness*, W.H. Freeman & Co., 1979.

The following article originally appeared as:

R.M. Karp, *Reducibility Among Combinatorial Problems*, Complexity of Computer Computations (R.E. Miller and J.W. Thatcher, eds.), Plenum Press, 1972, pp. 85–103.

REDUCIBILITY AMONG COMBINATORIAL PROBLEMS†

Richard M. Karp

University of California at Berkeley

Abstract: A large class of computational problems involve the determination of properties of graphs, digraphs, integers, arrays of integers, finite families of finite sets, boolean formulas and elements of other countable domains. Through simple encodings from such domains into the set of words over a finite alphabet these problems can be converted into language recognition problems, and we can inquire into their computational complexity. It is reasonable to consider such a problem satisfactorily solved when an algorithm for its solution is found which terminates within a number of steps bounded by a polynomial in the length of the input. We show that a large number of classic unsolved problems of covering, matching, packing, routing, assignment and sequencing are equivalent, in the sense that either each of them possesses a polynomial-bounded algorithm or none of them does.

1. INTRODUCTION

All the general methods presently known for computing the chromatic number of a graph, deciding whether a graph has a Hamilton circuit, or solving a system of linear inequalities in which the variables are constrained to be 0 or 1, require a combinatorial search for which the worst case time requirement grows exponentially with the length of the input. In this paper we give theorems which strongly suggest, but do not imply, that these problems, as well as many others, will remain intractable perpetually.

†This research was partially supported by National Science Foundation Grant GJ-474.

We are specifically interested in the existence of algorithms that are guaranteed to terminate in a number of steps bounded by a polynomial in the length of the input. We exhibit a class of well-known combinatorial problems, including those mentioned above, which are equivalent, in the sense that a polynomial-bounded algorithm for any one of them would effectively yield a polynomial-bounded algorithm for all. We also show that, if these problems do possess polynomial-bounded algorithms then all the problems in an unexpectedly wide class (roughly speaking, the class of problems solvable by polynomial-depth backtrack search) possess polynomial-bounded algorithms.

The following is a brief summary of the contents of the paper. For the sake of definiteness our technical development is carried out in terms of the recognition of languages by one-tape Turing machines, but any of a wide variety of other abstract models of computation would yield the same theory. Let Σ^* be the set of all finite strings of 0's and 1's. A subset of Σ^* is called a <u>language</u>. Let $\mathcal{P}$ be the class of languages recognizable in polynomial time by one-tape deterministic Turing machines, and let $\mathcal{NP}$ be the class of languages recognizable in polynomial time by one-tape nondeterministic Turing machines. Let Π be the class of functions from Σ^* into Σ^* computable in polynomial time by one-tape Turing machines. Let L and M be languages. We say that $L \propto M$ (L <u>is reducible to</u> M) if there is a function $f \in \Pi$ such that $f(x) \in M \Leftrightarrow x \in L$. If $M \in \mathcal{P}$ and $L \propto M$ then $L \in \mathcal{P}$. We call L and M equivalent if $L \propto M$ and $M \propto L$. Call L <u>(polynomial) complete</u> if $L \in \mathcal{NP}$ and every language in $\mathcal{NP}$ is reducible to L. Either all complete languages are in $\mathcal{P}$, or none of them are. The former alternative holds if and only if $\mathcal{P} = \mathcal{NP}$.

The main contribution of this paper is the demonstration that a large number of classic difficult computational problems, arising in fields such as mathematical programming, graph theory, combinatorics, computational logic and switching theory, are complete (and hence equivalent) when expressed in a natural way as language recognition problems.

This paper was stimulated by the work of Stephen Cook (1971), and rests on an important theorem which appears in his paper. The author also wishes to acknowledge the substantial contributions of Eugene Lawler and Robert Tarjan.

2. THE CLASS $\mathcal{P}$

There is a large class of important computational problems which involve the determination of properties of graphs, digraphs, integers, finite families of finite sets, boolean formulas and elements of other countable domains. It is a reasonable working hypothesis, championed originally by Jack Edmonds (1965) in connection with problems in graph theory and integer programming, and by now widely accepted, that such a problem can be regarded as tractable if and only if there is an algorithm for its solution whose running time is bounded by a polynomial in the size of the input. In this section we introduce and begin to investigate the class of problems solvable in polynomial time.

We begin by giving an extremely general definition of "deterministic algorithm", computing a function from a countable domain D into a countable range R.

For any finite alphabet A, let A^* be the set of finite strings of elements of A; for $x \in A^*$, let $\lg(x)$ denote the length of x.

A *deterministic algorithm* $\mathcal{A}$ is specified by:

a countable set D (the *domain*)
a countable set R (the *range*)
a finite alphabet Δ such that $\Delta^* \cap R = \phi$
an *encoding function* $E: D \to \Delta^*$
a *transition function* $\tau: \Delta^* \to \Delta^* \cup R$.

The *computation* of $\mathcal{A}$ on input $x \in D$ is the unique sequence $y_1, y_2, \ldots$ such that $y_1 = E(x)$, $y_{i+1} = \tau(y_i)$ for all i and, if the sequence is finite and ends with y_k, then $y_k \in R$. Any string occurring as an element of a computation is called an *instantaneous description*. If the computation of $\mathcal{A}$ on input x is finite and of length $t(x)$, then $t(x)$ is the *running time* of $\mathcal{A}$ on input x. $\mathcal{A}$ is *terminating* if all its computations are finite. A terminating algorithm $\mathcal{A}$ computes the function $f_{\mathcal{A}}: D \to R$ such that $f_{\mathcal{A}}(x)$ is the last element of the computation of $\mathcal{A}$ on x.

If $R = \{\text{ACCEPT}, \text{REJECT}\}$ then $\mathcal{A}$ is called a *recognition algorithm*. A recognition algorithm in which $D = \Sigma^*$ is called a *string recognition algorithm*. If $\mathcal{A}$ is a string recognition algorithm then the *language recognized by* $\mathcal{A}$ is $\{x \in \Sigma^* \mid f_{\mathcal{A}}(x) = \text{ACCEPT}\}$. If $D = R = \Sigma^*$ then $\mathcal{A}$ is called a *string mapping algorithm*. A terminating algorithm $\mathcal{A}$ with domain $D = \Sigma^*$ *operates in polynomial time* if there is a polynomial $p(\cdot)$ such that, for every $x \in \Sigma^*$, $t(x) \leq p(\lg(x))$.

To discuss algorithms in any practical context we must specialize the concept of deterministic algorithm. Various well known classes of string recognition algorithms (Markov algorithms, one-tape Turing machines, multitape and multihead Turing machines, random access machines, etc.) are delineated by restricting the functions E and τ to be of certain very simple types. These definitions are standard [Hopcroft & Ullman (1969)] and will not be repeated here. It is by now commonplace to observe that many such classes are equivalent in their capability to recognize languages; for each such class of algorithms, the class of languages recognized is the class of recursive languages. This invariance under changes in definition is part of the evidence that recursiveness is the correct technical formulation of the concept of decidability.

The class of languages recognizable by string recognition algorithms which operate in polynomial time is also invariant under a wide range of changes in the class of algorithms. For example, any language recognizable in time $p(\cdot)$ by a multihead or multitape Turing machine is recognizable in time $p^2(\cdot)$ by a one-tape Turing machine. Thus the class of languages recognizable in polynomial time by one-tape Turing machines is the same as the class recognizable by the ostensibly more powerful multihead or multitape Turing machines. Similar remarks apply to random access machines.

<u>Definition 1</u>. $\mathcal{P}$ is the class of languages recognizable by one-tape Turing machines which operate in polynomial time.

<u>Definition 2</u>. Π is the class of functions from Σ^* into Σ^* defined by one-tape Turing machines which operate in polynomial time.

The reader will not go wrong by identifying $\mathcal{P}$ with the class of languages recognizable by digital computers (with unbounded backup storage) which operate in polynomial time and Π with the class of string mappings performed in polynomial time by such computers.

<u>Remark</u>. If $f\colon \Sigma^* \to \Sigma^*$ is in Π then there is a polynomial $p(\cdot)$ such that $\lg(f(x)) \le p(\lg(x))$.

We next introduce a concept of reducibility which is of central importance in this paper.

<u>Definition 3</u>. Let L and M be languages. Then $L \propto M$ (L <u>is reducible to</u> M) if there is a function $f \in \Pi$ such that $f(x) \in M \Leftrightarrow x \in L$.

<u>Lemma 1</u>. If $L \propto M$ and $M \in \mathcal{P}$ then $L \in \mathcal{P}$.

<u>Proof</u>. The following is a polynomial-time bounded algorithm to decide if $x \in L$: compute $f(x)$; then test in polynomial time whether $f(x) \in M$.

We will be interested in the difficulty of recognizing subsets of countable domains other than Σ^*. Given such a domain D, there is usually a natural one-one encoding $e: D \to \Sigma^*$. For example we can represent a positive integer by the string of 0's and 1's comprising its binary representation, a 1-dimensional integer array as a list of integers, a matrix as a list of 1-dimensional arrays, etc.; and there are standard techniques for encoding lists into strings over a finite alphabet, and strings over an arbitrary finite alphabet as strings of 0's and 1's. Given such an encoding $e: D \to \Sigma^*$, we say that a set $T \subseteq D$ is recognizable in polynomial time if $e(T) \in \mathcal{P}$. Also, given sets $T \subseteq D$ and $U \subseteq D'$, and encoding functions $e: D \to \Sigma^*$ and $e': D' \to \Sigma^*$ we say $T \propto U$ if $e(T) \propto e'(U)$.

As a rule several natural encodings of a given domain are possible. For instance a graph can be represented by its adjacency matrix, by its incidence matrix, or by a list of unordered pairs of nodes, corresponding to the arcs. Given one of these representations, there remain a number of arbitrary decisions as to format and punctuation. Fortunately, it is almost always obvious that any two "reasonable" encodings e_0 and e_1 of a given problem are equivalent; i.e., $e_0(S) \in \mathcal{P} \Leftrightarrow e_1(S) \in \mathcal{P}$. One important exception concerns the representation of positive integers; we stipulate that a positive integer is encoded in a binary, rather than unary, representation. In view of the invariance of recognizability in polynomial time and reducibility under reasonable encodings, we discuss problems in terms of their original domains, without specifying an encoding into Σ^*.

We complete this section by listing a sampling of problems which are solvable in polynomial time. In the next section we examine a number of close relatives of these problems which are not known to be solvable in polynomial time. Appendix 1 establishes our notation.

Each problem is specified by giving (under the heading "INPUT") a generic element of its domain of definition and (under the heading "PROPERTY") the property which causes an input to be accepted.

SATISFIABILITY WITH AT MOST 2 LITERALS PER CLAUSE [Cook (1971)]
INPUT: Clauses $C_1, C_2, \ldots, C_p$, each containing at most 2 literals
PROPERTY: The conjunction of the given clauses is satisfiable; i.e., there is a set $S \subseteq \{x_1, x_2, \ldots, x_n, \bar{x}_1, \bar{x}_2, \ldots, \bar{x}_n\}$ such that

a) S does not contain a complementary pair of literals and
b) $S \cap C_k \neq \phi$, $k = 1,2,\ldots,p$.

MINIMUM SPANNING TREE [Kruskal (1956)]
INPUT: G, w, W
PROPERTY: There exists a spanning tree of weight $\leq W$.

SHORTEST PATH [Dijkstra (1959)]
INPUT: G, w, W, s, t
PROPERTY: There is a path between s and t of weight $\leq$ W.

MINIMUM CUT [Edmonds & Karp (1972)]
INPUT: G, w, W, s, t
PROPERTY: There is an s,t cut of weight $\leq$ W.

ARC COVER [Edmonds (1965)]
INPUT: G, k
PROPERTY: There is a set $Y \subseteq A$ such that $|Y| \leq k$ and every node is incident with an arc in Y.

ARC DELETION
INPUT: G, k
PROPERTY: There is a set of k arcs whose deletion breaks all cycles.

BIPARTITE MATCHING [Hall (1948)]
INPUT: $S \subseteq Z_p \times Z_p$
PROPERTY: There are p elements of S, no two of which are equal in either component.

SEQUENCING WITH DEADLINES
INPUT: $(T_1,\ldots,T_n) \in Z^n$, $(D_1,\ldots,D_n) \in Z^n$, k
PROPERTY: Starting at time 0, one can execute jobs 1,2,...,n, with execution times T_i and deadlines D_i, in some order such that not more than k jobs miss their deadlines.

SOLVABILITY OF LINEAR EQUATIONS
INPUT: (c_{ij}), (a_i)
PROPERTY: There exists a vector (y_j) such that, for each i, $\sum_j c_{ij}y_j = a_i$.

3. NONDETERMINISTIC ALGORITHMS AND COOK'S THEOREM

In this section we state an important theorem due to Cook (1971) which asserts that any language in a certain wide class NP is reducible to a specific set S, which corresponds to the problem of deciding whether a boolean formula in conjunctive normal form is satisfiable.

Let $P^{(2)}$ denote the class of subsets of $\Sigma^* \times \Sigma^*$ which are recognizable in polynomial time. Given $L^{(2)} \in P^{(2)}$ and a polynomial p, we define a language L as follows:

$$L = \{x \mid \text{there exists } y \text{ such that } \langle x,y\rangle \in L^{(2)} \text{ and } lg(y) \leq p(lg(x))\}.$$

We refer to L as the language derived from $L^{(2)}$ by p-bounded existential quantification.

Definition 4. NP is the set of languages derived from elements of $P^{(2)}$ by polynomial-bounded existential quantification.

There is an alternative characterization of NP in terms of nondeterministic Turing machines. A nondeterministic recognition algorithm A is specified by:

a countable set D (the domain)
a finite alphabet Δ such that $\Delta^* \cap \{ACCEPT, REJECT\} = \phi$
an encoding function $E: D \to \Delta^*$
a transition relation $\tau \subseteq \Delta^* \times (\Delta^* \cup \{ACCEPT, REJECT\})$

such that, for every $y_0 \in \Delta^*$, the set $\{\langle y_0, y\rangle \mid \langle y_0, y\rangle \in \tau\}$ has fewer than k_A elements, where k_A is a constant. A computation of A on input $x \in D$ is a sequence $y_1, y_2, \ldots$ such that $y_1 = E(x)$, $\langle y_i, y_{i+1}\rangle \in \tau$ for all i, and, if the sequence is finite and ends with y_k, then $y_k \in \{ACCEPT, REJECT\}$. A string $y \in \Delta^*$ which occurs in some computation is an instantaneous description. A finite computation ending in ACCEPT is an accepting computation. Input x is accepted if there is an accepting computation for x. If $D = \Sigma^*$ then A is a nondeterministic string recognition algorithm and we say that A operates in polynomial time if there is a polynomial $p(\cdot)$ such that, whenever A accepts x, there is an accepting computation for x of length $\leq p(lg(x))$.

A nondeterministic algorithm can be regarded as a process which, when confronted with a choice between (say) two alternatives, can create two copies of itself, and follow up the consequences of both courses of action. Repeated splitting may lead to an exponentially growing number of copies; the input is accepted if any sequence of choices leads to acceptance.

The nondeterministic 1-tape Turing machines, multitape Turing machines, random-access machines, etc. define classes of nondeterministic string recognition algorithms by restricting the encoding function E and transition relation τ to particularly simple forms. All these classes of algorithms, restricted to operate in polynomial time, define the same class of languages. Moreover, this class is NP.

Theorem 1. $L \in NP$ if and only if L is accepted by a nondeterministic Turing machine which operates in polynomial time.

Proof. $\Rightarrow$ Suppose $L \in NP$. Then, for some $L^{(2)} \in P^{(2)}$ and some polynomial p, L is obtained from $L^{(2)}$ by p-bounded existential quantification. We can construct a nondeterministic machine which first guesses the successive digits of a string y of length $\leq p(\lg(y))$ and then tests whether $\langle x,y\rangle \in L^{(2)}$. Such a machine clearly recognizes L in polynomial time.

$\Leftarrow$ Suppose L is accepted by a nondeterministic Turing machine T which operates in time p. Assume without loss of generality that, for any instantaneous description Z, there are at most two instantaneous descriptions that may follow Z (i.e., at most two primitive transitions are applicable). Then the sequence of choices of instantaneous descriptions made by T in a given computation can be encoded as a string y of 0's and 1's, such that $\lg(y) \leq p(\lg(x))$.

Thus we can construct a deterministic Turing machine T', with $\Sigma^* \times \Sigma^*$ as its domain of inputs, which, on input $\langle x,y\rangle$, simulates the action of T on input x with the sequence of choices y. Clearly T' operates in polynomial time, and L is obtained by polynomial bounded existential quantification from the set of pairs of strings accepted by T'.

The class NP is very extensive. Loosely, a recognition problem is in NP if and only if it can be solved by a backtrack search of polynomial bounded depth. A wide range of important computational problems which are not known to be in P are obviously in NP. For example, consider the problem of determining whether the nodes of a graph G can be colored with k colors so that no two adjacent nodes have the same color. A nondeterministic algorithm can simply guess an assignment of colors to the nodes and then check (in polynomial time) whether all pairs of adjacent nodes have distinct colors.

In view of the wide extent of NP, the following theorem due to Cook is remarkable. We define the satisfiability problem as follows:

SATISFIABILITY
INPUT: Clauses $C_1, C_2, \ldots, C_p$
PROPERTY: The conjunction of the given clauses is satisfiable; i.e., there is a set $S \subseteq \{x_1, x_2, \ldots, x_n; \bar{x}_1, \bar{x}_2, \ldots, \bar{x}_n\}$ such that
a) S does not contain a complementary pair of literals
and b) $S \cap C_k \neq \phi$, $k = 1,2,\ldots,p$.

Theorem 2 (Cook). If $L \in NP$ then $L \propto$ SATISFIABILITY.

The theorem stated by Cook (1971) uses a weaker notion of reducibility than the one used here, but Cook's proof supports the present statement.

Corollary 1. $P = NP \Leftrightarrow$ SATISFIABILITY $\in P$.

Proof. If SATISFIABILITY $\in P$ then, for each $L \in NP$, $L \in P$, since $L \propto$ SATISFIABILITY. If SATISFIABILITY $\notin P$, then, since clearly SATISFIABILITY $\in NP$, $P \neq NP$.

Remark. If $P = NP$ then NP is closed under complementation and polynomial-bounded existential quantification. Hence it is also closed under polynomial-bounded universal quantification. It follows that a polynomial-bounded analogue of Kleene's Arithmetic Hierarchy [Rogers (1967)] becomes trivial if $P = NP$.

Theorem 2 shows that, if there were a polynomial-time algorithm to decide membership in SATISFIABILITY then every problem solvable by a polynomial-depth backtrack search would also be solvable by a polynomial-time algorithm. This is strong circumstantial evidence that SATISFIABILITY $\notin P$.

4. COMPLETE PROBLEMS

The main object of this paper is to establish that a large number of important computational problems can play the role of SATISFIABILITY in Cook's theorem. Such problems will be called complete.

Definition 5. The language L is (polynomial) complete if

a) $L \in NP$

and b) SATISFIABILITY $\propto$ L.

Theorem 3. Either all complete languages are in P, or none of them are. The former alternative holds if and only if $P = NP$.

We can extend the concept of completeness to problems defined over countable domains other than Σ^*.

Definition 6. Let D be a countable domain, e a "standard" one-one encoding $e: D \to \Sigma^*$ and T a subset of D. Then T is complete if and only if $e(D)$ is complete.

Lemma 2. Let D and D' be countable domains, with one-one encoding functions e and e'. Let $T \subseteq D$ and $T' \subseteq D'$. Then $T \propto T'$ if there is a function $F: D \to D'$ such that

a) $F(x) \in T' \Leftrightarrow x \in T$

and b) there is a function $f \in \Pi$ such that $f(x) = e'(F(e^{-1}(x)))$ whenever $e'(F(e^{-1}(x)))$ is defined.

The rest of the paper is mainly devoted to the proof of the following theorem.

Main Theorem. All the problems on the following list are complete.

1. SATISFIABILITY
COMMENT: By duality, this problem is equivalent to determining whether a disjunctive normal form expression is a tautology.

2. 0-1 INTEGER PROGRAMMING
INPUT: integer matrix C and integer vector d
PROPERTY: There exists a 0-1 vector x such that $Cx = d$.

3. CLIQUE
INPUT: graph G, positive integer k
PROPERTY: G has a set of k mutually adjacent nodes.

4. SET PACKING
INPUT: Family of sets $\{S_j\}$, positive integer ℓ
PROPERTY: $\{S_j\}$ contains ℓ mutually disjoint sets.

5. NODE COVER
INPUT: graph G', positive integer ℓ
PROPERTY: There is a set $R \subseteq N'$ such that $|R| \leq \ell$ and every arc is incident with some node in R.

6. SET COVERING
INPUT: finite family of finite sets $\{S_j\}$, positive integer k
PROPERTY: There is a subfamily $\{T_h\} \subseteq \{S_j\}$ containing $\leq k$ sets such that $\cup T_h = \cup S_j$.

7. FEEDBACK NODE SET
INPUT: digraph H, positive integer k
PROPERTY: There is a set $R \subseteq V$ such that every (directed) cycle of H contains a node in R.

8. FEEDBACK ARC SET
INPUT: digraph H, positive integer k
PROPERTY: There is a set $S \subseteq E$ such that every (directed) cycle of H contains an arc in S.

9. DIRECTED HAMILTON CIRCUIT
INPUT: digraph H
PROPERTY: H has a directed cycle which includes each node exactly once.

10. UNDIRECTED HAMILTON CIRCUIT
INPUT: graph G
PROPERTY: G has a cycle which includes each node exactly once.

11. SATISFIABILITY WITH AT MOST 3 LITERALS PER CLAUSE
INPUT: Clauses $D_1, D_2, \ldots, D_r$, each consisting of at most 3 literals from the set $\{u_1, u_2, \ldots, u_m\} \cup \{\bar{u}_1, \bar{u}_2, \ldots, \bar{u}_m\}$
PROPERTY: The set $\{D_1, D_2, \ldots, D_r\}$ is satisfiable.

12. CHROMATIC NUMBER
INPUT: graph G, positive integer k
PROPERTY: There is a function $\phi: N \to Z_k$ such that, if u and v are adjacent, then $\phi(u) \neq \phi(v)$.

13. CLIQUE COVER
INPUT: graph G', positive integer ℓ
PROPERTY: N' is the union of ℓ or fewer cliques.

14. EXACT COVER
INPUT: family $\{S_j\}$ of subsets of a set $\{u_i, i = 1,2,\ldots,t\}$
PROPERTY: There is a subfamily $\{T_h\} \subseteq \{S_j\}$ such that the sets T_h are disjoint and $\cup T_h = \cup S_j = \{u_i, i = 1,2,\ldots,t\}$.

15. HITTING SET
INPUT: family $\{U_i\}$ of subsets of $\{s_j, j = 1,2,\ldots,r\}$
PROPERTY: There is a set W such that, for each i, $|W \cap U_i| = 1$.

16. STEINER TREE
INPUT: graph G, $R \subseteq N$, weighting function $w: A \to Z$, positive integer k
PROPERTY: G has a subtree of weight $\leq k$ containing the set of nodes in R.

17. 3-DIMENSIONAL MATCHING
INPUT: set $U \subseteq T \times T \times T$, where T is a finite set
PROPERTY: There is a set $W \subseteq U$ such that $|W| = |T|$ and no two elements of W agree in any coordinate.

18. KNAPSACK
INPUT: $(a_1, a_2, \ldots, a_r, b) \in Z^{n+1}$
PROPERTY: $\Sigma a_j x_j = b$ has a 0-1 solution.

19. JOB SEQUENCING
INPUT: "execution time vector" $(T_1, \ldots, T_p) \in Z^p$,
"deadline vector" $(D_1, \ldots, D_p) \in Z^p$
"penalty vector" $(P_1, \ldots, P_p) \in Z^p$
positive integer k
PROPERTY: There is a permutation π of $\{1,2,\ldots,p\}$ such that

$$\left(\sum_{j=1}^{p} [\text{if } T_{\pi(1)} + \cdots + T_{\pi(j)} > D_{\pi(j)} \text{ then } P_{\pi(j)} \text{ else } 0]\right) \leq k .$$

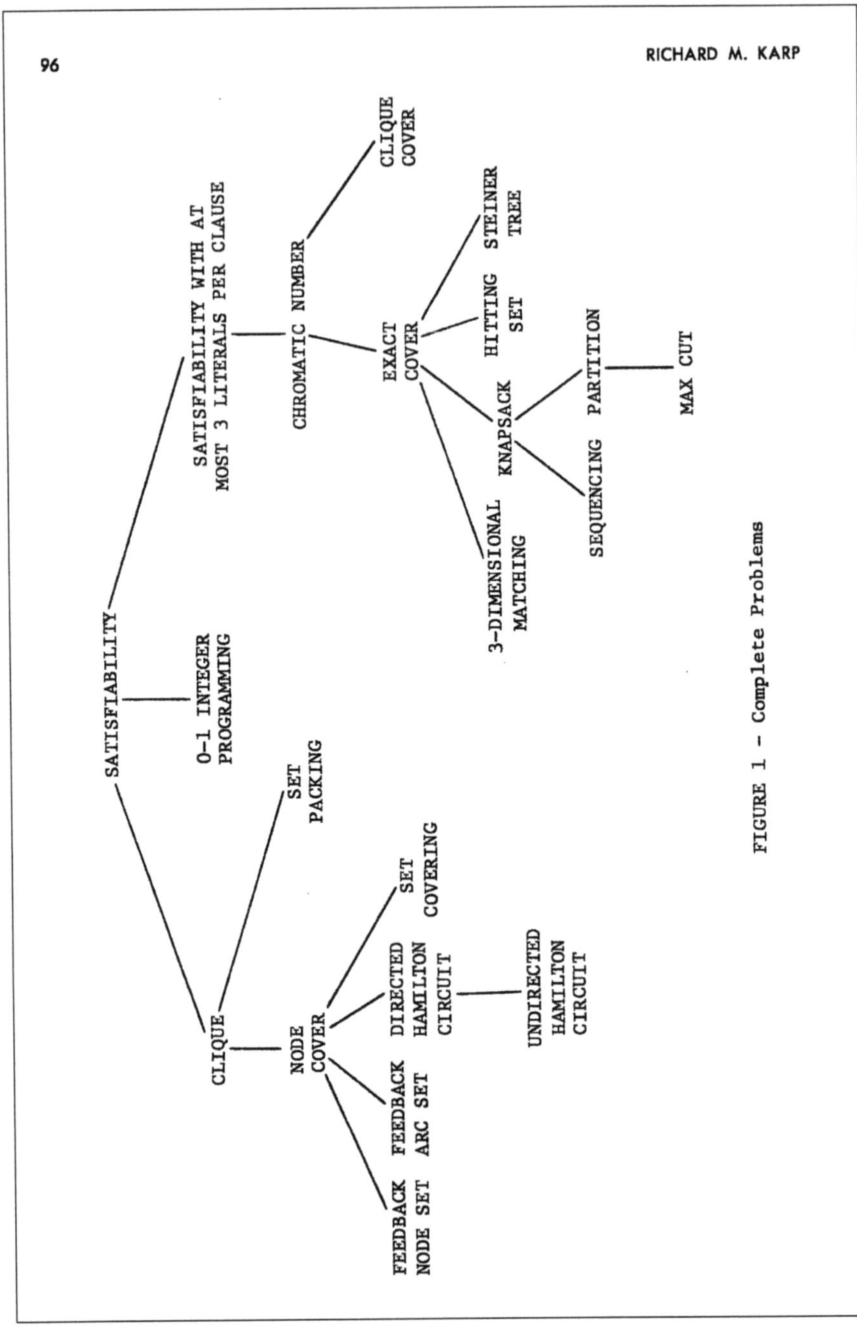

FIGURE 1 - Complete Problems

20. PARTITION
INPUT: $(c_1,c_2,\ldots,c_s) \in Z^s$
PROPERTY: There is a set $I \subseteq \{1,2,\ldots,s\}$ such that $\sum_{h\in I} c_h = \sum_{h\notin I} c_h$.

21. MAX CUT
INPUT: graph G, weighting function $w: A \rightarrow Z$, positive integer W
PROPERTY: There is a set $S \subseteq N$ such that

$$\sum_{\substack{\{u,v\}\in A\\ u\in S\\ v\notin S}} w(\{u,v\}) \geq W \quad .$$

It is clear that these problems (or, more precisely, their encodings into Σ^*), are all in NP. We proceed to give a series of explicit reductions, showing that SATISFIABILITY is reducible to each of the problems listed. Figure 1 shows the structure of the set of reductions. Each line in the figure indicates a reduction of the upper problem to the lower one.

To exhibit a reduction of a set $T \subseteq D$ to a set $T' \subseteq D'$, we specify a function $F: D \rightarrow D'$ which satisfies the conditions of Lemma 2. In each case, the reader should have little difficulty in verifying that F does satisfy these conditions.

SATISFIABILITY $\propto$ 0-1 INTEGER PROGRAMMING

$$c_{ij} = \begin{cases} 1 & \text{if } x_j \in C_i \\ -1 & \text{if } \bar{x}_j \in C_i \\ 0 & \text{otherwise} \end{cases} \qquad \begin{array}{l} i = 1,2,\ldots,p \\ j = 1,2,\ldots,n \end{array}$$

$b_i = 1 -$ (the number of complemented variables in C_i) ,
$i = 1,2,\ldots,p$.

SATISFIABILITY $\propto$ CLIQUE

$N = \{\langle\sigma,i\rangle \mid \sigma \text{ is a literal and occurs in } C_i\}$

$A = \{\{\langle\sigma,i\rangle,\langle\delta,j\rangle\} \mid i \neq j \text{ and } \sigma \neq \bar{\delta}\}$

$k = p$, the number of clauses.

CLIQUE $\propto$ SET PACKING

Assume $N = \{1,2,\ldots,n\}$. The elements of the sets $S_1,S_2,\ldots,S_n$ are those two-element sets of nodes $\{i,j\}$ not in A.

$S_i = \{\{i,j\} \mid \{i,j\} \notin A\}$, $i = 1,2,\ldots,n$

$\ell = k$.

CLIQUE $\propto$ NODE COVER

G' is the complement of G.
$\ell = |N| - k$

NODE COVER $\propto$ SET COVERING

Assume $N' = \{1,2,\ldots,n\}$. The elements are the arcs of G'. S_j is the set of arcs incident with node j. $k = \ell$.

NODE COVER $\propto$ FEEDBACK NODE SET

$V = N'$
$E = \{\langle u,v\rangle \mid \{u,v\} \in A'\}$
$k = \ell$

NODE COVER $\propto$ FEEDBACK ARC SET

$V = N' \times \{0,1\}$
$E = \{\langle\langle u,0\rangle,\langle u,1\rangle\rangle \mid u \in N'\} \cup \{\langle\langle u,1\rangle,\langle v,0\rangle\rangle \mid \{u,v\} \in A'\}$
$k = \ell$.

NODE COVER $\propto$ DIRECTED HAMILTON CIRCUIT

Without loss of generality assume $A' = Z_m$.

$$V = \{a_1,a_2,\ldots,a_\ell\} \cup \{\langle u,i,\alpha\rangle \mid u \in N' \text{ is incident with } i \in A' \text{ and } \alpha \in \{0,1\}\}$$

$$\begin{aligned} E = {} & \{\langle\langle u,i,0\rangle,\langle u,i,1\rangle\rangle \mid \langle u,i,0\rangle \in V\} \\ & \cup \{\langle\langle u,i,\alpha\rangle,\langle v,i,\alpha\rangle\rangle \mid i \in A',\ u \text{ and } v \text{ are incident with } i,\ \alpha \in \{0,1\}\} \\ & \cup \{\langle\langle u,i,1\rangle,\langle u,j,0\rangle\rangle \mid u \text{ is incident with } i \text{ and } j \text{ and } \not\exists h,\ i < h < j, \text{ such that } u \text{ is incident with } h\} \\ & \cup \{\langle\langle u,i,1\rangle,a_f\rangle \mid 1 \le f \le \ell \text{ and } \not\exists h > i \text{ such that } u \text{ is incident with } h\} \\ & \cup \{\langle a_f,\langle u,i,0\rangle\rangle \mid 1 \le f \le \ell \text{ and } \not\exists h < i \text{ such that } u \text{ is incident with } h\}. \end{aligned}$$

DIRECTED HAMILTON CIRCUIT $\propto$ UNDIRECTED HAMILTON CIRCUIT

$N = V \times \{0,1,2\}$
$A = \{\{\langle u,0\rangle,\langle u,1\rangle\},\{\langle u,1\rangle,\langle u,2\rangle\} \mid u \in V\}$
$\quad \cup \{\{\langle u,2\rangle,\langle v,0\rangle\} \mid \langle u,v\rangle \in E\}$

SATISFIABILITY $\propto$ SATISFIABILITY WITH AT MOST 3 LITERALS PER CLAUSE

Replace a clause $\sigma_1 \cup \sigma_2 \cup \cdots \cup \sigma_m$, where the σ_i are literals and $m > 3$, by

$$(\sigma_1 \cup \sigma_2 \cup u_1)(\sigma_3 \cup \cdots \cup \sigma_m \cup \bar{u}_1)(\bar{\sigma}_3 \cup u_1)\cdots(\bar{\sigma}_m \cup u_1) \quad ,$$

where u_1 is a new variable. Repeat this transformation until no clause has more than three literals.

SATISFIABILITY WITH AT MOST 3 LITERALS PER CLAUSE
$\propto$ CHROMATIC NUMBER

Assume without loss of generality that $m \geq 4$.

$$N = \{u_1,u_2,\ldots,u_m\} \cup \{\bar{u}_1,\bar{u}_2,\ldots,\bar{u}_m\} \cup \{v_1,v_2,\ldots,v_m\} \cup \{D_1,D_2,\ldots,D_r\}$$

$$A = \{\{u_i,\bar{u}_i\} \mid i=1,2,\ldots,n\} \cup \{\{v_i,v_j\} \mid i \neq j\} \cup \{\{v_i,x_j\} \mid i \neq j\} \cup \{\{v_i,\bar{x}_j\} \mid i \neq j\} \cup \{\{u_i,D_f\} \mid u_i \notin D_f\} \cup \{\{\bar{u}_i,D_f\} \mid \bar{u}_i \in D_f\}$$

$$k = r+1$$

CHROMATIC NUMBER $\propto$ CLIQUE COVER

G' is the complement of G
$\ell = k$.

CHROMATIC NUMBER $\propto$ EXACT COVER

The set of elements is

$$N \cup A \cup \{\langle u,e,f\rangle \mid u \text{ is incident with } e \text{ and } 1 \leq f \leq k\} .$$

The sets S_j are the following:

for each f, $1 \leq f \leq k$, and each $u \in N$,
$\{u\} \cup \{\langle u,e,f\rangle \mid e \text{ is incident with } u\}$;

for each $e \in A$ and each pair f_1, f_2 such that
$1 \leq f_1 \leq k$, $1 \leq f_2 \leq k$ and $f_1 \neq f_2$
$\{e\} \cup \{\langle u,e,f\rangle, f \neq f_1\} \cup \{\langle v,e,g\rangle \mid g \neq f_2\}$,

where u and v are the two nodes incident with e.

EXACT COVER $\propto$ HITTING SET

The hitting set problem has sets U_i and elements s_j, such that $s_j \in U_i \Leftrightarrow u_i \in S_j$.

EXACT COVER $\propto$ STEINER TREE

$N = \{n_0\} \cup \{S_j\} \cup \{u_i\}$
$R = \{n_0\} \cup \{u_i\}$
$A = \{\{n_0,S_j\}\} \cup \{\{S_j,u_i\} \mid u_i \in S_j\}$
$w(\{n_0,S_j\}) = |S_j|$
$w(\{S_j,u_i\}) = 0$
$k = |\{u_i\}|$.

EXACT COVER $\propto$ 3-DIMENSIONAL MATCHING

Without loss of generality assume $|S_j| \geq 2$ for each j. Let $T = \{\langle i,j\rangle \mid u_i \in S_j\}$. Let α be an arbitrary one-one function from $\{u_i\}$ into T. Let π: T → T be a permutation such that, for each fixed j, $\{<i,j>| \; u_i \in S_j\}$ is a cycle of π.

$$U = \{<\alpha(u_i),<i,j>,<i,j>>| \; <i,j> \in T\} \cup \{<\beta,\sigma,\pi(\sigma)>| \text{ for all } i, \beta \neq \alpha(u_i)\} \quad .$$

EXACT COVER ∝ KNAPSACK

Let $d = |\{S_j\}| + 1$. Let $e_{ji} = \begin{cases} 1 & \text{if } u_i \in S_j \\ 0 & \text{if } u_i \notin S_j \end{cases}$. Let

$r = |\{S_j\}|$, $a_j = \sum e_{ji} d^{i-1}$ and $b = \frac{d^t - 1}{d-1}$.

KNAPSACK ∝ SEQUENCING

$p = r$, $T_i = P_i = a_i$, $D_i = b$.

KNAPSACK ∝ PARTITION

$$\begin{aligned} s &= r+2 \\ c_i &= a_i \; , \quad i = 1,2,\ldots,r \\ c_{r+1} &= b+1 \\ c_{r+2} &= (\sum_{i=1}^{r} a_i) + 1 - b \end{aligned}$$

PARTITION ∝ MAX CUT

$$\begin{aligned} N &= \{1,2,\ldots,s\} \\ A &= \{\{i,j\}| \; i \in N, j \in N, i \neq j\} \\ w(\{i,j\}) &= c_i \cdot c_j \\ W &= \lceil \tfrac{1}{4} \Sigma c_i^2 \rceil \end{aligned}$$

Some of the reductions exhibited here did not originate with the present writer. Cook (1971) showed that SATISFIABILITY ∝ SATISFIABILITY WITH AT MOST 3 LITERALS PER CLAUSE. The reduction

SATISFIABILITY ∝ CLIQUE

is implicit in Cook (1970), and was also known to Raymond Reiter. The reduction

NODE COVER ∝ FEEDBACK NODE SET

was found by the Algorithms Seminar at the Cornell University Computer Science Department. The reduction

NODE COVER ∝ FEEDBACK ARC SET

was found by Lawler and the writer, and Lawler discovered the reduction

EXACT COVER ∝ 3-DIMENSIONAL MATCHING

The writer discovered that the exact cover problem was reducible to the directed traveling-salesman problem on a digraph in which the arcs have weight zero or one. Using refinements of the technique used in this construction, Tarjan showed that

EXACT COVER $\propto$ DIRECTED HAMILTON CIRCUIT

and, independently, Lawler showed that

NODE COVER $\propto$ DIRECTED HAMILTON CIRCUIT .

The reduction

DIRECTED HAMILTON CIRCUIT $\propto$ UNDIRECTED HAMILTON CIRCUIT

was pointed out by Tarjan.

Below we list three problems in automata theory and language theory to which every complete problem is reducible. These problems are not known to be complete, since their membership in *NP* is presently in doubt. The reader unacquainted with automata and language theory can find the necessary definitions in Hopcroft and Ullman (1969).

EQUIVALENCE OF REGULAR EXPRESSIONS
INPUT: A pair of regular expressions over the alphabet $\{0,1\}$
PROPERTY: The two expressions define the same language.

EQUIVALENCE OF NONDETERMINISTIC FINITE AUTOMATA
INPUT: A pair of nondeterministic finite automata with input alphabet $\{0,1\}$
PROPERTY: The two automata define the same language.

CONTEXT-SENSITIVE RECOGNITION
INPUT: A context-sensitive grammar Γ and a string x
PROPERTY: x is in the language generated by Γ.

First we show that

SATISFIABILITY WITH AT MOST 3 LITERALS PER CLAUSE
$\propto$ EQUIVALENCE OF REGULAR EXPRESSIONS .

The reduction is made in two stages. In the first stage we construct a pair of regular expressions over an alphabet $\Delta = \{u_1, u_2, \ldots, u_n, \bar{u}_1, \bar{u}_2, \ldots, \bar{u}_n\}$. We then convert these regular expressions to regular expressions over $\{0,1\}$.

The first regular expression is $\Delta^n\Delta^*$ (more exactly, Δ is written out as $(u_1+u_2+\cdots+u_n+\bar{u}_1+\cdots+\bar{u}_n)$, and Δ^n represents n copies of the expression for Δ concatenated together). The second regular expression is

$$\Delta^n\Delta^* \cup \bigcup_{i=1}^{n} (\Delta^*u_i\Delta^*\bar{u}_i\Delta^* \cup \Delta^*\bar{u}_i\Delta^*u_i\Delta^*) \cup \bigcup_{h=1}^{r} \theta(D_h)$$

where

$$\theta(D_h) = \begin{cases} \Delta^*\bar{\sigma}_1\Delta^* & \text{if } D_h = \sigma_1 \\ \Delta^*\bar{\sigma}_1\Delta^*\bar{\sigma}_2\Delta^* \cup \Delta^*\bar{\sigma}_2\Delta^*\bar{\sigma}_1\Delta^* & \text{if } D_h = \sigma_1 \cup \sigma_2 \\ \Delta^*\bar{\sigma}_1\Delta^*\bar{\sigma}_2\Delta^*\bar{\sigma}_3\Delta^* \cup \Delta^*\bar{\sigma}_1\Delta^*\bar{\sigma}_3\Delta^*\bar{\sigma}_2\Delta^* & \\ \quad \cup \Delta^*\bar{\sigma}_2\Delta^*\bar{\sigma}_1\Delta^*\bar{\sigma}_3\Delta^* \cup \Delta^*\bar{\sigma}_2\Delta^*\bar{\sigma}_3\Delta^*\bar{\sigma}_1\Delta^* & \\ \quad \cup \Delta^*\bar{\sigma}_3\Delta^*\bar{\sigma}_1\Delta^*\bar{\sigma}_2\Delta^* \cup \Delta^*\bar{\sigma}_3\Delta^*\bar{\sigma}_2\Delta^*\bar{\sigma}_1\Delta^* & \\ & \text{if } D_h = \sigma_1 \cup \sigma_2 \cup \sigma_3 . \end{cases}$$

Now let m be the least positive integer $\geq \log_2|\Delta|$, and let ϕ be a 1-1 function from Δ into $\{0,1\}^m$. Replace each regular expression by a regular expression over $\{0,1\}$, by making the substitution $a \to \phi(a)$ for each occurrence of each element of Δ.

EQUIVALENCE OF REGULAR EXPRESSIONS $\propto$ EQUIVALENCE OF NONDETERMINISTIC FINITE AUTOMATA

There are standard polynomial-time algorithms [Salomaa (1969)] to convert a regular expression to an equivalent nondeterministic automaton. Finally, we show that, for any $L \in NP$,

$$L \underset{=}{\propto} \text{CONTEXT-SENSITIVE RECOGNITION} \quad .$$

Suppose L is recognized in time $p(\ )$ by a nondeterministic Turing machine. Then the following language $\tilde{L}$ over the alphabet $\{0,1,\#\}$ is accepted by a nondeterministic linear bounded automaton which simulates the Turing machine:

$$\tilde{L} = \{\#^{p(\lg(x))}x\#^{p(\lg(x))} \mid x \in L\} \quad .$$

Hence $\tilde{L}$ is context-sensitive and has a context-sensitive grammar $\tilde{\Gamma}$. Thus $x \in L$ iff

$$\tilde{\Gamma}, \#^{p(\lg(x))}x\#^{p(\lg(x))}$$

is an acceptable input to CONTEXT-SENSITIVE RECOGNITION.

We conclude by listing the following important problems in NP which are not known to be complete.

GRAPH ISOMORPHISM
INPUT: graphs G and G'
PROPERTY: G is isomorphic to G'.

NONPRIMES
INPUT: positive integer k
PROPERTY: k is composite.

LINEAR INEQUALITIES
INPUT: integer matrix C, integer vector d
PROPERTY: $Cx \geq d$ has a rational solution.

APPENDIX I

Notation and Terminology Used in Problem Specification

PROPOSITIONAL CALCULUS

$x_1,x_2,\ldots,x_n$	$u_1,u_2,\ldots,u_m$	propositional variables
$\bar{x}_1,\bar{x}_2,\ldots,\bar{x}_n$	$\bar{u}_1,\bar{u}_2,\ldots,\bar{u}_m$	complements of propositional variables
σ,σ_i		literals
$C_1,C_2,\ldots,C_p$	$D_1,D_2,\ldots,D_r$	clauses

$C_k \subseteq \{x_1,x_2,\ldots,x_n,\bar{x}_1,\bar{x}_2,\ldots,\bar{x}_n\}$

$D_\ell \subseteq \{u_1,u_2,\ldots,u_m,\bar{u}_1,\bar{u}_2,\ldots,\bar{u}_m\}$

A clause contains no complementary pair of literals.

SCALARS, VECTORS, MATRICES

Z	the positive integers
Z^p	the set of p-tuples of positive integers
Z_p	the set $\{0,1,\ldots,p-1\}$
k,W	elements of Z
$\langle x,y\rangle$	the ordered pair $\langle x,y\rangle$
(a_i) (y_j) d	vectors with nonnegative integer components
(c_{ij}) C	matrices with integer components

GRAPHS AND DIGRAPHS

$G = (N,A)$ $G' = (N',A')$ finite graphs

N,N' sets of nodes A,A' sets of arcs

s,t,u,v nodes $e,\{u,v\}$ arcs

$(X,\bar{X}) = \{\{u,v\} \mid u \in X \text{ and } v \in \bar{X}\}$ cut

If $s \in X$ and $t \in \bar{X}$, $(X,\bar{X})$ is a s-t cut.

$w\colon A \to Z$ $w'\colon A' \to Z$ weight functions

The weight of a subgraph is the sum of the weights of its arcs.

$H = (V,E)$ digraph V set of nodes, E set of arcs

$e,\langle u,v\rangle$ arcs

SETS

ϕ	the empty set
$\lvert S\rvert$	the number of elements in the finite set S
$\{S_j\}$ $\{T_h\}$ $\{U_i\}$	finite families of finite sets

Chapter 9
Lagrangian Relaxation for Integer Programming

Arthur M. Geoffrion

Introduction by *Arthur M. Geoffrion*

It is a pleasure to write this commentary because it offers an opportunity to express my gratitude to several people who helped me in ways that turned out to be essential to the birth of [8]. They also had a good deal to do with shaping my early career and, consequently, much of what followed.

The immediate event that triggered my interest in this topic occurred early in 1971 in connection with a consulting project I was doing for Hunt-Wesson Foods (now part of ConAgra Foods) with my colleague Glenn Graves. It was a distribution system design problem: how many distribution centers should there be and where, how should plant outputs flow through the DCs to customers, and related questions. We had figured out how to solve this large-scale MILP problem optimally via Benders Decomposition, a method that had been known for about a decade but had not yet seen practical application to our knowledge. This involved repeatedly solving a large 0-1 integer linear programming master problem in alternation with as many pure classical transportation subproblems as there were commodity classes. The master problem was challenging, and one day Glenn, who did all the implementation, came up with a new way to calculate conditional "penalties" to help decide which variable to branch on in our LP-based branch-and-bound approach.

I regularly taught a doctoral course in those days that covered, *inter alia*, the main types of penalties used by branch-and-bound algorithms. But after studying the math that Glenn used to justify his, I didn't see a connection to any of the penalties I knew about. I did, however, notice that Glenn made use of a Lagrangean term, and I was very familiar with Lagrangeans owing to my earlier work on solving discrete optimization problems via Lagrange multipliers [2] and on duality theory in nonlinear programming [6]. It often happens that a mathematical result can be

Arthur M. Geoffrion
UCLA Anderson School of Management, Los Angeles, USA
e-mail: ageoffri@anderson.ucla.edu

M. Jünger et al. (eds.), *50 Years of Integer Programming 1958-2008*,
DOI 10.1007/978-3-540-68279-0_9, © Springer-Verlag Berlin Heidelberg 2010

derived in quite distinct ways, and so it was in this case: I found that not only Glenn's penalties, but several other kinds of penalties could be derived in a unified way as shown in Sec. 4 of [8], and that numerous special problem structures could be exploited to produce additional penalties. This pleased me greatly, because I had a passion for trying to unify and simplify results that others had derived from disparate viewpoints, especially in the context of exploiting special problem structure. At that point, I knew that I had to write this up.

Shortly it became clear that what I later dubbed *Lagrangean relaxation* was useful for exploiting various kinds of special structures of integer programming problems in other ways besides penalties. In particular, it can be used to tailor most of the main operations found in branch-and-bound algorithms as explained in Sec. 3 of [8]. It also rendered obsolete the need for so-called *surrogate constraints* as explained in Sec. 5, and it can be used to derive new cutting planes as explained in Sec. 6. Some basic theory of Lagrangean relaxation had to be filled in, the subject of Sec. 3, and this drew importantly on my earlier work on nonlinear duality. I had a working paper version of [8] by late 1971, and in late 1972 presented the main results at a symposium in Germany. When Glenn and I wrote up the work surrounding the Hunt-Wesson Foods project, we included a comment in Sec. 3.1 of [7] on the branching penalties used in our implementation.

To explain more fully where [8] came from, I should also explain how the triggering Hunt-Wesson project came about, especially since this was my first industrial consulting engagement since obtaining my Ph.D. 5 years earlier (how does one boot a consulting practice?), and I should comment on the prior research that supported [8] and the novel solution method used for the Hunt-Wesson problem. First a few words about the origin of the project.

A very senior UCLA colleague of mine, Professor Elwood Buffa, opened a door in 1970 that would change my life in unforeseen ways. A former doctoral student of his, Dr. William Taubert, was then a vice president of Hunt-Wesson Foods, which had been struggling for years to rationalize its network of distribution centers. El knew that I was working on large-scale optimization methods that might conceivably apply to such problems, but he couldn't have known whether I could adapt those methods successfully. Neither did I. With no prompting whatever, he decided to recommend me to Bill Taubert as a consultant. El didn't have to take that risk, nor did Bill in hiring me. If I failed—which my inexperience as a consultant and unfamiliarity with distribution systems should have made the safest bet—it would have been an embarrassment to El, Bill, and UCLA.

But a streak of good luck ensued, leading to a successful project at Hunt-Wesson Foods, to many more consulting engagements in what is now called supply chain management, to the founding of a consulting and software firm that celebrates its 30th anniversary this year (2008), to the discovery of several important research problems that would occupy most of the rest of my career, and to an appreciation for the synergies of research, practice, and teaching that has shaped my professional life, including my service to TIMS and INFORMS.

If fortune favors the prepared mind, mine must have been prepared by my previous work on topics that proved useful not only for the Hunt-Wesson Foods project and what followed from it, but also for the paper which this commentary introduces. Especially my work on integer programming (especially [3, 4]), nonlinear duality theory [6], and large-scale optimization methods (especially [5]). Most of that work came about because of another door opened for me by my dissertation advisor at Stanford University, Professor Harvey Wagner.

When I accepted a job at UCLA's business school in 1964, just prior to finishing my thesis, Harvey suggested that I would benefit from being a day-a-week consultant at RAND Corporation, just a few miles from UCLA. He arranged it with Dr. Murray Geisler, head of RAND's Logistics Department. At that time, RAND was not far past its prime as the greatest think tank in the world, including its astonishing role as the fertile spawning ground or incubator of such important OR methods as discrete event and Monte Carlo simulation, dynamic programming, game theory, parts of inventory and logistics theory, network flow theory, and mathematical programming—linear, quadratic, stochastic, and integer. RAND was also a major contributor to the very early history of artificial intelligence, digital computing, the Internet, both systems analysis and policy analysis, the U.S. space program, and much more besides. That day a week, which lasted fairly steadily until the early 1970s, was disproportionately important to my early research life.

I had fine operations research colleagues at UCLA, but none did research in optimization, whereas at RAND I could interact with many staff members and A-list consultants who did, including Robin Brooks, Eric Denardo, Ben Fox, Ray Fulkerson, Glenn Graves, Al Madansky, Harry Markowitz, Bob Thrall, and Philip Wolfe. Moreover, at RAND I had excellent computer programming and clerical/data services (they had an IBM 7044 when I arrived), a full-service publication department that professionally edited and widely disseminated most of my research papers on optimization, and a good library that would even translate Russian-language articles at my request. I was in heaven there, and could not overstate the advantages gained from RAND's infrastructure and my second set of colleagues there as I launched my career.

It was at RAND that, very early in 1965, Murray handed me a somewhat beat up copy of Egon Balas' additive algorithm paper prior to its publication [1] (written while Egon was still in Rumania), and asked me to take a look at it since it was creating a stir. Thus commenced my enduring interest in integer programming. I recast this work as LP-based implicit enumeration in a limited-circulation manuscript dated August 23, 1965, published internally at RAND in September 1967 and externally about two years later [4]. Murray quickly arranged for Richard Clasen—an important early figure in mathematical programming in his own right—to be assigned to me to implement my first 0-1 integer programming code, the RIP30C incarnation of which RAND distributed externally starting mid-1968. Murray also arranged for others to assist me with the extensive numerical experiments.

My debt to RAND goes beyond even what is mentioned above: as a hotbed of OR for many years, RAND's influence on nearby UCLA for more than a decade prior to my arrival helped to build and shape an OR group with a vitality and local culture that provided a comfortable home for my entire career. The group's founding in the early 1950s originated independently of RAND, but its frequent interactions with RAND staff and consultants in its early days were of incalculable value; there are records of visits in the 1950s by Kenneth Arrow, Richard Bellman, Abe Charnes, Bill Cooper, George Dantzig, Merrill Flood, Ray Fulkerson, Alan Manne, Harry Markowitz, Oscar Morgenstern, Lloyd Shapley, Andy Vaszonyi, and dozens of others. (As an aside, the phrase "management sciences" was coined during a conversation between Melvin Salveson, the former Tjalling Koopmans student who founded UCLA's OR group, and Merrill Flood of RAND in September, 1953, the same month when Mel hosted on campus the first pre-founding meeting—attended by many RAND OR people—of what became The Institute of Management Sciences (TIMS) three months later.) Some taught courses as lecturers, and some even joined the faculty. By the time of my arrival, these interactions had largely tailed off, but they left a palpable tradition of creativity and excellence in my group that inspired my best efforts as an impressionable young faculty member.

Let me summarize. The paper following this commentary did not appear out of nowhere. It was enabled by multiple gifts of wisdom and kindness toward me by Harvey Wagner, who taught me how to do research and arranged for me to consult at RAND; by Elwood Buffa, who dropped my first and all-important consulting job in my lap; by Murray Geisler, who turned my attention to integer programming and arranged generous assistance in support of my research; and by my early colleague/mentors at UCLA, Jim Jackson (an OR pioneer whose contributions included "Jackson networks") and Jacob Marschak (a world-class economist), who helped shape my understanding of what it means to be a professor, arranged for me to be supported from the outset on their research grants, and then helped me obtain my own grants (from NSF starting in 1970 and ONR starting in 1972). I will always be grateful to these people for the important roles they played in my professional life.

References

1. E. Balas, *An additive algorithm for solving linear programs with zero-one variables*, Operations Research 13 (1965) 517–546.
2. R. Brooks and A.M. Geoffrion, *Finding Everett's Lagrange multipliers by linear programming*, Operations Research 14 (1966) 1149–1153.
3. A.M. Geoffrion, *Integer programming by implicit enumeration and Balas' method*, Review of the Society for Industrial and Applied Mathematics 9 (1967) 178–190.
4. A.M. Geoffrion, *An improved implicit enumeration approach for integer programming*, Operations Research 17 (1969) 437–454.
5. A.M. Geoffrion, *Elements of large-scale mathematical programming*, Management Science 16 (1970) 652–691.
6. A.M. Geoffrion, *Duality in nonlinear programming*, SIAM Review 13 (1971) 1–37.
7. A.M. Geoffrion and G.W. Graves, *Multicommodity distribution system design by Benders decomposition*, Management Science 20 (1974) 822–844.
8. A.M. Geoffrion, *Lagrangean relaxation for integer programming*, Mathematical Programming Study 2 (1974) 82–114.

The following article originally appeared as:

A.M. Geoffrion, *Lagrangian Relaxation for Integer Programming*, Mathematical Programming Study 2 (1974) 82–114.

Mathematical Programming Study 2 (1974) 82–114. North-Holland Publishing Company

LAGRANGEAN RELAXATION FOR INTEGER PROGRAMMING *

A.M. GEOFFRION

University of California, Los Angeles, Calif., U.S.A.

Received 25 January 1974
Revised manuscript received 26 August 1974

Taking a set of "complicating" constraints of a general mixed integer program up into the objective function in a Lagrangean fashion (with fixed multipliers) yields a "Lagrangean relaxation" of the original program. This paper gives a systematic development of this simple bounding construct as a means of exploiting special problem structure. A general theory is developed and special emphasis is given to the application of Lagrangean relaxation in the context of LP-based branch-and-bound.

1. Introduction

The general integer linear programming problem can be written as

$$\text{(P)} \qquad \min_{x \geqq 0} \text{imize } c\,x, \quad \text{subject to } A\,x \geqq b, \quad B\,x \geqq d, \quad x_j \text{ integer}, \quad j \in I,$$

where b, c and d are vectors, A and B are matrices of conformable dimensions, and the index set I denotes the variables required to be integer. The reason for distinguishing two types of constraints is that the second of these, $B\,x \geqq d$, is supposed to have special structure.

We define the *Lagrangean relaxation* of (P) relative to $A\,x \geqq b$ and a conformable nonnegative vector λ to be

$$(\text{PR}_\lambda) \qquad \min_{x \geqq 0} \text{imize } c\,x + \lambda\,(b - A\,x), \quad \text{subject to } B\,x \geqq d, \quad x_j \text{ integer}, \quad j \in I.$$

* An earlier version of this paper was presented at the IBM International Symposium on Discrete Optimization, Wildbad, Germany, October 30–November 1, 1972. This research was supported by the Office of Naval Research under Contract Number N00014-69-A-0200-4042 and by the National Science Foundation under Grants GP-26294 and GP-36090X.

The fruitful application of (PR_λ) in specific cases requires judicious partitioning of the constraints into the two types $Ax \geqq b$ and $Bx \geqq d$, and an appropriate choice of $\lambda \geqq 0$.

Lagrangean relaxation has been used by Held and Karp [20,21] in their highly successful work on the traveling-salesman problem; by Fisher [6] in his promising algorithm for scheduling in the presence of resource constraints and in his efficient machine scheduling algorithm [7]; by Fisher and Schrage [9] in their proposed algorithm for scheduling hospital admissions; by Ross and Soland [26] in their remarkably efficient algorithm for the generalized assignment problem; and by Shapiro [27] and Fisher and Shapiro [10] in the context of a group theoretic approach to pure integer programming. See also [8]. Other authors have also made special application of Lagrangean relaxation ideas implicitly if not explicitly in their work. Not to be forgotten is the general relevance of the literature on Lagrangean methods for nonconvex optimization (e.g., [2, 5, 18]).

The purpose of this paper is to develop the theory and explore the usefulness of Lagrangean relaxation in the context of branch-and-bound or implicit enumeration methods for (P). In contrast with most of the references just cited, our emphasis is on LP-based branch-and-bound algorithms rather than those whose bounding constructs do not involve linear programming. This is not to deny the great value of non-LP-based techniques for special problems, but rather to stress the as yet untapped potential of Lagrangean relaxation as a means of making the most widely used general purpose approach more efficient for problems with special structure. The development is intended for use at two levels. Pedagogically it strives for a unified exposition of a number of old and new developments in integer programming. As a research effort it aims to develop what appears to be a potent general approach to the design of improved algorithms for special classes of integer programs. Although the algorithmic context of this paper is the branch-and-bound approach to integer linear programs, it is clear that these ideas can also be applied to other classes of algorithms and problems.

The paper is organized as follows. The basic results concerning the relation between (P), (PR_λ) and related problems are collected in Section 2. Lagrangean duality theory turns out to play a surprisingly major role. In Section 3, a generic LP-based branch-and-bound approach for (P) is reviewed, and the basic uses and strategies of Lagrangean relaxation in this context are described. Section 4 derives the standard penalties of Driebeek [4] and Tomlin [28] from the viewpoint of Lagrangean relaxation, and several new penalties are developed. In Section 5, the concept of surrogate constraints as developed by Glover [17] and the author [12] is shown to be subsumed by the Lagrangean relaxation viewpoint. Section 6 derives cutting-planes based on Lagrangean relaxation, including some which utilize the penalties of Section 4. Concluding comments are given in Section 7.

Three simple examples for the special constraints $Bx \geqq d$ will now be introduced. They will serve in the sequel to illustrate general ideas and to emphasize that Lagrangean relaxation is intended to be specialized to particular problem structures. The final subsection of this Introduction summarizes the special notations and assumptions commonly used in the sequel.

1.1. *Three examples*

The Lagrangean relaxation (PR_λ) must be much simpler to solve than (P) itself in order for it to yield any computational advantage. It should admit a closed form solution or be solvable by an efficient specialized algorithm. Thus the constraints $Bx \geqq d$ must possess considerable special structure. Three of the simplest possible examples of such structure are as follows. They will be referred to repeatedly in the sequel.

Example 1. The constraints $Bx \geqq d$ specify only upper bounds on some or all of the variables. For instance, in 0–1 programming problems the integer variables possess upper bounds of unity. It is easy to see that the optimal solution of (PR_λ) can be written down by inspection of the signs of the collected coefficient vector of x, namely $(c - \lambda A)$.

Example 2. The constraints $Bx \geqq d$ are as in Example 1 but also include some generalized upper bounding constraints of the form

$$\sum_{j \in J_k} x_j = 1, \qquad k = 1, 2, \ldots, K, \tag{1}$$

where $J_1, \ldots, J_K$ are disjoint subsets of I. Such constraints perform a "multiple choice" function. The optimal solution of (PR_λ) can again be written down by inspection, with a search for the smallest $(c - \lambda A)_j$ now being necessary over each subset J_k.

Example 3. The constraints $Bx \geqq d$ are as in Example 1 but also include some constraints of the form

$$\sum_{j \in J_k} \beta_{kj} x_j \leqq \beta_{kk} x_k, \qquad k = 1, \ldots, K, \tag{2}$$

where the K subsets $\{k, J_k\}$ are disjoint, $x_1, \ldots, x_K$ are 0–1 variables, the variables in J_k are continuous-valued, and all β coefficients are strictly positive. This type of constraint typically arises in location and expansion models. In the familiar capacitated plant location problem for example, x_k is 1 or 0, according to whether or not a plant of capacity β_{kk} is built at the kth site, x_j for $j \in J_k$ corresponds to the amounts shipped from plant site k to various destinations, and the β_{kj}'s are all unity. The Lagrangean relaxation (PR_λ) can be solved easily because it separates into K independent problems of the form

$$\begin{aligned} &\text{minimize } \sum_{j \in J_k} (c - \lambda A)_j x_j + (c - \lambda A)_k x_k, \\ &\text{subject to } \sum_{j \in J_k} \beta_{kj} x_j \leqq \beta_{kk} x_k, \\ &\qquad 0 \leqq x_j \leqq u_j, \qquad j \in J_k, \\ &\qquad x_k = 0 \text{ or } 1, \end{aligned} \tag{3^k_λ}$$

where u_j is the upper bound on variable x_j. If $x_k = 0$, it follows from the positivity of β_{kj} that $x_j = 0$ must hold for all $j \in J_k$. If $x_k = 1$, (3^k_λ) becomes a trivial "continuous knapsack problem" with bounded variables. The best of the solutions obtained under the two cases $x_k = 0$ and $x_k = 1$ yields the true optimal solution of (3^k_λ). From these K solutions one may directly assemble the optimal solution of (PR_λ).

These three examples are among the simplest types of special constraints $Bx \geqq d$ for which the associated Lagrangean relaxation can be optimized very efficiently. Whereas closed form solutions are available for these examples, other applications may call for specialized algorithms of a less trivial sort. In most practical applications of integer programming there are several obvious and tractable choices for the constraints to be designated as $Bx \geqq d$. In Held and Karp's excellent work on the traveling-salesman problem [20, 21], for example, (PR_λ) is a minimum spanning "1-tree" problem for which highly efficient algorithms are available. And in Fisher and Schrage's algorithm for hospital admissions scheduling [9], (PR_λ) separates into a relatively simple scheduling problem for each patient.

1.2. *Notation and assumptions*

Notation and terminology is generally standard and consistent with that of [14], a survey paper containing additional background material. However, *the reader should memorize the following peculiar notations:* if $(\cdot)$ is an optimization problem, then $v(\cdot)$ is its optimal value, $F(\cdot)$ is its set of feasible solutions, and $(\bar{\cdot})$ refers to the same problem with all integrality conditions on the variables dropped; the vector $\bar{\lambda}$ denotes an optimal multiplier vector (dual solution) associated with the constraints $Ax \geqq b$ for the ordinary linear program $(\bar{\mathrm{P}})$.

We adopt the convention that the optimal value of an infeasible optimization problem is $+\infty$ (resp. $-\infty$) if it is a minimizing (resp. maximizing) problem. The inner product of two vectors, be they row or column, is denoted simply by their juxtaposition.

Two benign assumptions are made throughout this paper in the interest of decluttering the exposition, except where explicitly stated to the contrary. The first is that the nonspecial constraints $Ax \geqq b$ are all inequality constraints. If some of these constraints were given as *equalities*, then the corresponding components of λ would not be required to be nonnegative. This is the only change required to accommodate equality constraints. The second assumption is that the special constraints $Bx \geqq d$ include upper bounds on all variables. This obviates the need for special treatment of the case where (P) or one of its relaxations has optimal value equal to $-\infty$, and also permits certain notational economies. This assumption is consistent with the vast majority of potential applications. It is a simple exercise to allow for its absence in all of the results to follow.

2. Theory of Lagrangean relaxation

The term *relaxation* is used in this paper in the following sense: a minimizing problem (Q) is said to be a relaxation of a minimizing problem (P) if $F(\mathrm{Q}) \supseteq F(\mathrm{P})$ and the objective function of (Q) is less than or equal to that of (P) on $F(\mathrm{P})$. Clearly (PR_λ) is a relaxation in this sense for all $\lambda \geqq 0$, for the extra Lagrangean term $\lambda(b - Ax)$ in the objective function of (PR_λ) must be nonpositive when $Ax \geqq b$ is satisfied. Notice that the common practice of relaxation by simply throwing away some of the constraints is equivalent to Lagrangean relaxation with $\lambda = 0$. Permitting $\lambda \neq 0$ (but always $\geqq 0$) allows the relaxation to be tighter.

The potential usefulness of any relaxation of (P), and of a Lagrangean relaxation in particular, is largely determined by how near its optimal value is to that of (P). This furnishes a criterion by which to measure the "quality" of a particular choice for λ. The ideal choice would be to take λ as an optimal solution of the (concave) program

$$\text{(D)} \qquad \underset{\lambda \geqq 0}{\text{maximize}}\ v(\text{PR}_\lambda),$$

which is designated by (D) because it coincides with the formal Lagrangean dual of (P) with respect to the constraints $A\,x \geqq b$ (see, e.g., [13]). This problem in turn is intimately linked to the following relaxation of (P):

$$\text{(P*)} \qquad \underset{x}{\text{minimize}}\ c\,x,$$

$$\text{subject to } A\,x \geqq b,$$
$$x \in \text{Co}\,\{x \geqq 0\colon B\,x \geqq d \text{ and } x_j \text{ integer}, j \in I\},$$

where Co denotes the convex hull of a set. It may be difficult to express the convex hull in (P*) as an explicit set of linear constraints, but in principle this is always possible and so (P*) may be regarded as a linear program. In fact, as we shall see, (P*) and (D) are essentially LP duals. An optimal multiplier vector corresponding to $A\,x \geqq b$ will be denoted by λ^* when (P*) has finite optimal value.

Theorem 1 describes some of the basic relationships between (P), (PR_λ), (D), (P*), and $(\bar{\text{P}})$ (the usual LP relaxation which drops the integrality requirements).

Theorem 1. (a)

(a) $$F(\bar{\text{P}}) \supseteq F(\text{P*}) \supseteq F(\text{P}), \qquad F(\text{PR}_\lambda) \supseteq F(\text{P}),$$
$$v(\bar{\text{P}}) \leqq v(\text{P*}) \leqq v(\text{P}), \qquad v(\text{PR}_\lambda) \leqq v(\text{P}) \quad \textit{for all } \lambda \geqq 0.$$

(b) *If* $(\bar{\text{P}})$ *is feasible, then* $v(\bar{\text{P}}) \leqq v(\text{PR}_{\bar{\lambda}})$.

(c) *If for a given* λ *a vector* x *satisfies the three conditions*

(i) x *is optimal in* (PR_λ),
(ii) $A\,x \geqq b$,
(iii) $\lambda\,(b - A\,x) = 0$,

then x *is an optimal solution of* (P). *If* x *satisfies* (i) *and* (ii) *but not* (iii), *then* x *is an* ε*-optimal solution of* (P) *with* $\varepsilon = \lambda\,(A\,x - b)$.

(d) *If* (P*) *is feasible, then*

$$v(\text{D}) \equiv \max_{\lambda \geqq 0} v(\text{PR}_\lambda) = v(\text{PR}_{\lambda^*}) = v(\text{P*}).$$

Proof. Parts (a) and (c) are very easy. Let $(\overline{\mathrm{P}})$ be feasible. Then it has finite optimal value, for $Bx \geqq d$ includes upper bounds on all variables, and

$$\begin{aligned} v(\overline{\mathrm{P}}) &= \max_{\lambda \geqq 0} v(\overline{\mathrm{PR}_\lambda}) && \text{(by the dual theorem of linear programming)[1]}, \\ &= v(\overline{\mathrm{PR}_{\bar\lambda}}) && \text{(by the definition of } \bar\lambda), \\ &\leqq v(\mathrm{PR}_{\bar\lambda}) && \text{(because } F(\overline{\mathrm{PR}_{\bar\lambda}}) \supseteq F(\mathrm{PR}_{\bar\lambda})). \end{aligned}$$

This proves part (b). An identical argument (the third portion is not needed) applied to (P*) yields the conclusion of part (d) if one uses the following observation in the obvious way:

$$v(\mathrm{PR}_\lambda) = [\min_x cx + \lambda(b - Ax),$$
$$\text{s.t. } x \in \mathrm{Co}\{x \geqq 0 \colon Bx \geqq d \text{ and } x_j \text{ integer}, j \in I\}],$$

which holds because the minimum value of a linear function over any compact set is not changed if the set is replaced by its convex hull.

A few comments are in order. Part (a) simply records the most obvious relations between (P) and its relaxations $(\overline{\mathrm{P}})$, (P*) and (PR_λ). Part (b) shows that $\bar\lambda$, an immediate by-product of optimally solving the standard LP relaxation $(\overline{\mathrm{P}})$, yields a Lagrangean relaxation that is at least as good as $(\overline{\mathrm{P}})$ itself (hopefully it will be better). Part (c) indicates the well-known conditions under which a solution of a Lagrangean relaxation is also optimal or near-optimal in (P). This is in recognition of the fact that Lagrangean relaxation is of interest not only for the lower bounds it yields on $v(\mathrm{P})$, but also for the possibility that it may actually yield an optimal or near-optimal solution of (P). It follows, incidentally, that (PR_λ) can yield in this manner a proven ε-optimal solution ($\varepsilon \geqq 0$) of (P) only if $v(\mathrm{PR}_\lambda) \geqq v(\mathrm{P}) - \varepsilon$. Thus the provable quality of the feasible solutions obtainable from Lagrangean relaxation by invoking part (c) is limited by the gap (if any) between $v(\mathrm{P})$ and $v(\mathrm{D})$. Part (d) establishes that Lagrangean relaxation can do as well as, but no better than (P*). Thus, the position of $v(\mathrm{P}^*)$ in the interval $[v(\overline{\mathrm{P}}), v(\mathrm{P})]$ is the question of central concern when analyzing the potential value of Lagrangean relaxation applied to a particular problem class.

[1] We have taken here the "partial" dual of $(\overline{\mathrm{P}})$ with respect to the constraints $Ax \geqq b$, rather than the "full" dual customarily used in linear programming. See [13] (especially Sec. 6.1) for an account of this generalization of the traditional duality theory. It is easily verified that $\bar\lambda$ is a bona fide optimal solution of the partial dual even though it may be defined in terms of the full dual.

It bears emphasis that the conclusion of part (d) is really a simple consequence of the fact that (P*) and (D) are essentially LP duals of one another. The true dual of (P*) when multipliers are introduced just for the $A\,x \geqq b$ constraints is

$$\underset{\lambda \geqq 0}{\text{maximize}}\ \Big[\min_{x}\ c\,x + \lambda\,(b - A\,x),$$
$$\text{s.t.}\quad x \in \mathrm{Co}\,\{x \geqq 0\colon B\,x \geqq d \text{ and } x_j \text{ integer}, j \in I\}\Big].$$

But, as observed in the proof of part (d), the maximand of this problem equals $v(\mathrm{PR}_\lambda)$, so (D) *must have the same optimal value and optimal solution set as the true dual of* (P*). Thus one may invoke most of the rich optimality/duality theory for linear programming to say much more about the relationship between (P*) and (D) than is said in Theorem 1 (d). For instance, one may assert when (P*) is feasible that the set of its optimal multipliers coincides with the set of optimal solutions of (D) and also with the negative of the set of subgradients at $y = 0$ of its (convex) b-perturbation function

$$\phi_b^*(y) \overset{\mathrm{d}}{=} \Big[\inf.\ c\,x,$$
$$\text{s.t.}\quad A\,x \geqq b - y,$$
$$x \in \mathrm{Co}\,\{x \geqq 0\colon B\,x \geqq d \text{ and } x_j \text{ integer}, j \in I\}\Big],$$

(see, e.g., [13, Th. 1 and 3]).

Theorem 1 leaves open two important questions: the relations between $v(\overline{\mathrm{P}})$ and $v(\mathrm{P}^*)$ and between $v(\mathrm{P}^*)$ and $v(\mathrm{P})$. These relations are taken up in the next two theorems.

Theorem 2 shows that $v(\overline{\mathrm{P}}) = v(\mathrm{P}^*)$ when the following holds:

Integrality Property. The optimal value of (PR_λ) is not altered by dropping the integrality conditions on its variables, i.e., $v(\mathrm{PR}_\lambda) = v(\overline{\mathrm{PR}_\lambda})$ for all $\lambda \geqq 0$.

Theorem 2. *Let* $(\overline{\mathrm{P}})$ *be feasible and* (PR_λ) *have the Integrality Property. Then* (P*) *is feasible and*

$$v(\overline{\mathrm{P}}) = v(\mathrm{PR}_{\bar{\lambda}}) = v(\mathrm{D}) = v(\mathrm{PR}_{\lambda_*}) = v(\mathrm{P}^*).$$

Proof. In view of Theorem 1 (b), (d), it is enough to show that $v(\overline{\mathrm{P}}) = v(\mathrm{P}^*)$. We have

$$v(\overline{\mathrm{P}}) = \max_{\lambda \geqq 0} v(\overline{\mathrm{PR}_\lambda}) \quad \text{(by duality)},$$
$$= \max_{\lambda \geqq 0} v(\mathrm{PR}_\lambda) \quad \text{(by the Integrality Property)},$$

$$= \max_{\lambda \geqq 0} [\min_x cx + \lambda(b - Ax),$$

$$\text{s.t.} \quad x \in \text{Co}\{x \geqq 0: Bx \geqq d \text{ and } x_j \text{ integer}, j \in I\}]$$

(by the observation used in the proof of Theorem 1 (d)),

$$= v(\text{P}^*) \qquad \text{(by duality).}$$

Notice that the feasibility of (P*) is a consequence of the fact that its dual has finite optimal value.

Thus Lagrangean relaxation can do no better than the standard LP relaxation ($\overline{\text{P}}$) when the constraint partition $Ax \geqq b$, $Bx \geqq d$ is such that the Integrality Property holds.[2] The best choice of λ for (PR_λ) is then $\bar{\lambda}$ from ($\overline{\text{P}}$). In this circumstance, Lagragean relaxation seems to be of questionable value unless a near-optimal solution of (D) can be found by specialized means more rapidly than ($\overline{\text{P}}$) can be solved by linear programming methods. Generally it is more promising to use Lagrangean relaxations for which the Integrality Property does not hold.

The Integrality Property clearly holds for Examples 1 and 2 (one may assume without loss of generality that the upper bounds on the integer variables are integers), but it does not hold for Example 3. The presence or absence of the Integrality Property is evident in many applications upon inspection of (PR_λ) in light of the special structure of the constraints $Bx \geqq d$. In other applications one may be able to appeal to the total unimodularity characterization of natural integer solutions of linear programming problems (e.g., [30]).

We now turn to the relationship between $v(\text{P}^*)$ [or $v(\text{D})$] and $v(\text{P})$. A sufficient condition for $v(\text{P}^*) = v(\text{P})$ obviously is

$$F(\text{P}^*) = \text{Co}[F(\text{P})],$$

but this is likely to be difficult to verify in specific cases because the "integer polyhedron" is a notoriously difficult object to study.

Most of what is known about the relationship in question is a consequence of the fact that (D) is the formal Lagrangean dual of (P) with respect to the constraints $Ax \geqq b$. Careful examination of Lagrangean duality theory shows that many of the results do not require convexity of the primal

[2] This fact has been noted by Nemhauser and Ullman [25] in the special context of Example 1.

problem. For instance, convexity is not used in the proofs of the key Lemmas 3, 4 and 5 of [13]. These results yield Theorem 3, which uses the following definitions. The *b-perturbation function* associated with (P) is defined as

$$\phi_b(y) \stackrel{\mathrm{d}}{=} [\inf_{x \geqq 0} c\,x, \quad \text{s.t. } A\,x \geqq b - y, \quad B\,x \geqq d, \quad x_j \text{ integer}, \quad j \in I].$$

A vector γ conformable with y is said to be a *global subgradient* of ϕ_b at $y = 0$ (assuming $\phi_b(0) \equiv v(\mathrm{P})$ is finite) if

$$\phi_b(y) \geqq v(\mathrm{P}) + \gamma\,y \quad \text{for all } y.$$

The adjective "global" is used to emphasize that the subgradient definition used here relates to a global rather than local aspect of ϕ_b (which is generally nonconvex).

Theorem 3. *Assume that* (P) *is feasible (and therefore has an optimal solution, since all variables are bounded).*

(a) *The following are equivalent:*

(1) $v(\mathrm{P}) = v(\mathrm{D})$.

(2) *There exists a global subgradient of* ϕ_b *at* $y = 0$.

(3) *There exists a pair* (x, λ) *satisfying* $\lambda \geqq 0$ *and conditions* (i), (ii) *and* (iii) *of Theorem* 1(c).

(b) *If* $v(\mathrm{P}) = v(\mathrm{D})$, *then each optimal solution of* (D) *is the negative of a global subgradient of* ϕ_b *at* $y = 0$ *and conversely, and any such solution* λ^* *yields the set of all optimal solutions of* (P) *as the vectors* x *which satisfy conditions* (i), (ii) *and* (iii) *of Theorem* 1 (c) *with* $\lambda = \lambda^*$.

The most interesting aspect of Theorem 3 is the criterion for the equality $v(\mathrm{P}) = v(\mathrm{D})$ in terms of the existence of a global subgradient of ϕ_b at the origin and the identification of these subgradients with the solutions of (D). The theorem also confirms that Lagrangean relaxation does indeed yield the optimal solutions of (P) when $v(\mathrm{P}) = v(\mathrm{D})$, via the optimality conditions of Theorem 1 (c).

The b-perturbation function ϕ_b thus emerges as a key object for study if one wishes to understand when $v(\mathrm{P}) = v(\mathrm{D})$ is likely to hold. What is known about ϕ_b? Clearly it is nonincreasing. It can also be shown to be lower semicontinuous. It is piecewise-linear and convex over any region where the optimal values of the integer variables stay constant, for in such a region the perturbed (P) reduces to a perturbed linear program. And ϕ_b is obviously bounded below by the piecewise-linear convex perturbation function ϕ_b^* defined earlier for (P*). In fact, this last observation can be strengthened to assert that ϕ_b^* is actually the *best* possible convex function which nowhere exceeds ϕ_b in value. This geometrically obvious but important result is stated more precisely as follows.

Theorem 4. *The b-perturbation function ϕ_b^* associated with* (P*) *is precisely the lower convex envelope of the b-perturbation function ϕ_b associated with* (P).

Proof. An alternative way of phrasing the result is to say that the epigraph of ϕ_b^* is the convex hull of the epigraph of ϕ_b; that is, $\text{Epi}\,[\phi_b^*] = \text{Co}\,\{\text{Epi}\,[\phi_b]\}$. Clearly,

$$\begin{aligned}\text{Epi}\,[\phi_b] &\overset{\text{d}}{=} \{(\mu, y): \mu \geqq \phi_b(y)\}\\ &= \{(\mu, y): \mu \geqq c\,x \text{ and } b - A\,x \leqq y \text{ for some } x \in X\},\end{aligned}$$

where

$$X \overset{\text{d}}{=} \{x \geqq 0: B\,x \geqq d \text{ and } x_j \text{ integer for } j \in I\},$$

and similarly for $\text{Epi}\,[\phi_b^*]$ with X replaced by $\text{Co}\,\{X\}$. Suppose that $(\bar{\mu}, \bar{y}) \in \text{Epi}\,[\phi_b^*]$. Then $\bar{\mu} \geqq c\,\bar{x}$ and $b - A\,\bar{x} \leqq \bar{y}$ for some $\bar{x} \in \text{Co}\,\{X\}$. Let $\bar{x} = \sum_h \theta_h x^h$, where $x^h \in X$, $\theta_h \geqq 0$ for all h and $\sum_h \theta_h = 1$. Define $\mu^h = c\,x^h$ and $y^h = b - A\,x^h$ for all h. Clearly $(\mu^h, y^h) \in \text{Epi}\,[\phi_b]$ for all h. Hence $\sum_h \theta_h(\mu^h, y^h) \in \text{Co}\,\{\text{Epi}\,[\phi_b]\}$. But

$$\begin{aligned}\sum_h \theta_h(\mu^h, y^h) &= \left(\textstyle\sum_h \theta_h \mu^h, \sum_h \theta_h y^h\right)\\ &= \left(\textstyle\sum_h \theta_h c\,x^h, \sum_h \theta_h(b - A\,x^h)\right) = (c\,\bar{x}, b - A\,\bar{x}) \leqq (\bar{\mu}, \bar{y})\end{aligned}$$

and so $(\bar{\mu}, \bar{y})$ must also be in $\text{Co}\,\{\text{Epi}\,[\phi_b]\}$. This shows that $\text{Epi}\,[\phi_b^*] \subseteq \text{Co}\,\{\text{Epi}\,[\phi_b]\}$. Now suppose $(\bar{\mu}, \bar{y}) \in \text{Co}\,\{\text{Epi}\,[\phi_b]\}$. Let $(\bar{\mu}, \bar{y}) = \sum_h \theta_h (\mu^h, y^h)$, where $(\mu^h, y^h) \in \text{Epi}\,[\phi_b]$, $\theta_h \geqq 0$ for all h and $\sum_h \theta_h = 1$. Let x^h be any point in X satisfying $\mu^h \geqq c\,x^h$ and $b - A\,x^h \leqq y^h$. Then

$$\sum_h \theta_h \mu^h \geqq \sum_h \theta_h c\,x^h = c\,\bar{x},$$

$$\sum_h \theta_h y^h \geqq \sum_h \theta_h(b - A\,x^h) = b - A\,\bar{x},$$

where $\bar{x} \stackrel{d}{=} \sum_h \theta_h x^h$. Thus $(\bar{\mu}, \bar{y}) \geqq (c\bar{x}, b - A\bar{x})$ with $\bar{x} \in \text{Co}\{X\}$, which shows that $(\bar{\mu}, \bar{y}) \in \text{Epi}[\phi_b^*]$. This completes the proof.

This is the central connection between (P) and (P*) – actually, between two parameterized families of problems of which (P) and (P*) are members of special significance. The duality gap (if any), $v(P) - v(D)$, is precisely equal to the difference between the b-perturbation function of (P) and its lower convex envelope, both evaluated at the origin. This characterization provides the basis for a qualitative understanding of duality gaps—and hence of the potential of Lagrangean relaxation—when applied to specific classes of problems with reference to salient characteristics of the data.

Some of these ideas are illustrated in Fig. 1 for a hypothetical mixed integer program with but a single A-type constraint (so that y is a scalar). Suppose that only two sets of values for the integer variables enter into an optimal solution of (P) as b varies. The piecewise-linear and convex b-perturbation functions for the two corresponding linear programs (with the integer variables fixed) are drawn as light lines. One of these linear programs becomes infeasible for $y \leqq y^1$, while the other becomes infeasible for $y \leqq y^2$. The pointwise minimum of these two functions is $\phi_b(y)$, which is superimposed as a heavy line. The lower convex envelope of $\phi_b(y)$, namely $\phi_b^*(y)$, is superimposed as a line with alternating dots and dashes. It is clear that there is no duality gap (difference between $\phi_b(y)$ and $\phi_b^*(y)$) for $y^2 \leqq y \leqq y^3$ or $y^4 \leqq y$. A global subgradient of ϕ_b at $y = 0$ will exist

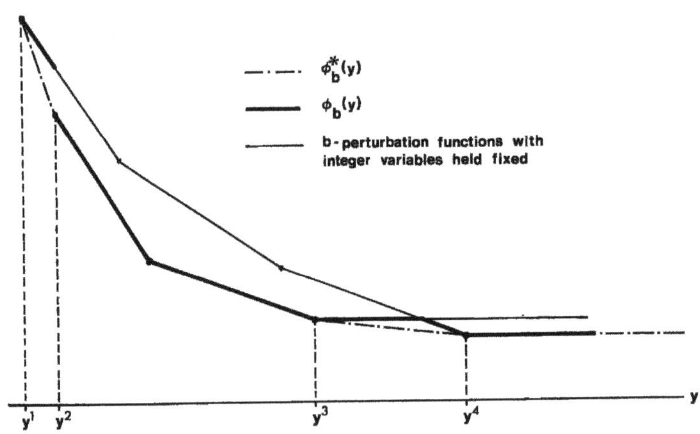

Fig. 1. Hypothetical illustration of b-perturbation functions.

(any subgradient of ϕ_b^* at $y = 0$ will do) if $y = 0$ falls in either of these intervals. If $y = 0$ falls between y^1 and y^2 or between y^3 and y^4, on the other hand, there will be a gap and no global subgradient of ϕ_b at $y = 0$ will exist.

We note in closing that the duality gap tends to be rather small for the class of problems with which we have numerical experience, namely capacitated facility location problems with additional constraints. The special constraints are as in Example 3. For four practical problems the values averaged as follows (after normalization via division by 0.01 v(P)):

$$\begin{aligned} v(\mathrm{P}) &= 100.00, \\ v(\mathrm{D}) &= 99.93, \\ v(\mathrm{PR}_{\bar{\lambda}}) &= 98.97, \\ v(\overline{\mathrm{P}}) &= 97.46. \end{aligned}$$

Notice that the duality gap is small by comparison with the gap between the integer problem and its usual LP relaxation, and that the LP multipliers $\bar{\lambda}$ yield a Lagrangean relaxation quite a bit better than the LP relaxation itself. See [15] for further details.

3. The use of Lagrangean relaxation in LP-based branch-and-bound

Virtually all of the current generally successful integer linear programming algorithms are of the branch-and-bound type with linear programming as the primary source of bounds [14]. This section and those to follow discuss the use of Lagrangean relaxation as a device for possibly improving the efficiency of such algorithms for special classes of problems.

A brief review of the usual LP-based branch-and-bound approach to (P) is necessary at this point. The terminology adopted is that of [14] which can be consulted for further details. At any given time there is a list of so-called *candidate problems*, each of which is simply (P) with certain additional "separation" constraints appended. The union of the feasible regions of the candidate problems constitutes a partition of the unenumerated portion of the feasible region of (P). There is also a number z^* representing the objective value of the *incumbent*, the best currently known feasible solution of (P) (initially z^* can be taken to be a suitably large number). The primary iterative step is to select one of the candidate problems, say (CP), and to examine it for the existence of a feasible solution of (P) with value better than z^*. The examination may be conclusive or inconclusive, depending on how much effort is expended; the usual practice involves solving the linear program $(\overline{\mathrm{CP}})$, which ignores all integrality conditions on the variables of (CP). A conclusive examination is one for which the outcome is

(i) that (CP) is infeasible (e.g. $(\overline{\text{CP}})$ is infeasible), or
(ii) that $v(\text{CP}) \geqq z^*$ (e.g. $v(\overline{\text{CP}}) \geqq z^*$), or
(iii) that $v(\text{CP}) < z^*$ and an optimal solution of (CP) is at hand (e.g. the optimal solution $\bar{x}$ of $(\overline{\text{CP}})$ happens to satisfy the integrality conditions); this solution replaces the current incumbent and z^* is updated.

Then (CP) is said to be *fathomed* and is deleted from the list of candidate problems. Otherwise, (CP) is not fathomed and must be separated into two or more simpler candidate (sub)problems to be added to the list. This is accomplished via mutually exclusive and exhaustive separation constraints. The usual practice (cf. [3]) is to select a particular *separation variable* $j_0 \in I$ and to invoke an interval dichotomy on its range. For instance, for $j_0 = 3$ one subproblem might receive the new constraint $x_3 \leqq 2$ and the other the new constraint $x_3 \geqq 3$. Candidate problems continue to be examined in this fashion, with fathoming or separation occurring each time, until the list of candidate problems is exhausted.

It should be evident that a Lagrangean relaxation of (CP), say (CPR_λ), is just as amenable as the usual linear programming relaxation $(\overline{\text{CP}})$ as a device for examining candidate problems: the infeasibility of (CPR_λ) implies that of (CP); $v(\text{CPR}_\lambda) \geqq z^*$ implies $v(\text{CP}) \geqq z^*$; and an optimal solution of (CPR_λ), say x^{R}, is optimal in (CP) if it is feasible in (CP) and satisfies complementary slackness (see Theorem 1 (c)). Note that if x^{R} is feasible in (CP) but does not satisfy complementary slackness, it may still improve on the incumbent, in which case it should be used to update the incumbent and z^* even though (CP) is not fathomed. In cases where x^{R} is not feasible in (CP) it may be worth trying to adjust it in some problem-specific manner so as to gain feasibility and, hopefully, to improve thereby on the incumbent. This is exactly the same tactic as is commonly used with $(\overline{\text{CP}})$ when the (fractional) LP solution is rounded to satisfy integrality in the hope of obtaining an improved feasible solution.

The usual linear programming relaxation $(\overline{\text{CP}})$ is also used commonly to derive conditional bounds for use in guiding separation, for tagging newly created candidate subproblems with lower bounds on their optimal value, and for reducing the range restrictions on integer variables without sacrificing optimality. Lagrangean relaxations of (CP) can be used for these same purposes. Suppose that some variable $j \in I$ has a fractional value in the LP solution $\bar{x}$ of $(\overline{\text{CP}})$. We are interested in lower bounds on $v(\text{CP} \mid x_j \leqq [\bar{x}_j])$ and $v(\text{CP} \mid x_j \geqq [\bar{x}_j] + 1)$, where " $|$ " signifies that the constraint following it is appended to the problem, and $[\bar{x}_j]$ stands for the integer part of $\bar{x}_j$. Such conditional bounds are given, respectively, by

$$v_D(j) \stackrel{d}{=} v(\text{CPR}_\lambda \mid x_j \leqq [\bar{x}_j]),$$
$$v_U(j) \stackrel{d}{=} v(\text{CPR}_\lambda \mid x_j \geqq [\bar{x}_j] + 1). \tag{4}$$

If $v_D(j) \geqq z^*$ holds, then the lower limit for x_j obviously can be tightened to $[\bar{x}_j] + 1$. Similarly, $v_U(j) \geqq z^*$ implies that the upper range restriction can be lowered to $[\bar{x}_j]$. It is even possible that both $v_D(j) \geqq z^*$ and $v_U(j) \geqq z^*$ hold, in which case it is clear that (CP) is fathomed. The bounds (4) can also be used to guide separation in the event that (CP) is not fathomed. Let $V_D(j)$ and $V_U(j)$ be computed for every $j \in I$ such that $\bar{x}_j$ is fractional. One appealing choice for the separation variable would be the one which maximizes the larger of $V_D(j)$ and $V_U(j)$ over all eligible j. Several successful integer programming codes have employed an analogous criterion based on $(\overline{\text{CP}})$ rather than (CPR$_\lambda$). Once a separation variable j_0 is selected, $V_D(j_0)$ and $V_U(j_0)$ yield lower bounds for future reference on the newly created candidate subproblems.

The computation of conditional bounds like (4) is taken up in more detail in Section 4. We note here only that the bounding problems have the same structure as (CPR$_\lambda$) since we have assumed that range restrictions on all variables are incorporated into the special constraints $Bx \geqq d$, just as (CPR$_\lambda$) will have the same structure as (PR$_\lambda$) if, as is usually the case, the separation constraints employed are simple range restrictions on the variables.

Thus we see that Lagrangean relaxation can be used for the standard branch-and-bound tasks of fathoming, generating improved feasible solutions, range reduction, and guiding separation. It can also be used to derive surrogate constraints and cutting-planes. These uses are taken up in Section 5 and 6.

We turn now to a discussion of the strategy questions which arise in connection with the use of (CPR$_\lambda$) as an adjunct to $(\overline{\text{CP}})$. The two main questions concern how λ is to be chosen and whether (CPR$_\lambda$) should be used before or after or even in place of $(\overline{\text{CP}})$. These questions cannot be answered definitively in general, but an obviously important consideration is whether or not the Integrality Property defined in Section 2 holds for the particular constraint partition under consideration.

Suppose the Integrality Property does hold. Then (CPR$_\lambda$) can be infeasible only if $(\overline{\text{CP}})$ is infeasible, and if $(\overline{\text{CP}})$ is feasible, then by Theorem 2 it must yield the best possible choice of λ for (CPR$_\lambda$) and $v(\text{CPR}_{\lambda*}) = v(\overline{\text{CP}})$. Thus (CPR$_\lambda$) cannot fathom (CP) by infeasibility or by value unless ($\overline{\text{CP}}$) would also do so. One can also show that at least one of the conditional bounds $V_D(j)$ and $V_U(j)$ must coincide with $v(\overline{\text{CP}})$ for each variable that is fractional in an optimal solution of ($\overline{\text{CP}}$) when the natural choice $\bar{\lambda}$ from ($\overline{\text{CP}}$) is used in (4). Moreover, *both* of the bounds coincide with $v(\overline{\text{CP}})$ in the special case of Examples 1 and 2 and perhaps in other cases as well. These facts argue against the use of a Lagrangean relaxation for which the Integrality Property holds. It has little to offer that cannot already be achieved by ($\overline{\text{CP}}$), though it may possibly prove to be more fruitful than ($\overline{\text{CP}}$) as a source of improved feasible solutions. It is important to recognize, however, that this negative conclusion rests on the implicit assumption that ($\overline{\text{CP}}$) is of manageable size as a linear program. If this is not the case, then (CPR$_\lambda$) may be a comparatively attractive computational alternative. A beautiful illustration is provided by Held and Karp's work on the traveling-salesman problem. Here ($\overline{\text{CP}}$) has such an enormous number of constraints that it is not practical to solve directly. Of course, the omission of ($\overline{\text{CP}}$) necessitates the introduction of some method for computing a near optimal λ (see below). And even if ($\overline{\text{CP}}$) *is* of manageable size it may still be sufficiently burdensome computationally that (CPR$_\lambda$) is attractive as a surrogate to be invoked *prior* to ($\overline{\text{CP}}$) during the examination of a candidate problem. The hope is that the Lagrangean relaxation will permit (CP) to be fathomed without having to resort to the more expensive linear program ($\overline{\text{CP}}$). The best choice for λ is likely to be a multiplier vector saved from the linear program corresponding to the prior candidate problem most closely related to the current one. Section 5 indicates how this tactic coincides in special cases with the use of surrogate constraints – a device which has proven quite effective computationally in some applications (cf. [14, Sec. 3.1.5]).

Now suppose that the Integrality Property does *not* hold. Then ($\overline{\text{CP}}$) does not necessarily yield the best choice for λ, and (CPR$_\lambda$) may succeed in fathoming where ($\overline{\text{CP}}$) fails. It makes strategic sense to invoke (CPR$_\lambda$) either before or after ($\overline{\text{CP}}$) or even in lieu of it, depending on the relative tightness and computational expense of the two relaxations. The most effective strategy also depends on the role played by ($\overline{\text{CP}}$) in generating the λ to be used by (CPR$_\lambda$), since ($\overline{\text{CP}}$) can be used to generate a starting (or even final) value of λ which can then be improved upon by some independent method. To indicate the possible methods for finding a suitable λ we shall consider for the sake of notational convenience the situation encountered before any branching has taken place. Then (CP) is (P) itself and (CPR_λ) is just (PR_λ). The general situation is entirely analogous.

There are two broad approaches to computing a suitable λ for (PR_λ): (sub)optimization of the concave Lagrangean dual problem (D) and (sub) optimization of the linear program (P*). The first approach yields λ directly, whereas the second yields λ indirectly as the multiplier vector associated with the $A\,x \geqq b$ constraints in (P*). The distinction should not be thought of as a rigid one; some methods can be described naturally from either viewpoint.

Consider the first approach. One of the most promising methods for seeking an optimal solution of (D) is via the Agmon–Motzkin–Schoenberg method as revived by Held and Karp [21]. See also the recent and extensive study of this method by Held, Wolfe and Crowder [22]. The idea is very simple. Let $\lambda^\nu \geqq 0$ be the current estimate of an optimal solution of (D) and let x^ν be an optimal solution of (PR_{λ^ν}). Then the new estimate is

$$\lambda^{\nu+1} = \max\,\{\lambda^\nu + \theta^\nu(b - A\,x^\nu), 0\},$$

where the max operator is applied component-wise and θ^ν is a positive step size satisfying certain requirements [22]. The vector $(b - A\,x^\nu)$ is a subgradient of $v(PR_\lambda)$ at $\lambda = \lambda^\nu$ but the sequence $\langle v(PR_{\lambda^\nu})\rangle$ is not necessarily monotone. Favorable computational experience has been reported for several different applications [8, 21, 22]. An alternative is to carry out an ascent method for (D); see [8, 10, 20]. Still another method is to optimize (D) by tangential approximation (outer linearization/relaxation) making use of the fact that the evaluation of $v(PR_\lambda)$ for a given λ yields a linear support at that point. The available evidence [20, 24] suggests that convergence is slow in some applications. A combination of ascent and tangential approximation is possible with the BOXSTEP method of Hogan, Marsten and Blankenship [23].

Consider now the indirect approach via (P*). Perhaps the most obvious method is to apply generalized programming (Dantzig–Wolfe decomposition, inner linearization/restriction) with the convex hull portion of the constraints of (P*) represented in terms of its extreme points. The column-generation problems are precisely of the form (PR_λ). Since this method is equivalent to the tangential approximation method for (D), however, its efficiency is suspect. Another possibility is to apply the primal-dual simplex method to (P*) with special provisions to accomodate the convex hull. This method, developed by Fisher and by Fisher and Shapiro, can also be interpreted as an ascent method for (D). Some encouraging computational experience has been reported [8]. In some applications the form of the constraints describing the convex hull in (P*) is known. Then it may be possible to apply the dual simplex method to (P*) with (most) violated constraints generated as needed. This is probably one of the best methods for obtaining a near-optimal λ fairly quickly when it applies. It has the added advantage of yielding valid constraints that may be appended to $(\overline{\mathrm{P}})$ to make it a tighter relaxation of (P).

Other specialized techniques, both exact and heuristic, can be devised for (D) or (P*) in particular applications.

4. Penalties

The so-called "penalty" concept in integer programming was propelled to prominence by Driebeek [4], although the essential notion was used earlier by Dakin [3] and Healy [19]. The original idea was to underestimate the amount by which the optimal value of the LP relaxation of the current candidate problem would increase if separation were carried out using a particular separation variable. The estimates of change, referred to as penalties, can be used to help guide separation and may also permit fathoming or range reduction. An important subsequent refinement of this original idea was the recognition that it is the candidate problems and subproblems themselves, and not their LP relaxations, which are central to the underlying enumerative process. Tomlin [28, 29] showed how to modify the penalty formulae so as to take at least partial account of the integrality conditions. The resulting penalties are underestimates of the difference between $v(\overline{\mathrm{CP}})$ and the optimal value of a candidate subproblem derived from (CP). See [14] for a discussion of current practice in the computation and use of penalties.

Lagrangean relaxation furnishes a convenient setting for deriving the simple and strengthened penalties alluded to above. This is done in subsection 4.1. More importantly, it leads naturally to extensions and specializations which do not follow as easily from the more traditional viewpoints. These are illustrated in subsections 4.2–4.4 for Examples 1–3. It is hoped that these improved penalties and their counterparts for other structures will add new vitality to the penalty concept by overcoming the limitations of standard penalties pointed out so clearly by Forrest, Hirst and Tomlin [11].

4.1. *Basic results: $Bx \geqq d$ vacuous*

The first task is to show how the formulae for simple and strengthened penalties are related to Lagrangean relaxation. This requires taking $\lambda = \bar{\lambda}$ and specializing $Bx \geqq d$ to be vacuous (in contrast to our usual convention, in this subsection, $Bx \geqq d$ will not include upper bounds on the variables). Define I_f to be the indices in I such that $\bar{x}_j$ is fractional ($\bar{x}$ is the optimal solution of $(\overline{\text{CP}})$). It is easy to verify that the objective function coefficient of $(\text{CPR}_{\bar{\lambda}})$ vanishes for all such $j \in I_f$, and hence for all such j we have

$$V_{\text{D}}(j) \stackrel{\text{d}}{=} v(\text{CPR}_{\bar{\lambda}} \mid x_j \leqq [\bar{x}_j]) = v(\overline{\text{CP}}),$$

$$V_{\text{U}}(j) \stackrel{\text{d}}{=} v(\text{CPR}_{\bar{\lambda}} \mid x_j \geqq [\bar{x}_j] + 1) = v(\overline{\text{CP}}).$$

Thus the Lagrangean relaxation $(\text{CPR}_{\bar{\lambda}})$ appears to yield zero "down" and "up" penalties for separation on x_j.

A simple remedy is to employ an alternative representation for x_j in terms of variables whose objective function coefficients in $(\text{CPR}_{\bar{\lambda}})$ do not vanish. Such a representation is available from the final tableau of the linear program $(\overline{\text{CP}})$ since $j \in I_f$ must be basic therein:

$$x_j = \bar{x}_j - \sum_{i \in N} \bar{a}_{ji} x_i,$$

where N is the set of nonbasic variables. The use of this representation in the definition of $V_{\text{D}}(j)$ and $V_{\text{U}}(j)$ leads to the following conditional bounds: for $j \in I_f$,

$$V_{\text{D}}^*(j) \stackrel{\text{d}}{=} v(\text{CPR}_{\bar{\lambda}} \mid \bar{x}_j - \textstyle\sum_{i \in N} \bar{a}_{ji} x_i \leqq [\bar{x}_j]),$$
$$V_{\text{U}}^*(j) \stackrel{\text{d}}{=} v(\text{CPR}_{\bar{\lambda}} \mid \bar{x}_j - \textstyle\sum_{i \in N} \bar{a}_{ji} x_i \geqq [\bar{x}_j] + 1). \tag{5}$$

Clearly,

$$V_{\text{D}}^*(j) \leqq v(\text{CP} \mid x_j \leqq [\bar{x}_j]),$$

$$V_{\text{U}}^*(j) \leqq v(\text{CP} \mid x_j \geqq [\bar{x}_j] + 1)$$

for all $j \in I_f$; that is, these conditional bounds really do underestimate the optimal value of the candidate problems that would result if (CP) were separated using x_j as the separation variable.

Unfortunately the computation of $V_{\text{D}}^*(j)$ and $V_{\text{U}}^*(j)$ may be onerous if $\bar{a}_{ji} \neq 0$ for some variables $i \in N \cap I$. The computation then requires solving a knapsack-type problem with some integer variables. Hence it is natural to think of estimating $V_D^*(j)$ and $V_U^*(j)$ from below by simply dropping all integrality conditions:

$$\begin{aligned} V_D^{*0}(j) &\overset{\mathrm{d}}{=} v(\overline{\mathrm{CPR}}_{\bar\lambda} \mid \bar{x}_j - \textstyle\sum_{i\in N} \bar{a}_{ji}x_i \leqq [\bar{x}_j]), \\ V_U^{*0}(j) &\overset{\mathrm{d}}{=} v(\overline{\mathrm{CPR}}_{\bar\lambda} \mid \bar{x}_j - \textstyle\sum_{i\in N} \bar{a}_{ji}x_i \geqq [\bar{x}_j] + 1). \end{aligned} \tag{6}$$

The computation of each of these conditional bounds merely requires minimizing a linear function with all nonnegative coefficients [$(c - \bar\lambda A) \geqq 0$, by duality] subject to a single linear constraint and $x \geqq 0$. This is sometimes referred to as a "continuous knapsack" type problem and it is easy to write down an explicit solution:

$$V_D^{*0}(j) = v(\overline{\mathrm{CP}}) + (\bar{x}_j - [\bar{x}_j]) \underset{i\in N:\, \bar{a}_{ji}>0}{\text{minimum}} \{(c - \bar\lambda A)_i/\bar{a}_{ji}\}, \tag{7}$$

$$V_U^{*0}(j) = v(\overline{\mathrm{CP}}) + ([\bar{x}_j] + 1 - \bar{x}_j) \underset{i\in N:\, \bar{a}_{ji}<0}{\text{minimum}} \{(c - \bar\lambda A)_i/(-\bar{a}_{ji})\}$$

(we have used the fact that $\bar\lambda b = v(\overline{\mathrm{CP}})$ by LP duality). These conditional bounds are identical to those associated with the simplest penalties mentioned earlier (cf. (5) in [4]).

The strengthened penalties of Tomlin can also be recovered from this viewpoint by retaining the condition that x_j must not be in the open interval (0, 1) for $j \in N \cap I$. Then we obtain

$$\begin{aligned} V_D^{*1}(j) &\overset{\mathrm{d}}{=} v(\overline{\mathrm{CPR}}_{\bar\lambda} \mid \bar{x}_j - \textstyle\sum_{i\in N} \bar{a}_{ji}x_i \leqq [\bar{x}_j] \text{ and } x_i \notin (0,1) \text{ for all } i \in N \cap I), \\ V_U^{*1}(j) &\overset{\mathrm{d}}{=} v(\overline{\mathrm{CPR}}_{\bar\lambda} \mid \bar{x}_j - \textstyle\sum_{i\in N} \bar{a}_{ji}x_i \geqq [\bar{x}_j] + 1 \text{ and } x_i \notin (0,1) \text{ for all } i \in N \cap I). \end{aligned} \tag{8}$$

It is not difficult to see that at most one variable need be at a positive level in an optimal solution of the modified continuous knapsack problems defined in (8). This observation leads to the explicit formulae:

$$\begin{aligned} V_D^{*1}(j) &= v(\overline{\mathrm{CP}}) + \underset{i\in N:\, \bar{a}_{ji}>0}{\text{minimum}} \begin{cases} (c - \bar\lambda A)_i (\bar{x}_j - [\bar{x}_j])/\bar{a}_{ji} & \text{if } i \notin I, \\ (c - \bar\lambda A)_i \max\{(\bar{x}_j - [\bar{x}_j])/\bar{a}_{ji}, 1\} & \text{if } i \in I, \end{cases} \\ V_U^{*1}(j) &= v(\overline{\mathrm{CP}}) + \underset{i\in N:\, \bar{a}_{ji}<0}{\text{minimum}} \begin{cases} (c - \bar\lambda A)_i ([\bar{x}_j] + 1 - \bar{x}_j)/(-\bar{a}_{ji}) & \text{if } i \notin I, \\ (c - \bar\lambda A)_i \max\{([\bar{x}_j] + 1 - \bar{x}_j)/(-\bar{a}_{ji}), 1\} & \text{if } i \in I. \end{cases} \end{aligned} \tag{9}$$

These formulae are identical with the strengthened penalties of Tomlin (cf. (10) and (11) of [28] or (3.5) and (3.6) of [29]). It is evident from the very definitions (5), (6) and (8) that

$$V_D^{*0}(j) \leqq V_D^{*1}(j) \leqq V_D^{*}(j),$$
$$V_U^{*0}(j) \leqq V_U^{*1}(j) \leqq V_U^{*}(j) \tag{10}$$

for all $j \in I_f$.

This completes the recovery of known formulae for Driebeek and Tomlin penalties for $j \in I_f$. Exactly the same type of analysis holds for penalties associated with *basic* variables of $I - I_f$. Such penalties are of interest as a means of obtaining tighter ranges on integer variables which happen to be naturally integer in the optimal solution of $(\overline{\text{CP}})$. For a *nonbasic* variable variable x_j in $I - I_f$ the quantity of interest is $v(\text{CPR}_\lambda \,|\, x_j \geqq 1)$; no alternative representation in terms of nonbasic variables is possible. Evidently,

$$v(\text{CPR}_{\bar{\lambda}} \,|\, x_j \geqq 1) = v(\overline{\text{CP}}) + (c - \bar{\lambda} A)_j. \tag{11}$$

Again this is a standard result.

Another technique for strengthening (6) makes use of the following elementary and well-known result.

Theorem 5. *Let* (IP) *be a minimizing integer linear program in which exactly one variable, say* x_h, *is declared to be integer-valued. Suppose that* $\bar{x}_h$, *the optimal level of* x_h *when* (IP) *is solved ignoring the integrality requirement, is fractional. Then the optimal value of* (IP) *is given by*

$$v(\text{IP}) = \min\{v(\overline{\text{IP}} \,|\, x_h = [\bar{x}_h]), v(\overline{\text{IP}} \,|\, x_h = [\bar{x}_h] + 1)\}.$$

The possibility that $(\overline{\text{IP}} \,|\, x_h = [\bar{x}_h])$ or $(\overline{\text{IP}} \,|\, x_h = [\bar{x}_h] + 1)$ or both are infeasible is not excluded (recall that our convention is to define a minimum over an empty set as $+\infty$).

Let $i_D(j)$ be the minimizing nonbasic i in the formula for $V_D^{*0}(j)$ given in (7). The index $i_U(j)$ is defined similarly. Then application of Theorem 5 in the obvious way permits the following improvement on (6) to be computed with only a little extra effort:

$$V_D^{*2}(j) \stackrel{\text{d}}{=} v(\overline{\text{CPR}}_{\bar{\lambda}} \,|\, \bar{x}_j - \textstyle\sum_{i \in N} \bar{a}_{ji} x_i \leqq [\bar{x}_j] \text{ and } x_{i_D(j)} \text{ integer}),$$
$$V_U^{*2}(j) \stackrel{\text{d}}{=} v(\overline{\text{CPR}}_{\bar{\lambda}} \,|\, \bar{x}_j - \textstyle\sum_{i \in N} \bar{a}_{ji} x_i \geqq [\bar{x}_j] + 1 \text{ and } x_{i_U(j)} \text{ integer}). \tag{12}$$

Neither (12) nor (8) necessarily dominates the other; one may verify the following relationship for $j \in I_f$:

$$(\bar{x}_j - [\bar{x}_j])/\bar{a}_{ji_D(j)} \left\{\begin{smallmatrix}\leqq\\ \geqq\end{smallmatrix}\right\} 1 \Rightarrow V_D^{*1}(j) \left\{\begin{smallmatrix}\geqq\\ \leqq\end{smallmatrix}\right\} V_D^{*2}(j),$$
$$([\bar{x}_j] + 1 - \bar{x}_j)/(-\bar{a}_{ji_U(j)}) \left\{\begin{smallmatrix}\leqq\\ \geqq\end{smallmatrix}\right\} 1 \Rightarrow V_U^{*1}(j) \left\{\begin{smallmatrix}\geqq\\ \leqq\end{smallmatrix}\right\} V_U^{*2}(j). \tag{13}$$

We are unable to supply a reference to the conditional bounds (12) in the published literature. However, Armstrong and Sinha [1] have independently and very recently proposed a precisely analogous strengthening of (8) for the mixed integer 0–1 case. They report favorable computational experience.

So far we have required $Bx \geqq d$ to be vacuous; that is, all upper bounds and other special constraints are treated as general A-type constraints. Analogs of the previous penalty results as well as new penalty results emerge easily by allowing $Bx \geqq d$ to be nonvacuous. This will now be illustrated for the three examples.

4.2. *Penalties for Example* 1

Example 1 differs from the previous development only in that $(\mathrm{CPR}_{\bar{\lambda}})$ now has upper-bounded variables. As indicated in Section 3, it can be shown that both $V_{\mathrm{D}}(j)$ and $V_{\mathrm{U}}(j)$ equal $v(\overline{\mathrm{CP}})$ for all $j \in I_f$ due to the vanishing of the corresponding objective function coefficients in $(\mathrm{CPR}_{\bar{\lambda}})$. The remedy for these vanishing penalties is again to invoke the representation for x_j which is available from the final LP tableau of $(\overline{\mathrm{CP}})$. This representation will be written as

$$x_j = \alpha_{j0} - \sum_{i \neq j} \alpha_{ji} x_i \quad \text{for } j \in I_f, \tag{14}$$

where, of course, many of the coefficients α_{ji} may be 0. The resulting strengthened conditional lower bounds on $v(\mathrm{CP} \mid x_j \leqq [\bar{x}_j])$ and $v(\mathrm{CP} \mid x_j \geqq [\bar{x}_j] + 1)$ for $j \in I_f$ are

$$\begin{aligned} V_{\mathrm{D}}^*(j) &\overset{\mathrm{d}}{=} v(\mathrm{CPR}_{\bar{\lambda}} \mid x_j = \alpha_{j0} - \textstyle\sum_{i \neq j} \alpha_{ji} x_i \leqq [\bar{x}_j]), \\ V_{\mathrm{U}}^*(j) &\overset{\mathrm{d}}{=} v(\mathrm{CPR}_{\bar{\lambda}} \mid x_j = \alpha_{j0} - \textstyle\sum_{i \neq j} \alpha_{ji} x_i \geqq [\bar{x}_j]) + 1). \end{aligned} \tag{15}$$

We have used the notations V_{D}^* and V_{U}^* as in (5) because (15) is an exact counterpart of (5). Like (5), (15) could be too expensive computationally because each estimate requires solving a knapsack-type problem in integer variables. The fact that the knapsack problem now has upper-bounded variables is a dubious advantage. The most easily computed lower approximation to (15) is obtained by dropping the integrality requirements as in (6):

$$\begin{aligned} V_{\mathrm{D}}^{*0}(j) &\overset{\mathrm{d}}{=} v(\overline{\mathrm{CPR}_{\bar{\lambda}}} \mid x_j = \alpha_{j0} - \textstyle\sum_{i \neq j} \alpha_{ji} x_i \leqq [\bar{x}_j]), \\ V_{\mathrm{U}}^{*0}(j) &\overset{\mathrm{d}}{=} v(\overline{\mathrm{CPR}_{\bar{\lambda}}} \mid x_j = \alpha_{j0} - \textstyle\sum_{i \neq j} \alpha_{ji} x_i \geqq [\bar{x}_j] + 1). \end{aligned} \tag{16}$$

The notations V_D^{*0} and V_U^{*0} have again been carried over. The differences

$$V_U^{*0}(j) - v(\overline{CP}), \qquad v_D^{*0}(j) = v(\overline{CP}) \tag{17}$$

are Driebeek-like up and down penalties for Example 1. The computation of (16) requires only slightly more effort than the computation of (6). A "continuous knapsack" problem with upper-bounded variables must now be solved. Explicit formulae for V_D^{*0} and V_U^{*0} are slightly more cumbersome than expression (7), but are easily programmed for a computer.

To strengthen (16) one may formally write the counterpart of (8), but unfortunately explicit calculation may be nearly as costly as that of (15) itself. This is because the upper bounds generally invalidate the key property of (8) that at most one variable need be at a positive level in an optimal solution of each associated optimization problem. Thus the strengthened penalties of Tomlin do not generalize usefully to $Bx \geqq d$ when it includes upper bounds on variables.

But the other technique based on Theorem 5 for strenthening $V_D^{*0}(j)$ and $V_U^{*0}(j)$ *does* generalize nicely. Let $i_D(j)$ and $i_U(j)$ be respectively the fractional-valued variables in the solutions of the optimizations corresponding to $V_D^{*0}(j)$ and $V_U^{*0}(j)$. It is easy to see that at most one variable need be fractional in each of these solutions; if none is, then that penalty cannot be strengthened by the present device. The strengthened conditional bounds analogous to (12) are:

$$\begin{aligned} V_D^{*2}(j) &\overset{d}{=} v(\overline{CPR_{\bar{\lambda}}} \mid x_j = \alpha_{j0} - \textstyle\sum_{i \neq j} \alpha_{ji} x_i \leqq [\bar{x}_j] \text{ and } x_{i_D(j)} \text{ integer}), \\ V_U^{*2}(j) &\overset{d}{=} v(\overline{CPR_{\bar{\lambda}}} \mid x_j = \alpha_{j0} - \textstyle\sum_{i \neq j} \alpha_{ji} x_i \\ &\qquad \geqq [\bar{x}_j] + 1 \text{ and } x_{i_U(j)} \text{ integer}). \end{aligned} \tag{18}$$

The required optimizations are inexpensive to carry out. Clearly,

$$\begin{aligned} V_D^{*0}(j) &\leqq V_D^{*2}(j) \leqq V_D^{*}(j) \leqq v(\mathrm{CP} \mid x_j \leqq [\bar{x}_j]), \\ V_U^{*0}(j) &\leqq V_U^{*2}(j) \leqq V_U^{*}(j) \leqq v(\mathrm{CP} \mid x_j \geqq [\bar{x}_j] + 1). \end{aligned} \tag{19}$$

Exactly the same types of penalties can be constructed for $j \in I - I_f$ when the objective function coefficient of x_j vanishes in $(\mathrm{CPR}_{\bar{\lambda}})$.

4.3. *Penalties for Example* 2

The development of penalties for Example 2 closely parallels that for Example 1. For $j \in I_f$, both up and down penalties again vanish, and it is necessary to use the final LP tableau representation of the form (14).[3] The resulting conditional bounds $V_D^*(j)$ and $V_U^*(j)$ defined in (15) may still be too expensive computationally to use in general, but the multiple choice constraints do tend to make the computation easier by comparison with Example 1. There are nontrivial situations where $V_D^*(j)$ and $V_U^*(j)$ can be computed relatively economically by a simple enumerative procedure. But in general one may have to fall back on the Driebeek-like penalties defined by (16). The required computations are no longer simple continuous knapsack problems with upper-bounded variables, but they can still be carried out efficiently by specialized techniques (e.g. by parametric optimization applied to the dual of $(\overline{\mathrm{CPR}_{\bar{x}}})$ with respect to the added constraint). Strengthening these penalties along the lines suggested by Tomlin as in (8) appears to be no easier in general than (15) itself. But again, as with Example 1, the strengthening of (18) based on Theorem 5 is attractive. The indices $i_D(j)$ and $i_U(j)$ may be selected to be any of the fractional-valued variables in the solutions of the optimizations corresponding to $V_D^{*0}(j)$ and $V_U^{*0}(j)$. The implementation of (18) on a computer is only slightly more expensive than that of (16). Naturally, (19) continues to hold. The reader should have no difficulty seeing what to do if penalties are desired for variables in $I - I_f$.

The special nature of the multiple choice constraints (1) makes it possible to define "cumulative" conditional bounds on the "upward" problems as follows:

$$V_U^{*0}(j;J_k) \stackrel{\mathrm{d}}{=} \max\{V_U^{*0}(j), V_D^{*0}(i) \quad \text{for } i \in \{J_k - j\}\} \tag{20}$$

where it is understood that j is in J_k in these definitions. That this provides true lower bounds on $v(\mathrm{CP} \mid x_j = 1)$ follows from the fact that $x_j = 1$ implies $x_i = 0$ for all $i \neq j$ in the same multiple choice set. Similar cumulative bounds hold for V^{*2} and V^*.

4.4. *Penalties for Example* 3

For Example 3 we must distinguish between the "switching" (x_k) versus the "nonswitching" variables in I_f. The up and down penalties associated with $V_D(j)$ and $V_U(j)$ are highly unlikely to vanish for the fractional switching variables. In fact, one can argue that they are likely to be quite large. Our experience with the practical facility location problems mentioned at the end of Section 2 has been that these penalties tend to be at least *an order of*

[3] Numbered displays from the discussion of Example 1 will be used here with the understanding that (CP), etc., have the structure of Example 2 rather than Example 1.

magnitude greater than the standard Tomlin penalties when $\bar{\lambda}$ is used, and yet take less time to compute [15]. For the nonswitching variables in I_f, however, it is easy to see that the naive penalties vanish and thus that alternative representations from the final LP tableau may be useful. The detailed discussion would be so close to that for Example 2 that it will not be given here.

5. Surrogate constraints

Consider the case where (P) is a pure 0–1 integer program with $Bx \geqq d$ consisting solely of unit upper bounds on all variables. The present author proposed [12] the use of "surrogate" constraints (after Glover [16]) of the form

$$cx + \lambda(b - Ax) < z^*, \tag{21}$$

with the prescription that $\lambda \geqq 0$ be chosen as the optimal dual vector corresponding to $Ax \geqq b$ in $(\bar{\text{P}})$ or some $(\overline{\text{CP}})$. Clearly such a constraint must be satisfied by every feasible solution to (P) with lower objective value than that of the incumbent. Two uses of this type of surrogate constraint were proposed in connection with the examination of a typical candidate problem: as a possible means of fathoming via the easy test

$$\underset{x=0,1}{\text{minimum}} \{cx + \lambda(b - Ax)\} \overset{?}{\geqq} z^* \tag{22}$$

and as a possible means of range reduction via the following easy tests applied to a typical (say the jth) variable:

$$\underset{x=0,1}{\text{minimum}} \{cx + \lambda(b - Ax) \text{ s.t. } x_j = 0\} \overset{?}{\geqq} z^*, \tag{23a}$$

$$\underset{x=0,1}{\text{minimum}} \{cx + \lambda(b - Ax) \text{ s.t. } x_j = 1\} \overset{?}{\geqq} z^*. \tag{23b}$$

If (22) holds, then (P) is fathomed. If (23a) [resp. (23b)] holds, then x_j must be 1 [resp. 0] in any feasible solution of (P) which is superior in value to the current incumbent. It is understood, naturally, that all separation constraints must also be honored in taking the minima in (22) and (23) when examining a candidate problem subsequent to (P). If all separation constraints involve only additional range restrictions on the variables, as is usually the case, then (22) and (23) remain computationally trivial.

It is easy to interpret (22) and (23) from the viewpoint of Lagrangean relaxation (remember that $Bx \geqq d$ consists of just the upper bound constraints $x_i \leqq 1$). Test (22) can be rewritten as

$$v(\mathrm{PR}_\lambda) \overset{?}{\geqq} z^*, \tag{24}$$

which is precisely the elementary fathoming criterion normally associated with (PR_λ). Similarly, (23) can be rewritten as

$$v(\mathrm{PR}_\lambda \mid x_j = 0) \overset{?}{\geqq} z^*, \tag{25a}$$

$$v(\mathrm{PR}_\lambda \mid x_j = 1) \overset{?}{\geqq} z^*. \tag{25b}$$

This is precisely the ordinary range reduction criterion described in Section 3. And the injunction to obtain λ from the usual linear programming relaxation is a consequence of Theorem 2, which implies that the strongest tests are obtained in this way.

Thus the surrogate constraint (21) and the tests based on it are seen to be completely subsumed by the simplest Lagrangean relaxation techniques for the special case of Example 1. Generalizations of (21)–(23) when (P) is not a pure 0–1 program or when $Bx \geqq d$ includes more than simple upper bounds can be obtained without difficulty. Some such generalizations were developed several years ago by this author in unpublished lecture notes and by Glover [17] using the surrogate constraint viewpoint, but in each case the same results may be obtained easily as special cases of more general and powerful results based on Lagrangean relaxation.

6. Cutting planes

For present purposes, a *cutting-plane* is any linear inequality which must be satisfied by all of the feasible solutions of a candidate problem but is violated by an optimal solution of its usual linear programming relaxation $(\overline{\mathrm{CP}})$. Appending cutting-planes to $(\overline{\mathrm{CP}})$ makes it a tighter relaxation of (CP) and thereby yields better bounds for use in a hybrid branch-and-bound algorithm (cf. [14, Sec. IV]). Cutting-planes may also, of course, be used in a purely cutting-plane approach.

This section explores the uses of Lagrangean relaxation as a source of cutting-planes. For notational convenience we only consider cuts relative to the initial candidate problem (P) itself. It is a simple matter to apply the ideas developed below to any candidate problem.

The simplest type of cutting-plane for (P) is (here $\lambda \geqq 0$)

$$v(\mathrm{PR}_\lambda) \leqq cx + \lambda(b - Ax). \tag{26}$$

A special case of this cut was proposed by Shapiro [27], who showed that it can be at least as strong as all of the cuts in a well-known group theoretic class. The validity of this constraint for any feasible solution of (P) follows from the definition of $v(\mathrm{PR}_\lambda)$ and the fact that the feasible region of (P) is contained in that of (PR_λ). It will be violated at $\bar{x}$, an optimal solution of $(\bar{\mathrm{P}})$, if $v(\mathrm{PR}_\lambda) > v(\bar{\mathrm{P}})$ holds, because $\lambda \geqq 0$ and $A\bar{x} \geqq b$ imply

$$v(\bar{\mathrm{P}}) = c\bar{x} \geqq c\bar{x} + \lambda(b - A\bar{x}).$$

The condition $v(\mathrm{PR}_\lambda) > v(\bar{\mathrm{P}})$ will hold when $v(\bar{\mathrm{P}}) < v(\mathrm{D})$ and λ is sufficiently near optimal in (D). Of course this condition is impossible when the Integrality Property holds; in fact, the Integrality Property implies that (26) cannot be violated by any solution of $(\bar{\mathrm{P}})$ whatever, because then

$$v(\mathrm{PR}_\lambda) = v(\overline{\mathrm{PR}}_\lambda) \leqq cx + \lambda(b - Ax)$$

for all x feasible in $(\overline{\mathrm{PR}}_\lambda)$ and thus for all x feasible in $(\bar{\mathrm{P}})$. Thus (26) can be a true cutting-plane only when the Integrality Property does not hold. Appending it to $(\bar{\mathrm{P}})$ must increase the optimal value of $(\bar{\mathrm{P}})$ at least to $v(\mathrm{PR}_\lambda)$ because (26) implies

$$cx \geqq v(\mathrm{PR}_\lambda) - \lambda(b - Ax) \geqq v(\mathrm{PR}_\lambda) \quad \text{for all } x \text{ feasible in } (\bar{\mathrm{P}}).$$

An improvement of (26) is obtained by replacing $v(\mathrm{PR}_\lambda)$ with $v(\mathrm{PR}_\lambda \mid x_{j_1}, \ldots, x_{j_p})$, which denotes the optimal value of (PR_λ) as a function of specified values for the distinguished *cut variables* $x_{j_1}, \ldots, x_{j_p}$. (If the values of the cut variables are such that no completion exists which is feasible in (PR_λ)—e.g., if an integer cut variable takes on a fractional value—then by convention, $v(\mathrm{PR}_\lambda \mid x_{j_1}, \ldots, x_{j_p})$ is defined to be $+\infty$ at such a point.) The constraint

$$v(\mathrm{PR}_\lambda \mid x_{j_1}, \ldots, x_{j_p}) \leqq cx + \lambda(b - Ax) \tag{27}$$

is valid by an argument similar to that for (26) and is uniformly at least as tight because

$$v(\mathrm{PR}_\lambda) \leqq v(\mathrm{PR}_\lambda \mid x'_{j_1}, \ldots, x'_{j_p}) \tag{28}$$

obviously holds for every feasible solution x' of (P). Strict inequality holds in (28) except when $x'_{j_1}, \ldots, x'_{j_p}$ happens to be part of an optimal solution of (PR_λ). This fact also renders (27) less susceptible to neutralization by the Integrality Property.

The difficulty with (27), of course, is that $v(\mathrm{PR}_\lambda \mid x_{j_1}, \ldots, x_{j_p})$ need not be a linear function. It depends upon the structure of (PR_λ) and the choice of cut variables. One source of nonlinearity has to do with the domain on which it is $+\infty$. Fortunately, (27) need only hold for *feasible* solutions of

(P), and so $v(\mathrm{PR}_\lambda \mid x_{j_1}, \ldots, x_{j_p})$ can be redefined arbitrarily wherever it is $+\infty$. It is clear that this redefinition should be linearly interpolative in nature. Of course, this still may not render (27) linear. It may be necessary to determine a linear lower bounding function $l_\lambda(x_{j_1}, \ldots, x_{j_p})$,

$$l_\lambda(x_{j_1}, \ldots, x_{j_p}) \leqq v(\mathrm{PR}_\lambda \mid x_{j_1}, \ldots, x_{j_p}) \quad \text{for all } x \text{ feasible in (P).} \tag{29}$$

Clearly, l_λ should be as "tight" as possible, especially in the vicinity of $\bar{x}$. Thus the linear constraint to be appended to $(\bar{\mathrm{P}})$ is of the form

$$l_\lambda(x_{j_1}, \ldots, x_{j_p}) \leqq c\,x + \lambda\,(b - A\,x), \tag{30}$$

where $(j_1, \ldots, j_P)$ is an arbitrary set of cut variable indices, $\lambda \geqq 0$, and (29) must hold.

The above ideas can be illustrated with reference to the three examples of Section 1. Examples 1 and 2 satisfy the Integrality Property and so constraint (26) cannot be violated at $\bar{x}$. Furthermore, it can be shown for these examples that no constraint of the form (30) can be violated at $\bar{x}$, no matter what λ or cut variables are chosen. Example 3, on the other hand, does lend itself to the derivation of useful cutting-planes. Cut (26) tends to be quite good, even when $\bar{\lambda}$, an immediate by-product of $(\bar{\mathrm{P}})$, is used. We see from the computational experience cited at the end of Section 2 that, in the four practical problems studied, a single cut of the form (26) with $\lambda = \bar{\lambda}$ raised the optimal value of $(\bar{\mathrm{P}})$ an average of at least 59.6% of the distance from $v(\bar{\mathrm{P}})$ to $v(\mathrm{P})$. If the effort to find an optimal λ were expended, (26) would raise the optimal value an average of at least 97.1% of the way to $v(\mathrm{P})$. It should also be noted that a cut of the form (30) is available as an immediate by-product of the evaluation of (PR_λ). Recall that (PR_λ) separates into independent subproblems of the form (3^k_λ):

$$\begin{aligned} v(\mathrm{PR}_\lambda) = \lambda\,b + \sum_{k=1}^{K} v(3^k_\lambda) \\ + \underset{x_j,\, j\in T}{\text{minimum}} \{\textstyle\sum_{j\in T} (c - \lambda\,A)\,x_j \text{ s.t. } 0 \leqq x_j \leqq u_j, \\ j \in T \text{ and } x_j \text{ integer, } j \in T \cap I\}, \end{aligned} \tag{31}$$

where T comprises the indices of all variables not appearing in any of the subproblems of type (3^k_λ). To evaluate $v(\mathrm{PR}_\lambda)$ one makes use of the fact that

$$v(3^k_\lambda) = \min\{v(3^k_\lambda \mid x_k = 0), v(3^k_\lambda \mid x_k = 1)\} \tag{32}$$

and of the fact that the last term involving $j \in T$ in (31) is trivially evaluated by inspection. Thus the binary variables x_k are obvious choices for cut variables. We have

$$v(\mathrm{PR}_\lambda \mid x_1, \ldots, x_K) = \mathrm{CON}_\lambda + \sum_{k=1}^{K} v(3_\lambda^k \mid x_k), \tag{33}$$

where the constant CON_λ equals the first and last terms of (31). The binary nature of the variables makes it easy to write down a linear function l_λ satisfying (29) with equality in this case:

$$\mathrm{CON}_\lambda + \sum_{k=1}^{K} v(3_\lambda^k \mid x_k = 1)\, x_k = v(\mathrm{PR}_\lambda \mid x_1, \ldots, x_K)$$

for all binary $(x_1, \ldots, x_K)$. Thus (30) becomes

$$\mathrm{CON}_\lambda + \sum_{k=1}^{K} v(3_\lambda^k \mid x_k = 1)\, x_k \leqq c\,x + \lambda\,(b - A\,x). \tag{34}$$

Our experience with the same four practical problems as mentioned above is that a single cut of this type raised $v(\bar{\mathrm{P}})$ an average of 69.7% of the way from $v(\bar{\mathrm{P}})$ to $v(\mathrm{P})$ when $\bar{\lambda}$ was used [15].

The derivation of a type (30) cut for Example 3 generalizes easily to the frequent situation where (PR_λ) separates into a number of independent subproblems involving 0–1 variables. Suppose

$$v(\mathrm{PR}_\lambda) = \lambda\,b + \sum_{k=1}^{P} v(\mathrm{PR}_\lambda^k),$$

where (PR_λ^k) involves the variables J_k ($J_1, \ldots, J_P$ is a mutually exclusive and exhaustive partition) among which is a 0–1 variable j_k. Suppose further that both $v(\mathrm{PR}_\lambda^k \mid x_{j_k} = 0)$ and $v(\mathrm{PR}_\lambda^k \mid x_{j_k} = 1)$ can be obtained inexpensively in the course of evaluating $v(\mathrm{PR}_\lambda^k)$. Then j_k is a natural choice for a cut variable and a type (30) constraint is

$$\lambda\,b + \sum_{k=1}^{P} v(\mathrm{PR}_\lambda^k \mid x_{j_k} = 0)(1 - x_{j_k}) + v(\mathrm{PR}_\lambda^k \mid x_{j_k} = 1)\, x_{j_k} \leqq$$
$$\leqq c\,x + \lambda\,(b - A\,x). \tag{35}$$

We have made use of the relations

$$v(\mathrm{PR}_\lambda \mid x_{j_1}, \ldots, x_{j_P}) = \lambda\,b + \sum_{k=1}^{P} v(\mathrm{PR}_\lambda^k \mid x_{j_k}),$$

$$v(\mathrm{PR}_\lambda^k \mid x_{j_k}) = v(\mathrm{PR}_\lambda^k \mid x_{j_k} = 0)(1 - x_{j_k}) + v(\mathrm{PR}_\lambda^k \mid x_{j_k} = 1)\, x_{j_k} \quad \text{for } x_{j_k} = 0,1.$$

The latter relation furnishes the required l_λ function with equality in (29). Constraint (35) is likely to improve on the counterpart of (26), namely

$$\lambda b + \sum_{k=1}^{P} v(\mathrm{PR}_\lambda^k) \leqq c\,x + \lambda(b - A\,x), \tag{36}$$

because

$$\begin{aligned} v(\mathrm{PR}_\lambda^k) &= \min\{v(\mathrm{PR}_\lambda^k \mid x_{j_k} = 0), v(\mathrm{PR}_\lambda^k \mid x_{j_k} = 1)\} \\ &\leqq v(\mathrm{PR}_\lambda^k \mid x_{j_k} = 0)(1 - x_{j_k}) \\ &\quad + v(\mathrm{PR}_\lambda^k \mid x_{j_k} = 1)\,x_{j_k} \quad \text{for } 0 \leqq x_{j_k} \leqq 1. \end{aligned}$$

It should also be pointed out that (35) and (36) can be decomposed into P component inequalities:

$$v(\mathrm{PR}_\lambda^k \mid x_{j_k} = 0)(1 - x_{j_k}) + v(\mathrm{PR}_\lambda^k \mid x_{j_k} = 1)\,x_{j_k} \leqq \sum_{j \in J_k} (c - \lambda\,A)_j x_j, \tag{35_k}$$

$$v(\mathrm{PR}_\lambda^k) \leqq \sum_{j \in J_k} (c - \lambda\,A)_j x_j. \tag{36_k}$$

The validity of (35_k) and (36_k) should be evident. Their sum over all k yields (35) and (36), respectively.

Other types of cutting-planes can be devised with the help of the penalty formulae of Section 4. In particular, useful cutting-planes for Examples 1 and 2 can be determined (recall that neither (26) nor (30) were useful in this context). Both the simple penalties based on (16) and the strengthened penalties based on (18) can be used to generate cuts violated by $\bar{x}$ so long as at least one of these penalties is nonzero. This may be done as follows. Consistency of notation requires that we let (CP) equal (P) when applying the results of Section 4.

Consider first the simple conditional bounds (16). Select any $j \in I_f$ such that at least one of the penalties is strictly positive and take this j to be the one cut variable. Clearly

$$\begin{aligned} & v(\mathrm{PR}_\lambda \mid x_j = \alpha_{j0} - \textstyle\sum_{i \neq j} \alpha_{ji} x_i) \leqq \\ & \leqq c\,x + \bar{\lambda} \leqq c\,x + \bar{\lambda}(b - A\,x) \quad \text{for all } x \text{ feasible in (P)}. \end{aligned} \tag{37}$$

The left-hand side of (37) is convex as a function of x_j, and thus the unique linear function passing through it at the points $[\bar{x}_j]$ and $[\bar{x}_j] + 1$ does not overestimate it for any integer value of x_j:

$$\begin{aligned} & V_{\mathrm{D}}^{*0}(j) + (V_{\mathrm{U}}^{*0}(j) - V_{\mathrm{D}}^{*0}(j))(x_j - [\bar{x}_j]) \leqq \\ & \leqq v(\overline{\mathrm{PR}}_{\bar{\lambda}} \mid x_j = \alpha_{j0} - \sum_{i \neq j} \alpha_{ji} x_i) \quad \text{for all integer } x_j. \end{aligned} \tag{38}$$

Together, (37) and (38) imply that

$$V_{\mathrm{D}}^{*0}(j) + (V_{\mathrm{U}}^{*0}(j) - V_{\mathrm{D}}^{*0}(j))(x_j - [\bar{x}_j]) \leqq c\,x + \bar{\lambda}(b - A\,x) \quad (39)$$

is a legitimate cut. Notice that if there are several $j \in I_f$ for which $V_{\mathrm{D}}^{*0}(j)$ and $V_{\mathrm{U}}^{*0}(j)$ are computed, then it is an easy matter to select j so as to yield the cut of type (39) which is most violated by $\bar{x}$.

Now consider the strengthened conditional bounds (18). An analog of inequality (37) holds, but the analog of (38) does not because of the added integrality requirement in (18). It appears necessary to require that $j \in I_f$ be a 0–1 variable if a cut is to be based on (18) with j as the single cut variable. Then

$$V_{\mathrm{D}}^{*2}(j) + (V_{\mathrm{U}}^{*2}(j) - V_{\mathrm{D}}^{*2}(j))\,x_j \leqq c\,x + \bar{\lambda}(b - A\,x) \quad (40)$$

is a legitimate cut. By (19), (40) is clearly a superior cut to (39). It is a simple matter to select j so as to yield the cut which is deepest at $\bar{x}$ among those of the form (40).

For Example 2 one should of course use in (39) and (40) the cumulative penalties defined in (20) in place of $V_{\mathrm{U}}^{*0}(j)$ or $V_{\mathrm{U}}^{*2}(j)$ if the necessary quantities are at hand. One may further improve cuts (39) and (40) when j is a multiple choice variable by using one of the obvious cuts

$$\sum_{j \in J_k} V_{\mathrm{U}}^{*0}(j; J_k)\,x_j \leqq c\,x + \bar{\lambda}(b - A\,x), \qquad k = 1, \ldots, K \quad (41)$$

or the still stronger cuts

$$\sum_{j \in J_k} V_{\mathrm{U}}^{*2}(j; J_k)\,x_j \leqq c\,x + \bar{\lambda}(b - A\,x), \qquad k = 1, \ldots, K. \quad (42)$$

Each of these cuts takes all of J_k as the set of cut variables. It is easy to verify that cuts of the form (41) [resp. (42)] are at least as strong as those of the form (39) [resp. (40)] for all x feasible in (P).

A cut similar to (41) was proposed by Healy [19]. To be precise, for the kth cut he omitted the term $\bar{\lambda}(b - A\,x)$ and used $V_{\mathrm{U}}^{*0}(j)$ as the coefficient of x_j, where $V_{\mathrm{U}}^{*0}(j)$ is computed with $B\,x \geqq d$ taken to consist of only the kth multiple choice constraint (no upper bounds or other multiple choice constraints are included). This cut is dominated by (41).

We note in closing that penalty-based cuts with more than one cut variable can often be obtained for Examples 1 and 2 and other structures by: (i) adding to $(\mathrm{PR}_{\bar{\lambda}})$ relations of the form (14) for any subset of j's in I_f so long as no variable appears with a nonzero coefficient in more than one of these relations, and then (ii) exploiting separability.

7. Conclusion

Lagrangean relaxation is a systematic exploitation of the formal Lagrangean dual problem in integer programming. This dual problem need not be solved optimally and need not be devoid of a duality gap in order to be useful. It provides a means for fathoming, range reduction, generating improved feasible solutions, and guiding separation (Sec. 3). It also provides new penalties (Sec. 4) and cutting-planes (Sec. 6) and supplants the narrower notion of surrogate constraints (Sec. 5). All of these functions usually can be tailored to the special structure of the particular problem class at hand, beginning with the judicious choice of the subset of constraints to play the role of $Bx \geqq d$. This has been carried out in detail for three of the simplest structures. Some of the uses of Lagrangean relaxation have been explored by other authors for several more complex structures [6], [7], [8], [9], [10], [20], [21], [26]. Yet it remains to work out the full import of Lagrangean relaxation even for these structures and for many others of importance. It is hoped that the framework of this paper will facilitate this effort and encourage new applications.

Acknowledgment

This paper has benefited from careful readings by Marshall L. Fisher and Roy E. Marsten.

References

[1] R.D. Armstrong and P. Sinha, "Improved penalty calculations for a mixed integer branch-and-bound algorithm", *Mathematical Programming* 6 (1974) 212–223.
[2] R. Brooks and A. Geoffrion, "Finding Everett's Lagrange multipliers by linear programming", *Operations Research* 14 (1966) 1149–1153.
[3] Dakin, R.J., "A tree search algorithm for mixed integer programming problems", *Computer Journal*, 8 (1965) 250–255.
[4] N.J. Driebeek, "An algorithm for the solution of mixed integer programming problems", *Management Science* 12 (1966) 576–587.
[5] Everett, H.M., "Generalized Lagrange multiplier method for solving problems of optimum allocation of resources", *Operations Research* 11 (1966) 399–417.
[6] M.L. Fisher, "Optimal solution of scheduling problems using Lagrange multipliers: Part I", *Operations Research* 21 (1973) 1114–1127.
[7] M.L. Fisher, "A dual algorithm for the one-machine scheduling problem", Graduate School of Business Rept., University of Chicago, Chicago, Ill. (1974).
[8] M.L. Fisher, W.D. Northup and J.F. Shapiro, "Using duality to solve discrete optimization problems: Theory and computational experience", Working Paper OR 030-74, Operations Research Center, M.I.T. (1974).

[9] M.L. Fisher and L. Schrage, "Using Lagrange multipliers to schedule elective hospital admissions", Working Paper, University of Chicago, Chicago, Ill. (1972).

[10] M.L. Fisher and J.F. Shapiro, "Constructive duality in integer programming", *SIAM Journal on Applied Mathematics*, to appear.

[11] J.J.H. Forrest, J.P.H. Hirst and J.A. Tomlin, "Practical solution of large mixed integer programming problems with UMPIRE", *Management Science* 20 (1974) 733–773.

[12] A.M. Geoffrion, "An improved implicit enumeration approach for integer programming", *Operations Research* 17 (1969) 437–454.

[13] A.M. Geoffrion, "Duality in nonlinear programming", *SIAM Review* 13 (1971) 1–37.

[14] A.M. Geoffrion and R.E. Marsten, "Integer programming algorithms: A framework and state-of-the-art survey", *Management Science* 18 (1972) 465–491.

[15] A.M. Geoffrion and R.D. McBride, "The capacitated facility location problem with additional constraints", paper presented to the Joint National Meeting of AIIE, ORSA, and TIMS, Atlantic City, November 8–10, 1972.

[16] F. Glover, "A multiphase-dual algorithm for the zero–one integer programming problem", *Operations Research* 13 (1965) 879–919.

[17] F. Glover, "Surrogate constraints", *Operations Research* 16 (1968) 741–749.

[18] H.J. Greenberg and T.C. Robbins, "Finding Everett's Lagrange multipliers by Generalized Linear Programming, Parts I, II, and III", Tech. Rept. CP-70008, Computer Science/Operations Research Center, Southern Methodist University, Dallas, Tex. revised (June 1972).

[19] W.C. Healy, Jr., "Multiple choice programming", *Operations Research* 12 (1964) 122–138.

[20] M. Held and R.M. Karp, "The traveling salesman problem and minimum spanning trees", *Operations Research* 18 (1970) 1138–1162.

[21] M. Held and R.M. Karp, "The traveling salesman problem and minimum spanning trees: Part II", *Mathematical Programming* 1 (1971) 6–25.

[22] M. Held, P. Wolfe and H.P. Crowder, "Validation of subgradient optimization", Mathematical Sciences Department, IBM Watson Research Center, Yorktown Heights, N.Y. (August 1973).

[23] W.W. Hogan, R.E. Marsten and J.W. Blankenship, "The BOXSTEP method for large scale optimization", Working Paper 660-73, Sloan School of Management, M.I.T. (December 1973).

[24] R.E. Marsten, private communication (August 22, 1973).

[25] G.L. Nemhauser and Z. Ullman, "A note on the generalized Lagrange multiplier solution to an integer programming problem", *Operations Research* 16 (1968) 450–452.

[26] G.T. Ross and R.M. Soland, "A branch and bound algorithm for the generalized assignment problem", *Mathematical Programming*, to appear.

[27] J.F. Shapiro, "Generalized Lagrange multipliers in integer programming", *Operations Research* 19 (1971) 68–76.

[28] J.A. Tomlin, "An improved branch and bound method for integer programming", *Operations Research* 19 (1971) 1070–1075.

[29] J.A. Tomlin, "Branch and bound methods for integer and non-convex programming", in: J. Abadie, ed., *Integer and nonlinear programming* (North-Holland, Amsterdam, 1970).

[30] A.F. Veinott and G.B. Dantzig, "Integral extreme points", *SIAM Review* 10 (1968) 371–372.

Chapter 10
Disjunctive Programming

Egon Balas

Introduction by *Egon Balas*

In April 1967 I and my family arrived into the US as fresh immigrants from behind the Iron Curtain. After a fruitful semester spent with George Dantzig's group in Stanford, I started working at CMU. My debut in integer programming and entry ticket into Academia was the additive algorithm for 0-1 programming [B65], an implicit enumeration procedure based on logical tests akin to what today goes under the name of constraint propagation. As it used only additions and comparisons, it was easy to implement and was highly popular for a while. However, I was aware of its limitations and soon after I joined CMU I started investigating cutting plane procedures, trying to use for this purpose the tools of convex analysis: support functions and their level sets, maximal convex extensions, polarity, etc. During the five years starting in 1969, I proposed a number of procedures based on the central idea of intersection cuts [2] (numbered references are to those at the end of the paper, whereas mnemonicized ones are to the ones listed at the end of this introduction): Given any convex set S containing the LP optimum of a mixed integer program (MIP) but containing no feasible integer point in its interior, one can generate a valid cut by intersecting the boundary of S with the extreme rays of the cone defined by the optimal solution to the linear programming relaxation of the MIP and taking the hyperplane defined by the intersection points as the cut. The search for the most appropriate sets S in this role has led to the concept of outer polars and related constructs [3, 6]. In our days, the idea of intersection cuts has been revived in the form of cutting planes from convex sets with lattice-free interiors, and is the object of numerous investigations (e.g., [ALWW07], [BC07], [CM07], [DW07]).

It was this line of research that has led to the idea of disjunctive programming, through a process outlined in section 1 of the paper below. Optimizing a function

Egon Balas
Carnegie Mellon University, Pittsburgh, USA
e-mail: eb17@andrew.cmu.edu

M. Jünger et al. (eds.), *50 Years of Integer Programming 1958-2008*,
DOI 10.1007/978-3-540-68279-0_10, © Springer-Verlag Berlin Heidelberg 2010

subject to a set of linear inequalities connected by conjunction or disjunction is a special type of nonconvex programming problem called disjunctive programming. Mixed 0-1 programming is its most important special case. More broadly speaking, disjunctive programming is optimization over a union of polyhedra. The basic document on disjunctive programming is the July 1974 technical report "Disjunctive Programming: Properties of the convex hull of feasible points," MSRR #348, referenced as [10] in the survey below, which however has not appeared in print until 24 years later, when it was published as an invited paper with a preface by Gérard Cornuéjols and Bill Pulleyblank [B98]. The reasons for this situation are complex. The 1974 paper was not rejected, but the report I received at the end of a very long refereeing process was asking for a revision that would have involved a major rewriting effort at a time when I was engaged in other, complementary research projects. As a result, when shortly thereafter I was invited by the late Peter Hammer to prepare an introductory survey of disjunctive programming for the upcoming 1977 ARIDAM conference in Vancouver, I decided to incorporate into that survey the main results of MSSR #348 and forego its publication as a stand alone paper. This is how this survey came about.

Looking back on that decision, I find that not much of substance was lost. Sections 2–6 of this survey adequately summarize the main findings of the 1974 paper, with one important exception, which I will now explain. MSRR #348 explored the basic properties of disjunctive programs from a polyhedral point of view. It gave two compact characterizations of the convex hull of a union P of polyhedra $P_i = \{x : A^i x \geq b^i\}$, $i \in Q$, in a higher dimensional space, a procedure we call today extended formulation. The first one describes conv$P$ as the set of points x satisfying $x = \sum(x^i : i \in Q)$ for some (x^i, x^i_0), $i \in Q$, such that $A^i x^i \geq b^i x^i_0, i \in Q$, $\sum(x^i_0 : i \in Q) = 1$, $x^i_0 \geq 0, i \in Q$. The second one describes the facets of conv$P$ as the vertices of the reverse polar of P, defined as $P^{\#} := \{y : yx \geq 1 \quad \forall x \in P\}$, shown to be the set of points y satisfying $y \geq u^i A^i, i \in Q$ for some $u^i \geq 0$ such that $u^i b^i \geq 1$, $i \in Q$. Both characterizations are linear in $|Q|$, the number of polyhedra in the union. In a certain sense the two characterizations are equivalent: projecting the higher-dimensional polyhedron of the convex hull characterization onto the x-space yields the set of all valid inequalities and conversely, taking the reverse polar of the reverse polar yields the convex hull. Thus in the survey paper I only included the second characterization, so that the first one did not appear in print until 1985 [B85]. But it is precisely this first characterization which has served as a prototype for the many extended formulations that have proved to be such prolific tools for polyhedral analysis of combinatorial problems, starting with the early studies of this type in the 1980's [BP83], [BP89], [BLP89], and continuing with a plethora of results, the more recent ones being exemplified by [PW06], [A06], [CW08]. It is also this first characterization that was generalized to nonlinear disjunctive programming [SM99] and was extensively used in the modeling of a variety of practical situations in industry.

Habent sua fata libelli, goes the Latin saying: *books have their own fate*. This apparently also applies to papers or theorems or discoveries. While the work on disjunctive programming, including the cutting planes that it entailed, stirred little if any enthusiasm at the time of its inception, about 15 years later when Sebastian Ceria, Gérard Cornuéjols and myself recast essentially the same results in a new framework which we called lift-and-project [BCC93], the reaction was quite different. This time our work was focused on algorithmic aspects, with the cutting planes generated in rounds and embedded into an enumerative framework (branch and cut), and was accompanied by the development of an efficient computer code (MIPO, developed by Sebastian) that was able to solve many problem instances that had been impervious to solution by branch-and-bound alone. Our interest in returning to the ideas of disjunctive programming was prompted by the exciting work of Lovász and Schrijver on matrix cones [LS91]. We discovered that a streamlined version of the Lovász-Schrijver procedure was isomorphic to the disjunctive programming procedure for generating the integer hull of a 0-1 program by starting with the higher dimensional representation and projecting it onto the original space. Thus the Lovász-Schrijver Theorem according to which n applications of this procedure (n being the number of 0-1 variables) yields the integer hull, follows directly from the sequential convexification procedure for facial disjunctive programs, of which 0-1 programs are a prime example (see Section 6 below). The reader will no doubt recognize in the linear program $P_1^*(g,\alpha_0)$ preceding Theorem 4.4 of section 4 below, the ancestor of the cut generating linear program of the lift-and-project (L&P) algorithm [BCC93], which is the specialization of $P_1^*(g,\alpha_0)$ to the case of the disjunction $x_k = 0$ or $x_k = 1$.

The computational success of L&P cuts triggered a strong revival of interest in cutting planes. Gérard and Sebastian soon discovered [BCCN96] that mixed integer Gomory (MIG) cuts, when used in the MIPO fashion, i.e., generated in rounds and embedded into a branch-and-bound framework, could also solve many of the problem instances unsolved at the time. Since the MIG cuts were easier to implement than the L&P ones, they were the first to find their way into the commercial codes. The combination of cutting planes with branch and bound played a central role in the revolution in the state of the art in integer programming that started in the mid-90s. The commercial implementation of lift-and-project cuts had to await the discovery of a method [BP03], [P03] for generating them directly from the LP simplex tableau, without explicit recourse to the higher dimensional cut generating linear program. Today, due to the efforts of Pierre Bonami [BB07], an open-source implementation is also publicly available [COIN-OR].

References

[A06] A. Atamtürk, *Strong formulations of robust mixed 0-1 programming*, Mathematical Programming 108 (2006) 235–250.

[ALWW07] K. Andersen, Q. Louveaux, R. Weismantel and L.A. Wolsey, *Inequalities from two rows of the simplex tableau*, Integer Programming and Combinatorial Optimization IPCO 12 (M. Fischetti and D.P. Williamson, eds.), Springer, 2007, pp. 1–16.

[B65] E. Balas, *An additive algorithm for solving linear programs in 0-1 variables*, Operations Research 13 (1965) 517–546.

[B85] E. Balas, *Disjunctive programming and a hierarchy of relaxations for discrete optimization problems*, SIAM Journal on Algebraic and Discrete Methods 6 (1985) 466–486.

[B98] E. Balas, *Disjunctive programming: Properties of the convex hull of feasible points*, Invited paper with a Foreword by G. Cornuéjols and G. Pulleyblank, Discrete Applied Mathematics 89 (1998) 1–44.

[BB07] E. Balas and P. Bonami, *New variants of lift-and-project cut generation from the LP tableau: Open source implementation and testing*, Integer Programming and Combinatorial Oprtimization IPCO 12 (M. Fischetti and D.P. Williamson, eds.), Springer, 2007, pp. 89–103.

[BC07] V. Borozan and G. Cornuéjols, *Minimal inequalities for integer constraints*, Technical Report, Tepper School, Carnegie Mellon University, 2007.

[BCC93] E. Balas, S. Ceria and G. Cornuéjols, *A lift-and-project cutting plane algorithm for mixed 0-1 programs*, Mathematical Programming 58 (1993) 295–324.

[BCCN96] E. Balas, S. Ceria, G. Cornuéjols and N. Natraj, *Gomory cuts revisited*, Operations Research Letters 19 (1996) 1–10.

[BLP89] M. Ball, W. Liu and W.R. Pulleyblank, *Two-terminal Steiner tree polyhedra*, Contributions to Operations Research and Economics, MIT Press, 1989, pp. 251–284.

[BP83] E. Balas and W.R. Pulleyblank, *The perfectly matchable subgraph polytope of a bipartite graph*, Networks 13 (1983) 495–516.

[BP89] E. Balas and W.R. Pulleyblank, *The perfectly matchable subgraph polytope of an arbitrary graph*, Combinatorica 9 (1989) 321–337.

[BP03] E. Balas and M. Perregaard, *A precise correspondence between lift-and-project cuts, simple disjunctive cuts, and mixed integer Gomory cuts for 0-1 programming*, Mathematical Programming 94 (2003) 221–245.

[CM07] G. Cornuéjols and F. Margot, *On the facets of mixed integer programs with two integer variables and two constraints*, Technical Report, Tepper School, Carnegie Mellon University, 2007.

[COIN-OR] http://www.coin-or.org

[CW08] M. Conforti and L.A. Wolsey, *Compact formulations as a union of polyhedra*, Mathematical Programming 114 (2008) 277–289.

[DW07] S.S. Dey and L.A. Wolsey, *Lifting integer variables in minimal inequalities corresponding to lattice-free triangles*, Integer Programming and Combinatorial Optimization IPCO 13 (A. Lodi, A. Panconesi, and G. Rinaldi, eds.), Springer, 2008, pp. 463–475.

[LS91] L. Lovász and A. Schrijver, *Cones of matrices and set functions and 0-1 optimization*, SIAM Journal of Optimization 1 (1991) 166–190.

[P03] M. Perregaard, *A Practical implementation of lift-and-project cuts: a computational exploration of lift-and-project with XPRESS-MP*, International Symposium on Mathematical Programming, Copenhagen, August 2003.

[PW06] Y. Pochet and L.A. Wolsey, *Production Planning by Mixed Integer Programming*, Springer, 2006.

[SM99] R. Stubbs and S. Mehrotra, *A branch and cut method for 0-1 mixed integer convex programming*, Mathematical Programming 86 (1999) 515–532.

The following article originally appeared as:

E. Balas, *Disjunctive Programming*, Discrete Optimization II (P.L. Hammer, E.L. Johnson, and B.H. Korte, eds.), Annals of Discrete Mathematics 5 (1979) 3–51.

Annals of Discrete Mathematics 5 (1979) 3–51

DISJUNCTIVE PROGRAMMING

Egon BALAS
Carnegie-Mellon University, Pittsburgh, PA 15213, U.S.A.

1. Introduction

This paper reviews some recent developments in the convex analysis approach to integer programming. A product of the last five years, these developments are based on viewing integer programs as disjunctive programs, i.e., linear programs with disjunctive constraints. Apart from the fact that this is the most natural and straightforward way of stating many problems involving logical conditions (dichotomies, implications, etc.), the disjunctive programming approach seems to be fruitful for zero-one programming both theoretically and practically. On the theoretical side, it provides neat structural characterizations which offer new insights. On the practical side, it produces a variety of cutting planes with desirable properties, and offers several ways of combining cutting planes with branch and bound.

The line of research that has led to the disjunctive programming approach originates in the work on intersection or convexity cuts by Young [40], Balas [2], Glover [23], Owen [36] and others (see also [13, 24, 42]). This geometrically motivated work can be described in terms of intersecting the edges of the cone originating at the linear programming optimum $\bar{x}$ with the boundary of some convex set S, whose interior contains $\bar{x}$ but no feasible integer points, and using the hyperplane defined by the intersection points as a cutting plane. An early forerunner of this kind of approach was a paper by Hoang Tuy [29].

In the early 70's, research on intersection or convexity cuts was pursued in two main directions. One, typified by [23, 18, 4], was aimed at obtaining stronger cuts by including into S some explicitly or implicitly enumerated feasible integer points. The other, initiated by [3], brought into play polarity, in particular outer polars (i.e., polars of the feasible set, scaled so as to contain all feasible 0–1 points in their boundary), and related concepts of convex analysis, like maximal convex extensions, support and gauge functions, etc. (see also [5, 15, 19]). Besides cutting planes, this second direction has also produced (see [5, 6]) a "constraint activating" method (computationally untested to date) based on the idea of "burying" the feasible set into the outer polar (without using cuts), by activating the problem constraints one by one, as needed. This research has yielded certain insights and produced reasonably strong cutting planes; but those procedures that were implemented (and, by the way, very few were) turned out to be computationally too expensive to be practically useful.

In 1973 Glover [25] discovered that intersection cuts derived from a convex polyhedron S can be strengthened by rotating the facets of S in certain ways, a procedure he called polyhedral annexation. This was an important step toward the development of the techniques discussed in this paper. The same conclusions were reached independently (and concomitantly) in a somewhat different context by Balas [7]. The new context was given by the recognition that intersection cuts could be viewed as derived from a disjunction. Indeed, requiring that no feasible integer point be contained in the interior of S, is the same as requiring every feasible integer point to satisfy at least one of the inequalities whose complements define the supporting halfspaces of S. This seemingly innocuous change of perspective proved to be extremely fruitful. For one thing, it led naturally and immediately to the consideration of disjunctive programs in their generality [8, 9], and hence to a characterization of all the valid inequalities for an integer program. By the same token, it offered the new possibility of generating cuts specially tailored for problems with a given structure. Besides, it offered a unified theoretical perspective on cutting planes and enumeration, as well as practical ways of combining the two approaches. Finally, it vastly simplified the proofs of many earlier results and opened the way to the subsequent developments to be discussed below.

Besides the above antecedents of the disjunctive programming approach there have been a few other papers concerned with linear (or nonlinear) programs with disjunctive constraints [26, 26a, 27]. The paper by Owen [36] deserves special mention, as the first occurrence of a cut with coefficients of different signs. However, these efforts were focused on special cases.

The main conceptual tool used in studying the structural properties of disjunctive programs is polarity. Besides the classical polar sets, we use a natural generalization of the latter which we call reverse polar sets [10]. This connects our research to the work of Fulkerson [20, 21, 22], whose blocking and antiblocking polyhedra are close relatives of the polar and reverse polar sets. There are also connections with the work of Tind [39] and Araoz [1]. One crucial difference in the way polars and reverse polars of a polyhedron are used in our work, versus that of the above mentioned authors, is the fact that we "dualize" the reverse polar $S^{\#}$ of a (disjunctive) set S, by representing it in terms of the inequalities (of the disjunction) defining S, rather than in terms of the points of S. It is precisely this element which leads to a linear programming characterization of the convex hull of feasible points of a disjunctive program.

Except for the specific applications described in Sections 7 and 8, which are new, the results reviewed here are from [8–12, 14]. We tried to make this review self-contained, giving complete proofs for most of the statements. Many of the results are illustrated on numerical examples. For further details and related developments the reader is referred to the above papers, as well as those of

Glover [25] and Jeroslow [32]. Related theoretical developments are also to be found in Jeroslow [30, 31], Blair and Jeroslow [17], Borwein [17a], Zemel [41], while some related computational procedures are discussed in Balas [12].

This paper is organized as follows. Section 2 introduces some basic concepts and terminology, and discusses ways of formulating disjunctive programs (DP). Section 3 contains the basic characterization of the family of valid inequalities for a DP, and discusses some of the implications of this general principle for deriving cutting planes. Section 4 extends the duality theorem of linear programming to DP. Section 5 discusses the basic properties of reverse polars and uses them to characterize the convex hull of a disjunctive set (the feasible set of a DP). It shows how the facets of the convex hull of the feasible set of a DP in n variables, defined by a disjunction with q terms, can be obtained from a linear program with $q(n+1)$ constraints.

In Section 6 we address the intriguing question of whether the convex hull of a disjunctive set can be generated sequentially, by imposing one by one the disjunctions of the conjunctive normal form, and producing at each step the convex hull of a disjunctive set with one elementary disjunction. We answer the question in the negative for the case of a general DP or a general integer program, but in the positive for a class of DP called facial, which subsumes the general (pure or mixed) zero-one program, the linear complementarity problem, the nonconvex (linearly constrained) quadratic program, etc.

The first 6 sections treat the integrality constraints of an (otherwise) linear program as disjunctions. When it comes to generating cutting planes from a particular noninteger simplex tableau, the disjunctions that can be used effectively are those involving the basic variables. Section 7 discusses a principle for strengthening cutting planes derived from a disjunction, by using the integrality conditions (if any) on the nonbasic variables. Section 8 illustrates on the case of multiple choice constraints how the procedures of Sections 3 and 7 can take advantage of problem structure. Section 9 discusses some ways in which disjunctive programming can be used to combine branch and bound with cutting planes. In particular, if LP_k, $k \in Q$, are the subproblems associated with the active nodes of the search tree at a given stage of a branch and bound procedure applied to a mixed integer program P, it is shown that a cut can be derived from the cost rows of the simplex tableaux associated with the problems LP_k, which provides a bound on the value of P, often considerably better than the one available from the branch and bound process. Finally, Section 10 deals with techniques for deriving disjunctions from conditional bounds. Cutting planes obtained from such disjunctions have been used with good results on large sparse set covering problems.

2. Linear programs with logical conditions

By disjunctive programming we mean linear programming with disjunctive constraints. Integer programs (pure and mixed) and a host of other nonconvex programming problems (the linear complementarity problem, the general quadratic program, separable nonlinear programs, bimatrix games, etc.) can be stated as linear programs with logical conditions. In the present context "logical conditions" means statements about linear inequalities, involving the operations "and" ($\wedge$, conjunction — sometimes denoted by juxtaposition), "or" ($\vee$, disjunction), "complement of" ($\neg$, negation). The operation "if $\cdots$ then" ($\Rightarrow$, implication) is known to be equivalent to a disjunction. The negation of a conjunctive set of inequalities is a disjunction whose terms are the same inequalities. The operation of conjunction applied to linear inequalities gives rise to (convex) polyhedral sets. The disjunctions are thus the crucial elements of a logical condition (the ones that make the constraint set nonconvex), and that is why we call this type of problem a *disjunctive program.*

A disjunctive program (DP) is then a problem of the form

$$\min\{cx \mid Ax \geqslant a_0,\ x \geqslant 0,\ x \in L\},$$

where A is an $m \times n$ matrix, a_0 is an m-vector, and L is a set of logical conditions. Since the latter can be expressed in many ways, there are many different forms in which a DP can be stated. Two of them are fundamental.

The constraint set of a DP (and the DP itself) is said to be in *disjunctive normal form* if it is defined by a disjunction whose terms do not contain further disjunctions; and in *conjunctive normal form,* if it is defined by a conjunction whose terms do not contain further conjunctions. The disjunctive normal form is then

$$\bigvee_{h \in Q} \begin{pmatrix} A^h x \geqslant a_0^h \\ x \geqslant 0 \end{pmatrix}, \tag{2.1}$$

while the conjunctive normal form is

$$\begin{aligned} &Ax \geqslant a_0, \qquad x \geqslant 0, \\ &\bigvee_{i \in Q_j} (d^i x \geqslant d_{i0}), \qquad j \in S \end{aligned} \tag{2.2}$$

or, alternatively,

$$\begin{pmatrix} Ax \geqslant a_0 \\ x \geqslant 0 \end{pmatrix} \wedge \left[\bigvee_{i \in Q_1} (d^i x \geqslant d_{i0})\right] \wedge \cdots \wedge \left[\bigvee_{i \in Q_{|S|}} (d^i x \geqslant d_{i0})\right]. \tag{2.2'}$$

Here each d^i is an n-vector and each d_{i0} a scalar, while the sets Q and Q_j, $j \in S$, may or may not be finite. The connection between the two forms is that each term of the disjunction (2.1) has, besides the $m+n$ inequalities of the system $Ax \geqslant a_0$, $x \geqslant 0$, precisely one inequality $d^i x \geqslant d_{i0}$, $i \in Q_j$, from each disjunction $j \in S$ of (2.2), and that all distinct systems $A^h x \geqslant a_0^h$, $x \geqslant 0$ with this property are present among the terms of (2.1); so that, if Q (and hence each Q_j, $j \in S$) is finite, then $|Q| = |\Pi_{j \in S} Q_j|$, where Π stands for cartesian product. Since the operations $\wedge$ and $\vee$ are distributive with respect to each other [i.e., if A, B, C are inequalities,

$A \wedge (B \vee C) = AB \vee AC$, and $A \vee (BC) = (A \vee B) \wedge (A \vee C)$], any logical condition involving these operations can be brought to any of the two fundamental forms, and each of the latter can be obtained from the other one.

We illustrate the meaning of these two forms on the case when the DP is a zero-one program in n variables. Then the disjunctive normal form (2.1) is

$$\bigvee_{h \in Q} (Ax \geq a_0, x \geq 0, x = x^h)$$

where $x^1, \ldots, x^{|Q|}$ is the set of all 0–1 points, and $|Q| = 2^n$; whereas the conjunctive normal form is

$$Ax \geq 0, \qquad x \geq 0, \qquad (x_j = 0) \vee (x_j = 1), \qquad j = 1, \ldots, n.$$

Once the inequalities occurring in the conjunctions and/or disjunctions of a DP are given, the disjunctive and conjunctive normal forms are unique. It is a fact of crucial practical importance, however, that the inequalities expressing the conditions of a given problem can be chosen in more than one way. For instance, the constraint set

$$\begin{aligned} &3x_1 + x_2 - 2x_3 + x_4 \leq 1, \\ &x_1 + x_2 + x_3 + x_4 \leq 1, \\ &x_j = 0 \text{ or } 1, \qquad j = 1, \ldots, 4, \end{aligned}$$

when put in disjunctive normal form, becomes a disjunction with $2^4 = 16$ terms; but the same constraint set can also be expressed as

$$\begin{aligned} &3x_1 + x_2 - 2x_3 + x_4 \leq 1, \\ &\bigvee_{i=1}^{4} \begin{pmatrix} x_i = 1 & \\ x_j = 0, & j \neq i \end{pmatrix} \vee (x_j = 0, \forall j), \end{aligned}$$

which gives rise to a disjunction with only 5 terms.

3. The basic principle of disjunctive programming

A constraint B is said to be a *consequence of*, or *implied by*, a constraint A, if every x that satisfies A also satisfies B. We are interested in the family of inequalities implied by the constraint set of a general disjunctive program (DP). All valid cutting planes for a DP belong of course to this family. On the other hand, the set of points satisfying all the inequalities in the family is precisely the convex hull of the set of feasible solutions to the DP. A characterization of this family is given in the next theorem, which is an easy but important generalization of a classical result. Let $x \in \mathbf{R}^n$, $\alpha \in \mathbf{R}^n$, $\alpha_0 \in \mathbf{R}$, $A^h \in \mathbf{R}^{m_h \times n}$, $a_0^h \in \mathbf{R}^{m_h}$, $h \in Q$ (not necessarily finite) and let a_j^h be the jth column of A^h, $h \in Q$, $j \in N = \{1, \ldots, n\}$.

Theorem 3.1. *The inequality* $\alpha x \geq \alpha_0$ *is a consequence of the constraint*

$$\bigvee_{h\in Q}\begin{pmatrix} A^h x \geq a_0^h \\ x \geq 0 \end{pmatrix} \tag{3.1}$$

if and only if there exists a set of $\theta^h \in \mathbf{R}^{m_h}$, $\theta^h \geq 0$, $h \in Q^*$, *satisfying*

$$\alpha \geq \theta^h A^h \quad \text{and} \quad \alpha_0 \leq \theta^h a_0^h, \quad \forall h \in Q^*, \tag{3.2}$$

where Q^* *is the set of those* $h \in Q$ *such that the system* $A^h x \geq a_0$, $x \geq 0$ *is consistent.*

Proof. $\alpha x \geq \alpha_0$ is a consequence of (3.1) if and only if it is a consequence of each term $h \in Q^*$ of (3.1). But according to a classical result on linear inequalities (see, for instance, Theorem 1.4.4 of [38], or Theorem 22.3 of [37]), this is the case if and only if the conditions stated in the theorem hold. □

Remark 3.1.1. If the ith inequality of a system $h \in Q^*$ of (3.1) is replaced by an equation, the ith component of θ^h is to be made unconstrained. If the variable x_j in (3.1) is allowed to be unconstrained, the jth inequality of each system $\alpha \geq \theta^h A^h$, $h \in Q^*$, is to be replaced by the corresponding equation in the "if" part of the statement.

With these changes, Theorem 3.1 remains true.

An alternative way of stating (3.2) is

$$\begin{aligned} &\alpha_j \geq \sup_{h\in Q^*} \theta^h a_j^h, \qquad j \in N, \\ &\alpha_0 \leq \inf_{h\in Q^*} \theta^h a_0^h. \end{aligned} \tag{3.3}$$

Since $Q^* \subseteq Q$, the if part of the Theorem remains of course valid if Q^* is replaced by Q.

Since (3.3) defines *all* the valid inequalities for (3.1), every valid cutting plane for a disjunctive program can be obtained from (3.3) by choosing suitable multipliers θ_i^h. If we think of (3.1) as being expressed in terms of the nonbasic variables of a basic optimal solution to the linear program associated with a DP, then a valid inequality $\alpha x \geq \alpha_0$ cuts off the optimal linear programming solution (corresponding to $x_j = 0$, $j \in N$) if and only if $\alpha_0 > 0$; hence α_0 will have to be fixed at a positive value. Inequalities with $\alpha_0 \leq 0$ may still cut off parts of the linear programming feasible set, but not the optimal solution $x = 0$.

The special case when each system $A^h x \geq a_0^h$, $h \in Q$, consists of the single inequality $a^h x \geq a_{h0}$ (a^h a vector, a_{h0} a positive scalar) deserves special mention. In this case, choosing multipliers $\theta^h = 1/a_{h0}$, $h \in Q$, we obtain the inequality

$$\sum_{j\in J}\left(\max_{h\in Q} a_j^h / a_{h0}\right) x_j \geq 1, \tag{3.4}$$

which (for Q finite) is Owen's cut [36]. It can also be viewed as a slightly improved version of the intersection cut from the convex set

$$S=\{x \mid a^h x \leq a_{h0},\ h \in Q\},$$

which has the same coefficients as (3.4) except for those (if any) $j \in J$ such that $a_j^h < 0$, $\forall h \in Q$. For the latter, the intersection cut from S has zero coefficients whereas the corresponding coefficients of (3.4) are negative.

Whenever all the coefficients of (3.4) are positive (in terms of intersection cuts, this corresponds to the case when S is bounded), (3.4) is the strongest inequality implied by the disjunction $\vee_{h \in Q}(a^h x \geq a_{h0})$; in the presence of negative coefficients, however, (3.4) can sometimes be further strengthened.

Due to the generality of the family of inequalities defined by (3.3), not only can the earlier cuts of the literature be easily recovered by an appropriate choice of the multipliers θ^h (see [8] for details), but putting them in the form (3.3) indicates, by the same token, ways in which they can be strengthened by appropriate changes in the multipliers.

A significant new feature of the cutting planes defined by (3.3) consists in the fact that they can have coefficients of different signs. The classical cutting planes, as well as the early intersection/convexity cuts and the group theoretic cutting planes (including those corresponding to facets of the corner polyhedron), are all restricted to positive coefficients (when stated in the form $\geq$, in terms of the nonbasic variables of the tableau from which they were derived). This important limitation, which tends to produce degeneracy in dual cutting plane algorithms, can often be overcome in the case of the cuts obtained from (3.3) by an appropriate choice of multipliers.

Another important feature of the principle expressed in Theorem 3.1 for generating cutting planes is the fact that in formulating a given integer program as a disjunctive program, one can take advantage of any particular structure the problem may have. In Section 7 we will illustrate this on some frequently occurring structures. We finish this section by an example of a cut for a general mixed integer program.

Example 3.1. Consider the mixed integer program whose constraint set is

$$x_1 = 0.2 + 0.4(-x_3) + 1.3(-x_4) - 0.01(-x_5) + 0.07(-x_6)$$
$$x_2 = 0.9 - 0.3(-x_3) + 0.4(-x_4) - 0.04(-x_5) + 0.1(-x_6)$$
$$x_j \geq 0, \quad j = 1, \ldots, 6, \quad x_j \text{ integer}, \quad j = 1, \ldots, 4.$$

This problem is taken from Johnson's paper [35], which also lists six cutting planes derived from the extreme valid inequalities for the associated group problem:

$$0.75x_3+0.875x_4+0.0125x_5+0.35x_6\geq 1,$$
$$0.778x_3+0.444x_4+0.40x_5+0.111x_6\geq 1,$$
$$0.333x_3+0.667x_4+0.033x_5+0.35x_6\geq 1,$$
$$0.50x_3+x_4+0.40x_5+0.25x_6\geq 1,$$
$$0.444x_3+0.333x_4+0.055x_5+0.478x_6\geq 1,$$
$$0.394x_3+0.636x_4+0.346x_5+0.155x_6\geq 1.$$

The first two of these inequalities are the mixed-integer Gomory cuts derived from the row of x_1 and x_2 respectively. To show how they can be improved, we first derive them as they are. To do this, for a row of the form

$$x_i=a_{i0}+\sum_{j\in J}a_{ij}(-x_j),$$

with x_j integer-constrained for $j\in J_1$, continuous for $j\in J_2$, one defines $f_{ij}=a_{ij}-[a_{ij}]$, $j\in J\cup\{0\}$, and $\varphi_{i0}=f_{i0}$,

$$\varphi_{ij}=\begin{cases} f_{ij}, & j\in J_1^+=\{j\in J_1 \mid f_{i0}\geq f_{ij}\},\\ f_{ij}-1, & j\in J_1^-=\{j\in J_1 \mid f_{i0}<f_{ij}\},\\ a_{ij}, & j\in J_2.\end{cases}$$

Then every x which satisfies the above equation and the integrality constraints on x_j, $j\in J_1\cup\{i\}$, also satisfies the condition

$$y_i=\varphi_{i0}+\sum_{j\in J}\varphi_{ij}(-x_j),\qquad y_i \text{ integer.}$$

For the two equations of the example, the resulting conditions are

$$y_1=0.2-0.6(-x_3)-0.7(-x_4)-0.01(-x_5)+0.07(-x_6),\qquad y_1 \text{ integer,}$$
$$y_2=0.9+0.7(-x_3)+0.4(-x_4)-0.04(-x_5)+0.1(-x_6),\qquad y_2 \text{ integer.}$$

Since each y_i is integer-constrained, they have to satisfy the disjunction $y_i\leq 0\vee y_i\geq 1$. Applying the above theorem with multipliers $\theta_i=1/a_{i0}$ then gives the cut (3.4) which in the two cases $i=1,2$ is

$$\frac{0.6}{0.8}x_3+\frac{0.7}{0.8}x_4+\frac{0.01}{0.8}x_5+\frac{0.07}{0.2}x_6\geq 1,$$

and

$$\frac{0.7}{0.9}x_3+\frac{0.4}{0.9}x_4+\frac{0.04}{0.1}x_5+\frac{0.1}{0.9}x_6\geq 1.$$

These are precisely the first two inequalities of the above list.

Since all cuts discussed here are stated in the form ≥ 1, the smaller the jth coefficient, the stronger is the cut in the direction j. We would thus like to reduce the size of the coefficients as much as possible.

Now suppose that instead of $y_1 \leq 0 \vee y_1 \geq 1$, we use the disjunction

$$\{y_1 \leq 0\} \vee \begin{Bmatrix} y_1 \geq 1 \\ x_1 \geq 0 \end{Bmatrix},$$

which of course is also satisfied by every feasible x.

Then, applying Theorem 3.1 with multipliers 5, 5 and 15 for $y_1 \leq 0$, $y_1 \geq 1$ and $x_1 \geq 0$ respectively, we obtain the cut whose coefficients are

$$\max\left\{\frac{5\times(-0.6)}{5\times 0.2}, \frac{5\times 0.6+15\times(-0.4)}{5\times 0.8+15\times(-0.2)}\right\} = -3,$$

$$\max\left\{\frac{5\times(-0.7)}{5\times 0.2}, \frac{5\times 0.7+15\times(-1.3)}{5\times 0.8+15\times(-0.2)}\right\} = -3.5,$$

$$\max\left\{\frac{5\times(-0.01)}{5\times 0.2}, \frac{5\times 0.01+15\times 0.01}{5\times 0.8+15\times(-0.2)}\right\} = 0.2,$$

$$\max\left\{\frac{5\times 0.07}{5\times 0.2}, \frac{5\times(-0.07)+15\times(-0.07)}{5\times 0.8+15\times(-0.2)}\right\} = 0.35,$$

that is

$$-3x_3 - 3.5x_4 + 0.2x_5 + 0.35x_6 \geq 1.$$

The sum of coefficients on the left hand side has been reduced from 1.9875 to -5.95.

Similarly, for the second cut, if instead of $y_2 \leq 0 \vee y_2 \geq 1$ we use the disjunction

$$\begin{Bmatrix} y_2 \leq 0 \\ x_1 \geq 0 \end{Bmatrix} \vee \{y_2 \geq 1\},$$

with multipliers 10, 40 and 10 for $y_2 \leq 0$, $x_1 \geq 0$ and $y_2 \geq 1$ respectively, we obtain the cut

$$-7x_3 - 4x_4 + 0.4x_5 - x_6 \geq 1.$$

Here the sum of left hand side coefficients has been reduced from 1.733 to -11.6.

4. Duality

In this section we state a duality theorem for disjunctive programs, which generalizes to this class of problems the duality theorem of linear programming.

Consider the disjunctive program

$$z_0 = \min cx,$$

$$\text{(P)} \qquad \bigvee_{h \in Q} \left\{ \begin{array}{c} A^h x \geqslant b^h \\ x \geqslant 0 \end{array} \right\},$$

where A^h is a matrix and b^h a vector, $\forall h \in Q$.

We define the *dual* of (P) to be the problem

$$w_0 = \max w,$$

$$\text{(D)} \qquad \bigwedge_{h \in Q} \left\{ \begin{array}{r} w - u^h b^h \leqslant 0 \\ u^h A^h \leqslant c \\ u^h \geqslant 0 \end{array} \right\}.$$

The constraint set of (D) requires each u^h, $h \in Q$, to satisfy the corresponding bracketed system, and w to satisfy each of them.

Let

$$X_h = \{x \mid A^h x \geqslant b^h, x \geqslant 0\}, \qquad \bar{X}_h = \{x \mid A^h x \geqslant 0, x \geqslant 0\};$$

$$U_h = \{u^h \mid u^h A^h \leqslant c, u^h \geqslant 0\}, \qquad \bar{U}_h = \{u^h \mid u^h A^h \leqslant 0, u^h \geqslant 0\}.$$

Further, let

$$Q^* = \{h \in Q \mid X_h \neq \emptyset\}, \qquad Q^{**} = \{h \in Q \mid U_h \neq \emptyset\}.$$

We will assume the following

Regularity condition:

$$(Q^* \neq \emptyset, Q \setminus Q^{**} \neq \emptyset) \Rightarrow Q^* \setminus Q^{**} \neq \emptyset;$$

i.e., if (P) is feasible and (D) is infeasible, then there exists $h \in Q$ such that $X_h \neq \emptyset$, $U_h = \emptyset$.

Theorem 4.1. *Assume that* (P) *and* (D) *satisfy the regularity condition. Then exactly one of the following two situations holds.*

(1) *Both problems are feasible; each has an optimal solution and* $z_0 = w_0$.

(2) *One of the problems is infeasible; the other one either is infeasible or has no finite optimum.*

Proof. (1) Assume that both (P) and (D) are feasible. If (P) has no finite minimum, then there exists $h \in Q$ such that $\bar{X}_h \neq \emptyset$ and $\bar{x} \in \bar{X}_h$ such that $c\bar{x} < 0$. But then $U_h = \emptyset$, i.e., (D) is infeasible; a contradiction.

Thus (P) has an optimal solution, say $\bar{x}$. Then the inequality $cx \geqslant z_0$ is a consequence of the constraint set of (P); i.e., $x \in X_h$ implies $cx \geqslant z_0$, $\forall h \in Q$. But then for all $h \in Q^*$, there exists $u^h \in U_h$ such that $u^h b^h \geqslant z_0$. Further, since (D) is feasible, for each $h \in Q \setminus Q^*$ there exists $\hat{u}^h \in U_h$; and since $X_h = \emptyset$ (for $h \in Q \setminus Q^*$), there also exists $\bar{u}^h \in \bar{U}_h$ such that $\bar{u}^h b^h > 0$, $\forall h \in Q \setminus Q^*$. But then, defining

$$u^h(\lambda) = \hat{u}^h + \lambda \bar{u}^h, \quad h \in Q \setminus Q^*,$$

for λ sufficiently large, $u^h(\lambda) \in U_h$, $u^h(\lambda)b^h \geqslant z_0$, $\forall h \in Q \setminus Q^*$.

Hence for all $h \in Q$, there exist vectors u^h satisfying the constraints of (D) for $w = z_0$. To show that this is the maximal value of w, we note that since $\bar{x}$ is optimal for (P), there exists $h \in Q$ such that

$$c\bar{x} = \min\{cx \mid x \in X_h\}.$$

But then by linear programming duality,

$$\begin{aligned} c\bar{x} &= \max\{u^h b^h \mid u^h \in U_h\} \\ &= \max\{w \mid w - u^h b^h \leqslant 0,\ u^h \in U_h\} \\ &\geqslant \max\left\{w \mid \bigwedge_{h \in Q} (w - u^h b^h \leqslant 0,\ u^h \in U_h)\right\} \end{aligned}$$

i.e., $w \leqslant z_0$, and hence the maximum value of w is $w_0 = z_0$.

(2) Assume that at least one of (P) and (D) is infeasible. If (P) is infeasible, $X_h = \emptyset$, $\forall h \in Q$; hence for all $h \in Q$, there exists $\bar{u}^h \in \bar{U}_h$ such that $\bar{u}^h b^h > 0$.

If (D) is also infeasible, we are done. Otherwise, for each $h \in Q$ there exists $\hat{u} \in U_h$. But then defining

$$u^h(\lambda) = \hat{u}^h + \lambda \bar{u}^h, \qquad h \in Q,$$

$u^h(\lambda) \in U_h$, $h \in Q$, for all $\lambda > 0$, and since $\bar{u}^h b^h > 0$, $\forall h \in Q$, w can be made arbitrarily large by increasing λ; i.e., (D) has no finite optimum.

Conversely, if (D) is infeasible, then either (P) is infeasible and we are done, or else, from the regularity condition, $Q^* \setminus Q^{**} \neq \emptyset$; and for $h \in Q^* \setminus Q^{**}$ there exists $\hat{x} \in X_h$ and $\bar{x} \in \bar{X}_h$ such that $c\bar{x} < 0$. But then

$$x(\mu) = \hat{x} + \mu \bar{x}$$

is a feasible solution to (P) for any $\mu > 0$, and since $c\bar{x} < 0$, z can be made arbitrarily small by increasing μ; i.e., (P) has no finite optimum. □

The above theorem asserts that either situation 1 or situation 2 holds for (P) and (D) *if* the regularity condition is satisfied. The following Corollary shows that the condition is not only sufficient but also necessary.

Corollary 4.1.1. *If the regularity condition does not hold, then if* (P) *is feasible and* (D) *is infeasible,* (P) *has a finite minimum (i.e., there is a "duality gap").*

Proof. Let (P) be feasible, (D) infeasible, and $Q^* \setminus Q^{**} = \emptyset$, i.e., for every $h \in Q^*$, let $U_h \neq \emptyset$. Then for each $h \in Q^*$, $\min\{cx \mid x \in X_h\}$ is finite, hence (P) has a finite minimum. □

Remark. The theorem remains true if some of the variables of (P) [of (D)] are unconstrained, and the corresponding constraints of (D) [of (P)] are equalities.

The regularity condition can be expected to hold in all but some rather peculiar situations. In linear programming duality, the case when both the primal and the dual problem is infeasible only occurs for problems whose coefficient matrix A has the very special property that there exists $x\neq 0$, $u\neq 0$, satisfying the homogeneous system

$$\begin{aligned} &Ax\geqslant 0, \qquad x\geqslant 0;\\ &uA\leqslant 0, \qquad u\geqslant 0. \end{aligned}$$

In this context, our regularity condition requires that, if the primal problem is feasible and the dual is infeasible, then at least one of the matrices A^h whose associated sets U_h are infeasible, should not have the above mentioned special property.

Though most problems satisfy this requirement, nevertheless there *are* situations when the regularity condition breaks down, as illustrated by the following example.

Consider the disjunctive program

$$\text{(P)}\qquad \begin{aligned} &\min -x_1-2x_2,\\ &\left\{\begin{array}{c} -x_1+x_2\geqslant 0\\ -x_1-x_2\geqslant -2\\ x_1, x_2\geqslant 0 \end{array}\right\} \vee \left\{\begin{array}{c} -x_1+x_2\geqslant 0\\ x_1-x_2\geqslant 1\\ x_1, x_2\geqslant 0 \end{array}\right\} \end{aligned}$$

and its dual

$$\text{(D)}\qquad \begin{array}{llll} \max w, & & & \\ w & +2u_2^1 & & \leqslant 0,\\ & -u_1^1-\ u_2^1 & & \leqslant -1,\\ & u_1^1-\ u_2^1 & & \leqslant -2,\\ w & & -u_2^2 & \leqslant 0,\\ & & -u_1^2+u_2^2 & \leqslant -1,\\ & & u_1^2-u_2^2 & \leqslant -2,\\ & & u_i^k\geqslant 0, & i=1,2; \quad k=1,2. \end{array}$$

The primal problem (P) has an optimal solution $\bar{x}=(0,2)$, with $c\bar{x}=-4$; whereas the dual problem (D) is infeasible. This is due to the fact that $Q^*=\{1\}$, $Q\setminus Q^{**}=\{2\}$ and $X_2=\emptyset$, $U_2=\emptyset$, i.e., $Q^*\setminus Q^{**}=\emptyset$, hence the regularity condition is violated. Here

$$X_2=\left\{x\in\mathbf{R}_+^2 \,\middle|\, \begin{array}{c}-x_1+x_2\geqslant 0\\ x_1-x_2\geqslant 1\end{array}\right\},\qquad U_2=\left\{u\in\mathbf{R}_+^2 \,\middle|\, \begin{array}{c}-u_1^2+u_2^2\leqslant -1\\ u_1^2-u_2^2\leqslant -2\end{array}\right\}.$$

5. The convex hull of a disjunctive set

Having described the family of all valid inequalities, one is of course interested in identifying the strongest ones among the latter, i.e., the facets of the convex hull of feasible points of a disjunctive program.

If we denote the feasible set of a DP by

$$F=\left\{x\in\mathbf{R}^n \,\middle|\, \bigvee_{h\in Q}\begin{pmatrix}A^h x\geqslant a_0^h\\ x\geqslant 0\end{pmatrix}\right\},$$

then for a given scalar α_0, the family of inequalities $\alpha x\geqslant\alpha_0$ satisfied by all $x\in F$, i.e., the family of valid inequalities for the DP, is obviously isomorphic to the family of vectors $\alpha\in F^{\#}_{(\alpha_0)}$, where

$$F^{\#}_{(\alpha_0)}=\{y\in\mathbf{R}^n \mid yx\geqslant\alpha_0,\ \forall x\in F\},$$

in the sense that $\alpha x\geqslant\alpha_0$ is a valid inequality if and only if $\alpha\in F^{\#}_{(\alpha_0)}$.

In view of its relationship with ordinary polar sets, we call $F^{\#}_{(\alpha_0)}$ the *reverse polar* of F. Indeed, the ordinary polar set of F is

$$F^{\circ}=\{y\in\mathbf{R}^n \mid yx\leqslant 1,\ \forall x\in F\},$$

and if we denote by $F^{\circ}_{(\alpha_0)}$ the polar of F scaled by α_0, (i.e., the set obtained by replacing 1 with α_0 in F), then $F^{\#}_{(\alpha_0)}=-F^{\circ}_{(-\alpha_0)}$.

The size (as opposed to the sign) of α_0 is of no interest here. Therefore we will distinguish only between the 3 cases $\alpha_0>0$ (or $\alpha_0=1$), $\alpha_0=0$ and $\alpha_0<0$ (or $\alpha_0=-1$). (When the sign of α_0 makes no difference or is clear from the context, we will simply write $F^{\#}$.) For $\alpha_0\leqslant 0$, as mentioned above, $F^{\#}_{(\alpha_0)}$ is (the negative of) an ordinary polar set, whose properties are described in the literature. The most interesting case for us, however, is $\alpha_0>1$, since this is the only case when the inequality $\alpha x\geqslant\alpha_0$ cuts off the point $x=0$. This is why we need the concept of reverse polars.

For an arbitrary set $S\subseteq\mathbf{R}^n$ we will denote by cl S, conv S, cone S, int S and dim S, the closure, the convex hull, the conical hull, the interior and the dimension of S, respectively. For a polyhedral set $S\subseteq\mathbf{R}^n$ we will denote by vert S and dir S the set of vertices (extreme points) and the set of extreme direction vectors of S, respectively. For definitions and background material on these and related concepts (including ordinary polar sets), the reader is referred to [37] or [38] (see also [28]).

In [10] we showed that while some of the basic properties of polar sets carry over to reverse polars, others can only be recovered in a modified form. In the first category we mention (a) $(\lambda S)^{\#}=(1/\lambda)S^{\#}$; (b) $S\subseteq T\Rightarrow S^{\#}\supseteq T^{\#}$; (c) $(S\cup T)^{\#}=S^{\#}\cap T^{\#}$, properties which follow from the definitions. In the second one we state a few theorems, which are from [10] (see also [1]).

Theorem 5.1. (i) *If* $\alpha_0\leqslant 0$, *then* $S^{\#}\neq\emptyset$ *and*

$$0\in\operatorname{int}\operatorname{cl}\operatorname{conv} S\Leftrightarrow S^{\#}\ \textit{is bounded}$$

(ii) *If* $\alpha_0 > 0$, *then*

$$0 \in \text{cl conv } S \Leftrightarrow S^{\#} = \emptyset \Leftrightarrow S^{\#} \textit{ is bounded.}$$

Proof. (i) follows from the corresponding property of the ordinary polar S° of S and the fact that $S^{\#}_{(\alpha_0)} = -S^{\circ}_{(\alpha_0)}$.

(ii) For $\alpha_0 > 0$, if $S^{\#} \neq \emptyset$ there exists $y \in \mathbf{R}^n$ such that $xy \geqslant \alpha_0$, $\forall x \in \text{cl conv } S$. But $0 \cdot y < \alpha_0$, hence $0 \notin \text{cl conv } S$. Thus, if $0 \in \text{cl conv } S$, then $S^{\#} = \emptyset$; and hence $S^{\#}$ is bounded. Conversely, if $0 \notin \text{cl conv } S$, there exists a hyperplane $ax = \alpha_0$ separating 0 from cl conv S, i.e., such that ($\alpha_0 > 0$ and) $ax \geqslant \alpha_0$, $\forall x \in \text{cl conv } S$; which implies $a \in S^{\#}$, i.e., $S^{\#} \neq \emptyset$. It also implies $\lambda a \in S^{\#}$, $\forall \lambda > 1$, i.e., $S^{\#}$ is unbounded. □

From this point on, we restrict our attention to sets S whose convex hull is polyhedral and pointed. For the disjunctive set F, this condition is satisfied if Q is finite. Most of the results carry over to the general case, but proofs are simpler with the above assumptions.

Theorem 5.2. *If* cl conv S *is polyhedral, so is* $S^{\#}$.

Proof. Let $u_1, \ldots, u_p$ and $v_1, \ldots, v_q$ be the vertices and extreme direction vectors, respectively, of cl conv S. Then for every $y \in S$ there exist scalars $\lambda_i \geqslant 0$, $i = 1, \ldots, p$, $\mu_j \geqslant 0$, $j = 1, \ldots, q$, with $\sum_{i=1}^{p} \lambda_i = 1$, such that

$$y = \sum_{i=1}^{p} u_i \lambda_i + \sum_{j=1}^{q} v_j \mu_j,$$

and it can easily be seen that for arbitrary x, $xy \geqslant \alpha_0$, $\forall y \in S$, if and only if $xu_i \geqslant \alpha_0$, $i = 1, \ldots, p$, and $xv_j \geqslant 0$, $j = 1, \ldots, q$. Thus

$$S^{\#} = \left\{ x \in R^n \,\middle|\, \begin{matrix} xu_i \geqslant \alpha_0, & i = 1, \ldots, p \\ xv_j \geqslant 0\ , & j = 1, \ldots, q \end{matrix} \right\},$$

i.e., $S^{\#}$ is polyhedral. □

The next result describes the crucial involutory property of polar and reverse polar sets.

Theorem 5.3. *Assume* $S^{\#} \neq \emptyset$. *Then*

$$S^{\#\#} = \begin{cases} \text{cl conv } S + \text{cl cone } S, & \text{if } \alpha_0 > 0, \\ \text{cl cone } S, & \text{if } \alpha_0 = 0, \\ \text{cl conv } (S \cup \{0\}). & \text{if } \alpha_0 < 0. \end{cases}$$

Proof. $S^{\#\#} = \{x \in \mathbf{R}^n \mid xy \geqslant \alpha_0, \forall y \in S^{\#}\}$

$$= \left\{ x \in \mathbf{R}^n \,\middle|\, \begin{matrix} xy \geqslant \alpha_0, & \text{for all } y \text{ s.t.} \\ u_i y \geqslant \alpha_0, & i = 1, \ldots, p \\ v_i y \geqslant 0, & i = 1, \ldots, q \end{matrix} \right\}.$$

But $xy \geqslant \alpha_0$ is a consequence of the system $u_i y \geqslant \alpha_0$, $i=1,\dots,p$ and $v_i y \geqslant 0$, $i=1,\dots,q$ (consistent, since $S^{\#} \neq \emptyset$), if and only if there exists a set of $\theta_i \geqslant 0$, $i=1,\dots,p$, $\sigma_i \geqslant 0$, $i=1,\dots,q$, such that

$$x = \sum_{i=1}^{p} \theta_i u_i + \sum_{i=1}^{q} \sigma_i v_i, \tag{5.1}$$

with

$$\sum_{i=1}^{p} \theta_i \alpha_0 \geqslant \alpha_0.$$

Since $S^{\#}$ is polyhedral, so is $S^{\#\#}$. thus $S^{\#\#}$ is the closed set of points $x \in \mathbf{R}^n$ of the form (5.1) with $\theta_i \geqslant 0$, $i=1,\dots,p$ $\sigma_i \geqslant 0$, $i=1,\dots,q$, and

$$\sum_{i=1}^{p} \theta_i \begin{cases} \geqslant 1, & \text{if } \alpha_0 > 0, \\ \geqslant 0, & \text{if } \alpha_0 = 0, \\ \leqslant 1, & \text{if } \alpha_0 < 0. \end{cases}$$

But these are precisely the expressions for the three sets claimed in the theorem to be equal to $S^{\#\#}$ in the respective cases. □

Corollary 5.3.1. cl conv $S = S^{\#\#}_{(1)} \cap S^{\#\#}_{(-1)}$.

Proof. Follows immediately from the proof of Theorem 5.3, where $x \in S^{\#\#}_{(1)} \cap S^{\#\#}_{(-1)}$ corresponds to $\sum_{i=1}^{p} \theta_i = 1$. □

Example 5.1. Consider the disjunctive set

$$F = \left\{ x \in \mathbf{R}^2 \;\middle|\; \begin{array}{rr} -x_1 & -x_2 \geqslant -2 \\ x_1 & \geqslant 0 \\ & x_2 \geqslant 0 \\ \multicolumn{2}{c}{x_1 \geqslant 1 \vee x_2 \geqslant 1} \end{array} \right\}$$

illustrated in Fig. 1(a). Its reverse polars for $\alpha_0 = 1$ and $\alpha_0 = -1$ are the sets

$$F^{\#}_{(1)} = \left\{ y \in \mathbf{R}^2 \;\middle|\; \begin{array}{rr} 2y_1 & \geqslant 1 \\ & 2y_2 \geqslant 1 \\ y_1 & \geqslant 1 \\ & y_2 \geqslant 1 \end{array} \right\} = \left\{ y \in \mathbf{R}^2 \;\middle|\; \begin{array}{rr} y_1 & \geqslant 1 \\ & y_2 \geqslant 1 \end{array} \right\}$$

and

$$F^{\#}_{(-1)} = \left\{ y \in \mathbf{R}^2 \;\middle|\; \begin{array}{rr} 2y_1 & \geqslant -1 \\ & 2y_2 \geqslant -1 \\ y_1 & \geqslant -1 \\ & y_2 \geqslant -1 \end{array} \right\} = \left\{ y \in \mathbf{R}^2 \;\middle|\; \begin{array}{rr} 2y_1 & \geqslant -1 \\ & 2y_2 \geqslant -1 \end{array} \right\}$$

shown in Fig. 1(b) and (c).

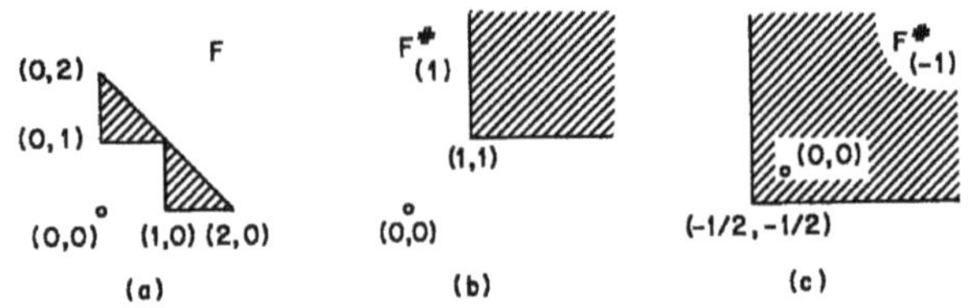

Fig. 1.

Finally, the sets $F^{\#\#}$ corresponding to $\alpha_0=1$ and $\alpha_0=-1$ (shown in Fig. 2(a), (b)) are

$$F^{\#\#}_{(1)}=\left\{x\in\mathbf{R}^2\,\middle|\,\begin{array}{l} x_1+x_2\geqslant 1\\ x_1 \qquad\;\; \geqslant 0\\ \qquad\; x_2\geqslant 0\end{array}\right\},\qquad F^{\#\#}_{(-1)}=\left\{x\in\mathbf{R}^2\,\middle|\,\begin{array}{l} -x_1-x_2\geqslant -2\\ \;\; x_1 \qquad\quad \geqslant 0\\ \qquad\quad\; x_2\geqslant 0\end{array}\right\}$$

and their intersection (shown in Fig. 2(c)) is

$$F^{\#\#}_{(1)}\cap F^{\#\#}_{(-1)}=\text{cl conf } F=\left\{x\in\mathbf{R}^2\,\middle|\,\begin{array}{l} -x_1-x_2\geqslant -2\\ \;\; x_1+x_2\geqslant 1\\ \;\; x_1 \qquad\;\; \geqslant 0\\ \qquad\;\; x_2\geqslant 0\end{array}\right\}.$$

The next theorem is needed to prove some other essential properties of reverse polars.

Theorem 5.4. $S^{\#\#\#}=S^{\#}$.

Proof. If $\alpha_0\leqslant 0$, this follows from the corresponding property of ordinary polars. If $\alpha_0>0$ and $0\in \text{cl conv } S$, then $S^{\#}=\emptyset$, $S^{\#\#}=\mathbf{R}^n$, and $S^{\#\#\#}=\emptyset=S^{\#}$. Finally, if $\alpha_0>0$ and $0\notin \text{cl conv } S$, then

$$\begin{aligned} S^{\#\#\#}&=(\text{cl conv } S+\text{cl cone } S)^{\#} \qquad \text{(from Theorem 5.3)}\\ &=\{y\in\mathbf{R}^n \mid xy\geqslant\alpha_0, \forall x\in(\text{cl conv } S+\text{cl cone } S)\}\\ &=\{y\in\mathbf{R}^n \mid xy\geqslant\alpha_0, \forall x\in S\}=S^{\#}. \quad\square\end{aligned}$$

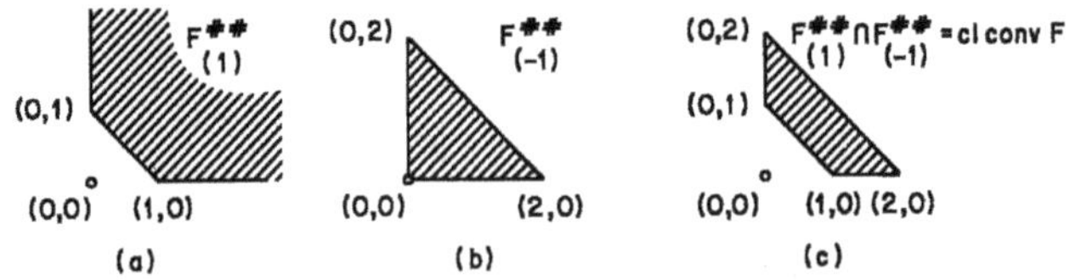

Fig. 2.

The above results can be used to characterize the facets of cl conv S. To simplify the exposition, we will assume that S is a full-dimensional pointed polyhedron, which implies that $S^{\#\#}$ is also full-dimensional. For the general case the reader is referred to [10].

We recall that an inequality $\pi x \geqslant \pi_0$ defines a facet of a convex polyhedral set C if $\pi x \geqslant \pi_0$, $\forall x \in C$, and $\pi x = \pi_0$ for exactly d affinely independent points x of C, where $d = \dim C$. The facet of C defined by $\pi x \geqslant \pi_0$ is then $\{x \in C \mid \pi x = \pi_0\}$; but as customary in the literature, for the sake of brevity we call $\pi x \geqslant \pi_0$ itself a facet.

We proceed in two steps, the first of which concerns the facets of $S^{\#\#}$.

Theorem 5.5. *Let* $\dim S = n$, *and* $\alpha_0 \neq 0$. *Then* $\alpha x \geqslant \alpha_0$ *is a facet of* $S^{\#\#}$ *if and only if* $\alpha \neq 0$ *is a vertex of* $S^{\#}$.

Proof. From Theorem 5.4, $\alpha \in \mathbf{R}^n$ is a vertex of $S^{\#}$ if and only if

$$\alpha \in S^{\#} = \left\{ y \in \mathbf{R}^n \,\middle|\, \begin{array}{ll} uy \geqslant \alpha_0, & \forall u \in \text{vert } S^{\#\#} \\ vy \geqslant 0, & \forall v \in \text{dir } S^{\#\#} \end{array} \right\}$$

and α satisfies with equality a subset of rank n of the system defining $S^{\#}$. Further, $\alpha \neq 0$ if and only if this subset of inequalities is not homogeneous (i.e., at least one right hand side coefficient is $\alpha_0 \neq 0$).

On the other hand, $\alpha x \geqslant \alpha_0$ is a facet of $S^{\#\#}$ if and only if (i) $\alpha x \geqslant \alpha_0$, $\forall x \in S^{\#\#}$, i.e., $\alpha \in S^{\#}$; and (ii) $\alpha x = \alpha_0$ for exactly n affinely independent points of $S^{\#\#}$. But (ii) holds if and only if $\alpha u = \alpha_0$ for r vertices u of $S^{\#\#}$, and $\alpha v = 0$ for s extreme direction vectors v of $S^{\#\#}$, with $r \geqslant 1$ (since $\alpha \neq 0$) and $r + s \geqslant n$, such that the system of these equations is of rank n.

Thus the two sets of conditions (for $\alpha x \geqslant \alpha_0$ to be a facet of $S^{\#\#}$ and for α to be a vertex of $S^{\#}$) are identical. □

By arguments similar to the above proof one shows that $\alpha x \geqslant 0$ is a facet of $S^{\#\#}$ if and only if α is an extreme direction vector of $S^{\#}$. Unlike for $\alpha_0 \neq 0$, the homogeneous inequality $\alpha x \geqslant 0$ is a facet of $S^{\#\#}_{(1)}$ if and only if it is also a facet of $S^{\#\#}_{(-1)}$ [of $S^{\#\#}_{(0)}$], due to the fact that every extreme direction vector of $S^{\#}_{(1)}$ is also an extreme direction vector of $S^{\#}_{(-1)}$ [of $S^{\#}_{(0)}$], and vice-versa.

Theorem 5.6. *Let* $\dim S = n$, *and* $\alpha_0 \neq 0$. *Then* $\alpha x \geqslant \alpha_0$ *is a facet of* cl conv S *if and only if it is a facet of* $S^{\#\#}_{(\alpha_0)}$.

Proof. (i) If $\alpha_0 > 0$, the halfspace $\alpha x \geqslant \alpha_0$ contains cl conv S if and only if it contains cl conv S + cl cone S. If $\alpha_0 < 0$, the halfspace $\alpha x \geqslant \alpha_0$ contains cl conv S if and only if it contains cl conv $(S \cup \{0\})$. From Theorem 4.3, in both cases $\alpha x \geqslant \alpha_0$ is a supporting halfspace for cl conv S if and only if it is a supporting halfspace for $S^{\#\#}_{(\alpha_0)}$.

(ii) Next we show that

$$\{x \in \text{cl conv } S \mid \alpha x = \alpha_0\} = \{x \in S^{\#\#}_{(\alpha_0)} \mid \alpha x = \alpha_0\}.$$

The relation $\subseteq$ follows from $\text{cl conv } S \subseteq S^{\#\#}_{(\alpha_0)}$ (Theorem 5.3). To show the converse, assume it to be false, and let $x \in S^{\#\#}_{(\alpha_0)} \setminus \text{cl conv } S$ satisfy $\alpha x = \alpha_0$. From Theorem 5.3, $x = \lambda u$ for some $u \in \text{cl conv } S$, and $\lambda > 1$ if $\alpha_0 > 0$, $0 < \lambda < 1$ if $\alpha_0 < 0$. In each case, $\alpha x = \alpha_0$ implies $\alpha u = (1/\lambda)\alpha x < \alpha_0$ for some $u \in \text{cl conv } S \subseteq S^{\#\#}_{(\alpha_0)}$, contrary to the assumption that $\alpha x \geq \alpha_0$, $\forall x \in S^{\#\#}_{(\alpha_0)}$. □

By an argument similar to the above proof one can show that if $\alpha x \geq 0$ is a facet of cl conv S, then it is a facet of $S^{\#\#}_{(\alpha_0)}$ for both $\alpha_0 = 1$ and $\alpha_0 = -1$. The converse, however, is not true, i.e., $\alpha x \geq 0$ can be a facet of both $S^{\#\#}_{(1)}$ and $S^{\#\#}_{(-1)}$, without being a facet of $S^{\#\#}_{(1)} \cap S^{\#\#}_{(-1)}$.

We are now ready to characterize the facets of the convex hull of the disjunctive set

$$F = \left\{x \in \mathbf{R}^n \;\middle|\; \bigvee_{h \in Q} \begin{pmatrix} A^h x \geq a_0^h \\ x \geq 0 \end{pmatrix}\right\},$$

where Q is assumed to be finite, and F to be full-dimensional (for the general case see [10]).

Theorem 5.7. *$\alpha x \geq \alpha_0$, with $\alpha_0 \neq 0$, is a facet of* cl conv F *if and only if $\alpha \neq 0$ is a vertex of the polyhedron*

$$F^{\#}_{(\alpha_0)} = \left\{ y \in \mathbf{R}^n \;\middle|\; \begin{array}{l} y \geq \theta^h A^h, \quad h \in Q^* \\ \text{for some } \theta^h \geq 0, \quad h \in Q^* \\ \text{satisfying } \theta^h a_0^h \geq \alpha_0 \end{array} \right\},$$

where Q^* is the set of those $h \in Q$ for which the system $A^h x \geq a_0^h$, $x \geq 0$, is consistent.

Proof. From Theorem 3.1, the set $F^{\#}_{(\alpha_0)} = \{y \in \mathbf{R}^n \mid xy \geq \alpha_0, \forall x \in F\}$ is of the form claimed above. The rest is a direct application of Theorems 5.5 and 5.6. □

As for the case $\alpha_0 = 0$, from the above comments it follows that if $\alpha x \geq 0$ is a facet of cl conv F, then $\alpha \neq 0$ is an extreme direction vector of $F^{\#}_{(\alpha_0)}$ for all α_0. The converse is not true, but if $\alpha \neq 0$ is an extreme direction vector of $F^{\#}_{(\alpha_0)}$ for some α_0 (hence for all α_0), then $\alpha x \geq 0$ is either a facet of cl conv F, or the intersection of two facets, $\alpha^1 x \geq \alpha_0^1$ and $\alpha^2 x \geq \alpha_0^2$, with $\alpha^1 + \alpha^2 = \alpha$ and $\alpha_0^1 = -\alpha_0^2 \neq 0$ (see [10] for the details).

Since the facets of cl conv F are vertices (in the nonhomogeneous case) or extreme direction vectors (in the homogeneous case) of the convex polyhedron $F^{\#}$, they can be found by maximizing or minimizing some properly chosen linear function on $F^{\#}$, i.e., by solving a linear program of the form

$$\begin{aligned}&\min gy,\\ &\left.\begin{aligned} y-\theta^h A^h &\geq 0,\\ \theta^h a_0^h &\geq \alpha_0,\\ \theta^h &\geq 0\end{aligned}\right\} h\in Q^*,\end{aligned} \qquad P_1^*(g,\alpha_0)$$

or its dual

$$\begin{aligned}&\max \sum_{h\in Q^*}\alpha_0\xi_0^h,\\ &\left.\begin{aligned}\sum_{h\in Q^*}\xi^h &= g,\\ a_0^h\xi_0^h - A^h\xi^h &\leq 0,\\ \xi_0^h\geq 0,\ \xi^h &\geq 0.\end{aligned}\right\} h\in Q^*.\end{aligned} \qquad P_2^*(g,\alpha_0)$$

From Theorem 5.1, if $\alpha_0\leq 0$ then $F^{\#}_{(\alpha_0)}\neq\emptyset$, i.e., $P_1^*(g,\alpha_0)$ is always feasible; whereas if $\alpha_0>0$, then $P_1^*(g,\alpha_0)$ is feasible if and only if $0\notin \operatorname{cl}\operatorname{conv} F$. This latter condition expresses the obvious fact that an inequality which cuts off the origin can only be derived from a disjunction which itself cuts off the origin.

Two problems arise in connection with the use of the above linear programs to generate facets of cl conv F. The first one is that sometimes only Q is known, but Q^* is not. This can be taken care of by working with Q rather than Q^*. Let $P_k(g,\alpha_0)$ denote the problem obtained by replacing Q^* with Q in $P_k^*(g,\alpha_0)$, $k=1,2$. It was shown in [10], that if $P_2(g,\alpha_0)$ has an optimal solution $\bar{\xi}$ such that

$$(\bar{\xi}_0^h=0,\ \bar{\xi}^h\neq 0)\Rightarrow h\in Q^*,$$

then every optimal solution of $P_1(g,\alpha_0)$ is an optimal solution of $P_1^*(g,\alpha_0)$. Thus, one can start by solving $P_2(g,\alpha_0)$. If the above condition is violated for some $h\in Q\setminus Q^*$, then h can be removed from Q and $P_2(g,\alpha_0)$ solved for Q redefined in this way. When necessary, this procedure can be repeated.

The second problem is that, since the facets of cl conv F of primary interest are the nonhomogeneous ones (in particular those with $\alpha_0>0$, since they cut off the origin), one would like to identify the class of vectors g for which $P_1^*(g,\alpha_0)$ has a finite minimum. It was shown in [10], that $P_1^*(g,\alpha_0)$ has a finite minimum if and only if $\lambda g\in\operatorname{cl}\operatorname{conv} F$ for some $\lambda>0$; and that, should g satisfy this condition, $\alpha x\geq\alpha_0$ is a facet of cl conv F (where F is again assumed full-dimensional) if and only if $\alpha=\bar{y}$ for every optimal solution $(\bar{y},\bar{\theta})$ to $P_1^*(g,\alpha_0)$.

As a result of these characterizations, facets of the convex hull of the feasible set F can be computed by solving the linear program $P_1(g,\alpha_0)$ or its dual. If the disjunction defining F has many terms, like in the case when F comes from the disjunctive programming formulation of a 0–1 program with a sizeable number of

0–1 conditions, $P_1(g, \alpha_0)$ is too large to be worth solving. If, however, F is made to correspond to a relaxation of the original zero-one program, involving zero-one conditions for only a few well chosen variables, then $P_1(g, \alpha_0)$ or its dual is practically solvable and provides the strongest possible cuts obtainable from those particular zero-one conditions.

On the other hand, since the constraint set of $P_2(g, \alpha_0)$ consists of $|Q|$ more or less loosely connected subsystems, one is tempted to try to approximate an optimal solution to $P_2(g, \alpha_0)$ — and thereby to $P_1(g, \alpha_0)$ — by solving the subsystems independently. Early computational experience indicates that these approximations are quite good.

We now give a numerical example for a facet calculation.

Example 5.2. Find all those facets of cl conv F which cut off the origin (i.e., all facets of the form $\alpha x \geq 1$), where $F \subset \mathbf{R}^2$ is the disjunctive set

$$F = F_1 \vee F_2 \vee F_3 \vee F_4,$$

with

$$\begin{aligned}
F_1 &= \{x \mid -x_1 + 2x_2 \geq 6,\ 0 \leq x_1 \leq 1,\ x_2 \geq 0\},\\
F_2 &= \{x \mid 4x_1 + 2x_2 \geq 11,\ 1 \leq x_1 \leq 2.5,\ x_2 \geq 0\},\\
F_3 &= \{x \mid -x_1 + x_2 \geq -2, 2.5 \leq x_1 \leq 4, x_2 \geq 0\},\\
F_4 &= \{x \mid x_1 + x_2 \geq 6, 4 \leq x_1 \leq 6, x_2 \geq 0\},
\end{aligned}$$

(see Fig. 3).

After removing some redundant constraints, F can be restated as the set of those $x \in \mathbf{R}^2_+$ satisfying

$$\{-x_1 + 2x_2 \geq 6\} \vee \left\{\begin{array}{l} 4x_1 + 2x_2 \geq 11 \\ -x_1 + \ \ x_2 \geq -2 \end{array}\right\} \vee \{x_1 + x_2 \geq 6\}$$

and the corresponding problem $P_1(g, 1)$ is

$$\begin{array}{llllll}
\min g_1 y_1 + g_2 y_2 & & & & & \\
y_1 & & +\theta_1^1 & & & \geq 0 \\
 & y_2 & -2\theta_1^1 & & & \geq 0 \\
y_1 & & & -4\theta_1^2 + \theta_2^2 & & \geq 0 \\
 & y_2 & & -2\theta_1^2 - \theta_2^2 & & \geq 0 \\
y_1 & & & & -\theta_1^3 & \geq 0 \\
 & y_2 & & & -\theta_1^3 & \geq 0 \\
 & & 6\theta_1^1 & & & \geq 1 \\
 & & & 11\theta_1^2 - 2\theta_2^2 & & \geq 1 \\
 & & & & 6\theta_1^3 & \geq 1 \\
 & & \theta_1^1,\ \theta_1^2,\ \theta_2^2,\ \theta_1^3 \geq 0. & & &
\end{array}$$

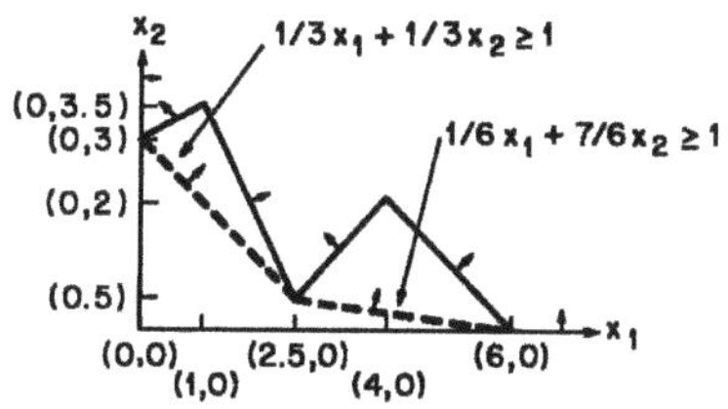

Fig. 3.

Solving this linear program for $g=(1, 1)$, yields the optimal points

$$(y;\theta)=(\tfrac{1}{3}, \tfrac{1}{3}; \tfrac{1}{6}, \tfrac{1}{9}, \tfrac{1}{9}, \tfrac{1}{6}), \quad \text{and} \quad (y;\theta)=(\tfrac{1}{3}, \tfrac{1}{3}; \tfrac{1}{6}, \tfrac{1}{9}, \tfrac{1}{9}, \tfrac{1}{3}),$$

which have the same y-component: $(\frac{1}{3}, \frac{1}{3})$. These points are optimal (and the associated y is unique) for all $g>0$ such that $g_1<5g_2$. For $g_1=5g_2$, in addition to the above points, which are still optimal, the points

$$(y;\theta)=(\tfrac{1}{6}, \tfrac{7}{6}; \tfrac{1}{6}, \tfrac{2}{9}, \tfrac{13}{18}, \tfrac{1}{6}) \quad \text{and} \quad (y;\theta)=(\tfrac{1}{6}, \tfrac{7}{6}; \tfrac{7}{12}, \tfrac{2}{9}, \tfrac{13}{18}, \tfrac{1}{6}),$$

which again have the same y-component $y=(\frac{1}{6}, \frac{7}{6})$, also become optimal; and they are the only optimal solutions for all $g>0$ such that $g_1>5g_2$.

We have thus found that the convex hull of F has two facets which cut off the origin, corresponding to the two vertices $y^1=(\frac{1}{3}, \frac{1}{3})$ and $y^2=(\frac{1}{6}, \frac{7}{6})$ of $F^{\#}_{(1)}$.

$$\tfrac{1}{3}x_1+\tfrac{1}{3}x_2\geqslant 1,$$
$$\tfrac{1}{6}x_1+\tfrac{7}{6}x_2\geqslant 1.$$

6. Facial disjunctive programs

In this section we discuss the following problem [10]. Given a disjunctive program in the conjunctive normal form (2.2), is it possible to generate the convex hull of feasible points by imposing the disjunctions $j\in S$ one by one, at each step calculating a "partial" convex hull, i.e., the convex hull of the set defined by the inequalities generated earlier, plus one of the disjunctions?

For instance, in the case of an integer program, is it possible to generate the convex hull of feasible points by first producing all the facets of the convex hull of points satisfying the linear inequalities, plus the integrality condition on, say, x_1; then adding all these facet-inequalities to the constraint set and generating the facets of the convex hull of points satisfying this amended set of inequalities, plus the integrality condition on x_2; etc.? The question has obvious practical importance, since calculating facets of the convex hull of points satisfying one disjunction is a considerably easier task, as shown in the previous section, then calculating facets of the convex hull of the full disjunctive set.

The answer to the above question is negative in general, but positive for a very important class of disjunctive programs, which we term facial. The class includes (pure or mixed) 0–1 programs.

The fact that in the general case the above procedure does not produce the convex hull of the feasible points can be illustrated on the following 2-variable problem.

Example 6.1. Given the set

$$F_0 = \{x \in \mathbf{R}^2 \mid -2x_1 + 2x_2 \leqslant 1,\ 2x_1 - 2x_2 \leqslant 1,\ 0 \leqslant x_1 \leqslant 2,\ 0 \leqslant x_2 \leqslant 2\}$$

find $F = \text{cl conv}\,(F_0 \cap \{x \mid x_1, x_2 \text{ integer}\})$. Denoting

$$F_1 = \text{cl conv}\,(F_0 \cap \{x \mid x_1 \text{ integer}\}), \qquad F_2 = \text{cl conv}\,(F_1 \cap \{x \mid x_2 \text{ integer}\}),$$

the question is whether $F_2 = F$. As shown in Fig. 4, the answer is no, since

$$F_2 = \left\{ x \;\middle|\; \begin{array}{l} 2x_1 - x_2 \geqslant 0 \\ -2x_1 + 3x_2 \geqslant 0 \\ -2x_1 + x_2 \geqslant -2 \\ 2x_1 - 3x_2 \geqslant -2 \end{array} \right\}, \quad \text{while} \quad F = \left\{ x \;\middle|\; \begin{array}{l} -x_1 + x_2 = 0 \\ 0 \leqslant x_1 \leqslant 2 \\ 0 \leqslant x_2 \leqslant 2 \end{array} \right\}.$$

If the order in which the integrality constraints are imposed is reversed, the outcome remains the same.

Consider a disjunctive program stated in the conjunctive normal form (2.2), and denote

$$F_0 = \{x \in \mathbf{R}^n \mid Ax \geqslant a_0,\ x \geqslant 0\}.$$

The disjunctive program (and its constraint set) is called *facial* if every

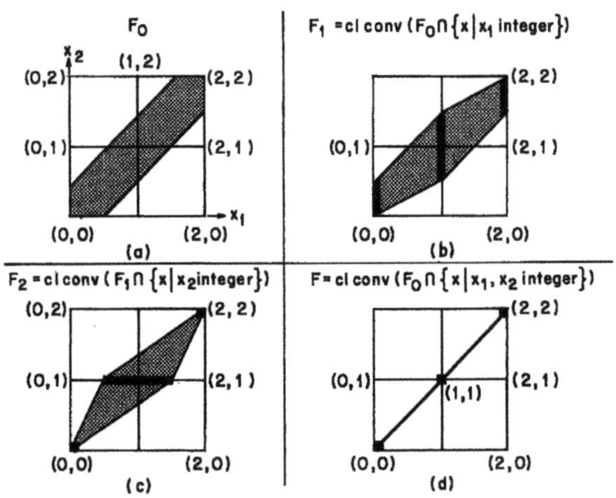

Fig. 4.

inequality $d^i x \geq d_{i0}$ that appears in a disjunction of (2.2), defines a face of F_0; i.e., if for all $i \in Q_j$, $j \in S$, the set

$$F_0 \cap \{x \mid d^i x \geq d_{i0}\}$$

is a face of F_0. (A face of a polyhedron P is the intersection of P with some of its boundary planes.)

Clearly, DP is facial if and only if for every $i \in Q_j$, $j \in S$, $F_0 \subseteq \{x \mid d^i x \leq d_{i0}\}$.

The class of disjunctive programs that have the facial property includes the most important cases of disjunctive programming, like the 0–1 programming (pure or mixed), nonconvex quadratic programming, separable programming, the linear complementarity problem, etc.; but not the general integer programming problem. In all the above mentioned cases the inequalities $d^i x \geq d_{i0}$ of each disjunction actually define facets, i.e., $(d-1)$-dimensional faces of F_0, where d is the dimension of F_0.

Another property that we need is the boundedness of F_0. Since this can always be achieved, if necessary, by regularizing F_0, its assumption does not represent a practical limitation.

Theorem 6.1. *Let the constraint set of a* DP *be represented as*

$$F = \left\{x \in F_0 \;\middle|\; \bigvee_{i \in Q_j} (d^i x \geq d_{i0}),\, j \in S\right\},$$

where

$$F_0 = \{x \in \mathbf{R}^n \mid Ax \geq a_0,\ x \geq 0\}.$$

For an arbitrary ordering of S, define recursively

$$F_j = \operatorname{conv}\left[\bigvee_{i \in Q_j} (F_{j-1} \cap \{x \mid d^i x \geq d_{i0}\})\right], \quad j = 1, \ldots, |S|.$$

If F is facial and F_0 is bounded, then

$$F_{|S|} = \operatorname{conv} F.$$

The proof of this theorem [10] uses the following auxiliary result.

Lemma 6.1.1. *Let $P_1, \ldots, P_r$ be a finite set of polytopes (bounded polyhedra), and $P = \bigcup_{h=1}^r P_h$.*

Let $H^+ = \{x \in \mathbf{R}^n \mid d^i x \leq d_{i0}\}$ be an arbitrary halfspace, and $H^- = \{x \in \mathbf{R}^n \mid d^i x \geq d_{i0}\}$. If $P \subseteq H^+$, then

$$H^- \cap \operatorname{conv} P = \operatorname{conv}(H^- \cap P).$$

Proof. Let $H^- \cap \operatorname{conv} P \neq \emptyset$ (otherwise the Lemma holds trivially). Clearly, $(H^- \cap P) \subseteq (H^- \cap \operatorname{conv} P)$, and therefore

$$\operatorname{conv}(H^- \cap P) \subseteq \operatorname{conv}(H^- \cap \operatorname{conv} P) = H^- \cap \operatorname{conv} P.$$

To prove $\supseteq$, let $u_1, \ldots, u_p$ be the vertices of all the polytopes P_h, $h = 1, \ldots, r$. Obviously, p is finite, conv P is closed, and vert conv $P \subseteq (\bigcup_{h=1}^{r} \text{vert } P_h)$. Then

$$x \in H^- \cap \text{conv } P \Rightarrow d^i x \geq d_{i0} \quad \text{and} \quad x = \sum_{k=1}^{p} \lambda_k u_k,$$

with

$$\sum_{k=1}^{p} \lambda_k = 1, \quad \lambda_k \geq 0, \qquad k = 1, \ldots, p.$$

Further, $P \subseteq H^+$ implies $d^i u_k \leq d_{i0}$, $k = 1, \ldots, p$. We claim that in the above expression for x, if $\lambda_k > 0$ then $d^i u_k \geq d_{i0}$. To show this, we assume there exist $\lambda_k > 0$ such that $d^i u_k < d_{i0}$. Then

$$d^i x = d^i\Big(\sum_{k=1}^{p} \lambda_k u_k\Big) < d_{i0}\Big(\sum_{k=1}^{p} \lambda_k\Big) = d_{i0},$$

a contradiction. Hence x is the convex combination of points $u^k \in H^- \cap P$, or $x \in \text{conv}(H^- \cap P)$. □

Another relation to be used in the proof of the theorem, is the fact that for arbitrary $S_1, S_2 \subseteq \mathbf{R}^n$,

$$\text{conv}(\text{conv } S_1 \cup \text{conv } S_2) = \text{conv}(S_1 \cup S_2).$$

Proof of Theorem 6.1. For $j = 1$, the statement is true from the definitions and the obvious relation

$$\bigvee_{i \in Q_1} (F_0 \cap \{x \mid d^i x \geq d_{i0}\}) = \Big\{x \in F_0 \,\Big|\, \bigvee_{i \in Q_1} (d^i x \geq d_{i0})\Big\}. \tag{6.1}$$

To prove the theorem by induction on j, suppose the statement is true for $j = 1, \ldots, k$. Then

$$F_{k+1} = \text{conv}\Big[\bigvee_{i \in Q_{k+1}} (F_k \cap \{x \mid d^i x \geq d_{i0}\})\Big] \text{ (by definition)}$$

$$= \text{conv}\Big[\bigvee_{i \in Q_{k+1}} \Big(\Big\{x \mid d^i x \geq d_{i0}\Big\} \cap \text{conv}\Big\{x \in F_0 \,\Big|\, \bigvee_{i \in Q_j} (d^i x \geq d_{i0}),\ j = 1, \ldots, k\Big\}\Big)\Big]$$

(from the assumption)

$$= \text{conv}\Big[\bigvee_{i \in Q_{k+1}} \text{conv}\Big(\{x \mid d^i x \geq d_{i0}\} \cap \Big\{x \in F_0 \,\Big|\, \bigvee_{i \in Q_j} (d^i x \geq d_{i0}),\ j = 1, \ldots, k\Big\}\Big)\Big]$$

(from Lemma 6.1.1, since DP is facial, hence $F_0 \subseteq \{x \mid d^i x \leq d_{i0}\}$, and F_0 is bounded.)

$$= \text{conv}\Big[\Big(\bigvee_{i \in Q_{k+1}} \{x \mid d^i x \geq d_{i0}\}\Big) \cap \Big\{x \in F_0 \,\Big|\, \bigvee_{i \in Q_j} (d^i x \geq d_{i0}),\ j = 1, \ldots, k\Big\}\Big]$$

(from (6.1) and the distributivity of $\cap$ with respect to $\bigvee$)

$$= \operatorname{conv}\left\{x \in F_0 \,\middle|\, \bigvee_{i \in Q_j} (d^i x \geq d_{i0}),\ j = 1, \ldots, k+1\right\},$$

i.e., the statement is also true for $j = k+1$. □

Theorem 6.1 implies that for a bounded facial disjunctive program with feasible set F, the convex hull of F can be generated in $|S|$ stages, (where S is as in (2.2)), by generating at each stage a "partial" convex hull, namely the convex hull of a disjunctive program with only one disjunction.

In terms of a 0–1 program, for instance, the above result means that the problem

$$\min\{cx \mid Ax \geq b,\ 0 \leq x \leq e,\ x_j = 0 \text{ or } 1,\ j = 1, \ldots, n\},$$

where $e = (1, \ldots, 1)$, is equivalent to (has the same convex hull of its feasible points, as)

$$\min\{cx \mid Ax \geq b,\ 0 \leq x \leq e,\ \alpha^i x \geq \alpha_{i0},\ i \in H_1,\ x_j = 0 \text{ or } 1,\ j = 2, \ldots, n\} \tag{6.2}$$

where $\alpha^i x \geq \alpha_{i0}$, $i \in H_1$, are the facets of

$$F_1 = \operatorname{conv}\{x \mid Ax \geq b,\ 0 \leq x \leq e,\ x_1 = 0 \text{ or } 1\}.$$

In other words, x_1 is guaranteed to be integer-valued in a solution of (6.2) although the condition $x_1 = 0$ or 1 is not present among the constraints of (6.2). A 0–1 program in n variables can thus be replaced by one in $n-1$ variables at the cost of introducing new linear inequalities. The inequalities themselves are not expensive to generate, since the disjunction that gives rise to them ($x_1 = 0 \vee x_1 = 1$) has only two terms. The difficulty lies rather in the number of facets that one would have to generate, were one to use this approach for solving 0–1 programs. However, by using some information as to which inequalities (facets of a "partial" convex hull) are likely to be binding at the optimum, one might be able to make the above approach efficient by generating only a few facets of the "partial" convex hull at each iteration. This question requires further investigation. For additional results on facial disjunctive programs see [33, 34].

7. Disjunctive programs with explicit integrality constraints

The theory reviewed in the previous sections derives cutting planes from disjunctions. In this context, 0–1 or integrality conditions are viewed as disjunctions, and the disjunction to be used for deriving a cut usually applies to the basic variables.

In this section we discuss a principle for strengthening cutting planes derived from disjunctions in the case when, besides the disjunction which applies to the basic variables, there are also integrality constraints on some of the nonbasic variables. In [14] we first proved this principle for arbitrary cuts, by using subadditive functions, then applied it to cuts from disjunctions. Here we prove the principle directly for the latter case, without recourse to concepts outside the framework of disjunctive programming.

Let a DP be stated in the disjunctive normal form (2.1), and assume in addition that some components of x are integer-constrained. In order for the principle that we are going to discuss to be applicable, it is necessary for each $A^h x$, $h \in Q$, to have a lower bound, say b_0^h. With these additional features, and denoting by J the index set for the components of x ($|J| = n$), the constraint set of the DP can be stated as

$$A^i x \geqslant b_0^i, \qquad i \in Q, \tag{7.1}$$

$$x \geqslant 0,$$

$$\bigvee_{i \in Q} (A^i x \geqslant a_0^i), \tag{7.2}$$

and

$$x_j \text{ integer}, \qquad j \in J_1 \subset J, \tag{7.3}$$

where

$$a_0^i \geqslant b_0^i \qquad i \in Q. \tag{7.4}$$

Let $Q = \{1, \ldots, q\}$, and let a_j^i stand for the j-th column of A^i, $j \in J$, $i \in Q$.

Theorem 7.1. [14] *Define*

$$M = \left\{ m \in \mathbf{R}^q \,\middle|\, \sum_{i \in Q} m_i \geqslant 0,\ m_i \text{ integer},\ i \in Q \right\}. \tag{7.5}$$

Then every $x \in \mathbf{R}^n$ *that satisfies* (7.1), (7.2), (7.3), *also satisfies the inequality*

$$\sum_{j \in J} \alpha_j x_j \geqslant \alpha_0, \tag{7.6}$$

where

$$\alpha_j = \begin{cases} \inf\limits_{m \in M} \max\limits_{i \in Q} \theta^i [a_j^i + m_i (a_0^i - b_0^i)], & j \in J_1 \\ \max\limits_{i \in Q} \theta^i a_j^i, & j \in J \setminus J_1 = J_2 \end{cases} \tag{7.7}$$

and

$$\alpha_0 = \min_{i \in Q} \theta^i a_0^i. \tag{7.8}$$

To prove this theorem we will use the following auxiliary result.

Lemma 7.1. *Let* $m_j \in M$, $m_j = (m_{ij})$, $j \in J_1$. *Then for every* $x \in \mathbf{R}^n$ *satisfying* (7.3) *and* $x \geqslant 0$, *either*

$$\sum_{j \in J_1} m_{ij} x_j = 0, \qquad \forall i \in Q \tag{7.9}$$

or

$$\bigvee_{i \in Q} \left(\sum_{j \in J_1} m_{ij} x_j \geq 1 \right) \tag{7.10}$$

holds.

Proof. If the statement is false, there exists $\bar{x} \geq 0$ satisfying (7.3) and such that

$$\sum_{i \in Q} \sum_{j \in J_1} m_{ij} \bar{x}_j < 0.$$

On the other hand, from $\bar{x} \geq 0$ and the definition of M,

$$\sum_{i \in Q} \sum_{j \in J_1} m_{ij} \bar{x}_j \geq 0,$$

a contradiction. □

Proof of Theorem 7.1. We first show that every x which satisfies (7.1), (7.2) and (7.3), also satisfies

$$\bigvee_{i \in Q} \left[\sum_{j \in J_1} [a_j^i + m_{ij}(a_0^i - b_0^i)] x_j + \sum_{j \in J_2} a_j^i x_j \geq a_0^i \right] \tag{7.2'}$$

for any set of $m_j \in M$, $j \in J_1$. To see this, write (7.2′) as

$$\bigvee_{i \in Q} \left[\sum_{j \in J} a_j^i x_j + (a_0^i - b_0^i) \sum_{j \in J_1} m_{ij} x_j \geq a_0^i \right]. \tag{7.2''}$$

From Lemma 7.1, either (7.9) or (7.10) holds for every $x \geq 0$ satisfying (7.3). If (7.9) holds, then (7.2″) is the same as (7.2) which holds by assumption. If (7.10) holds, there exists $k \in Q$ such that $\sum_{j \in J_1} m_{kj} x_j = 1 + \lambda$ for some $\lambda \geq 0$. But then the kth term of (7.2″) becomes

$$\sum_{j \in J} a_j^k x_j \geq b_0^k - \lambda (a_0^k - b_0^k)$$

which is satisfied since $\lambda(a_0^k - b_0^k) \geq 0$ and x satisfies (7.1). This proves that every feasible x satisfies (7.2′).

Applying to (7.2′) Theorem 3.1 then produces the cut (7.6) with coefficients defined by (7.7), (7.8). Taking the infimum over M is justified by the fact that (7.6) is valid with α_0 as in (7.8), α_j as in (7.7) for $j \in J_2$, and

$$\alpha_j = \max_{i \in Q} \theta^i [a_j^i + m_{ij}(a_0^i - b_0^i)]$$

for $j \in J_1$, for arbitrary $m_j \in M$. □

Corollary 7.1.1. [14]. *Let the vectors* σ^i, $i \in Q$, *satisfy*

$$\sigma^i (a_0^i - b_0^i) = 1, \qquad \sigma^i a_0^i > 0. \tag{7.11}$$

Then every $x \in \mathbf{R}^n$ *that satisfies* (7.1), (7.2) *and* (7.3), *also satisfies*

$$\sum_{j \in J} \beta_j x_j \geq 1, \tag{7.6$'$}$$

where

$$\beta_j = \begin{cases} \min\limits_{m \in M} \max\limits_{i \in Q} \dfrac{\sigma^i a_j^i + m_i}{\sigma^i a_0^i}, & j \in J_1, \\ \max\limits_{i \in Q} \dfrac{\sigma^i a_j^i}{\sigma^i a_0^i}, & j \in J_2. \end{cases} \tag{7.7$'$}$$

Proof. Given any σ^i, $i \in Q$, satisfying (7.11), if we apply Theorem 7.1 by setting $\theta^i = (\sigma^i / \sigma^i a_0^i)$, $i \in Q$, in (7.7) and (7.8), we obtain the cut (7.6′), with β_j defined by (7.7′), $j \in J$. □

Note that the cut-strengthening procedure of Theorem 7.1 requires, in order to be applicable, the existence of lower bounds on each component of $A^i x$, $\forall i \in Q$. This is a genuine restriction, but one that is satisfied in many practical instances. Thus, if x is the vector of nonbasic variables associated with a given basis, assuming that $A^i x$ is bounded below for each $i \in Q$ amounts to assuming that the basic variables are bounded below and/or above. In the case of a 0–1 program, for instance, such bounds not only exist but are quite tight.

Example 7.1. Consider again the mixed-integer program of Example 3.1 (taken from [35]), and assume this time that x_1 and x_2 are 0–1 variables rather than just integer constrained, i.e., let the constraint set of the problem be given by

$$x_1 = 0.2 + 0.4(-x_3) + 1.3(-x_4) - 0.01(-x_5) + 0.07(-x_6)$$
$$x_2 = 0.9 - 0.3(-x_3) + 0.4(-x_4) - 0.04(-x_5) + 0.1(-x_6)$$
$$x_j \geq 0,\ j = 1, \ldots, 6; \quad x_j = 0 \text{ or } 1,\ j = 1, 2; \quad x_j \text{ integer},\ j = 3, 4.$$

This change does not affect the Gomory cuts or the cuts obtainable from extreme valid inequalities for the group problem, which remain the same as listed in Example 3.1.

Now let us derive a cut, strengthened by the above procedure, from the disjunction

$$\left\{ \begin{matrix} x_1 \geq 0 \\ -x_2 \geq 0 \end{matrix} \right\} \vee \{x_2 \geq 1\}.$$

Since x_1, $-x_2$ and x_2 are bounded below by 0, -1 and 0 respectively, we have

$$a_0^1 = \begin{pmatrix} -0.2 \\ 0.9 \end{pmatrix}, \quad b_0^1 = \begin{pmatrix} -0.2 \\ -0.1 \end{pmatrix}; \quad a_0^2 = 0.1, \quad b_0^2 = -0.9.$$

Applying Corollary 7.1.1, we choose $\sigma^1=(4,1)$, $\sigma^2=1$, which is easily seen to satisfy (7.11). Since Q has only 2 elements, the set M of (7.5) becomes

$$M=\{m=(m_1, m_2) \mid m_1+m_2\geqslant 0;\ m_1,\ m_2 \text{ integer}\}$$

and, since at the optimum we may assume equality,

$$M=\{m=(m_1, -m_1) \mid m_1 \text{ integer}\}.$$

The coefficients defined by (7.7′) become

$$\beta_3=\min_{m_1\,\text{integer}}\max\left\{\frac{4\times(-0.4)+1\times(-0.3)+m_1}{4\times(-0.2)+1\times 0.9},\ \frac{1\times 0.3-m_1}{1\times 0.1}\right\}=-7 \quad (\text{with } m_1=1),$$

$$\beta_4=\min_{m_1\,\text{integer}}\max\left\{\frac{4\times(-1.3)+1\times 0.4+m_1}{4\times(-0.2)+1\times 0.9},\ \frac{1\times(-0.4)-m_1}{1\times 0.1}\right\}=-24 \ (\text{with } m_1=2),$$

$$\beta_5=\max\left\{\frac{4\times 0.01+1\times(-0.04)}{4\times(-0.2)+1\times 0.9},\ \frac{1\times 0.04}{1\times 0.1}\right\}=0.4,$$

$$\beta_6=\max\left\{\frac{4\times(-0.07)+1\times 0.1}{4\times(-0.2)+1\times 0.9},\ \frac{1\times(-0.1)}{1\times 0.1}\right\}=-1,$$

and the cut is

$$-7x_3-24x_4+0.4x_5-x_6\geqslant 1,$$

which has a smaller coefficient for x_4 (and hence is stronger) than the cut derived in Example 3.1.

In the above example, the integers m_i were chosen by inspection. Derivation of an optimal set of m_i requires the solution of a special type of optimization problem. Two efficient algorithms are available [14] for doing this when the multipliers σ^i are fixed. Overall optimization would of course require the simultaneous choice of the σ^i and the m_i, but a good method for doing that is not yet available.

The following algorithm, which is one of the two procedures given in [14], works with fixed σ^i, $i\in Q$. It first finds optimal noninteger values for the m_i, $i\in Q$, and rounds them down to produce an initial set of integer values. The optimal integer values, and the corresponding value of β_j, are then found by applying an iterative step k times, where $k\leqslant|Q|-1$, $|Q|$ being the number of terms in the disjunction from which the cut is derived.

Algorithm for calculating β_j, $j\in J_1$, of (7.7′)

Denote

$$\alpha_i=\sigma^i a_j^i, \qquad \lambda_i=(\sigma^i a_0^i)^{-1}, \tag{7.12}$$

and

$$\gamma = \sum_{i \in Q} \alpha_i \Big/ \sum_{i \in Q} \frac{1}{\lambda_i}. \tag{7.13}$$

Calculate

$$m_i^* = \frac{\gamma}{\lambda_i} - \alpha_i, \quad i \in Q, \tag{7.14}$$

set $m_i = [m_i^*]$, $i \in Q$, define $k = -\sum_{i \in Q}[m_i^*]$, and apply k times the following

Iterative Step. Find

$$\lambda_s(\alpha_s + m_s + 1) = \min_{i \in Q} \lambda_i(\alpha_i + m_i + 1)$$

and set

$$m_s \leftarrow m_s + 1, \qquad m_i \leftarrow m_i, \qquad i \in Q \setminus \{s\}.$$

This algorithm was shown in [14] to find an optimal set of m_i (and the associated value of β_j) in k steps, where $k = -\sum_{i \in Q}[m_i^*] \leqslant |Q| - 1$.

Example 7.2. Consider the integer program with the constraint set

$$\begin{aligned}
&x_1 = \tfrac{1}{6} + \tfrac{7}{6}(-x_5) - \tfrac{2}{6}(-x_6) + \tfrac{5}{6}(-x_7),\\
&x_2 = \tfrac{2}{6} + \tfrac{1}{6}(-x_5) + \tfrac{1}{6}(-x_6) - \tfrac{1}{6}(-x_7),\\
&x_3 = \tfrac{3}{6} - \tfrac{2}{6}(-x_5) + \tfrac{4}{6}(-x_6) - \tfrac{1}{6}(-x_7),\\
&x_4 = \tfrac{1}{6} + \tfrac{4}{6}(-x_5) + \tfrac{5}{6}(-x_6) - \tfrac{1}{6}(-x_7),\\
&x_1 + x_2 + x_3 + x_4 \geqslant 1,\\
&x_j = 0 \text{ or } 1, j = 1, \dots, 4; \qquad x_j \geqslant 0 \text{ integer}, \ j = 5, 6, 7.
\end{aligned}$$

We wish to generate a strengthened cut from the disjunction

$$x_1 \geqslant 1 \vee x_2 \geqslant 1 \vee x_3 \geqslant 1 \vee x_4 \geqslant 1.$$

If we apply Theorem 3.1 without strengthening and choose $\theta^i = (1 - a_{i0})^{-1}$, $i = 1, 2, 3, 4$, we obtain the cut

$$\tfrac{2}{3}x_5 + \tfrac{2}{3}x_6 + \tfrac{1}{3}x_7 \geqslant 1,$$

whose jth coefficient is

$$\alpha_j = \max_{i \in \{1,2,3,4\}} \frac{-a_{ij}}{1 - a_{i0}}.$$

To apply the strengthening procedure, we note that each x_j, $j = 1, 2, 3, 4$ is bounded below by 0. Using $\sigma^i = 1$ (which satisfies (6.11) since $\sigma^i(a_0^i - b_0^i) = \sigma^i[1 - a_{i0} - (-a_{i0})] = 1$, $i = 1, 2, 3, 4$, and $\sigma^i a_0^i = \sigma^i(1 - a_{i0}) > 0$, we obtain

$$\beta_5 = \min_{m \in M} \max\{\tfrac{6}{5}(-\tfrac{7}{6} + m_1), \tfrac{6}{4}(-\tfrac{1}{6} + m_2), \tfrac{6}{3}(\tfrac{2}{6} + m_3), \tfrac{6}{5}(-\tfrac{4}{6} + m_4)\},$$

$$\beta_6 = \min_{m \in M} \max\{\tfrac{6}{5}(\tfrac{2}{6} + m_1), \tfrac{6}{4}(-\tfrac{1}{6} + m_2), \tfrac{6}{3}(-\tfrac{4}{6} + m_3), \tfrac{6}{5}(-\tfrac{5}{6} + m_4)\},$$

$$\beta_7 = \min_{m \in M} \max\{\tfrac{6}{5}(-\tfrac{5}{6} + m_1), \tfrac{6}{4}(\tfrac{1}{6} + m_2), \tfrac{6}{3}(\tfrac{1}{6} + m_3), \tfrac{6}{5}(\tfrac{1}{6} + m_4)\}.$$

Next we apply the above Algorithm for calculating β_j:

For $j = 5$: $\gamma = -\tfrac{10}{17}$; $m_1^* = \tfrac{23}{34}$, $m_2^* = -\tfrac{23}{102}$, $m_3^* = -\tfrac{32}{51}$, $m_4^* = \tfrac{11}{51}$.

Thus our starting values are $[m_1^*] = 0$, $[m_2^*] = -1$, $[m_3^*] = -1$, $[m_4^*] = 0$. Since $k = -(-1) - (-1) = 2$, the Iterative step is applied twice:

1. $\min\{-\tfrac{1}{5}, -\tfrac{1}{4}, \tfrac{2}{3}, \tfrac{2}{5}\} = -\tfrac{1}{4}$, $s = 2$; $m_1 = 0$, $m_2 = -1 + 1 = 0$, $m_3 = -1$, $m_4 = 0$.
2. $\min\{-\tfrac{1}{5}, \tfrac{5}{4}, \tfrac{2}{3}, \tfrac{2}{5}\} = -\tfrac{1}{5}$, $s = 1$; $m_1 = 1$, $m_2 = 0$, $m_3 = -1$, $m_4 = 0$.

These are the optimal m_i, and

$$\beta_5 = \max\{-\tfrac{1}{5}, -\tfrac{1}{4}, -\tfrac{4}{3}, -\tfrac{4}{5}\} = -\tfrac{1}{5}.$$

For $j = 6$; $\gamma = -\tfrac{8}{17}$, $[m_1^*] = -1$, $[m_2^*] = -1$, $[m_3^*] = 0$, $[m_4^*] = 0$; $k = 2$.

1. $\min\{\tfrac{2}{5}, -\tfrac{1}{4}, \tfrac{2}{3}, \tfrac{1}{5}\} = -\tfrac{1}{4}$, $s = 2$; $m_1 = -1$, $m_2 = 0$, $m_3 = 0$, $m_4 = 0$.
2. $\min\{\tfrac{2}{5}, \tfrac{5}{4}, \tfrac{2}{3}, \tfrac{1}{5}\} = \tfrac{1}{5}$, $s = 4$; $m = -1$, $m = 0$, $m = 0$, $m = 1$.

$\beta_6 = \max\{-\tfrac{4}{5}, -\tfrac{1}{4}, -\tfrac{4}{3}, \tfrac{1}{5}\} = \tfrac{1}{5}$.

For $j = 7$: $\gamma = -\tfrac{2}{17}$, $[m_1^*] = 0$, $[m_2^*] = -1$, $[m_3^*] = -1$, $[m_4^*] = -1$; $k = 3$.

1. $\min\{\tfrac{1}{5}, \tfrac{1}{4}, \tfrac{1}{3}, \tfrac{1}{5}\} = \tfrac{1}{5}$, $s = 1$; $m_1 = 1$, $m_2 = -1$, $m_3 = -1$, $m_4 = -1$;
2. $\min\{\tfrac{7}{5}, \tfrac{1}{4}, \tfrac{1}{3}, \tfrac{1}{5}\} = \tfrac{1}{5}$, $s = 4$; $m_1 = 1$, $m_2 = -1$, $m_3 = -1$, $m_4 = 0$;
3. $\min\{\tfrac{7}{5}, \tfrac{1}{4}, \tfrac{1}{3}, \tfrac{7}{5}\} = \tfrac{1}{4}$, $s = 2$; $m_1 = 1$, $m_2 = 0$, $m_3 = -1$, $m_4 = 0$.

$\beta_7 = \max\{\tfrac{1}{5}, \tfrac{1}{4}, -\tfrac{5}{3}, \tfrac{1}{5}\} = \tfrac{1}{4}$.

Thus the strengthened cut is

$$-\tfrac{1}{5}x_5 + \tfrac{1}{5}x_6 + \tfrac{1}{4}x_7 \geqslant 1.$$

The frequently occurring situation, when $|Q| = 2$, deserves special mention. In this case the coefficients β_j, $j \in J_1$, are given by

$$\beta_j = \min\{\lambda_1(\alpha_1 + \langle m_0^* \rangle), \lambda_2(\alpha_2 - [m_0^*])\}, \tag{7.15}$$

where

$$m_0^* = \frac{\lambda_2\alpha_2 - \lambda_1\alpha_1}{\lambda_1 + \lambda_2}, \tag{7.16}$$

with α_i, λ_i, $i=1, 2$, defined by (7.12), and $\langle m_0^* \rangle$ = the smallest integer $\geq m_0^*$. The optimal value of $m_1 = -m_2$ is either $\langle m_0^* \rangle$ or $[m_0^*]$, according to whether the minimum in (7.15) is attained for the first or the second term.

The strengthening procedure discussed in this section produces the seemingly paradoxical situation that weakening a disjunction by adding a new term to it, may result in a strengthening of the cut derived from the disjunction; or, conversely, dropping a term from a disjunction may lead to a weakening of the inequality derived from the disjunction. For instance, if the disjunction used in Example 7.2 is replaced by the stronger one

$$x_1 \geq 1 \vee x_2 \geq 1 \vee x_3 \geq 1,$$

then the cut obtained by the strengthening procedure is

$$-\tfrac{1}{5}x_5 + \tfrac{2}{5}x_6 + \tfrac{1}{4}x_7 \geq 1,$$

which is weaker than the cut of the example, since the coefficient of x_6 is $\frac{2}{5}$ instead of $\frac{1}{5}$. The explanation of this strange phenomenon is to be sought in the fact that the strengthening procedure uses the lower bounds on each term of the disjunction. In Example 7.2, besides the disjunction $x_1 \geq 1 \vee x_2 \geq 1 \vee x_3 \geq 1 \vee x_4 \geq 1$, the procedure also uses the information that $x_i \geq 0$, $i=1, 2, 3, 4$. When the above disjunction is strengthened by omitting the term $x_4 \geq 1$, the procedure does not any longer use the information that $x_4 \geq 0$.

8. Some frequently occurring disjunctions

As mentioned earlier, one of the main advantages of the disjunctive programming approach is that it can make full use of the special structure inherent in many combinatorial problems. In [8, 9] cutting planes are derived from the logical conditions of the set partitioning problem, the linear complementarity problem, the two forms of representation for nonconvex separable programs, etc. More general complementarity problems are discussed in [33, 34]. Here we illustrate the procedure on the frequently occurring logical condition (where $x_i \geq 0$ integer, $i \in Q$)

$$\sum_{i \in Q} x_i = 1, \tag{8.1}$$

often called a multiple choice constraint.

If all the problem constraints are of this form, we have a set partitioning problem. But the cut that we derive uses only one condition (8.1), so it applies to arbitrary integer programs with at least one equation (8.1). It also applies to mixed-integer programs, provided the variables x_i, $i \in Q$, are integer-constrained. Here we discuss the all integer case, to simplify the notation.

Let I and J be the index sets for the basic and nonbasic variables in a basic feasible noninteger solution of the form

$$x_i = a_{i0} + \sum_{j \in J} a_{ij}(-x_j), \qquad i \in I. \tag{8.2}$$

In [8], [9], several cutting planes are derived from the disjunction

$$\bigvee_{i \in Q} \begin{pmatrix} x_i = 1 \\ x_h = 0, \ \forall h \in Q \setminus \{i\} \end{pmatrix}.$$

Here we describe another cut, which in most cases turns out to be stronger than those mentioned above. It is derived from the disjunction

$$\sum_{i \in Q_1} x_i = 0 \vee \sum_{i \in Q_2} x_i = 0, \tag{8.3}$$

clearly valid for any partition (Q_1, Q_2) of Q in the sense of being satisfied by every integer x satisfying (8.1). We require only that

$$\{i \in Q_k \mid a_{i0} > 0\} \neq \emptyset, \quad k = 1, 2. \tag{8.4}$$

Denoting

$$\beta_j^k = \sum_{i \in I \cap Q_k} a_{ij}, \qquad k = 1, 2; j \in J \cup \{0\}, \tag{8.5}$$

(8.3) can be written as

$$\begin{Bmatrix} \sum_{j \in J} \beta_j^1 x_j \geq \beta_0^1 \\ \sum_{j \in J \cap Q_1} x_j = 0 \end{Bmatrix} \vee \begin{Bmatrix} \sum_{j \in J} \beta_j^2 x_j \geq \beta_0^2 \\ \sum_{j \in J \cap Q_2} x_j = 0 \end{Bmatrix},$$

which implies the disjunction

$$\sum_{j \in J \setminus Q_1} \beta_j^1 x_j \geq \beta_0^1 \vee \sum_{j \in J \setminus Q_2} \beta_j^2 x_j \geq \beta_0^2, \tag{$*$}$$

with $\beta_0^k > 0$, $k = 1, 2$. Note that once the sets $I \cap Q_k$, $k = 1, 2$, are chosen, the sets $J \cap Q_1$ and $J \cap Q_2$ can be "optimized," in the sense of putting an index $j \in J \cap Q$ into Q_1 if $(\beta_j^1/\beta_0^1) \geq (\beta_j^2/\beta_0^2)$, and in Q_2 otherwise. Using this device while applying Theorem 3.1 to $(*)$ with multipliers $\theta^k = 1/\beta_0^k$, $k = 1, 2$, we obtain the cut

$$\sum \beta_j x_j \geq 1, \tag{8.6}$$

with coefficients

$$\beta_j = \begin{cases} \max\left\{\dfrac{\beta_j^1}{\beta_0^1}, \dfrac{\beta_j^2}{\beta_0^2}\right\}, & j \in J \setminus Q, \\ \max\left\{0, \min\left\{\dfrac{\beta_j^1}{\beta_0^1}, \dfrac{\beta_j^2}{\beta_0^2}\right\}\right\}, & j \in J \cap Q. \end{cases} \tag{8.7}$$

We now apply the strengthening procedure of Section 7 to the coefficients β_j, $j \in J\backslash Q$ (the coefficients indexed by $J \cap Q$ can usually not be further strengthened). This, as mentioned earlier, assumes that all x_j, $j \in J\backslash Q$, are integer constrained. A lower bound on $\sum_{j\in J\backslash Q_k} \beta_j^k x_j$ is $\beta_0^k - 1$, for $k = 1, 2$, since

$$\sum_{i\in I\cap Q_k} x_i = \beta_0^k + \sum_{j\in J\backslash Q_k} \beta_j^k(-x_j) \leq 1, \qquad k = 1, 2.$$

The multipliers $\sigma^k = 1$, $k = 1, 2$, satisfy condition (7.11) of Corollary 7.1.1, since

$$\sigma^k[\beta_0^k - (\beta_0^k - 1)] = 1, \qquad \sigma^k \beta_0^k > 0, \qquad k = 1, 2,$$

and thus the jth coefficient of the strengthened cut becomes (Corollary 7.1.1)

$$\beta_j = \min_{m\in M} \max_{k\in\{1,2\}} \left\{\frac{\beta_j^k + m_k}{\beta_0^k}\right\},$$

with M defined by (7.5). Applying the closed form solution (7.15) to the minimization problem involved in calculating β_j (in the special case of a disjunction with only two terms), we obtain

$$\alpha_1 = \sigma^k \beta_j^k = \beta_j^k, \qquad \lambda_k = (\sigma^k \beta_0^k)^{-1} = \frac{1}{\beta_0^k}, \qquad k = 1, 2,$$

and hence

$$\beta_j = \min\left\{\frac{\beta_j^1 + \langle m_0^*\rangle}{\beta_0^1}, \frac{\beta_j^2 - [m_0^*]}{\beta_0^2}\right\}$$

where

$$m_0^* = \frac{\beta_j^2 \beta_0^1 - \beta_j^1 \beta_0^2}{\beta_0^1 + \beta_0^2}.$$

We have thus proved

Theorem 8.1. *If* (8.2) *is a basic feasible noninteger solution of the linear program associated with an integer program whose variables have to satisfy* (8.1), *then for any partition* $(I \cap Q_1, I \cap Q_2)$ *of the set* $I \cap Q$ *satisfying* (8.4), *the inequality* (8.6) *is a valid cut, with coefficients*

$$\beta_j = \begin{cases} \max\left\{0, \min\left\{\dfrac{\beta_j^1}{\beta_0^1}, \dfrac{\beta_j^2}{\beta_0^2}\right\}\right\}, & j \in J \cap Q, \\[2ex] \min\left\{\dfrac{\beta_j^1 + \langle m_0^*\rangle}{\beta_0^1}, \dfrac{\beta_j^2 - [m_0^*]}{\beta_0^2}\right\}, & j \in J \setminus Q, \end{cases} \tag{8.9}$$

where the β_j^k, $k = 1, 2$, $j \in J$, are defined by (8.5) and m_0^* is given by (8.8).

We illustrate this cut on a set partitioning problem, which is a special case of Theorem.

Example 8.1. Consider the set partitioning problem whose cost vector is $c = (5, 4, 3, 2, 2, 3, 1, 1, 1, 0)$ and whose coefficient matrix is given in Table 1.

Table 1

	1	2	3	4	5	6	7	8	9	10
1	1					1	1		1	
2	1	1								1
3		1	1			1	1	1		
4		1	1	1			1			
5			1	1	1	1		1	1	

The linear programming optimum is obtained for $x_4 = x_8 = x_9 = 1/3$, $x_7 = 2/3$, $x_{10} = 1$, and $x_j = 0$ for all other j. The associated system (8.2) is shown in the form of a simplex tableau in Table 2 (artificial variables have been removed).

Table 2

	1	$-x_1$	$-x_6$	$-x_3$	$-x_5$	$-x_2$
x_0	-2	5	2	1	1	3
x_{10}	1	1	0	0	0	1
x_8	$\frac{1}{3}$	$-\frac{1}{3}$	$\frac{2}{3}$	$\frac{2}{3}$	$\frac{1}{3}$	$\frac{1}{3}$
x_7	$\frac{2}{3}$	$\frac{1}{3}$	$\frac{1}{3}$	$\frac{1}{3}$	$-\frac{1}{3}$	$\frac{2}{3}$
x_4	$\frac{1}{3}$	$-\frac{1}{3}$	$-\frac{2}{3}$	$\frac{2}{3}$	$-\frac{1}{3}$	$\frac{1}{3}$
x_9	$\frac{1}{3}$	$\frac{2}{3}$	$\frac{2}{3}$	$-\frac{1}{3}$	$\frac{1}{3}$	$-\frac{2}{3}$

We choose the disjunction corresponding to row 5 of the matrix shown in Table 1, which is of the form (8.1), with $Q = \{3, 4, 5, 6, 8, 9\}$.

We define $I \cap Q_1 = \{4, 8\}$, $I \cap Q_2 = \{9\}$, and we obtain the coefficients shown in Table 3.

Table 3

	$j=1$	$j=6$	$j=3$	$j=5$	$j=2$	$j=0$
β_j^1	$-\frac{2}{3}$	0	$\frac{4}{3}$	0	$\frac{2}{3}$	$\frac{2}{3}$
β_j^2	$\frac{2}{3}$	$\frac{2}{3}$	$-\frac{1}{3}$	$\frac{1}{3}$	$-\frac{2}{3}$	$\frac{1}{3}$

Since $J \cap Q = \{3, 5, 6\}$, we need the values m_0^* for $j \in J \setminus Q = \{1, 2\}$. They are $m_0^*(1) = \frac{2}{3}$, $m_0^*(2) = -\frac{2}{3}$. Hence

$$\beta_1 = \min\left\{\frac{-\frac{2}{3}+1}{\frac{2}{3}}, \frac{\frac{2}{3}-0}{\frac{1}{3}}\right\} = \frac{1}{2},$$

$$\beta_6 = \min\left\{0, \max\left\{\frac{0}{\frac{2}{3}}, \frac{\frac{2}{3}}{\frac{1}{3}}\right\}\right\} = 0,$$

$$\beta_3 = \min\left\{0, \max\left\{\frac{\frac{4}{3}}{\frac{2}{3}}, \frac{-\frac{1}{3}}{\frac{1}{3}}\right\}\right\} = 0,$$

$$\beta_5 = \min\left\{0, \max\left\{\frac{0}{\frac{2}{3}}, \frac{-\frac{2}{3}}{\frac{1}{3}}\right\}\right\} = 0,$$

$$\beta_2 = \min\left\{\frac{\frac{2}{3}+0}{\frac{2}{3}}, \frac{-\frac{2}{3}-(-1)}{\frac{1}{3}}\right\} = 1,$$

and we obtain the cut

$$\tfrac{1}{2}x_1 + x_2 \geq 1,$$

or

	1	$-x_1$	$-x_6$	$-x_3$	$-x_5$	$-x_2$
s	-1	$-\frac{1}{2}$	0	0	0	-1

which is considerably stronger than the traditional cuts that one can derive from Table 2, and it actually implies that the nonbasic variable x_2 has to be 1 in any integer solution.

Dual cutting plane methods have been found reasonably successful on set partitioning problems. Using stronger cuts can only enhance the efficiency of such methods, since the computational cost of the cut (8.9) is quite modest.

9. Combining cutting planes with branch and bound

The disjunctive programming approach offers various ways of combining branch and bound with cutting planes, some of which are currently the object of computational testing. Here we discuss one feature which seems to us crucial.

For any problem P, let $v(P)$ denote the value of P (i.e., of an optimal solution to P).

Suppose we are using branch and bound to solve a mixed-integer 0–1 program P, stated as a maximization problem. If $\{P_i\}_{i\in Q}$ is the set of active subproblems (active nodes of the search tree) at a given stage of the procedure and $\bar{v}(P_i)$ is the available upper bound on $v(P_i)$ (assume, for the sake of simplicity, that $\bar{v}(P_i) = v(\mathrm{LP}_i)$, where LP_i is the linear program associated with P_i), then $\max_{i\in Q} \bar{v}(P_i)$ is an upper bound on $v(P)$. Also, $\underline{v}(P)$, the value of the best integer solution found up to the stage we are considering, is of course a lower bound on $v(P)$; i.e., at any stage in the procedure,

$$\underline{v}(P) \leq v(P) \leq \max_{i\in Q} \bar{v}(P_i). \tag{9.1}$$

Hence the importance of finding good bounds for both sides of (9.1).

It is a crucial feature of the approach reviewed here that it can be used to derive a cutting plane from the optimal simplex tableaux associated with the subproblems LP_i, $i \in Q$, which provides an upper bound on $v(P)$ at least as good as, and often better than $\max_{i\in Q} \bar{v}(P_i)$.

Let the linear program LP associated with the mixed-integer 0–1 program P have an optimal solution of the form

$$x_h = a_{h0} + \sum_{j\in J} a_{hj}(-x_j), \quad h \in I \cup \{0\}, \tag{9.2}$$

where I and J are the index sets for the basic and nonbasic variables respectively, and let I_1 and J_1 be the respective index sets for the integer constrained variables. Here $a_{h0} \geq 0$, $h \in I$, and $a_{h0} \leq 1$, $h \in I_1$. Further, since P is a maximization problem and the solution (9.2) is optimal, $a_{0j} \geq 0$, $j \in J$.

Now let $\{P_k\}_{k \in Q}$ be the set of active subproblems, and for $k \in Q$, let the optimal solution to LP_k, the linear program associated with P_k, be of the form

$$x_h = a^k_{h0} + \sum_{j \in J^k} a^k_{hj}(-x_j), \qquad h \in I^k \cup \{0\}, \tag{9.3$_k$}$$

where I^k, J^k are defined with respect to LP_k the same way as I, J with respect to LP. Again $a^k_{0j} \geq 0$, $\forall j \in J^k$, since each LP_k is a maximization problem.

In order to derive a valid cutting plane from $(9.3)_k$, $k \in Q$, we view the branching process as the imposition of the disjunction

$$\bigvee_{k \in Q} \begin{pmatrix} D^k x \geq d^k_0 \\ Ax \geq b \\ x \geq 0 \end{pmatrix}, \tag{9.4}$$

where $Ax \geq b$ stands for the system

$$\sum_{j \in J} (-a_{hj}) x_j \geq -a_{h0}, \qquad h \in I,$$

$$\sum_{j \in J} a_{hj} x_j \geq a_{h0} - 1, \qquad h \in I_1,$$

expressing the conditions $x_h \geq 0$, $h \in I$, $x_h \leq 1$, $h \in I_1$, while each $D^k x \geq d^k_0$ is composed of inequalities of the form

$$\sum_{j \in J} a_{ij} x_j \geq a_{i0}$$

or

$$\sum_{j \in J} (-a_{ij}) x_j \geq 1 - a_{i0},$$

corresponding to the conditions $x_i \leq 0$ or $x_i \geq 1$ for some $i \in I_1$ whose totality, together with $Ax \geq b$, $x \geq 0$, defines P_k.

Now consider the cut derived from (9.4) on the basis of Theorem 3.1, with the optimal dual variables (obtained by solving LP_k) used as the multipliers θ^k, $k \in Q$. By optimal dual variables we mean those obtained by minimizing the sum of $a_{0j} x_j$, $j \in J$, subject to the kth term of (9.4). If for $k \in Q$ we denote by (u^k, v^k) the optimal dual vector associated with the kth term of (9.4), and

$$\alpha^k = u^k D^k + v^k A, \qquad \alpha^k_0 = u^k d^k_0 + v^k b, \tag{9.5}$$

and if $\alpha^k_0 > 0$, $k \in Q$, then according to Theorem 3.1, the inequality

$$\sum_{j \in J} \left(\max_{k \in Q} \frac{\alpha^k_j}{\alpha^k_0} \right) x_j \geq 1 \tag{9.6}$$

is satisfied by every x satisfying (9.4), i.e., by every feasible integer solution. The condition $\alpha_0^k>0$, $k\in Q$, amounts to requiring that $v(\mathrm{LP}_k)<v(\mathrm{LP})$, i.e., that the "branching constraints" $D^kx\geq d_0^k$ force v(LP) strictly below v(LP), $\forall k\in Q$. This is a necessary and sufficient condition for the procedure discussed here to be applicable. Should the condition not be satisfied for some $k\in Q$, one can use a different objective function for LP_k than for the rest of the subproblems — but we omit discussing this case here. Note that, since $b\leq 0$, $v^kb\leq 0$, $\forall k\in Q$, and thus $\alpha_0^k>0$ implies $u^kd_0^k>0$, $\forall k\in Q$.

Since the multipliers (u^k, v^k) are optimal solutions to the linear programs dual to LP_k, $k\in Q$, they maximize the right hand side coefficient α_0^k of each inequality $\alpha^kx\geq\alpha_0^k$ underlying the cut (9.6) subject to the condition that $\alpha_j^k\leq a_{0j}$, $\forall j\in J$.

We now proceed to strengthen the inequality (9.6) via the procedure of Section 7. To do this, we have to derive lower bounds on α^kx, $k\in Q$. We have

$$\begin{aligned}\alpha^kx-\alpha_0^k&=u^k(D^kx-d_0^k)+v^k(Ax-b)\\&\geq u^k(D^kx-d_0^k)\geq -u^ke,\end{aligned}$$

where $e=(1,\ldots,1)$. The first inequality holds since $Ax-b\geq 0$ for all x satisfying (9.4), while the second one follows from the fact that each inequality of the system $D^kx-d_0^k\geq 0$ is either of the form $-x_i\geq 0$ or of the form $x_i-1\geq 0$, with $i\in I_1$, and in both cases -1 is a lower bound on the value of the left hand side. Thus

$$\alpha^kx\geq\alpha_0^k-u^ke,\qquad k\in Q \tag{9.7}$$

holds for every x satisfying (9.4). Note that $u^kd_0^k>0$ implies $u^k\neq 0$ and hence $u^k\geq 0$ implies $u^ke>0$, $k\in Q$.

We now apply Corollary 7.1.1 to the system

$$\begin{aligned}&\alpha^kx\geq\alpha_0^k-u^ke,\qquad k\in Q,\\&x\geq 0,\\&\bigvee_{k\in Q}(\alpha^kx\geq\alpha_0^k),\\&x_j \text{ integer},\qquad j\in J_1.\end{aligned} \tag{9.8}$$

We choose $\sigma^k=1/u^ke$, $k\in Q$, which satisfies condition (7.11) of the Corollary:

$$(1/u^ke)[\alpha_0^k-(\alpha_0^k-u^ke)]=1,\qquad (1/u^ke)\alpha_0^k>0.$$

The strengthened cut is then

$$\sum_{j\in J}\beta_jx_j\geq 1, \tag{9.9}$$

with

$$\beta_j=\begin{cases}\min\limits_{m\in M}\max\limits_{k\in Q}\dfrac{(\alpha_j^k/u^ke)+m_k}{\alpha_0^k/u^ke}, & j\in J_1,\\[2ex] \max\limits_{k\in Q}\dfrac{\alpha_j^k}{\alpha_0^k}, & j\in J\setminus J_1,\end{cases} \tag{9.10}$$

where

$$M=\left\{m \in \mathbf{R}^{|Q|} \,\middle|\, \sum_{k\in Q} m_k \geq 0,\ m_k \text{ integer},\ k \in Q\right\}. \tag{9.11}$$

The values of α_j^k, α_0^k and $u^k e$ needed for computing the cut coefficients, are readily available from the cost row of the simplex tableaux associated with the optimal solutions to LP and LP$_k$, $k \in Q$. If the latter are represented in the form (9.2) and (9.3)$_k$ respectively, and if d_j^k and a_j denote the jth column of D^k and A of (9.4), while S_k is the row index set for D^k, $k \in Q$, we have for all $k \in Q$,

$$\begin{aligned} a_{0j}^k &= a_{0j} - u^k d_j^k - v^k a_j \\ &= a_{0j} - \alpha_j^k, \qquad j \in J^k \cap J, \end{aligned}$$

and

$$a_{0j}^k = 0 + u_j^k, \qquad j \in J^k \cap S_k,$$

since the indices $j \in S_k$ correspond to the slack variables of the system $-D^k x \leq -d_0^k$, whose costs are 0 (note that $S_k \cap J = \emptyset$ by definition). Further, for $j \in J \setminus J^k = J \cap I^k$ the reduced cost of x_j in LP$_k$ is 0, hence for all $k \in Q$

$$\begin{aligned} 0 &= a_{0j} - u^k d_j^k - v^k a_j \\ &= a_{0j} - \alpha_j^k. \end{aligned}$$

Finally,

$$\begin{aligned} a_{00}^k &= a_{00} - u^k d_0^k - v^k b \\ &= a_{00} - \alpha_0^k, \qquad \forall k \in Q. \end{aligned}$$

From the above expressions we then have for $k \in Q$,

$$\alpha_j^k = \begin{cases} a_{0j} - a_{0j}^k, & j \in J \cap J^k, \\ a_{0j}, & j \in J \setminus J^k, \end{cases} \tag{9.12}$$

$$\alpha_0^k = a_{00} - a_{00}^k,$$

and

$$u^k e = \sum_{i \in J^k \cap S_k} a_{0i}^k \tag{9.13}$$

(since $u_i^k = 0$, $\forall i \in S_k \cap I^k$).

The representation (9.3)$_k$ of the optimal solution to LP$_k$ assumes that the slack variable $j \in S_k$ of each "branching constraint" $x_i \leq 0$ or $x_i \geq 1$ that is tight at the optimum, is among the nonbasic variables with $a_{0j} > 0$. If one prefers instead to replace these slacks with the corresponding structural variables x_i and regard the latter as "fixed" at 0 or 1, and if F_k denotes the index set of the variables fixed in LP$_k$, the reduced costs a_{0i}^k, $i \in J^k \cap F_k$ are then the same, except for their signs, as a_{0j}^k, $j \in J^k \cap S_k$, and the only change required in the expressions derived above is to replace (9.13) by

$$u^k e = \sum_{i \in J^k \cap F_k} |a_{0i}^k|. \tag{9.13'}$$

Of course, in order to calculate a cut of the type discussed here, one needs the reduced costs a_{0j}^k for both the free and the fixed variables.

We have thus proved the following result.

Theorem 9.1. *If* LP *and* LP_k, $k \in Q$, *have optimal solutions of the form* (9.2) *and* $(9.3)_k$ *respectively, with* $a_{00} > a_{00}^k$, $k \in Q$, *then every feasible integer solution satisfies the inequality* (9.9), *with coefficients defined by* (9.10), (9.11), (9.12) *and* (9.13) [*or* (9.13′)].

In the special case when $|Q| = 2$ and LP_1, LP_2 are obtained from LP by imposing $x_i \leq 0$ and $x_i \geq 1$ respectively (for some $i \in I_1$ such that $0 < a_{i0} < 1$), the definition of β_j for $j \in J_1$ becomes

$$\beta_j = \min\left\{\frac{(\alpha_j^1/u_i^1) + \langle m_0^* \rangle}{\alpha_0^1/u_i^1}, \frac{(\alpha_j^2/u_i^2) - [m_0^*]}{\alpha_0^2/u_i^2}\right\} \tag{9.10'}$$

with

$$m_0^* = \frac{\alpha_j^2 \alpha_0^1 - \alpha_j^1 \alpha_0^2}{u_i^2 \alpha_0^1 + u_i^1 \alpha_0^2}. \tag{9.14}$$

We now state the property of the cut (9.9) mentioned at the beginning of this section.

Corollary 9.1.1. *Adding the cut* (9.9) *to the constraints of* LP *and performing one dual simplex pivot in the cut row reduces the objective function value from* a_{00} *to* $\bar{a}_{00}$ *such that*

$$\bar{a}_{00} \leq \max_{k \in Q} a_{00}^k. \tag{9.15}$$

Proof. For each $k \in Q$, $a_{00}^k = a_{00} - \alpha_0^k$. Now suppose (9.15) is false, i.e., $\bar{a}_{00} > a_{00}^k$, $\forall k \in Q$, Then

$$\bar{a}_{00} = a_{00} - \min_{j \in J \mid \beta_j > 0} a_{0j}/\beta_j > a_{00} - \alpha_0^k, \quad \forall k \in Q,$$

and hence for all $k \in Q$,

$$\begin{aligned} \alpha_0^k &> \min_{j \in J \mid \beta_j > 0} a_{0j}/\beta_j \\ &\geq \min_{j \in J \mid \alpha_j^s > 0} a_{0j}(\alpha_0^s/\alpha_j^s), \end{aligned} \tag{9.16}$$

where

$$\frac{\alpha_j^s}{\alpha_0^s} = \max_{h \in Q} \frac{\alpha_j^h}{\alpha_0^h}. \tag{9.17}$$

The second inequality in (9.16) holds since the cut (9.9) is a strengthened version of (9.6), in the sense that

$$\beta_j \leq \max_{h \in Q} \frac{\alpha_j^h}{\alpha_0^h}, \quad j \in J.$$

Now suppose the minimum in the second inequality of (9.16) is attained for $j = t$. Since (9.16) holds for all $k \in Q$, we then have

$$\alpha_0^s > a_{0t}(\alpha_0^s/\alpha_t^s)$$

or (since $\alpha_t^s > 0$, $\alpha_0^s > 0$), $\alpha_t^s > a_{0t}$. But this contradicts the relation $\alpha_t^s \leq a_{0t}$ implied by (9.12) □

Note that the Corollary remains true if the strenghtened cut (9.9) is replaced by its weaker counterpart (9.6). In that case, however, (9.15) holds with equality. The remarkable fact about the property stated in the Corollary is that by using the strengthened cut (9.9) one often has (9.15) satisfied as strict inequality. More precisely, we have the following

Remark 9.1. (9.15) holds as strict inequality if $\beta_t < \alpha_t^s/\alpha_0^s$, where s is defined by (9.17), and t by

$$a_{0t}(\alpha_0^s/\alpha_t^s) = \min_{j \in J | \alpha_j^s > 0} a_{0j}(\alpha_0^s/\alpha_t^s).$$

Note that for (9.15) to hold as strict inequality the pivot discussed in the Corollary need not occur on a "strengthened" cut coefficient. All that is needed, is that the coefficient on which the pivot *would* occur in case the unstrengthened cut were used, should be "strengthened" i.e., reduced by the strengthening procedure).

The significance of the cut of Theorem 9.1 is that it concentrates in the form of a single inequality much of the information generated by the branch and bound procedure up to the point where it is derived, and thus makes it possible, if one so wishes, to start a new tree search while preserving a good deal of information about the earlier one.

Theorem 9.1 and its Corollary are stated for (pure or mixed) 0–1 programs; but the 0–1 property (as opposed to integrality) is only used for the derivation of the lower bounds on $D^k x \geq d_0^k$, $k \in Q$; hence it only involves those variables on which branching has occured. The results are therefore valid for mixed-integer programs with some 0–1 variables, provided the strengthening procedure is only used on cuts derived from branching on 0–1 variables.

Example 9.1. Consider the problem in the variables $x_j \geq 0$, $j = 1, \ldots, 6$; $x_j = 0$ or 1, $j = 1$, 4; x_j integer, $j = 2$, 3, whose linear programming relaxation has the optimal solution shown in Table 4.

Table 4

	1	$-x_3$	$-x_4$	$-x_5$	$-x_6$
x_0	1.1	2.0	0.2	0.05	1.17
x_1	0.2	0.4	1.3	−0.01	0.07
x_2	0.9	−0.3	0.4	−0.04	0.1

If we solve the problem by branch and bound, using the rules of always selecting for branching (a) LP_k with the largest a_{00}^k, and (b) x_i with the largest max {up penalty, down penalty}, we generate the search tree shown in Fig. 5. The optimal solution is $x = (0, 2, 1, 0, 20, 0)$, with value -1.9, found at node 6. To prove optimality, we had to generate two more nodes, i.e., the total number of nodes generated (apart from the starting node) is 8.

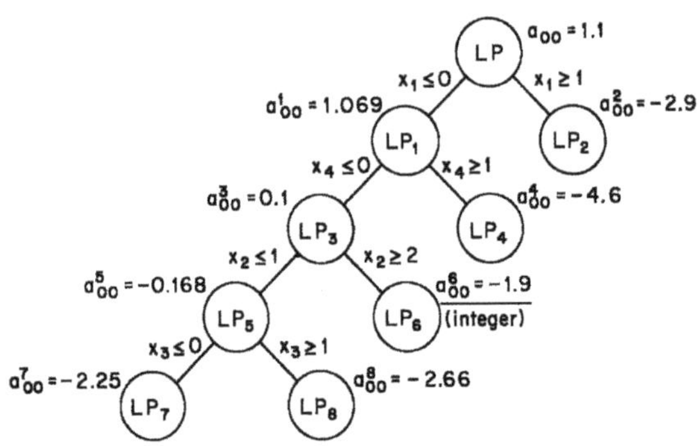

Fig. 5.

Suppose now that after generating the first four nodes, we wish to use the available information to derive a cut. At that point there are three active nodes, associated with LP_k, for $k = 2, 3, 4$. The corresponding reduced cost coefficients a_{0j}^k, $j \in J^k$, are shown in Table 5. The slack variables of the "branching constraints" $x_1 \leqslant 0$, $x_1 \geqslant 1$, $x_4 \leqslant 0$, and $x_4 \geqslant 1$ are denoted by x_7, x_8, x_9 and x_{10} respectively.

Table 5

k	2				3				4			
J^k	3	4	8	6	7	9	5	6	3	10	1	6
a_{0j}^k	4.0	6.7	5.0	1.52	5.0	6.3	0.1	0.82	4.0	6.7	5.0	1.52
a_{00}^k	−2.9				0.1				−4.6			

The coefficients α_j^k, α_0^k and $u^k e$, extracted from Table 9.2, and the cost row of Table 4, are as follows:

$k=2$: $\alpha_0^2 = a_{00} - a_{00}^2 = 4$; $u^2 e = u_1^2 = a_{08}^2 = 5$; $\alpha_3^2 = a_{03} - a_{03}^2 = -2$,
$\alpha_4^2 = a_{04} - a_{04}^2 = -6.5$, $\alpha_5^2 = a_{05} = 0.05$, $\alpha_6^2 = a_{06} - a_{06}^2 = -0.35$;

$k=3$: $\alpha_0^3 = 1$; $u^3 e = u_7^3 + u_9^3 = 11.3$; $\alpha_3^3 = 2$, $\alpha_4^3 = 0.2$, $\alpha_5^3 = -0.05$, $\alpha_6^3 = 0.35$;

$k=4$: $\alpha_0^4 = 5.7$; $u^4 e = u_{10}^4 = 6.7$; $\alpha_3^4 = -2$, $\alpha_4^4 = 0.2$, $\alpha_5^4 = 0.05$, $\alpha_6^4 = -0.35$.

The coefficients of the strengthened cut are shown below, where $M = \{m \in \mathbf{R}^3 \mid m_1 + m_2 + m_3 \geq 0,\ m_i \text{ integer},\ i = 1, 2, 3\}$.

$$\beta_3 = \min_{m \in M} \max \left\{ \frac{(-2/5) + m_1}{4/5}, \frac{(2/11.3) + m_2}{1/11.3}, \frac{(-2/6.7) + m_3}{5.7/6.7} \right\} = 0.75,$$

with $m = (1, -1, 0)$.

$$\beta_4 = \min_{m \in M} \max \left\{ \frac{(-6.5/5) + m_1}{4/5}, \frac{(0.2/11.3) + m_2}{1/11.3}, \frac{(0.2/6.7) + m_3}{5.7/6.7} \right\} = 0.035,$$

with $m = (1, -1, 0)$.

$$\beta_5 = \max \left\{ \frac{0.05}{4}, \frac{-0.05}{1}, \frac{0.05}{5.7} \right\} = 0.0125,$$

$$\beta_6 = \max \left\{ \frac{-0.35}{4}, \frac{0.35}{1}, \frac{-0.35}{5.7} \right\} = 0.35.$$

Adding to the optimal tableau of LP the cut

$$0.75x_3 + 0.035x_4 + 0.0125x_5 + 0.35x_6 \geq 1$$

produces Table 6 and the two pivots shown in Tables 6 and 7 produce the optimal Tableau 8. Thus no further branching is required.

Table 6

	1	$-x_3$	$-x_4$	$-x_5$	$-x_6$
x_0	1.1	2.0	0.2	0.05	1.17
x_1	0.2	0.4	1.3	−0.01	0.07
x_2	0.9	−0.3	0.4	−0.04	0.1
s	−1.0	[−0.75]	−0.035	−0.0125	−0.35

Table 7

	1	$-s$	$-x_4$	$-x_5$	$-x_6$
x_0	−1.57	2.67	0.106	0.01675	1.117
x_1	−0.333	0.533	1.281	[−0.01675]	−0.117
x_2	1.3	−0.4	0.414	−0.035	0.240
x_3	1.33	−1.33	0.047	0.01675	0.467

Table 8

	1	$-s$	$-x_4$	$-x_5$	$-x_6$
x_0	-1.9	2.687	1.387	1.0	1.0
x_5	20.0	-31.8	-73.0	-60.0	7.0
x_3	2.0			-2.0	
x_3	1.0			1.0	

10. Disjunctions from conditional bounds

In solving pure or mixed integer programs by branch and bound, the most widely used rule for breaking up the feasible set is to choose an integer-constrained variable x_i whose value a_{i0} at the linear programming optimum is noninteger, and to impose the disjunction $(x_i \leqslant [a_{i0}]) \vee (x_i \geqslant [a_{i0}]+1)$. It has been observed, however, that in the presence of multiple choice constraints, i.e., of constraints of the form

$$\sum_{i \in Q} x_i = 1,$$

it is more efficient to use a disjunction of the form

$$\Big(\sum_{i \in Q_1} x_i = 0\Big) \vee \Big(\sum_{i \in Q_2} x_i = 0\Big),$$

where $Q_1 \cup Q_2 = Q$, $Q_1 \cap Q_2 = \emptyset$, and Q_1 and Q_2 are about equal in size.

This is just one example of a situation where it is possible to branch so as to fix the values of several variables on each branch. The circumstance that makes this possible in the above instance is the presence of the rather tight multiple choice constraint. More generally, a tightly constrained feasible set makes it possible to derive disjunctions stronger than the usual dichotomy on a single variable. On the other hand, the feasible set of any integer program becomes more or less tightly constrained after the discovery of a "good" solution (in particular, of an optimal solution), provided that one restricts it to those solutions better than the current best. Such a "tightly constrained" state of the feasible set can be expressed in the form of an inequality

$$\pi x \leqslant \pi_0,$$

with $\pi \geqslant 0$, and with a relatively small $\pi_0 > 0$. One way of doing this, if the problem is of the form

$$\min\{cx \mid Ax \geqslant b,\ x \geqslant 0,\ x_j \text{ integer},\ j \in N\}, \tag{P}$$

with c integer, and if z_U is the value of the current best integer solution, is to find a set of multipliers u such that

$$uA \leqslant c, \qquad u \geqslant 0,$$

and define

$$\pi = c - uA, \qquad \pi_0 = z_U - ub - 1.$$

Then multiplying $Ax \geq b$ by $-u$ and adding the resulting inequality, $-uAx \leq -ub$, to $cx \leq z_U - 1$, yields the inequality

$$\pi x \leq \pi_0,$$

satisfied by every feasible integer x such that $cx < z_U$. Here $\pi \geq 0$, $\pi_0 > 0$, and the size of π_0 depends on the gap between the upper bound z_U and the lower bound ub on the value of (P).

Now suppose we have such an inequality $\pi x \leq \pi_0$. Without loss of generality, we may assume that $\pi_j > 0$, $\forall j$ (by simply deleting the zero components). Then the following statement holds [12].

Theorem 10.1. *Let* $\pi \in \mathbf{R}^n$, $\pi_0 \in \mathbf{R}$, $(\pi, \pi_0) > 0$, $N = \{1, \ldots, n\}$, *and for* $i = 1, \ldots, p$, $1 \leq p \leq n$, let $Q_i \subseteq N$, $Q_i \neq \emptyset$, *with*

$$\pi_{j(i)} = \min_{j \in Q_i} \pi_j.$$

If the sets Q_i, $i = 1, \ldots, p$, *satisfy the conditions*

$$\sum_{i \mid j \in Q_i} \pi_{j(i)} \leq \pi_j, \quad j \in N, \tag{10.1}$$

and

$$\sum_{i=1}^{p} \pi_{j(i)} > \pi_0, \tag{10.2}$$

then every integer vector $x \geq 0$ *such that* $\pi x \leq \pi_0$, *satisfies the disjunction*

$$\bigvee_{i=1}^{p} (x_j = 0, j \in Q_i). \tag{10.3}$$

Proof. Every integer $x \geq 0$ which violates (10.3) satisfies

$$\sum_{j \in Q_i} x_j \geq 1, \qquad i = 1, \ldots, p. \tag{10.4}$$

Multiplying by $\pi_{j(i)}$ the ith inequality of (10.4) and adding up the resulting inequalities, one obtains

$$\sum_{j \in \bigcup_{i=1}^{p} Q_i} \left(\sum_{i \mid j \in Q_i} \pi_{j(i)} \right) x_j \geq \sum_{i=1}^{p} \pi_{j(i)}. \tag{10.5}$$

Further,

$$\sum_{j\in N}\pi_j x_j \geqslant \sum_{j\in\bigcup_{i=1}^{p}Q_i}\pi_j x_j$$

$$\geqslant \sum_{j\in\bigcup_{i=1}^{p}Q_i}\Big(\sum_{i\,|\,j\in Q_i}\pi_{j(i)}\Big)x_j \quad \text{[from (10.1)]}$$

$$\geqslant \sum_{i=1}^{p}\pi_{j(i)} \quad \text{[from (10.5)]}$$

$$> \pi_0 \quad \text{[from (10.2)].}$$

Hence every integer $x \geqslant 0$ that violates the disjunction (10.3) also violates the inequality $\pi x \leqslant \pi_0$. □

One way of looking at Theorem 10.1 is as follows. Suppose the constraints of an integer program which include the inequality $\pi x \leqslant \pi_0$, were amended by the additional constraints (10.4).

From the proof of the Theorem, these inequalities give a lower bound on πx which exceeds π_0; this contradiction then produces the disjunction (10.3). Since the inequalities (10.4) are not actually part of the problem, we call the bound on πx derived from them a *conditional* bound, and the disjunction obtained from such bounds, a disjunction from conditional bounds.

Example 10.1. The inequality

$$9x_1+8x_2+8x_3+7x_4+7x_5+6x_6+6x_7+5x_8+5x_9+5x_{10}+4x_{11}$$
$$+4x_{12}+3x_{13}+3x_{14}+3x_{15}+2x_{16}+2x_{17}\leqslant 10,$$

together with the condition $x \geqslant 0$, x_j integer $\forall j$, implies the disjunction

$$(x_j=0, j=1,2,3,4,5,6,7)\vee(x_j=0, j=1,8,9,10,11,12,13,14)\vee$$
$$\vee(x_j=0, j=2,3,8,9,10,15,16,17).$$

Indeed,

$$\pi_{j(1)}=\min\{9,8,8,7,7,6,6\} \quad =6,$$
$$\pi_{j(2)}=\min\{9,5,5,5,4,4,3,3\}=3,$$
$$\pi_{j(3)}=\min\{8,8,5,5,5,3,2,2\}=2,$$

and (10.1), (10.2) are satisfied, since $6+3+2>10$, while $6+3\leqslant 9 (j=1)$, $6+2\leqslant 8 (j=2,3)$, $6\leqslant 7 (j=4,5)$, $6\leqslant 6 (j=6,7)$, $3+2\leqslant 5 (j=8,9,10)$, $3\leqslant 4 (j=11,12)$, $3\leqslant 3 (j=13,14)$, $2\leqslant 3 (j=15)$, $2\leqslant 2 (j=16,17)$.

Next we outline a procedure [12] based on Theorem 10.1 for systematically generating disjunctions of the type (10.3) from an inequality $\pi x \leqslant \pi_0$, with $\pi_j > 0$, $\forall j$.

1. Choose some $S \subset N$ such that

$$\sum_{j\in S}\pi_j > \pi_0$$

but

$$\sum_{j \in T} \pi_j \leqslant \pi_0$$

for all $T \subset S$, $T \neq S$. Order $S = \{j(1), \ldots, j(p)\}$ according to decreasing values of $\pi_{j(i)}$ and go to 2.

2. Set

$$Q_1 = \{j \in N \mid \pi_j \geqslant \pi_{j(1)}\}$$

and define recursively

$$Q_i = \left\{j \in N \mid \pi_j \geqslant \pi_{j(i)} + \sum_{k=1}^{i-1} \pi_{j(k)} \delta_j^k\right\}, \qquad i = 2, \ldots, p$$

where $\delta_j^k = 1$ if $j \in Q_k$, $\delta_j^k = 0$ otherwise. The sets Q_i, $i = 1, \ldots, p$, obtained in this way, satisfy (10.1), (10.2).

In the above example, $S = \{7, 14, 17\}$. If, on the other hand, one uses $S = \{5, 12\}$ (which is also admissible, since $\pi_5 + \pi_{12} = 7 + 4 > 10$), one obtains the disjunction

$$(x_j = 0, j = 1, \ldots, 5) \vee (x_j = 0,\ j = 6, \ldots, 12).$$

A disjunction of the form (10.3) can be used to partition the feasible set into p subproblems, the kth one of which is constrained by

$$\sum_{j \in Q_i} x_j \geqslant 1, \quad i = 1, \ldots, k-1; \qquad x_j = 0, \quad \forall j \in Q_k.$$

Another way of using a disjunction of the form (10.3) is to derive cutting planes. This has been explored in the context of set covering problems [12] and found to yield good results. In particular, let $A = (a_{ij})$ be a 0–1 matrix and consider the set covering problem

$$\min \{cx \mid Ax \geqslant e, x_j = 0 \text{ or } 1, j \in N\}, \tag{SC}$$

where e is the vector of 1's of appropriate dimension. For every row of A, let

$$N_i = \{j \in N \mid a_{ij} = 1\}.$$

Now suppose a prime cover (a basic feasible integer solution) $\bar{x}$ is known; then $z_U = c\bar{x}$ is an upper bound on the value of the optimum. If $\bar{u}$ is any feasible solution to the dual of the linear programming relaxation of (SC), i.e., any vector satisfying $\bar{u}A \leqslant c$, $\bar{u} \geqslant 0$, then setting $\pi = c - \bar{u}A$ and $\pi_0 = z_U - \bar{u}e - 1$ one obtains an inequality $\pi x \leqslant \pi_0$ which is satisfied by every integer solution better than $\bar{x}$, and which can therefore be used to derive a disjunction of form (10.3).

Suppose this is done, and a disjunction (10.3) is at hand, known to be satisfied by every feasible integer x better than the current best. Then for each $i \in \{1, \ldots, p\}$, one chooses a row $h(i)$ of A, such that $N_{h(i)} \cap Q_i$ is "large"—or, conversely, $N_{h(i)} \setminus Q_i$ is "small." Clearly (and this is true for any choice of the indices $h(i)$), the disjunction (10.3) implies

$$\bigvee_{i=1}^{p}\left(\sum_{j\in N_{h(i)}\setminus Q_i} x_j \geq 1\right), \tag{10.6}$$

which in turn implies the inequality

$$\sum_{j\in W} x_j \geq 1, \tag{10.7}$$

where

$$W = \bigcup_{i=1}^{p} [N_{h(i)} \setminus Q_i].$$

The class of cutting planes (10.7) obtained in this way was shown in [12] to include as a proper subclass the Bellmore–Ratliff inequalities [16] derived from involutory bases. An all integer algorithm which uses these cutting planes was implemented and tested with good results on set covering problems with up to 1,000 variables (see [12] for details).

References

[1] J. A. Araoz Durand, Polyhedral neopolarities, Ph.D. Dissertation, University of Waterloo (November 1973).

[2] E. Balas, Intersection cuts — A new type of cutting planes for integer programming, Operations Res. 19 (1971) 19–39.

[3] E. Balas, Integer programming and convex analysis: Intersection cuts from outer polars, Math. Programming 2 (1972) 330–382.

[4] E. Balas, Ranking the facets of the octahedron, Discrete Math. 2 (1972) 1–15.

[5] E. Balas, Nonconvex quadratic programming via generalized polars, SIAM J. Appl. Math. 28 (1975) 335–349.

[6] E. Balas, A constraint-activating outer polar method for pure and mixed integer 0–1 programs, in: P. L. Hammer and G. T. Zoutendijk, eds., Mathematical Programming in Theory and Practice, (North-Holland, Amsterdam, 1974 275–310.

[7] E. Balas, On the use of intersection cuts in branch and bound. Paper presented at the 8th International Symposium on Mathematical Programming, Stanford, CA(August 27–31, 1973).

[8] E. Balas, Intersection cuts from disjunctive constraints, MSRR No. 330, Carnegie–Mellon University, First draft (August 1973); Revised and expanded (February 1974).

[9] E. Balas, Disjunctive programming: Cutting planes from logical conditions, in: O. L. Mangasarian, R. R. Meyer and S. M. Robinson, eds., Nonlinear Programming 2 (Academic Press, New York, 1975) 279–312.

[10] E. Balas, Disjunctive programming: Properties of the convex hull of feasible points, MSSR No. 348, Carnegie–Mellon University (July 1974).

[11] E. Balas, A Note on duality in disjunctive programming, J. Optimization Theory Appl. 15 (1977).

[12] E. Balas, Set covering with cutting planes from conditional bounds, MSRR No. 399. Carnegie–Mellon University (July 1976).

[13] E. Balas, V. J. Bowman, F. Glover and D. Sommer, An intersection cut from the dual of the unit hypercube, Operations Res. 19 (1971) 40–44.

[14] E. Balas and R. Jeroslow, Strengthening cuts for mixed integer programs, MSRR No. 359, Carnegie–Mellon University (February 1975).

[15] E. Balas and A. Zoltners, Intersection cuts from outer polars of truncated cubes, Naval Res. Logist. Quart. 22 (1975) 477–496.
[16] M. Bellmore and H.D. Ratliff, Set covering and involutory bases, Management Sci. 18 (1971) 194–206.
[17] C.E. Blair and R.G. Jeroslow, A converse for disjunctive constraints, MSRR No. 393, Carnegie–Mellon University (June 1976).
[17a] J. Borwein, A strong duality theory for the minimum of a family of convex programs, Dalhousie University, Halifax (June 1978).
[18] C.A. Burdet, Enumerative inequalities in integer programming, Math. Programming 2 (1972) 32–64.
[19] C.A. Burdet, Polaroids: A new tool in nonconvex and integer programming, Naval Res. Logist. Quart. 20 (1973) 13–24.
[20] D.R. Fulkerson, Blocking Polyhedra, in: B. Harris, ed., Graph Theory and Its Applications (Academic Press, New York, 1972) 93–112.
[21] D.R. Fulkerson, Anti-blocking polyhedra, J. Combinatorial Theory 12 (1972) 50–71.
[22] D.R. Fulkerson, Blocking and anti-blocking pairs of polyhedra, Math. Programming 1 (1971) 168–194.
[23] F. Glover, Convexity cuts and cut search, Operations Res. 21 (1973) 123–134.
[24] F. Glover, Convexity cuts for multiple choice problems, Discrete Math. 6 (1973) 221–234.
[25] F. Glover, Polyhedral annexation in mixed integer programming. MSRS 73–9, University of Colorado (August 1973). Revised and published as: Polyhedral annexation in mixed integer and combinatorial programming, Math. Programming 9 (1975) 161–188.
[26] F. Glover and D. Klingman, The generalized lattice point problem, Operations Res. 21 (1973) 141–156.
[26a] F. Glover, D. Klingman and J. Stutz, The disjunctive facet problem: formulation and solution techniques, Operations Res. 22 (1974) 582–601.
[27] J.J. Greenberg and W.P. Pierskalla, Stability theorems for infinitely constrained mathematical programs, J. Optimization Theory Appl. 16 (1975) 409–428.
[28] B. Grünbaum, Convex Polytopes (J. Wiley, New York, 1967).
[29] Hoang Tuy, Concave programming under linear constraints (Russian), Doklady Akademii Narck SSSR (1964). English translation in Soviet Math. Dokl. (1964) 1437–1440.
[30] R.G. Jeroslow, The principles of cutting plane theory: part I, Carnegie–Mellon University (February 1974).
[31] R.G. Jeroslow, Cutting planes for relaxations of integer programs, MSRR No. 347, Carnegie–Mellon University (July 1974).
[32] R.G. Jeroslow, Cutting plane theory: Disjunctive methods, Ann. Discrete Math., vol. 1: Studies in Integer Programming, (1977) 293–330.
[33] R.G. Jeroslow, Cutting planes for complementarity constraints, SIAM J. Control 16 (1978).
[34] R.G. Jeroslow, A cutting plane game and its algorithms, CORE Discussion Paper 7724 (June 1977).
[35] E.L. Johnson, The group problem for mixed integer programming, Math. Programming Study 2 (1974) 137–179.
[36] G. Owen, Cutting planes for programs with disjunctive constraints, J. Optimization Theory Appl. 11 (1973) 49–55.
[37] R.T. Rockafellar, Convex Analysis (Princeton University Press, Princeton, N.J, 1970).
[38] J. Stoer and C. Witzgall, Convexity and Optimization in Finite Dimensions I (Springer, Berlin, 1970).
[39] J. Tind, Blocking and anti-blocking sets, Math. Programming 6 (1974) 157–166.
[40] R.D. Young, Hypercylindrically deduced cuts in zero-one integer programs, Operations Res. 19 (1971) 1393–1405.
[41] E. Zemel, On the facial structure of 0–1 polytopes, Ph.D. Dissertation, GSIA, Carnegie–Mellon University (May 1976).
[42] P. Zwart, Intersection cuts for separable programming, Washington University, St. Louis (January 1972).

The following article originally appeared as:

E. Balas, *Erratum to: Disjunctive Programming*, Discrete Applied Mathematics 5 (1983) 247–248.

Discrete Applied Mathematics 5 (1983) 247–248
North-Holland Publishing Company 247

ERRATUM

To: E. Balas, Disjunctive Programming, in: P.L. Hammer, E.L. Johnson and B.H. Korte, eds., Discrete Optimization II, Ann. Discrete Math. 5 (North-Holland, Amsterdam, 1979) 3–51

E. BALAS

Carnegie-Mellon University, Pittsburgh, PA 15213, USA

Received 18 May 1982

On page 36, lines 4–5, we assert that for $k=1,2$, $\beta_0^k - 1$ is a lower bound on $\sum_{j\in J\setminus Q_k} \beta_j^k x_j$, since

$$\sum_{i\in I\cap Q_k} x_i = \beta_0^k + \sum_{j\in J\setminus Q_k} \beta_j^k(-x_j) \le 1, \quad k=1,2.$$

This is not true in general. Instead, from the definitions it follows that for $k=1,2$,

$$\beta_0^k + \sum_{j\in J\setminus Q_k} \beta_j^k(-x_j) = \sum_{i\in I\cap Q_k} x_i + \sum_{j\in J\cap Q_k} \beta_j^k x_j,$$

and thus an upper bound on the value of the expression on the lefthand side is given by

$$\begin{aligned}\gamma_0^k &= \max\left\{ \sum_{i\in I\cap Q_k} x_i + \sum_{j\in J\cap Q_k} \beta_j^k x_j \;\middle|\; \sum_{j\in Q} x_j \le 1 \right\}\\ &= \max\left\{1, \max_{j\in J\cap Q_k} \beta_j^k\right\}.\end{aligned}$$

Thus the lower bound on $\sum_{j\in J\setminus Q_k} \beta_j^k x_j$ to be used in the derivation of the cut of Theorem 8.1 is $\beta_0^k - \gamma_0^k$ rather than $\beta_0^k - 1$. Accordingly, the multipliers $\sigma^k = 1$, $k=1,2$, should be replaced by $\sigma^k = 1/\gamma_0^k$, $k=1,2$, and the coefficients β_j^k, β_0^k, should be replaced everywhere on page 36 by β_j^k/γ_0^k and β_0^k/γ_0^k, respectively. Thus the correct version of (8.8) and (8.9) is

$$m_0^* = \frac{\beta_j^2\beta_0^1 - \beta_j^1\beta_0^2}{\beta_0^1\gamma_0^2 + \beta_0^2\gamma_0^1} \tag{8.8}$$

and

$$\beta_j = \begin{cases} \max\left\{0, \min\left\{\dfrac{\beta_j^1}{\beta_0^1}, \dfrac{\beta_j^2}{\beta_0^2}\right\}\right\}, & j\in J\cap Q,\\[2ex] \min\left\{\dfrac{(\beta_j^1/\gamma_0^1) + \langle m_0^*\rangle}{\beta_0^1/\gamma_0^1}, \dfrac{(\beta_j^2/\gamma_0^2) - [m_0^*]}{\beta_0^2/\gamma_0^2}\right\}, & j\in J\setminus Q, \end{cases} \tag{8.9}$$

respectively.

0166-218X/83/0000–0000/$03.00 © 1983 North-Holland

Example 8.1, which illustrates Theorem 8.1, contains some typographical errors. In the expressions for β_6, β_3 and β_5 on page 37, min and max should be interchanged. Besides, the entries a_{46} and a_{45} of Table 2 (also on page 37) should read $-\frac{1}{3}$ and $\frac{1}{3}$ instead of $-\frac{2}{3}$ and $-\frac{1}{3}$ respectively. As a result of this and the change in (8.8) and (8.9), the values of $m_0^*(1)$ and $m_0^*(2)$ are $\frac{3}{5}$ and $-\frac{3}{5}$, instead of $\frac{2}{3}$ and $-\frac{2}{3}$, respectively, and the cut becomes

$$x_1 + x_2 + x_5 + \tfrac{1}{2}x_6 \geq 1.$$

Part II
From the Beginnings to the State-of-the-Art

The celebration of 50 Years of Integer Programming started with three invited talks by Gérard Cornuéjols, William Cook, and Laurence Wolsey on integer programming and combinatorial optimization from the beginnings to the state-of-the-art. While their original lectures can be enjoyed on the DVD-Video that accompanies this book, the written versions of their presentations are presented here. Michele Conforti and Giacomo Zambelli have joined Gérard Cornuéjols as co-authors and François Vanderbeck has joined Laurence Wolsey as a co-author.

Chapter 11
Polyhedral Approaches to Mixed Integer Linear Programming

Michele Conforti, Gérard Cornuéjols,* and Giacomo Zambelli

Abstract This survey presents tools from polyhedral theory that are used in integer programming. It applies them to the study of valid inequalities for mixed integer linear sets, such as Gomory's mixed integer cuts.

11.1 Introduction

11.1.1 Mixed integer linear programming

In this tutorial we consider *mixed integer linear programs*. These are problems of the form

$$\begin{aligned} \max \quad & cx + hy \\ & Ax + Gy \leqslant b \\ & x \geqslant 0 \text{ integral} \\ & y \geqslant 0, \end{aligned} \tag{11.1}$$

Michele Conforti
Università di Padova, Italy
e-mail: conforti@math.unipd.it

Gérard Cornuéjols
Carnegie Mellon University, Pittsburgh, USA, and Université d'Aix-Marseille, France
e-mail: gc0v@andrew.cmu.edu

Giacomo Zambelli
Università di Padova, Italy
e-mail: giacomo@math.unipd.it

* Supported by NSF grant CMMI0653419, ONR grant N00014-09-1-0133 and ANR grant ANR06-BLAN-0375.

M. Jünger et al. (eds.), *50 Years of Integer Programming 1958-2008*,
DOI 10.1007/978-3-540-68279-0_11, © Springer-Verlag Berlin Heidelberg 2010

where the data are the row vectors $c \in \mathbb{Q}^n$, $h \in \mathbb{Q}^p$, the matrices $A \in \mathbb{Q}^{m\times n}$, $G \in \mathbb{Q}^{m\times p}$ and the column vector $b \in \mathbb{Q}^m$; and the variables are the column vectors $x \in \mathbb{R}^n$ and $y \in \mathbb{R}^p$. The set S of feasible solutions to (11.1) is called a *mixed integer linear set* when $p \geqslant 1$ and a *pure integer linear set* when $p = 0$.

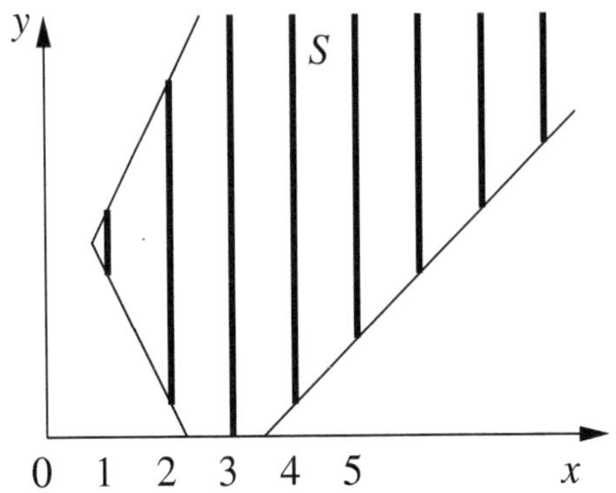

Fig. 11.1 A mixed integer set

The polyhedral approach is a powerful tool for solving mixed integer linear programs (11.1). This is the topic of this tutorial.

11.1.2 Historical perspective

Babylonian tablets show that mathematicians were already solving systems of linear equations over 3000 years ago. The eighth book of the Chinese *Nine Books of Arithmetic*, written over 2000 years ago, describes the Gaussian elimination method. In 1809, Gauss [29] used this method in his work and presented it as a "standard technique". Surprisingly, the method was subsequently named after him.

A major breakthrough occurred when mathematicians started analyzing systems of linear *inequalities*. This is a fertile ground for beautiful theories. In 1826 Fourier [28] gave an algorithm for solving such systems by eliminating variables one at a time. Other important contributions are due to Farkas [26] and Minkowski [39]. Systems of linear inequalities define polyhedra and it is natural to optimize a linear function over them. This is the topic of *linear programming*, arguably one of the greatest successes of computational mathematics in the twentieth century. The *simplex method*, developed by Dantzig [20] in 1951, is currently used to solve large-scale applications in all sorts of human endeavors. It is often desirable to find *integer solutions* to linear programs. This is the topic of this tutorial. The first algorithm for solving (11.1) in the pure integer case was discovered in 1958 by Gomory [31].

When considering algorithmic questions, a fundamental issue is the increase in computing time when the size of the problem instance increases. In the 1960s Edmonds [23] was one of the pioneers in stressing the importance of *polynomial-time* algorithms. These are algorithms whose computing time is bounded by a polynomial function of the instance size. In particular Edmonds [24] pointed out that, by being a bit careful with the intermediate numbers that are generated, the Gaussian elimination method can be turned into a polynomial-time algorithm. The existence of a polynomial-time algorithm for linear programming remained a challenge for many years. This question was resolved positively by Khachiyan [34] in 1979, and later by Karmarkar [33] using a totally different algorithm. Both algorithms were (and still are) very influential, each in its own way. In integer programming, Lenstra [36] found a polynomial-time algorithm when the number of variables is fixed. Although integer programming is NP-hard in general, the polyhedral approach has proved successful in practice. It can be traced back to the work of Dantzig, Fulkerson and Johnson [21]. Research is currently very active in this area. Also very promising are non-polyhedral approximations that can be computed in polynomial-time, such as semidefinite relaxations (Lovász and Schrijver [37], Goemans and Williamson [30]).

In the next section, we motivate the polyhedral approach by presenting a cutting plane method for solving mixed integer linear programs (11.1).

11.1.3 Cutting plane methods

Solving a mixed integer linear program (MILP) such as (11.1) is NP-hard (Cook [16]). One approach that has been quite successful in practice is based on an idea that is commonly used in computational mathematics: Find a relaxation that is easier to compute and gives a tight approximation. We focus on *linear programming (LP) relaxations*.

Given a mixed integer linear set $S := \{(x,y) \in \mathbb{Z}^n_+ \times \mathbb{R}^p_+ : Ax + Gy \leqslant b\}$, a linear programming relaxation of S is a set $P' = \{(x,y) : A'x + G'y \leqslant b'\}$ that contains S. Why LP relaxations? Mainly for two reasons: As mentioned already, there are efficient practical algorithms for solving linear programs [20], [33]. Second, one can generate a sequence of LP relaxations that provide increasingly tight approximations of the set S.

For a mixed integer set S, there is a *natural LP relaxation*:

$$P_0 := \{(x,y) : Ax + Gy \leqslant b,\ x \geqslant 0,\ y \geqslant 0\}$$

which is obtained from the system that defines S by discarding the integrality requirement on the vector x.

Let (x^0, y^0) be an optimal solution and z^0 the value of the linear program

$$\max\{cx + hy : (x,y) \in P_0\} \tag{11.2}$$

whose constraint set is the natural linear programming relaxation of S. We will assume that we have a linear programming solver at our disposal, thus (x^0, y^0) and z^0

are available to us. Let z^* be the optimal value of (11.1). Since $S \subseteq P_0$, it follows that $z^0 \geqslant z^*$. Furthermore if x^0 is an integral vector, then $(x^0, y^0) \in S$, $z^* = z^0$ and the MILP (11.1) is solved.

A strategy for dealing with an optimal solution (x^0, y^0) of (11.2) that is not in S is to find an inequality $\alpha x + \gamma y \leqslant \beta$ that is satisfied by every point in S and such that $\alpha x^0 + \gamma y^0 > \beta$. The existence of such an inequality is guaranteed when (x^0, y^0) is an optimal *basic* solution of (11.2), which can be found by standard LP algorithms.

An inequality $\alpha x + \gamma y \leqslant \beta$ that is satisfied by every point in S is a *valid inequality* for S. If such an inequality is violated by (x^0, y^0), it is a *cutting plane* separating (x^0, y^0) from S.

Define now

$$P_1 := P_0 \cap \{(x,y) : \alpha x + \gamma y \leqslant \beta\}.$$

Since $S \subseteq P_1 \subset P_0$, a linear programming relaxation for MILP based on P_1 is stronger than the natural relaxation based on P_0 in the sense that the solution (x^0, y^0) of the natural LP relaxation does not belong to P_1. So the optimal value of the LP

$$\max\{cx + hy : (x,y) \in P_1\}$$

is at least as good an approximation of the value z^* as z_0. The recursive application of this idea leads to the *cutting plane approach*:

Starting with $i = 0$, repeat:

Recursive Step: Solve the linear program $\max\{cx + hy : (x,y) \in P_i\}$. If the associated optimal basic solution (x^i, y^i) belongs to S , Stop.

Otherwise solve the following separation problem:

Find a cutting plane $\alpha x + \gamma y \leqslant \beta$ separating (x^i, y^i) from S.

Set $P_{i+1} := P_i \cap \{(x,y) : \alpha x + \gamma y \leqslant \beta\}$ and repeat the recursive step.

If (x^i, y^i) is not in S, there are infinitely many cutting planes separating (x^i, y^i) from S. So the separation problem that is recursively solved in the above algorithm has many solutions. There is usually a tradeoff between the running time of a separation procedure and the quality of the cutting planes it produces. We will discuss several possibilities in this survey.

For now, we illustrate the cutting plane approach on a two variable example (see Figure 11.2):

$$\begin{aligned} \max z = \; & 11x_1 + 4.2x_2 \\ & -x_1 + x_2 \leq 2 \\ & 8x_1 + 2x_2 \leq 17 \\ & x_1, x_2 \geq 0 \text{ integer.} \end{aligned} \tag{11.3}$$

We first add slack variables x_3 and x_4 to turn the inequality constraints into equalities. The problem becomes:

$$\begin{aligned} z - 11x_1 - 4.2x_2 &= 0 \\ -x_1 + x_2 + x_3 &= 2 \\ 8x_1 + 2x_2 + x_4 &= 17 \\ x_1, x_2, x_3, x_4 &\geq 0 \text{ integer.} \end{aligned}$$

Solving the LP relaxation, we get the optimal tableau:

$$\begin{aligned} z + 1.16x_3 + 1.52x_4 &= 28.16 \\ x_2 + 0.8x_3 + 0.1x_4 &= 3.3 \\ x_1 - 0.2x_3 + 0.1x_4 &= 1.3 \\ x_1, x_2, x_3, x_4 &\geq 0. \end{aligned}$$

The corresponding basic solution is $x_3 = x_4 = 0$, $x_1 = 1.3$, $x_2 = 3.3$ with objective value $z = 28.16$. Since the values of x_1 and x_2 are not integer, this is not a solution of (11.3). We can generate a cut from the constraint $x_2 + 0.8x_3 + 0.1x_4 = 3.3$ using the following reasoning. Since x_2 is an integer variable, we have

$$0.8x_3 + 0.1x_4 = 0.3 + k \quad \text{where } k \in \mathbb{Z}.$$

Since the left-hand-side is nonnegative, we must have $k \geqslant 0$, which implies

$$0.8x_3 + 0.1x_4 \geqslant 0.3.$$

This is the famous Gomory fractional cut [31]. Note that it cuts off the above fractional LP solution $x_3 = x_4 = 0$.

Since $x_3 = 2 + x_1 - x_2$ and $x_4 = 17 - 8x_1 - 2x_2$, we can express Gomory's fractional cut in the space (x_1, x_2). This yields $x_2 \leqslant 3$ (see Figure 11.2).

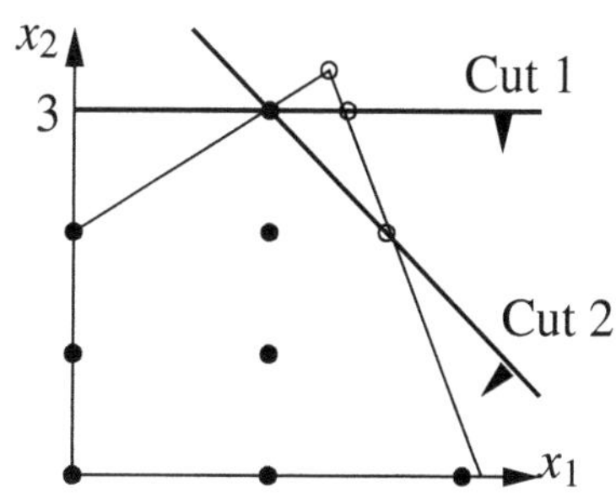

Fig. 11.2 The first two cuts in the cutting plane algorithm

Adding this cut to the linear programming relaxation, we get the following formulation.

$$\begin{aligned} \max\ 11x_1 + 4.2x_2 & \\ -x_1 + x_2 &\leq 2 \\ 8x_1 + 2x_2 &\leq 17 \\ x_2 &\leq 3 \\ x_1, x_2 &\geq 0. \end{aligned}$$

Solving this linear program, we find the basic solution $x_1 = 1.375$, $x_2 = 3$ with value $z = 27.725$. From the optimal tableau, we can generate a new fractional cut:

$$x_1 + x_2 \leqslant 4.$$

Adding this cut to the LP, we find a new LP optimum $x_1 = 1.5$, $x_2 = 2.5$ with value $z = 27$. Two more iterations are needed to obtain the optimal integer solution $x_1 = 1$, $x_2 = 3$ with value $z = 23.6$.

11.2 Polyhedra and the fundamental theorem of integer programming

A *polyhedron* in $\mathbb{R}^n$ is a set of the form $P := \{x \in \mathbb{R}^n : Ax \leqslant b\}$ where A is a real matrix and b a real vector. If A and b have rational entries, P is a *rational polyhedron*. A polyhedron of the form $\{x \in \mathbb{R}^n : Ax \leqslant 0\}$ is called a *polyhedral cone*. Note that a polyhedral cone is always nonempty since it contains the null vector 0.

For $S \subseteq \mathbb{R}^n$, the *convex hull* of S is the set $\text{conv}(S) := \{x \in \mathbb{R}^n : x = \sum_{i=1}^k \lambda_i x^i$ where $k \geqslant 1$, $\lambda \in \mathbb{R}^k_+, \sum_{i=1}^k \lambda_i = 1$ and $x^1, \ldots, x^k \in S\}$. This is the smallest convex set that contains S. Note that $\text{conv}(\emptyset) = \emptyset$. The convex hull of a finite set of points in $\mathbb{R}^n$ is called a *polytope*. It will sometimes be useful to work with $\overline{\text{conv}}(S)$, the closure of $\text{conv}(S)$, which is the smallest closed convex set that contains S. The *conic hull* of a nonempty set $S \subseteq \mathbb{R}^n$ is $\text{cone}(S) := \{x \in \mathbb{R}^n : x = \sum_{i=1}^k \lambda_i x^i$ where $k \geqslant 1$, $\lambda \in \mathbb{R}^k_+$ and $x^1, \ldots, x^k \in S\}$. If S is a finite set, $\text{cone}(S)$ is said to be *finitely generated*. It will be convenient to define $\text{cone}(\emptyset) = \{0\}$.

Given a cone C and $r \in C \setminus \{0\}$, the set $\text{cone}(r) = \{\lambda r : \lambda \geqslant 0\}$ is called a *ray* of C. Since $\text{cone}(\lambda r) = \text{cone}(r)$ for every $\lambda > 0$, we will sometimes simply refer to a vector $r \in C$ as a ray, to denote the corresponding ray $\text{cone}(r)$. So when we say that r and r' are distinct rays, we mean $\text{cone}(r) \neq \text{cone}(r')$. We say that $r \in C \setminus \{0\}$ is an *extreme ray* if there are no distinct rays r^1 and r^2 such that $r = r^1 + r^2$. We say that C is *pointed* if, for every $r \in C \setminus \{0\}$, $-r \notin C$. One can verify that *if C is a finitely generated pointed cone, then C is generated by its extreme rays.*

An important theorem, due to Minkowski and Weyl, states that every polyhedron P can be written as the sum of a polytope Q and a finitely generated cone C. Here $Q + C := \{x \in \mathbb{R}^n : x = y + z$ for some $y \in Q$ and $z \in C\}$. Note that $P = \emptyset$ if and only if $Q = \emptyset$. If $P := \{x \in \mathbb{R}^n : Ax \leqslant b\}$ is nonempty, then C is the cone $\{x \in \mathbb{R}^n : Ax \leqslant 0\}$, which is called the *recession cone* of P and denoted by $\text{rec}(P)$. We will prove this theorem in Section 11.2.3.

Using the Minkowski-Weyl theorem, we will prove the fundamental theorem of integer programming, due to Meyer [38]:

Given rational matrices A and G and a rational vector b, let $P := \{(x,y) : Ax + Gy \leqslant b\}$ and $S := \{(x,y) \in P : x \text{ integral}\}$. Then there exist rational matrices A', G' and a rational vector b' such that $\text{conv}(S) = \{(x,y) : A'x + G'y \leqslant b'\}$.

In other words, when P is a rational polyhedron, then the convex hull of S is also a rational polyhedron. Note that, if matrices A, G are not rational, then $\text{conv}(S)$ may not be a polyhedron as shown by the example $S := \{x \in \mathbb{Z}^2 : 1 \leqslant x_2 \leqslant \sqrt{2}x_1\}$. In this case, infinitely many inequalities are required to describe $\text{conv}(S)$ by a system of linear inequalities. In this survey we assume that the data A, G, b are rational. Meyer's theorem is the theoretical underpinning of the polyhedral approach to integer programming. Indeed it shows that (11.1), the problem of optimizing a linear function over a mixed integer set S, is equivalent to solving a linear program. The main difficulty is that the polyhedron conv(S) is not known explicitly. In the later sections of the tutorial, we will address the constructive aspects of conv(S).

11.2.1 Farkas' lemma and linear programming duality

The following is a fundamental fact in linear algebra:

Theorem 11.1. *A system of linear equations $Ax = b$ is infeasible if and only if the system $uA = 0,\ ub < 0$ is feasible.*

A constructive proof of Theorem 11.1 is straightforward using Gaussian elimination on the system $Ax = b$. Furthermore, one can decide in polynomial time which of the two systems is feasible, and find a solution, again using Gaussian elimination (Edmonds [24], see e.g. Schrijver [45] p. 27). Farkas' lemma [26] provides a certificate of the solvability of a system of linear *inequalities* $Ax \leq b$ in the same spirit as Theorem 11.1. However its proof is not as straightforward. This is not surprising since checking feasibility of a system of linear inequalities is a linear programming problem. In fact, Farkas' lemma can be used to derive the strong duality theorem of linear programming, as we will show later in this section. We first give a proof of Farkas' lemma based on Theorem 11.1, following Conforti, Di Summa and Zambelli [15]. A linear system $a_i x = b_i,\ i = 1, \ldots, p,\ \ a_i x \leqslant b_i,\ i = p+1, \ldots, m$ is *minimally infeasible* if it has no solution but each of the m linear systems obtained by removing a single constraint has a solution.

Theorem 11.2 (Farkas' lemma). *The system of linear inequalities $Ax \leqslant b$ is infeasible if and only if the system $uA = 0,\ ub < 0,\ u \geqslant 0$ is feasible.*

Proof. Assume $uA = 0,\ ub < 0,\ u \geqslant 0$ is feasible. Then $0 = uAx \leqslant ub < 0$ for any x satisfying $Ax \leqslant b$. It follows that $Ax \leqslant b$ is infeasible and this proves the "if" part.

Now we prove the "only if" part. Let $A'x \leqslant b'$ be an infeasible system. Let a_i, $i \in R$, denote the rows of A'. Remove inequalities from the system $A'x \leqslant b'$ until it becomes minimally infeasible. Let $Ax \leqslant b$ be the resulting system and let $M \subseteq R$ index the rows of A. We will show that there exists $u \geqslant 0$ such that $uA = 0$ and $ub < 0$. Setting $u_i = 0$ for all $i \in R \setminus M$, this will show that there exists $u \geqslant 0$ such that $uA' = 0$ and $ub' < 0$, proving the "only if" part. Given $S \subseteq M$, define $\bar{S} := M \setminus S$.

Claim. *For every $S \subseteq M$ the system $a_i x = b_i,\ i \in S,\ \ a_i x \leqslant b_i,\ i \in \bar{S}$ is minimally infeasible.*

The proof is by induction on $|S|$. The claim is true when $S = \emptyset$. Consider $S \subseteq M$ with $|S| \geqslant 1$ and assume by induction that the claim holds for any set of cardinality smaller than $|S|$.

The system $a_i x = b_i$, $i \in S$, $a_i x \leqslant b_i$, $i \in \bar{S}$ is infeasible. Let $k \in M$. By the induction hypothesis applied to $S \setminus \{k\}$, the system

$$a_i x = b_i,\ i \in S \setminus \{k\},\ \ a_i x \leqslant b_i,\ i \in \bar{S} \tag{11.4}$$

is feasible for any $k \in S$. Therefore, to prove the claim, we only need to show that

$$a_i x = b_i,\ i \in S,\ \ a_i x \leqslant b_i,\ i \in \bar{S} \setminus \{k\} \tag{11.5}$$

is feasible for any $k \in \bar{S}$. Let $h \in S$. By induction, there exist x^h and x^k such that

$$a_i x^h = b_i,\ i \in S \setminus \{h\},\ \ a_i x^h \leqslant b_i,\ i \in \bar{S}$$

and

$$a_i x^k = b_i,\ i \in S \setminus \{h\},\ \ a_i x^k \leqslant b_i,\ i \in \bar{S} \cup \{h\} \setminus \{k\}.$$

Notice that $a_h x^k \leqslant b_h$ and $a_h x^h > b_h$, so, setting $\alpha = a_h x^h - b_h > 0$ and $\beta = b_h - a_h x^k \geqslant 0$, the vector $y = \frac{\alpha}{\alpha+\beta} x^k + \frac{\beta}{\alpha+\beta} x^h$ is a solution for (11.5). This proves the claim.

Now, since $a_i x \leqslant b_i$, $i \in M$, is infeasible, then clearly $a_i x = b_i$, $i \in M$, is infeasible and by Theorem 11.1, there exists a vector u such that $uA = 0$, $ub < 0$. The lemma holds if $u \geqslant 0$. So suppose $u_1 < 0$. According to the Claim, there is a vector x^* such that $a_i x^* = b_i$, $i \in M \setminus \{1\}$. Since $Ax \leq b$ is an infeasible system, $a_1 x^* > b_1$. This (together with $u_1 < 0$) shows $u(Ax^* - b) < 0$, contradicting

$$u(Ax^* - b) = (uA)x^* - ub > 0$$

where the inequality follows from $uA = 0$ and $ub < 0$. □

Equivalently, Farkas' lemma can be written as:

Corollary 11.1. *The system of linear inequalities $Ax \leqslant b$ is feasible if and only if $ub \geq 0$ for every vector u satisfying $uA = 0$, $u \geqslant 0$.*

Farkas' lemma can also be restated as follows:

Theorem 11.3. *The system $Ax = b$, $x \geqslant 0$ is feasible if and only if $ub \geq 0$ for every u satisfying $uA \geqslant 0$.*

Proof. The system $Ax = b$, $x \geqslant 0$ is equivalent to $Ax \leqslant b$, $-Ax \leqslant -b$, $-x \leqslant 0$. The theorem follows by applying Corollary 11.1 to this latter system. □

The following is a more general, yet still equivalent, form of Farkas' lemma.

Theorem 11.4. *The system $Ax + By \leq f$, $Cx + Dy = g$, $x \geqslant 0$ is feasible if and only if $uf + vg \geq 0$ for every (u, v) satisfying $uA + vC \geqslant 0$, $uB + vD = 0$, $u \geq 0$.*

Checking the feasibility of a system of linear inequalities can be done in polynomial time (Khachiyan [34], Karmarkar [33]). Next we derive the fundamental theorem of Linear Programming from Farkas' lemma.

Theorem 11.5 (Linear Programming Duality). *Let*

$$P := \{x : Ax \leq b\} \quad \text{and} \quad D := \{u : uA = c, \ u \geq 0\}.$$

If P and D are both nonempty, then

$$\max\{cx : x \in P\} = \min\{ub : u \in D\}. \tag{11.6}$$

Proof. For $\bar{x} \in P$ and $\bar{u} \in D$ we have that $c\bar{x} = \bar{u}A\bar{x} \leq \bar{u}b$. Therefore "max $\leq$ min" always holds. To prove equality, we need to prove that the system

$$-cx + b^T u^T \leq 0, \, Ax \leq b, \, A^T u^T = c^T, \, u^T \geq 0 \tag{11.7}$$

is feasible. By Farkas' lemma (Theorem 11.4), (11.7) is feasible if and only $\mu b + \nu c^T \geq 0$ for every vector (λ, μ, ν) satisfying

$$\mu A - \lambda c = 0, \quad \nu A^T + \lambda b^T \geq 0, \quad \lambda, \mu \geq 0.$$

If $\lambda > 0$, then $\mu b = \lambda^{-1} \lambda b^T \mu^T \geq -\lambda^{-1} \nu A^T \mu^T = -\lambda^{-1} \nu c^T \lambda = -\nu c^T$.
If $\lambda = 0$, let $\bar{x} \in P$ and $\bar{u} \in D$. Then $\mu b \geq \mu A\bar{x} = 0 \geq -\nu A^T u^T = -\nu c^T$.
Therefore in both cases we have $\mu b + \nu c^T \geq 0$ and the proof is complete. □

Theorem 11.6 (Complementary slackness). *Let X^{opt} and U^{opt} be subsets of $P := \{x : Ax \leqslant b\}$ and $D := \{u : uA = c, \ u \geqslant 0\}$ respectively. Define $I = \{i : u_i > 0$ for some $u \in U^{\text{opt}}\}$.*

Then X^{opt} and U^{opt} are the sets of optimal solutions of the pair of dual LPs

$$\max\{cx : x \in P\} \quad \text{and} \quad \min\{ub : u \in D\}$$

if and only if

$$X^{\text{opt}} = \{x : a_i x = b_i, i \in I, \, a_i x \leq b_i, i \notin I\}.$$

Proof. By Theorem 11.5, the sets X^{opt} and U^{opt} are the sets of optimal solutions of the pair of dual LPs above if and only if $cx = ub$ for every $x \in X^{\text{opt}}$, $u \in U^{\text{opt}}$. For every $x \in P$ and $u \in D$, $cx = uAx \leqslant ub$, hence equality holds throughout if and only if $a_i x = b_i$ for every $i \in I$ and every $x \in X^{\text{opt}}$. □

Here is another well-known consequence of Farkas' lemma:

Remark 11.1. Let $P := \{x : Ax \leq b\}$ and $D := \{u : uA = c, \ u \geq 0\}$. If $D = \emptyset$ and $P \neq \emptyset$, then $\max\{cx : x \in P\}$ is unbounded.

Proof. Since $D = \emptyset$, by Farkas' lemma, the system $Ay \leq 0, \ cy > 0$ is feasible: Let $\bar{y}$ be a solution of this system and $\bar{x} \in P$. Then $\bar{x} + \lambda \bar{y} \in P$ for every $\lambda \geq 0$. Since $c\bar{y} > 0$, $\max\{cx : x \in P\}$ is unbounded. □

11.2.2 Carathéodory's theorem

Theorem 11.7 (Carathéodory's theorem). *If the linear system $Ax = b, x \geqslant 0$ is feasible, then it has a solution $\bar{x}$ where the columns of A corresponding to the positive entries of $\bar{x}$ are linearly independent.*

Proof. Let $\bar{x}$ be a solution with the minimum number of positive entries and let $\bar{A}$ be the column submatrix of A corresponding to the positive entries of $\bar{x}$. If the columns of $\bar{A}$ are linearly dependent, there exists a vector $y \neq 0$ such that $Ay = 0$ and $y_j = 0$ whenever $\bar{x}_j = 0$. We assume w.l.o.g. that y has at least one positive component. Let $\lambda = \min_{y_j>0} \frac{\bar{x}_j}{y_j}$, let j^* be an index for which this minimum is attained and define $x^* = \bar{x} - \lambda y$. Clearly $Ax^* = b$. By the choice of λ, $x^* \geqslant 0$ and $x^*_j = 0$ whenever $\bar{x}_j = 0$. Furthermore $x^*_{j^*} = 0$ while $\bar{x}_{j^*} > 0$, a contradiction to our choice of $\bar{x}$. □

The next theorem combines Carathéodory's Theorem 11.7 with a strengthening of Farkas' lemma.

Theorem 11.8. *For a linear system $Ax = b, x \geqslant 0$ exactly one of the following two alternatives holds:*

- *$Ax = b, x \geqslant 0$ admits a solution $\bar{x}$ where the columns of A, corresponding to the positive entries of $\bar{x}$, are linearly independent.*
- *There is a vector u such that $uA \geqslant 0$, $ub < 0$ and there is a column submatrix A^0 of A such that $uA^0 = 0$ and $\text{rank}(A^0) = \text{rank}(A|b) - 1$.*

Proof. If $Ax = b, x \geqslant 0$ admits a solution, then Theorem 11.7 shows that the first outcome holds. So we assume that $Ax = b$, $x \geqslant 0$ is infeasible and we show that the second outcome holds.
If $\text{rank}(A) = \text{rank}(A|b) - 1$, then by standard linear algebra $Ax = b$ is infeasible, so by Theorem 11.1 there exists a vector u such that $uA = 0$, $ub < 0$ and the theorem obviously holds in this case.
So we consider the case $\text{rank}(A) = \text{rank}(A|b)$ and we can assume that A has full row-rank m. By Theorem 11.3, there exists a vector u such that $uA \geqslant 0$, $ub < 0$. Among such vectors, choose u such that the column submatrix A^0 of A where $uA^0 = 0$ has maximum rank. Suppose by contradiction that $\text{rank}(A^0) \leqslant m-2$. Then $\text{rank}(A^0|A\mathbf{1}) \leqslant m-1$ where $\mathbf{1}$ denotes the vector of all 1s. So there exists a vector $v \neq 0$ such that $vA^0 = 0$, $vA\mathbf{1} = 0$, and we may choose v such that $vb \geqslant 0$.
Let J be the set of column-indices of $A = (a^1, \dots, a^n)$ and J^0 be the subset of J, corresponding to the column indices of A^0. Since $v \neq 0$ and A has full row-rank, there is an index j such that $va^j \neq 0$. Since $vA\mathbf{1} = 0$, then there is an index j such that $va^j > 0$. (Note that such an index j is in $J \setminus J^0$).
Let $\lambda = \min_{j \in J: va^j > 0} \frac{ua^j}{va^j}$ and let j^* be an index for which this minimum is attained. Then by the choice of λ, $(u - \lambda v)A \geqslant 0$ and $(u - \lambda v)b < 0$. Furthermore $(u - \lambda v)A^0 = 0$ and $(u - \lambda v)a^{j^*} = 0$. Since $vA^0 = 0$ and $va^{j^*} > 0$, then $A^0 x = a^{j^*}$ is infeasible by Theorem 11.1. This implies $\text{rank}(A^0|a^{j^*}) = \text{rank}(A^0) + 1$. Therefore $u - \lambda v$ contradicts the choice of u. □

11.2.3 The theorem of Minkowski-Weyl

We first present the Minkowski-Weyl theorem for cones.

Theorem 11.9 (Minkowski-Weyl theorem for cones). *For a set $C \subseteq \mathbb{R}^n$, the two following conditions are equivalent:*

1. *There is a matrix A such that $C = \{x \in \mathbb{R}^n : Ax \geq 0\}$.*
2. *There is a matrix R such that $C = \{x \in \mathbb{R}^n : x = R\mu,\ \mu \geqslant 0\}$.*

Theorem 11.9 states that a cone is polyhedral if and only if it is finitely generated. The columns of R are the *generators* of the cone C.

A pair of two matrices (A,R) that satisfy $\{x \in \mathbb{R}^n : Ax \geq 0\} = \{x \in \mathbb{R}^n : x = R\mu,\ \mu \geqslant 0\}$ will be called an *MW-pair*. Theorem 11.9 states that, for every matrix A, there exists a matrix R such that (A,R) is an MW-pair and, conversely, for every matrix R, there exists a matrix A such that (A,R) is an MW-pair. The next lemma shows that any one of these two statements implies the other.

Lemma 11.1. *A pair of matrices (A,R) is an MW-pair if and only if (R^T, A^T) is an MW-pair.*

Proof. We only need to show that if (A,R) is an MW-pair, then (R^T, A^T) is an MW-pair. Assume (A,R) is an MW-pair. By Farkas' lemma (Theorem 11.3), $\{x \in \mathbb{R}^n : x = R\mu, \mu \geqslant 0\} = \{x \in \mathbb{R}^n : x^T y \geq 0$ for every y satisfying $R^T y \geq 0\}$. Therefore

$$\{x : Ax \geq 0\} = \{x : x^T y \geq 0 \text{ for every } y \text{ such that } R^T y \geq 0\}. \tag{11.8}$$

We need to show that $\{y : R^T y \geq 0\} = \{y : y = A^T \nu,\ \nu \geq 0\}$.
We first show $\{y : R^T y \geq 0\} \subseteq \{y : y = A^T \nu,\ \nu \geq 0\}$. Let $\bar{y}$ satisfying $R^T \bar{y} \geq 0$. By (11.8), inequality $\bar{y}^T x \geq 0$ is valid for $Ax \geq 0$. By Farkas' Lemma (Theorem 11.3), it follows that there exists a vector ν such that $\bar{y} = A^T \nu$ and $\nu \geqslant 0$.
We show the reverse inclusion. Given $\bar{y}$ such that $\bar{y} = A^T \nu$ for some $\nu \geq 0$, $R^T \bar{y} = R^T A^T \nu \geq 0$, where the inequality follows from the assumption that (A,R) is an MW-pair, because AR has nonnonnegative entries (since $ARe^j \geq 0$ for the unit vectors e^j). □

Proof of Theorem 11.9
By Lemma 11.1 it is enough to prove that 2 implies 1.
Let $r^1, \ldots, r^k$ be the columns of R. We may assume that the vectors $r^1, \ldots, r^k$ span $\mathbb{R}^n$, else all these vectors satisfy an equation $dr = 0$ where $d \neq 0$ and one variable can be eliminated (i.e., the dimension can be reduced). Now consider the half spaces $\{x \in \mathbb{R}^n : ax \geq 0\}$ that contain $\{r^1, \ldots, r^k\}$ such that the hyperplane $\{x \in \mathbb{R}^n : ax = 0\}$ contains $n-1$ linearly independent vectors among $r^1, \ldots, r^k$. Since these vectors span $\mathbb{R}^n$, there is a *finite* number of such half-spaces. (In fact, at most $\binom{k}{n-1}$). Let A be the matrix whose rows contain the incidence vector a of all such subspaces and consider the cones

$$C_A := \{x \in \mathbb{R}^n : Ax \geq 0\}, \quad C_R := \{x \in \mathbb{R}^n : x = R\mu, \mu \geqslant 0\}.$$

Since every inequality of the system $Ax \geq 0$ is valid for $r^1, \ldots, r^k$, we have $C_R \subseteq C_A$.

Let $\bar{x} \notin C_R$. Then the system $\bar{x} = R\mu, \mu \geq 0$ is infeasible. By Theorem 11.8, there exists u such that $uR \geqslant 0$, $u\bar{x} < 0$ and there exists a column submatrix R^0 of R such that $uR^0 = 0$ and $\text{rank}(R^0) = \text{rank}(R|\bar{x}) - 1 = n - 1$ (because $r^1, \ldots, r^k$ span $\mathbb{R}^n$). Therefore u is one of the vectors a from $Ax \geq 0$. But then $a\bar{x} < 0$, i.e., $\bar{x} \notin C_A$. So C_R coincides with C_A and the theorem follows. □

Remark 11.2. The proof of Theorem 11.9 shows that, if A is a rational matrix, there exists a rational matrix R such that (A, R) is an MW-pair.

We now present the Minkowski-Weyl theorem for polyhedra. Given subsets V, R of $\mathbb{R}^n$, the *Minkowski sum* of V, R is the set:

$$V + R := \{x \in \mathbb{R}^n : \text{ there exist } v \in V,\ r \in R \text{ such that } x = v + r\}.$$

If one of V, R is empty, the Minkowski sum of V, R is empty.

Theorem 11.10 (Minkowski-Weyl theorem for polyhedra). *For a subset P of $\mathbb{R}^n$, the following two conditions are equivalent:*

1. *P is a polyhedron, i.e., there is a matrix A and a vector b such that $P = \{x \in \mathbb{R}^n : Ax \leqslant b\}$.*
2. *There exist vectors $v^1, \ldots, v^p$, $r^1, \ldots, r^q$ such that*

$$P = \text{conv}(v^1, \ldots, v^p) + \text{cone}(r^1, \ldots, r^q).$$

Proof. We show that 1 implies 2. Consider the polyhedral cone $C_P := \{(x, y) \in \mathbb{R}^{n+1} : by - Ax \geqslant 0, y \geqslant 0\}$. Notice that $P = \{x : (x, 1) \in C_P\}$. By Theorem 11.9, the cone C_P is finitely generated. Since $y \geqslant 0$ for every vector $(x, y) \in C_P$, we can assume that $y = 0$ or 1 for all the rays that generate C_P. That is,

$$C_P = \text{cone}\left\{ \begin{pmatrix} v^1 \\ 1 \end{pmatrix}, \ldots, \begin{pmatrix} v^p \\ 1 \end{pmatrix}, \begin{pmatrix} r^1 \\ 0 \end{pmatrix}, \ldots, \begin{pmatrix} r^q \\ 0 \end{pmatrix} \right\}.$$

Therefore $P = \text{conv}\{v^1, \ldots, v^p\} + \text{cone}\{r^1, \ldots, r^q\}$.

The converse statement follows analogously from Theorem 11.9. □

Corollary 11.2 (Minkowski-Weyl theorem for polytopes). *For a set $P \subseteq \mathbb{R}^n$, the following two conditions are equivalent:*

1. *P is bounded and there is a matrix A and a vector b such that $P = \{x \in \mathbb{R}^n : Ax \leqslant b\}$.*
2. *There is a finite set of vectors $v^1, \ldots, v^p$ such that $P = \text{conv}(v^1, \ldots, v^p)$.*

For a matrix $V := (v^1, \ldots, v^p)$ whose columns are the vectors v^j, it will be convenient to denote by $\text{conv}(V)$ the set $\text{conv}(v^1, \ldots, v^p)$. Similarly $\text{cone}(V)$ will denote the set $\text{cone}(v^1, \ldots, v^p)$.

11.2.4 Projections

Let $P \subseteq \mathbb{R}^{n+p}$ where $(x,y) \in P$ will be interpreted as meaning $x \in \mathbb{R}^n$ and $y \in \mathbb{R}^p$. The *projection* of P onto the x-space $\mathbb{R}^n$ is

$$\text{proj}_x(P) := \{x \in \mathbb{R}^n \ : \ \exists y \in \mathbb{R}^p \text{ with } (x,y) \in P\}.$$

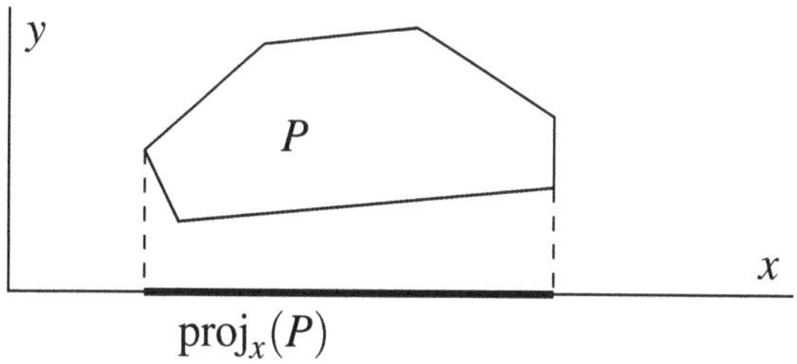

Fig. 11.3 Projection

Theorem 11.11. *Let $P := \{(x,y) \in \mathbb{R}^n \times \mathbb{R}^p : Ax + Gy \leqslant b\}$. Then $\text{proj}_x(P) = \{x \in \mathbb{R}^n : v^t(b - Ax) \geqslant 0$ for all $t \in T\}$ where $\{v^t\}_{t \in T}$ is the set of extreme rays of $Q := \{v \in \mathbb{R}^m : vG = 0, v \geqslant 0\}$.*

Proof. Let $x \in \mathbb{R}^n$. By Farkas's Lemma, $Gy \leqslant b - Ax$ has a solution y if and only if $v^t(b - Ax) \geqslant 0$ for all $v \in Q$. Since $v \geqslant 0$ for every $v \in Q$, then Q is pointed, hence it is generated by its extreme rays, and the statement follows. □

Remark 11.3. It follows from Theorem 11.11 that the projection of a polyhedron is again a polyhedron. If a polyhedron is rational, then its projection is also a rational polyhedron.

Remark 11.4. Variants of Theorem 11.11 can be proved similarly:
If $y \geqslant 0$ in P, then the relevant cone Q is $\{v : vG \geqslant 0, \ v \geqslant 0\}$.
If $y \geqslant 0$ and $Ax + Gy = b$ in P, the relevant cone is $\{v : vG \geqslant 0\}$ with v unrestricted in sign.

Enumerating the extreme rays of Q may not be an easy task in applications. Another way of obtaining the projection of P is to eliminate the variables y_i one at a time (Fourier elimination procedure):

Consider a polyhedron $P \subseteq \mathbb{R}^{n+1}$ defined by the system of inequalities:

$$\sum_{j=1}^{n} a_{ij}x_j + g_i z \leqslant b_i \quad \text{for } i \in I. \tag{11.9}$$

Let $I^0 = \{i \in I : g_i = 0\}$, $I^+ = \{i \in I : g_i > 0\}$, $I^- = \{i \in I : g_i < 0\}$. The Fourier procedure eliminates the variable z as follows: It keeps the inequalities of (11.9) in I^0, and it combines each pair of inequalities $i \in I^+$ and $l \in I^-$ to eliminate z.

Theorem 11.12. *The system of* $|I^0|+|I^+||I^-|$ *inequalities obtained by the Fourier procedure is the projection* $\text{proj}_x(P)$ *of* P *in the x-space* $\mathbb{R}^n$.

We leave the proof as a (fairly easy) exercise to the reader.

11.2.5 The fundamental theorem for MILP

Let $S := \{(x,y) : Ax+Gy \leqslant b,\ x \text{ integral}\}$ be a mixed integer set, where matrices A, G and vector b have rational entries. Meyer [38] proved that $\text{conv}(S)$ is a rational polyhedron, i.e., there exist rational matrices A' and G' and a rational vector b' such that $\text{conv}(S) = \{(x,y) : A'x+G'y \leqslant b'\}$.

Note first that if S contains finitely many vectors (for instance this happens when S is the set of integral vectors contained in a polytope), the above result follows from Corollary 11.2, without the need for the assumption that A, G and b have rational entries.

Theorem 11.13 (Meyer [38]). *Let* $P := \{(x,y) : Ax+Gy \leqslant b\}$ *be a rational polyhedron and let* $S := \{(x,y) \in P : x \text{ integral}\}$. *Then* $\text{conv}(S)$ *is a rational polyhedron. Furthermore, if* S *is nonempty, the recession cones of* $\text{conv}(S)$ *and* P *coincide.*

Proof. If $S = \emptyset$, the theorem is obvious. We assume that S and P are both nonempty. By the Minkowski-Weyl theorem (Theorem 11.10), $P = \text{conv}(V) + \text{cone}(R)$ where $V = (v^1,\dots,v^p)$, $R = (r^1,\dots,r^q)$. Since P is a rational polyhedron, by Remark 11.2 we can assume that V is a rational matrix and R is an integral matrix.
Consider the following truncation of P

$$T := \{(x,y) : (x,y) = \sum_{i=1}^{p} \lambda_i v^i + \sum_{j=1}^{q} \mu_j r^j,\ \sum_{i=1}^{p} \lambda_i = 1, \lambda \geq 0,\ 0 \leq \mu \leq \mathbf{1}\}.$$

T is the projection of a rational polyhedron and therefore it is a rational polyhedron by Remark 11.3. Furthermore T is bounded, which implies that it is a rational polytope by Corollary 11.2. Let $T_I := \{(x,y) \in T : x \text{ integral}\}$.

Claim. $\text{conv}(T_I)$ *is a rational polytope.*

Since T is a polytope, the set $X := \{x :$ there exist y such that $(x,y) \in T_I\}$ is a finite set. For fixed $\bar{x} \in X$, the set $T_{\bar{x}} := \{(\bar{x},y) : (\bar{x},y) \in T_I\}$ is a rational polytope and therefore by Corollary 11.2, $T_{\bar{x}} = \text{conv}(V_{\bar{x}})$ for some rational matrix $V_{\bar{x}}$. Since X is a finite set, there is a rational matrix V_{T_I} which contains all the columns of all the matrices $V_{\bar{x}}$, for $\bar{x} \in X$. Therefore $\text{conv}(T_I) = \text{conv}(V_{T_I})$ and this proves the claim.

A point $(\bar{x},\bar{y})$ belongs to S if and only if $\bar{x}$ is integral and there exist multipliers $\lambda \geqslant 0$, $\sum_{i=1}^{p} \lambda_i = 1$ and $\mu \geq 0$ such that

$$(\bar{x},\bar{y}) = \sum_{i=1}^{p} \lambda_i v^i + \sum_{j=1}^{q} (\mu_j - \lfloor \mu_j \rfloor) r^j + \sum_{j=1}^{q} \lfloor \mu_j \rfloor r^j.$$

The point $\sum_{i=1}^{p} \lambda_i v^i + \sum_{j=1}^{q} (\mu_j - \lfloor \mu_j \rfloor) r^j$ belongs to T and therefore it also belongs to T_I since $\bar{x}$ and $\sum_{j=1}^{q} \lfloor \mu_j \rfloor r^j$ are integral vectors. Thus

$$S = T_I + R_I \tag{11.10}$$

where R_I is the set of integral conic combinations of $r^1, \ldots, r^q$.
This shows in particular that T_I is nonempty. Equation (11.10) implies that $\text{conv}(S) = \text{conv}(T_I) + \text{cone}(R)$. By the above claim, $\text{conv}(T_I)$ is a rational polytope. Thus $\text{conv}(S) = \text{conv}(T_I) + \text{cone}(R)$ is a rational polyhedron having the same recession cone (namely $\text{cone}(R)$) as P. □

Remark 11.5. In Theorem 11.13:

- If matrices A, G are not rational, then $\text{conv}(S)$ may not be a polyhedron. One such example is the set $S := \{x \in \mathbb{Z}^2 : 1 \leqslant x_2 \leqslant \sqrt{2}x_1\}$.
- If A, G are rational matrices but b is not rational, then $\text{conv}(S)$ is a polyhedron that has the same recession cone as P. However $\text{conv}(S)$ is not always a rational polyhedron. (This can be inferred from the above proof).

11.2.6 Valid inequalities

An inequality $cx \leq \delta$ is *valid* for the set $P \subseteq \mathbb{R}^n$ if $cx \leq \delta$ is satisfied by every point in P.

Theorem 11.14. *Let $P := \{x : Ax \leq b\}$ be a nonempty polyhedron. An inequality $cx \leq \delta$ is valid for P if and only if the system $uA = c, ub \leq \delta, u \geq 0$ is feasible.*

Proof. Consider the linear program $\max\{cx : x \in P\}$. Since $P \neq \emptyset$ and $cx \leq \delta$ is a valid inequality for P, the above program admits a finite optimum and its value is $\delta' \leq \delta$.
By Remark 11.1, the set $D = \{u : uA = c, u \geq 0\}$ is nonempty. Therefore by Theorem 11.5, δ' is the common value of the equation (11.6). Thus an optimum solution u of $\min\{ub : u \in D\}$ satisfies $uA = c$, $ub \leqslant \delta$, $u \geqslant 0$.

Conversely, assume $uA = c$, $ub \leqslant \delta$, $u \geqslant 0$ is feasible. Then, for all $x \in P$, we have $cx = uAx \leqslant ub \leqslant \delta$. This shows that $cx \leqslant \delta$ is valid for P. □

11.2.7 Facets

Let $P := \{x \in \mathbb{R}^n : Ax \leqslant b\}$ be a polyhedron. A *face* of P is a set of the form

$$F := P \cap \{x \in \mathbb{R}^n : cx = \delta\}$$

where $cx \leqslant \delta$ is a valid inequality for P (the inequality is said to *define* the face F). A face is itself a polyhedron since it is the intersection of the polyhedron P with another polyhedron (the hyperplane $cx = \delta$). A face of P is said *proper* if it is nonempty and properly contained in P. Maximal proper faces of P are called *facets*.

Theorem 11.15. *Let $P := \{x \in \mathbb{R}^n : Ax \leqslant b\}$ be a nonempty polyhedron. Let M be the set of row indices of A and let $I \subseteq M$. The set*

$$F_I := \{x \in \mathbb{R}^n : a_i x = b_i, i \in I,\ a_i x \leq b_i, i \in M \setminus I\}.$$

is a face of P. Conversely, if F is a nonempty face of P, then $F = F_I$ for some $I \subseteq M$.

Proof. Let u be a vector such that $u_i > 0, i \in I,\ u_i = 0, i \in M \setminus I$, and let $c := uA$, $\delta := ub$. Then, given $x \in P$, clearly x satisfies $cx = \delta$ if and only if it satisfies $a_i x = b_i$, $i \in I$. Thus $F_I = P \cap \{x \in \mathbb{R}^n :\ cx = \delta\}$, so F_I is a face.

Conversely, let $F := \{x \in P : cx = \delta\}$ be a nonempty face of P. Then F is the set of optimal solutions of the LP $\max\{cx :\ x \in P\}$. Let I be the set of indices defined as in Theorem 11.6. Then by Theorem 11.6, $F = F_I$. □

Theorem 11.15 has important consequences:

- The number of faces of P is finite.
- The intersection of faces of P is a face of P.
- If F is a face of P and F' is a face of F, then F' is also a face of P.
- If F_1, F_2 are faces of P, there is a unique minimal face F of P that contains both F_1 and F_2 (The system of equalities that defines F is the intersection of the systems of equalities that define F_1 and F_2).

Furthermore, by Theorem 11.15, we can express a face F of P as follows. Let $A_F^= x \leqslant b_F^=$ be the set of all the inequalities $a_i x \leqslant b_i$ in $Ax \leqslant b$ such that $F \subseteq \{x \in \mathbb{R}^n : a_i x = b_i\}$. Then

$$F = P \cap \{x \in \mathbb{R}^n :\ A_F^= x = b_F^=\}.$$

An inequality $a_i x \leqslant b_i$ from $Ax \leqslant b$ such that $a_i x = b_i$ for all $x \in P$ is called an *implicit equality* of P. Let us partition the inequalities $Ax \leqslant b$ defining P into the implicit equalities $A^= x \leqslant b^=$ and the rest $A^< x \leqslant b^<$ (either of these two families of inequalities could be empty). Thus $P = \{x \in \mathbb{R}^n :\ A^= x = b^=,\ A^< x \leqslant b^<\}$ and for each inequality $a_i x \leqslant b_i$ in $A^< x \leqslant b^<$, there exists $\bar{x} \in P$ such that $a_i \bar{x} < b_i$.

Remark 11.6. P contains a point $\bar{x}$ such that $A^= \bar{x} = b^=$, $A^< \bar{x} < b^<$.

Proof. Indeed, for every inequality $a_i x \leqslant b_i$ in $A^< x \leqslant b^<$, there is a point $x^i \in P$ such that $a_i x^i < b_i$. Let r be the number of these points. Then $\bar{x} := \frac{1}{r} \sum_{i=1}^r x^i$ satisfies the statement. □

An inequality $a_i x \leq b_i$ of the system $Ax \leq b$ is *redundant* if $a_i x \leq b_i$ is a valid inequality for the system obtained from $Ax \leq b$ by removing the inequality $a_i x \leq b_i$. Theorem 11.14 provides a characterization of redundant inequalities.

Let $I^<$ denote the set of indices of the rows of $A^< x \leq b^<$. For every $i \in I^<$, denote by $A_{-i}^< x \leqslant b_{-i}^<$ the system obtained from $A^< x \leq b^<$ by removing the inequality $a_i^< x \leq b_i^<$.

Lemma 11.2. *Assume that $A^{<}x \le b^{<}$ contains no redundant inequality. Then for every $i \in I^{<}$, the polyhedron P contains a point x^i satisfying*

$$A^{=}x = b^{=},\ A^{<}_{-i}x < b^{<}_{-i},\ a^{<}_i x = b^{<}_i.$$

Proof. Let $i \in I^{<}$. Since no inequality in $A^{<}x \le b^{<}$ is redundant, the system:

$$A^{=}x = b^{=},\ A^{<}_{-i}x \le b^{<}_{-i},\ a^{<}_i x > b^{<}_i$$

is feasible. Let $\bar{x}^i$ be a point satisfying this system. By Remark 11.6, there is a point $\bar{x}$ satisfying $A^{=}x = b^{=}, A^{<}x < b^{<}$. Then a point on the segment having $\bar{x}$ and $\bar{x}^i$ as endpoints satisfies the above property. □

Theorem 11.16. *Let $P \subseteq \mathbb{R}^n$ be a polyhedron. Partition the inequalities defining P into the implicit equalities and the rest $P = \{x \in \mathbb{R}^n : A^{=}x = b^{=}, A^{<}x \le b^{<}\}$. Assume that $A^{<}x \le b^{<}$ contains no redundant inequality. Then a face of P is a facet if and only if it is of the form*

$$F_i := \{x \in P, a^{<}_i x = \beta^{<}_i\},\ \text{for some } i \in I^{<}.$$

Proof. Let F be a facet of P. Since no inequality in $A^{<}x \le b^{<}$ in an implicit equality for P, by Theorem 11.15 and the maximality of F, we have that $F = F_i$, for some $i \in I^{<}$.
Conversely, we show that for every $i \in I^{<}$, the set F_i is a facet. By Lemma 11.2, F_i is a proper face of P and it is not contained in any other face F_j, $j \in I^{<}$. Since all the facets of P are of the form F_j for $j \in I^{<}$, F_i is a proper face of P that is maximal with respect to inclusion, i.e., F_i a facet of P. □

This result states that, if a polyhedron in $\mathbb{R}^n$ has m facets, any representation by a system of linear inequalities in $\mathbb{R}^n$ contains at least m inequalities. In integer linear programming, we often consider polyhedra that are given implicitly as $\text{conv}(S)$. It is not unusual for such polyhedra to have a number of facets that is exponential in the size of the input. Thus their representation by linear inequalities in $\mathbb{R}^n$ is large. In some cases, there is a way to get around this difficulty: a polyhedron with a large number of facets can sometimes be obtained as the projection of a polyhedron with a small number of facets. We illustrate this idea in the next section.

11.3 Union of polyhedra

In this section, we prove a result of Balas [2, 3] about the union of k polyhedra. Consider k polyhedra $P_i := \{x \in \mathbb{R}^n : A_i x \leqslant b^i\}$, $i = 1, \dots, k$ and their union $\cup_{i=1}^k P_i$. We will show that $\overline{\text{conv}}(\cup_{i=1}^k P_i)$, the smallest closed convex set that contains $\cup_{i=1}^k P_i$, is a polyhedron. Furthermore we will show that this polyhedron can be obtained as the projection onto $\mathbb{R}^n$ of a polyhedron with polynomially many variables and constraints in a higher-dimensional space.

The closure is needed as shown by the following example: P_1 consists of a single point, and P_2 is a line that does not contain the point P_1 (see Figure 11.4). Let P_3 denote the line going through P_1 that is parallel to P_2. It is easy to verify that $\overline{\text{conv}}(P_1 \cup P_2) = \text{conv}(P_2 \cup P_3)$ and that $\text{conv}(P_1 \cup P_2) = \text{conv}(P_2 \cup P_3) \setminus (P_3 \setminus P_1)$ (indeed, a point x^* in $P_3 \setminus P_1$ is not in $\text{conv}(P_1 \cup P_2)$, but there is a sequence of points $x^k \in \text{conv}(P_1 \cup P_2)$ that converges to x^*). Here $\text{conv}(P_1 \cup P_2)$ is not a closed set, and therefore it is not a polyhedron.

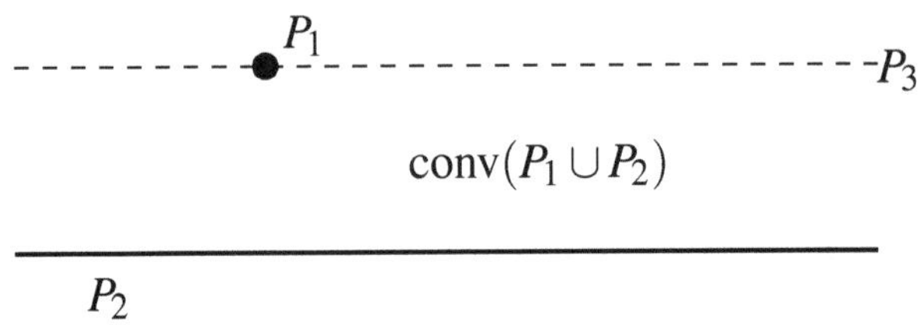

Fig. 11.4 $\overline{\text{conv}}(P_1 \cup P_2) \neq \text{conv}(P_1 \cup P_2)$

We recall that, by Minkowski-Weil's Theorem 11.10, $P_i = Q_i + C_i$, where Q_i is a polytope and C_i a finitely generated cone.

Theorem 11.17. *Let* $P_i = Q_i + C_i$ *be nonempty polyhedra* $i = 1, \dots, k$. *Then* $Q = \text{conv}(\cup_{i=1}^k Q_i)$ *is a polytope,* $C = \text{conv}(\cup_{i=1}^k C_i)$ *is a finitely generated cone, and* $\overline{\text{conv}}(\cup_{i=1}^k P_i)$ *is the polyhedron*

$$\overline{\text{conv}}(\cup_{i=1}^k P_i) = Q + C.$$

Proof. We first show that Q is a polytope and C a finitely generated cone. For $i = 1, \dots, k$, let V_i and C_i be finite sets in $\mathbb{R}^n$ such that $Q_i = \text{conv}(V_i)$ and $C_i = \text{cone}(R_i)$. Then it is straightforward to show that $Q = \text{conv}(\cup_{i=1}^k V_i)$ and $C = \text{cone}(\cup_{i=1}^k R_i)$, thus Q is a polytope and C a finitely generated cone.

We show $\overline{\text{conv}}(\cup_{i=1}^k P_i) \subseteq Q + C$.
We just need to show $\text{conv}(\cup_{i=1}^k P_i) \subseteq Q + C$, because $Q + C$ is a polyhedron, and so it is closed. Let $x \in \text{conv}(\cup_{i=1}^k P_i)$. Then x is a convex combination of a finite number of points in $\cup_{i=1}^k P_i$. Since P_i is convex, we can write x as a convex combination of points $z^i \in P_i$, say $x = \sum_{i=1}^k y_i z^i$ where $y_i \geqslant 0$ for $i = 1, \dots, k$ and $\sum_{i=1}^k y_i = 1$. Since $P_i = Q_i + C_i$, then $z^i = w^i + x^i$ where $w^i \in Q_i$, $x^i \in C_i$, thus $x = \sum_{i=1}^k y_i w^i + \sum_{i=1}^k y_i x^i$, so $x \in Q + C$ since $\sum_{i=1}^k y_i w^i \in Q$ and $\sum_{i=1}^k y_i x^i \in C$.

We show $Q + C \subseteq \overline{\text{conv}}(\cup_{i=1}^k P_i)$.
Let $x \in Q + C$. Then $x = \sum_{i=1}^k y_i w^i + \sum_{i=1}^k x^i$ where $w^i \in Q_i$, $y_i \geqslant 0$, $x^i \in C_i$ for $i = 1, \dots, k$, and $\sum_{i=1}^k y_i = 1$.

Define $I := \{i : y_i > 0\}$ and consider the point

$$x^{\varepsilon} := \sum_{i \in I}(y_i - \frac{k}{|I|}\varepsilon)w^i + \sum_{i=1}^{k} \varepsilon(w^i + \frac{1}{\varepsilon}x^i)$$

for $\varepsilon > 0$ small enough so that $y_i - \frac{k}{|I|}\varepsilon \geqslant 0$ for all $i \in I$.
Notice that $x^{\varepsilon} \in \text{conv}(\cup_{i=1}^{k} P_i)$ since $\sum_{i \in I}(y_i - \frac{k}{|I|}\varepsilon) + \sum_{i=1}^{k} \varepsilon = 1$ and $w^i + \frac{1}{\varepsilon}x^i \in P_i$.
Since $\lim_{\varepsilon \to 0^+} x^{\varepsilon} = x$, we have $x \in \overline{\text{conv}}(\cup_{i=1}^{k} P_i)$. □

Corollary 11.3. *If $P_1, \ldots, P_k$ are polyhedra with the same recession cone, then the convex hull of their union* $\text{conv}(\cup_{i=1}^{k} P_i)$ *is a polyhedron.*

Proof. We leave it as an exercise to the reader to check how the last part of the proof of Theorem 11.17 simplifies to show $Q + C \subseteq \text{conv}(\cup_{i=1}^{k} P_i)$. □

Although $\overline{\text{conv}}(\cup_{i=1}^{k} P_i)$ may have exponentially many facets, Balas [2, 3] proved that it is the projection of a higher-dimensional polyhedron Y with a polynomial size representation:

$$Y := \begin{cases} A_i x^i \leqslant b^i y_i \\ \sum x^i = x \\ \sum y_i = 1 \\ \quad y_i \geqslant 0 \;\; \text{for } i = 1, \ldots, k. \end{cases} \tag{11.11}$$

In this formulation, x^i is a vector in $\mathbb{R}^n$ and y_i is a scalar, for $i = 1, \ldots, k$. The vector $x \in \mathbb{R}^n$ corresponds to the original space onto which Y is projected. Thus, the polyhedron Y is defined in $\mathbb{R}^{kn+n+k}$. A formulation with a polynomial number of variables and constraints is said to be *compact*. The gist of Balas's result is that $\overline{\text{conv}}(\cup_{i=1}^{k} P_i)$ has a compact representation with respect to the systems $A_i x \leqslant b^i$, $i = 1, \ldots, k$. This fact will be fundamental in the development of this survey.

Since it is convenient to consider also the case where some of the systems $A_i x \leqslant b^i$ are infeasible, we need a condition that is typically satisfied in the context of integer programming.

Given the polyhedra $P_i := \{x \in \mathbb{R}^n : A_i x \leqslant b^i\}$, $i = 1, \ldots, k$, let $C_i := \{x \in \mathbb{R}^n : A_i x \leqslant 0\}$. So C_i is the recession cone of P_i when $P_i \neq \emptyset$.

Cone Condition: *If $\cup P_i \neq \emptyset$, then $C_j \subseteq \text{conv}(\cup_{i: P_i \neq \emptyset} C_i)$ for $j = 1, \ldots, k$.*

By Minkowski-Weil's Theorem 11.10, if $P_i \neq \emptyset$ then $P_i = Q_i + C_i$ for some polytope Q_i. If we let $Q_i = \emptyset$ whenever $P_i = \emptyset$, then $P_i = Q_i + C_i$, $i = 1 \ldots, k$.

Theorem 11.18 (Balas [2, 3]). *Consider k polyhedra $P_i := \{x \in \mathbb{R}^n : A_i x \leqslant b^i\}$ and let Y be the polyhedron defined in (11.11). Then*

$$\text{proj}_x Y = Q + C,$$

where $Q = \text{conv}(\cup_{i=1}^{k} Q_i)$ and $C = \text{conv}(\cup_{i=1}^{k} C_i)$.
Furthermore, if the Cone Condition is satisfied, then

$$\text{proj}_x Y = \overline{\text{conv}}(\cup_{i=1}^{k} P_i).$$

Proof. Notice that, if the Cone Condition is satisfied, then $C = \text{conv}(\cup_{i:P_i\neq\emptyset}C_i)$, therefore by Theorem 11.17 $Q+C = \overline{\text{conv}}(\cup_{i=1}^k P_i)$. Thus, if $\text{proj}_x Y = Q+C$, then $\text{proj}_x Y = \overline{\text{conv}}(\cup_{i=1}^k P_i)$. So we only need to show $\text{proj}_x Y = Q+C$.

(a) $Q+C \subseteq \text{proj}_x Y$.

The result holds trivially when $\cup_{i=1}^k P_i = \emptyset$, so we assume $\cup_{i=1}^k P_i \neq \emptyset$. Without loss of generality, $P_1,\dots,P_h$ are nonempty and $P_{h+1},\dots,P_k$ are empty, where $1 \leqslant h \leqslant k$.

Let $x \in Q+C$. Then there exist $w^i \in Q^i$, $i = 1,\dots,h$, and $z^i \in C_i$, $i = 1,\dots,k$, such that $x = \sum_{i=1}^h y_i w^i + \sum_{i=1}^k z^i$, where $y_i \geqslant 0$ for $i = 1,\dots,h$ and $\sum_{i=1}^h y_i = 1$. Let $x^i = y_i w^i + z^i$ for $i = 1,\dots,h$ and $y_i = 0$, $x^i = z^i$ for $i = h+1,\dots,k$. Then $A_i x^i \leqslant b^i y_i$ for $i = 1,\dots k$ and $x = \sum_{i=1}^k x^i$. This shows that $x \in \text{proj}_x Y$ and therefore (a) holds.

(b) $\text{proj}_x Y \subseteq Q+C$.

The result holds if $Y = \emptyset$, so we assume $Y \neq \emptyset$. Let $x \in \text{proj}_x Y$. By the definition of projection, there exist $x^1,\dots,x^k,y$ such that $x = \sum_{i=1}^k x^i$ where $Ax^i \leqslant b^i y_i$, $\sum y_i = 1$, $y \geqslant 0$. Let $I := \{i : y_i > 0\}$.

For $i \in I$, let $z^i := \frac{x^i}{y_i}$. Then $z^i \in P_i$. Since $P_i = Q_i + C_i$, we can write $z^i = w^i + y^i r^i$ where $w^i \in Q_i$ and $r^i \in C_i$.

For $i \notin I$, we have $A_i x^i \leqslant 0$, that is $x^i \in C_i$. Let $r^i = x^i$ if $i \notin I$. Then

$$x = \sum_{i\in I} y_i z^i + \sum_{i\notin I} x^i = \underbrace{\sum_{i\in I} y_i w^i}_{\in Q} + \underbrace{\sum_{i=1}^{k} r^i}_{\in C} \in Q+C. \qquad \square$$

Remark 11.7. The Cone Condition assumption in Theorem 11.18 is necessary as shown by the following example (see Figure 11.5): $P_1 := \{x \in \mathbb{R}^2 : 0 \leqslant x \leqslant 1\}$ and $P_2 := \{x \in \mathbb{R}^2 : x_1 \leqslant 0, x_1 \geqslant 1\}$. Note that $P_2 = \emptyset$ and $C_2 = \{x \in \mathbb{R}^2 : x_1 = 0\}$. So in this case $\text{proj}_x Y = P_1 + C_2 = \{x \in \mathbb{R}^2 : 0 \leqslant x_1 \leqslant 1\}$, which is different from $\overline{\text{conv}}(P_1 \cup P_2) = P_1$.

Remark 11.8. The Cone Condition assumption in Theorem 11.18 is automatically satisfied if
(i) $C_i = \{0\}$ whenever $P_i = \emptyset$, or
(ii) all the cones C_i are identical.
For example (i) holds when all the P_is are nonempty, or when $C_i = \{0\}$ for all i.

11.4 Split disjunctions

Let $P := \{(x,y) \in \mathbb{R}^n \times \mathbb{R}^p : Ax + Gy \leqslant b\}$ and let $S := P \cap (\mathbb{Z}^n \times \mathbb{R}^p)$. For $\pi \in \mathbb{Z}^n$ and $\pi_0 \in \mathbb{Z}$, define

$$\begin{aligned} \Pi_1 &= P \cap \{(x,y) : \pi x \leqslant \pi_0\} \\ \Pi_2 &= P \cap \{(x,y) : \pi x \geqslant \pi_0 + 1\}. \end{aligned} \tag{11.12}$$

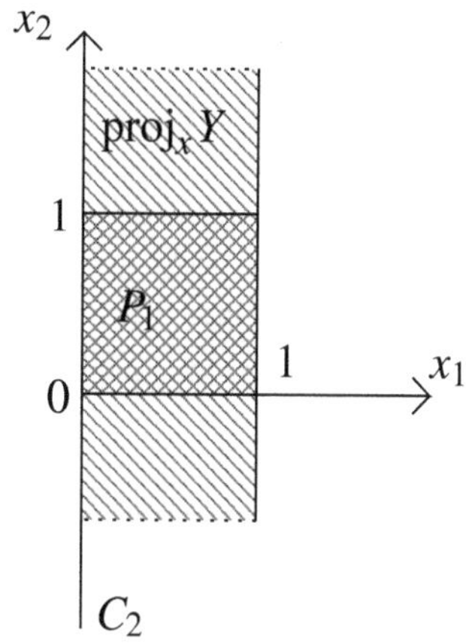

Fig. 11.5 $\overline{\mathrm{conv}}(P_1 \cup P_2) \neq \mathrm{proj}_x Y$

Clearly $S \subseteq \Pi_1 \cup \Pi_2$ and therefore $\mathrm{conv}(S) \subseteq \mathrm{conv}(\Pi_1 \cup \Pi_2)$. We call this latter set $P^{(\pi,\pi_0)}$.

Theorem 11.19. *$P^{(\pi,\pi_0)}$ is a polyhedron.*

Proof. It follows from Theorem 11.17 that $\overline{P^{(\pi,\pi_0)}}$ is a polyhedron, thus we only need to show that $P^{(\pi,\pi_0)}$ is closed. This is obvious if at least one of Π_1 and Π_2 is empty, so we assume $\Pi_1, \Pi_2 \neq \emptyset$. Because $\Pi_1 \cup \Pi_2 \subset P^{(\pi,\pi_0)} \subseteq P$, $\Pi_1 = P^{(\pi,\pi_0)} \cap \{(x,y) : \pi x \leqslant \pi_0\}$ and $\Pi_2 = P^{(\pi,\pi_0)} \cap \{(x,y) : \pi x \geqslant \pi_0 + 1\}$. Thus $P^{(\pi,\pi_0)} = \Pi_1 \cup \Pi_2 \cup \Pi$, where $\Pi = P^{(\pi,\pi_0)} \cap \{(x,y) : \pi_0 \leqslant \pi x \leqslant \pi_0 + 1\}$. By definition, Π_1 and Π_2 are polyhedra, so they are closed. We show that Π is a polyhedron, thus $P^{(\pi,\pi_0)}$ is closed because it is the union of a finite number of closed sets.

Let $P_1 := \Pi_1 \cap \{(x,y) : \pi x = \pi_0\}$ and $P_2 := \Pi_2 \cap \{(x,y) : \pi x = \pi_0 + 1\}$. Notice that P_1 and P_2 have the same recession cone $C := \{(x,y) : Ax + Gy \leqslant 0, \pi x = 0\}$, thus, by Corollary 11.3, $\mathrm{conv}(P_1 \cup P_2)$ is a polyhedron.

We show that $\Pi = \mathrm{conv}(P_1 \cup P_2)$, thus showing that Π is a polyhedron.

The inclusion $\Pi \supseteq \mathrm{conv}(P_1 \cup P_2)$ comes from the definition. We prove $\Pi \subseteq \mathrm{conv}(P_1 \cup P_2)$. If $(\bar{x},\bar{y}) \in \Pi$, then there exist $(x',y') \in \Pi_1$, $(x'',y'') \in \Pi_2$ such that $(\bar{x},\bar{y})$ is contained in the line segment L joining (x',y') and (x'',y''). Since $\pi_0 \leqslant \pi\bar{x} \leqslant \pi_0 + 1$, L intersects $\{(x,y) : \pi x = \pi_0\}$ in a point $(\bar{x}',\bar{y}') \in \Pi_1$, and $\{(x,y) : \pi x = \pi_0 + 1\}$ in a point $(\bar{x}'',\bar{y}'') \in \Pi_2$. Furthermore, $(\bar{x},\bar{y})$ is contained in the line segment joining $(\bar{x}',\bar{y}')$ and $(\bar{x}'',\bar{y}'')$. Thus $(\bar{x},\bar{y}) \in \mathrm{conv}(P_1 \cup P_2)$. □

Lemma 11.3. *The polyhedra Π_1 and Π_2 satisfy the Cone Condition of Theorem 11.18.*

Proof. The conditions of Theorem 11.18 trivially hold when Π_1 and Π_2 are either both empty or both nonempty.

Assume now that $\Pi_1 = \emptyset$ and $\Pi_2 \neq \emptyset$. For the conditions of Theorem 11.18 to hold, we need to show that $C_1 \subseteq C_2$ where $C_1 := \{(x,y) : Ax+Gy \leqslant 0, \pi x \leqslant 0\}$ and $C_2 := \{(x,y) : Ax+Gy \leqslant 0, \pi x \geqslant 0\}$.

We claim that $C_1 = \{(x,y) : Ax+Gy \leqslant 0, \pi x = 0\}$. Suppose not. Then there exists $(\bar{x},\bar{y})$ such that $A\bar{x}+G\bar{y} \leqslant 0$ and $\pi\bar{x} < 0$. Since $\Pi_2 \neq \emptyset$, there exists $(\hat{x},\hat{y})$ such that $A\hat{x}+G\hat{y} \leqslant b$. Now consider the point $(x^\lambda,y^\lambda) = (\hat{x},\hat{y}) + \lambda(\bar{x},\bar{y})$. We have $Ax^\lambda + Gy^\lambda \leqslant b$ and, for $\lambda \geqslant \frac{\pi\hat{x}-\pi_0}{|\pi\bar{x}|}$, we have $\pi x^\lambda \leqslant \pi_0$. But then (x^λ,y^λ) is in Π_1, contradicting the assumption that $\Pi_1 = \emptyset$. This proves the claim.

The claim implies $C_1 \subseteq C_2$. □

Thus, by Theorem 11.18, $P^{(\pi,\pi_0)}$ has the following extended formulation, with additional vectors of variables (x^1,y^1), (x^2,y^2) and variable λ.

$$\begin{aligned} Ax^1+Gy^1 &\leqslant \lambda b \\ \pi x^1 &\leqslant \lambda\pi_0 \\ Ax^2+Gy^2 &\leqslant (1-\lambda)b \\ \pi x^2 &\geqslant (1-\lambda)(\pi_0+1) \\ x^1+x^2 &= x \\ y^1+y^2 &= y \\ 0 \leqslant \lambda &\leqslant \mathbf{1}. \end{aligned} \tag{11.13}$$

If the system $Ax+Gy \leqslant b$ has m constraints, the extended formulation (11.13) has $2m+n+p+4$ constraints. By contrast, a formulation of $P^{(\pi,\pi_0)}$ in the original space (x,y) may be considerably more complicated, as $P^{(\pi,\pi_0)}$ may have a large number of facets (Recall from Section 11.2.7 that any description of a polyhedron must have at least one inequality for each facet).

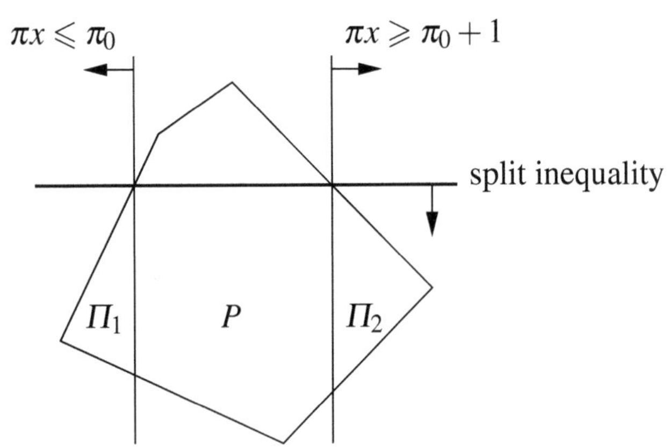

Fig. 11.6 A split inequality

A *split* is a disjunction $\pi x \leqslant \pi_0$ or $\pi x \geqslant \pi_0+1$ where $\pi \in \mathbb{Z}^n$ and $\pi_0 \in \mathbb{Z}$. We will also say that (π,π_0) defines a split. An inequality $cx+hy \leqslant c_0$ is called a *split inequality* [17] if it is valid for some polyhedron $P^{(\pi,\pi_0)}$ where $(\pi,\pi_0) \in \mathbb{Z}^n \times \mathbb{Z}$ defines a split (see Figure 11.6).

The *split closure* P^{Split} of P is

$$\bigcap_{(\pi,\pi_0)\in\mathbb{Z}^n\times\mathbb{Z}} P^{(\pi,\pi_0)}. \tag{11.14}$$

Clearly $S \subseteq P^{\text{Split}} \subseteq P$. In general P^{Split} provides a tighter relaxation of S than P. Although each of the sets $P^{(\pi,\pi_0)}$ is a polyhedron, it is not clear however that P^{Split} is a polyhedral set, for it is the intersection of infinitely many of them. This will be discussed in Section 11.6.

11.4.1 One-side splits, Chvátal inequalities

Given a polyhedron $P := \{(x,y) \in \mathbb{R}^n \times \mathbb{R}^p : \; Ax + Gy \leqslant b\}$, we consider the mixed integer set $S = P \cap (\mathbb{Z}^n \times \mathbb{R}^p)$. Let $\pi \in \mathbb{Z}^n$ and $z := \max\{\pi x : (x,y) \in P\}$. A split defined by $(\pi,\pi_0) \in \mathbb{Z}^n \times \mathbb{Z}$ is a *one-side split* for P if

$$\pi_0 \leq z < \pi_0 + 1. \tag{11.15}$$

This is equivalent to:

$$\Pi_1 \subseteq P \text{ and } \Pi_2 = \emptyset$$

where Π_1 and Π_2 are the polyhedra defined in (11.12). Note that, if (π,π_0) defines a one-side split, then the polyhedron $P^{(\pi,\pi_0)}$ can be easily described in its original space, for $P^{(\pi,\pi_0)} = \Pi_1 = \{(x,y) : \; Ax + Gy \leqslant b, \; \pi x \leq \pi_0\}$. The inequality $\pi x \leq \pi_0$ is valid for S. It is called a *Chvátal inequality*.

The *Chvátal closure* P^{Ch} of P is

$$\bigcap_{(\pi,\pi_0)\in\mathbb{Z}^n\times\mathbb{Z} \text{ defines a one-side split}} P^{(\pi,\pi_0)}. \tag{11.16}$$

Equivalently, P^{Ch} is the set of vectors (x,y) that satisfy the system $Ax + Gy \leqslant b$ and all the Chvátal inequalities. Note that $S \subseteq P^{\text{Split}} \subseteq P^{\text{Ch}} \subseteq P$.

Notice that inequality $\pi x \leq \pi_0$ satisfies (11.15) if and only if $\lfloor z \rfloor = \pi_0$. Since π is an integral vector and $\pi x \leq z$ is a valid inequality for P, by Theorem 11.14, $\pi x \leq \pi_0$ is a Chvátal inequality if and only if the system:

$$u \geq 0, \; uA = \pi \in \mathbb{Z}^n, \; uG = 0, \; ub \leqslant z \tag{11.17}$$

is feasible. By Theorem 11.11, condition $u \geq 0$, $uG = 0$ shows that $\pi x \leq z$ is valid for the projection $\text{proj}_x(P)$ of P onto the x-space.

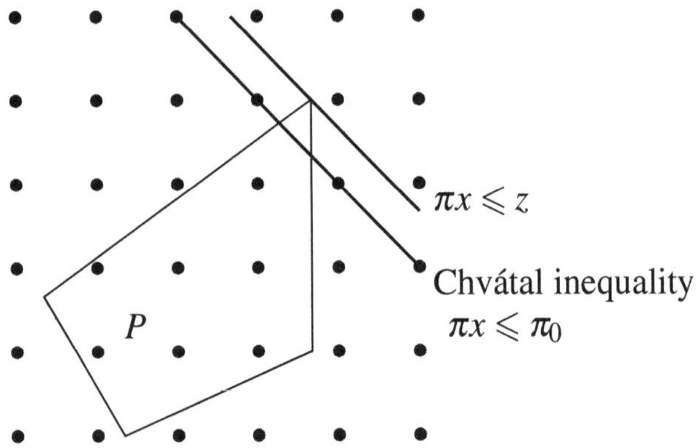

Fig. 11.7 Chvátal inequality

We can assume that u is chosen so that the coefficients of π are relatively prime: If not, let m be the G.C.D. of the components of π: The inequality $\frac{\pi}{m}x \le \lfloor \frac{\pi_0}{m} \rfloor$ is a Chvátal inequality that dominates $\pi x \le \pi_0$. The equation $\pi x = \pi_0 = \lfloor z \rfloor$ admits infinitely many integral solutions, while the equation $\pi x = \alpha$, has obviously no integral solution for $\lfloor z \rfloor < \alpha \le z$.

Therefore the Chvátal closure is obtained by taking any rational inequality $\pi x \le z$ that is valid for the projection $\text{proj}_x(P)$ of P (which is generated without loss of generality by a vector u such that $u \ge 0$, $uA = \pi \in \mathbb{Z}^n$, $uG = 0$ such that the coefficients of π are relatively prime) and then tightening the inequality by decreasing the right-hand side until an integral point (and hence infinitely many integral points) are encountered. See Figure 11.7.

11.5 Gomory's mixed-integer inequalities

We consider a polyhedron $P := \{(x,y) \in \mathbb{R}^n_+ \times \mathbb{R}^p_+ : Ax + Gy \le b\}$ and the set $S := P \cap (\mathbb{Z}^n_+ \times \mathbb{R}^p_+)$. Note that, here, P is defined by a system of inequalities together with nonnegativity constraints on the variables, and this is important in this section. By standard linear algebra, any system of linear inequalities can be converted into a system of the above type (variables that are unrestricted in sign can be replaced by the difference of two nonnegative variables, etc.).

We consider the following equality form of the system defining P:

$$Ax + Gy + Is = b, \; x, y, s \ge 0. \tag{11.18}$$

For $\lambda \in \mathbb{Q}^m$, we consider the equation $\lambda Ax + \lambda Gy + \lambda Is = \lambda b$, which we denote by

$$\sum_{j=1}^{n} a_j^\lambda x_j + \sum_{j=1}^{p} g_j^\lambda y_j + \sum_{i=1}^{m} \lambda_i s_i = \beta^\lambda. \tag{11.19}$$

Let $f_0 = \beta^\lambda - \lfloor \beta^\lambda \rfloor$ and $f_j = a_j^\lambda - \lfloor a_j^\lambda \rfloor$. We consider the following *Gomory mixed-integer (GMI) inequality* [32] :

$$\sum_{j=1}^{n} \left(\lfloor a_j^\lambda \rfloor + \frac{(f_j - f_0)^+}{1-f_0}\right)x_j + \frac{1}{1-f_0}\left(\sum_{j:g_j^\lambda<0} g_j^\lambda y_j + \sum_{i:\lambda_i<0} \lambda_i s_i\right) \leqslant \lfloor \beta^\lambda \rfloor \tag{11.20}$$

where $(f_j - f_0)^+ = \max\{f_j - f_0,\ 0\}$. By substituting $s = b - (Ax + Gy)$ we get an inequality in the (x,y)-space.

We denote by $(\pi^\lambda, \pi_0^\lambda)$ the vector in $\mathbb{Z}^n \times \mathbb{Z}$ defining the following split:

$$\text{Either } \sum_{f_j \le f_0} \lfloor a_j^\lambda \rfloor x_j + \sum_{f_j > f_0} \lceil a_j^\lambda \rceil x_j \le \lfloor \beta^\lambda \rfloor \tag{11.21}$$

$$\text{or } \sum_{f_j \le f_0} \lfloor a_j^\lambda \rfloor x_j + \sum_{f_j > f_0} \lceil a_j^\lambda \rceil x_j \ge \lfloor \beta^\lambda \rfloor + 1; \tag{11.22}$$

where (11.21) is $\pi^\lambda x \leqslant \pi_0^\lambda$ while (11.22) is $\pi^\lambda x \geqslant \pi_0^\lambda + 1$.

Lemma 11.4. *Inequality* (11.20) *is a split inequality, valid for* $P^{(\pi^\lambda, \pi_0^\lambda)}$.

Proof. Consider the sets

$$\Pi_1 = P \cap \{(x,y) :\ \pi^\lambda x \leqslant \pi_0^\lambda\}, \quad \Pi_2 = P \cap \{(x,y) :\ \pi^\lambda x \geqslant \pi_0^\lambda + 1\}.$$

To prove that inequality (11.20) is a split inequality, we will show that it is valid for $\Pi_1 \cup \Pi_2$.

Since the constraints $x \ge 0$, $y \ge 0$, $s \ge 0$ are valid for Π_1 and Π_2, it is enough to show that each of Π_1 and Π_2 admits a valid inequality $ax + gy + ls \le \lfloor \beta^\lambda \rfloor$ whose coefficients are greater than or equal to the corresponding coefficients in inequality (11.20). (This is where the nonnegativity plays a role).

Since inequality $\pi^\lambda x \leqslant \pi_0^\lambda$ has this property, inequality (11.20) is valid for Π_1. Inequality (11.20) is valid for Π_2 since it is implied by the inequality

$$\frac{1}{1-f_0}(11.19) - \frac{f_0}{1-f_0}(11.22).$$

□

A consequence of the above lemma is that the GMI inequalities (11.20) are valid for S.

Lemma 11.5. *Chvátal inequalities are GMI inequalities.*

Proof. For $\lambda \ge 0$ such that $\lambda A \in \mathbb{Z}^n$ and $\lambda G = 0$ the Chvátal inequality (11.17) coincides with the inequality given by the formula (11.20). □

The *Gomory mixed integer closure* P^{GMI} of P is the set of points in P that satisfy all the GMI inequalities (11.20). It follows from Lemmas 11.4 and 11.5 that $P^{\text{Split}} \subseteq P^{\text{GMI}} \subseteq P^{\text{Ch}}$. In light of the particular derivation of the GMI inequalities, it may appear that the first containment can be strict. This is not the case: In the next section we show that P^{Split} coincides with P^{GMI}.

11.5.1 Equivalence of split closure and Gomory mixed integer closure

In this section we will need the following.

Lemma 11.6. *Let $P := \{x \in \mathbb{R}^n : Ax \leqslant b\}$ be a polyhedron and let $\Pi := P \cap \{x : \pi x \leqslant \pi_0\}$. If $\Pi \neq \emptyset$ and $\alpha x \leqslant \beta$ is a valid inequality for Π, then there exists a scalar $\lambda \in \mathbb{R}_+$ such that*

$$\alpha x - \lambda(\pi x - \pi_0) \leqslant \beta$$

is valid for P.

Proof. By Theorem 11.14, since $\Pi \neq \emptyset$, there exist $u \geqslant 0$, $\lambda \geqslant 0$ such that

$$\alpha = uA + \lambda \pi \quad \text{and} \quad \beta \geqslant ub + \lambda \pi_0.$$

Since $uAx \leqslant ub$ is valid for P, so is $uAx \leqslant \beta - \lambda \pi_0$. Since $uAx = \alpha x - \lambda \pi x$, the inequality $\alpha x - \lambda(\pi x - \pi_0) \leqslant \beta$ is valid for P. □

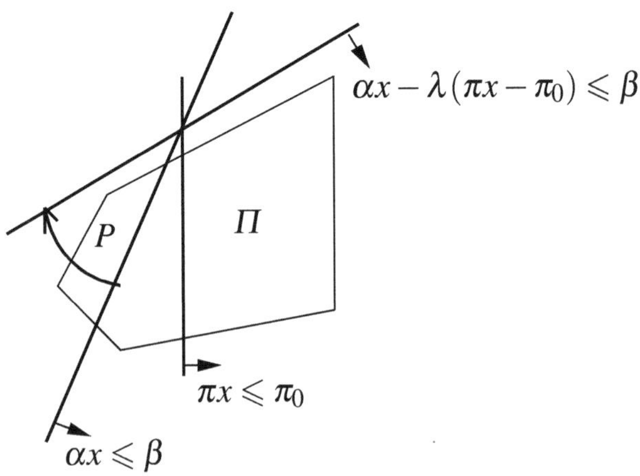

Fig. 11.8 Illustration of Lemma 11.6

Remark 11.9. The assumption $\Pi \neq \emptyset$ is necessary in Lemma 11.6, as shown by the following example: $P := \{x \in \mathbb{R}^2 : x_1 \geqslant 0, x_2 \geqslant 0\}$ and $\Pi := P \cap \{x : x_2 \leqslant -1\}$.

Thus Π is empty. The inequality $x_1 \leqslant 1$ is valid for Π but there is no scalar λ such that $x_1 - \lambda(x_2+1) \leqslant 1$ is valid for P.

Theorem 11.20 (Nemhauser and Wolsey [40]). *Let $P := \{(x,y) \in \mathbb{R}^n_+ \times \mathbb{R}^p_+ : Ax + Gy \leqslant b\}$ be a polyhedron and let $S := P \cap (\mathbb{Z}^n_+ \times \mathbb{R}^p_+)$. Then P^{Split} coincides with P^{GMI}.*

Proof. (Cornuéjols and Li [18]) Lemma 11.4 shows that $P^{\text{Split}} \subseteq P^{\text{GMI}}$. To prove the reverse inclusion, we show that every split inequality is a GMI inequality.

We assume that the constraints $x \geqslant 0$ and $y \geqslant 0$ are part of the system $Ax + Gy \leqslant b$ that defines P. Let $cx + hy \leqslant c_0$ be a split inequality. Let $(\pi, \pi_0) \in \mathbb{Z}^n \times \mathbb{Z}$ define the split disjunction used in deriving this inequality and let Π_1, Π_2 be the corresponding intersections with P as defined in (11.12).

First assume that one of Π_1, Π_2 is empty. Then the inequality $cx + hy \leqslant c_0$ is a Chvátal inequality and by Lemma 11.5 it is also a GMI inequality.

We now assume that both Π_1, Π_2 are nonempty. By Lemma 11.6, there exist $\alpha, \beta \in \mathbb{R}_+$ such that

$$cx + hy - \alpha(\pi x - \pi_0) \leqslant c_0 \text{ and} \tag{11.23}$$

$$cx + hy + \beta(\pi x - (\pi_0 + 1)) \leqslant c_0 \tag{11.24}$$

are both valid for P. We can assume $\alpha > 0$ and $\beta > 0$ since, otherwise, $cx + hy \leqslant c_0$ is valid for P and therefore also for P^{GMI}. We now introduce slack variables s_1 and s_2 in (11.23) and (11.24) respectively and subtract (11.23) from (11.24). We obtain

$$(\alpha+\beta)\pi x + s_2 - s_1 = (\alpha+\beta)\pi_0 + \beta.$$

Dividing by $\alpha + \beta$ we get

$$\pi x + \frac{s_2}{\alpha+\beta} - \frac{s_1}{\alpha+\beta} = \pi_0 + \frac{\beta}{\alpha+\beta}. \tag{11.25}$$

We now derive the GMI inequality associated with equation (11.25). Note that the fractional part of the right-hand side is $\frac{\beta}{\alpha+\beta}$ and that the continuous variable s_2 has a positive coefficient while s_1 has a negative coefficient. So the GMI inequality is

$$\pi x + \frac{1}{\alpha} s_1 \leq \pi_0.$$

We now use (11.23) to eliminate s_1 to get the GMI inequality in the space of the x, y variables. The resulting inequality is $cx + hy \leqslant c_0$ and therefore $cx + hy \leqslant c_0$ is a GMI inequality. □

11.6 Polyhedrality of closures

In this section we show that the GMI closure (or equivalently, the split closure) of a rational polyhedron is a rational polyhedron. This result is due to Cook, Kannan,

and Schrijver [17]. Simpler proofs appear in [1], [22] and [47]. In this section, we follow the approach of Dash, Günlük and Lodi [22]. The idea is to prove that a finite number of splits are sufficient to generate P^{Split} defined in (11.14). The result follows, since then P^{Split} is the intersection of a finite number of polyhedra and therefore is a polyhedron.
For this idea to work, it is fundamental to assume that the polyhedra that we deal with are rational. We first illustrate this in an easy case: The Chvátal closure of a pure integer set.

11.6.1 The Chvátal closure of a pure integer set

We consider here pure integer sets of the type $S := P \cap \mathbb{Z}^n$, where $P := \{x \in \mathbb{R}^n : Ax \leqslant b\}$ is a rational polyhedron. Therefore we can assume that A, b have integral entries.
In the pure integer case, a Chvátal inequality $\pi x \leq \pi_0$ is derived from a vector u satisfying:

$$u \geq 0,\ uA = \pi \in \mathbb{Z}^n,\ \lfloor ub \rfloor = \pi_0. \tag{11.26}$$

Lemma 11.7. *Let $(\pi, \pi_0) \in \mathbb{Z}^{n+1}$ and $u \in \mathbb{Q}$ satisfying (11.26) such that $\pi x \leqslant \pi_0$ is not valid for P and not redundant for P^{Ch}. Then $u < \mathbf{1}$.*

Proof. Suppose u does not satisfy $u < \mathbf{1}$. We will show that $\pi x \leqslant \pi_0$ is the sum of a Chvátal inequality $\pi^1 x \leqslant \pi_0^1$ and an inequality $\pi^2 x \leqslant \pi_0^2$ valid for P such that $\pi^2 \neq 0$. This contradicts the assumption that $\pi x \leqslant \pi_0$ is not valid for P and not redundant for P^{Ch}.

Let $f = u - \lfloor u \rfloor$, and let $\pi^1 = fA$, $\pi_0^1 = \lfloor fb \rfloor$, $\pi^2 = \lfloor u \rfloor A$, $\pi_0^2 = \lfloor \lfloor u \rfloor b \rfloor$. Since A is an integral matrix, π^2 is an integral vector. Since π is an integral vector, $\pi^1 = \pi - \pi^2$ is integral as well.

Since b is integral, $\pi_0^2 = \lfloor u \rfloor b$, therefore $\pi^2 x \leq \pi_0^2$ is valid for P. Furthermore, $\pi x \leqslant \pi_0$ is the sum of $\pi^1 x \leqslant \pi_0^1$ and $\pi^2 x \leqslant \pi_0^2$. Since $\pi x \leqslant \pi_0$ is not valid for P, then $f \neq 0$. Since u does not satisfy $u < \mathbf{1}$, then $\lfloor u \rfloor \neq 0$, thus $\pi^2 \neq 0$. □

Theorem 11.21 (Chvátal [13]). *P^{Ch} is a rational polyhedron.*

Proof. By Lemma 11.7, any irredundant valid inequality for P^{Ch} that is not valid for P is of the form $(uA)x \leqslant \lfloor ub \rfloor$ for some u satisfying $uA \in \mathbb{Z}^n$, $0 \leqslant u < \mathbf{1}$. Since $\{uA \in \mathbb{R}^n : 0 \leqslant u < \mathbf{1}\}$ is bounded, $\{uA \in \mathbb{Z}^n : 0 \leqslant u < \mathbf{1}\}$ is finite. Thus there is only a finite number of such inequalities, hence P^{Ch} is a polyhedron. □

11.6.2 The split closure of a mixed integer set

This is more tricky, uses the fact that $P^{\text{Split}} = P^{\text{GMI}}$ (Theorem 11.20), but the idea is the same.

Throughout this section, $P := \{(x,y) \in \mathbb{R}^n_+ \times \mathbb{R}^p_+ : Ax + Gy \le b\}$ is a rational polyhedron, and $S := P \cap (\mathbb{Z}^n_+ \times \mathbb{R}^p_+)$. We will also assume that A, G, b are integral. We let $s = b - (Ax + Gy)$ be the slacks of $Ax + Gy \leqslant b$.
Recall from Section 11.5 that a vector $\lambda \in \mathbb{Q}^m$ yields the GMI inequality:

$$\sum_{j=1}^{n}(\lfloor a_j^\lambda \rfloor + \frac{(f_j - f_0)^+}{1-f_0})x_j + \frac{1}{1-f_0}\sum_{j:g_j^\lambda<0} g_j^\lambda y_j + \sum_{i:\lambda_i<0} +\lambda_i s_i \leqslant \lfloor \beta^\lambda \rfloor$$

which is valid for S. We denote it by $GM(\lambda)$.

Given $\lambda \in \mathbb{Q}^m$ and the corresponding GMI inequality $GM(\lambda)$, we consider the following partitions of $M := \{1,\dots,m\}$ and $P := \{1,\dots,p\}$:

$$M^+ := \{i \in M : \lambda_i \ge 0\} \quad M^- := \{i \in M : \lambda_i < 0\}$$

$$P^+ := \{j \in P : g_j^\lambda \ge 0\} \quad P^- := \{j \in P : g_j^\lambda < 0\}$$

and the following cone:

$$C_\lambda := \{\mu \in \mathbb{R}^m : g_j^\mu \ge 0, j \in P^+,\ g_j^\mu \le 0, j \in P^-,\ \mu_i \ge 0, i \in M^+,\ \mu_i \le 0, i \in M^-\} \tag{11.27}$$

Lemma 11.8. *Let $\lambda \in \mathbb{Q}^m$. Let $\lambda^1, \lambda^2 \in C_\lambda$ where $\lambda = \lambda^1 + \lambda^2$ and $\lambda^2 \in \mathbb{Z}^m \setminus \{0\}$. Then*

$$GM(\lambda) = GM(\lambda^1) + GM(\lambda^2)$$

and either $GM(\lambda)$ is valid for P or it is redundant for P^{GMI}.

Proof. Since $\lambda^2 b$ is an integer, $f_0 = 0$ and $GM(\lambda^2)$ is the following inequality:

$$\sum_{j=1}^{n} a_j^{\lambda^2} x_j + \sum_{j:g_j^{\lambda^2}<0} g_j^{\lambda^2} y_j + \sum_{i:\lambda_i^2<0} +\lambda_i^2 s_i \leqslant \beta^{\lambda^2}.$$

Therefore $GM(\lambda^2)$ is implied by the following inequalities, valid for P:

$$\sum_{j=1}^{n} a_j^{\lambda^2} x_j + \sum_{j=1}^{p} g_j^{\lambda^2} y_j + \sum_{i=1}^{m} \lambda_i^2 s_i = \beta^{\lambda^2}, \quad s \ge 0, \quad y \ge 0.$$

Hence $GM(\lambda^2)$ is valid for P. Moreover, since A, G, b are integral, all coefficients of $GM(\lambda^2)$ are integral.

Since λ^1, $\lambda^2 \in C_\lambda$, $g_j^\lambda < 0$ implies $g_j^{\lambda^1} \leqslant 0$ and $g_j^{\lambda^2} \leqslant 0$; and $\lambda_i < 0$ implies $\lambda_i^1 \leqslant 0$ and $\lambda_i^2 \leqslant 0$. This shows that $GM(\lambda) = GM(\lambda^1) + GM(\lambda^2)$. Thus, since $GM(\lambda^2)$ is valid for P, either $GM(\lambda)$ is valid for P, or it is redundant for P^{GMI}. □

Let Δ be the largest of the absolute values of the determinants of the square submatrices of G.

Lemma 11.9. *Let $\lambda \in \mathbb{Q}^m$ such that $GM(\lambda)$ is not valid for P and it is irredundant for P^{GMI}. Then*

$$-m\Delta \le \lambda_i \le m\Delta, \quad i = 1,\ldots,m. \tag{11.28}$$

Proof. We will show that if λ does not satisfy (11.28), then there exist $\lambda^1, \lambda^2 \in C_\lambda$ such that $\lambda = \lambda^1 + \lambda^2$ and $\lambda^2 \in \mathbb{Z}^m \setminus \{0\}$. This will prove the lemma since then, by Lemma 11.8, either $GM(\lambda)$ is valid for P or it is redundant for P^{GMI}, a contradiction.

Assume λ violates (11.28). Let $r^1,\ldots,r^q$ be a set of vectors generating C_λ. By Remark 11.2, we can choose $r^1,\ldots,r^q$ integral, and by standard linear algebra we can choose them so that $-\Delta\mathbf{1} \le r^t \le \Delta\mathbf{1}$, $t = 1,\ldots,q$ (we leave this as an exercise). Since $\lambda \in C_\lambda$, by Carathéodory's Theorem 11.7, $\lambda = \sum_{t=1}^q \nu_t r^t$, and at most m of the ν_t are positive, while the others are 0. Let

$$\lambda^1 = \sum_{t=1}^q (\nu_t - \lfloor \nu_t \rfloor) r^t, \quad \lambda^2 = \sum_{t=1}^q \lfloor \nu_t \rfloor r^t.$$

Clearly $\lambda^1, \lambda^2 \in C_\lambda$ and $\lambda = \lambda^1 + \lambda^2$. Since $r^1,\ldots,r^q$ are integral vectors, λ^2 is integral. We show that $\lambda^2 \ne 0$. Since at most m of the ν_t are positive, and by definition $-\Delta\mathbf{1} \le r^t \le \Delta\mathbf{1}$, $t = 1,\ldots,q$, then $-\Delta\mathbf{m} \le \lambda^1 \le \Delta\mathbf{m}$. Thus $\lambda^2 \ne 0$, as λ violates (11.28). □

Theorem 11.22 (Cook, Kannan, and Schrijver [17]). *$P^{\mathrm{GMI}} (= P^{\mathrm{Split}})$ is a rational polyhedron.*

Proof. By Lemma 11.4, for every $\lambda \in \mathbb{Q}^m$, $GM(\lambda)$ is a split inequality valid for $P^{(\pi^\lambda,\pi_0^\lambda)}$, where the split $(\pi^\lambda, \pi_0^\lambda) \in Z^{n+1}$ is defined by (11.21)–(11.22). By Lemma 11.9, if $GM(\lambda)$ is irredundant for P^{GMI} and not valid for P, then λ satisfies (11.28). By Theorem 11.20, $P^{\mathrm{GMI}} = P^{\mathrm{Split}}$, thus any inequality valid for $P^{(\pi^\lambda,\pi_0^\lambda)}$ is valid for P^{GMI}. Therefore

$$P^{\mathrm{GMI}} = \bigcap_{\substack{(\pi^\lambda,\pi_0^\lambda)\in\mathbb{Z}^{n+1} \text{ s.t.} \\ \lambda \text{ satisfies (11.28)}}} P^{(\pi^\lambda,\pi_0^\lambda)}.$$

Since the set $\{\lambda \in \mathbb{R}^m : \lambda \text{ satisifes } (11.28)\}$ is bounded, the set $\{(\pi^\lambda, \pi_0^\lambda) \in \mathbb{Z}^{n+1} : \lambda \text{ satisifes } (11.28)\}$ is finite. Therefore P^{GMI} is the intersection of a finite number of polyhedra, hence it is a polyhedron. □

A natural question is whether one can optimize a linear function over P^{GMI} in polynomial time. It turns out that this problem is NP-hard (Caprara and Letchford [12], Cornuéjols and Li [19]). Equivalently, given a point $(\bar{x},\bar{y}) \in P$, it is NP-hard to find a GMI inequality that cuts off $(\bar{x},\bar{y})$ or show that none exists. A similar NP-hardness result was proved earlier by Eisenbrand [25] for the Chvátal closure P^{Ch}.

Note that this is in contrast with the problem of finding a GMI inequality that cuts off a *basic* solution $(\bar{x},\bar{y}) \in P \setminus S$. Indeed, any row of the simplex tableau where $\bar{x}_j$ is fractional generates a GMI inequality that cuts off $(\bar{x},\bar{y})$.

Although it is NP-hard to optimize over the Chvátal closure, there are empirical results on its strength. For 24 instances from the MIPLIB library [8] (all the pure integer programs in MIPLIB 3 with nonzero integrality gap), Fischetti and Lodi [27] found that the Chvátal closure closes at least 63 % of the integrality gap on average (The *integrality gap* is the difference between the values of the objective function when optimized over $\text{conv}(S)$ and over P respectively). Bonami, Cornuéjols, Dash, Fischetti, and Lodi [10] found that the Chvátal closure closes at least 29 % of the integrality gap on average on the remaining 41 MIPLIB instances (all the MIPLIB 3 instances that have at least one continuous variable and nonzero integrality gap).

The split closure and the GMI closure are identical. How tight is it in practice? Balas and Saxena [7] addressed this question by formulating the separation problem for the split closure as a parametric mixed integer linear program with a single parameter in the objective function and the right hand side. They found that the split closure closes 72 % of the integrality gap on average on the MIPLIB instances. This experiment shows that the split closure is surprisingly strong. Finding deep split inequalities efficiently remains a challenging practical issue.

11.7 Lift-and-project

In this section, we consider mixed 0,1 linear programs. These are mixed integer linear programs where the integer variables are only allowed to take the values 0 or 1. It will be convenient to write mixed 0,1 linear programs in the form

$$\begin{aligned} \min\ & cx \\ & Ax \geqslant b \\ & x_j \in \{0,1\} \text{ for } j = 1,\dots,n \\ & x_j \geqslant 0 \qquad \text{for } j = n+1,\dots,n+p, \end{aligned}$$

where the matrix $A \in \mathbb{Q}^{m\times(n+p)}$, the row vector $c \in \mathbb{Q}^{n+p}$ and the column vector $b \in \mathbb{Q}^m$ are data, and $x \in \mathbb{R}^{n+p}$ is a column vector of variables.

Consider the polyhedron $P := \{x \in \mathbb{R}^{n+p}_+ : Ax \geqslant b\}$ and the mixed 0,1 linear set $S := \{x \in \{0,1\}^n \times \mathbb{R}^p_+ : Ax \geqslant b\}$. Without loss of generality, throughout this section we assume that the constraints $Ax \geqslant b$ include $-x_j \geqslant -1$ for $j = 1,\dots,n$, but not $x \geqslant 0$.

Balas, Ceria and Cornuéjols [4] study the following "lift-and-project" relaxation for S: given an index $j \in \{1,\dots,n\}$, let

$$P_j = \text{conv}\{(Ax \geqslant b, x \geqslant 0, x_j = 0) \cup (Ax \geqslant b, x \geqslant 0, x_j = 1)\}.$$

Clearly $S \subseteq P_j \subseteq P$, so P_j is a relaxation of S tighter than P, and by definition it is the tightest possible among the relaxations that ignore the integrality of all the variables x_i for $i \neq j$.

The set $\bigcap_{j=1}^n P_j$ is called the *lift-and-project closure*. It is a better approximation of $\text{conv}(S)$ than P:

$$\operatorname{conv}(S) \subseteq \bigcap_{j=1}^{n} P_j \subseteq P.$$

How much better is it in practice? Bonami and Minoux [11] performed computational experiments (see also Bonami's dissertation [9]). On 35 mixed 0,1 linear programs from MIPLIB, they found that the lift-and-project closure reduces the integrality gap by 37 % on average.

11.7.1 Lift-and-project cuts

Optimizing a linear function over P_j amounts to solving a linear program. In fact, it is possible to express P_j using Theorem 11.18 and then projecting onto the x-space. P_j is the convex hull of the union of two polyhedra:

$$\begin{array}{rcr} Ax \geqslant b & & Ax \geqslant b \\ x \geqslant 0 & \text{and} & x \geqslant 0 \\ -x_j \geqslant 0 & & x_j \geqslant 1 \end{array}$$

By Theorem 11.18,

$$P_j = \operatorname{proj}_x \left\{ \begin{array}{rl} Ax^0 & \geqslant by_0 \\ -x_j^0 & \geqslant 0 \\ Ax^1 & \geqslant by_1 \\ x_j^1 & \geqslant y_1 \\ x^0 + x^1 & = x \\ y_0 + y_1 & = 1 \\ x, x^0, x^1, y_0, y_1 & \geqslant 0. \end{array} \right.$$

Let e_j denote the j-th unit vector. Using the projection theorem (Theorem 11.11), we get that P_j is defined by the inequalities $\alpha x \geqslant \beta$ such that

$$\begin{array}{cccccccl} \alpha & & -uA & +u_0 e_j & & & \geqslant 0 \\ \alpha & & & & -vA & -v_0 e_j & \geqslant 0 \\ & \beta & -ub & & & & \leqslant 0 \\ & \beta & & & -vb & -v_0 & \leqslant 0 \\ & & u, & u_0, & v, & v_0 & \geqslant 0. \end{array} \tag{11.29}$$

The inequality $\alpha x \geqslant \beta$ is called a *lift-and-project* inequality. Clearly lift-and-project inequalities are special type of split inequalities, relative to splits of the type $x_j \leqslant 0$ or $x_j \geqslant 1$.

Given a fractional point $\bar{x}$, we can determine if there exists a lift-and-project inequality $\alpha x \geqslant \beta$ valid for P_j that cuts off $\bar{x}$. In fact, this problem amounts to finding $(\alpha, \beta, u, u_0, v, v_0)$ satisfying (11.29) such that $\alpha\bar{x} - \beta < 0$. In order to find a "best" cut in cone (11.29), one usually adds a normalization constraint to truncate the cone. We then obtain the following *cut generating LP*:

$$
\begin{array}{llllll}
\min \alpha\bar{x} - \beta & & & & & \\
\alpha & -uA & +u_0 e_j & & & \geqslant 0 \\
\alpha & & & -vA & -v_0 e_j & \geqslant 0 \\
\beta & -ub & & & & \leqslant 0 \\
\beta & & & -vb & -v_0 & \leqslant 0 \\
 & \sum_{i=1}^m u_i & +u_0 & +\sum_{i=1}^m v_i & +v_0 & = 1 \\
 & u, & u_0, & v, & v_0 & \geqslant 0.
\end{array}
\tag{11.30}
$$

Balas and Perregaard [6] give a precise correspondence between the basic feasible solutions of (11.30) and the basic solutions (possibly infeasible) of the usual LP relaxation

$$(\mathrm{R}) \quad \min\{cx : Ax \geqslant b, x \geqslant 0\}.$$

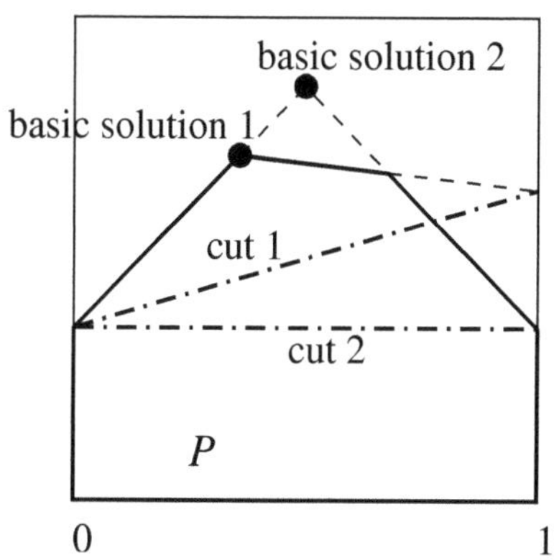

Fig. 11.9 Correspondence between basic solutions and lift-and-project cuts

A geometric view of this correspondence may be helpful: The $n+p$ extreme rays emanating from a basic solution of (R) intersect the hyperplanes $x_j = 0$ and $x_j = 1$ in $n+p$ points (some of these points may be at infinity). These points uniquely determine a hyperplane $\alpha x = \beta$ where (α, β) are associated with a basic feasible solution of the cut generating LP (11.30). For example, in Figure 11.9, cut 1 corresponds to the basic solution 1 of (R) and cut 2 corresponds to the basic (infeasible) solution 2 of (R).

Using the correspondence, Balas and Perregaard [6] show how simplex pivots in the cut generating LP (11.30) can be mimicked by pivots in (R). The major practical consequence is that the cut generating LP (11.30) need not be formulated and solved explicitly. A sequence of increasingly deep lift-and-project cuts can be computed by pivoting directly in (R). We elaborate on these pivoting rules in section 11.7.3.

11.7.2 Strengthened lift-and-project cuts

Again we consider the mixed 0,1 linear set $S := \{x \in \{0,1\}^n \times \mathbb{R}^p_+ : Ax \geqslant b\}$. We assume that the constraints $Ax \geqslant b$ contain $-x_j \geqslant -1$ for $j = 1, \dots, n$, but not $x \geqslant 0$. The cut generating LP (11.30) produces a lift-and-project inequality $\alpha x \geqslant \beta$ that is valid for P_j. The derivation only uses the integrality of variable x_j, not of the variables x_k for $k = 1, \dots, n$ and $k \neq j$. Balas and Jeroslow [5] found a simple way to use the integrality of the other variables to strengthen the lift-and-project cut. This strengthening has the nice property that it is straightforward to implement once the cut generating LP (11.30) has been solved.

Note that, given u, u_0, v, v_0, the optimal values of α_k and β in (11.30) are:

$$\alpha_k = \begin{cases} \max(ua^k, va^k) & \text{for } k \neq j, \\ \max(ua^j - u_0, va^j + v_0) & \text{for } k = j, \end{cases} \tag{11.31}$$

where a^k denotes the k-th column of A, and

$$\beta = \min(ub, vb + v_0).$$

To strengthen the inequality $\alpha x \geqslant \beta$, one can try to decrease the coefficients α_k. Balas and Jeroslow [5] found a way to do just that by using the integrality of the variables x_k for $k = 1, \dots, n$.

Theorem 11.23 (Balas and Jeroslow [5]). *Let $\bar{x}$ satisfy $Ax \geqslant b, x \geqslant 0$. Given an optimal solution u, u_0, v, v_0 of the cut generating LP (11.30), define $m_k = \frac{va^k - ua^k}{u_0 + v_0}$,*

$$\alpha_k = \begin{cases} \min(ua^k + u_0\lceil m_k \rceil, va^k - v_0\lfloor m_k \rfloor) & \text{for } k = 1, \dots, n, \\ \max(ua^k, va^k) & \text{for } k = n+1, \dots, n+p, \end{cases}$$

and $\beta = \min(ub, vb + v_0)$. Then the inequality $\alpha x \geqslant \beta$ is valid for $\operatorname{conv}(S)$.

Proof. For $\pi \in \mathbb{Z}^n$, the following disjunction is valid for $\operatorname{conv}(S)$:

$$\text{either } \sum_{k=1}^{n} \pi_k x_k \geqslant 0 \quad \text{or} \quad -\sum_{k=1}^{n} \pi_k x_k \geqslant 1.$$

Let us repeat the derivation of (11.30) with this disjunction in place of $-x_j \geqslant 0$ or $x_j \geqslant 1$ as before. We consider the union of

$$\begin{array}{rcl} Ax \geqslant b & & Ax \geqslant b \\ x \geqslant 0 & \text{and} & x \geqslant 0 \\ \sum_{k=1}^{n} \pi_k x_k \geqslant 0 & & -\sum_{k=1}^{n} \pi_k x_k \geqslant 1. \end{array}$$

Using Theorem 11.18 and the projection theorem (Theorem 11.11), we get that any inequality $\alpha x \geqslant \beta$ that satisfies

$$\begin{array}{llllll}
\alpha & -uA & -u_0(\sum_{k=1}^n \pi_k e_k) & & & \geqslant 0 \\
\alpha & & & -vA & +v_0(\sum_{k=1}^n \pi_k e_k) & \geqslant 0 \\
\beta & -ub & & & & \leqslant 0 \\
\beta & & & -vb & -v_0 & \leqslant 0 \\
& u, & u_0, & v, & v_0 & \geqslant 0
\end{array}$$

is valid for $\mathrm{conv}(S)$. We can choose u, u_0, v, v_0 to be an optimal solution of the original cut generating LP (11.30). This implies that, for $k = 1, \ldots, n$, we can choose $\alpha_k = \max(ua^k + u_0\pi_k, va^k - v_0\pi_k)$. Smaller coefficients α_k produce stronger inequalities since the variables are nonnegative. What is the best choice of $\pi_k \in \mathbb{Z}$ to get a small α_k? It is obtained by equating $ua^k + u_0\pi_k$ and $va^k - v_0\pi_k$, which yields the value m_k in the statement of the theorem (both u_0 and v_0 are strictly positive since otherwise $\alpha x \geqslant \beta$ is valid for P, contradicting that it is a cut for $\bar{x}$), and then rounding this value m_k either up or down since π_k must be integer. The best choice is the minimum stated in the theorem. □

Bonami and Minoux [11] found that applying the Balas-Jeroslow strengthening step improves the average gap closed by an additional 8 %, as compared to the lift-and-project closure, on the 35 MIPLIB instances in their experiment. Specifically, the integrality gap closed goes from 37 % to 45 %. The time to perform the strengthening step is negligible.

11.7.3 Improving mixed integer Gomory cuts by lift-and-project

In this section we discuss the correspondence between basic feasible solutions of the cut generating LP (11.30) and basic solutions (possibly infeasible) of the usual LP relaxation (R) introduced in Section 11.7.1. The simplex tableaux of (11.30) and (R) will be referred to as *large* and *small* respectively.

Let

$$x_j = a_{j0} - \sum_{h \in J} a_{jh} x_h \tag{11.32}$$

be a row of the small optimal simplex tableau such that $0 < a_{j0} < 1$. The GMI cut from this row is equivalent to the strengthened lift-and-project cut from some basic feasible solution of (11.30), where index j in (11.30) is the same as in (11.32). To identify this solution, partition J into subsets M_1 and M_2, such that $h \in M_1$ if $a_{jh} < 0$, and $h \in M_2$ if $a_{jh} > 0$ ($h \in J$ such that $a_{jh} = 0$ can go into either subset). Then eliminating α, β from (11.30), the n columns indexed by $M_1 \cup M_2$ together with the two columns indexed by u_0 and v_0 define a feasible basis of the resulting system of $n+2$ equations. The strengthened lift-and-project cut associated with this basic feasible solution to (11.30) is equivalent to the GMI cut from (11.32).

To evaluate the GMI cut generated from the small simplex tableau (11.32) as a lift-and-project cut, we calculate the reduced costs in the large tableau of the nonbasic variables of the above solution to (11.30). Each row x_i of the small tableau corresponds to a pair of columns of the large tableau, associated with variables u_i

and v_i. The reduced costs $r(u_i)$, $r(v_i)$ of these variables in the large tableau are known simple functions of the entries a_{ih} and a_{jh}, for $h \in J$, of rows j and i of the small tableau. If they are all nonnegative, the current large tableau is optimal, hence the GMI cut from (11.32) cannot be improved. Otherwise, the cut can be improved by executing a pivot in a row i of the small tableau, such that $r(u_i) < 0$ or $r(v_i) < 0$.

To identify the nonbasic variable x_k to replace x_i in the basis of the small tableau, we calculate for each $h \in J$ the objective function value $f(a_{ih})$ of (11.30) resulting from the corresponding exchange in the large tableau. This value is a known simple function of the ratio a_{jh}/a_{ih} and of the coefficients of rows j and i of the small tableau. Any column h for which $f(a_{ih}) < 0$ is a candidate for an improving pivot, and the most negative value indicates the best column k.

Executing the pivot in the small tableau that exchanges x_i for x_k yields a new simplex tableau (whose solution is typically infeasible), whose j-th row (the same j as before!) is of the form

$$x_j = a_{j0} + ta_{i0} - \sum_{h \in J \cup i \setminus k} (a_{jh} + ta_{ih})x_h, \tag{11.33}$$

with $t := a_{jk}/a_{ik}$. The GMI cut from (11.33) is then stronger than the one from (11.32), in the sense that it cuts off the LP optimum of (R) by a larger amount.

These steps can then be repeated with (11.33) replacing (11.32) for as long as improvements are possible.

Practical experience shows that in about three quarters of the cases GMI cuts from the optimal simplex tableau can be improved by the pivoting procedure described above. On the other hand, improvements beyond 10 pivots are not frequent, and beyond 20 pivots they are very rare.

This procedure was extensively tested and has been incorporated into the mixed integer module of XPRESS, with computational results reported in [42].

11.7.4 Sequential convexification

Theorem 11.24 (Balas [2]). $P_n(P_{n-1}(\ldots P_2(P_1)\ldots)) = \text{conv}(S)$.

Before proving Theorem 11.24, we need a lemma. Let $H \subseteq \mathbb{R}^n$ be a hyperplane and $S \subseteq \mathbb{R}^n$. In general, $\text{conv}(S) \cap H \neq \text{conv}(S \cap H)$, as shown by the example where S consists of two points not in H but the line segment connecting them intersects H. The following lemma shows that equality holds when S lies entirely in one of the closed half spaces defined by the hyperplane H (see Figure 11.10).

Lemma 11.10. *Let* $H := \{x \in \mathbb{R}^n : ax = b\}$ *be a hyperplane and* $S \subseteq \{x : ax \leqslant b\}$. *Then* $\text{conv}(S) \cap H = \text{conv}(S \cap H)$.

Proof. Clearly $\text{conv}(S \cap H) \subseteq \text{conv}(S)$ and $\text{conv}(S \cap H) \subseteq H$ so $\text{conv}(S \cap H) \subseteq \text{conv}(S) \cap H$.

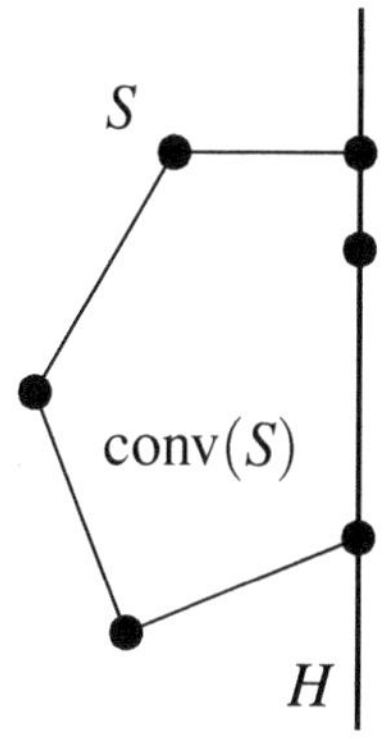

Fig. 11.10 Illustration of Lemma 11.10

We show $\operatorname{conv}(S) \cap H \subseteq \operatorname{conv}(S \cap H)$. Let $x \in \operatorname{conv}(S) \cap H$. This means $ax = b$ and $x = \sum_{i=1}^{k} \lambda_i x^i$ where $x^1, \ldots, x^k \in S$, $\lambda \geqslant 0$ and $\sum_{i=1}^{k} \lambda_i = 1$. We have

$$b = ax = \sum_{i=1}^{k} \lambda_i a x^i \leqslant \sum_{i=1}^{k} \lambda_i b = b \tag{11.34}$$

where the inequality follows from $ax^i \leqslant b$ and $\lambda_i \geqslant 0$. Relation (11.34) implies that these inequalities are in fact equations, i.e., $ax^i = b$ for $i = 1, \ldots, k$. Therefore $x^i \in S \cap H$. This implies $x \in \operatorname{conv}(S \cap H)$. □

Proof of Theorem 11.24. By induction. Let $S_t := \{x \in \{0,1\}^t \times \mathbb{R}_+^{n-t+p} : Ax \geqslant b\}$. We want to show $P_t(P_{t-1}(\ldots P_2(P_1)\ldots)) = \operatorname{conv}(S_t)$. By definition, this is true for $t = 1$, so consider $t \geqslant 2$. Suppose that this is true for $t-1$. By the induction hypothesis we have

$$\begin{aligned} P_t(P_{t-1}(\ldots P_2(P_1)\ldots)) &= P_t(\operatorname{conv}(S_{t-1})) \\ &= \operatorname{conv}(\operatorname{conv}(S_{t-1}) \cap \{x_t = 0\}) \cup (\operatorname{conv}(S_{t-1}) \cap \{x_t = 1\}). \end{aligned}$$

By Lemma 11.10, $\operatorname{conv}(S_{t-1}) \cap \{x_t = 0\} = \operatorname{conv}(S_{t-1} \cap \{x_t = 0\})$ and $\operatorname{conv}(S_{t-1}) \cap \{x_t = 1\} = \operatorname{conv}(S_{t-1} \cap \{x_t = 1\})$. Thus

$$\begin{aligned} P_t(P_{t-1}(\ldots P_2(P_1)\ldots)) &= \operatorname{conv}((S_{t-1} \cap \{x_t = 0\}) \cup (S_{t-1} \cap \{x_t = 1\})) \\ &= \operatorname{conv}(S_t). \end{aligned}$$

□

11.8 Rank

11.8.1 Chvátal rank

In this section, we consider a pure integer set $S := P \cap \mathbb{Z}^n$ where $P := \{x \in \mathbb{R}^n : Ax \leqslant b\}$ is a rational polyhedron. We denote conv(S) by P_I. The Chvátal closure P^{Ch} introduced in Section 11.4.1 will be denoted by $P^{(1)}$ in this section. We can iterate the closure process to obtain the Chvátal closure of $P^{(1)}$. We denote by $P^{(2)}$ this second Chvátal closure. Iteratively, we define the tth Chvátal closure of P to be the Chvátal closure of $P^{(t-1)}$, for $t \geqslant 2$ integer. An inequality that is valid for $P^{(t)}$ but not $P^{(t-1)}$ is said to have *Chvátal rank* t. Are there inequalities of arbitrary large Chvátal rank or is there a value t after which $P^{(t)} = P^{(t+1)}$? The main result of this section is that the second statement is the correct one. In fact, we will prove that there exists a finite t such that $P^{(t)} = P_I$. Therefore, every valid inequality for $P_I := \text{conv}(S)$ has a bounded Chvátal rank. This result for the pure integer case is in contrast with the situation for the mixed case, as we will see in the next section.

We will need the following theorem, whose proof can be found in textbooks such as [45].

Theorem 11.25 (Integer Farkas Lemma or Kronecker Approximation Theorem). *Let A be a rational matrix and b a rational vector. The system $Ax = b$ has an integral solution if and only if for every rational vector u such that uA is integral, ub is an integer.*

Given a set $P \subset \mathbb{R}^n$, we denote by aff(P) the *affine hull* of P, that is the minimal affine subspace of $\mathbb{R}^n$ containing P.

Lemma 11.11. *Let $P \subseteq \mathbb{R}^n$ be a nonempty rational polyhedron such that* $\text{aff}(P) \cap \mathbb{Z}^n \neq \emptyset$. *If $P_I = \emptyset$, then* $\dim(\text{rec}(P)) < \dim(P)$.

Proof. Let $d = \dim(P) = \dim(\text{aff}(P))$. Suppose, by contradiction, that $P_I = \emptyset$ and there are d linearly independent integral vectors $r^1, \ldots, r^d \in \text{rec}(P)$. Let $z \in P$. Since $\text{aff}(P) \cap \mathbb{Z}^n \neq \emptyset$, and $z + r^1, \ldots, z + r^d$ is a basis of aff(P), there exist $\mu_1, \ldots, \mu_d$ such that $z + \sum_{i=1}^d \mu_i r^i \in \mathbb{Z}^n$. Thus $z + \sum_{i=1}^d (\mu_i - \lfloor \mu_i \rfloor) r^i$ is an integral point in P, contradicting the fact that $P_I = \emptyset$. □

A consequence of the above lemma is that every rational polyhedron having full-dimensional recession cone contains an integer point.

Lemma 11.12. *Let $P \subseteq \mathbb{R}^n$ be a rational polyhedron such that* $\text{aff}(P) \cap \mathbb{Z}^n \neq \emptyset$. *Then* $P_I = \{x : Ax \leqslant b\} \cap \text{aff}(P)$ *for some integral A and b such that, for every row a_i of A,*

1. *a_i is not orthogonal to* aff(P);
2. *there exists $d_i \in \mathbb{R}$ such that $a_i x \leqslant d_i$ is valid for P.*

Proof. Assume first $P_I \neq \emptyset$. Then clearly there exist an integral matrix A and an integral vector b such that $P_I = \{x : Ax \leqslant b\} \cap \text{aff}(P)$ and no row of A is orthogonal

to aff(P). We prove 2): Since rec$(P_I) = $ rec(P) by Theorem 11.13, for every row a_i, $d_i = \max\{a_i x : x \in P\}$ is finite, thus $a_i x \leqslant d_i$ is valid for P.
Assume now $P_I = \emptyset$. By standard linear algebra, aff$(P) = z + L$ where $z \in P$ and L is a linear subspace of $\mathbb{R}^n$ such that $\dim(L) = \dim(P)$. Notice that rec$(P) \subset L$. By Lemma 11.11, $\dim(\text{rec}(P)) < \dim(P)$, thus there exists an integral $a \in L$ such that a is orthogonal to rec(P). Thus both $u = \max\{ax : x \in P\}$ and $l = \min\{ax : x \in P\}$ are finite, hence $P_I = \{x : ax \leqslant -1, -ax \leqslant 0\} = \emptyset$, $a, -a$ are not orthogonal to aff(P), and $ax \leqslant u$, $-ax \leqslant -l$ are valid for P. □

Lemma 11.13. *Let P be a rational polyhedron and F a nonempty face of P. Then $F^{(s)} = P^{(s)} \cap F$ for every $s \in \mathbb{Z}_+$.*

Proof. It suffices to show that $F^{(1)} = P^{(1)} \cap F$. This is a consequence of the following statement, which we prove next:

If $cx \le \lfloor d \rfloor$ is a Chvátal inequality for F, there is a Chvátal inequality $c^* x \le \lfloor d^* \rfloor$ for P such that $F \cap \{x : cx \le \lfloor d \rfloor\} = F \cap \{x : c^* x \le \lfloor d^* \rfloor\}$.

Since P is rational, by Theorem 11.15, we can write P as $\{x : A'x \le b', A''x \le b''\}$, where A', A'', b', b'' are integral, so that $F = \{x : A'x \le b', A''x = b''\}$. We can assume that $d = \max\{cx : x \in F\}$. By the duality theorem 11.5 there exist vectors y', y'' such that

$$y'A' + y''A'' = c,\ y'b' + y''b'' = d,\ y' \ge 0.$$

Note that y'' is unrestricted in sign. To obtain a Chvátal inequality for P, we have to use nonnegative multipliers. Define c^* and d^* as:

$$c^* = y'A' + (y'' - \lfloor y'' \rfloor)A'',\ d^* = y'b' + (y'' - \lfloor y'' \rfloor)b''.$$

The multipliers y' and $y'' - \lfloor y'' \rfloor$ are nonnegative. We have $c^* = c - (\lfloor y'' \rfloor)A'')$, $d^* = d - (\lfloor y'' \rfloor)b'')$. Since A'' is an integral matrix and b'', c are integral vectors, then c^* is integral and $\lfloor d \rfloor = \lfloor d^* \rfloor - (\lfloor y'' \rfloor)b''$. So $c^* x \le \lfloor d^* \rfloor$ is a Chvátal inequality for P and $F \cap \{x : c^* x \le \lfloor d^* \rfloor\} = F \cap \{x : \lfloor y'' \rfloor A'' x = \lfloor y'' \rfloor b'',\ c^* x \le \lfloor d^* \rfloor\} F \cap \{x : cx \le \lfloor d \rfloor\}$. □

Theorem 11.26 (Chvátal [13], Schrijver [44]). *Let P be a rational polyhedron. Then there exists $t \in \mathbb{Z}_+$ such that $P^{(t)} = P_I$.*

Proof. The proof is by induction on $d = \dim(P)$, the cases $d = -1$, $d = 0$ being trivial. If aff$(P) \cap \mathbb{Z}^n = \emptyset$, by Theorem 11.25 there exists an integral vector a and a scalar $d \notin \mathbb{Z}$ such that $P \subseteq \{x : ax = d\}$, hence $P_I = \emptyset = \{x : ax \leqslant \lfloor d \rfloor, -ax \leqslant -\lceil d \rceil\} = P^{(1)}$. Therefore we may assume aff$(P) \cap \mathbb{Z}^n \neq \emptyset$. By Lemma 11.12, $P_I = \{x : Ax \leqslant b\} \cap$ aff(P) for some integral A and b such that, for every row a_i of A, a_i is not orthogonal to aff(P) and $a_i x \leqslant d_i$ is valid for P for some $d_i \in \mathbb{R}$.
We only need to show that, for any row a_i of A, there exists a nonnegative integer t such that the inequality $a_i x \leqslant b_i$ is valid for $P^{(t)}$. Suppose not, then, since $a_i x \leqslant d_i$ is valid for P, there exists an integer $d > b_i$ and $r \in \mathbb{Z}_+$ such that, for every $s \geqslant r$, $a_i x \leqslant d$ is valid for $P^{(s)}$ but $a_i x \leqslant d - 1$ is not valid for $P^{(s)}$. Then $F = P^{(r)} \cap \{x : a_i x = d\}$ is a face of $P^{(r)}$ and $F_I = \emptyset$. Since a_i is not orthogonal to aff(P), $\dim(F) <$

$\dim(P)$, therefore, by induction, there exists h such that $F^{(h)} = \emptyset$. By Lemma 11.13, $F^{(h)} = P^{(r+h)} \cap F$, hence $a_i x < d$ for every $x \in P^{(r+h)}$, therefore $a_i x \leqslant \beta - 1$ is valid for $P^{(r+h+1)}$, contradicting the choice of d and r. □

11.8.2 Split rank

Let $P := \{(x,y) \in \mathbb{R}^n \times \mathbb{R}^p : Ax + Gy \leqslant b\}$ and let $S := P \cap (\mathbb{Z}^n \times \mathbb{R}^p)$. In this section, we denote the split closure P^{Split} of P by P^1.

For $k \geqslant 2$, P^k denotes the split closure relative to P^{k-1} and it is called the k-th split closure relative to P. It follows from Theorem 11.22 that P^k is a polyhedron. Unlike for the pure integer case, there is in general no finite r such that $P^r = \text{conv}(S)$ in the mixed integer case, as shown by the following example [17].

Example 11.1. Let $S := \{(x,y) \in \mathbb{Z}^2_+ \times \mathbb{R}_+ : x_1 \geqslant y,\ x_2 \geqslant y,\ x_1 + x_2 + 2y \leqslant 2\}$. Starting from $P := \{(x_1,x_2,y) \in \mathbb{R}^3_+ : x_1 \geqslant y,\ x_2 \geqslant y,\ x_1 + x_2 + 2y \leqslant 2\}$, we claim that there is no finite r such that $P^r = \text{conv}(S)$.

To see this, note that P is a simplex with vertices $O = (0,0,0)$, $A = (2,0,0)$, $B = (0,2,0)$ and $C = (\frac{1}{2},\frac{1}{2},\frac{1}{2})$ (see Figure11.11). S is contained in the plane $y = 0$. More generally, consider a simplex P with vertices O,A,B and $C = (\frac{1}{2},\frac{1}{2},t)$ with $t > 0$. Let $C_1 = C$, let C_2 be the point on the edge from C to A with coordinate $x_1 = 1$ and C_3 the point on the edge from C to B with coordinate $x_2 = 1$. Observe that no split disjunction removes all three points C_1, C_2, C_3. Let Q_i be the intersection of all split inequalities that do not cut off C_i. All split inequalities belong to at least one of these three sets, thus $P^1 = Q_1 \cap Q_2 \cap Q_3$. Let S_i be the simplex with vertices O,A,B,C_i. Clearly, $S_i \subseteq Q_i$. Thus $S_1 \cap S_2 \cap S_3 \subseteq P^1$. It is easy to verify that $(\frac{1}{2},\frac{1}{2},\frac{t}{3}) \in S_i$ for $i = 1,2$ and 3. Thus $(\frac{1}{2},\frac{1}{2},\frac{t}{3}) \in P^1$. By induction, $(\frac{1}{2},\frac{1}{2},\frac{t}{3^k}) \in P^k$.

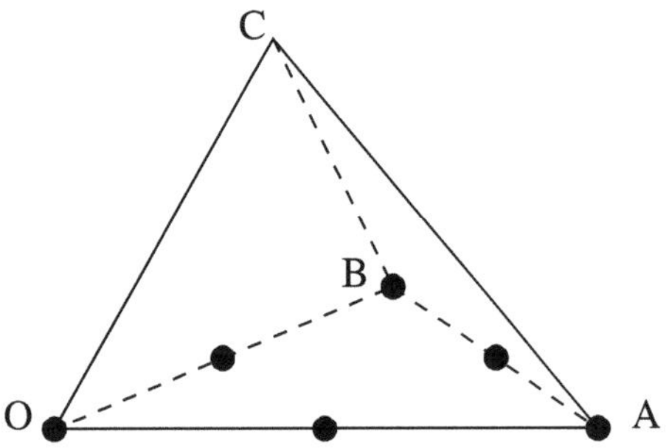

Fig. 11.11 Example showing that the split rank can be unbounded

Remark 11.10. For mixed 0,1 programs, Theorem 11.24 implies that $P^n = \text{conv}(S)$.

Example 11.2. Cornuéjols and Li [18] observed that the n-th split closure is needed for 0,1 programs, i.e., there are examples where $P^k \neq \text{conv}(S)$ for all $k < n$. They use the following well-known polytope studied by Chvátal, Cook, and Hartmann [14]:

$$P_{\text{CCH}} := \{x \in [0,1]^n : \sum_{j \in J} x_j + \sum_{j \notin J} (1 - x_j) \geqslant \tfrac{1}{2}, \text{ for all } J \subseteq \{1, 2, \cdots, n\}\}$$

Let F_j be the set of all vectors $x \in \mathbb{R}^n$ such that j components of x are $\frac{1}{2}$ and each of the remaining $n - j$ components are equal to 0 or 1. The polytope P_{CCH} is the convex hull of F_1.

Lemma 11.14. *If a polyhedron $P \subseteq \mathbb{R}^n$ contains F_j, then its split closure P^1 contains F_{j+1}.*

Proof. It suffices to show that, for every $(\pi, \pi_0) \in \mathbb{Z}^n \times \mathbb{Z}$, the polyhedron $\Pi = \text{conv}((P \cap \{x : \pi x \leqslant \pi_0\}) \cup (P \cap \{x : \pi x \geqslant \pi_0 + 1\}))$ contains F_{j+1}. Let $v \in F_{j+1}$ and assume w.l.o.g. that the first $j+1$ elements of v are equal to $\frac{1}{2}$. If $\pi v \in \mathbb{Z}$, then clearly $v \in \Pi$. If $\pi v \notin \mathbb{Z}$, then at least one of the first $j+1$ components of π is nonzero. Assume w.l.o.g. that $\pi_1 > 0$. Let $v_1, v_2 \in F_j$ be equal to v except for the first component which is 0 and 1 respectively. Notice that $v = \frac{v_1 + v_2}{2}$. Clearly, each of the intervals $[\pi v_1, \pi v]$ and $[\pi v, \pi v_2]$ contains an integer. Since πx is a continuous function, there are points $\tilde{v}_1$ on the line segment $\text{conv}(v, v_1)$ and $\tilde{v}_2$ on the line segment $\text{conv}(v, v_2)$ with $\pi \tilde{v}_1 \in \mathbb{Z}$ and $\pi \tilde{v}_2 \in \mathbb{Z}$. This means that $\tilde{v}_1$ and $\tilde{v}_2$ are in Π. Since $v \in \text{conv}(\tilde{v}_1, \tilde{v}_2)$, this implies $v \in \Pi$. □

Starting from $P = P_{\text{CCH}}$ and applying the lemma recursively, it follows that the $(n-1)$-st split closure relative to P_{CCH} contains F_n, which is nonempty. Since $\text{conv}(P_{\text{CCH}} \cap \{0,1\}^n)$ is empty, the n-th split closure is needed to obtain $\text{conv}(P_{\text{CCH}} \cap \{0,1\}^n)$.

End of Example 11.2.

Remark 11.11. In view of Example 11.1 showing that no bound may exist on the split rank when the integer variables are general, and Remark 11.10 showing that the rank is always bounded when they are 0,1 valued, one is tempted to convert general integer variables into 0,1 variables. For a bounded integer variable $0 \leqslant x \leqslant u$, there are several natural transformations:

(i) a binary expansion of x (see Owen and Mehrotra [41]);

(ii) $x = \sum_{i=1}^{u} i z_i$, $\sum z_i \leqslant 1$, $z_i \in \{0,1\}$ (see Sherali and Adams [46] and Köppe, Louveaux and Weismantel [35]);

(iii) $x = \sum_{i=1}^{u} z_i$, $z_i \leqslant z_{i-1}$, $z_i \in \{0,1\}$ (see Roy [43]).

More studies are needed to determine whether any practical benefit can be gained from such transformations.

References

1. K. Andersen, G. Cornuéjols and Y. Li, *Split closure and intersection cuts*, Mathematical Programming 102 (2005) 457–493.
2. E. Balas, *Disjunctive programming: properties of the convex hull of feasible points*, GSIA Management Science Research Report MSRR 348, Carnegie Mellon University (1974), published as invited paper in Discrete Applied Mathematics 89 (1998) 1–44.
3. E. Balas, *Disjunctive programming and a hierarchy of relaxations for discrete optimization problems*, SIAM Journal on Algebraic and Discrete Methods 6 (1985) 466–486.
4. E. Balas, S. Ceria, and G. Cornuéjols, *A lift-and-project cutting plane algorithm for mixed 0-1 programs*, Mathematical Programming 58 (1993) 295–324.
5. E. Balas and R. Jeroslow, *Strengthening cuts for mixed integer programs*, European Journal of Operations Research 4 (1980) 224–234.
6. E. Balas and M. Perregaard, *A Precise correspondence between lift-and-project cuts, simple disjunctive cuts and mixed integer Gomory cuts for 0-1 programming*, Mathematical Programming 94 (2003) 221–245.
7. E. Balas and A. Saxena, *Optimizing over the split closure*, Mathematical Programming 113 (2008) 219–240.
8. R.E. Bixby, S. Ceria, C.M. McZeal, and M.W.P. Savelsbergh, *An updated mixed integer programming library: MIPLIB 3.0*, Optima 58 (1998) 12–15.
9. P. Bonami, *Etude et mise en oeuvre d'approches polyédriques pour la résolution de programmes en nombres entiers ou mixtes généraux*, PhD Thesis, Université de Paris 6 (2003).
10. P. Bonami, G. Cornuéjols, S. Dash, M. Fischetti, and A. Lodi, *Projected Chvátal-Gomory cuts for mixed integer linear programs*, Mathematical Programming 113 (2008) 241–257.
11. P. Bonami and M. Minoux, *Using rank-1 lift-and-project closures to generate cuts for 0–1 MIPs, a computational investigation*, Discrete Optimization 2 (2005) 288–307.
12. A. Caprara and A.N. Letchford, *On the separation of split cuts and related inequalities*, Mathematical Programming 94 (2003) 279–294.
13. V. Chvátal, *Edmonds polytopes and a hierarchy of combinatorial optimization*, Discrete Mathematics 4 (1973) 305–337.
14. V. Chvátal, W. Cook, and M. Hartmann, *On cutting-plane proofs in combinatorial optimization*, Linear Algebra and its Applications 114/115 (1989) 455–499.
15. M. Conforti, M. Di Summa, and G. Zambelli, *Minimally infeasible set partitioning problems with balanced constraints*, Mathematics of Operations Research 32 (2007) 497–507.
16. S.A. Cook, *The complexity of theorem-proving procedures*, Proceedings of the Third Annual ACM Symposium on the Theory of Computing (Shaker Heights, Ohio 1971), ACM, New York, 1971, pp. 151–158.
17. W. Cook, R. Kannan, and A. Schrijver, *Chvátal closures for mixed integer programming problems*, Mathematical Programming 47 (1990) 155–174.
18. G. Cornuéjols and Y. Li, *On the rank of mixed 0,1 polyhedra*, Mathematical Programming 91 (2002) 391–397.
19. G. Cornuéjols and Y. Li, *A connection between cutting plane theory and the geometry of numbers*, Mathematical Programming 93 (2002) 123–127.
20. G.B. Dantzig, *Maximization of a linear function of variables subject to linear inequalities*, in: Activity Analysis of Production and allocation (T.C. Koopmans, ed.), Wiley N.Y. (1951) 339–347.
21. G. Dantzig, R. Fulkerson, and S. Johnson, *Solution of a large-scale traveling-salesman problem*, Operations Research 2 (1954) 393–410.
22. S. Dash, O. Günlük and A. Lodi, *MIR closures of polyhedral sets*, Mathematical Programming 121 (2010) 33–60.
23. J. Edmonds, *Paths, trees, and flowers*, Canadian Journal of Mathematics 17 (1965) 449–467.
24. J. Edmonds, *Systems of distinct representatives and linear algebra*, Journal of Research of the National Bureau of Standards B 71 (1967) 241–245.

25. F. Eisenbrand, *On the membership problem for the elementary closure of a polyhedron*, Combinatorica 19 (1999) 297–300.
26. Gy. Farkas, *On the applications of the mechanical principle of Fourier*, Mathematikai és Természettudományi Értesotö 12 (1894) 457–472.
27. M. Fischetti and A. Lodi, *Optimizing over the first Chvátal closure*, Mathematical Programming 110 (2007) 3–20.
28. J.B.J. Fourier, *Solution d'une question particulière du calcul des inégalités*, Nouveau Bulletin des Sciences par la Société Philomatique de Paris (1826) 317–319.
29. C.F. Gauss, *Theoria Motus Corporum Coelestium in Sectionibus Conicis Solem Ambientium*, F. Perthes & J.H. Besser, Hamburg, 1809.
30. M.X. Goemans and D.P. Williamson, *Improved approximation algorithms for maximum cut and satisfiability problems using semidefinite programming*, Journal of the ACM 42 (1995) 1115–1145.
31. R.E. Gomory, *Outline of an algorithm for integer solutions to linear programs*, Bulletin of the American Mathematical Society 64 (1958) 275–278.
32. R.E. Gomory, *An algorithm for integer solutions to linear programs*, Recent Advances in Mathematical Programming (R.L. Graves and P. Wolfe, eds.), McGraw-Hill, New York,1963, pp. 269–302.
33. N. Karmarkar, *A new polynomial-time algorithm for linear programming*, Combinatorica 4 (1984) 373–395.
34. L.G. Khachiyan, *A polynomial algorithm in linear programming*, Soviet Mathematics Doklady 20 (1979) 191–194.
35. M. Köppe, Q. Louveaux, and R. Weismantel, *Intermediate integer programming representations using value disjunctions*, Discrete Optimization 5 (2008) 293–313.
36. H.W. Lenstra, *Integer programming with a fixed number of variables*, Mathematics of Operations Research 8 (1983) 538–548.
37. L. Lovász and A. Schrijver, *Cones of matrices and set-functions and 0-1 optimization*, SIAM Journal of Optimization 1 (1991) 166–190.
38. R.R. Meyer, *On the existence of optimal solutions to integer and mixed integer programming problems*, Mathematical Programming 7 (1974) 223–235.
39. H. Minkowski, *Geometrie der Zahlen (Erste Lieferung)*, Teubner, Leipzig, 1896.
40. G.L. Nemhauser and L.A. Wolsey, *A recursive procedure to generate all cuts for 0-1 mixed integer programs*, Mathematical Programming 46 (1990) 379–390.
41. J.H. Owen and S. Mehrotra, *On the value of binary expansions for general mixed-integer linear programs*, Operations Research 50 (2002) 810–819.
42. M. Perregaard, *A practical implementation of lift-and-project cuts: A computational exploration of lift-and-project cuts with XPRESS-MP*, 18th ISMP, Copenhagen, 2003.
43. J.-S. Roy, *"Binarize and project" to generate cuts for general mixed-integer programs*, Algorithmic Operations Research 2 (2007) 37–51.
44. A. Schrijver, *On cutting planes*, Annals of Discrete Mathematics 9 (1980) 291–296.
45. A. Schrijver, *Theory of Linear and Integer Programming*, Wiley, New York, 1986.
46. H. Sherali and W. Adams, *A reformulation-linearization technique for solving discrete and continuous nonconvex problems*, Kluwer Academic Publishers, Dordrecht, 1998.
47. J.P. Vielma, *A constructive characterization of the split closure of a mixed integer linear program*, Operations Research Letters 35 (2007) 29–35.

Chapter 12
Fifty-Plus Years of Combinatorial Integer Programming

William Cook

Abstract Throughout the history of integer programming, the field has been guided by research into solution approaches to combinatorial problems. We discuss some of the highlights and defining moments of this area.

12.1 Combinatorial integer programming

Integer-programming models arise naturally in optimization problems over combinatorial structures, most notably in problems on graphs and general set systems. The translation from combinatorics to the language of integer programming is often straightforward, but the new rendering typically suggests direct lines of attack via linear programming.

As an example, consider the stable-set problem in graphs. Given a graph $G = (V,E)$ with vertices V and edges E, a *stable set* of G is a subset $S \subseteq V$ such that no two vertices in S are joined by an edge. The *stable-set problem* is to find a maximum-cardinality stable set. To formulate this as an integer-programming (IP) problem, consider a vector of variables $x = (x_v : v \in V)$ and identify a set $U \subseteq V$ with its characteristic vector $\bar{x}$, defined as $\bar{x}_v = 1$ if $v \in U$ and $\bar{x}_v = 0$ otherwise. For $e \in E$ write $e = (u,v)$, where u and v are the ends of the edge. The stable-set problem is equivalent to the IP model

$$\begin{aligned} \max \sum (x_v : v \in V) \\ x_u + x_v \leqslant 1,\ \forall\, e = (u,v) \in E, \\ x_v \geqslant 0,\ \forall\, v \in V, \\ x_v \text{ integer},\ \forall\, v \in V. \end{aligned} \tag{12.1}$$

William Cook
School of Industrial and Systems Engineering, Georgia Institute of Technology, Atlanta, USA
e-mail: `bico@isye.gatech.edu`

M. Jünger et al. (eds.), *50 Years of Integer Programming 1958-2008*,
DOI 10.1007/978-3-540-68279-0_12, © Springer-Verlag Berlin Heidelberg 2010

To express this model in matrix notation, let A denote the edge-vertex incidence matrix of G, that is, A has rows indexed by E, columns indexed by V, and for each $e \in E$ and $v \in V$, entry $A_{ev} = 1$ if v is an end of e and $A_{ev} = 0$ otherwise. Letting $\mathbf{0}$ and $\mathbf{1}$ denote the vectors of all zeros and all ones, respectively, problem (12.1) can be written as

$$\max(\mathbf{1}^T x : Ax \leqslant \mathbf{1},\ x \geqslant \mathbf{0},\ x \text{ integer}). \tag{12.2}$$

In a similar fashion, the *vertex-cover problem* can be modeled as

$$\min(\mathbf{1}^T x : Ax \geqslant \mathbf{1},\ x \geqslant \mathbf{0},\ x \text{ integer}). \tag{12.3}$$

This later problem asks for a minimum-cardinality set $C \subseteq V$ such that every edge in E has at least one of its ends in C.

By dropping the integrality constraints on the variables, we obtain linear programming (LP) relaxations for the IP models. From these relaxations we get the LP dual

$$\min(y^T \mathbf{1} : y^T A \geqslant \mathbf{1}^T,\ y \geqslant \mathbf{0}) \tag{12.4}$$

for the stable-set problem and the LP dual

$$\max(y^T \mathbf{1} : y^T A \leqslant \mathbf{1}^T,\ y \geqslant \mathbf{0}) \tag{12.5}$$

for the vertex-cover problem. Solutions to (12.4) and (12.5) give upper bounds and lower bounds for the two combinatorial problems, respectively, through the weak LP-duality theorem. When these bounds are unsatisfactory in a given application, IP techniques can be employed to improve the relaxations and reduce the gap between the cardinality of a stable set or vertex cover and the value of the corresponding dual LP. The primary tool for obtaining such an improvement is to add, to the LP relaxation, inequalities that are satisfied by all integer solutions, but violated by an optimal solution to the LP problem. This is known as the *cutting-plane method* and it was first proposed in the context of a combinatorial problem, as we discuss in the next section.

The use of cutting planes is a practical step to improve a given model, but the LP-duality framework can also be a powerful tool in proving theorems in combinatorics. The idea is to formulate a relaxation such that the LP optimum can always be attained by an integer-valued vector. Such a relaxation gives a characterization of optimal solutions to the combinatorial problem. If it can be shown that the dual LP also always has integer solutions, then strong LP duality provides a form of combinatorial min-max theorem. Such statements are among the most beautiful results in combinatorics.

The pursuit of LP-based proofs of min-max theorems often involves showing that a polyhedron P, defined as the solution set to a system of linear inequalities $Ax \leqslant b$, has the property that each of its vertices is integer valued. In this case P is said to be an *integer polyhedron*. If for each integer objective vector w the dual LP $\min(y^T b : y^T A = w^T, y \geqslant \mathbf{0})$ has an integer optimal solution, then the system $Ax \leqslant b$ is called *totally dual integral* (TDI). Such systems are often the final goal of research efforts, since they translate to min-max results for the weighted version of the combinatorial problem, where w provides the weights and the integer dual solutions correspond to combinatorial structures.

The study of integer polyhedra and totally dual integral systems for combinatorial problems is known as *polyhedral combinatorics*. The area we are calling *combinatorial integer programming* includes both the application of IP techniques to solve particular instances of possibly difficult problems, as well as the development of general methods and examples in polyhedral combinatorics.

In this paper we give a brief history of combinatorial IP. It is not our intent to be comprehensive in any form, but rather to touch on some of the highlights in the development and growth of the field, with particular emphasis on breakthroughs in combinatorial methods that have led to successful techniques for general integer programming. For a marvelously complete account of the history of the wider field of combinatorial optimization, the reader is directed to the work of Schrijver [135]. We refer the reader also to the reprints of classical papers contained in this volume, and to the fascinating historical perspectives offered by the authors of these papers in their newly written introductions.

12.2 The TSP in the 1950s

The birth of combinatorial integer programming occurred in the spring of 1954, pre-dating the start of general integer-programming research by several years. The event was described as follows in the popular journal *Newsweek*, July 26, 1954.

> Finding the shortest route for a traveling salesman—starting from a given city, visiting each of a series of other cities, and then returning to his original point of departure—is more than an after-dinner teaser. For years it has baffled not only goods- and salesman-routing businessmen but mathematicians as well. If a drummer visits 50 cities, for example, he has 10^{62} (62 zeros) possible itineraries. No electronic computer in existence could sort out such a large number of routes and find the shortest.
>
> Three Rand Corp. mathematicians, using Rand McNally road-map distances between the District of Columbia and major cities in each of the 48 states, have finally produced a solution. By an ingenious application of linear programming—a mathematical tool recently used to solve production-scheduling problems—it took only a few weeks for the California experts to calculate "by hand" the shortest route to cover the 49 cities: 12,345 miles.

The California experts were George Dantzig, Ray Fulkerson, and Selmer Johnson, part of an exceptionally strong and influential center for the new field of mathematical programming, housed at the RAND Corporation in Santa Monica. Dantzig et al. took up the computational challenge of the traveling salesman problem (TSP), solving a 49-city instance with hand-only computations. Along the way they set the stage for the study of integer programming.

The traveling salesman problem

Before going into the details of the RAND team's work, it is appropriate to consider the research environment where conditions were set for their breakthrough study. The starting point of the discussion is the TSP itself and how the problem came to the prominent role it has played in the history of integer programming. Concerning this issue, Dantzig et al. [29] write the following.

> The origin of this problem is somewhat obscure. It appears to have been discussed informally among mathematicians at mathematics meetings for many years. Surprisingly little in the way of results has appeared in the mathematical literature. It may be that the minimal-distance tour problem was stimulated by the so-called Hamiltonian game which is concerned with finding the number of tours possible over a specified network. The latter problem is cited by some as the origin of group theory and has some connections with the famous Four-Color Conjecture. Merrill Flood (Columbia University) should be credited with stimulating interest in the traveling-salesman problem in many quarters. As early as 1937, he tried to obtain near optimal solutions in reference to routing of school buses. Both Flood and A. W. Tucker (Princeton University) recall that they heard the problem first in a seminar talk by Hassler Whitney at Princeton in 1934 (although Whitney, recently queried, does not seem to recall the problem).

This brief summary of TSP history is expanded in Hoffman and Wolfe [81] and Schrijver [135, 136]. In these works it is noted that Karl Menger described a geometric variant of the TSP in a record of a mathematics colloquium held in Vienna on February 5, 1930 [113]. Schrijver [135, 136] also points out that Menger and Whitney met at Harvard University in 1930–31, during a semester-long visit by Menger. This Menger-Whitney interaction supports the idea of a possible connection between Menger's Vienna colloquium and Whitney's Princeton seminar.

It remains a question whether Whitney did in fact discuss the TSP at Princeton. There unfortunately is not an accessible record at Princeton University covering the seminars delivered in the Department of Mathematics in the 1930s. The Pusey Library at Harvard University does, however, contain an archive of 3.9 cubic feet of Whitney's papers, and within the collection there is a set of handwritten notes that appear to be preparation for a seminar by Whitney, written sometime in the years shortly after 1930. The notes give an introduction to graph theory, including the following paragraph.

> A similar, but much more difficult problem is the following. Can we trace a simple closed curve in a graph through each vertex exactly once? This corresponds to the following problem. Given a set of countries, is it possible to travel through them in such a way that at the end of the trip we have visited each country exactly once?

This is an unusual example for the *Hamilton-circuit problem*, and clearly not a far step away from the TSP. The geographic aspect also matches well with Flood's recollection of the "'48-states problem' of Hassler Whitney" in a 1984 interview [45] with Albert Tucker.

There is not a record of TSP research, under the TSP name, in the late 1930s and into the 1940s, but by the end the 1940s it had become a known challenge in Princeton and RAND, supported by the interest of Merrill Flood. On the Princeton side, Harold Kuhn writes the following in a recent email letter [95].

> The traveling salesman problem was known by name around Fine Hall by 1949. For instance, it was one of a number of problems for which the RAND corporation offered a money prize. I believe that the list was posted on a bulletin board in Fine Hall in the academic year 1948–49.

At the RAND Corporation, Julia Robinson published a research report in 1949 that appears to be the first reference to the TSP by name [132]. Interestingly, Robinson formulates the problem as finding "the shortest route for a salesman starting from Washington, visiting all the state capitals and then returning to Washington."

Heller and Kuhn

Robinson's work considers an LP approach to the TSP, treating a variation of the *assignment problem*. This problem asks for an optimal assignment of n workers to n tasks, where the quality of assigning worker i to task j is specified by a weight w_{ij}. The goal is to maximize the total weight of the n assignments. A feasible solution to the problem can be viewed as a graph having vertices labeled 1 up to n, with the assignment of worker i to task j indicated by an edge directed from i to j. The assignment solution gives a collection of disjoint directed circuits meeting every vertex. The connection to the TSP is clear: a TSP solution is the special case when the assignment yields a single circuit containing all vertices.

The assignment problem is a member of a more general class called *transportation problems*. Efforts to solve instances from this class played a prominent role in the early history of linear programming. Julia Robinson's research in this area is mentioned in the following quote from Dantzig et al. [29].

> The relations between the traveling-salesman problem and the transportation problem appear to have been first explored by M. Flood, J. Robinson, T. C. Koopmans, M. Beckmann, and latter by I. Heller and H. Kuhn.

Robinson's paper begins an LP line of attack, but it is the work of Isidor Heller and Harold Kuhn that appears to have had the most influence on the computational study of Dantzig et al. Both Heller [72] and Kuhn [91] began investigations of linear descriptions of the convex hull of TSP tours, considering tours as characteristic vectors of their edge sets. Their approach aims at a direct LP formulation of the TSP.

In notes from a George Dantzig Memorial Lecture delivered in 2008 [94], Kuhn writes the following concerning his TSP study.

> My efforts centered around the formulation of the traveling salesman problem as a linear program that has as feasible solutions the convex hull of the "tours". A tour is defined as a 0-1 matrix that presents a permutation that is a single cycle. For example, with 5 cities, there are 24 tours that are the extreme points of an 11 dimensional feasible set. In the summer of 1953, I found that this convex polyhedron has 390 faces, a very discouraging fact.
>
> I had a number of contacts with George throughout the summer discussing this and other problems. And I know that George attended my lecture at the end of the summer (as did Selmer Johnson, Ray Fulkerson, and Alan Hoffman). We were both keenly aware of the fact that, although the complete set of faces (or constraints) in the linear programming formulation of the Traveling Salesman Problem was enormous, if you could find an optimal

solution to a relaxed problem with a subset of the faces that is a tour, then you had solved the underlying Traveling Salesman Problem.

Fig. 12.1 Harold Kuhn, 1961. Photograph courtesy of Harold Kuhn.

Kuhn presented his TSP work at the Sixth Symposium in Applied Mathematics, held in Santa Monica on August 26–28, 1953. His lecture is titled "The traveling salesman problem" in the conference program [2], and Dantzig, Fulkerson, and Johnson are listed as participants of the meeting. Kuhn [95] notes that his Santa Monica lecture included the following points.

1. A statement of the TSP as an LP and a clear statement that if you solved over a subset of the constraints and obtained a tour then the tour was optimal.
2. Results of a shooting experiment on the five-city TSP polytope to explore the distribution of its faces. (This experiment is described in [55] and in [93].)
3. The fact that in the five-city polytope any two vertices are contained in a face of dimension one, that is, the polytope is neighborly, and that the six-city polytope does not have this property.
4. An account of constructions of classes of faces whose number grows exponentially with the number of cities.

Concerning this last point, Kuhn writes that the number of faces "was so large that it discouraged me from pursuing this direction of research."

Kuhn's work was inspired in part by a study of the five-city TSP polytope by Heller, which was also carried out in 1953. Heller presented his research in a lecture "On the problem of shortest path between points" [72] at the Summer Meeting of the

American Mathematical Society, held in Kingston, Ontario, August 31 to September 5, 1953. Dantzig again participated in the meeting, while Fulkerson and Johnson are not listed as participants [3].

The studies of Heller and Kuhn conclude with the fact that the natural LP model of the TSP necessarily contains far too many inequalities for any solver to handle directly. Undeterred, Dantzig et al. saw this as an opportunity to demonstrate the versatility of the simplex algorithm.

The cutting-plane method

The approach adopted by the RAND team is laid out in a preliminary version [28] of their paper. In the following quote, C_1 denotes the solution set of the LP relaxation, T_n denotes the convex hull of all tours through n cities, and d_{ij} is the cost of travel between city i and city j.

> What we do is this: Pick a tour x which looks good, and consider it as an extreme point of C_1; use the simplex algorithm to move to an adjacent extreme point e in C_1 which gives a smaller value of the functional; either e is a tour, in which case start again with this new tour, or there exists a hyperplane separating e from the convex of tours; in the latter case cut down C_1 by one such hyperplane that passes through x, obtaining a new convex C_2 with x as an extreme point. Starting with x again, repeat the process until a tour $\hat{x}$ and a convex $C_m \supset T_n$ are obtained over which $\hat{x}$ gives a minimum of $\sum d_{ij}x_{ij}$.

The process clearly applies to problems beyond the TSP, and it is known today as the *primal cutting-plane method*; see, for example, Letchford and Lodi [102].

The published version of the paper excludes a general description of their method, relying on a sequence of five-city and six-city examples to convey the idea. There was a four-month gap between the release of the preliminary report and the submission of their paper, and the authors appear to have changed their minds as to how best to describe the methodology. Among other things, the preliminary report also contains a discussion of the convex hull of tours, similar in style to the modern treatment of the TSP polytope. Regarding this, Fulkerson writes in a September 2, 1954, letter to *Operations Research* editor George Shortly: "In an effort to keep the version submitted for publication elementary, we avoid going into these matters in any detail." It is a pity this choice was made, but it is not a surprising decision given the nature of operations research literature at the time.

The LP relaxation adopted by Dantzig et al. has a variable x_{ij} for each unordered pair of cities (i,j). It is convenient to describe this model in terms of a complete graph $G=(V,E)$, denoting variable x_{ij} as x_e, where e is the edge having ends i and j. The initial relaxation consists of the *degree equations*

$$\sum(x_e : v \text{ is an end of } e) = 2 \quad \text{for all cities } v \tag{12.6}$$

together with the restriction $x_e \geqslant 0$ for all $e \in E$. A ready supply of potential cutting planes is derived from the observation that every proper subset of k cities can contain at most $k-1$ edges in a tour. The corresponding *subtour constraints* are

$$\sum(x_e : \text{both ends of } e \text{ are in } S) \leqslant |S|-1 \quad \text{for all } S \subseteq V, S \neq V. \tag{12.7}$$

These inequalities are called "loop conditions" in [29] and they are the first line-of-defense in the RAND computations.

The published descriptions of the small TSP examples in [29] focus on the integration of the simplex algorithm and the cutting-plane method, suggesting how Dantzig et al. were able to handle the 49-city LP relaxation with hand-only calculations. For this large TSP, using LP duality, they present a succinct proof that their method produced an optimal tour. The final LP relaxation contains a set of 23 subtour inequalities (of which 16 have the form $x_e \leqslant 1$ for edges $e \in E$, that is, the set S has only two cities) and two additional inequalities. The second of these two non-subtour cutting planes points to Irving Glicksberg as an unsung hero in the TSP effort, cited in the footnote: "We are indebted to I. Glicksberg of Rand for pointing out relations of this kind to us."

Fig. 12.2 Irving Glicksberg, 1978. Photograph copyright Mathematisches Forschungsintitut Oberwolfach.

It should be noted that the Dantzig et al. study considers the symmetric version of the TSP, where the travel cost between city i and city j is the same as the cost between city j and city i. This differs from the Heller and Kuhn studies, where the directed version of the problem is considered. This point generated some discussion among TSP researchers. In his September 2, 1954, letter to George Shortley, Fulkerson writes the following.

> The assumption $d_{ij} = d_{ji}$ certainly seems to be of some importance, although we are not sure that it is crucial. (Dr. I. Heller, who has done considerable research on the problem, feels that the symmetry assumption, which permits representing the convex of tours in a different space, is of the utmost importance.) It is true, as the referee says, that the loop conditions and combinatorial analysis can be used for directed tours as well, and some work should be done along these lines. (The fact that the analogues of the loop conditions are faces of the convex of directed tours has been known for a couple of years.) However, if one has a symmetric problem, much is gained by using undirected tours. This is probably due to two

> things: (1) The simplex algorithm of linear programming becomes especially easy, and (2) there is some reason to believe that the convex of undirected tours may have significantly fewer faces than the directed tours.

This point is also brought up in a letter from Fulkerson to Heller, dated March 11, 1954.

> I read your abstracts "On the problem of shortest path between points" in the November issue of the Bulletin of the American Mathematical Society with much interest. If it is not too much trouble, I would greatly appreciate it if you would send me more details of your results.
>
> Recently, G. Dantzig, S. Johnson, and I have been working on computational aspects of the problem via linear programming techniques even though we don't know, of course, all the faces of the convex C_n of tours for general n. The methods we have been using seem hopeful, however; in particular, an optimal tour has been found by hand computation for a large scale problem using 48 cities, rather quickly. We have found it convenient in translating Dantzig's simplex algorithm in terms of the map of points, to identify tours which differ only in direction of traversal. For example, C_5 can be characterized by a system of 25 hyperplanes in 10 dimensional space. We don't know very much about C_n in general, but thought we might learn more from reading your papers, if they are available.

Similar requests for polyhedral results were sent from Dantzig to Kuhn (March 11, 1954) and from Dantzig to Tucker (March 25, 1954). It is clear that Dantzig et al. were actively seeking more information on the facial structure of the TSP polytope, to better equip their cutting-plane method. This is a topic that was taken up in force two decades later, as we describe in Section 12.6.

Reduced-cost fixing and branch-and-bound algorithms

In the 1954 reports and in a follow-up 1959 paper [30], Dantzig et al. assert the effectiveness of a method for transforming the TSP into a problem on a sparse graph. This transformation is known as *reduced-cost fixing* and it was described for the first time in this TSP work. The idea is the following. When a minimization LP problem with nonnegative variables is solved by the simplex method, the objective function is rewritten in the form $z = z_o + \sum(\bar{c}_j x_j : j = 1, \ldots, m)$ such that z_o is a constant, $\bar{c}_j \geqslant 0$ for all j, and a solution vector x^* is found such that $x_j^* = 0$ for all j such that $\bar{c}_j$ is positive. If the variables are required to take on integer values, then any x_j such that $z_o + \bar{c}_j$ is greater than the cost of a known feasible integer solution can be set to the value 0, and eliminated from the problem. This process is used in modern integer-programming solvers, reducing the problem space in a preliminary step to an enumeration phase.

The RAND team's implementation of this idea is more subtle, since they do not actually solve the LP relaxation in their primal cutting-plane method, carrying out only single pivots of the simplex algorithm. Nevertheless, they show that reduced-cost fixing can be accomplished, taking advantage of the fact that the variables in the TSP relaxation are bounded between 0 and 1. The explicit variable bounds allow one to obtain a bound on the objective value even in the case when some of the $\bar{c}_j$ values are negative, and again variables can be eliminated. In the following comment on this process from Dantzig et al. [29], E denotes a value such that variables x_j with $\bar{c}_j > E$ can be eliminated.

> During the early stages of the computation, E may be quite large and very few links can be dropped by this rule; however, in the latter stages often so many links are eliminated that one can list all possible tours that use the remaining admissible links.

A general method for carrying out this enumeration of tours is not given, but in [30] an example is used to describe a possible scheme, relying on forbidding subtours and vertices of degree three when growing a tour in the sparse edge set.

The enumeration aspect of the Dantzig et al. work has not been followed up to any large degree in modern computational studies of the TSP, but it was pursued in the late 1950s in various combinatorial approaches by Frederick Bock [14], G. A. Croes [24], and others. These studies, in turn, contributed to the development of the *branch-and-bound* algorithm, where the set of solutions is split into two or more subsets (the branching step), a lower-bounding method is applied separately to each of the subsets (the bounding step), and the process is applied repeatedly to the resulting subproblems (growing a search tree).

The first full-version of the branch-and-bound method may be the TSP algorithm described in the 1958 Ph.D. thesis of Willard Eastman [32]. In Eastman's algorithm, the lower bound is provided by the solution of a variant of the assignment problem. In his branching step, a subtour having k edges in the assignment solution is chosen, and k subproblems are created by setting to 0, one at a time, each of the variables corresponding to the edges in the subtour. Eastman carried out his method on a ten-city TSP; an illustration of part of his search tree is given in Figure 12.3.

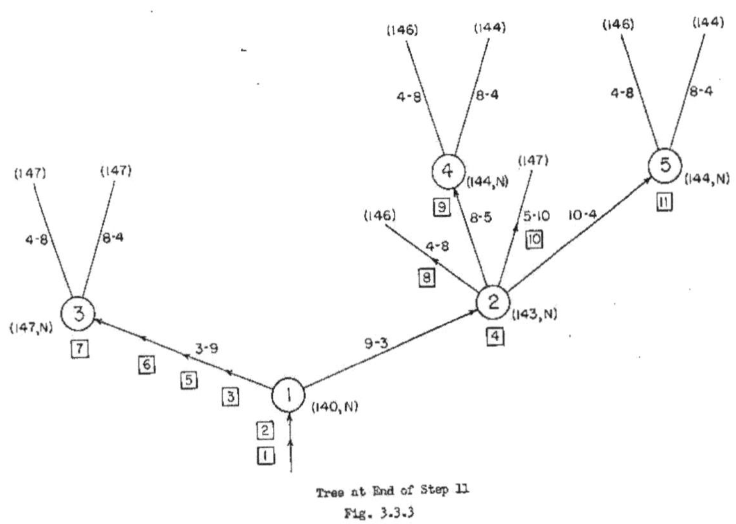

Fig. 12.3 Branch-and-bound search tree from W.L. Eastman's 1958 Ph.D. thesis.

The TSP-driven branch-and-bound research had a great impact on the practical solution of general integer-programming instances, starting with the important 1960 paper of Land and Doig [98]. We close this section by noting that the name "branch-and-bound" first appeared several years after Land and Doig, when the method was again applied to the TSP in the 1963 work of Little, Murty, Sweeney, and Karel [104].

12.3 Proving theorems with linear-programming duality

In the preface to a collection of his papers [114], Alan Hoffman thanks Harold Kuhn and David Gale: "in fond recollection of the early '50s, when we taught each other to use the ostensibly practical subject of linear programming to prove aesthetic combinatorial theorems that were ostentatiously useless." The work of these gentlemen and their colleagues set basic research directions that formed a roadmap for the early development of polyhedral combinatorics. The topics considered include the assignment problem by Kuhn [92], systems of distinct representatives by Hoffman and Kuhn [78], bipartite matching by Hoffman [74], network flows by Lester Ford and Ray Fulkerson [46] and David Gale [51], and partially ordered sets by Dantzig and Hoffman [31]. This work was carried out in an incredibly active span of years in the mid-1950s. A nice overview of the activity can be seen in the volume *Linear Inequalities and Related Systems*, edited in 1956 by Kuhn and Tucker [96]. Leafing through the pages of the book, it is striking how many household names appear among the authors. The volume also contains a bibliography of 289 books and papers covering research on systems of linear inequalities, with the majority written after 1950.

An important general concept that came out of this body of work is the notion of totally unimodularity, introduced by Hoffman and Joseph Kruskal [77]. A matrix is called *totally unimodular* if each of its subdeteriminants is 0, 1, or -1. The well-known Hoffman-Kruskal result states that an integral matrix A is totally unimodular if and only if for each integral vector b the set $\{x : Ax \leqslant b, x \geqslant \mathbf{0}\}$ is an integer polyhedron.

The following illustration of this concept is adopted from Hoffman's short survey paper "Linear programming" in *Applied Mechanics Reviews*, 1956 [75]. Consider a *bipartite* graph $G = (V,E)$. By definition, V can be written as the disjoint union of sets U and W such that each $e \in E$ has one end in U and one end in W. An inductive proof shows that the edge-vertex incidence matrix A of such a graph is totally unimodular. It follows that both sides of the LP-duality equation

$$\max(\mathbf{1}^T x : Ax \leqslant \mathbf{1},\ x \geqslant \mathbf{0}) = \min(y^T \mathbf{1} : y^T A \geqslant \mathbf{1}^T,\ y \geqslant \mathbf{0})$$

are attained by integer solutions, $\bar{x}$ and $\bar{y}$, assuming that the optima exist. Note that $\bar{x}$ is the incidence vector of a stable set of G, while $\bar{y}$ is the incidence vector of a set of edges F such that each vertex in V meets at least one edge in F, that is,

Fig. 12.4 Esther and Alan Hoffman, Washington D.C., 1951. Photograph courtesy of Alan Hoffman.

Fig. 12.5 Alan Hoffman, 2000. Photograph by Sue Clites.

F is an *edge cover*. Also, the optimality condition is satisfied as long as G has no isolated vertices. We conclude that for such a graph, the maximum cardinality of a stable set is equal to the minimum cardinality of an edge cover. Perhaps not "ostentatiously useless", but a pretty min-max result nonetheless. Hoffman [114] writes the following concerning his joint work with Kruskal.

> In this paper the concept (not the name) of total unimodularity was shown to be a neat explanation (via Cramer's rule) of the fact that some linear programming problems have all their vertices integral. I do not think this paper would have been accepted for publication if we had not fancied it up with a supçon of generalization: the main idea is too obvious and folklorish. And we also thought that we introduced a new class of matrices with the "unimodular property", but Jack Edmonds later found that our new class wasn't really new after all. It is nevertheless true that totally unimodular matrices (as Berge christened them), and unimodular matrices generally, are key to understanding how linear programming duality underlies a wide variety of extremal combinatorial analysis.

Indeed, total unimodularity provides a unifying theme for combinatorial min-max theorems, and it remains a fundamental tool in polyhedral combinatorics.

12.4 Cutting-plane computation

Returning to the cutting-plane method, the publication of the 1954 TSP paper did not begin an immediate revolution in the practical solution of integer-programming problems. Within the RAND Corporation, however, the cutting-plane strategy was explored as a computational tool in the years following the Dantzig-Fulkerson-Johnson success.

Markowitz and Manne

An important contribution in this effort is documented in the paper "On the solution of discrete programming problems" by Harry Markowitz and Alan Manne [111], first published as a RAND research paper in 1956 [110]. Markowitz and Manne formulate a general mixed-integer-programming model and describe, in abstract terms, how it can be solved with a variant of the cutting-plane method. They introduce their procedure as follows [111].

> We do not present an automatic algorithm. We present, rather, a general approach susceptible to a number of variations depending on the problem and the judgment of the user. The approach is of little or no purely mathematical interest. Its only recommendation consists of a few empirical observations: When applied to very small discrete problems (with a few thousand a priori possibilities) it has produced and confirmed the answer almost immediately. Its application to two moderate-size problems is described subsequently. There is a danger, of course, in generalizing from so few observations. They provide encouragement, rather than proof.
>
> Our procedure for handling discrete problems was suggested by that employed in the solution of the 'traveling-salesman' problem by Dantzig, Fulkerson, and Johnson.

Fig. 12.6 Harry Markowitz, 2000. Photograph by Sue Clites.

Despite these modest words, the Markowitz-Manne approach is an interesting variation of the method used in the TSP work. The new ideas are to (1) allow cuts that possibly remove integer solutions that are known to have objective value no better than a previously computed solution, (2) use linear constraints to partition the feasible region, allowing the cutting-plane method to be applied independently to each of the subregions, and (3) allow the simplex algorithm to compute optimal LP solutions to the relaxations, rather than carrying out single simplex pivots, thus obtaining bounds to measure the quality of previously computed solutions. The second point is a clear precursor to the modern branch-and-cut version of the branch-and-bound algorithm, and the third point is the adoption of the now common "dual" cutting-plane method.

Markowitz and Manne begin their presentation with a description of the primal cutting-plane approach of Dantzig et al. and then lay out a general step of the partition+relaxation strategy. Details of a possible implementation of the abstract ideas are provided through two examples, one in production scheduling and one in air transportation. The step-by-step elaboration of the technique on these problems provides great insight into the practical application of LP arguments in integer programming.

The following simple remark concludes the Markowitz-Manne paper [111].

> The solutions to the two examples presented here, along with those to the traveling-salesman problem, suggest that the human being with simple aids can frequently produce solutions with near-optimum payoffs.

It is interesting to see these famous researchers (Markowitz was awarded a Noble Prize in 1990) getting their hands dirty with detailed integer-programming calculations. One must imagine that the members of the RAND Corporation were a driven group of problem solvers, using the focus of real computations to guide their research.

Dantzig in 1957

Foremost among these IP problem solvers is undoubtedly George Dantzig. He returns to the cutting-plane method in a 1957 paper [27], summarizing some of his work following the TSP study. Here Dantzig also describes the "dual" cutting-plane approach, considering cuts that remove fractional optimal LP solutions.

> The linear programming approach consists in putting such additional linear-inequality constraints on the system that the fractional extreme points of C where the total value of z is maximized will be excluded, while the set of extreme points of the convex hull C^* of admissible solutions will be unaltered. The procedure would be straightforward except that the rules for generating the *complete set* of additional constraints is not known. For practical problems, however, rules for generating a partial set of constraints is often sufficient to yield the required solution.

This methodology is applied to an example of the *knapsack problem*, that is, an IP model where the feasible region is determined by a single inequality constraint and all variables are restricted to take on 0 or 1 values. In the knapsack metaphor, the coefficients in the inequality are the weights of the objects and the right-hand-side is the knapsack's capacity. To run his cutting-plane approach, Dantzig considers what are now known as *cover inequalities*, expressing that the sum of k variables can be no more than $k-1$ if the weight of the corresponding k objects exceeds the capacity of the knapsack. In the following quote from [27], condition "(14)" refers to such a cutting plane, form "(11)" refers to the objective function, and condition "(9)" refers to the single knapsack constraint.

> Form (11) is maximized under conditions (9) and $0 \leqslant x_j \leqslant 1$, but with the constraint (14) added. Again a new fractional extreme point may turn up for the new convex C, and it will be necessary again to seek a condition that will exclude it. For the most part the conditions added will be other partial sums of the x_j similar to (14). However, at times more subtle relations will be required until an extreme point is obtained that is admissible.
>
> Since the discovery of these more subtle relations is more an art than a science, the reader may dismiss the whole approach as worthless. However, experiments with many problems by the author and others indicate that very often a practical problem can be solved using only such obvious supplementary conditions as (14).

The use of the phrase "many problems" suggests that the cutting-plane method was indeed in use, at least at the RAND Corporation.

Gomory's IP algorithm

A common thread in the discussions of Dantzig, Fulkerson, and Johnson [29], Markowitz and Manne [111], and Dantzig [27] is the need for creativity in the discovery of inequalities to add as cutting planes, with appeals to Irving Glicksberg, to a "human being", and to "more of an art than a science", respectively. Such creativity would limit the automation of the procedure on the class of digital computers that was becoming available. This subject was addressed by Princeton researcher Ralph Gomory, with the publication of his stunning four-page paper [54] in 1958.

Fig. 12.7 George Dantzig, Ralph Gomory, and Ellis Johnson, 1982. Photograph by Sue Clites.

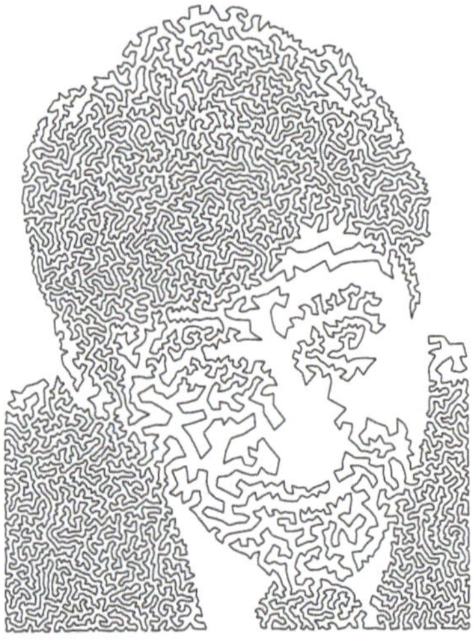

Fig. 12.8 Ralph Gomory as a TSP tour. Image by Robert Bosch, December 2007.

It is the purpose of this note to outline a finite algorithm for obtaining integer solutions to linear programs. The algorithm has been programmed successfully on an E101 computer and used to run off the integer solution to small (seven or less variables) linear programs completely automatically.

The algorithm closely resembles the procedures already used by Dantzig, Fulkerson, and Johnson, and Markowitz and Manne to obtain solutions to discrete variable programming problems. Their procedure is essentially this. Given the linear program, first maximize the objective function using the simplex method, then examine the solution. If the solution is not in integers, ingenuity is used to formulate a new constraint that can be shown to be satisfied by the still unknown integer solution but not by the noninteger solution already attained. This additional constraint is added to the original ones, the solution already attained becomes nonfeasible, and a new maximum satisfying the new constraint is sought. This process is repeated until an integer maximum is obtained, or until some argument shows that a nearby integer point is optimal. What has been needed to transform this procedure into an algorithm is a systematic method for generating the new constraints. A proof that the method will give the integer solution in a finite number of steps is also important. This note will describe an automatic method of generating new constraints. The proof of the finiteness of the process will be given separately.

Fig. 12.9 Magazine advertisement for the Burroughs E101 computer, 1959.

The great importance of Gomory's algorithm is covered in detail in other parts of this volume; we focus here only on its connections to developments in combinatorial integer programming.

In one direction, the connection to combinatorial IP is not as strong as one might guess. Indeed, in his 1991 paper "Early integer programming" [56], Gomory describes how he became aware of the existing cutting-plane research only after the main details of his procedure had been worked out. Nonetheless, the history of success with the cutting-plane method in combinatorial integer programming likely played a major role in the acceptance of Gomory's algorithm as a viable technique for solving general IP problems.

In the other direction, an initial wave of combinatorial projects took the direct approach of formulating IP models and turning Gomory loose on small instances. Representative papers at this kind are those by Lambert [97] and Miller, Tucker, and Zemlin [116], where IP formulations of the TSP are presented together with reports of solutions to instances having five cities and four cities, respectively. Rapid growth in the size of the IP formulations in these studies limits the applicability of the methodology.

An interesting hybrid approach was explored several years later by Glenn Martin, described in an unpublished manuscript from 1966 [112]. Martin considers the TSP, but he begins with a simple LP relaxation consisting of the degree equations and a subtour constraint for the ends of the cheapest edge incident to each city. He applies Gomory's algorithm to obtain an integer optimal solution x^* to the relaxation. If x^* is a tour, then it is an optimal solution to the TSP; otherwise he adds, by hand, subtour inequalities violated by x^* and applies Gomory again. Using three rounds of the procedure, Martin repeated the Dantzig-Fulkerson-Johnson feat of solving the 49-city USA instance. This effective approach is considered in further studies by Takis Miliotis [115] in the mid-1970s.

12.5 Jack Edmonds, polynomial-time algorithms, and polyhedral combinatorics

The work of Gomory centers on the automation of the cutting-plane procedure, making it suitable for implementation on a digital computer. In 1960, a branch-and-bound alternative was proposed by Ailsa Land and Alison Doig [98], working at the London School of Economics. In the following quote, these two authors comment on their IP algorithm.

> Until recently there was no general automatic routine for solving such problems, as opposed to procedures for proving the optimality of conjectured solutions, and the work here is intended to fill the gap.

Fig. 12.10 Ailsa Land, Banff, 1977. Photograph courtesy of Ailsa Land.

Fig. 12.11 Alison Doig, *The Sun*, October 21, 1965. Courtesy of Alison (Doig) Harcourt.

Variations of their approach became the dominant practical method for the solution of IP instances. Concerning this, Ailsa Land and Susan Powell [99] make the following remark in a 2007 paper.

> While branch and bound began to be built into computer codes, the cutting plane approach was obviously more elegant, and we spent a great deal of time experimenting with it. ... Work was done, but it was not published because as a method to solve problems branch and bound resoundingly won.

They go on to write: "It is gratifying that the combination, 'branch and cut', is now often successful in dealing with real problems."

The importance of the automatic nature of the Gomory and Land-Doig algorithms cannot be disputed, but a critical theoretical question remained. The algorithms were shown to be finite, but this in itself is not a substantial issue for the problem class. Consider, for example, the TSP, where it is obvious that one can simply list all possible tours. That this approach is not an acceptable solution is suggested already by Karl Menger [113], in his initial description of the problem.

> This problem can naturally be solved using a finite number of trials. Rules which reduce the number of trials below the number of permutations of the given point set are not known.

What is sought is an algorithm that is efficient, not just finite.

A search for a better-than-finite algorithm for the assignment problem was a focus of early mathematical-programming research in the United States, starting with a 1951 lecture of John von Neumann at Princeton. Two years later, an efficient solution method was famously developed by Harold Kuhn [92], armed with a copy of Jenő Egerváry's 1931 paper [44] and a large Hungarian-English dictionary.

Kuhn's [93] personal account of the events leading up to this work is delightful to read, as is Schrijver's [135] thorough description of the history of assignment-problem computation and algorithms.

An analysis of Kuhn's Hungarian algorithm appeared in a 1957 paper of James Munkres [117], showing that it can be implemented to run in time polynomial in n, the number of vertices. This notion of polynomial time did not immediately become a standard means for evaluating algorithms, however. In particular, the criterion was not used in the discussions of the finite algorithms for integer programming.

Jack Edmonds took charge of this issue, several years later, dramatically bringing the notion of polynomial-time algorithms and good characterizations into the hearts and minds of the research community. His efforts of persuasion began at a workshop in the summer of 1961, held at the RAND Corporation. Edmonds, working at the National Bureau of Standards in Washington, D.C., joined a group of young researchers invited to take part in the workshop together with leading figures in the field, including Dantzig, Fulkerson, Hoffman and others. His RAND lecture, and a 1963 research paper [33, 35], concerned the problem of finding optimal matchings in a general graph. Edmonds [35] writes the following.

> I am claiming, as a mathematical result, the existence of a *good* algorithm for finding a maximum cardinality matching in a graph.
>
> There is an obvious finite algorithm, but that algorithm increases in difficulty exponentially with the size of the graph. It is by no means obvious whether *or not* there exists an algorithm whose difficulty increases only algebraically with the size of the graph.

Not only did this paper of Edmonds establish the basis for complexity theory, the technique he employed opened up the world of polyhedral combinatorics beyond unimodularity. The linear constraints present in the natural formulation of the matching problem do not define an integer polyhedron. Edmonds nonetheless provides a simple description of a full set of inequalities defining the convex hull of the integer points in the relaxation.

A paper of Gomory [55] has a fascinating section covering the discussion that took place after Gomory's TSP lecture at the IBM Scientific Computing Symposium on Combinatorial Problems, March 16–18, 1964. This record includes the following remarks of Edmonds, in response to a comment of Harold Kuhn.

> The algorithm I had in mind is one I introduced in a paper submitted to the Canadian Journal of Mathematics. This algorithm depends crucially on what amounts to knowing all the bounding inequalities of the associated convex polyhedron—and, as I said, there are many of them. The point is that the inequalities are known by an easily verifiable characterization rather than by an exhausting listing—so their number is not important.
>
> This sort of thing should be expected for a class of extremum problems with combinatorially special structure. For the traveling salesman problem, the vertices of the associated polyhedron have a simple characterization despite their number—so might the bounding inequalities have a simple characterization despite their number. At least we should hope they have, because finding a really good traveling salesman algorithm is undoubtedly equivalent to finding such a characterization.

The thesis of Edmonds was clear: the existence of polynomial-time algorithms goes hand-in-hand with polyhedral characterizations.

Fig. 12.12 Jack Edmonds, 2009. Photograph by Marc Uetz.

The application of Edmonds' thesis to matching problems begins, for a graph $G=(V,E)$, with the simple inequalities

$$\begin{aligned} \Sigma(x_e : e \text{ meets vertex } v) \leqslant 1 \quad & \text{for all vertices } v \in V, \\ x_e \geqslant 0 \quad & \text{for all edges } e \in E. \end{aligned} \tag{12.8}$$

The polyhedron P defined by this system has as vertices each incidence vector of a matching of G, but it may have non-integer vertices as well. Consider, for example, three edges f, g, and h that form a triangle in G. Setting $\bar{x}_f = \bar{x}_g = \bar{x}_h = 1/2$ and $\bar{x}_e = 0$ for all other edges e gives a vertex $\bar{x}$ of P. Such half-integer vertices can be cut off from P by the addition of inequalities

$$\Sigma(x_e : e \text{ has both ends in } S) \leqslant (|S|-1)/2$$

for each set $S \subseteq V$ of odd cardinality. Edmonds calls these constraints *blossom inequalities*. His remarkable theorem is that adding these inequalities to (12.8) gives a defining system for the convex hull of matchings. Edmonds' proof is via a polynomial-time algorithm that constructs a matching and a corresponding dual solution that together satisfy the LP-duality equation.

The Chvátal closure

The method of Edmonds considers the full set of blossom inequalities in a single stroke, rather than introducing them one at a time in a cutting-plane implementation. An exquisite theory considering waves of inequalities was developed by Vašek Chvátal [17], summarized by the famous slogan

combinatorics = number theory + linear programming

from his paper "Edmonds polytopes and a hierarchy of combinatorial problems", published in 1973. The waves considered by Chvátal are the following. Given a

Fig. 12.13 Vašek Chvátal. Photograph by Adrian Bondy. All rights reserved.

polyhedron P and an inequality $c^T x \leqslant \delta$, with c integral, satisfied by each of its members, each integer vector in P also satisfies $c^T x \leqslant \lfloor \delta \rfloor$, where $\lfloor \delta \rfloor$ denotes δ rounded down to the nearest integer. Let P' denote the members of P that satisfy all such inequalities. Chvátal called P' the "elementary closure" of P; nowadays it is referred to as the *Chvátal closure*. The main result of [17] is that for any bounded polyhedron P, a finite number of applications of the closure operation results in the convex hull of the integer points in P. Thus combinatorial theorems can be proved by repeatedly rounding down inequalities obtained as linear combinations of previously derived inequalities. This theory can be interpreted either in terms of *cutting-plane proofs* [20, 21] or geometrically as the *Chvátal rank* of polyhedra [133]; it provides an important connection between the polyhedral methods of Edmonds and the IP algorithm of Gomory.

Polyhedral combinatorics in the 1970s

Edmonds himself followed the matching breakthrough with a series of results, applying his polyhedral methods to spanning trees [39], branchings [37], matroid intersection [40], submodular functions [38], and, together with Ellis Johnson, the Chinese postman problem [36, 42, 43]. His leadership and amazing research moved polyhedral combinatorics into high gear. Highlights of the maturing field in the 1970s include the following projects.

Fig. 12.14 Bernhard Korte and László Lovász, 1982. Photograph courtesy of the Research Institute for Discrete Mathematics, University of Bonn.

- Ray Fulkerson [48, 49] develops his theory of blocking and anti-blocking polyhedra.
- László Lovász [105, 106] proves the weak perfect-graph conjecture.
- Egon Balas and Manfred Padberg [4, 5] study set-covering problems.
- Manfred Padberg [122], George Nemhauser and Leslie Trotter [118, 119], Vašek Chvátal [19], and Laurence Wolsey [143] study the stable-set polytope.
- Jack Edmonds and Rick Giles [41] and Alan Hoffman [76] show that total dual integrality implies primal integrality, that is, if $Ax \leqslant b$ is TDI and b is integer, then $P = \{x : Ax \leqslant b\}$ is an integer polyhedron. Further properties of TDI systems are investigated by Rick Giles and William Pulleyblank [52] and Alexander Schrijver [134].
- William Pulleyblank and Jack Edmonds [130, 129] describe the facet-defining inequalities of the matching polytope.
- Paul Seymour [138] provides a deep characterization of a certain combinatorial class of integer polyhedra and TDI systems, receiving a Fulkerson Prize in 1979.

- Jack Edmonds and Rick Giles [41] propose a general LP framework that includes a min-max theorem for directed cuts in graphs proved by Cláudio Lucchesi and Daniel Younger [109]. Other LP-based min-max frameworks are described by András Frank [47] and Alan Hoffman and Donald Schwartz [80]
- William Cunningham and Alfred Marsh, III [26] show that the blossom system for matchings is TDI. Further studies of the blossom system are made by Alan Hoffman and Rosa Oppenheim [79] and Alexander Schrijver and Paul Seymour [137].
- Paul Seymour's [139] decomposition theorem for regular matroids yields a polynomial-time algorithm to test if a matrix is totally unimodular.

This period was a golden era for polyhedral combinatorics, attracting great talent to the field and establishing a standard of quality and elegance.

Fig. 12.15 James Ho, Ellis Johnson, George Nemhauser, Jack Edmonds, and George Dantzig, 1985. Photograph courtesy of George Nemhauser.

12.6 Progress in the solution of the TSP

On the computational side, the TSP continued to lead the way in studies of combinatorial IP methods. TSP research in the 1960s and 1970s was dominated first by the work of Michael Held and Richard Karp, and later by the return of the cutting-plane method.

Dynamic programming

The straightforward enumeration algorithm for the TSP, listing all tours and choosing the cheapest, solves an n-city instance in time proportional to $n!$. Analysis of the cutting-plane method has not improved this result—the number of cuts needed in a worst-case example cannot be easily estimated. A breakthrough occurred in 1962, however, using an alternative algorithmic technique called *dynamic programming*. This method was shown by Held and Karp [69] to solve any instance of the TSP in time proportional to $n^2 2^n$.

The general approach was layed out in Richard Bellman's book *Dynamic Programming*, published by Princeton University Press in 1957 [7]. Bellman was another prominent member of the mathematical-programming group at the RAND Corporation, where he introduced the dynamic-programming technique in a 1953 technical report [6]. The subject goes well beyond its application to IP problems, encompassing general models in multistage decision making.

At roughly the same time as the Held and Karp study, dynamic programming for the TSP was also proposed by Bellman [8, 9] and R. H. Gonzales [57]. The idea used in all three projects is to build a TSP tour by computing optimal paths visiting subsets of cities, gradually increasing the number of cities covered in each path. The $n^2 2^n$ worst-case bound for the method is significantly better than $n!$, although it is still far from a practical result for instances of the size tackled by Dantzig et al. with cutting planes.

The Held and Karp paper [69] includes a computational study, giving very fast solutions for examples having 13 or fewer cities, and good approximate solutions for larger problems. An IBM press release from January 2, 1964, describes the availability of the TSP code as follows.

> IBM mathematicians (left to right) Michael Held, Richard Shareshian and Richard M. Karp review the manual describing a new computer program which provides business and industry with a practical scientific method for handling a wide variety of complex scheduling tasks. The program, available to users of the IBM 7090 and 7094 data processing systems, consists of a set of 4,500 instructions which tell the computer what to do with data fed into it. It grew out of the trio's efforts to find solutions for a classic mathematical problem—the "Traveling Salesman" problem—which has long defied solution by man, or by the fastest computers he uses.

The accompanying photograph of Held, Karp, and Shareshian is shown in Figure 12.16.

An effective branch-and-bound algorithm

Held and Karp's $n^2 2^n$ result is to this day the best known bound for general TSP algorithms. The dynamic-programming approach does not, however, extend to a practical method for large-scale instances, and the 49-city USA solution remained a computational record throughout the 1960s. At the end of the decade Held and Karp struck again, this time with a branch-and-bound method that succeeded in pushing

Fig. 12.16 Michael Held, Richard Shareshian, and Richard Karp, 1964. Photograph courtesy of IBM Corporate Archives.

the limits of TSP computation. Richard Karp made the following remarks on this joint work in his 1985 Turing Award Lecture [86].

> After a long series of unsuccessful experiments, Held and I stumbled upon a powerful method of computing lower bounds. This bounding technique allowed us to prune the search severely, so that we were able to solve problems with as many as 65 cities. I don't think any of my theoretical results have provided as great a thrill as the sight of the numbers pouring out of the computer on the night Held and I first tested our bounding method.

The Held-Karp bounding technique relies on an iterative approach for obtaining a good approximation to the value of the LP relaxation consisting of the degree equations and all subtour constraints, avoiding the simplex method and the cutting-plane approach. Each step of the bounding algorithm computes an optimal spanning tree for a problem obtained by deleting a single city; the value of the tree plus the cost of the cheapest two edges meeting the deleted city is a lower bound on the cost of any tour. The edge costs are adjusted after each step, according to the shape of the resulting tree. The form of the cost adjustment is to add or subtract, for each city v, a fixed value δ_v from the cost of each edge meeting v. This adjustment does not alter the TSP, but it can change the spanning tree and the implied lower bound. The spanning-tree polyhedron result of Edmonds [39] connects the subtour constraints with the tree computations, and the edge-cost adjustment accounts for dual variables on the degree equations, in a technique known as Lagrangian relaxation.

The tenacity of Held and Karp set a new standard in TSP computation. Using a computer implementation written together with Linda Ibrahim, their branch-and-bound algorithm solved a set of instances having up to 64 cities [71].

Implementing the cutting-plane method

Despite the great early success of the cutting-plane method, the approach was not really pursued as a TSP tool after the solution of the 49-city problem. Except for the publication of a RAND report by John Robacker in 1955, describing a set of tests on nine-city TSP instances, no further computations were reported with the Dantzig-Fulkerson-Johnson technique in the decade following their published result. This point is brought up in a 1964 lecture by Gomory [55].

> I do not see why this particular approach stopped where it did. It should be possible to use the same approach today, but it an algorithmic manner. We no longer have to be artistic about generating the separating hyperplanes or cuts, since this is now done automatically in integer programming. It seems likely that one can get over the difficulties of maintaining the basis as well. So it should be possible to do the whole thing now systematically. This is an approach one might not expect to work, but we already know that it does.

Saman Hong, supervised by Mandell Bellmore, responded to this call in his Ph.D. work at The Johns Hopkins University. His thesis *A Linear Programming Approach for the Traveling Salesman Problem* appeared in 1972, and reports the first fully automatic version of the cutting-plane method for the TSP.

Fig. 12.17 Saman Hong, 1971. Photograph courtesy of Saman Hong.

Hong's work uses subtour constraints together with a version of Edmonds' blossom inequalities, embedded in a combined branch-and-bound and cutting-plane approach, now called *branch-and-cut*. His computational results are modest, solving instances with up to 25 cities, but he opened the way for a renewed attack with TSP cuts.

Following Hong, the team of Martin Grötschel and Manfred Padberg took up the study of TSP cutting planes, combining to push all aspects of the technology. The pair started their effort in 1974, working at Bernhard Korte's Institut für Ökonometrie und Operations Research at the University of Bonn. They focused on the study of structural properties of the TSP polytope, including an important proof that a generalization of the *comb inequalities*, introduced by Vašek Chvátal [18], are facet defining.

Their work set the stage for a big push in TSP computation, beginning with a study by Grötschel using an instance consisting of 120 cities from West Germany. Grötschel [59] describes his method as follows.

> After every LP-run we represented the optimal solution graphically by hand on a map. In the beginning a plotter was used, but as the number of different fractional components of the solutions increased there were not enough symbols to distinguish them and the plottings became too cluttered. Using the graphical representation of the optimal solution we looked for subtour elimination constraints and comb inequalities to cut off the present solution and added them to the present constraint system. Drawing and searching took from 30 man-minutes in the beginning up to 3 man-hours after the last runs.

After thirteen rounds of the procedure, an optimal solution to the 120-city TSP was found. This work was carried out in 1975, and it is first described in Grötschel's 1977 Ph.D. thesis [58].

Manfred Padberg commented on this successful computation in his 2007 paper [123], and notes that a study together with Hong was begun hot on the heels of Grötschel's project.

> To my pleasant surprise, Martin included numerical experiments in his dissertation; he solved a 120-city traveling salesman problem to optimality which was a world record. In early 1975 I met Saman Hong, a Korean 1972 Johns Hopkins' Ph.D., in New York. We started a project to solve symmetric TSPs, by implementing an exact arithmetic, primal simplex algorithm using Don Goldfarb's "steepest edge" criteria for pivot column selection with automatic constraint generation.

The joint work of Padberg and Hong [124] was a computational success, automating the primal cutting-plane algorithm, solving instances with up to 75 cities, and computing good lower bounds on other instances. The largest example treated in the study is a 318-city instance considered earlier by Shen Lin and Brian Kernighan [103]. This instance arose in an application to guide a drilling machine through a set of holes; it is treated as a *Hamiltonian-path* problem, asking for a single path joining specified starting and ending points and covering all other cities.

Padberg was not satisfied with the good approximations for the large instances, and several years after the completion of his study with Hong he continued his pursuit for a solution to the Lin-Kernighan example. Padberg [123] comments on this effort in the following passage.

> Some day in early 1979 I approached Harlan Crowder of IBM Research with a proposal to push the exact solvability of TSPs up a notch, just like he had done earlier with Mike

Held and Phil Wolfe. He was all for it and we sat down to discuss what had to be done. It seemed feasible and so we did it. It took maybe a couple of months to string it all together. Harlan had other duties as well and I was back teaching at NYU. One evening we had it all together and submitted a computer run for the 318-city symmetric TSP. We figured it would take hours to solve and went to the "Side Door", a restaurant not far from IBM Research, to have dinner. On the way back we discussed all kinds of "bells and whistles" we might want to add to the program in case of a failure. When we got to IBM Research and checked the Computer Room for output it was there. The program proclaimed optimality of the solution it had found in under 6 minutes of computation time!

Fig. 12.18 Ellis Johnson, Tito Ciriani, Manfred Padberg, Mario Lucertini, Harlan Crowder (sitting), 1982. Photograph courtesy of Manfred Padberg.

The Crowder-Padberg study [25] concluded with the solution to the 318-city instance and a large collection of smaller examples. This was a triumph of the cutting-plane method and the long-term research efforts of Grötschel and Padberg.

12.7 Widening the field of application in the 1980s

In the twenty-five years following Dantzig-Fulkerson-Johnson, the few computational studies with the cutting-plane method focused on efforts at solving instances of the TSP. The landscape changed dramatically, however, in the 1980s. By the end of the decade the technique was in wide use, covering numerous fundamental models as well as applied problems from business and industry.

The transition away from the confines of the TSP was led by a focused effort of Martin Grötschel, Michael Jünger, and Gerhard Reinelt, with their study of the linear-ordering problem [61, 62, 63, 84, 131]. Over a period of four years, the team managed to duplicate the scope of research that had evolved for the TSP over the previous two and half decades. In a single project, Grötschel et al. formulated an

LP relaxation, carried out a polyhedral study, developed a class of potential cutting planes, created efficient separation algorithms to produce cuts, implemented a branch-and-cut framework, gathered problem data, and carried out a large-scale computational test. This work set a standard for future studies and demonstrated that the algorithmic success with the TSP was not an isolated event.

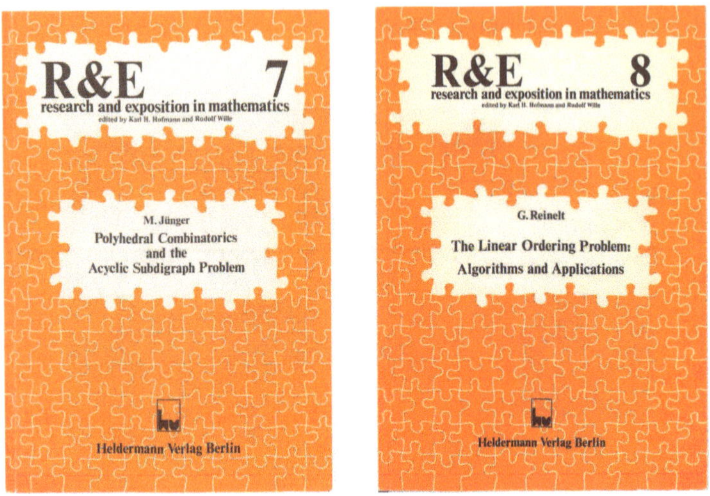

Fig. 12.19 Ph.D. theses of Michael Jünger and Gerhard Reinelt, 1985.

Fig. 12.20 Michael Jünger, Martin Grötschel, Jeff Edmonds, Yoshiko Wakabayashi, Mario Sakumoto, and Gerhard Reinelt, providing input to order a selection of beers, 1984. Photograph courtesy of Gerhard Reinelt.

The linear-ordering problem

Fresh from theoretical and computational work with the TSP, Grötschel began a study with Ph.D. students Jünger and Reinelt, aiming to apply the TSP lessons to a new problem area. The team began their work in 1981 at the University of Bonn, moving in 1983 to the University of Augsburg. The *linear-ordering problem* they consider is defined as follows. Let $D = (V, E)$ be a complete directed graph with weights $(w_e : e \in E)$ on the edges. The problem is to find an ordering of the vertices $v_1, v_2, \ldots, v_n$ that maximizes the sum of the weights of the edges that are directed consistent with the ordering, that is, from a vertex lower in the order to a vertex higher in the order. The ordering can be thought of as a ranking of the vertices, with the weight w_e, for directed edge $e = (u, v)$, giving the payoff for ranking u before v. The problem was formulated by Bernhard Korte and Walter Oberhofer [89, 90] in the late 1960s; applications are described in Jan Karel Lenstra's 1977 thesis [101]. The Grötschel et al. study provides a complete package of tools for solving real-world instances arising in a variety of settings.

In the years following the linear-ordering project, cutting-plane research entered a period of rapid growth, with studies covering a wide range of models. A 1994 survey paper by Jünger, Reinelt, and Stephan Thienel [85] lists twenty-one cutting-plane projects carried out by various research teams.

The breadth of this cutting-plane work was aided by the development of high quality of LP solvers such as Robert Bixby's CPLEX code, making it much easier to experiment with combinatorial IP techniques. Bixby [13] describes this point as follows.

> What was needed was a numerically robust code that was also flexible enough to be embedded in these integer programming applications. It had to be a code that made it easy to handle the kinds of operations that arose in a context in which it was natural to begin with a model instantiated in one form followed by a sequence of problem modifications (such as row and column additions and deletions and variable fixings) interspersed with resolves. These needs were among the fundamental motivations behind the development of the callable-library version of the CPLEX code.

The growth and availability of such flexible LP codes went hand-in-hand with the expansion of the cutting-plane method.

Advancing the TSP

During these expansion years, the TSP was certainly not left behind. Indeed, two large-scale computational projects were initiated by Padberg and Giovanni Rinaldi [126, 127, 128] and by Grötschel and Olaf Holland [82, 60]. The important goal of these studies was to assess whether the performance of the cutting-plane method could be substantially enhanced by digging deeper into the polyhedral structure of problem classes and by considering more sophisticated computational tools available in branch-and-cut implementations. The spectacular success of the two studies demonstrates that this is indeed the case. Among the computational achievements were the solution of a 666-city world TSP instance by Grötschel and Holland and the solution of a 2,392-city circuit-board drilling problem by Padberg and Rinaldi. The term "branch-and-cut" was first used in the Padberg-Rinaldi study, and their work introduced important components that continue to be used in modern implementations of the solution scheme.

Fig. 12.21 Giovanni Rinaldi, Michele Conforti, Monique Laurent, M.R. Rao, and Manfred Padberg, 1985. Photograph courtesy of Monique Laurent.

12.8 Optimization ≡ Separation

The most widely circulated news event in the history of mathematical programming occurred in the summer of 1979, when Leonid Khachiyan [88] published a polynomial-time algorithm for solving linear-programming problems. The story was covered on the front page of the *New York Times* and in newspapers around the world. Khachiyan's work made use of the ellipsoid method for convex programming developed by David Yudin and Arkadi Nemirovski [144], and the papers [88] and [144] were jointly awarded a Fulkerson Prize in 1982.

In the introduction to the current paper, we took care to state that LP duality is used to obtain bounds on combinatorial problems, rather than assuming one could actually solve a given relaxation. With Khachiyan's result this is no longer an issue in theoretical work, since the ellipsoid method can deliver the required optimal LP value. The real power of the result goes well beyond this, however, as discovered by Martin Grötschel, László Lovász, and Alexander Schrijver [64], Richard Karp and Christos Papadimitriou [87], and Manfred Padberg and M. R. Rao [125]. Each of these teams showed that subject to some modest technical conditions, the ellipsoid

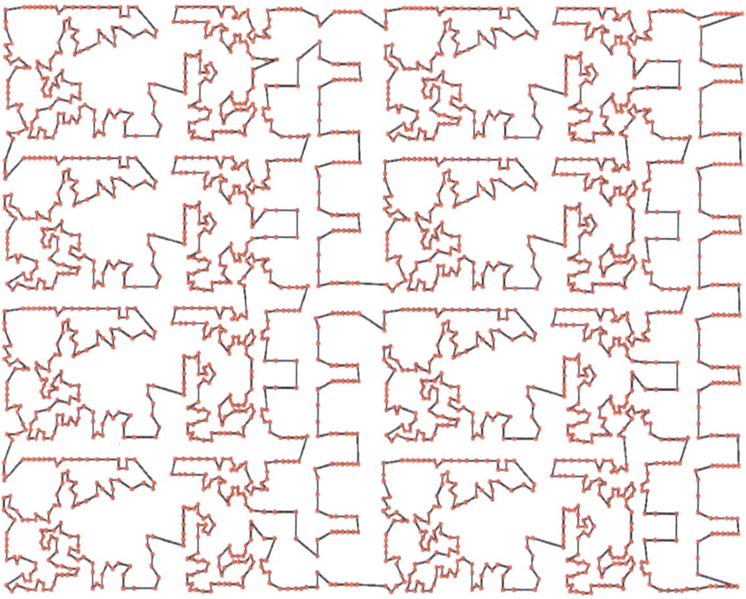

Fig. 12.22 Padberg and Rinaldi's 2,392-city TSP.

method can be used to prove that the problem of optimizing a linear function over a rational polyhedron P is polynomial-time equivalent to the separation problem for P, that is, given a vector $\bar{x}$, either show that $\bar{x} \in P$ or find a linear inequality $c^T x \leqslant \beta$ that is satisfied by all points in P but violated by $\bar{x}$. A particularly sharp version of this equivalence is derived in the Grötschel et al. paper [64], and it too was awarded a Fulkerson Prize in 1982.

The optimization $\equiv$ separation result has wide-reaching applications in combinatorial integer programming, giving a precise algorithmic realization of Jack Edmonds' appeal for linking polynomial-time algorithms and polyhedral descriptions. Recall the quote of Edmonds concerning a characterization of the TSP polytope: "finding a really good traveling salesman algorithm is undoubtedly equivalent to finding such a characterization." As with many other aspects of combinatorial optimization, Edmonds' insight was right on the money. The ellipsoid method tells us that what is needed to solve the TSP is a polyhedral characterization yielding a polynomial-time separation algorithm. The study of separation algorithms for classes of combinatorial problems, and for particular classes of inequalities, is now a standard part of the field, in both practical and theoretical research.

The optimization $\equiv$ separation paradigm is the central theme of a beautiful monograph *Geometric Algorithms and Combinatorial Optimization* by Grötschel et al. [66], published in 1988. This work intertwines combinatorics, polyhedra, and the geometry of numbers, to produce polynomial-time algorithms and deep insights into a host of combinatorial optimization problems. The monograph is on a very short list of must-read books for any student of integer programming.

Fig. 12.23 Alexander Schrijver, László Lovász, and Martin Grötschel, Amsterdam, 1991. Photograph courtesy of Martin Grötschel.

The discovery and elaboration of optimization ≡ separation is a crowning achievement in combinatorial integer programming and it might well be viewed as marking the end of the initial development phase of the field.

12.9 State of the art

The applied and theoretical work of the 1980s brought to fruition the early visions of Dantzig-Fulkerson-Johnson, Edmonds, Gomory, Hoffman, Kuhn, and others. The accomplishments of that decade set the stage for the now mature field of combinatorial integer programming, where deep theoretical questions and ever more complex practical computations drive the growth of the discipline.

At this point in the narrative we cannot hope to do justice to the wide range of activities being carried out by the research community, and we refer the reader to state-of-the-art surveys included in this volume. We limit our discussion to several highlights that are representative of the overall advancement of the field.

Balanced matrices and perfect graphs

Claude Berge, a pioneer in both graph theory and optimization, was the catalyst for two major results in the late 1990s and early 2000s. Both studies concern the integrality of polyhedra and are thus central to the theme of combinatorial integer programming.

The first of the two results is the 1999 publication of an algorithm for decomposing and recognizing balanced matrices by Michele Conforti, Gérard Cornuéjols, and M. R. Rao [22]. Berge [11, 12] introduced this class of $0/1$ matrices in the early 1970s, generalizing the notion of a bipartite graph. Matrix A is called *balanced* if it does not contain a square submatrix of odd order having exactly two ones in every row and exactly two ones in every column. If A is balanced, then both $Ax \leqslant \mathbf{1}, x \geqslant \mathbf{0}$ and $Ax \geqslant \mathbf{1}, x \geqslant \mathbf{0}$ are totally dual integral systems [50]. The definition of the class tells us which matrices are not balanced. The achievement of Conforti et al. was to answer the other natural question, showing which matrices are in fact balanced. Their study was awarded a Fulkerson Prize in 2000.

Fig. 12.24 W. Cunningham, A. Schrijver, M. Laurent, B. Gamble, F. B. Shepherd, D. Williamson, A. Hoffman, C. De Simone, D. Shmoys, J. Geelen, J. Kleinberg, S. Fekete, M. Goemans, M. Conforti, G. Cornuéjols, and A. Gerards. Bellairs Research Institute, March 1995. Photograph courtesy of David Williamson.

The second major result was the proof of Berge's [10] strong perfect-graph conjecture by Maria Chudnovsky, Neil Robertson, Paul Seymour, and Robin Thomas [16]. Perfection is defined in terms of the stable sets and cliques of a graph. A clique in $G = (V, E)$ is a set $C \subseteq V$ such that each pair of vertices in C is joined by an edge in E; the clique-covering number of G is the minimum number of cliques covering all vertices, that is, each $v \in V$ is a member of a least one of the cliques. If G has a stable set of cardinality k, then the clique-covering number of G must be at least k.

Graph G if called *perfect* if for every induced subgraph H of G the clique-covering number of H is equal to the cardinality of its largest stable set. It follows that if we let A denote the clique-vertex incidence matrix of G, then G is perfect if and only if for each $0/1$ vector $w = (w_v : v \in V)$ the optimal values in the LP duality equation

$$\max(w^T x : Ax \leqslant \mathbf{1}, x \geqslant \mathbf{0}) = \min(y^T \mathbf{1} : y^T A \geqslant w^T, y \geqslant \mathbf{0})$$

are attained by $0/1$ vectors. This is just a reinterpretation of the definition, but it hints at the result that G is perfect if and only if $\{x : Ax \leqslant \mathbf{1}, x \geqslant \mathbf{0}\}$ is an integer polyhedron [49, 105, 19]. This close connection with integer programming is one of the drivers of the great interest in the study of perfection.

The strong perfect-graph conjecture states that a graph is perfect if and only if neither it or its complement contains as an induced subgraph an odd circuit having at least five edges. The simplicity of this possible characterization drew considerable attention in the forty years after Berge proposed the problem in 1960, leading to a great body of work in the literature. This research culminated in the Chudnovsky et al. proof, announced in May 2002, just one month before Berge passed away. Their published version runs 179 pages in the *Annals of Mathematics* and it is one of the great achievements in graph theory and polyhedral combinatorics. An interesting account of the steps and missteps along the way to the final proof can be found in Seymour [140].

Fig. 12.25 Robin Thomas, Paul Seymour, Neil Robertson and Maria Chudnovsky signing copies of their proof of the strong perfect graph conjecture, November 1, 2002. Photograph courtesy of the American Institute of Mathematics.

With the conjecture now a theorem, the characterization was used to obtain a polynomial-time recognition algorithm for perfect graphs [15]. Thus three important classes of matrices, totally unimodular, balanced, and perfect, all are recognizable in polynomial time.

This area of polyhedral combinatorics is well-developed, but many interesting open questions remain. A nice treatment can be found in the book of Cornuéjols [23], including offers of cash rewards of $5,000 for each of eighteen conjectures. Of this potential $90,000, the teams Chudnovsky-Seymour and Chudnovsky-Robertson-Seymour-Thomas have thus far collected $25,000, leaving plenty of money on the table for future work.

Semidefinite programming

One of the important applications of the ellipsoid method in the Grötschel et al. [64, 65, 66] study is the development of a polynomial-time algorithm to find a maximum-weight stable set in a perfect graph. Building on earlier work of Lovász [107] concerning the Shannon capacity of graphs, Grötschel et al. consider a convex relaxation of the stable-set problem involving a matrix of variables that must be symmetric and positive semidefinite. The optimization $\equiv$ separation theory provides a polynomial-time algorithm to optimize over this non-polyhedral set, and in the case of perfect graphs the relaxation coincides with the stable-set polytope.

The semidefinite-relaxation idea was further developed by Lovász and Schrijver [108] in a hierarchy of relaxations for $0/1$ integer-programming problems. This framework was shown to have particularly interesting consequences for the study of stable-set polytopes beyond the class of perfect graphs.

The Lovász-Schrijver study, in turn, generated interest in the model of semidefinite programming (SDP). Here linear programming is extended by replacing the standard vector of variables by a symmetric and positive semidefinite matrix. The interest in SDP models was heightened by two additional developments in the early 1990s. First, Farid Alizadeh [1] and Yurii Nesterov and Arkadi Nemirovski [120, 121] showed that LP interior-point methods could be extended to semidefinite programming, providing an efficient practical tool for solving these models. A nice description of this work can be found in Todd [141]. Second, Michel Goemans and David Williamson [53] utilized SDP relaxations in their breakthrough result on the max-cut problem in graphs, yielding a strong new approximation algorithm; their result was awarded a Fulkerson Prize in 2000.

With these applications and algorithms in place, the past decade has seen the SDP area grow by leaps and bounds. Henry Wolkowicz [142] lists 1,060 references in an online SDP bibliography; IP-related SDP work is covered in surveys by Christoph Helmberg [73] and Monique Laurent and Franz Rendl [100]. SDP methods are now an exciting tool in the study of approximation algorithms and in the study of lower bounds for combinatorial optimization problems.

Schrijver's Meisterwerk

A milestone in combinatorial integer programming was reached in 2003 with the publication of Alexander Schrijver's three-volume monograph *Combinatorial Optimization: Polyhedra and Efficiency* [135]. The breadth and depth of coverage in the monograph is breathtaking, as are Schrijver's historical treatments. The volumes total 1,881 pages and include over 4,000 references. The importance of this work can hardly be overstated. Schrijver's beautiful scholarly writing has defined the field, giving combinatorial optimization an equal footing with much older, more established areas of applied mathematics. His monograph received the 2004 Lanchester Prize.

The following quote from Schrijver's preface emphasizes the role of polyhedral methods in the broad study of combinatorial optimization.

> Pioneered by the work of Jack Edmonds, polyhedral combinatorics has proved to be a most powerful, coherent, and unifying tool throughout combinatorial optimization. Not only has it led to efficient (that is, polynomial-time) algorithms, but also, conversely, efficient algorithms often imply polyhedral characterizations and related min-max theorems. It makes the two sides closely intertwined.

This connection will undoubtedly continue, advancing both combinatorial optimization and general integer programming.

Fig. 12.26 Alexander Schrijver, 2007. Photograph copyright Wim Klerkx, Hollandse Hoogte.

References

1. F. Alizadeh, *Interior point methods in semidefinite programming with applications to combinatorial optimization*, SIAM Journal on Optimization 5 (1995) 13–51.
2. American Mathematical Society, The Sixth Symposium in Applied Mathematics, Bulletin of the American Mathematical Society 59 (1953) 513–514.
3. American Mathematical Society, The Summer Meeting in Kingston, Bulletin of the American Mathematical Society 59 (1953) 515–568.
4. E. Balas and M.W. Padberg, *On the set-covering problem*, Operations Research 20 (1972) 1152–1161.
5. E. Balas and M.W. Padberg, *On the set-covering problem: II. An algorithm for set partitioning*, Operations Research 23 (1975) 74–90.
6. R. Bellman, *An introduction to the theory of dynamic programming*, Research Report R-245, RAND Corporation, Santa Monica, California, USA, 1953.
7. R. Bellman, *Dynamic Programming*, Princeton University Press, Princeton, New Jersey, USA, 1957.
8. R. Bellman, *Combinatorial processes and dynamic programming*, Combinatorial Analysis (R. Bellman and M. Hall, Jr., eds.), American Mathematical Society, Providence, Rhode Island, USA, 1960, pp. 217–249.
9. R. Bellman, *Dynamic programming treatment of the travelling salesman problem*, Journal of the Association for Computing Machinery 9 (1962) 61–63.
10. C. Berge, *Färbung von Graphen, deren sämtliche bzw. deren ungerade Kreise starr sind (Zusammenfassung)*, Wiss. Z. Martin-Luther-Univ. Halle-Wittenberg Math.-Natur. Reihe 10 (1961) 114–115.
11. C. Berge, *Sur certains hypergraphes généralisant les graphes bipartis*, Combinatorial Theory and its Applications I (P. Erdős, A. Rẽnyi, and V. Sós, eds.), Colloq. Math. Soc. János Bolyai 4 (1970) 119–133.
12. C. Berge, *Balanced matrices*, Mathematical Programming 2 (1972) 19–31.
13. R.E. Bixby, *Solving real-world linear programs: A decade and more of progress*, Operations Research 50 (2002) 3–15.
14. F. Bock, *An algorithm for solving "traveling-salesman" and related network optimization problems*, Research Report, Armour Research Foundation, Presented at the Operations Research Society of America Fourteenth National Meeting, St. Louis, October 24, 1958.
15. M. Chudnovsky, G. Cornuéjols, X. Liu, P.D. Seymour, and K. Vušković, *Recognizing Berge graphs*, Combinatorica 25 (2005) 143–186.
16. M. Chudnovsky, N. Robertson, P. Seymour, and R. Thomas, *The strong perfect graph theorem*, Annals of Mathematics 164 (2006) 51–229.
17. V. Chvátal, *Edmonds polytopes and a hierarchy of combinatorial problems*, Discrete Mathematics 4 (1973) 305–337.
18. V. Chvátal, *Edmonds polytopes and weakly hamiltonian graphs*, Mathematical Programming 5 (1973) 29–40.
19. V. Chvátal, *On certain polyhedra associated with graphs*, Journal of Combinatorial Theory B 18 (1975) 138–154.
20. V. Chvátal, *Cutting planes in combinatorics*, European Journal of Combinatorics 6 (1985) 217–226.
21. V. Chvátal, W. Cook, and M. Hartmann, *On cutting-plane proofs in combinatorial optimization*, Linear Algebra and Its Applications 114/115 (1989) 455–499.
22. M. Conforti, G. Cornuéjols, and M.R. Rao, *Decomposition of balanced matrices*, Journal of Combinatorial Theory, Series B 77 (1999) 292–406.
23. G. Cornuéjols, *Combinatorial Optimization: Packing and Covering*, SIAM, Philadelphia, USA,2001.
24. G.A. Croes, *A method for solving traveling-salesman problems*, Operations Research 6 (1958) 791–812.

25. H. Crowder and M.W. Padberg, *Solving large-scale symmetric travelling salesman problems to optimality*, Management Science 26 (1980) 495–509.
26. W.H. Cunningham and A.B. Marsh, III, *A primal algorithm for optimum matching*, Mathematical Programming Study 8 (1978) 50–72.
27. G.B. Dantzig, *Discrete-variable extremum problems*, Operations Research 5 (1957) 266–277.
28. G.B. Dantzig, D.R. Fulkerson, and S.M. Johnson, *Solution of a large scale traveling salesman problem*, Technical Report P-510, RAND Corporation, Santa Monica, California, USA, 1954.
29. G.B. Dantzig, D.R. Fulkerson, and S.M. Johnson, *Solution of a large-scale traveling-salesman problem*, Operations Research 2 (1954) 393–410.
30. G.B. Dantzig, D.R. Fulkerson, and S.M. Johnson, *On a linear-programming, combinatorial approach to the traveling-salesman problem*, Operations Research 7 (1959) 58–66.
31. G.B. Dantzig and A.J. Hoffman, *Dilworth's theorem on partially ordered sets*, Linear Inequalities and Related Systems (H.W. Kuhn and A.W. Tucker, eds.), Princeton University Press, Princeton, New Jersey, USA, 1956, pp. 207–214.
32. W.L. Eastman, *Linear Programming with Pattern Constraints*, Ph.D. Thesis, Department of Economics, Harvard University, Cambridge, Massachusetts, USA, 1958.
33. J. Edmonds, *Paths, trees, and flowers*, Working paper, National Bureau of Standards and Princeton University, February 1963.
34. J. Edmonds, *Maximum matching and a polyhedron with 0,1-vertices*, Journal of Research of the National Bureau of Standards 69B (1965) 125–130.
35. J. Edmonds, *Paths, trees, and flowers*, Canadian Journal of Mathematics 17 (1965) 449–467.
36. J. Edmonds, *The Chinese postman's problem*, Bulletin of the Operations Research Society of America 13 (1965) B-73.
37. J. Edmonds, *Optimum branchings*, Journal of Research National Bureau of Standards 71B (1967) 233–240.
38. J. Edmonds, *Submodular functions, matroids, and certain polyhedra*, Combinatorial Structures and Their Applications (R. Guy, H. Hanani, N. Sauer, and J. Schönheim, eds.), Gordon and Breach, New York, USA, 1970, pp. 69–87.
39. J. Edmonds, *Matroids and the greedy algorithm*, Mathematical Programming 1 (1971) 127–136.
40. J. Edmonds, *Matroid intersection*, Discrete Optimization I (P.L. Hammer, E.L. Johnson, and B.H. Korte, eds.), North-Holland, 1979, pp. 39–49.
41. J. Edmonds and R. Giles, *A min-max relation for submodular functions on a graph*, Studies in Integer Programming (P.L. Hammer, E.L. Johnson, B.H. Korte, and G.L. Nemhauser, eds.), Annals of Discrete Mathematics 1 (1977) 185–204.
42. J. Edmonds and E.L. Johnson, *Matching, a well-solved class of integer linear programs*, Combinatorial Structures and Their Applications (R. Guy, H. Hanani, N. Sauer, and J. Schönheim, eds.), Gordon and Breach, New York, USA, 1970, pp. 89–92.
43. J. Edmonds and E.L. Johnson, *Matching, Euler tours, and the Chinese postman*, Mathematical Programming 5 (1973) 88–124.
44. J. Egerváry, *Matrixok kombinatorius tulajonságairól*, Matematikai és Fizikai Lapok 38 (1931) 16–28.
45. M.M. Flood, Merrill Flood (with Albert Tucker), Interview of Merrill Flood in San Francisco on 14 May 1984, The Princeton Mathematics Community in the 1930s, Transcript Number 11 (PMC11), Princeton University. (1984).
46. L.R. Ford, Jr. and D.R. Fulkerson, *Maximal flow through a network*, Canadian Journal of Mathematics 8 (1956) 399–404.
47. A. Frank, *Kernel systems of directed graphs*, Acta Scientiarum Mathematicarum [Szeged] 41 (1979) 63–76.
48. D.R. Fulkerson, *Blocking and anti-blocking pairs of polyhedra*, Mathematical Programming 1 (1971) 168–194.
49. D.R. Fulkerson, *Anti-blocking polyhedra*, Journal of Combinatorial Theory, Series B 12 (1972) 50–71.

50. D.R. Fulkerson, A.J. Hoffman, and R. Oppenheim, *On balanced matrices*, Mathematical Programming Study 1 (1974) 120–132.
51. D. Gale, *A theorem on flows in networks*, Pacific Journal of Mathematics 7 (1957) 1073–1082.
52. F.R. Giles and W.R. Pulleyblank, *Total dual integrality and integer polyhedra*, Linear Algebra and Its Applications 25 (1979) 191–196.
53. M. Goemans and D. Williamson, *Improved approximation algorithms for maximum cut and satisfiability problems*, Journal of the Association of Computing Machinery 42 (1995) 1115–1145.
54. R.E. Gomory, *Outline of an algorithm for integer solutions to linear programs*, Bulletin of the American Mathematical Society 64 (1958) 275–278.
55. R.E. Gomory, *The traveling salesman problem*, Proceedings of the IBM Scientific Computing Symposium on Combinatorial Problems, IBM, White Plains, New York, USA, 1966, pp. 93–121.
56. R.E. Gomory, *Early integer programming*, History of Mathematical Programming—A Collection of Personal Reminiscences (J.K. Lenstra, A.H.G. Rinnooy Kan, and A. Schrijver, eds.), North-Holland, 1991, pp. 55–61.
57. R.H. Gonzales, *Solution to the traveling salesman problem by dynamic programming on the hypercube*, Technical Report Number 18, Operations Research Center, Massachusetts Institute of Technology, Cambridge, Massachusetts, USA, 1962.
58. M. Grötschel, *Polyedrische Charakterisierungen kombinatorischer Optimierungsprobleme*, Anton Hain Verlag, Meisenheim/Glan, Germany, 1977.
59. M. Grötschel, *On the symmetric travelling salesman problem: Solution of a 120-city problem*, Mathematical Programming Study 12 (1980) 61–77.
60. M. Grötschel and O. Holland, *Solution of large-scale symmetric travelling salesman problems*, Mathematical Programming 51 (1991) 141–202.
61. M. Grötschel, M. Jünger, and G. Reinelt, *A cutting plane algorithm for the linear ordering problem*, Operations Research 32 (1984) 1195–1220.
62. M. Grötschel, M. Jünger, and G. Reinelt, *On the acyclic subgraph polytope*, Mathematical Programming 33 (1985) 28–42.
63. M. Grötschel, M. Jünger, and G. Reinelt, *Facets of the linear ordering polytope*, Mathematical Programming 33 (1985) 43–60.
64. M. Grötschel, L. Lovász, and A. Schrijver, *The ellipsoid method and its consequences in combinatorial optimization*, Combinatorica 1 (1981) 169–197.
65. M. Grötschel, L. Lovász, and A. Schrijver, *Polynomial algorithms for perfect graphs*, Topics on Perfect Graphs (C. Berge and V. Chvátal, eds.), Annals of Discrete Mathematics 21 (1984) 325–356.
66. M. Grötschel, L. Lovász, and A. Schrijver, *Geometric Algorithms and Combinatorial Optimization*, Springer. (1988).
67. M. Grötschel and G.L. Nemhauser, *George Dantzig's contributions to integer programming*, Discrete Optimization 5 (2008) 168–173.
68. M. Grötschel and W.R. Pulleyblank, *Clique tree inequalities and the symmetric travelling salesman problem*, Mathematics of Operations Research 11 (1986) 537–569.
69. M. Held and R.M. Karp, *A dynamic programming approach to sequencing problems*, Journal of the Society of Industrial and Applied Mathematics 10 (1962) 196–210.
70. M. Held and R.M. Karp, *The traveling-salesman problem and minimum spanning trees*, Operations Research 18 (1970) 1138–1162.
71. M. Held and R.M. Karp, *The traveling-salesman problem and minimum spanning trees: Part II*, Mathematical Programming 1 (1971) 6–25.
72. I. Heller, *On the problem of the shortest path between points*, I. Abstract 664t, Bulletin of the American Mathematical Society 59 (1953) 551.
73. C. Helmberg, *Semidefinite Programming for Combinatorial Optimization*, Habilitationsschrift, TU Berlin, ZIB-Report ZR-00-34, Konrad-Zuse-Zentrum Berlin, 2000.
74. A.J. Hoffman, *Generalization of a theorem of König*, Journal of the Washington Academy of Sciences 46 (1956) 211–212.

75. A.J. Hoffman, *Linear programming*, Applied Mechanics Reviews 9 (1956) 185–187.
76. A.J. Hoffman, *A generalization of max flow-min cut*, Mathematical Programming 6 (1974) 352–359.
77. A.J. Hoffman and J.B. Kruskal, *Integral boundary points of convex polyhedra*, Linear Inequalities and Related Systems (H.W. Kuhn and A.W. Tucker, eds.), Princeton University Press, Princeton, New Jersey, USA, 1956, pp. 223–246.
78. A.J. Hoffman and H.W. Kuhn, *Systems of distinct representatives and linear programming*, The American Mathematical Monthly 63 (1956) 455–460.
79. A.J. Hoffman and R. Oppenheim, *Local unimodularity in the matching polytope*, Algorithmic Aspects of Combinatorics (B.R. Alspach, P. Hell, and D.J. Miller, eds.), North-Holland, 1978, pp. 201–209.
80. A.J. Hoffman and D.E. Schwartz, *On lattice polyhedra*, Combinatorics Volume I (A. Hajnal and V.T. Sós, eds.), North-Holland, 1978, pp. 593–598.
81. A.J. Hoffman and P. Wolfe, *History*, The Traveling Salesman Problem (E.L. Lawler, J.K. Lenstra, A.H.G. Rinnooy Kan, and D.B. Shmoys, eds.), John Wiley & Sons, Chichester, UK, 1985, pp. 1–15.
82. O.A. Holland, *Schnittebenenverfahren für Travelling-Salesman und verwandte Probleme*, Ph.D. Thesis, Universität Bonn, Bonn, Germany, 1987.
83. S. Hong, *A Linear Programming Approach for the Traveling Salesman Problem*, Ph.D. Thesis, Johns Hopkins University, Baltimore, Maryland, USA, 1972.
84. M. Jünger, *Polyhedral Combinatorics and the Acyclic Subdigraph Problem*, Heldermann Verlag, Berlin, Germany, 1985.
85. M. Jünger, G. Reinelt, and S. Thienel, *Practical problem solving with cutting plane algorithms*, Combinatorial Optimization (W. Cook, L. Lovász, and P. Seymour, eds.), DIMACS Series in Discrete Mathematics and Theoretical Computer Science 20, American Mathematical Society, Providence, Rhode Island, USA, 1995, pp. 111–152.
86. R.M. Karp, *Combinatorics, complexity, and randomness*, Communications of the ACM 29 (1986) 98–109.
87. R.M. Karp and C.H. Papadimitriou, *On linear characterization of combinatorial optimization problems*, SIAM Journal on Computing 11 (1982) 620–632.
88. L.G. Khachiyan, *A polynomial algorithm in linear programming*, Soviet Mathematics Doklady 20 (1979) 191–194.
89. B. Korte and W. Oberhofer, *Zwei Algorithmen zur Lösung eines komplexen Reihenfolgeproblems*, Mathematical Methods of Operations Research 12 (1968) 217–231.
90. B. Korte and W. Oberhofer, *Zur Triangulation von Input-Output-Matrizen*, Jahrbücher für Nationalökonomie und Statistik 182 (1969) 398–433.
91. H.W. Kuhn, *On certain convex polyhedra*, Abstract 799t, Bulletin of the American Mathematical Society 61 (1955) 557–558.
92. H.W. Kuhn, *The Hungarian method for the assignment problem*, Naval Research Logistics Quarterly 2 (1955) 83–97.
93. H.W. Kuhn, *On the origin of the Hungarian method*, History of Mathematical Programming—A Collection of Personal Reminiscences (J.K. Lenstra, A.H.G. Rinnooy Kan, and A. Schrijver, eds.), North-Holland, 1991, pp. 77–81.
94. H.W. Kuhn, *57 years of close encounters with George*, George Dantzig Memorial Site, INFORMS, 2008, available at http://www2.informs.org/History/dantzig/articles_kuhn.html.
95. H.W. Kuhn, Email letter sent on December 15, 2008.
96. H.W. Kuhn and A.W. Tucker, eds. *Linear Inequalities and Related Systems*, Princeton University Press, Princeton, New Jersey, USA, 1956.
97. F. Lambert, *The traveling-salesman problem*, Cahiers du Centre de Recherche Opérationelle 2 (1960) 180–191.
98. A.H. Land and A.G. Doig, *An automatic method of solving discrete programming problems*, Econometrica 28 (1960) 497–520.
99. A.H. Land and S. Powell, *A survey of the operational use of ILP models*, History of Integer Programming: Distinguished Personal Notes and Reminiscences (K. Spielberg and M. Guignard-Spielberg, eds.), Annals of Operations Research 149 (2007) 147–156.

100. M. Laurent and F. Rendl, *Semidefinite programming and integer programming*, Handbook on Discrete Optimization (K. Aardal, G.L. Nemhauser, and R. Weismantel, eds.), Elsevier, 2005, pp. 393–514.
101. J.K. Lenstra, *Sequencing by Enumerative Methods*, Mathematical Centre Tracts 69, Mathematisch Centrum, Amsterdam, 1977.
102. A.N. Letchford and A. Lodi, *Primal cutting plane algorithms revisited*, Mathematical Methods of Operations Research 56 (2002) 67–81.
103. S. Lin and B.W. Kernighan, *An effective heuristic algorithm for the traveling-salesman problem*, Operations Research 21 (1973) 498–516.
104. J.D. Little, K.G. Murty, D.W. Sweeney, and C. Karel, *An algorithm for the traveling salesman problem*, Operations Research 11 (1963) 972–989.
105. L. Lovász, *A characterization of perfect graphs*, Journal of Combinatorial Theory, Series B 13 (1972) 95–98.
106. L. Lovász, *Normal hypergraphs and the perfect graph conjecture*, Discrete Mathematics 2 (1972) 253–267.
107. L. Lovász, *On the Shannon capacity of graphs*, IEEE Transactions on Information Theory 25 (1979) 1–7.
108. L. Lovász and A. Schrijver, *Cones of matrices and set-functions, and 0-1 optimization*, SIAM Journal on Optimization 1 (1991) 166–190.
109. C.L. Lucchesi and D.H. Younger, *A minimax theorem for directed graphs*, The Journal of the London Mathematical Society 17 (1978) 369–374.
110. A.S. Manne and H.M. Markowitz, *On the solution of discrete programming problems*, Technical Report P-711, RAND Corporation, Santa Monica, California, USA, 1956.
111. H.M. Markowitz and A.S. Manne, *On the solution of discrete programming problems*, Econometrica 25 (1957) 84–110.
112. G.T. Martin, *Solving the traveling salesman problem by integer linear programming*, Technical Report, C-E-I-R, New York, USA, 1966.
113. K. Menger, *Bericht über ein mathematisches Kolloquium*, Monatshefte für Mathematik und Physik 38 (1931) 17–38.
114. C.A. Micchelli, *Selected Papers of Alan Hoffman with Commentary*, World Scientific Publishing Company, 2003.
115. P. Miliotis, *Using cutting planes to solve the symmetric travelling salesman problem*, Mathematical Programming 15 (1978) 177–188.
116. C.E. Miller, A.W. Tucker, and R.A. Zemlin, *Integer programming formulation of traveling salesman problems*, Journal of the Association for Computing Machinery 7 (1960) 326–329.
117. J. Munkres, *Algorithms for the assignment and transportation problems*, Journal of the Society for Industrial and Applied Mathematics 5 (1957) 32–33.
118. G.L. Nemhauser and L.E. Trotter, Jr., *Properties of vertex packing and independence system polyhedra*, Mathematical Programming 6 (1974) 48–61.
119. G.L. Nemhauser and L.E. Trotter, Jr., *Vertex packings: Structural properties and algorithms*, Mathematical Programming 8 (1975) 232–248.
120. Y. Nesterov and A. Nemirovski, *Conic formulation of a convex programming problem and duality*, Optimization Methods and Software 1 (1992) 95–115.
121. Y. Nesterov and A. Nemirovski, *Interior Point Polynomial Algorithms in Convex Programming*, SIAM, Philadelphia, USA, 1994.
122. M.W. Padberg, *On the facial structure of set packing polyhedra*, Mathematical Programming 5 (1973) 199–215.
123. M.W. Padberg, *Mixed-integer programming — 1968 and thereafter*, History of Integer Programming: Distinguished Personal Notes and Reminiscences (K. Spielberg and M. Guignard-Spielberg, eds.), Annals of Operations Research 149 (2007) 147–156.
124. M.W. Padberg and S. Hong, *On the symmetric travelling salesman problem: A computational study*, Mathematical Programming Study 12 (1980) 78–107.
125. M.W. Padberg and M.R. Rao, *The Russian method for linear programming III: Bounded integer programming*, Research Report 81-39, New York University, Graduate School of Business Administration, 1981.

126. M.W. Padberg and G. Rinaldi, *Optimization of a 532-city symmetric traveling salesman problem by branch and cut*, Operations Research Letters 6 (1987) 1–7.
127. M.W. Padberg and G. Rinaldi, *Facet identification for the symmetric traveling salesman polytope*, Mathematical Programming 47 (1990) 219–257.
128. M.W. Padberg and G. Rinaldi, *A branch-and-cut algorithm for the resolution of large-scale symmetric traveling salesman problems*, SIAM Review 33 (1991) 60–100.
129. W.R. Pulleyblank, *Faces of Matching Polyhedra*, Ph.D. Thesis, Department of Combinatorics and Optimization, University of Waterloo, Waterloo, Ontario, Canada, 1973.
130. W.R. Pulleyblank and J. Edmonds, *Facets of 1-matching polyhedra*, Hypergraph Seminar (C. Berge and D. Ray-Chaudhuri, eds.), Springer, 1974, pp. 214–242.
131. G. Reinelt, *The Linear Ordering Problem: Algorithms and Applications*, Heldermann Verlag, Berlin, Germany, 1985.
132. J. Robinson, *On the Hamiltonian game (a traveling salesman problem)*, Research Memorandum RM-303, RAND Corporation, Santa Monica, California, USA, 1949.
133. A. Schrijver, *On cutting planes*, Combinatorics 79 Part II (M. Deza and I.G. Rosenberg, eds.), Annals of Discrete Mathematics 9, North-Holland, 1980, pp. 291–296.
134. A. Schrijver, *On total dual integrality*, Linear Algebra and Its Applications 38 (1981) 27–32.
135. A. Schrijver, *Combinatorial Optimization: Polyhedra and Efficiency*, Springer, Berlin, Germany, 2003.
136. A. Schrijver, *On the history of combinatorial optimization (till 1960)*, Handbook of Discrete Optimization (K. Aardal, G.L. Nemhauser, and R. Weismantel, eds.), Elsevier, 2005, pp. 1–68.
137. A. Schrijver and P.D. Seymour, *A proof of total dual integrality of matching polyhedra*, Report ZN 79/77, Stichting Mathematisch Centrum, Amsterdam, 1977.
138. P.D. Seymour, *The matroids with the max-flow min-cut property*, Journal of Combinatorial Theory, Series B 23 (1977) 289–295.
139. P.D. Seymour, *Decomposition of regular matroids*, Journal of Combinatorial Theory, Series B 28 (1980) 305–359.
140. P.D. Seymour, *How the proof of the strong perfect graph conjecture was found*, Gazette des Mathematiciens 109 (2006) 69–83.
141. M. Todd, *Semidefinite optimization*, Acta Numerica 10 (2001) 515–560.
142. H. Wolkowicz, *Semidefinite and cone programming bibliography/comments/abstracts*, 2008, see http://orion.uwaterloo.ca/~hwolkowi/henry/book/fronthandbk.d/handbooksdp.html.
143. L.A. Wolsey, *Further facet generating procedures for vertex packing polytopes*, Mathematical Programming 11 (1976) 158–163.
144. D.B. Yudin and A S. Nemirovski, *Evaluation of the informational complexity of mathematical programming problems*, Ékonomica i Matematicheskie Metody 12 (1976) 128–142.

Chapter 13
Reformulation and Decomposition of Integer Programs

François Vanderbeck and Laurence A. Wolsey

Abstract We examine ways to reformulate integer and mixed integer programs. Typically, but not exclusively, one reformulates so as to obtain stronger linear programming relaxations, and hence better bounds for use in a branch-and-bound based algorithm. First we cover reformulations based on decomposition, such as Lagrangean relaxation, the Dantzig-Wolfe reformulation and the resulting column generation and branch-and-price algorithms. This is followed by an examination of Benders' type algorithms based on projection. Finally we discuss extended formulations involving additional variables that are based on problem structure. These can often be used to provide strengthened a priori formulations. Reformulations obtained by adding cutting planes in the original variables are not treated here.

13.1 Introduction

Integer linear programs (IPs) and mixed integer linear programs (MIPs) are often difficult to solve, even though the state-of-the-art mixed integer programming solvers are in many cases remarkably effective, and have improved radically in the last ten years. These solvers typically use branch-and-cut involving cutting planes to obtain improved linear programming bounds and branching to carry out implicit enumeration of the solutions. However these systems essentially ignore problem structure.

The goal in this chapter is to show the numerous ways in which, given an initial formulation of an IP, *problem structure* can be used to obtain improved prob-

François Vanderbeck
Institut de Mathématiques de Bordeaux (IMB) and INRIA, Université de Bordeaux, France
e-mail: `fv@math.u-bordeaux1.fr`

Laurence A. Wolsey
Center for Operations Research and Econometrics, Université Catholique de Louvain, Belgium
e-mail: `laurence.wolsey@uclouvain.be`

M. Jünger et al. (eds.), *50 Years of Integer Programming 1958-2008*,
DOI 10.1007/978-3-540-68279-0_13,

lem formulations and more effective algorithms that take the structure into account. One common way to obtain reformulations is by adding valid inequalities (cutting planes) in the original variables. This topic is treated in considerable detail in Chapter 11. Here we consider other possibilities. The general motivation is to obtain a reformulation for which the optimal value of the linear programming relaxation is closer to the optimal value of the IP than that of the original formulation and that is computationally tractable.

One approach is to introduce new variables so as to better model the structure of the problem—the resulting *extended formulations* will be studied in detail. Introducing new variables typically permits one to model some combinatorial structure more precisely and to induce integrality through tighter linear constraints linking the variables. One such extended formulation is provided by the classical Minkowski representation of a polyhedron in terms of its extreme points and extreme rays. An alternative is to develop reformulations based on projection onto a subset of the variables, based on Farkas' lemma and/or Fourier-Motzkin elimination. Projection allows one to reduce the number of variables so that calculations are typically faster: thus for a mixed integer program one might project onto the integer variables, and for an extended formulation giving an improved bound one might project so as to obtain the tightened bound while working in the space of the original variables.

There are also other reasons leading us to look at alternative formulations. One might be to treat or eliminate symmetry among solutions (see Chapter 17), another might be to obtain variables that are more effective as branching variables, or variables for which one can develop effective valid inequalities.

Reformulations often rely on a decomposition of the problem. Given a hard integer program (IP) in the form

$$\min\{cx : x \in X\} \quad \text{where } X = \{x \in \mathbb{Z}^n_+ : Ax \geqslant a\},$$

one typical way to obtain a set with structure is to *decompose* X into two (or more) sets $X = Y \cap Z$, where one or both of the sets Y, Z has *structure* and is a candidate for reformulation. In addition reformulations often require specific solution methods: the reformulation may involve a very large number of variables and/or constraints, in which case it becomes necessary to develop algorithms that treat the corresponding columns or rows implicitly, Dantzig-Wolfe decomposition and Benders' decomposition being the two classical examples.

The contents of this chapter are as follows. In Section 13.2 we introduce the different concepts used later. We give definitions and simple examples of polyhedra, formulations, extended formulations and reformulations obtained by projection. We discuss how decomposition can be used to obtain simpler sets, and what we mean by a set with structure.

In Section 13.3 we consider reformulations that are appropriate when the optimization problem over a "simpler" set Z, obtained by dropping some "hard" constraints, is relatively easy to solve. In particular we consider the Lagrangean dual approach to obtain tight bounds and related algorithms, and the Dantzig-Wolfe reformulation whose linear programming relaxation gives an identical bound. The basic column generation algorithm to solve the linear programming relaxation of the Dantzig-Wolfe reformulation is presented, as well as its integration into a branch-and-bound algorithm to solve the integer problem. In Section 13.4 we consider formulations and algorithms based on projection, in particular Benders' reformulation. Projection typically leads to formulations with a very large number of constraints, so here the algorithms rely on cut generation.

The reformulations in Sections 13.3 and 13.4 are generic. In Section 13.5 we consider sets with more structure for which it is possible to obtain interesting extended formulations. In many cases optimization over the sets is polynomially solvable. We show extended formulations a) based on variable splitting such as the multi-commodity reformulation of single source fixed charge network flow problems, b) for sets over which one can optimize by dynamic programming, c) for sets in the form of disjunctions, and d) for a variety of other sets with structure.

In Section 13.6 we discuss hybrid reformulations and algorithms; for example if $X = Y \cap Z$ and both sets have some special structure, we might wish to combine a (large) extended formulation for Y with a (large) cutting plane description for Z. Section 13.7 consists of historical notes as well as a few references concerning recent theoretical and computational developments.

13.2 Polyhedra, reformulation and decomposition

13.2.1 Introduction

Given a problem that has been formulated as a linear integer program, we are interested in finding reformulations (alternative problem descriptions) that are more effective in one way or another. We present some basic results about polyhedra, and give definitions of formulations and extended formulations, with a couple of examples to show how reformulations arise. Finally we discuss how decomposition leads one to simpler subsets, and indicate how their structure can be exploited to provide reformulations and possibly specialized algorithms.

Throughout we assume that our objective is to solve the integer program

$$\text{(IP)} \qquad \min\{cx : x \in X\}$$

where $X \subseteq \mathbb{Z}^n$ is a discrete solution set that can be modeled as the set of integer points satisfying a set of linear inequalities

$$X = P \cap \mathbb{Z}^n \text{ with } P = \{x \in \mathbb{R}^n_+ : Ax \geqslant a\},$$

or the mixed integer program

$$\text{(MIP)} \qquad \min\{cx + hy : (x,y) \in X^M\}$$

where $X^M \subseteq \mathbb{Z}^n \times \mathbb{R}^p$ is given in the form

$$X^M = P^M \cap (\mathbb{Z}^n \times \mathbb{R}^p) \text{ with } P^M = \{(x,y) \in \mathbb{R}^n_+ \times \mathbb{R}^p_+ : Gx + Hy \geqslant b\}.$$

P and P^M will be referred to as the initial formulations of X and X^M respectively. For simplicity, results are presented for the integer set X, unless the presence of continuous variables y is important.

13.2.2 Polyhedra and reformulation

Here we study the feasible solutions sets X and X^M arising in IP and MIP respectively. Throughout we will use the term *reformulation* informally to mean any alternative description of problems IP or MIP.

Definition 13.1. A *polyhedron* $P \subseteq \mathbb{R}^n$ is the intersection of a finite number of half-spaces. In other words there exists $A \in \mathbb{R}^{m \times n}$ and $a \in \mathbb{R}^m$ such that $P = \{x \in \mathbb{R}^n : Ax \geqslant a\}$.

Definition 13.2. A polyhedron P is a *formulation* for X if $X = P \cap \mathbb{Z}^n$.

Sets such as X have many formulations. If P^1, P^2 are two formulations for X with $P^1 \subset P^2$, we say that P^1 is a *stronger* formulation than P^2 because

$$z(c) = \min\{cx : x \in X\} \geqslant \min\{cx : x \in P^1\} \geqslant \min\{cx : x \in P^2\} \quad \forall c \in \mathbb{R}^n$$

and thus the lower bound on $z(c)$ provided by the linear programming relaxation with formulation P^1 is always greater than or equal to that provided by P^2.

Definition 13.3. Given $X \subseteq \mathbb{R}^n$, the *convex hull of* X, denoted $\text{conv}(X)$, is the smallest closed convex set containing X.

The convex hull of an integer set X (or a mixed integer set X^M defined by rational data) is a polyhedron. Thus the strongest possible formulation is provided by $\text{conv}(X)$ because $z(c) = \min\{cx : x \in \text{conv}(X)\}$.

Given an initial formulation P of X, one classical way to obtain a stronger formulation is to add valid inequalities (cutting planes) in the x variables so as to obtain a better approximation to $\text{conv}(X)$. This is discussed in Chapter 11. The main concepts presented in this chapter, extended formulations and projection, are now defined.

Definition 13.4. An *extended formulation for a polyhedron* $P \subseteq \mathbb{R}^n$ is a polyhedron $Q = \{(x,w) \in \mathbb{R}^{n+p} : Gx + Hw \geqslant d\}$ such that $P = \text{proj}_x(Q)$.

Definition 13.5. Given a set $U \subseteq \mathbb{R}^n \times \mathbb{R}^p$, the *projection* of U on the first n variables, $x = (x_1, \ldots, x_n)$, is the set

$$\text{proj}_x(U) = \{x \in \mathbb{R}^n : \exists\, w \in \mathbb{R}^p \text{ with } (x,w) \in U\}.$$

Minkowski's representation of a polyhedron in terms of its extreme points and extreme rays gives an extended formulation that can be useful for both linear and integer programs.

Definition 13.6. Given a non-empty polyhedron $P \subseteq \mathbb{R}^n$,

i) $x \in P$ is an *extreme point of* P if $x = \lambda x^1 + (1-\lambda)x^2$, $0 < \lambda < 1$, $x^1, x^2 \in P$ implies that $x = x^1 = x^2$.
ii) r is a *ray* of P if $r \neq 0$ and $x \in P$ implies $x + \mu r \in P$ for all $\mu \in \mathbb{R}^1_+$.
iii) r is an *extreme ray of* P if r is a ray of P and $r = \mu_1 r^1 + \mu_2 r^2$, $\mu_i > 0$ $(i = 1,2)$, r^1, r^2 rays of P implies $r^1 = \alpha r^2$ for some $\alpha > 0$.

From now on we assume that $rank(A) = n$ which is necessary for P to have extreme points.

Theorem 13.1 (Minkowski). *Every polyhedron* $P = \{x \in \mathbb{R}^n : Ax \geqslant a\}$ *can be represented in the form*

$$P = \{x \in \mathbb{R}^n : x = \sum_{g \in G} \lambda_g x^g + \sum_{r \in R} \mu_r v^r,\ \sum_{g \in G} \lambda_g = 1, \lambda \in \mathbb{R}^{|G|}_+, \mu \in \mathbb{R}^{|R|}_+\}$$

where $\{x^g\}_{g \in G}$ *are the extreme points of* P *and* $\{v^r\}_{r \in R}$ *the extreme rays of* P.

Example 1 *The polyhedron*

$$P = \{x \in \mathbb{R}^2_+ : 4x_1 + 12x_2 \geqslant 33, 3x_1 - x_2 \geqslant -1, x_1 - 4x_2 \geqslant -23\}$$

has the extended formulation

$$\begin{aligned} Q = \{(x,\lambda,\mu) \in \mathbb{R}^2 \times \mathbb{R}^3_+ \times \mathbb{R}^2_+ :& \\ x = \begin{pmatrix} \frac{33}{4} \\ 0 \end{pmatrix} \lambda_1 + \begin{pmatrix} \frac{21}{40} \\ \frac{103}{40} \end{pmatrix} \lambda_2 + \begin{pmatrix} \frac{19}{11} \\ \frac{68}{11} \end{pmatrix} \lambda_3 + \begin{pmatrix} 1 \\ 0 \end{pmatrix} \mu_1 + \begin{pmatrix} 4 \\ 1 \end{pmatrix} \mu_2,& \\ \lambda_1 + \lambda_2 + \lambda_3 = 1\},& \end{aligned}$$

see Figure 13.1.

The concept of extended formulation for a polyhedron generalizes to sets X of integer points, and in particular one can apply Definition 13.4 to $\text{conv}(X)$.

Definition 13.7. An *extended formulation for an IP set* $X \subseteq \mathbb{Z}^n$ is a polyhedron $Q \subseteq \mathbb{R}^{n+p}$ such that $X = \text{proj}_x(Q) \cap \mathbb{Z}^n$.

Minkowski's Theorem (Theorem 13.1) obviously provides an extended formulation for X. Specifically take

$$Q = \{(x,\lambda,\mu) \in \mathbb{R}^n \times \mathbb{R}^{|G|}_+ \times \mathbb{R}^{|R|}_+ : x = \sum_{g \in G} \lambda_g x^g + \sum_{r \in R} \mu_r v^r,\ \sum_{g \in G} \lambda_g = 1\}$$

where $\{x^g\}_{g \in G}$ are the extreme points and $\{v^r\}_{r \in R}$ the extreme rays of $\text{conv}(X)$.

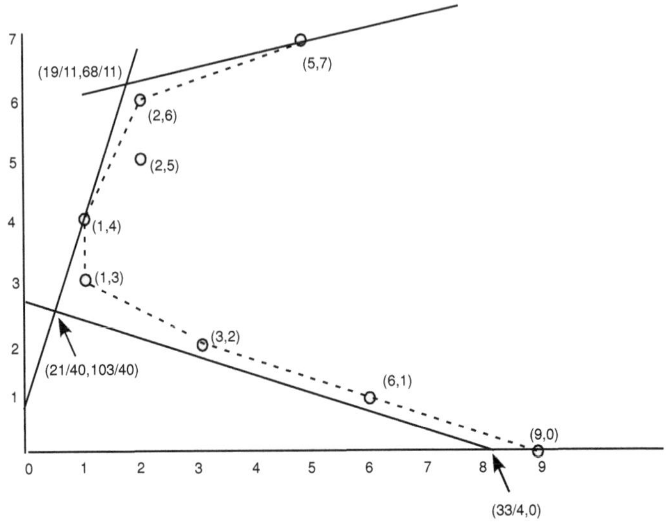

Fig. 13.1 Extreme Points and Rays of P and $\text{conv}(P \cap \mathbb{Z}^n)$

Definition 13.8. An extended formulation $Q \subseteq \mathbb{R}^{n+p}$ for an IP set $X \subseteq \mathbb{Z}^n$ is *tight* if $\text{proj}_x(Q) = \text{conv}(X)$.

An extended formulation $Q \subseteq \mathbb{R}^{n+p}$ for an IP set $X = P \cap \mathbb{Z}^n$ is *compact* if the length of the description of Q is polynomial in the length of the description of X (i.e., the length of the description of the initial formulation P of X).

In general the number of extreme points and extreme rays of $\text{conv}(X)$ is not polynomial in the length of the description of X, so the extended formulation provided by Minkowski's Theorem is not compact. Similarly the number of inequalities in the x variables required to describe $\text{conv}(X)$ is usually not polynomial in the length of the description of X.

In the framework of integer programs one also encounters more general reformulations in which some of the additional variables are required to be integer, replacing the integrality constraints on some of the original variables. It may then be possible to drop the original variables.

Definition 13.9. An *extended IP-formulation for an IP set* $X \subseteq \mathbb{Z}^n$ is a set $Q_I = \{(x, w^1, w^2) \in \mathbb{R}^n \times \mathbb{Z}^{p_1} \times \mathbb{R}^{p_2} : Gx + H^1 w^1 + H^2 w^2 \geqslant b\}$ such that $X = \text{proj}_x Q_I$.

There is a somewhat similar result to Minkowski's theorem concerning an extended IP-formulation. Again we assume rationality of the data in the case of mixed integer sets.

Theorem 13.2. *Every IP set* $X = \{x \in \mathbb{Z}^n : Ax \geqslant a\}$ *can be represented in the form* $X = \text{proj}_x(Q_I)$, *where*

$$Q_I = \{(x,\lambda,\mu) \in \mathbb{R}^n \times \mathbb{Z}_+^{|G|} \times \mathbb{Z}_+^{|R|} : x = \sum_{g\in G} \lambda_g x^g + \sum_{r\in R} \mu_r v^r,\ \sum_{g\in G} \lambda_g = 1\},$$

$\{x^g\}_{g\in G}$ is a finite set of integer points in X, and $\{v^r\}_{r\in R}$ are the extreme rays (scaled to be integer) of $\text{conv}(X)$.

Note that when X is bounded, all the points of X must be included in the set $\{x^g\}_{g\in G}$ and $R = \emptyset$. When X is unbounded, the set $\{x^g\}_{g\in G}$ includes all of the extreme points of $\text{conv}(X)$ and typically other points, see Example 2 below.

Theorem 13.2 provides an example of a common situation with extended IP-formulations in which there is a linear transformation $x = Tw$ linking all (or some) of the original x variables and the additional variables w. In such cases IP can be reformulated in terms of the additional variables in the form

$$\min\{cTw : ATw \geqslant a, w \in W\},$$

where the set W provides an appropriate representation of the integrality of the original x variables.

Example 2 *The set of integer points $X = P \cap \mathbb{Z}^2$ where*

$$P = \{x \in \mathbb{R}_+^2 : 4x_1 + 12x_2 \geqslant 33, 3x_1 - x_2 \geqslant -1, x_1 - 4x_2 \geqslant -23\}$$

has an extended IP-formulation, based on Theorem 13.2:

$$Q = \{(x,\lambda,\mu) \in \mathbb{R}^2 \times \mathbb{Z}_+^6 \times \mathbb{Z}_+^2 : x = \begin{pmatrix} 9 \\ 0 \end{pmatrix} \lambda_1 + \begin{pmatrix} 3 \\ 2 \end{pmatrix} \lambda_2 + \begin{pmatrix} 1 \\ 3 \end{pmatrix} \lambda_3 + \begin{pmatrix} 1 \\ 4 \end{pmatrix} \lambda_4 +$$
$$\begin{pmatrix} 2 \\ 6 \end{pmatrix} \lambda_5 + \begin{pmatrix} 5 \\ 7 \end{pmatrix} \lambda_6 + \begin{pmatrix} 2 \\ 5 \end{pmatrix} \lambda_7 + \begin{pmatrix} 6 \\ 1 \end{pmatrix} \lambda_8 + \begin{pmatrix} 1 \\ 0 \end{pmatrix} \mu_1 + \begin{pmatrix} 4 \\ 1 \end{pmatrix} \mu_2,\ \sum_{p=1}^{6} \lambda_p = 1\}.$$

Here the points $(2,5)^T$ and $(6,1)^T$ are not extreme points of $\text{conv}(X)$. *However they cannot be obtained as an integer combination of the extreme points and rays of* $\text{conv}(X)$, *so they are necessary for this description. See Figure 13.1.*

Given an IP set X or a MIP set X^M, an alternative is to concentrate on a subset of the more important variables (for instance the integer variables in an MIP). Here projection is the natural operation and the lemma of Farkas a basic tool. From now on, we typically assume that all the variables x or (x,y) encountered in IP or MIP are non-negative.

Lemma 13.1 (Farkas [36]). *Given $A \in \mathbb{R}^{m\times n}$ and $a \in \mathbb{R}^m$, the polyhedron $\{x \in \mathbb{R}_+^n : Ax \geqslant a\} \neq \emptyset$ if and only if $va \leqslant 0$ for all $v \in \mathbb{R}_+^m$ such that $vA \leqslant 0$.*

This immediately gives a characterization of the projection of a polyhedron. Specifically if $Q = \{(x,w) \in \mathbb{R}_+^n \times \mathbb{R}_+^p : Gx + Hw \geqslant d\}$, it follows from the definition that $x \in \text{proj}_x(Q)$ if and only if $Q(x) = \{w \in \mathbb{R}_+^p : Hw \geqslant d - Gx\}$ is nonempty. Now Farkas' Lemma, with $A = H$ and $a = d - Gx$, gives:

Theorem 13.3 (Projection). *Let* $Q = \{(x,w) \in \mathbb{R}^n \times \mathbb{R}^p_+ : Gx + Hw \geqslant d\}$. *Then*

$$\begin{aligned}\text{proj}_x(Q) &= \{x \in \mathbb{R}^n : v(d - Gx) \leqslant 0 \; \forall \, v \in V\} \\ &= \{x \in \mathbb{R}^n : v^j(d - Gx) \leqslant 0 \text{ for } j = 1, \ldots, J\}\end{aligned}$$

where $V = \{v \in \mathbb{R}^m_+ : vH \leqslant 0\}$ and $\{v^j\}_{j=1}^J$ are the extreme rays of V.

Example 3 *Given the polyhedron* $Q = \{(x,y) \in \mathbb{R}^2_+ \times \mathbb{R}^3_+ :$

$$\begin{aligned}-2x_1 - 3x_2 - \;\; 4y_1 + \;\; y_2 - 4y_3 &\geqslant \;\; -9 \\ -7x_1 - 5x_2 - 12y_1 - 2y_2 + 4y_3 &\geqslant -11\},\end{aligned}$$

we have that $V = \{v \in \mathbb{R}^2_+ : -4v_1 - 12v_2 \leqslant 0, v_1 - 2v_2 \leqslant 0, -4v_1 + 4v_2 \leqslant 0\}$. *The extreme rays are* $v^1 = (1,1)^T$ *and* $v^2 = (2,1)^T$. *From Theorem 13.3, one obtains*

$$proj_x(Q) = \{x \in \mathbb{R}^2_+ : 9x_1 + 8x_2 \leqslant 20, 11x_1 + 11x_2 \leqslant 29\}.$$

The classical application of this approach is to reformulate mixed integer programs.

Now we illustrate by example the sort of reformulations that can arise using additional variables and projection for a problem with special structure.

Example 4 Formulations of the Directed Steiner Tree Problem

Given a digraph $D = (V,A)$ *with costs* $c \in \mathbb{R}^{|A|}_+$, *a root* $r \in V$ *and a set* $T \subseteq V \setminus \{r\}$ *of terminals, the problem is to find a minimum cost subgraph containing a directed path from* r *to each node in* T.

One way to formulate this problem is to construct a subgraph in which one requires $|T|$ *units to flow out from node* r *and one unit to flow into every node of* T. *This leads one to introduce the variables:*

$x_{ij} = 1$ *if arc* (i,j) *forms part of the subgraph and* $x_{ij} = 0$ *otherwise, and* y_{ij} *is the flow in arc* (i,j). *The resulting MIP formulation is*

$$\min \sum_{(i,j) \in A} c_{ij} x_{ij}$$

$$-\sum_{j \in V^+(r)} y_{rj} = -|T| \qquad (13.1)$$

$$-\sum_{j \in V^+(i)} y_{ij} + \sum_{j \in V^-(i)} y_{ji} = 1 \qquad \text{for } i \in T \qquad (13.2)$$

$$-\sum_{j \in V^+(i)} y_{ij} + \sum_{j \in V^-(i)} y_{ji} = 0 \qquad \text{for } i \in V \setminus (T \cup \{r\}) \qquad (13.3)$$

$$y_{ij} \leqslant |T| x_{ij} \qquad \text{for } (i,j) \in A \qquad (13.4)$$

$$y \in \mathbb{R}^{|A|}_+, \; x \in \{0,1\}^{|A|},$$

where $V^+(i) = \{j : (i,j) \in A\}$ *and* $V^-(i) = \{j : (j,i) \in A\}$, *(13.1) indicates that* $|T|$ *units flow out from node* r, *(13.2) that a net flow of one unit arrives at each node* $i \in T$, *(13.3) that there is conservation of flow at the remaining nodes and (13.4) that the flow on each arc does not exceed* $|T|$ *and is only positive if the arc has been installed.*

This problem has special network structure that we now exploit.

Multicommodity flow variables

To obtain an extended formulation, consider the flow directed towards node k as a separate commodity for each node $k \in T$. Then w^k_{ij} denotes the flow in arc (i,j) of commodity k with destination $k \in T$. The resulting extended formulation is:

$$\min \sum_{(i,j)\in A} c_{ij}x_{ij}$$

$$-\sum_{j\in V^+(r)} w^k_{rj} = -1 \qquad \text{for } k \in T \tag{13.5}$$

$$-\sum_{j\in V^+(i)} w^k_{ij} + \sum_{j\in V^-(i)} w^k_{ji} = 0 \qquad \text{for } i \in V\setminus\{r,k\},\ k \in T \tag{13.6}$$

$$-\sum_{j\in V^+(k)} w^k_{kj} + \sum_{j\in V^-(k)} w^k_{jk} = 1 \qquad \text{for } k \in T \tag{13.7}$$

$$w^k_{ij} \leqslant x_{ij} \qquad \text{for } (i,j) \in A,\ k \in T \tag{13.8}$$

$$w \in \mathbb{R}^{|T|\times|A|}_+,\ x \in [0,1]^{|A|}.$$

Constraints (13.5)–(13.7) are flow conservation constraints and (13.8) variable upper bound constraints for each commodity. The constraints $y_{ij} = \sum_{k\in T} w^k_{ij}\ \ (i,j) \in A$ provide the link between the original flow variables y and the new multi-commodity flow variables w, but the y variables are unnecessary as there are no costs on the flows.

The main interest of such an extended formulation is that the value of its linear programming relaxation is considerably stronger than that of the original formulation because the relationship between the flow variables y_{ij} or w^k_{ij} and the arc selection variables x_{ij} is more accurately represented by (13.8) than by (13.4).

Projection onto the binary arc variables

It is well-known (from the max flow/min cut theorem) that one can send flow of one unit from r to k in a network (V,A) with capacities if and only if the capacity of each cut separating r and k is at least one. Considering the arc capacities to be x_{ij}, this immediately validates the following formulation in the arc variables x. Equivalently one can apply Theorem 13.3 to the extended formulation $Q = \{(x,w) \in [0,1]^{|A|} \times \mathbb{R}^{|T|\times|A|}_+$ satisfying (13.5)–(13.8)$\}$ and project out the w variables. In both cases one obtains the formulation:

$$\min \sum_{(i,j)\in A} c_{ij}x_{ij}$$

$$\sum_{(i,j)\in\delta^+(U)} x_{ij} \geqslant 1 \qquad \textit{for } U \subseteq V \textit{ with } r \in U,\ T \setminus U \neq \emptyset$$

$$x \in \{0,1\}^{|A|},$$

where $\delta^+(U) = \{(i,j) \in A : i \in U, j \notin U\}$ *is the directed cut set consisting of arcs with their tails in* U *and their heads in* $V \setminus U$.

The potential interest of this reformulation is that the number of variables required is as small as possible and the value of the linear programming relaxation is the same as that of the multi-commodity extended formulation. In Section 13.5 we will consider the more general problem in which there are also costs on the flow variables y_{ij}.

13.2.3 Decomposition

When optimizing over the feasible set X of IP is too difficult, we need to address the question of how to "decompose" X so as to arrive at one or more sets with structure, and also indicate what we mean by "structure".

We first present three ways of *decomposing*.

1. *Intersections*. $X = Y \cap Z$. Now if the set Z has structure, we can consider reformulations for the set Z. More generally, one might have $X = X^1 \cap \cdots \cap X^K$ where several of the sets X^k have structure. Another important variant is that in which $X = Y \cap Z$ and Z itself decomposes into sets Z^k each with distinct variables, namely $Z = Z^1 \times \cdots \times Z^K$.
2. *Unions (or Disjunctions)*. $X = Y \cup Z$ where Z has structure. Again one might have $X = X^1 \cup \cdots \cup X^K$ where several of the sets X^k have structure.
3. *Variable Fixing*. Suppose that $X \subset \mathbb{Z}^n \times \mathbb{R}^p$. For fixed values $\bar{x}$, let $Z(\bar{x}) = \{(x,y) \in X : x = \bar{x}\}$. This is of interest if $Z(\bar{x})$ has structure for all relevant values of $\bar{x}$. Again an important case is that in which $Z(\bar{x})$ decomposes into sets with distinct variables, i.e., $Z(\bar{x}) = Z^1(\bar{x}^1) \times \cdots \times Z^K(\bar{x}^K)$ and each set $Z^k(\bar{x}^k)$ just involves the variables y^k, where $y = (y^1, \ldots, y^K)$.

Now we indicate in what circumstances we say that the set Z obtained above has *structure*.

i) Either there is a polynomial algorithm for the *optimization* problem $\min\{cx : x \in Z\}$, denoted $\mathrm{OPT}(Z,c)$, or $\mathrm{OPT}(Z,c)$ can be solved rapidly in practice. Based on decomposition by intersection, ways to reformulate and exploit such sets are the subject of the next section.
ii) There is a polynomial algorithm for the *separation problem*, $\mathrm{SEP}(Z,x^*)$, defined as follows:
Given the set $Z \subseteq \mathbb{R}^n$ and $x^* \in \mathbb{R}^n$, is $x^* \in \mathrm{conv}(Z)$? If not, find a valid inequality

$\pi x \geqslant \pi_0$ for Z cutting off x^* (i.e., $\pi x \geqslant \pi_0$ for all $x \in Z$ and $\pi x^* < \pi_0$). More generally there is a polyhedron P' (often $P' = \text{conv}(Z')$ where $Z \subseteq Z'$) for which there is a separation algorithm (exact or heuristic) that can be solved rapidly in practice.
Such sets are amenable to reformulation by the addition of cutting planes. A special case of this type, treated in Section 13.4, is that in which the set $Z(\bar{x})$, obtained by variable fixing, has structure of type i). Combined with projection, this leads to reformulations and algorithms in the space of the x variables.

iii) Set Z has specific structure that can be exploited by introducing new variables that better describe the integrality of the variables. Examples of sets with interesting extended formulations include network design problems with 0-1 variables to indicate which arcs are open, such as the Steiner tree problem in Example 4, and scheduling problems in which it is useful to model start times in detail. Problems that can be solved by dynamic programming and problems of optimizing over sets defined by disjunctions are also candidates for reformulation through the introduction of new variables. Extended formulations for a wide variety of such problems are presented in Section 13.5.

13.3 Price or constraint decomposition

Consider a (minimization) problem of the form

$$\text{(IP)} \qquad z = \min\{cx : x \in X\}$$

that is difficult, but with the property that a subset of the constraints of X defines a set Z ($X \subset Z$) over which optimization is "relatively easy". More specifically,

$$\text{(IP)} \qquad z = \min\{cx : \underbrace{Dx \geqslant d, Bx \geqslant b, x \in \mathbb{Z}^n_+}_{x \in X}\} \tag{13.9}$$

where the constraints $Dx \geqslant d$ represent "complicating constraints" that define the integer set $Y = \{x \in \mathbb{Z}^n_+ : Dx \geqslant d\}$, while the constraints $Bx \geqslant b$ define a set $Z = \{x \in \mathbb{Z}^n_+ : Bx \geqslant b\}$ that is "tractable", meaning that $\min\{cx : x \in Z\}$ can be solved rapidly in practice.

Here we examine how one's ability to optimize over the simpler set Z can be exploited to produce dual bounds by relaxing the complicating constraints and penalizing their violation in the objective function (a procedure called Lagrangean relaxation). The prices associated to each constraint placed in the objective function are called Lagrange multipliers or dual variables, and the aim is to choose the prices to try to enforce satisfaction of the complicating constraints $Dx \geqslant d$. An alternative is to view the problem of optimizing over X as that of selecting a solution from the set Z that also satisfies the constraints defining Y. This leads to the so-called Dantzig-Wolfe reformulation in which variables are associated to the points of the set Z as specified in Theorems 13.1 or 13.2. The LP solution to this reformulation provides a dual bound that is typically tighter than that of the LP relaxation of the original formulation of X and is equal to the best bound that can be derived by Lagrangean relaxation of the constraints $Dx \geqslant d$. This will be demonstrated below.

In many applications of interest $Bx \geqslant b$ has *block diagonal* structure: i.e., $Z = Z^1 \times \cdots \times Z^K$ in which case the integer program takes the form

$$(\mathrm{IP_{BD}}) \qquad \min\{\sum_{k=1}^{K} c^k x^k : (x^1,\ldots,x^K) \in Y,\ x^k \in Z^k \text{ for } k = 1,\ldots,K\}$$

and can be written explicitly as:

$$\begin{array}{lllllll}
(\mathrm{IP_{BD}}) & \min & c^1x^1 & + c^2x^2 & + \cdots + & c^K x^K & \\
 & & D^1x^1 & + D^2x^2 & + \cdots + & D^K x^K & \geqslant d \\
 & & B^1x^1 & & & & \geqslant b^1 \\
 & & & B^2x^2 & & & \geqslant b^2 \\
 & & & & \ddots & & \geqslant \vdots \\
 & & & & & B^K x^K & \geqslant b^K \\
 & & \multicolumn{4}{l}{x^1 \in \mathbb{Z}^{n_1}_+,\ x^2 \in \mathbb{Z}^{n_2}_+, \ldots, x^K \in \mathbb{Z}^{n_K}_+.} &
\end{array}$$

Here relaxing the constraints $\sum_{k=1}^{K} D^k x^k \geq d$ allow one to decompose the problem into K smaller size optimization problems: $\min\{c^k x^k : x^k \in Z^k\}$.

Another important special case is the *identical sub-problem* case in which $D^k = D$, $B^k = B$, $c^k = c$, $Z^k = Z^*$ for all k. In this case the "complicating" constraints only depend on the aggregate variables

$$y = \sum_{k=1}^{K} x^k\,, \tag{13.10}$$

so the complicating constraints correspond to a set of the form $Y = \{y \in \mathbb{Z}^n_+ : Dy \geqslant d\}$. The problem can now be written as:

$$(\mathrm{IP_{IS}}) \qquad \min\{cy : Dy \geqslant d, y = \sum_{k=1}^{K} x^k,\ x^k \in Z^* \text{ for } k = 1,\ldots,K\}\,. \tag{13.11}$$

Example 5 (The bin packing problem)
Given an unlimited supply of bins of capacity 1 and a set of items indexed by $i = 1,\ldots,n$ *of size* $s_i \in (0,1]$*, the problem is to find the minimum number of bins that are needed in order to pack all the items. Let* K *be an upper bound on the number of bins that are needed (*$K = n$*, or* K *is the value of any feasible solution). A direct IP formulation is*

$$\min \sum_{k=1}^{K} u_k \tag{13.12}$$

$$\sum_{k=1}^{K} x_{ik} = 1 \quad \text{for } i = 1,\dots,n \tag{13.13}$$

$$\sum_i s_i x_{ik} \leqslant u_k \quad \text{for } k = 1,\dots,K \tag{13.14}$$

$$x \in \{0,1\}^{nK} \tag{13.15}$$

$$u \in \{0,1\}^{K} \tag{13.16}$$

where $u_k = 1$ *if bin k is used and* $x_{ik} = 1$ *if the item of size i is placed in bin k. This is a natural candidate for price decomposition. Without the constraints (13.13), the problem that remains decomposes into K identical knapsack problems.*

In this section,

i) we review the Lagrangean relaxation and Dantzig-Wolfe reformulation approaches, showing the links between them and the fact that both provide the same dual bound;
ii) we then discuss algorithms to compute this dual bound: sub-gradient methods and the column generation procedure, as well as stabilization techniques that are used to improve convergence, and
iii) we consider the combination of column generation with branch-and-bound to solve problems to integer optimality: deriving branching schemes when using a Dantzig-Wolfe reformulation can be nontrivial in the case of a block diagonal structure with identical sub-problems.
For simplicity, most of these developments are presented for the case of a single subsystem involving only bounded integer variables. However the developments easily extend to the case of a mixed integer or unbounded subsystem Z, or to a subsystem with *block diagonal* structure. The case where these blocks are identical will be discussed separately. The economic interpretation of the algorithms reviewed here will justify the use of the terminology "price decomposition".

13.3.1 Lagrangean relaxation and the Lagrangean dual

The Lagrangean relaxation approach to a problem IP with the structure outlined above consists of turning the "difficult" constraints $Dx \geqslant d$ into constraints that can be violated at a price π, while keeping the remaining constraints describing the set $Z = \{x \in \mathbb{Z}^n_+ : Bx \geqslant b\}$. This gives rise to the so-called *Lagrangean sub-problem*:

$$L(\pi) = \min_x \{cx + \pi(d - Dx) : \; Bx \geqslant b, x \in \mathbb{Z}^n_+\} \tag{13.17}$$

that by assumption is relatively tractable. For any non-negative penalty vector $\pi \geqslant 0$, the dual function $L(\pi)$ defines a dual (lower) bound on the optimal value z of IP: indeed the optimal solution x^* of IP satisfies $cx^* \geqslant cx^* + \pi(d - Dx^*) \geqslant L(\pi)$ (the first inequality results from x^* being feasible for IP and $\pi \geqslant 0$ and the second because x^* is feasible in (13.17)). The problem of maximizing this bound over the set of admissible dual vectors is known as the *Lagrangean dual*:

$$\text{(LD)} \qquad z_{\text{LD}} = \max_{\pi \geqslant 0} L(\pi) = \max_{\pi \geqslant 0} \min_{x \in Z} \{cx + \pi(d - Dx)\}. \tag{13.18}$$

We now reformulate the Lagrangean dual as a linear program, assuming that the constraint set Z is non-empty and bounded. The Lagrangean sub-problem achieves its optimum at an extreme point x^t of $\text{conv}(Z)$, so one can write

$$z_{\text{LD}} = \max_{\pi \geqslant 0} \min_{t=1,\ldots,T} \{cx^t + \pi(d - Dx^t)\}, \tag{13.19}$$

where $\{x^t\}_{t=1,\ldots,T}$ is the set of extreme points of $\text{conv}(Z)$, or alternatively $\{x^t\}_{t=1,\ldots,T}$ is the set of all points of Z. Introducing an additional variable σ representing a lower bound on the $(c - \pi D)x^t$ values, we can now rewrite LD as the linear program:

$$z_{\text{LD}} = \max\ \pi d + \sigma \tag{13.20}$$

$$\pi D x^t + \sigma \leqslant cx^t \qquad \text{for } t = 1, \ldots, T \tag{13.21}$$

$$\pi \geqslant 0, \sigma \in \mathbb{R}^1. \tag{13.22}$$

Taking its linear programming dual gives:

$$z_{\text{LD}} = \min \sum_{t=1}^{T} (cx^t)\lambda_t \tag{13.23}$$

$$\sum_{t=1}^{T} (Dx^t)\lambda_t \geqslant d \tag{13.24}$$

$$\sum_{t=1}^{T} \lambda_t = 1 \tag{13.25}$$

$$\lambda \in \mathbb{R}_+^T. \tag{13.26}$$

From formulation (13.23)–(13.26), one easily derives the following result.

Theorem 13.4 (Lagrangean duality).

$$z_{\text{LD}} = \min\{cx : Dx \geqslant d, x \in \text{conv}(Z)\}. \tag{13.27}$$

Indeed, by definition of the set of points $\{x^t\}_{t=1}^T$, $\text{conv}(Z) = \{x = \sum_{t=1}^T x^t \lambda_t : \sum_{t=1}^T \lambda_t = 1,\ \lambda_t \geqslant 0\ t = 1, \ldots, T\}$. Thus, the value of the Lagrangean dual is equal to the value of the linear program obtained by minimizing cx over the intersection of the "complicating" constraints $Dx \geqslant d$ with the convex hull of the "tractable" set Z.

Example 6 (Lagrangean relaxation for the bin packing problem).
Continuing Example 5, consider an instance of the bin packing problem with $n = 5$ *items and size vector* $s = (\frac{1}{6}, \frac{2}{6}, \frac{2}{6}, \frac{3}{6}, \frac{4}{6})$. *Dualizing the constraints (13.13), the Lagrangean subproblem (13.17) takes the form:* $\min\{\sum_{k=1}^{K} u_k - \sum_{i=1}^{n} \pi_i(1 - \sum_{k=1}^{K} x_{ik}) : (13.14) - (13.16\}$. *Arbitrarily taking dual variables* $\pi = (\frac{1}{3}, \frac{1}{3}, \frac{1}{3}, \frac{1}{2}, \frac{1}{2})$ *and using the fact that this problem splits up into an identical knapsack problem for each k, the Lagrangean sub-problem becomes:*

$$
\begin{aligned}
L(\pi) = \sum_{i=1}^{5} \pi_i + K \min(u - \frac{1}{3}x_1 - \frac{1}{3}x_2 - \frac{1}{3}x_3 - \frac{1}{2}x_4 - \frac{1}{2}x_5) \\
\frac{1}{6}x_1 + \frac{2}{6}x_2 + \frac{2}{6}x_3 + \frac{3}{6}x_4 + \frac{4}{6}x_5 \leqslant u \\
x \in \{0,1\}^5, \; u \in \{0,1\}.
\end{aligned}
$$

The optimal solution is $x = (1,1,0,1,0), u = 1$. *For* $K = n$ *(a trivial solution is to put each item in a separate bin), the resulting lower bound is* $\frac{12}{6} - \frac{5}{6} = \frac{7}{6}$. *The best Lagrangean dual bound* $z_{\text{LD}} = 2$ *is attained for* $\pi = (0,0,0,1,1)$, $x = (0,0,0,0,1)$ *and* $u = 1$.

13.3.2 Dantzig-Wolfe reformulations

Here we consider two closely related extended formulations for problem IP: $\min\{cx : Dx \geqslant d, x \in Z\}$, and then we consider the values of the corresponding linear programming relaxations.

We continue to assume that Z is bounded. The Dantzig-Wolfe reformulation resulting from Theorem 13.1 (called the convexification approach) takes the form:

$$
\text{(DWc)} \quad z^{\text{DWc}} = \min \sum_{g \in G^c} (cx^g)\lambda_g \tag{13.28}
$$

$$
\sum_{g \in G^c} (Dx^g)\lambda_g \geqslant d \tag{13.29}
$$

$$
\sum_{g \in G^c} \lambda_g = 1 \tag{13.30}
$$

$$
x = \sum_{g \in G^c} x^g \lambda_g \in \mathbb{Z}^n \tag{13.31}
$$

$$
\lambda \in \mathbb{R}_+^{|G^c|} \tag{13.32}
$$

where $\{x^g\}_{g \in G^c}$ are the extreme points of $\text{conv}(Z)$.

The Dantzig-Wolfe reformulation resulting from Theorem 13.2 (called the discretization approach) is

$$\text{(DWd)} \quad z^{\text{DWd}} = \min \sum_{g \in G^d} (cx^g)\lambda_g \tag{13.33}$$

$$\sum_{g \in G^d} (Dx^g)\lambda_g \geqslant d \tag{13.34}$$

$$\sum_{g \in G^d} \lambda_g = 1 \tag{13.35}$$

$$\lambda \in \{0,1\}^{|G^d|} \tag{13.36}$$

where $\{x^g\}_{g \in G^d}$ are all the points of Z.

As pointed out above, the extreme points of $\text{conv}(Z)$ are in general a strict subset of the points of Z ($G^c \subseteq G^d$). Note however that the distinction between the two approaches disappears when considering the LP relaxations of the Dantzig-Wolfe reformulations: both sets allow one to model $\text{conv}(Z)$ and they provide a dual bound that is equal to the value of the Lagrangean dual.

Observation 1

i) The linear program (13.23)–(13.26) is precisely the linear programming relaxation of DWc.

ii) It is identical to the linear programming relaxations of DWd (any point of Z can be obtained as a convex combination of extreme points of $\text{conv}(Z)$*). Hence*

$$z_{\text{LP}}^{\text{DWc}} = z_{\text{LP}}^{\text{DWd}} = \min\{cx : Dx \geqslant d, x \in \text{conv}(Z)\} = z_{\text{LD}},$$

where $z_{\text{LP}}^{\text{DWc}}$ *and* $z_{\text{LP}}^{\text{DWd}}$ *denote the values of the LP relaxations of DWc and DWd respectively.*

In addition there is no difference between DWc and DWd when $Z \subset \{0,1\}^n$ as every point $x \in Z$ is an extreme point of $\text{conv}(Z)$. In other words

$$x = \sum_{g \in G^C} x^g \lambda_g \in \{0,1\}^n \text{ in DWc if and only if } \lambda \in \{0,1\}^{|G^d|} \text{ in DWd.}$$

To terminate this subsection we examine the form DWd takes when there is block diagonal structure. Specifically the multi-block Dantzig-Wolfe reformulation is:

$$\min\left\{ \sum_{k=1}^{K} \sum_{g \in G_k^d} (cx^g)\lambda_{kg} : \sum_{k=1}^{K} \sum_{g \in G_k^d} (Dx^g)\lambda_{kg} \geqslant d, \right. \tag{13.37}$$
$$\left. \sum_{g \in G_k^d} \lambda_{kg} = 1 \text{ for } k = 1,\ldots,K,\ \lambda \in \{0,1\}^{\sum_k |G_k^d|} \right\}.$$

where $Z^k = \{x^g\}_{g \in G_k^d}$ for all k and $x^k = \sum_{g \in G_k^d} x^g \lambda_{kg} \in Z^k$.

Identical subproblems

When the subproblems are identical for $k = 1,\ldots,K$, the above model admits many different representations of the same solution: any permutation of the k indices defines a symmetric solution. To avoid this symmetry, it is normal to introduce the aggregate variables $v_g = \sum_{k=1}^{K} \lambda_{kg}$. Defining $Z^* = Z^1 = \cdots = Z^K$ and $Z^* = \{x^g\}_{g\in G^*}$, one obtains the reformulation:

$$\text{(DWad)} \quad \min \sum_{g\in G^*} (cx^g)v_g \tag{13.38}$$

$$\sum_{g\in G^*} (Dx^g)v_g \geqslant d \tag{13.39}$$

$$\sum_{g\in G^*} v_g = K \tag{13.40}$$

$$v \in \mathbb{Z}_+^{|G^*|}, \tag{13.41}$$

where $v_g \in \mathbb{Z}_+$ is the number of copies of x^g used in the solution. The projection of reformulation solution v into the original variable space will only provide the aggregate variables y defined in (13.10):

$$y = \sum_{g\in G^*} x^g v_g\,. \tag{13.42}$$

Example 7 (The cutting stock problem)
An unlimited number of strips of length L are available. Given $d \in \mathbb{Z}_+^n$ and $s \in \mathbb{R}_+^n$, the problem is to obtain d_i strips of length s_i for $i = 1,\ldots,n$ by cutting up the smallest possible number of strips of length L.

Here $Z^* = \{x \in \mathbb{Z}_+^n : \sum_{i=1}^n s_i x_i \leqslant L\}$, each point x^g of Z^* corresponds to a cutting pattern, $D = I$ and $c = 1$, so one obtains directly the formulation

$$\min\left\{\sum_{g\in G^*} v_g : \sum_{g\in G^*} (x^g)v_g \geqslant d, v \in Z_+^{|G^*|}\right\}$$

in the form DWad, without the cardinality constraint (13.40). The bin packing problem is the special case in which $d_i = 1$ for all i and each cutting pattern contains each strip length at most once.

To complete the picture we describe how to solve the linear programming relaxation of the Dantzig-Wolfe reformulation in the next subsection and how to use this reformulation in a branch-and-bound approach to find an optimal integer solution (subsection 13.3.5).

13.3.3 Solving the Dantzig-Wolfe relaxation by column generation

Here we consider how to compute the dual bound provided by the "Dantzig-Wolfe relaxation" using column generation. Alternative ways to compute this dual bound are then discussed in the next subsection.

Consider the linear relaxation of DWc given in (13.28)–(13.32) or DWd given in (13.33)–(13.36) which are equivalent as noted in Observation 1. This LP is traditionally called the (Dantzig-Wolfe) master problem (MLP). It has a very large number of variables that will be introduced dynamically in the course of the optimization by the revised simplex method. We assume that Z is a bounded integer set. Let $\{x^g\}_{g\in G}$ be either the extreme points of $\text{conv}(Z)$ or all the points of Z. Suppose that, at iteration t of the simplex algorithm, only a subset of points $\{x^g\}_{g\in G^t}$ with $G^t \subset G$ are known. They give rise to the *restricted master linear program*:

$$\text{(RMLP)} \qquad z^{\text{RMLP}} = \min \sum_{g\in G^t} (cx^g)\lambda_g \tag{13.43}$$

$$\sum_{g\in G^t} (Dx^g)\lambda_g \geqslant d \tag{13.44}$$

$$\sum_{g\in G^t} \lambda_g = 1 \tag{13.45}$$

$$\lambda \in \mathbb{R}_+^{|G^t|}.$$

The dual of RMLP takes the form:

$$\max \ \pi d + \sigma \tag{13.46}$$

$$\pi D x^g + \sigma \leqslant cx^g \quad \text{for } g \in G^t \tag{13.47}$$

$$\pi \geqslant 0, \ \sigma \in \mathbb{R}^1. \tag{13.48}$$

Let λ' and (π', σ') represent the primal and the dual solutions of the restricted master program RMLP respectively.

The column generation algorithm follows directly from the following simple observations exploiting both primal and dual representations of the master problem.

Observation 2

i) *Given a current dual solution (π', σ'), the reduced cost of the column associated to solution x^g is $cx^g - \pi' Dx^g - \sigma'$.*
ii) $\zeta = \min_{g\in G}(cx^g - \pi' Dx^g) = \min_{x\in Z}(c - \pi' D)x$. *Thus, instead of examining the reduced costs of the huge number of columns, pricing can be carried out implicitly by solving a single integer program over the set Z.*
iii) *The solution value of the restricted Master problem $z^{\text{RMLP}} = \sum_{g\in G^t}(cx^g)\lambda'_g = \pi' d + \sigma'$ gives an upper bound on z_{MLP}. MLP is solved when $\zeta - \sigma' = 0$, i.e., when there is no column with negative reduced cost.*
iv) *The pricing problem defined in ii) is equivalent to the Lagrangean sub-problem given in (13.17); hence, each pricing step provides a Lagrangean dual bound.*

v) *For another view point on iv), note that the dual solution π' of RMLP, completed by ζ, forms a feasible solution (π', ζ) for the dual of MLP:*

$$\{\max \pi d + \sigma : \pi D x^g + \sigma \leqslant c x^g \text{ for } g \in G; \pi \geqslant 0, \sigma \in \mathbb{R}^1\},$$

and therefore $\pi' d + \zeta$ gives a lower bound on z^{MLP}.

vi) *If the solution λ' to RMLP is integer, the corresponding value of z^{RMLP} provides a valid primal (upper) bound for problem IP.*

Point ii) is crucial as our motivation for the Dantzig-Wolfe reformulation was the assumption that solving an optimization problem over Z is relatively tractable. Point vi) highlights a strong point of the column generation approach: it may produce primal integer solutions in the course of the solution of MLP.

Column Generation Algorithm for a master program of the form (13.23)–(13.26):

i) Initialize primal and dual bounds $PB = +\infty$, $DB = -\infty$. Generate a subset of points x^g so that RMLP is feasible. (Master feasibility can be achieved using artificial columns. It is standard to combine Phases 1 and 2 of the simplex method to eliminate these artificial columns from the LP solution).
ii) Iteration t:
   a) Solve RMLP over the current set of columns $\{x^g\}_{g \in G^t}$; record the primal solution λ^t and the dual solution (π^t, σ^t).
   b) Check whether λ^t defines an integer solution of IP; if so update PB. If $PB = DB$, stop.
   c) Solve the pricing problem

$$(SP^t) \quad \zeta^t = \min\{(c - \pi^t D)x : x \in Z\}.$$

   Let x^t be an optimal solution.
   If $\zeta^t - \sigma^t = 0$, set $DB = z^{\text{RMLP}}$ and stop; the Dantzig-Wolfe master problem MLP is solved.
   Otherwise, add x^t to G^t and include the associated column in RMLP (its reduced cost is $\zeta^t - \sigma^t < 0$).
   d) Compute the dual bound: $L(\pi^t) = \pi^t d + \zeta^t$; update $DB = \max\{DB, L(\pi^t)\}$. If $PB = DB$, stop.

iii) Increment t and return to ii).

When problem IP has a block diagonal structure with the kth subproblem having optimal value ζ^k, the corresponding upper bounds on the unrestricted master LP value z^{MLP} are of the form $\pi' d + \sum_{k=1}^K \sigma'_k$ and the lower bounds of the form $\pi' d + \sum_{k=1}^K \zeta^k$. When the K subsystems are identical these bounds take the form $\pi' d + K\sigma'$ and $\pi' d + K\zeta$ respectively. The typical behavior of these upper and lower bounds in the course of the column generation algorithm is illustrated in Figure 13.2. Example 8 demonstrates the column generation procedure on an instance of the bin packing problem.

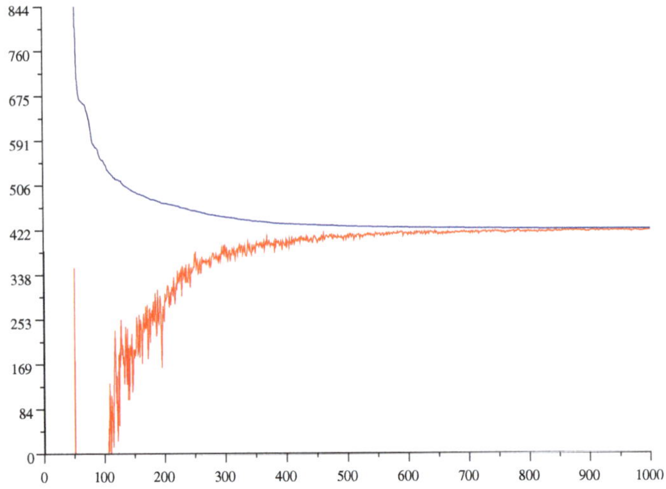

Fig. 13.2 Convergence of the column generation algorithm

Example 8 (Column generation for the bin packing problem)
Consider the same instance as in Example 6 with $n = 5$ items and size vector $s = (\frac{1}{6}, \frac{2}{6}, \frac{2}{6}, \frac{3}{6}, \frac{4}{6})$. Initialize the restricted master RMLP with the trivial packings in which each item is in a separate bin. The initial restricted master then takes the form:

$$\min v_1 + v_2 + v_3 + v_4 + v_5$$

$$\begin{pmatrix} 1 & 0 & 0 & 0 & 0 \\ 0 & 1 & 0 & 0 & 0 \\ 0 & 0 & 1 & 0 & 0 \\ 0 & 0 & 0 & 1 & 0 \\ 0 & 0 & 0 & 0 & 1 \end{pmatrix} \begin{pmatrix} v_1 \\ v_2 \\ v_3 \\ v_4 \\ v_5 \end{pmatrix} \geqslant \begin{pmatrix} 1 \\ 1 \\ 1 \\ 1 \\ 1 \end{pmatrix}, v \in \mathbb{R}^5_+$$

Its optimal value is $Z = 5$ with dual solution $\pi = (1,1,1,1,1)$. The column generation sub-problem is

$$\zeta = 1 - \max\{x_1 + x_2 + x_3 + x_4 + x_5 : x_1 + 2x_2 + 2x_3 + 3x_4 + 4x_5 \leqslant 6,\ x \in \{0,1\}^5\}.$$

The optimal solution of the knapsack problem is $x^6 = (1,1,1,0,0)$ with value 3, which gives the lower bound $L(\pi) = \sum_i \pi_i + K\,(1-3) = -5$ (with $K = 5$). x^6 is added to the restricted master with associated variable v_6. The successive iterations give

t	Z^t	*master sol.*	π^t	$L(\pi^t)$	PB	x^t
5	5	$v_1 = v_2 = v_3 = v_4 = v_5 = 1$	$(1,1,1,1,1)$	-5	5	$(1,1,1,0,0)$
6	3	$v_4 = v_5 = v_6 = 1,$	$(0,0,1,1,1)$	-2	3	$(0,0,1,1,0)$
7	3	$v_1 = v_4 = v_5 = 1$	$(0,1,0,1,1)$	-2	3	$(0,1,0,1,0)$
8	3	$v_1 = v_6 = v_7 = v_8 = \frac{1}{2}, v_5 = 1$	$(1,0,0,1,1)$	-2	3	$(1,0,0,0,1)$
9	2.5	$v_6 = v_7 = v_8 = \frac{1}{2}, v_9 = 1$	$(0,\frac{1}{2},\frac{1}{2},\frac{1}{2},1)$	0	3	$(0,1,0,0,1)$
10	2.33	$v_6 = v_8 = v_{10} = \frac{1}{3}, v_7 = v_9 = \frac{2}{3}$	$(\frac{1}{3},\frac{1}{3},\frac{1}{3},\frac{2}{3},\frac{2}{3})$	$\frac{2}{3}$	3	$(1,1,0,1,0)$
11	2.25	$v_6 = v_{11} = \frac{1}{4}, v_9 = v_{10} = \frac{1}{2}, v_7 = \frac{3}{4}$	$(\frac{1}{4},\frac{1}{4},\frac{1}{2},\frac{1}{2},\frac{3}{4})$	$\frac{4}{3}$	3	$(0,0,1,0,1)$
12	2	$v_{11} = v_{12} = 1$	$(0,0,0,1,1)$	2	2	$(0,0,0,0,1)$

In this example, the master problem has an optimal solution that is integer, so this is an optimal solution of the bin packing problem (the column generation procedure ends with $PB = DB$).

The column generation algorithm has an appealing economic interpretation, derived directly from linear programming duality. Dantzig-Wolfe decomposition can be viewed as a procedure for decentralizing the decision-making process. The master problem plays the role of the coordinator setting prices that serve as incentives to meet the global constraints $\sum_k Dx^k \geqslant d$. These prices are submitted to the subdivisions. Each independent subdivision uses these prices to evaluate the profitability of its activities ($x^k \in Z^k$) and returns an interesting business proposal (with negative reduced cost). The procedure iterates until no more improving proposals can be generated, and the given prices are optimal.

13.3.4 Alternative methods for solving the Lagrangean dual

By Observation 1, the above column generation algorithm solves the Lagrangean dual $z_{\mathrm{LD}} = \max_{\pi \geqslant 0} L(\pi)$. Alternatives to the column generation approach to solving the Lagrangean dual can be related to the different formulations of the problem: its max-min form (13.19) or the dual linear program (13.20)–(13.22). The dual point of view is particularly important in the analysis of the convergence of methods for solving the Lagrangean dual: convergence is driven by the successive dual solutions, even for the column generation procedure. Dual analysis has inspired enhanced column generation algorithms making use of so-called stabilization techniques. A better theoretical convergence rate can only be achieved by using non-linear programming techniques such as the bundle method. On the other hand, simpler methods (such as the sub-gradient algorithm), whose convergence in practice is worse than that of the standard column generation approach, remain useful because of their easy implementation and their ability to cope with large size problems.

Here we review some of the classical alternative approaches to solving the Lagrangean dual arising from the different formulations given in Section 13.3.1.

Note that $L(\pi) = \min_{g \in G}(c - \pi D)x^g + \pi d$ is a piecewise affine concave function of π, as illustrated in Figure 13.3. Solving the Lagrangean dual requires the

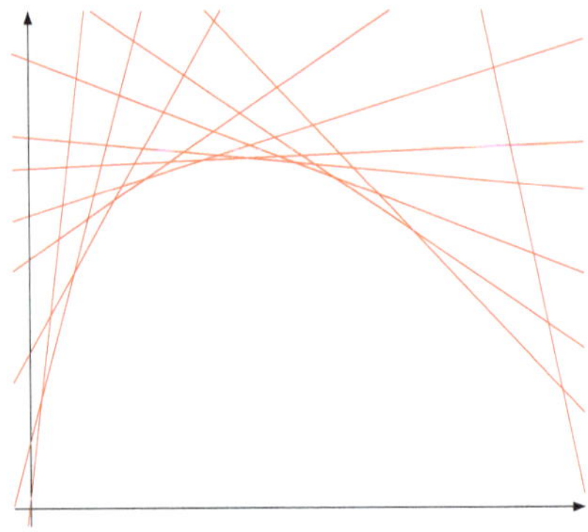

Fig. 13.3 The Lagrangean dual function $L(\pi)$ seen as a piecewise affine concave function; we assume $\pi \in \mathbb{R}^1$ in this representation; each segment/hyperplane is defined by a vector x^t.

maximization of this non-differentiable concave function. A simple method for this is:

The sub-gradient algorithm (for solving the Lagrangean dual in its form (13.19)):

i) Initialize $\pi^0 = 0$, $t = 1$.
ii) Iteration t,

a) Solve the Lagrangean subproblem (13.17) to obtain the dual bound $L(\pi^t) = \min\{cx + \pi^t(d - Dx)\}$ *and an optimal solution* x^t.
b) Compute the violation $(d - Dx^t)$ *of the dualized constraints; this provides a "sub-gradient" that can be used as a "potential direction of ascent" to modify the dual variables.*
c) Update the dual solution by making a small step in the direction of the sub-gradient

$$\pi^{t+1} = \max\{0, \pi^t + \varepsilon_t(d - Dx^t)\}$$

where ε_t *is an appropriately chosen step-size.*

iii) If $t < \tau$*, increment* t *and return to ii).*

Central to this approach is the simple dual price updating rule of step ii.c). The rule leads to an increase in the prices associated with violated constraints and a decrease for non-tight constraints. Observe, however, that it ignores all previously generated points x^g for $g = 1, \ldots, t-1$ when updating π. Not surprisingly this can result in poor performance. Moreover, the convergence of the algorithm is quite sensitive to the selection of the step size (choosing ε_t too large leads to oscillations and possible divergence, choosing it too small leads to slow convergence or convergence to a non-optimal point). It is usual to use a normalized step size: $\varepsilon_t = \frac{\alpha_t}{\|d - Dx^t\|}$. Standard choices are:

i) $\alpha_t = C(PB - L(\pi^t))$ with $C \in (0,2)$, where the primal bound PB acts as an overestimate of the unknown Lagrangean dual value z_{LD}, so the step size reduces as one gets closer to the optimal value z_{LD};
ii) the α_t form a geometric series: $\alpha_t = C\rho^t$ with $\rho \in (0,1)$ and $C > 0$;
iii) the α_t form a divergent series: $\alpha^t \to 0$ and $\sum_t \alpha^t \to \infty$; for instance, take $\alpha_t = \frac{1}{t}$.

Convergence is guaranteed for i) if PB is replaced by a lower bound on z_{LD} and for ii) if C and ρ are sufficiently large. Step size iii) is always convergent, but convergence is very slow because of the divergent sequence. Parameter τ in step iii) of the algorithm allows one to limit the number of iterations. Another standard heuristic termination rule is to stop when the dual bound $DB = \max_t L(\pi^t)$ has not improved for several iterations.

The sub-gradient approach can be used as a heuristic to produce a candidate solution for the primal problem (13.27). However it is not guaranteed to satisfy constraints $Dx \geqslant d$ while the primal solution of (13.23)–(13.26) does. The candidate, denoted $\hat{x}$, is obtained as a convex combination of previously generated points x^g for $g = 1,\dots,t$. Possible choices of updating rules are:

i) $\hat{x} = \sum_{g=1}^{t} x^g \lambda_g$ where $\lambda_g = \frac{\alpha_g}{\sum_{g=1}^{t} \alpha_g}$, or
ii) $\hat{x} = \alpha\hat{x} + (1-\alpha)x^t$ with $\alpha \in (0,1)$.

The latter rule is of interest because it puts more weight on the points x^t generated most recently. Using step size iii), the theory predicts the convergence of $\hat{x}$ towards an optimal solution to (13.27). In practice however, one would first check whether $\hat{x}$ verifies $Dx \geqslant d$ and if so record the associated value as an upper bound on z_{LD} that can be helpful in monitoring convergence (although there is no monotonic convergence of these upper bounds as in Figure 13.2). If furthermore $\hat{x}$ verifies the integrality conditions, then it defines a primal bound PB.

The *volume algorithm* is a variant of the sub-gradient method in which one uses the information of all the previously generated Lagrangean subproblem solutions to estimate both primal and dual solutions to (13.23)–(13.26), thus providing better stopping criteria. At each iteration,

i) the estimate of a primal solution is updated using: $\hat{x} = \eta\hat{x} + (1-\eta)x^t$ with a suitable $\eta \in (0,1)$;
ii) the dual solution estimate $\hat{\pi}$ is defined by the price vector that has generated the best dual bound so far: $\hat{\pi} = \operatorname{argmax}_{g=1,\dots,t} L(\pi^g)$;
iii) the "direction of ascent" is defined by the violation $(d - D\hat{x})$ of the dualized constraint by the primal solution estimate $\hat{x}$ instead of using the latest Lagrangean sub-problem solution x^t;
iv) the dual price updating rule consists in taking a step from $\hat{\pi}$ instead of π^t: $\pi^{t+1} = \max\{0, \hat{\pi} + \varepsilon_t(d - D\hat{x})\}$.

The method is inspired by the conjugate gradient method. It is equivalent to making a suitable correction v^t in the dual price updating direction $\pi^{t+1} = \max\{0, \pi^t + \varepsilon_t(d - Dx^t) + v^t\}$. The name *volume* refers to the underlying theory saying that the weight $(1-\eta)\eta^{g-1}$ of the gth solution x^g in the primal solution estimate $\hat{x}$ approximates the volume under the hyperplane $\pi D x^t + \sigma = cx^g$ in the dual polyhedron of

Figure 13.3 augmented by the constraint $\sigma \geqslant \hat{\pi}d$. The algorithm stops when primal feasibility is almost reached: $\|(d - D\hat{x})\| \leqslant \varepsilon$ and the duality gap is small enough: $\|c\hat{x} - \hat{\pi}d\| \leqslant \varepsilon$. The implementation of the method is as simple as that of the sub-gradient algorithm, while its convergence performance is typically better.

The linear programming representation (13.20)–(13.22) of the Lagrangean dual suggests the use of a cutting plane procedure to dynamically introduce the constraints associated with the different points x^g. This procedure is a standard non-linear programming approach to maximize a concave non-differentiable function, known as *Kelley's cutting plane algorithm.* It is identical to the above column generation procedure but seen in the dual space: point x^g defines a violated cut for (13.20)–(13.22) if and only if it defines a negative reduced cost column for (13.23)–(13.26).

The convergence of the basic column generation algorithm (or its dual counterpart) suffers several drawbacks, as illustrated in Figure 13.2: i) during the initial stages, when few points x^g are available, primal and dual bounds are very weak and ineffective, ii) convergence can be slow with very little progress made in improving the bounds, iii) the dual bounds can behave erratically as π jumps from one extreme point solution to another at successive iterations, and iv) the upper bounds z^{RMLP} can remain stuck at the same value due to degeneracy (iterating between alternative solutions of the same value).

Efforts have been made to construct more sophisticated and robust algorithms. They combine several mechanisms:

i) proper initialization (warm start): what is essential is to have meaningful dual solutions π from the outset (using a dual heuristic or a rich initial set of points x^g, produced for instance by the sub-gradient method);
ii) stabilization techniques that penalize deviations of the dual solutions from a *stability center* $\hat{\pi}$, defined as the dual solution providing the best dual bound so far: the dual problem becomes

$$\max_{\pi \geqslant 0} \{L(\pi) + S(\pi - \hat{\pi})\},$$

where S is a penalty function that increases as π moves away from $\hat{\pi}$;
iii) smoothing techniques that moderate the current dual solution based on previous iterates: the price vector sent to the subproblem is

$$\overline{\pi}^t = \alpha \overline{\pi}^{t-1} + (1-\alpha)\pi^t,$$

where π^t is the current dual solution of RMLP, $\alpha \in (0,1)$ is a smoothing parameter, and $\overline{\pi}^{t-1}$ is the smoothed price of the previous iterate.
iv) an interior point approach providing dual solutions corresponding to points in the center of the face of optimal solutions of RMLP as opposed to the extreme points generated by simplex-based algorithms;
v) reformulation strategies to avoid degeneracy or symmetries. For instance, when the MLP is a set covering problem, a dynamic row aggregation and disaggregation procedure allows one to control degeneracy and to reduce the number

of iterations. Another approach consists in adding valid dual cuts in (13.20)–(13.22) to break dual symmetries. These mechanisms can be combined into hybrid methods. For instance, combining ii) and iii) by smoothing around a stability center:

$$\overline{\pi}^t = \alpha\hat{\pi} + (1-\alpha)\pi^t \,. \tag{13.49}$$

Stabilization techniques differ essentially in the choice of the penalty function. Several typical penalty functions are pictured in Figure 13.4 for a 1-dimensional vector π. When S is a piecewise linear function, the modified dual problem can still be formulated as a linear program (with artificial variables). For instance, to model a boxstep penalty function $S(\pi_i) = 0$ if $\pi \in [0, \overline{\pi}_i]$ and $-\infty$ otherwise (for $\overline{\pi}_i = 2 * \hat{\pi}_i$), the master program (13.23)–(13.26) is augmented with artificial columns ρ_i for $i = 1, \ldots m$, whose costs are defined by the upper bounds $\overline{\pi}_i$ on the the dual prices. The resulting primal-dual pair of augmented formulations of the master are:

$$\begin{array}{llll}
\min \sum_{t=1}^{T}(cx^t)\lambda_t + \sum_i \overline{\pi}_i\rho_i & & \max \sum_i \pi_i d_i + \sigma & \\
\sum_{t=1}^{T}(D_i x^t)\lambda_t + \rho_i \geqslant d_i \text{ for all } i & & \sum_i \pi_i D_i x^t + \sigma \leqslant cx^t \text{ for all } t & \\
\sum_{t=1}^{T}\lambda_t = 1 & & \pi_i \leqslant \overline{\pi}_i \text{ for all } i & \\
\lambda \in \mathbb{R}^T_+,\ \rho \in \mathbb{R}^m_+ & & \pi \geqslant 0,\ \sigma \in \mathbb{R}^1. &
\end{array} \tag{13.50}$$

Properly setting the parameters that define this stabilization function may require difficult experimental tuning.

In theory the convergence rates of all the LP-based methods (with or without piece-wise linear penalty functions) are the same (although LP stabilization helps in practice). However using a quadratic penalty allows one to benefit from the quadratic convergence rate of Newton's method to get an improved theoretical convergence rate. The *bundle* method consists in choosing the penalty function $S = \frac{\|\pi-\hat{\pi}\|^2}{\eta}$ where η is a parameter that is dynamically adjusted to help convergence. (In the case of equality constraints $Dx = d$, the bundle method has an intuitive interpretation in the primal space: solving the penalized dual is equivalent to solving the augmented Lagrangean subproblem: $\min\{cx + \hat{\pi}(d - Dx) + \eta\|d - Dx\|^2 : x \in \text{conv}(Z)\}$.) The method calls for the solution of a quadratic program at each iteration (the dual restricted master involves the maximization of a concave objective under linear constraints). Experimentally use of the bundle method leads to a drastic reduction in the number of iterations for some applications. The extra computing time in solving the quadratic master is often minor.

Interior-point based solution approaches such as the Analytic Center Method (ACCPM) can also be shown theoretically to have a better rate of convergence. Even smoothing techniques can benefit from theoretical analysis: using rule (13.49), one can show that at each iteration either the dual bound is strictly improved, or the column generated based on the smoothed prices $\overline{\pi}^t$ has a strictly negative reduced cost for the original prices π^t.

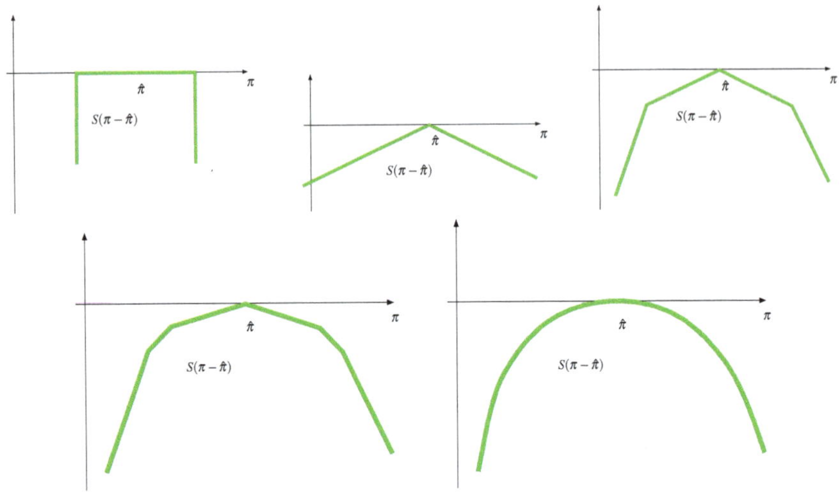

Fig. 13.4 Examples of penalty functions: the boxstep; three piece-wice linear penalty functions; the quadratic penalty of the bundle method.

In practice, each of the above enhancement techniques has been shown to significantly reduce the number of iterations in certain applications. However there may be overheads that make each iteration slightly more time consuming. Another factor in assessing the impact of the enhanced techniques is the time required by the pricing subproblem solver: it has been observed that stabilized, smoothed or centered dual prices π can make the pricing problem harder to solve in practice. Thus the benefits from using stabilization techniques are context dependent.

13.3.5 Optimal integer solutions: branch-and-price

To solve problem IP based on its Dantzig-Wolfe reformulation, one must combine column generation with branch-and-bound; the resulting algorithm is known as *branch-and-price* or *IP column generation*. The issues are how to select branching constraints and how to carry out pricing (solve the resulting subproblem(s)) after adding these constraints. Note that a standard branching scheme consisting in imposing a disjunctive constraint on a variable λ_g of the Dantzig-Wolfe reformulation that is currently fractional is not advisable. First, it induces an unbalanced enumeration tree: rounding down a λ_g variable is weakly constraining, while rounding it up is considerably more constraining, especially when the corresponding bounds are 0 and 1 respectively. Second, on the down branch it is difficult to impose an upper bound on a λ_g variable: the associated column is likely to be returned as the solution of the pricing problem unless one specifically excludes it from the sub-problem solution set (essentially adding the constraint $x \neq x^g$ in the sub-problem which destroys its structure), or one computes the next best column. The alternative is to attempt to express branching restrictions in terms of the variables of the original formulation. In general, deriving an appropriate branching scheme in a column generation context can be non-trivial, especially when tackling problems with identical subsystems.

Below we start by considering the case of a single subsystem. The branching schemes developed for this case already indicate some of the issues and extend directly to the case with multiple but distinct subsystems. We will then consider the case of a set partitioning master program with multiple identical subsystems in 0-1 variables. In this case, a classical approach is the Ryan and Foster branching scheme. We place it in the context of alternative schemes. From this discussion, we indicate the basic ideas for dealing with the general case. In particular, we outline a general branching and pricing scheme that is guaranteed to produce a finite branching tree and to maintain the structure of the pricing problem when the set Z is bounded.

13.3.5.1 Branch-and-price with a single or multiple distinct subsystems

We describe the algorithm for a single subsystem, which extends to the case of distinct subsystems. We suppose that λ^* is an optimal solution of the Dantzig-Wolfe linear programming relaxation.

i) **Integrality Test**. If λ^* is integer, or more generally if $x^* = \sum_{g \in G} x^g \lambda_g^* \in \mathbb{Z}^n$, stop. x^* is an optimal solution of IP.
ii) **Branching.** Select a variable x_j for which $x_j^* = \sum_{g \in G} x_j^g \lambda_g^* \notin \mathbb{Z}$. Separate into two subproblems with feasible regions $X \cap \{x : x_j \leqslant \lfloor x_j^* \rfloor\}$ and $X \cap \{x : x_j \geqslant \lceil x_j^* \rceil\}$.
Let us consider just the up-branch (U); the down-branch is treated similarly. The new IP for which we wish to derive a lower bound is the problem:

$$z^U = \min\{cx : Dx \geqslant d, x \in Z, x_j \geqslant \lceil x_j^* \rceil\}.$$

There are now two options, depending whether the new constraint is treated as a complicating constraint, or becomes part of the "tractable" subproblem.

Option 1. The branching constraint is dualized as a "difficult" constraint: $Y_1^U = \{x \in \mathbb{Z}^n : Dx \geqslant d, x_j \geqslant \lceil x_j^* \rceil\}$ and $Z_1^U = Z$.

iii) **Solving the new MLP**: The resulting linear program is

$$
\begin{aligned}
(\text{MLP}_1) \qquad z^{\text{MLP}_1} = \min & \sum_{g\in G} (cx^g)\lambda_g \\
& \sum_{g\in G} (Dx^g)\lambda_g \geqslant d \\
& \sum_{g\in G} x_j^g \lambda_g \geqslant \lceil x_j^* \rceil \\
& \sum_{g\in G} \lambda_g = 1 \\
& \lambda \in \mathbb{R}_+^{|G|},
\end{aligned}
$$

where $\{x^g\}_{g\in G}$ is the set of points of Z.

iv) **Solving the new subproblem**. Suppose that an optimal dual solution after iteration t is $(\pi^t, \mu^t, \sigma^t) \in \mathbb{R}_+^m \times \mathbb{R}_+^1 \times \mathbb{R}^1$. The subproblem now takes the form:

$$
(\text{SP}_1^t) \qquad \zeta_1^t = \min\{(c - \pi^t D)x - \mu^t x_j : x \in Z\}.
$$

Option 2. The branching constraint is enforced in the sub-problem: $Y_2^U = Y$ and $Z_2^U = Z \cap \{x_j \geqslant \lceil x_j^* \rceil\}$.

iii) **Solving the new MLP**: The resulting linear program is

$$
\begin{aligned}
(\text{MLP}_2) \qquad z^{\text{MLP}_2} = \min & \sum_{g\in G_2^U} (cx^g)\lambda_g \\
& \sum_{g\in G_2^U} (Dx^g)\lambda_g \geqslant d \\
& \sum_{g\in G_2^U} \lambda_g = 1 \\
& \lambda \in \mathbb{R}_+^{|G_2^U|},
\end{aligned}
$$

where $\{x^g\}_{g\in G_2^U}$ is the set of points of Z_2^U.

iv) **Solving the new subproblem**. Suppose that an optimal dual solution after iteration t is $(\pi^t, \sigma^t) \in \mathbb{R}_+^m \times \mathbb{R}^1$. The subproblem now takes the form:

$$
(\text{SP}_2^t) \qquad \zeta_2^t = \min\{(c - \pi^t D)x : x \in Z \cap \{x : x_j \geqslant \lceil x_j^* \rceil\}\}.
$$

Note that, with Option 2, branching on $x_j \geqslant \lceil x_j^* \rceil$ on the up-branch can be viewed as partitioning the set Z into two sets $Z \setminus Z_2^U$ and Z_2^U: adding the constraint $\sum_{g\in G_2^U} \lambda_g = 1$ is equivalent to adding $\sum_{g\in G\setminus G_2^U} \lambda_g = 0$ and thus the columns of $Z \setminus Z_2^U$ are removed from the master.

Both Options 1 and 2 have certain advantages and disadvantages:

- *Strength of the linear programming bound*

$$\begin{aligned} z^{\mathrm{MLP}_1} &= \min\{cx : Dx \geqslant d, x \in \mathrm{conv}(Z), x_j \geqslant \lceil x_j^* \rceil\} \\ \leqslant z^{\mathrm{MLP}_2} &= \min\{cx : Dx \geqslant d, x \in \mathrm{conv}(Z \cap \{x : x_j \geqslant \lceil x_j^* \rceil\})\}, \end{aligned}$$

  so Option 2 potentially leads to better bounds.
- *Complexity of the subproblem*
  For Option 1 the subproblem is unchanged, whereas for Option 2 the subproblem may remain tractable, or it may become more complicated if the addition of bounds on the variables makes it harder to solve.
- *Getting Integer Solutions*
  If an optimal solution x^* of IP is not an extreme point of conv(Z), there is no chance that x^* will ever be obtained as an optimal solution of the subproblem under Option 1. Under Option 2, because of the addition of the bound constraints, one can eventually generate a column $x^g = x^*$ in the interior of $\mathrm{conv}(Z)$.

The above pros and cons suggest that Option 2 may be preferable if the modified subproblem remains tractable.

In the above we only consider branching at the root node and the modifications to the column generation procedure after adding a single branching constraint. The two options can be used throughout the branch-and-price tree, adding a new lower or upper bound on a variable on each branch. Both schemes also extend to mixed integer programs in which case branching is carried out only on the integer variables.

13.3.5.2 Branch-and-price with identical subsystems

In the case of identical subsystems the Dantzig-Wolfe reformulation is given by DWad (13.38)–(13.41). Here the model variables result from an aggregation: $\nu_g = \sum_{k=1}^{K} \lambda_{kg}$ with $\sum_{g \in G} \nu_g = K$. Hence, there is no direct mapping back to the original distinct subsystem variables $(x^1, \ldots, x^K)$. The projection (13.42) of reformulation solution ν into the original variable space will only provide the aggregate variables y defined in (13.10). The *"Integrality Test"* needs to be adapted. Moreover, branching on a single component of y is typically not enough to eliminate a fractional solution. In particular, the Option 1 scheme typically does not suffice because one may have $y_j^* = \sum_{g \in G} x_j^g \lambda_g^* \in \mathbb{Z}$ for all j even though the current master solution does not provide an optimal integer solution to the original problem. The extension consists in defining branching entities involving more than one variable x_j of the original formulation. This can be interpreted as defining auxiliary variables on which to branch. The branching constraint can then either go in the master (as in Option 1) or be enforced in the pricing problem (as in Option 2), which amounts to branching on appropriately chosen subsets $\hat{Z} \subset Z$.

First, we provide an *"Integrality Test"* although its definition is not unique.

Integrality Test. *Sort the columns x^g with $\nu_g^* > 0$ in lexicographic order. Disaggregate ν into λ variables using the recursive rule:*

$$\lambda_{kg}^* = \min\{1, \nu_g - \sum_{\kappa=1}^{k-1} \lambda_{\kappa g}^*, (k - \sum_{\gamma \prec g} \nu_g^*)^+\} \quad \text{for } g \in G,\ k = 1, \ldots, K, \tag{13.51}$$

where $g_1 \prec g_2$ if g_1 precedes g_2 in the lexicographic order. For all k, let $(x^k)^* = \sum_{g \in G^c} x^g \lambda_{kg}^*$. If $x^* \in \mathbb{Z}^{Kn}$, stop. x^* is a feasible solution of IP.

Note that if ν^* is integer, the point x^* obtained by the above mapping will be integer. In general x^* can be integer even when ν^* is not. However, when $Z \subset \{0,1\}^n$, ν^* is integer if and only if x^* is integer.

Let us now discuss Branching. We first treat the special case of (13.11) in which the master problem is a set partitioning problem. Then we present briefly possible extensions applicable to the general case.

The Set Partitioning Case

For many applications with identical binary subsystems, one has $Z \subseteq \{0,1\}^n$, $D = I, d = (1, \ldots, 1)$, and the master takes the form of:

$$\min\{\sum_g (c x^g)\, \nu_g : \sum_g x_j^g\, \nu_g = 1\ \forall j,\ \sum_g \nu_g = K,\ \nu_g \in \{0,1\}^{|G|}\}. \tag{13.52}$$

One example is the bin packing problem of Example 8 in which Z is the set of solutions of a 0-1 knapsack problem. Another is the graph (vertex) coloring problem in which columns correspond to node subsets that can receive the same color and Z is the set of stable sets of the graph.

Assume that the solution to the master LP is fractional with $\nu^* \notin \{0,1\}^{|G|}$. Branching on a single component y_j is not an option. Indeed, if $\hat{G} = \{g : x_j^g = 1\}$, $y_j^* = \sum_{g \in G} x_j^g \nu_g^* = \sum_{g \in \hat{G}} \nu_g^* = 1$ for any master LP solution. However there must exist a pair of coordinates i and j such that

$$w_{ij}^* = \sum_{g: x_i^g = 1, x_j^g = 1} \nu_g^* = \alpha \quad \text{with } 0 < \alpha < 1,$$

so that one can branch on the disjunctive constraint:

$$(w_{ij} = \sum_{g: x_i^g = 1, x_j^g = 1} \nu_g = 0) \quad \text{or} \quad (w_{ij} = \sum_{g: x_i^g = 1, x_j^g = 1} \nu_g = 1),$$

where $w_{ij} = \sum_k x_i^k x_j^k$ is interpreted as an auxiliary variable indicating whether or not components i and j are in the same subset of the partition.

We present three ways to handle the branching constraint, numbered 3, 4 and 5 to distinguish them from the Options 1 and 2 above. They are illustrated on the up-branch $w_{ij} = \sum_{g: x_i^g = 1, x_j^g = 1} \nu_g = 1$.

Option 3. The branching constraint is dualized as a "difficult" constraint: $Y_3^U = \{x \in \mathbb{Z}^n : Dx \geqslant d, w_{ij} \geqslant 1\}$ and $Z_3^U = Z$. Then the master includes the constraint

$\sum_{g:x_i^g=1,x_j^g=1} \nu_g \geqslant 1$ with associated dual variable μ and the pricing subproblem needs to be amended to correctly model the reduced costs of a column; it takes the form:

$$\zeta_3 = \min\{(c - \pi D)x - \mu w_{ij} : x \in Z, w_{ij} \leqslant x_i, w_{ij} \leqslant x_j, w_{ij} \in \{0,1\}\}.$$

If one wishes to enforce branching directly in the pricing subproblem, note that one cannot simply set $w_{ij} = 1$ in the subproblem because this branching constraint must only be satisfied by one of the K subproblem solutions. Instead one must restrict the subproblem to $\hat{Z}$ in such a way that any linear combination of its solutions $x \in \hat{Z}$ satisfies $w_{ij} = \sum_{g \in \hat{G}:x_i^g=1,x_j^g=1} \nu_g = 1$. This can be done through options 4 or 5:

Option 4. Let $Y_4^U = \{x \in \mathbb{Z}^n : Dx \geqslant d\}$ and $\hat{Z} = Z_4^U = Z \cap \{x_i = x_j\}$. The combination of this restriction on the solution set with the set partitioning constraints $\sum_{g \in \hat{G}:x_i^g=1} \nu_g = 1$ and $\sum_{g \in \hat{G}:x_j^g=1} \nu_g = 1$ results in the output: $\sum_{g \in \hat{G}:x_i^g=1,x_j^g=1} \nu_g = 1$. With this option the master is unchanged, while the pricing subproblem is:

$$\zeta_4 = \min\{(c - \pi D)x : x \in Z, x_i = x_j\}.$$

Option 5. Here on the up branch one works with two different subproblems: one whose solutions have $w_{ij} = 1$ and the other whose solutions have $w_{ij} = 0$. Let $Y_5^U = \{x \in \mathbb{Z}^n : Dx \geqslant d\}$ and $\hat{Z} = Z_{5A}^U \cup Z_{5B}^U$ with $Z_{5A}^U = Z \cap \{x_i = x_j = 0\}$ and $Z_{5B}^U = Z \cap \{x_i = x_j = 1\}$. Then, in the master program the convexity constraint $\sum_{g \in G} \nu_g = K$ is replaced by $\sum_{g \in G_{5A}^U} \nu_g = K - 1$ and $\sum_{g \in G_{5B}^U} \nu_g = 1$, and there are two pricing subproblems, one over set Z_{5A}^U and one over set Z_{5B}^U:

$$\zeta_{5A} = \min\{(c - \pi D)x : x \in Z, x_i = x_j = 0\}$$

and

$$\zeta_{5B} = \min\{(c - \pi D)x : x \in Z, x_i = x_j = 1\}.$$

Option 3 can be seen as an extension of Option 1. Option 4 is known in the literature as the Ryan and Foster branching scheme. Option 5 can be seen as an extension of Option 2. The analysis of the advantages and disadvantages of Options 3, 4 and 5 provides a slightly different picture from the comparison of Options 1 and 2:

- *Strength of the linear programming bound*

$$\begin{aligned} z^{\mathrm{MLP}_3} &= \min\{cx : Dx \geqslant d, x \in \mathrm{conv}(Z)^K, w_{ij} \geqslant 1\} \\ \leqslant z^{\mathrm{MLP}_4} &= \min\{cx : Dx \geqslant d, x \in \mathrm{conv}(Z_2^U)^K\}, \\ \leqslant z^{\mathrm{MLP}_5} &= \min\{cx : Dx \geqslant d, x \in (\mathrm{conv}(Z_{5A}^U)^{K-1} \times \mathrm{conv}(Z_{5B}^U))\}, \end{aligned}$$

- *Complexity of the subproblem*
  The three options assume a change of structure in the subproblem (even Option 3). The Option 5 modifications of fixing some of the subproblem variables are the least significant.

- *Getting Integer Solutions*
  Both Option 4 and 5 allow one to generate a column $x^g = x^*$ in the interior of $\text{conv}(Z)$, but Option 5 is better in this regard.

The down-branch can be treated similarly: $Y_3^D = \{x \in \mathbb{Z}^n : Dx \geqslant d, w_{ij} = 0\}$, $Z_4^D = Z \cap \{x_i + x_j \leqslant 1\}$, $Z_{5A}^D = Z \cap \{x_i = 0\}$ and $Z_{5B}^D = Z \cap \{x_i = 1, x_j = 0\}$.

Note that the pricing problem modifications are easy to handle in some application while they make the pricing problem harder in others. The Option 3 modifications affect the cost structure in a way that is not amenable to standard pricing problem solvers in both of our examples: bin packing and vertex coloring. The Option 4 modifications do not affect the structure of the stable set sub-problem for the vertex coloring problem: addition of the inequality $x_i + x_j \leqslant 1$ on the down-branch amounts to adding an edge in the graph, while adding $x_i = x_j$ in the up-branch amounts to aggregating the two nodes—contracting an edge. However, for the bin packing application, a constraint of the form $x_i + x_j \leqslant 1$ in the down-branch destroys the knapsack problem structure, so that a standard special purpose knapsack solver can no longer be used, while the up-branch can be handled by the aggregation of items. The Option 5 modifications are easily handled by preprocessing for both the bin packing and vertex coloring problems.

The General Case with Identical Subsystems

For the general case, such as the cutting stock problem of Example 7, the Master LP relaxation is

$$\min\{\sum_{g \in G} (cx^g)v_g : \sum_{g \in G} (Dx^g)v_g \geqslant d, \sum_{g \in G} v_g = K, v \in \mathbb{R}_+^{|G|}\}.$$

If its solution v does not pass the *"Integrality Test"*, one must apply an ad-hoc branching scheme. The possible choices can be understood as extensions of the schemes discussed in Options 1 to 5.

Option 1. Branching on the aggregate variables y does not guarantee the elimination of all fractional solutions. As we have seen in the set partitioning case, no fractional solutions can be eliminated in this way. However for the general case, in some (if not all) fractional solutions, there exists a coordinate i for which $y_i = \sum_{g \in G} x_i^g v_g = \alpha \notin \mathbb{Z}$. Then one can create two branches

$$\sum_{g \in G} x_i^g v_g \leqslant \lfloor \alpha \rfloor \text{ and } \sum_{g \in G} x_i^g v_g \geqslant \lceil \alpha \rceil.$$

This additional constraint in the master does not change the structure of the pricing problem that becomes

$$\zeta = \min\{(c - \pi D)x - \mu_i x_i : x \in Z\}$$

where μ_i (resp. $-\mu_i$) is the dual variable associated to up-branch (resp. down-branch) constraint.

Options 3 and 4. If the original variables do not offer a large enough spectrum of branching objects (i.e., if the integrality of the aggregate y_i value does not yield an integer solution x to the original problem), one can call on an extended formulation, introducing auxiliary integer variables. Then one can branch on the auxiliary variables, either by dualizing the branching constraint in the master (Option 3) or, when possible, by enforcing it in the subproblem (Option 4). A natural approach is to exploit the extended formulation that is implicit to the solution of the pricing problem. For example, in the vehicle routing problem, solutions are the incidence vectors of the nodes in a route, whereas the edges defining the routes implicitly define the costs of the route; branching on the aggregated edge variables summed over all the vehicles allows one to eliminate all fractional solutions. For the cutting stock problem, solving the knapsack subproblem by dynamic programming amounts to searching for a longest path in a pseudo-polynomial size network whose nodes represent capacity consumption levels (see Section 13.5.4). Branching on the associated edge flows in this network permits one to eliminate all fractional solutions.

Options 2 and 5. For a general integer problem, a generalization of the Option 2 approach is to look for a pair consisting of an index j and an integer bound l_j for which $\sum_{g:x_j^g \geqslant l_j} \nu_g = \alpha \notin \mathbb{Z}$, and then create the two branches:

$$\sum_{g \in \hat{G}} \nu_g \geqslant \lceil \alpha \rceil \text{ or } \sum_{g \in G \setminus \hat{G}} \nu_g \geqslant K - \lfloor \alpha \rfloor \tag{13.53}$$

where $\hat{Z} = Z \cap \{x_j \geqslant l_j\} = \{x^g\}_{g \in \hat{G}}$. Then pricing is carried out independently over the two sets $\hat{Z}$ and $Z \setminus \hat{Z} = Z \cap \{x_j \leqslant l_j - 1\}$ on both branches. As in the set partitioning special case, one may have to consider sets $\hat{Z}$ defined by more than a single *component bound*. It is easy to show that if a solution ν does not pass the *"Integrality Test"* there must exists a branching set $\hat{Z} = Z \cap \{sx \geqslant l\}$, where $l \in \mathbb{Z}^n$ is a vector of bounds and $s \in \{-1,1\}^n$ defines the sign of each component bound, such that $\sum_{g:x^g \in \hat{Z}} \nu_g = \alpha \notin \mathbb{Z}$. Then, branching takes a form generalizing (13.53) and pricing is carried out independently for $\hat{Z}$ and its complementary sets: the technicalities are beyond the scope of this chapter (see the references provided in Section 13.7); in particular, to avoid the proliferation of the number of cases to consider when pricing, it is important to chose a branching set $\hat{Z}$ that is either a subset of a previously defined branching set or lies in the complement of all previously defined branching sets.

Option 1 can always be tried as a first attempt to eliminate a fractional solution. Although easy to implement, the resulting branching can be weak (low improvement in the dual bound). Options 3 and 4 are application specific schemes (whether the branching constraint can be enforced in the subproblem and whether this modifies its structure are very much dependent on the application). By comparison Option 5 is a generic scheme that can be applied to all applications for which adding bounds on the subproblem variables does not impair its solution (i.e., it works if Z is bounded). Typically it provides the strongest dual bound improvement.

13.3.6 Practical aspects

In developing a branch-and-price algorithm, there are many practical issues such as a proper initialization of the restricted master program, stabilization of the column generation procedure (as discussed in Section 13.3.4), early termination of the master LPs, adapting primal heuristics and preprocessing techniques to a column generation context, combining column and cut generation, and branching strategies. Note that the branching schemes of Section 13.3.5 must be understood as default schemes that are called upon after using possible branching on constraint strategies that can yield a more balanced search tree.

Initialization is traditionally carried out by running a primal heuristic and using the heuristic solutions as an initial set of columns. Another classical option is to run a sub-gradient or a volume algorithm to obtain an initial bundle of columns before going into the more computationally intensive LP based column generation procedure. An alternative is to run a dual heuristic to estimate the dual prices. These estimates are then used to define the cost of the artificial columns associated with each of the master constraints as presented in (13.50).

The column generation approach is often used in primal heuristics. A branch-and-price algorithm can be turned into a heuristic by solving the pricing problem heuristically and carrying out partial branching. A classical heuristic consists in solving the integer master program restricted to the columns generated at the root node using a standard MIP solver (hoping that this integer program is feasible). Another common approach is to apply iterative rounding of the master LP solution, which corresponds to plunging depth-first into the branch-and-price tree (partial backtracking yields diversification in this primal search). The branching scheme underlying such a rounding procedure is simpler than for exact branch-and-price (for instance one can branch directly on the master variables as only one branch is explored).

13.4 Resource or variable decomposition

The "classical" problem tackled by resource decomposition is the mixed integer program

$$
\begin{aligned}
(\text{MIP}) \qquad z^{\text{MIP}} = \min\, & cx + hy \\
& Gx + Hy \geqslant d \\
& x \in Z^n,\ y \in \mathbb{R}^p_+
\end{aligned}
$$

where the integer variables x are seen as the "important" decision variables (possibly representing the main investment decisions). One approach is then to decompose the optimization in two stages: first choosing x and then computing the associated optimal y. A feedback loop allowing one to adjust the x solution after obtaining pricing information from the optimization of y makes the Benders' approach different from simple hierarchical optimization.

In this section we first derive the Benders' reformulation in the space of the x variables and show how it can be solved using branch-and-cut. We then consider the case in which the y variables are integer variables, as well as the case with block diagonal structure in which the subproblem obtained when the x variables are fixed decomposes, and finally we discuss one computational aspect.

13.4.1 Benders' reformulation

The approach here is to rewrite MIP as a linear integer program just in the space of the integer variables x. First we rewrite the problem as

$$z^{\text{MIP}} = \min\{cx + \phi(x) : x \in \text{proj}_x(Q) \cap \mathbb{Z}^n\}$$

where

$$Q = \{(x,y) \in \mathbb{R}^n \times \mathbb{R}^p_+ : Gx + Hy \geqslant d\}$$

and

$$\phi(x) = \min\{hy : Hy \geqslant d - Gx, y \in \mathbb{R}^p_+\}$$

is the second stage problem that remains once the important variables have been fixed in the first stage. This can in turn be written as

$$z^{\text{MIP}} = \min\{cx + \sigma : x \in \text{proj}_x(Q) \cap \mathbb{Z}^n, (\sigma, x) \in P_\phi\}$$

where $P_\phi = \{(\sigma,x) : \sigma \geqslant \phi(x)\}$. Note that when x yields a feasible second stage problem, i.e., $x \in \text{proj}_x(Q)$, P_ϕ can be described by linear inequalities. By LP duality, $\phi(x) = \max\{u(d - Gx) : uH \leqslant h, u \in \mathbb{R}^m_+\} = \max_{t=1,\dots,T} u^t(d - Gx)$ where $\{u^t\}_{t=1}^T$ are the extreme points of $U = \{u \in \mathbb{R}^m_+ : uH \leqslant h\}$. In addition a polyhedral description of $\text{proj}_x(Q)$ is given by Theorem 13.3. Thus we obtain the reformulation:

$$\begin{aligned} \text{(RMIP)} \quad z^{\text{MIP}} = \min\ & cx + \sigma \\ & u^t(d - Gx) \leq \sigma \qquad \text{for } t = 1,\dots,T \\ & v^r(d - Gx) \leqslant 0 \qquad \text{for } r = 1,\dots,R \\ & x \in \mathbb{Z}^n,\ \sigma \in \mathbb{R}^1, \end{aligned}$$

where $\{u^t\}_{t=1}^T$ and $\{v^r\}_{r=1}^R$ are the extreme points and extreme rays of U respectively.

RMIP is a linear integer program with a very large (typically exponential) number of constraints. With modern mixed integer programming software, the natural way to solve such a problem is by branch-and-cut.

Specifically at each node of the enumeration tree, a linear programming relaxation is solved starting with a subset of the constraints of RMIP. If this linear program is infeasible, RMIP at that node is infeasible, and the node can be pruned. Otherwise if (σ^*, x^*) is the current linear programming solution, violated constraints are found by solving the linear programming separation problem

$$\phi(x^*) = \min\{hy : Hy \geqslant d - Gx^*, y \in \mathbb{R}^p_+\}, \tag{13.54}$$

or its dual $\max\{u(d - Gx^*) : uH \leqslant h, u \in \mathbb{R}^m_+\}$. There are three possibilities:

i) The linear programming separation problem (13.54) is infeasible and one obtains a new extreme ray v^r with $v^r(d - Gx^*) > 0$. (An extreme ray is obtained as the dual solution on termination of the simplex algorithm). The violated constraint $v^r(d - Gx) \leqslant 0$, called a *feasibility cut*, is added, and one iterates.
ii) ii) The linear programming separation subproblem is feasible, and one obtains a new dual extreme point u^t with $\phi(x^*) = u^t(d - Gx^*) > \sigma^*$. The violated constraint $\sigma \geqslant u^t(d - Gx)$, called an *optimality cut*, is added, and one iterates.
iii) The linear programming separation subproblem is feasible with optimal value $\phi(x^*) = \sigma^*$. Then (x^*, σ^*) is a solution to the linear programming relaxation of RMIP and the node is solved.

Example 9 *Consider the mixed integer program*

$$\begin{array}{rl} \min & -4x_1 - 7x_2 - 2y_1 - 0.25y_2 + 0.5y_3 \\ & -2x_1 - 3x_2 - 4y_1 + y_2 - 4y_3 \geqslant -9 \\ & -7x_1 - 5x_2 - 12y_1 - 2y_2 + 4y_3 \geqslant -11 \\ & x \leqslant 3,\ x \in \mathbb{Z}^2_+,\ y \in \mathbb{R}^3_+ \end{array}$$

where the feasible region is similar to that of Example 3.

The extreme rays $v^1 = (1,1)^T, v^2 = (2,1)^T$ of the feasible region of the dual $U = \{u \in \mathbb{R}^2_+ : -4u_1 - 12u_2 \leqslant -2, u_1 - 2u_2 \leqslant -0.25, -4u_1 + 4u_2 \leqslant 0.5\}$ were calculated in Example 3. The extreme points are $u^1 = (1/32, 5/32), u^2 = (1/20, 3/10)$, so the resulting complete reformulation RMIP *is:*

$$\begin{array}{rl} \min & \sigma - 4x_1 - 7x_2 \\ & -9x_1 - 8x_2 \geqslant -20 \\ & -11x_1 - 11x_2 \geqslant -29 \\ & \sigma - 1.15625x_1 - 0.875x_2 \geqslant -2 \\ & \sigma - 1.15x_1 - 0.9x_2 \geqslant -2.1 \\ & x \leqslant 3,\ x \in \mathbb{Z}^2_+,\ \sigma \in \mathbb{R}^1. \end{array}$$

Now we assume that the extreme points and rays of U are not known, and the problem is to be solved by branch-and-cut. One starts at the initial node 0 with only the bound constraints $0 \leqslant x \leqslant 3$ and dynamically adds Benders' cuts during branch-and-cut. We further assume that a lower bound of -100 on the optimal value of $\phi(x)$ is given.

Node 0. Iteration 1. *Solve the Master linear program:*

$$\begin{aligned}\zeta = \min \sigma - 4x_1 - 7x_2 \\ \sigma \geqslant -100 \\ x_1 \leqslant 3,\ x_2 \leqslant 3,\ x \in \mathbb{R}^2_+,\ \sigma \in \mathbb{R}^1.\end{aligned}$$

Solution of the LP Master $\zeta = -133, x = (3,3), \sigma = -100.$
Solve the separation linear program

$$\begin{array}{rl} \min & -2y_1 - 0.25y_2 + 0.5y_3 \\ & -4y_1 + y_2 - 4y_3 \geqslant -9+15 \\ & -12y_1 - 2y_2 + 4y_3 \geqslant -11+36 \\ & y \in \mathbb{R}^3_+. \end{array}$$

The ray $v = (1,1)$ *shows that the separation LP is infeasible. The corresponding feasibility cut* $-9x_1 - 8x_2 \geqslant -20$ *is added to the Master LP.*

Node 0. Iteration 2.
Solution of the LP Master: $\zeta = -117.5, x = (0,2.5), \sigma = -100.$
Solution of the Separation LP: $\phi(x) = 3/16 > \sigma$. $u = (1/32, 5/32)$. *The corresponding optimality cut* $\sigma - 1.15625x_1 - 0.875x_2 \geqslant -2$ *is added to the Master LP.*

Node 0. Iteration 3.
Solution of the LP Master: $\zeta = -17\frac{5}{16}, x = (0,2.5), \sigma = \frac{3}{16}.$
Solution of the Separation LP: $\phi(x) = \sigma$. *The LP at node 0 is solved.*
Create node 1 by branching on $x_2 \leqslant 2$ *and node 2 by branching on* $x_2 \geqslant 3$, *see Figure 13.5.*

Node 1. Iteration 1
The constraint $x_2 \leqslant 2$ *is added to the Master LP of Node 0, Iteration 3.*
Solution of the LP Master: $\zeta = -15.514, x = (4/9,2), \sigma = 0.264.$
Solution of the Separation LP: $\phi(x) = \sigma$. *The LP at node 1 is solved.*
Create node 3 by branching on $x_1 \leqslant 0$ *and node 4 by branching on* $x_1 \geqslant 1$.

Node 3. Iteration 1
The constraint $x_1 \leqslant 0$ *is added to the Master LP of Node 1, Iteration 1.*
Solution of the LP Master: $\zeta = -14.25, x = (0,2), \sigma = -0.25.$
Solution of the Separation LP: $\phi(x) = \sigma$. *The LP at node 3 is solved. The solution is integer. The value* -14.25 *and the solution* $x = (0,2), y = (0.25,0,0.5)$ *are stored. The node is pruned by optimality.*

Node 4. Iteration 1
The constraint $x_1 \geqslant 1$ *is added to the Master LP of Node 1, Iteration 1.*
Solution of the LP Master: $\zeta = -13.26$. *The node is pruned by bound.*

Node 2. Iteration 1
The constraint $x_2 \geqslant 3$ *is added to the Master LP of Node 0, Iteration 3.*

The LP Master is infeasible. The node is pruned by infeasibility.

All nodes have been pruned. The search is complete. The optimal solution is $x = (0,2), y = (0.25, 0, 0.5)$ *with value* -14.25. *The branch-and-cut tree is shown in Figure 13.5.*

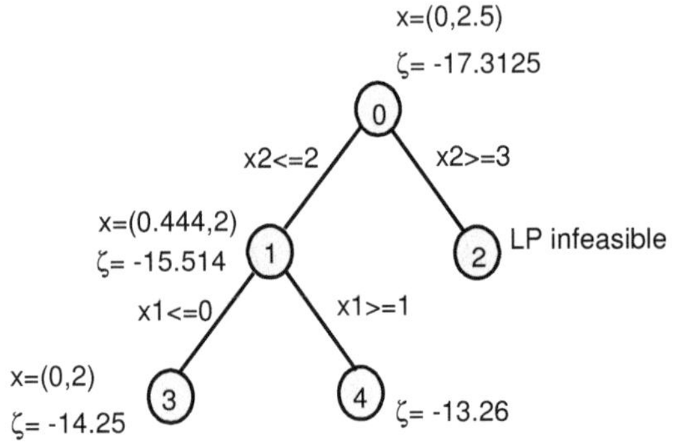

Fig. 13.5 Branch-and-Cut Tree for Benders' Approach

13.4.2 Benders with integer subproblems

The Benders' approach has also been found useful in tackling integer programming models of the form

$$\min\{cx + hy : Gx + Hy \geqslant d,\ x \in \{0,1\}^n,\ y \in Y \subseteq \mathbb{Z}^p\},$$

where the x variables are 0-1 and represent the "strategic decisions", and the y variables are also integer. Once the x variables are fixed, there remains a difficult combinatorial problem to find the best corresponding y in the second stage. Typical examples are vehicle routing (or multi-machine scheduling) in which the x variables may be an assignment of clients to vehicles (or jobs to machines) and the y variables describe the feasible tours of each vehicle (or the sequence of jobs on each machine).

As before one can design a Benders' reformulation and branch-and-cut algorithm in the (σ, x) variables:

$$z^{\text{MIP}} = \min\{cx + \sigma, \sigma \geqslant \phi(x), x \in \mathbb{Z}^n,\ \sigma \in \mathbb{R}^1\},$$

where $\phi(x) = \infty$ when $x \notin \text{proj}_x(Q)$. However the separation subproblem is no longer a linear program, but the integer program:

$$\phi(x) = \min\{hy : Hy \geqslant d - Gx, y \in Y\}. \tag{13.55}$$

Now one cannot easily derive a polyhedral description of the projection into the x-space as in the continuous subproblem case. The combinatorial subproblem (13.55) must be solved repeatedly at each branch-and-bound node. It is often tackled by constraint programming techniques, especially when it reduces to a feasibility problem (in many applications $h = 0$).

A naive variant of the algorithm presented in Section 13.4.1 is to solve the master problem to integer optimality before calling the second stage problem: one only calls the separation algorithm when RMIP has an integer solution $x^* \in \{0,1\}^n$. The separation is typically easier to solve in this case. This approach is often used when the subproblem is handled by constraint programming. There are three possible outcomes:

i) The separation subproblem is infeasible for the point $x^* \in \{0,1\}^n$, and one can add the infeasibility cut

$$\sum_{j:x_j^*=0} x_j + \sum_{j:x_j^*=1} (1 - x_j) \geqslant 1 \tag{13.56}$$

that cuts off the point x^*.

ii) The separation subproblem is feasible for x^*, but $\phi(x^*) > \sigma^*$. One can add the optimality cut

$$\sigma \geqslant \phi(x^*) - (\phi(x^*) - M)\Big(\sum_{j:x_j^*=0} x_j + \sum_{j:x_j^*=1} (1 - x_j)\Big)$$

that cuts off the point (σ^*, x^*), where M is a lower bound on ϕ.

iii) The separation subproblem is feasible for x^*, and $\phi(x^*) = \sigma^* = hy^*$. Now (x^*, y^*) is a feasible solution with value $cx^* + \phi(x^*)$. The node can be pruned by optimality.

This naive version has to be improved to have any chance of working in practice (for instance, in some applications one can add certain valid inequalities in the x variables a priori). In particular it is important to find inequalities that cut off more than just the point x^*. One case in which a slightly stronger inequality can be generated is that in which $x^* \in \{0,1\}$ infeasible implies x infeasible whenever $x \geqslant x^*$. In this case one searches for a minimal infeasible solution $\tilde{x} \leqslant x^*$ and the infeasibility cut (13.56) is replaced by the inequality:

$$\sum_{j:\tilde{x}_j=1} (1 - x_j) \geqslant 1$$

stating that in any feasible solution at least one variable with $\tilde{x}_j = 1$ must be set to zero.

Finally note that one can also work with a relaxation of (13.55) as any feasibility cut or optimality cut that is valid for the relaxation is valid for (13.55).

13.4.3 Block diagonal structure

In many applications MIP has block diagonal structure of the form

$$\begin{array}{llllll} \min cx + & h^1y^1 & + h^2y^2 & + \cdots & + h^Ky^K & \\ G^1x + & H^1y^1 & & & & \geqslant d^1 \\ G^2x + & & H^2y^2 & & & \geqslant d^2 \\ \ddots & & & \ddots & & \geqslant \vdots \\ G^Kx + & & & & H^Ky^K & \geqslant d^K \\ & x \in X, \; y^k \in Z^k \text{ for } k = 1,\ldots,K & & & & \end{array}$$

Here the second stage subproblem breaks up into K subproblems

$$\zeta^k = \min\{h^ky^k : H^ky^k \geqslant d^k - G^kx, y^k \in Z^k\} \text{ for } k = 1,\ldots,K.$$

One important and well-known case is that of two-stage stochastic linear and integer programming, where x represent the first stage decisions (discrete or otherwise). Then depending on a discrete probability distribution, one observes the random variables involving one or more elements of (G^k, H^k, h^k, d^k) with probability p_k before taking an optimal second stage decision y^k. Note that all the subproblems will have a similar structure in the relatively common situation in which the matrices H^k, G^k are independent of k.

We now consider an example in which all the costs are restricted to the x variables, but the subproblems are hard combinatorial problems.

Example 10 (Multi-Machine Job Assignment Problem)
There are K machines and n jobs. Each job j has a release date r_j and a due date d_j. The processing time of job j on machine k is p^k_j and the associated processing cost is c^k_j. The problem is to assign each job to one machine so that the jobs on each machine can be scheduled without preemption while respecting the release and due dates, and the sum of the assignment costs are minimized.

Letting $x^k_j = 1$ if job j is assigned to machine k, the problem can be written as

$$z^{\text{MIP}} = \min\Big\{\sum_{k=1}^{K}\sum_{j=1}^{n} c^k_j x^k_j : \sum_{k=1}^{K} x^k_j = 1 \text{ for } j = 1,\ldots,n, \; x^k \in Z^k \text{ for } k = 1,\ldots,K\Big\},$$

where $x^k \in Z^k$ if and only if the set $S^k = \{j : x^k_j = 1\}$ of jobs can be scheduled on machine k. The set Z^k can be represented as a linear integer program, but the feasibility problem for each machine is well-solved in practice by the "Cumulative Constraint" from Constraint Programming. Given a proposed assignment x^*, one calls the Cumulative Constraint *in turn for each of the K subproblems. Either x^* is a feasible assignment, or one or more infeasibility cuts*

$$\sum_{j \in S^k} x^k_j \leqslant |S^k| - 1,$$

are added (involving as small as possible a set S^k of infeasible jobs). Note that as the costs are limited to the x variables, there are no optimality cuts for this problem. Results are also significantly improved by the a priori addition of valid inequalities in the x_i^k variables.

13.4.4 Computational aspects

Much recent research has shown the importance of normalization in generating cutting planes, and Benders' algorithm is no exception. Returning to the algorithm outlined in Subsection 13.4.1, given (x^*, σ^*), a violated feasibility or optimality cut is generated if and only if there is no feasible point (x^*, y^*) attaining the present lower bound $cx^* + \sigma^*$, or equivalently the set

$$\{y \in \mathbb{R}^p_+ : Hy \geqslant d - Gx^*, hy \leqslant \sigma^*\} = \emptyset.$$

By Farkas' Lemma this holds if and only if

$$\{(u, u_0) \in \mathbb{R}^m_+ \times \mathbb{R}^1_+ : u(d - Gx^*) - u_0\sigma^* > 0, uH - u_0 h \leqslant 0\} \neq \emptyset.$$

Taking the normalization $\sum_{i=1}^m u_i + u_0 = 1$ motivated by the aim of generating a minimal infeasible subsystem of inequalities and also the fact that this normalization has been effective for other problems, the earlier separation problem (13.54) can be replaced by the linear program:

$$\begin{aligned} \zeta = \max\ & u(d - Gx^*) - u_0\sigma^* \\ & uH - u_0 h \leqslant 0 \\ & \sum_{i=1}^m u_i + u_0 = 1 \\ & u \in \mathbb{R}^m_+, u_0 \in \mathbb{R}^1_+. \end{aligned}$$

Now if $\zeta > 0$, the inequality $u(d - Gx) \leqslant u_0\sigma$ is violated by ζ. Note that this is a feasibility cut when $u_0 = 0$ and an optimality cut when $u_0 > 0$. A recent computational study has shown that Benders' algorithm is significantly more effective and requires far fewer iterations when this normalized separation problem is used.

13.5 Extended formulations: problem specific approaches

We now consider the use and derivation of extended formulations based on explicit problem structure in more detail.

Typically we again have a decomposition $X = Y \cap Z$ of the feasible region, and Z has some specific structure that we wish to exploit. In nearly all such cases a minimal inequality description of $\text{conv}(Z)$ in the original space of variables requires a very large number of constraints. However there is the possibility that one can find a compact extended formulation that is tight or at least considerably stronger than the initial formulation for Z. This section is mainly about such reformulations.

First it is natural to ask when there is hope of finding such a compact and tight extended formulation for Z. An important indication is given by the "Polynomial Equivalence of Optimization and Separation". Informally it states that, subject to certain technical conditions:

A family of problems $\min\{cx : x \in Z \subseteq \mathbb{Z}^n\}$ is polynomially solvable if and only if for all instances Z there is a polynomial separation algorithm for $\text{conv}(Z)$.

Assuming P $\neq$ NP, this tells us that a tight and compact extended formulation can only exist for a problem for which the optimization/separation problem is in P. However it gives no guarantee of the existence of such a formulation.

Below we briefly discuss ways in which "relatively compact" extended formulations can be used. Then we look at different ways to derive extended formulations. We have attempted to classify them according to the method of derivation. In particular we consider extended formulations based on variable splitting, dynamic programming algorithms, unions of polyhedra, explicit convex hull descriptions or the associated separation problem, and finally a couple of miscellaneous extended IP-formulations are presented.

13.5.1 Using compact extended formulations

Here we consider briefly different ways to make use of extended formulations that are compact or of "reasonable size".

Intersection

Given an initial formulation P of X, the decomposition $X = Y \cap Z$ and an extended formulation Q for Z, then $Q' = P \cap Q$ is an extended formulation for X. Assuming that Q is compact, one simple option is to feed the reformulated problem

$$\max\{cx + 0w : (x,w) \in Q', x \in \mathbb{Z}^n\}$$

to an MIP solver. Alternatively one might also try to improve the formulation of Y and combine this with the extended formulation Q so as to produce an even stronger reformulation, see Section 13.6.

Projection

Again given the decomposition $X = Y \cap Z$ and an extended formulation Q for Z, one may wish to avoid explicit introduction of the new variables $w \in \mathbb{R}^p$. One possibility is to use linear programming to provide a separation algorithm for $\text{proj}_x(Q)$.

Separation Algorithm

Given $Q = \{(x,w) \in \mathbb{R}^n_+ \times \mathbb{R}^p_+ : Gx + Hw \geqslant d\}$ and $x^* \in \mathbb{R}^n_+$,

i) check whether $Q(x^*) = \{w \in \mathbb{R}^p : Hw \geqslant d - Gx^*\} \neq \emptyset$. This can be tested by linear programming.
ii) If $Q(x^*) \neq \emptyset$, then $x^* \in \text{proj}_x(Q)$. Stop.
iii) If $Q(x^*) = \emptyset$, then by Farkas' lemma there exists $v^* \in V = \{v \in \mathbb{R}^m_+ : vH \leqslant 0\}$ with $v^*(d - Gx^*) > 0$ (v^* is obtained as a dual solution of the linear program used in i)). Then $v^*Gx \geqslant v^*d$ is a valid inequality for $\text{proj}_x(Q)$ cutting off x^*.

Note that the Minkowski non-compact extended formulation of Z (see Section 13.2) can be used in a similar manner to provide a separation algorithm for $\text{conv}(Z)$. However in this case a column generation approach (or some alternative) must be used, and the resulting column generation subproblem is the optimization problem over Z.

Inequality representation of $\text{proj}_x(Q)$

One can sometimes obtain an explicit polyhedral description of $\text{proj}_x(Q)$ by way of linear inequalities. In the simple cases the projection can be obtained directly from inspection of Q. Otherwise given $Q = \{(x,w) \in \mathbb{R}^n_+ \times \mathbb{R}^p_+ : Gx + Hw \geqslant d\}$, one may be able to describe all the extreme rays $\{v^1, \ldots, v^T\}$ of $V = \{v \in \mathbb{R}^m_+ : vH \leqslant 0\}$. This immediately gives the polyhedral description $\{x \in \mathbb{R}^n_+ : v^tGx \geqslant v^td,\ t = 1, \ldots, T\}$ of $\text{proj}_x(Q)$. Alternatively, a systematic method of projecting out variables one at a time, known as "Fourier-Motzkin elimination", can be used to eliminate the w variables in certain cases.

13.5.2 Variable splitting I: multi-commodity extended formulations

Using a multi-commodity extended formulation of the flows as for the directed Steiner tree problem presented in Example 4 is an example of variable splitting. Here we consider a more general fixed charge network flow problem, and present two further applications to the asymmetric traveling salesman problem and a lot-sizing problem.

Single-source fixed charge network flows

Given a directed graph or network $D = (V,A)$, a root $r \in V$, a vector $b \in \mathbb{R}^{|V|}$ of demands with $b_r < 0$, $b_v \geq 0$ for all $v \in V \setminus \{r\}$, unit flow costs $c \in \mathbb{R}^{|A|}$ and fixed costs $q \in \mathbb{R}^{|A|}_+$ for the use of an arc, the problem is to find a feasible flow that minimizes the sum of all the flow and fixed costs. This can be formulated as the mixed integer program:

$$
\begin{aligned}
\min \sum_{(u,v)\in A} & (q_{uv}x_{uv} + c_{uv}y_{uv}) & \\
\sum_{u\in\delta^-(v)} y_{uv} - \sum_{u\in\delta^+(v)} y_{vu} &= b_v & \text{for } v \in V \\
y_{uv} &\leq |b_r| x_{uv} & \text{for } (u,v) \in A \\
y \in \mathbb{R}^{|A|}_+, \; x &\in [0,1]^{|A|}. &
\end{aligned}
$$

The linear programming relaxation of this model does not provide good bounds because, when $y_{uv} > 0$ for some arc (u,v), one typically has $y_{uv} \ll |b_r|$. Thus $x_{uv} = \frac{y_{uv}}{|b_r|} \ll 1$, which means that the fixed cost term $q_{uv}x_{uv}$ seriously underestimates the correct fixed cost q_{uv}. One way to improve the formulation is to use a *multi-commodity* reformulation.

Let $T = \{v \in V \setminus \{r\} : b_v > 0\}$ be the set of terminals, or commodities. We now treat flow with destination $t \in T$ as a distinct commodity and define the variable w^t_{uv} to be the flow in arc (u,v) with destination $t \in T$. The resulting reformulation is

$$\min\{qx + cy : (x,y,w) \in Q, \; x \in \mathbb{Z}^{|A|}\},$$

where Q is the polyhedron

$$
\begin{aligned}
\sum_j w^t_{jr} - \sum_j w^t_{rj} &= -b_t & \text{for } t \in T \\
\sum_j w^t_{jv} - \sum_j w^t_{vj} &= 0 & \text{for } v \in V \setminus \{r,t\}, \; t \in T \\
\sum_j w^t_{jt} - \sum_j w^t_{tj} &= b_t & \text{for } t \in T \\
w^t_{ij} &\leq b_t x_{ij} & \text{for } (i,j) \in A, \; t \in T \\
y_{ij} &= \sum_{t\in T} w^t_{ij} & \text{for } (i,j) \in A \qquad (13.57) \\
y \in \mathbb{R}^{|A|}_+, \; w \in \mathbb{R}^{|A|\cdot|T|}_+, \; x &\in [0,1]^{|A|}. &
\end{aligned}
$$

Note that now the bound on the flow on the decision variable x_{ij} is $x_{ij} \geqslant \max_{t\in T} \frac{w^t_{ij}}{b_t}$. Again considering the linear programming relaxation, it is often the case that $w^t_{ij} = b_t$ for some commodity t, and this forces $x_{ij} = 1$, so that in this case the evaluation of the fixed cost for the arc (i,j) is exact.

For the special case of the directed Steiner tree problem introduced in Section 13.2.2, we noted that projection of the above formulation leads us to the reformulation $\min\{qx : x \in P', \, x \in \mathbb{Z}^n\}$ where P' is the polyhedron

$$\{x \in [0,1]^{|A|} : \sum_{i \in U, j \in V \setminus U} x_{ij} \geq 1, \text{ for } U \subseteq V \text{ with } r \in U, t \in T \cap (V \setminus U)\}.$$

As P' has an exponential number of constraints, one can use the max-flow/min-cut theorem to provide a polynomial separation algorithm for the polyhedron P'. Note that this is exactly the Benders' separation problem. For this special case, the linear programming relaxation has an optimal solution that solves the original problem in certain cases, in particular when the network is Series Parallel, or when $T = V \setminus \{r\}$ (minimum weight spanning tree) or $|T| = 2$ (shortest path).

More generally network design problems, in which the first stage variables are the choice of open arcs (or the multiples of capacity installed) and the second stage variables are the resulting flows, are often treated by Benders' approach.

TSP and sub-tour polytope: a three-index flow reformulation

It is well known and follows directly from the last reformulation that the asymmetric traveling salesman problem $(ATSP)$ can be written as the integer program:

$$\min \sum c_{ij} x_{ij} \tag{13.58}$$

$$\sum_j x_{ij} = 1 \qquad \text{for } i \in V \tag{13.59}$$

$$\sum_i x_{ij} = 1 \qquad \text{for } j \in V \tag{13.60}$$

$$\sum_{i \in U} \sum_{j \in V \setminus U} x_{ij} \geq 1 \qquad \text{for } U \subset V \text{ with } \phi \subset U \tag{13.61}$$

$$x \in \{0,1\}^{|A|}, \tag{13.62}$$

where $x_{ij} = 1$ if arc (i,j) lies on the tour. Let $Z = \{x \in \mathbb{Z}^{|A|}$ satisfying (13.61) and (13.62)$\}$. To model these connectivity constraints one can again use multi-commodity flows to ensure that one unit can flow from some root node $r \in V$ to every other node. This leads to the extended formulation Q for $\text{conv}(Z)$:

$$\sum_j w^t_{rj} - \sum_j w^t_{jr} = 1 \qquad \text{for } t \in V \setminus \{r\}$$

$$\sum_j w^t_{ij} - \sum_j w^t_{ji} = 0 \qquad \text{for } i \in V \setminus \{r,t\}, \, t \in V \setminus \{r\}$$

$$w^t_{ij} \leq x_{ij} \qquad \text{for } (i,j) \in A, \, t \in V \setminus \{r\}$$

$$x \in [0,1]^{|A|}, \, w \in [0,1]^{|A|(|V|-1)}$$

where w^t_{ij} is the flow in (i,j) from node r to node t. Here Q is a tight and compact extended formulation for Z.

For the symmetric traveling salesman problem on an undirected graph $G = (V,E)$, one can also make use of this reformulation by setting $y_e = x_{ij} + x_{ji}$, and adding $w^t_{ij} + w^{t'}_{ji} \leq y_e$ for all $(i,j) \in E$, $t,t' \in T$. Conversely it can be shown that projection onto the edge variables y gives back the well-known sub-tour elimination constraints $\sum_{e \in E(S)} y_e \leqslant |S| - 1$, where $E(S) = \{e = (i,j) \in E : i,j \in S\}$.

Uncapacitated lot-sizing

The uncapacitated lot-sizing problem involves time periods $t = 1,\ldots,n$, demands d_t in period t, production costs p_t, a set-up or fixed production cost q_t and a unit (end-of-period) storage cost h_t.

Letting x_t, s_t be the production and end-stock in period t, and $y_t \in \{0,1\}$ indicate if there is a set-up or not, a natural formulation as an MIP is given by:

$$\min \sum_{t=1}^{n} p_t x_t + \sum_{t=0}^{n} h_t s_t + \sum_{t=1}^{n} q_t y_t$$

$$s_{t-1} + x_t = d_t + s_t \qquad \text{for } t = 1,\ldots,n \tag{13.63}$$

$$x_t \leq M y_t \qquad \text{for } t = 1,\ldots,n \tag{13.64}$$

$$x \in \mathbb{R}^n_+,\ s \in \mathbb{R}^{n+1}_+,\ y \in \{0,1\}^n \tag{13.65}$$

with feasible region $X^{\text{LS-U}}$. We also use the notation $d_{ut} \equiv \sum_{j=u}^{t} d_j$

For this problem various polynomial algorithms are known, as well as a complete description of the convex hull of solutions given by an exponential number of facet-defining inequalities.

As this problem can be viewed as a special case of the fixed charge network flow problem, it is easy to add an additional subscript to the production and stock variables indicating the period t in which the units will be used to satisfy the demand.

Rescaling the resulting production variable, one can define new variables w_{ut} to be the fraction of the demand in period t satisfied by production in period u. This leads immediately to the following extended formulation $Q^{\text{LS-U}}$ for $X^{\text{LS-U}}$

$$\sum_{u=1}^{t} w_{ut} = 1 \qquad \text{for } t = 1,\ldots,n \tag{13.66}$$

$$w_{ut} \leq y_u \qquad \text{for } 1 \leq u \leq t \leq n \text{ with } d_{ut} > 0 \tag{13.67}$$

$$w \in \mathbb{R}^{(n-1)n/2}_+,\ y \in [0,1]^n \tag{13.68}$$

$$x_u = \sum_{t=u}^{n} d_t w_{ut} \qquad \text{for } u = 1,\ldots,n \tag{13.69}$$

$$s_t = \sum_{u \leq t} \sum_{t < \ell} d_\ell w_{u\ell} \qquad \text{for } t = 1,\ldots,n. \tag{13.70}$$

It can be shown that $\text{proj}_{x,s,y}(Q) = \text{conv}(X^{\text{LS-U}})$. It follows that the linear program

$$\min\{px + hs + qy, (x,s,y,w) \in Q^{\text{LS-U}}\}$$

has an optimal solution that solves the lot-sizing problem. Note that this formulation can also be obtained from the complete multi-commodity reformulation by elimination of the multi-commodity stock variables.

13.5.3 Variable splitting II

Here we present other reformulations obtained by variable splitting. Given an integer variable x with $0 \leqslant x \leqslant C$, it is possible to model it with binary variables, either with a so-called unary expansion:

$$x = \sum_{q=0}^{C} q z_q, \; \sum_{q=0}^{C} z_q = 1, z \in \{0,1\}^{C+1},$$

or with a binary expansion

$$x = \sum_{p=0}^{P} 2^p w_p \leqslant C, \; w \in \{0,1\}^{P+1},$$

where $P = \log_2 \lfloor C \rfloor$.

Time-indexed formulation

Machine scheduling problems are traditionally modeled using variables representing the starting time (or completion time) of the jobs. However, when using these variables, sequencing constraints (enforcing that a machine can only process one job at a time) are not easily modeled as linear mixed integer programs. Consider a single machine scheduling problem, in which there are n jobs with processing times p_j, release dates r_j and deadlines d_j for job j. Let the variable $y_j \in \mathbb{R}^1_+$ represent the start-time of job j, with $r_j \leqslant y_j \leqslant d_j - p_j$ for all j. These variables must satisfy the disjunctive constraints

$$y_j \geqslant y_i + p_i, \text{ or } y_i \geqslant y_j + p_j \text{ for } i \neq j$$

which are often modeled in mixed integer programming by the introduction of so-called big M constraints of the form $y_j \geqslant y_i + p_i - M(1 - \delta_{ij})$, where the 0-1 variable $\delta_{ij} = 1$ if job i precedes j.

Time-indexed variables, based on the unary decomposition of the y variables, allow one to build a linear IP-reformulation avoiding the big M constraints. Assuming integer processing times p_j, one can discretize the time horizon into T periods. One can then introduce new variables w_t^j where $w_t^j = 1$ if job j starts at the beginning of the interval $[t-1,t]$, and $w_t^j = 0$ otherwise. Then one obtains the IP-reformulation

$$\sum_{t=1}^{T} w_t^j = 1 \qquad \text{for } j = 1,\ldots,n$$

$$\sum_{j=1}^{n} \sum_{u=t-p_j+1}^{t} w_u^j \leqslant 1 \qquad \text{for } t = 1,\ldots,T-p_j+1,\ j = 1,\ldots,n$$

$$w_t^j \in \{0,1\} \qquad \text{for } t = r_j,\ldots,d_j - p_j + 1,\ j = 1,\ldots,n$$

where the first constraint ensures that each job j is started once, the second that at most one job is on the machine in each period, the range of definition of the variables handles the release and due dates, and the original variables are obtained by setting $y_j = \sum_t (t-1) w_t^j$.

Many different objective functions and constraints, such as precedence constraints, are easily handled using such time-indexed variables. Though pseudo-polynomial in size, the linear programming relaxation of this extended IP-formulation typically provides a much stronger bound than that of a big-M formulation in the (y,δ) variables.

Capacity-indexed variables

In capacitated vehicle routing problems with integral demands d_i and a vehicle capacity C, it has been proposed to apply variable splitting to the arc indicator variables. Specifically if $x^a = 1$ indicates that an arc a forms part of a vehicle route, $w_q^a = 1$ indicates that $a = (i,j)$ forms part of the route and the total load of the vehicle while traversing arc a is q. Now as a quantity d_i is delivered to client i, one must have

$$\sum_{a \in \delta^-(i)} w_q^a = \sum_{a \in \delta^+(i)} w_{q-d_i}^a \text{ for } d_i \leqslant q \leqslant C$$

and flow conservation becomes:

$$\sum_{q=0}^{C} \sum_{a \in \delta^-(i)} q w_q^a - \sum_{q=0}^{C} \sum_{a \in \delta^+(i)} q w_q^a = d_i \text{ for } i \in V.$$

Summing over $S \subset V$ and defining aggregate variables $y_q^-(S) = \sum_{a \in \delta^-(S)} w_q^a$ and $y_q^+(S) = \sum_{a \in \delta^+(S)} w_q^a$, one obtains integer knapsack sets

$$\sum_{q=0}^{C} q y_q^-(S) - \sum_{q=0}^{C} q y_q^+(S) = \sum_{i \in S} d_i,\ y_q^-(S),\ y_q^+(S) \in \mathbb{Z}_+^{C+1}$$

for which a variety of cutting planes can be generated. Here $x^a = \sum_q w_q^a$ provides the link to the original arc variables.

Fractionality-indexed variables and network dual MIPs

A network dual set is a mixed integer set in which all the constraints have two non-zero entries of $+1$ and -1 respectively. Thus we consider the set

$$X^{\text{ND}} = \{x \in \mathbb{R}^n : x_i - x_j \geqslant b_{ij} \text{ for } i, j \in N,\ x_i \in \mathbb{Z}^1 \text{ for } i \in I \subset N\}$$

where $N = \{1, \dots, n\}$. Such sets have been studied recently motivated by research on lot-sizing problems.

For the presentation here, we assume that each right-hand side value b_{ij} is a multiple of $\frac{1}{K}$, so we can write $b_{ij} = \lfloor b_{ij} \rfloor + \frac{h_{ij}}{K}$ with $h_{ij} \in \mathbb{Z}^1_+$ and $h_{ij} \in \{0, 1, \dots, K-1\}$. As a consequence of this assumption, one can assume that $Kx_i \in \mathbb{Z}^1$ for all i.

Following the idea of a unary expansion, we can write

$$Kx_i = K\lfloor x_i \rfloor + \sum_{h=0}^{K-1} hz_h,\ \sum_{h=0}^{K-1} z_h = 1,\ z \in \mathbb{Z}^K_+.$$

This in turn can be rewritten as

$$\begin{aligned} Kx_i &= \lfloor x_i \rfloor + (\lfloor x_i \rfloor + z_{K-1}) + (\lfloor x_i \rfloor + z_{K-2} + z_{K-1}) + \cdots + (\lfloor x_i \rfloor + z_1 + \cdots + z_{K-1}) \\ &= \sum_{h=0}^{K-1} (\lfloor x_i \rfloor + \sum_{j=K-h}^{K-1} z_j) \\ &= \sum_{h=0}^{K-1} w_i^h \end{aligned}$$

where $w_i^h = \lfloor x_i \rfloor$ if $x_i - \lfloor x_i \rfloor < \frac{K-h}{K}$ and $w_i^h = \lceil x_i \rceil$ if $x_i - \lfloor x_i \rfloor \geqslant \frac{K-h}{K}$.

With these variables, one obtains the extended formulation

$$x_i = \frac{1}{K} \sum_{h=0}^{K-1} w_i^h \ i \in N \tag{13.71}$$

$$w_i^t - w_j^{f(t)} \geqslant \lfloor b_{ij} \rfloor \qquad \text{for } t = 0, \dots, K - h_{ij} - 1,\ i, j \in N \tag{13.72}$$

$$w_i^t - w_j^{f(t)} \geqslant \lfloor b_{ij} \rfloor + 1 \qquad \text{for } t = K - h_{ij}, \dots, K-1,\ i, j \in N \tag{13.73}$$

$$x_i = w_i^h \qquad \text{for } h = 0, \dots, K-1,\ i \in I, \tag{13.74}$$

where $f(t) = t + h_{ij} \pmod K$. For the integer variables x_i with $i \in I$, one can use (13.74) to eliminate the corresponding w variables. The important observation is that this reformulation again has network dual structure, but with an integer right hand side. Thus the corresponding matrix is totally unimodular and the extremal solutions are integer. So it provides a tight and compact extended formulation for X^{ND}.

We now indicate briefly how network dual sets arise in lot-sizing problems.

Example 11 *Consider the set*

$$s_{k-1} + \sum_{u=k}^{t} C y_u + r_t \geqslant \sum_{u=k}^{t} d_u \qquad \text{for } 1 \leqslant k \leqslant t \leqslant n \tag{13.75}$$

$$s \in \mathbb{R}_+^{n+1}, r \in \mathbb{R}_+^n, y \in [0,1]^n, \tag{13.76}$$

known as the constant capacity Wagner-Whitin relaxation with backlogging, where s_t, y_t are the same stock and set-up variables introduced earlier for the lot-sizing problem, and r_t represents the backlog/shortage at the end of period t.

Introducing new variables: $z_t = \sum_{u=1}^t y_u$, $\sigma_{k-1} = -(s_{k-1} - Cz_{k-1} + \sum_{u=1}^{k-1} d_u)/C$ and $\rho_t = (r_t + Cz_t - \sum_{u=1}^t d_u)/C$, constraint (13.75) after division by C can be written as $\rho_t - \sigma_{t-1} \geqslant 0$, $\frac{1}{C} s_{k-1} \geqslant 0$ becomes $z_{k-1} - \sigma_{k-1} \geqslant (\sum_{u=1}^{k-1} d_u)/C$, $\frac{1}{C} r_t \geqslant 0$ becomes $\rho_t - z_t \geqslant -(\sum_{u=1}^t d_u)/C$, and $0 \leqslant y_t \leqslant 1$ becomes $0 \leqslant z_t - z_{t-1} \leqslant 1$.

Thus one obtains the reformulation:

$$\begin{aligned}
\rho_t - \sigma_{k-1} &\geqslant 0 && \text{for } 1 \leqslant k \leqslant t \leqslant n\\
z_{k-1} - \sigma_{k-1} &\geqslant \Big(\sum_{u=1}^{k-1} d_u\Big)/C && \text{for } k = 1, \dots, n\\
\rho_t - z_t &\geqslant -\Big(\sum_{u=1}^{t} d_u\Big)/C && \text{for } t = 1, \dots, n\\
-z_t + z_{t-1} &\geqslant -1 && \text{for } t = 1, \dots, n\\
z_t - z_{t-1} &\geqslant 0 && \text{for } t = 1, \dots, n\\
\rho, \sigma \in \mathbb{R}^n, z &\in \mathbb{Z}^n, &&
\end{aligned}$$

which is precisely a network dual MIP.

More generally when the b_t take arbitrary values, the extended formulation (13.71)–(13.74) can always be reduced to a size that is polynomial in F, the number of distinct fractional values taken by the continuous variables in the extreme point solutions. For the lot-sizing set (13.75)–(13.76), F is $\Theta(n^2)$, corresponding to the values 0 and $\sum_{u=k}^t d_u/C \pmod 1$, so that the extended formulation is both tight and compact.

13.5.4 Reformulations based on dynamic programming

In many cases, solving a problem by dynamic programming can be interpreted as transforming it to a shortest or longest path problem (in an appropriate network of possibly very large size). It is then natural to look for a reformulation as a network flow problem. More generally, a dynamic programming recursion can often be written as a linear program, and the dual of this linear program provides an extended formulation in which the variables indicate which terms are tight in the dynamic programming recursion. We demonstrate this with two examples, the first of which illustrates the simple case in which the dynamic program corresponds to a longest path algorithm.

The integer knapsack problem

Consider the integer knapsack problem:

$$z = \max\{\sum_{j=1}^{n} c_j x_j : \sum_{j=1}^{n} a_j x_j = b, x \in \mathbb{Z}_+^n\}$$

with $\{a_j\}_{j=1}^n$, b positive integers. (The standard inequality knapsack problem is obtained by taking $a_n = 1$ and $c_n = 0$). It is well-known that the dynamic programming recursion:

$$G(t) = \max_{j=1,\ldots,n: t-a_j \geqslant 0} \{c_j + G(t-a_j)\}$$

with $G(0) = 0$, can be used to find $z = G(b)$ and then the corresponding optimal solution. One can convert the recursion into a linear program in which the values $G(t)$ for $t = 0,\ldots,b$ are the variables:

$$\begin{aligned}
&\min G(b) \\
&G(t) - G(t-a_j) \geqslant c_j \qquad \text{for } t = a_j,\ldots,b,\ j = 1,\ldots,n \\
&G(0) = 0.
\end{aligned}$$

Defining dual variables $w_{j,t-a_j}$ for all t, j with $t - a_j \geqslant 0$, the linear programming dual is

$$\begin{aligned}
\max \sum_{j=1}^{n} \sum_{t=0}^{b-a_j} c_j w_{jt} & \\
\sum_j w_{jt} = +1 & \qquad \text{for } t = 0 \\
-\sum_j w_{j,t-a_j} + \sum_j w_{jt} = 0 & \qquad \text{for } t = 1,\ldots,b-1 \qquad (13.77) \\
-\sum_j w_{j,t-a_j} = -1 & \qquad \text{for } t = b \\
w_{jt} \geqslant 0 & \qquad \text{for } t = 0,1,\ldots,b-a_j,\ j = 1,\ldots,n.
\end{aligned}$$

The resulting problem can be viewed as a longest path problem in a network $D = (V,A)$ with nodes $V = \{0,1,\ldots,b\}$ and arcs $(t, t+a_j) \in A$ for all $t \in \{0,1,\ldots,b-a_j\}$

with weight c_j for all j. Any path from 0 to b corresponds to a feasible solution of the knapsack problem. Adding the equations $x_j = \sum_{t=0}^{b-a_j} w_{jt}$ that count the number of times j-type arcs are used, one has that the polyhedron is a tight extended formulation for $Z = \{x \in \mathbb{Z}^n_+ : \sum_{j=1}^n a_j x_j = b\}$.

An instance of the network corresponding to this extended formulation is shown in Figure 13.5.4.

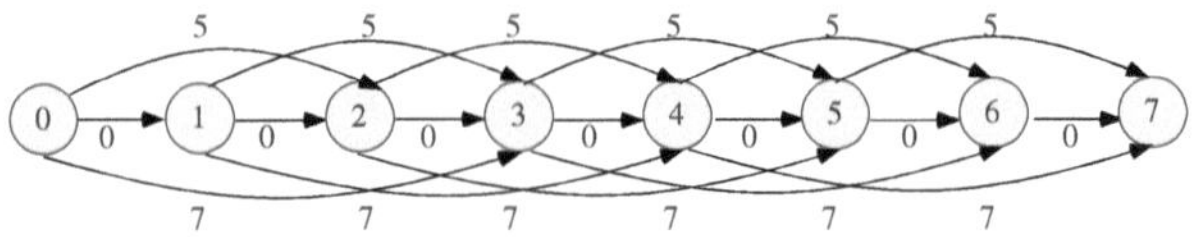

Fig. 13.6 Knapsack Longest Path: $a = (2,3,1), b = 7, c = (5,7,0)$

For this instance, the optimal linear programming solution $x_1 = \frac{7}{2}, x_2 = x_3 = 0$ is not integral and provides an upper bound on z of 17.5. The linear programming relaxation of the extended formulation has an optimal solution $w_0^1 = w_2^1 = w_4^2 = 1$, $w_t^j = 0$ otherwise, giving the optimal solution $x_1 = 2, x_2 = 1$ of value 17.

Optimal cardinality constrained subtrees of a tree

The second example involves a somewhat different dynamic program. One is given a rooted directed tree $T = (V,A)$ with node weights $c \in \mathbb{R}^{|V|}$. Node 1 is the root. The problem is to find an optimal rooted subtree with 1 as the root containing at most K nodes. A natural IP formulation is given by

$$\max\left\{\sum_{v\in V} c_v x_v : x_{p(v)} \geqslant x_v \text{ for } v \in V, \sum_{v\in V} x_v \leqslant K, x \in \{0,1\}^{|V|}\right\},$$

where $x_v = 1$ if v forms part of the subtree, $p(v)$ is the predecessor of v on the path from v to the root and $x_{p(1)} = 1$ by definition. For simplicity, we suppose that it is a binary tree and the left and right sons of node v are the nodes $2v$ and $2v+1$ respectively.

Let $H(v,k)$ denote the maximum weight subtree with at most k nodes rooted at v. The dynamic programming recursion is:

$$H(v,k) = \max\{H(v,k-1), c_v + \max_{t=0,\dots,k-1} [H(2v,t) + H(2v+1,k-1-t)]\},$$

where the first term in the maximization can be dropped for $v \neq 1$. Replacing the max by appropriate inequalities and taking the optimal value $H(1,K)$ as the objective function leads to the linear program:

$$\begin{aligned}
&\min H(1,K) \\
H(1,k) - H(1,k-1) &\geqslant 0 && \text{for } k = 1,\ldots,K \\
H(v,k) - H(2v,t) - H(2v+1,k-1-t) &\geqslant c_v && \text{for } v \in V,\ 0 \leqslant t < k \leqslant K \\
H(v,k) &\geqslant 0 && \text{for } v \in V,\ k = 0,\ldots,K.
\end{aligned}$$

Taking $y_{1,k}$ and $w_{v,k,t,k-1-t}$ as dual variables, we obtain

$$\begin{aligned}
\max \sum_{v\in V} c_v \sum_{k=1}^{K}\sum_{t=0}^{k-1} w_{v,k,t,k-1-t} \\
\sum_t w_{1,K,t,K-1-t} + y_{1,K} &\leqslant 1 \\
\sum_t w_{1,k,t,K-1-t} + y_{1,k} - y_{1,k+1} &\leqslant 1 && \text{for } k = 1,\ldots,K-1 \\
\sum_{t=0}^{k-1} w_{v,k,t,k-1-t} - \sum_{\kappa>k} w_{p(v),\kappa,k,\kappa-1-k} &\leqslant 0 && \text{for } v > 1 \text{ even},\ k = 1,\ldots,K \\
\sum_{t=0}^{k-1} w_{v,k,t,k-1-t} - \sum_{\kappa>k} w_{p(v),\kappa,\kappa-1-k,k} &\leqslant 0 && \text{for } v > 1 \text{ odd},\ k = 1,\ldots,K \\
w,y &\geqslant 0.
\end{aligned}$$

where $p(v) = \lfloor \frac{k}{2} \rfloor$. Here $w_{v,k,t,k-1-t} = 1$ means that the optimal tree contains a subtree rooted at v containing k nodes with t (resp $k-1-t$) nodes in the subtrees rooted in its left (resp. right) successors, and $y_{1k} = 1$ indicates that $H(1,k) = H(1,k-1)$. Setting $x_v = \sum_{k=1}^{K}\sum_{t=0}^{k-1} w_{v,k,t,k-1-t}$ allows us to complete the extended formulation.

13.5.5 The union of polyhedra

One of the very basic polyhedral results of relevance to integer programming concerns the union of polyhedra. Assume $P = \text{conv}(P^1 \cup \cdots \cup P^K)$ where $P^k = \{x \in R^n : A^k x \leq b^k\}$ and $C^k = \{x \in \mathbb{R}^n : A^k x \leq 0\}$ is the recession cone of P^k for all k.

Theorem 13.5 (Balas). *If $P^k \neq \phi$ and $C = C^k$ for $k = 1,\ldots,K$, then*

$$\begin{aligned}
\text{conv}(\cup_{k=1}^{K} P^k) = \text{proj}_x\{(x,w,\delta) \in \mathbb{R}^n \times \mathbb{R}^{nK} \times \mathbb{R}^K_+ : x = \textstyle\sum_{k=1}^{K} w^k, \\
A^k w^k \leq b^k \delta^k \ \textit{for } k = 1,\ldots,K,\ \textstyle\sum_{k=1}^{K} \delta^k = 1\}.
\end{aligned}$$

Disjunctions arise frequently in integer programming. Given a 0-1 set $X = P \cap \mathbb{Z}^n$ where $P = \{x \in \mathbb{R}^n : Ax \leq b,\ 0 \leqslant x \leqslant 1\}$ it is natural to select some variable j and consider the disjunction

$$P = P_j^0 \cup P_j^1 \text{ where } P_j^i = \{x \in P : x_j = i\} \text{ for } i = 0,1.$$

One use of extended formulations is to give tightened formulations that are then projected back into the original space. One example using the above disjunction is the lift-and-project approach presented in Chapter 11. Here we consider situations in which a problem becomes easy when the value of one variable is fixed. Then, if one can describe the convex hull of solutions when this variable is fixed, an extended formulation is obtained for the original set by taking the convex hull of the union of the convex hulls.

$1-k$ **configurations**

A $1-k$ configuration is a special 0-1 knapsack set of the form

$$Y = \Big\{(x_0,x) \in \{0,1\}^{n+1} : kx_0 + \sum_{j=1}^{n} x_j \leqslant n\Big\}.$$

To describe its convex hull $O(n^k)$ valid inequalities are needed. Now observe that $Y = Y^0 \cup Y^1$ where $Y^0 = \{x \in \{0,1\}^{n+1} : x_0 = 0\}$ and $Y^1 = \{x \in \{0,1\}^{n+1} : x_0 = 1, \sum_{j=1}^{n} x_j \leqslant n-k\}$. To obtain the convex hulls of Y^0 and Y^1, it suffices to drop the integrality constraints in their initial descriptions. Theorem 13.5 then gives the extended formulation Q:

$$\begin{aligned}
x_j &= x_{j0} + x_{j1} && \text{for } j = 0,\dots,n\\
x_{00} &= 0,\ 0 \leqslant x_{j0} \leqslant \delta_0 && \text{for } j = 1,\dots,n\\
x_{01} &= \delta_1,\ 0 \leqslant x_{j1} \leqslant \delta_1 && \text{for } j = 1,\dots,n\\
\sum_{j=1}^{n} x_{j1} &\leqslant (n-k)\delta_1\\
\delta_0 + \delta_1 &= 1,\ \delta \in \mathbb{R}^2_+.
\end{aligned}$$

After renaming x_{j1} as w_j, and replacing δ_1 by x_0 and x_{j0} by $x_j - w_j$ for $j = 1,\dots,n$, the resulting tight extended formulation is:

$$\begin{aligned}
0 \leqslant x_j - w_j &\leqslant 1 - x_0 && \text{for } j = 1,\dots,n\\
0 \leqslant w_j &\leqslant x_0 && \text{for } j = 1,\dots,n\\
\sum_{j=1}^{n} w_j &\leqslant (n-k)x_0\\
x \in [0,1]^{n+1},\ & w \in [0,1]^n.
\end{aligned}$$

Circular ones matrices

Consider the set $X = \{x \in \{0,1\}^n : Ax \leqslant b\}$ where A is a *circular ones* matrix, i.e, each row is either of the form

$$0\,0\,0\,1\,1\,1\,1\,0\,0$$

with 0's followed by 1's followed by 0's, or of the form

$$1\,1\,0\,0\,1\,1\,1\,1\,1$$

with 1's followed by 0's followed by 1's.

Let $P^k = \{x \in [0,1]^n : Ax \leqslant b, \sum_{j=1}^n x_j = k\}$ for $k = 0,\ldots,n$. Observe first that subtracting a row of the second type from a row of all 1's gives a row of the first type. Secondly a 0-1 matrix with only rows of the first type is known as a *consecutive 1's matrix*, and is known to be totally unimodular. It follows that $P^k = \text{conv}(P^k \cap \mathbb{Z}^n)$ and

$$\text{conv}(X) = \text{conv}(\cup_{k=0}^n P^k),$$

so a tight extended formulation is obtained immediately from Theorem 13.5.

13.5.6 From polyhedra and separation to extended formulations

Given the set $X \subseteq \mathbb{Z}^n$, suppose that a family of valid inequalities for X is known. This family explicitly or implicitly describes a polyhedron P containing the feasible region X. A first possibility is that the inequalities directly suggest an extended formulation.

Uncapacitated lot-sizing

Let $X^{\text{LS-U}}$ be as described in (13.63)–(13.65). It has been shown that every non-trivial facet-defining inequality for $\text{conv}(X^{\text{LS-U}})$ is of the form

$$\sum_{j \in S} x_j + \sum_{j \in L \setminus S} d_{jl} y_j \geqslant d_{1l} \tag{13.78}$$

where $L = \{1,\ldots,l\}$, $S \subseteq L$, $l = 1,\ldots,n$ and $d_{ut} \equiv \sum_{j=u}^t d_j$.

Let $\mu_{jl} = \min\{x_j, d_{jl}y_j\}$ for $1 \leqslant j \leqslant l \leqslant n$. One sees that (13.78) is satisfied for all S if and only if $\sum_{j=1}^l \min\{x_j, d_{jl}y_j\} \geqslant d_{1l}$. It follows immediately that a tight and compact extended formulation is given by the polyhedron consisting of the original constraints (13.63)–(13.65) less the integrality constraints, plus the constraints

$$\sum_{j=1}^{l} \mu_{jl} \geqslant d_{1l} \qquad \text{for } l = 1,\ldots,n$$

$$\mu_{jl} \leqslant x_j \qquad \text{for } 1 \leq j \leq l \leq n$$

$$\mu_{jl} \leqslant d_{jl} y_j \qquad \text{for } 1 \leq j \leq l \leq n.$$

A second possibility is that the separation problem for P can be formulated as an optimization problem that can be reduced to a linear program. Specifically suppose that $P = \{x \in \mathbb{R}^n : \pi^t x \geqslant \pi_0^t,\ t = 1, \dots, T\}$. Now $x^* \in P$ if and only if $\zeta \geqslant 0$ where $\zeta = \min_{t=1,\dots,T}(\pi^t x^* - \pi_0^t)$. Suppose now that the latter can be reformulated as a linear program:

$$\zeta = \min_{w}\{gx^* + hw - d_0 : Gx^* + Hw \geqslant d, w \in \mathbb{R}_+^p\}.$$

By LP duality, $\zeta \geqslant 0$ if and only if there exists a dual feasible solution with a non-negative value, namely

$$\{u \in \mathbb{R}^p : ud - uGx^* \geqslant d_0 - gx^*, uH \leqslant h, u \in \mathbb{R}_+^m\} \neq \emptyset.$$

Finally letting x vary, this gives us an extended formulation

$$Q = \{(x,u) \in \mathbb{R}^n \times \mathbb{R}^p : ud - uGx \geqslant d_0 - gx, uH \leqslant h, u \in \mathbb{R}_+^m\}$$

for which $P = \mathrm{proj}_x(Q)$.

Subtour elimination constraints

Consider the relaxation of the set of forests or symmetric traveling salesman tours consisting of the set Y defined by the exponential family of subtour elimination constraints. Specially set $Z = \cap_{k=1}^K Z^k$ where $Z^k = P_Z^k \cap \mathbb{Z}^{|E|}$ and

$$P_Z^k = \Big\{x \in [0,1]^{|E|} : \sum_{e \in E(S)} x_e \leqslant |S| - 1 \text{ for } S \subseteq V \text{ with } k \in S\Big\}.$$

Now consider the separation problem for $x^* \in [0,1]^{|E|}$. One sees that $x^* \in P_Z^k$ if and only if

$$\max_{S: k \in S \subseteq V} \Big\{ \sum_{e \in E(S)} x_e^* - |S \setminus \{k\}| \Big\} \leqslant 0.$$

Letting $v_j = 1$ if $j \in S$ and $u_e = 1$ if $e = (i,j) \in E(S)$, this optimization problem can be formulated as the IP

$$\zeta = \max \sum_{e \in E} x_e^* u_e - \sum_{j \in V \setminus \{k\}} v_j \qquad (13.79)$$

$$u_e \leqslant v_i, u_e \leqslant v_j \qquad \text{for } e = (i,j) \in E \qquad (13.80)$$

$$u_e \geqslant v_i + v_j - 1 \qquad \text{for } e = (i,j) \in E \qquad (13.81)$$

$$u \in \{0,1\}^{|E|},\ v \in \{0,1\}^{|V|},\ v_k = 1. \qquad (13.82)$$

It can then easily be shown that the constraints (13.81) can be dropped, and in addition that the integrality and bounds can be relaxed. It follows that $\zeta \leqslant 0$ if and only if $\eta \leqslant 0$ where

$$\begin{aligned}\eta = \max \sum_{e\in E} x_e^* u_e - \sum_{j\in V\setminus\{k\}} v_j & \\ u_e \leqslant v_i, u_e \leqslant v_j & \quad \text{for } e=(i,j)\in E \\ u \in \mathbb{R}^{|E|},\ v \in \mathbb{R}_+^{|V|}. &\end{aligned}$$

In this last linear program, either $\eta = 0$ or it is unbounded, so the dual of this linear program is feasible if and only if $\eta \leqslant 0$. In other words $x^* \in [0,1]^{|E|}$ is in P_Z^k if and only if $Q^k(x^*) \neq \emptyset$, where $Q^k(x)$ is the polyhedron

$$\begin{aligned} w_{ijk} + w_{jik} = x_e & \quad \text{for } e=(i,j)\in E \\ \sum_{j:j<i} w_{jik} + \sum_{j:j>i} w_{ijk} \leqslant 1 & \quad \text{for } i \neq k \\ \sum_{j:j<i} w_{jik} + \sum_{j:j>i} w_{ijk} \leqslant 0 & \quad \text{for } i = k \\ x \in \mathbb{R}^{|E|}, w_{ijk},\ w_{jik} \geqslant 0 & \quad \text{for } e=(i,j)\in E. \end{aligned}$$

13.5.7 Miscellaneous reformulations

There are several other reasons that might lead one to try an alterative formulation. An important one, already discussed in Section 13.3, is the problem of symmetry. A second is to find good branching directions for use in the context of branch-and-bound and branch-and-cut, and a third as before is to derive stronger linear programming bounds.

Symmetry-breaking in vertex coloring

Given a graph $G = (V,E)$ with $|V| = n$ and $|E| = m$, the textbook formulation for vertex coloring is based on the variables:
$y_k = 1$ if color k is used
$x_{ik} = 1$ if vertex i receives color k, where $k = 1,\ldots,K$ are the permissible colors.

This leads to the formulation:

$$\begin{aligned} \min \sum_k y_k & \\ \sum_k x_{ik} = 1 & \quad \text{for } i \in V \\ x_{ik} + x_{jk} \leqslant y_k & \quad \text{for } (i,j)\in E,\ k=1,\ldots,K \\ x_{ik} \leqslant y_k & \quad \text{for } i \in V,\ k=1,\ldots,K \\ x \in \{0,1\}^{|V|\times K},\ y \in \{0,1\}^K. & \end{aligned}$$

Clearly given any coloring, any permutation of the colors leads to essentially the same solution independently of the structure of the graph. To avoid this symmetry and also to tighten the formulation, it suffices to observe that, given any feasible coloring, each stable set can be assigned the color of its node of minimum index. Hence one can eliminate all variables x_{ik} with $k > i$, and also eliminate y_k by setting $y_k = x_{kk}$. Note that a similar approach applies for the bin packing problem of Example 5.

Boolean reformulation: 0-1 knapsack

Given two 0-1 knapsack sets of the form

$$X^i = \left\{x \in \{0,1\}^n : \sum_{j=1}^{n} a^i_j x_j \leqslant a^i_0\right\} \text{ for } i = 1,2$$

with $\{a^i_j\}$ positive integers, it is natural to ask when $X^1 = X^2$, or the two sets are equal. In particular one might be interested in finding the set of integer coefficients for which the right-hand side value a^i_0 or the sum of the weights $\sum_{j=1}^{n} a^i_j$ is minimum. It also seems likely that the corresponding formulation P_{X^i} is typically tighter when the coefficients are smaller.

Example 12 *Consider the knapsack set*

$$X = P^1 \cap \mathbb{Z}^n \text{ where } P^1 = \{x \in [0,1]^5 : 97x_1 + 65x_2 + 47x_3 + 46x_4 + 25x_5 \leqslant 136\}.$$

It can be verified that X can also be expressed as

$$X = P^2 \cap \mathbb{Z}^n \text{ where } P^2 = \{x \in [0,1]^5 : 5x_1 + 3x_2 + 3x_3 + 2x_4 + 1x_5 \leqslant 6\}$$

and this is the reformulation with integer coefficients with the minimum possible right hand-side value.

In addition it is easy to check that the extreme points of P^2 all lie in P^1 and thus $P^2 \subset P^1$.

Improved branching variables for an equality integer program

Consider the set

$$X = \{x \in \mathbb{Z}^n_+ : Ax = b\}$$

with $A \in Z^{m\times n}$ and $b \in \mathbb{Z}^m$. "Integer programming in a fixed number of variables is polynomially solvable" is one of the most fundamental results in integer programming. Lattice reformulations and the calculation of a reduced basis of a lattice play an important role in the proof of this result. Here we indicate briefly how a lattice reformulation can be used as a heuristic to look for effective branching variables. See the references cited in Section 13.7 for the appropriate lattice definitions.

Suppose that $x^0 \in \mathbb{Z}^n$ with $Ax^0 = b$, then X can be rewritten as $X = \{x \in \mathbb{Z}^n_+ : x = y + x^0, Ay = 0\}$. Now given a matrix $T \in \mathbb{Z}^{n\times(n-m)}$ such that $\{y \in \mathbb{Z}^n : Ay = 0\} = \{y \in \mathbb{Z}^n : y = Tw, w \in \mathbb{Z}^{n-m}\}$, then $X = \text{proj}_x(W)$ where

$$W = \{(x,w) \in \mathbb{R}^n_+ \times \mathbb{Z}^{n-m} : x = x^0 + Tw\}.$$

Here the extended IP-formulation does not provide tighter bounds. However it is possible to find an appropriate matrix T in polynomial time using a "reduced basis" algorithm, and for certain instances the new integer variables w are much more effective variables for branching than the original variables x.

Example 13 *Consider the set $X = \{x \in \mathbb{Z}^5_+ : ax = b\}$ where*

$$a = (11737, 7263, 9086, 32560, 20823), \quad b = 639253.$$

This has the extended formulation

$$\begin{pmatrix} x_1 \\ x_2 \\ x_3 \\ x_4 \\ x_5 \end{pmatrix} = \begin{pmatrix} 28 \\ 51 \\ -40 \\ 17 \\ -12 \end{pmatrix} + \begin{pmatrix} -1 & -1 & 7 & 239 \\ 0 & 0 & -11 & 616 \\ -1 & 0 & -10 & -445 \\ 0 & 1 & 4 & 33 \\ 1 & -1 & -2 & -207 \end{pmatrix} w, \quad x \in \mathbb{R}^5_+, \ w \in \mathbb{Z}^4.$$

Here branching on w_4, it is easily verified that $X = \emptyset$, whereas this is very hard to detect when branching on the x variables. In fact that the best MIP solvers all require millions of nodes to prove infeasibility for this tiny instance when using the original formulation.

13.5.8 Existence of polynomial size extended formulations

Yannakakis has shown that for the perfect matching polytope there is no extended formulation that is "symmetric" in a very general sense. This includes formulations in which one chooses a root, such as the extended formulation for the subtour polytope in Subsection 13.5.2. Thus it appears very unlikely that every family of IPs: $\min\{cx : x \in X\}$ that is polynomially solvable has a polynomial size extended formulation whose projection in the original variables provides $\text{conv}(X)$. It remains a major challenge to discover necessary and/or sufficient conditions for the existence of polynomial size extended formulations for such problems.

On the other hand it has very recently been shown that for the 0-1 knapsack problem $z = \min\{cx : ax \geqslant b, x \in \{0,1\}^n\}$, given any $\varepsilon > 0$, there exists a polynomial size extended formulation based on disjunctions for which the value z_{LP} of the linear programming relaxation is such that $z \leqslant (1+\varepsilon)z_{LP}$.

13.6 Hybrid algorithms and stronger dual bounds

Here we consider ways to obtain stronger dual bounds for the problem $z = \min\{cx : x \in Y \cap Z\}$ by using properties of both the sets Y and Z. Thus we assume as before that optimizing over Z is relatively easy, and now we assume also that we can either optimize over Y relatively easily, or that we have a cut generation routine for Y or some polyhedron P_Y containing $\text{conv}(Y)$.

13.6.1 *Lagrangean decomposition or price-and-price*

Here we assume that we can optimize efficiently over the set Z and also over the set Y. We reformulate IP by duplicating the x variables giving the new formulation:

$$\begin{aligned} &\min cy \\ &y - z = 0 \\ &y \in Y \\ &z \in Z. \end{aligned}$$

Applying Lagrangean relaxation, the subproblem with dual variables $u \in \mathbb{R}^n$ gives two subproblems $\min\{(c-u)y : y \in Y\}$ and $\min\{uz : z \in Z\}$, and by Theorem 13.4 the value of the resulting Lagrangean dual is $\min\{cx : x \in \text{conv}(Y) \cap \text{conv}(Z)\}$. This model can be solved either by dual methods such as a basic subgradient approach, or by a column generation approach (called Price-and-Price in this context).

In the latter case, *the restricted master problem at iteration t* is constructed from a set $\{y^i\}_{i \in I^{t-1}}$ of extreme points of $\text{conv}(Y)$ and a set $\{(z^j)\}_{j \in J^{t-1}}$ of extreme points of $\text{conv}(Z)$ giving the linear program:

$$\begin{aligned} \text{(RMPP)} \qquad & \min\ cx \\ & x - \sum_{i \in I^{t-1}} \lambda_i y^i = 0 \\ & \sum_{i \in I^{t-1}} \lambda_i = 1 \\ & x - \sum_{j \in J^{t-1}} \beta_j z^j = 0 \\ & \sum_{j \in J^{t-1}} \beta_j = 1 \\ & \lambda \in \mathbb{R}_+^{I^{t-1}},\ \beta \in \mathbb{R}_+^{J^{t-1}}, \end{aligned}$$

where the x variables can be easily eliminated. If (π, π_0, μ, μ_0) are optimal dual variables, one can solve the two pricing subproblems

$$\zeta^1 = \min\{\pi x - \pi_0,\ x \in Y\}$$

and

$$\zeta^2 = \min\{\mu x - \mu_0, x \in Z\}.$$

If $\zeta^1 < 0$ or $\zeta^2 < 0$, then the corresponding optimal solution provides a new column to be added, and one updates RMPP. If $\zeta^1 = \zeta^2 = 0$, the algorithm terminates. In practice, convergence (and dual instability) require an even more careful treatment in price-and-price than in branch-and-price.

13.6.2 Cut-and-price

Here we assume that we can optimize efficiently over the set $Z = \{x \in \mathbb{Z}^n_+ : Bx \geqslant b\}$ and that there is a cut generation algorithm for $Y = \{x \in \mathbb{Z}^n_+ : Dx \geqslant d\}$, or more realistically for some polyhedron P_Y containing $\mathrm{conv}(Y)$.

The restricted master problem at iteration t.
This problem is constructed from a set $\{x^i\}_{i \in I^{t-1}}$ of extreme points of $\mathrm{conv}(Z)$ and a set $\{(\alpha^j, \alpha_0^j)\}_{j \in J^{t-1}}$ of valid inequalities for P_Y (or Y), including the constraints $Dx \geqslant d$, giving the linear program:

$$\begin{aligned}
\text{(RMCP)} \qquad & \min cx \\
& x - \sum_{i \in I^{t-1}} \lambda_i x^i = 0 \\
& \sum_{i \in I^{t-1}} \lambda_i = 1 \\
& \sum_{j \in J^{t-1}} \alpha^j x \geqslant \alpha_0^j \qquad \text{for } j \in J^{t-1} \\
& \lambda \in \mathbb{R}^{I^{t-1}}_+ ,
\end{aligned}$$

Let (x, λ) be a primal optimal solution and $(\pi, \pi_0, \mu) \in \mathbb{R}^n \times \mathbb{R}^1 \times \mathbb{R}^{|J^{t-1}|}_+$ a dual optimal solution. Here again, one can eliminate the x variables, observing that $\pi = c - \sum_{j \in J^{t-1}} \mu_j^t \alpha^j$ from dual feasibility.

The order in which the two subproblems are solved below is arbitrary. We have chosen to give priority to column generation.

The Optimization Subproblem – Adding Columns.
Solve $\zeta^t = \min\{\pi x - \pi_0 : x \in Z\}$ with solution x^t.
If $\zeta^t < 0$, the column corresponding to x^t has negative reduced cost. Set $I^t = I^{t-1} \cup \{t\}$, set $t \leftarrow t+1$, and reoptimize RMCP.

Otherwise go to the (Constraint) Separation Subproblem.

The Separation Subproblem – Adding Constraints.
Solve the separation problem to see if the point $x = \sum_{i \in I_{t-1}} \lambda_i x^i$ can be cut off.
If a cut (α^t, α_0^t) is generated, set $J^t = J^{t-1} \cup \{t\}$, set $t \leftarrow t+1$, and reoptimize RMCP. Otherwise stop.

On termination $x = \sum_{i \in I^{t-1}} \lambda_i x^i \in P_Y \cap \text{conv}(Z)$. If the separation routine is exact for $\text{conv}(Y)$, the optimal value on termination is $\min\{cx : x \in \text{conv}(Y) \cap \text{conv}(Z)\}$ as with the other hybrid approaches.

Example 14 (The Vehicle Routing Problem)
Given a fleet of K identical vehicles of capacity C, and clients with demands d_i for $i = 1, \ldots, n$, the problem is to determine a delivery route for each vehicle starting and ending at the depot, so that the demand of each client is satisfied by exactly one vehicle, the total amount delivered by a vehicle does not exceed its capacity and the total travel costs are minimized. Consider a complete graph $H = (V, E)$, where the nodes $V = \{0, \ldots, n+1\}$ correspond to departure from the depot (node 0), the n customers and arrival at the depot (node $n+1$). The travel cost on edge e is c_e.

One possibility is to formulate the problem with K distinct vehicles based on the variables x_e^k such that $x_e^k = 1$ if edge e is traversed by vehicle k. However as the vehicles are identical, one can attempt to build a formulation using the variables x_e specifying the number of vehicles traversing edge e. Note that $x_e \in \{0,1\}$ for all e. This leads to a standard formulation

$$\min \sum_{e \in E} c_e x_e \tag{13.83}$$

$$\sum_{e \in \delta(i)} x_e = 2 \qquad \text{for } i \in V \setminus \{0, n+1\} \tag{13.84}$$

$$\sum_{e \in \delta(i)} x_e = K \qquad \text{for } i \in \{0, n+1\} \tag{13.85}$$

$$\sum_{e \in \delta(S)} x_e \geqslant 2B(S) \qquad \text{for } S \subseteq V \setminus \{0, n+1\} \tag{13.86}$$

$$x \in \{0,1\}^{|E|}, \tag{13.87}$$

where $B(S)$ denotes the minimum number of vehicles required to visit the set S of clients. The value of $B(S)$ is in fact the solution of a bin-packing problem, but a valid formulation is obtained if one ensures that the number of vehicles traveling through S is sufficient to satisfy the sum of the demands, i.e., $\sum_{e \in \delta(S)} x_e \geqslant 2\,(\sum_{i \in S} d_i)/C$.

On the other hand the price decomposition approach leads to an extended formulation in which one must select K feasible routes in such a way that each client is visited exactly once, leading to the master problem

$$\min\Big\{\sum_{g \in G}\big(\sum_{e} c_e x_e^g\big)\lambda_g : \sum_{g \in G}\big(\sum_{e \in \delta(i)} x_e^g\big)\lambda_g = 2 \ \text{ for } i \in V \setminus \{0, n+1\}, \tag{13.88}$$
$$\sum_{g \in G} \lambda_g = K,\ \lambda \in \{0,1\}^{|G|}\Big\}$$

where $Z = \{x^g\}_{g\in G}$ is the set of edge incidence vectors of feasible routes.

Unfortunately optimizing over this set Z is a hard problem that is not tractable in practice. This suggests using a relaxation of the set Z in which feasible routes are replaced by "q-routes", where a q-route *is a walk beginning at node 0 and ending at node $n+1$ (possibly visiting some client nodes more than once) for which the sum of the demands at the nodes visited does not exceed the capacity. It is easily seen that if the union of K q-routes satisfies the degree constraints (13.84)–(13.85), then one has K feasible routes. However, in the LP relaxation of (13.88), inequalities (13.86) are useful cuts. Thus, a hybrid cut-and-price approach can be implemented where the master is*

$$
\begin{aligned}
\min \sum_{e\in E} c_e\, x_e & \\
x \text{ satisfies } (13.84)(13.86) & \\
x_e = \sum_{p\in P} q_e^p \lambda_p \qquad & \text{for } e \in E \\
\sum_{p\in P} \lambda_p = K, & \\
x \in \mathbb{R}^{|E|},\ \lambda \in \{0,1\}^P &
\end{aligned}
$$

in a form ready to be tackled by a cut-and-price algorithm. The degree constraints are kept throughout, the constraints (13.86) are generated by cutting planes, and the q-routes are generated by column generation. Branching is dealt with by branching on the original x_e variables.

In practice one may choose to eliminate the original x_e variables by substitution, the cut generation problem is tackled using a heuristic because the calculation of the exact bin-packing value $B(S)$ is hard. Cuts of the form (13.86) can be generated by identifying small sets S that require more than one vehicle, or else inequalities are generated in which $B(S)$ is replaced by a lower bound $(\sum_{i\in S} d_i)/C$ or $\lceil(\sum_{i\in S} d_i)/C\rceil$. The separation problem for the inequalities with right hand side $(\sum_{i\in S} d_i)/C$ is solvable by maximum flow algorithms. For the column generation problem, a dynamic programming algorithm is used to find q-routes of minimum reduced cost.

13.7 Notes

Here we present notes providing some basic historical references, some references for results or applications mentioned in the chapter, and a few recent references concerning interesting extensions or examples of the ideas presented in the different sections.

13.7.1 Polyhedra

The result (Theorem 13.1) that every polyhedron is finitely generated by extreme points and extreme rays is due to Minkowski [73] and its converse, Theorem 13.3, to Weyl [95]. Meyer [72] showed that for integer programs and mixed integer programs with rational data the convex hull of solutions is a polyhedron. Theorem 13.2 on the representation of integer sets is proved in Giles and Pulleyblank [47]. For Farkas' lemma, see [36], and the earlier work of Fourier [40, 41].

13.7.2 Dantzig-Wolfe and price decomposition

The first use of an optimization subproblem to price out an exponential number of non-basic variables can be found in a paper of Ford and Fulkerson [39] on multi-commodity flows. Specifically they used a path-flow formulation, and then using the LP dual variables on the arcs, they solved shortest path problems for each commodity to find a path with negative reduced cost to enter the basis. This was closely followed by the Dantzig-Wolfe decomposition algorithm [22]. The first applications to discrete problems were the two papers on the cutting stock problem of Gilmore and Gomory [48, 49], introduced in Example 7, in which the subproblem was a knapsack problem. Eisenbrand and Shmonin [30] have recently shown that an optimal solution of the cutting stock problem is of polynomial size. Another early application was the model of Dzielinski and Gomory [28] on multi-item lot-sizing in which the subproblem was a single item lot-sizing problem.

Lagrangean relaxation

Early work showing the effectiveness of Lagrange multipliers in optimization can be found in Everett [35]. The first demonstration of the effectiveness of Lagrangean relaxation were the seminal papers of Held and Karp [54, 55] on the symmetric traveling salesman problem, based on the 1-tree relaxation that can be solved by a greedy algorithm. The survey of Geoffrion [45] clarified the properties of Lagrangean relaxation as applied to integer programs, including the integrality property, and Fisher [38] was one of several researchers to popularize the approach. See also Lemaréchal [65]

Later dual heuristics, or approximate algorithms for the Lagrangean dual, were proposed by numerous authors, including Bilde and Krarup [12] and Erlenkotter [33] for uncapacitated facility location, Wong [97] for directed Steiner trees and Balakrishnan, Magnanti and Wong [2] for multicommodity uncapacitated fixed charge network flows.

Solving the Lagrangean dual

The subgradient algorithm was proposed in Uzawa [86], Ermolev [34] and Polyak [78]. For early applications to integer programming, see Held and Karp [54, 55] and Held et al. [56]. Its variant, the volume algorithm, is due to Barahona and Anbil [5]. The cutting plane algorithm applied to the LP form of the Lagrangean dual is known as the method of Kelley [60] or Cheney-Goldstein [18]. It is the equivalent of the column generation approach but carried out in the dual space. The piecewise linear stabilization of column generation is studied in du Merle et al. [27] and Ben Amor et al. [8]. Stabilization based on smoothing dual prices was introduced by Neame [74] (using a convex combination of the current master dual solution and that of the previous iterate) and Wenges [94] (using a convex combination of the current dual solution and the dual solution that yielded the best Lagrangean bound). Recently Pessoa et al [76] have proved that at each iteration either the column generated with the smoothed prices has a strictly negative reduced cost for the restricted master, or one gets a strictly improving dual bound and a new associated stability center. Rousseau et al. [82] consider interior point stabilization.

The Bundle method, in which a quadratic term is introduced in the restricted master dual problem to penalize the deviation from a stability center, was developed by Lemaréchal [63], see also [64, 61]. There has been a large amount of research on such methods in the last few years. In many cases, and particular for very large problems in which the column generation approach is much too slow, the proximal bundle method has been effective. See Borndorfer et al. [13, 14] for applications to vehicle and duty scheduling in public transport and airline crew scheduling. Bundle's numerical performance is compared to LP based column generation in [16], and many references can be found in the thesis of Weider [93].

The analytic center cutting plane method (ACCPM) is due to Goffin and Vial [50].

Branching and column generation

For some of the first successful applications of integer programming column generation to routing problems, see Desrochers, Soumis et al. [26, 24] and Desrochers and Soumis [25]. See Soumis [84] for an annotated bibliography. The branching rule of Ryan and Foster appears in [83]. Vanderbeck and Wolsey [91, 89] discuss different branching strategies (extending the scheme of Ryan and Foster to cases where the master is not a set partitioning problem) and their inherent difficulties. Villeneuve et al. [92] suggest that one can always proceed by using standard branching in an "original" formulation and re-apply Dantzig-Wolfe reformulation to the problem augmented with branching constraints, but this leads to problems of symmetry in the case of multiple identical subproblems. Examples of branching on auxiliary variables, implicitly using an extended formulation as presented in Options 3 and 4 can be found in Belov et al. [7], Campêlo et al. [17] and Carvalho [23]. Elhallaoui et al. [31] consider the dynamic aggregation of set partitioning constraints. The scheme presented in Option 2 and its extension presented in Option 5 has been proposed as a generic all-purpose scheme by Vanderbeck [90] (although it normally assumes a bounded subproblem, it is can also be used in some application specific contexts in which the subproblem is unbounded).

13.7.3 Resource decomposition

The resource decomposition approach that became known as Benders' algorithm was proposed by Benders [9]. Geoffrion [43] produced the first important surveys on different ways to create decomposition algorithms, as well as an extension to nonlinear programs [44]. Geoffrion and Graves [46] reported a successful application of Benders' algorithm to a large distribution problem. Magnanti and Wong [68] studied ways to obtain strong Benders cuts. Since branch-and-cut algorithms became a practical possibility, this allows one to solve the Benders' reformulation directly by solving LP subproblems to generate cuts at the nodes rather than having to solve an integer program at each iteration, as proposed originally. Applications of Benders' algorithm to two stage stochastic programs are numerous, see for example Van Slyke and Wets [87]. The case with integer variables at both stages was treated by Laporte and Louveaux [62] among others. The multi-machine job assignment problem was first treated by Jain and Grossman [57]. The importance of normalization and the computational effectiveness of using a modified linear program to solve the separation problem is demonstrated in Fischetti et al. [37].

13.7.4 Extended formulations

Apart from Minkowski's representation of a polyhedron, extended formulations were not considered systematically as a tool for modeling integer programs until the 70's.

Grötschel, Lovasz and Schrijver's paper on the equivalence of optimization and separation [52] implies that, unless $P = NP$, one can only hope to find tight and compact extended formulations for integer programs if the corresponding optimization problem is polynomially solvable. Balas and Pulleyblank [4] gave an extended formulation for the perfectly matchable subgraph polytope of a bipartite graph and extended formulations have been proposed for a variety of combinatorial optimization problems in the last twenty years.

Variable splitting I: multi-commodity extended formulations

Rardin and Choe [81] explored the effectiveness of multi-commodity reformulations, and Wong [96] showed that the multi-commodity reformulation gives the spanning tree polytope. For the Steiner problem on series parallel graphs, see Prodon et al. [80]. Bilde and Krarup [11] showed that the extended facility location reformulation for uncapacitated lot-sizing was integral, and later Eppen and Martin [32] proposed an alternative formulation. The book of Pochet and Wolsey [77] contains numerous reformulations for different single and multi-item lot-sizing problems.

Variable splitting II

Pritsker et al. [79] contains one of the first uses of a time-indexed formulation for a scheduling problem. Gouveia [51] demonstrates the use of capacity indexed variables. The reformulation of network dual MIPs was studied in Conforti et al. [19], and the specific formulation proposed here is from Conforti et al. [21]. The first compact extended formulation for the constant capacity Wagner-Whitin relaxation with backlogging is due to Van Vyve [88].

Extended formulations based on dynamic programming

Martin [69] and Eppen and Martin [32] show how dynamic programs can be used to derive extended formulations. The longest/shortest path formulations for knapsack problems were known in the early 70's and probably date from the work of Gilmore and Gomory [48] on knapsack functions or Gomory on group problems. For dynamic programs that are not of the shortest path type, see Martin et al. [71]. The cardinality constrained problem is a natural generalization of the problem of finding an optimal subtree of a tree.

The union of polyhedra

The characterization of the convex hull of the union of polyhedra is due to Balas [3]. Recently Conforti and Wolsey [20] show how the union of polyhedra can be used to develop compact and tight extended formulations for several problems whose complexity was not previously known.

$1-k$ configurations are studied by Padberg [75]. Circular ones matrices are treated in Bartholdi et al. [6], see also Eisenbrand et al. [29].

From polyhedra and separation to extended formulations

Martin [70] demonstrates how LP separation algorithms can lead to extended formulations.

Miscellaneous

Equivalent knapsack problems are studied in Bradley et al. [15]. The polynomiality of IP with a fixed number of variables is due to H.W. Lenstra, Jr., [67] and the lattice reformulation demonstrated in the example was proposed by Aardal and A.K. Lenstra [1]. See Lenstra, Lenstra and Lovász [66] for properties of reduced bases and a polynomial algorithm to compute a reduced basis.

Existence of polynomial size extended formulations

Yannakakis [98] presents lower bounds on the size of an extended formulation for a given class of problems, and shows that even though weighted matching is polynomially solvable, it is most unlikely that there is a tight and compact extended formulation. The existence of polynomial size extended formulations approximating the convex hull of the 0-1 knapsack polytope is from Bienstock and McClosky [10].

13.7.5 Hybrid algorithms and stronger dual bounds

For Lagrangean decomposition, see Jornsten and Nasberg [59] and Guignard and Kim [53]. For cut-and-price, recent papers include Fukasawa et al. [42] on vehicle routing and Ochoa et al. [85] on capacitated spanning trees. In the latter paper use was also made of the capacity-indexed variables from subsection 13.5.3. Jans and Degraeve [58] combine an extended formulation and column generation for a multi-item lot-sizing problem.

References

1. K. Aardal and A.K. Lenstra, *Hard equality constrained integer knapsacks, Erratum:* Mathematics of Operations Research *31, 2006, page 846*, Mathematics of Operations Research 29 (2004) 724–738.
2. A. Balakrishnan, T.L. Magnanti, and R.T. Wong, *A dual ascent procedure for large-scale uncapacitated network design*, Operations Research 37 (1989) 716–740.
3. E. Balas, *Disjunctive programming: properties of the convex hull of feasible points*, originally as GSIA Management Science Research Report MSRR 348, Carnegie Mellon University,1974, Discrete Applied Mathematics 89 (1998) 1–44.
4. E. Balas and W.R. Pulleyblank, *The perfectly matchable subgraph polytope of a bipartite graph*, Networks 13 (1983) 495–516.
5. F. Barahona and R. Anbil, *The volume algorithm: Producing primal solutions with a subgradient method*, Mathematical Programming 87 (2000) 385–399.
6. J.J. Bartholdi, J.B. Orlin, and H. Ratliff, *Cyclic scheduling via integer programs with circular ones*, Mathematical Programming 28 (1980) 1074–1085.
7. G. Belov, A.N. Letchford, and E. Uchoa, *A node-flow model for the 1D stock cutting: robust branch-cut-and-price*, Tech. report, University of Lancaster, 2005.

8. H. Ben Amor, J. Desrosiers, and A. Frangioni, *On the choice of explicit stabilizing terms in column generation*, Discrete Applied Mathematics 157 (2009) 1167–1184.
9. J.F. Benders, *Partitioning procedures for solving mixed variables programming problems*, Numerische Mathematik 4 (1962) 238–252.
10. D. Bienstock and B. McClosky, *Tightening simple mixed-integer sets with guaranteed bounds*, Tech. report, Columbia University, New York, July 2008.
11. O. Bilde and J. Krarup, *Plant location, set covering and economic lot sizes: An* $O(mn)$ *algorithm for structured problems*, Optimierung bei Graphentheoretischen und Ganzzahligen Probleme (L. Collatz et al., ed.), Birkhauser Verlag, Basel, 1977, pp. 155–180.
12. O. Bilde and J. Krarup, *Sharp lower bounds and efficient algorithms for the simple plant location problem*, Annals of Discrete Mathematics 1 (1977) 79–97.
13. R. Borndörfer, A. Löbel, and S. Weider, *A bundle method for integrated multi-depot vehicle and duty scheduling in public transit*, ZIB Report 04-14, Konrad-Zuse Zentrum, Berlin, 2004.
14. R. Borndörfer, U. Schelten, T. Schlechter, and S. Weider, *A column generation approach to airline crew scheduling*, ZIB Report 05-37, Konrad-Zuse Zentrum, Berlin, 2005.
15. G.H. Bradley, P.L. Hammer, and L.A. Wolsey, *Coefficent reduction for inequalities in 0-1 variables*, Mathematical Programming 7 (1974) 263–282.
16. O. Briant, C. Lemaréchal, Ph. Meurdesoif, S. Michel, N. Perrot, and F. Vanderbeck, *Comparison of bundle and classical column generation*, Mathematical Programming 113 (2008) 299–344.
17. M. Campêlo, V. Campos, and R. Corréa, *On the asymmetric representatives formulation for the vertex coloring problem*, Notes in Discrete Mathematics 19 (2005) 337–343.
18. E. Cheney and A. Goldstein, *Newton's method for convex programming and Tchebycheff approximations*, Numerische Mathematik 1 (1959) 253–268.
19. M. Conforti, M. Di Summa, F. Eisenbrand, and L.A. Wolsey, *Network formulations of mixed integer programs*, Mathematics of Operations Research 34 (2009) 194–209.
20. M. Conforti and L.A. Wolsey, *Compact formulations as a union of polyhedra*, Mathematical Programming 114 (2008) 277–289.
21. M. Conforti, L.A. Wolsey, and G. Zambelli, *Projecting an extended formulation for mixed integer covers on bipartite graphs*, Tech. report, University of Padua, November 2008.
22. G.B. Dantzig and P. Wolfe, *Decomposition principle for linear programs*, Operations Research 8 (1960) 101–111.
23. J.V. de Carvalho, *Exact solution of bin packing problems using column generation and branch-and-bound*, Annals of Opererations Research 86 (1999) 629–659.
24. J. Desrosiers, Y. Dumas, M.M. Solomon, and F. Soumis, *Time constrained routing and scheduling*, Network Routing (C.L. Monma M.O. Ball, T.L. Magnanti and G.L. Nemhauser, eds.), Handbooks in Operations Research and Management Science, Vol. 8, Elsevier, 1995.
25. J. Desrosiers and F. Soumis, *A column generation approach to the urban transit crew scheduling problem*, Transportation Science 23 (1989) 1–13.
26. J. Desrosiers, F. Soumis, and M. Desrochers, *Routing with time windows by column generation*, Networks 14 (1984) 545–565.
27. O. Du Merle, D. Villeneuve, J. Desrosiers, and P. Hansen, *Stabilized column generation*, Discrete Mathematics 194 (1999) 229–237.
28. B. Dzielinski and R. Gomory, *Optimal programming of lot-sizes, inventories and labor allocations*, Management Science 11 (1965) 874–890.
29. F. Eisenbrand, G. Oriolo, G. Stauffer, and P. Ventura, *Circular ones matrices and the stable set polytope of quasi-line graphs*, Integer Programming and Combinatorial Optimization, IPCO 2005 (M. Jünger and V. Kaibel, eds.), Lecture Notes in Computer Science 3509, Springer, 2005, pp. 291–305.
30. F. Eisenbrand and G. Shmonin, *Carathéodory bounds for integer cones*, Operations Research Letters 34 (2006) 564–568.
31. I. Elhallaoui, D. Villeneuve, F. Soumis, and G. Desaulniers, *Dynamic aggregation of set-partitioning constraints in column generation*, Operations Research 53 (2005) 632–645.
32. G.D. Eppen and R.K. Martin, *Solving multi-item capacitated lot-sizing problems using variable definition*, Operations Research 35 (1987) 832–848.

33. D. Erlenkotter, *A dual-based procedure for uncapacitated facility location*, Operations Research 26 (1978) 992–1009.
34. Y.M. Ermolèv, *Methods of solution of nonlinear extremal problems*, Kibernetica 2 (1966) 1–17.
35. H. Everett III, *Generalized lagrange multiplier method for solving problems of optimal allocation of resources*, Operations Research 11 (1963) 399–417.
36. Gy. Farkas, *On the applications of the mechanical principle of Fourier*, Mathematikai és Természettudományi Értesotö 12 (1894) 457–472.
37. M. Fischetti, D. Salvagnin, and A. Zanette, *Minimal infeasible subsystems and Benders' cuts*, Mathematical Programming to appear (2009).
38. M.L. Fisher, *The lagrangean relaxation method for solving integer programming problems*, Management Science 27 (1981) 1–18.
39. L.R. Ford, Jr. and D.R. Fulkerson, *A suggested computation for maximal multi-commodity network flows*, Management Science 5 (1958) 97–101.
40. J.B.J. Fourier, *Solution d'une question particulière du calcul des inégalités*, Nouveau Bulletin des Sciences par la Société Philomatique de Paris (1826) 317–319.
41. J.B.J. Fourier, from 1824, republished as Second extrait in oeuvres de fourier, tome ii (G. Darboux, ed.), Gauthier-Villars, Paris, 1890, see D.A. Kohler, Translation of a report by Fourier on his work on linear inequalities, Opsearch 10 (1973) 38–42.
42. R. Fukosawa, H. Longo, J. Lysgaard, M. Reis, E. Uchoa, and R.F. Werneck, *Robust branch-and-cut-and-price for the capacitated vehicle routing problem*, Mathematical Programming 106 (2006) 491–511.
43. A.M. Geoffrion, *Elements of large scale mathematical programming I and II*, Management Science 16 (1970) 652–691.
44. A.M. Geoffrion, *Generalized Benders' decomposition*, Journal of Optimization Theory and Applications 10 (1972) 237–260.
45. A.M. Geoffrion, *Lagrangean relaxation for integer programming*, Mathematical Programming Study 2 (1974) 82–114.
46. A.M. Geoffrion and G.W. Graves, *Multicommodity distribution design by Benders' decomposition*, Management Science 20 (1974) 822–844.
47. R. Giles and W.R. Pulleyblank, *Total dual integrality and integral polyhedra*, Linear algebra and its applications 25 (1979) 191–196.
48. P.C. Gilmore and R.E. Gomory, *A linear programming approach to the cutting stock problem*, Operations Research 9 (1961) 849–859.
49. P.C. Gilmore and R.E. Gomory, *A linear programming approach to the cutting stock problem: Part ii*, Operations Research 11 (1963) 863–888.
50. J.-L. Goffin and J.-P. Vial, *Convex non-differentiable optimization: a survey focused on the analytic center cutting plane method*, Optimization Methods and Software 17 (2002) 805–867.
51. L. Gouveia, *A 2n constraint formulation for the capacitated minimal spanning tree problem*, Operations Research 43 (1995) 130–141.
52. M. Grötschel, L. Lovász, and A. Schrijver, *The ellipsoid method and its consequences in combinatorial optimization*, Combinatorica 1 (1981) 169–197.
53. M. Guignard and S. Kim, *Lagrangean decomposition for integer programming: Theory and applications*, RAIRO 21 (1987) 307–323.
54. M. Held and R.M. Karp, *The traveling salesman problem and minimum spanning trees*, Operations Research 18 (1970) 1138–1162.
55. M. Held and R.M. Karp, *The traveling salesman problem and minimum spanning trees: Part II*, Mathematical Programming 1 (1971) 6–25.
56. M. Held, P. Wolfe, and H.P. Crowder, *Validation of subgradient optimization*, Mathematical Programming 6 (1974) 62–88.
57. V. Jain and I.E. Grossman, *Algorithms for hybrid milp/clp models for a class of optimization problems*, INFORMS J. Computing 13 (2001) 258–276.
58. R. Jans and Z. Degraeve, *Improved lower bounds for the capacitated lot sizing problem with set-up times*, Operations Research Letters 32 (2004) 185–195.

59. K. Jornsten and M. Nasberg, *A new Lagrangian relaxation approach to the generalized assignment problem*, European Journal of Operational Research 27 (1986) 313–323.
60. J.E. Kelley, *The cutting plane method for solving convex programs*, SIAM Journal 8 (1960) 703–712.
61. K.C. Kiwiel, *An aggregate subgradient method for nonsmooth convex minimization*, Mathematical Programming 27 (1983) 320–341.
62. G. Laporte and F.V. Louveaux, *The integer L-shaped method for stochastic integer programs with complete recourse*, Operations Research Letters 13 (1993) 133–142.
63. C. Lemaréchal, *An algorithm for minimizing convex functions*, Information Processing '74 (J.L. Rosenfeld, ed.), North Holland, 1974, pp. 552–556.
64. C. Lemaréchal, *Nonsmooth optimization and descent methods*, Tech. report, IIASA, 1978.
65. C. Lemaréchal, *Lagrangean relaxation*, Computational Combinatorial Optimization (M. Jünger and D. Naddef, eds.), Lecture Notes in Computer Science 2241, Springer, 2001, pp. 112–156.
66. A.K. Lenstra, H.W. Lenstra, Jr., and L. Lovász, *Factoring polynomials with rational coefficients*, Mathematische Annalen 261 (1982) 515–534.
67. H.W. Lenstra, Jr., *Integer programming with a fixed number of variables*, Mathematics of Operations Research 8 (1983) 538–547.
68. T.L. Magnanti and R.T. Wong, *Accelerated Benders' decomposition: Algorithmic enhancement and model selection criteria*, Operations Research 29 (1981) 464–484.
69. R.K. Martin, *Generating alternative mixed integer programming models using variable definition*, Operations Research 35 (1987) 820–831.
70. R.K. Martin, *Using separation algorithms to generate mixed integer model reformulations*, Operations Research Letters 10 (1991) 119–128.
71. R.K. Martin, R.L. Rardin, and B.A. Campbell, *Polyhedral characterization of discrete dynamic programming*, Operations Research 38 (1990) 127–138.
72. R.R. Meyer, *On the existence of optimal solutions to integer and mixed integer programming problems*, Mathematical Programming 7 (1974) 223–235.
73. H. Minkowski, *Geometrie der Zahlen (erste Lieferung)*, Teubner, Leipzig, 1986.
74. P.J. Neame, *Nonsmooth dual methods in integer programing*, Ph.D. thesis, Depart. of Math. and Statistics, The University of Melbourne, 1999.
75. M.W. Padberg, *$(1,k)$-configurations and facets for packing problems*, Mathematical Programming 18 (1980) 94–99.
76. A. Pessoa, E. Uchoa, M. Poggi de Aragao, and R. Rodrigues, *Algorithms over arc-time indexed formulations for single and parallel machine scheduling problems*, Tech. report, Rio de Janeiro, 2009.
77. Y. Pochet and L.A. Wolsey, *Production planning by mixed-integer programming*, Springer Series in Operations Research and Financial Engineering, Springer, New York, 2006.
78. B.T. Polyak, *A general method for solving extremum problems*, Soviet Mathematic Doklady 8 (1967) 593–597.
79. A.A.B. Pritsker, L.J. Watters, and P.J. Wolfe, *Multiproject scheduling with limited resources: a zero-one programming approach*, Management Science 16 (1969) 93–108.
80. A. Prodon, T.M. Liebling, and H. Gröflin, *Steiner's problem on 2-trees*, Tech. Report RO 850351, Département de Mathématiques, Ecole Polytechnique Fédérale de Lausanne, 1985.
81. R.L. Rardin and U. Choe, *Tighter relaxations of fixed charge network flow problems*, Tech. Report report J-79-18, School of Industrial and Systems Engineering, Georgia Institute of Technology, 1979.
82. L.-M. Rousseau, M. Gendreau, and D. Feillet, *Interior point stabilization for column generation*, Tech. report, University de Montreal, 2003.
83. D.M. Ryan and B.A. Foster, *An integer programming approach to scheduling*, Computer Scheduling of Public Transport Urban Passenger Vehicle and Crew Scheduling (A. Wren, ed.), North-Holland, Amsterdam, 1981, pp. 269–280.
84. F. Soumis, *Decomposition and column generation*, Annotated Bibliographies in Combinatorial Optimization (F. Maffioli M. Dell'Amico and S. Martello, eds.), Wiley, Chichester, 1997, pp. 115–126.

85. E. Uchoa, R. Fukasawa, J. Lysgaard, A. Pessoa, M.P. Aragao, and D. Andrade, *Robust branch-and-cut-and-price for the capacitated minimum spanning tree problem over an extended formulation*, Mathematical Programming 112 (2008) 443–472.
86. H. Uzawa, *Iterative methods for concave programming*, Studies in Linear and Nonlinear Programming (K. Arrow, L. Hurwicz, and H. Uzawa, eds.), Stanford University Press, 1959.
87. R.M. Van Slyke and R. Wets, *L-shaped linear programs with applications to optimal control and stochastic programming*, SIAM J. of Applied Mathematics 17 (1969) 638–663.
88. M. Van Vyve, *Linear programming extended formulations for the single-item lot-sizing problem with backlogging and constant capacity*, Mathematical Programming 108 (2006) 53–78.
89. F. Vanderbeck, *On Dantzig-Wolfe decomposition in integer programming and ways to perform branching in a branch-and-price algorithm*, Operations Research 48 (2000) 111–128.
90. F. Vanderbeck, *Branching in branch-and-price: a generic scheme*, Research Report Inria-00311274, University Bordeaux I and INRIA, 2006, revised 2008.
91. F. Vanderbeck and L.A. Wolsey, *An exact algorithm for IP column generation*, Operations Research Letters 19 (1996) 151–159.
92. D. Villeneuve, J. Desrosiers, M.E. Lübbecke, and F. Soumis, *On compact formulations for integer programs solved by column generation*, Annals of Operations Research 139 (2006) 375–388.
93. S. Weider, *Integration of vehicle and duty scheduling in public transport*, Ph.D. thesis, Faculty of Mathematics and Sciences, The Technical University, Berlin, 2007.
94. P. Wentges, *Weighted dantzig-wolfe decomposition for linear mixed-integer programming*, International Transactions on Operational Research 4 (1997) 151–162.
95. H. Weyl, *The elementary theory of convex polyhedra*, Contributions to the Theory of Games I (H.W. Kuhn and A.W. Tucker, eds.), Princeton University Press, Princton N.J, translated from 1935 original in German, 1950, pp. 3–18.
96. R.T. Wong, *Integer programming formulations of the traveling salesman problem*, Proceedings of IEEE International Conference on Circuits and Computers, 1980, pp. 149–152.
97. R.T. Wong, *Dual ascent approach for Steiner tree problems on directed graphs*, Mathematical Programming 28 (1984) 271–287.
98. M. Yannakakis, *Expressing combinatorial optimization problems by linear programs*, Journal of Computer and System Sciences 43 (1991) 441–466.

Part III
Current Topics

Six survey talks on current hot topics were given at the 12th Combinatorial Optimization Workshop, Aussois, France, 7–11 January 2008, in the days following the celebration of 50 Years of Integer Programming 1958–2008. The speakers were Fritz Eisenbrand, Andrea Lodi, François Margot, Franz Rendl, Jean-Philippe P. Richard, and Robert Weismantel. For the written versions, Robert Weismantel has been joined by the co-authors Raymond Hemmecke, Matthias Köppe, and Jon Lee, and Jean-Philippe P. Richard has been joind by the co-author Santanu S. Dey.

Chapter 14
Integer Programming and Algorithmic Geometry of Numbers
A tutorial

Friedrich Eisenbrand

Abstract This chapter surveys a selection of results from the interplay of integer programming and the geometry of numbers. Apart from being a survey, the text is also intended as an entry point into the field. I therefore added exercises at the end of each section to invite the reader to delve deeper into the presented material.

14.1 Lattices, integer programming and the geometry of numbers

The central objects of study of the *geometry of numbers* are lattices. A *lattice* is a set $\Lambda = \{y \in \mathbb{R}^n : y = Ax,\ x \in \mathbb{Z}^n\}$, where $A \in \mathbb{R}^{n\times n}$ is a nonsingular matrix. We say that the lattice is *generated by* A and write $\Lambda = \Lambda(A)$. The matrix A is called a *basis* of the lattice Λ. If A is a rational matrix, i.e., $A \in \mathbb{Q}^{n\times n}$, then Λ is a *rational lattice*.

A very important problem, which has also received a lot of attention in computer science and optimization, is the *shortest vector problem* with respect to the ℓ_p-norm for some $p \in \mathbb{N}_+ \cup \{\infty\}$. It is as follows.

> Given a rational lattice-basis $A \in \mathbb{Q}^{n\times n}$, compute a nonzero vector $v \in \Lambda(A)$ with minimal norm $\|v\|_p$.

If the norm is not specified, we implicitly assume the ℓ_2-norm and denote the length of a shortest vector w.r.t. the ℓ_2-norm as $SV(\Lambda)$. The shortest vector problem can (in fixed dimension) be solved efficiently with *lattice basis reduction*. In varying dimension the shortest vector problem is NP-hard under randomized reductions [2]. Still, the fastest algorithms [43, 3] for this problem rely on basis reduction.

Friedrich Eisenbrand
Department of Mathematics, EPFL, Lausanne, Switzerland
e-mail: friedrich.eisenbrand@epfl.ch

M. Jünger et al. (eds.), *50 Years of Integer Programming 1958-2008*,
DOI 10.1007/978-3-540-68279-0_14, © Springer-Verlag Berlin Heidelberg 2010

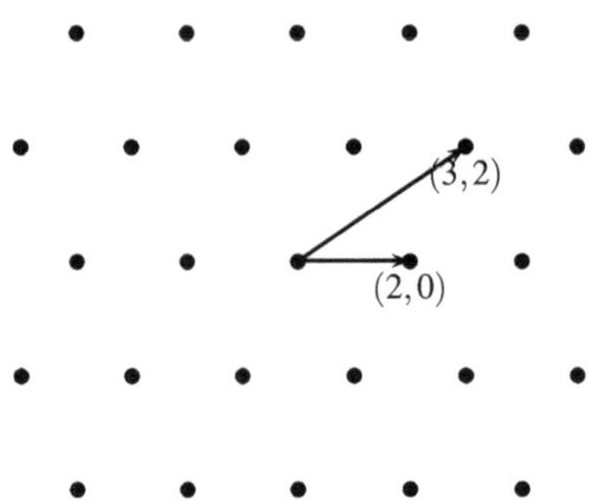

Fig. 14.1 The lattice generated by $(2,0)$ and $(3,2)$.

Lenstra [56] has shown that *integer programming* can be solved in polynomial time, if the dimension is fixed. His algorithm is based on lattice basis reduction. Why do lattices and the algorithmic geometry of numbers come into play in integer programming?

The *greatest common divisor* of two integers $a_0, a_1 \in \mathbb{Z}$ where not both a_0 and a_1 are zero, is the largest integer that divides both a_0 and a_1. It is denoted by $\gcd(a_0, a_1)$. The greatest common divisor can be efficiently computed with the *Euclidean algorithm*, see, e.g., [51, 1]. It computes the remainder sequence $a_0, a_1, \ldots, a_{k-1}, a_k \in \mathbb{N}_+$, where a_i, $i \geqslant 2$ is given by $a_{i-2} = a_{i-1} q_{i-1} + a_i$, $q_i \in \mathbb{N}$, $0 < a_i < a_{i-1}$, and a_k divides a_{k-1} exactly. Then $a_k = \gcd(a_0, a_1)$. The Euclidean algorithm can be interpreted as a reduction algorithm and we will see that the 2-dimensional basis reduction algorithm of Lagrange (see Section 14.2) works along the same lines.

Now, the connection between integer linear programming with algorithmic number theory reveals itself already in the following theorem, which is proved at the beginning of every course on elementary number theory, see, e.g., [65].

Theorem 14.1. *Let $a, b \in \mathbb{Z}$ be integers that are not both equal to zero. The greatest common divisor* $\gcd(a,b)$ *is the smallest integer in the set*

$$\{ax + by \colon x, y \in \mathbb{Z}, ax + by \geqslant 1\}. \tag{14.1}$$

The problem to find the minimum in (14.1) can be modeled as an integer program in two variables and one constraint.

$$\begin{aligned} \min \quad & ax + by \\ & ax + by \geqslant 1 \\ & x, y \in \mathbb{Z}. \end{aligned}$$

This already explains why efficient methods for integer programming with a fixed number of variables incorporate *reduction techniques*, which in a basic form appear already in the Euclidean algorithm. Such reduction techniques are in the focus of this chapter.

14.2 Informal introduction to basis reduction

Informally, *lattice basis reduction* is about transforming a lattice basis B into a lattice basis B' that generates the same lattice, $\Lambda(B) = \Lambda(B')$ and from which a shortest vector can (in fixed dimension) be easily determined. Before we make this more precise, we have to understand what valid transformations are, i.e., when $\Lambda(B) = \Lambda(B')$ holds. Recall that an integer matrix $U \in \mathbb{Z}^{n\times n}$ is called *unimodular* if $\det(U) = \pm 1$. Thus $U \in \mathbb{Z}^{n\times n}$ is unimodular if and only if U is non-singular and U^{-1} is a matrix with integer entries, see exercise 1.

Lemma 14.1. *Let $B, B' \in \mathbb{R}^{n\times n}$ be two rational non-singular matrices. One has $\Lambda(B) = \Lambda(B')$ if and only if there exists a unimodular matrix $U \in \mathbb{Z}^{n\times n}$ with $B' = B \cdot U$.*

Proof. Suppose $\Lambda(B) = \Lambda(B')$. Then, every column of B is in $\Lambda(B')$ which implies that there exists an integral matrix $U \in \mathbb{Z}^{n\times n}$ with $B = B' \cdot U$. Similarly, there exists an integral matrix $V \in \mathbb{Z}^{n\times n}$ with $B' = B \cdot V$. From this it follows that $B = B \cdot V \cdot U$ and since B is non-singular, this implies that $V \cdot U = I_n$, where I_n is the $n \times n$ identity matrix. This implies that U is unimodular since $1 = \det(V \cdot U) = \det(V) \cdot \det(U)$ and since both $\det(V)$ and $\det(U)$ are integers, one has $\det(U) = \pm 1$. On the other hand, if $B' = B \cdot U$ with an integral U, then $\Lambda(B') \subseteq \Lambda(B)$. If U is unimodular, then $B = B' \cdot U^{-1}$ and U^{-1} is integral, which implies $\Lambda(B) \subseteq \Lambda(B')$. □

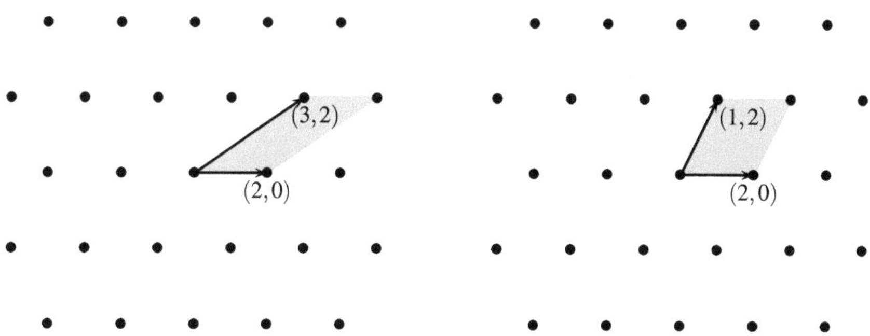

Fig. 14.2 A lattice, two different bases and the lattice determinant, which is depicted as the volume of the parallelepipeds defined by the bases respectively.

Lemma 14.1 implies that the absolute value $|\det(B)|$ of the determinant of a basis B of Λ is an invariant of Λ. This value is called the *determinant* of Λ. The set $\Pi(B) = \{B\lambda \colon \lambda \in \mathbb{R}^n, 0 \leqslant \lambda < 1\}$ is called the *parallelepiped* spanned by the columns of B. The volume of this parallelepiped is the absolute value of the determinant of B. Thus the lattice determinant is the volume of the parallelepiped defined by the basis elements of any basis, see Figure 14.2.

The result of the multiplication of B by a unimodular matrix U can also be obtained by a sequence of *elementary column operations* on B:

i) Swap of two columns.
ii) Multiplication of a column with -1.
iii) Addition of an *integral* multiple of one column to *another column*.

Now that we know which operations to apply, let us consider an example of lattice basis reduction. Suppose that we want to determine the shortest vector in the lattice which is generated by the following two column-vectors v_1 and v_2.

It is difficult to guess what the shortest vector of this lattice could be. We subtract v_1 from v_2 and replace v_2 with the outcome of this operation. The result is a new basis.

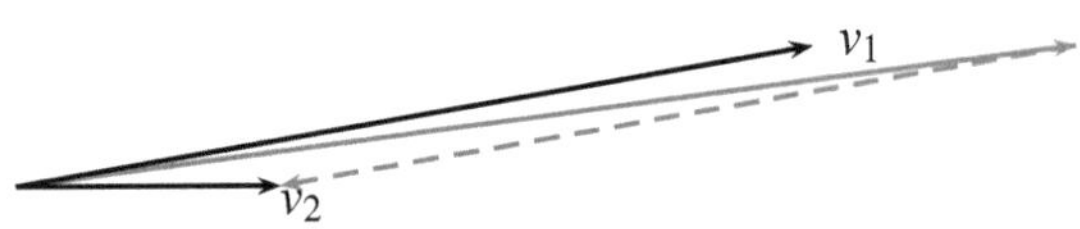

Still, it is not easy to determine the shortest vector. Next we subtract 3-times v_2 from v_1 and replace v_1 with the outcome of this operation.

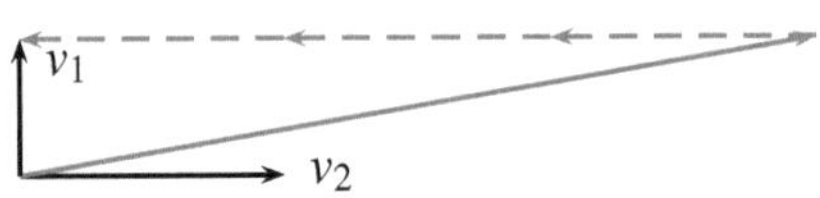

Now the shortest vector reveals itself easily. Since the obtained basis of our lattice consists of two orthogonal vectors, the shortest vector of the lattice is the shortest vector of the basis itself. In fact, we have traced above the reduction algorithm of Lagrange which was also described by Gauß [29].

Intuitively it seems clear that the shortest vector problem should still be easy, if the basis is *almost orthogonal*. We will deliver a precise definition of this in Section 14.5, where we describe the LLL-algorithm.

Exercises

1) Prove that $U \in \mathbb{Z}^{n \times n}$ is unimodular if and only if U is non-singular and U^{-1} is a matrix with integer entries.
2) Let $B \in \mathbb{Q}^{n \times n}$ be a lattice basis that consists of pairwise orthogonal vectors. Prove that the shortest vector of $\Lambda(B)$ is the shortest column vector of B.

3) Let $\Lambda, \Lambda' \subseteq \mathbb{Z}^n$ be lattices and suppose that $\Lambda \supseteq \Lambda'$. Show that $\det(\Lambda)$ divides $\det(\Lambda')$.
4) Consider three points $v_1, v_2, v_3 \in \mathbb{Z}^2$ that are not co-linear, i.e., there is no line containing all three points. Show that the triangle spanned by v_1, v_2 and v_3 does not contain an integer point apart from v_1, v_2 and v_3 itself, if and only if the matrix $(v_2 - v_1, v_3 - v_2)$ is unimodular.
   A similar statement cannot be made in $\mathbb{R}^3$. Provide an example of linearly independent integer vectors $v_1, v_2, v_3 \in \mathbb{Z}^3$ such that the simplex $\operatorname{conv}\{0, v_1, v_2, v_3\}$ does not contain an integer point apart from $0, v_1, v_2$ and v_3 and $\det(v_1, v_2, v_3) \neq \pm 1$.
5) (Picks formula) The convex hull $P = \operatorname{conv}\{v_1, \dots, v_n\}$ of integer points $v_i \in \mathbb{Z}^2$, $i = 1, \dots, n$ is a convex lattice polygon. Let A, I and B be the area, number of integer points in the interior and boundary of P respectively. Prove Picks formula
$$A = I + B/2 - 1.$$
   Hint: Exercise 4).

14.3 The Hermite normal form

The section is devoted to the algorithmic solution of the following problem

> Given a rational matrix $A \in \mathbb{Q}^{m \times n}$ of full row-rank and a rational vector $b \in \mathbb{Q}^m$, decide whether there exists an integral vector $x \in \mathbb{Z}^n$ with $Ax = b$.

This problem is solved with the Hermite normal form of a rational matrix which is central to this chapter. We will also see that the set $\{Ax \colon x \in \mathbb{Z}^n\}$ is also a lattice, namely the lattice that is generated by the Hermite normal form of A.

We motivate the Hermite normal form first via certificates of unsolvability of rational linear equations. Suppose we are given a system $Ax = b$ of linear equations over the reals. How can one certify that this system does not have a solution? The following well known theorem from linear algebra provides a certificate.

Theorem 14.2. *Let $A \in \mathbb{R}^{m \times n}$ be a matrix and $b \in \mathbb{R}^m$ be a vector. The system $Ax = b$ does not have a solution $x \in \mathbb{R}^n$ if and only if there exists a $\lambda \in \mathbb{R}^m$ with $\lambda^T A = 0$ and $\lambda^T b \neq 0$.*

Proof. If x is a solution to the system, then $\lambda^T A = 0$ implies $\lambda^T b = \lambda^T A x = 0^T x = 0$.

On the other hand, if $Ax = b$ does not have a solution then b is not in the vector space
$$Z = \{z \in \mathbb{R}^m \colon z = Ax, x \in \mathbb{R}^n\}.$$
The vector space Z is the kernel of a matrix $C \in \mathbb{R}^{k \times m}$,
$$Z = \{z \in \mathbb{R}^m \colon Cz = 0\}.$$

We have $CA = 0$ and $Cb \neq 0$. We can choose a y with $y^T Cb \neq 0$. We then have $y^T CA = 0$, so $\lambda = y^T C$ serves our purpose. □

Now let $A \in \mathbb{Q}^{m\times n}$ be a rational matrix and let $b \in \mathbb{Q}^m$ be a rational vector. What happens if we ask, whether a system $Ax = b$ has an integer solution $x \in \mathbb{Z}^n$?

Clearly $Ax = b$ has an integer solution if and only if $(\alpha \cdot A)x = \alpha \cdot b$ has an integer solution for $\alpha \in \mathbb{R} - \{0\}$. This means that we can multiply A and b with the least common multiple of the denominators of their entries and obtain an integral matrix and an integral vector.

The simplest nontrivial system of rational equations over the integers is one equation and two variables. This case was already considered by Gauß [29], who mentions the following theorem. Recall the notation $u \mid v$ which stands for *u divides v* for integers u and v.

Theorem 14.3. *Let a,b and c be integers where a or b are nonzero. The system*

$$ax + by = c \tag{14.2}$$

has a solution with integers x and y if and only if $\gcd(a,b) \mid c$.

Proof. Let d be a common divisor of a and b, i.e., $d \cdot a' = a$ and $d \cdot b' = b$ with integers a' and b'. If $ax + by = c$, then

$$d(a'x + b'y) = c$$

which implies that $d \mid c$. Thus if (14.2) has an integral solution, then $\gcd(a,b) \mid c$.

On the other hand let $\gcd(a,b) \cdot k = c$ for some integer k. Recall that there exist integers x' and y' with $ax' + by' = \gcd(a,b)$. An integral solution of (14.2) is then $x = kx'$ and $y = ky'$. □

In other words, (14.2) does not have an integral solution, if and only if there exists an $\alpha \in \mathbb{R}$ such that $\alpha \cdot (a,b) \in \mathbb{Z}^2$ and $\alpha \cdot c \notin \mathbb{Z}$. It turns out that this simple observation can be generalized.

Theorem 14.4. *Let $A \in \mathbb{Q}^{m\times n}$ and $b \in \mathbb{Q}^m$. The system*

$$Ax = b \tag{14.3}$$

does not have an integral solution, if and only if there exists a $\lambda \in \mathbb{R}^m$ with $\lambda^T A \in \mathbb{Z}^n$ and $\lambda^T b \notin \mathbb{Z}$.

To prove this theorem, we now introduce the Hermite normal form. A rational matrix $A \in \mathbb{Q}^{m\times n}$ of full row rank is said to be in *Hermite normal form (HNF)* if it has the form $[B \mid 0]$, where B is a nonsingular, lower triangular matrix with non-negative entries, in which each row has a unique maximal entry, located on the diagonal. The following is an example of a matrix in HNF

$$\begin{pmatrix} 2 & 0 & 0 \\ 1 & 3 & 0 \end{pmatrix}$$

Theorem 14.5. *Each rational matrix $A \in \mathbb{Q}^{m\times n}$ of full row-rank can be brought into Hermite normal form by a finite series of elementary column operations.*

Proof. By scaling all entries of A with a positive common multiple α of the denominators, we obtain an integral matrix A'. If $[H' \mid 0]$ is a HNF of A', then $(1/\alpha)[H' \mid 0]$ is an HNF of A. Therefore we can assume without loss of generality that A is integral.

Algorithm 14.1 computes the Hermite normal form and a corresponding unimodular matrix of an integral matrix with full row rank with the use of the extended Euclidean algorithm. Here the term "update columns i and j" with a 2×2-matrix $\left(\begin{smallmatrix} a & b \\ c & d\end{smallmatrix}\right)$ means that the new columns c_i and c_j result from their old values by multiplying (c_i, c_j) with $\left(\begin{smallmatrix} a & b \\ c & d\end{smallmatrix}\right)$, i.e., $(c_i, c_j) := (c_i, c_j)\left(\begin{smallmatrix} a & b \\ c & d\end{smallmatrix}\right)$.

The function $\mathbf{exggT}(a,b)$ implements the *extended Euclidean algorithm*, which computes the triple (g,x,y), where $g = \gcd(a,b)$ and x and y are integers with $g = xa + yb$, see, e.g., [51]. Since the matrix $\begin{pmatrix} x & -b/g \\ y & -a/g\end{pmatrix}$ is unimodular, the updating step in the algorithm corresponds to a sequence of elementary column operations.

Algorithm 14.1 HNF of A

Input: $A \in \mathbb{Z}^{m\times n}$ full row rank
Return: $[H \mid 0] \in \mathbb{Z}^{m\times n}$ HNF of A, unimodular matrix $U \in \mathbb{Z}^{n\times n}$ with $A \cdot U = [H \mid 0]$

```
H ← A
U ← I_n
for i = 1 to m do
   for j = i+1 to n do
      if H_{i,j} ≠ 0 then
         (g,x,y) = exggT(H_{i,i}, H_{i,j})
         update columns i and j of H and U with ( x  -H_{i,j}/g ; y  H_{i,i}/g )
      end if
   end for
   for j = 1 to i-1 do
      H_{i,j} = q · H_{i,i} + r (division with remainder, r non-negative)
      update columns j and i of H and U with ( 1 0 ; -q 1 ) {reduce entries to the left of diagonal
      element H_{i,i}}
   end for
end for
```

□

It is an easy exercise to show that the following invariant holds after each "update columns" operation of Algorithm 14.1

$$A \cdot U = H.$$

Let us trace the algorithm, when it is executed on the matrix $\left(\begin{smallmatrix} 2 & 3 & 4 \\ 2 & 4 & 6\end{smallmatrix}\right)$. The greatest common divisor of $(2,3)$ is $1 = (-1)\cdot 2 + 1 \cdot 3$. Thus we update column 1 and 2 with the matrix $\left(\begin{smallmatrix} -1 & -3 \\ 1 & 2\end{smallmatrix}\right)$, obtaining $\left(\begin{smallmatrix} 1 & 0 & 4 \\ 2 & 2 & 6\end{smallmatrix}\right)$. The transforming matrix U yields $\begin{pmatrix} -1 & -3 & 0 \\ 1 & 2 & 0 \\ 0 & 0 & 1\end{pmatrix}$. Eliminating 4 yields $H = \left(\begin{smallmatrix} 1 & 0 & 0 \\ 2 & 2 & -2\end{smallmatrix}\right)$ and $U = \begin{pmatrix} -1 & -3 & 4 \\ 1 & 2 & -4 \\ 0 & 0 & 1\end{pmatrix}$.

Eliminating -2 yields $H = \left(\begin{smallmatrix} 1 & 0 & 0 \\ 2 & 2 & 0 \end{smallmatrix}\right)$, $U = \left(\begin{smallmatrix} -1 & -4 & 1 \\ 1 & 4 & -2 \\ 0 & -1 & 1 \end{smallmatrix}\right)$

Now reducing 2 in the lower left corner yields the Hermite normal form $H = \left(\begin{smallmatrix} 1 & 0 & 0 \\ 0 & 2 & 0 \end{smallmatrix}\right)$ and the unimodular matrix transformation matrix $U = \left(\begin{smallmatrix} 3 & -4 & 1 \\ -3 & 4 & -2 \\ 1 & -1 & 1 \end{smallmatrix}\right)$ with $A \cdot U = H$.

Proof (of Theorem 14.4). Suppose $Ax = b$ has an integral solution $x \in \mathbb{Z}^n$. Then with $\lambda^T A \in \mathbb{Z}^n$ one has $\lambda^T b = \lambda^T Ax \in \mathbb{Z}$.

Suppose now that $Ax = b$ does not have an integral solution. If $Ax = b$ is not solvable in $\mathbb{R}^n$, then there exists a $\lambda \in \mathbb{R}^m$ with $\lambda^T A = 0$ and $\lambda^T b \neq 0$ and thus a λ with $\lambda^T A = 0$ and $\lambda^T b = 1/2$. Thus we can assume that $Ax = b$ has a solution in $\mathbb{R}^n$ and therefore we can assume that A has full row-rank.

There exists a unimodular matrix $U \in \mathbb{Z}^{n \times n}$ such that $A \cdot U = [B \mid 0]$ is in HNF. Thus $B^{-1}A$ is an integral matrix. We claim that $B^{-1}b$ is not an integral vector and thus that there exists a row λ^T of B^{-1} with $\lambda^T A \in \mathbb{Z}^n$ and $\lambda^T b \notin \mathbb{Z}$.

If $B^{-1}b$ is integral, then

$$A \cdot U \begin{pmatrix} B^{-1}b \\ 0 \end{pmatrix} = [B \mid 0] \begin{pmatrix} B^{-1}b \\ 0 \end{pmatrix} = b$$

shows that $x = U \cdot \begin{pmatrix} B^{-1}b \\ 0 \end{pmatrix}$ is an integral solution of $Ax = b$, which is a contradiction. □

The next theorem shows that the HNF is unique. For this we extend the notation $\Lambda(A) = \{Ax \colon x \in \mathbb{Z}^n\}$ also for the case where $A \in \mathbb{Q}^{m \times n}$ is a rational matrix of full row-rank which is not necessarily a basis. Such a set is also a lattice, since $\Lambda(A) = \Lambda(B)$, where $[B \mid 0]$ is the HNF of A and B is nonsingular.

Theorem 14.6. *Let A and A' be integral matrices of full row-rank, with HNF $[B \mid 0]$ and $[B' \mid 0]$ respectively. Then $\Lambda(A) = \Lambda(A')$ if and only if $B = B'$.*

Proof. If $B = B'$, the clearly $\Lambda(A) = \Lambda(A')$.

It remains to prove that if B and B' are different, then $\Lambda(B) \neq \Lambda(B')$. So assume that B and B' nevertheless generate the same lattice Λ. Let i be the smallest row-index such that row i of B differs from row i of B'. Without loss of generality assume that $B_{i,j} > B'_{i,j}$ holds for some $1 \leqslant j < i$. Observe that the i-th components of elements of Λ whose first $i-1$ components are 0, are integral multiples of $B_{i,i}$. Notice that the first $i-1$ components of the vector $\beta = B^{(j)} - B'^{(j)} \in \Lambda$ are 0, where $B^{(j)}$ and $B'^{(j)}$ denote the j-th column of B and B' respectively. But the i-th component of β is strictly between 0 and $B_{i,i}$. This is a contradiction. □

Corollary 14.1. *The Hermite normal form of a rational matrix with full row-rank is unique.*

Clearly, Algorithm 14.1 requires $O(m \cdot n)$ extended gcd-computations and $O(m^2 \cdot n)$ arithmetic operations. But is it a polynomial algorithm? This cannot be argued, since the binary encoding length of the numbers could grow exponentially. The *size* of an integer z is the number of bits which are required to encode z. We define

$\text{size}(z) = 1 + \lceil \log_2(|z|+1) \rceil$. Likewise, the size of a matrix $A \in \mathbb{Z}^{m \times n}$ is the number of bits needed to encode A, i.e., $\text{size}(A) = mn + \sum_{i,j} \text{size}(a_{ij})$, see [75, p. 29]. In fact Fang and Havas [23] provide examples of pivoting schemes, where the size of the numbers in the intermediate steps of the HNF-algorithm above grows exponentially.

Nevertheless, it can be shown that the HNF of a rational matrix can be computed in polynomial time, see [47, 75]. The key to Schrijver's method [75] is the following lemma.

Lemma 14.2. *Let $A \in \mathbb{Z}^{m \times n}$ be a matrix of full row rank and let $d = \det(\Lambda(A))$ be the lattice determinant of $\Lambda(A)$ and let D be a multiple of d. Then*

$$\Lambda(A) = \Lambda([A \mid D \cdot I_m]).$$

Proof. Clearly $\Lambda(A) \subseteq \Lambda([A \mid D \cdot I_m])$. Let Λ denote $\Lambda(A)$. For the reverse inclusion we simply have to show that $D \cdot e_i \in \Lambda$ for each unit vector e_i, $i = 1, \ldots, m$. Let $B \in \mathbb{Z}^{m \times m}$ be a basis of Λ and let $\widetilde{B} \in \mathbb{Z}^{m \times m}$ be the adjoint of B. All column vectors of $B \cdot \widetilde{B}$ are elements of Λ. But Cramers rule says $B^{-1} = (1/\det(B))\widetilde{B}$. Therefore $B \cdot \widetilde{B} = \det(B) \cdot I_m$. Since $d = |\det(B)|$ we see that $d \cdot e_i \in \Lambda$, for each unit vector e_i, $i = 1, \ldots, n$. Since $d \mid D$ we also have that $D \cdot e_i \in \Lambda$. □

Corollary 14.2. *Let $[H \mid 0]$ be the HNF of $A \in \mathbb{Z}^{m \times n}$, where A has full row-rank and let $d = \det(\Lambda(A))$ be the determinant of $\Lambda(A)$. Let $[H' \mid 0]$ be the HNF of $[A \mid d \cdot I_m] \in \mathbb{Z}^{m \times (n+m)}$. Then $H' = H$.*

Theorem 14.7. *There exists a polynomial time algorithm which computes the HNF of a rational matrix of full row-rank.*

Proof. Since we can scale A with the product of the denominators of A we can assume without loss of generality that A is an integral matrix. We start by identifying m linearly independent columns of A and by computing $D = |\det(A)|$. This can be done with Gaussian elimination in polynomial time [17]. The fact that the encoding length of D is polynomial follows from the Hadamard bound (14.9) which we discuss later. Exercise 14.2.3) shows that $D \mid \det(\Lambda(A))$.

We compute the HNF of $[A \mid D \cdot I_m]$ as in Algorithm 14.1, but keeping numbers in a row reduced modulo D until we eliminate D in this row stemming from $D \cdot I_m$.

The important observation is this. The first i rows of H remain unchanged after the i-th run through the first **for** loop. The remaining rows of H can be kept reduced modulo D, since the last $m - i$ columns of H are of the form $\binom{0}{d \cdot I_{m-i}}$. This procedure requires a polynomial amount of extended-gcd computations and arithmetic operations on numbers of size at most $\text{size}(D)$. □

Theorem 14.8. *There exists a polynomial algorithm which computes the HNF $[H \mid 0]$ of a rational matrix $A \in \mathbb{Q}^{m \times n}$ and the corresponding unimodular matrix $U \in \mathbb{Z}^{n \times n}$ with $A \cdot U = [H \mid 0]$.*

Proof. Select m linearly independent columns of A using Gaussian elimination. Assume without loss of generality that those are the first m columns. The matrix $[H \mid 0]$ is the HNF of A if and only if the HNF of $A' = \begin{bmatrix} A \\ 0 \mid I_{n-m} \end{bmatrix}$ is of the form $H' = \begin{bmatrix} H \mid 0 \\ B \end{bmatrix}$ with some matrix B. Compute H' with the polynomial modification of Algorithm 14.1. We have $A' \cdot U = H'$ and since A' is nonsingular we have $U = H' \cdot A'^{-1}$. Clearly $A \cdot U = [H \mid 0]$. □

The proof of the next theorem is left as an excise.

Theorem 14.9. *There exists a polynomial algorithm for the following problem*

> *Given an integral matrix $A \in \mathbb{Z}^{m \times n}$ and an integral vector $b \in \mathbb{Z}^m$, compute a solution $\hat{x} \in \mathbb{Z}^n$ of $Ax = b$ or establish that no such $\hat{x}$ exists.*

Notes

Showing that Gaussian elimination is in fact a polynomial-time algorithm is not trivial. Edmonds [17] has shown that, if one keeps the numerators and denominators in the intermediate steps of the algorithm gcd-free, then the size of the numbers remains polynomial. Von zur Gathen and Sieveking [81] showed that systems of rational equations over the integers can be solved in polynomial time.

In fact, there is an arsenal of ways to do linear algebra in polynomial time, one of them being a modular method again. If one wants to compute the determinant of an integral matrix $A \in \mathbb{Z}^{n \times n}$ for example, one could apply the Hadamard bound (14.9) to compute a $D \in \mathbb{Z}$ with $2 \cdot |\det(A)| < D$. If we now know a number $x \in \mathbb{Z}, 0 \leqslant x < D$ with $x \equiv \det(A) \pmod{D}$, then we can retrieve $\det(A)$ from x. In fact, if $x \leqslant D/2$, then $\det(A) = x$ and if $x > D/2$, then $\det(A) = x - D$. One computes with the sieve of Erathostenes the first k prime numbers $p_1, \ldots, p_k$ such that $p_1 \cdots p_k \geqslant D$ holds. The prime-number theorem guarantees that the *value* of p_k is polynomial in the encoding length of D. One can then compute the value $\det(A) \pmod{p_i}$ for each p_i with Gaussian elimination in the field $\mathbb{Z}_{p_i}$ and reconstruct x with the Chinese remainder theorem, see, e.g.,, [80].

Exercises

1) Compute by hand the HNF of $\begin{pmatrix} 3 & 5 & 7 & 2 \\ 2 & 4 & 9 & 3 \end{pmatrix}$.
2) Write a computer program which implements Algorithm 14.1.
3) Prove Theorem 14.9.
4) Let $D \in \mathbb{N}_+$ be an integer which is not necessarily prime and let $A \in \mathbb{Z}^{m \times n}$ and $b \in \mathbb{Z}^m$ be an integer matrix and vector respectively. Show that there is a polynomial algorithm which computes a solution to the system

$$Ax = b \pmod{D}, x \in \mathbb{Z}^n$$

   or asserts that the system does not have a solution.
5) Describe a unimodular matrix G such that the step "update columns" j and i of H and U with $\begin{pmatrix} 1 & 0 \\ -q & 1 \end{pmatrix}$ in Algorithm 14.1 corresponds to the multiplication of H and U with G from the right.

6) Modify Algorithm 14.1 such that it detects on the fly whether A has full row-rank and possibly determines a row which is in the span of the other rows.

14.4 Minkowski's theorem

At the end of the 19-th century, Minkowski opened the stage for the geometry of numbers with his famous book *Geometrie der Zahlen* [61]. One of his main results is described in this section. He used geometric methods to prove upper bounds on numbers which are representable by positive definite quadratic forms. In the setting of lattices his result means that a lattice $\Lambda \subseteq \mathbb{R}^n$ contains a nonzero lattice point whose norm is bounded roughly by $\sqrt{\frac{2}{\pi e} n} \det(\Lambda)^{1/n}$. The simplicity and elegance of his approach is stunning. His result was preceded by a sequence of bounds which we briefly discuss. Lagrange and Gauß [29] proved that a 2-dimensional lattice has a nonzero lattice point whose norm is bounded by $\sqrt{4/3}\det(\Lambda)^{1/2}$. In his *Disquisitiones Arithmeticae* [29] Gauß proved that a 3-dimensional lattice has a nonzero lattice vector of norm bounded by $\sqrt{4/3}\det(\Lambda)^{1/3}$. Hermite [37] generalized this result to arbitrary dimension and provided the upper bound $(4/3)^{(n-1)/4}\det(\Lambda)^{1/n}$. All these results were algorithmic, in the sense that they provided algorithms computing nonzero lattice vectors achieving these bounds. It is remarkable that these algorithms run in polynomial time, if the dimension is fixed. The bound obtained by Minkowski is much stronger than the one of Hermite. It is however not known how to compute a nonzero lattice vector satisfying this bound in polynomial time, if the dimension is not fixed.

A *convex body* is a compact and full-dimensional convex set $K \subseteq \mathbb{R}^n$. In its simplest form, Minkowski's theorem is as follows.

Theorem 14.10. *Let $K \subseteq \mathbb{R}^n$ be a convex body which is symmetric around the origin ($x \in K$ implies $-x \in K$). If* $\operatorname{vol}(K) \geqslant 2^n$*, then K contains a nonzero integral vector $v \in \mathbb{Z}^n \setminus \{0\}$.*

Proof. We first argue that it suffices to assume that $\operatorname{vol}(K) > 2^n$ holds. If $\operatorname{vol}(K) = 2^n$, then, since K is compact, there exists an $\varepsilon > 0$ such that the distance to K of each integer point that is not contained in K is at least ε. This means that there exists a constant $\delta > 0$ such that the sets $(1+\delta)\cdot K$ and K contain the same integer points. The volume of $(1+\delta)\cdot K$ however is strictly larger than 2^n.

Suppose now that the theorem does not hold. Consider the set $S = 1/2\cdot K$ and the translates of S with integer vectors

$$S+v,\, v \in \mathbb{Z}^n.$$

If $S+v_1 \cap S+v_2 \neq \emptyset$ for some integral vectors $v_1 \neq v_2$, then there exist $k_1, k_2 \in K$ such that

$$0 \neq v_1 - v_2 = 1/2(k_2 - k_1).$$

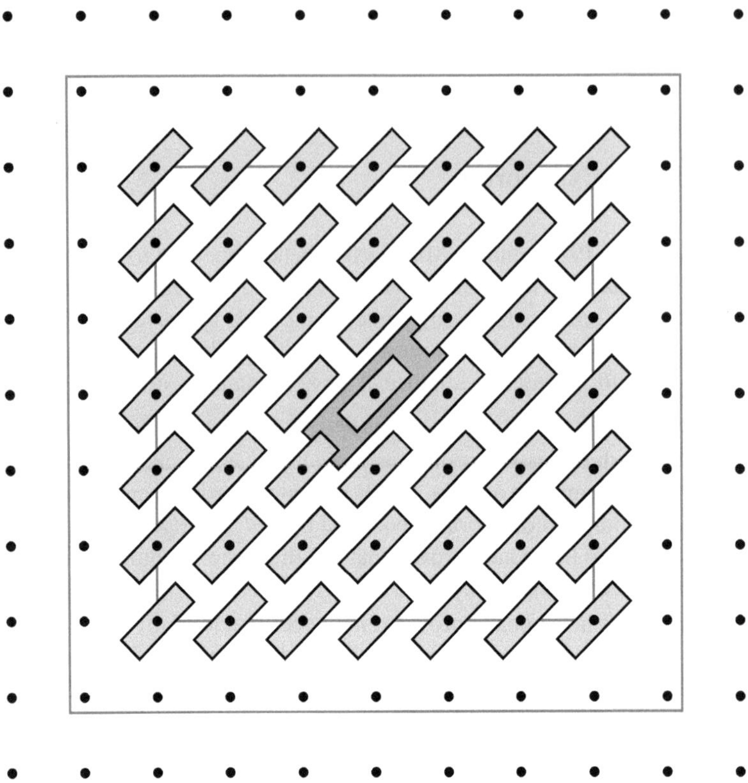

Fig. 14.3 An illustration of the proof of Minkowski's theorem, where the dark rectangle stands for K and the light rectangles for the sets $S+v$. The inner square represents the points $x \in \mathbb{R}^2$ with $\|x\|_\infty \leqslant M$ and the outer square represents the set (14.5), being the points $x \in \mathbb{R}^2$ with $\|x\|_\infty \leqslant M+D$.

Due to the symmetry of K around the origin we have $-k_1 \in K$ and due to the convexity of K we have $1/2(k_2 - k_1) \in K$, which is a nonzero integral point. This shows that the translates $S+v$, $v \in \mathbb{Z}^n$ do not intersect.

Now we consider the volume

$$V_M = \mathrm{vol}\left(\bigcup_{\substack{v \in \mathbb{Z}^n \\ \|v\|_\infty \leqslant M}} S+v \right)$$

for some $M \in \mathbb{N}$. Since the $S+v$ do not intersect we have

$$\begin{aligned} V_M &= \sum_{v \in \mathbb{Z}^n, \|v\|_\infty \leqslant M} \mathrm{vol}(S) \\ &= (2 \cdot M + 1)^n \cdot \mathrm{vol}(S). \end{aligned} \tag{14.4}$$

Since S is bounded, S has finite diameter $D \in R$. This means that the union

$$\bigcup_{\substack{v\in\mathbb{Z}^n\\ \|v\|_\infty\leqslant M}} S+v$$

is a subset of

$$\{x \in \mathbb{R}^n : \|x\|_\infty \leqslant M+D\}. \tag{14.5}$$

The volume of the set in (14.5) is $2^n \cdot (M+D)^n$. Therefore the inequality

$$2^n \cdot (M+D)^n \geqslant (2\cdot M+1)^n \cdot \operatorname{vol}(S)$$

must hold. As M tends to infinity, the expression

$$\frac{(2\cdot M+2\cdot D)^n}{(2\cdot M+1)^n}$$

tends to one. This is a contradiction, since $\operatorname{vol}(S)>1$. □

Minkowski's theorem has also a version for general lattices. Let $\Lambda(B) \subseteq \mathbb{R}^n$ be a lattice and $K \subseteq \mathbb{R}^n$ be a convex set which is symmetric around the origin with $\operatorname{vol}(K) \geqslant 2^n \det(\Lambda)$. The mapping $\phi(x) = B^{-1}x$ maps Λ to $\mathbb{Z}^n$ and $\phi(K)$ is a convex body which is symmetric around the origin. The volume of $\phi(K)$ is $\operatorname{vol}(\phi(K)) = (1/\det(B))\operatorname{vol}(K)$ and thus $\operatorname{vol}(\phi(K)) \geqslant 2^n$. Theorem 14.10 implies that $\phi(K)$ contains a nonzero integer vector or equivalently K contains a nonzero lattice vector from Λ.

Theorem 14.11 (Minkowski's convex body theorem [61]). *Let $\Lambda \subseteq \mathbb{R}^n$ be a lattice and let $K \subseteq \mathbb{R}^n$ be a convex body of volume $\operatorname{vol}(K) \geqslant 2^n \det(\Lambda)$ that is symmetric about the origin. Then K contains a nonzero lattice point.*

As announced in the introduction of this section, Minkowski's theorem can be used to derive an upper bound on the length of a shortest vector of a lattice Λ in terms of the determinant of Λ. Let V_n be the volume of the n-dimensional unit ball. By scaling the unit ball with $\alpha \in \mathbb{R}_+$ one obtains a ball of volume $\alpha^n \cdot V_n$. This is greater or equal to $2^n \det(\Lambda)$ for $\alpha \geqslant 2\cdot \sqrt[n]{\det(\Lambda)/V_n}$. This has the following consequence.

Theorem 14.12. *A lattice $\Lambda \subseteq \mathbb{R}^n$ has a nonzero lattice point of length less than or equal to $2\cdot \sqrt[n]{\det(\Lambda)/V_n}$.*

The formula for V_n is

$$V_n = \frac{\pi^{\lfloor n/2\rfloor} 2^{\lceil n/2\rceil}}{\prod_{0\leqslant 2i \leqslant n}(n-2i)}.$$

Using Stirling's formula ($n! \approx \sqrt{2\pi n}\left(\frac{n}{e}\right)^n$) one sees that this is roughly $\left(\frac{2\pi e}{n}\right)^{n/2}$. The bound of Theorem 14.12 is thus roughly $\sqrt{\frac{2}{\pi e}n}\det(\Lambda)^{1/n}$.

Exercises

1) Show that a lattice Λ has a nonzero lattice point v with $\|v\|_\infty \leqslant \sqrt[n]{\det(\Lambda)}$.
2) Show that Minkowski's theorem holds also for convex sets that are full-dimensional and closed, i.e., the boundedness condition is not really necessary.
3) Let $K \subseteq \mathbb{R}^n$ be a convex body of volume $\operatorname{vol}(K) \geqslant k \cdot 2^n$. Show that K contains at least $2 \cdot k$ nonzero integer points.
4) (Two squares theorem) In this exercise you will prove that a prime number p with $p \equiv 1 \pmod 4$ can be written as the sum of two square numbers $p = a^2 + b^2$ for $a, b \in \mathbb{N}$.

   a) Show that the equation $q^2 \equiv -1 \pmod p$ has a solution.
   b) Consider the lattice Λ generated by $\left(\begin{smallmatrix} 1 & 0 \\ q & p \end{smallmatrix}\right)$ and the disk of radius $\sqrt{p \cdot 2 - \varepsilon}$ around 0 for a small $\varepsilon > 0$.
      i) Show that $\|v\|^2$ is divisible by p for each $v \in \Lambda$.
      ii) Show that there exists a $v \in \Lambda \setminus \{0\}$ with $\|v\|^2 = p$.
      iii) Conclude that p is the sum of two squares.

   Hints: Wilson's theorem, see, e.g., [65] states $(p-1)! \equiv -1 \pmod p$. If $q^2 \equiv -1$ does not have a solution, then $\mathbb{Z}_p^*$ can be paired (perfectly matched) into $(p-1)/2$ pairs, where the product of each pair is congruent to -1. But $(p-1)/2$ is even ($p \equiv 1 \pmod 4$), implying $(p-1)! \equiv 1 \pmod p$, contradicting Wilson's theorem.

14.5 The LLL algorithm

Recall the intuition from the introductory section on basis reduction. We want to transform a lattice basis into an equivalent one that is almost orthogonal. The hope is that a shortest vector can be determined from such a basis more easily. For a clear understanding of what almost orthogonal should mean we first recall the Gram-Schmidt procedure.

Let $U \subseteq \mathbb{R}^n$ be a subspace of $\mathbb{R}^n$ and let $v \in \mathbb{R}^n$. The *projection* of v *onto the orthogonal complement of* U is a vector $v^* = v - h$, where $h \in U$ and $\langle v - h, u \rangle = 0$ for each $u \in U$. If $U = \langle x_1, \ldots, x_k \rangle$ is the subspace generated by $x_1, \ldots, x_k$, then $v - h$ is also called the projection of v onto the orthogonal complement of $x_1, \ldots, x_k$.

The Gram-Schmidt procedure computes a set of vectors $b_1^*, \ldots, b_n^*$ such that the following conditions hold.

i) The vectors $b_1, \ldots, b_k$ span the same subspace as $b_1^*, \ldots, b_k^*$ for each $k = 1, \ldots, n$.

ii) The vectors $b_1^*, \ldots, b_n^*$ are pairwise orthogonal.

The procedure is described below. It is easy to see that i) and ii) hold for B^*.

The Gram-Schmidt orthogonalization procedure decomposes the matrix $B = (b_1, \ldots, b_n)$ into

$$B = B^* \cdot R \tag{14.6}$$

Fig. 14.4 The projection b_2^* of b_2 onto to the orthogonal complement of the subspace generated by b_1.

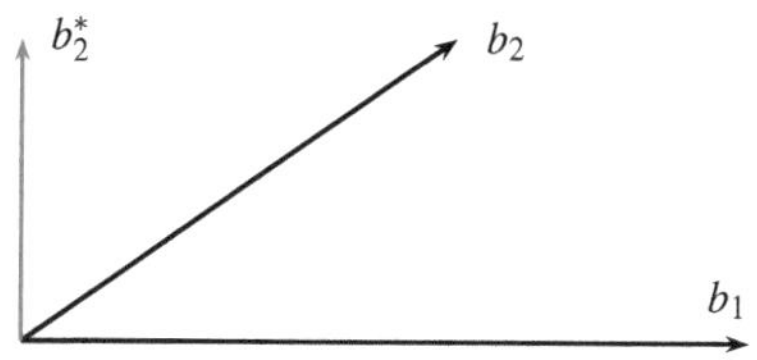

Algorithm 14.2 Gram-Schmidt orthogonalization

Input: $B = (b_1, \ldots, b_n) \in \mathbb{R}^{n \times n}$ nonsingular
Return: $B^* = (b_1^*, \ldots, b_n^*) \in \mathbb{R}^{n \times n}$ satisfying conditions i) and ii)
  $b_1^* \leftarrow b_1$
  for $j = 2, \ldots, k$ **do**
    $b_j^* \leftarrow b_j - \sum_{i=1}^{j-1} \mu_{ji} b_i^*$, where $\mu_{ji} = \langle b_j, b_i^* \rangle / \|b_i^*\|^2$
  end for

where B^* is a matrix with pairwise orthogonal columns $b_1^*, \ldots, b_n^*$ and

$$R = \begin{pmatrix} 1 & & \mu \\ & \ddots & \\ 0 & & 1 \end{pmatrix}$$

is an upper triangular matrix with diagonal elements 1. The decomposition (14.6) is the *Gram-Schmidt orthogonalization (GSO)* of B.

From the Gram-Schmidt orthogonalization procedure, we can also deduce the so-called Hadamard bound. Notice that

$$\det(B) = \det(B^* \cdot R) = \det(B^*). \tag{14.7}$$

Since $\det(B^* \cdot B^{*T}) = \|b_1^*\|^2 \cdots \|b_n^*\|^2$ and since $\det(B^* \cdot B^{*T}) = \det(B^*)^2$ we conclude that

$$|\det(B)| = \prod_{i=1}^{n} \|b_i^*\|. \tag{14.8}$$

Since we have $\|b_i\| \geqslant \|b_i^*\|$ for $i = 1, \ldots, n$ we obtain the *Hadamard bound*

$$|\det(B)| \leqslant \prod_{i=1}^{n} \|b_i\|. \tag{14.9}$$

We can use equation (14.9) now to measure how far a lattice basis B deviates from being orthogonal. The *orthogonality defect* of the lattice basis B is the number γ such that

$$\gamma \cdot |\det(B)| = \prod_{i=1}^{n} \|b_i\| \tag{14.10}$$

holds. The columns of B are pairwise orthogonal if and only if $\gamma = 1$. With the next theorem, we connect the computation of a shortest vector to this orthogonality defect. It implies that a shortest vector can be found among $(2 \cdot \gamma + 1)^n$ candidates.

Theorem 14.13. *Let $B \in \mathbb{Q}^{n \times n}$ be a lattice basis with orthogonality defect γ. A shortest non-zero vector $v \in \Lambda(B)$ is of the form*

$$v = \sum_{i=1}^{n} x_i \cdot b_i, \text{ with } x_i \in \mathbb{Z}, -\gamma \leqslant x_i \leqslant \gamma. \tag{14.11}$$

Proof. Since $\|b_j\| \geqslant \|b_j^*\|$ for $j = 1, \ldots, n$ and since

$$\|b_1\| \cdots \|b_n\| = \gamma \cdot \|b_1^*\| \cdots \|b_n^*\|,$$

we can conclude that

$$\|b_n\| \leqslant \gamma \cdot \|b_n^*\| \tag{14.12}$$

holds. Consider now $v = B \cdot x$ in (14.11). Since $B \cdot x = B^* \cdot R \cdot x$, where B^* is the Gram-Schmidt orthogonalization of B, and since the last component of $R \cdot x$ is equal to x_n, we conclude that

$$\|B \cdot x\| \geqslant |x_n| \cdot \|b_n^*\| \geqslant (|x_n|/\gamma)\|b_n\|. \tag{14.13}$$

From this we can conclude that, if v is a shortest vector, then x_n satisfies $-\gamma \leqslant x_n \leqslant \gamma$.

The orthogonality defect is invariant under the permutation of columns of B. Therefore each vector in the basis can play the role of being the last vector in the above argument. This implies that each component x_i in (14.11) satisfies $-\gamma \leqslant x_i \leqslant \gamma$ which implies the assertion. □

This implies that the shortest vector problem can be solved in fixed dimension, if we can compute a lattice basis B' of $\Lambda(B)$ whose orthogonality defect is bounded by a constant, depending on the dimension only. Therefore we call a lattice basis $B \in \mathbb{Q}^{n \times n}$ *reduced* if its orthogonality defect is bounded by a constant γ_n, depending only on the dimension n.

The LLL-algorithm[55] is an algorithm which reduces a lattice basis in polynomial time. The orthogonality defect of an LLL-reduced basis can be bounded by $2^{n(n-1)/4}$. Consequently, the shortest vector problem can be solved in time $2^{O(n^3)}$ times a polynomial in the binary encoding length of the input.

Normalization

Let $B = B^* \cdot R$ be the GSO of B. If R would be the identity matrix, then B would be B^* and thus orthogonal. The first step of the LLL-algorithm is *normalization*: One applies elementary column operations to transform the upper triangular matrix

R into a matrix that is as close as possible to the identity matrix. Normalization is in the literature also sometimes referred to as *size reduction*.

Let r_{ij} be the j-th entry of the i-th row of R. By subtracting $\lfloor r_{ij} \rceil$ times the i-th column of R from the j-th column, the new entry r'_{ij} at position ij will satisfy $-1/2 < r'_{ij} \leqslant 1/2$. Notice that the entries in a row below the i-th row of R remain unchanged. Thus working our way from the last to the first row, we obtain a basis $B' = B^* \cdot R'$ with

$$-1/2 < r'_{ij} \leqslant 1/2, \text{ for } 1 \leqslant i < j \leqslant n. \tag{14.14}$$

This procedure is called a *normalization step*.

Swapping

The LLL-algorithm iterates normalization and swapping steps. More precisely it normalizes the basis and then searches for two consecutive basis elements which should be swapped. This its continued, until a certain condition holds.

Algorithm 14.3 LLL algorithm

Repeat the following two steps, as long as there exists a j, $1 \leqslant j \leqslant n-1$ with

$$\|b^*_{j+1} + \mu_{j+1,j} b^*_j\|^2 < 3/4 \, \|b^*_j\|^2 : \tag{14.15}$$

Normalize B
Swap b_j and b_{j+1}

Notice that the condition (14.15) is invariant under normalization, since B^* is left untouched by the normalization procedure. Let us shed some light on the swapping operation and on this condition (14.15) that has to hold when swapping is applied. The vector $b^*_{j+1} + \mu_{j+1,j} b^*_j$ is the new j-th vector of B^* after the swap because

$$b^*_{j+1} = b_{j+1} - \sum_{i=1}^{j} \mu_{j+1,i} b^*_i. \tag{14.16}$$

Thus the vector $b^*_{j+1} + \mu_{j+1j} b^*_j$ is the projection of b_{j+1} onto the orthogonal complement of $b_1, \ldots, b_{j-1}$.

The condition (14.15) ensures that the norm of this new j-th column has decreased by a factor of $3/4$ at least. Since the vectors b^*_μ for $\mu \neq j, j+1$ remain the same (see exercise 14.5.2), the only side effect is an increase of the norm of the $j+1$-st column of B^*. The rest of the GSO remains unchanged.

More precisely, if the norm j-th column decreases by a factor of α then the $j+1$-st column increases by a factor of $1/\alpha$ since the product of the norms of the columns of B^* is equal to $|\det(B)|$ which is left invariant by a swap of columns in B.

Analysis

The above observation allows us now to show that the algorithm runs in polynomial time. The *potential* of a lattice basis B is defined as

$$\phi(B) = \|b_1^*\|^{2n} \|b_2^*\|^{2(n-1)} \|b_3^*\|^{2(n-2)} \cdots \|b_n^*\|^2. \tag{14.17}$$

A normalization step does not change the potential, since B^* remains unchanged. How does a swap affect the potential? Let B' be the basis after a swap operation, where the j-th and $j+1$-st column are swapped. We have (see exercise 2)

$$\begin{aligned} \frac{\phi(B)}{\phi(B')} &= \frac{\|b_j^*\|^{2\cdot(n-j+1)} \|b_{j+1}^*\|^{2(n-j)}}{\|b_j'^*\|^{2\cdot(n-j+1)} \|b_{j+1}'^*\|^{2(n-j)}} \\ &= \frac{\|b_j^*\|^{2\cdot(n-j)} \|b_{j+1}^*\|^{2(n-j)}}{\|b_j'^*\|^{2\cdot(n-j)} \|b_{j+1}'^*\|^{2(n-j)}} \cdot \frac{\|b_j^*\|^2}{\|b_j'^*\|^2} \\ &= \frac{\|b_j^*\|^2}{\|b_j'^*\|^2} \\ &\geqslant \tfrac{4}{3}. \end{aligned} \tag{14.18}$$

This shows that the potential drops at every iteration by at least $3/4$. Next we show that the potential of a lattice basis $B \in \mathbb{Z}^{n \times n}$ is an integer.

Let B_i be the matrix consisting of the first i columns of B. Then we have

$$\det(B_i^T \cdot B_i) = \|b_1^*\|^2 \cdots \|b_i^*\|^2 \in \mathbb{N}. \tag{14.19}$$

Consequently we have

$$\phi(B) = \prod_{i=1}^{n} \det(B_i^T \cdot B_i) \in \mathbb{N}. \tag{14.20}$$

This shows that the potential is bounded from below by 1 and the algorithm terminates in $O(\log \phi(B))$ steps. Clearly $\phi(B) \leqslant \det(B)^{2\cdot n} \leqslant (\sqrt{n}M)^{2\cdot n^2}$, where M is an upper bound on the absolute value of an entry in B.

We thus have the following theorem.

Theorem 14.14. *The LLL-algorithm terminates in $O(n^2(\log n + s))$ iterations, where s is the largest binary encoding length of a coefficient of $B \in \mathbb{Z}^{n \times n}$.*

In order to conclude that the LLL-algorithm runs in polynomial time, we also have to bound the binary encoding length of the numbers which occur in the intermediate steps of the algorithm. This is very important but a bit tedious and we don't do it here and refer, for example to [55]. We conclude with the following theorem.

Theorem 14.15. *The LLL-algorithm runs in polynomial time in the input encoding length of the initial lattice basis.*

Orthogonality defect and approximation of the shortest vector

The next theorem bounds the length of a shortest vector from below by means of the GSO. Here, $SV(\Lambda)$ denotes the length of a shortest nonzero vector of Λ.

Theorem 14.16. *Let B be a lattice basis and let $B^* = (b_1^*, \ldots, b_n^*)$ be its Gram-Schmidt orthogonalization, then $SV(\Lambda(B)) \geqslant \min_{i=1,\ldots,n} \|b_i^*\|_2$.*

Proof. Let $0 \neq v \in \Lambda$, then $v = Bx$ and let $k \leqslant n$ be the largest index with $x_k \neq 0$. With $B = B^* \cdot R$ we have

$$\begin{aligned} v &= B^* \cdot R \cdot x \\ &= x_k b_k^* + \sum_{i=1}^{k-1} \lambda_i b_i^*, \text{ for some } \lambda_i \in \mathbb{R}. \end{aligned}$$

This implies $\|v\| \geqslant |x_k| \, \|b_k^*\|$ and the theorem follows. □

With this lower bound, we can show that the first vector of an LLL-reduced basis is an approximation of the shortest vector, which is exponential in the dimension only.

Upon termination of the LLL-algorithm we have that each $\mu_{j+1,j}^2 \leqslant 1/4$. Since also $\|b_{j+1}^* + \mu_{j+1,j} b_j^*\|^2 \geqslant 3/4 \, \|b_j^*\|^2$ and since $\|b_{j+1}^* + \mu_{j+1,j} b_j^*\|^2 = \|b_{j+1}^*\|^2 + \mu_{j+1,j}^2 \|b_j^*\|^2$ we have

$$\|b_{j+1}^*\|^2 \geqslant 1/2 \|b_j^*\|^2 \tag{14.21}$$

for each $j = 1, \ldots, n-1$. Thus it follows by induction that

$$\|b_j^*\|^2 \geqslant 2^{i-j} \|b_i^*\|^2, \text{ for } 1 \leqslant i < j \leqslant n. \tag{14.22}$$

From this we can conclude that

$$\|b_1\|^2 = \|b_1^*\|^2 \leqslant 2^{n-1} SV(\Lambda(B)). \tag{14.23}$$

Also since

$$b_j = b_j^* + \sum_{i=1}^{j-1} \mu_{ji} b_i^* \tag{14.24}$$

and since each μ_{ji} has absolute value less that $1/2$ we have

$$\|b_j\|^2 \leqslant \|b_j^*\|^2 + 1/4 \sum_{i=1}^{j-1} \|b_i^*\| \tag{14.25}$$

and by applying (14.22) we obtain

$$\|b_j\|^2 \leqslant \|b_j^*\|^2 (1 + 1/4 \sum_{i=1}^{j-1} 2^{j-i}) \leqslant 2^{j-1} \|b_j^*\|^2. \tag{14.26}$$

This implies the following for the orthogonality defect

$$\|b_1\| \cdots \|b_n\| \leqslant 2^{n(n-1)/4} \|b_1^*\| \cdots \|b_n^*\| = 2^{n(n-1)/4} |\det(B)|. \tag{14.27}$$

Together with Theorem 14.13 we thus have the next theorem.

Theorem 14.17. *A shortest vector of an integral lattice can be computed in time $2^{O(n^3)}$ times a polynomial in the length of the input encoding.*

Notes

The LLL-algorithm requires $O(n^5 \log B)$ arithmetic operations on rational numbers of size $O(n + \log B)$. Here B is an upper bound on the norm of the basis vectors. Schnorr [72] presented an floating-point variant of the LLL algorithm which requires $O(n^4 \log B)$ arithmetic operations on rationals of size $O(n + \log B)$. Improvements on the algorithm itself and its analysis were given by Kaltofen [40] and Storjohann [78] among others. Using naive arithmetic, the analysis of the LLL-algorithm presented here amounts to a bit-complexity of $O(n^4 \log B (n + \log B)^2)$. Recently, Nguyen and Stehlé [63] have presented a floating-point variant of the LLL-algorithm which requires $O(n^5 (n + \log B) \log B)$ bit-operations. In fixed dimension, this matches the bit-complexity of the Euclidean algorithm. The greatest common divisor of two s-bit integers can be computed with $O(M(s) \log s)$ bit-operations [73], where $M(s)$ is the number of bit-operations, which are required for the multiplication of two s-bit numbers. For a long time, the fastest method for integer multiplication was the one by Schönhage and Strassen [74], requiring $O(s \log s \log \log s)$ bit-operations. Martin Fürer [26] improved this complexity recently to $O(s \log s \cdot 2^{O(\log^* s)})$. It is an interesting open problem, whether a shortest vector in fixed dimension can be computed with $O(M(s) \log s)$ bit-operations. Eisenbrand and Rote [20] showed that one can compute a shortest vector in fixed dimension n using $O(M(s) \log^{n-1} s)$ bit-operations. Recently Gama and Nguyen [27] proved that using a shortest vector oracle in dimension k, one can compute a $((1+\varepsilon)\gamma_k)^{(n-k)/(k-1)}$ approximation of the shortest vector, where γ_k is the so-called *Hermite constant.*

Exercises

1. Prove that $\|b_i\| \geqslant \|b_i^*\|$ holds for each $i = 1, \ldots, n$ for the GSO of B.
2. Let
$$B = (b_1, \ldots, b_{i-1}, b_i, b_{i+1}, b_{i+2}, \ldots, b_n)$$
and
$$C = (b_1, \ldots, b_{i-1}, b_{i+1}, b_i, b_{i+2}, \ldots, b_n)$$
be two lattice bases. Notice that C originates from B via swapping the i-th and $i+1$-st column. Prove that B^* and C^* only differ in the i-th and $i+1$-st column. Show further that $\|b_i^*\| \cdot \|b_{i+1}^*\| = \|c_i^*\| \cdot \|c_{i+1}^*\|$ holds.

3. Let $B \in \mathbb{R}^{n \times n}$ be a matrix. Prove that $|\det(B)| \leqslant (\sqrt{n}M)^n$, where M is an upper bound on the absolute values of the entries in B. *Hint: Hadamard bound!*
4. Estimate the total number of arithmetic operations which are performed by the LLL-algorithm in terms of $\phi(B)$.
5. Let α be a fixed constant and let $\Lambda = \Lambda(A)$, $A \in \mathbb{Q}^{n \times n}$ be a rational lattice in fixed dimension n.

   a) Prove that one can enumerate all vectors $v \in \Lambda(A)$ with $\|v\| \leqslant \alpha \cdot SV(\Lambda)$ in polynomial time.
   b) Let $\|\cdot\|$ be any fixed norm. Show that a shortest vector of Λ w.r.t. $\|\cdot\|$ can be computed in polynomial time, if $\|v\|$ can be evaluated in polynomial time in the binary input encoding of v for any $v \in \Lambda$.

14.6 Kannan's shortest vector algorithm

As we have mentioned above, one can compute a shortest vector of a lattice that is represented by a LLL-reduced basis $b_1, \ldots, b_n$ in $2^{O(n^3)}$ steps via enumerating the candidates $\sum_{j=1}^{n} \lambda_j b_j$, where $|\lambda_j| \leqslant 2^{n(n-1)/4}$ and choosing the shortest nonzero vector from this set.

Kannan [42, 43] provided an algorithm for the shortest vector problem, whose dependence on the dimension is $2^{O(n \log n)}$. Helfrich [36] improved Kannan's algorithm. Recently, Ajtai, Kumar and Sivakumar [3] presented a randomized algorithm for the shortest vector problem, with an expected dependence of $2^{O(n)}$ which is the subject of the next section. In this section, we describe Kannan's algorithm.

Korkine-Zolotareff reduction

A lattice basis $b_1, \ldots, b_n$ is *Korkine-Zolotareff reduced*, or *K-Z reduced* for short, if the following conditions hold.

i) The vector b_1 is a shortest vector of the lattice generated by $b_1, \ldots, b_n$.
ii) The numbers μ_{jk} in the Gram-Schmidt orthogonalization of $b_1, \ldots, b_n$ satisfy $|\mu_{jk}| \leqslant 1/2$.
iii) If $b_2', \ldots, b_n'$ denote the projections of $b_2, \ldots, b_n$ onto the orthogonal complement of the space generated by b_1, then $b_2', \ldots, b_n'$ is Korkine-Zolotareff reduced.

A two-dimensional lattice basis that is K-Z reduced is also called *Gauß* reduced, see [29]. The algorithm of Kannan computes a Korkine-Zolotareff reduced basis in dimension n by first computing a partially Korkine-Zolotareff reduced lattice basis, from which a shortest vector is among $2^{O(n \log n)}$ candidates. The basis is partially Korkine-Zolotareff reduced with the help of an algorithm for Korkine-Zolotareff reduction in dimension $n-1$.

With a shortest vector at hand, one can then compute a fully K-Z reduced basis by K-Z reducing the projection along the orthogonal complement of this shortest vector. A lattice basis $b_1,\dots,b_n$ is *partially Korkine-Zolotareff reduced* or *partially K-Z reduced* for short, if it satisfies the following properties.

1. If $b'_2,\dots,b'_n$ denotes the projection of $b_2,\dots,b_n$ onto the orthogonal complement of the space generated by b_1, then $b'_2,\dots,b'_n$ is Korkine-Zolotareff reduced.
2. The numbers μ_{jk} in the Gram-Schmidt orthogonalization of $b_1,\dots,b_n$ satisfy $|\mu_{jk}| \leqslant 1/2$.
3. $\|b'_2\| \geqslant 1/2\,\|b_1\|$.

Notice that, once Conditions 1 and 3 hold, Condition 2 can be satisfied via a normalization step. Normalization does not destroy Conditions 1 and 3. Condition 1 can be satisfied by applying Kannan's algorithm for full K-Z reduction to $b'_2,\dots,b'_n$, and applying the transformation to the original vectors $b_2,\dots,b_n$. Then if Condition 3 is not satisfied, then Helfrich [36] has proposed to replace b_1 and b_2 with the *Gauß-reduction* of this pair, or equivalently its K-Z reduction. Clearly, if b_1,b_2 is Gauß-reduced, which means that $\|b_1\| \leqslant \|b_2\|$ and the angle enclosed by b_1 and b_2 is at least 60° and at most 120°, then Condition 3 holds.

The following algorithm computes a partially K-Z reduced basis from a given input basis $b_1,\dots,b_n$. It uses as a subroutine an algorithm to K-Z reduce the lattice basis $b'_2,\dots,b'_n$.

Algorithm 14.4 Partial K-Z reduction

1. Apply the LLL-algorithm to $b_1,\dots,b_n$.
2. K-Z reduce $b'_2,\dots,b'_n$ and apply the corresponding transformation to $b_2,\dots,b_n$.
3. Perform normalization step on $b_1,\dots,b_n$.
4. If $\|b'_2\| < 1/2\,\|b_1\|$, then replace b_1,b_2 by its Gauß reduction and go to Step 2.

We show in a moment that we can extract a shortest vector from a partially K-Z reduced basis in $2^{O(n\log n)}$ steps, but before, we analyze the running time of the algorithm.

Theorem 14.18 ([36]). *Step 4 of Algorithm 14.4 is executed at most* $\log n + 6$ *times.*

Proof. Let v be a shortest vector and let $b_1,\dots,b_n$ be the lattice basis immediately before Step 4 of Algorithm 14.4 and let $b'_2,\dots,b'_n$ denote the projection of $b_2,\dots,b_n$ onto the orthogonal complement of b_1.

If Step 4 is executed, then v is not equal to b_1. Then clearly, the projection of v onto the orthogonal complement of b_1 is nonzero. Since $b'_2,\dots,b'_n$ is K-Z reduced it follows that $\|v\| \geqslant \|b'_2\|$ holds. Denote the Gauß reduction of b_1,b_2 by $\widetilde{b_1},\widetilde{b_2}$. The determinant of $\Lambda(b_1,b_2)$ is equal to $\|b_1\|\,\|b'_2\|$. After the Gauß reduction in Step 4, we have therefore

$$\|\widetilde{b}_1\| \leqslant 2\sqrt{\|b_1\|\,\|b_2'\|} \tag{14.28}$$

$$\leqslant 2\sqrt{\|b_1\|\,\|v\|}. \tag{14.29}$$

Dividing this inequality by $\|v\|$ gives

$$\frac{\|\widetilde{b}_1\|}{\|v\|} \leqslant 2\sqrt{\frac{\|b_1\|}{\|v\|}}.$$

Thus, if $b_1^{(i)}$ denotes the first basis vector after the i-th execution of Step 4, one has

$$\frac{\|b_1^{(i)}\|}{\|v\|} \leqslant 4\left(\frac{\|b_1^{(0)}\|}{\|v\|}\right)^{(1/2)^i}. \tag{14.30}$$

Since we start with a LLL-reduced basis, we know that $\|b_1^{(0)}\|/\|v\| \leqslant 2^{(n-1)/2}$ holds, and consequently that $\|b_1^{(\log n)}\|/\|v\| \leqslant 8$. Each further Gauß reduction decreases the length of the first basis vector by at least $3/4$. Therefore the number of runs through Step 4 is bounded by $\log n + 6$. □

Extracting a shortest vector

We now argue that with such a partially K-Z reduced basis $b_1,\ldots,b_n$ at hand, one only needs to check $n^{O(n)}$ candidates for the shortest vector. Let $v = \sum_{j=1}^n \lambda_j b_j$ be a shortest vector. After rewriting each b_j in terms of the Gram-Schmidt orthogonalization one obtains

$$\begin{aligned} v &= \sum_{j=1}^{n}\sum_{k=1}^{j}(\lambda_j \mu_{jk} b_k^*) \\ &= \sum_{k=1}^{n}(\sum_{j=k}^{n}\lambda_j \mu_{jk})b_k^*, \end{aligned}$$

where the μ_{jk} are as in Algorithm 14.2.

The length of v satisfies

$$\|v\|^2 = \sum_{k=1}^{n}(\sum_{j=k}^{n}(\lambda_j\mu_{jk}))^2\|b_k^*\|^2. \tag{14.31}$$

Consider the coefficient $c_n = |\lambda_n \mu_{nn}| = |\lambda_n|$ of $\|b_n^*\|$ in (14.31). We can bound this absolute value by $|\lambda_n| \leqslant \|v\|/\|b_n^*\| \leqslant \|b_1\|/\|b_n^*\|$. This leaves us $1+2\|b_1\|/\|b_n^*\|$ possibilities for λ_n. Suppose now that we picked $\lambda_n,\ldots,\lambda_{j+1}$ and inspect the coefficient c_j of $\|b_j^*\|$ in (14.31), which is

$$\begin{aligned} c_j &= |\sum_{k=j}^{n} (\lambda_k \mu_{kj})| \\ &= |\lambda_j + \sum_{k=j+1}^{n} (\lambda_k \mu_{kj})|. \end{aligned}$$

Since the inequality $c_j \leqslant \|b_1\|/\|b_j^*\|$ must hold, this leaves only $1+2\|b_1\|/\|b_j^*\|$ possibilities to pick λ_j. Thus by choosing the coefficients $\lambda_n,\dots,\lambda_1$ in this order, one has at most $\prod_{j=1}^n (1+2\|b_1\|/\|b_j^*\|)$ candidates.

Suppose $\|b_j^*\| > \|b_1\|$ for some j. Then b_j can never have a nonzero coefficient λ_j in a shortest vector representation $v = \sum_{j=1}^n \lambda_j b_j$. Because in that case, v has a nonzero component in its projection to the orthogonal complement of $b_1\mathbb{R}+\ldots+b_{i-1}\mathbb{R}$ and since $b_2',\dots,b_n'$ is K-Z reduced, this implies that $\|v\| \geqslant \|b_j^*\| > \|b_1\|$, which is impossible. Thus we can assume that $\|b_j^*\| \leqslant \|b_1\|$ holds for all $j=1,\dots,n$. Otherwise, b_j can be discarded. Therefore the number of candidates N for the tuples $(\lambda_1,\dots,\lambda_n)$ satisfies

$$\begin{aligned} N &\leqslant \prod_{j=1}^{n} (1+2\|b_1\|/\|b_j^*\|) \\ &\leqslant \prod_{j=1}^{n} (3\|b_1\|/\|b_j^*\|) \\ &= 3^n \|b_1\|^n/\det(\Lambda). \end{aligned}$$

Next we give an upper bound for $\|b_1\|$. If b_1 is a shortest vector, then Minkowski's theorem, (Theorem 14.11) guarantees that $\|b_1\| \leqslant \sqrt{n}\det(\Lambda)^{1/n}$ holds. If b_1 is not a shortest vector, then the shortest vector v has a nonzero projection onto the orthogonal complement of $b_1\mathbb{R}$. Since $b_2',\dots,b_n'$ is K-Z reduced, this implies that $\|v\| \geqslant \|b_2'\| \geqslant 1/2\|b_1\|$, since the basis is partially K-Z reduced. In any case we have $\|b_1\| \leqslant 2\sqrt{n}\det(\Lambda)^{1/n}$ and thus that $N \leqslant 6^n n^{n/2}$.

Now it is clear how to compute a K-Z reduced basis and thus a shortest vector. With an algorithm for K-Z reduction in dimension $n-1$, one uses Algorithm 14.4 to partially K-Z reduce the basis and then one checks all possible candidates for a shortest vector. Then one performs K-Z reduction on the basis for the projection onto the orthogonal complement of the shortest vector. Kannan [43] has shown that this procedure for K-Z reduction requires $n^{O(n)}\varphi$ operations, where φ is the binary encoding length of the initial basis and where the operands during the execution of the algorithm have at most $O(n^2\varphi)$ bits.

Theorem 14.19 ([43]). *Let B be a lattice basis of binary encoding length φ. There exists an algorithm which computes a K-Z reduced basis and requires $n^{O(n)} \cdot p(\varphi)$ arithmetic operations on rationals of size $O(n^2\varphi)$, where $p(\cdot)$ is a fixed polynomial.*

Notes

Kannan [44] also developed an algorithm for the *closest vector problem* whose running time is $2^{O(n\log n)}$ times a polynomial in the encoding length of the input. Here, one is given a rational vector $x \in \mathbb{Q}^n$ and a lattice $\Lambda \subseteq \mathbb{Q}^n$. The task is to compute a lattice point $v \in \Lambda$ which is closest to x, i.e., minimizing $\|v-x\|$. Kannan also showed that the integer programming feasibility problem can be solved within this complexity bound. Furthermore he showed that one can compute an approximate solution of the closest vector problem with a polynomial number of queries to an oracle which solves the shortest vector problem of an $n+1$-dimensional lattice. Blömer [9] showed that there exists an algorithm for the closest vector problem which runs in time $n!$ times a polynomial in the input encoding length. This means an exponential improvement over Kannan's algorithm [44] and its subsequent improvement by Helfrich [36]. Hanrot and Stehlé [33] improved this further to $n^{0.5 \cdot n+o(n)}$ and showed that the shortest vector problem can be solved in time $n^{0.184 \cdot n+o(n)}$ times a polynomial in the input length. Stehlé and Pujol [66] improve the arithmetic complexity of Kannan's algorithm.

Exercises

1. Prove that a 2-dimensional lattice basis is Gauß reduced if and only if it is LLL-reduced.
2. Prove that the shortest vector can be extracted in time $2^{O(n^2)}$ out of a LLL-reduced basis by adapting the arguments given in this section.

14.7 A randomized simply exponential algorithm for shortest vector

Ajtai, Kumar and Sivakumar [3] described a randomized method which outputs a shortest vector of a lattice with very high probability and has running time $2^{O(n)}$ times a polynomial in the binary input encoding length of the lattice basis. We follow the description of their algorithm in Oded Regev's excellent lecture notes [68]. At first sight, the algorithm is quite different from the shortest-vector algorithms that are based on reduction and candidate-trial, as Kannan's algorithm.

Our task is to compute the shortest vector of a lattice $\Lambda(B)$, for a nonsingular $B \in \mathbb{Q}^{n\times n}$. We can assume that $2 \leqslant SV(\Lambda) < 3$ holds, see exercise 14.7.1. The idea is to sample points in $\mathbb{R}^n$ and to translate these samples into lattice points. How is this translation done? A point $x \in \mathbb{R}^n$ can be represented as a linear combination of the basis vectors

$$x = \sum_{i=1}^{n} \lambda_i b_i, \tag{14.32}$$

for some $\lambda_i \in \mathbb{R}$. If we round all λ_i down, we obtain a lattice point

$$\sum_{i=1}^{n} \lfloor \lambda_i \rfloor \cdot b_i \in \Lambda(B). \tag{14.33}$$

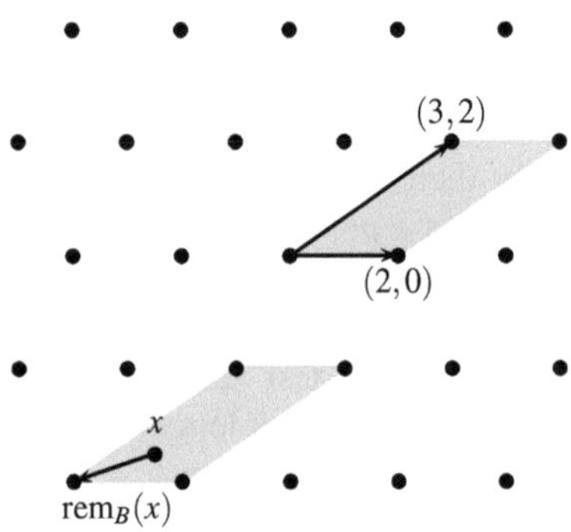

Fig. 14.5 The remainder of a point w.r.t. a basis B.

The *remainder* of x modulo the *basis* B is the vector

$$\mathrm{rem}_B(x) = \sum_{i=1}^{n} (\lambda_i - \lfloor \lambda_i \rfloor) b_i.$$

The lattice point in (14.33) is $x - \mathrm{rem}_B(x)$ and we associate it to x. Clearly one has

$$\mathrm{rem}_B(x) = \mathrm{rem}_B(x+v) \tag{14.34}$$

for each $v \in \Lambda(B)$.

Sieving

What we discussed above suggests the following approach. We sample a number of points $x_1, \ldots, x_k$ uniformly at random from $B_2(0)$, the ball of radius 2 around 0, and compute their remainders $y_i = \mathrm{rem}_B(x_i)$. The norms of these remainders are bounded by $R = \sum_{i=1}^n \|b_i\|$.

If we sample enough points (and the next lemma provides a bound on k), then there will be two such sampled points x_i, x_j such that their remainders y_i and y_j have distance at most $R/2$. The length of the lattice vector $x_i - y_i - (x_j - y_j)$ is then bounded by $R/2 + 4$.

Lemma 14.3 (Sieving Lemma). *Let $y_1, \ldots, y_k \in \mathbb{R}^n$ be points contained in the ball $B_R(0)$. There is a polynomial algorithm that computes a mapping $\tau : \{1, \ldots, k\} \to \{1, \ldots, k\}$ with the following properties:*

i) The cardinality of the image of τ is bounded by 5^n.
ii) For each $i = 1,\dots,k$ one has $\|y_i - y_{\tau(i)}\| \leqslant R/2$.

In other words, we can identify $\leqslant 5^n$ *centers* among the $y_1,\dots,y_k$ such that the balls of radius $R/2$ around these centers cover all of the remaining y_i. The mapping τ associates the y_i to a center of a ball that contains y_i.

Proof (of Lemma 14.3). We describe the simple procedure. In the beginning, all points are colored black. We iterate the following steps until there are no black points left: Choose any black point y_i and define $\tau(j) = i$, for each j such that y_j is black and $\|y_i - y_j\| \leqslant R/2$ holds. Color all points at distance at most $R/2$ from y_i red.

Call the points y_i with $\tau(i) = i$ *centers*. The number of centers is the size of the image of τ. Two centers have distance at least $R/2$ from each other. This means that the balls of radius $R/4$ around the centers do not intersect. On the other hand, these balls all fit into the ball of radius $R + R/4$. This implies that the number of centers is bounded by

$$\mathrm{vol}(B_{5\cdot R/4}(0))/\mathrm{vol}((B_{R/4}(0)) = 5^n,$$

and the assertion follows. □

The randomized algorithm for shortest vector

The sieving lemma provides a procedure to generate lattice vectors whose length is roughly $R/2$. This idea is now iterated in the random sampling algorithm for shortest vector.

Algorithm 14.5 Randomized algorithm for shortest vector

1. $R_0 \leftarrow \sum_{i=1}^n \|b_i\|$
   Choose $N = 2^{8n} \log R_0$ points $x_1,\dots,x_N$ uniformly at random from $B_2(0)$
   Initialize a set of tuples $L = \{(x_i, y_i) : y_i = \mathrm{rem}_B(x_i), i = 1,\dots,N\}$
   $R \leftarrow R_0$
2. **While** $R > 6$ **do**
   Apply the sieving algorithm to the vectors y_i for each $(x_i, y_i) \in L$
   Remove from L all tuples (x_i, y_i), where y_i is a center of the sieving procedure
   Replace each of the remaining (x_j, y_j) with $(x_j, y_j - (y_{\tau(j)} - x_{\tau(j)}))$
   $R \leftarrow R/2 + 2$.
3. For any two pairs (x_i, y_i), (x_j, y_j) in L compute the difference $x_i - y_i - (x_j - y_j)$ and output the shortest nonzero vector among these differences

---

Lemma 14.4. *At the beginning of each iteration of the* **while** *loop one has for each $(x_i, y_i) \in L$*

i) $x_i - y_i \in \Lambda(B)$
ii) $\|y_i\| \leqslant R$.

Proof. At the beginning of the first iteration, these invariants hold, as we discussed above.

We now show that these conditions hold after the instructions of the while-loop have been executed, if they were true at the beginning of that iteration. We only need to consider a tuple (x_j, y_j), where y_j is not a center in the sieving procedure. Let y_i be the center of y_j, i.e., $\tau(j) = i$. Since $x_j - y_j \in \Lambda(B)$ and $x_i - y_i \in \Lambda(B)$ we conclude that $x_j - y_j - (x_i - y_i) \in \Lambda(B)$, which implies condition i).

Since y_i is the center of y_j we have $\|y_j - y_i\| \leqslant R/2$. From this we conclude

$$\|y_j - (y_i - x_i)\| \leqslant \|y_j - y_i\| + \|x_i\| \leqslant R/2 + 2.$$

□

Lemma 14.4 and Exercise 14.7.4 imply that the algorithm runs in time $2^{O(n)}$ times a polynomial in the input encoding of the lattice basis B. Furthermore, at the end of the algorithm, there are at least

$$2^{8n} \log R_0 - 5^n \log R_0 \geqslant 2^{7n} \log R_0 \tag{14.35}$$

tuples (x_i, y_i) in L. The lattice vectors $x_i - y_i$ are short, i.e., $\|x_i - y_i\| \leqslant 2 + 6 = 8$. But how can we be sure that they are not all zero? Exercise 3 shows that, even when L is initialized, roughly half of these lattice vectors could be zero.

It turns out that the following observation that follows from (14.34) is crucial.

> Consider a tuple (x_i, y_i) in L before Step 3 of the algorithm. The algorithm would behave just the same until this point if x_i was replaced by $x_i + v_i$ after the initialization (Step 1), where $v_i \in \Lambda(B)$ is an arbitrary lattice vector.

When do we have to really *know* a point x_i and not just $y_i = \text{rem}_B(x_i)$? The value of x_i is needed only when y_i was a center of the sieving procedure and tuples (x_j, y_j) are replaced by $(x_j, y_j - (y_i - x_i))$. Now we shed some light on why we sample the points $x_1, \ldots, x_N$ from $B_2(0)$ and not from a ball of other radius. This is connected to the fact that $2 \leqslant SV(\Lambda(B)) < 3$ and this crucial observation from above.

Let $v \in \Lambda(B)$ be a shortest vector. Consider the set $C_1 = B_2(0) \cap B_2(v)$ and the set $C_2 = B_2(0) \cap B_2(-v)$.

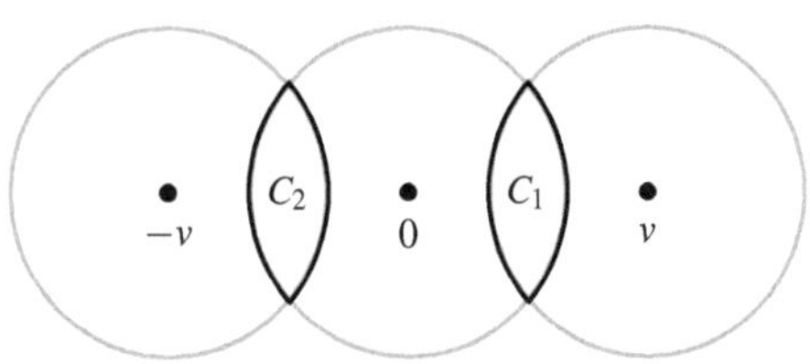

Fig. 14.6 The sets C_1 and C_2.

The mapping

$$f(x) = \begin{cases} x - v, & \text{if } x \in C_1 \\ x + v, & \text{if } x \in C_2 \\ x, & \text{if } x \in B_2(0) \backslash (C_1 \cup C_2) \end{cases} \tag{14.36}$$

is a bijection of $B_2(0)$ which maps C_1 to C_2 and vice-versa. A point $x \in B_2(0)$ which is chosen uniformly at random, remains to be chosen uniformly at random after we toss a fair coin and map x to $f(x)$ in the case that the coin shows heads.

This shows, together with the observation above, that we could, just before we enter Step 3 of the algorithm, toss a fair coin for each x_i with $(x_i, y_i) \in L$ and replace x_i by $f(x_i)$ if the coin shows head.

The next theorem implies that a shortest vector of $\Lambda(B)$ is returned with high probability.

Theorem 14.20. *At the end of Step 2 there exists with high probability a subset $L' \subseteq L$ of size 2^n which satisfies the following conditions*

a) $x_i - y_i = x_j - y_j$ *for each* $(x_i, y_i), (x_j, y_j) \in L'$.
b) $x_i \in C_1 \cup C_2$ *for each* $(x_i, y_i) \in L'$.

Before we prove the theorem we derive the main result of this section.

Theorem 14.21. *The random sampling algorithm outputs a shortest vector with high probability.*

Proof. Consider the set L' from Theorem 14.20. For each $(x_i, y_i) \in L'$, we toss a coin and replace x_i with $f(x_i)$ if the coin shows head. If we consider a difference

$$x_i - y_i - (x_j - y_j), \text{ where } (x_i, y_i) \neq (x_j, y_j) \in L'$$

then the probability that this difference becomes $\pm v$ is exactly $1/2$. (One coin shows head and coin shows tail). Since the cardinality of L' is 2^n with high probability, the algorithm will output a shortest vector of $\Lambda(B)$. □

We now prove Theorem 14.20. Recall that the set L contains at least $2^{7 \cdot n}$ tuples at the end of Step 2. The next lemma shows that the expected amount of these tuples, whose first component belongs to $C_1 \cup C_2$ is at least $2^{5 \cdot n + 1}$.

Lemma 14.5. *Consider the sets C_1 and C_2 described above. One has*

$$\text{vol}(C_1)/\text{vol}(B_2(0)) = \text{vol}(C_2)/\text{vol}(B_2(0)) \geqslant 2^{-2n}. \tag{14.37}$$

Lemma 14.5 shows that the expected number of tuples (x_i, y_i) with $x_i \in C_1 \cap C_2$ is $2^{6n+1} \log R_0$. The probability that this number of tuples is smaller than $2^{6n} \log R_0$ at the end of Step 1 is exponentially small.

Since we delete in each iteration 5^n tuples and perform at most $\log R_0$ many iterations, L contains at the end of Step 2, 2^{5n} tuples (x_i, y_i) with $x_i \in C_1 \cup C_2$ with high probability. Notice that at this point, a difference $x_i - y_i$ is a vector, whose length is bounded by 8.

Lemma 14.6.

$$|\Lambda(B) \cap B_8(0)| \leqslant 2^{4n}. \tag{14.38}$$

Lemma 14.6 and the previous discussion show that there exists a lattice vector $w \in \Lambda(B)$ such that $x_i - y_i = w$ and $x_i \in C_1 \cup C_2$ for at least 2^n of these tuples. This shows Theorem 14.20.

Notes

Ajtai et al. [4] also showed that there is a randomized algorithm to compute K-Z reduced bases in time $2^{O(n)}$. Together with the block-reduction algorithm of Schnorr [72], this implies the existence of a randomized polynomial time approximation algorithm for shortest vector, with an approximation factor of $2^{n \log\log n / \log n}$. Schnorr [72] previously showed that one approximate the shortest vector within a factor of $2^{n \log\log^2 n / \log n}$.

We did not mention how the sampling of points in $B_2(0)$ is done. A more general procedure for this task (sampling in convex bodies) was presented by Dyer, Kannan and Frieze [16]. Here the authors show how to sample points whose statistical distance to the uniform distribution is exponentially small. This serves the purpose of algorithm 14.5, see also exercise 7.

Blömer and Naewe [10] modified the sampling procedure described above to compute shortest vectors which are outside a given subspace. Under some length conditions on the target vector, they achieve a simply exponential algorithm for closest vector.

The shortest vector problem is also very interesting from the viewpoint of computational complexity. Van Emde Boas [22] proved that the shortest vector problem with respect to the ℓ_∞ norm is NP-hard, and he conjectured that it is NP-hard with respect to the Euclidean norm. In the same paper he proved that the closest vector problem is NP-hard for any ℓ_p norm. Ajtai [2] proved that the shortest vector problem with respect to the ℓ_2-norm is NP-hard for randomized problem reductions. This means that the reduction makes use of results of a probabilistic algorithm. Ajtai also showed that approximating the length of a shortest vector in a given lattice within a factor $1 + 1/2^{n^c}$ is NP-hard for some constant c. The non-approximability factor was improved to $(1 + 1/n^\varepsilon)$ by Cai and Nerurkar [12]. Micciancio [60] improved this factor substantially by showing that it is NP-hard to approximate the shortest vector in a given lattice within any constant factor less that $\sqrt{2}$ for randomized problem reductions, and that the same result holds for deterministic problem reductions (the "normal" type of reductions used in an NP-hardness proof) under the condition that a certain number theoretic conjecture holds. Goldreich and Goldwasser [30] proved that it is not NP-hard to approximate the shortest vector, or the closest vector, within a factor $\sqrt{n}$ unless the polynomial-time hierarchy collapses. Dinur [15] showed that it is NP-hard to approximate shortest vectors w.r.t. ℓ_∞ within almost polynomial factors. Khot [50] showed that the shortest vector problem w.r.t. the ℓ_p-norm is hard to approximate within any constant if $p > 1$. Using norm-embeddings and a reduction to the ℓ_2 norm, Regev and Rosen [69] showed that this also holds for $p = 1$ and $p = \infty$. The currently strongest inapproximability result for shortest vector is due to Haviv and Regev [34].

Exercises

1. Justify the assumption $2 \leqslant SV(\Lambda) \leqslant 3$. *Hint: Suppose that $B \in \mathbb{Z}^{n \times n}$ and consider the lattices $\Lambda\left((3/2)^i/\|b_1\| \cdot B\right)$.*
2. Show that $\|\text{rem}_B(x)\|$ is bounded by $\sum_{i=1}^n \|b_i\|$.
3. Suppose that x is chosen uniformly at random from $B_2(0)$. Show that the probability that
$$x - \text{rem}_B(x) = 0 \tag{14.39}$$
holds, is at most $1/2$. For each fixed n, provide a family of lattice bases B_k^n with $2 \leqslant SV(\Lambda(B_k^n)) < 3$ such that the probability for the event (14.39) tends to $1/2$ for $k \to \infty$.
4. Prove that the number of iterations through the **while** loop of algorithm 14.5 is bounded by $\log R_0 - 1$.
5. Prove Lemma 14.5.
6. Prove Lemma 14.6. *Hint: Packing argument!*
7. How many bits does one have to know of each sampled point x_i in Algorithm 14.5? Show that the number of bits which are necessary to represent the numbers in the intermediate steps of Algorithm 14.5 is polynomial in the input encoding of the lattice basis.

8*. A prominent research question is the one whether there exists a deterministic algorithm for shortest vector with running time $2^{O(n)}$ times a polynomial in the input encoding.

9*. Is there a (randomized) $2^{O(n)}$-algorithm which computes a shortest nonnegative lattice vector, i.e., a $v \in \Lambda - \{0\}$ with $v \geqslant 0$ and $\|v\|$ minimal?

14.8 Integer programming in fixed dimension

In this section, we will describe Lenstra's algorithm [56] which solves the following integer feasibility problem.

> Given $A \in \mathbb{Z}^{m \times n}$ and $b \in \mathbb{Z}^m$, decide whether there exists an integer point that satisfies the inequality system $Ax \leqslant b$.

The algorithm runs in polynomial time if the dimension n is fixed. The integer feasibility problem is in NP [11]. In other words, if there exists an integer solution of $Ax \leqslant b$, then there exists one, whose size is polynomial in the size of the largest absolute value of a coefficient of the system $Ax \leqslant b$ and n, see, e.g., [75].

We start with the flatness theorem, which is the key concept of any polynomial time algorithm for integer programming in fixed dimension. The flatness theorem shows that the feasibility problem can be decided with a variant of branch-and-bound and in particular that the feasibility problem can be solved in polynomial time, if the dimension n is fixed.

Let $K \subseteq \mathbb{R}^n$ be a nonempty closed subset of $\mathbb{R}^n$ and let $d \in \mathbb{R}^n$ be a vector. The *width* of K *along* d is the number

$$w_d(K) = \max\{d^T x \colon x \in K\} - \min\{d^T x \colon x \in K\}.$$

If the maximum or minimum does not exist, then $w_d(K)$ is defined to be infinity.

Suppose now that $K \subseteq \mathbb{R}^n$ is a convex body which does not contain an integer point. The *flatness theorem*, attributed to Khinchin 1948, ensures that there exists a nonzero integer vector $d \in \mathbb{Z}^n$ such that $w_d(K)$ is bounded by a constant. The *width* of P is defined as

$$w(P) = \min_{d \in \mathbb{Z}^n \setminus \{0\}} w_d(P).$$

A $d \in \mathbb{Z}^n \setminus \{0\}$ which minimizes $w_d(P)$ is called a *flat direction* of P.

Theorem 14.22 (Khinchine's flatness theorem [49]). *Let $K \subseteq \mathbb{R}^n$ be a convex body. Either K contains an integer point, or $w(K) \leqslant \omega(n)$, where $\omega(n)$ is a constant depending on the dimension only.*

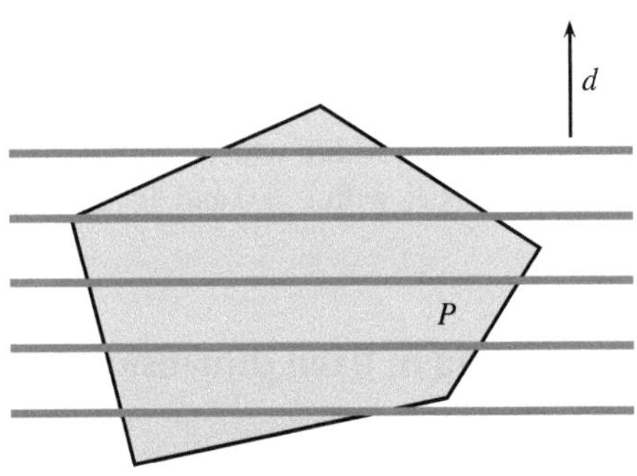

Fig. 14.7 Branching on a flat direction. The vector $d \neq 0$ is integral and the grey lines represent the lines $d^T x = \delta$ where $\delta \in \mathbb{Z}$ and $\max_{x \in P} d^T x \geqslant \delta \geqslant \min_{x \in P} d^T x$. The feasibility of P can be checked by checking the feasibility of the five 1-dimensional polytopes which are obtained from the intersection of P with the gray lines.

If $w_d(P) > \omega(n)$ for a flat direction d of P, then the flatness theorem guarantees the existence of an integer point in P. If, on the other hand, $w_d(P) \leqslant \omega(n)$, then an integer point $x^* \in P \cap \mathbb{Z}^n$ must lie on one of the constant number of hyperplanes

$$c^T x = \delta, \text{ where } \delta \in \mathbb{Z}, \text{ and } \min\{c^T x \colon x \in P\} \leqslant \delta \leqslant \max\{c^T x \colon x \in P\}.$$

Later on in this section, we prove the flatness theorem. If the reader has understood how the flatness theorem is proved, he can also solve exercises 6 and 7. These exercises ask you to prove that the width and a flat direction of a rational polyhedron in fixed dimension can be computed in polynomial time.

Algorithm 14.6 Lenstra's algorithm

Input: $A \in \mathbb{Z}^{m \times n}$ and $b \in \mathbb{Z}^m$ with $P = \{x \in \mathbb{R}^n \colon Ax \leqslant b\}$ full-dimensional
Return: "Yes" if P is integer feasible and "No" otherwise.

```
Compute w(P) and a flat direction d of P
if w(P) > ω(n) then
   return "Yes"
end if
for δ = ⌈min_{x∈P} d^T x⌉, ..., ⌊max_{x∈P} d^T x⌋ do
   Compute C ∈ Z^{m×(n−1)} and f ∈ Z^m such that Cx ⩽ f is integer feasible if and only if Ax ⩽
   b, d^T x = δ is integer feasible (See exercise 2)
   Recursively solve the integer feasibility problem for Cx ⩽ f
end for
```

Exercises 2, 6 and 7 show that Lenstra's algorithm runs in polynomial time in the input encoding length.

Theorem 14.23. *Lenstra's algorithm runs in polynomial time in the input encoding length if the dimension is fixed.*

Proving the flatness theorem

Let us first discuss how to compute a flat direction of an ellipsoid. An ellipsoid is the image $f(B^n)$ of the n-dimensional unit ball $B^n = \{x \in \mathbb{R}^n \colon \|x\| \leqslant 1\}$ under an affine map $f(x) = A^{-1}x + a$, where $A \in \mathbb{R}^{n \times n}$ is a non-singular matrix, and $a \in \mathbb{R}^n$ is a vector. This ellipsoid can also be described as $E(A,a) = \{x \in \mathbb{R}^n \colon \|A(x-a)\| \leqslant 1\}$.

Consider now an ellipsoid $E(A,a)$ and a direction $d \in \mathbb{Z}^n \setminus \{0\}$. We first want to compute the width of $E(A,a)$ along d and the discuss how to find an integral vector $d \neq 0$ along which the width is minimal.

Since the width is invariant under translation, we can assume that $a = 0$ holds. The width of $E(A,0)$ along d is

$$\max_{x_1,x_2 \in E(A,0)} d^T(x_1 - x_2). \tag{14.40}$$

We have $d^T x_1 - d^T x_2 = d^T A^{-1}(Ax_1 - Ax_2)$ and Ax_1 and Ax_2 are contained in the unit ball if and only if $x_1, x_2 \in E(A,0)$. Consequently (14.40) is simply $2 \cdot \|d^T A^{-1}\|$.

Keep in mind that we want to compute a flat direction of $E(A,0)$ or, in other words, that we are interested in an integral d such that $2\cdot\|d^TA^{-1}\|$ is as small as possible. This is a shortest vector problem in the lattice $\Lambda(A^{-1^T})$.

The *dual lattice* Λ^* of a lattice $\Lambda\subseteq\mathbb{R}^n$ is defined as

$$\Lambda^*=\{x\in\mathbb{R}^n\colon x^Ty\in\mathbb{Z}\text{ for all }y\in\Lambda\}.$$

Exercise 5 shows that $\Lambda^*(A)=\Lambda(A^{-1^T})$ holds. Our discussion above implies the following theorem.

Theorem 14.24. *A nonzero vector $d\in\mathbb{Z}^n$ is a flat direction of the ellipsoid $E(A,a)$ if and only if $A^{-1^T}d$ is a shortest nonzero vector of $\Lambda^*(A)$. A flat direction of an ellipsoid can be computed in polynomial time in fixed dimension. If the dimension is varying, one can compute an integer vector $d\in\mathbb{Z}^n\setminus\{0\}$ in polynomial time such that $w_d(E(A,a))\leqslant 2^{n-1}w(E(A,a))$ holds.*

The *covering radius* $\mu(\Lambda)$ of a lattice $\Lambda\in\mathbb{R}^n$ is the smallest number α such that the balls of radius α centered at the lattice points cover $\mathbb{R}^n$. Alternatively, the covering radius is the largest distance of a point $x\in\mathbb{R}^n$ to the lattice Λ.

The *packing radius* $\rho(\Lambda)$ of Λ is the largest number β such that the balls of radius β around the lattice points do not properly intersect. Alternatively, the packing radius is $SV(\Lambda)/2$.

We are now very close to proving the flatness theorem for ellipsoids. What if $E(A,a)$ does not contain an integer point? Inspecting the definition of $E(A,a)=\{x\in\mathbb{R}^n\colon\|A(x-a)\|\leqslant 1\}$ we see that this is the case if and only if the distance of Aa to the lattice $\Lambda(A)$ is larger than 1, i.e., $\mu(\Lambda(A))>1$. If we can now infer from this that the packing radius of $\Lambda^*(A)$ is bounded by a constant, depending only on the dimension, we are done, since the width of $E(A,a)$ is exactly $2\cdot SV(\Lambda^*(A))=4\cdot\rho(\Lambda^*(A))$. This is established in the next theorem via basis reduction.

Theorem 14.25. *Let $\Lambda\subseteq\mathbb{R}^n$ be a lattice and Λ^* be its dual. One has*

$$\mu(\Lambda)\cdot\rho(\Lambda^*)\leqslant f(n),$$

where $f(n)$ is a constant depending only on n which satisfies $f(n)\leqslant n/4\cdot 2^{n(n-1)/4}$.

Proof. Suppose that $B=(b_1,\dots,b_n)\in\mathbb{Q}^{n\times n}$ is a basis of the rational lattice $\Lambda(B)$ with orthogonality defect γ, i.e., one has

$$\|b_1\|\cdots\|b_n\|=\gamma\cdot|\det(B)|. \tag{14.41}$$

Assume that the longest basis vector is b_n, in other words that $\|b_n\|\geqslant\|b_j\|$ for $j=1,\dots,n-1$. This assumption can be made without loss of generality, since the orthogonality defect is invariant under swapping of columns.

Let $u\in\mathbb{R}^n$ be a point whose distance to Λ is $\mu(\Lambda)$. Since B is a basis of $\mathbb{R}^n$ we can write $u=\sum_{i=1}^n\lambda_ib_i$, with $\lambda_i\in\mathbb{R}$. The vector $v=\sum_{i=1}^n\lfloor\lambda_i\rceil b_i$ belongs to the lattice $\Lambda(B)$, where $\lfloor\lambda_i\rceil$ denotes the closest integer to λ_i. Clearly $\|v-u\|\geqslant\mu(\Lambda)$.

We also have

$$\begin{aligned}\|v-u\| &= \|\textstyle\sum_{i=1}^n (\lfloor\lambda_i\rceil - \lambda_i) b_i\| \\ &\leqslant \textstyle\sum_{i=1}^n \|(\lfloor\lambda_i\rceil - \lambda_i) b_i\| \\ &\leqslant \tfrac{1}{2}\textstyle\sum_{i=1}^n \|b_i\| \\ &\leqslant \tfrac{n}{2}\|b_n\|,\end{aligned} \tag{14.42}$$

where the last inequality in (14.42) follows from the fact that the last basis vector b_n is the longest one in the basis. Since $\|v-u\| \geqslant \mu(\Lambda)$ we therefore have

$$\|b_n\| \geqslant 2\mu(\Lambda)/n. \tag{14.43}$$

Let $B = B^* \cdot R$ be the GSO of B. The following facts (see exercise 14.5.1)

$$\begin{aligned}&\|b_1\|\cdots\|b_n\| = \gamma \cdot \|b_1^*\|\cdots\|b_n^*\| \\ &\|b_j\| \geqslant \|b_j^*\|, \quad j = 1,\ldots,n.\end{aligned}$$

imply $\|b_n\| \leqslant \gamma \cdot \|b_n^*\|$, which together with (14.43) implies

$$\|b_n^*\| \geqslant 2\mu(\Lambda)/(n\cdot\gamma).$$

Now b_n^* is orthogonal to the vectors b_j^*, $j = 1,\ldots,n-1$. Since R is an upper triangular matrix with only 1's on its diagonal, it follows that $(b_n^*)^T B^* \cdot R = (0,\ldots,0,\|b_n^*\|^2)$ from which we can conclude that

$$d = b_n^*/\|b_n^*\|^2 \in \Lambda^*(B)^{\cdot}$$

The norm of d is the reciprocal of the norm of b_n^* and thus satisfies

$$\|d\| \leqslant n\cdot\gamma/(2\mu(\Lambda)).$$

This bounds the length of a shortest vector of $\Lambda^*(B)$ and consequently one has $\rho(\Lambda^*) \leqslant n\cdot\gamma/(4\mu(\Lambda))$ implying

$$\rho(\Lambda^*)\mu(\Lambda) \leqslant n\cdot\gamma/4.$$

By using the LLL-bound (14.27) on γ we obtain

$$\rho(\Lambda^*)\mu(\Lambda) \leqslant (n/4)2^{n(n-1)/4}.$$

□

Theorem 14.26 (Flatness theorem for ellipsoids). *Let $E(A,a) \subseteq \mathbb{R}^n$ be an ellipsoid which does not contain an integer point, then $w(E(A,a)) \leqslant 4\cdot f(n)$.*

Proof. If $E(A,a)$ does not contain an integer point, then the covering radius of $\Lambda(A)$ is at least one. Since $\mu(\Lambda(A))\cdot\rho(\Lambda^*(A)) \leqslant f(n)$ it follows thus that $\rho(\Lambda^*(A)) \leqslant f(n)$ and consequently that $w(E(A,a)) \leqslant 4\cdot f(n)$. □

The flatness theorem 14.22 for general convex bodies $K \subseteq \mathbb{R}^n$ follows from the fact that K can be well approximated by an ellipsoid. John [39] showed that there

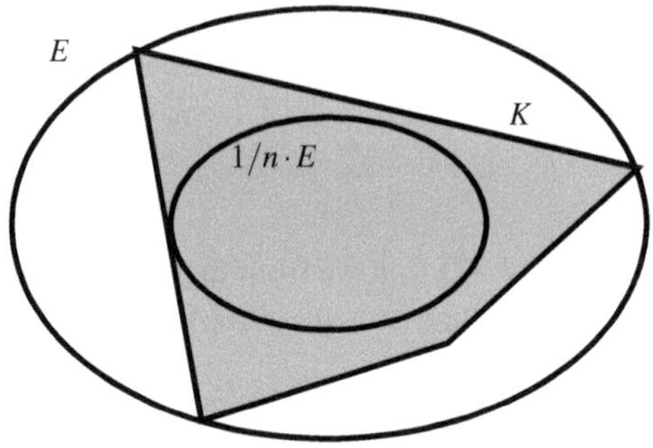

Fig. 14.8 A John-pair of ellipsoids.

exists an ellipsoid E containing K such that if E is centrally scaled by $1/n$, then it is contained in K. In other words one has

$$E/n \subseteq K \subseteq E.$$

This implies the flatness theorem with $\omega(n) \leqslant 4 \cdot n \cdot f(n)$.

Notes

The bound $f(n) \leqslant (n/4)2^{n(n-1)/4}$ of Theorem 14.25 is far from optimal. Lagarias, Lenstra and Schnorr [54] proved that $f(n) = O(n^{3/2})$. Banaszcyk [5] proved that $f(n) = O(n)$ holds.

Grötschel, Lovász and Schrijver [32] have shown that there exists a polynomial time algorithm which computes an ellipsoid E such that $1/(n(n+1)) \cdot E \subseteq K \subseteq E$ holds, if K is a convex body which is equipped with a *weak separation oracle*. Nesterov and Nemirovski [62] describe a polynomial algorithm which computes E such that $((1+\varepsilon)n)^{-1} E \subseteq P \subseteq E$ holds, where $P \subseteq \mathbb{R}^n$ is a full-dimensional polytope. Here $\varepsilon > 0$ is a parameter of the algorithm and the algorithm runs in polynomial time in the binary encoding length of P (rational data) and the binary encoding length of ε. A faster algorithm was provided by Khachiyan [48].

Finding a maximum volume ellipsoid inscribed in a polyhedron is a convex programming problem of *LP-type*. This means that such an ellipsoid can be computed in linear time in the RAM-model of computation if the dimension is fixed, see [59]. In the RAM-model, all numbers have input length one and basic arithmetic operations take constant time. The binary encoding length of the numbers in the input do not contribute to the input length. In fact those could even be real numbers. Gärtner [28] provided a subexponential algorithm for problems of LP-type in varying dimension, and thus also a subexponential algorithm to compute the maximum volume ellipsoid of a polytope. His algorithm is a generalization of the linear programming algorithm of Matoušek, Sharir and Welzl [59].

If the dimension is fixed, then using the algorithms for the largest enclosed ellipsoid by Matoušek et al. [59] Lenstra's algorithm requires $O(m \cdot s)$ arithmetic operations on rational numbers of size $O(s)$, where s is the largest binary encoding length of a coefficient of $Ax \leqslant b$. If the number of constraints m is fixed, then this matches the running time of the Euclidean algorithm, see, e.g., [51]. If one uses the floating point variant of the LLL-algorithm of Nguyen and Stehlé [64], then the bit-complexity for fixed dimension and number of constraints is $O(s^2)$ which also matches the bit-complexity of the Euclidean algorithm.

Exercises

1. Show how to compute a feasible point $x \in \mathbb{Z}^n$ of $Ax \leqslant b$ if it exists, by using a polynomial number of calls to an oracle solving an integer feasibility problem on an input which is of polynomial size in size(A), size(b) and n.
2. Consider a system of inequalities
$$Ax \leqslant b, d^T x = \beta, \tag{14.44}$$
where $A \in \mathbb{Z}^{m \times n}$, $b \in \mathbb{Z}^m$, $d \in \mathbb{Z}^n$ and $\beta \in \mathbb{Z}$.
Show how to reduce the feasibility problem of (14.44) to a feasibility problem in dimension $n-1$ involving m inequalities. This new system should be of polynomial size in the size of the system (14.44).
Hint: Hermite normal form
3. Consider the polyhedron $P = \{x \in \mathbb{R}^n : Ax \leqslant b\}$, where $A \in \mathbb{Z}^{m \times n}$ and $b \in \mathbb{Z}^m$. Suppose that P has a flat direction whose first component is nonzero. Formulate a mixed integer program to compute the width and a flat direction of P. Conclude that the encoding length of a flat direction is polynomial in the encoding length of $Ax \leqslant b$.
4. Let $d \in \mathbb{R}^n \setminus \{0\}$ and let B^n be the n-dimensional unit ball. Show that $w_d(B^n) = 2 \cdot \|d\|$.
5. Prove that the dual lattice of $\Lambda(A)$ is $\Lambda((A^{-1})^T)$, i.e., $\Lambda^*(A) = \Lambda((A^T)^{-1})$.
6. Let $A \in \mathbb{Q}^{n \times n}$ be a nonsingular rational matrix and $a \in \mathbb{Q}^n$ be a rational vector. Show that there exists a polynomial algorithm which outputs either an integer vector in $E(A,a)$, or a nonzero integer vector $d \in \mathbb{Z}^n$ with $w_d(E(A,a)) \leqslant n \cdot 2^{n(n-1)/4}$.
7. Let $P = \{x \in \mathbb{R}^n : Ax \leqslant b\}$ be a rational polyhedron in fixed dimension n. Show that the width and a flat direction of P can be computed in polynomial time. *Hint: Exercise 5 of Section 14.5.*

14.9 The integer linear optimization problem

In this section we want to consider the integer optimization problem in fixed dimension

$$\max\{c^T x\colon Ax \leqslant b, x \in \mathbb{Z}^n\}. \tag{14.45}$$

In the analysis of the algorithm that follows, we use the parameters m and s, where m is the number of inequalities of the system $Ax \leqslant b$ and s is an upper bound on the binary encoding length of a coefficient of A, b and c.

The *greatest common divisor* of two integers a and b can be computed with the Euclidean algorithm with $O(s)$ arithmetic operations, where s is an upper bound on the binary encoding length of the integers a and b. On the other hand we have the following well known formula

$$\gcd(a,b) = \min\{ax_1 + bx_2\colon ax_1 + bx_2 \geqslant 1, x_1, x_2 \in \mathbb{Z}\}. \tag{14.46}$$

This implies that the greatest common divisor can be computed with an algorithm for the integer optimization problem in dimension 2 with one constraint.

The integer optimization problem can be reduced to the integer feasibility problem with binary search. One has a complexity of $O(m+s)$ for the integer feasibility problem in fixed dimension. This follows from an analysis of Lenstra's algorithm in combination with an efficient algorithm to compute a Löwner-John ellipsoid, see [59, 82]. With binary search for an optimal point, one obtains a running time of $O(m \cdot s + s^2)$. If, in addition to the dimension, also the number of constraints is fixed, this results in an $O(s^2)$ algorithm for the integer optimization problem, which is in contrast to the linear running time of the Euclidean algorithm.

Clarkson [13] has shown that the integer optimization problem with m constraints can be solved with an expected number of $O(m)$ arithmetic operations and $O(\log m)$ calls to an oracle solving the integer optimization problem on a constant size subset of the input constraints. Therefore we concentrate now on the integer optimization problem with a fixed number of constraints. In this section we outline an algorithm which solves the integer optimization problem in fixed dimension with a fixed number of constraints with $O(s)$ arithmetic operations on rational numbers of size $O(s)$. The algorithm relies on the LLL-algorithm.

The first step is to reduce the integer optimization problem over a full-dimensional polytope with a fixed number of facets to a disjunction of integer optimization problems over a constant number of *two-layer simplices*. A two-layer simplex is a full-dimensional simplex, whose vertices can be partitioned into two sets V and W, such that the objective function of the elements in each of the sets V and W agree, i.e., for all $v_1, v_2 \in V$ one has $c^T v_1 = c^T v_2$ and for all $w_1, w_2 \in W$ one has $c^T w_1 = c^T w_2$.

How can one reduce the integer optimization problem over a polytope P to a sequence of integer optimization problems over two-layer simplices? Simply consider the hyperplanes $c^T x = c^T v$ for each vertex v of P. If the number of constraints defining P is fixed, then these hyperplanes partition P into a constant number of polytopes, whose vertices can be grouped into two groups, according to their objective function value. Thus we can assume that the vertices of P can be partitioned into two sets V and W, such that the objective function values of the elements in each of the sets V and W agree. Carathéodory's theorem, see Schrijver [75, p. 94], implies that P is covered by the simplices that are spanned by the vertices of P. These simplices are two-layer simplices. Therefore, the integer optimization problem in fixed dimension with a fixed number of constraints can be reduced to a constant number of integer optimization problems over a two-layer simplex by applying a constant number of arithmetic operations.

The key idea is then to let the objective function slide into the two-layer simplex, until the width of the truncated simplex exceeds the flatness bound. In this way, one can be sure that the optimum of the integer optimization problem lies in the truncation, which is still flat. Thereby one has reduced the *integer optimization problem* in dimension n to a constant number of *integer optimization problems* in dimension $n-1$ and binary search can be avoided.

How do we determine a parameter π such that the truncated two-layer simplex $\Sigma \cap (c^T x \geqslant \pi)$ just exceeds the flatness bound? We explain the idea with the help of the 3-dimensional example in Figure 14.9.

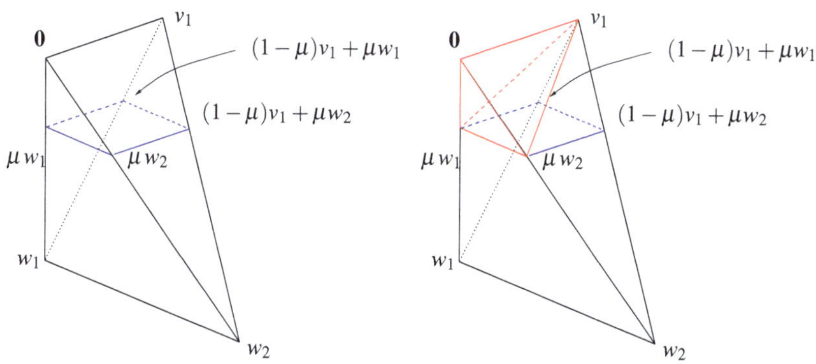

Fig. 14.9 Reduction to the parametric lattice width problem.

Here we have a two-layer simplex Σ in 3-space. The set V consists of the points $\mathbf{0}$ and v_1 and W consists of w_1 and w_2. The objective is to find a highest point in the vertical direction. The picture on the left describes a particular point in time, where the objective function slid into Σ. So we consider the truncation $\Sigma \cap (c^T x \geqslant \pi)$ for some $\pi \geqslant c^T w_1$. This truncation is the convex hull of the points

$$\mathbf{0}, v_1, \mu w_1, \mu w_2, (1-\mu)v_1 + \mu w_1, (1-\mu)v_1 + \mu w_2, \tag{14.47}$$

where $\mu = \pi / c^T w_1$. Now consider the simplex $\Sigma_{V,\mu W}$, which is spanned by the points $\mathbf{0}, v_1, \mu w_1, \mu w_2$. This simplex is depicted on the right in Figure 14.9. If this simplex is scaled by 2, then it contains the truncation $\Sigma \cap (c^T x \geqslant \pi)$. This is easy to see, since the scaled simplex contains the points $2(1-\mu)v_1$, $2\mu w_1$ and $2\mu w_2$. So we have the condition $\Sigma_{V,\mu W} \subseteq \Sigma \cap (c^T x \geqslant \pi) \subseteq 2\Sigma_{V,\mu W}$. From this we can infer the important observation

$$w(\Sigma_{V,\mu W}) \leqslant w(\Sigma \cap (c^T x \geqslant \pi)) \leqslant 2\, w(\Sigma_{V,\mu W}). \tag{14.48}$$

This means that we essentially determine the correct π by determining a $\mu \geqslant 0$, such that the width of the simplex $\Sigma_{V,\mu W}$ just exceeds the flatness bound. The width of $\Sigma_{V,\mu W}$ is roughly (up to a constant factor) the length of the shortest vector of the lattice $L(A_\mu)$, where A_μ is the matrix

$$A_\mu = \begin{pmatrix} \mu w_1^T \\ \mu w_2^T \\ v_1 \end{pmatrix}.$$

Thus we have to find a parameter μ, such that the shortest vector of $L(A_\mu)$ is sandwiched between $\omega(n)+1$ and $\gamma \cdot (\omega(n)+1)$ for some constant γ. This problem can be understood as a *parametric shortest vector* problem.

To describe this problem let us introduce some notation. We define for an $n \times n$-matrix $A = (a_{ij})_{\forall i,j}$, the matrix $A^{\mu,k} = (a_{ij})^{\mu,k}_{\forall i,j}$, as

$$a_{ij}^{\mu,k} = \begin{cases} \mu \cdot a_{ij}, & \text{if } i \leqslant k, \\ a_{ij}, & \text{otherwise.} \end{cases} \tag{14.49}$$

In other words, the matrix $A^{\mu,k}$ results from A by scaling the first k rows with μ. The parametric shortest vector problem is now defined as follows.

> Given a nonsingular matrix $A \in \mathbb{Z}^{n\times n}$ and some $U \in \mathbb{N}$, find a parameter $p \in \mathbb{N}$ such that $U \leqslant \mathrm{SV}(L(A^{p,k})) \leqslant 2^{n+1/2} \cdot U$ or assert that $\mathrm{SV}(L) > U$.

It turns out that the parametric shortest vector problem can be solved in linear time when the dimension is fixed with a cascaded LLL-algorithm. From this, it follows that the integer optimization problem in fixed dimension with a fixed number of constraints can be solved in linear time. Together with Clarkson's result we obtain.

Theorem 14.27 ([18]). *The integer optimization problem* (14.45) *can be solved with an expected number of* $O(m + s\log m)$ *arithmetic operations on rationals of size* $O(s)$.

Notes

A polynomial time algorithm for the *two-variable integer programming problem* was presented by Hirschberg and Wong [38] and Kannan [46] for special cases and by Scarf [70, 71] for the general case. Then, Lenstra [56] proved that integer programming in arbitrary fixed dimension can be solved in polynomial time. Afterwards, various authors were looking for faster algorithms for the two-dimensional case [24, 84, 41]. A linear-time algorithm in the arithmetic model was presented by Eisenbrand and Laue [19].

Exercises[1]

1*) Can one solve an integer program in fixed dimension with a fixed number of constraints in quadratic time in the bit-model of complexity ?
Recent results on the bit-complexity of the LLL-algorithm [63] could help answering this question

2*) Is the shortest vector problem solvable in time $O(M(s)\log s)$ if the dimension is fixed?
See also the Notes section of Chapter 14.5.

14.10 Diophantine approximation and strongly polynomial algorithms

In this section we review a very important appliction of the LLL-algorithm in optimization, namely the rounding algorithm of Frank and Tardos [25]. The input of a combinatorial optimization problem usually consists of a combinatorial structure, like a graph or a subset system and some numbers, which reflect weights on the edges or costs of subsets or alike. An algorithm is *strongly polynomial* if

A) it consist only of basic arithmetic operations $+,-,*,/$ and comparisons $<,>,=$;

B) the number of such basic operations which are carried out is polynomial in the input length, where the numbers in the input have length one;

C) the encoding lengths of the numbers in the intermediate steps of the algorithm are polynomial in the binary encoding length of the input, where numbers in the input account with their binary encoding length.

An algorithm is weakly polynomial if it satisfies A) and C) but instead of B) satisfies the weaker condition that the number of basic operations is bounded in the binary encoding length of the input.

Algorithms which are based on *bit-scaling* of the involved numbers usually have the property to be weakly polynomial only. It is an outstanding open problem, whether linear programming, which can be solved in weakly polynomial time, can also be solved in strongly polynomial time.

A *0/1-optimization problem* is an integer program, where the integer variables are restricted to be either 0 or 1. The result that we survey now shows a facet of the tremendous importance of the LLL-algorithm in the area of algorithms and complexity. In fact it holds in greater generality, see [25, 31]. We treat a less general version here, to keep the exposition as simple as possible.

Theorem 14.28 ([25]). *If a 0/1 optimization problem can be solved in weakly polynomial time, then it can be solved in strongly polynomial time.*

[1] Those are not really "exercises" but rather research questions.

The starting point of the method leading to this result is Dirichlet's theorem. Suppose that you have an n-dimensional vector $\alpha \in \mathbb{R}^n$ which is represented by n real numbers $\alpha_1, \dots, \alpha_n \in \mathbb{R}$ and you are interested in a vector $p \in \mathbb{Z}^n$ with integer components of small absolute value and which encloses a small angle with α. Such a p could be a candidate to replace the linear objective function $\alpha^T x$ with the linear objective function $p^T x$, hoping that the optimal 0/1-solutions are the same.

Theorem 14.29 (Dirichlet). *Let $Q \geqslant 1$ be a real number and let $\alpha_1, \dots, \alpha_n \in \mathbb{R}$. There exists an integer q and integers $p_1, \dots, p_n$ with*

i) $1 \leqslant q \leqslant Q^n$ *and*
ii) $|q \cdot \alpha_i - p_i| \leqslant 1/Q$ *for* $i = 1, \dots, n$.

Proof. Consider the lattice Λ which is generated by the following matrix

$$\begin{pmatrix} 1 & & & & \alpha_1 \\ & 1 & & & \alpha_2 \\ & & \ddots & & \vdots \\ & & & 1 & \alpha_n \\ & & & & \frac{1}{Q^{n+1}} \end{pmatrix}. \tag{14.50}$$

The determinant of Λ is $1/Q^{n+1}$. The cube $K = \{x \in \mathbb{R}^{n+1} : -1/Q \leqslant x \leqslant 1/Q$ is symmetric around 0 and has volume $2^{n+1} \cdot 1/Q^{n+1}$. Minkowski's theorem implies that this set contains a nonzero lattice point [2]

$$\begin{pmatrix} -p_1 + q \cdot \alpha_1 \\ \vdots \\ -p_n + q \cdot \alpha_n \\ q/Q^{n+1} \end{pmatrix}.$$

We can assume q to be positive, since the negative of this vector is also contained in K. The integers p_i, $i = 1, \dots, n$ and q are as it is stated in the theorem. □

If Q in the Dirichlet theorem is a positive integer, then one can replace condition ii) by the slightly stronger condition

ii') $|q \cdot \alpha_i - p_i| < 1/Q$ for $i = 1, \dots, n$,

see exercise 1).

Suppose now that we want to solve a 0/1 optimization problem

$$\max_{x \in \mathscr{F}} \alpha^T x \tag{14.51}$$

[2] Because, if this cube would not contain a nonzero lattice point, then one could scale it with a factor larger than one and still it would not contain a nonzero lattice point. The volume of this scaled cube is strictly larger than $2^{n+1} \cdot 1/Q^{n+1}$. This would contradict Minkowski's theorem.

where $\mathscr{F} \subseteq \{0,1\}^n$. The set $\mathscr{F}$ could, for example, be the characteristic vectors of matchings of a graph $G = (V,E)$ or characteristic vectors of other combinatorial structures.

Our goal is to replace α by an integer vector v whose binary encoding length is polynomial in the dimension n such that the optimal points in $\mathscr{F}$ w.r.t. the objective functions $\alpha^T x$ and $v^T x$ are the same.

To this end, consider an optimal solution $\bar{x}$ of problem (14.51), i.e., an $\bar{x}$ with

$$\alpha^T(\bar{x} - x) \geqslant 0 \text{ for all } x \in \mathscr{F}.$$

The points $\bar{x} - x$ above have components in $\{0, \pm 1\}$. Our integer vector $v \in \mathbb{Z}^n$ will have the following property.

For each $y \in \{0, \pm 1\}^n$ one has

$$\alpha^T y \geqslant 0 \text{ if and only if } v^T y \geqslant 0.$$

If this holds, then we can safely replace α by v in (14.51).

Let us assume for a moment that Dirichlet's theorem has an efficient implementation, i.e., that the q and p_i satisfying conditions i) and ii) can be computed in polynomial time. Assume further that the largest absolute value of a component in α is one, or in other words that $\|\alpha\|_\infty = 1$. If this would not hold, then we could simply scale α by $1/\|\alpha\|_\infty$. Set Q in Dirichlet's theorem to $Q := n$ and let q and p be the resulting integer and integer vector respectively.

How large are the absolute values of the components of p? Since $1 \leqslant q \leqslant n^n$ (i) and $|q\alpha_i - p_i| < 1/n$ (ii') it follows that $\|p\|_\infty \leqslant q\|\alpha\|_\infty \leqslant n^n$. The binary encoding length of p is therefore polynomial in n, which is one of the objectives that we want an approximation of α to satisfy.

How close comes p in terms of being a valid replacement v of α? The next lemma shows, that this vector comes very close already to what we want. Let $\operatorname{sign}(x)$ denote the sign of a real number x, i.e.,

$$\operatorname{sign}(x) = \begin{cases} 1, & \text{if } x > 0, \\ -1, & \text{if } x < 0, \\ 0, & \text{if } x = 0. \end{cases}$$

Lemma 14.7. *Let $y \in \{0, \pm 1\}^n$ be a vector. If $\operatorname{sign}(p^T y) \neq 0$, then $\operatorname{sign}(p^T y) = \operatorname{sign}(\alpha^T y)$.*

Proof. Suppose that $p^T y > 0$. Since p and y are integer vectors, it follows that $p^T y \geqslant 1$ holds. Property ii) of the Dirichlet theorem implies $|q\alpha_i - p_i| < 1/n$ for each i. Thus $|q\alpha^T y - p^T y| = |(q\alpha - p)^T y| < n/n = 1$. In other words, the distance of $p^T y$ to $q\alpha^T y$ is less than one, implying that $q\alpha^T y > 0$ and consequently $\alpha^T y > 0$.

The case $p^T y < 0$ is analogous. □

If $\operatorname{sign}(p^T y) = 0$, then we cannot infer anything about $\operatorname{sign}(\alpha^T y)$. The idea is now to apply recursion. If $\operatorname{sign}(p^T y) = 0$, then clearly

$$\operatorname{sign}(\alpha^T y) = \operatorname{sign}\left((q\cdot\alpha - p)^T y\right)$$

Since one component of α is 1, say component i, we clearly have $p_i = q$ and thus the vector $q\cdot\alpha - p$ has at least one more zero component than α. Let v' be an integer vector which has been recursively computed and satisfies for each $y \in \{0,\pm 1\}^n$ $\operatorname{sign}(v'^T y) = \operatorname{sign}\left((q\cdot\alpha - p)^T y\right)$.

Let M be a large weight. The above discussion shows that

$$\forall y \in \{0,\pm 1\}^n\colon\ \operatorname{sign}(\alpha^T y) = \operatorname{sign}\left((M\cdot p + v')^T y\right).$$

Thus we set $v := M\cdot p + v'$.

Let us deduce a bound on M. This number M has to be large enough so that $M\cdot p^T y$ dominates the sign of $(M\cdot p + v')^T y$ if $p^T y \neq 0$. This clearly holds if M is at least as large as the absolute value of $v'^T y$. To this end, let $t(n)$ be an upper bound on the largest absolute value of a component of v which is constructed by the above described recursive procedure. The absolute value of $v'^T y$ is then bounded by $n\cdot t(n-1)$. Thus we set M to this value. How large is $\|v\|_\infty$? For this we have the recursive formula

$$t(n) \leqslant n\cdot t(n-1)\cdot n^n + t(n-1),$$

which shows a crude bound of $t(n) \leqslant (n+1)^{n^2+n}$. The binary encoding length of v is thus $O(n^2 \log n)$.

What about our assumption from the beginning that Dirichlet's theorem has an efficient implementation? In fact, it is an outstanding open problem, whether the q and the p_i as in Dirichlet's theorem can be found in polynomial time. However, we have the LLL-algorithm. The proofs of the next two lemmas are left as an exercise.

Lemma 14.8. *Let $Q \geqslant 1$ be an integer and $\alpha_1,\dots,\alpha_n \in \mathbb{Q}$ be rational numbers. There exists a polynomial time algorithm that computes an integer q and integers $p_1,\dots,p_n$ with*

I) $1 \leqslant q \leqslant \sqrt{n+1}2^{(n-1)/2}\cdot Q^n$ and
II) $|q\cdot\alpha_i - p_i| < \sqrt{n+1}2^{(n-1)/2}\cdot 1/Q$ for $i = 1,\dots,n$.

Lemma 14.9. *Let $Q \geqslant 1$ be an integer and $\alpha_1,\dots,\alpha_n \in \mathbb{Q}$ be rationals numbers. There exists a polynomial algorithm that computes an integer q and integers $p_1,\dots,p_n$ with*

a) $1 \leqslant q \leqslant 2^{n^2} Q^n$ and
b) $|q\cdot\alpha_i - p_i| < 1/Q$ for $i = 1,\dots,n$.

If we replace the guarantees of Dirichlet's theorem with the ones obtained in Lemma 14.9, then the binary encoding length of v is still polynomial in n. This means that we have an algorithm to round an objective function vector $\alpha \in \mathbb{Q}^n$ to an equivalent integer vector $v \in \mathbb{Z}^n$. The only trouble is that the LLL-algorithm is not *strongly polynomial* and our goal is to develop a strongly polynomial time algorithm. This is finally achieved by rounding the α_i first individually before the algorithm behind Lemma 14.9 is applied. The rounding operation $\lfloor\cdot\rfloor$ is not part of the standard set of operations on which a strongly polynomial time algorithm can rely. Let $\beta \in \mathbb{R}$ be a real number with $0 < \beta \leqslant 1$ and let $K \geqslant 2$ be a positive integer. In exercise 4, you are asked to compute $\lfloor K \cdot \beta \rfloor$ with a linear (in the binary encoding length of K) number of elementary operations $(+,-,*,/,<,>,=)$.

Theorem 14.30. *There exists a strongly polynomial algorithm that, given* n real numbers $\alpha_1,\ldots,\alpha_n$ *with* $\|\alpha\|_\infty = 1$, *computes integers* $p_1,\ldots,p_n$ *and* q *with*

$$|q\cdot\alpha_i - p_i| < 1/n,\ i=1,\ldots,n,\ \text{and}\ 1 \leqslant q \leqslant 2^{n^2+n}n^n.$$

Proof. For each i compute:

$$\alpha_i' = \frac{\lfloor \alpha_i 2^{n^2+n+1} n^{n+1} \rfloor}{2^{n^2+n+1} n^{n+1}}$$

in strongly polynomial time (see exercise 4) and run the algorithm behind Theorem 14.9 on input α' and $Q' = 2\cdot n$. This algorithm returns q and p.

We now have to show the following claim.

$$|q\cdot\alpha_i - p_i| \leqslant 1/n,\ i=1,\ldots,n,\ \text{and}\ q \leqslant 2^{n^2+n}n^n. \tag{14.52}$$

Clearly one has $1 \leqslant q \leqslant 2^{n^2}(2n)^n = 2^{n^2+n}n^n$. On the other hand we have

$$\begin{aligned} |q\alpha_i - p_i| &\leqslant |q\alpha_i - q\alpha_i'| + |q\alpha_i' - p_i| \\ &< \frac{2^{n^2+n}n^n}{2^{n^2+n+1}n^{n+1}} + 1/(2n) \\ &= 1/n. \end{aligned}$$

which shows the claim. □

If we adapt the argument from before with the new parameters of Theorem 14.30 we obtain the result of Frank and Tardos [25].

Theorem 14.31. *There exists a strongly polynomial algorithm that, given* $\alpha \in \mathbb{R}^n$, *computes a* $v \in \mathbb{Z}^n$ *whose binary encoding length is polynomial in* n *such that* $\operatorname{sign}(v^T y) = \operatorname{sign}(\alpha^T y)$ *for all* $y \in \{0,\pm 1\}^n$.

Consequently, a $0/1$-optimization problem can be solved in polynomial time if and only if it can be solved in strongly polynomial time.

Notes

The result of Tardos and Frank [25] is more general than for the case restricted to $0/1$-problems described above. If one wants to optimize a linear function over the convex hull of a set of rational points, then the linear function can be replaced by an integer linear objective function, whose binary encoding length is polynomial in the largest binary encoding length of a point in the set, see also [58, 31]. Tardos [79] has shown that a linear program $\max\{c^T x\colon Ax \leqslant b\}$ can be solved within a number of elementary operations bounded by a polynomial in the binary encoding length of A. This can also be shown with a rounding argument, see [31]. Diophantine approximation is also discussed in a recent survey of Lenstra [57].

Exercises

1. Show that condition ii) can be strengthened to $|q \cdot \alpha_i - p_i| < 1/Q$ for $i = 1, \ldots, n$.
2. Prove Lemma 14.8.
   Hint: The shortest vector of the lattice generated by the matrix (14.50) *has* ℓ_2*-norm at most* $\sqrt{n+1}/Q$ *and the LLL-algorithm finds a vector* x *which is a* $2^{(n-1)/2}$*-approximation of this shortest vector. Bound the* ℓ_∞*-norm of* x.
3. Prove Lemma 14.9.
   Hint: Set $Q' := \lceil \sqrt{n+1} \cdot 2^{(n-1)/2} \cdot Q \rceil$ *and apply Lemma 14.8 using* α *and* Q'. *The bound a) holds for all but a finite number of* n. *For this fixed number of* n *one can afford to find a shortest vector w.r.t. the* ℓ_∞*-norm (see exercise 14.5.5).*
4. Let $\beta \in \mathbb{R}$ be a real number with $0 < \beta \leqslant 1$ and let $K \geqslant 2$ be a positive integer. Show how to compute $\lfloor K \cdot \beta \rfloor$ with a linear (in the binary encoding length of K) number of elementary RAM operations $(+, -, *, /, <, >, =)$.
5. Show that for a 0/1-optimization problem (14.51) there exists a $v \in \mathbb{Z}^n$ with $\|v\|_\infty = 2^{O(n \log n)}$ such that an $x \in \mathscr{F}$ is optimal w.r.t. $\alpha^T x$ if and only if x is optimal w.r.t. $v^T x$.
   Hint: Use linear programming duality and the Hadamard bound.

14.11 Parametric integer programming

The *Frobenius problem* is as follows. Given n integers $a_1, \ldots, a_n$ whose greatest common divisor is one, compute the largest $t \in \mathbb{N}$ which cannot be written as

$$x_1 \cdot a_1 + \cdots + x_n \cdot a_n = t, \quad x_1, \ldots, x_n \in \mathbb{N}_0.$$

The Frobenius problem is NP-complete [67]. Kannan [45] showed that it can be solved in polynomial time, if n is fixed. In fact, Kannan showed something more general. The Frobenius problem can be solved with binary search, if the validity of the following *forall/exist*-statement can be decided in polynomial time.

$$\forall y \in \mathbb{Z},\, y \geqslant N \quad \exists x_1, \ldots, x_n \in \mathbb{N}_0 \quad : \quad y = x_1 \cdot a_1 + \cdots + x_n a_n \tag{14.53}$$

Kannan's result is a polynomial time procedure to decide forall/exist statements of the following form, where $Q \subseteq \mathbb{R}^m$ is a polyhedron, $A \in \mathbb{Z}^{m\times n}$ is a matrix and $t \in \mathbb{N}$ is an integer.

$$\forall b \in \left(Q \cap (\mathbb{R}^{m-t} \times \mathbb{Z}^t)\right) \quad Ax \leqslant b \text{ is IP-feasible.} \tag{14.54}$$

More precisely, he showed that such statements can be decided in polynomial time, if n, t and the affine dimension of Q are fixed.

Theorem 14.32 ([45]). *If n,t and the affine dimension of Q are fixed, then $\forall\exists$-statements can be decided in polynomial time.*

In this section we review the basic ideas of his procedure and the generalization of Eisenbrand and Shmonin [21], which only assumes n and t to be fixed whereas the dimension of Q can vary.

Theorem 14.33 ([21]). *If n,t are fixed, then $\forall\exists$-statements can be decided in polynomial time.*

Suppose that the width of a polyhedron $P \subseteq \mathbb{R}^n$ is the width along the integer direction c, i.e., $w(P) = w_c(P)$ with $c \in \mathbb{Z}^n$. Let $\beta = \min\{c^T x\colon x \in P\}$. If $w(P) > \omega(n)$, then we can scale and translate P, to obtain a polyhedron P' which is sandwiched between the hyperplanes $c^T x = \beta$ and $c^T x = \beta + \omega(n)$, see figure 14.10.

Theorem 14.34. *If $w(P) > \omega(n)$, then there exists an integer point in*

$$P \cap \left(\beta \leqslant c^T x \leqslant \beta + \omega(n)\right).$$

The width of P' is exactly $\omega(n)$ and by the flatness theorem (Theorem 14.22), P' contains an integer point. From this it follows that there exists an integer point on one of the constant number of lower dimensional polyhedra

$$P \cap c^T x = \lceil \beta \rceil + i, \text{ for } i = 0, \ldots, \omega(n). \tag{14.55}$$

Theorem 14.35 (Strengthening of flatness theorem). *Let $P \subseteq \mathbb{R}^n$ be a polyhedron with $w(P) = w_c(P)$ for an integer vector $c \in \mathbb{Z}^n \setminus \{0\}$. The polyhedron P contains an integer point if and only if at least one of the polyhedra*

$$P \cap \left(c^T x = \lceil \beta \rceil + i\right) \quad i = 0, \ldots, \omega(n)$$

contains an integer point.

We continue with a description of Kannan's ideas in dimension 2. Suppose that $A \in \mathbb{Z}^{m\times 2}$ and $Q \subseteq \mathbb{R}^m$ and that we wish to decide the $\forall\exists$-statement (14.54) with $t = 0$. We then would have to decide, whether there exists a $b \in Q$ such that $Ax \leqslant b$ is *integer infeasible*. We denote the polyhedron $\{x \in \mathbb{R}^2\colon Ax \leqslant b\}$ by P_b. Let us make the following simplifying assumptions.

I) The flat direction of P_b is always the first unit vector e_1.

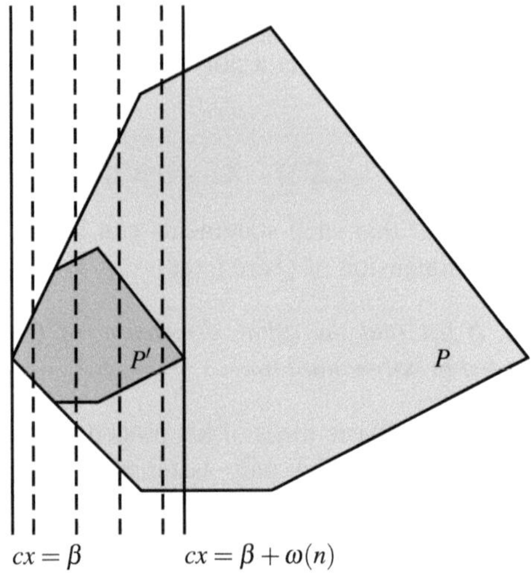

Fig. 14.10 An illustration of the strengthening of the flatness theorem. There must be an integer point on one of the dashed lines $c^T x = \lceil \beta \rceil + i$, $i = 1, \dots, \omega(n)$.

II) The optimum basis of the linear program $\min\{e_1^T x \colon x \in \mathbb{R}^2, Ax \leqslant b\}$ is the same for all $b \in Q$.

It follows from II) that the optimum solution of the linear program $\min\{e_1^T x \colon x \in P_b\}$ is of the form $G \cdot b$ for a fixed matrix $G \in \mathbb{R}^{2\times 2}$.

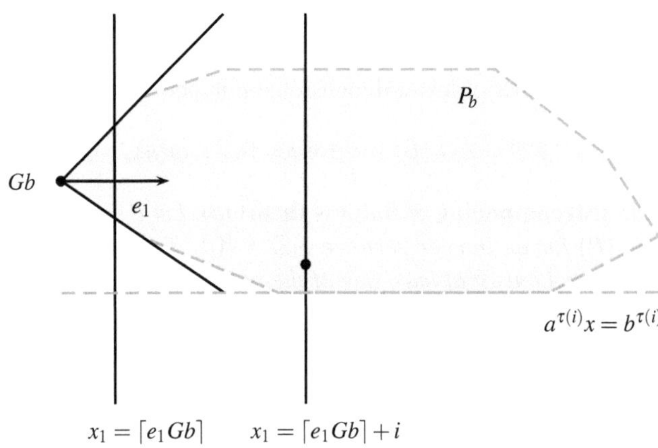

Fig. 14.11 Illustration in dimension 2.

The strengthening of the flatness theorem tells that P_b contains an integral point if and only if there exists an integral point on the lines $x_1 = \lceil e_1 Gb \rceil + j$ for $j = 0, \dots, \omega(2)$. The intersections of these lines with the polyhedron P_b are 1-

dimensional polyhedra. Some of the constraints $ax \leqslant b$ of $Ax \leqslant b$ are "pointing upwards", i.e., $ae_2 < 0$, where e_2 is the second unit-vector. Let $a_1x_1 + a_2x_2 \leqslant b$ be a constraint pointing upwards such that the intersection point $(\lceil e_1Gb \rceil + j, y)$ of $a_1x_1 + a_2x_2 = \beta$ with the line $x_1 = \lceil e_1Gb \rceil + j$ has the largest second component. The line $x_1 = \lceil e_1Gb \rceil + j$ contains an integral point in P_b if and only if $(\lceil e_1Gb \rceil + j, \lceil y \rceil)$ is contained in P_b. This point is illustrated in Figure 14.11. We assume that this "highest upwards constraint" for the line $x_1 = z + i$ is always fixed, lets say it is the constraint $a^{\tau(i)}x \leqslant b^{\tau(i)}$ for a fixed mapping $\tau : \{0, \ldots, \omega(2)\} \rightarrow \{1, \ldots, m\}$. We make this assumption explicit.

III) The highest intersection point of a line $a^jx = b^j$ with the line $x_1 = z + i$ is the intersection point of the line $a^{\tau(i)}x = b^{\tau(i)}$, where $\tau : \{0, \ldots, \omega(2)\} \rightarrow \{1, \ldots, m\}$ is a fixed mapping.

We now formulate a mixed-integer program from which we can extract the integer solution of P_b on the line $x_1 = z + i$, assuming that I-III) holds.

$$\begin{aligned} (Gb)_1 \leqslant z &< (Gb)_1 + 1 \\ x_1^i &= z + i \\ y = (b^{\tau(i)} &- a_1^{\tau(i)} x_1)/a_2^{\tau(i)} \\ y \leqslant x_2^i &< y + 1 \\ x_1^i, x_2^i&, z \in \mathbb{Z}. \end{aligned} \tag{14.56}$$

For a given $b \in Q$, the line $x_1 = z + i$ contains an integer point in P_b if and only if $x^i = (x_1^i, x_2^i) \in P_b$.

Recall that we want to decide the forall/exist-statement (14.54)) by searching a $b \in Q$ such that P_b does not contain an integer point. In other words, we are looking for a $b \in Q$ for which each of the x^i for $i = 0, \ldots, \omega(2)$ violates some constraint of $Ax \leqslant b$. To do that, we again fix a mapping $\mu : \{0, \ldots, \omega(2)\} \rightarrow \{1, \ldots, m\}$, where $a^{\mu(i)}x \leqslant b^{\mu(i)}$ is supposed to be the constraint of $Ax \leqslant b$ which is violated by x^i. The number of such mappings μ is $m^{\omega(2)+1}$, which is polynomial in m, since $\omega(2)$ is a constant.

We enumerate all these mappings μ and append to each of the $\omega(2) + 1$ MIP's (14.56) the constraint

$$a^{\mu(i)}x^i > b^{\mu(i)}. \tag{14.57}$$

The conjunction of all these MIP's is a MIP with a constant number of integer variables, which can be solved with Lenstra's algorithm [56] in polynomial time. This shows that we can decide forall/exist statements, given that the conditions I-III) hold.

We justify condition I, at the heart of which lies the following theorem.

Theorem 14.36. *Let n be fixed and $A \in \mathbb{Z}^{m \times n}$ be a matrix of full column rank. There exist a set $D \subseteq \mathbb{Z}^n \setminus \{0\}$ of polynomial (in the encoding length of A) cardinality such that for each $b \in \mathbb{R}^m$ there exists a $d \in D$ with $w(P_b) = w_d(P_b)$. Such a set D can be computed in polynomial time.*

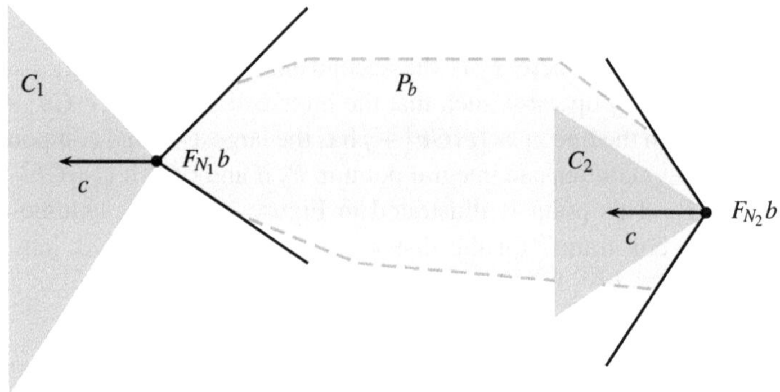

Fig. 14.12 The flat direction c and the two cones C_1 and C_2.

Proof. Suppose that P_b is non-empty and let $c \in \mathbb{Z}^n$ be its width direction. Then there are two bases N_1 and N_2 such that

$$\max\{cx : Ax \leqslant b\} = cF_{N_1}b \quad \text{and} \quad \min\{cx : Ax \leqslant b\} = cF_{N_2}b \tag{14.58}$$

and c belongs to the cones C_1 and C_2 defined by the optimal bases with respect to $\max\{c^Tx\colon x \in P_b\}$ and $\min\{c^Tx\colon x \in P_b\}$, see Figure 14.12. Here F_{N_1} and F_{N_2} are the inverses of the basis matrices of N_1 and N_2 respectively.

In fact, equations (14.58) hold for any vector c in $C_1 \cap C_2$. Thus, the lattice width of P_b is equal to the optimum value of the following optimization problem:

$$\min\{c(F_{N_1} - F_{N_2})b : c \in C_1 \cap C_2 \cap \mathbb{Z}^n \setminus \{0\}\}. \tag{14.59}$$

The latter can be viewed as an integer programming problem. Indeed, the cones C_1 and C_2 can be represented by some systems of linear inequalities, say $cD_1 \leqslant 0$ and $cD_2 \leqslant 0$, respectively, where $D_1, D_2 \in \mathbb{Z}^{n\times n}$. The minimum (14.59) is taken over all integral vectors c satisfying $cD_1 \leqslant 0$ and $cD_2 \leqslant 0$, except the origin. Since both cones C_1 and C_2 are simplicial, i.e., generated by n linearly independent vectors, the origin is a vertex of $C_1 \cap C_2$ and therefore can be cut off by a single inequality, for example, $cD_1\mathbf{1} \leqslant -1$, where $\mathbf{1}$ denotes the n-dimensional all-one vector. It is important that all other integral vectors c in $C_1 \cap C_2$ satisfy this inequality and therefore remain feasible. Thus, the problem (14.59) can be rewritten as

$$\min\{c(F_{N_1} - F_{N_2})b : cD_1 \leqslant 0, cD_2 \leqslant 0, cD_1\mathbf{1} \leqslant -1\, c \in \mathbb{Z}^n\}.$$

For a given b, this is an integer programming problem. Therefore, the optimum value of (14.59) is attained at some vertex of the integer hull of the underlying polyhedron

$$\{c : cD_1 \leqslant 0, cD_2 \leqslant 0, cD_1\mathbf{1} \leqslant -1\} \tag{14.60}$$

Shevchenko [76] and Hayes and Larman [35] proved that the number of vertices of the integer hull of a rational polyhedron is polynomial in fixed dimension. Tight bounds for this number were presented by Cook et al. [14] and Bárány et al. [6]. □

To justify conditions I,II) and III) one then partitions the parameter polyhedron Q into a polynomial number of polyhedra such that for those b in one partition, the width direction is always invariant. Via a unimodular transformation one can assume that this flat direction is the first unit vector, see exercises 1 and 2) for further details.

Notes

The problem to decide forall/exist statements belongs to the second level of the polynomial hierarchy and is Π_2^p-complete, see [77, 83]. Barvinok and Woods [7] extended the counting algorithm of Barvinok [8] such that it computes a short rational generating function for an integer projection of the set of integer points in a polytope. The algorithm is based on Kannan's partitioning lemma. The variant of the partitioning theorem which we described was used and further refined for in implementation of the algorithm [7] by Köppe et al. [52]. Köppe and Verdoolaege [53] presented an algorithm which computes a formula for the number of integer points in a parameterized polyhedron.

Exercises

1. Consider the polyhedra $P_b = \{x \in \mathbb{R}^n : Ax \leqslant b\}$ for $b \in Q$ and let $d \in \mathbb{Z}^n - \{0\}$, where $A \in \mathbb{Z}^{m \times n}$ has full column rank and n is fixed. Write down a polynomial number of inequalities which describe those $b \in Q$ such the $w(P_b) = w_d(P_b)$. *Hint: Associate to each $d \in D$ from Theorem 14.36 an "entering basis" and "leaving basis".*
2. Write down a linear program which partitions Q further into parameters such that condition III) holds.

3*. I s there a polynomial algorithm which, given a matrix $A \in \mathbb{Z}^{m \times n}$, where n is fixed and a polyhedron $Q \subseteq \mathbb{R}^m$ determines a $b \in Q$ such that the number of integer points in P_b is minimal?

Acknowledgments

I am grateful to Damien Stehlé for numerous comments and suggestions which helped me a lot to improve this manuscript. I also want to thank Johannes Blömer for several discussions on shortest and closest vectors.

References

1. A.V. Aho, J.E. Hopcroft, and J.D. Ullman, *The design and analysis of computer algorithms*, Addison-Wesley, Reading, 1974.
2. M. Ajtai, *The shortest vector problem in L2 is NP-hard for randomized reductions*, Proceedings of the 30th Annual ACM Symposium on Theory of Computing (STOC-98) (New York), ACM Press, 1998, pp. 10–19.
3. M. Ajtai, R. Kumar, and D. Sivakumar, *A sieve algorithm for the shortest lattice vector problem*, Proceedings of the thirty-third annual ACM symposium on Theory of computing, ACM Press, 2001, pp. 601–610.
4. M. Ajtai, R. Kumar, and D. Sivakumar, *Sampling short lattice vectors and the closest lattice vector problem*, Proceedings of the 17th IEEE Annual Conference on Computational Complexity (Montreal, Canada), 2002, pp. 53–67.
5. W. Banaszczyk, *Inequalities for convex bodies and polar reciprocal lattices in* R^n. *II. Application of K-convexity*, Discrete Comput. Geom. 16 (1996) 305–311.
6. I. Bárány, R. Howe, and L. Lovász, *On integer points in polyhedra: A lower bound*, Combinatorica 12 (1992) 135–142.
7. A. Barvinok and K. Woods, *Short rational generating functions for lattice point problems*, Journal of the American Mathematical Society 16 (2003) 957–979.
8. A. Barvinok, *A polynomial time algorithm for counting integral points in polyhedra when the dimension is fixed*, Mathematics of Operations Research 19 (1994) 769–779.
9. J. Blömer, *Closest vectors, successive minima, and dual HKZ-bases of lattices*, Automata, Languages and Programming, 27th International Colloquium, ICALP 2000, Geneva, Switzerland, July 9-15, 2000, Proceedings (U. Montanari, J. D. P. Rolim, and E. Welzl, eds.), Lecture Notes in Computer Science 1853, Springer, 2000, pp. 248–259.
10. J. Blömer and S. Naewe, *Sampling methods for shortest vectors, closest vectors and successive minima*, Proceedings of the 34th International Colloquium on Automata, Languages and Programming, ICALP 2007 (L. Arge, C. Cachin, T. Jurdzinski, and A. Tarlecki, eds.), Lecture Notes in Computer Science 4596, Springer, 2007, pp. 65–77.
11. I. Borosh and L.B. Treybig, *Bounds on positive integral solutions of linear diophantine equations*, Proceedings of the American Mathematical Society 55 (1976) 299–304.
12. J.-Y. Cai and A.P. Nerurkar, *Approximating the svp to within a factor* $(1+1/dim^{\varepsilon})$ *is NP-hard under randomized reductions*, Proceedings of the 38th IEEE Conference on Computational Complexity (Pittsburgh), IEEE Computer Society Press, 1998, pp. 46–55.
13. K.L. Clarkson, *Las Vegas algorithms for linear and integer programming when the dimension is small*, Journal of the Association for Computing Machinery 42 (1995) 488–499.
14. W. Cook, M.E. Hartmann, R. Kannan, and C. McDiarmid, *On integer points in polyhedra*, Combinatorica 12 (1992) 27–37.
15. I. Dinur, *Approximating* SVP_∞ *to within almost-polynomial factors is NP-hard*, Theoretical Computer Science 285 (2002) 55–71.
16. M. Dyer, A. Frieze, and R. Kannan, *A random polynomial-time algorithm for approximating the volume of convex bodies*, Journal of the ACM 38 (1991) 1–17.
17. J. Edmonds, *Systems of distinct representatives and linear algebra*, Journal of Reseach of the National Bureau of Standards 71B (1967) 241–245.
18. F. Eisenbrand, *Fast integer programming in fixed dimension*, Proceedings of the 11th Annual European Symposium on Algorithms, ESA' 2003 (G. Di Battista and U. Zwick, eds.), Lecture Notes in Computer Science 2832, Springer, 2003, pp. 196–207.
19. F. Eisenbrand and S. Laue, *A linear algorithm for integer programming in the plane*, Mathematical Programming 102 (2005) 249–259.
20. F. Eisenbrand and G. Rote, *Fast reduction of ternary quadratic forms*, Cryptography and Lattices Conference, CALC 2001 (J. Silverman, ed.), Lecture Notes in Computer Science 2146, Springer, 2001, pp. 32–44.
21. F. Eisenbrand and G. Shmonin, *Parametric integer programming in fixed dimension*, Mathematics of Operations Research (2008), to appear.

22. P. van Emde Boas, *Another NP-complete partition problem and the complexity of computing short vectors in a lattice*, Technical Report MI-UvA-81-04, Mathematical Institute, University of Amsterdam, Amsterdam, 1981.
23. X.G. Fang and G. Havas, *On the worst-case complexity of integer Gaussian elimination*, Proceedings of the 1997 International Symposium on Symbolic and Algebraic Computation (Kihei, HI) (New York), ACM, 1997, pp. 28–31.
24. S.D. Feit, *A fast algorithm for the two-variable integer programming problem*, Journal of the Association for Computing Machinery 31 (1984) 99–113.
25. A. Frank and É. Tardos, *An application of simultaneous Diophantine approximation in combinatorial optimization*, Combinatorica 7 (1987) 49–65.
26. M. Fürer, *Faster integer multiplication*, Proceedings of the 39th Annual ACM Symposium on Theory of Computing, STOC 2007 (D.S. Johnson and U. Feige, eds.), ACM, 2007, pp. 57–66.
27. N. Gama and P.Q. Nguyen, *Finding short lattice vectors within mordell's inequality*, Proceedings of the 40th Annual ACM Symposium on Theory of Computing, Victoria, British Columbia, Canada, STOC 2008 (R.E. Ladner and C. Dwork, eds.), ACM, 2008, pp. 207–216.
28. B. Gärtner, *A subexponential algorithm for abstract optimization problems*, SIAM Journal on Computing 24 (1995) 1018–1035.
29. C.F. Gauß, *Disquisitiones arithmeticae*, Gerh. Fleischer Iun., 1801.
30. O. Goldreich and S. Goldwasser, *On the limits of non-approximability of lattice problems*, Proceedings of the 30th Annual ACM Symposium on Theory of Computing (New York), ACM Press, 1998, pp. 1–9.
31. M. Grötschel, L. Lovász, and A. Schrijver, *Geometric algorithms and combinatorial optimization*, Algorithms and Combinatorics, Vol. 2, Springer, 1988.
32. M. Grötschel, L. Lovász, and A. Schrijver, *Geometric methods in combinatorial optimization*, Progress in Combinatorial Optimization (W.R. Pulleyblank, ed.), Academic Press, Toronto, 1984, pp. 167–183.
33. G. Hanrot and D. Stehlé, *Improved analysis of Kannan's shortest lattice vector algorithm (extended abstract)*, Advances in cryptology—CRYPTO 2007, Lecture Notes in Comput. Science 4622, Springer, Berlin, 2007, pp. 170–186.
34. I. Haviv and O. Regev, *Tensor-based hardness of the shortest vector problem to within almost polynomial factors*, STOC'07—Proceedings of the 39th Annual ACM Symposium on Theory of Computing, ACM, New York, 2007, pp. 469–477.
35. A.C. Hayes and D.G. Larman, *The vertices of the knapsack polytope*, Discrete Applied Mathematics 6 (1983) 135–138.
36. B. Helfrich, *Algorithms to construct Minkowski reduced and Hermite reduced lattice bases*, Theoretical Computer Science 41 (1985) 125–139.
37. C. Hermite, *Extraits de lettres de M. Ch. Hermite à M. Jacobi sur différents objets de la théorie des nombres*, Journal für die reine und angewandte Mathematik 40 (1850).
38. D.S. Hirschberg and C.K. Wong, *A polynomial algorithm for the knapsack problem in two variables*, Journal of the Association for Computing Machinery 23 (1976) 147–154.
39. F. John, *Extremum problems with inequalities as subsidiary conditions*, Studies and Essays Presented to R. Courant on his 60th Birthday, January 8, 1948, Interscience Publishers, Inc., New York, N. Y., 1948, pp. 187–204.
40. E. Kaltofen, *On the complexity of finding short vectors in integer lattices*, Computer Algebra: Proceedings of EUROCAL '1983 (J. VanHulzen, ed.), Lecture Notes in Computer Science 162, Springer, 1983, pp. 236–244.
41. N. Kanamaru, T. Nishizeki, and T. Asano, *Efficient enumeration of grid points in a convex polygon and its application to integer programming*, International Journal of Computational Geometry & Applications 4 (1994) 69–85.
42. R. Kannan, *Improved algorithms for integer programming and related problems*, Proceedings of the 15th Annual ACM Symposium on Theory of Computing (New York), ACM Press, 1983, pp. 193–206.
43. R. Kannan, *Minkowski's convex body theorem and integer programming*, Mathematics of Operations Research 12 (1987) 415–440.

44. R. Kannan, *Minkowski's convex body theorem and integer programming*, Mathematics of Operations Research 12 (1987) 415–440.
45. R. Kannan, *Lattice translates of a polytope and the Frobenius problem*, Combinatorica 12 (1992) 161–177.
46. R. Kannan, *A polynomial algorithm for the two-variable integer programming problem*, Journal of the Association for Computing Machinery 27 (1980) 118–122.
47. R. Kannan and A. Bachem, *Polynomial algorithms for computing the Smith and Hermite normal forms of an integer matrix*, SIAM Journal on Computing 8 (1979) 499–507.
48. L.G. Khachiyan and M.J. Todd, *On the complexity of approximating the maximal inscribed ellipsoid for a polytope*, Mathematical Programming 61 (1993) 137–159.
49. A. Khinchine, *A quantitative formulation of Kronecker's theory of approximation (in russian)*, Izvestiya Akademii Nauk SSR Seriya Matematika 12 (1948) 113–122.
50. S. Khot, *Hardness of approximating the shortest vector problem in high l_p norms*, Journal of Computer and System Sciences 72 (2006) 206–219.
51. D. Knuth, *The art of computer programming*, Vol. 2, Addison-Wesley, 1969.
52. M. Köppe, S. Verdoolaege, and K. Woods, *An implementation of the barvinok–woods integer projection algorithm*, Technical report, Katholieke Universiteit Leuven, 2008.
53. M. Köppe and S. Verdoolaege, *Computing parametric rational generating functions with a primal Barvinok algorithm*, Electronic Journal of Combinatorics 15:#R16 (2008).
54. J. Lagarias, H. Lenstra, and C. Schnorr, *Korkin-zolotarev bases and successive minima of a lattice and its reciprocal lattice*, Combinatorica 10 (1990) 333–348.
55. A.K. Lenstra, H.W. Lenstra, and L. Lovász, *Factoring polynomials with rational coefficients*, Mathematische Annalen 261 (1982) 515–534.
56. H.W. Lenstra, *Integer programming with a fixed number of variables*, Mathematics of Operations Research 8 (1983) 538–548.
57. H.W. Lenstra, Jr., *Lattices*, Algorithmic number theory: lattices, number fields, curves and cryptography, Math. Sci. Res. Inst. Publ., Cambridge Univ. Press, Cambridge, 2008, pp. 127–181.
58. L. Lovász, *An algorithmic theory of numbers, graphs and convexity*, SIAM, 1986.
59. J. Matoušek, M. Sharir, and E. Welzl, *A subexponential bound for linear programming*, Algorithmica 16 (1996) 498–516.
60. D. Micciancio, *The shortest vector in a lattice is hard to approximate to within some constant*, Proceedings of the 39th Annual Symposium on Foundations of Computer Science (Los Alamitos, CA), IEEE Computer Society, 1998, pp. 92–98.
61. H. Minkowski, *Geometrie der Zahlen*, Teubner, Leipzig, 1896.
62. Y.E. Nesterov and A.S. Nemirovski, *Self–concordant functions and polynomial–time methods in convex programming*, Technical report, Central Economic and Mathematical Institute, USSR Academy of Science, Moscow, USSR, 1989.
63. P.Q. Nguyen and D. Stehlé, *Floating-point LLL revisited*, Proceedings of the 24th Annual International Conference on the Theory and Applications of Cryptographic Techniques (EUROCRYPT 05) (R. Cramer, ed.), Lecture Notes in Computer Science 3494, Springer, 2005, pp. 215–233.
64. P.Q. Nguyen and D. Stehlé, *Floating-point LLL revisited*, Advances in cryptology—EUROCRYPT 2005, Lecture Notes in Computer Science 3494, Springer, 2005, pp. 215–233.
65. I. Niven, H.S. Zuckerman, and H.L. Montgomery, *An introduction to the theory of numbers, fifth edition*, Wiley, 1991.
66. X. Pujol and D. Stehlé, *Rigorous and efficient short lattice vectors enumeration*, 14th International Conference on the Theory and Application of Cryptology and Information Security (ASIACRYPT 2008) 2008, pp. 390–405.
67. J.L. Ramírez-Alfonsín, *Complexity of the Frobenius problem*, Combinatorica 16 (1996) 143–147.
68. O. Regev, *Lattices in computer science*, Lecture notes, Tel Aviv University, 2004, http://www.cs.tau.ac.il/~odedr/teaching/lattices_fall_2004/index.html.

69. O. Regev and R. Rosen, *Lattice problems and norm embeddings*, STOC'06: Proceedings of the 38th Annual ACM Symposium on Theory of Computing, ACM, New York, 2006, pp. 447–456.
70. H.E. Scarf, *Production sets with indivisibilities. Part I: generalities*, Econometrica 49 (1981) 1–32.
71. H.E. Scarf, *Production sets with indivisibilities. Part II: The case of two activities*, Econometrica 49 (1981) 395–423.
72. C.-P. Schnorr, *A hierarchy of polynomial time lattice basis reduction algorithms*, Theoretical Computer Science 53 (1987) 201–224.
73. A. Schönhage, *Schnelle Berechnung von Kettenbruchentwicklungen*, Acta Informatica 1 (1971) 139–144.
74. A. Schönhage and V. Strassen, *Schnelle Multiplikation grosser Zahlen*, Computing 7 (1971) 281–292.
75. A. Schrijver, *Theory of linear and integer programming*, John Wiley & Sons, 1986.
76. V.N. Shevchenko, *On the number of extreme points in integer programming*, Kibernetika (1981) 133–134 (Russian).
77. L.J. Stockmeyer, *The polynomial-time hierarchy*, Theoretical Computer Science 3 (1976) 1–22.
78. A. Storjohann, *Faster algorithms for integer lattice reduction*, Technical Report 249, Department of Computer Science, ETH Zürich, 1996.
79. É. Tardos, *A strongly polynomial algorithm to solve combinatorial linear programs*, Operations Research 34 (1986) 250–256.
80. J. von zur Gathen and J. Gerhard, *Modern computer algebra*, Cambridge University Press, 1999.
81. J. von zur Gathen and M. Sieveking, *Weitere zum Erfüllungsproblem polynomial äquivalente kombinatorische Aufgaben*, Komplexität von Entscheidungsproblemen: ein Seminar (E. Specker and V. Strassen, eds.), Lecture Notes in Computer Science 43, Springer, 1976, pp. 49–71.
82. E. Welzl, *Smallest enclosing disks (balls and ellipsoids)*, New results and new trends in computer science (Graz, 1991), Lecture Notes in Computer Science 555, Springer, Berlin, 1991, pp. 359–370.
83. C. Wrathall, *Complete sets and the polynomial-time hierarchy*, Theoretical Computer Science 3 (1976) 23–33.
84. L.Y. Zamanskij and V.D. Cherkasskij, *A formula for determining the number of integral points on a straight line and its application*, Ehkon. Mat. Metody 20 (1984) 1132–1138, (in Russian).

Chapter 15
Nonlinear Integer Programming

Raymond Hemmecke, Matthias Köppe, Jon Lee, and Robert Weismantel

Abstract Research efforts of the past fifty years have led to a development of linear integer programming as a mature discipline of mathematical optimization. Such a level of maturity has not been reached when one considers nonlinear systems subject to integrality requirements for the variables. This chapter is dedicated to this topic. The primary goal is a study of a simple version of general nonlinear integer problems, where all constraints are still linear. Our focus is on the computational complexity of the problem, which varies significantly with the type of nonlinear objective function in combination with the underlying combinatorial structure. Numerous boundary cases of complexity emerge, which sometimes surprisingly lead even to polynomial time algorithms. We also cover recent successful approaches for more general classes of problems. Though no positive theoretical efficiency results are available, nor are they likely to ever be available, these seem to be the currently most successful and interesting approaches for solving practical problems. It is our belief that the study of algorithms motivated by theoretical considerations and those motivated by our desire to solve practical instances should and do inform one another. So it is with this viewpoint that we present the subject, and it is in this direction that we hope to spark further research.

Raymond Hemmecke
FMA/IMO, Otto-von-Guericke-Universität Magdeburg, Germany
e-mail: hemmecke@imo.math.uni-magdeburg.de

Matthias Köppe
Department of Mathematics, University of California, Davis, USA
e-mail: mkoeppe@math.ucdavis.edu

Jon Lee
IBM T.J. Watson Research Center, Yorktown Heights, New York, USA
e-mail: jonlee@us.ibm.com

Robert Weismantel
FMA/IMO, Otto-von-Guericke-Universität Magdeburg, Germany
e-mail: weismant@imo.math.uni-magdeburg.de

M. Jünger et al. (eds.), *50 Years of Integer Programming 1958-2008*,
DOI 10.1007/978-3-540-68279-0_15, © Springer-Verlag Berlin Heidelberg 2010

15.1 Overview

In the past decade, *nonlinear* integer programming has gained a lot of mindshare. Obviously many important applications demand that we be able to handle nonlinear objective functions and constraints. Traditionally, nonlinear mixed-integer programs have been handled in the context of the field of global optimization, where the main focus is on numerical algorithms to solve nonlinear continuous optimization problems and where integrality constraints were considered as an afterthought, using branch-and-bound over the integer variables. In the past few years, however, researchers from the field of integer programming have increasingly studied nonlinear mixed-integer programs from their point of view. Nevertheless, this is generally considered a very young field, and most of the problems and methods are not as well-understood or stable as in the case of linear mixed-integer programs.

Any contemporary review of nonlinear mixed-integer programming will therefore be relatively short-lived. For this reason, our primary focus is on a classification of nonlinear mixed-integer problems from the point of view of computational complexity, presenting theory and algorithms for the efficiently solvable cases. The hope is that at least this part of the chapter will still be valuable in the next few decades. However, we also cover recent successful approaches for more general classes of problems. Though no positive theoretical efficiency results are available — nor are they likely to ever be available, these seem to be the currently most successful and interesting approaches for solving practical problems. It is our belief that the study of algorithms motivated by theoretical considerations and those motivated by our desire to solve practical instances should and do inform one another. So it is with this viewpoint that we present the subject, and it is in this direction that we hope to spark further research.

Let us however also remark that the selection of the material that we discuss in this chapter is subjective. There are topics that some researchers associate with "nonlinear integer programming" that are not covered here. Among them are pseudo-Boolean optimization, max-cut and quadratic assignment as well as general 0/1 polynomial programming. There is no doubt that these topics are interesting, but, in order to keep this chapter focused, we refrain from going into these topics. Instead we refer the interested reader to the references [56] on max-cut, [33] for recent advances in general 0/1 polynomial programming, and the excellent surveys [30] on pseudo-Boolean optimization and [104, 35] on the quadratic assignment problem.

A general model of mixed-integer programming could be written as

$$\begin{aligned} \text{max/min} \quad & f(x_1,\dots,x_n) \\ \text{s.t.} \quad & g_1(x_1,\dots,x_n) \leqslant 0 \\ & \vdots \\ & g_m(x_1,\dots,x_n) \leqslant 0 \\ & \mathbf{x} \in \mathbb{R}^{n_1} \times \mathbb{Z}^{n_2}, \end{aligned} \tag{15.1}$$

where $f, g_1, \ldots, g_m \colon \mathbb{R}^n \to \mathbb{R}$ are arbitrary nonlinear functions. However, in parts of the chapter, we study a rather restricted model of nonlinear integer programming, where the nonlinearity is confined to the objective function, i.e., the following model:

$$\begin{aligned} \text{max/min} \quad & f(x_1, \ldots, x_n) \\ \text{subject to} \quad & A\mathbf{x} \leqslant \mathbf{b} \\ & \mathbf{x} \in \mathbb{R}^{n_1} \times \mathbb{Z}^{n_2}, \end{aligned} \tag{15.2}$$

where A is a rational matrix and $\mathbf{b}$ is a rational vector. It is clear that this model is still NP-hard, and that it is much more expressive and much harder to solve than integer linear programs.

We start out with a few fundamental hardness results that help to get a picture of the complexity situation of the problem.

Even in the pure continuous case, nonlinear optimization is known to be hard.

Theorem 15.1. *Pure continuous polynomial optimization over polytopes ($n_2 = 0$) in varying dimension is NP-hard. Moreover, there does not exist a fully polynomial time approximation scheme (FPTAS) (unless* P = NP*).*

Indeed the max-cut problem can be modeled as minimizing a quadratic form over the cube $[-1,1]^n$, and thus inapproximability follows from a result by Håstad [66]. On the other hand, pure continuous polynomial optimization problems over polytopes ($n_2 = 0$) can be solved in polynomial time when the dimension n_1 is fixed. This follows from a much more general result on the computational complexity of approximating the solutions to general algebraic formulae over the real numbers by Renegar [112].

However, as soon as we add just two integer variables, we get a hard problem again:

Theorem 15.2. *The problem of minimizing a degree-*4 *polynomial over the lattice points of a convex polygon is NP-hard.*

This is based on the NP-completeness of the problem whether there exists a positive integer $x < c$ with $x^2 \equiv a \pmod{b}$; see [54, 45].

We also get hardness results that are much worse than just NP-hardness. The negative solution of Hilbert's tenth problem by Matiyasevich [96, 97], based on earlier work by Davis, Putnam, and Robinson, implies that nonlinear integer programming is *incomputable*, i.e., there cannot exist any general algorithm. (It is clear that for cases where finite bounds for all variables are known, an algorithm trivially exists.) Due to a later strengthening of the negative result by Matiyasevich (published in [77]), there also cannot exist any such general algorithm for even a small fixed number of integer variables; see [45].

Theorem 15.3. *The problem of minimizing a linear form over polynomial constraints in at most 10 integer variables is not computable by a recursive function.*

Another consequence, as shown by Jeroslow [76], is that even integer quadratic programming is incomputable.

Theorem 15.4. *The problem of minimizing a linear form over quadratic constraints in integer variables is not computable by a recursive function.*

How do we get positive complexity results and practical methods? One way is to consider subclasses of the objective functions and constraints. First of all, for the problem of *concave minimization* or *convex maximization* which we study in Section 15.2, we can make use of the property that optimal solutions can always be found on the boundary (actually on the set of vertices) of the convex hull of the feasible solutions. On the other hand, as in the pure continuous case, *convex minimization*, which we address in Section 15.3), is much easier to handle, from both theoretical and practical viewpoints, than the general case. Next, in Section 15.4, we study the general case of polynomial optimization, as well as practical approaches for handling the important case of quadratic functions. Finally, in Section 15.5, we briefly describe the practical approach of global optimization.

For each of these subclasses covered in Sections 15.2–15.5, we discuss positive complexity results, such as *polynomiality results in fixed dimension*, if available (Sections 15.2.1, 15.3.1, 15.4.1), including some *boundary cases of complexity* in Sections 15.2.2, 15.3.2, and 15.5.2, and discuss *practical algorithms* (Sections 15.2.3, 15.3.3, 15.4.2, 15.4.3, 15.5.1).

We end the chapter with conclusions (Section 15.6), including a table that summarizes the complexity situation of the problem (Table 15.1).

15.2 Convex integer maximization

15.2.1 Fixed dimension

Maximizing a convex function over the integer points in a polytope in fixed dimension can be done in polynomial time. To see this, note that the optimal value is taken on at a vertex of the convex hull of all feasible integer points. But when the dimension is fixed, there is only a polynomial number of vertices, as Cook et al. [39] showed.

Theorem 15.5. *Let* $P = \{\mathbf{x} \in \mathbb{R}^n : A\mathbf{x} \leqslant \mathbf{b}\}$ *be a rational polyhedron with* $A \in \mathbb{Q}^{m\times n}$ *and let* ϕ *be the largest binary encoding size of any of the rows of the system* $A\mathbf{x} \leqslant \mathbf{b}$. *Let* $P^{\mathrm{I}} = \operatorname{conv}(P \cap \mathbb{Z}^n)$ *be the integer hull of* P. *Then the number of vertices of* P^{I} *is at most* $2m^n(6n^2\phi)^{n-1}$.

Moreover, Hartmann [65] gave an algorithm for enumerating all the vertices, which runs in polynomial time in fixed dimension.

By using Hartmann's algorithm, we can therefore compute all the vertices of the integer hull P^{I}, evaluate the convex objective function on each of them and pick the best. This simple method already provides a polynomial-time algorithm.

15.2.2 Boundary cases of complexity

In the past fifteen years algebraic geometry and commutative algebra tools have shown their exciting potential to study problems in integer optimization (see [20, 131] and references therein). The presentation in this section is based on the papers [47, 102].

The first key lemma, extending results of [102] for combinatorial optimization, shows that when a suitable geometric condition holds, it is possible to efficiently reduce the convex integer maximization problem to the solution of polynomially many linear integer programming counterparts. As we will see, this condition holds naturally for a broad class of problems in variable dimension. To state this result, we need the following terminology. A *direction* of an edge e (i.e., a one-dimensional face) of a polyhedron P is any nonzero scalar multiple of $\mathbf{u}-\mathbf{v}$ with $\mathbf{u},\mathbf{v}$ any two distinct points in e. A set of vectors *covers all edge-directions of* P if it contains a direction of each edge of P. A *linear integer programming oracle* for matrix $A \in \mathbb{Z}^{m\times n}$ and vector $\mathbf{b} \in \mathbb{Z}^m$ is one that, queried on $\mathbf{w} \in \mathbb{Z}^n$, solves the linear integer program $\max\{\mathbf{w}^\top\mathbf{x} : A\mathbf{x} = \mathbf{b},\ \mathbf{x} \in \mathbb{N}^n\}$, that is, either returns an optimal solution $\mathbf{x} \in \mathbb{N}^n$, or asserts that the program is infeasible, or asserts that the objective function $\mathbf{w}^\top\mathbf{x}$ is unbounded.

Lemma 15.1. *For any fixed d there is a strongly polynomial oracle-time algorithm that, given any vectors $\mathbf{w}_1,\dots,\mathbf{w}_d \in \mathbb{Z}^n$, matrix $A \in \mathbb{Z}^{m\times n}$ and vector $\mathbf{b} \in \mathbb{Z}^m$ endowed with a linear integer programming oracle, finite set $E \subset \mathbb{Z}^n$ covering all edge-directions of the polyhedron* $\operatorname{conv}\{\mathbf{x} \in \mathbb{N}^n : A\mathbf{x} = \mathbf{b}\}$, *and convex functional $c : \mathbb{R}^d \longrightarrow \mathbb{R}$ presented by a comparison oracle, solves the convex integer program*

$$\max\left\{c(\mathbf{w}_1^\top\mathbf{x},\dots,\mathbf{w}_d^\top\mathbf{x}) :\ A\mathbf{x} = \mathbf{b},\ \mathbf{x} \in \mathbb{N}^n\right\}.$$

Here, *solving* the program means that the algorithm either returns an optimal solution $\mathbf{x} \in \mathbb{N}^n$, or asserts the problem is infeasible, or asserts the polyhedron $\{\mathbf{x} \in \mathbb{R}^n_+ : A\mathbf{x} = \mathbf{b}\}$ is unbounded; and *strongly polynomial oracle-time* means that the number of arithmetic operations and calls to the oracles are polynomially bounded in m and n, and the size of the numbers occurring throughout the algorithm is polynomially bounded in the size of the input (which is the number of bits in the binary representation of the entries of $\mathbf{w}_1,\dots,\mathbf{w}_d,A,\mathbf{b},E$).

Let us outline the main ideas behind the proof to Lemma 15.1, and let us point out the difficulties that one has to overcome. Given data for a convex integer maximization problem $\max\{c(\mathbf{w}_1^\top\mathbf{x},\dots,\mathbf{w}_d^\top\mathbf{x}) :\ A\mathbf{x} = \mathbf{b},\ \mathbf{x} \in \mathbb{N}^n\}$, consider the polyhedron $P := \operatorname{conv}\{\mathbf{x} \in \mathbb{N}^n : A\mathbf{x} = \mathbf{b}\} \subseteq \mathbb{R}^n$ and its linear transformation $Q := \{(\mathbf{w}_1^\top\mathbf{x},\dots,\mathbf{w}_d^\top\mathbf{x}) :\ \mathbf{x} \in P\} \subseteq \mathbb{R}^d$. Note that P has typically exponentially many vertices and is not accessible computationally. Note also that, because c is convex, there is an optimal solution $\mathbf{x}$ whose image $(\mathbf{w}_1^\top\mathbf{x},\dots,\mathbf{w}_d^\top\mathbf{x})$ is a vertex of Q. So an important ingredient in the solution is to construct the vertices of Q. Unfortunately, Q may also have exponentially many vertices even though it lives in a space $\mathbb{R}^d$ of fixed dimension. However, it can be shown that, when the number of *edge-directions* of P is polynomial, the number of vertices of Q is polynomial. Nonetheless, even in this case, it is not possible to construct these vertices directly, because the number of vertices of P may still be exponential. This difficulty can finally be overcome by using a suitable *zonotope*. See [47, 102] for more details.

Let us now apply Lemma 15.1 to a broad (in fact, *universal*) class of convex integer maximization problems. Lemma 15.1 implies that these problems can be solved in polynomial time. Given an $(r+s)\times t$ matrix A, let A_1 be its $r\times t$ sub-matrix consisting of the first r rows and let A_2 be its $s\times t$ sub-matrix consisting of the last s rows. Define the *n-fold matrix* of A to be the following $(r+ns)\times nt$ matrix,

$$A^{(n)} := (\mathbf{1}_n \otimes A_1) \oplus (I_n \otimes A_2) = \begin{pmatrix} A_1 & A_1 & A_1 & \cdots & A_1 \\ A_2 & 0 & 0 & \cdots & 0 \\ 0 & A_2 & 0 & \cdots & 0 \\ \vdots & \vdots & \ddots & \vdots & \vdots \\ 0 & 0 & 0 & \cdots & A_2 \end{pmatrix}.$$

Note that $A^{(n)}$ depends on r and s: these will be indicated by referring to A as an "$(r+s)\times t$ matrix."

We establish the following theorem, which asserts that convex integer maximization over n-fold systems of a fixed matrix A, in variable dimension nt, are solvable in polynomial time.

Theorem 15.6. *For any fixed positive integer d and fixed $(r+s)\times t$ integer matrix A there is a polynomial oracle-time algorithm that, given n, vectors $\mathbf{w}_1,\ldots,\mathbf{w}_d \in \mathbb{Z}^{nt}$ and $\mathbf{b}\in\mathbb{Z}^{r+ns}$, and convex function $c:\mathbb{R}^d \longrightarrow \mathbb{R}$ presented by a comparison oracle, solves the convex n-fold integer maximization problem*

$$\max\left\{c(\mathbf{w}_1^\top \mathbf{x},\ldots,\mathbf{w}_d^\top \mathbf{x}) : A^{(n)}\mathbf{x}=\mathbf{b},\ \mathbf{x}\in\mathbb{N}^{nt}\right\}.$$

The equations defined by an n-fold matrix have the following, perhaps more illuminating, interpretation: splitting the variable vector and the right-hand side vector into components of suitable sizes, $\mathbf{x} = (\mathbf{x}^1,\ldots,\mathbf{x}^n)$ and $\mathbf{b} = (\mathbf{b}^0,\mathbf{b}^1,\ldots,\mathbf{b}^n)$, where $\mathbf{b}^0 \in \mathbb{Z}^r$ and $\mathbf{x}^k \in \mathbb{N}^t$ and $\mathbf{b}^k \in \mathbb{Z}^s$ for $k = 1,\ldots,n$, the equations become $A_1(\sum_{k=1}^n \mathbf{x}^k) = \mathbf{b}^0$ and $A_2\mathbf{x}^k = \mathbf{b}^k$ for $k = 1,\ldots,n$. Thus, each component $\mathbf{x}^k$ satisfies a system of constraints defined by A_2 with its own right-hand side $\mathbf{b}^k$, and the sum $\sum_{k=1}^n \mathbf{x}^k$ obeys constraints determined by A_1 and $\mathbf{b}^0$ restricting the "common resources shared by all components".

Theorem 15.6 has various applications, including multiway transportation problems, packing problems, vector partitioning and clustering. For example, we have the following corollary providing the first polynomial time solution of convex 3-way transportation.

Corollary 15.1 (Convex 3-way transportation). *For any fixed d,p,q there is a polynomial oracle-time algorithm that, given n, arrays $\mathbf{w}_1,\ldots,\mathbf{w}_d \in \mathbb{Z}^{p\times q\times n}$, $\mathbf{u}\in\mathbb{Z}^{p\times q}$, $\mathbf{v}\in\mathbb{Z}^{p\times n}$, $\mathbf{z}\in\mathbb{Z}^{q\times n}$, and convex $c:\mathbb{R}^d \longrightarrow \mathbb{R}$ presented by comparison oracle, solves the convex integer 3-way transportation problem*

$$\max\{c(\mathbf{w}_1^\top \mathbf{x},\ldots,\mathbf{w}_d^\top \mathbf{x}) : \mathbf{x}\in\mathbb{N}^{p\times q\times n},\ \sum_i x_{i,j,k}=z_{j,k},\ \sum_j x_{i,j,k}=v_{i,k},\ \sum_k x_{i,j,k}=u_{i,j}\}.$$

Note that in contrast, if the dimensions of two sides of the tables are variable, say, q and n, then already the standard *linear* integer 3-way transportation problem over such tables is NP-hard, see [41, 42, 43].

In order to prove Theorem 15.6, we need to recall some definitions. The *Graver basis* of an integer matrix $A\in\mathbb{Z}^{m\times n}$, introduced in [60], is a canonical finite set $\mathscr{G}(A)$ that can be defined as follows. For each of the 2^n orthants $\mathscr{O}_j$ of $\mathbb{R}^n$ let H_j denote the inclusion-minimal Hilbert basis of the pointed rational polyhedral cone $\ker(A)\cap\mathscr{O}_j$. Then the Graver basis $\mathscr{G}(A)$ is defined to be the union $\mathscr{G}(A)=\cup_{i=1}^{2^n}H_j\setminus\{\mathbf{0}\}$ over all these 2^n Hilbert bases. By this definition, the Graver basis $\mathscr{G}(A)$ has a nice representation property: every $\mathbf{z}\in\ker(A)\cap\mathbb{Z}^n$ can be written as a sign-compatible nonnegative integer linear combination $\mathbf{z}=\sum_i\alpha_i\mathbf{g}_i$ of Graver basis elements $\mathbf{g}_i\in\mathscr{G}(A)$. This follows from the simple observation that $\mathbf{z}$ has to belong to some orthant $\mathscr{O}_j$ of $\mathbb{R}^n$ and thus it can be represented as a sign-compatible nonnegative integer linear combination of elements in $H_j\subseteq\mathscr{G}(A)$. For more details on Graver bases and the currently fastest procedure for computing them see [125, 68, 1].

Graver bases have another nice property: They contain all edge directions in the integer hulls within the polytopes $P_{\mathbf{b}}=\{\mathbf{x}: A\mathbf{x}=\mathbf{b},\mathbf{x}\geqslant\mathbf{0}\}$ as $\mathbf{b}$ is varying. We include a direct proof here.

Lemma 15.2. *For every matrix $A\in\mathbb{Z}^{m\times n}$ and every vector $\mathbf{b}\in\mathbb{N}^m$, the Graver basis $\mathscr{G}(A)$ of A covers all edge-directions of the polyhedron* $\mathrm{conv}\{\mathbf{x}\in\mathbb{N}^n : A\mathbf{x}=\mathbf{b}\}$ *defined by A and* $\mathbf{b}$.

Proof. Consider any edge e of $P:=\mathrm{conv}\{\mathbf{x}\in\mathbb{N}^n : A\mathbf{x}=\mathbf{b}\}$ and pick two distinct points $\mathbf{u},\mathbf{v}\in e\cap\mathbb{N}^n$. Then $\mathbf{g}:=\mathbf{u}-\mathbf{v}$ is in $\ker(A)\cap\mathbb{Z}^n\setminus\{\mathbf{0}\}$ and hence, by the representation property of the Graver basis $\mathscr{G}(A)$, $\mathbf{g}$ can be written as a finite *sign-compatible* sum $\mathbf{g}=\sum\mathbf{g}^i$ with $\mathbf{g}^i\in\mathscr{G}(A)$ for all i. Now, we claim that $\mathbf{u}-\mathbf{g}^i\in P$ for all i. To see this, note first that $\mathbf{g}^i\in\mathscr{G}(A)\subseteq\ker(A)$ implies $A\mathbf{g}^i=\mathbf{0}$ and hence $A(\mathbf{u}-\mathbf{g}^i)=A\mathbf{u}=\mathbf{b}$; and second, note that $\mathbf{u}-\mathbf{g}^i\geqslant\mathbf{0}$: indeed, if $g^i_j\leqslant 0$ then $u_j-g^i_j\geqslant u_j\geqslant 0$; and if $g^i_j>0$ then sign-compatibility implies $g^i_j\leqslant g_j$ and therefore $u_j-g^i_j\geqslant u_j-g_j=v_j\geqslant 0$.

Now let $\mathbf{w}\in\mathbb{R}^n$ be a linear functional uniquely maximized over P at the edge e. Then for all i, as just proved, $\mathbf{u}-\mathbf{g}^i\in P$ and hence $\mathbf{w}^\top\mathbf{g}^i\geqslant 0$. But $\sum\mathbf{w}^\top\mathbf{g}^i=\mathbf{w}^\top\mathbf{g}=\mathbf{w}^\top\mathbf{u}-\mathbf{w}^\top\mathbf{v}=0$, implying that in fact, for all i, we have $\mathbf{w}^\top\mathbf{g}^i=0$ and therefore $\mathbf{u}-\mathbf{g}^i\in e$. This implies that each $\mathbf{g}^i$ is a direction of the edge e (in fact, moreover, all $\mathbf{g}^i$ are the same, so $\mathbf{g}$ is a multiple of some Graver basis element).

In Section 15.3.2, we show that for fixed matrix A, the size of the Graver basis of $A^{(n)}$ increases only polynomially in n implying Theorem 15.14 that states that certain integer convex n-fold *minimization* problems can be solved in polynomial time when the matrix A is kept fixed. As a special case, this implies that the integer *linear* n-fold *maximization* problem can be solved in polynomial time when the matrix A is kept fixed. Finally, combining these results with Lemmas 15.1 and 15.2, we can now prove Theorem 15.6.

Proof (of Theorem 15.6). The algorithm underlying Proposition 15.14 provides a polynomial time realization of a linear integer programming oracle for $A^{(n)}$ and $\mathbf{b}$. The algorithm underlying Proposition 15.2 allows to compute the Graver basis $\mathscr{G}(A^{(n)})$ in time polynomial in the input. By Lemma 15.2, this set $E := \mathscr{G}(A^{(n)})$ covers all edge-directions of the polyhedron $\operatorname{conv}\{\mathbf{x} \in \mathbb{N}^{nt} : A^{(n)}\mathbf{x} = \mathbf{b}\}$ underlying the convex integer program. Thus, the hypothesis of Lemma 15.1 is satisfied and hence the algorithm underlying Lemma 15.1 can be used to solve the convex integer maximization problem in polynomial time.

15.2.3 Reduction to linear integer programming

In this section it is our goal to develop a basic understanding about when a discrete polynomial programming problem can be tackled with classical linear integer optimization techniques. We begin to study polyhedra related to polynomial integer programming. The presented approach applies to problems of the kind

$$\max\{f(\mathbf{x}) : A\mathbf{x} = \mathbf{b},\ \mathbf{x} \in \mathbb{Z}^n_+\}$$

with convex polynomial function f, as well as to models such as

$$\max\{\mathbf{c}^\top \mathbf{x} : \mathbf{x} \in K_I\}, \tag{15.3}$$

where K_I denotes the set of all integer points of a compact *basic-closed* semi-algebraic set K described by a finite number of polynomial inequalities, i. e.,

$$K = \{\mathbf{x} \in \mathbb{R}^n : p_i(\mathbf{x}) \leqslant 0,\ i \in M,\ \mathbf{l} \leqslant \mathbf{x} \leqslant \mathbf{u}\}.$$

We assume that $p_i \in \mathbb{Z}[\mathbf{x}] := \mathbb{Z}[x_1, \dots, x_n]$, for all $i \in M = \{1, \dots, m\}$, and $\mathbf{l}, \mathbf{u} \in \mathbb{Z}^n$.

One natural idea is to derive a linear description of the convex hull of K_I. Unfortunately, $\operatorname{conv}(K_I)$ might contain interior integer points that are not elements of K, see Figure 15.1. On the other hand, if a description of $\operatorname{conv}(K_I)$ is at hand, then the mathematical optimization problem of solving (15.3) can be reduced to optimizing the linear objective function $\mathbf{c}^\top \mathbf{x}$ over the polyhedron $\operatorname{conv}(K_I)$. This is our first topic of interest. In what follows we denote for a set D the projection of D to a set of variables x by the symbol D_x.

Definition 15.1. For polynomials $p_1, \dots, p_m : \mathbb{Z}^n \to \mathbb{Z}$ we define the polyhedron associated with the vector $\mathbf{p}(\mathbf{x}) = (p_1(\mathbf{x}), \dots, p_m(\mathbf{x}))$ of polynomials as

$$P_{\mathbf{p}} = \operatorname{conv}(\{(\mathbf{x}, \mathbf{p}(\mathbf{x})) \in \mathbb{R}^{n+m} : \mathbf{x} \in [\mathbf{l}, \mathbf{u}] \cap \mathbb{Z}^n\}).$$

The significance of the results below is that they allow us to reformulate the non-linear integer optimization problem $\max\{f(\mathbf{x}) : A\mathbf{x} = \mathbf{b},\ \mathbf{x} \in \mathbb{Z}^n_+\}$ as the optimization problem $\min\{\pi : A\mathbf{x} = \mathbf{b},\ f(\mathbf{x}) \leqslant \pi,\ \mathbf{x} \in \mathbb{Z}^n_+\}$. This in turn has the same objective function value as the linear integer program: $\min\{\pi : A\mathbf{x} = \mathbf{b},\ (\mathbf{x}, \pi) \in P_f,\ \mathbf{x} \in \mathbb{Z}^n_+\}$.

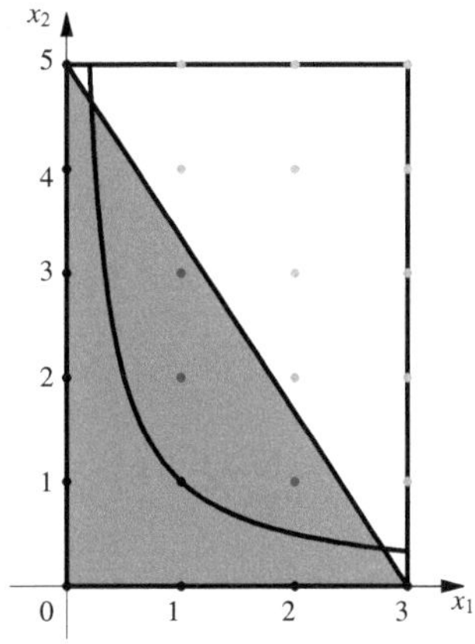

Fig. 15.1 Let $K = \{\mathbf{x} \in \mathbb{R}^2 \mid x_1 x_2 - 1 \leqslant 0,\ 0 \leqslant x_1 \leqslant 3,\ 0 \leqslant x_2 \leqslant 5\}$. The point $(1,2)$ is contained in $\operatorname{conv}(K_I)$. But it violates the constraint $x_1 x_2 - 1 \leqslant 0$.

In this situation, an H-description or V-description of the polyhedron P_f is sufficient to reduce the original nonlinear optimization problem to a linear integer problem.

Proposition 15.1. *For a vector of polynomials* $\mathbf{p} \in \mathbb{Z}[\mathbf{x}]^m$, *let*

$$K_I = \{\mathbf{x} \in \mathbb{Z}^n : \mathbf{p}(\mathbf{x}) \leqslant \mathbf{0},\ \mathbf{l} \leqslant \mathbf{x} \leqslant \mathbf{u}\}.$$

Then,

$$\operatorname{conv}(K_I) \subseteq \left(P_{\mathbf{p}} \cap \{(\mathbf{x}, \boldsymbol{\pi}) \in \mathbb{R}^{n+m} : \boldsymbol{\pi} \leqslant \mathbf{0}\}\right)_{\mathbf{x}}. \tag{15.4}$$

Proof. It suffices to show that $K_I \subseteq \left(P_{\mathbf{p}} \cap \{(\mathbf{x}, \boldsymbol{\pi}) \in \mathbb{R}^{n+m} : \boldsymbol{\pi} \leqslant \mathbf{0}\}\right)_{\mathbf{x}}$. Let us consider $\mathbf{x} \in K_I \subseteq [\mathbf{l}, \mathbf{u}] \cap \mathbb{Z}^n$. By definition, $(\mathbf{x}, \mathbf{p}(\mathbf{x})) \in P_{\mathbf{p}}$. Moreover, we have $\mathbf{p}(\mathbf{x}) \leqslant \mathbf{0}$. This implies $(\mathbf{x}, \mathbf{p}(\mathbf{x})) \in \left(P_{\mathbf{p}} \cap \{(\mathbf{x}, \boldsymbol{\pi}) \in \mathbb{R}^{n+m} : \boldsymbol{\pi} \leqslant \mathbf{0}\}\right)$, and thus,

$$\mathbf{x} \in \left(P_{\mathbf{p}} \cap \{(\mathbf{x}, \boldsymbol{\pi}) \in \mathbb{R}^{n+m} : \boldsymbol{\pi} \leqslant \mathbf{0}\}\right)_{\mathbf{x}}.$$

Note that even in the univariate case, equality in Formula (15.4) of Proposition 15.1 does not always hold. For instance, if $K_I = \{x \in \mathbb{Z} : x^2 - 5 \leqslant 0,\ -3 \leqslant x \leqslant 5\}$, then

$$\operatorname{conv}(K_I) = [-2, 2] \neq [-2.2, 2.2] \subseteq \left(P_{x^2} \cap \{(x, \pi) \in \mathbb{R}^2 : \pi - 5 \leqslant 0\}\right)_x.$$

Although even in very simple cases the sets $\operatorname{conv}(K_I)$ and $\left(P_{\mathbf{p}} \cap \{(\mathbf{x}, \boldsymbol{\pi}) : \boldsymbol{\pi} \leqslant \mathbf{0}\}\right)_{\mathbf{x}}$ differ, it is still possible that the integer points in K_I and $\left(P_{\mathbf{p}} \cap \{(\mathbf{x}, \boldsymbol{\pi}) : \boldsymbol{\pi} \leqslant \mathbf{0}\}\right)_{\mathbf{x}}$ are equal. In our example with $K_I = \{x \in \mathbb{Z} : x^2 - 5 \leqslant 0,\ -3 \leqslant x \leqslant 5\}$, we then obtain,

$$K_I = \{-2, -1, 0, 1, 2\} = [-2.2, 2.2] \cap \mathbb{Z}.$$

Of course, for any $\mathbf{p} \in \mathbb{Z}[\mathbf{x}]^m$ we have that

$$K_I = \{\mathbf{x} \in \mathbb{Z}^n : \mathbf{p}(\mathbf{x}) \leqslant \mathbf{0},\, \mathbf{l} \leqslant \mathbf{x} \leqslant \mathbf{u}\} \subseteq \left(P_{\mathbf{p}} \cap \{(\mathbf{x}, \boldsymbol{\pi}) : \boldsymbol{\pi} \leqslant \mathbf{0}\}\right)_{\mathbf{x}} \cap \mathbb{Z}^n. \qquad (15.5)$$

The key question here is when equality holds in Formula (15.5).

Theorem 15.7. *Let* $\mathbf{p} \in \mathbb{Z}[\mathbf{x}]^m$ *and* $K_I = \{\mathbf{x} \in \mathbb{Z}^n : \mathbf{p}(\mathbf{x}) \leqslant \mathbf{0},\, \mathbf{l} \leqslant \mathbf{x} \leqslant \mathbf{u}\}$. *Then,*

$$K_I \;=\; \left(P_{\mathbf{p}} \cap \{(\mathbf{x}, \boldsymbol{\pi}) : \boldsymbol{\pi} \leqslant \mathbf{0}\}\right)_{\mathbf{x}} \cap \mathbb{Z}^n$$

holds if every polynomial $p' \in \{p_i : i = 1, \ldots, m\}$ *satisfies the condition*

$$p'\Big(\sum_{\mathbf{k}} \lambda_{\mathbf{k}} \mathbf{k}\Big) - \sum_{\mathbf{k}} \lambda_{\mathbf{k}} p'(\mathbf{k}) \quad < \quad 1, \qquad (15.6)$$

for all $\lambda_{\mathbf{k}} \geqslant 0$, $\mathbf{k} \in [\mathbf{l}, \mathbf{u}] \cap \mathbb{Z}^n$, $\sum_{\mathbf{k}} \lambda_{\mathbf{k}} = 1$ *and* $\sum_{\mathbf{k}} \lambda_{\mathbf{k}} \mathbf{k} \in \mathbb{Z}^n$.

Proof. Using (15.5), we have to show that $\left(P_{\mathbf{p}} \cap \{(\mathbf{x}, \boldsymbol{\pi}) : \boldsymbol{\pi} \leqslant \mathbf{0}\}\right)_{\mathbf{x}} \cap \mathbb{Z}^n \subseteq K_I$ if all p_i, $i \in \{1, \ldots, m\}$, satisfy (15.6). Let $\mathbf{x} \in \left(P_{\mathbf{p}} \cap \{(\mathbf{x}, \boldsymbol{\pi}) \in \mathbb{R}^{m+n} : \boldsymbol{\pi} \leqslant \mathbf{0}\}\right)_{\mathbf{x}} \cap \mathbb{Z}^n$. Then, there exists a $\boldsymbol{\pi} \in \mathbb{R}^m$ such that

$$(\mathbf{x}, \boldsymbol{\pi}) \in P_{\mathbf{p}} \cap \{(\mathbf{x}, \boldsymbol{\pi}) \in \mathbb{R}^{n+m} : \boldsymbol{\pi} \leqslant 0\}.$$

By definition, $\boldsymbol{\pi} \leqslant \mathbf{0}$. Furthermore, there must exist nonnegative real numbers $\lambda_{\mathbf{k}} \geqslant 0$, $\mathbf{k} \in [\mathbf{l}, \mathbf{u}] \cap \mathbb{Z}^n$, such that $\sum_{\mathbf{k}} \lambda_{\mathbf{k}} = 1$ and $(\mathbf{x}, \boldsymbol{\pi}) = \sum_{\mathbf{k}} \lambda_{\mathbf{k}} (\mathbf{k}, \mathbf{p}(\mathbf{k}))$. Suppose that there exists an index i_0 such that the inequality $p_{i_0}(\mathbf{x}) \leqslant 0$ is violated. The fact that $p_{i_0} \in \mathbb{Z}[\mathbf{x}]$ and $\mathbf{x} \in \mathbb{Z}^n$, implies that $p_{i_0}(\mathbf{x}) \geqslant 1$. Thus, we obtain

$$\sum_{\mathbf{k}} \lambda_{\mathbf{k}} p_{i_0}(\mathbf{k}) = \pi_{i_0} \leqslant 0 < 1 \leqslant p_{i_0}(\mathbf{x}) = p_{i_0}\Big(\sum_{\mathbf{k}} \lambda_{\mathbf{k}} \mathbf{k}\Big),$$

or equivalently, $p_{i_0}\left(\sum_{\mathbf{k}} \lambda_{\mathbf{k}} \mathbf{k}\right) - \sum_{\mathbf{k}} \lambda_{\mathbf{k}} p_{i_0}(\mathbf{k}) \geqslant 1$. Because this is a contradiction to our assumption, we have that $p_i(\mathbf{x}) \leqslant 0$ for all i. Hence, $\mathbf{x} \in K_I$. This proves the claim.

The next example illustrates the statement of Theorem 15.7.

Example 15.1. Let $p \in \mathbb{Z}[\mathbf{x}]$, $\mathbf{x} \mapsto p(\mathbf{x}) := 3x_1^2 + 2x_2^2 - 19$. We consider the semi-algebraic set

$$K = \{\mathbf{x} \in \mathbb{R}^2 \mid p(\mathbf{x}) \leqslant 0,\ 0 \leqslant x_1 \leqslant 3,\ 0 \leqslant x_2 \leqslant 3\} \text{ and } K_I = K \cap \mathbb{Z}^2.$$

It turns out that the convex hull of K_I is described by $x_1 + x_2 \leqslant 3$, $0 \leqslant x_1 \leqslant 2$ and $0 \leqslant x_2$. Notice that the polynomial p is convex. This condition ensures that p satisfies Equation (15.6) of Theorem 15.7. We obtain in this case

$$\begin{aligned}\left(P_p \cap \{(\mathbf{x}, \pi) \in \mathbb{R}^3 : \pi \leqslant 0\}\right)_{\mathbf{x}} = \{\, \mathbf{x} \in \mathbb{R}^2 :\ & 9x_1 + 6x_2 \leqslant 29, \\ & 3x_1 + 10x_2 \leqslant 31, \\ & 9x_1 + 10x_2 \leqslant 37, \\ & 15x_1 + 2x_2 \leqslant 37, \\ & 15x_1 + 6x_2 \leqslant 41, \\ & -x_1 \leqslant 0,\ 0 \leqslant x_2 \leqslant 3\}.\end{aligned}$$

The sets K, $\operatorname{conv}(K_I)$ and $\left(P_p \cap \{(\mathbf{x},\pi) \in \mathbb{R}^3 : \pi \leqslant 0\}\right)_{\mathbf{x}}$ are illustrated in Figure 15.2. Note that here $K_I = \left(P_p \cap \{(\mathbf{x},\pi) : \pi \leqslant 0\}\right)_{\mathbf{x}} \cap \mathbb{Z}^2$.

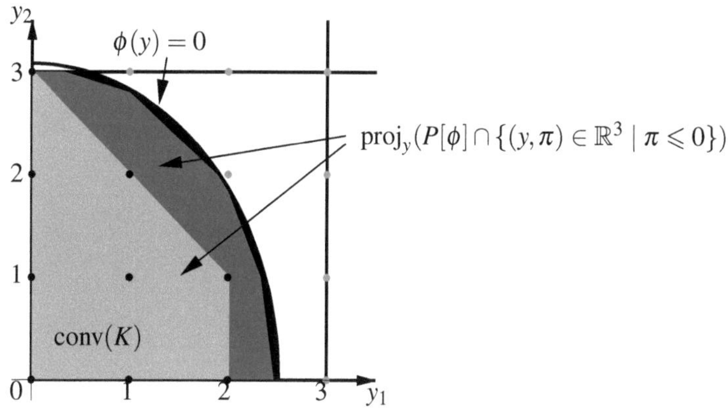

Fig. 15.2 Illustration of Theorem 15.7 in Example 15.1.

Next we introduce a large class of nonlinear functions for which one can ensure that equality holds in Formula (15.5).

Definition 15.2 (Integer-Convex Polynomials). A polynomial $f \in \mathbb{Z}[\mathbf{x}]$ is said to be *integer-convex* on $[\mathbf{l},\mathbf{u}] \subseteq \mathbb{R}^n$, if for any finite subset of nonnegative real numbers $\{\lambda_{\mathbf{k}}\}_{\mathbf{k}\in[\mathbf{l},\mathbf{u}]\cap\mathbb{Z}^n} \subseteq \mathbb{R}_+$ with $\sum_{\mathbf{k}\in[\mathbf{l},\mathbf{u}]\cap\mathbb{Z}^n} \lambda_{\mathbf{k}} = 1$ and $\sum_{\mathbf{k}\in[\mathbf{l},\mathbf{u}]\cap\mathbb{Z}^n} \lambda_{\mathbf{k}}\mathbf{k} \in [\mathbf{l},\mathbf{u}] \cap \mathbb{Z}^n$, the following inequality holds:

$$f\Big(\sum_{\mathbf{k}\in[\mathbf{l},\mathbf{u}]\cap\mathbb{Z}^n} \lambda_{\mathbf{k}}\mathbf{k}\Big) \leqslant \sum_{\mathbf{k}\in[\mathbf{l},\mathbf{u}]\cap\mathbb{Z}^n} \lambda_{\mathbf{k}} f(\mathbf{k}). \tag{15.7}$$

If (15.7) holds strictly for all $\{\lambda_{\mathbf{k}}\}_{\mathbf{k}\in[\mathbf{l},\mathbf{u}]\cap\mathbb{Z}^n} \subseteq \mathbb{R}_+$, and $\mathbf{x} \in [\mathbf{l},\mathbf{u}] \cap \mathbb{Z}^n$ such that $\sum_{\mathbf{k}} \lambda_{\mathbf{k}} = 1$, $\mathbf{x} = \sum_{\mathbf{k}} \lambda_{\mathbf{k}}\mathbf{k}$, and $\lambda_{\mathbf{x}} < 1$, then the polynomial f is called *strictly integer-convex* on $[\mathbf{l},\mathbf{u}]$.

By definition, a (strictly) convex polynomial is (strictly) integer-convex. Conversely, a (strictly) integer-convex polynomial is not necessarily (strictly) convex. Figure 15.3 gives an example.

Integer convexity is inherited under taking conic combinations and applying a composition rule.

(a) For any finite number of integer-convex polynomials $f_s \in \mathbb{Z}[\mathbf{x}]$, $s \in \{1,...,t\}$, on $[\mathbf{l},\mathbf{u}]$, and nonnegative integers $a_s \in \mathbb{Z}_+$, $s \in \{1,\ldots,t\}$, the polynomial $f \in \mathbb{Z}[\mathbf{x}]$, $\mathbf{x} \mapsto f(\mathbf{x}) := \sum_{s=1}^t a_s f_s(\mathbf{x})$, is integer-convex on $[\mathbf{l},\mathbf{u}]$.
(b) Let $\mathbf{l},\mathbf{u} \in \mathbb{Z}^n$ and let $h \in \mathbb{Z}[\mathbf{x}]$, $\mathbf{x} \mapsto h(\mathbf{x}) := \mathbf{c}^\top\mathbf{x} + \gamma$, be a linear function. Setting $W = \left\{\mathbf{c}^\top\mathbf{x} + \gamma : \mathbf{x} \in [\mathbf{l},\mathbf{u}]\right\}$, for every integer-convex univariate polynomial $q \in \mathbb{Z}[\mathbf{w}]$, the function $p \in \mathbb{Z}[\mathbf{x}]$, $\mathbf{x} \mapsto p(\mathbf{x}) := q(h(\mathbf{x}))$ is integer-convex on $[\mathbf{l},\mathbf{u}]$.

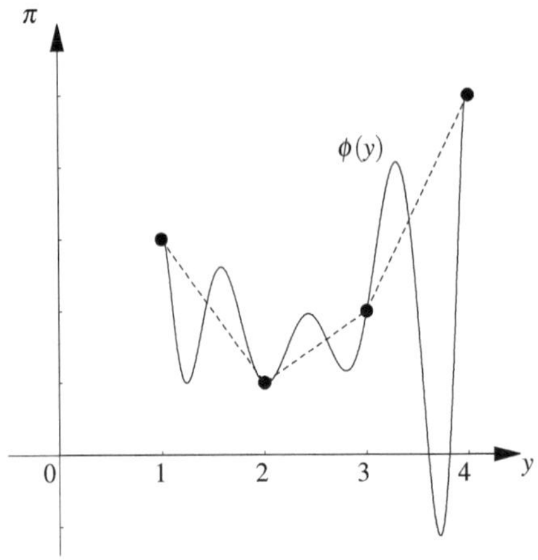

Fig. 15.3 The graph of an integer-convex polynomial p on $[1,4]$.

Indeed, integer-convex polynomial functions capture a lot of combinatorial structure. In particular, we can characterize the set of all vertices in an associated polyhedron. Most importantly, if f is integer-convex on $[\mathbf{l},\mathbf{u}]$, then this ensures that for any integer point $\mathbf{x} \in [\mathbf{l},\mathbf{u}]$ the value $f(\mathbf{x})$ is not underestimated by all $\pi \in \mathbb{R}$ with $(\mathbf{x},\pi) \in P_f$, where P_f is the polytope associated with the graph of the polynomial $f \in \mathbb{Z}[\mathbf{x}]$.

Theorem 15.8. *For a polynomial $f \in \mathbb{Z}[\mathbf{x}]$ and $\mathbf{l},\mathbf{u} \in \mathbb{Z}^n$, $\mathbf{l}+\mathbf{1} < \mathbf{u}$, let*

$$P_f = \operatorname{conv}(\{(\mathbf{x}, f(\mathbf{x})) \in \mathbb{Z}^{n+1} \mid \mathbf{x} \in [\mathbf{l},\mathbf{u}] \cap \mathbb{Z}^n\}).$$

Then f to be integer-convex on $[\mathbf{l},\mathbf{u}]$ is equivalent to the condition that for all $(\mathbf{x},\pi) \in P_f$, $\mathbf{x} \in \mathbb{Z}^n$ we have that $f(\mathbf{x}) \leqslant \pi$. Moreover, if f is strictly integer-convex on $[\mathbf{l},\mathbf{u}]$, then for every $\mathbf{x} \in [\mathbf{l},\mathbf{u}] \cap \mathbb{Z}^n$, the point $(\mathbf{x}, f(\mathbf{x}))$ is a vertex of P_f.

Proof. First let us assume that f is integer-convex on $[\mathbf{l},\mathbf{u}]$. Let $(\mathbf{x},\pi) \in P_f$ such that $\mathbf{x} \in \mathbb{Z}^n$. Then, there exist nonnegative real numbers $\{\lambda_\mathbf{k}\}_{\mathbf{k}\in[\mathbf{l},\mathbf{u}]\cap\mathbb{Z}^n} \subseteq \mathbb{R}_+$, $\sum_\mathbf{k} \lambda_\mathbf{k} = 1$, such that $(\mathbf{x},\pi) = \sum_\mathbf{k} \lambda_\mathbf{k} (\mathbf{k}, f(\mathbf{k}))$. It follows that

$$f(\mathbf{x}) = f\Big(\sum_\mathbf{k} \lambda_\mathbf{k}\mathbf{k}\Big) \leqslant \sum_\mathbf{k} \lambda_\mathbf{k} f(\mathbf{k}) = \pi.$$

Next we assume that f is not integer-convex on $[\mathbf{l},\mathbf{u}]$. Then, there exists a subset of nonnegative real numbers $\{\lambda_\mathbf{k}\}_{\mathbf{k}\in[\mathbf{l},\mathbf{u}]\cap\mathbb{Z}^n} \subseteq \mathbb{R}_+$ with $\sum_\mathbf{k} \lambda_\mathbf{k} = 1$ such that

$$\mathbf{x} := \sum_\mathbf{k} \lambda_\mathbf{k}\mathbf{k} \in [\mathbf{l},\mathbf{u}] \cap \mathbb{Z}^n \text{ and } \pi := \sum_\mathbf{k} \lambda_\mathbf{k} f(\mathbf{k}) < f\Big(\sum_\mathbf{k} \lambda_\mathbf{k}\mathbf{k}\Big) = f(\mathbf{x}).$$

But then, $(\mathbf{x},\pi) = \sum_\mathbf{k} \lambda_\mathbf{k}(\mathbf{k}, f(\mathbf{k})) \in P_f$ violates the inequality $f(\mathbf{x}) \leqslant \pi$. This is a contradiction to the assumption.

If f is strictly integer-convex on $[\mathbf{l},\mathbf{u}]$, then for each $\mathbf{x} \in [\mathbf{l},\mathbf{u}] \cap \mathbb{Z}^n$, we have that

$$f(\mathbf{x}) < \sum_{\mathbf{k}\in[\mathbf{l},\mathbf{u}]\cap\mathbb{Z}^n\setminus\{\mathbf{x}\}} \lambda_{\mathbf{k}} f(\mathbf{k}),$$

for all $\lambda_{\mathbf{k}} \in \mathbb{R}_+$, $\mathbf{k} \in [\mathbf{l},\mathbf{u}] \cap \mathbb{Z}^n \setminus \{\mathbf{x}\}$, with $\mathbf{x} = \sum_{\mathbf{k}} \lambda_{\mathbf{k}}\mathbf{k}$ and $\sum_{\mathbf{k}} \lambda_{\mathbf{k}} = 1$. Thus, every point $\big(\mathbf{x}, f(\mathbf{x})\big)$, $\mathbf{x} \in [\mathbf{l},\mathbf{u}] \cap \mathbb{Z}^n$, is a vertex of P_f.

15.3 Convex integer minimization

The complexity of the case of convex integer minimization is set apart from the general case of integer polynomial optimization by the existence of bounding results for the coordinates of optimal solutions. Once a finite bound can be computed, it is clear that an algorithm for minimization exists. Thus the fundamental incomputability result for integer polynomial optimization (Theorem 15.3) does not apply to the case of convex integer minimization.

The first bounds for the optimal integer solutions to convex minimization problems were proved by [79, 126]. We present the sharpened bound that was obtained by [12, 11] for the more general case of quasi-convex polynomials. This bound is a consequence of an efficient theory of quantifier elimination over the reals; see [111].

Theorem 15.9. *Let $f, g_1, \dots, g_m \in \mathbb{Z}[x_1, \dots, x_n]$ be quasi-convex polynomials of degree at most $d \geqslant 2$, whose coefficients have a binary encoding length of at most ℓ. Let*

$$F = \{\mathbf{x} \in \mathbb{R}^n : g_i(\mathbf{x}) \leqslant 0 \quad \text{for } i = 1, \dots, m\}$$

be the (continuous) feasible region. If the integer minimization problem $\min\{f(\mathbf{x}) : \mathbf{x} \in F \cap \mathbb{Z}^n\}$ *is bounded, there exists a radius $R \in \mathbb{Z}_+$ of binary encoding length at most $(md)^{O(n)}\ell$ such that*

$$\min\{f(\mathbf{x}) : \mathbf{x} \in F \cap \mathbb{Z}^n\} = \min\{f(\mathbf{x}) : \mathbf{x} \in F \cap \mathbb{Z}^n,\ \|\mathbf{x}\| \leqslant R\}.$$

15.3.1 Fixed dimension

In fixed dimension, the problem of convex integer minimization can be solved using variants of Lenstra's algorithm [88] for integer programming. Indeed, when the dimension n is fixed, the bound R given by Theorem 15.9 has a binary encoding size that is bounded polynomially by the input data. Thus, a Lenstra-type algorithm can be started with a "small" (polynomial-size) initial outer ellipsoid that includes a bounded part of the feasible region containing an optimal integer solution.

The first algorithm of this kind for convex integer minimization was announced by Khachiyan [79]. In the following we present the variant of Lenstra's algorithm due to Heinz [67], which seems to yield the best known complexity bound for the problem. The complexity result is the following.

Theorem 15.10. *Let $f, g_1, \dots, g_m \in \mathbb{Z}[x_1, \dots, x_n]$ be quasi-convex polynomials of degree at most $d \geqslant 2$, whose coefficients have a binary encoding length of at most ℓ. There exists an algorithm running in time $m\ell^{O(1)} d^{O(n)} 2^{O(n^3)}$ that computes a minimizer $\mathbf{x}^* \in \mathbb{Z}^n$ of the problem* (15.1) *or reports that no minimizer exists. If the algorithm outputs a minimizer $\mathbf{x}^*$, its binary encoding size is $\ell d^{O(n)}$.*

A complexity result of greater generality was presented by Khachiyan and Porkolab [80]. It covers the case of minimization of convex polynomials over the integer points in convex semialgebraic sets given by *arbitrary* (not necessarily quasi-convex) polynomials.

Theorem 15.11. *Let $Y \subseteq \mathbb{R}^k$ be a convex set given by*

$$Y = \{\mathbf{y} \in \mathbb{R}^k : Q_1 \mathbf{x}^1 \in \mathbb{R}^{n_1} : \ \cdots \ Q_\omega \mathbf{x}^\omega \in \mathbb{R}^{n_\omega} : P(\mathbf{y}, \mathbf{x}^1, \dots, \mathbf{x}^\omega)\}$$

with quantifiers $Q_i \in \{\exists, \forall\}$, where P is a Boolean combination of polynomial inequalities

$$g_i(\mathbf{y}, \mathbf{x}^1, \dots, \mathbf{x}^\omega) \leqslant 0, \quad i = 1, \dots, m$$

with degrees at most $d \geqslant 2$ and coefficients of binary encoding size at most ℓ. There exists an algorithm for solving the problem $\min\{y_k : \mathbf{y} \in Y \cap \mathbb{Z}^k\}$ in time $\ell^{O(1)}(md)^{O(k^4)\prod_{i=1}^{\omega} O(n_i)}$.

When the dimension $k + \sum_{i=1}^{\omega} n_i$ is fixed, the algorithm runs in polynomial time. For the case of convex minimization where the feasible region is described by convex polynomials, the complexity bound of Theorem 15.11, however, translates to $\ell^{O(1)} m^{O(n^2)} d^{O(n^4)}$, which is worse than the bound of Theorem 15.10 [67].

In the remainder of this subsection, we describe the ingredients of the variant of Lenstra's algorithm due to Heinz. The algorithm starts out by "rounding" the feasible region, by applying the shallow-cut ellipsoid method to find proportional inscribed and circumscribed ellipsoids. It is well-known [61] that the shallow-cut ellipsoid method only needs an initial circumscribed ellipsoid that is "small enough" (of polynomial binary encoding size – this follows from Theorem 15.9) and an implementation of a *shallow separation oracle*, which we describe below.

For a positive-definite matrix A we denote by $\mathscr{E}(A, \hat{\mathbf{x}})$ the ellipsoid $\{\mathbf{x} \in \mathbb{R}^n : (\mathbf{x} - \hat{\mathbf{x}})^\top A (\mathbf{x} - \hat{\mathbf{x}}) \leqslant 1\}$.

Lemma 15.3 (Shallow separation oracle). *Let $g_0, \dots, g_{m+1} \in \mathbb{Z}[\mathbf{x}]$ be quasi-convex polynomials of degree at most d, the binary encoding sizes of whose coefficients are at most r. Let the (continuous) feasible region $F = \{\mathbf{x} \in \mathbb{R}^n : g_i(\mathbf{x}) < 0\}$ be contained in the ellipsoid $\mathscr{E}(A, \hat{\mathbf{x}})$, where A and $\hat{\mathbf{x}}$ have binary encoding size at most ℓ. There exists an algorithm with running time $m(lnr)^{O(1)} d^{O(n)}$ that outputs*

(a) *"true" if*

$$\mathscr{E}((n+1)^{-3} A, \hat{\mathbf{x}}) \subseteq F \subseteq \mathscr{E}(A, \hat{\mathbf{x}}); \tag{15.8}$$

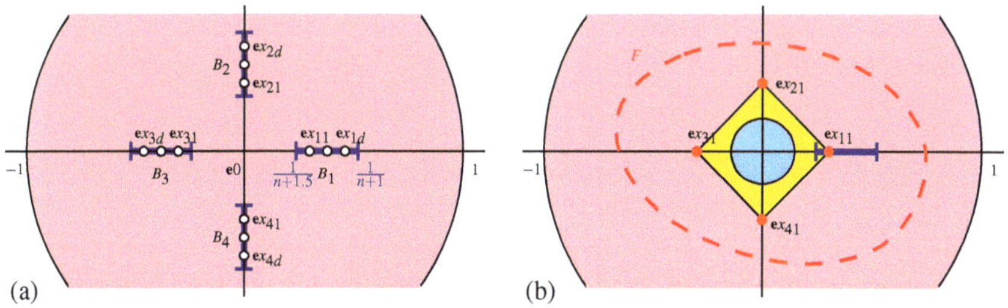

Fig. 15.4 The implementation of the shallow separation oracle. **(a)** Test points $\mathbf{x}_{ij}$ in the circumscribed ball $\mathscr{E}(1,\mathbf{0})$. **(b)** Case I: All test points $\mathbf{x}_{i1}$ are (continuously) feasible; so their convex hull (a cross-polytope) and its inscribed ball $\mathscr{E}((n+1)^{-3},\mathbf{0})$ are contained in the (continuous) feasible region F.

(b) *otherwise, a vector* $\mathbf{c} \in \mathbb{Q}^n \setminus \{\mathbf{0}\}$ *of binary encoding length* $(l+r)(dn)^{\mathrm{O}(1)}$ *with*

$$F \subseteq \mathscr{E}(A,\hat{\mathbf{x}}) \cap \left\{ \mathbf{x} \in \mathbb{R}^n : \mathbf{c}^\top(\mathbf{x}-\hat{\mathbf{x}}) \leqslant \tfrac{1}{n+1}(\mathbf{c}^\top A\mathbf{c})^{1/2} \right\}. \tag{15.9}$$

Proof. We give a simplified sketch of the proof, without hard complexity estimates. By applying an affine transformation to $F \subseteq \mathscr{E}(A,\hat{\mathbf{x}})$, we can assume that F is contained in the unit ball $\mathscr{E}(I,\mathbf{0})$. Let us denote as usual by $\mathbf{e}_1,\dots,\mathbf{e}_n$ the unit vectors and by $\mathbf{e}_{n+1},\dots,\mathbf{e}_{2n}$ their negatives. The algorithm first constructs numbers $\lambda_{i1},\dots,\lambda_{id} > 0$ with

$$\frac{1}{n+\frac{3}{2}} < \lambda_{i1} < \cdots < \lambda_{id} < \frac{1}{n+1} \tag{15.10}$$

and the corresponding point sets $B_i = \{\mathbf{x}_{ij} := \lambda_{ij}\mathbf{e}_i : j = 1,\dots,d\}$, (Figure 15.4 (a)). The choice of the bounds (15.10) for λ_{ij} will ensure that we either find a large enough inscribed ball for (a) or a deep enough cut for (b). Then the algorithm determines the (continuous) feasibility of the center $\mathbf{0}$ and the $2n$ innermost points $\mathbf{x}_{i,1}$.

Case I. If $\mathbf{x}_{i,1} \in F$, $i = 1,\dots,2n$, then the cross-polytope $\operatorname{conv}\{\mathbf{x}_{i,1} : i = 1,\dots,2n\}$ is contained in F; see Figure 15.4 (b). An easy calculation shows that the ball $\mathscr{E}((n+1)^{-3},\mathbf{0})$ is contained in the cross-polytope and thus in F; see Figure 15.4. Hence the condition in (a) is satisfied and the algorithm outputs "true".

Case II. We now discuss the case when the center $\mathbf{0}$ violates a polynomial inequality $g_0(\mathbf{x}) < 0$ (say). Let $F_0 = \{\mathbf{x} \in \mathbb{R}^n : g_0(\mathbf{x}) < 0\} \supseteq F$. Due to convexity of F_0, for all $i = 1,\dots,n$, one set of each pair $B_i \cap F_0$ and $B_{n+i} \cap F_0$ must be empty; see Figure 15.5 (a). Without loss of generality, let us assume $B_{n+i} \cap F_0 = \emptyset$ for all i. We can determine whether an n-variate polynomial function of known maximum degree d is constant by evaluating it on $(d+1)^n$ suitable points (this is a consequence of the Fundamental Theorem of Algebra). For our case of quasi-convex polynomials, this can be improved; indeed, it suffices to test whether the gradient ∇g_0 vanishes on the nd points in the set $B_1 \cup \cdots \cup B_n$. If it does, we know that g_0 is constant, thus $F = \emptyset$, and so can we return an arbitrary vector $\mathbf{c}$. Otherwise, there is a point $\mathbf{x}_{ij} \in B_i$ with $\mathbf{c} := \nabla g_0(\mathbf{x}_{ij}) \neq \mathbf{0}$; we return this vector as the desired normal vector of a shallow

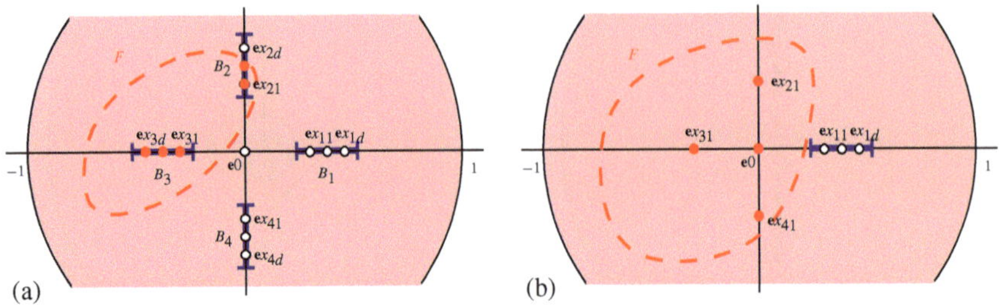

Fig. 15.5 The implementation of the shallow separation oracle. **(a)** Case II: The center $\mathbf{0}$ violates a polynomial inequality $g_0(\mathbf{x}) < 0$ (say). Due to convexity, for all $i = 1, \ldots, n$, one set of each pair $B_i \cap F$ and $B_{n+i} \cap F$ must be empty. **(b)** Case III: A test point $\mathbf{x}_{k1}$ is infeasible, as it violates an inequality $g_0(\mathbf{x}) < 0$ (say). However, the center $\mathbf{0}$ is feasible at least for this inequality.

cut. Due to the choice of λ_{ij} as a number smaller than $\frac{1}{n+1}$, the cut is deep enough into the ellipsoid $\mathscr{E}(A, \hat{\mathbf{x}})$, so that (15.9) holds.

Case III. The remaining case to discuss is when $\mathbf{0} \in F$ but there exists a $k \in \{1, \ldots, 2n\}$ with $\mathbf{x}_{k,1} \notin F$. Without loss of generality, let $k = 1$, and let $\mathbf{x}_{1,1}$ violate the polynomial inequality $g_0(\mathbf{x}) < 0$, i.e., $g_0(\mathbf{x}_{1,1}) \geqslant 0$; see Figure 15.5 (b). We consider the univariate polynomial $\phi(\lambda) = g_0(\lambda \mathbf{e}_i)$. We have $\phi(0) = g_0(\mathbf{0}) < 0$ and $\phi(\lambda_{1,1}) \geqslant 0$, so ϕ is not constant. Because ϕ has degree at most d, its derivative ϕ' has degree at most $d-1$, so ϕ' has at most $d-1$ roots. Thus, for at least one of the d different values $\lambda_{1,1}, \ldots, \lambda_{1,d}$, say $\lambda_{1,j}$, we must have $\phi'(\lambda_{1,j}) \neq 0$. This implies that $\mathbf{c} := \nabla g_0(\mathbf{x}_{1,j}) \neq \mathbf{0}$. By convexity, we have $\mathbf{x}_{1,j} \notin F$, so we can use $\mathbf{c}$ as the normal vector of a shallow cut.

By using this oracle in the shallow-cut ellipsoid method, one obtains the following result.

Corollary 15.2. *Let $g_0, \ldots, g_m \in \mathbb{Z}[\mathbf{x}]$ be quasi-convex polynomials of degree at most $d \geqslant 2$. Let the (continuous) feasible region $F = \{\mathbf{x} \in \mathbb{R}^n : g_i(\mathbf{x}) \leqslant 0\}$ be contained in the ellipsoid $\mathscr{E}(A_0, \mathbf{0})$, given by the positive-definite matrix $A_0 \in \mathbb{Q}^{n\times n}$. Let $\varepsilon \in \mathbb{Q}_{>0}$ be given. Let the entries of A_0 and the coefficients of all monomials of $g_0, \ldots, g_m$ have binary encoding size at most ℓ.*

There exists an algorithm with running time $m(\ell n)^{O(1)} d^{O(n)}$ that computes a positive-definite matrix $A \in \mathbb{Q}^{n\times n}$ and a point $\hat{\mathbf{x}} \in \mathbb{Q}^n$ with

(a) *either* $\mathscr{E}((n+1)^{-3}A, \hat{\mathbf{x}}) \subseteq F \subseteq \mathscr{E}(A, \hat{\mathbf{x}})$
(b) *or* $F \subseteq \mathscr{E}(A, \hat{\mathbf{x}})$ *and* $\operatorname{vol} \mathscr{E}(A, \hat{\mathbf{x}}) < \varepsilon$.

Finally, there is a lower bound for the volume of a continuous feasible region F that can contain an integer point.

Lemma 15.4. *Under the assumptions of Theorem 15.2, if $F \cap \mathbb{Z}^n \neq \emptyset$, then there exists an $\varepsilon \in \mathbb{Q}_{>0}$ of binary encoding size $\ell(dn)^{O(1)}$ with* $\operatorname{vol} F > \varepsilon$.

On the basis of these results, one obtains a Lenstra-type algorithm for the decision version of the convex integer minimization problem with the desired complexity. By applying binary search, the optimization problem can be solved, which provides a proof of Theorem 15.10.

15.3.2 Boundary cases of complexity

In this section we present an optimality certificate for problems of the form

$$\min\{f(\mathbf{x}) : A\mathbf{x} = \mathbf{b}, \mathbf{l} \leqslant \mathbf{x} \leqslant \mathbf{u}, \mathbf{x} \in \mathbb{Z}^n\},$$

where $A \in \mathbb{Z}^{d\times n}$, $\mathbf{b} \in \mathbb{Z}^d$, $\mathbf{l}, \mathbf{u} \in \mathbb{Z}^n$, and where $f : \mathbb{R}^n \to \mathbb{R}$ is a separable convex function, that is, $f(\mathbf{x}) = \sum_{i=1}^{n} f_i(x_i)$ with convex functions $f_i : \mathbb{R} \to \mathbb{R}$, $i = 1, \ldots, n$. This certificate then immediately leads us to a oracle-polynomial time algorithm to solve the separable convex integer minimization problem at hand. Applied to separable convex n-fold integer minimization problems, this gives a polynomial time algorithm for their solution [70].

For the construction of the optimality certificate, we exploit a nice super-additivity property of separable convex functions.

Lemma 15.5. *Let $f : \mathbb{R}^n \to \mathbb{R}$ be a separable convex function and let $\mathbf{h}_1, \ldots \mathbf{h}_k \in \mathbb{R}^n$ belong to a common orthant of $\mathbb{R}^n$, that is, they all have the same sign pattern from $\{\geqslant 0, \leqslant 0\}^n$. Then, for any $\mathbf{x} \in \mathbb{R}^n$ we have*

$$f\Big(\mathbf{x} + \sum_{i=1}^{k} \mathbf{h}_i\Big) - f(\mathbf{x}) \geqslant \sum_{i=1}^{k} [f(\mathbf{x} + \mathbf{h}_i) - f(\mathbf{x})].$$

Proof. The claim is easy to show for $n = 1$ by induction. If, w.l.o.g., $h_2 \geqslant h_1 \geqslant 0$ then convexity of f implies $[f(x+h_1+h_2) - f(x+h_2)]/h_1 \geqslant [f(x+h_1) - f(x)]/h_1$, and thus $f(x+h_1+h_2) - f(x) \geqslant [f(x+h_2) - f(x)] + [f(x+h_1) - f(x)]$. The claim for general n then follows from the separability of f by adding the superadditivity relations of each one-parametric convex summand of f.

A crucial role in the following theorem is again played by the Graver basis $\mathscr{G}(A)$ of A. Let us remind the reader that the Graver basis $\mathscr{G}(A)$ has a nice representation property due to its definition: every $\mathbf{z} \in \ker(A) \cap \mathbb{Z}^n$ can be written as a sign-compatible nonnegative integer linear combination $\mathbf{z} = \sum_i \alpha_i \mathbf{g}_i$ of Graver basis elements $\mathbf{g}_i \in \mathscr{G}(A)$. This followed from the simple observation that $\mathbf{z}$ has to belong to some orthant $\mathscr{O}_j$ of $\mathbb{R}^n$ and thus it can be represented as a sign-compatible nonnegative integer linear combination of elements in $H_j \subseteq \mathscr{G}(A)$ belonging to this orthant. Note that by the integer Carathéodory property of Hilbert bases, at most $2 \cdot \dim(\ker(A)) - 2$ vectors are needed in such a representation [119]. It is precisely this simple representation property of $\mathscr{G}(A)$ combined with the superadditivity of the separable convex function f that turns $\mathscr{G}(A)$ into an optimality certificate for $\min\{f(\mathbf{x}) : A\mathbf{x} = \mathbf{b}, \mathbf{l} \leqslant \mathbf{x} \leqslant \mathbf{u}, \mathbf{x} \in \mathbb{Z}^n\}$.

Theorem 15.12. *Let $f : \mathbb{R}^n \to \mathbb{R}$ be a separable convex function given by a comparison oracle that when queried on $\mathbf{x}, \mathbf{y} \in \mathbb{Z}^n$ decides whether $f(\mathbf{x}) < f(\mathbf{y})$, $f(\mathbf{x}) = f(\mathbf{y})$, or $f(\mathbf{x}) > f(\mathbf{y})$. Then $\mathbf{x}_0$ is an optimal feasible solution to $\min\{f(\mathbf{x}) : A\mathbf{x} = \mathbf{b}, \mathbf{l} \leqslant \mathbf{x} \leqslant \mathbf{u}, \mathbf{x} \in \mathbb{Z}^n\}$ if and only if for all $\mathbf{g} \in \mathscr{G}(A)$ the vector $\mathbf{x}_0 + \mathbf{g}$ is not feasible or $f(\mathbf{x}_0 + \mathbf{g}) \geqslant f(\mathbf{x}_0)$.*

Proof. Assume that $\mathbf{x}_0$ is not optimal and let $\mathbf{x}_{\min}$ be an optimal solution to the given problem. Then $\mathbf{x}_{\min} - \mathbf{x}_0 \in \ker(A)$ and thus it can be written as a sign-compatible nonnegative integer linear combination $\mathbf{x}_{\min} - \mathbf{x}_0 = \sum_i \alpha_i \mathbf{g}_i$ of Graver basis elements $\mathbf{g}_i \in \mathscr{G}(A)$. We show that one of the $\mathbf{g}_i$ must be an improving vector, that is, for some $\mathbf{g}_i$ we have that $\mathbf{x}_0 + \mathbf{g}_i$ is feasible and $f(\mathbf{x}_0 + \mathbf{g}_i) < f(\mathbf{x}_0)$.

For all i, the vector $\mathbf{g}_i$ has the same sign-pattern as $\mathbf{x}_{\min} - \mathbf{x}_0$ and it is now easy to check that the coordinates of $\mathbf{x}_0 + \mathbf{g}_i$ lie between the corresponding coordinates of $\mathbf{x}_0$ and $\mathbf{x}_{\min}$. This implies in particular $\mathbf{l} \leqslant \mathbf{x}_0 + \mathbf{g}_i \leqslant \mathbf{u}$. Because $\mathbf{g}_i \in \ker(A)$, we also have $A(\mathbf{x}_0 + \mathbf{g}_i) = \mathbf{b}$ for all i. Consequently, for all i the vector $\mathbf{x}_0 + \mathbf{g}_i$ would be a feasible solution. It remains to show that one of these vectors has a strictly smaller objective value than $\mathbf{x}_0$.

Due to the superadditivity from Lemma 15.5, we have

$$0 \geqslant f(\mathbf{x}_{\min}) - f(\mathbf{x}_0) = f\Big(\mathbf{x}_0 + \sum_{i=1}^{2n-2} \alpha_i \mathbf{g}_i\Big) - f(\mathbf{x}_0) \geqslant \sum_{i=1}^{k} \alpha_i [f(\mathbf{x}_0 + \mathbf{g}_i) - f(\mathbf{x}_0)].$$

Thus, at least one of the summands $f(\mathbf{x}_0 + \mathbf{g}_i) - f(\mathbf{x}_0)$ must be negative and we have found an improving vector for $\mathbf{z}_0$ in $\mathscr{G}(A)$.

We now turn this optimality certificate into a polynomial oracle-time algorithm to solve the separable convex integer minimization problem $\min\{f(\mathbf{x}) : A\mathbf{x} = \mathbf{b}, \mathbf{l} \leqslant \mathbf{x} \leqslant \mathbf{u}, \mathbf{x} \in \mathbb{Z}^n\}$. For this, we call $\alpha\mathbf{g}$ a *greedy* Graver improving vector if $\mathbf{x}_0 + \alpha\mathbf{g}$ is feasible and such that $f(\mathbf{x}_0 + \alpha\mathbf{g})$ is minimal among all such choices of $\alpha \in \mathbb{Z}_+$ and $\mathbf{g} \in \mathscr{G}(A)$. Then the following result holds.

Theorem 15.13. *Let $f : \mathbb{R}^n \to \mathbb{R}$ be a separable convex function given by a comparison oracle and assume that $|f(\mathbf{x})| < M$ for all $\mathbf{x} \in \{\mathbf{x} : A\mathbf{x} = \mathbf{b}, \mathbf{l} \leqslant \mathbf{x} \leqslant \mathbf{u}, \mathbf{x} \in \mathbb{Z}^n\}$. Then any feasible solution $\mathbf{x}_0$ to $\min\{f(\mathbf{x}) : A\mathbf{x} = \mathbf{b}, \mathbf{l} \leqslant \mathbf{x} \leqslant \mathbf{u}, \mathbf{x} \in \mathbb{Z}^n\}$ can be augmented to optimality by a number of greedy Graver augmentation steps that is polynomially bounded in the encoding lengths of A, $\mathbf{b}$, $\mathbf{l}$, $\mathbf{u}$, M, and $\mathbf{x}_0$.*

Proof. Assume that $\mathbf{x}_0$ is not optimal and let $\mathbf{x}_{\min}$ be an optimal solution to the given problem. Then $\mathbf{x}_{\min} - \mathbf{x}_0 \in \ker(A)$ and thus it can be written as a sign-compatible nonnegative integer linear combination $\mathbf{x}_{\min} - \mathbf{x}_0 = \sum_i \alpha_i \mathbf{g}_i$ of at most $2n-2$ Graver basis elements $\mathbf{g}_i \in \mathscr{G}(A)$. As in the proof of Theorem 15.12, sign-compatibility implies that for all i the coordinates of $\mathbf{x}_0 + \alpha_i \mathbf{g}_i$ lie between the corresponding coordinates of $\mathbf{x}_0$ and $\mathbf{x}_{\min}$. Consequently, we have $\mathbf{l} \leqslant \mathbf{x}_0 + \alpha_i \mathbf{g}_i \leqslant \mathbf{u}$. Because $\mathbf{g}_i \in \ker(A)$, we also have $A(\mathbf{x}_0 + \alpha_i \mathbf{g}_i) = \mathbf{b}$ for all i. Consequently, for all i the vector $\mathbf{x}_0 + \alpha_i \mathbf{g}_i$ would be a feasible solution.

Due to the superadditivity from Lemma 15.5, we have

$$0 \geqslant f(\mathbf{x}_{\min}) - f(\mathbf{x}_0) = f\Big(\mathbf{x}_0 + \sum_{i=1}^{2n-2} \alpha_i \mathbf{g}_i\Big) - f(\mathbf{x}_0) \geqslant \sum_{i=1}^{k} [f(\mathbf{x}_0 + \alpha_i \mathbf{g}_i) - f(\mathbf{x}_0)].$$

Thus, at least one of the summands $f(\mathbf{x}_0 + \alpha_i \mathbf{g}_i) - f(\mathbf{x}_0)$ must be smaller than $\frac{1}{2n-2}[f(\mathbf{x}_{\min}) - f(\mathbf{x}_0)]$, giving an improvement that is at least $\frac{1}{2n-2}$ times the maximal possible improvement $f(\mathbf{x}_{\min}) - f(\mathbf{x}_0)$. Such a geometric improvement, however, implies that the optimum is reached in a number of greedy augmentation steps which is polynomial in the encoding lengths of A, $\mathbf{b}$, $\mathbf{l}$, $\mathbf{u}$, M, and $\mathbf{x}_0$ [4].

Thus, once we have a polynomial size Graver basis, we get a polynomial time algorithm to solve the convex integer minimization problem at hand.

For this, let us consider again n-fold systems (introduced in Section 15.2.2). Two nice stabilization results established by Hoşten and Sullivant [72] and Santos and Sturmfels [114] immediately imply that if A_1 and A_2 are kept fixed, then the size of the Graver basis increases only polynomially in the number n of copies of A_1 and A_2.

Proposition 15.2. *For any fixed $(r+s) \times t$ integer matrix A there is a polynomial time algorithm that, given any n, computes the Graver basis $\mathscr{G}(A^{(n)})$ of the n-fold matrix $A^{(n)} = (\mathbf{1}_n \otimes A_1) \oplus (I_n \otimes A_2)$.*

Combining this proposition with Theorem 15.13, we get the following nice result from [70].

Theorem 15.14. *Let A be a fixed integer $(r+s) \times t$ matrix and let $f : \mathbb{R}^{nt} \to \mathbb{R}$ be any separable convex function given by a comparison oracle. Then there is a polynomial time algorithm that, given n, a right-hand side vector $\mathbf{b} \in \mathbb{Z}^{r+ns}$ and some bound $|f(\mathbf{x})| < M$ on f over the feasible region, solves the n-fold convex integer programming problem*

$$\min\{f(\mathbf{x}) : \; A^{(n)}\mathbf{x} = \mathbf{b}, \; \mathbf{x} \in \mathbb{N}^{nt}\}.$$

Note that by applying an approach similar to Phase I of the simplex method one can also compute an initial feasible solution x_0 to the n-fold integer program in polynomial time based on greedy Graver basis directions [47, 68].

We wish to point out that the presented approach can be generalized to the mixed-integer situation and also to more general objective functions that satisfy a certain superadditivity/subadditivity condition, see [69, 87] for more details. Note that for mixed-integer convex problems one may only expect an approximation result, as there need not exist a rational optimum. In fact, already a mixed-integer greedy augmentation vector can be computed only approximately. Nonetheless, the technical difficulties when adjusting the proofs for the pure integer case to the mixed-integer situation can be overcome [69]. It should be noted, however, that the Graver basis of n-fold matrices does not show a stability result similar to the pure integer case as presented in [72, 114]. Thus, we do not get a nice polynomial time algorithm for solving mixed-integer convex n-fold problems.

15.3.3 Practical algorithms

In this section, the methods that we look at, aimed at formulations having convex continuous relaxations, are driven by O.R./engineering approaches, transporting and motivated by successful mixed-integer linear programming technology and smooth continuous nonlinear programming technology. In Section 15.3.3.1 we discuss general algorithms that make few assumptions beyond those that are typical for convex continuous nonlinear programming. In Section 15.3.3.2 we present some more specialized techniques aimed at convex quadratics.

15.3.3.1 General algorithms

Practical, broadly applicable approaches to general mixed-integer nonlinear programs are aimed at problems involving convex minimization over a convex set with some additional integrality restriction. Additionally, for the sake of obtaining well-behaved continuous relaxations, a certain amount of smoothness is usually assumed. Thus, in this section, the model that we focus on is

$$\begin{aligned} \min \quad & f(\mathbf{x},\mathbf{y}) \\ \text{s.t.} \quad & g(\mathbf{x},\mathbf{y}) \leqslant \mathbf{0} \\ & \mathbf{l} \leq \mathbf{y} \leq \mathbf{u} \\ & \mathbf{x} \in \mathbb{R}^{n_1},\ \mathbf{y} \in \mathbb{Z}^{n_2}, \end{aligned} \tag{P[$\mathbf{l},\mathbf{u}$]}$$

where $f\colon \mathbb{R}^n \to \mathbb{R}$ and $g\colon \mathbb{R}^n \to \mathbb{R}^m$ are twice continuously-differentiable convex functions, $\mathbf{l} \in (\mathbb{Z} \cup \{-\infty\})^{n_2}$, $\mathbf{u} \in (\mathbb{Z} \cup \{+\infty\})^{n_2}$, and $\mathbf{l} \leq \mathbf{u}$. It is also helpful to assume that the feasible region of the relaxation of (P[$\mathbf{l},\mathbf{u}$]) obtained by replacing $\mathbf{y} \in \mathbb{Z}^{n_2}$ with $\mathbf{y} \in \mathbb{R}^{n_2}$ is bounded. We denote this continuous relaxation by ($\mathrm{P}_{\mathbb{R}}$[$\mathbf{l},\mathbf{u}$]).

To describe the various algorithmic approaches, it is helpful to define some related subproblems of (P[$\mathbf{l},\mathbf{u}$]) and associated relaxations. Our notation is already designed for this. For vector $\mathbf{l}' \in (\mathbb{Z} \cup \{-\infty\})^{n_2}$ and $\mathbf{u}' \in (\mathbb{Z} \cup \{+\infty\})^{n_2}$, with $\mathbf{l} \leq \mathbf{l}' \leq \mathbf{u}' \leq \mathbf{u}$, we have the *subproblem* (P[$\mathbf{l}',\mathbf{u}'$]) and its associated continuous relaxation ($\mathrm{P}_{\mathbb{R}}$[$\mathbf{l}',\mathbf{u}'$]).

Already, we can see how the family of relaxations ($\mathrm{P}_{\mathbb{R}}$[$\mathbf{l}',\mathbf{u}'$]) leads to the obvious extension of the Branch-and-Bound Algorithm of mixed-integer linear programming. Indeed, this approach was experimented with in [64]. The *Branch-and-Bound Algorithm* for mixed-integer nonlinear programming has been implemented as `MINLP-BB` [89], with continuous nonlinear-programming subproblem relaxations solved with the active-set solver `filterSQP` and also as `SBB`, with associated subproblems solved with any of CONOPT, SNOPT and MINOS. Moreover, Branch-and-Bound is available as an algorithmic option in the actively developed code Bonmin [22, 24, 26], which can be used as a callable library, as a stand-alone solver, via the modeling languages AMPL and GAMS, and is available in source-code form, under the Common Public License, from COIN-OR [29], available for running on NEOS [28]. By default, relaxations of subproblems are solved with the interior-point solver Ipopt (whose availability options include all of those for Bonmin), though there is also an interface to filterSQP. The Branch-and-Bound Algorithm in Bonmin includes effective strong branching and SOS branching. It can also be used as a robust heuristic on problems for which the relaxation $(\mathrm{P}_{\mathbb{R}})$ does not have a convex feasible region, by setting negative "cutoff gaps".

Another type of algorithmic approach emphasizes continuous nonlinear programming over just the continuous variables of the formulation. For fixed $\bar{\mathbf{y}} \in \mathbb{Z}^{n_2}$, we define

$$\begin{array}{rl} \min & f(\mathbf{x},\mathbf{y}) \\ \text{s.t.} & g(\mathbf{x},\mathbf{y}) \leqslant \mathbf{0} \\ & \mathbf{y} = \bar{\mathbf{y}} \\ & \mathbf{x} \in \mathbb{R}^{n_1}. \end{array} \tag{$\mathrm{P}^{\bar{\mathbf{y}}}$}$$

Clearly any feasible solution to such a continuous nonlinear-programming subproblem $(\mathrm{P}^{\bar{\mathbf{y}}})$ yields an upper bound on the optimal value of $(\mathrm{P}[\mathbf{l},\mathbf{u}])$. When $(\mathrm{P}^{\bar{\mathbf{y}}})$ is infeasible, we may consider the continuous nonlinear-programming feasibility subproblem

$$\begin{array}{rl} \min & \sum_{i=1}^{m} w_i \\ \text{s.t.} & g(\mathbf{x},\mathbf{y}) \leqslant \mathbf{w} \\ & \mathbf{y} = \bar{\mathbf{y}} \\ & \mathbf{x} \in \mathbb{R}^{n_1} \\ & \mathbf{w} \in \mathbb{R}^{m}_{+}. \end{array} \tag{$\mathrm{F}^{\bar{\mathbf{y}}}$}$$

If we can find a way to couple the solution of upper-bounding problems $(\mathrm{P}^{\bar{\mathbf{y}}})$ (and the closely related feasibility subproblems $(\mathrm{F}^{\bar{\mathbf{y}}})$) with a lower-bounding procedure exploiting the convexity assumptions, then we can hope to build an iterative procedure that will converge to a global optimum of $(\mathrm{P}[\mathbf{l},\mathbf{u}])$. Indeed, such a procedure is the *Outer-Approximation (OA) Algorithm* [49, 50]. Toward this end, for a finite set of "linearization points"

$$\mathscr{K} := \{(\mathbf{x}^k \in \mathbb{R}^{n_1}, \mathbf{y}^k \in \mathbb{R}^{n_2}) \ : \ k = 1,\ldots,K\},$$

we define the mixed-integer *linear* programming relaxation

$$
\begin{aligned}
\min\quad & z \\
\text{s.t.}\quad & \nabla f(\mathbf{x}^k,\mathbf{y}^k)^\top \begin{pmatrix}\mathbf{x}-\mathbf{x}^k\\ \mathbf{y}-\mathbf{y}^k\end{pmatrix} + f(\mathbf{x}^k,\mathbf{y}^k) \leqslant z, \quad \forall(\mathbf{x}^k,\mathbf{y}^k)\in\mathscr{K} \\
& \nabla g(\mathbf{x}^k,\mathbf{y}^k)^\top \begin{pmatrix}\mathbf{x}-\mathbf{x}^k\\ \mathbf{y}-\mathbf{y}^k\end{pmatrix} + g(\mathbf{x}^k,\mathbf{y}^k) \leqslant 0, \quad \forall(\mathbf{x}^k,\mathbf{y}^k)\in\mathscr{K} \\
& \mathbf{x}\in\mathbb{R}^{n_1} \\
& \mathbf{y}\in\mathbb{R}^{n_2}, \quad \mathbf{l}\le\mathbf{y}\le\mathbf{u} \\
& z\in\mathbb{R}.
\end{aligned}
\tag{$\mathrm{P}^{\mathscr{K}}[\mathbf{l},\mathbf{u}]$}
$$

We are now able to concisely state the basic OA Algorithm 15.1.

Algorithm 15.1 OA Algorithm

Input: The mixed-integer nonlinear program $(\mathrm{P}[\mathbf{l},\mathbf{u}])$.
Output: An optimal solution $(\mathbf{x}^*,\mathbf{y}^*)$.

1. Solve the nonlinear-programming relaxation $(\mathrm{P}_{\mathbb{R}})$, let $(\mathbf{x}^1,\mathbf{y}^1)$ be an optimal solution, and let $K := 1$, so that initially we have $\mathscr{K} = \{(\mathbf{x}^1,\mathbf{y}^1)\}$.
2. Solve the mixed-integer linear programming relaxation $(\mathrm{P}^{\mathscr{K}}[\mathbf{l},\mathbf{u}])$, and let $(\mathbf{x}^*,\mathbf{y}^*,z^*)$ be an optimal solution. If $(\mathbf{x}^*,\mathbf{y}^*,z^*)$ corresponds to a feasible solution of $(\mathrm{P}[\mathbf{l},\mathbf{u}])$ (i.e., if $f(\mathbf{x}^*,\mathbf{y}^*)\le z^*$ and $g(\mathbf{x}^*,\mathbf{y}^*)\le\mathbf{0}$), then STOP (with the optimal solution $(\mathbf{x}^*,\mathbf{y}^*)$ of $(\mathrm{P}[\mathbf{l},\mathbf{u}])$).
3. Solve the continuous nonlinear-programming subproblem $(\mathrm{P}^{\bar{\mathbf{y}}^*})$.
   i. Either a feasible solution $(\mathbf{x}^*,\mathbf{y}^*,z^*)$ is obtained,
   ii. or $(\mathrm{P}^{\bar{\mathbf{y}}^*})$ is infeasible, in which case we solve the nonlinear-programming feasibility subproblem $(\mathrm{F}^{\bar{\mathbf{y}}})$, and let its solution be $(\mathbf{x}^*,\mathbf{y}^*,\mathbf{u}^*)$.
4. In either case, we augment the set $\mathscr{K}$ of linearization points, by letting $K := K+1$ and $(\mathbf{x}^K,\mathbf{y}^K) := (\mathbf{x}^*,\mathbf{y}^*)$.
5. GOTO 2.

Each iteration of Steps 3–4 generate a linear cut that can improve the mixed-integer linear programming relaxation $(\mathrm{P}^{\mathscr{K}}[\mathbf{l},\mathbf{u}])$ that is repeatedly solved in Step 2. So clearly the sequence of optimal objective values for $(\mathrm{P}^{\mathscr{K}}[\mathbf{l},\mathbf{u}])$ obtained in Step 2 corresponds to a nondecreasing sequence of lower bounds on the optimum value of $(\mathrm{P}[\mathbf{l},\mathbf{u}])$. Moreover each linear cut returned from Steps 3–4 cuts off the previous solution of $(\mathrm{P}^{\mathscr{K}}[\mathbf{l},\mathbf{u}])$ from Step 2. A precise proof of convergence (see for example [22]) uses these simple observations, but it also requires an additional assumption that is standard in continuous nonlinear programming (i.e., a "constraint qualification").

Implementations of OA include `DICOPT` [40] which can be used with either of the mixed-integer linear programs codes `Cplex` and `Xpress-MP`, in conjunction with any of the continuous nonlinear programming codes `CONOPT`, `SNOPT` and `MINOS` and is available with `GAMS`. Additionally `Bonmin` has OA as an algorithmic option, which can use `Cplex` or the COIN-OR code `Cbc` as its mixed-integer linear programming solver, and `Ipopt` or `FilterSQP` as its continuous nonlinear programming solver.

Generalized Benders Decomposition [55] is a technique that is closely related to and substantially predates the OA Algorithm. In fact, one can regard OA as a proper strengthening of Generalized Benders Decomposition (see [49, 50]), so as a practical tool, we view it as superseded by OA.

Substantially postdating the development of the OA Algorithm is the simpler and closely related *Extended Cutting Plane (ECP) Algorithm* introduced in [135]. The original ECP Algorithm is a straightforward generalization of *Kelley's Cutting-Plane Algorithm* [78] for convex continuous nonlinear programming (which predates the development of the OA Algorithm). Subsequently, the ECP Algorithm has been enhanced and further developed (see, for example [136, 134]) to handle, for example, even pseudo-convex functions.

The motivation for the ECP Algorithm is that continuous nonlinear programs are expensive to solve, and that all the associated solutions give us are further linearization points for $(\mathrm{P}^{\mathscr{K}}[\mathbf{l},\mathbf{u}])$. So the ECP Algorithm dispenses altogether with the solution of continuous nonlinear programs. Rather, in the most rudimentary version, after each solution of the mixed-integer linear program $(\mathrm{P}^{\mathscr{K}}[\mathbf{l},\mathbf{u}])$, the most violated constraint (i.e, of $f(\mathbf{x}^*,\mathbf{y}^*) \leq z$ and $g(\mathbf{x}^*,\mathbf{y}^*) \leq \mathbf{0}$) is linearized and appended to $(\mathrm{P}^{\mathscr{K}}[\mathbf{l},\mathbf{u}])$. This simple iteration is enough to easily establish convergence (see [135]). It should be noted that for the case in which there are no integer-constrained variables, then at each step $(\mathrm{P}^{\mathscr{K}}[\mathbf{l},\mathbf{u}])$ is just a continuous linear program and we exactly recover Kelley's Cutting-Plane Algorithm for convex continuous nonlinear programming.

It is interesting to note that Kelley, in his seminal paper [78], already considered application of his approach to *integer* nonlinear programs. In fact, Kelley cited Gomory's seminal work on integer programming [59, 58] which was also taking place in the same time period, and he discussed how the approaches could be integrated.

Of course, many practical improvements can be made to the rudimentary ECP Algorithm. For example, more constraints can be linearized at each iteration. An implementation of the ECP Algorithm is the code `Alpha-ECP` (see [134]) which uses `Cplex` as its mixed-integer linear programming solver and is available with `GAMS`. The general experience is that for mildly nonlinear problems, an ECP Algorithm can outperform an OA Algorithm. But for a highly nonlinear problem, the performance of the ECP Algorithm is limited by the performance of Kelley's Cutting-Plane Algorithm, which can be quite poor on highly-nonlinear purely continuous problems. In such cases, it is typically better to use an OA Algorithm, which will handle the nonlinearity in a more sophisticated manner.

In considering again the performance of an OA Algorithm on a mixed-integer nonlinear program $(\mathrm{P}[\mathbf{l},\mathbf{u}])$, rather than the convex continuous nonlinear programming problems $(\mathrm{P}^{\bar{\mathbf{y}}^*})$ and $(\mathrm{F}^{\bar{\mathbf{y}}})$ being too time consuming to solve (which led us to the ECP Algorithm), it can be the case that solution of the mixed-integer linear programming problems $(\mathrm{P}^{\mathscr{K}}[\mathbf{l},\mathbf{u}])$ dominates the running time. Such a situation led to the *Quesada-Grossmann Branch-and-Cut Algorithm* [108]. The viewpoint is that the mixed-integer linear programming problems $(\mathrm{P}^{\mathscr{K}}[\mathbf{l},\mathbf{u}])$ are solved by a Branch-

and-Bound or Branch-and-Cut Algorithm. During the solution of the mixed-integer linear programming problem $(\mathrm{P}^{\mathscr{K}}[\mathbf{l},\mathbf{u}])$, whenever a new solution is found (i.e., one that has the variables $\mathbf{y}$ integer), we interrupt the solution process for $(\mathrm{P}^{\mathscr{K}}[\mathbf{l},\mathbf{u}])$, and we solve the convex continuous nonlinear programming problems $(\mathrm{P}^{\bar{\mathbf{y}}^*})$ to derive new outer approximation cuts that are appended to mixed-integer linear programming problem $(\mathrm{P}^{\mathscr{K}}[\mathbf{l},\mathbf{u}])$. We then continue with the solution process for $(\mathrm{P}^{\mathscr{K}}[\mathbf{l},\mathbf{u}])$. The Quesada-Grossmann Branch-and-Cut Algorithm is available as an option in `Bonmin`.

Finally, it is clear that the essential scheme of the Quesada-Grossmann Branch-and-Cut Algorithm admits enormous flexibility. The *Hybrid Algorithm* [22] incorporates two important enhancements.

First, we can seek to further improve the linearization $(\mathrm{P}^{\mathscr{K}}[\mathbf{l},\mathbf{u}])$ by solving convex continuous nonlinear programming problems at additional nodes of the mixed-integer linear programming Branch-and-Cut tree for $(\mathrm{P}^{\mathscr{K}}[\mathbf{l},\mathbf{u}])$ — that is, not just when solutions are found having $\mathbf{y}$ integer. In particular, at any node $(\mathrm{P}^{\mathscr{K}}[\mathbf{l}',\mathbf{u}'])$ of the mixed-integer linear programming Branch-and-Cut tree, we can solve the associated convex continuous nonlinear programming subproblem $(\mathrm{P}_{\mathbb{R}}[\mathbf{l}',\mathbf{u}'])$: Then, if in the solution $(\mathbf{x}^*,\mathbf{y}^*)$ we have that $\mathbf{y}^*$ is integer, we may update the incumbent and fathom the node; otherwise, we append $(\mathbf{x}^*,\mathbf{y}^*)$ to the set $\mathscr{K}$ of linearization points. In the extreme case, if we solve these continuous nonlinear programming subproblems at every node, we essentially have the Branch-and-Bound Algorithm for mixed-integer nonlinear programming.

A second enhancement is based on working harder to find a solution $(\mathbf{x}^*,\mathbf{y}^*)$ with $\mathbf{y}^*$ integer at selected nodes $(\mathrm{P}^{\mathscr{K}}[\mathbf{l}',\mathbf{u}'])$ of the mixed-integer linear programming Branch-and-Cut tree. The idea is that at a node $(\mathrm{P}^{\mathscr{K}}[\mathbf{l}',\mathbf{u}'])$, we perform a time-limited mixed-integer linear programming Branch-and-Bound Algorithm. If we are successful, then we will have found a solution to the node with $(\mathbf{x}^*,\mathbf{y}^*)$ with $\mathbf{y}^*$ integer, and then we perform an OA iteration (i.e., Steps 3-4) on $(\mathrm{P}[\mathbf{l}',\mathbf{u}'])$ which will improve the linearization $(\mathrm{P}^{\mathscr{K}}[\mathbf{l}',\mathbf{u}'])$. We can then repeat this until we have solved the mixed-integer nonlinear program $(\mathrm{P}[\mathbf{l}',\mathbf{u}'])$ associated with the node. If we do this without time limit at the root node $(\mathrm{P}[\mathbf{l},\mathbf{u}])$, then the entire procedure reduces to the OA Algorithm. The Hybrid Algorithm was developed for and first made available as part of `Bonmin`.

`FilMint` [2] is another successful modern code, also based on enhancing the general framework of the Quesada-Grossmann Branch-and-Cut Algorithm. The main additional innovation introduced with `FilMint` is the idea of using ECP cuts rather than only performing OA iterations for getting cuts to improve the linearizations $(\mathrm{P}^{\mathscr{K}}[\mathbf{l}',\mathbf{u}'])$. Subsequently, this feature was also added to `Bonmin`. `FilMint` was put together from the continuous nonlinear programming active-set code `FilterSQP`, and the mixed-integer linear programming code `MINTO`. Currently, `FilMint` is only generally available via NEOS [51].

It is worth mentioning that just as for mixed-integer linear programming, effective heuristics can and should be used to provide good upper bounds quickly. This can markedly improve the performance of any of the algorithms described above. Some examples of work in this direction are [27] and [23].

15.3.3.2 Convex quadratics and second-order cone programming

Though we will not go into any details, there is considerable algorithmic work and associated software that seeks to leverage more specialized (but still rather general and powerful) nonlinear models and existing convex continuous nonlinear-programming algorithms for the associated relaxations. In this direction, recent work has focused on conic programming relaxations (in particular, the semi-definite and second-order cones). On the software side, we point to work on the binary quadratic and max-cut problems (via semi-definite relaxation) [109, 110] with the code `BiqMac` [21]. We also note that `Cplex` (v11) has a capability aimed at solving mixed-integer quadratically-constrained programs that have a convex continuous relaxation.

One important direction for approaching quadratic models is at the modeling level. This is particulary useful for the convex case, where there is a strong and appealing relationship between quadratically constrained programming and second-order cone programming (SOCP). A *second-order cone constraint* is one that expresses that the Euclidean norm of an affine function should be no more than another affine function. An *SOCP* problem consists of minimizing a linear function over a finite set of second-order cone constraints. Our interest in SOCP stems from the fact that (continuous) convex quadratically constrained programming problems can be reformulated as SOCP problems (see [93]). The appeal is that very efficient interior-point algorithms have been developed for solving SOCP problems (see [57], for example), and there is considerable mature software available that has functionality for efficient handling of SOCP problems; see, for example: `SDPT3` [118] (GNU GPL open-source license; Matlab), `SeDuMi` [120] (GNU GPL open-source license; Matlab), `LOQO` [94] (proprietary; C library with interfaces to AMPL and Matlab), `MOSEK` [100] (proprietary; C library with interface to Matlab), `Cplex` [73] (proprietary; C library). Note also that `MOSEK` and `Cplex` can handle integer variables as well; one can expect that the approaches essentially marry specialized SOCP solvers with Branch-and-Bound and/or Outer-Approximation Algorithms. Further branch-and-cut methods for mixed-integer SOCP, employing linear and convex quadratic cuts [37] and a careful treatment of the non-differentiability inherent in the SOCP constraints, have recently been proposed [48].

Also in this vein is recent work by Günlük and Linderoth [62, 63]. Among other things, they demonstrated that many practical mixed-integer quadratically constrained programming formulations have substructures that admit extended formulations that can be easily strengthened as certain integer SOCP problems. This approach is well known in the mixed-integer linear programming literature. Let

$$Q := \{w \in \mathbb{R}, \mathbf{x} \in \mathbb{R}^n_+, \mathbf{z} \in \{0,1\}^n \ : \ w \geq \sum_{i=1}^{n} r_i x_i^2,\, u_i z_i \geq x_i \geq l_i z_i,\ i = 1,2,\dots,n\},$$

where $r_i \in \mathbb{R}_+$ and $u_i, l_i \in \mathbb{R}$ for all $i = 1,2,\dots,n$. The set Q appears in several formulations as a substructure. Consider the following extended formulation of Q

$$\bar{Q} := \{w \in \mathbb{R},\ \mathbf{x} \in \mathbb{R}^n, \mathbf{y} \in \mathbb{R}^n, \mathbf{z} \in \mathbb{R}^n : w \geq \sum_i r_i y_i, (x_i, y_i, z_i) \in S_i, i = 1, 2, \ldots, n\},$$

where

$$S_i := \{(x_i, y_i, z_i) \in \mathbb{R}^2 \times \{0,1\}\ :\ y_i \geq x_i^2,\ u_i z_i \geq x_i \geq l_i z_i,\ x_i \geq 0\},$$

and $u_i, l_i \in \mathbb{R}$. The convex hull of each S_i has the form

$$S_i^c\ :=\ \{(x_i, y_i, z_i) \in \mathbb{R}^3\ :\ y_i z_i \geq x_i^2,\ u_i z_i \geq x_i \geq l_i z_i,\ 1 \geq z_i \geq 0,\ x_i, y_i \geq 0\}$$

(see [36, 62, 63, 130]). Note that $x_i^2 - y_i z_i$ is *not* a convex function, but nonetheless S_i^c is a convex set. Finally, we can state the result of [62, 63], which also follows from a more general result of [71], that the convex hull of the extended formulation $\bar{Q}$ has the form

$$\bar{Q}^c := \{w \in \mathbb{R}, \mathbf{x} \in \mathbb{R}^n, \mathbf{y} \in \mathbb{R}^n, \mathbf{z} \in \mathbb{R}^n :\geq \sum_{i=1}^n r_i y_i, (x_i, y_i, z_i) \in S_i^c,\ i = 1, 2, \ldots, n\}.$$

Note that all of the nonlinear constraints describing the S_i^c and $\bar{Q}^c$ are rather simple quadratic constraints. Generally, it is well known that even the "restricted hyperbolic constraint"

$$y_i z_i \geq \sum_{k=1}^n x_k^2\ ,\ \mathbf{x} \in \mathbb{R}^n,\ y_i \geq 0,\ z_i \geq 0$$

(more general than the nonconvexity in S_i^c) can be reformulated as the second-order cone constraint

$$\left\| \begin{pmatrix} 2\mathbf{x} \\ y_i - z_i \end{pmatrix} \right\|_2 \leq y_i + z_i\ .$$

In this subsection, in the interest of concreteness and brevity, we have focused our attention on convex quadratics and second-order cone programming. However, it should be noted that a related approach, with broader applicability (to all convex objective functions) is presented in [52], and a computational comparison is available in [53]. Also, it is relevant that many convex non-quadratic functions are representable as second-order cone programs (see [5, 16]).

15.4 Polynomial optimization

In this section, we focus our attention on the study of optimization models involving polynomials only, but without any assumptions on convexity or concavity. It is worth emphasizing the fundamental result of Jeroslow (Theorem 15.4) that even pure integer quadratically constrained programming is *undecidable*. One can however avoid this daunting pitfall by bounding the variables, and this is in fact an assumption that we should and do make for the purpose of designing practical approaches. From the point of view of most applications that we are aware of, this is a very reasonable assumption. We must be cognizant of the fact that the geometry of even quadratics on boxes is daunting from the point of view of mathematical programming; see Figure 15.6. Specifically, the convex envelope of the graph of the product x_1x_2 on a box deviates badly from the graph, so relaxation-based methods are intrinsically handicapped. It is easy to see, for example, that for $\delta_1, \delta_2 > 0$, x_1x_2 is strictly convex on the line segment joining $(0,0)$ and (δ_1, δ_2); while x_1x_2 is strictly concave on the line segment joining $(\delta_1, 0)$ and $(0, \delta_2)$.

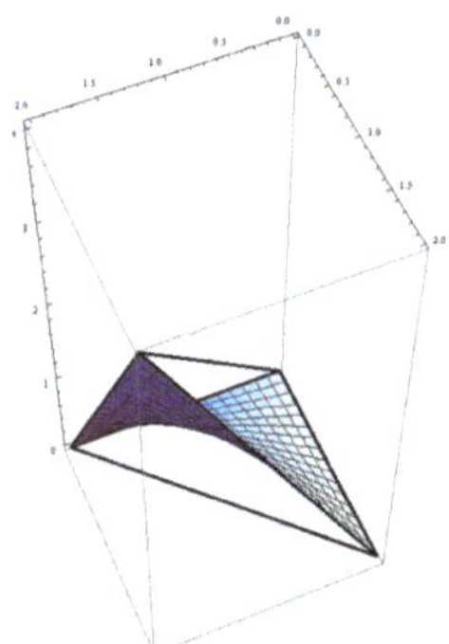

Fig. 15.6 Tetrahedral convex envelope of the graph of the product x_1x_2 on a box.

Despite these difficulties, we have positive results. In Section 15.4.1, the highlight is a *fully polynomial time approximation scheme (FPTAS)* for problems involving maximization of a polynomial in fixed dimension, over the mixed-integer points in a polytope. In Section 15.4.2, we broaden our focus to allow feasible regions defined by inequalities in polynomials (i.e., semi-algebraic sets). In this setting, we do not present (nor could we expect) complexity results as strong as for linear constraints, but rather we show how tools of semi-definite programming are being developed to provide, in a systematic manner, strengthened relaxations. Finally, in Section 15.4.3, we describe recent computational advances in the special case of semi-algebraic programming for which all of the functions are quadratic — i.e., *mixed-integer quadratically constrained programming* (MIQCP).

15.4.1 Fixed dimension and linear constraints: An FPTAS

As we pointed out in the introduction (Theorem 15.2), optimizing degree-4 polynomials over problems with two integer variables is already a hard problem. Thus, even when we fix the dimension, we cannot get a polynomial-time algorithm for solving the optimization problem. The best we can hope for, even when the number of both the continuous and the integer variables is fixed, is an approximation result.

Definition 15.3 (FPTAS).

(a) An algorithm $\mathscr{A}$ is an ε-*approximation algorithm* for a maximization problem with optimal cost $f_{\max}$, if for each instance of the problem of encoding length n, $\mathscr{A}$ runs in polynomial time in n and returns a feasible solution with cost $f_{\mathscr{A}}$, such that $f_{\mathscr{A}} \geqslant (1-\varepsilon) \cdot f_{\max}$.
(b) A family $\{\mathscr{A}_\varepsilon\}_\varepsilon$ of ε-approximation algorithms is a *fully polynomial time approximation scheme (FPTAS)* if the running time of $\mathscr{A}_\varepsilon$ is polynomial in the encoding size of the instance and $1/\varepsilon$.

Indeed it is possible to obtain an FPTAS for general polynomial optimization of mixed-integer feasible sets in polytopes [45, 44, 46]. To explain the method of the FPTAS, we need to review the theory of *short rational generating functions* pioneered by Barvinok [13, 14]. The FPTAS itself appears in Section 15.4.1.3.

15.4.1.1 Introduction to rational generating functions

We explain the theory on a simple, one-dimensional example. Let us consider the set S of integers in the interval $P = [0,\dots,n]$; see the top of Figure 15.7 (a). We associate with S the polynomial $g(S;z) = z^0 + z^1 + \cdots + z^{n-1} + z^n$; i.e., every integer $\alpha \in S$ corresponds to a monomial z^α with coefficient 1 in the polynomial $g(S;z)$. This polynomial is called the *generating function* of S (or of P). From the viewpoint of computational complexity, this generating function is of exponential size (in the encoding length of n), just as an explicit list of all the integers $0, 1, \dots, n-1, n$ would be. However, we can observe that $g(S;z)$ is a finite geometric series, so there exists a simple summation formula that expresses it in a much more compact way:

$$g(S;z) = z^0 + z^1 + \cdots + z^{n-1} + z^n = \frac{1-z^{n+1}}{1-z}. \tag{15.11}$$

The “long” polynomial has a “short” representation as a rational function. The encoding length of this new formula is *linear* in the encoding length of n.

Suppose now someone presents to us a finite set S of integers as a generating function $g(S;z)$. Can we decide whether the set is nonempty? In fact, we can do something much stronger even – we can *count* the integers in the set S, simply by evaluating at $g(S;z)$ at $z=1$. On our example we have $|S| = g(S;1) = 1^0 + 1^1 + \cdots + 1^{n-1} + 1^n = n+1$. We can do the same on the shorter, rational-function formula if we are careful with the (removable) singularity $z=1$. We just compute the limit

$$|S| = \lim_{z\to 1} g(S;z) = \lim_{z\to 1} \frac{1-z^{n+1}}{1-z} = \lim_{z\to 1} \frac{-(n+1)z^n}{-1} = n+1$$

using the Bernoulli–l’Hôpital rule. Note that we have avoided to carry out a polynomial division, which would have given us the long polynomial again.

The summation formula (15.11) can also be written in a slightly different way:

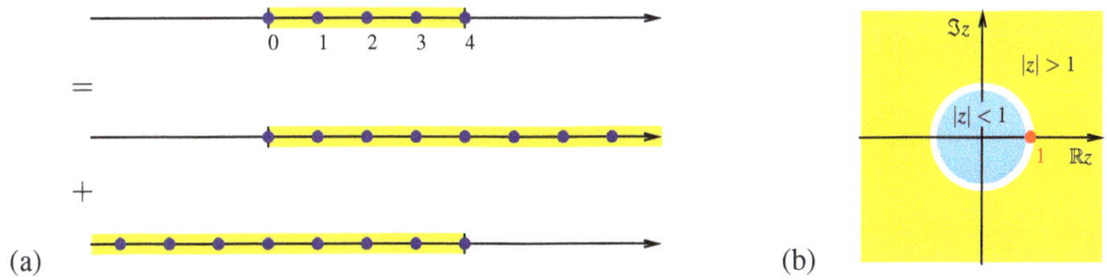

Fig. 15.7 **(a)** One-dimensional Brion theorem. **(b)** The domains of convergence of the Laurent series.

$$g(S;z) = \frac{1}{1-z} - \frac{z^{n+1}}{1-z} = \frac{1}{1-z} + \frac{z^n}{1-z^{-1}} \tag{15.12}$$

Each of the two summands on the right-hand side can be viewed as the summation formula of an infinite geometric series:

$$g_1(z) = \frac{1}{1-z} = z^0 + z^1 + z^2 + \dots, \tag{15.13a}$$

$$g_2(z) = \frac{z^n}{1-z^{-1}} = z^n + z^{n-1} + z^{n-2} + \dots. \tag{15.13b}$$

The two summands have a geometrical interpretation. If we view each geometric series as the generating function of an (infinite) lattice point set, we arrive at the picture shown in Figure 15.7. We remark that all integer points in the interval $[0,n]$ are covered twice, and also all integer points outside the interval are covered once. This phenomenon is due to the *one-to-many correspondence* of rational functions to their Laurent series. When we consider Laurent series of the function $g_1(z)$ about $z=0$, the pole $z=1$ splits the complex plane into two domains of convergence (Figure 15.7): For $|z|<1$, the power series $z^0+z^1+z^2+\dots$ converges to $g_1(z)$. As a matter of fact, it converges absolutely and uniformly on every compact subset of the open circle $\{z \in \mathbb{C} : |z| < 1\}$. For $|z|>1$, however, the series diverges. On the other hand, the Laurent series $-z^{-1}-z^{-2}-z^{-3}-\dots$ converges (absolutely and compact-uniformly) on the open circular ring $\{z \in \mathbb{C} : |z| > 1\}$ to the function $g_1(z)$, whereas it diverges for $|z|<1$. The same holds for $g_2(z)$. Altogether we have:

$$g_1(z) = \begin{cases} z^0 + z^1 + z^2 + \dots, & \text{for } |z| < 1, \\ -z^{-1} - z^{-2} - z^{-3} - \dots, & \text{for } |z| > 1, \end{cases} \tag{15.14}$$

$$g_2(z) = \begin{cases} -z^{n+1} - z^{n+2} - z^{n+3} - \dots, & \text{for } |z| < 1, \\ z^n + z^{n-1} + z^{n-2} + \dots, & \text{for } |z| > 1. \end{cases} \tag{15.15}$$

We can now see that the phenomenon we observed in formula (15.13) and Figure 15.7 is due to the fact that we had picked two Laurent series for the summands $g_1(z)$ and $g_2(z)$ that do not have a common domain of convergence; the situation of

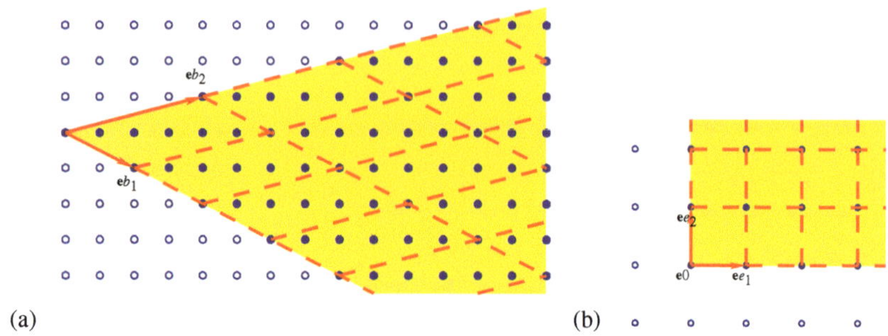

Fig. 15.8 **(a)** Tiling a rational two-dimensional cone with copies of the fundamental parallelepiped. **(b)** The semigroup $S \subseteq \mathbb{Z}^2$ generated by $\mathbf{b}_1$ and $\mathbf{b}_2$ is a linear image of $\mathbb{Z}^2_+$.

formula (15.13) and Figure 15.7 appears again in the d-dimensional case as *Brion's Theorem.*

Let us now consider a *two-dimensional* cone C spanned by the vectors $\mathbf{b}_1 = (\alpha, -1)$ and $\mathbf{b}_2 = (\beta, 1)$; see Figure 15.8 for an example with $\alpha = 2$ and $\beta = 4$. We would like to write down a generating function for the integer points in this cone. We apparently need a generalization of the geometric series, of which we made use in the one-dimensional case. The key observation now is that using copies of the half-open *fundamental parallelepiped*, $\Pi = \{ \lambda_1 \mathbf{b}_1 + \lambda_2 \mathbf{b}_2 : \lambda_1 \in [0,1), \lambda_2 \in [0,1) \}$, the cone can be *tiled*:

$$C = \bigcup_{\mathbf{s} \in S} (\mathbf{s} + \Pi), \text{ where } S = \{ \mu_1 \mathbf{b}_1 + \mu_2 \mathbf{b}_2 : (\mu_1, \mu_2) \in \mathbb{Z}^2_+ \} \tag{15.16}$$

(a disjoint union). Because we have chosen *integral* generators $\mathbf{b}_1, \mathbf{b}_2$, the integer points are "the same" in each copy of the fundamental parallelepiped. Therefore, also the integer points of C can be tiled by copies of $\Pi \cap \mathbb{Z}^2$; on the other hand, we can see $C \cap \mathbb{Z}^2$ as a finite disjoint union of copies of S, shifted by the integer points of Π:

$$C \cap \mathbb{Z}^2 = \bigcup_{\mathbf{s} \in S} (\mathbf{s} + (\Pi \cap \mathbb{Z}^2)) = \bigcup_{\mathbf{x} \in \Pi \cap \mathbb{Z}^2} (\mathbf{x} + S). \tag{15.17}$$

The set S is just the image of $\mathbb{Z}^2_+$ under the matrix $(\mathbf{b}_1, \mathbf{b}_2) = \left(\begin{smallmatrix} \alpha & \beta \\ -1 & 1 \end{smallmatrix}\right)$; cf. Figure 15.8. Now $\mathbb{Z}^2_+$ is the direct product of $\mathbb{Z}_+$ with itself, whose generating function is the geometric series $g(\mathbb{Z}_+; z) = z^0 + z^1 + z^2 + z^3 + \cdots = \frac{1}{1-z}$. We thus obtain the generating function as a product, $g(\mathbb{Z}^2_+; z_1, z_2) = \frac{1}{1-z_1} \cdot \frac{1}{1-z_2}$. Applying the linear transformation $(\mathbf{b}_1, \mathbf{b}_2)$,

$$g(S; z_1, z_2) = \frac{1}{(1 - z_1^{\alpha} z_2^{-1})(1 - z_1^{\beta} z_2^{1})}.$$

From (15.17) it is now clear that $g(C; z_1, z_2) = \sum_{\mathbf{x} \in \Pi \cap \mathbb{Z}^2} z_1^{x_1} z_2^{x_2} g(S; z_1, z_2)$; the multiplication with the monomial $z_1^{x_1} z_2^{x_2}$ corresponds to the shifting of the set S by the vector (x_1, x_2). In our example, obviously $\Pi \cap \mathbb{Z}^2 = \{ (i, 0) : i = 0, \ldots, \alpha + \beta - 1 \}$.

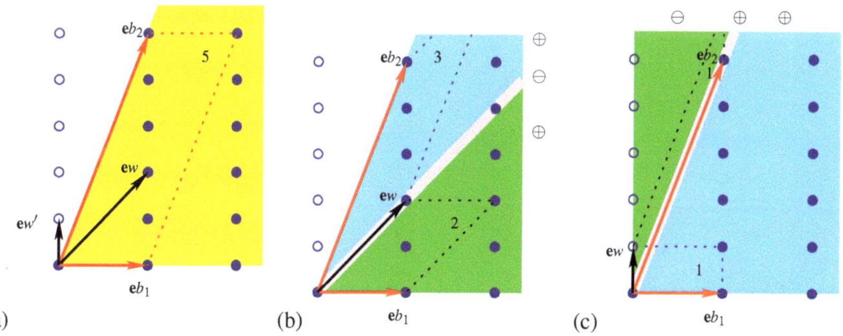

Fig. 15.9 **(a)** A cone of index 5 generated by $\mathbf{b}^1$ and $\mathbf{b}^2$. **(b)** A triangulation of the cone into the two cones spanned by $\{\mathbf{b}^1, \mathbf{w}\}$ and $\{\mathbf{b}^2, \mathbf{w}\}$, having an index of 2 and 3, respectively. We have the inclusion-exclusion formula $g(\text{cone}\{\mathbf{b}_1, \mathbf{b}_2\}; \mathbf{z}) = g(\text{cone}\{\mathbf{b}_1, \mathbf{w}\}; \mathbf{z}) + g(\text{cone}\{\mathbf{b}_2, \mathbf{w}\}; \mathbf{z}) - g(\text{cone}\{\mathbf{w}\}; \mathbf{z})$; here the one-dimensional cone spanned by $\mathbf{w}$ needed to be subtracted. **(c)** A signed decomposition into the two unimodular cones spanned by $\{\mathbf{b}^1, \mathbf{w}'\}$ and $\{\mathbf{b}^2, \mathbf{w}'\}$. We have the inclusion-exclusion formula $g(\text{cone}\{\mathbf{b}_1, \mathbf{b}_2\}; \mathbf{z}) = g(\text{cone}\{\mathbf{b}_1, \mathbf{w}'\}; \mathbf{z}) - g(\text{cone}\{\mathbf{b}_2, \mathbf{w}'\}; \mathbf{z}) + g(\text{cone}\{\mathbf{w}'\}; \mathbf{z})$.

Thus

$$g(C; z_1, z_2) = \frac{z_1^0 + z_1^1 + \cdots + z_1^{\alpha+\beta-2} + z_1^{\alpha+\beta-1}}{(1 - z_1^{\alpha} z_2^{-1})(1 - z_1^{\beta} z_2^{1})}.$$

Unfortunately, this formula has an exponential size as the numerator contains $\alpha + \beta$ summands. To make the formula shorter, we need to recursively break the cone into "smaller" cones, each of which have a much shorter formula. We have observed that the length of the formula is determined by the number of integer points in the fundamental parallelepiped, the *index* of the cone. *Triangulations* usually do not help to reduce the index significantly, as another two-dimensional example shows. Consider the cone C' generated by $\mathbf{b}_1 = (1, 0)$ and $\mathbf{b}_2 = (1, \alpha)$; see Figure 15.9. We have $\Pi' \cap \mathbb{Z}^2 = \{(0,0)\} \cup \{\,(1,i) : i = 1, \ldots, \alpha - 1\,\}$, so the rational generating function would have α summands in the numerator, and thus have exponential size. Every attempt to use triangulations to reduce the size of the formula fails in this example. The choice of an interior vector $\mathbf{w}$ in Figure 15.9, for instance, splits the cone of index 5 into two cones of index 2 and 3, respectively – and also a one-dimensional cone. Indeed, every possible triangulation of C' into unimodular cones contains at least α two-dimensional cones! The important new idea by Barvinok was to use so-called *signed decompositions* in addition to triangulations in order to reduce the index of a cone. In our example, we can choose the vector $\mathbf{w} = (0, 1)$ *from the outside of the cone* to define cones $C_1 = \text{cone}\{\mathbf{b}_1, \mathbf{w}\}$ and $C_2 = \text{cone}\{\mathbf{w}, \mathbf{b}_2\}$; see Figure 15.9. Using these cones, we have the inclusion-exclusion formula

$$g(C'; z_1, z_2) = g(C_1; z_1, z_2) - g(C_2; z_1, z_2) + g(C_1 \cap C_2; z_1, z_2).$$

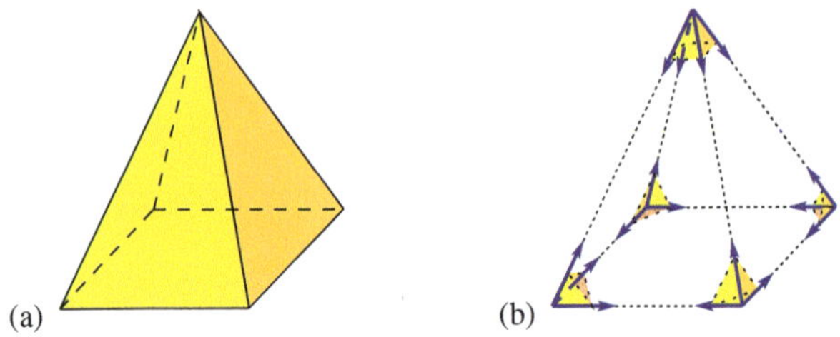

Fig. 15.10 Brion's theorem, expressing the generating function of a polyhedron to those of the supporting cones of all vertices.

It turns out that all cones C_1 and C_2 are unimodular, and we obtain the rational generating function by summing up those of the subcones,

$$g(C';z_1,z_2) = \frac{1}{(1-z_1)(1-z_2)} - \frac{1}{(1-z_1^1 z_2^\alpha)(1-z_2)} + \frac{1}{1-z_1 z_2^\alpha}.$$

15.4.1.2 Barvinok's algorithm for short rational generating functions

We now present the general definitions and results. Let $P \subseteq \mathbb{R}^d$ be a rational polyhedron. We first define its *generating function* as the *formal Laurent series* $\tilde{g}(P;\mathbf{z}) = \sum_{\boldsymbol{\alpha}\in P\cap\mathbb{Z}^d} \mathbf{z}^{\boldsymbol{\alpha}} \in \mathbb{Z}[[z_1,\ldots,z_d,z_1^{-1},\ldots,z_d^{-1}]]$, i.e., without any consideration of convergence properties. (A formal *power* series is not enough because monomials with negative exponents can appear.) As we remarked above, this encoding of a set of lattice points does not give an immediate benefit in terms of complexity. We will get short formulas only when we can identify the Laurent series with certain rational functions. Now if P is a polytope, then $\tilde{g}(P;\mathbf{z})$ is a *Laurent polynomial* (i.e., a *finite* sum of monomials with positive or negative integer exponents), so it can be naturally identified with a rational function $g(P;\mathbf{z})$. Convergence comes into play whenever P is not bounded, since then $\tilde{g}(P;\mathbf{z})$ can be an infinite formal sum. We first consider a *pointed* polyhedron P, i.e., P does not contain a straight line.

Theorem 15.15. *Let $P \subseteq \mathbb{R}^d$ be a pointed rational polyhedron. Then there exists a non-empty open subset $U \subseteq \mathbb{C}^d$ such that the series $\tilde{g}(P;\mathbf{z})$ converges absolutely and uniformly on every compact subset of U to a rational function $g(P;\mathbf{z}) \in \mathbb{Q}(z_1,\ldots,z_d)$.*

Finally, when P contains an integer point and also a straight line, there does not exist any point $\mathbf{z} \in \mathbb{C}^d$ where the series $\tilde{g}(P;\mathbf{z})$ converges absolutely. In this case we set $g(P;\mathbf{z}) = 0$; this turns out to be a consistent choice (making the map $P \mapsto g(P;\mathbf{z})$ a *valuation*, i.e., a finitely additive measure). The rational function $g(P;\mathbf{z}) \in \mathbb{Q}(z_1,\ldots,z_d)$ defined as described above is called the *rational generating function* of $P \cap \mathbb{Z}^d$.

Theorem 15.16 (Brion [31]). *Let P be a rational polyhedron and $V(P)$ be the set of vertices of P. Then, $g(P;\mathbf{z}) = \sum_{\mathbf{v}\in V(P)} g(C_P(\mathbf{v});\mathbf{z})$, where $C_P(\mathbf{v}) = \mathbf{v} + \operatorname{cone}(P - \mathbf{v})$ is the* supporting cone *of the vertex $\mathbf{v}$; see Figure 15.10.*

We remark that in the case of a non-pointed polyhedron P, i.e., a polyhedron that has no vertices because it contains a straight line, both sides of the equation are zero.

Barvinok's algorithm computes the rational generating function of a polyhedron P as follows. By Brion's theorem, the rational generating function of a polyhedron can be expressed as the sum of the rational generating functions of the supporting cones of its vertices. Every supporting cone $\mathbf{v}_i + C_i$ can be triangulated to obtain simplicial cones $\mathbf{v}_i + C_{ij}$. If the dimension is fixed, these polyhedral computations all run in polynomial time.

Now let K be one of these simplicial cones, whose *basis vectors* $\mathbf{b}_1,\dots,\mathbf{b}_d$ (i.e., primitive representatives of its extreme rays) are the columns of some matrix $B \in \mathbb{Z}^{d\times d}$; then the index of K is $|\det B|$. Barvinok's algorithm now computes a *signed decomposition* of K to produce simplicial cones with smaller index. To this end, it constructs a vector $\mathbf{w} = \alpha_1\mathbf{b}_1 + \cdots + \alpha_d\mathbf{b}_d \in \mathbb{Z}^d \setminus \{\mathbf{0}\}$ with $|\alpha_i| \leqslant |\det B|^{-1/d}$. The existence of such a vector follows from Minkowski's first theorem, and it can be constructed in polynomial time using integer programming or lattice basis reduction followed by enumeration. The cone is then decomposed into cones spanned by d vectors from the set $\{\mathbf{b}_1,\dots,\mathbf{b}_d,\mathbf{w}\}$; each of the resulting cones then has an index at most $(\operatorname{ind} K)^{(d-1)/d}$. In general, these cones form a signed decomposition of K; only if $\mathbf{w}$ lies inside K, they form a triangulation (see Figure 15.9). The resulting cones and their intersecting proper faces (arising in an inclusion-exclusion formula) are recursively processed, until cones of low index (for instance unimodular cones) are obtained. Finally, for a unimodular cone $\mathbf{v} + B\mathbb{R}^d_+$, the rational generating function is $\mathbf{z}^{\mathbf{a}}/\prod_{j=1}^d (1-\mathbf{z}^{\mathbf{b}_j})$, where $\mathbf{a}$ is the unique integer point in the fundamental parallelepiped. We summarize Barvinok's algorithm as Algorithm 15.2.

Algorithm 15.2 Barvinok's Algorithm

Input: A polyhedron $P \subset \mathbb{R}^d$ given by rational inequalities.
Output: The rational generating function for $P \cap \mathbb{Z}^d$ in the form

$$g_P(\mathbf{z}) = \sum_{i\in I} \varepsilon_i \frac{\mathbf{z}^{\mathbf{a}_i}}{\prod_{j=1}^d (1-\mathbf{z}^{\mathbf{b}_{ij}})} \tag{15.18}$$

where $\varepsilon_i \in \{\pm 1\}$, $\mathbf{a}_i \in \mathbb{Z}^d$, and $\mathbf{b}_{ij} \in \mathbb{Z}^d$.

1. Compute all vertices $\mathbf{v}_i$ and corresponding supporting cones C_i of P.
2. Triangulate C_i into simplicial cones C_{ij}, keeping track of all the intersecting proper faces.
3. Apply signed decomposition to the cones $\mathbf{v}_i + C_{ij}$ to obtain unimodular cones $\mathbf{v}_i + C_{ijl}$, keeping track of all the intersecting proper faces.
4. Compute the unique integer point $\mathbf{a}_i$ in the fundamental parallelepiped of every resulting cone $\mathbf{v}_i + C_{ijl}$.
5. Write down the formula (15.18).

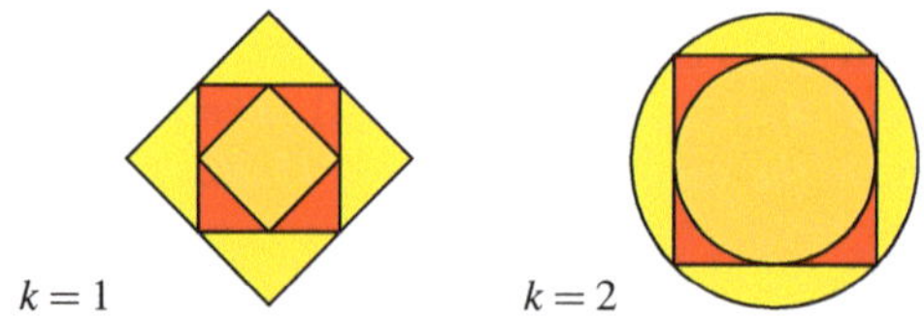

Fig. 15.11 Approximation properties of ℓ_k-norms.

We remark that it is possible to avoid computations with the intersecting proper faces of cones (step 2 of the algorithm) entirely, using techniques such as polarization, irrational decomposition [81], or half-open decomposition [32, 82].

Due to the descent of the indices in the signed decomposition procedure, the depth of the decomposition tree is at most $\left\lfloor 1+\frac{\log_2 \log_2 D}{\log_2 \frac{d}{d-1}}\right\rfloor$, where $D=|\det B|$. Because at each decomposition step at most $\mathrm{O}(2^d)$ cones are created and the depth of the tree is doubly logarithmic in the index of the input cone, Barvinok could obtain a polynomiality result *in fixed dimension*:

Theorem 15.17 (Barvinok [13]). *Let d be fixed. There exists a polynomial-time algorithm for computing the rational generating function* (15.18) *of a polyhedron $P \subseteq \mathbb{R}^d$ given by rational inequalities.*

15.4.1.3 The FPTAS for polynomial optimization

We now describe the fully polynomial-time approximation scheme, which appeared in [45, 44, 46]. It makes use of the elementary relation

$$\max\{s_1,\dots,s_N\} = \lim_{k\to\infty} \sqrt[k]{s_1^k+\cdots+s_N^k}, \tag{15.19}$$

which holds for any finite set $S=\{s_1,\dots,s_N\}$ of non-negative real numbers. This relation can be viewed as an approximation result for ℓ_k-norms. Now if P is a polytope and f is an objective function non-negative on $P\cap\mathbb{Z}^d$, let $\mathbf{x}^1,\dots,\mathbf{x}^N$ denote all the feasible integer solutions in $P\cap\mathbb{Z}^d$ and collect their objective function values $s_i=f(\mathbf{x}^i)$ in a vector $\mathbf{s}\in\mathbb{Q}^N$. Then, comparing the unit balls of the ℓ_k-norm and the ℓ_∞-norm (Figure 15.11), we get the relation

$$L_k := N^{-1/k}\|\mathbf{s}\|_k \leqslant \|\mathbf{s}\|_\infty \leqslant \|\mathbf{s}\|_k =: U_k.$$

Thus, for obtaining a good approximation of the maximum, it suffices to solve a summation problem of the polynomial function $h=f^k$ on $P\cap\mathbb{Z}^d$ for a value of k that is large enough. Indeed, for $k=\lceil (1+1/\varepsilon)\log N\rceil$, we obtain $U_k-L_k\leqslant \varepsilon f(\mathbf{x}^{\max})$. On the other hand, this choice of k is polynomial in the input size (because $1/\varepsilon$ is encoded in unary in the input, and $\log N$ is bounded by a polynomial in the binary encoding size of the polytope P). Hence, when the dimension d is fixed, we can expand the polynomial function f^k as a list of monomials in polynomial time.

Solving the summation problem can be accomplished using short rational generating functions as follows. Let $g(P;\mathbf{z})$ be the rational generating function of $P\cap\mathbb{Z}^d$, computed using Barvinok's algorithm. By symbolically applying differential operators to $g(P;\mathbf{z})$, we can compute a short rational function representation of the Laurent polynomial $g(P,h;\mathbf{z}) = \sum_{\boldsymbol{\alpha}\in P\cap\mathbb{Z}^d} h(\boldsymbol{\alpha})\mathbf{z}^{\boldsymbol{\alpha}}$, where each monomial $\mathbf{z}^{\boldsymbol{\alpha}}$ corresponding to an integer point $\boldsymbol{\alpha}\in P\cap\mathbb{Z}^d$ has a coefficient that is the value $h(\boldsymbol{\alpha})$. To illustrate this, consider again the generating function of the interval $P=[0,4]$,

$$g_P(z) = z^0+z^1+z^2+z^3+z^4 = \frac{1}{1-z} - \frac{z^5}{1-z}.$$

We now apply the differential operator $z\frac{\mathrm{d}}{\mathrm{d}z}$ and obtain

$$\left(z\frac{\mathrm{d}}{\mathrm{d}z}\right) g_P(z) = 1z^1+2z^2+3z^3+4z^4 = \frac{1}{(1-z)^2} - \frac{-4z^5+5z^4}{(1-z)^2}.$$

Applying the same differential operator again, we obtain

$$\left(z\frac{\mathrm{d}}{\mathrm{d}z}\right)\left(z\frac{\mathrm{d}}{\mathrm{d}z}\right) g_P(z) = 1z^1+4z^2+9z^3+16z^4 = \frac{z+z^2}{(1-z)^3} - \frac{25z^5-39z^6+16z^7}{(1-z)^3}.$$

We have thus evaluated the monomial function $h(\alpha)=\alpha^2$ for $\alpha=0,\dots,4$; the results appear as the coefficients of the respective monomials. The same works for several variables, using the partial differential operators $z_i\frac{\partial}{\partial z_i}$ for $i=1,\dots,d$. In fixed dimension, the size of the rational function expressions occuring in the symbolic calculation can be bounded polynomially. Thus one obtains the following result.

Theorem 15.18.

(a) Let $h(x_1,\dots,x_d)=\sum_{\boldsymbol{\beta}} c_{\boldsymbol{\beta}}\mathbf{x}^{\boldsymbol{\beta}}\in\mathbb{Q}[x_1,\dots,x_d]$ be a polynomial. Define the differential operator

$$D_h = h\left(z_1\frac{\partial}{\partial z_1},\dots,z_d\frac{\partial}{\partial z_d}\right) = \sum_{\boldsymbol{\beta}} c_{\boldsymbol{\beta}}\left(z_1\frac{\partial}{\partial z_1}\right)^{\beta_1}\cdots\left(z_d\frac{\partial}{\partial z_d}\right)^{\beta_d}.$$

Then D_h maps the generating function $g(P;\mathbf{z})=\sum_{\boldsymbol{\alpha}\in P\cap\mathbb{Z}^d}\mathbf{z}^{\boldsymbol{\alpha}}$ to the weighted generating function $(D_h g)(\mathbf{z}) = g(P,h;\mathbf{z}) = \sum_{\boldsymbol{\alpha}\in P\cap\mathbb{Z}^d} h(\boldsymbol{\alpha})\mathbf{z}^{\boldsymbol{\alpha}}$.

(b) Let the dimension d be fixed. Let $g(P;\mathbf{z})$ be the Barvinok representation of the generating function $\sum_{\boldsymbol{\alpha}\in P\cap\mathbb{Z}^d}\mathbf{z}^{\boldsymbol{\alpha}}$ of $P\cap\mathbb{Z}^d$. Let $h\in\mathbb{Q}[x_1,\dots,x_d]$ be a polynomial, given as a list of monomials with rational coefficients $c_{\boldsymbol{\beta}}$ encoded in binary and exponents $\boldsymbol{\beta}$ encoded in unary. We can compute in polynomial time a Barvinok representation $g(P,h;\mathbf{z})$ for the weighted generating function $\sum_{\boldsymbol{\alpha}\in P\cap\mathbb{Z}^d} h(\boldsymbol{\alpha})\mathbf{z}^{\boldsymbol{\alpha}}$.

Thus, we can implement the polynomial time algorithm 15.3.

Taking the discussion of the convergence of the bounds into consideration, one obtains the following result.

Algorithm 15.3

Input: A rational convex polytope $P \subset \mathbb{R}^d$; a polynomial objective function $f \in \mathbb{Q}[x_1,\dots,x_d]$ that is non-negative over $P \cap \mathbb{Z}^d$, given as a list of monomials with rational coefficients $c_{\boldsymbol{\beta}}$ encoded in binary and exponents $\boldsymbol{\beta}$ encoded in unary; an index k, encoded in unary.

Output: A lower bound L_k and an upper bound U_k for the maximal function value f^* of f over $P \cap \mathbb{Z}^d$. The bounds L_k form a nondecreasing, the bounds U_k a nonincreasing sequence of bounds that both reach f^* in a finite number of steps.

1. Compute a short rational function expression for the generating function $g(P;\mathbf{z}) = \sum_{\boldsymbol{\alpha} \in P \cap \mathbb{Z}^d} \mathbf{z}^{\boldsymbol{\alpha}}$. Using residue techniques, compute $|P \cap \mathbb{Z}^d| = g(P;\mathbf{1})$ from $g(P;\mathbf{z})$.
2. Compute the polynomial f^k from f.
3. From the rational function $g(P;\mathbf{z})$ compute the rational function representation of $g(P, f^k;\mathbf{z})$ of $\sum_{\boldsymbol{\alpha} \in P \cap \mathbb{Z}^d} f^k(\boldsymbol{\alpha})\mathbf{z}^{\boldsymbol{\alpha}}$ by Theorem 15.18. Using residue techniques, compute

$$L_k := \left\lceil \sqrt[k]{g(P, f^k;\mathbf{1})/g(P;\mathbf{e1})} \right\rceil \quad \text{and} \quad U_k := \left\lfloor \sqrt[k]{g(P, f^k;\mathbf{1})} \right\rfloor.$$

Theorem 15.19 (Fully polynomial-time approximation scheme). *Let the dimension d be fixed. Let $P \subset \mathbb{R}^d$ be a rational convex polytope. Let f be a polynomial with rational coefficients that is non-negative on $P \cap \mathbb{Z}^d$, given as a list of monomials with rational coefficients $c_{\boldsymbol{\beta}}$ encoded in binary and exponents $\boldsymbol{\beta}$ encoded in unary.*

(i) Algorithm 15.3 computes the bounds L_k, U_k in time polynomial in k, the input size of P and f, and the total degree D. The bounds satisfy the following inequality:

$$U_k - L_k \leqslant f^* \cdot \left(\sqrt[k]{|P \cap \mathbb{Z}^d|} - 1\right).$$

(ii) For $k = (1 + 1/\varepsilon)\log(|P \cap \mathbb{Z}^d|)$ (a number bounded by a polynomial in the input size), L_k is a $(1-\varepsilon)$-approximation to the optimal value f^* and it can be computed in time polynomial in the input size, the total degree D, and $1/\varepsilon$. Similarly, U_k gives a $(1+\varepsilon)$-approximation to f^*.

(iii) With the same complexity, by iterated bisection of P, we can also find a feasible solution $\mathbf{x}_\varepsilon \in P \cap \mathbb{Z}^d$ with $\left|f(\mathbf{x}_\varepsilon) - f^*\right| \leqslant \varepsilon f^*$.

The mixed-integer case can be handled by *discretization* of the continuous variables. We illustrate on an example that one needs to be careful to pick a sequence of discretizations that actually converges. Consider the mixed-integer *linear* optimization problem depicted in Figure 15.12, whose feasible region consists of the point $(\frac{1}{2},1)$ and the segment $\{(x,0) : x \in [0,1]\}$. The unique optimal solution is $x = \frac{1}{2}$, $z = 1$. Now consider the sequence of grid approximations where $x \in \frac{1}{m}\mathbb{Z}_{\geqslant 0}$. For even m, the unique optimal solution to the grid approximation is $x = \frac{1}{2}$, $z = 1$. However, for odd m, the unique optimal solution is $x = 0$, $z = 0$. Thus the full sequence of the optimal solutions to the grid approximations does not converge because it has two limit points; see Figure 15.12.

To handle polynomial objective functions that take arbitrary (positive and negative) values on the feasible region, one can shift the objective function by a constant

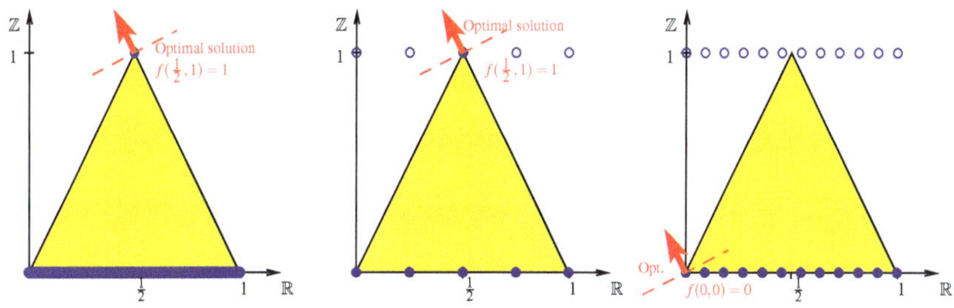

Fig. 15.12 A mixed-integer linear optimization problem and a sequence of optimal solutions to grid problems with two limit points, for even m and for odd m.

that is large enough. Then, to obtain a strong approximation result, one iteratively reduces the constant by a factor. Altogether we have the following result.

Theorem 15.20 (Fully polynomial-time approximation schemes). *Let the dimension $n = n_1 + n_2$ be fixed. Let an optimization problem* (15.2) *of a polynomial function f over the mixed-integer points of a polytope P and an error bound ε be given, where*

(I_1) f is given as a list of monomials with rational coefficients $c_{\boldsymbol{\beta}}$ encoded in binary and exponents $\boldsymbol{\beta}$ encoded in unary,
(I_2) P is given by rational inequalities in binary encoding,
(I_3) the rational number $\frac{1}{\varepsilon}$ is given in unary encoding.

(a) There exists a fully polynomial time approximation scheme (FPTAS) for the maximization *problem for all polynomial functions $f(\mathbf{x},\mathbf{z})$ that are* non-negative *on the feasible region. That is, there exists a polynomial-time algorithm that, given the above data, computes a feasible solution $(\mathbf{x}_\varepsilon,\mathbf{z}_\varepsilon) \in P \cap (\mathbb{R}^{n_1} \times \mathbb{Z}^{n_2})$ with*

$$\left|f(\mathbf{x}_\varepsilon,\mathbf{z}_\varepsilon) - f(\mathbf{x}_{\max},\mathbf{z}_{\max})\right| \leqslant \varepsilon f(\mathbf{x}_{\max},\mathbf{z}_{\max}).$$

(b) There exists a polynomial-time algorithm that, given the above data, computes a feasible solution $(\mathbf{x}_\varepsilon,\mathbf{z}_\varepsilon) \in P \cap (\mathbb{R}^{n_1} \times \mathbb{Z}^{n_2})$ with

$$\left|f(\mathbf{x}_\varepsilon,\mathbf{z}_\varepsilon) - f(\mathbf{x}_{\max},\mathbf{z}_{\max})\right| \leqslant \varepsilon\left|f(\mathbf{x}_{\max},\mathbf{z}_{\max}) - f(\mathbf{x}_{\min},\mathbf{z}_{\min})\right|.$$

15.4.2 Semi-algebraic sets and SOS programming

In this section we use results from algebraic geometry over the reals to provide a convergent (and in the case of binary optimization, finite) sequence of semi-definite relaxations for the general polynomial optimization problem over semi-algebraic sets:

$$\begin{aligned} Z^* = \text{minimize } & f(\mathbf{x}) \\ \text{s.t. } & g_i(\mathbf{x}) \geqslant \mathbf{0}, \quad i = 1, \dots, m, \\ & \mathbf{x} \in \mathbb{R}^n, \end{aligned} \tag{15.20}$$

where $f, g_i \in \mathbb{R}[\mathbf{x}]$ are polynomials defined as:

$$f(\mathbf{x}) = \sum_{\boldsymbol{\alpha} \in \mathbb{Z}^n_+} f_{\boldsymbol{\alpha}} \mathbf{x}^{\boldsymbol{\alpha}}, \qquad g_i(\mathbf{x}) = \sum_{\boldsymbol{\alpha} \in \mathbb{Z}^n_+} g_{i,\boldsymbol{\alpha}} \mathbf{x}^{\boldsymbol{\alpha}},$$

where there are only finitely many nonzero coefficients $f_{\boldsymbol{\alpha}}$ and $g_{i,\boldsymbol{\alpha}}$. Moreover, let $K = \{\mathbf{x} \in \mathbb{R}^n : g_i(\mathbf{x}) \geqslant 0,\ i = 1, \dots, m\}$ denote the set of feasible solutions. Note that problem (15.20) can model binary optimization, by taking $f(\mathbf{x}) = \mathbf{c}^\top \mathbf{x}$, and taking as the polynomials $g_i(\mathbf{x})$, $\mathbf{a}_i^\top \mathbf{x} - b_i$, $x_j^2 - x_j$ and $-x_j^2 + x_j$ (to model $x_j^2 - x_j = 0$). Problem (15.20) can also model bounded integer optimization (using for example the equation $(x_j - l_j)(x_j - l_j + 1) \cdot \ldots \cdot (x_j - u_j) = 0$ to model $l_j \leqslant x_j \leqslant u_j$), as well as bounded mixed-integer nonlinear optimization. Problem (15.20) can be written as:

$$\begin{aligned} \text{maximize } & \gamma \\ \text{s.t. } & f(\mathbf{x}) - \gamma \geqslant 0, \ \ \forall \mathbf{x} \in K. \end{aligned} \tag{15.21}$$

This leads us to consider conditions for polynomials to be nonnegative over a set K.

Definition 15.4. Let $p \in \mathbb{R}[\mathbf{x}]$ where $\mathbf{x} = (x_1, \dots, x_n)^\top$. The polynomial p is called *sos* (sum of squares), if there exist polynomials $h_1, \dots, h_k \in \mathbb{R}[\mathbf{x}]$ such that $p = \sum_{i=1}^k h_i^2$.

Clearly, in multiple dimensions if a polynomial can be written as a sum of squares of other polynomials, then it is nonnegative. However, is it possible for a polynomial in higher dimensions to be nonnegative without being a sum of squares? The answer is yes. The most well-known example is probably the Motzkin-polynomial $M(x,y,z) = x^4y^2 + x^2y^4 + z^6 - 3x^2y^2z^2$, which is nonnegative without being a sum of squares of polynomials.

The following theorem establishes a certificate of positivity of a polynomial on the set K, under a certain assumption on K.

Theorem 15.21 ([107] [75]). *Suppose that the set K is compact and there exists a polynomial $h(\mathbf{x})$ of the form*

$$h(\mathbf{x}) = h_0(\mathbf{x}) + \sum_{i=1}^m h_i(\mathbf{x}) g_i(\mathbf{x}),$$

such that $\{\mathbf{x} \in \mathbb{R}^n : h(\mathbf{x}) \geqslant 0\}$ is compact and $h_i(\mathbf{x})$, $i = 0, 1, \dots, m$, are polynomials that have a sum of squares representation. Then, if the polynomial g is strictly positive over K, then there exist $p_i \in \mathbb{R}[\mathbf{x}]$, $i = 0, 1, \dots, m$, that are sums of squares such that

$$g(\mathbf{x}) = p_0(\mathbf{x}) + \sum_{i=1}^m p_i(\mathbf{x}) g_i(\mathbf{x}). \tag{15.22}$$

Note that the number of terms in Equation (15.22) is linear. While the assumption of Theorem 15.21 may seem restrictive, it is satisfied in several cases:

(a) For binary optimization problems, that is, when K includes the inequalities $x_j^2 \geqslant x_j$ and $x_j \geqslant x_j^2$ for all $j = 1, \dots, n$.
(b) If all the g_j's are linear, i.e., K is a polyhedron.
(c) If there is one polynomial g_k such that the set $\{\mathbf{x} \in \mathbb{R}^n : g_k(\mathbf{x}) \geqslant 0\}$ is compact.

More generally, one way to ensure that the assumption of Theorem 15.21 holds is to add to K the extra quadratic constraint $g_{m+1}(\mathbf{x}) = a^2 - \|\mathbf{x}\|^2 \geqslant 0$ for some a sufficiently large. It is also important to emphasize that we do not assume that K is convex. Notice that it may even be disconnected.

Let us now investigate algorithmically when a polynomial is a sum of squares. As we will see this question is strongly connected to semi-definite optimization. The idea of using semi-definite optimization for solving optimization problems over polynomials is due to [121] and further expanded in [83] and [105]. We consider the vector

$$\mathbf{v}_d(\mathbf{x}) = (\mathbf{x}^{\boldsymbol{\alpha}})_{|\boldsymbol{\alpha}| \leqslant d} = (1, x_1, \dots, x_n, x_1^2, x_1x_2, \dots, x_{n-1}x_n, x_n^2, \dots, x_1^d, \dots, x_n^d)^\top,$$

of all the monomials $\mathbf{x}^{\boldsymbol{\alpha}}$ of degree less than or equal to d, which has dimension $s = \sum_{i=0}^{d} \binom{n}{i} = \binom{n+d}{d}$.

Proposition 15.3 ([38]). *The polynomial $g(\mathbf{x})$ of degree $2d$ has a sum of squares decomposition if and only if there exists a positive semi-definite matrix Q for which $g(\mathbf{x}) = \mathbf{v}_d(\mathbf{x})^\top Q \mathbf{v}_d(\mathbf{x})$.*

Proof. Suppose there exists an $s \times s$ matrix $Q \succeq 0$ for which $g(\mathbf{x}) = \mathbf{v}_d(\mathbf{x})^\top Q \mathbf{v}_d(\mathbf{x})$. Then $Q = HH^\top$ for some $s \times k$ matrix H, and thus,

$$g(\mathbf{x}) = \mathbf{v}_d(\mathbf{x})^\top HH^\top \mathbf{v}_d(\mathbf{x}) = \sum_{i=1}^{k} (H^\top \mathbf{v}_d(\mathbf{x}))_i^2.$$

Because $(H^\top \mathbf{v}_d(\mathbf{x}))_i$ is a polynomial, then $g(\mathbf{x})$ is expressed as a sum of squares of the polynomials $(H^\top \mathbf{v}_d(\mathbf{x}))_i$.

Conversely, suppose that $g(\mathbf{x})$ has a sum of squares decomposition $g(\mathbf{x}) = \sum_{i=1}^{\ell} h_i(\mathbf{x})^2$. Let $\mathbf{h}_i$ be the vector of coefficients of the polynomial $h_i(\mathbf{x})$, i.e., $h_i(\mathbf{x}) = \mathbf{h}_i^\top \mathbf{v}_d(\mathbf{x})$. Thus,

$$g(\mathbf{x}) = \sum_{i=1}^{\ell} \mathbf{v}_d(\mathbf{x})^\top \mathbf{h}_i \mathbf{h}_i^\top \mathbf{v}_d(\mathbf{x}) = \mathbf{v}_d(\mathbf{x})^\top Q \mathbf{v}_d(\mathbf{x}),$$

with $Q = \sum_{i=1}^{\ell} \mathbf{h}_i \mathbf{h}_i^\top \succeq 0$, and the proposition follows.

Proposition 15.3 leads to an algorithm. Given a polynomial $f(\mathbf{x}) \in \mathbb{R}[x_1, \dots, x_n]$ of degree $2d$. In order to compute the minimum value $f^* = \min\{f(\mathbf{x}) : \mathbf{x} \in \mathbb{R}^n\}$ we introduce an artificial variable λ and determine

$$\max\{\lambda : \lambda \in \mathbb{R},\ f(\mathbf{x}) - \lambda \geqslant 0\}.$$

With the developments above, we realize that we can determine a lower bound for f^* by computing the value

$$p^{sos} = \max\{\lambda : \lambda \in \mathbb{R},\ f(\mathbf{x}) - \lambda \text{ is sos}\} \leqslant f^*.$$

The latter task can be accomplished by setting up a semi-definite program. In fact, if we denote by $f_{\boldsymbol{\alpha}}$ the coefficient of the monomial $\mathbf{x}^{\boldsymbol{\alpha}}$ in the polynomial f, then $f(\mathbf{x}) - \lambda$ is sos if and only if there exists an $s \times s$ matrix $Q \succeq 0$ for which $f(\mathbf{x}) - \lambda = \mathbf{v}_d(\mathbf{x})^\top Q \mathbf{v}_d(\mathbf{x})$. Now we can compare the coefficients on both sides of the latter equation. This leads to the SOS-program

$$\begin{aligned} p^{sos} = \max\ & \lambda \\ \text{s.t.}\ & f_{\mathbf{0}} - \lambda = Q_{\mathbf{0},\mathbf{0}} \\ & \sum_{\boldsymbol{\beta},\boldsymbol{\gamma},\ \boldsymbol{\beta}+\boldsymbol{\gamma}=\boldsymbol{\alpha}} Q_{\boldsymbol{\beta},\boldsymbol{\gamma}} = f_{\boldsymbol{\alpha}} \\ & Q = (Q_{\boldsymbol{\beta},\boldsymbol{\gamma}})_{\boldsymbol{\beta},\boldsymbol{\gamma}} \succeq 0. \end{aligned}$$

In a similar vein, Theorem 15.21 and Proposition 15.3 jointly imply that we can use semi-definite optimization to provide a sequence of semi-definite relaxations for the optimization problem (15.21). Assuming that the set K satisfies the assumption of Theorem 15.21, then if $f(\mathbf{x}) - \gamma > 0$ for all $\mathbf{x} \in K$, then

$$f(\mathbf{x}) - \gamma = p_0(\mathbf{x}) + \sum_{i=1}^{m} p_i(\mathbf{x}) g_i(\mathbf{x}), \tag{15.23}$$

where $p_i(\mathbf{x})$, $i = 0, 1, \ldots, m$ have a sum of squares representation. Theorem 15.21 does not specify the degree of the polynomials $p_i(\mathbf{x})$. Thus, we select a bound $2d$ on the degree of $p_i(\mathbf{x})$, and we apply Proposition 15.3 to each of the polynomials $p_i(\mathbf{x})$, that is, $p_i(\mathbf{x})$ is a sum of squares if and only if $p_i(\mathbf{x}) = \mathbf{v}_d(\mathbf{x})^\top Q_i \mathbf{v}_d(\mathbf{x})$ with $Q_i \succeq 0$, $i = 0, 1, \ldots, m$. Substituting to (15.23), we obtain that γ, $Q_i, i = 0, 1, \ldots, m$, satisfy linear equations that we denote as $L(\gamma, Q_0, Q_1, \ldots, Q_m) = 0$. Thus, we can find a lower bound to problem (15.20) by solving the semi-definite optimization problem

$$\begin{aligned} Z_d = \max\ & \gamma \\ \text{s.t.}\ & L(\gamma, Q_0, Q_1, \ldots, Q_m) = 0, \\ & Q_i \succeq 0,\ i = 0, 1, \ldots, m. \end{aligned} \tag{15.24}$$

Problem (15.24) involves semi-definite optimization over $m+1$ $s \times s$ matrices. From the construction we get the relation $Z_d \leqslant Z^*$. It turns out that as d increases, Z_d converges to Z^*. Moreover, for binary optimization, there exists a finite d for which $Z_d = Z^*$ [84].

Problem (15.24) provides a systematic way to find convergent semi-definite relaxations to problem (15.20). While the approach is both general (it applies to very general nonconvex problems including nonlinear mixed-integer optimization problems) and insightful from a theoretical point of view, it is only practical for values of $d = 1,2$, as large scale semi-definite optimization problems cannot be solved in practice. In many situations, however, Z_1 or Z_2 provide strong bounds. Let us consider an example.

Example 15.2. Let us minimize $f(x_1,x_2) = 2x_1^4 + 2x_1^3x_2 - x_1^2x_2^2 + 5x_2^4$ over $\mathbb{R}^2$. We attempt to write

$$\begin{aligned} f(x_1,x_2) &= 2x_1^4 + 2x_1^3x_2 - x_1^2x_2^2 + 5x_2^4 \\ &= \begin{pmatrix} x_1^2 \\ x_2^2 \\ x_1x_2 \end{pmatrix}^\top \begin{bmatrix} q_{11} & q_{12} & q_{13} \\ q_{12} & q_{22} & q_{23} \\ q_{13} & q_{23} & q_{33} \end{bmatrix} \begin{pmatrix} x_1^2 \\ x_2^2 \\ x_1x_2 \end{pmatrix} \\ &= q_{11}x_1^4 + q_{22}x_2^4 + (q_{13} + 2q_{12})x_1^2x_2^2 + 2q_{13}x_1^3x_2 + 2q_{23}x_1x_2^3. \end{aligned}$$

In order to have an identity, we obtain

$$q_{11} = 2,\ q_{22} = 5,\ q_{33} + 2q_{12} = -1,\ 2q_{13} = 2,\ q_{23} = 0.$$

Using semi-definite optimization, we find a particular solution such that $Q \succeq 0$ is given by

$$Q = \begin{bmatrix} 2 & -3 & 1 \\ -3 & 5 & 0 \\ 1 & 0 & 5 \end{bmatrix} = HH^\top, \qquad H = \frac{1}{\sqrt{2}} \begin{bmatrix} 2 & 0 \\ -3 & 1 \\ 1 & 3 \end{bmatrix}.$$

It follows that $f(x_1,x_2) = \frac{1}{2}(2x_1^2 - 3x_2^2 + x_1x_2)^2 + \frac{1}{2}(x_2^2 + 3x_1x_2)^2$, and thus the optimal solution value is $\gamma^* = 0$ and the optimal solution is $x_1^* = x_2^* = 0$.

15.4.3 Quadratic functions

In this section, we focus on instances of polynomial programming where the functions are all quadratic. The specific form of the *mixed-integer quadratically constrained programming* problem that we consider is

$$\begin{aligned} \min \quad & q_0(\mathbf{x}) \\ \text{s.t.} \quad & q(\mathbf{x}) \leqslant 0 \\ & \mathbf{l} \leq \mathbf{x} \leq \mathbf{u} \qquad\qquad (\text{MIQCP}[\mathbf{l},\mathbf{u}]) \\ & x_i \in \mathbb{R}, \text{ for } i = 1,\ldots,k, \\ & x_i \in \mathbb{Z}, \text{ for } i = k+1,\ldots,n, \end{aligned}$$

where $q_0\colon \mathbb{R}^n \to \mathbb{R}$ and $q\colon \mathbb{R}^n \to \mathbb{R}^m$ are quadratic, $\mathbf{l},\mathbf{u} \in \mathbb{Z}^n$, and $\mathbf{l} \leq \mathbf{u}$. We denote the continuous relaxation by $(\text{MIQCP}_{\mathbb{R}}[\mathbf{l},\mathbf{u}])$. We emphasize that we are not generally making any convexity/concavity assumptions on the quadratic functions q_i, so when we do require any such assumptions we will state so explicitly.

Of course one can write a binary constraint $y_i \in \{0,1\}$ as the (nonconvex) quadratic inequality $y_i(1-y_i) \leq 0$ in the bound-constrained variable $0 \leq y_i \leq 1$. So, in this way, the case of binary variables y_i can be seen as the special case of (MIQCP[$\mathbf{l},\mathbf{u}$]) with no discrete variables (i.e., $k=n$). So, in a sense, the topic of *mixed-binary quadratically constrained programming* can be seen as a special case of *(purely continuous) quadratically constrained programming.* We are not saying that it is necessarily useful to do this from a computational viewpoint, but it makes it clear that the scope of even the purely continuous quadratic model includes quadratic models having both binary and continuous variables, and in particular mixed-$\{0,1\}$ linear programming.

In addition to the natural mathematical interest in studying mixed-integer quadratically constrained programming, there is a wealth of applications that have motivated the development of practical approaches; for example: Trimloss problems (see [85], for example), portfolio optimization (see [25], for example), Max-Cut and other binary quadratic models (see [109, 110] and the references therein).

In the remainder of this section, we describe some recent work on practical computational approaches to nonconvex quadratic optimization models. Rather than attempt a detailed survey, our goal is to present a few recent and promising techniques. One could regard these techniques as belonging more to the field of global optimization, but in Section 15.5 we present material on global optimization aimed at more general unstructured nonlinear integer programming problems.

15.4.3.1 Disjunctive programming

It is not surprising that integer variables in a mixed-integer *nonlinear* program can be treated with disjunctive programming [8, 10]. A corresponding branch-and-cut method was first described in [124] in the context of 0/1 mixed convex programming.

Here we describe an intriguing result from [115, 116, 117], which shows that one can also make useful disjunctions from nonconvex quadratic functions in a mixed-integer quadratically-constrained programming problem or even in a purely continuous quadratically-constrained programming problem. The starting point for this approach is that we can take a quadratic form $\mathbf{x}^\top A_i \mathbf{x}$ in $\mathbf{x} \in \mathbb{R}^n$, and rewrite it via an extended formulation as the linear form $\langle A_i, \mathbf{X}\rangle$, using the matrix variable $\mathbf{X} \in \mathbb{R}^{n\times n}$, and the nonlinear equation $\mathbf{X} = \mathbf{x}\mathbf{x}^\top$. The standard approach is to relax $\mathbf{X} = \mathbf{x}\mathbf{x}^\top$ to the convex inequality $\mathbf{X} \succeq \mathbf{x}\mathbf{x}^\top$. But the approach of [115, 116, 117] involves working with the nonconvex inequality $\mathbf{X} \preceq \mathbf{x}\mathbf{x}^\top$. This basic idea is as follows. Let $\mathbf{v} \in \mathbb{R}^n$ be arbitrary (for now). We have the equation

$$\langle \mathbf{v}\mathbf{v}^\top, \mathbf{X}\rangle = \langle \mathbf{v}\mathbf{v}^\top, \mathbf{x}\mathbf{x}^\top\rangle = (\mathbf{v}^\top \mathbf{x})^2 ,$$

which we relax as the *concave* inequality

$$(\mathbf{v}^\top \mathbf{x})^2 \geq \langle \mathbf{v}\mathbf{v}^\top, \mathbf{X}\rangle. \qquad (\Omega)$$

If we have a point $(\hat{\mathbf{x}}, \hat{\mathbf{X}})$ that satisfies the convex inequality $\mathbf{X} \succeq \mathbf{x}\mathbf{x}^\top$, but for which $\hat{\mathbf{X}} \neq \hat{\mathbf{x}}\hat{\mathbf{x}}^\top$, then it is the case that $\hat{\mathbf{X}} - \hat{\mathbf{x}}\hat{\mathbf{x}}^\top$ has a positive eigenvalue λ. Let $\mathbf{v}$ denote a unit-length eigenvector belonging to λ. Then

$$\begin{aligned} \lambda &= \lambda \|\mathbf{v}\|_2^2 \\ &= \langle \mathbf{v}\mathbf{v}^\top, \hat{\mathbf{X}} - \hat{\mathbf{x}}\hat{\mathbf{x}}^\top \rangle. \end{aligned}$$

So, $\lambda > 0$ if and only if $(\mathbf{v}^\top \hat{\mathbf{x}})^2 < \langle \mathbf{v}\mathbf{v}^\top, \hat{\mathbf{X}} \rangle$. That is, every positive eigenvalue of $\hat{\mathbf{X}} - \hat{\mathbf{x}}\hat{\mathbf{x}}^\top$ yields an inequality of the form (Ω) that is violated by $(\hat{\mathbf{x}}, \hat{\mathbf{X}})$. Next, we make a disjunction on this violated nonconvex inequality (Ω). First, we choose a suitable polyhedral relaxation $\mathscr{P}$ of the feasible region, and we let $[\eta_L, \eta_U]$ be the range of $\mathbf{v}^\top \mathbf{x}$ as $(\mathbf{x}, \mathbf{X})$ varies over the relaxation $\mathscr{P}$. Next, we choose a value $\theta \in (\eta_L, \eta_U)$ (e.g., the midpoint), and we get the polyhedral disjunction:

$$\left\{ (\mathbf{x}, \mathbf{X}) \in \mathscr{P} : \begin{array}{l} \eta_L(\mathbf{v}) \leq \mathbf{v}^\top \mathbf{x} \leq \theta \\ (\mathbf{v}^\top \mathbf{x})(\eta_L(\mathbf{v}) + \theta) \; - \; \theta \eta_L(\mathbf{v}) \geq \langle \mathbf{v}\mathbf{v}^\top, \mathbf{X} \rangle \end{array} \right\}$$

or

$$\left\{ (\mathbf{x}, \mathbf{X}) \in \mathscr{P} : \begin{array}{l} \theta \leq \mathbf{v}^\top \mathbf{x} \leq \eta_U(\mathbf{v}) \\ (\mathbf{v}^\top \mathbf{x})(\eta_U(\mathbf{v}) + \theta) \; - \; \theta \eta_U(\mathbf{v}) \geq \langle \mathbf{v}\mathbf{v}^\top, \mathbf{X} \rangle. \end{array} \right\}.$$

Notice that the second part of the first (resp., second) half of the disjunction corresponds to a secant inequality over the interval between the point θ and the lower (resp., upper) bound for $\mathbf{v}^\top \mathbf{x}$. Finally, we use the linear-programming technology of ordinary disjunctive programming to separate, via a linear inequality, the point $(\hat{\mathbf{x}}, \hat{\mathbf{X}})$ from the convex closure of the two halves of the disjunction. Details and extensions of this idea appear in [115, 116, 117].

15.4.3.2 Branch and cut

A branch-and-cut scheme for optimization of a nonconvex quadratic form over a box was recently developed by Vandenbussche and Nemhauser [133, 132]. They use a formulation of Balas via linear programming with complementarity conditions, based on the necessary optimality conditions of continuous quadratic programming (see [9]). Specifically, they consider the problem

$$\begin{array}{ll} \min & \frac{1}{2}\mathbf{x}^\top Q\mathbf{x} + \mathbf{c}^\top \mathbf{x} \\ \text{s.t.} & \mathbf{x} \in [0,1]^n, \end{array} \qquad (\text{BoxQP}[Q,\mathbf{c}])$$

where Q is an $n \times n$ symmetric, non positive semi-definite matrix, and $\mathbf{c} \in \mathbb{R}^n$. The KKT necessary optimality conditions for (BoxQP$[Q,\mathbf{c}]$) are

$$\mathbf{y} - Q\mathbf{x} - \mathbf{z} = \mathbf{c}, \tag{15.25}$$
$$\mathbf{y}^\top(\mathbf{1} - \mathbf{x}) = 0 \tag{15.26}$$
$$\mathbf{z}^\top\mathbf{x} = 0 \tag{15.27}$$
$$\mathbf{x} \in [0,1]^n \tag{15.28}$$
$$\mathbf{y},\ \mathbf{z} \in \mathbb{R}^n_+ \,. \tag{15.29}$$

Vandenbussche and Nemhauser, appealing to a result of Balas, define $\mathscr{P}(Q,\mathbf{c})$ as the polyhedron defined as the *convex hull* of solutions to (15.25–15.29), and they work with the reformulation of (BoxQP) as the linear program

$$\begin{array}{ll} \min & \frac{1}{2}\mathbf{c}^\top\mathbf{x} + \frac{1}{2}\mathbf{1}^\top\mathbf{y} \\ \text{s.t.} & (\mathbf{x},\mathbf{y},\mathbf{z}) \in \mathscr{P}(Q,\mathbf{c}). \end{array} \tag{Balas[$Q,\mathbf{c}$]}$$

The main tactic of Vandenbussche and Nemhauser is to develop cutting planes for $\mathscr{P}(Q,c)$.

Burer and Vandenbussche pursue a similar direction, but they allow general polyhedral constraints and employ semi-definite relaxations [34].

15.4.3.3 Branch and bound

Linderoth also looks at quadratically-constrained programs that are not convex [92]. He develops a novel method for repeatedly partitioning the continuous feasible region into the Cartesian product of triangles and rectangles. What is particularly interesting is that to do this effectively, Linderoth develops convex envelopes of bilinear functions over rectangles and triangles (also see Anstreicher and Burer's paper [7]), and then he demonstrates that these envelopes involve hyperbolic constraints which can be reformulated as the second-order cone constraints. It is interesting to compare this with the similar use of second-order cone constraints for convex quadratics (see Section 15.3.3.2).

One can view the technique of Linderoth as being a specialized "Spatial Branch-and-Bound Algorithm." In Section 15.5 we will describe the Spatial Branch-and-Bound Algorithm for global optimization in its full generality.

15.5 Global optimization

In the present section we take up the subject of global optimization of rather general nonlinear functions. This is an enormous subject, and so we will point to just a couple of directions that we view as promising. On the practical side, in Section 15.5.1 we describe the Spatial Branch-and-Bound Algorithm which is one of the most successful computational approaches in this area. In Section 15.5.2, from the viewpoint of complexity theory, with a goal of trying to elucidate the boundary between tractable and intractable, we describe some very recent work on global optimization of a very general class of nonlinear functions over an independence system.

15.5.1 Spatial Branch-and-Bound

In this section we address methods for global optimization of rather general mixed-integer nonlinear programs having non-convex relaxations. Again, to have any hope at all, we assume that the variables are bounded. There is a very large body of work on solution techniques in this space. We will not attempt to make any kind of detailed survey. Rather we will describe one very successful methodology, namely the *Spatial Branch-and-Bound Algorithm.* We refer to [123, 128] and the references therein.

The Spatial Branch-and-Bound Algorithm for mixed-integer nonlinear programming has many similarities to the ordinary branch-and-bound employed for the solution of mixed-integer linear programs, but there are many additional wrinkles. Moreover, the techniques can be integrated. In what follows, we will concentrate on how continuous nonlinearities are handled. We leave it to the reader to see how these techniques would be integrated with the associated techniques for mixed-integer linear programs.

One main difference with the mixed-integer linear case is that all nonlinear functions in a problem instance are symbolically and recursively decomposed via simple operators, until we arrive at simple functions. The simple operators should be in a limited library. For example: sum, product, quotient, exponentiation, power, logarithm, sine, cosine, absolute value. Such a decomposition is usually represented via a collection of rooted directed acyclic graphs. At each root is a nonlinear function occurring in the problem formulation. Leaves are constants, affine functions and atomic variables. Each non-leaf node is thought of as an auxiliary variable and also as representing a simple operator, and its children are the arguments of that operator.

An inequality constraint in the problem formulation can be thought of as a bounding interval on a root. In addition, the objective function is associated with a root, and so lower and upper bounds on the optimal objective value can also be thought of as a bounding interval on a root. Simple bounds on a variable in the problem formulation can be thought of as a bounding interval on a leaf. In this way, we have an extended-variable reformulation of the given problem.

Bounds are propagated up and down each such rooted directed acyclic graph via interval arithmetic and a library of convex envelopes or at least linear convex relaxations of the graphs of simple nonlinear operators acting on one or two variables on simple domains (intervals for univariate operators and simple polygons for bivariate operators). So, in this way, we have a now tractable convex or even linear relaxation of the extended-variable reformulation, and this is used to get a lower bound on the minimum objective value.

The deficiency in our relaxation is localized to the graphs of simple functions that are only approximated by convex sets. We can seek to improve bounds by branching on the interval for a variable and reconvexifying on the subproblems. For example, we may have a variable v that is a convex function f in a variable w on an interval $[l,u]$. Then the convex envelope of the graph $G[l,u] := \{(v,w) \; : \; v = f(w)\}$ is precisely

$$\tilde{G}[l,u] := \left\{(v,w) \; : \; f(w) \leq v \leq f(l) + \left(\frac{f(u)-f(l)}{u-l}\right)(w-l)\right\}.$$

We may find that at a solution of the relaxation, the values of the variables (u,v), say $(\hat{v},\hat{w}) \in \tilde{G}[l,u]$, are far from $G[l,u]$. In such a case, we can *branch* by choosing a point $b \in [l,u]$ (perhaps at or near $\hat{v}$), and forming two subproblems in which the bounding interval for v is amended as $[l,b]$ in one and $[b,u]$ in the other. The value in branching on such a continuous variable is that we now reconvexify on the subintervals, effectively replacing $\tilde{G}[l,u]$ with the smaller set $\tilde{G}[l,b] \cup \tilde{G}[b,u]$. In particular, if we did choose $b = \hat{v}$, then $(\hat{v},\hat{w}) \notin \tilde{G}[l,b] \cup \tilde{G}[l,b]$, and so the algorithm makes some progress. We note that a lot of work has gone into good branching strategies (see [15] for example).

Finally, a good Spatial Branch-and-Bound procedure should have an effective strategy for finding good feasible solutions, so as to improve the objective upper bound (for minimization problems). A good rudimentary strategy is to take the solution of a relaxation as a starting point for a continuous nonlinear-programming solver aimed at finding a locally-optimal solution of the continuous relaxation (of either the original or extended-variable formulation). Then if a feasible solution to this relaxation is obtained and if it happens to have integer values for the appropriate variables, then we have an opportunity to update the objective value upper bound. Alternatively, one can use a solver aimed mainly at mixed-integer nonlinear programs having *convex* relaxation as a heuristic also from such a starting point. In fact, the Branch-and-Bound Algorithm in `Bonmin` has options aimed at giving good solutions from such a starting point, even for non-convex problems.

The Spatial Branch-and-Bound Algorithm relies on the rapid and tight convexification of simple functions on simple domains. Therefore, considerable work has gone into developing closed-form expressions for such envelopes. This type of work has paralleled some research in mixed-integer linear programming that has focused on determining convex hulls for simple constraints. Useful results include: univariate functions [3, 122, 36], univariate monomials of odd degree [90, 91], bilinear functions [6, 98], trilinear functions [99], so-called $(n-1)$-convex functions [74], and fractional terms [127]. Further relevant work includes algorithms exploiting variable transformations and appropriate convex envelopes and relaxations. For example, for the case of "signomials" (i.e., terms of the form $a_0 x_1^{a_1} x_2^{a_2} \cdots x_n^{a_n}$, with $a_i \in \mathbb{R}$), see [106].

We do not make any attempt to exhaustively review available software for global optimization. Rather we just mention that state-of-the-art codes implementing a

Spatial Branch-and-Bound Algorithm include `Baron` [113, 128, 129] and the new open-source code `Couenne` [15].

15.5.2 Boundary cases of complexity

Now, we shift our attention back to the viewpoint of complexity theory. Our goal is to sample a bit of the recent work that is aimed at revealing the boundary between tractable and intractable instances of nonlinear discrete optimization problems. We describe some very recent work on global optimization of a very general class of nonlinear functions over an independence system (see [86]). Other work in this vein includes [17, 19].

Specifically, we consider the problem of optimizing a nonlinear objective function over a weighted independence system presented by a linear-optimization oracle. While this problem is generally intractable, we are able to provide a polynomial-time algorithm that determines an "r-best" solution for nonlinear functions of the total weight of an independent set, where r is a constant that depends on certain Frobenius numbers of the individual weights and is independent of the size of the ground set.

An *independence system* is a nonempty set of vectors $S \subseteq \{0,1\}^n$ with the property that $\mathbf{x} \in \{0,1\}^n$, $\mathbf{x} \leqslant \mathbf{y} \in S$ implies $\mathbf{x} \in S$. The general nonlinear optimization problem over a multiply-weighted independence system is as follows. Given an independence system $S \subseteq \{0,1\}^n$, weight vectors $\mathbf{w}_1, \ldots, \mathbf{w}_d \in \mathbb{Z}^n$, and a function $f : \mathbb{Z}^d \to \mathbb{R}$, find $\mathbf{x} \in S$ minimizing the objective $f(\mathbf{w}_1^\top \mathbf{x}, \ldots, \mathbf{w}_d^\top \mathbf{x})$.

The representation of the objective in the above composite form has several advantages. First, for $d > 1$, it can naturally be interpreted as *multi-criteria optimization*: the d given weight vectors $\mathbf{w}_1, \ldots, \mathbf{w}_d$ represent d different criteria, where the value of $\mathbf{x} \in S$ under criterion i is its i-th total weight $\mathbf{w}_i^\top \mathbf{x}$ and the objective is to minimize the "balancing" $f(\mathbf{w}_1^\top \mathbf{x}, \ldots, \mathbf{w}_d^\top \mathbf{x})$ of the d given criteria by the given function f. Second, it allows us to classify nonlinear optimization problems into a hierarchy of increasing generality and complexity: at the bottom lies standard linear optimization, recovered with $d = 1$ and f the identity on $\mathbb{Z}$; and at the top lies the problem of maximizing an arbitrary function, which is typically intractable, arising with $d = n$ and $\mathbf{w}_i = \mathbf{1}_i$ the i-th standard unit vector in $\mathbb{Z}^n$ for all i.

The computational complexity of the problem depends on the number d of weight vectors, on the weights $w_{i,j}$, on the type of function f and its presentation, and on the type of independence system S and its presentation. For example, when S is a *matroid*, the problem can be solved in polynomial time for any fixed d, any $\{0,1,\ldots,p\}$-valued weights $w_{i,j}$ with p fixed, and any function f presented by a *comparison oracle*, even when S is presented by a mere *membership oracle*, see [17]. Also, for example, when S consists of the *matchings* in a given bipartite graph G, the problem can be solved in polynomial time for any fixed d, any weights $w_{i,j}$ presented in unary, and any *concave* function f, see [18]; but on the other hand, for *convex* f, already with fixed $d = 2$ and $\{0,1\}$-valued weights $w_{i,j}$, the problem includes as a special case the *exact matching problem* whose complexity is long open [101, 103].

In view of the difficulty of the problem already for $d = 2$, we take a first step and concentrate on *nonlinear optimization over a (singly) weighted independence system*, that is, with $d = 1$, single weight vector $\mathbf{w} = (w_1, \ldots, w_n)$, and univariate function $f : \mathbb{Z} \to \mathbb{R}$. The function f can be arbitrary and is presented by a *comparison oracle* that, queried on $\mathbf{x}, \mathbf{y} \in \mathbb{Z}$, asserts whether or not $f(\mathbf{x}) \leqslant f(\mathbf{y})$. The weights w_j take on values in a p-tuple $\mathbf{a} = (a_1, \ldots, a_p)$ of positive integers. Without loss of generality we assume that $\mathbf{a} = (a_1, \ldots, a_p)$ is *primitive*, by which we mean that the a_i are distinct positive integers whose greatest common divisor $\gcd(\mathbf{a}) := \gcd(a_1, \ldots, a_p)$ is 1. The independence system S is presented by a *linear-optimization oracle* that, queried on vector $\mathbf{c} \in \mathbb{Z}^n$, returns an element $\mathbf{x} \in S$ that maximizes the linear function $\mathbf{c}^\top \mathbf{x} = \sum_{j=1}^n c_j x_j$. It turns out that this problem is already quite intriguing, and so we settle for an approximative solution in the following sense, that is interesting in its own right. For a nonnegative integer r, we say that $\mathbf{x}^* \in S$ is an *r-best solution* to the optimization problem over S if there are at most r better objective values attained by feasible solutions. In particular, a 0-best solution is optimal. Recall that the *Frobenius number* of a primitive $\mathbf{a}$ is the largest integer $\mathrm{F}(\mathbf{a})$ that is not expressible as a nonnegative integer combination of the a_i. We prove the following theorem.

Theorem 15.22. *For every primitive p-tuple* $\mathbf{a} = (a_1, \ldots, a_p)$*, there is a constant* $r(\mathbf{a})$ *and an algorithm that, given any independence system* $S \subseteq \{0,1\}^n$ *presented by a linear-optimization oracle, weight vector* $\mathbf{w} \in \{a_1, \ldots, a_p\}^n$*, and function* $f : \mathbb{Z} \to \mathbb{R}$ *presented by a comparison oracle, provides an* $r(\mathbf{a})$*-best solution to the nonlinear problem* $\min\{f(\mathbf{w}^\top \mathbf{x}) \ : \ \mathbf{x} \in S\}$*, in time polynomial in n. Moreover:*

1. *If a_i divides a_{i+1} for $i = 1, \ldots, p-1$, then the algorithm provides an optimal solution.*
2. *For $p = 2$, that is, for $\mathbf{a} = (a_1, a_2)$, the algorithm provides an* $\mathrm{F}(\mathbf{a})$*-best solution.*

In fact, we give an explicit upper bound on $r(\mathbf{a})$ in terms of the Frobenius numbers of certain subtuples derived from $\mathbf{a}$. An interesting special case is that of $\mathbf{a} = (2,3)$. Because $\mathrm{F}(2,3) = 1$, the solution provided by our algorithm in that case is either optimal or second best.

The proof of Theorem 15.22 is pretty technical, so we only outline the main ideas. Below we present a naïve solution strategy that does not directly lead to a good approximation. However, this naïve approach is used as a basic building block. One partitions the independence system into suitable pieces, to each of which a suitable refinement of the naïve strategy is applied separately. Considering the monoid generated by $\{a_1, \ldots, a_p\}$ allows one to show that the refined naïve strategy applied to each piece gives a good approximation within that piece. In this way, the approximation quality $r(\mathbf{a})$ can be bounded as follows, establishing a proof to Theorem 15.22.

Lemma 15.6. *Let* $\mathbf{a} = (a_1, \ldots, a_p)$ *be any primitive p-tuple. Then the following hold:*

1. *An upper bound on* $r(\mathbf{a})$ *is given by* $r(\mathbf{a}) \leqslant (2\max(\mathbf{a}))^p$.

2. *For divisible* $\mathbf{a}$, *we have* $r(\mathbf{a}) = 0$.
3. *For* $p = 2$, *that is, for* $\mathbf{a} = (a_1, a_2)$, *we have* $r(\mathbf{a}) = \mathrm{F}(\mathbf{a})$.

Before we continue, let us fix some notation. The *indicator* of a subset $J \subseteq N$ is the vector $\mathbf{1}_J := \sum_{j \in J} \mathbf{1}_j \in \{0,1\}^n$, so that $\mathrm{supp}(\mathbf{1}_J) = J$. Unless otherwise specified, $\mathbf{x}$ denotes an element of $\{0,1\}^n$ and $\boldsymbol{\lambda}, \boldsymbol{\tau}, \boldsymbol{\nu}$ denote elements of $\mathbb{Z}_+^p$. Throughout, $\mathbf{a} = (a_1, \ldots, a_p)$ is a *primitive* p-tuple. We will be working with weights taking values in $\mathbf{a}$, that is, vectors $\mathbf{w} \in \{a_1, \ldots, a_p\}^n$. With such a weight vector w being clear from the context, we let $N_i := \{j \in N \ : \ w_j = a_i\}$ for $i = 1, \ldots, p$, so that $N = \biguplus_{i=1}^p N_i$. For $\mathbf{x} \in \{0,1\}^n$ we let $\lambda_i(\mathbf{x}) := |\mathrm{supp}(\mathbf{x}) \cap N_i|$ for $i = 1, \ldots, p$, and $\boldsymbol{\lambda}(\mathbf{x}) := (\lambda_1(\mathbf{x}), \ldots, \lambda_p(\mathbf{x}))$, so that $\mathbf{w}^\top \mathbf{x} = \boldsymbol{\lambda}(\mathbf{x})^\top \mathbf{a}$. For integers $z, s \in \mathbb{Z}$ and a set of integers $Z \subseteq \mathbb{Z}$, we define $z + sZ := \{z + sx \ : \ x \in Z\}$.

Let us now present the naïve strategy to solve the univariate nonlinear problem $\min\{f(\mathbf{w}^\top \mathbf{x}) \ : \ \mathbf{x} \in S\}$. Consider a set $S \subseteq \{0,1\}^n$, weight vector $\mathbf{w} \in \{a_1, \ldots, a_p\}^n$, and function $f : \mathbb{Z} \to \mathbb{R}$ presented by a comparison oracle. Define the *image* of S under $\mathbf{w}$ to be the set of values $\mathbf{w}^\top \mathbf{x}$ taken by elements of S; we denote it by $\mathbf{w} \cdot S$.

We point out the following simple observation.

Proposition 15.4. *A necessary condition for any algorithm to find an r-best solution to the problem* $\min\{f(\mathbf{w}^\top \mathbf{x}) \ : \ x \in S\}$, *where the function f is presented by a comparison oracle only, is that it computes all but at most r values of the image* $\mathbf{w} \cdot S$ *of S under* $\mathbf{w}$.

Note that this necessary condition is also sufficient for computing the *objective value* $f(\mathbf{w}^\top \mathbf{x}^*)$ of an r-best solution, but not for computing an actual r-best solution $\mathbf{x}^* \in S$, which may be harder. Any point $\bar{\mathbf{x}}$ attaining $\max\{\mathbf{w}^\top \mathbf{x} \ : \ \mathbf{x} \in S\}$ provides an approximation of the image given by

$$\{\mathbf{w}^\top \mathbf{x} \ : \ \mathbf{x} \leqslant \bar{\mathbf{x}}\} \subseteq \mathbf{w} \cdot S \subseteq \{0, 1, \ldots, \mathbf{w}^\top \bar{\mathbf{x}}\}\,. \tag{15.30}$$

This suggests the following natural naïve strategy for finding an approximative solution to the optimization problem over an independence system S that is presented by a linear-optimization oracle.

(Naïve Strategy)
input Independence system $S \subseteq \{0,1\}^n$ presented by a linear-optimization oracle, $f : \mathbb{Z} \to \mathbb{R}$ presented by a comparison oracle, and $\mathbf{w} \in \{a_1, \ldots, a_p\}^n$
obtain $\bar{\mathbf{x}}$ attaining $\max\{\mathbf{w}^\top \mathbf{x} \ : \ \mathbf{x} \in S\}$ using the linear-optimization oracle for S
output $\mathbf{x}^*$ as one attaining $\min\{f(\mathbf{w}^\top \mathbf{x}) \ : \ \mathbf{x} \leqslant \bar{\mathbf{x}}\}$ using the algorithm of Lemma 15.7 below.

Unfortunately, as the next example shows, the number of values of the image that are missing from the approximating set on the left-hand side of equation (15.30) cannot generally be bounded by any constant. So by Proposition 15.4, this strategy cannot be used *as is* to obtain a provably good approximation.

Example 15.3. Let $\mathbf{a} := (1,2)$, $n := 4m$, $\mathbf{y} := \sum_{i=1}^{2m} \mathbf{1}_i$, $\mathbf{z} := \sum_{i=2m+1}^{4m} \mathbf{1}_i$, and $\mathbf{w} := \mathbf{y} + 2\mathbf{z}$, i.e.,

$$\mathbf{y} = (1,\ldots,1,0,\ldots,0), \quad \mathbf{z} = (0,\ldots,0,1,\ldots,1), \quad \mathbf{w} = (1,\ldots,1,2,\ldots,2),$$

define f on $\mathbb{Z}$ by

$$f(k) := \begin{cases} k, & k \text{ odd}, \\ 2m, & k \text{ even}, \end{cases}$$

and let S be the independence system

$$S := \{\mathbf{x} \in \{0,1\}^n \, : \, \mathbf{x} \leqslant \mathbf{y}\} \cup \{\mathbf{x} \in \{0,1\}^n \, : \, \mathbf{x} \leqslant \mathbf{z}\}.$$

Then the unique optimal solution of the linear-objective problem $\max\{\mathbf{w}^\top \mathbf{x} \, : \, \mathbf{x} \in S\}$ is $\bar{\mathbf{x}} := \mathbf{z}$, with $\mathbf{w}^\top \bar{\mathbf{x}} = 4m$, and therefore

$$\begin{aligned} \{\mathbf{w}^\top \mathbf{x} \, : \, \mathbf{x} \leqslant \bar{\mathbf{x}}\} &= \{2i \, : \, i = 0,1,\ldots,2m\} \text{ and} \\ \mathbf{w} \cdot S &= \{i \, : \, i = 0,1,\ldots,2m\} \cup \{2i \, : \, i = 0,1,\ldots,2m\}. \end{aligned}$$

So all m odd values (i.e., $1,3,\ldots,2m-1$) in the image $\mathbf{w} \cdot S$ are missing from the approximating set $\{\mathbf{w}^\top \mathbf{x} \, : \, \mathbf{x} \leqslant \bar{\mathbf{x}}\}$ on the left-hand side of (15.30), and $\mathbf{x}^*$ attaining $\min\{f(\mathbf{w}^\top \mathbf{x}) \, : \, \mathbf{x} \leqslant \bar{\mathbf{x}}\}$ output by the above strategy has objective value $f(\mathbf{w}^\top \mathbf{x}^*) = 2m$, while there are $m = \frac{n}{4}$ better objective values (i.e., $1,3,\ldots,2m-1$) attainable by feasible points (e.g., $\sum_{i=1}^{k} \mathbf{1}_i$, for $k = 1,3,\ldots,2m-1$).

Nonetheless, a more sophisticated refinement of the naïve strategy, applied repeatedly to several suitably chosen subsets of S rather than S itself, will lead to a good approximation. Note that the naïve strategy can be efficiently implemented as follows.

Lemma 15.7. *For every fixed p-tuple* $\mathbf{a}$*, there is a polynomial-time algorithm that, given univariate function* $f : \mathbb{Z} \to \mathbb{R}$ *presented by a comparison oracle, weight vector* $\mathbf{w} \in \{a_1,\ldots,a_p\}^n$*, and* $\bar{\mathbf{x}} \in \{0,1\}^n$*, solves* $\min\{f(\mathbf{w}^\top \mathbf{x}) \, : \, \mathbf{x} \leqslant \bar{\mathbf{x}}\}$.

Proof. Consider the following algorithm:

---

input function $f : \mathbb{Z} \to \mathbb{R}$ presented by a comparison oracle, $\mathbf{w} \in \{a_1,\ldots,a_p\}^n$ and $\bar{\mathbf{x}} \in \{0,1\}^n$
let $N_i := \{j : w_j = a_i\}$ and $\tau_i := \lambda_i(\bar{\mathbf{x}}) = |\mathrm{supp}(\bar{\mathbf{x}}) \cap N_i|$, $i = 1,\ldots,p$.
For *every choice of* $\boldsymbol{\nu} = (\nu_1,\ldots,\nu_p) \leqslant (\tau_1,\ldots,\tau_p) = \boldsymbol{\tau}$
determine some $\mathbf{x}_{\boldsymbol{\nu}} \leqslant \bar{\mathbf{x}}$ with $\lambda_i(\mathbf{x}_{\boldsymbol{\nu}}) = |\mathrm{supp}(\mathbf{x}_{\boldsymbol{\nu}}) \cap N_i| = \nu_i$, $i = 1,\ldots,p$.
output $\mathbf{x}^*$ as one minimizing $f(\mathbf{w}^\top \mathbf{x})$ among the $\mathbf{x}_{\boldsymbol{\nu}}$ by using the comparison oracle of f.

As the value $\mathbf{w}^\top \mathbf{x}$ depends only on the cardinalities $|\mathrm{supp}(\mathbf{x}) \cap N_i|$, $i = 1,\ldots,p$, it is clear that

$$\{\mathbf{w}^\top \mathbf{x} \, : \, \mathbf{x} \leqslant \bar{\mathbf{x}}\} = \{\mathbf{w}^\top \mathbf{x}_{\boldsymbol{\nu}} \, : \, \boldsymbol{\nu} \leqslant \boldsymbol{\tau}\}.$$

Clearly, for each choice $\boldsymbol{\nu} \leqslant \boldsymbol{\tau}$ it is easy to determine some $\mathbf{x}_{\boldsymbol{\nu}} \leqslant \bar{\mathbf{x}}$ by zeroing out suitable entries of $\bar{\mathbf{x}}$. The number of choices $\boldsymbol{\nu} \leqslant \boldsymbol{\tau}$ and hence of loop iterations and

Objective function	Constraints: Linear	Constraints: Convex Polynomial	Constraints: Arbitrary Polynomial
Linear	**Polynomial-time in fixed dimension:** – Lenstra's algorithm [88] – Generalized basis reduction, Lovász–Scarf [95] – Short rational generating functions, Barvinok [13]	**Polynomial-time in fixed dimension:** Lenstra-type algorithms **(15.3.1)** – Khachiyan–Porkolab [80] – Heinz [67]	**Incomputable:** Hilbert's 10th problem, Matiyasevich [96] **(15.1)**, even for: – quadratic constraints, Jeroslow [76] – fixed dimension 10, Matiyasevich [77]
Convex max **(15.2)**	**Polynomial-time in fixed dimension:** Cook et al. [39] **(15.2.1)**		**Incomputable (15.1)**
Convex min **(15.3)**	**Polynomial-time in fixed dimension:** Lenstra-type algorithms: Khachiyan–Porkolab [80], Heinz [67] **(15.3.1)**		**Incomputable (15.1)**
Arbitrary Polynomial **(15.4)**	**NP-hard, inapproximable,** even for quadratic forms over hypercubes: MAX-CUT, Håstad [66] **(15.1)** **NP-hard,** even for fixed dimension 2, degree 4 **(15.1)** **FPTAS in fixed dimension:** Short rational generating functions, De Loera et al. [45] **(15.4.1)**		**Incomputable (15.1)**

Table 15.1 Computational complexity and algorithms for nonlinear integer optimization.

comparison-oracle queries of f to determine $\mathbf{x}^*$ is

$$\prod_{i=1}^{p}(\tau_i+1) \leqslant (n+1)^p.$$

15.6 Conclusions

In this chapter, we hope to have succeeded in reviewing mixed-integer nonlinear programming from two important viewpoints.

We have reviewed the computational complexity of several important classes of mixed-integer nonlinear programs. Some of the negative complexity results (incomputability, NP-hardness) that appeared in Section 15.1 have been supplemented by polynomiality or approximability results in fixed dimension. Table 15.1 gives a summary. In addition to that, and not shown in the table, we have explored the boundary between tractable and intractable problems, by highlighting interesting cases in varying dimension where still polynomiality results can be obtained.

Additionally, we have reviewed a selection of practical algorithms that seem to have the greatest potential from today's point of view. Many of these algorithms, at their core, are aimed at *integer convex minimization*. Here we have nonlinear branch-and-bound, outer approximation, the Quesada–Grossman algorithm, hybrid algorithms, and generalized Benders decomposition. As we have reported, such approaches can be specialized and enhanced for problems with SDP constraints, SOCP constraints, and (convex) quadratics. For *integer polynomial programming* (without convexity assumptions), the toolbox of Positivstellensätze and SOS programming is available. For the case of *quadratics* (without convexity assumptions), specialized versions of disjunctive programming, branch-and-cut, and branch-and-bound have been devised. Finally, for general *global optimization*, spatial branch-and-bound is available as a technique, which relies heavily on convexification methods.

It is our hope that, by presenting these two viewpoints to the interested reader, this chapter will help to create a synergy between both viewpoints in the near future. May this lead to a better understanding of the field, and to much better algorithms than what we have today!

Acknowledgments

We would like to thank our coauthors, Jesús De Loera and Shmuel Onn, for their permission to base the presentation of some of the material in this chapter on our joint papers [47, 87, 86].

References

1. 4ti2 team, *4ti2 – a software package for algebraic, geometric and combinatorial problems on linear spaces*, Available at http://www.4ti2.de.
2. K. Abhishek, S. Leyffer, and J.T. Linderoth, *Filmint: An outer-approximation-based solver for nonlinear mixed integer programs*, Preprint ANL/MCS-P1374-0906, 2006.
3. C.S. Adjiman, *Global optimization techniques for process systems engineering*, Ph.D. thesis, Princeton University, June 1998.
4. R.K. Ahuja, T.L. Magnanti, and J.B. Orlin, *Network flows: Theory, algorithms, and applications*, Prentice-Hall, Inc., New Jersey, 1993.
5. S. Aktürk, A. Atamtürk, and S. Gürel, *A strong conic quadratic reformulation for machine-job assignment with controllable processing times*, Operations Research Letters 37 (2009) 187–191.
6. F.A. Al-Khayyal and J.E. Falk, *Jointly constrained biconvex programming*, Mathematics of Operations Research 8 (1983) 273–286.

7. K. Anstreicher and S. Burer, *Computable representations for convex hulls of low-dimensional quadratic forms*, Technical report, Department of Management Sciences, University of Iowa, 2007.
8. E. Balas, *Disjunctive programming: Properties of the convex hull of feasible points*, MSRR No. 330, Carnegie Mellon University, Pittsburgh, 1974.
9. E. Balas, *Nonconvex quadratic programming via generalized polars*, SIAM Journal on Applied Mathematics 28 (1975) 335–349.
10. E. Balas, *Disjunctive programming: Properties of the convex hull of feasible points*, Discrete Applied Mathematics 89 (1998) 3–44.
11. B. Bank, J. Heintz, T. Krick, R. Mandel, and P. Solernó, *Une borne optimale pour la programmation entiére quasi-convexe*, Bull. Soc. math. France 121 (1993) 299–314.
12. B. Bank, J. Heintz, T. Krick, R. Mandel, and P. Solernó, *A geometrical bound for integer programming with polynomial constraints*, Fundamentals of Computation Theory, Lecture Notes In Computer Science 529, Springer, 1991, pp. 121–125.
13. A.I. Barvinok, *A polynomial time algorithm for counting integral points in polyhedra when the dimension is fixed*, Mathematics of Operations Research 19 (1994) 769–779.
14. A.I. Barvinok and J.E. Pommersheim, *An algorithmic theory of lattice points in polyhedra*, New Perspectives in Algebraic Combinatorics (L.J. Billera, A. Björner, C. Greene, R.E. Simion, and R.P. Stanley eds.), Math. Sci. Res. Inst. Publ., Vol. 38, Cambridge Univ. Press, Cambridge, 1999, pp. 91–147.
15. P. Belotti, J. Lee, L. Liberti, F. Margot, and A. Wächter, *Branching and bounds tightening techniques for non-convex MINLP*, IBM Research Report RC24620, 2008.
16. A. Ben-Tal and A. Nemirovski, *Lectures on modern convex optimization: Analysis, algorithms, and engineering applications*, MPS-SIAM Series on Optimization, SIAM, Philadelphia, USA, 2001.
17. Y. Berstein, J. Lee, H. Maruri-Aguilar, S. Onn, E. Riccomagno, R. Weismantel, and H. Wynn, *Nonlinear matroid optimization and experimental design*, SIAM Journal on Discrete Mathematics 22 (2008) 901–919.
18. Y. Berstein and S. Onn, *Nonlinear bipartite matching*, Discrete Optimization 5 (2008) 53–65.
19. Y. Berstein, J. Lee, S. Onn, and R. Weismantel, *Nonlinear optimization for matroid intersection and extensions*, IBM Research Report RC24610, 2008.
20. D. Bertsimas and R. Weismantel, *Optimization over integers*, Dynamic Ideas, Belmont, Ma., 2005.
21. Biq-Mac Solver - Binary quadratic and Max cut Solver, biqmac.uni-klu.ac.at, 2006.
22. P. Bonami, L. Biegler, A. Conn, G. Cornuéjols, I. Grossmann, C. Laird, J. Lee, A. Lodi, F. Margot, N. Sawaya, and A. Wächter, *An algorithmic framework for convex mixed integer nonlinear programs*, Discrete Optimization 5 (2008) 186–204.
23. P. Bonami, G. Cornuéjols, A. Lodi, and F. Margot, *A feasibility pump for mixed integer nonlinear programs*, Mathematical Programming 119 (2009) 331–352.
24. P. Bonami and J. Lee, *Bonmin users' manual*, Technical report, June 2006.
25. P. Bonami and M.A. Lejeune, *An exact solution approach for integer constrained portfolio optimization problems under stochastic constraints*, Operations Research 57 (2009) 650–670.
26. P. Bonami, J. Forrest, J. Lee, and A. Wächter, *Rapid development of an MINLP solver with COIN-OR*, Optima 75 (2007) 1–5.
27. P. Bonami and J.P.M. Gonçalves, *Primal heuristics for mixed integer nonlinear programs*, IBM Research Report RC24639, 2008.
28. Bonmin, neos.mcs.anl.gov/neos/solvers/minco:Bonmin/AMPL.html.
29. Bonmin, projects.coin-or.org/Bonmin, v. 0.99.
30. E. Boros and P.L. Hammer, *Pseudo-Boolean optimization*, Discrete Applied Mathematics 123 (2002) 155–225.
31. M. Brion, *Points entiers dans les polyédres convexes*, Ann. Sci. École Norm. Sup. 21 (1988) 653–663.
32. M. Brion and M. Vergne, *Residue formulae, vector partition functions and lattice points in rational polytopes*, J. Amer. Math. Soc. 10 (1997) 797–833.

33. C. Buchheim and G. Rinaldi, *Efficient reduction of polynomial zero-one optimization to the quadratic case*, SIAM Journal on Optimization 18 (2007) 1398–1413.
34. S. Burer and D. Vandenbussche, *A finite branch-and-bound algorithm for nonconvex quadratic programming via semidefinite relaxations*, Mathematical Programming 113 (2008) 259–282.
35. R.E. Burkard, E. Çela, and L. Pitsoulis, *The quadratic assignment problem*, Handbook of Combinatorial Optimization (Dordrecht), Computer-aided chemical engineering, Kluwer Academic Publishers, 1998, pp. 241–339.
36. S. Ceria and J. Soares, *Convex programming for disjunctive convex optimization*, Mathematical Programming 86 (1999) 595–614.
37. M.T. Çezik and G. Iyengar, *Cuts for mixed 0-1 conic programming*, Mathematical Programming 104 (2005) 179–202.
38. M.D. Choi, T.Y. Lam, and B. Reznick, *Sums of squares of real polynomials*, Proceedings of symposia in pure mathematics 58 (1995) 103–126.
39. W.J. Cook, M.E. Hartmann, R. Kannan, and C. McDiarmid, *On integer points in polyhedra*, Combinatorica 12 (1992) 27–37.
40. GAMS Development Corp., *DICOPT*, www.gams.com/dd/docs/solvers/dicopt.pdf.
41. J.A. De Loera and S. Onn, *The complexity of three-way statistical tables*, SIAM Journal of Computing 33 (2004) 819–836.
42. J.A. De Loera and S. Onn, *All linear and integer programs are slim 3-way transportation programs*, SIAM Journal of Optimization 17 (2006) 806–821.
43. J.A. De Loera and S. Onn, *Markov bases of three-way tables are arbitrarily complicated*, Journal of Symbolic Computation 41 (2006) 173–181.
44. J.A. De Loera, R. Hemmecke, M. Köppe, and R. Weismantel, *FPTAS for mixed-integer polynomial optimization with a fixed number of variables*, 17th ACM-SIAM Symposium on Discrete Algorithms, 2006, pp. 743–748.
45. J.A. De Loera, R. Hemmecke, M. Köppe, and R. Weismantel, *Integer polynomial optimization in fixed dimension*, Mathematics of Operations Research 31 (2006) 147–153.
46. J.A. De Loera, R. Hemmecke, M. Köppe, and R. Weismantel, *FPTAS for optimizing polynomials over the mixed-integer points of polytopes in fixed dimension*, Mathematical Programming 118 (2008) 273–290.
47. J.A. De Loera, R. Hemmecke, S. Onn, and R. Weismantel, *N-fold integer programming*, Discrete Optimization 5 (2008), 231–241.
48. S. Drewes and S. Ulbrich, *Mixed integer second order cone programming*, IMA Hot Topics Workshop, Mixed-Integer Nonlinear Optimization: Algorithmic Advances and Applications, November 17–21, 2008.
49. M.A. Duran and I.E. Grossmann, *An outer-approximation algorithm for a class of mixed-integer nonlinear programs*, Mathematical Programming 36 (1986) 307–339.
50. M.A. Duran and I.E. Grossmann, *Erratum: "An outer-approximation algorithm for a class of mixed-integer nonlinear programs" [Mathematical Programming 36 (1986) 307–339]*, Mathematical Programming 39 (1987) 337.
51. FilMINT, www-neos.mcs.anl.gov/neos/solvers/minco:FilMINT/AMPL.html.
52. A. Frangioni and C. Gentile, *Perspective cuts for a class of convex 0-1 mixed integer programs*, Mathematical Programmming 106 (2006) 225–236.
53. A. Frangioni and C. Gentile, *A computational comparison of reformulations of the perspective relaxation: SOCP vs. cutting planes*, Operations Research Letters 37 (2009) 206–210.
54. M.R. Garey and D.S. Johnson, *Computers and intractability: A guide to the theory of NP-completeness*, W.H. Freeman and Company, New York, NY, 1979.
55. A.M. Geoffrion, *Generalized Benders decomposition*, J. Optimization Theory Appl. 10 (1972) 237–260.
56. M.X. Goemans and D.P. Williamson, *Improved approximation algorithms for maximum cut and satisfiability problems using semidefinite programming*, Journal of the ACM 42 (1995) 1115–1145.

57. D. Goldfarb, S.C. Liu, and S.Y. Wang, *A logarithmic barrier function algorithm for quadratically constrained convex quadratic programming*, SIAM Journal on Optimization 1 (1991) 252–267.
58. R.E. Gomory, *An algorithm for integer solutions to linear programs*, Princeton IBM Mathematics Research Project, Technical Report No. 1, Princeton University, November 17, 1958.
59. R.E. Gomory, *Outline of an algorithm for integer solutions to linear programs*, Bulletin of the American Mathematical Society 64 (1958) 275–278.
60. J.E. Graver, *On the foundations of linear and integer linear programming I*, Mathematical Programming 8 (1975) 207–226.
61. M. Grötschel, L. Lovász, and A. Schrijver, *Geometric algorithms and combinatorial optimization*, Springer, 1988.
62. O. Günlük and J. Linderoth, *Perspective relaxation of mixed integer nonlinear programs with indicator variables*, Integer Programming and Combinatorial Optimization 2008 – Bertinoro, Italy (A. Lodi, A. Panconesi, and G. Rinaldi, eds.), Lecture Notes in Computer Science 5035, Springer, 2008, pp. 1–16.
63. O. Günlük and J. Linderoth, *Perspective reformulations of mixed integer nonlinear programs with indicator variables*, Optimization Technical Report, ISyE Department, University of Wisconsin-Madison, June 20, 2008.
64. O.K. Gupta and A. Ravindran, *Branch and bound experiments in convex nonlinear integer programming*, Management Sci. 31 (1985) 1533–1546.
65. M.E. Hartmann, *Cutting planes and the complexity of the integer hull*, Phd thesis, Cornell University, Department of Operations Research and Industrial Engineering, Ithaca, NY, 1989.
66. J. Håstad, *Some optimal inapproximability results*, Proceedings of the 29th Symposium on the Theory of Computing (STOC), ACM, 1997, pp. 1–10.
67. S. Heinz, *Complexity of integer quasiconvex polynomial optimization*, Journal of Complexity 21 (2005) 543–556.
68. R. Hemmecke, *On the positive sum property and the computation of Graver test sets*, Mathematical Programming 96 (2003) 247–269.
69. R. Hemmecke, M. Köppe, and R. Weismantel, *Oracle-polynomial time convex mixed-integer minimization*, Manuscript, 2008.
70. R. Hemmecke, S. Onn, and R. Weismantel, *A polynomial oracle-time algorithm for convex integer minimization*, Manuscript, 2008.
71. J.-B. Hiriart-Urruty and C. Lemaréchal, *Convex analysis and minimization algorithms ii: Advanced theory and bundle methods.*, Grundlehren der Mathematischen Wissenschaften 306, Springer, 1993.
72. S. Hoşten and S. Sullivant, *A finiteness theorem for Markov bases of hierarchical models*, Journal of Combinatorial Theory Ser. A 114 (2007) 311–321.
73. Ilog-Cplex, www.ilog.com/products/cplex, v. 10.1.
74. M. Jach, D. Michaels, and R. Weismantel, *The convex envelope of (n-1)-convex functions*, SIAM Journal on Optimization 19 (2008), 1451–1466.
75. T. Jacobi and A. Prestel, *Distinguished representations of strictly positive polynomials*, Journal für die Reine und Angewandte Mathematik 532 (2001) 223–235.
76. R.G. Jeroslow, *There cannot be any algorithm for integer programming with quadratic constraints*, Operations Research 21 (1973) 221–224.
77. J.P. Jones, *Universal diophantine equation*, Journal of Symbolic Logic 47 (1982) 403–410.
78. J.E. Kelley, Jr., *The cutting-plane method for solving convex programs*, Journal of the Society for Industrial and Applied Mathematics 8 (1960) 703–712.
79. L.G. Khachiyan, *Convexity and complexity in polynomial programming*, Proceedings of the International Congress of Mathematicians, August 16–24, 1983, Warszawa (New York) (Zbigniew Ciesielski and Czesław Olech, eds.), North-Holland, 1984, pp. 1569–1577.
80. L.G. Khachiyan and L. Porkolab, *Integer optimization on convex semialgebraic sets.*, Discrete and Computational Geometry 23 (2000) 207–224.
81. M. Köppe, *A primal Barvinok algorithm based on irrational decompositions*, SIAM Journal on Discrete Mathematics 21 (2007) 220–236.

82. M. Köppe and S. Verdoolaege, *Computing parametric rational generating functions with a primal Barvinok algorithm*, The Electronic Journal of Combinatorics 15 (2008) 1–19, #R16.
83. J.B. Lasserre, *Global optimization with polynomials and the problem of moments*, SIAM Journal on Optimization 11 (2001) 796–817.
84. M. Laurent, *A comparison of the Sherali–Adams, Lovász–Schrijver and Lasserre relaxations for 0-1 programming*, Mathematics of Operations Research 28 (2003) 470–496.
85. J. Lee, *In situ column generation for a cutting-stock problem*, Computers & Operations Research 34 (2007) 2345–2358.
86. J. Lee, S. Onn, and R. Weismantel, *Nonlinear optimization over a weighted independence system*, IBM Research Report RC24513, 2008.
87. J. Lee, S. Onn, and R. Weismantel, *On test sets for nonlinear integer maximization*, Operations Research Letters 36 (2008) 439–443.
88. H.W. Lenstra, *Integer programming with a fixed number of variables*, Mathematics of Operations Research 8 (1983) 538–548.
89. S. Leyffer, *User manual for MINLP_BB*, Technical report, University of Dundee, UK, March 1999.
90. L. Liberti, *Comparison of convex relaxations for monomials of odd degree*, Optimization and Optimal Control (I. Tseveendorj, P.M. Pardalos, and R. Enkhbat, eds.), World Scientific, 2003.
91. L. Liberti and C.C. Pantelides, *Convex envelopes of monomials of odd degree*, Journal of Global Optimization 25 (2003) 157–168.
92. J. Linderoth, *A simplicial branch-and-bound algorithm for solving quadratically constrained quadratic programs*, Mathematical Programming 103 (2005) 251–282.
93. M. Sousa Lobo, L. Vandenberghe, S. Boyd, and H. Lebret, *Applications of second-order cone programming*, Linear Algebra Appl. 284 (1998) 193–228, ILAS Symposium on Fast Algorithms for Control, Signals and Image Processing (Winnipeg, MB, 1997).
94. LOQO, www.princeton.edu/~rvdb, v. 4.05.
95. L. Lovász and H.E. Scarf, *The generalized basis reduction algorithm*, Mathematics of Operations Research 17 (1992) 751–764.
96. Y.V. Matiyasevich, *Enumerable sets are diophantine*, Doklady Akademii Nauk SSSR 191 (1970) 279–282, (Russian); English translation, Soviet Mathematics Doklady 11 (1970) 354–357.
97. Y.V. Matiyasevich, *Hilbert's tenth problem*, The MIT Press, Cambridge, MA, USA, 1993.
98. G.P. McCormick, *Computability of global solutions to factorable nonconvex programs: Part i — convex underestimating problems*, Mathematical Programming 10 (1976) 146–175.
99. C.A. Meyer and C.A. Floudas, *Trilinear monomials with mixed sign domains: Facets of the convex and concave envelopes*, Journal of Global Optimization 29 (2004) 125–155.
100. MOSEK, www.mosek.com, v. 5.0.
101. K. Mulmuley, U.V. Vazirani, and V.V. Vazirani, *Matching is as easy as matrix inversion*, Combinatorica 7 (1987) 105–113.
102. S. Onn and U.G. Rothblum, *Convex combinatorial optimization*, Disc. Comp. Geom. 32 (2004) 549–566.
103. C.H. Papadimitriou and M. Yanakakis, *The complexity of restricted spanning tree problems*, Journal of the Association for Computing Machinery 29 (1982) 285–309.
104. P.M. Pardalos, F. Rendl, and H. Wolkowicz, *The quadratic assignment problem: A survey and recent developments.*, Quadratic Assignment and Related Problems (P.M. Pardalos and H. Wolkowicz, eds.), DIMACS Series in Discrete Mathematics and Theoretical Computer Science, vol. 16, American Mathematical Society, DIMACS Workshop May 20–21, 1993 1994, pp. 1–42.
105. P.A. Parrilo, *Semidefinite programming relaxations for semialgebraic problems*, Mathematical Programming 96 (2003) 293–320.
106. R. Pörn, K.-M. Björk, and T. Westerlund, *Global solution of optimization problems with signomial parts*, Discrete Optimization 5 (2008) 108–120.
107. M. Putinar, *Positive polynomials on compact semi-algebraic sets*, Indiana University Mathematics Journal 42 (1993) 969–984.

108. I. Quesada and I.E. Grossmann, *An LP/NLP based branch and bound algorithm for convex MINLP optimization problems*, Computers & Chemical Engineering 16 (1992) 937–947.
109. F. Rendl, G. Rinaldi, and A. Wiegele, *A branch and bound algorithm for max-cut based on combining semidefinite and polyhedral relaxations*, Integer Programming and Combinatorial Optimization 2007 – Ithaca, New York (M. Fischetti and D.P. Williamson, eds.), Lecture Notes in Computer Science 4513, Springer, 2007, pp. 295–309.
110. F. Rendl, G. Rinaldi, and A. Wiegele, *Solving max-cut to optimality by intersecting semidefinite and polyhedral relaxations*, Technical report, Alpen-Adria-Universität Klagenfurt, Inst. f. Mathematik, 2008.
111. J. Renegar, *On the computational complexity and geometry of the first-order theory of the reals. part III: Quantifier elimination*, Journal of Symbolic Computation 13 (1992) 329–352.
112. J. Renegar, *On the computational complexity of approximating solutions for real algebraic formulae*, SIAM Journal on Computing 21 (1992) 1008–1025.
113. N.V. Sahinidis, *BARON: A general purpose global optimization software package*, Journal of Global Optimization 8 (1996) 201–205.
114. F. Santos and B. Sturmfels, *Higher Lawrence configurations*, Journal of Combinatorial Theory Ser. A 103 (2003) 151–164.
115. A. Saxena, P. Bonami, and J. Lee, *Disjunctive cuts for non-convex mixed integer quadratically constrained programs*, Integer Programming and Combinatorial Optimization 2008 – Bertinoro, Italy (A. Lodi, A. Panconesi, and G. Rinaldi, eds.), Lecture Notes in Computer Science 5035, Springer, 2008, pp. 17–33.
116. A. Saxena, P. Bonami, and J. Lee, *Convex relaxations of non-convex mixed integer quadratically constrained programs: Extended formulations*, IBM Research Report RC24621, 2008.
117. A. Saxena, P. Bonami, and J. Lee, *Convex relaxations of non-convex mixed integer quadratically constrained programs: Projected formulations*, IBM Research Report RC24695, 2008.
118. SDPT3, www.math.nus.edu.sg/~mattohkc/sdpt3.html, v. 4.0 (beta).
119. A. Sebö, *Hilbert bases, Caratheodory's Theorem and combinatorial optimization*, Proceedings of the IPCO conference, Waterloo, Canada, 1990, pp. 431–455.
120. SeDuMi, sedumi.mcmaster.ca, v. 1.1.
121. N.Z. Shor, *An approach to obtaining global extremums in polynomial mathematical programming*, Kibernetika 52 (1987) 102–106.
122. E.M.B. Smith, *On the optimal design of continuous processes*, Ph.D. thesis, Imperial College of Science, Technology and Medicine, University of London, Oct. 1996.
123. E.M.B. Smith and C.C. Pantelides, *A symbolic reformulation/spatial branch-and-bound algorithm for the global optimisation of nonconvex MINLPs*, Computers & Chemical Engineering 23 (1999) 457–478.
124. R.A. Stubbs and S. Mehrotra, *A branch-and-cut method for 0-1 mixed convex programming*, Mathematical Programming 86 (1999) 515–532.
125. B. Sturmfels, *Gröbner bases and convex polytopes*, American Mathematical Society, Providence, RI, 1996.
126. S.P. Tarasov and L.G. Khachiyan, *Bounds of solutions and algorithmic complexity of systems of convex diophantine inequalities*, Soviet Math. Doklady 22 (1980) 700–704.
127. M. Tawarmalani and N. Sahinidis, *Convex extensions and envelopes of semi-continuous functions*, Mathematical Programming 93 (2002) 247–263.
128. M. Tawarmalani and N.V. Sahinidis, *Convexification and global optimization in continuous and mixed-integer nonlinear programming: Theory, algorithms, software and applications*, Nonconvex Optimization and Its Applications, vol. 65, Kluwer Academic Publishers, Dordrecht, 2002.
129. M. Tawarmalani and N.V. Sahinidis, *Global optimization of mixed-integer nonlinear programs: A theoretical and computational study*, Mathematical Programming 99 (2004) 563–591.
130. M. Tawarmalani and N.V. Sahinidis, *Semidefinite Relaxations of Fractional Programs via Novel Convexification Techniques*, Journal of Global Optimization 20 (2001) 137–158.
131. R.R. Thomas, *Algebraic methods in integer programming*, Encyclopedia of Optimization (C. Floudas and P. Pardalos, eds.), Kluwer Academic Publishers, Dordrecht, 2001.

132. D. Vandenbussche and G.L. Nemhauser, *A branch-and-cut algorithm for nonconvex quadratic programs with box constraints*, Mathematical Programming 102 (2005) 559–575.
133. D. Vandenbussche and G.L. Nemhauser, *A polyhedral study of nonconvex quadratic programs with box constraints*, Mathematical Programming 102 (2005) 531–557.
134. T. Westerlund and K. Lundqvist, *Alpha-ECP, version 5.101: An interactive MINLP-solver based on the extended cutting plane method*, Technical Report 01-178-A, Process Design Laboratory at Abo Akademi University, Updated version of 2005-10-21.
135. T. Westerlund and F. Pettersson, *An extended cutting plane method for solving convex MINLP problems*, Computers and Chemical Engineering 19(Suppl.) (1995) S131–S136.
136. T. Westerlund and R. Pörn, *Solving pseudo-convex mixed integer optimization problems by cutting plane techniques*, Optimization and Engineering 3 (2002) 253–280.

Chapter 16
Mixed Integer Programming Computation

Andrea Lodi

In memory of my friend and colleague Lorenzo Brunetta (1966–2008).

Abstract The first 50 years of Integer and Mixed-Integer Programming have taken us to a very stable paradigm for solving problems in a reliable and effective way. We run over these 50 exciting years by showing some crucial milestones and we highlight the building blocks that are making nowadays solvers effective from both a performance and an application viewpoint. Finally, we show that a lot of work must still be done for improving the solvers and extending their modeling capability.

16.1 Introduction

We consider a general *Mixed Integer Linear Program* (MIP) in the form

$$\min\{c^T x : Ax \geq b,\ x \geq 0,\ x_j \in \mathbb{Z}\ \forall j \in I\} \tag{16.1}$$

where we do not assume that the matrix A has any special structure. Thus, the algorithmic approach relies on the iterative solution, through general-purpose techniques, of the *Linear Programming* (LP) relaxation

$$\min\{c^T x : Ax \geq b,\ x \geq 0\}, \tag{16.2}$$

i.e., the same as problem (16.1) above but the integrality requirement on the x variables in the set I has been dropped. We denote an optimal solution of problem (16.2) as x^*. The reason for dropping such constraints is that MIP is NP-hard while LP is polynomially solvable and general-purpose techniques for its solution are efficient in practice.

Andrea Lodi
DEIS, University of Bologna, Italy
e-mail: andrea.lodi@unibo.it

M. Jünger et al. (eds.), *50 Years of Integer Programming 1958-2008*,
DOI 10.1007/978-3-540-68279-0_16,

Due to space limitations, this chapter does not cover LP state-of-the-art, while it covers the basic characteristics and components of current, commercial and non-commercial, MIP solvers. However, Bixby et al. [27] report that in 2004 an LP was solved, by Cplex 8, a million times faster than it was by Cplex 1 in 1990, three orders of magnitudes due to hardware and to software improvements, respectively. This gives a clear indication of how much LP technology has been and is important for MIP development.

Roughly speaking, using the LP computation as a tool, MIP solvers integrate the *branch-and-bound* and the *cutting plane* algorithms through variations of the general *branch-and-cut* scheme proposed by Padberg and Rinaldi [76, 77] in the context of the *Traveling Salesman Problem* (TSP). The reader is referred to Chapter 12 for a detailed account on the history of the branch-and-cut framework which is, in turn, tightly connected to the history of the TSP and other combinatorial optimization problems like the *Linear Ordering*, see Grötschel, Jünger and Reinelt [57].

The branch-and-bound algorithm, Land and Doig [63]. In its basic version the branch-and-bound algorithm iteratively partitions the solution space into sub-MIPs (the children nodes) which have the same theoretical complexity of the originating MIP (the father node, or the root node if it is the initial MIP). Usually, for MIP solvers the branching creates two children by using the rounding of the solution of the LP relaxation value of a fractional variable, say x_j, constrained to be integral

$$x_j \leqslant \lfloor x_j^* \rfloor \quad \text{or} \quad x_j \geqslant \lfloor x_j^* \rfloor + 1. \tag{16.3}$$

The two children above are often referred to as *left* (or "down") branch and *right* (or "up") branch, respectively. On each of the sub-MIPs the integrality requirement on the variables $x_j, \forall j \in I$ is relaxed and the LP relaxation is solved. Despite the theoretical complexity, the sub-MIPs become smaller and smaller due to the partition mechanism (basically some of the decisions are taken) and eventually the LP relaxation is directly integral for all the variables in I. In addition, the LP relaxation is solved at every node to decide if the node itself is worthwhile to be further partitioned: if the LP relaxation value is already not smaller than the best feasible solution encountered so far, called *incumbent*, the node can safely be fathomed because none of its children will yield a better solution than the incumbent. Finally, a node is also fathomed if its LP relaxation is infeasible.

The cutting plane algorithm, Gomory [55]. Any MIP can be solved without branching by simply finding its "right" linear programming description, more precisely, the *convex hull* of its (mixed-)integer solutions. In order to do that, one has to iteratively solve the so called *separation problem*

> Given a feasible solution x^* of the LP relaxation (16.2) which is not feasible for the MIP (16.1), find a linear inequality $\alpha^T x \geqslant \alpha_0$ which is valid for (16.1), i.e., satisfied by all feasible solutions $\bar{x}$ of the system (16.1), while it is violated by x^*, i.e., $\alpha^T x^* < \alpha_0$.

Any inequality solving the separation problem is called a *cutting plane* (or a *cut*, for short) and has the effect of tightening the LP relaxation to better approximate the convex hull.

Gomory [55] has given an algorithm that converges in a finite number of iterations for pure *Integer Linear Programming* (IP)[1] with integer data. Such an algorithm solves the separation problem above in an efficient and elegant manner in the special case in which x^* is an optimal basis of the LP relaxation. No algorithm of this kind is known for MIPs, that being one of the most intriguing open questions in the area (see e.g., Cook, Kannan and Schrijver [37]). An important step in this direction is the work of Jörg [60].

The idea behind integrating the two algorithms above is that LP relaxations (16.2) do not naturally well approximate, in general, the convex hull of (mixed-)integer solutions of MIPs (16.1). Thus, some extra work to devise a better approximation by tightening any relaxation with additional linear inequalities (cutting planes) increases the chances that fewer nodes in the search tree are needed. On the other hand, pure cutting plane algorithms show, in general, a slow convergence and the addition of too many cuts can lead to very large LPs which in turn present numerical difficulties for the solvers. The branch-and-cut algorithm has been proven to be very effective initially for combinatorial optimization problems (like TSP) with special-purpose cuts based on a polyhedral analysis and later on in the general MIP context.

The chapter is organized as follows.

In the first part, Section 16.2, we discuss the evolution of MIP solvers having in mind both a performance perspective and a modeling/application viewpoint. At the beginning of the section we present some important MIP milestones with no aim of being exhaustive with respect to algorithms and software. We then go into the details of the basic components of MIP codes in Section 16.2.1. Finally, in Section 16.2.2 we describe some important tools that allow a relevant degree of flexibility in the development of MIP-based applications.

In the second part, Section 16.3, we discuss the challenges for the next generation MIP solvers by first presenting a list of difficult MIP classes on which better performance/strategies would be extremely beneficial. We again present the content first from a performance angle, Section 16.3.1, and second with a modeling perspective, Section 16.3.2.

Some conclusions are drawn in Section 16.4.

16.2 MIP evolution

Despite quite some work on basically all aspects of IP and in particular on cutting planes, the early general-purpose MIP solvers were mainly concerned with developing a fast and reliable LP machinery used within good branch-and-bound schemes.

There are at least two remarkable exceptions to this trend. The first is a paper published in 1983 by Crowder, Johnson and Padberg [42] which describes the implementation of a general-purpose code for pure 0-1 IPs, called PIPX, that used

[1] IPs are the special case of MIPs where all variables belong to I, i.e., are constrained to be integer.

the IBM linear programming system MPSX/370 and the IBM integer programming system MIP/370 as building blocks. The authors were able to solve ten instances obtained from real-world industrial applications. Those instances were very large for that time, up to 2,756 binary variables and 755 constraints. Some cutting planes were integrated in a good branch-and-bound framework (with some preprocessing to reduce coefficients of variables) and in particular *cover inequalities* which will be discussed in Section 16.2.1.2 below.

The second exception is a paper published in 1987 by Van Roy and Wolsey [87] in which the authors describe a system called MPSARX that solved mixed 0-1 industrial instances arising from applications like production planning, material requirement planning, multilevel distribution planning, etc. Also in this case, the instances were very large for the time, up to 2,800 variables and 1,500 constraints. The system acted in two phases: in phase one some preprocessing was performed and, more importantly, cutting planes were generated. Such a phase was called reformulation in [87] and performed by a system called ARX. The second phase was the solution of the new model by the mathematical programming system called SCICONIC/VM.

After the success[2] of the two early MIP solvers [42, 87] and, mainly, the formalization of the branch-and-cut algorithm in the TSP context [77], a major step was required to prove that cutting plane generation in conjunction with branching could work in general, i.e., without exploiting the structure of the underlying polyhedron, in particular in the cutting plane separation.

The proof of concept came through two papers both published in 1996. In [19], Balas, Ceria and Cornuéjols showed that a branch-and-cut algorithm based on general-purpose *disjunctive* cuts (see Balas [18]) could lead to very effective results for 0-1 MIPs. More precisely, the cutting planes used are the so called *lift-and-project* cuts in which the separation problem amounts to solving an LP in an extended space and the computational results in [19] showed a very good and stable behavior. In addition, these results showed that the quality and the speed of LP solvers in the late nineties allowed the solution of LPs as an additional building block of the overall algorithm. Such an outcome was not granted and LP rapid evolution showed to have reached a very stable state.

In [20], Balas, Ceria, Cornuéjols and Natraj revisited the classical Gomory mixed-integer cuts [56] by overruling the common sense that these cuts had a purely theoretical interest. In fact, the results in [20] showed that, embedded in a branch-and-cut framework, Gomory mixed-integer cuts are a fundamental tool for the solution of 0-1 MIPs. Such an improvement was mainly obtained by separating and adding groups (denoted as "rounds") of cuts[3] instead of one cut at a time. It turns out that Gomory mixed-integer cuts are one of the most crucial ingredients for the success of current MIP solvers.

---

[2] The paper by Crowder, Johnson and Padberg [42] won the "Frederick W. Lanchester Prize" awarded by INFORMS in 1983 while the paper by Van Roy and Wolsey [87] won the "Beale-Orchard-Hays" Prize awarded by MPS in 1988.

[3] In principle, one Gomory mixed-integer cut can be separated from each tableau row where an integer variable is non-zero and fractional.

The two papers above have played a central role in the move to the current generation of MIP solvers and the fact that the use of cutting planes, and in particular within a branch-and-cut framework, has been a breakthrough is shown by the following very meaningful experiment due to Achterberg and Bixby [8]. On a testbed of 1,734 MIP instances all Cplex versions beginning with Cplex 1.2, the first having MIP capability, have been extensively compared. The results of this experiment are shown in Tables 16.1 and 16.2. Specifically, Table 16.1 reports the evolution of Cplex by comparing each version with the most current one, version 11.0. The first column of the table indicates the version while the second recalls the year in which the version has been released. The third and fourth columns count the number of times each version is better and worse with respect to Cplex 11.0, respectively: computing times within 10% are considered equivalent, i.e., counted neither better nor worse. Finally, the last column gives the computing time as geometric mean again normalized with respect to Cplex 11.0. A time limit of 30,000 CPU seconds for each instance was provided on a cluster of Intel Xeon machines 3 GHz. Besides

Cplex version	year	better	worse	time
11.0	2007	–	–	1.00
10.0	2005	201	650	1.91
9.0	2003	142	793	2.73
8.0	2002	117	856	3.56
7.1	2001	63	930	4.59
6.5	1999	71	997	**7.47**
6.0	1998	55	1060	21.30
5.0	1997	45	1069	22.57
4.0	1995	37	1089	26.29
3.0	1994	34	1107	34.63
2.1	1993	13	1137	56.16
1.2	1991	17	1132	67.90

Table 16.1 Computing times for 12 Cplex versions: normalization with respect to Cplex 11.0.

the comparison with the current Cplex version, the most interesting way of interpreting the results reported in the table is by looking at the trend the columns show: together with a very stable decreasing of the computing time from version 1.2 up, it is clear that the biggest step forward in a version-to-version scale has been made with Cplex 6.5 which is the first having full cutting plane capability, and in particular Gomory mixed-integer cuts. Indeed, the geometric mean of the computing times drops from 21.30 to 7.47 going from version 6.0 to 6.5, by far the biggest decrease.

The trend is confirmed by the numbers reported in Table 16.2. On a slightly larger set of 1,852 MIPs (including some models in which older versions encountered numerical troubles), the table highlights the version-to-version improvement in the number of solved problems. Besides the first two columns which report the same information as Table 16.1, the third and fourth column report the number and percentage of problems solved to proven optimality, respectively. Finally, the last

Cplex version	year	# optimal	% optimal	v-to-v % improvement
11.0	2007	1,243	67.1%	7.8%
10.0	2005	1,099	59.3%	3.5%
9.0	2003	1,035	55.9%	2.6%
8.0	2002	987	53.3%	2.5%
7.1	2001	941	50.8%	4.3%
6.5	1999	861	46.5%	**13.4%**
6.0	1998	613	33.1%	1.0%
5.0	1997	595	32.1%	1.8%
4.0	1995	561	30.3%	4.4%
3.0	1994	479	25.9%	6.2%
2.1	1993	365	19.7%	4.7%
1.2	1991	278	15.0%	—

Table 16.2 Version-to-version comparison on 12 Cplex versions with respect to the number of solved problems.

column gives precisely the version-to-version improvement in the percentage of problems optimally solved.

Just to give an "absolute" flavor of the improvement due to cuts, we consider instance `p2756`, the largest instance in the seminal paper by Crowder, Johnson and Padberg [42]. If the cut generation phase is disabled Cplex 11 is able to solve the problem by exploring 3,415,408 branch-and-bound nodes which are necessary to close the percentage gap of 13.5 of the root node relaxation. In the default version, i.e., with cutting planes, Cplex 11 solves the problem within 11 nodes. The root relaxation improved—among others—by 22 Gomory mixed-integer cuts and 23 cover inequalities has a percentage gap of only 0.2.

We end this section by recognizing the crucial impact had on the MIP development by two factors. On the one side, the creation of archives of instances used to compare algorithmic ideas and software implementations through benchmarking has contributed to define a common methodology. The last in order of time of such libraries is the MIPLIB 2003 maintained at ZIB [11]. Figure 16.1 (courtesy of ZIB) shows that every library tends rapidly to be crunched and the community is probably ready for new challenging instances.

On the other hand, the availability of commercial and non-commercial MIP solvers has created over the past decades and continues to create nowadays a very fruitful competition that stimulates the entire MIP community to go "larger" and "quicker".

16.2.1 A performance perspective

The current generation of MIP solvers incorporates key ideas continuously developed during the first 50 years of Integer Programming. In the next sections we will

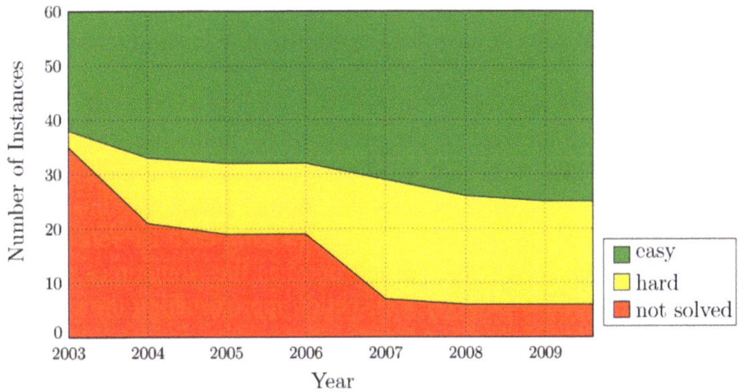

Fig. 16.1 Comparison of the number of solved MIPLIB 2003 instances at the beginning of each year. "Easy" means, that the instance could be solved within one hour using a commercial MIP-solver, "hard" stands for instances, that were solved, but not in the previous conditions.

discuss these main ingredients[4]. As it will be clear case by case, most of the times these ideas have been developed by studying special-purpose, often, very small substructures (see e.g., Wolsey [89]) and then extended and generalized to MIP. This seems to be a key feature of MIP.

16.2.1.1 Presolving

In the presolving (often called preprocessing) phase the solver tries to detect certain changes in the input that will probably lead to a better performance of the solution process. This is generally done without "changing" the set of optimal solutions of the problem at hand[5] and it affects two main situations.

On the one side, it is often the case that MIP models, in particular those originating from real-world applications and created by using modeling languages, contain some "garbage", i.e., irrelevant or redundant information that tend to slow down the solution process forcing the solver to perform useless operations. More precisely, there are two types of sources of inefficiency: first, the models are unnecessary large and thus harder to manage. This is the case in which there are redundant constraints or variables which are already fixed and nevertheless appear in the model as additional constraints. Second, the variable bounds can be unnecessary large or the

---

[4] Each of the ingredients is treated in a somehow concise way with the aim of giving an overview, discussing the main connections and providing useful pointers. For many more details, the reader is referred to Part II of the excellent Ph.D. dissertation of Achterberg [6].

[5] In fact, the set of optimal solutions might change due to presolve in case, for example, of symmetry breaking reductions, see Chapter 17.

constraints could have been written in a loose way, for example with coefficients weaker than they could possibly be.

Thus, modern MIP solvers have the capability of "cleaning up" the models so as to create a presolved instance associated with the original one on which the MIP technology is then applied. With respect to the two issues above, MIP presolve is able to reduce the size of the problems by removing such redundancies and it generally provides tools that, exploiting the information on the integrality of the variables in set I, strengthen variables' bound and constraints' coefficients. If the first improvement has only the effect of making each LP relaxation smaller and then quicker to be solved, the second one has the, sometimes crucial, effect of making such relaxations stronger, i.e., better approximating the convex hull of (mixed-)integer solutions.

On the other hand, more sophisticated presolve mechanisms are also able to discover important implications and sub-structures that might be of fundamental importance later on in the computation for both branching purposes and cutting plane generation. As an example, the presolve phase determines the clique table or conflict graph, i.e., groups of binary variables such that no more than one can be non-zero at the same time. The conflict graph is then fundamental to separate *clique inequalities* (see Johnson and Padberg [61]) which are written as

$$\sum_{j \in Q} x_j \leqslant 1 \tag{16.4}$$

where Q denotes a subset of (indices of) binary variables with the property, stated above, that at most one of them can be non-zero.

The most fundamental work about presolving is the one of Savelsbergh [85] to which the interested reader is referred to.

16.2.1.2 Cutting plane generation

The importance of cutting planes has been already discussed extensively. The arsenal of separation algorithms has been continuously enlarged over the years, thus the main issue, discussed in Section 16.3.1.3, is probably how to select cuts instead of how to generate them.

A large group of cuts, with strong relationship among each other, includes Chvátal-Gomory cuts [55, 34], Gomory mixed-integer cuts [56], mixed-integer rounding cuts [72, 46], $\{0, \frac{1}{2}\}$ cuts [31], lift-and-project cuts [18, 19] and split cuts [37, 21]. Essentially, all these inequalities are obtained by applying a disjunctive argument on an integer or mixed-integer set of a single constraint often derived by aggregating many others. This group of cuts is presented in a brilliant and unified way by Cornuéjols [38] and the reader is referred to that paper and to Chapter 11.

Besides the group above and clique inequalities already discussed in the previous section, two more classes of very useful cuts are generally used within a MIP solver, namely *cover inequalities*[6] and *flow cover inequalities*, which are briefly discussed in the following.

Cover cuts. Somehow similarly to the clique inequalities, cover constraints define a property on a set of binary variables. More precisely, given a knapsack kind constraint in the form $\alpha^T x \leqslant \alpha_0$ where we assume that $\alpha \in \mathbb{Z}_+^{|V|}$, $\alpha_0 \in \mathbb{Z}_+$ and V is a subset of (indices of) binary variables, a set $Q \subseteq V$ is called a *cover* if $\sum_{j \in Q} \alpha_j > \alpha_0$. Moreover, the cover is said to be minimal if $\sum_{j \in Q \setminus \{\ell\}} \alpha_j \leqslant \alpha_0$ for all $\ell \in Q$. In other words, Q is a set of binary variables which cannot be all together non-zero at the same time. In the light of the definition above, the simplest version of a cover inequality is

$$\sum_{j \in Q} x_j \leqslant |Q| - 1. \tag{16.5}$$

The amount of work devoted to cover cuts is huge starting with the seminal work of Balas [17] and Wolsey [88]. In particular, cover cuts do not define facets, i.e., polyhedral faces of dimension one less of the dimension of the associated polyhedron[7], but can be lifted (see Padberg [75]) to become facet defining.

Flow cover cuts. The polyhedral investigation of the so called *0-1 single node flow problem* is at the base of the definition of flow cover cuts by Padberg, Van Roy and Wolsey [78]. Essentially, this is a fixed charge problem with arcs incoming and outgoing to a single node: to each of these arcs is associated a continuous variables measuring the flow on the arc and upper bounded by the arc capacity, if the flow over the arc is non-zero a fixed cost must be paid and this is triggered by a binary variable associated with the same arc. A flow balance on the node must also be satisfied. Flow cover cuts can then be devised as mixed-integer rounding inequalities (see Marchand [67]) and then strengthened by lifting (see Gu, Nemhauser and Savelsbergh [58]). The interested reader is also referred to Louveaux and Wolsey [65] for a general discussion on these mixed-integer sets.

Finally, it is worth noting that despite the origin of the flow cover cuts is a specific polyhedral context, they are useful and applied in general within MIP solvers. The reason is that it is easy to aggregate variables (and constraints) of a MIP in order to derive a mixed-integer set like the 0-1 single node flow set (see e.g., [67]).

We end the section by noting that the separation of all mentioned cuts (including cliques) except lift-and-project cuts is NP-hard for a general x^*. Note that this is not in contrast with the fact that one can separate, e.g., Gomory mixed-integer cuts by a polynomial (very cheap) procedure [55, 56] once the fractional solution x^* to be cut off is a vertex of the continuous relaxation.

---

[6] Recall that cover cuts were already used in the pioneering works [42, 87], see Section 16.2 above.

[7] For a detailed presentation of polyhedral basic concepts as dimensions, faces, facets, etc. the reader is referred for example to Papadimitriou and Steiglitz [79] and Nemhauser and Wolsey [71].

16.2.1.3 Sophisticated branching strategies

The branching mechanism introduced in Section 16.1 requires to take two independent and important decisions at any step: node selection and variable selection. We will analyze them separately in the following by putting more attention on the latter.

Node Selection. This is very classical: one extreme is the so called *best-bound first* strategy in which one always considers the most promising node, i.e., the one with the smallest LP value, while the other extreme is *depth first* where one goes deeper and deeper in the tree and starts backtracking only once a node is fathomed, i.e., it is either (mixed-)integer feasible, or LP infeasible or it has a lower bound not better (smaller) than the incumbent. The pros and cons of each strategy are well known: the former explores less nodes but generally maintains a larger tree in terms of memory while the latter can explode in terms of nodes and it can, in the case some bad decisions are taken at the beginning, explore useless portions of the tree itself. All other techniques, more or less sophisticated, are basically hybrids around these two ideas, like interleaving best-bound and diving[8] in an appropriate way.

Variable Selection. The variable selection problem is the one of deciding how to partition the current node, i.e., on which variable to branch on in order to create the two children. For a long time, a classical choice has been branching on the most fractional variable, i.e., in the 0-1 case the closest to 0.5. That rule has been computationally shown by Achterberg, Koch and Martin [10] to be worse than a complete random choice. However, it is of course very easy to evaluate. In order to devise stronger criteria one has to do much more work. The extreme is the so called *strong branching* technique (see e.g., Applegate, Bixby, Chvátal and Cook [16] and Linderoth and Savelsbergh [64]). In its full version, at any node one has to tentatively branch on each candidate fractional variable and select the one on which the increase in the bound on the left branch times the one on the right branch is the maximum. Of course, this is generally unpractical but its computational effort can be easily limited in two ways: on the one side, one can define a much smaller candidate set of variables to branch on and, on the other hand, can limit to a fixed (small) amount the number of Simplex pivots to be performed in the variable evaluation. Another sophisticated technique is *pseudocost branching* which goes back to Benichou, Gauthier, Girodet and Hentges [24] and keeps a history of the success (in terms of the change in the LP relaxation value) of the branchings already performed on each variable as an indication of the quality of the variable itself. The most recent effective and sophisticated method introduced by Achterberg, Koch and Martin [10] is called *reliability branching* and it integrates strong and pseudocost branchings. The idea is to define a reliability threshold, i.e., a level below which the information of the pseudocosts is not considered accurate enough and some strong branching is performed. Such a threshold is mainly associated with the number of previous branching decisions that involved the variable.

---

[8] A *dive* in the tree is a sequence of branchings without backtracking.

We end the section by noting that until now the methods proposed in research papers and implemented by MIP solvers to do enumeration have been rather structured, very relying on the (almost always binary) tree paradigm which has proven to be stable and effective. Few attempts of seriously revisiting the tree framework have been made, one notable exception being the work of Chvátal called *resolution search* [35]. In that paper, the idea is to detect conflicts, i.e., sequences of variable fixings that yield an infeasible subproblem, making each sequence minimal and use the information to construct an enumeration strategy. A similar concept of conflicts has been used by Achterberg [5, 6] to do propagation in MIP and it has been known in the Artificial Intelligence and SATisfiability (SAT) communities with the name of "no good recording" (see e.g., Stallman and Sussman [86]) since the seventies.

16.2.1.4 Primal heuristics

Primal heuristics are a relatively "fresh" component of MIP solvers in the sense that in the last few years the capability of the solvers to quickly find in the search tree very high quality solutions has improved dramatically. On the other hand, it is also true that rounding heuristics, going from easy to complex, have been part of the arsenal of the MIP solvers since almost the beginning. Thus, one could say that the continuous hybridization between rounding and diving techniques with local search has been the way to obtain the recent brilliant results.

Following the structure of Chapter 9 of [6], we distinguish among the following types of heuristics.

Rounding heuristics. Starting from a feasible solution of the continuous relaxation (generally the optimal one) one rounds up or down the fractional values of the variables in I in the attempt to produce a (mixed-)integer solution still satisfying the linear constraints. This can be done in a straightforward and very fast manner or in a more and more complex way, i.e., by allowing some backtracking mechanism in the case a (partial) mixed-integer assignment becomes infeasible.

Diving heuristics. With respect to a rounding heuristic, the diving is performed in an iterative way: after rounding one or more variables the LP relaxation is solved again and the process is iterated. In this category, Achterberg [6] distinguishes between "hard" rounding in which the variables are really fixed to a specified value and "soft" rounding in which the effect is obtained implicitly by changing the objective function coefficient of the variables to be "rounded" in an appropriate way. An algorithm falling in the latter category is the *feasibility pump* heuristic proposed for 0-1 MIPs by Fischetti, Glover and Lodi [49] and later extended to general MIPs by Bertacco, Fischetti and Lodi [26] and Achterberg and Berthold [7]. The idea of the algorithm is to alternatively satisfy either the linear constraints (by solving LP relaxations) or integer constraints (by applying rounding). The LP relaxations are obtained by replacing the original objective function with one taking into account the distance with respect to the current (mixed-)integer rounded solution. If the trajectory of the rounded solutions intersects the one of the LP (neighborhood) relaxations, then a feasible solution with respect to both the linear constraints and the integer ones has been found. Such an algorithmic framework has been recently extended to convex mixed-integer non-linear programming by Bonami, Cornuéjols, Lodi and Margot [29].

Improving heuristics. The heuristics in this category are designed to improve the quality of the current incumbent solution, i.e., they exploit that solution or a group of solutions so as to get a better one. These heuristics are usually local search methods and we have recently seen a variety of algorithms which solve sub-MIPs for exploring the neighborhood of the incumbent or of a set of solutions. When these sub-MIPs are solved in a general-purpose fashion, i.e., through a nested call to a MIP solver, this is referred to as MIPping and will be the content of Section 16.2.1.5 below. Improvement heuristics of this type have connections to the so called *metaheuristic* techniques [54] which have proven very successful on hard and large combinatorial problems in the nineties. Techniques from metaheuristics have been now incorporated in the MIP solvers, see e.g., [50, 82], and as a result MIP solvers have now improved their ability of quickly finding very good feasible solutions and can be seen as competitive heuristic techniques if used in a truncated way, i.e., with either a time or node limit.

Before closing this section, it is interesting to note that Achterberg [6] has shown that the impact of heuristics is not dramatic in terms of ability of the MIP solver to prove optimality in a (much) quicker way[9]. However, the psychological impact for the user who sees a high quality feasible solution provided very early is huge.

16.2.1.5 MIPping

MIP computation has reached such an effective and stable quality to allow the solution of sub-MIPs during the solution of a MIP itself, the *MIPping* approach (see Fischetti, Lodi and Salvagnin [52] for a survey). In other words, optimization sub-problems are formulated as general-purpose MIPs and solved, often heuristically, by using MIP solvers, i.e., without taking advantage of any special structure. The novelty is not in solving optimization sub-problems having different levels of complexity, but in using a general-purpose MIP solver for handling them. In some sense such a technique, originally proposed by Fischetti and Lodi [50] for modeling large neighborhoods of 0-1 solutions, shows the high quality of the current generation of MIP solvers like the solution of large LPs in the cutting plane generation phase [19] showed more than ten years ago the effectiveness of the LP phase.

After the original paper on the primal heuristic side, the MIPping technique has been successfully applied in the cutting plane generation phase [51], for accelerating Benders decomposition [80], to detect dominated nodes in the search tree [84] and again to devise very high quality primal solutions [45], just to cite a few applications.

---

[9] This issue will be discussed in more detail in Section 16.3.1.5.

16.2.2 A modeling/application perspective

Solving a MIP to optimality is only one aspect of using a MIP solver for applications, sometimes not the most important one. Nowadays MIP solvers include useful tools for complex algorithmic design (i.e., one can use the blocks of a solver to devise a specialized algorithm) and data and model analysis. In this way, several additional types of users besides the ones that want to use an MIP solver as a black-box are taken into account: (a) the academic user who wants to develop a sophisticated branch-and-cut or hybrid algorithm, (b) the practitioner who likes to play with parameters and use a combination of his/her favorite techniques, and, finally, (c) the freshman who might need help with modeling and solution. Some of these tools are discussed in the following.

Automatic tuning of the parameters. The number of parameters, corresponding to different algorithmic options such as LP methods, branching strategies, cutting level etc. gives the user a wide degree of flexibility but makes, at the same time, the hand-tuning of parameters fairly complex. This is the reason for attempting to supporting such a tuning in an automatic way: the tuning tool analyzes a model or a group of models and suggests a suite of parameter settings that provide better performance than the default parameter settings. In the academic context, automatic tuning has been studied by Baz, Brooks, Gosavi and Hunsaker [22]. (Tool mainly intended for users of type (c) above.)

Multiple solutions. The capability of finding not only one among all the optimal solutions of a MIP but many of them and all those with a prefixed quality is a fundamental option for application-oriented solutions in which the availability of a set of solutions instead of a single one allows flexibility and support for decision making. From a theoretical viewpoint such a problem has been studied by Danna, Fenelon, Gu and Wunderling [44] who also describe an implementation within Cplex. (Tool intended for all users, mainly (b) and (c).)

Detection of sources of infeasibility. Real-world models are often over constrained and sources of infeasibility must be removed so as to allow a solution. Sometimes such a detection helps the user to better understand his/her own model and to make a step forward in the direction of solving the problem even before any optimal/heuristic solution has been computed. From a theoretical viewpoint finding the minimal set of constraints that once removed make an LP feasible is NP-hard in its own, thus MIP solvers incorporate heuristics following the work of the academic community [12, 33]. (Tool intended for all users, mainly (b) and (c).)

Callbacks. Any additional knowledge of the user with respect to the problem at hand, in particular if such a knowledge is hard to automatically discover by the MIP solver, should be used in the design of the solution algorithm. Callbacks are pieces of code that allow the flexibility of accommodating the user code for specific cutting plane generation, primal heuristics, branching strategies, etc. Moreover, several of these callbacks allow recovering information from the system in the solution phase so as to favor a better understanding of the algorithm evolution. (Tool mainly intended for users (a) and (b).)

16.3 MIP challenges

Overall, a big challenge from both performance and modeling viewpoints is *accuracy* which is a new issue, or more precisely, an old issue that has become very important after realizing that MIP solvers can now really solve interesting problems. As an example, accuracy in computation and benchmarking are the topics of the issue 77 of *Optima*: `http://www.mathprog.org/?nav=optima_newsletter`.

Accuracy checks are needed in basically all parts of a MIP solver but maybe the most crucial part is cutting plane generation. Sophisticated cutting plane procedures challenge the floating-point arithmetic of the solvers heavily, the danger being the generation of invalid cuts, i.e., linear inequalities cutting off (mixed-)integer feasible solutions, possibly the optimal one. This issue has been recently investigated in two papers. Margot [68] studied a methodology for testing accuracy and strength of cut generators based on random dives of the search tree by recording a well-chosen set of indicators. Cook, Dash, Fukasawa and Goycoolea [36] proposed a method that weakens slightly the Gomory mixed-integer cuts but makes them "safe" with respect to accumulation of floating-point errors.

Both papers above investigate the accuracy issue of cutting plane generation mainly from a research viewpoint. At the moment, the standard method of dealing with inaccuracy of cuts in the solvers is discarding "suspicious" or "dangerous" cuts without doing extra work to either test their correctness or separating them in a proven careful way. Among other indicators, cutting planes with high rank[10] are considered suspicious in the sense that they might have accumulated relevant floating-point errors while dense inequalities are said to be dangerous since they generally slow down the LP machinery.

The cutting plane accuracy case is indeed rather typical also with respect to some of the issues that will be discussed in the next sections. Nowadays both commercial and non-commercial MIP solvers are very reliable and the methodology they use is well understood. Thus, a criterion for deciding if a new idea coming from academia can make it "inside" a MIP solver is its capability of being integrated in such a methodology in a smooth way. Thus, cutting planes potentially very effective but numerically "unsafe" are discarded.

Before getting into the details of performance and modeling aspects, we would like to briefly give a non-exhaustive list of difficult classes of MIPs. The improvement on dealing with any of these classes would represent a solid progress for the MIP methodology.

Badly modeled MIPs. MIP difficulties often come from bad modeling. This might happen for two main reasons.

---

[10] The rank of the inequalities in the original formulation is 0 while every inequality obtained as combination of two or more inequalities of rank 0 has rank 1 and so on. The reader is referred to Chvátal [34] for an accurate definition of the rank of an inequality and its implications for IP.

On the one side, the user might have not paid enough attention to the modeling phase and used the MIP methodology too optimistically. Common mistakes[11] are, for example, variables unnecessarily unbounded or free, very small and/or very large coefficients, unnecessarily large bigM values, etc. These mistakes cause numerical difficulties for the solver and sometimes forbid the sophisticated and effective techniques discussed in the previous sections to be applied successfully.

On the other hand, the MIP modeling capability might not be sufficient to carefully express the real problem at hand. This is the case, for example, of models having too complicated disjunctive constraints or of problems being naturally nonlinear on which classical linearization tools are not effective. An example of the latter case will be discussed in Section 16.3.2.2.

Large-size MIPs. The size of the model still matters. On the one side, the inherent structure of a model is the most important factor determining its practical hardness for a MIP solver. Indeed, cases in which small MIPs are hard to solve are no surprise (see e.g., Salazar-González [83]). On the other hand, very large models involving several hundreds of thousands of variables and constraints tend to generally challenge MIP solvers a lot. This is mainly due to two reasons. First, solving very large LP relaxations can be slow. Second, a MIP of very large size is often associated with complex real-world systems involving heterogeneous constraints, generally resulting in weak LP relaxations.

Chapter 13 of this book is devoted to decomposition and reformulation which are very useful techniques in this cases. However, from an application point of view, some of the tools discussed in Section 16.2.2, as the automatic tuning and the callbacks, are extremely useful to deal with very large models, often from a heuristic point of view.

Market Split-type MIPs. Cornuéjols and Dawande [39] formulated a class of small 0-1 IPs that were very difficult for branch-and-cut solvers and problems in such a class are referred to as Market Split (or sometimes Market Share) instances. That seminal paper brought the attention of the mathematical programming community essentially to the problem of solving a system of diophantine equations with upper and lower bounds on the variables. Related to this class are all those problems which contain (hard) knapsack equality constraints which might involve huge coefficients. The methodology which has been developed over the years to deal with these difficult instances involves *lattice basis reduction* [2, 1, 3] which is quite hard to integrate within a MIP solver. Indeed, in spite of the improvement in the MIP solvers in the last decade, those problems remain hard for MIP.

MIPs from Scheduling. Scheduling applications are among the most important ones in Operations Research. Many problems of that kind can be formulated as MIPs, in general by using disjunctive constraints and bigM formulations. However, it is rather disappointing to admit that state-of-the-art MIP technology and solvers are only very rarely the best way to go for solving scheduling problems, at least in their pure form. For large scale applications in which the scheduling part is only

[11] Note that with the word "mistake" we are indicating trivial untidiness that any user could have avoided and not sophisticated modeling decisions which might require a high degree of mathematical and optimization knowledge.

one component, then MIP might still be effective. Of course, a natural (hard) question is if there exist good MIP models for Scheduling or we should better look for extending modeling capability in different directions by hybridizing the solvers[12]. This issue is discussed in Section 16.3.2.

16.3.1 A performance perspective

The performance of MIP solvers can/must be improved in many different directions. In the following section we will briefly discuss our favorite ones.

16.3.1.1 Branching versus cutting

Karamanov and Cornuéjols [62] experimented with branching on Gomory mixed-integer disjunctions obtained by the optimal basis of the LP relaxation. More precisely, from each simplex tableau row i in which an integer variable x_ℓ is basic at the fractional value x_ℓ^*, say $\bar{a}_i^T x = x_\ell^*$, instead of deriving the associated Gomory mixed-integer cut, they consider the corresponding disjunction

$$\pi^T x \leqslant \lfloor x_\ell^* \rfloor \quad \text{or} \quad \pi^T x \geqslant \lfloor x_\ell^* \rfloor + 1, \tag{16.6}$$

where

$$\pi_j = \begin{cases} \lfloor \bar{a}_{ij} \rfloor, & \text{if } j \in I, j \neq \ell, \bar{a}_{ij} - \lfloor \bar{a}_{ij} \rfloor \leqslant x_\ell^* - \lfloor x_\ell^* \rfloor, \\ \lceil \bar{a}_{ij} \rceil, & \text{if } j \in I, j \neq \ell, \bar{a}_{ij} - \lfloor \bar{a}_{ij} \rfloor > x_\ell^* - \lfloor x_\ell^* \rfloor, \\ 1, & \text{if } j = \ell, \\ 0, & \text{otherwise.} \end{cases}$$

The best disjunction according to a heuristic measure is selected for branching.

In Figure 16.2 we give an intuition of why such an approach can give some advantages with respect to cutting. The left-hand side of the figure shows that the intersection cut (see Balas [18]) $\beta^T x \leqslant \beta_0$ derived from the associated disjunction can be weaker than the two branchings in isolation since its intersections with the two polyhedra, obtained by using the branching conditions, is above the two new optimal LP relaxation solutions $x^{*,1}$ and $x^{*,2}$. The situation is even more extreme on the right-hand side of the figure where the left polyhedron obtained by imposing the condition $\pi^T x \leqslant \pi_0$ is empty while the associated intersection cut does not take any advantage on that.

In the line of the work in [62], Cornuéjols, Liberti and Nannicini [40] try to improve the disjunction to be used for branching by row manipulations in the spirit of the reduce-and-split approach for cuts proposed by Andersen, Cornuéjols and Li [13].

[12] Obviously, this question is tightly connected to the "bad modeling" point above.

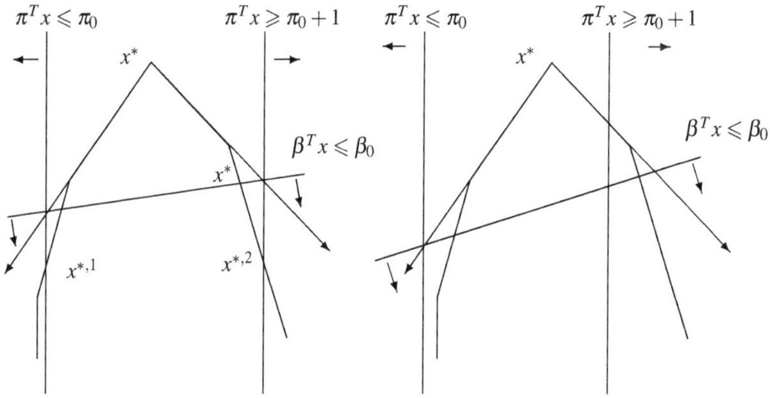

Fig. 16.2 Branching on Gomory mixed-integer disjunctions (the notation is $\pi_0 = \lfloor x_\ell^* \rfloor$).

Branching on general disjunctions has been also used in context less related to cutting plane theory. A pioneering work in this direction is the one of Owen and Mehrotra [74] where the general disjunctions to branch on are generated by a search heuristic in the neighborhood containing all disjunctions with coefficients in $\{-1,0,1\}$ on the integer variables with fractional values at the current node. Moreover, branching on appropriate disjunctions has recently been proposed in the context of highly symmetric MIPs by Ostrowsky, Linderoth, Rossi and Smriglio [73] (see also Chapter 17). Finally, Bertacco [25] has experimented using the MIPping approach for devising general disjunctions either associated with "thin" directions of the polyhedron[13] or resulting in the maximization of the corresponding lower bound. A very similar approach has been followed recently by Mahajan and Ralphs [66].

The real impact of branching on disjunctions is however still to be asserted.

Before ending the section, we would like to note that the work in [62] highlighted the possibility of using some cutting plane theory in the branching context, where, as already discussed, branching on variables is by far predominant with respect to anything else. Overall, it seems that a tighter coordination of the two most fundamental ingredients of the MIP solvers, branching and cutting, can lead to strong improvements.

16.3.1.2 Dealing with general integer and continuous variables

A very important class of MIPs is 0-1 IPs, for instance for the most famous combinatorial optimization problem, i.e., the TSP. Many of the sophisticated techniques discussed in the previous sections have been originally proposed for this class and eventually extended to general MIPs.

[13] Such thin directions are used in the context of lattice basis reduction to reformulate the problem, see also the following section.

As an example, branching on variables is particularly natural and effective in the 0-1 case: the two children created by branching according to (16.3) correspond to fixing the associated variable and are simply obtained by changing the upper and lower bound for the left and right child, respectively. In the case of general integer variables, however, the classical variable branching does not fix the associated variable anymore in the children nodes and in particular when these variables have a large difference between lower and upper bound, branching (16.3) turns out to be less effective.

Another example is in the context of primal heuristics. The techniques discussed in Section 16.2.1.4 basically assume that, once the integer variables have been fixed, what is left is simply solving an LP to find the good value for the rest, i.e., the continuous variables. Of course, this has a single positive outcome over three: the LP can be feasible and luckily provide a solution of "good" quality. In fact, it might be either infeasible (worst case) or provide a solution of "bad" quality, basically useless.

The two examples above suggest that very little has been tailored to deal with both general integer and continuous variables. That might also be a reason for which some of the difficult MIPs discussed in Section 16.3 are hard. Indeed, the relative weakness of variable branching is a reason for which MIP solvers are not performing well on solving systems of diophantine equations in the general form, i.e., not 0-1. Instead, basis reduction methods essentially look for a direction in which the underlying polyhedron is "thin" and a reformulation of the problem exploiting such a direction makes variable branching much more effective. Moreover, such a lack might partially explain the difficulty with Scheduling models since in many of them continuous variables play a central role while they are put aside by disjunctive formulations. For example, Scheduling problems with precedences are modeled with binary variables and bigM constraints which generally lead to weak LP relaxations while the inherent structure is captured by the continuous variables corresponding to the starting time of the jobs.

It is worth mentioning that an exception to this trend concerns the cutting plane phase. Although most of the cuts discussed in Section 16.2.1 are essentially based on disjunctions on the integer part of the model, the continuous variables are lifted with sophisticated techniques and special mixed-integer structures are explicitly exploited (see e.g., Wolsey [89]).

Overall, we feel that a urgent MIP challenge is dealing with general integer and continuous variables with special-purpose, sophisticated, techniques.

16.3.1.3 Cutting planes exploitation

As shown in Section 16.2, cutting plane generation has been a key step for the success of MIP solvers and their capability of being effective for a wide variety of problems. However, the most natural question is: are we using cuts in the best possible way?

The answer is, in our view, rather negative. Fundamental questions about the use of cutting planes remain open and the fact that we do not understand many issues possibly reduces what we could really gain from them. Among these questions, we like to mention the following ones.

Accuracy. We have already pointed out that floating-point inaccuracy in the cut generation can lead to troubles for MIP solvers and that the common reaction is defining confidence threshold values for a cut to be safe. This is only a heuristic action, probably very conservative too. The role of the rank and other aspects such as normalizations, i.e., conditions to truncate the cone associated with disjunctive cutting planes, are not fully understood and much more can be done in this direction, some attempts being reported in Fischetti, Lodi and Tramontani [53].

Cut selection and interaction. The number of procedures we currently have for generating cuts is relevant and so is the number of cuts we could generate at any round. Thus, one of the most relevant issues is which cuts should be selected and added to the current LP relaxation. In other words, cuts should be able to interact in a profitable way together so as to have an overall "cooperative behavior". Results in this direction have been reported by Andreello, Caprara and Fischetti [15].

Cut strength. The strength of a cut is generally measured by its violation. However, scaling and other factors can make such a measure highly misleading. If this is the case, cut selection is very hard and, for example, separating over a larger class of cuts is not necessarily a good idea with respect to preferring a smaller (thus, theoretically weaker) sub-class. An example of such a counter-intuitive behavior is the separation of Chvátal-Gomory cuts instead of their stronger mixed integer rounding version as experienced by Dash, Günlük and Lodi [46]: separating over the sub-class as a heuristic mixed integer rounding procedure helped accelerating the cutting plane convergence dramatically[14].

We close this section by mentioning one of the most exciting and potentially helpful challenges in the cutting plane context even if this is not strictly related to cut exploitation. A recent series of papers [14, 48, 41, 47] has brought the attention of the community to the possibility of generating cuts using more than one row of the simplex tableau at a time. In particular, this is based on characterizing lattice-free triangles (and lattice-free bodies in general) instead of simply split disjunctions as in the case of the single row tableau cuts discussed in Section 16.2.1.2. (For cuts from more tableau rows the reader is also referred to Chapter 19.)

16.3.1.4 Symmetry

Symmetric MIP formulations are among the most difficult ones to deal with. Examples of difficult symmetric MIPs are those arising in coloring, scheduling and covering applications, just to mention a few. In the recent years some special tech-

---

[14] Note that such an outcome might be also related to the stable behavior of Chvátal-Gomory cuts with respect to Gomory mixed-integer or mixed integer rounding cuts due to their numerical accuracy (see e.g., [51, 91]).

niques have been incorporated into the MIP solvers but there is room for improvement.

Chapter 17 of this book is entirely devoted to symmetry.

16.3.1.5 Code complexity or performance variability

The interaction of key ingredients of a MIP solver presented in Sections 16.2.1.2–16.2.1.5 has side effects: sometimes positive and sometimes negative ones. A positive example is any improvement in LP performance. It explicitly speeds up node throughput, and implicitly helps because one can now do more strong branching and MIPping in the same amount of computing time. On the negative side, finding a (near-)optimal solution very early in the search tree explicitly improves the quality of the primal bound but might sometimes hurt in proving optimality, or at least it does not help as expected.

The latter is an example of what we call performance variability studied by Danna [43]: some good features that might not be always helpful because the complexity of the code creates unexpected (negative) interactions. A partial explanation of this special case is that the number of fathomed nodes increases due to the fact of having a very tight primal bound, thus reducing the amount of information one collects from solving relaxations and then resulting in a less precise guidance in the branching decisions.

A deeper understanding through sophisticated testing techniques of these performance effects would be highly beneficial for the design of more robust MIP solvers, where more robust also means that among the same class of instances a solver performs somehow homogeneously. Some steps in this direction has been done by Hooker [59] and McGeoch [69] but such work is not specific for MIP.

Another negative example can be possibly derived from a very recent work of Zanette, Fischetti and Balas [91]. These authors show that a careful implementation of the original cutting plane algorithm of Gomory [55] can be surprisingly effective. In particular, they notice that such an algorithm does not suffer from the classical issues always documented as the quick weakening of the quality of cuts after few iterations because of numerical difficulties such as huge determinants, saturation, etc. The analysis shows that one crucial point is that the Simplex algorithm (the one generally used to solve the LP relaxations) tends to recover primal feasibility after the addition of cuts through (very) few dual pivots with the effect of two consecutive optimal bases being very "close". This behavior is of course good since the algorithm is fast but comparing it with its *lexicographic* version[15] one can notice that the much higher effectiveness (and lack of numerical difficulties) of the cutting planes in the latter seems to be associated to consecutive optimal bases being

[15] The lexicographic dual simplex is a generalized version of the simplex algorithm where, instead of considering the minimization of the objective function, viewed without loss of generality as an additional integer variable $x_0 = c^T x$, one is interested in the lexicographical minimization of the entire solution vector $(x_0, x_1, \dots, x_n)$, where $(x_0, x_1, \dots, x_n) >_{LEX} (y_0, y_1, \dots, y_n)$ means that there exists an index k such that $x_i = y_i$ for all $i = 1, \dots, k-1$ and $x_k > y_k$.

very “far apart”. On the one side, this behavior might be interpreted as a lack of understanding (still) of Gomory mixed-integer cuts. On the other hand, it shows that recovering feasibility quickly (certainly a good feature) might hurt the quality (for cutting) of the bases, thus creating a negative effect overall.

We close this section by formulating a very intriguing research question associated with the first negative example of side effects above (Wolsey [90]). Besides avoiding good primal solutions hurting the optimality proof, how can one use them to have instead a strong speed up? This seems to be a very hard topic whose impact might, however, be crucial for radical steps forward.

16.3.2 A modeling perspective

Of course, new challenges in modeling and application viewpoints involve the development of additional tools in the spirit of the ones described in Section 16.2.2. Among all possible, our favorite would be a tool for detecting minimal sources of numerical instability, i.e., providing the user a set of few columns which make the LP difficult to solve because they are ill-conditioned (Rothberg [81]).

However, the main challenge from an application viewpoint seems to be *dissemination*. More precisely, an interesting direction is extending the current modeling (and solving) capability by taking into account (classes of) non-linear constraints within the MIP framework, i.e., while keeping the structure and infra-structure that have made MIP solvers very successful and reliable. Both commercial and non-commercial solvers are currently following such a direction in rather different ways.

Of course, one difficulty in this context is related to the modeling language one decides to use for expressing the non-linear constraints. Such a problem does not exist in the pure MIP case because the two formats MPS and LP commonly used have been proven robust for the purposes of the problems at hand.

Nevertheless, two successful non-commercial stories are described in the next two sections. (The reader is also referred to Chapter 15 for both theoretical and practical aspects of Nonlinear Integer Programming.)

16.3.2.1 SCIP

The solver SCIP (Solving Constraint Integer Programs, Achterberg [6]) provides within the framework of a MIP solver a tight integration of *Constraint Programming* (CP) and *SATisfiability* techniques and, of course, MIP methodology. SCIP is currently the non-commercial MIP solver with the best performance[16] but in our view its main characteristic is that it can handle, in principle, arbitrary (non-linear) constraints by domain propagation.

[16] Note that SCIP must rely on an external LP solver.

A Chip verification example. Recently, such a tight integration of techniques and capability to handle non-linear constraints has been successfully applied to the property checking problem for System-on-Chip design by Achterberg, Brinkmann and Wedler [9]. Without going into the details of the application, current MIP and CP solvers were considered not promising for the control part of chip verification but at the same time SAT solvers have troubles in the case of verification of arithmetic circuits. The algorithm in [9] presents a unified decision procedure that tackles the property checking problem at the Register Transfer (RT) level. For each RT operation, a specific domain propagation algorithm is applied, using both, bit- and word-level representations. In addition, conflict clauses are learned by analyzing infeasible LPs and deductions, and by employing reverse propagation.

In summary, using the MIP framework an algorithm based on SCIP proved successful on a rather complex and heterogeneous application in which the components which are hard for MIP are handled by domain propagation and SAT techniques.

16.3.2.2 Bonmin

The solver Bonmin (Basic Open-source Nonlinear Mixed INteger programming, Bonami et al. [28]) has been developed for Convex Mixed-Integer Non-Linear Programs (MINLP) within the framework of the MIP solver Cbc [32] by essentially replacing everything which was peculiar for the linear case with a non-linear counterpart but keeping the structure as unchanged as possible[17].

A network design example in the water distribution. The solver Bonmin has been recently applied successfully by Bragalli et al. [30] to a water network design problem in which one has to select among a discrete set the diameter of the pipe to install on the arcs. The model does not have special difficulties besides the so called Hazen-Williams equation modeling pressure loss in water pipes. However, such an equation is very "nasty", i.e., the model is highly non-linear and non-convex. In this context Bonmin only behaves as a heuristic but can be tailored a bit to avoid a too early termination.

A classical MIP model from the eighties linearizes such an equation using the so-called *Special Ordered Sets* of type 2 (SOS2) introduced by Beale and Tomlin [23]. Specifically, by defining a set of variables to be an SOS2, a MIP solver automatically imposes that at most 2 such variables can take a non-zero value, and that they must be consecutive. However, Cplex 10.2 does not find any feasible solution for the instance `fossolo` (see Figure 16.3) in 2 days of CPU time (!!) while Bonmin finds a very accurate one in seconds. Moreover, using the diameters computed by Bonmin, the MIP computation is unable to certify the solution to be feasible even by allowing 1,000 linearization points. In other words, besides the fact that Bonmin was lucky, i.e., heuristically, effective for such a non-convex problem, it is easy to see that the attempt of solving such a problem with a pure MIP solver was a very bad

[17] A very similar approach using the MIP solver MINTO [70] as starting point has been followed by Abhishek, Leyffer and Linderoth and ended up with a Convex Mixed-Integer Non-Linear solver called FilMINT [4].

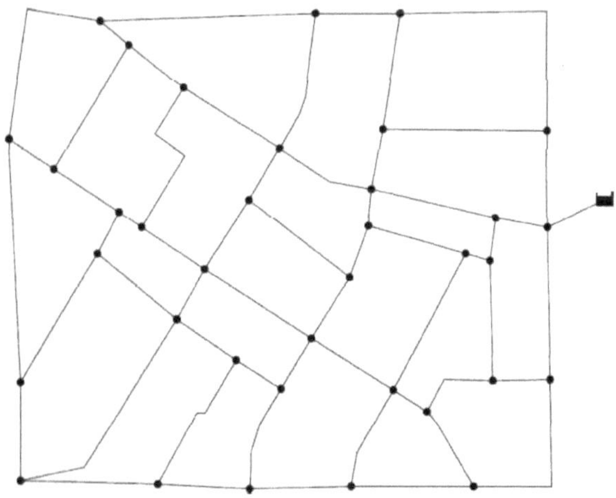

Fig. 16.3 A network design example in the water distribution, instance `fossolo`.

idea because even when all true decisions were taken (the diameters) the problem remained hard.

16.4 Conclusions

We have looked back at the first 50 years of MIP by showing some milestones that have made possible the current state-of-the-art, i.e., the ability of solving in a reliable and effective way many pure and real-world problems. We have also shown that it is not at all the end of the story. Indeed, we do not fully understand even basic components as remarkably shown by Zanette, Balas and Fischetti [91] in the most recent IPCO 2008 meeting which implements for the first time (!) the approach considered by the editors of the present book as the beginning of IP, i.e., Gomory's 1958 paper [55], and demonstrates that it had been somehow overlooked. In addition, there are still very difficult classes of MIPs on which the current solvers are not effective. One of the reasons can be found in the intrinsic weakness of MIP formulations when facing some constraints that can be, otherwise, naturally stated in non-linear or combinatorial fashion. We call for a dissemination of the MIP framework in the direction of increasing its modeling and solving capability.

Acknowledgements The writing of this chapter as well as the talk delivered in the Aussois meeting celebrating the 50 Years of Integer Programming have been challenging. I would like to thank the editors for challenging me and a number of friends and colleagues for very beneficial discussions on the subject and comments on preliminary versions of this chapter. Among them, Ed Klotz, Matteo Fischetti, Egon Balas, Claudia D'Ambrosio and Andrea Tramontani. A special thank goes to Bob Bixby and Tobias Achterberg for sharing with me the results of the experiments on the

Cplex evolution presented in Section 16.2. I am also indebted to Tobias after having read his remarkable thesis that was very helpful in the preparation of this chapter. A warm thank goes to Emilie Danna with whom I share the interest for performance variability and branching issues. Emilie is also trying to teach me some statistics. Finally, I am indebted to two anonymous referees for very useful comments and a careful reading.

References

1. K. Aardal, R.E. Bixby, C.A.J. Hurkens A.K. Lenstra, and J.W. Smeltink, *Market split and basis reduction: Towards a solution of the Cornuéjols-Dawande instances*, INFORMS Journal on Computing 12 (2000) 192–202.
2. K. Aardal, C.A.J. Hurkens, and A.K. Lenstra, *Solving a system of diophantine equations with lower and upper bounds on the variables*, Mathematics of Operations Research 25 (2000) 427–442.
3. K. Aardal and A.K. Lenstra, *Hard equality constrained integer knapsacks*, Mathematics of Operations Research 29 (2004) 724–738.
4. K. Abhishek, S. Leyffer, and J.T. Linderoth, *FilMINT: An outer-approximation-based solver for nonlinear mixed integer programs*, Preprint ANL/MCS-P1374-0906, Mathematics and Computer Science Division, Argonne National Lab, 2006.
5. T. Achterberg, *Conflict analysis in mixed integer programming*, Discrete Optimization 4 (2007) 4–20.
6. T. Achterberg, *Constraint integer programming*, Ph.D. thesis, ZIB, Berlin, 2007.
7. T. Achterberg and T. Berthold, *Improving the feasibility pump*, Discrete Optimization 4 (2007) 77–86.
8. T. Achterberg and R.E. Bixby, Personal communication, 2008.
9. T. Achterberg, R. Brinkmann, and M. Wedler, *Property checking with constraint integer programming*, Tech. Report 07-37, ZIB, Berlin, 2007.
10. T. Achterberg, T. Koch, and A. Martin, *Branching rules revisited*, Operations Research Letters 33 (2005) 42–54.
11. T. Achterberg, T. Koch, and A. Martin, *MIPLIB 2003*, Operations Research Letters 34 (2006) 361–372, see http://miplib.zib.de.
12. E. Amaldi, M.E. Pfetsch, and L.E. Trotter Jr., *On the maximum feasible subsystem problem, IISs, and IIS-hypergraphs*, Mathematical Programming 95 (2003) 533–554.
13. K. Andersen, G. Cornuéjols, and Y. Li, *Reduce-and-split cuts: Improving the performance of mixed integer Gomory cuts*, Management Science 51 (2005) 1720–1732.
14. K. Andersen, Q. Louveaux, R. Weismantel, and L.A. Wolsey, *Inequalities from two rows of a simplex tableau*, Integer Programming and Combinatorial Optimization IPCO 2007 (M. Fischetti and D.P. Williamson, eds.), Lecture Notes in Computer Science 4513, Springer-Verlag, 2007, pp. 1–15.
15. G. Andreello, A. Caprara, and M. Fischetti, *Embedding cuts in a branch and cut framework: a computational study with* $\{0, \frac{1}{2}\}$*-cuts*, INFORMS Journal on Computing 19 (2007) 229–238.
16. D. Applegate, R.E. Bixby, V. Chvátal, and W.J. Cook, *The traveling salesman problem. A computational study*, Princeton University Press, 2007.
17. E. Balas, *Facets of the knapsack polytope*, Mathematical Programming 8 (1975) 146–164.
18. E. Balas, *Disjunctive programming*, Annals of Discrete Mathematics 5 (1979) 3–51.
19. E. Balas, S. Ceria, and G. Cornuéjols, *Mixed 0-1 programming by lift-and-project in a branch-and-cut framework*, Management Science 42 (1996) 1229–1246.
20. E. Balas, S. Ceria, G. Cornuéjols, and N. Natraj, *Gomory cuts revisited*, Operations Research Letters 19 (1996) 1–9.
21. E. Balas and A. Saxena, *Optimizing over the split closure*, Mathematical Programming 113 (2008) 219–240.

22. M. Baz, J.P. Brooks, A. Gosavi, and B. Hunsaker, *Automated tuning of optimization software parameters*, Tech. Report 2007-7, University of Pittsburgh, 2007.
23. E.M.L. Beale and J.A. Tomlin, *Special facilities in a general mathematical programming system for non-convex problems using ordered sets of variables*, OR 69. Proceedings of the Fifth International Conference on Operational Research (J. Lawrence, ed.), Tavistock Publications, 1970, pp. 447–454.
24. M. Benichou, J.M. Gauthier, P. Girodet, and G. Hentges, *Experiments in mixed-integer programming*, Mathematical Programming 1 (1971) 76–94.
25. L. Bertacco, *Exact and heuristic methods for mixed integer linear programs*, Ph.D. thesis, Università degli Studi di Padova, 2006.
26. L. Bertacco, M. Fischetti, and A. Lodi, *A feasibility pump heuristic for general mixed-integer problems*, Discrete Optimization 4 (2007) 63–76.
27. R.E. Bixby, M. Fenelon, Z. Gu, E. Rothberg, and R. Wunderling, *Mixed-integer programming: A progress report*, The Sharpest Cut: The Impact of Manfred Padberg and his Work (M. Grötschel, ed.), MPS-SIAM Series on Optimization, 2004, pp. 309–325.
28. P. Bonami, L.T. Biegler, A.R. Conn, G. Cornuéjols, I.E. Grossmann, C.D. Laird, J. Lee, A. Lodi, F. Margot, N. Sawaya, and A.Wächter, *An algorithmic framework for convex mixed integer nonlinear programs*, Discrete Optimization 5 (2008) 186–204.
29. P. Bonami, G. Cornuéjols, A. Lodi, and F. Margot, *A feasibility pump for mixed integer nonlinear programs*, Mathematical Programming 119 (2009) 331–352.
30. C. Bragalli, C. D'Ambrosio, J. Lee, A. Lodi, and P. Toth, *Water network design by MINLP*, Tech. Report RC24495, IBM, 2008.
31. A. Caprara and M. Fischetti, $\{0, \frac{1}{2}\}$ *Chvátal-Gomory cuts*, Mathematical Programming 74 (1996) 221–235.
32. Cbc, https://projects.coin-or.org/Cbc.
33. J.W. Chinneck, *Fast heuristics for the maximum feasible subsystem problem*, INFORMS Journal on Computing 13 (2001) 210–223.
34. V. Chvátal, *Edmonds polytopes and a hierarchy of combinatorial problems*, Discrete Mathematics 4 (1973) 305–337.
35. V. Chvátal, *Resolution search*, Discrete Applied Mathematics 73 (1997) 81–99.
36. W.J. Cook, S. Dash, R. Fukasawa, and M. Goycoolea, *Numerically accurate Gomory mixed-integer cuts*, Tech. report, School of Industrial and Systems Engineering, Georgia Tech, 2007, http://mgoycool.uai.cl/papers/cdfg08_ijoc.pdf.
37. W.J. Cook, R. Kannan, and A. Schrijver, *Chvátal closures for mixed integer programming problems*, Mathematical Programming 47 (1990) 155–174.
38. G. Cornuéjols, *Valid inequalities for mixed integer linear programs*, Mathematical Programming 112 (2008) 3–44.
39. G. Cornuéjols and M. Dawande, *A class of hard small 0-1 programs*, INFORMS Journal on Computing 11 (1999) 205–210.
40. G. Cornuéjols, L. Liberti, and G. Nannicini, *Improved strategies for branching on general disjunctions*, Tech. report, LIX, École Polytechnique, Optimization Online, paper 2071, 2008.
41. G. Cornuéjols and F. Margot, *On the facets of mixed integer programs with two integer variables and two constraints*, Mathematical Programming 120 (2009) 429–456.
42. H. Crowder, E. Johnson, and M.W. Padberg, *Solving large scale zero-one linear programming problem*, Operations Research 31 (1983) 803–834.
43. E. Danna, *Performance variability in mixed integer programming*, Workshop on Mixed Integer Programming, Columbia University, New York, 2008, see http://coral.ie.lehigh.edu/mip-2008/abstracts.html#Danna.
44. E. Danna, M. Fenelon, Z. Gu, and R.Wunderling, *Generating multiple solutions for mixed integer programming problems*, Integer Programming and Combinatorial Optimization IPCO 2007 (M. Fischetti and D.P. Williamson, eds.), Lecture Notes in Computer Science 4513, Springer-Verlag, 2007, pp. 280–294.
45. E. Danna, E. Rothberg, and C. Le Pape, *Exploiting relaxation induced neighborhoods to improve MIP solutions*, Mathematical Programming 102 (2005) 71–90.

46. S. Dash, O. Günlük, and A. Lodi, *MIR closures of polyhedral sets*, Mathematical Programming 121 (2010) 33–60.
47. S. Dey and L.A. Wolsey, *Lifting integer variables in minimal inequalities corresponding to lattice-free triangles*, Integer Programming and Combinatorial Optimization IPCO 2008 (A. Lodi, A. Panconesi, and G. Rinaldi, eds.), Lecture Notes in Computer Science, 5035, Springer-Verlag, 2008, pp. 463–475.
48. D.G. Espinoza, *Computing with multi-row gomory cuts*, Integer Programming and Combinatorial Optimization IPCO 2008 (A. Lodi, A. Panconesi, and G. Rinaldi, eds.), Lecture Notes in Computer Science 5035, Springer-Verlag, 2008, pp. 214–224.
49. M. Fischetti, F. Glover, and A. Lodi, *The feasibility pump*, Mathematical Programming 104 (2005) 91–104.
50. M. Fischetti and A. Lodi, *Local branching*, Mathematical Programming 98 (2002) 23–47.
51. M. Fischetti and A. Lodi, *Optimizing over the first Chvátal closure*, Mathematical Programming 110 (2007) 3–20.
52. M. Fischetti, A. Lodi, and D. Salvagnin, *Just MIP it!*, MATHEURISTICS: Hybridizing metaheuristics and mathematical programming (V. Maniezzo, T. Stützle, and S. Voss, eds.), Operations Research/Computer Science Interfaces Series, Springer, 2009.
53. M. Fischetti, A. Lodi, and A. Tramontani, *On the separation of disjunctive cuts*, Mathematical Programming, DOI 10.1007/s10107-009-0300-y, 2010.
54. F.W. Glover and G.A. Kochenberger (eds.), *Handbook of metaheuristics*, Kluwer Academic Publishers, Dordrecht, The Netherlands, 2003.
55. R.E. Gomory, *Outline of an algorithm for integer solutions to linear programs*, Bulletin of the American Mathematical Society 64 (1958) 275–278.
56. R.E. Gomory, *An algorithm for the mixed integer problem*, Tech. Report RM-2597, The Rand Corporation, 1960.
57. M. Grötschel, M. Jünger, and G. Reinelt, *A cutting plane algorithm for the linear ordering problem*, Operations Research 32 (1984) 1195–1220.
58. Z. Gu, G.L. Nemhauser, and M.W.P. Savelsbergh, *Mixed flow covers for mixed 0-1 integer programs*, Mathematical Programming 85 (1999) 439–467.
59. J.N. Hooker, *Needed: An empirical science of algorithms*, Operations Research 42 (1994) 201–212.
60. M. Jörg, *k-disjunctive cuts and cutting plane algorithms for general mixed integer linear programs*, Ph.D. thesis, Technische Universität München, 2008.
61. E.L. Johnson and M.W. Padberg, *Degree-two inequalities, clique facets and biperfect graphs*, Annals of Discrete Mathematics 16 (1982) 169–187.
62. M. Karamanov and G. Cornuéjols, *Branching on general disjunctions*, Tech. report, Tepper School of Business, Carnegie Mellon University, 2005, revised 2008.
63. A.H. Land and A.G. Doig, *An automatic method of solving discrete programming problems*, Econometrica 28 (1960) 497–520.
64. J.T. Linderoth and M.W.P. Savelsbergh, *A computational study of search strategies for mixed integer programming*, INFORMS Journal on Computing 11 (1999) 173–187.
65. Q. Louveaux and L.A. Wolsey, *Lifting, superadditivity, mixed integer rounding and single node flow sets revisited*, 4OR 1 (2003) 173–207.
66. A. Mahajan and T.K. Ralphs, *Experiments with branching using general disjunctions*, Tech. Report COR@L Lab, Lehigh University, 2008.
67. H. Marchand, *A polyhedral study of the mixed knapsack set and its use to solve mixed integer programs*, Ph.D. thesis, Université Catholique de Louvain, 1998.
68. F. Margot, *Testing cut generators for mixed-integer linear programming*, Tech. Report E-43, Tepper School of Business, Carnegie Mellon University, 2007.
69. C.C. McGeogh, *Experimental analysis of algorithms*, Notices of the American Mathematical Association 48 (2001) 304–311.
70. MINTO, http://coral.ie.lehigh.edu/minto/.
71. G.L. Nemhauser and L.A. Wolsey, *Integer and combinatorial optimization*, Wiley-Interscience, New York, 1988.

72. G.L. Nemhauser and L.A. Wolsey, *A recursive procedure to generate all cuts for 0-1 mixed integer programs*, Mathematical Programming 46 (1990) 379–390.
73. J. Ostrowsky, J. Linderoth, F. Rossi, and S. Smriglio, *Constraint orbital branching*, Integer Programming and Combinatorial Optimization IPCO 2008 (A. Lodi, A. Panconesi, and G. Rinaldi, eds.), Lecture Notes in Computer Science 5035, Springer-Verlag, 2008, pp. 225–239.
74. J. Owen and S. Mehrotra, *Experimental results on using general disjunctions in branch-and-bound for general-integer linear program*, Computational Optimization and Applications 20 (2001) 159–170.
75. M.W. Padberg, *A note on 0-1 programming*, Operations Research 23 (1975) 833–837.
76. M.W. Padberg and G. Rinaldi, *Optimization of a 532-city symmetric traveling salesman problem by branch and cut*, Operations Research Letters 6 (1987) 1–7.
77. M.W. Padberg and G. Rinaldi, *A branch and cut algorithm for the resolution of large-scale symmetric traveling salesmen problems*, SIAM Review 33 (1991) 60–100.
78. M.W. Padberg, T.J. Van Roy, and L.A. Wolsey, *Valid inequalities for fixed charge problems*, Operations Research 33 (1985) 842–861.
79. C.H. Papadimitriou and K. Steiglitz, *Combinatorial optimization: Algorithms and complexity*, Prentice-Hall, 1982.
80. W. Rei, J.-F. Cordeau, M. Gendreau, and P. Soriano, *Accelerating Benders decomposition by local branching*, Tech. Report C7PQMR PO2006-02-X, C.R.T., Montréal, 2006.
81. E. Rothberg, Personal communication, 2007.
82. E. Rothberg, *An evolutionary algorithm for polishing mixed integer programming solutions*, INFORMS Journal on Computing 19 (2007) 534–541.
83. J.J. Salazar González, *Difficult tiny MIPs arising from an application in commutative algebra*, Poster presentation, MIP, Berkeley, 2009.
84. D. Salvagnin, *A dominance procedure for integer programming*, Master's thesis, University of Padova, October 2005.
85. M.P.W. Savelsbergh, *Preprocessing and probing techniques for mixed integer programming problems*, ORSA Journal on Computing 6 (1994) 445–454.
86. R.M. Stallman and G.J. Sussman, *Forward reasoning and dependency directed backtracking in a system for computer-aided circuit analysis*, Artificial Intelligence 9 (1977) 135–196.
87. T.J. Van Roy and L.A. Wolsey, *Solving mixed integer programming problems using automatic reformulation*, Operations Research 35 (1987) 45–57.
88. L.A. Wolsey, *Facets for a linear inequality in 0-1 variables*, Mathematical Programming 8 (1975) 165–178.
89. L.A. Wolsey, *Strong formulations for mixed integer programs: Valid inequalities and extended formulations*, Mathematical Programming 97 (2003) 423–447.
90. L.A. Wolsey, Personal communication, 2005.
91. A. Zanette, M. Fischetti, and E. Balas, *Can pure cutting plane algorithms work?*, Integer Programming and Combinatorial Optimization IPCO 2008 (A. Lodi, A. Panconesi, and G. Rinaldi, eds.), Lecture Notes in Computer Science 5035, Springer-Verlag, 2008, pp. 416–434.

Chapter 17
Symmetry in Integer Linear Programming

François Margot*

Abstract An integer linear program (ILP) is symmetric if its variables can be permuted without changing the structure of the problem. Areas where symmetric ILPs arise range from applied settings (scheduling on identical machines), to combinatorics (code construction), and to statistics (statistical designs construction). Relatively small symmetric ILPs are extremely difficult to solve using branch-and-cut codes oblivious to the symmetry in the problem. This paper reviews techniques developed to take advantage of the symmetry in an ILP during its solution. It also surveys related topics, such as symmetry detection, polyhedral studies of symmetric ILPs, and enumeration of all non isomorphic optimal solutions.

17.1 Introduction

An integer linear program (ILP) is *symmetric* if its variables can be permuted without changing the structure of the problem. Symmetric ILPs frequently appear when formulating classical problems in combinatorics or optimization. For example, graph coloring, scheduling of jobs on parallel identical machines, covering design or codes construction are problems involving symmetries. Additional real world examples can be found in [107, 108, 109]. Even for relatively modestly sized problems, ILPs with large symmetry groups are difficult to solve using traditional branch-and-bound or branch-and-cut algorithms. (We assume that the reader is familiar with these procedures, as excellent introductions can be found in [38, 59, 90, 117].) The trouble comes from the fact that many subproblems in the enumeration tree are isomorphic, forcing a wasteful duplication of effort.

François Margot
Tepper School of Business, Carnegie Mellon University, Pittsburgh, USA
e-mail: `fmargot@andrew.cmu.edu`

* Supported by ONR grant N00014-03-1-0133.

M. Jünger et al. (eds.), *50 Years of Integer Programming 1958-2008*, DOI 10.1007/978-3-540-68279-0_17, © Springer-Verlag Berlin Heidelberg 2010

Even fairly small symmetric ILPs might be difficult to solve using state-of-the-art ILP solvers. Table 17.1 gives characteristics of a few problems. The first three problems are covering, packing, and orthogonal array construction problems (see [15] for details), the next two are covering problems (see [74] for details), *cod93* is a binary error-correcting code construction (see [74] for details), and *STS81* is a covering problem with a constraint matrix corresponding to a Steiner triple system (see [74] for details). Finally, *codbt24* is the problem of constructing a binary-ternary covering code of radius one with two binary entries and four ternary ones [9, 28]: For integers $a, b \geq 0$, let W be the set of words of length $a+b$ where the first a entries in a word take values in $\{0,1\}$ and the last b entries take values in $\{0,1,2\}$. The problem *codbtab* is then the problem of finding a minimum cardinality set $C \subseteq W$ such that for each $w \in W$, there exists $c \in C$ such that the Hamming distance between w and c is at most one. (An ILP formulation of each problem listed in this paper is available [71].)

Despite the rather small number of variables and small integrality gap (except for *cod93* and *STS81* where the gap is quite large; $OA_2(6,3,3,2)$ has no gap at all; it amounts to prove that no integer solution with value 54 exists), these problems are challenging for state-of-the-art codes. Using one of the leading commercial codes available today, finding an optimal solution (when it exists) is relatively easy. Proving its optimality, however, is much harder. Setting the upper bound value to the optimal value (except for the infeasible problem $OA_2(6,3,3,2)$), only problems $CA_7(7,2,4,7)$ (6 minutes) and $OA_2(6,3,3,2)$ (19 hours) can be solved in less than 24 hours. On the other hand, a code taking advantage of the symmetry in the problem (the code of [75], described in Section 17.9) is able to solve most of them in under a minute. The exceptions are $PA_7(7,2,4,7)$ (1 hour and 30 minutes), *cov*1174 (1 hour and 15 minutes), and *codbt24* (10 hours) [2].

Problem	n	Opt	LP	$\|G\|$
$CA_7(7,2,4,7)$	128	113	112	645,120
$PA_7(7,2,4,7)$	128	−108	−112	645,120
$OA_2(6,3,3,2)$	729	–	54	33,592,320
*cov*1054	252	51	50	3,628,800
*cov*1174	330	17	15.71	39,916,800
cod93	512	−40	−51.20	185,794,560
codbt24	324	36	29.45	248,832
*STS*81	81	61	27	1,965,150,720

Table 17.1 Problem characteristics; n is the number of variables, Opt is the optimal value, LP is the value of the LP relaxation of the initial formulation, and $|G|$ is the number of permutations in the symmetry group.

[2] The machine used is a 64 bits `Monarch Empro 4-Way Tower Server` with four `AMD Opteron 852 2.6GHz` processors, each with eight `DDR-400` SDRAM of 2 `GB` and running `Linux Fedora 9`. The compiler is `g++ version 4.3.0 20080428 (Red Hat 4.3.0-8)`. Results are obtained using only one processor.

Various techniques for dealing with symmetric problems have been studied by several research communities, yielding similar approaches. However, since the favorite algorithmic tools available to each community are different, slightly different algorithms have been developed. In this paper, we survey some of the approaches that have been developed in the Mathematical Programming community for solving symmetric ILPs. Limited comparisons and pointers to similar approaches developed in the Computer Science and Constraint Programming communities are given. A recent survey of techniques developed in the Constraint Programming Community to solve symmetric problems is [49].

The main approaches to deal with symmetries that are discussed here are perturbation (Section 17.4), fixing variables (Section 17.5), symmetry breaking inequalities (Section 17.8), and pruning of the enumeration tree (Section 17.9). The remainder of the paper covers related topics, such as detecting symmetries (Section 17.3), symmetric polyhedra (Section 17.6), partitioning problems (Section 17.7), enumeration of all non isomorphic solutions (Section 17.11), further developments using isomorphism pruning (Section 17.12), choice of formulation (Section 17.13), and the use of additional symmetries (Section 17.14). Finally, basic definitions and notation are covered in Section 17.2, and a very brief introduction to computational group representation is given in Section 17.10.

17.2 Preliminaries

In this section, basic definitions and notation are presented. The reader is invited to use standard references on permutation groups [20, 52, 102] to complement the extremely succinct material given here.

An unordered set containing elements $e_1,\ldots,e_n$ is denoted by the notation $\{e_1,\ldots,e_n\}$ while an ordered set of the same elements is denoted by $(e_1,\ldots,e_n)$.

Let Π^n be the set of all permutations of the ground set $I^n = \{1,\ldots,n\}$. Π^n is known as the *symmetric group* of I^n. A permutation $\pi \in \Pi^n$ is represented by an n-vector, with π_i being the image of i under π. The permutation π such that $\pi_i = i$ for $i = 1,\ldots$ is the *identity* permutation and is denoted by I.

If v is an n-vector and $\pi \in \Pi^n$, let $w = \pi(v)$ denote the vector w obtained by permuting the coordinates of v according to π, i.e.,

$$w_{\pi_i} = v_i, \text{ for all } i \in I^n.$$

The *composition* of two permutations $\pi^1, \pi^2 \in \Pi^n$, written $\pi^1 \cdot \pi^2$, is defined as the permutation $h = \pi^1(\pi^2)$. The composition of permutations is associative, i.e., for any $\pi^1, \pi^2, \pi^3 \in G$, we have $(\pi^1 \cdot \pi^2) \cdot \pi^3 = \pi^1 \cdot (\pi^2 \cdot \pi^3)$. The neutral element for the composition is the identity permutation, i.e., for any $\pi \in \Pi^n$, we have $\pi \cdot I = I \cdot \pi = \pi$.

A subset $G \subseteq \Pi^n$ with the composition operation defined above is a *group*, if it contains the identity permutation I and satisfies the following properties:

(i) For any $g^1, g^2 \in G$, we have $g^1 \cdot g^2 \in G$;
(ii) For any permutation $g \in G$, there exists an inverse permutation $g^{-1} \in G$ such that $g \cdot g^{-1} = g^{-1} \cdot g = I$.

If G is a group, the permutation g^{-1} of point (ii) above is unique. The number of permutations in G, denoted by $|G|$, is the *order* of the group. A group is finite if its order is finite. All groups considered in this paper are finite. A *subgroup* of a group G is any subset of G that is a group. To simplify the notation, we make no difference between a set $S \subseteq I^n$ and its characteristic vector. Hence $\pi(S)$ is the subset of I^n containing π_i for all $i \in S$.

Let $K = \{0, 1, \ldots, k\}$ for some positive integer value k. We consider an integer linear program of the form

$$\begin{aligned} \min \quad & c^T x \\ & Ax \geqslant b, \\ & x \in K^n, \end{aligned} \tag{17.1}$$

where A is an $m \times n$ matrix, b is an m-vector, c is an n-vector, and all three have rational entries. Let Q be the set of all feasible solutions of ILP (17.1). Note that sometimes not all n variables are requested to be integer, or some variables have slightly different bounds. These are small extensions of the model, and most results and algorithms can be adapted to handle them. However, symmetry between continuous variables is not always exploited.

The *symmetry group* G of ILP (17.1) is the set of all permutations π of the n variables mapping Q on itself and mapping each feasible solution on a feasible solution having the same value, i.e.,

$$G = \{\pi \in \Pi^n : c^T \bar{x} = c^T \pi(\bar{x}) \text{ and } \pi(\bar{x}) \in Q \text{ for all } \bar{x} \in Q\}.$$

Example 17.1. The following ILP with four binary variables is used to illustrate several definitions.

$$\begin{array}{llllll} \min & x_1 & +x_2 & +x_3 & +x_4 & \\ & x_1 & +x_2 & & & \geq 1 \\ & & x_2 & +x_3 & & \geq 1 \\ & & & x_3 & +x_4 & \geq 1 \\ & x_1 & & & +x_4 & \geq 1 \\ & x_1, & x_2, & x_3, & x_4 & \in \{0,1\}. \end{array} \tag{17.2}$$

The set Q contains 7 solutions: $(1,0,1,0)$, $(0,1,0,1)$, $(1,1,1,0)$, $(1,1,0,1)$, $(1,0,1,1)$, $(0,1,1,1)$, and $(1,1,1,1)$. The symmetry group G of ILP (17.2) contains eight permutations: the identity permutation I, $(2,3,4,1)$, $(3,4,1,2)$, $(4,1,2,3)$, $(3,2,1,4)$, $(4,3,2,1)$, $(1,4,3,2)$, and $(2,1,4,3)$. □

It is straightforward to check that G is indeed a group. Note that in most situations G is not known, but a subgroup G' of G is. All the results in this paper hold if G is replaced by G', but it should be expected that symmetry removal obtained with G' is weaker than the one obtained with G.

The *orbit* of any $v \in \mathbb{R}^n$ under G is

$$\text{orb}(v,G) = \{w \in \mathbb{R}^n : w = g(v) \text{ for some } g \in G\}\,.$$

The *stabilizer* of any $v \in \mathbb{R}^n$ in G is the subgroup of G given by

$$\text{stab}(v,G) = \{g \in G : g(v) = v\}\,.$$

Given a set $H = \{g^1,\dots,g^s\} \subseteq \Pi^n$, the collection of all $g \in \Pi^n$ that can be obtained by composing the given permutations in an arbitrary way (including using any of them more than once) is the group G *generated by* H. The permutations in H are *generators* of G. Any subgroup of Π^n can be generated by a set of $O(n^2)$ generators.

Given a group $G \subseteq \Pi^n$ and a nonempty set $J \subseteq \{1,\dots,n\}$, the *restriction* of G to J is the set of all permutations of the elements in J that are induced by permutations in $\text{stab}(J,G)$. The restriction of G to J is a group.

A vector $v \in \mathbb{R}^n$ is *lexicographically smaller* (resp., *lexicographically larger*) than a vector $w \in \mathbb{R}^n$ if for some $1 \le p \le n$ we have $v_i = w_i$ for $i = 1,\dots,p-1$ and $v_p < w_p$ (resp., $v_p > w_p$). This is denoted by

$$v <_L w \quad (\text{resp., } v >_L w).$$

Given a group $G \subseteq \Pi^n$, a vector v is *lexicomin* (resp., *lexicomax*) in its orbit under G if there is no $g \in G$ with $g(v) <_L v$ (resp., $g(v) >_L v$).

Example 1 (cont.). The two permutations $(2,3,4,1)$ and $(1,4,3,2)$ form a set of generators for G. The orbit of the vector $v = (0,1,0,1)$ contains only vector $(1,0,1,0)$ in addition to v itself. Vector v is lexicomin in its orbit under G. The stabilizer of v is the subgroup G' of G containing four permutations: I, $(3,4,1,2)$, $(3,2,1,4)$, and $(1,4,3,2)$. The restriction of G to $J = \{2,4\}$ contains only the identity and the permutation swapping 2 and 4 as $\text{stab}(J,G) = \text{stab}(v,G) = G'$. □

17.3 Detecting symmetries

In most cases where a symmetric ILP occurs, the fact that symmetries are present is known to the modeler. However, sometimes only a subgroup G' of the true symmetry group G is known. Devising techniques to compute generators of the symmetry group of an ILP is thus of practical importance.

The main difficulty to face is that G is implicitly defined in term of the feasible set of the ILP. By definition, an ILP with n variables and an empty feasible set has Π^n for its symmetry group. As deciding if the feasible set of an ILP is empty or not is an NP-complete problem, it is rather easy to show that deciding if $G = \Pi^n$ is also an NP-complete problem. Indeed, adding two integer variables $0 \le y_1, y_2 \le k$ and the constraint $y_1 + y_2 = 1$ to the ILP, we get that the symmetry group of the modified ILP is Π^{n+2} if and only if the original one is infeasible.

While this leaves little hope to find a polynomial-time algorithm for computing generators of G, the situation is in fact much worse. The implicit definition of G makes it difficult to design practical algorithms for this task, even worst-case exponential-time ones. One practical algorithm computes a subgroup G^{LP} of G defined as the symmetry group of the LP relaxation: Consider an ILP of the form (17.1) with n variables and m constraints. For a permutation $\pi \in \Pi^n$ and a permutation $\sigma \in \Pi^m$ of the m rows, let $A(\pi, \sigma)$ be the matrix obtained from A by permuting its columns according to π and its rows according to σ. The subgroup G^{LP} is given by

$$G^{LP} = \{\pi \in \Pi^n : \pi(c) = c \text{ and } \exists\, \sigma \in \Pi^m \text{ with } \sigma(b) = b,\ A(\pi, \sigma) = A\}\ .$$

Example 1 (cont.). For ILP (17.2), we have $G^{LP} = G$, as permuting both the variables and constraints according to any $\pi \in G$ yields an ILP identical to ILP(17.2).

However, consider the ILP obtained by adding to ILP (17.2) the inequality

$$-2x_1 - x_2 - 2x_3 - 2x_4 \geq -6\ . \tag{17.3}$$

Let H be its symmetry group and H^{LP} be the symmetry group of its LP relaxation. One can check that the only feasible solution of ILP (17.2) that becomes infeasible is $(1,1,1,1)$. It follows that H is identical to G. However, adding (17.3) destroys in the LP relaxation the symmetry between x_2 and the other variables. As a result, we have that H^{LP} contains only I and $(3,2,1,4)$ and $H \neq H^{LP}$. □

In the case where A is a $(0,1)$-matrix, it is possible to cast the computation of generators of G^{LP} as computing generators of the automorphism group of a bipartite graph with colored vertices. (The automorphism group of a graph is the set of all permutations of its nodes that maps adjacent nodes to adjacent nodes.) Indeed, the matrix

$$\left(\begin{array}{c|c} 0 & A^T \\ \hline A & 0 \end{array}\right)$$

can be seen as the adjacency matrix of a bipartite graph H having one vertex for each column of A and one for each row of A, the two sets of vertices forming the two sides of the bipartition. Two column-vertices (resp., row-vertices) have the same color if and only if their corresponding objective coefficients (resp., right hand side coefficients) are identical. Any automorphism of H that fixes both sides of the bipartition and the color classes induces a permutation of the columns of A that is in G^{LP}. Any permutation of G^{LP} can be extended to an automorphism of H that fixes the bipartition and color classes.

When A is not a $(0,1)$-matrix, it is possible to modify the above construction to get the correct result. Details can be found in [103]. Note that mapping the instance of a problem to a colored graph such that color preserving automorphisms of the graph correspond to symmetries of the problem is standard procedure. For example, see [1, 98, 99] for satisfiability problems and [95] for Constraint Programming in general.

The computational complexity status of computing generators of the automorphism group of a graph is an open problem as it is equivalent to the complexity status of the famous Graph Isomorphism problem (see [69] for a detailed discussion of complexity of problems related to permutation groups). However, efficient algorithms (such as `nauty` [77], `Saucy` [32], as part of `MAGMA` [10] or `GAP` [113]) are available and perform satisfactorily in many instances.

Another track is to formulate the problem of computing permutations in G^{LP} as an ILP [66], although this approach seems unlikely to be competitive when the order of G^{LP} is large.

While working with G^{LP} instead of G might result in a loss of efficiency, most applications have a known LP formulation for which G^{LP} is either a large subgroup of G or G itself, except for infeasible problems. Usually, the symmetry group used is either built from the modeler knowledge or by computing G^{LP} for some ILP formulation.

17.4 Perturbation

One of the first ideas that comes to mind when facing a symmetric problem is to perturb it slightly to destroy the symmetry or to capture some of the symmetry in the problem. For example, adding a small random perturbation to the coefficients c_i for $i = 1, \ldots, n$ easily destroys the symmetry in the problem. This, in general, does not help much and can even be counter-productive. The main reason is that when ILP (17.1) is infeasible, the same amount of work has to be done regardless of the objective function. In addition, even if the ILP is feasible, once the algorithm has found an optimal solution, the remainder of the computation can be seen as solving an infeasible ILP. Destroying symmetry by perturbation of the objective function thus achieves very little and using symmetry information in an efficient way while solving the problem is a far superior alternative.

The same is true for the "lexicographic" perturbation $c_i = 2^i$ for $i = 1, \ldots, n$ that can be used for certain binary problems, with the additional caveat that this transformation is numerically unstable and can only be used for very small problems.

Using perturbation is sometimes helpful when trying to find a good solution, but using symmetry information is a superior approach when dealing with an infeasible problem or when proving optimality of a solution.

17.5 Fixing variables

Another simple idea to reduce symmetry in an ILP is to fix variables. While this could be considered a special case of symmetry breaking inequalities (a topic covered in Section 17.8), we treat it separately as it can easily be combined with other techniques for dealing with symmetric ILPs when it is used as a preprocessing step.

Let ILP be a particular instance of ILP (17.1) with symmetry group G. Suppose that it is known that, for some $t \geq 1$, some index set $F = \{i_1, \ldots, i_t\}$ and some values $\{\bar{x}_{i_1}, \ldots, \bar{x}_{i_t}\}$, ILP has an optimal solution with $x_i = \bar{x}_i$ for all $i \in F$. Let the *fixed ILP* (FILP) be obtained by adding to ILP the constraints $x_i = \bar{x}_i$ for all $i \in F$. Let the *reduced ILP* (RILP) be obtained from ILP by substituting x_i by $\bar{x}_i$ for all $i \in F$. Let G^F (resp., G^R) be the symmetry group of FILP (resp., RILP). Note that FILP is an ILP with n variables, while RILP has $n - t$ variables.

Example 17.2. Consider the ILP with six binary variables

$$\begin{array}{rrrrrrrcr} \min & x_1 & +x_2 & +x_3 & +x_4 & -x_5 & -x_6 & & \\ & x_1 & +x_2 & & & -x_5 & -x_6 & \geq & -1 \\ & & x_2 & +x_3 & & & -x_6 & \geq & 0 \\ & & & x_3 & +x_4 & -x_5 & -x_6 & \geq & -1 \\ & x_1 & & & +x_4 & -x_5 & & \geq & 0 \\ & x_1 & +x_2 & & & +x_5 & & \geq & 1 \\ & & & x_3 & +x_4 & & +x_6 & \geq & 1 \\ & x_1, & x_2, & x_3, & x_4 & x_5, & x_6 & \in & \{0,1\}. \end{array} \tag{17.4}$$

Its feasible set contains 36 solutions, 9 with $(x_5,x_6) = (0,0)$, 10 with $(x_5,x_6) = (1,0)$, 10 with $(x_5,x_6) = (0,1)$, and 7 with $(x_5,x_6) = (1,1)$. The symmetry group G of ILP (17.4) contains only two permutations: I and $(3,4,1,2,6,5)$.

A simple analysis of ILP (17.4) proves that there exists an optimal solution with $x_5 = 1$ and $x_6 = 1$. Adding these two constraints to the ILP yields FILP and its symmetry group G^F is G. Substituting x_5 and x_6, we get RILP which is exactly ILP (17.2) with a symmetry group G^R containing eight permutations generated by $\{(2,3,4,1),(1,4,3,2)\}$. □

In this section, we discuss properties that might indicate which of the formulations ILP, FILP or RILP should be used. This is of course a difficult question and only partial answers or rules of thumb can be given.

Let v be the n-vector with $v_i = \bar{x}_i$ for all $i \in F$ and $v_i = -1$ otherwise. Assuming that in FILP there are at least two possible values for variable x_i for all $i \notin F$, we have that $G^F = \text{stab}(v,G)$. It follows that $|G^F| \leq |G|$. On the other hand, G^R contains a subgroup G' that is the restriction of $\text{stab}(v,G)$ to the variables indices in the complement of F. As shown in Example 17.1 at the end of Section 17.2, the order of a restriction of a group can be smaller than the order of the group itself and thus we might have $|G^R| < |\text{stab}(v,G)| = G^F$. However, as shown in Example 17.2, it is also possible to have $|G^R| > |G|$. As a result, there is no general relation between $|G^R|$ and either $|G^F|$ or $|G|$.

Which of the three ILP formulations to use depends on the solution algorithm $\mathbb{A}$. If $\mathbb{A}$ is a branch-and-bound algorithm oblivious to symmetry, regardless of the sizes of the symmetry groups, using either FILP or RILP produces similar results, and this should not be worse than using ILP. (A very simple illustration can be found in [57].) On the other hand, if $\mathbb{A}$ uses the symmetry group of the problem, the situation is not so clear cut. If $\mathbb{A}$ is efficient in using the symmetry group, it might be better to solve ILP than FILP or RILP. In the remainder of the section, we give two examples where this happens. In general, however, it should be expected that solving FILP or RILP is more efficient, in particular if the number of fixed variables is large.

An example where solving ILP is easier than solving FILP is for an ILP formulation to solve the classical edge coloring problem for a graph H with maximum vertex degree Δ. We want to decide if a coloring of the edges of H with Δ colors exists or not, such that any two edges sharing an endpoint receive distinct colors. A simple ILP formulation for edge coloring (EC) uses Δ binary variables x_{ej} for $j = 1, \ldots, \Delta$ for each edge e in the graph with the meaning that $x_{ej} = 1$ if and only if e receives color j. Constraints are simply

$$\sum_{j=1}^{\Delta} x_{ej} = 1, \quad \text{for each edge } e \in E(H), \tag{17.5}$$

$$\sum_{e \in \delta(v)} x_{ej} \leq 1, \quad \text{for all } v \in V(H), \text{ for all } j = 1, \ldots, \Delta, \tag{17.6}$$

where $\delta(v)$ is the set of edges incident with vertex v. The symmetry group G of EC is the product of the automorphism group of the graph H with the symmetric group on the Δ colors.

Obviously, permuting the Δ colors in any feasible solution of EC yields another feasible solution. For a vertex v of maximum degree Δ, all edges in $\delta(v)$ must receive Δ distinct colors. As a result, it is valid to fix the colors on $\delta(v)$ to any valid coloring. This will break the symmetry between the colors, yielding the FEC formulation. The symmetry group G^F contains all permutations of G fixing edges in $\delta(v)$ and their colors. Using this ILP formulation for coloring the edges of a clique on 9 nodes and fixing the colors as described above results in a solution time orders of magnitude larger for FEC than for EC for algorithms of [75].

Going back to the general case, it is also sometimes the case that solving ILP is better than solving RILP. An example from [74] is the code construction problem *cod93* listed in Table 17.1. It has 512 variables, a group with order 185,794,560, an optimal value of -40 and an LP relaxation value of -51.20. By fixing a few variables and substituting them in the formulation, one obtains an equivalent problem *cod93r* that has 466 variables, a group order of 362,880, an optimal value of -39 and an LP relaxation value of -47.00. Yet, several algorithms using the symmetry in the problem require a smaller enumeration tree to prove that no solution of value smaller than -40 exists in *cod93* than to prove that no solution of value smaller than -39 exists in *cod93r*.

If ILP is binary and an algorithm based on pruning of the enumeration tree (see Section 17.9) is used, it can be shown that fixing a set F of variables to value 1 and then use the algorithm on FILP is never superior to using the algorithm on the original ILP and branching first on the variables in F, creating only one subproblem corresponding to the fixing. A similar result for non binary ILPs can be stated provided that the variables and the values to which they are fixed satisfy a technical condition. It follows that for these algorithms, solving FILP is not a good idea. A similar result holds for comparing ILP with RILP: the latter might be preferable only if $|G^R| > |\mathrm{stab}(v, G)|$.

17.6 Symmetric polyhedra and related topics

The definition of a symmetric ILP given in Section 17.2 involves the objective function c. If c in (17.1) is replaced by the zero vector, the symmetry group G of the corresponding ILP is the *symmetry group of the polyhedron* corresponding to the convex hull of the characteristic vectors of the solutions of the problem. (Note that this group should not be confused with the group of the geometric symmetries of the polyhedron; only symmetries permuting space coordinates are considered here.) Many combinatorial polyhedra have large symmetry groups. Just to cite one example, the polytope associated with the Traveling Salesman Problem (TSP) [3, 63] on the complete graph on n nodes has a symmetry group of order $n!$. This statement might be a little bit misleading, since there are many ILP formulations of the TSP, some of them having less symmetry than others. We are talking here about the most studied formulation using exclusively binary edge variables x_{ij} for all $i, j = 1, \ldots, n$ and $i < j$. This polytope has received a lot of attention and has been the focus of intense computational studies in the last decades with impressive results [3]. However, in these studies, the topic of symmetry is rarely considered, as the objective function used in most instances essentially destroys the symmetry.

Similarly, many polyhedra related to combinatorial problems have large symmetry groups, but studies of their facial structure rarely rely on this knowledge. There are several polytopes closely linked to permutations or permutation groups: The *permutahedron* is the convex hull of all permutations of the entries of the n-vector $(1, 2, 3, \ldots, n)$. Its complete linear description is known [6]. A generalization of this polytope is the *permutahedron of a poset*, the convex hull of all permutations π such that if $i < j$ is the poset, then $\pi_i < \pi_j$. Its complete linear description is known for some classes of posets [4]. The *assignment* polytope is the convex hull of all binary $n \times n$ matrices with exactly one nonzero entry per row and per column. These matrices are in bijection with permutations of n elements: For a matrix M and permutation π, entry $M_{ij} = 1$ if and only if $\pi_i = j$. A complete linear description of the assignment polytope is known [84]. In [13, 14], the *permutation* polytope is studied. This polytope is the convex hull of the vertices of the assignment polytope corresponding to permutations that are in a given group G. While a complete linear description of the permutation polytope is given in [14], that paper also proves that deciding if one of its inequalities is violated by a given matrix $\overline{M}$ is an NP-complete problem.

Let Q be the feasible set of an ILP with variables x and let P be the convex hull of Q. The action of altering the ILP by adding to it a number of variables y, adding a number of constraints and modifying the original constraints such that the projection of the resulting feasible set on the space of the x variables remains P is known as a *lifting* P' of P. If Q has a symmetry group G, a *symmetric lifting* of P is a lifting P' of P such that G is the restriction of the symmetry group of P' to the x variables. Two important results of Yannakakis [118] are that neither the matching polytope nor the TSP polytope have a symmetric lifting of subexponential size.

One notable exception where the symmetry group G of the polyhedron plays a central role in the study of its facial structure is the problem of obtaining its complete linear description by enumeration. Typically, extreme points of the polyhedron are partitioned into orbits under G and then, for one vertex v in each orbit the facets incident to v are described. These algorithms are based on clever enumeration procedures using the symmetry group of the polyhedron and can be carried out for problems of small dimension. For example, a complete linear description is known for the TSP polytope on a complete graph with up to 10 nodes [27], the Linear Ordering polytope with up to 8 items [27], the Cut polytope on a complete undirected graph with up to 9 nodes [27], the Metric cone and Metric polytope on a complete graph with up to 8 nodes [35, 36].

All enumeration algorithms are limited in the size of instances they can tackle. They might give hints on the form of the complete linear description for all instances but, most of the time, classes of facet defining inequalities for large instances are not facet defining in smaller ones. The search for facet defining inequalities for symmetric polytopes thus requires some effort. One could hope that some generic process could be used to generate strong valid (let alone facet defining) inequalities using the symmetry group of the polytope, but no such process seems to be known. Most derivations of valid inequalities are problem specific and only use the existence of the symmetry group in a limited way. For many symmetric ILPs used to test the existence of combinatorial objects (such as problems similar to those listed in Table 17.1) the gap between the optimal value of the ILP formulation and its LP relaxation is large. The generation of strong valid inequalities for these problems has received little attention. One recent paper [70] does this for the very hard problem *codbt06*, better known as the Football Pool problem [28, 53] or as the construction of an optimal ternary covering code of length 6. This problem is still open despite extensive efforts for solving it [67, 86].

One main tenet of polyhedral combinatorics is that the knowledge of families of facets of the convex hull P of the feasible set of an ILP is useful for solving the problem, even if the complete linear description of P is not known. If the number of facets in a family F is exponential in the size of the problem encoding, F can still be used in an efficient way providing that it has an efficient *separation algorithm*. Such an algorithm takes a point $\bar{x}$ as input and either outputs one inequality corresponding to a facet in F violated by $\bar{x}$ or guarantees that all such inequalities are satisfied by $\bar{x}$.

Many papers describe facets and separation algorithms for specific symmetric polyhedra. For a symmetric problem with symmetry group G, any valid inequality for P of the form $ax \leq a_0$ generates a collection of "symmetric" inequalities of the form $g(a) \cdot x \leq a_0$ for all $g \in G$. This naturally leads to look for a separation algorithm for this class of inequalities, i.e., an algorithm for the following problem:

LINEAR OPTIMIZATION UNDER SYMMETRY

Input: A group G permuting the elements in I^n given by a set of t generators, a vector $a \in \mathbb{R}^n$ and a point $\bar{x} \in \mathbb{R}^n$;
Output: A permutation $g \in G$ maximizing $g(a) \cdot \bar{x}$.

In practice, the typical situation is to look for the most violated inequality in a class of facets under all possible permutations in G, and numerous examples where this can be done efficiently are known. For example, virtually all exact separation algorithms for facets of the TSP polytope described in [3] fall into this category. Nevertheless, it is straightforward to show that the above problem is NP-hard, using a polynomial transformation from the following problem (a variant of *Problem 5* of [14], variant shown there to be NP-hard):

MAXIMUM OVERLAP UNDER SYMMETRY

Input: A group G permuting the elements in I^n given by a set of t generators, and two functions $\phi_1, \phi_2 : I^n \to \{0,1\}$;
Output: A permutation $g \in G$ maximizing $|\{i \in I^n : \phi_1(i) = \phi_2(g(i))\}|$.

Indeed, define $a_i := \phi_2(i)$ and $\bar{x}_i := \phi_1(i)$ for all $i \in I^n$. For $g \in G$, $k, \ell = 0, 1$ and $s = 1, 2$, define

$$\begin{aligned} y^g_{k\ell} &= |\{i \in I^n : \phi_1(i) = k,\ \phi_2(g(i)) = \ell\}| \\ z_s &= |\{i \in I^n : \phi_s(i) = 1\}| \,. \end{aligned}$$

The first problem asks for a permutation $g \in G$ maximizing y^g_{11} while the second problem asks for maximizing $y^g_{11} + y^g_{00}$. But as $y^g_{11} + y^g_{10} = z_1$ and $y^g_{10} + y^g_{00} = n - z_2$, we have $y^g_{11} + y^g_{00} = 2\, y^g_{11} + n - z_1 - z_2$. As the above transformation is polynomial in the size of the instance of the second problem, it is a polynomial time reduction from the second problem to the first one.

17.7 Partitioning problems

Several classes of symmetric problems arising in practice are of the partitioning type: Given a set S of s elements, find a partition of S into at most t subsets with each of the subsets having to meet the same requirements. There is immediately a symmetry between the subsets in the partition. A typical ILP formulation for such a problem uses binary variables x_{ij} for all $i = 1, \ldots, s$ and $j = 1, \ldots, t$ with the meaning

$$x_{ij} = \begin{cases} 1, & \text{if } i \text{ is assigned to subset } j, \\ 0, & \text{otherwise.} \end{cases} \tag{17.7}$$

The problem formulation might use additional variables y_i for $i = 1, \ldots, n_y$ with $y_i \in \mathbb{Z}$ for $i \in Y \subseteq \{1, \ldots n_y\}$. The problem can then be written as

$$\begin{aligned} &\min c^{xT} x + c^{yT} y \\ &A^x\, x + A^y\, y \geqslant b \\ &\quad \sum_{j=1}^{t} x_{ij} = 1, \text{ for all } i = 1,\dots,s, \\ &\quad x_{ij} \in \{0,1\}, \text{ for all } i = 1,\dots,s,\ j = 1,\dots,t, \\ &\qquad y_i \in \mathbb{Z}, \text{ for all } i \in Y, \end{aligned} \tag{17.8}$$

where A^x is an $m \times (s \cdot t)$ matrix, A^y is an $m \times n_y$ matrix, c^x is an $(s \cdot t)$-vector, c^y is an n_y-vector, and b is an m-vector and all these matrices and vectors are rational. In addition, we assume that any permutation of the t subsets can be extended to a permutation in the symmetry group of the ILP. In other words, if h is any permutation of I^t, then there exists a permutation h' of I^{n_y} and

$$(x_{11},\dots,x_{s1},x_{12},\dots,x_{s2},\dots,x_{1t},\dots,x_{st},y_1,\dots,y_{n_y})$$

is feasible if and only if

$$(x_{1h(1)},\dots,x_{sh(1)},x_{1h(2)},\dots,x_{sh(2)},\dots,x_{1h(t)},\dots,x_{sh(t)},y_{h'(1)},\dots,y_{h'(n_y)})$$

is and both solutions have the same objective value. We also assume that the vector c^x is symmetric with respect to the t subsets, i.e., writing $c(x_{ij})$ for the entry of c^x corresponding to variable x_{ij}, we assume that $c(x_{ij}) = c(x_{ij'})$ for all $j, j' = 1,\dots,t$.

A few examples of problems fitting this model are bin packing, cutting stock, scheduling on identical machines, graph coloring (either vertex-coloring or edge-coloring), and graph partitioning. Many other practical problems featuring partitions into interchangeable subsets fit this model too (see [108] for examples).

Note that, in addition to the symmetry between the subsets, it is possible that some symmetry between the elements also occurs. Such symmetry occurs for example for graph coloring with a graph with a nontrivial automorphism group, or for bin packing with multiple items having identical dimensions. The material in this section concentrates on dealing with the symmetry between subsets only.

17.7.1 Dantzig-Wolfe decomposition

To address the symmetry between the subsets of the partition, a Dantzig-Wolfe decomposition approach can be tried: Given the collection of all binary s-vectors z^ℓ for $\ell = 1,\dots,u$ corresponding to the characteristic vector of a subset of the partition, the above problem can sometimes be reformulated with variables λ^ℓ for $\ell = 1,\dots,u$ and objective vector c^λ with $c^\lambda_\ell = \sum_{i=1}^{s} c(x_{i1})\, z^\ell_i$:

$$
\begin{aligned}
\min c^{\lambda^T} \lambda + c^{y^T} y \\
A^{\lambda}\, \lambda + A^{y'}\, y \geqslant b', \\
\sum_{\ell=1}^{u} z_i^{\ell}\, \lambda^{\ell} = 1, & \text{ for all } i = 1,\ldots,s, \\
\sum_{\ell=1}^{u} \lambda^{\ell} = t, \\
\lambda^{\ell} \in \{0,1\}, & \text{ for } \ell = 1,\ldots,u, \\
y_i \in \mathbb{Z}, & \text{ for all } i \in Y.
\end{aligned}
$$

This reformulation makes the symmetry between the subsets of the partition disappear, as it only asks for t of the given z^{ℓ} vectors that are disjoint and cover all items in S, without paying attention to their ordering. The disadvantage is of course the large number of variables. This type of formulation requires column generation procedures, as soon as the size of the problem is non-trivial. A pricing problem for vectors z^{ℓ} not included in the formulation then has to be solved.

Examples of successful applications of this approach can be found for the cutting stock problem [114, 115], scheduling [34, 37, 7, 116], edge coloring [83], vertex coloring [79], and graph partitioning [80]. See section 13.3 for a more general presentation and additional examples.

17.7.2 Partitioning orbitope

Consider the polytope obtained by taking the convex hull of the feasible solutions to the partitioning part of ILP (17.8), namely:

$$\sum_{j=1}^{t} x_{ij} = 1, \text{ for all } i = 1,\ldots,s, \tag{17.9}$$

$$x_{ij} \in \{0,1\}, \text{ for all } i = 1,\ldots,s,\ j = 1,\ldots,t. \tag{17.10}$$

Assuming that this system is part of a larger problem that has a symmetry between the sets in the partition, one can try to remove that symmetry from the problem by partitioning the feasible set of (17.9)–(17.10) in equivalence classes under that symmetry and by selecting one representative of each class. One natural way to do this is to arrange variables x_{ij} for $i = 1,\ldots,s$, $j = 1,\ldots,t$ in a matrix X with x_{ij} being the entry in row i and column j of X. A matrix $\overline{X}$ is *feasible* if its entries $\bar{x}_{ij}$ satisfy the constraints in (17.9)-(17.10). The symmetry group G^C considered here is the group that permutes the columns of $\overline{X}$ in any possible way. For column j of $\overline{X}$, define its *value* v_j as

$$v_j = \sum_{i=1}^{s} 2^{s-i} \cdot \bar{x}_{ij} \; .$$

A matrix $\overline{X}$ is then a representative of its equivalence class under G^C if and only if its columns are ordered in non-increasing order of their values. The *partitioning orbitope*, introduced in [61], is the polytope obtained as the convex hull of these representative matrices. Its complete linear description is known and is based on *shifted column inequalities* (SCI). An SCI has a *bar* formed by variables $\{x_{\bar{i}\bar{j}}, x_{\bar{i}\bar{j}+1}, \ldots, x_{\bar{i}t}\}$ for some $2 \leq \bar{i} \leq s$, $2 \leq \bar{j} \leq \min\{\bar{i}, t\}$ and a *shifted column* (SC) consisting of $\bar{i} - \bar{j} + 1$ variables, exactly one on each of the diagonals $x_{d+1,1}, x_{d+2,2}, \ldots, x_{d+\bar{j}-1,\bar{j}-1}$ for $d = 0, \ldots, \bar{i} - \bar{j}$ and with the condition that the variable selected in diagonal d has a column index no larger than the one of the variable selected in diagonal $d+1$ for $d = 0, \ldots, \bar{i} - \bar{j} - 1$ (see Figure 17.1). The SCI with bar B and shifted column S is then $x(B) - x(S) \leq 0$, using the notation $x(A) = \sum_{x_{ij} \in A} x_{ij}$.

Theorem 17.1. *[61] A linear description of the the partitioning orbitope is given by*

$$\sum_{j=1}^{t} x_{ij} = 1, \text{ for all } i = 1, \ldots, s, \tag{17.11}$$

$$x(B) - x(S) \leq 0, \text{ for all SCI with } B \text{ its bar and } S \text{ its SC}, \tag{17.12}$$

$$x_{ij} = 0, \text{ for } i = 1, \ldots, s,\ j = i+1, \ldots t, \tag{17.13}$$

$$x_{ij} \geq 0, \text{ for } i = 1, \ldots, s,\ j = 1, \ldots t. \tag{17.14}$$

The *packing orbitope* is obtained by replacing in (17.9) the equality sign by $\leq$. A result similar to Theorem 17.1 is given in [61] for the packing orbitope.

While the formulation of Theorem 17.1 has an exponential number of constraints, efficient separation algorithms are given in [61]. Empirical results using this formulation and further development of fixing algorithms based on it can be found in [60].

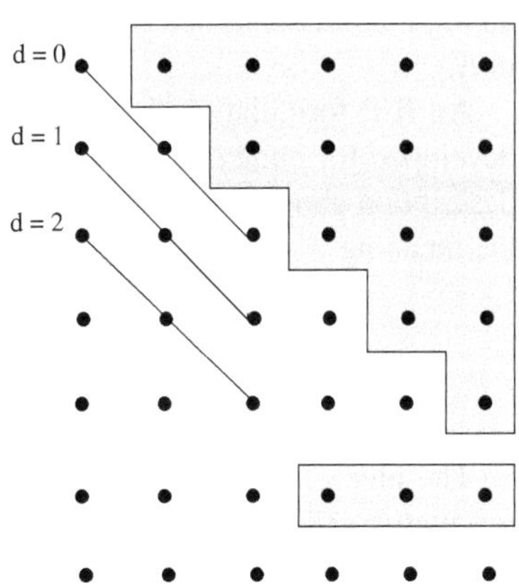

Fig. 17.1 Graphic representation of an SCI with $s = 7$, $t = 6$, $\bar{i} = 6$, and $\bar{j} = 4$. Entries in the shaded rectangle form the bar B and have coefficients 1; exactly one entry in each of the three diagonal segments has a coefficient -1; these three entries form the shifted column S. All other entries have coefficient 0, including entries in the top right shaded part corresponding to constraints (17.13).

Note that the ordering of the objects in S, or equivalently the ordering of the rows in the matrix X might have a big influence on the efficiency derived from the partitioning orbitope. An extreme example is solving an edge coloring problem on a graph H with maximum vertex degree Δ using the ILP formulation described in Section 17.5. If the edges corresponding to the first Δ rows of X are edges adjacent to a vertex v of maximum degree in H, the constraints (17.5)–(17.6) together with (17.13)–(17.14) immediately imply $x_{ii} = 1$ for all $i = 1, \ldots, \Delta$. As a result, all SCI inequalities are satisfied and the gain obtained by using the partitioning orbitope amounts to fixing the colors on the edges adjacent to v. It is likely, however, that a different ordering of the rows of X allows to derive more strength from the SCI. The effect of different orders of the elements in S when using the orbitope has not been investigated.

It could be the case that the partitioning orbitope is particularly useful when it is not clear how to construct a set S of t objects, no two of which can be in the same subset. This is the case for the graph partitioning problem used in the computational experiments of [60].

Alternative integer linear descriptions for the partitioning orbitope are sometimes used and examples can be found in [81, 92]. Generalizing Theorem 17.1 to the case where the right hand side of (17.9) is larger than 1 seems difficult, but would have practical implications.

17.7.3 Asymmetric representatives

Consider the partitioning problem where the goal is to minimize the number of subsets required to partition the s elements, with the additional constraint that the elements in any set U_d from a given list $U = \{U_1, \ldots, U_w\}$ cannot all be assigned to the same subset of the partition. (We assume that $|U_d| \geq 2$ for all $d = 1, \ldots, w$ since otherwise the problem is infeasible.) For example, the problem of coloring the nodes of a graph $H = (V, E)$ with the minimum number of colors fits this description, taking V as elements and U as the list of all pairs of elements corresponding to edges in E.

An ILP formulation for this problem (named *Asymmetric Representatives* and developed for node coloring [23, 24, 25] but that can be generalized to handle the description above), uses binary variables z_{ij} for all $i = 1, \ldots, s$ and $j = 1, \ldots, s$ with the meaning

$$z_{ij} = \begin{cases} 1, & \text{if } i \text{ is the representative of } j, \\ 0, & \text{otherwise.} \end{cases} \tag{17.15}$$

The idea of the formulation is that each element selects an element as its representative, all elements selecting the same representative forming a subset of the partition. Representative elements are elements i with $z_{ii} = 1$. The formulation is as follows.

$$\min \sum_{i=1}^{s} z_{ii}$$

$$\sum_{i=1}^{s} z_{ij} = 1, \text{ for all } j = 1,\dots,s, \tag{17.16}$$

$$\sum_{j \in U_d} z_{ij} \le (|U_d| - 1) \cdot z_{ii}, \text{ for all } d = 1,\dots,w,\ i = 1,\dots,s, \tag{17.17}$$

$$z_{ij} \in \{0,1\}, \text{ for all } i = 1,\dots,s,\ j = 1,\dots,s. \tag{17.18}$$

Inequalities (17.16) force each element to select exactly one representative, while an inequality (17.17) prevents the elements in set U_d to all select i as representative and requires that $z_{ii} = 1$ if node i is used as representative of one of the elements in U_d.

This formulation might or might not have symmetries, but the symmetry between the subsets in the partition is destroyed. Nevertheless, this formulation still has several equivalent solutions: for a given partition of the elements, all elements in a subset of the partition can select any of the elements in the subset to get a feasible solution with the same objective value. The formulation can thus also impose that $z_{ij} = 0$ for all $i > j$. (A more general formulation is given in [25], using a partial order on the elements instead of the total order used here.) For coloring the nodes of a graph $H = (V,E)$, the formulation simplifies to

$$\min \sum_{i=1}^{s} z_{ii}$$

$$\sum_{(i,j) \notin E} z_{ij} = 1, \text{ for all } j = 1,\dots,s, \tag{17.19}$$

$$z_{ij} + z_{ik} \le z_{ii}, \text{ for all distinct } i,j,k,\ (i,j),(i,k) \notin E,\ (j,k) \in E, \tag{17.20}$$

$$z_{ij} \in \{0,1\}, \text{ for all } i = 1,\dots,s, j = 1,\dots,s, \tag{17.21}$$

$$z_{ij} = 0, \text{ for all } i = 1,\dots,s,\ j = 1,\dots,s,\ i > j, \tag{17.22}$$

$$z_{ij} = 0, \text{ for all } (i,j) \in E. \tag{17.23}$$

Polyhedral results for this formulation can be found in [23, 24, 25] and [22] gives computational results. Investigation of a similar formulation for the stable set problem is available in [21].

17.8 Symmetry breaking inequalities

A natural way to get rid of symmetry in a problem is to add symmetry breaking inequalities. There are two main ways to do so. The first one, using *dynamic sym-*

metry breaking inequalities, is to generate inequalities during the solution process. These inequalities might be invalid for the initial formulation but, due to the development of the enumeration, it is guaranteed that adding them does not prevent the discovery of an optimal solution. The second one, using *static symmetry breaking inequalities*, is to add inequalities to the initial formulation (explicitly or implicitly), cutting some of the symmetric solutions while keeping at least one optimal solution. An example of static symmetry breaking inequalities are the inequalities describing the partitioning orbitope of Section 17.7.2.

17.8.1 Dynamic symmetry breaking inequalities

Several examples of static symmetry breaking inequalities can be found in the Mathematical Programming literature. On the other hand, dynamic symmetry breaking has been investigated mostly in the Constraint Programming literature [43, 45, 50, 91, 94, 96, 97]. The main reason for this is that constraint programming can express constraints in a form different than linear inequalities [111]. Indeed, when the ILP is not binary, some of the constraints described below simply cannot be expressed by linear inequalities. For binary problems, however, an example are the *isomorphism inequalities* of [73].

The basis of most dynamic symmetry breaking inequalities is that if a node a of the enumeration tree has a subset of variables $(x_{i_1},\dots,x_{i_k})$ fixed respectively to some values $(v_{i_1},\dots,v_{i_k})$, then for any $g \in G$ adding an inequality that cuts all solutions with $(x_{g(i_1)},\dots,x_{g(i_k)})$ fixed respectively to values $(v_{i_1},\dots,v_{i_k})$ can be added at some nodes of the enumeration tree. However, this inequality is not valid for the initial formulation and can only be added during the enumeration.

The drawback of this approach is either the huge number of constraints to handle [50], or the choice of a subset of constraints to use [31, 94, 96], or the design of a separation algorithm. In [45], an implementation based on the computational group theory package `GAP` [113] is presented.

17.8.2 Static symmetry breaking inequalities

The most general description of static symmetry breaking inequalities for the symmetry group G of ILP (17.1) is probably the following. A *fundamental region* for G is a closed set F in $\mathbb{R}^n$ such that:

(i) $g(int(F)) \cap int(F) = \emptyset$, for all $g \in G$, $g \neq I$,
(ii) $\cup_{g \in G}\, g(F) = \mathbb{R}^n$,

where $int(F)$ denotes the interior of F. Observe that (i) forces F to be not too large, while (ii) implies that F contains at least one optimal solution of ILP (17.1). Indeed, if x^* is an optimal solution, (ii) guarantees that $g(F)$ contains x^* for some $g \in G$ or, equivalently, that $g^{-1}(x^*) \in F$. We thus get:

Theorem 17.2. *Let G be the symmetry group for ILP (17.1) and let F be a fundamental region for G. Then an optimal solution to ILP (17.1) can be found by optimizing over the intersection of the feasible set of ILP (17.1) with F.*

It turns out that finding a linear description of a fundamental region for G is quite easy, at least in theory. The following result can be found in [52].

Theorem 17.3. *Let G be the symmetry group of ILP (17.1) and let $\bar{x} \in \mathbb{R}^n$ such that $g(\bar{x}) \neq \bar{x}$ for all $g \in G, g \neq I$. Then*

$$F = \{x \in \mathbb{R}^n : (g(\bar{x}) - \bar{x}) \cdot x \leq 0, \textit{ for all } g \in G,\ g \neq I\} \tag{17.24}$$

is a fundamental region for G.

The obvious practical weakness of this result is the huge number of inequalities (one for each permutation $g \in G$, except the identity) in this description with many of them not being facet defining for F. Another weakness, shared by almost all practical methods using static symmetry breaking inequalities, is that several isomorphic solutions might still be present in the boundary of F. In practice, relatively simple sets of static symmetry breaking inequalities are used, and most of them can be derived using a weak version of Theorem 17.2, as given in the next corollary.

Corollary 17.1. *Theorem 17.2 remains true when the fundamental region F is replaced by the region obtained from Theorem 17.3 by relaxing its statement in either of the following ways (or both):*

(i) Inequalities in (17.24) are written only for a subset of permutations in G.
(ii) The condition that $g(\bar{x}) \neq \bar{x}$ for all $g \in G,\ g \neq I$ is removed.

Proof. The proof of (i) is immediate. For (ii), let F be the feasible set defined by (17.24) for $\bar{x}$. For any real number $\varepsilon > 0$, define $\bar{x}(\varepsilon) \in \mathbb{R}^n$ such that $\bar{x}(\varepsilon)_i = \bar{x}_i + \varepsilon^i$ for $i = 1, \ldots, n$. For $\varepsilon > 0$ small enough, all components of $\bar{x}(\varepsilon)$ are distinct and thus, using (17.24), it defines a fundamental region F_ε for G. Let Q be the set of feasible solutions to ILP (17.1) and let $z \in Q - F$. We now show that there exists $\varepsilon(z) > 0$, such that $z \notin F_\varepsilon$ for all $0 < \varepsilon < \varepsilon(z)$. This implies that F contains all the feasible points from a fundamental region, yielding the result. Define

$$h_z(\varepsilon) = \max_{g \in G} \{(g(\bar{x}_\varepsilon) - \bar{x}_\varepsilon) \cdot z\} .$$

Observe that $h_z(\varepsilon)$ is the maximum of a finite number of polynomials in ε of degree n and thus is a continuous function in ε. As $z \notin F$, we have $h_z(0) > 0$, implying that there exists $\varepsilon(z) > 0$ such that $h_z(\varepsilon) > 0$ for all $0 < \varepsilon < \varepsilon(z)$. As Q is a finite set, for $0 < \delta < \min\{\varepsilon(z) : z \in Q\}$ we have $F_\delta \cap Q \subseteq F \cap Q$. □

We list a few applications of Theorem 17.2 or Corollary 17.1 below.

(i) The ILP has integer variables $0 \le x_i \le k$ for $i = 1,\dots,n$ and G restricted on these variables contains all permutations of I^n.

Add inequalities

$$x_1 \ge x_2 \ge \dots \ge x_n \, .$$

This result is widely known and used routinely. It can be obtained using Corollary 17.1 with $\bar{x}$ defined by $\bar{x}_i = i$ for $i = 1,\dots,n$ and observing that the $n-1$ inequalities given above dominate the remainder of the inequalities obtained from the theorem.

(ii) The ILP has integer variables $0 \le x_{ij} \le k$ for $i = 1,\dots,s$ and $j = 1,\dots,t$ and G, when restricted to these variables, contains all permutations of the first indices and all the permutations of the second indices. In other words, when arranging the variables x_{ij} in a two dimensional matrix X as in Section 17.7.2, all permutations of the rows of X and all the permutations of the columns of X can be extended to permutations in G. Add inequalities expressing that the columns of X must be in non-increasing lexicographic order and that the rows of X must be in non-increasing lexicographic order [40, 41]. This is the *Lex*^{2} symmetry breaking set, following the terminology of [96]. This result can be obtained from Theorem 17.2 using a matrix $\overline{X}$ with $\overline{x}_{ij} = (k+1)^{s \cdot t - (i-1) \cdot t - j}$ for $i = 1,\dots,s$ and $j = 1,\dots,t$. For example, for $s = 3, t = 4$ and $k = 1$ we get

$$\overline{X} = \begin{pmatrix} 2048 & 1024 & 512 & 256 \\ 128 & 64 & 32 & 16 \\ 8 & 4 & 2 & 1 \end{pmatrix} .$$

Using only the permutations in G that either swap two adjacent columns or swap two adjacent rows, we get the *Lex*^{2} set of constraints. For example, using the permutation swapping the first two columns, we get the inequality

$$-1024\, x_{11} - 64\, x_{21} - 4\, x_{31} + 1024\, x_{12} + 64\, x_{22} + 4\, x_{32} \le 0 \qquad (17.25)$$

implying the non-increasing lexicographic ordering of these columns. Similarly, using the permutation swapping the first two rows, we get the inequality

$$\begin{aligned} &- 1920\, x_{11} - 960\, x_{12} - 480\, x_{13} - 240\, x_{14} \\ &+ 1920\, x_{21} + 960\, x_{22} + 480\, x_{23} + 240\, x_{24} \le 0 \end{aligned} \qquad (17.26)$$

implying the non-increasing lexicographic ordering of these rows. Of course, using inequalities (17.25) or (17.26) is not advisable in practice when s or t is large, due to the numerical instability introduced by large coefficients. This is an example where Constraint Programming is a more flexible framework than Mathematical Programming to handle symmetry, as the lexicographic orderings on the rows and columns do not have to be expressed as linear inequalities.

However, in the case where $k = 1$ and exactly one entry in each row must have value 1, we can do much better than using the linear inequalities above: To enforce the non-increasing lexicographic ordering of the columns, we can use the Shifted Column inequalities of Section 17.7.2 and to enforce the non-increasing lexicographic ordering of the rows, we can use the inequalities

$$x_{ip} \le \sum_{j=p}^{t} x_{i+1,j}, \quad \text{for } i = 1,\ldots,s-1,\ p = 1\ldots,t.$$

Note that [41] shows that when $\overline{X}$ must have exactly one 1 per row, the Lex^2 set of constraints can be reinforced by adding the constraints that the sum of the entries in each column is also non-increasing, i.e., adding the constraints

$$\sum_{i=1}^{s} x_{ij} \ge \sum_{i=1}^{s} x_{i,j+1}, \quad \text{for } j = 1\ldots,t-1.$$

This result is of course implied by the inequalities obtained from Theorem 17.2 (it is implied by the lexicographic ordering of the columns and of the rows) but it does not seem easy to derive it algebraically directly from the inequalities obtained from the theorem.

(iii) The generalization of (ii) where the matrix X is d-dimensional with $d \ge 3$ and where any permutation of indices along any dimension of the matrix can be extended to a permutation in G can be handled similarly to (ii). For $i = 1,\ldots,d$, define $X_{i,\ell}$ as the $(d-1)$-dimensional matrix obtained from X by selecting all entries whose ith index is equal to ℓ.

Imposing, for each $i = 1,\ldots,d$, and each possible value of ℓ that the entries in $X_{i,\ell}$ are lexicographically not smaller than entries in $X_{i,\ell+1}$ is valid [40]. This result can be derived from Theorem 17.2 similarly to (ii). A weaker result for the case where entries in X are binary, imposing only that the sum of the entries in $X_{i,\ell}$ is not smaller than the sum of the entries in $X_{i,\ell+1}$ is given in [101].

(iv) The matrix X is an $s \times t$ matrix and any permutation of the columns of X can be extended to a permutation in G. A lexicographic ordering on the columns of X can be imposed. This can be done using the inequalities

$$\sum_{i=1}^{s} (k+1)^{s-i} \cdot x_{ij} \ge \sum_{i=1}^{s} (k+1)^{s-i} \cdot x_{i,j+1}, \quad \text{for } j = 1,\ldots,t-1,$$

but this is not numerically very stable when s is large. Note that this is identical to case (i) applied to the variables corresponding to $\sum_{i=1}^{s} (k+1)^{s-i} \cdot x_{ij}$ for $j = 1,\ldots,t$. Application of this idea can be found [33] for solving a layout problem and in [56] for solving lot-sizing problems. Of course, in special cases, the partitioning orbitope of Section 17.7.2 for example, better inequalities can be used. Weaker conditions have also been tested on practical problems (see [56, 93, 108] for examples and comparisons) such as the following three possibilities: First,

$$\sum_{i=1}^{s} x_{ij} \ge \sum_{i=1}^{s} x_{i,j+1}, \quad \text{for } j = 1,\ldots,t-1.$$

This can be obtained from Corollary 17.1 using $\bar{x}_{ij} = j$ for $i = 1,\ldots,s$, $j = 1\ldots,t$. Second,

$$\sum_{i=1}^{s} i \cdot x_{ij} \geq \sum_{i=1}^{s} i \cdot x_{i,j+1}, \quad \text{for } j = 1,\ldots,t-1,$$

This can be obtained from Corollary 17.1 using $\bar{x}_{ij} = i \cdot j$ for $i = 1,\ldots,s$, $j = 1\ldots,t$. Finally,

$$\sum_{i=1}^{s} i^2 \cdot x_{ij} \geq \sum_{i=1}^{s} i^2 \cdot x_{i,j+1}, \quad \text{for } j = 1,\ldots,t-1.$$

This can be obtained from Corollary 17.1 using $\bar{x}_{ij} = i^2 \cdot j$ for $i = 1,\ldots,s$, $j = 1\ldots,t$.

It should be noted that the impact of different sets of static symmetry breaking inequalities is difficult to estimate. In most cases, only empirical evaluation of specific implementations for specific classes of problems are available. Very little is known about desirable properties of such a set. Discussion related to the choice of $\bar{x}$ and separation of the inequalities in (17.24) when G is the symmetric group Π^n or a cyclic group can be found in [44]. In [66], a ranking of sets of static symmetry breaking inequalities is introduced. It is based on the maximum number of points in the orbit of an extreme point of the polytope that are cut by the set of inequalities.

The discussion of efficiency of different sets of static symmetry breaking inequalities when embedded in a branch-and-bound algorithm is muddled by the interaction between the set of inequalities and valid choices for branching decisions. Some experiments have been made [56, 107, 108, 109], but no definite answer is available. There are simply too many variables to consider: problem classes, formulation choice, large or small group order, choice of algorithm, coupling with other symmetry breaking techniques, etc.

Deciding with confidence beforehand that using a given set of dynamic symmetry breaking inequalities is better or worse than using a given set of static symmetry breaking inequalities is extremely difficult. One general rule of thumb is that for problems with a symmetry group of order up to a few thousands of permutations, dynamic symmetry breaking might be very effective. However, when the order of the symmetry groups is larger than, say, a million, dynamic symmetry breaking inequalities can be effective, but only when coupled with other symmetry breaking techniques. Some limited comparisons are reported in [91, 97].

17.9 Pruning the enumeration tree

A special case of static symmetry breaking inequalities for ILP (17.1) is obtained from Theorem 17.2 and Theorem 17.3 using $\bar{x}_i = (k+1)^{n-i}$, for $= 1,\ldots,n$ in the latter. Then, a vector $z \in \{0,\ldots,k\}^n$ is in the fundamental region F if and only if $(g(\bar{x}) - \bar{x}) \cdot z \leq 0$ for all $g \in G$, which is equivalent to

$$\max\{g(\bar{x}) \cdot z : g \in G\} \leq \bar{x} \cdot z \quad \text{and to} \quad \max\{\bar{x} \cdot g(z) : g \in G\} \leq \bar{x} \cdot z.$$

This last expression is equivalent to say that z is in F if and only if z is lexicomax in its orbit under G, due to the particular choice of $\bar{x}$. This is hardly a surprising or difficult result, and restricting the search for lexicomax (or lexicomin) solutions in their orbit has been used routinely (see [19, 100, 104, 112], just to name a few, more general expositions and applications can be found in [62, 78]). In [31], in the setting of clausal propositional logic, predicates similar to the constraints above are derived. Results on reducing the number of constraints that must be included to break all symmetries is also discussed. However, the number of constraints that must be included is, in general, too large for this approach to work. Experiments with a small subset of the constraints that does not break all symmetries have been tried [1, 31, 94, 96]. Moreover, even if it were possible to add all these inequalities, the resulting feasible set rarely is an integral polytope, implying that some work is left do be done to solve the problem. An alternative is to handle these constraints by pruning the enumeration tree: Node a of the enumeration tree is pruned if it can be shown that none of the solutions in the subtree rooted at a is lexicomax in its orbit.

A direct use of this idea leads to algorithms that fix (or build) an order on the variables and where lexicomax solutions with respect to that order are sought. These algorithms are presented in Section 17.9.1. However, it is possible to relax the need of an order on the variables. The resulting algorithms are covered in Section 17.9.2. Both types of algorithms are similar and it might help to study first algorithms working with a fixed order on the variables, as their description is a little bit simpler.

We focus on three different algorithms for tackling problems with arbitrary symmetry groups. These algorithms were developed independently of each other and look different, although they all use the same basic principles. These algorithms are: Symmetry Backtracking Search (`SBS`) [11, 12], Symmetry Breaking via Dominance Detection (`SBDD`) [39, 46, 96], and Isomorphism Pruning (`IsoP`) [74, 75]. Algorithms for handling special symmetry groups (such as product of several disjoint symmetric groups) have been studied too [42].

Related but simpler and less efficient algorithms for solving quadratic assignment problems can be found in [8, 76]. They essentially just avoid the creation of isomorphic subproblems from the same parent node. Nevertheless, this simple operation makes a difference when solving difficult benchmark quadratic assignment problems [2].

A few definitions are necessary to help in the description of the algorithms.

Let a be a node of the enumeration tree T and, for $i = 1, \dots, n$, let D_i^a be the possible values for variable x_i at node a. Let ILP^a be the ILP (17.1) with the additional constraints $x_i \in D_i^a$ for $i = 1, \dots, n$. The *path of a* in T is the path P_a from the root node of T to a. A node a_1 is a *son* of node a if aa_1 is an edge of T and the path of a_1 goes through a. See Figure 17.2.

We consider the following three branching rules at node a:

(i) *Partitioning*: Select a variable x_i with $|D_i^a| \geq 2$. Partition D_i^a into $2 \leq \ell \leq |D_i^a|$ non-empty sets $D_i^a(1), \dots, D_i^a(\ell)$. Create the ILP for son a_j for $j = 1, \dots, \ell$ by replacing $x_i \in D_i^a$ in ILP^a by $x_i \in D_i^a(j)$. To avoid cumbersome notation, we assume that the partition of D_i^a satisfies that, for all $1 \leq j \leq \ell - 1$, if $t \in D_i^a(j)$ and $t' \in D_i^a(j+1)$ then $t < t'$.

(ii) *Splitting*: Select a variable x_i such that $D_i^a = \{v_1, \dots, v_\ell\}$ with $\ell \geq 2$ and $v_j < v_{j+1}$ for $j = 1, \dots, \ell - 1$. Create the ILP for son a_j for $j = 1, \dots, \ell$, by replacing $x_i \in D_i^a$ in ILP^a by $x_i = v_j$.

(iii) *Minimum Index Splitting*: Similar to (ii), the only difference is that i must be the smallest index with $|D_i^a| \geq 2$.

It should be clear that (ii) is more restrictive than (i) and that (iii) is more restrictive than (ii). Note also that if one chooses $\ell = |D_i^a|$ in (i), the resulting rule is (ii). In this section, we assume implicitly that one of these three rules is used, with an arbitrary rule for selecting the branching variable for Partitioning and Splitting. Suppose that a branching rule is fixed and let T be the enumeration tree obtained following that rule, pruning nodes only when they are infeasible (pruned nodes are included in T). Tree T is called the *full enumeration tree* for the selected branching rule. Note that if a is a feasible leaf of T then $|D_i^a| = 1$ for all $i = 1, \dots, n$. Feasible leaves of T are in bijection with the feasible solutions of the ILP.

17.9.1 Pruning with a fixed order on the variables

In this section, we assume that a total order on the variables is fixed from the beginning of the algorithm. To simplify notation and without loss of generality we take this order as the natural order defined by the indices on the variables, with x_1 being the first variable in the order.

Note that the material in this section can be adapted so that the order used is built during the execution of the algorithm: The algorithm works with a partial order that is refined during the execution. The initial partial order is an order where all variables have the same priority. Then, each time an operation is performed, it must be valid for an extension of the current partial order and the partial order is modified to include the corresponding constraints. However, the resulting algorithm can be

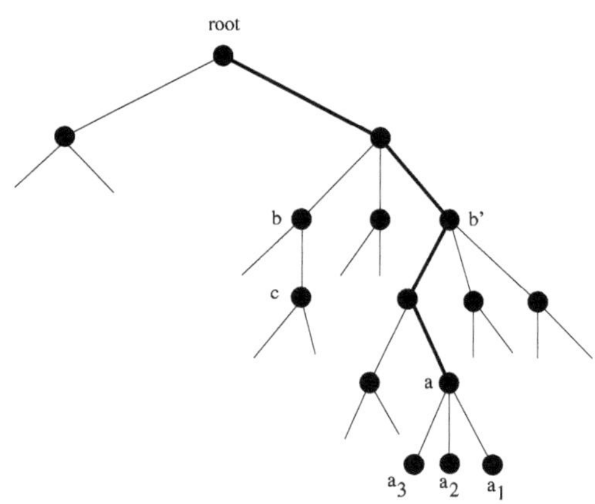

Fig. 17.2 Node a and the ordering of its sons; path P_a (in bold); b to the immediate left of P_a; c to the left of P_a.

seen as a constrained version of the algorithms presented in Section 17.9.2 and as such does not deserve much attention.

We know that node a of the enumeration tree can be pruned if none of the optimal solutions in the subtree rooted at a can be lexicomax in its orbit. It is possible to identify situations were this holds using the following extension of lexicographic ordering.

Given two nodes a and b of T, we say that $D^b = (D_1^b, \ldots, D_n^b)$ is *lexico-set larger* than $D^a = (D_1^a, \ldots, D_n^a)$, written $D^b >_{Ls} D^a$, if and only for some $t \in \{1, \ldots, n\}$ we have $\min\{j : j \in D_i^b\} \geq \max\{j : j \in D_i^a\}$ for $i = 1, \ldots, t-1$ and $\min\{j : j \in D_t^b\} > \max\{j : j \in D_t^a\}$. By extension, the definition applies to any pair of n-vectors of subsets of I^n. We write $g(D^a)$ for the vector of subsets obtained by permuting the entries of D^a according to g.

Example 17.3. Consider $D_1^a = \{0,1\}$, $D_2^a = \{0\}$, $D_3^a = \{1\}$, $D_4^a = \{1\}$ and $D_1^b = \{1\}$, $D_2^b = \{1\}$, $D_3^b = \{0\}$, $D_4^b = \{0,1\}$. We have $D^b >_{Ls} D^a$ and for $g = (3,4,2,1)$, we have $g(D^a) = D^b$. □

We then have:

Theorem 17.4. *Let a be a node of the full enumeration tree T. If there exists $g \in G$ such that $g(D^a) >_{Ls} D^a$ then node a can be pruned.*

The pruning done by Theorem 17.4 is called *isomorphism pruning* (*IP*). It is immediate that if IP is able to prune node a, it is also able to prune all sons d of a, as $D_i^d \subseteq D_i^a$ for all $i = 1, \ldots, n$ and thus $g(D^d) >_{Ls} D^d$ if $g(D^a) >_{Ls} D^a$. Nodes of T that are not pruned thus form a subtree of T containing the root of T as well as all solutions that are lexicomax in their orbit. The validity of the pruning follows.

If we assume that at the root r of T we have $D_i^r = \{0, \ldots, k\}$ for all $i = 1, \ldots, n$, IP is virtually useless until branching on variable x_1 has occurred. In general, IP is more efficient when the domains D_i^a for $i = 1, \ldots, t$ are small and t is large. As a consequence, most algorithms using IP also use the Minimum Index Splitting rule [19, 26, 72, 73, 74, 75, 96, 100].

Having the branching order essentially fixed from the beginning is a minor restriction when pure backtracking enumeration algorithms are used, but it is potentially a major drawback when using branch-and-bound algorithms or other domain reduction techniques, as it is well known that a clever branching variable choice can reduce the enumeration tree drastically. Nevertheless, algorithms based on IP and a fixed ordering of the variables have been shown, on many instances, to be orders of magnitude faster than a branch-and-bound algorithm oblivious to existing symmetries. It should be noted, however, that different orderings of the variables produce wildly different performance, transforming a problem that can be solved in seconds into one that is essentially impossible to solve. For many problems, finding a "reasonable" ordering of the variables is not too difficult. However, as mentioned in Section 17.5, for proving that no proper coloring of the edges of a clique on 9 nodes with 8 colors exists, two "reasonable" ordering of the variables yield running times that are orders of magnitude apart.

17.9.1.1 Additional domain reduction or additional inequalities

Some care must be taken when additional inequalities or variable domain reduction techniques are used together with IP (or symmetry breaking inequalities, but we focus on IP here). It is valid to use either, provided that it can be shown that at least one optimal solution x^* that is lexicomax in its orbit under G remains feasible. This is usually quite difficult to prove, but the following are examples where this is possible:

(i) ILP cutting planes: Add any inequality to ILP^a that is valid for the convex hull of the integer solutions of ILP^a; indeed, as no integer point is cut by these, all optimal solutions that are lexicomax in their orbit are kept. This implies that any of the standard cutting plane generators for ILP (Gomory, Cover, Knapsack, etc.) can be used.

(ii) Strict exclusion algorithms: Excluding value v from D_i^a for some i is valid if it is known that no optimal solution $\bar{x}$ of ILP, valid for ILP^a and lexicomax in its orbit, has $\bar{x}_i = v$. This implies that many of the usual techniques for excluding values for variables can be used. It is however necessary to stress that merely guaranteeing that there is an optimal solution $\bar{x}$ of ILP^a with $\bar{x}_i \neq v$ is not enough for having the right of excluding value v from D_i^a.

(iii) Strict exclusion algorithms working under symmetry: Similar to (ii) with the additional constraint that if $x_i \neq v$ is produced by the algorithm for ILP^a then, for any $g \in G$, it can produce that $x_{g(i)} \neq v$ in the ILP obtained by permuting the variables in ILP^a according to g. This requirement prevents the use of exclusion algorithms that use information related to lexicographic order. It is, however, usually met by typical exclusion algorithms that are used in ILP. The interest of imposing the constraint of working under symmetry on the exclusion algorithms is explained later in this section and also in Section 17.10.

(iv) IP exclusion: Suppose at node a, for some $g \in G$, and for some $t \in \{1, \ldots, n\}$, we have $\min\{j : j \in D_{g(i)}^a\} = \max\{j : j \in D_i^a\}$ for $i = 1, \ldots, t$ and $\max\{j : j \in D_{g(t)}^a\} > \max\{j : j \in D_t^a\}$. Notice that any feasible solution $\bar{x}$ for ILP^a with $\bar{x}_{g(t)} = v$ for $v = \max\{j : j \in D_{g(t)}^a\}$ is not lexicomax in its orbit, as $g^{-1}(\bar{x})$ is lexicographically larger. It is thus valid to exclude v from $D_{g(t)}^a$.

(v) Orbit exclusion: Suppose at node a, for some $p \geq 1$, we have $|D_i^a| = 1$ for $i = 1, \ldots, p$ and $|D_{p+1}^a| > 1$. Let v be the vector defined by $v_i = D_i^a$ for $i = 1, \ldots, p$ and $v_i = -1$ otherwise. Let $\mathbb{O}$ be an orbit under the stabilizer of v in G. Let D be the intersection of the domains D_i^a for all $i \in \mathbb{O}$. Set $D_i^a = D$ for all $i \in \mathbb{O}$.

Note that (ii) and (iii) were introduced in [74, 75] and that particular cases of (iv) have been used in [11, 12] as well as many papers in the Constraint Programming literature (this is a filtering algorithm), as well as under the name *0-fixing* in [74, 75]. Observe that (iv) does not fit the conditions of (iii). It is important to understand also that (v) is valid only if the other exclusion algorithms used satisfy (iii) or (iv). (v) was introduced in [75] under the name *orbit fixing* and a generalization of (v) is also given there.

17.9.2 Pruning without a fixed order of the variables

The practical need, mentioned in Section 17.9.1, of branching using the Minimum Index Splitting to perform IP can however be relaxed. A few definitions are needed before continuing the presentation.

Graphically, the full enumeration tree T is drawn with the root at the top and the sons a_j of a for $j = 1, \ldots, \ell$ are drawn from right to left below a, starting with a_1 (see Figure 17.2). We say that $a_{j'}$ is to the left of a_j if $j < j'$. A node b is *to the immediate left of* P_a if b is the son of a node $c \in P_a - a$ and b is to the left of the son of c that is part of P_a. A node c is *to the left of* P_a if c is in the subtree rooted at a node to the immediate left of P_a.

If we assume that the domains of the variables are modified only by the branching operation, we have [39]:

Theorem 17.5. *Let a be a node of the full enumeration tree T. If there exists a node b to the immediate left of the path of a and a permutation $g \in G$ such that $D^a_{g(i)} \subseteq D^b_i$ for $i = 1, \ldots, n$, then a can be pruned.*

Proof. All optimal solutions of the ILP are feasible leaves of T. Since T is drawn according to the convention above, the orbit $\mathbb{O}$ under G of any optimal solution has one solution s^* that is to the left of the path to any $s \in \mathbb{O} - s^*$. We claim that no node t in the path to s^* is pruned by the pruning of the statement. Indeed, if t is pruned, then there exists b to the left of the path of t and $g \in G$ with $D^t_{g(i)} \subseteq D^b_i$ for $i = 1, \ldots, n$. But then $g(s^*)$ is a feasible leaf of T to the left of the path to s^*, a contradiction with the definition of s^*. □

To showcase the connection between Theorem 17.5 and the lexicographic pruning of Section 17.9.1, suppose that the branching rule used to produce T in Theorem 17.5 is the Minimum Index Splitting rule. The pruning made by Theorem 17.5 is then equivalent to the one done by Theorem 17.4.

One major difference, however, between the two theorems is that Theorem 17.4 has the same efficiency when additional exclusion of values are done whereas, as stated, Theorem 17.5 might miss some pruning due to shrinkage of some domains in b. This motivates the tracking of branching decisions that have been made on the path of a. Let o^a be the *order list* at a, where o^a is simply the ordered list of the indices of the variables used as branching variables on the path of a. In this section, we assume that the Splitting branching rule is used to simplify the presentation, but the algorithms can handle the Partitioning branching rule.

The definition of lexico-set larger given in Section 17.9.1 can be extended so that comparisons are made according to an ordered list of indices: Let o be an ordered list of p indices with o_i denoting the ith element in the list. We say that, with respect to o, $D^b = (D^b_1, \ldots, D^b_n)$ is *lexico-set larger* than $D^a = (D^a_1, \ldots, D^a_n)$, written $D^b >_{os} D^a$, if and only for some $t \in \{1, \ldots, p\}$ we have $\min\{j : j \in D^b_{o_i}\} \geq \max\{j : j \in D^a_{o_i}\}$ for $i = 1, \ldots, t-1$ and $\min\{j : j \in D^b_{o_t}\} > \max\{j : j \in D^a_{o_t}\}$.

We then have [11, 12]:

Theorem 17.6. *Let a be a node of the full enumeration tree T obtained using the Splitting branching rule and let o^a be the order list at a. If there exists $g \in G$ such that $g(D^a) >_{o^a s} D^a$ then node a can be pruned.*

17.9.2.1 Additional domain reduction or additional inequalities

As in Section 17.9.1, it is possible to couple the pruning of Theorem 17.6 with domain reduction techniques and cutting planes, with restrictions similar to those listed in Section 17.9.1.1.

It is easy to overlook that two techniques conflict with each other. An example reported in the literature [96] is the attempt to use jointly the pruning of Theorem 17.6 with the *Lex*^{2} symmetry breaking constraints described in Section 17.8.2.

While this is perfectly valid if Theorem 17.4 or Theorem 17.6 is used with an ordering of the variable for which lexicomax solutions satisfy the *Lex*^{2} constraints, it might fail when Theorem 17.6 is used with an arbitrary order.

17.10 Group representation and operations

While Theorem 17.5 and Theorem 17.6 form the basis for pruning algorithms, they are only existence results. For a practical implementation, we need an algorithm for checking if there exists $g \in G$ satisfying the statement. The implementation of `SBS` from [11, 12] and implementation of `IsoP` from [74, 75] work with an arbitrary group given by a collection of generators. The implementations of `SBDD` from [39, 96] use problem specific algorithms or work only for very simple groups (typically symmetric groups). The `SBDD` implementation of [46] uses an approach similar to [11, 12], but few details are available, preventing a finer comparison with the implementations of `SBS` and `IsoP`. This section covers the basics for handling computational group operations needed in a branch-and-bound algorithm and points differences between the implementations of [11, 12] and [74, 75][3].

The group representation and algorithms are based on the *Schreier-Sims* representation of G, a tool widely used in computational group theory [16, 17, 18, 19, 54, 62, 64, 65]. The reader is referred to [16, 17, 55, 106] for a comprehensive overview of the field.

Let $G_0 = G$ and $G_i = \text{stab}(i, G_{i-1})$ for $i = 1, \dots, n$. Observe that $G_0, G_1, \dots, G_n$ are nested subgroups of G. For $t = 1, \dots, n$, let $\text{orb}(t, G_{t-1}) = \{j_1, \dots, j_p\}$ be the orbit of t under G_{t-1}. Then for each $1 \le i \le p$, let h_{t,j_i} be any permutation in G_{t-1} sending t on j_i, i.e., $h_{t,j_i}[t] = j_i$. Let $U_t = \{h_{t,j_1}, \dots, h_{t,j_p}\}$. Note that U_t is never empty as $\text{orb}(t, G_{t-1})$ always contains t.

Arrange the permutations in the sets U_t, $t = 1, \dots, n$ in an $n \times n$ table T, with

[3] The algorithms of [74, 75] assume a fixed order of the variables, but as pointed out first by Ostrowski [87] and as the presentation in this paper shows, this requirement can be waived.

$$T_{t,j} = \begin{cases} h_{t,j}, & \text{if } j \in \text{orb}(t, G_{t-1}), \\ \emptyset, & \text{otherwise.} \end{cases}$$

Example 17.4. As observed in Example 17.1, the symmetry group G of ILP (17.2) is $\{I, (2,3,4,1), (3,4,1,2), (4,1,2,3), (3,2,1,4), (4,3,2,1), (1,4,3,2), (2,1,4,3)\}$.

We have $G_0 := G$ and $\text{orb}(1, G_0) = \{1,2,3,4\}$ with $h_{1,1} = I$, $h_{1,2} = (2,3,4,1)$, $h_{1,3} = (3,4,1,2)$, and $h_{1,4} = (4,1,2,3)$. Then $G_1 := \text{stab}(1, G_0) = \{I, (1,4,3,2)\}$ and $\text{orb}(2, G_1) = \{2,4\}$ with $h_{2,2} = I$ and $h_{2,4} = (1,4,3,2)$.

And finally we obtain, $G_2 := \text{stab}(2, G_1) = \{I\}$, $G_3 := \text{stab}(3, G_2) = \{I\}$ and $G_4 := \text{stab}(4, G_3) = \{I\}$. The corresponding table T is:

	1	2	3	4
1	I	$h_{1,2}$	$h_{1,3}$	$h_{1,4}$
2		I		$h_{2,4}$
3			I	
4				I

□

The table T is called the Schreier-Sims representation of G. It is possible to make a small generalization of the presentation by ordering the points of the ground set in an arbitrary order β, called the *base* of the table. In that case, the subgroups $G(\beta)_t$ for $t = 1, \ldots, n$ are defined as the stabilizer of β_t in $G(\beta)_{t-1}$, with $G(\beta)_0 = G$. The corresponding table is denoted by $T(\beta)$. Row t of $T(\beta)$ corresponds to the element t, $U(\beta)_t$ is the set of non-empty entries in row t of $T(\beta)$ and $J(\beta)_t$ denotes the corresponding set of indices $\{j \in I^n : T(\beta)[t,j] \neq \emptyset\}$, also called the *basic orbit* of t in T (following the terminology of [65]). When the base β is fixed, we sometimes drop the qualifier (β) in these symbols, but from now on each table T is defined with respect to a base.

The most interesting property of this representation of G is that each $g \in G$ can be uniquely written as

$$g = g_1 \cdot g_2 \cdot \ldots \cdot g_n \tag{17.27}$$

with $g_i \in U_i$ for $i = 1, \ldots n$. Hence the permutations in the table form a set of generators of G. It is called a strong set of generators, since the equation (17.27) shows that $g \in G$ can be expressed as a product of at most n permutations in the set.

For a permutation $g \in G$, it is easy to find the n permutations $g_1, \ldots, g_n$ of (17.27): the permutations $g_2, \ldots, g_n$ all stabilize element 1, forcing g_1 to be $T[1, g(1)]$. Then, as $g_3, \ldots, g_n$ all stabilize element 2, we must have $(g_1 \cdot g_2)(2) = g(2)$, i.e., $g_2(2) = (g_1^{-1} \cdot g)(2)$ and thus $g_2 = T[2, (g_1^{-1} \cdot g)(2)]$. With a similar reasoning we obtain $g_3, \ldots, g_n$.

Algorithms for creating the table $T(\beta)$ and for changing the base β of the representation can be found in [16, 18, 19, 54, 55, 62, 64, 65, 106]. For a group G given by a set T of generators, algorithms for creating the table and with worst-case running time in $O(n^6 + n^2 \cdot |T|)$ [62] have been devised. Faster but more complex algorithms are also known [5, 55, 58, 105, 106]. The complexity of the algorithm of Jerrum [58] is in $O(n^5 + n^2 \cdot |T|)$ and the one of Babai et al. [5] is in $O(n^4 \cdot \log_c n + n^2 \cdot |T|)$ where c is a constant and one from [106] is $O(n^2 \cdot \log^3 |G| + |T| \cdot n^2 \cdot \log |G|)$. Since we might assume that the permutation group is given by a set of strong generators, the speed of the algorithm for finding the representation of the group is not particularly relevant here. Note also that the cardinality of the ground set of the groups that are usually of interest are small (for computational group theory standards, at least) and that the simpler algorithms perform satisfactorily in the large majority of the cases.

Algorithms for changing the base of the table can be found in [11, 19, 30, 55, 62, 106] with worst-case running time up to $O(n^6)$, while more complex algorithms run in almost linear time. An algorithm with worst case complexity in $O(n^6)$ or even $O(n^4)$ might seem impractical for values of $n \geq 100$. It turns out that the complexity bounds given above are very pessimistic and are usually attained for the symmetric group on n elements. The amount of time spent in the algorithms dealing with the group operations for the applications of [74, 75] stays below 10% of the total cpu time.

Although the algorithms are described here for a 2-dimensional table T, a more space efficient implementation uses a vector of ordered lists instead, as most entries in the table are usually empty. For example, when solving the covering design problem *cov1054* mentioned in Section 17.1, $n = 252$, the group has order $10! = 3,628,800$, but the number of entries in the table is, on average, 550. It is worth noting that most of the cpu time used by the group algorithms is spent multiplying permutations. Speedup may be obtained in some cases by keeping permutations in product form (see [55, 106] for details).

Property (17.27) is the corner stone of the algorithm testing the existence of $g \in G$ satisfying Theorem 17.6. To simplify the notation, assume that the Minimum Index Splitting rule is used, and thus at node a, for some $p \geq 1$, we have $o_i^a = i$ and $x_i = v_i$ for $i = 1, \dots, p$.

We use a backtracking algorithm to construct g, if it exists: For $v = 1, \dots, k$, let F_v^a be the set of indices of variables that have value v at a, i.e., $F_v^a = \{i \in I^n : D_i^a = \{v\}\}$.

0) Let $g_0 := I$, $i = 1$ and T be a Schreier-Sims table for G with base $(1, 2, \dots, n)$ and initialize the sets F_v^a for $v = 0, \dots, k$.
1) Let v be the value that x_i takes at a.
2) If $g_{i-1}^{-1}(i) \in F_w^a$ for some $w > v$ then STOP.
3) For all $j \in g_{i-1}(F_v^a)$ do
    - 3.1) Let $h_i := T[i, j]$.
    - 3.2) If $h_i \neq \emptyset$ then
        - 3.2.1) Remove index $g_{i-1}^{-1}(j)$ from F_v^a.
        - 3.2.2) $g_i := h_i^{-1} \cdot g_{i-1}$.
        - 3.2.3) $i := i + 1$; If $i \leq p$ then go to step 1).

If the algorithm terminates in step 2), then g_{i-1} is a permutation showing that a can be pruned, according to Theorem 17.6. Although this algorithm is essentially the one used in implementations, important variations occur. The implementations of SBS of [11, 12] and of IsoP of [74, 75] differ in modifications reducing the pruning that is done to improve running time. Another important difference is that in [11, 12], the algorithm is called before the branching variable is chosen while in [74, 75] it is called to weed out all values in the domain of the selected branching variable that would lead to a node that could be pruned by isomorphism. Advantages of the former is that information about the orbits of variables can be used to select the branching variable (experiments with different selection rules are described in [88]), while the latter leverages the fact that the operations performed by the algorithm at different sons of a node are very similar and can be collapsed efficiently in one application of the algorithm as described below.

Note that in both implementations of [11, 12] and [74, 75], the sets F_v^a contain only the indices in $\{1,\ldots,p\}$. The motivation for this choice is firstly the orbit exclusion algorithm of Section 17.9.1.1. Secondly, the fact that the stabilizer used in the orbit exclusion algorithm is a group means that orbit computations can be performed by computing generators of that group.

This last point is the approach taken in [11, 12], where a clever shortcut is used: when step 3.2.2) is reached with $i = p$, orbits are updated to reflect the effect of g_p and backtracking can be made directly to the smallest index i in step 3.2.2) for which $h_i \neq I$. This potentially speeds up the execution, but might miss some pruning that could be obtained from step 2.

In [74, 75] the algorithm is used before branching on variable x_{p+1}. In other words, at node a, assuming that more than one value is in D_{p+1}^a, the index $p+1$ would appear in F_v^a for all $v \in D_{p+1}^a$. The algorithm is then modified as follows:

(i) In step 2), if $g_{i-1}^{-1}(i) = p+1$, then instead of stopping, the value w is removed from D_{p+1}^a and $p+1$ is removed from F_w^a. This is an application of the domain reduction (iv) of Section 17.9.1.1.
(ii) If the stopping criterion is met in step 2), but $p+1$ was the index $g_{i-1}^{-1}(j)$ in an earlier step 3.2.1) used to build the current permutation g_{i-1}, then we can backtrack to that step 3.2.1), remove the value v from D_{p+1}^a, remove $p+1$ from F_v^a and continue. This is also an application of (iv) of Section 17.9.1.1.
(iii) In step 3.2.1), if $g_{i-1}^{-1}(j) = p+1$, then $p+1$ is removed from all the sets F_w^a.

Using these modifications, the values in D_{p+1}^a at termination of the algorithm are the values that need to be used to create sons by splitting. Of course, if that domain is empty then a is pruned.

Another modification used in [74, 75] is to ignore variables that have been set to 0 by branching. Suppose that variables that have been set to positive values by branching at a are $(i_1,\ldots,i_t)$ and assume that the base β of T starts with these elements in that order. The above algorithm is then run with $i := i_1$ and instead of incrementing i by one, it skips from i_j to i_{j+1}. The justification for this modification is that if $x_{\bar{i}} = 0$ is set for some $\bar{i}$, and a permutation g is obtained in Step 2) of the algorithm, then the minimum value in the domain of $j = g^{-1}(\bar{i})$ cannot be smaller than 0. Using the original algorithm, we could exclude from D_j^a all values larger than 0 and continue. This reduction is possibly missed by the modified algorithm, but it remains correct.

The importance of this modification is on display when solving a problem where the variables taking a positive value in any optimal solution are a small subset of the variables. (This is a usual feature of combinatorial problems that are typically solved by ILP.) The depth of the backtracking of the modified algorithm is then much smaller, with a significant impact on the running time.

Unfortunately, there is no direct comparison available between the implementations of [11, 12] and [74, 75], as the only published results for the former are for the well-known *Queens problem*: On an $n \times n$ chessboard, place n queens that do not attack each other. It would be foolish to draw conclusions on implementations of algorithms designed to handle groups with large orders based only on results on an application where the group order is 8. In Section 17.13, a comparison (hardly an ideal one but a little bit more meaningful) between implementations of `SBDD` and `IsoP` is given.

17.11 Enumerating all non-isomorphic solutions

One important application of the pruning algorithms of Section 17.9 is their use for enumerating all non-isomorphic solutions to a symmetric ILP. Pruning is the only technique that can reliably list only non-isomorphic solutions, as, in general, symmetry breaking inequalities are either too expensive to use or do not remove all symmetry from the problem. There is interest emanating from the Combinatorics and Statistics communities (among others) for enumerating all non-isomorphic graphs or matrices with certain properties. For example, covering designs [51, 82] or balanced incomplete block designs [29] have a long history. For small values of the parameters, complete or implicit enumeration algorithms were used to generate solutions. Enumerating all non-isomorphic solution is interesting, since these objects are used to build other mathematical objects or used to test conjectures. A few of the enumeration results obtained by pruning algorithms are available in [15, 26, 68, 73, 75, 78, 94, 96, 104]. Note that the solutions obtained by a pruning algorithm are non-isomorphic solutions with respect to the symmetry group G used by the algorithm. It might happen that G is only a subgroup of the symmetry group of the feasible set of the problem and thus that isomorphic solutions remain in the output.

Enumerating all non-isomorphic solutions also provides a powerful debugging tool. Replicating enumeration results on problems having thousands of non-isomorphic solutions is a much better indication that an algorithm is correct than when it is run to find an optimal solution. Indeed, in symmetric problems, a faulty algorithm can still find optimal solutions with surprising ease, whereas it is more likely that at least one of the non-isomorphic optimal solutions is missed or that two isomorphic solutions appear in the output when enumerating all non-isomorphic solutions.

Some empirical results for enumeration of all non-isomorphic solution to a combinatorial problem are given in Section 17.13.

17.12 Furthering the reach of isomorphism pruning

Isomorphism pruning and other techniques for handling symmetries in ILP help push the boundaries of the problems that can be solved routinely. However, many symmetric ILPs have a parametrized formulation yielding a never ending stream of challenging problems. When trying to solve an ILP, it is possible to use some kind of partial isomorphism pruning in order to speed up the process. For example, skipping the isomorphism pruning at deep level in the tree is usually beneficial as illustrated on a few examples in [96]. Another idea is to use the orbits of the variables not yet fixed to some values for branching variable selection [88].

Another recent development is the idea of branching on constraints using orbits. Given a branching corresponding to a valid disjunction $a \cdot x \leq b$ or $a \cdot x \geq b+1$ one son is created with the first constraint, while the second one is generated using $g(a) \cdot x \geq b+1$ for all $g \in G$. Further, when the disjunction is chosen carefully, it is sometimes possible to enumerate all non-isomorphic solution to $a \cdot x = t$ for some values of t and use these solutions for solving the original problem. This technique was pioneered by Östergård and Blass [85] in the combinatorics community and used for improving the lower bound for the Football Pool problem using integer programming [67, 86]. In [89], Ostrowski et al. formalize and generalize the technique and apply it successfully to solve Steiner Triple Systems and Covering Design problems that were previously out of reach.

A special situation of practical interest occurs when the symmetry group G is the product of a nontrivial group with a symmetric group. This is the typical situation for partitioning problems of the form considered in Section 17.7.2 when symmetry between the elements is also present. This happens for example for graph coloring problems where the automorphism group of the graph is non trivial. As symmetric groups are the most challenging ones to handle using the algorithms mentioned in Section 17.10 while they can be handled easily without complex representation, in [75] hybrid algorithms are used where the symmetric groups are handled directly. Improvement in reported running times are significant.

17.13 Choice of formulation

When dealing with a symmetric problem, the choice of the formulation might have a big impact on the performances of a solution technique. This is well known for ILPs, and it is even more acute when dealing with symmetries. While tightness of linear relaxation, number of variables, and constraint types can be used as guides for choosing an ILP formulation, things are obscured when dealing with a symmetric ILP, as different variable choices might yield very different symmetry groups. One would expect that fewer variables and a symmetry group with smaller order is better, but as described in Section 17.5 things are not so simple. In [110], several formulations for constructing a particular class of combinatorial designs are compared.

Just to have a concrete example illustrating how alternative formulations can make a big difference, consider the problem of constructing Balanced Incomplete Block Designs (BIBD). A BIBD is a binary matrix with given dimensions $b \times v$ with exactly k ones per row and constant dot product of value λ between any two of its columns (see [29] for background material). The results of [46, 96] are obtained using a nonlinear formulation with $b \cdot v$ binary variables, one for each entry in the matrix. This yields a formulation where the symmetry group has order $v! \cdot b!$. It is not convenient to use these variables to get an ILP formulation. An alternative linear formulation can be built from a list $\{R_1, \ldots, R_q\}$ of all possible binary rows of length v and having exactly k ones. Define one integer variable x_i taking values between 0 and λ, indicating how many times row R_i appears in the solution, for $i = 1, \ldots, q$. Constraints are that, for each pair j_1, j_2 of columns, the sum of all variables corresponding to rows having 1s in columns j_1 and j_2 should be exactly λ. This yields a formulation with $\frac{v!}{(v-k)! \cdot k!}$ variables, $\frac{v(v-1)}{2}$ constraints and a symmetry group of order $v!$.

For many instances, the latter formulation seems much simpler to solve than the former. The following table compares results for three codes for solving BIBD problems. The comparison is far from ideal, since the three codes are run on different machines and the formulations are not identical. Both `GAP-ECLIPSE` [46] and `SBDD+STAB` [96] use the first formulation, while `IsoP` [75] uses the second one. It is clear that `SBDD+STAB` outperforms `GAP-ECLIPSE`, but this should not be too surprising as `SBDD+STAB` uses the particular (extremely simple) structure of the symmetry group, while `GAP-ECLIPSE` is a code that can be used with any group. Nevertheless, the difference in running time on the third problem is quite stunning, possibly due to different branching variable choices. On the other hand, on hard problems, `IsoP` seems significantly faster than `SBDD+STAB`, even after discounting for the difference in machine speeds. It is likely that part of the difference can be attributed to the formulation `IsoP` uses.

v	k	λ	# solutions	`GAP-ECLIPSE`	`SBDD+STAB`	`IsoP`
11	5	2	1	19	0	1
13	4	1	1	42	0	10
13	3	1	2	59,344	0	1
7	3	8	5,413		21,302	115
9	3	3	22,521		34,077	1,147
15	3	1	80		2,522	161
13	4	2	2,461		18,496	5,142
11	5	4	4,393		83,307	3,384

Table 17.2 Comparison of formulations for BIBD problems. Number of non isomorphic solutions and times (rounded down) in seconds for enumerating them. Empty entries indicate that the corresponding result is not available. Times for `GAP-ECLIPSE` are from [46], obtained on a 2.6GHz Pentium IV processor. Times for `SBDD+STAB` are from [96], obtained a 1.4GHz Pentium Mobile laptop running Windows XP. Times for `IsoP` are obtained on the machine mentioned in Section 17.1.

17.14 Exploiting additional symmetries

So far, only symmetries of the original ILP (17.1) were considered. However, while solving the ILP by branch-and-bound, it is sometimes the case that, at node a of the enumeration tree, additional symmetries exist in ILP^a, as seen in Example 17.2 in Section 17.5. These symmetries can be used when solving ILP^a, provided that they can be identified. This last point is a big hurdle to clear. As pointed out in Section 17.3, automatic symmetry detection is a difficult problem in itself. Few papers attempt to use an automatic symmetry detection algorithm at nodes of the tree, and when they do, results are unconvincing [88], as the time spend in searching for additional symmetries is not compensated by a commensurate reduction in the size of the enumeration tree. Successful exploitation of additional symmetries are limited to cases where problem specific rules for generating the symmetries are designed from the start [47]. Development of theory and algorithms for exploiting additional symmetries can be found in [48].

References

1. F.A. Aloul, A. Ramani, I.L. Markov, and K.A. Sakallah, *Solving difficult instances of Boolean satisfiability in the presence of symmetry*, IEEE Transactions on CAD 22 (2003) 1117–1137.
2. K.M. Anstreicher, *Recent advances in the solution of quadratic assignment problems*, Mathematical Programming 97 (2003) 27–42.
3. D.L. Applegate, R.E. Bixby, V. Chvátal, and W.J. Cook, *The Traveling Salesman Problem, A Computational Study*, Princeton, 2006.
4. A. von Arnim, R. Schrader, and Y. Wang, *The permutahedron of N-sparse posets*, Mathematical Programming 75 (1996) 1–18.
5. L. Babai, E.M. Luks, and Á. Seress, *Fast management of permutation groups I*, SIAM Journal on Computing 26 (1997) 1310–1342.
6. E. Balas, *A linear characterization of permutation vectors*, Management Science Research Report 364, Carnegie Mellon University, Pittsburgh, PA, 1975.
7. C. Barnhart, E.L. Johnson, G.L. Nemhauser, M.W.P. Savelsbergh, and P.H. Vance, *Branch-and-price: Column generation for solving huge integer programs*, Operations Research 46 (1998) 316–329.
8. M.S. Bazaraa and O. Kirca, *A branch-and-bound based heuristic for solving the quadratic assignment problem*, Naval Research Logistics Quarterly 30 (1983) 287–304.
9. R. Bertolo, P. Östergård, and W.D. Weakley, *An updated table of binary/ternary mixed covering codes*, Journal of Combinatorial Designs 12 (2004) 157–176.
10. W. Bosma, J. Cannon, and C. Playoust, *The Magma algebra system. I. The User Language*, Journal of Symbolic Computations 24 (1997) 235–265.
11. C.A. Brown, L. Finkelstein, and P.W. Purdom, *Backtrack searching in the presence of symmetry*, Lecture Notes in Computer Science 357, Springer, 1989, pp. 99–110.
12. C.A. Brown, L. Finkelstein, and P.W. Purdom, *Backtrack Searching in the Presence of Symmetry*, Nordic Journal of Computing 3 (1996) 203–219.
13. C. Buchheim and M. Jünger, *Detecting symmetries by branch&cut*, Mathematical Programming 98 (2003) 369–384.
14. C. Buchheim and M. Jünger, *Linear optimization over permutation groups*, Discrete Optimization 2 (2005) 308–319.
15. D.A. Bulutoglu and F. Margot, *Classification of orthogonal arrays by integer programming*, Journal of Statistical Planning and Inference 138 (2008) 654–666.

16. G. Butler, *Computing in permutation and matrix groups II: Backtrack algorithm*, Mathematics of Computation 39 (1982) 671–680.
17. G. Butler, *Fundamental Algorithms for Permutation Groups*, Lecture Notes in Computer Science 559, Springer, 1991.
18. G. Butler and J.J. Cannon, *Computing in permutation and matrix groups I: Normal closure, commutator subgroups, series*, Mathematics of Computation 39 (1982) 663–670.
19. G. Butler and W.H. Lam, *A general backtrack algorithm for the isomorphism problem of combinatorial objects*, Journal of Symbolic Computation 1 (1985) 363–381.
20. P.J. Cameron, *Permutation Groups*, London Mathematical Society, Student Text 45, Cambridge University Press,1999.
21. M. Campêlo and R.C. Corrêa, *A Lagrangian decomposition for the maximum stable set problem*, Working Paper (2008), Universidade Federal do Ceará, Brazil.
22. M. Campêlo, V.A. Campos, and R.C. Corrêa, *Um algoritmo de Planos-de-Corte para o número cromático fracionário de um grafo*, Pesquisa Operational 29 (2009) 179–193.
23. M. Campêlo, R. Corrêa, and Y. Frota, *Cliques, holes and the vertex coloring polytope*, Information Processing Letters 89 (2004) 159–164.
24. M. Campêlo, V. Campos, and R. Corrêa, *On the asymmetric representatives formulation for the vertex coloring problem*, Electronic Notes in Discrete Mathematics 19 (2005) 337–343.
25. M. Campêlo, V. Campos, and R. Corrêa, *On the asymmetric representatives formulation for the vertex coloring problem*, Discrete Applied Mathematics 156 (2008) 1097–1111.
26. R.D. Cameron, C.J. Colbourn, R.C. Read, and N.C. Wormald, *Cataloguing the graphs on 10 vertices*, Journal of Graph Theory 9 (1985) 551–562.
27. T. Christof and G. Reinelt, *Decomposition and parallelization techniques for enumerating the facets of combinatorial polytopes*, International Journal on Computational Geometry and Applications 11 (2001) 423–437.
28. G. Cohen, I. Honkala, S. Litsyn, and A. Lobstein, *Covering Codes*, North Holland, 1997.
29. C.J. Colbourn and J.H. Dinitz (eds.): *The CRC Handbook of Combinatorial Designs*, CRC Press, 2007.
30. G. Cooperman, L. Finkelstein, and N. Sarawagi, *A random base change algorithm for permutation groups*, Proceedings of the International Symposium on Symbolic and Algebraic Computations – ISSAC 90, ACM Press, 1990, pp. 161–168.
31. J. Crawford, M.L. Ginsberg, E. Luks, and A. Roy, *Symmetry-breaking predicates for search problems*, KR'96: Principles of Knowledge Representation and Reasoning (L.C. Aiello, J. Doyle, and S. Shapiro, eds.), 1996, pp. 148–159.
32. P. Darga, M.H. Liffiton, K.A. Sakallah, and I.L. Markov, *Exploiting structure in symmetry generation for CNF*, Proceedings of the 41st Design Automation Conference, San Diego 2004, pp. 530–534.
33. Z. Degraeve, W. Gochet, and R. Jans, *Alternative formulations for a layout problem in the fashion industry*, European Journal of Operational Research 143 (2002) 80–93.
34. M. Desrochers and F. Soumis, *A column generation approach to the urban transit crew scheduling problem*, Transportation Science 23 (1989) 1–13.
35. A. Deza, K. Fukuda, T. Mizutani, and C. Vo, *On the face lattice of the metric polytope*, Discrete and Computational Geometry: Japanese Conference (Tokyo, 2002), Lecture Notes in Computer Science 2866, Springer, 2003, pp. 118–128.
36. A. Deza, K. Fukuda, D. Pasechnik, and M. Sato, *On the skeleton of the metric polytope*, Discrete and Computational Geometry: Japanese Conference (Tokyo, 2000), in Lecture Notes in Computer Science 2098, Springer, 2001, pp. 125–136.
37. Y. Dumas, M. Desrochers, and F. Soumis, *The pickup and delivery problem with time windows*, European Journal of Operations Research 54 (1991) 7–22.
38. M. Elf, C. Gutwenger, M. Jünger, and G. Rinaldi, *Branch-and-cut algorithms for combinatorial optimization and their implementation in ABACUS*, in [59] (2001) 155–222.
39. T. Fahle, S. Shamberger, and M. Sellmann, *Symmetry breaking*, Proc. 7th International Conference on Principles and Practice of Constraint Programming – CP 2001, Lecture Notes in Computer Science 2239, Springer, 2001, pp. 93–107.

40. P. Flener, A. Frisch, B. Hnich, Z. Kiziltan, I. Miguel, J. Pearson, and T. Walsh, *Symmetry in matrix models*, working paper APES-30-2001, 2001.
41. P. Flener, A. Frisch, B. Hnich, Z. Kiziltan, I. Miguel, J. Pearson, and T. Walsh, *Breaking row and column symmetries in matrix models*, Proc. 8th International Conference on Principles and Practice of Constraint Programming – CP 2002, Lecture Notes in Computer Science 2470, Springer, 2002, pp. 462–476.
42. P. Flener, J. Pearson, M. Sellmann, P. van Hentenryck, and M. Ågren, *Dynamic structural symmetry breaking for constraint satisfaction problems*, DOI 10.1007/s10601-008-9059-7, Constraints 14 (2009).
43. F. Focacci and M. Milano, *Global cut framework for removing symmetries*, Proc. 7th International Conference on Principles and Practice of Constraint Programming – CP 2001, Lecture Notes in Computer Science 2239, Springer, 2001, pp. 77–92.
44. E.J. Friedman, *Fundamental domains for integer programs with symmetries*, Proceedings of COCOA 2007, Lecture Notes in Computer Science 4616, 2007, pp. 146–153.
45. I.P. Gent, W. Harvey, and T. Kelsey, *Groups and constraints: Symmetry breaking during search*, Proc. 8th International Conference on Principles and Practice of Constraint Programming – CP 2002, Lecture Notes in Computer Science 2470, Springer, 2002, pp. 415–430.
46. I.P. Gent, W. Harvey, T. Kelsey, and S. Linton, *Generic SBDD using computational group theory*, Proc. 9th International Conference on Principles and Practice of Constraint Programming – CP 2003, Lecture Notes in Computer Science 2833, Springer, 2003, pp. 333–347.
47. I.P. Gent, T. Kelsey, S. Linton, I. McDonald, I. Miguel, and B. Smith, *Conditional symmetry breaking*, Proc. 11th International Conference on Principles and Practice of Constraint Programming, Lecture Notes in Computer Science 3709, 2005, pp. 333–347.
48. I.P. Gent, T. Kelsey, S.T. Linton, J. Pearson, and C.M. Roney-Dougal, *Groupoids and conditional symmetry*, Proc. 13th International Conference on Principles and Practice of Constraint Programming, Lecture Notes in Computer Science 4741, 2007, pp. 823–830.
49. I.P. Gent, K.E. Petrie, and J.-F. Puget, *Symmetry in constraint programming*, Handbook of Constraint Programming (F. Rossi, P. van Beek, and T. Walsh eds.), Elsevier, 2006, pp. 329–376.
50. I.P. Gent and B.M. Smith, *Symmetry breaking in constraint programming*, Proceedings of ECAI-2002, IOS Press, 2002, pp. 599–603.
51. D.M. Gordon and D.R. Stinson, *Coverings*, The CRC Handbook of Combinatorial Designs (C.J. Colbourn and J.H. Dinitz, eds.), CRC Press, 2007, pp. 365–372.
52. L.C. Grove and C.T. Benson, *Finite Reflection Groups*, Springer, 1985.
53. H. Hämäläinen, I. Honkala, S. Litsyn, and P. Östergård, *Football pools–A game for mathematicians*, American Mathematical Monthly 102 (1995) 579–588.
54. C.M. Hoffman, *Group-Theoretic Algorithms and Graph Isomorphism*, Lecture Notes in Computer Science 136, Springer, 1982.
55. D.F. Holt, B. Eick, and E.A. O'Brien, *Handbook of Computational Group Theory*, Chapman & Hall/CRC, 2004.
56. R. Jans, *Solving lotsizing problems on parallel identical machines using symmetry breaking constraints*, INFORMS Journal on Computing 21 (2009) 123–136.
57. R. Jans and Z. Degraeve, *A note on a symmetrical set covering problem: The lottery problem*, European Journal of Operational Research 186 (2008) 104–110.
58. M. Jerrum, *A compact representation for permutation groups*, Journal of Algorithms 7 (1986) 60–78.
59. M. Jünger and D. Naddef (eds.), *Computational Combinatorial Optimization*, Lecture Notes in Computer Science 2241, Springer, 2001.
60. V. Kaibel, M. Peinhardt, and M.E. Pfetsch, *Orbitopal fixing*, Proceedings of the 12th International Integer Programming and Combinatorial Optimization Conference (M. Fischetti and D.P. Williamson, eds.), Lecture Notes in Computer Science 4513, Springer, 2007, pp. 74–88.
61. V. Kaibel and M.E. Pfetsch, *Packing and partitioning orbitopes*, Mathematical Programming 114 (2008) 1–36.
62. D.L. Kreher and D.R. Stinson, *Combinatorial Algorithms, Generation, Enumeration, and Search*, CRC Press, 1999.

63. E.L. Lawler, J.K. Lenstra, A.H.G. Rinnooy Kan, and D.B. Shmoys, *The Traveling Salesman Problem*, Wiley, 1985.
64. J.S. Leon, *On an algorithm for finding a base and a strong generating set for a group given by generating permutations*, Mathematics of Computation 35 (1980) 941–974.
65. J.S. Leon, *Computing automorphism groups of combinatorial objects*, Computational Group Theory (M.D. Atkinson, ed.), Academic Press, 1984, pp. 321–335.
66. L. Liberti, *Automatic generation of symmetry-breaking constraints*, Proceedings of COCOA 2008, Lecture Notes in Computer Science 5165, 2008, pp. 328–338.
67. J. Linderoth, F. Margot, and G. Thain, *Improving bounds on the football pool problem via symmetry: Reduction and high-throughput computing*, to appear in INFORMS Journal on Computing, 2009.
68. C. Luetolf and F. Margot, *A catalog of minimally nonideal matrices*, Mathematical Methods of Operations Research 47 (1998) 221–241.
69. E. Luks, *Permutation groups and polynomial-time computation*, Groups and Computation (L. Finkelstein and W. Kantor, eds.), DIMACS Series in Discrete Mathematics and Theoretical Computer Science 11 (1993) 139–175.
70. J.L. Marenco and P.A. Rey, *The football pool polytope*, Electronic Notes in Discrete Mathematics 30 (2008) 75–80.
71. http://wpweb2.tepper.cmu.edu/fmargot/index.html
72. F. Margot, *Pruning by isomorphism in branch-and-cut*, Mathematical Programming 94 (2002) 71–90.
73. F. Margot, *Small covering designs by branch-and-cut*, Mathematical Programming 94 (2003) 207–220.
74. F. Margot, *Exploiting orbits in symmetric ILP*, Mathematical Programming 98 (2003) 3–21.
75. F. Margot, *Symmetric ILP: Coloring and small integers*, Discrete Optimization 4 (2007) 40–62.
76. T. Mautor and C. Roucairol, *A new exact algorithm for the solution of quadratic assignment problems*, Discrete Applied Mathematics 55 (1994) 281–293.
77. B.D. McKay, *Nauty User's Guide (Version 2.2)*, Computer Science Department, Australian National University, Canberra.
78. B.D. McKay, *Isomorph-free exhaustive generation*, Journal of Algorithms 26 (1998) 306–324.
79. A. Mehrotra and M.A. Trick, *A column generation approach for graph coloring*, INFORMS Journal on Computing 8 (1996) 344–354.
80. A. Mehrotra and M.A. Trick, *Cliques and clustering: A combinatorial approach*, Operations Research Letters 22 (1998) 1–12.
81. I. Méndez-Dìaz and P. Zabala, *A branch-and-cut algorithm for graph coloring*, Discrete Applied Mathematics 154 (2006) 826–847.
82. W.H. Mills and R.C. Mullin, *Coverings and packings*, Contemporary Design Theory: A Collection of Surveys (J.H. Dinitz and D.R. Stinson, eds.), Wiley, 1992, pp. 371–399.
83. G.L. Nemhauser and S. Park, *A polyhedral approach to edge colouring*, Operations Research Letters 10 (1991) 315–322.
84. G.L. Nemhauser and L.A. Wolsey, *Integer and Combinatorial Optimization*, Wiley, 1988.
85. P. Östergård and W. Blass, *On the size of optimal binary codes of length 9 and covering radius 1*, IEEE Transactions on Information Theory 47 (2001) 2556–2557.
86. P. Östergård and A. Wassermann, *A new lower bound for the football pool problem for six matches*, Journal of Combinatorial Theory Ser. A 99 (2002) 175–179.
87. J. Ostrowski, Personal communication, 2007.
88. J. Ostrowski, J. Linderoth, F. Rossi, and S. Smirglio, *Orbital branching*, IPCO 2007: The Twelfth Conference on Integer Programming and Combinatorial Optimization, Lecture Notes in Computer Science 4513, Springer, 2007, pp. 104–118.
89. J. Ostrowski, J. Linderoth, F. Rossi, and S. Smirglio, *Constraint orbital branching*, IPCO 2008: The Thirteenth Conference on Integer Programming and Combinatorial Optimization, Lecture Notes in Computer Science 5035, Springer, 2008, pp. 225–239.

90. M.W. Padberg and G. Rinaldi, *A branch-and-cut algorithm for the resolution of large scale symmetric travelling salesman problems*, SIAM Review 33 (1991) 60–100.
91. K.E. Petrie and B.M. Smith, *Comparison of symmetry breaking methods in constraint programming*, In Proceedings of SymCon05, 2005.
92. F. Plastria, *Formulating logical implications in combinatorial optimisation*, European Journal of Operational Research 140 (2002) 338–353.
93. J.-F. Puget, *On the satisfiability of symmetrical constrainted satisfaction problems*, Proceedings of the 7th International Symposium on Methodologies for Intelligent Systems, Lecture Notes in Artificial Intelligence 689, Springer, 1993, pp. 350–361.
94. J.-F. Puget, *Symmetry breaking using stabilizers*, Proc. 9th International Conference on Principles and Practice of Constraint Programming – CP 2003, Lecture Notes in Computer Science 2833, Springer, 2003, pp. 585–599.
95. J.-F. Puget, *Automatic detection of variable and value symmetries*, Proc. 11th International Conference on Principles and Practice of Constraint Programming – CP 2005, Lecture Notes in Computer Science 3709, Springer, 2005, pp. 475–489.
96. J.-F. Puget, *Symmetry breaking revisited*, Constraints 10 (2005) 23–46.
97. J.-F. Puget, *A comparison of SBDS and dynamic lex constraints*, In Proceeding of SymCon06, 2006, pp. 56–60.
98. A. Ramani, F.A. Aloul, I.L. Markov, and K.A. Sakallah, *Breaking instance-independent symmetries in exact graph coloring*, Journal of Artificial Intelligence Research 26 (2006) 191–224.
99. A. Ramani and I.L. Markov, *Automatically exploiting symmetries in constraint programming*, CSCLP 2004 (B. Faltings et al., eds.), Lecture Notes in Artificial Intelligence 3419, Springer, 2005, pp. 98–112.
100. R.C. Read, *Every one a winner or how to avoid isomorphism search when cataloguing combinatorial configurations*, Annals of Discrete Mathematics 2 (1978) 107–120.
101. P.A. Rey, *Eliminating redundant solutions of some symmetric combinatorial integer programs*, Electronic Notes in Discrete Mathematics 18 (2004) 201–206.
102. J.J. Rotman, *An Introduction to the Theory of Groups*, 4th edition, Springer, 1994.
103. D. Salvagnin, *A Dominance Procedure for Integer Programming*, Master's thesis, University of Padova, 2005.
104. E. Seah and D.R. Stinson, *An enumeration of non-isomorphic one-factorizations and Howell designs for the graph K_{10} minus a one-factor*, Ars Combinatorica 21 (1986) 145–161.
105. Á. Seress, *Nearly linear time algorithms for permutation groups: An interplay between theory and practice*, Acta Applicandae Mathematicae 52 (1998) 183–207.
106. Á. Seress, *Permutation Group Algorithms*, Cambridge Tracts in Mathematics 152, Cambridge University Press, 2003.
107. H.D. Sherali, B.M.P. Fraticelli, and R.D. Meller, *Enhanced model formulations for optimal facility layout*, Operations Research 51 (2003) 629–644.
108. H.D. Sherali and J.C. Smith, *Improving discrete model representations via symmetry considerations*, Management Science 47 (2001) 1396–1407.
109. H.D. Sherali, J.C. Smith, and Y. Lee, *Enhanced model representations for an intra-ring synchronous optical network design problem allowing demand splitting*, INFORMS Journal on Computing 12 (2000) 284–298.
110. B. Smith, *Reducing symmetry in a combinatorial design problem*, Proc. 8th International Conference on Principles and Practice of Constraint Programming – CP 2002, Lecture Notes in Computer Science 2470, Springer, 2002, pp. 207–213.
111. B.M. Smith, S.C. Brailsford, P.M. Hubbard, and H.P. Williams, *The progressive party problem: Integer linear programming and constraint programming compared*, Constraints 1 (1996) 119–138.
112. R.G. Stanton and J.A. Bates, *A computer search for B-coverings*, in Combinatorial Mathematics VII (R.W. Robinson, G.W. Southern, and W.D. Wallis, eds.), Lecture Notes in Computer Science 829, Springer, 1980, pp. 37–50.
113. The GAP Group: *GAP – Groups, Algorithms, and Programming, Version 4.4.10 (2007)*, http://www.gap-system.org.

114. P.H. Vance, *Branch-and-price algorithms for the one-dimensional cutting stock problem*, Computational Optimization and Applications 9 (1998) 111–228.
115. P.H. Vance, C. Barnhart, E.L. Johnson, and G.L. Nemhauser, *Solving binary cutting stock problems by column generation and branch-and-bound*, Computational Optimization and Applications 3 (1994) 111–130.
116. P.H. Vance, C. Barnhart, E.L. Johnson, and G.L. Nemhauser, *Airline crew scheduling: A new formulation and decomposition algorithm*, Operations Research 45 (1997) 188–200.
117. L.A. Wolsey, *Integer Programming*, Wiley, 1998.
118. M. Yannakakis, *Expressing combinatorial optimization problems by linear programs*, Journal of Computer and System Sciences 43 (1991) 441–466.

Chapter 18
Semidefinite Relaxations for Integer Programming

Franz Rendl

Abstract We survey some recent developments in the area of semidefinite optimization applied to integer programming. After recalling some generic modeling techniques to obtain semidefinite relaxations for NP-hard problems, we look at the theoretical power of semidefinite optimization in the context of the Max-Cut and the Coloring Problem. In the second part, we consider algorithmic questions related to semidefinite optimization, and point to some recent ideas to handle large scale problems. The survey is concluded with some more advanced modeling techniques, based on matrix relaxations leading to copositive matrices.

18.1 Introduction

Looking back at fifty years of integer programming, there is wide agreement that *Polyhedral Combinatorics* is a major ingredient to approach NP-hard integer optimization problems. Having at least a partial description of the convex hull of all feasible solutions of an integer program can be exploited by the strong algorithmic machinery available to solve linear programming problems, notably the Simplex method. First systematic attempts to use polyhedral techniques to solve 0/1 integer optimization go back to [10].

We consider an abstract combinatorial optimization problem (COP) given as follows. Let E be a finite set and let $\mathscr{F}$ be a (finite) family of subsets of E. The family $\mathscr{F}$ denotes the set of feasible solutions of (COP). Each $e \in E$ has a given integer cost c_e. We define the cost $c(F)$ of $F \in \mathscr{F}$ to be

$$c(F) := \sum_{e \in F} c_e.$$

Franz Rendl
Department of Mathematics, Alpen-Adria Universität, Klagenfurt, Austria
e-mail: `franz.rendl@uni-klu.ac.at`

M. Jünger et al. (eds.), *50 Years of Integer Programming 1958-2008*,
DOI 10.1007/978-3-540-68279-0_18, © Springer-Verlag Berlin Heidelberg 2010

The problem (COP) now consists in finding a feasible solution F of minimum cost:

$$\text{(COP)} \quad z^* = \min\{c(F) : F \in \mathscr{F}\}.$$

The traveling salesman problem (TSP) could be modeled with E being the edge set of the underlying graph G. An edge set F is in $\mathscr{F}$ exactly if it is the edge set of a Hamiltonian cycle in G.

By assigning to each $F \in \mathscr{F}$ a characteristic vector $x_F \in \{0,1\}^n$ with $(x_F)_e = 1$ if and only if $e \in F$, we can write (COP) as a linear program as follows. Let $P := \text{conv}\{x_F : F \in \mathscr{F}\}$ denote the convex hull of the incidence vectors of feasible solutions. Then it is clear that

$$z^* = \min\{c^T x_F : F \in \mathscr{F}\} = \min\{c^T x : x \in P\}.$$

This is the basic principle underlying the polyhedral approach to solve combinatorial optimization problems.

As an example, consider $\mathscr{F}$ to be the set of all permutation matrices. Birkhoff's theorem states that the convex hull of permutation matrices is the polyhedron of doubly stochastic matrices:

$$\text{conv}\{X : X \in \Pi\} = \Omega.$$

For notation, see the end of this section. Hence the combinatorial problem of finding a permutation ϕ minimizing $\sum_i c_{i,\phi(i)}$ can be solved through the linear program

$$\min\{\langle C, X\rangle : X \in \Omega\}.$$

The practical difficulty lies in the fact that in general the polyhedron P is not easily available. The use of a computationally tractable partial description of P in combination with systematic enumeration, like Branch and Bound, has led to quite successful solution methods for a variety of combinatorial optimization problems like the TSP, see e.g., [47]. It turned out however, that for some prominent NP-hard problems like Stable-Set or Max-Cut, this polyhedral approach was not as successful as one might have hoped in view of the results for TSP, see e.g., [5].

This motivated the study of more powerful approximation techniques for (COP). One such possibility consists in studying matrix liftings of the form

$$\mathscr{M} := \text{conv}\{x_F x_F^T : F \in \mathscr{F}\}, \tag{18.1}$$

see e.g., [53, 69, 70]. Any relaxation based on $\mathscr{M}$ lies in the space $\mathscr{S}_n$ of symmetric matrices, rather than $\mathbb{R}^n$, the "natural" space of (COP). The modeling power of using $\mathscr{M}$ comes from the fact that any quadratic constraint on $x \in P$, such as $x_i x_j = 0$, translates into a linear constraint on $X \in \mathscr{M}$, such as $x_{ij} = 0$. Moreover, any $X \in \mathscr{M}$ is positive semidefinite.

Polyhedral implications of working with $\mathscr{M}$ instead of P are investigated e.g., in [53, 4, 69] under the key words "lift and project". Using the condition that $\mathscr{M}$ is contained in the cone

$$\mathscr{S}_n^+ := \{X \in \mathscr{S}_n : X \succeq 0\}$$

of positive semidefinite matrices leads to semidefinite relaxations which are a generalization of linear programs. Formally, a semidefinite program, SDP for short, is defined by the data $C, A_1, \ldots, A_m \in \mathscr{S}_n$ and $b \in \mathbb{R}^m$ and consists of the following

$$\text{(SDP)} \quad z_p = \inf\{\langle C, X\rangle : \langle A_i, X\rangle = b_i,\ i = 1, \ldots, m,\ \ X \succeq 0\}. \tag{18.2}$$

In this chapter we will consider relaxations of integer programs which are based on SDP, rather than purely polyhedral combinatorics. We first recall some basics about SDP in Section 18.2. In Section 18.3 various modeling ideas are described which all lead to SDP relaxations. SDP as generalization of LP often provides tighter approximations, at increased computational cost. The theoretical power of SDP to approximate some NP-hard optimization problems is presented in Section 18.4. The hyperplane rounding idea of Goemans and Williamson [20] turns out to be a generic method to generate provably good feasible solutions for a variety of problems. In Section 18.5 we turn to algorithms for solving SDP. While interior-point based algorithms are still the method of choice, it turns out that large scale problems are beyond the scope of these algorithms and alternatives, capable of handling an arbitrary number of constraints at least approximately, are necessary. Several of these are discussed in 18.5. Finally, we touch some more recent modeling techniques which go beyond SDP in Section 18.6.

This article lines up with several survey papers devoted to the connection between semidefinite optimization and integer programming. The interested reader is referred to [46] for an extensive summary on the topic covering the development until 2003. The surveys by Lovász [52], Goemans [19] and Helmberg [27] all focus on the same topic, but also reflect the scientific interests and preferences of the respective authors. The present paper is no exception to this principle. The material selected, and also omitted, reflects the author's subjective view on the subject. Since the scientific area covered here is still under rapid development, the present survey is far from complete. Some discussion and pointers to material not covered will be given at the end of each chapter.

We introduce some notation. The vector of all ones is denoted e, and $J = ee^T$ is the all-ones matrix. The identity matrix is denoted by $I = (e_1, \ldots, e_n)$. Thus the e_i represent the standard unit vectors. For $i \neq j$ we define $E_{ij} = e_i e_j^T + e_j e_i^T$.

The set of symmetric matrices is denoted by $\mathscr{S}$ and $\mathscr{S}^+$ denotes the closed convex cone of semidefinite matrices. Its interior is the set of definite matrices, denoted $\mathscr{S}^{++}$. We also use the cone $\mathscr{C}$ of *copositive matrices*, given by $\mathscr{C} = \{X : v^T X v \geqslant 0 \; \forall \text{ vectors } v \geqslant 0\}$. Its dual cone $\mathscr{C}^*$ consists of the set of *completely positive matrices*, $\mathscr{C}^* = \{X : X = VV^T,\ V \geqslant 0 \text{ is } n \times k\}$. The standard simplex is $\Delta = \{x : x \geqslant 0,\ e^T x = 1\}$. The convex hull of a set S is denoted by $\operatorname{conv}(S)$. We denote by Π the set of permutation matrices and by Ω the set of doubly stochastic matrices, $\Omega = \{X : Xe = X^T e = e, X \geqslant 0\}$. For $a, b \in \mathbb{R}^n$, $\langle a, b\rangle_+$ denotes the maximal scalar product of a and b, if the entries in a and b can be permuted arbitrarily.

It is given e.g., by sorting the entries in both a and b in nondecreasing order. The minimal scalar product $\langle a,b\rangle_-$ is defined analogously.

If A and B are matrices of the same order, then the *Hadamard* or elementwise product is $C = A \circ B$ with $c_{ij} = a_{ij}b_{ij}$. The Kronecker product (or tensor product) $A \otimes B$ of two matrices A and B consists of the matrix of all pairwise products of elements from A and B. Formally, if $A = (a_{ij})$ is $m \times n$ and B is $p \times q$, then

$$A \otimes B = \begin{pmatrix} a_{11}B & \dots & a_{1n}B \\ \vdots & \ddots & \vdots \\ a_{m1}B & \dots & a_{mn}B \end{pmatrix}.$$

The operator $\operatorname{Diag} : \mathbb{R}^n \mapsto \mathscr{S}_n$ maps a vector y to the diagonal matrix $\operatorname{Diag}(y)$, its adjoint mapping $\operatorname{diag}(X)$ extracts the main diagonal from the matrix X.

If G is a graph, and $S \subseteq V(G)$ is a subset of its vertices, we denote by $\delta(S)$ the set of all edges ij such that $i \in S$, $j \notin S$. The neighbourhood $N(i)$ of vertex $i \in V(G)$ is the set of all vertices, adjacent to $i, N(i) = \{j : ij \in E(G)\}$. The complement graph $\overline{G}$ of a graph G has edges $ij \in E(\overline{G})$ whenever $i \neq j$ and $ij \notin E(G)$.

18.2 Basics on semidefinite optimization

Problem (18.2) is a convex optimization problem, because a linear function is optimized over the convex set

$$F_p := \{X : A(X) = b,\ X \succeq 0\}.$$

The linear operator $A(X)$ maps matrices into $\mathbb{R}^m$ and has $A(X)_i = \langle A_i, X\rangle$. Its adjoint A^T, defined through the adjoint identity

$$y^T A(X) = \langle A^T(y), X\rangle$$

is given by $A^T(y) = \sum_i y_i A_i$. The problem (SDP) as a convex problem possesses a dual, which is most conveniently derived through the Lagrangian

$$L(X,y) = \langle C,X\rangle + y^T(b - A(X)) = b^T y + \langle C - A^T(y), X\rangle \tag{18.3}$$

and the Minimax inequality

$$\inf_{u \in U} \sup_{v \in V} f(u,v) \geqslant \sup_{v \in V} \inf_{u \in U} f(u,v),$$

which holds for any function $f : U \times V \mapsto \mathbb{R}$. To get the dual, we first observe

$$\sup_y L(X,y) = \begin{cases} \langle C,X\rangle, & \text{if } A(X) = b, \\ +\infty, & \text{otherwise.} \end{cases}$$

To see what happens to $\inf_{X\succeq 0} L(X,y)$ we recall Fejer's theorem.

Theorem 18.1. *The matrix $M \in \mathscr{S}^+$ if and only if $\langle M,X\rangle \geqslant 0\ \forall X \in \mathscr{S}^+$.*

In the language of convex analysis, this translates into the fact that the cone dual to the semidefinite matrices is again the cone of semidefinite matrices. Recall that the dual cone K^* of the cone $K \subseteq \mathbb{R}^d$ is by definition

$$K^* := \{y \in \mathbb{R}^d : \langle y,x\rangle \geqslant 0\ \forall x \in K\}.$$

Now if $C - A^T(y) \succeq 0$, then by Fejer's theorem $\inf_{X\succeq 0} \langle C - A^T(y), X\rangle = 0$. On the other hand, if $C - A^T(y) \notin \mathscr{S}^+$, then there must exist some $X \succeq 0$ such that $\langle C - A^T(y), X\rangle < 0$. We conclude

$$\inf_{X\succeq 0} L(X,y) = \begin{cases} b^T y, & \text{if } C - A^T(y) \succeq 0, \\ -\infty, & \text{otherwise.} \end{cases} \tag{18.4}$$

Therefore (18.2) is equivalent to $z_p = \inf_{X\succeq 0} \sup_y L(X,y)$ and the Minimax inequality implies

$$z_p \geqslant \sup_y \inf_{X\succeq 0} L(X,y) =: z_d.$$

The problem defining z_d can be rewritten using (18.4) to yield

$$z_d = \sup\ b^T y \text{ such that } C - A^T(y) \succeq 0. \tag{18.5}$$

The last problem is again a semidefinite program, which is usually called the dual of (18.2). In contrast to linear programming, strong duality ($z_p = z_d$) does not hold in general. Moreover, attainment of sup and inf only holds under some additional conditions. The following condition, often called Slater's constraint qualification, insures $z_p = z_d$. Problem (18.2) satisfies the *Slater constraint qualification* if there exists a positive definite matrix $X \succ 0$ such that $A(X) = b$. Such matrices are often called *strictly feasible*.

Theorem 18.2. *(see e.g., Duffin [14]) If (18.2) satisfies the Slater condition and z_p is finite, then the dual problem is feasible, the dual supremum is attained, and $z_p = z_d$.*

In most of the semidefinite relaxations considered in this chapter, both the primal and the dual problem satisfy the Slater condition, therefore we have the following situation. Suppose

$$\exists X \succ 0 \text{ such that } A(X) = b \text{ and } \exists y \text{ with } C - A^T(y) \succ 0. \tag{18.6}$$

Then (X,y,Z) is optimal for (18.2) and (18.5) if and only if

$$A(X) = b,\ X \succeq 0,\ C - A^T(y) = Z \succeq 0,\ \langle Z,X\rangle = 0. \tag{18.7}$$

Note that in the dual problem, we have introduced for notational and algorithmic convenience the slack matrix Z in the formulation. The condition $\langle Z,X\rangle = 0$ is a consequence of $0 = z_p - z_d = \langle C,X\rangle - b^T y = \langle C - A^T(y), X\rangle$.

18.3 Modeling with semidefinite programs

The basic idea to come to semidefinite relaxations of (COP), and more generally integer programs, consists in working with $\mathscr{M}$ from (18.1), instead of P. Going from $\mathbb{R}^n$ to the space of symmetric matrices $\mathscr{S}_n$ allows to replace quadratic constraints and cost functions by linear ones. As a first example we consider a semidefinite relaxation of quadratic 0/1 optimization.

18.3.1 Quadratic 0/1 optimization

For given $Q \in \mathscr{S}_n$ we consider

$$\text{(QP)} \qquad \min_{x \in \{0,1\}^n} x^T Q x. \tag{18.8}$$

This problem is equivalent to Max-Cut, see e.g., [25, 5], and therefore NP-hard. We may assume without loss of generality that there is no additional linear term $c^T x$ in the objective function (18.8), because $x_i^2 = x_i$ allows us to add c to the main diagonal of Q. The cost function can be rewritten as

$$x^T Q x = \langle Q, xx^T \rangle.$$

Following the linearization idea we introduce a matrix variable X taking the role of xx^T. Since $x_i^2 = x_i$, the main diagonal of X is equal to x. The nonconvex constraint $X - xx^T = 0$ is relaxed to $X - xx^T \succeq 0$. The Schur-complement lemma shows that this set is convex.

Lemma 18.1. *Let* $M = \begin{pmatrix} A & B \\ B^T & C \end{pmatrix}$ *be symmetric and* A *invertible. Then* $M \succeq 0$ *if and only if* $A \succeq 0$ *and* $C - B^T A^{-1} B \succeq 0$.

Proof. This well known fact follows from the similarity transformation

$$\begin{pmatrix} I & 0 \\ -B^T A^{-1} & I \end{pmatrix} \begin{pmatrix} A & B \\ B^T & C \end{pmatrix} \begin{pmatrix} I & 0 \\ -B^T A^{-1} & I \end{pmatrix}^T = \begin{pmatrix} A & 0 \\ 0 & C - B^T A^{-1} B \end{pmatrix}.$$

□

Hence we get the following SDP relaxation of (QP):

$$\min \langle Q, X \rangle \text{ such that } X - xx^T \succeq 0, \ \operatorname{diag}(X) = x. \tag{18.9}$$

We use the Schur-complement lemma to replace the quadratic condition $X - xx^T \succeq 0$ by $\begin{pmatrix} 1 & x^T \\ x & X \end{pmatrix} \succeq 0$.

18.3.2 Max-Cut and graph bisection

Let A be the (weighted) adjacency matrix of a graph G on n vertices. A basic graph optimization problem consists in separating the vertex set V of G into two sets S and $V \setminus S$ such that

$$\sum_{ij \in \delta(S)} a_{ij}$$

is optimized. In case of Max-Cut one has

$$(\mathrm{MC}) \qquad z_{\mathrm{mc}} = \max_{S \subseteq V} \sum_{ij \in \delta(S)} a_{ij}. \tag{18.10}$$

The *Bisection Problem* has an additional input parameter s, specifying $|S| = s$, and is defined as follows.

$$(\mathrm{BS}) \qquad z_{\mathrm{bs}} = \min_{S \subseteq V, |S| = s} \sum_{ij \in \delta(S)} a_{ij} \tag{18.11}$$

The special case $s = \frac{n}{2}$ is often called Equicut. Kernighan and Lin [41] investigated various local improvement heuristics for this problem, see also [38]. All these bisection problems are well known to be NP-hard. A simple integer program for these problems can be obtained as follows. Subsets $S \subseteq V$ are modeled by vectors $x \in \{-1,1\}^n$ with $x_i = 1$ exactly if $i \in S$. Then $e = ij \in \delta(S)$ precisely if $x_i x_j = -1$. A convenient way to model the objective function makes use of the *Laplacian*, associated to A, which is defined as follows.

$$L = L_A = \mathrm{Diag}(Ae) - A \tag{18.12}$$

A simple calculation shows that

$$\sum_{ij \in \delta(S)} a_{ij} = \sum_{ij \in E} a_{ij} \frac{1 - x_i x_j}{2} = \frac{1}{4} x^T L x, \tag{18.13}$$

if $x \in \{-1,1\}^n$ represents S. Hence we have

$$\begin{aligned} z_{\mathrm{mc}} &= \max_{x \in \{-1,1\}^n} \frac{1}{4} x^T L x, \\ z_{\mathrm{bs}} &= \min_{x \in \{-1,1\}^n, e^T x = n - 2s} \frac{1}{4} x^T L x. \end{aligned}$$

The linearization idea gives again $x^T L x = \langle L, xx^T \rangle$, and we introduce $X = xx^T$ and now have $\mathrm{diag}(X) = e$. The cardinality constraint is easily incorporated using

$$|e^T x| = n - 2s \text{ if and only if } \langle J, xx^T \rangle = (n - 2s)^2.$$

Hence we get the SDP relaxation of Max-Cut as

$$z_{\mathrm{mc,sdp}} = \max \frac{1}{4}\langle L, X\rangle \text{ such that } \operatorname{diag}(X) = e,\ X \succeq 0, \tag{18.14}$$

and for bisection as

$$z_{\mathrm{bs,sdp}} = \min \frac{1}{4}\langle L, X\rangle \text{ such that } \operatorname{diag}(X) = e,\ \langle J, X\rangle = (n-2s)^2,\ X \succeq 0.$$

Quadratic 0/1 optimization can linearly (and bijectively) be mapped to optimization in $-1/1$ variables, showing the claimed equivalence between Max-Cut and quadratic 0/1 optimization. One may therefore wonder whether there is also some relation between the two relaxations (18.9) and (18.14). It is not too difficult to verify that these relaxations are in fact also equivalent, see e.g., [31, 45]. Helmberg [26] provides an explicit transformation between these problems, which preserves structure, like sparsity. The equivalence of Max-Cut and quadratic 0/1 optimization was already pointed out by Hammer, see [25].

18.3.3 Stable sets, cliques and the Lovász theta function

Perhaps the earliest use of SDP to get relaxations of NP-hard problems goes back to a fundamental paper by Lovász in 1979, see [51]. Consider the stable set problem in an unweighted graph G. We recall that a set $S \subseteq V(G)$ is stable if the vertices in S are pairwise nonadjacent. Consequently, S forms a complete subgraph in the complement graph $\overline{G}$ of G. The stable set problem asks to find a stable set of maximum cardinality in G. The cardinality of a largest stable set in G, denoted $\alpha(G)$, is called the *stability number of* G. Modeling stable sets by their characteristic vectors, we get

$$(\mathrm{STAB}) \quad \alpha(G) = \max\{e^T x :\ x_i x_j = 0\ \forall ij \in E(G),\ x \in \{0,1\}^n\}. \tag{18.15}$$

Following our linearization idea for 0/1 optimization, we obtain the following SDP relaxation, which was studied in some detail in [53].

$$\alpha(G) \leqslant \vartheta_1(G) = \max\{e^T x :\ X - xx^T \succeq 0,\ \operatorname{diag}(X) = x,\ x_{ij} = 0\ \forall ij \in E\}. \tag{18.16}$$

Note in particular that the quadratic equations in x are now linear in X. This model can in fact be simplified by eliminating x. Supppose x is the characteristic vector of a (nonempty) stable set. Then

$$X = \frac{1}{x^T x} xx^T$$

satisfies the following conditions:

$$X \succeq 0,\ \operatorname{tr}(X) = 1,\ x_{ij} = 0\ \forall ij \in E,\ \operatorname{rank}(X) = 1.$$

Moreover

$$\langle J, X\rangle = \frac{1}{x^T x}(e^T x)^2 = e^T x,$$

because $e^T x = x^T x$. The following result is well known.

Lemma 18.2.

$$\alpha(G) = \max\{\langle J, X\rangle : X \succeq 0,\ \mathrm{tr}(X) = 1,\ x_{ij} = 0\ \forall ij \in E,\ \mathrm{rank}(X) = 1\}. \quad (18.17)$$

Proof. Feasibility of X implies $X = aa^T$ where the vector a is nonzero only on some stable set S, because $x_{ij} = a_i a_j = 0\ \forall ij \in E$. Looking at the nonzero submatrix $a_S a_S^T$ indexed by S, the cost function becomes $(e^T a_S)^2$, which is maximal exactly if a_S is parallel to the all ones vector e. Hence a is (a multiple of) the characteristic vector of S and the maximization forces S to be a maximum stable set. □

Leaving out the (nonconvex) rank condition, we obtain another SDP relaxation for $\alpha(G)$:

$$\alpha(G) \leqslant \vartheta_2(G) = \max\{\langle J, X\rangle : \mathrm{tr}(X) = 1,\ x_{ij} = 0\ \forall ij \in E,\ X \succeq 0\}.$$

This is the relaxation proposed by Lovász in [51]. Lovász and Schrijver [53] show that $\vartheta_1(G) = \vartheta_2(G)$ and this function is usually called the *Lovász theta function* $\vartheta(G)$.

Let us introduce the linear operator $A_G : \mathscr{S}_n \mapsto \mathbb{R}^E$ associated to the edge set E of G,

$$A_G(X)_{ij} = \langle X, E_{ij}\rangle = x_{ij} + x_{ji} = 2x_{ij}\ \forall ij \in E$$

and its adjoint

$$A_G^T(y) = \sum_{ij \in E} y_{ij} E_{ij}.$$

Using these operators, we get the following primal-dual pair of semidefinite programs for $\vartheta(G)$.

$$\vartheta(G) = \max\{\langle J, X\rangle : \mathrm{tr}(X) = 1,\ A_G(X) = 0,\ X \succeq 0\} \quad (18.18)$$

$$= \min\{t : tI + A_G^T(y) \succeq J\}.$$

Strong duality is justified by the observation that $\frac{1}{n}I$ is strictly feasible for the maximization, and setting $y = 0$ and $t = n+1$ gives a strictly feasible point for the minimization problem. We will now use the adjoint A^T to express symmetric matrices in the following way. Let $Z \in \mathscr{S}_n$, then

$$Z = \mathrm{Diag}(y) + A_G^T(u) + A_{\overline{G}}^T(v),$$

if we define

$$y_i = z_{ii},\ u_{ij} = z_{ij}\ \forall ij \in E(G),\ v_{ij} = z_{ij}\ \forall ij \in E(\overline{G}),$$

Suppose that X is feasible for (18.18). Then

$$X = \mathrm{Diag}(x) + A_{\overline{G}}^T(\xi),$$

if we set $x_i = x_{ii}$ and $\xi_{ij} = x_{ij}$ $\forall ij \in E(\overline{G})$. By construction $x_{ij} = 0$ $\forall ij \in E(G)$. We substitute for X and obtain

$$\vartheta(G) = \max\{e^T x + 2e^T \xi : e^T x = 1,\ \mathrm{Diag}(x) + A_{\overline{G}}^T(\xi) \succeq 0\}. \tag{18.19}$$

Both models are mathematically equivalent, but from a computational point of view (18.18) is preferable in case $|E|$ is small. If $|E|$ is large (G is dense), then the second formulation (18.19) is more efficient, because in this case the number of variables $|E(\overline{G})|$ is small. Computational experience with both models is given in [15]. It turns out that $\vartheta(G)$ can be computed for graphs with $n \leqslant 200$ in acceptable time using interior-point methods and the sparse or the dense model depending on $|E(G)|$. From [15] it is also clear that interior-point methods cannot handle graphs with $n \geqslant 200$ and $|E(G)| \approx \frac{n^2}{4}$, which is the worst case in terms of computational effort for both models, see also Table 18.2 below.

18.3.4 Chromatic number

A k-coloring of a graph G is a k-partition $(V_1, \ldots, V_k)$ of the vertex set $V(G)$ such that each V_i is a stable set in G. The *chromatic number* $\chi(G)$ is the smallest value k, such that G has a k-coloring. Let us encode k-partitions of $V(G)$ by the characteristic vectors s_i for V_i, thus

$$s_i \in \{0,1\}^n \text{ and } (s_i)_u = 1 \text{ if } u \in V_i \text{ and } 0 \text{ otherwise.} \tag{18.20}$$

The partition property implies that $s_i \neq 0$ and that $\sum_i s_i = e$. Let $n_i := |V_i|$ denote the cardinality of V_i. We call the matrix

$$M = \sum_i s_i s_i^T \tag{18.21}$$

the *partition matrix* associated to the k-partition $(V_1, \ldots, V_k)$. Note that $\mathrm{rank}(M) = k$. It is clear from the definition that for any k-partition matrix M there exists a permutation matrix $P \in \Pi$ such that

$$M = P \begin{pmatrix} J_{n_1} & 0 & \ldots & 0 \\ 0 & J_{n_2} & \ddots & \vdots \\ \vdots & \ddots & \ddots & 0 \\ 0 & \ldots & 0 & J_{n_k} \end{pmatrix} P^T.$$

Partition matrices have the following characterization.

Lemma 18.3. *Let M be a symmetric 0/1 matrix. Then M is a k-partition matrix if and only if*

$$\operatorname{diag}(M) = e, \ \operatorname{rank}(M) = k, \ M \succeq 0. \tag{18.22}$$

Proof. Suppose that M is a k-partition matrix. Then this matrix obviously satisfies (18.22). Conversely let M satisfy (18.22). We need to show that M is (after appropriate renumbering) a direct sum of all ones blocks. Thus we need to show that $m_{ij} = m_{ik} = 1$ implies $m_{jk} = 1$. If this is not the case then the submatrix of M, indexed by i, j, k is

$$\begin{pmatrix} 1 & 1 & 1 \\ 1 & 1 & 0 \\ 1 & 0 & 1 \end{pmatrix},$$

which has negative determinant, contradicting $M \succeq 0$. Therefore M is indeed a direct sum of all ones blocks, and the rank condition shows that there must be k such blocks. □

This leads to the following somewhat nonstandard way to define $\chi(G)$.

$$\chi(G) = \min\{\operatorname{rank}(M) : \ m_{ij} \in \{0,1\}, \ \operatorname{diag}(M) = e, \ m_{ij} = 0 \ \forall ij \in E, \ M \succeq 0\}.$$

It turns out that in fact we can get rid of both the rank and the semidefiniteness condition on M by introducing another semidefiniteness constraint.

Lemma 18.4. *Let M be a symmetric 0/1 matrix. Then M is a k-partition matrix if and only if*

$$\operatorname{diag}(M) = e, \ (tM - J \succeq 0 \iff t \geqslant k). \tag{18.23}$$

This result is implicitly proved in many sources, see e.g., [51]. A simple argument can be found also in [15]. Here we only take a closer look at how the semidefiniteness condition comes about. Looking at principal submatrices of $tM - J$, we first observe that any minor with two rows from the same partition block is singular, because it contains two identical rows. Hence nonsingular principal submatrices can have at most one row from each partition block. Therefore these must be of the form

$$tI_s - J_s$$

where $s \leqslant k$ denotes the order of the submatrix. Clearly, this matrix is semidefinite if and only if $t \geqslant s$, and since s could be as large as k we get the desired condition.

Using this result we can express the chromatic number as the optimal solution of the following SDP in binary variables:

$$\chi(G) = \min\{t : \ tM - J \succeq 0, \ \operatorname{diag}(M) = e, \ m_{ij} = 0 \ \forall ij \in E, m_{ij} \in \{0,1\}\}. \tag{18.24}$$

To get a tractable relaxation, we leave out the 0/1 condition and parametrize the matrix tM as $tM = tI + A_{\overline{G}}^T(y)$. We get the following SDP as lower bound on $\chi(G)$:

$$\chi(G) \geqslant \vartheta(\overline{G}) = \min\{t : \ tI + A_{\overline{G}}^T(y) \succeq J\}. \tag{18.25}$$

Comparing with the minimization problem in (18.18), we see that the above SDP is indeed the dual applied to the complement graph. Hence we have shown the sandwich theorem of Lovász [51].

Theorem 18.3.

$$\alpha(\overline{G}) \leqslant \vartheta(\overline{G}) \leqslant \chi(G).$$

The notation for ϑ is not uniform in the literature. If one starts from the clique number of a graph G, it seems more natural to denote by $\vartheta(G)$ what we denote by $\vartheta(\overline{G})$.

The problem (18.25) can be rephrased in the following way: Minimize z_{11} where the matrix $Z = (z_{ij}) \succeq 0, z_{ii} = z_{11}\ \forall i,\ z_{ij} = -1\ \forall ij \in E(G)$. This follows simply from $Z = tI + A_{\overline{G}}^T(y) - J$. This problem in turn can be rewritten as

$$\min\{\lambda : V \succeq 0, v_{ii} = 1\ \forall i,\ v_{ij} = \lambda\ \forall ij \in E(G)\}. \tag{18.26}$$

The optimal value ϑ in (18.25) gives $\lambda = -\frac{1}{\vartheta - 1}$. This last model will be used later on to analyze algorithms for graph coloring, see [39].

18.3.5 General graph partition

The concept of partition matrices from (18.23) in the context of coloring can also be recycled to model the following general partition problems. Given the weighted adjacency matrix A of a graph G on n nodes and an integer $2 \leqslant k \leqslant n$, the general *k-partition problem* consists in finding a k-partition $(S_1, \ldots, S_k)$ of $V(G)$ with $|S_i| \geqslant 1$ such that the total weight of edges joining different partition blocks is optimized. In case of maximization, this problem also runs under the name of *Max-k-Cut*. Max-Cut, see the previous sections, corresponds to the special case $k = 2$. We represent partitions again through the respective characteristic vectors s_i, see (18.20), which we collect in the matrix $S = (s_1, \ldots, s_k)$. The characteristic vector $s_i \in \{0,1\}^n$ for the partition block S_i allows us to express the weight of the edges in $\delta(S_i)$ using the Laplacian L, see (18.12) as follows,

$$\sum_{rs \in \delta(S_i)} a_{rs} = s_i^T L s_i.$$

Therefore

$$\frac{1}{2}\langle S, LS\rangle = \frac{1}{2}\sum_i s_i^T L s_i$$

gives the total weight of all edges joining different partition blocks. We obtain a semidefinite relaxation of Max-k-Cut using (18.23), once $\langle S, LS\rangle = \langle L, SS^T\rangle$ is used to replace SS^T by a new matrix variable. For notational convenience we introduce the matrix variable $Y = \frac{1}{k-1}(kSS^T - J)$. (Note that SS^T takes the role of M in (18.23) and that $Le = 0$ implies $\langle L, J\rangle = 0$.)

$$\max\{\frac{k-1}{2k}\langle L,Y\rangle : \operatorname{diag}(Y)=e,\ Y\succeq 0,\ y_{ij}\geqslant -\frac{1}{k-1}\ \forall i\neq j\}. \tag{18.27}$$

The sign constraints follow from $SS^T \geqslant 0$. This model has been investigated theoretically by Frieze and Jerrum, see [18] and also by DeKlerk et al. [16].

A more restrictive version of k-partition, called *k-Equicut*, asks for k-partitions with $|S_i| = \frac{n}{k}\ \forall i$. In this case the objective is to minimize the total weight of edges cut. A feasible partition $(S_1,\dots,S_k)$ for this problem implies the following additional constraint

$$S^T e = \frac{n}{k} e$$

which together with $Se = e$ (note the varying dimension of e) gives

$$SS^T e = \frac{n}{k} e,$$

hence $\frac{n}{k}$ is eigenvalue of SS^T with eigenvector e, for any feasible partition matrix S. Setting $X = SS^T$, this gives the following relaxation of k-Equicut.

$$\min\{\frac{1}{2}\langle L,X\rangle : \operatorname{diag}(X)=e,\ Xe=\frac{n}{k}e,\ X\succeq 0,\ X\geqslant 0\}. \tag{18.28}$$

We note that this model does not contain the condition $kX \succeq J \succeq 0$, but only the weaker $X \succeq 0$. The following simple observation explains why.

Lemma 18.5. *Let $A \in \mathscr{S}_n^+$ and suppose $Av = \lambda v$ holds with $\|v\| = 1$. Then*

$$A - \lambda vv^T \succeq 0.$$

Proof. The spectral decomposition theorem for A yields $A = \lambda vv^T + \sum_i \lambda_i v_i v_i^T$ where we identified one of the eigenvalues of A as λ. $A \succeq 0$ implies $\lambda_i \geqslant 0$ hence $A - \lambda vv^T = \sum_i \lambda_i v_i v_i^T \succeq 0$. □

This justifies the semidefiniteness constraint $X \succeq 0$ in (18.28). It is also interesting to note that the sign constraint $X \geqslant 0$ together with the eigenvector condition $Xe = \frac{n}{k}e$ implies that

$$\frac{n}{k}I - X \succeq 0$$

because in this case $\|X\|_1 = \frac{n}{k}$. Therefore, any eigenvalue λ of X satisfies $|\lambda| \leqslant \frac{n}{k}$. In summary, the constraints in (18.28) imply

$$\frac{n}{k}I \succeq X \succeq \frac{1}{k}J.$$

It should be observed that both SDP relaxations for partitioning are formulated in the space of symmetric $n\times n$ matrices, but the models include roughly $\frac{n^2}{2}$ sign constraints, which we will later see to be a computational challenge. Further details in this direction can be found e.g., in [40] and [50].

We close this section with a bound on general k-partition, proposed by Donath and Hoffman in 1973, see [13]. In view of the results from the subsequent Section 18.3.7 on connections between eigenvalue optimization and semidefinite programming, this result may well be the first SDP bound of a combinatorial optimization problem, but in disguised form, see below.

We consider k-partition, where the cardinalities m_i of S_i are also specified through integers

$$m_1 \geqslant \dots m_k \geqslant 1 \text{ and } \sum_i m_i = n.$$

Minimizing the weight cut by S is equivalent to maximizing the weight not cut by S, which is equal to $\frac{1}{2}\langle S, AS\rangle$.

The starting point in [13] is the observation that any feasible partition matrix S satisfies

$$S^T S = \operatorname{diag}(m_1, \dots, m_k) =: M.$$

In words, the columns of S are pairwise orthogonal. An upper bound on the total weight not cut is therefore given by

$$\max_{S^T S = M} \langle S, AS\rangle = \max_{Y^T Y = I_k} \langle Y, AYM\rangle = \sum_{i=1}^{k} m_i \lambda_i(A), \tag{18.29}$$

where $\lambda_1(A) \geqslant \dots \geqslant \lambda_n(A)$ are the eigenvalues of A in nonincreasing order.

The first equality follows from the substitution $S = YM^{\frac{1}{2}}$ and the second from Theorem 18.4 below. The last problem can in fact be expressed as the optimal solution of a semidefinite program, see Section 18.3.7. Further details and extensions on using this approach are contained e.g., in [66, 40].

18.3.6 *Generic cutting planes*

In the previous sections we saw SDP relaxations of a variety of graph optimization problems. The matrix dimension typically was $n = |V(G)|$ and the number of (primal) constraints was n in case of (18.14) and (18.9). The SDP model of $\vartheta(G)$ can have up to $\frac{n^2}{4}$ equations, depending on $|E(G)|$, and which of the two computational models (18.18) and (18.19) is used. The graph partition models (18.27) and (18.28) both have roughly $\frac{n^2}{2}$ sign constraints. All these models can of course be tightened by adding further constraints valid for all points in $\mathscr{M}$.

In the following we explain a generic way to tighten these semidefinite relaxations by adding further valid constraints. The class of *hypermetric inequalities* provides a huge source of additional cutting planes, some of which also have an explicit combinatorial interpretation. To explain this class we concentrate first on the $-1/1$ model and the related relaxation (18.14).

Suppose b is an integer vector and $b^T e = 1$. Then $b^T x$ is odd $\forall x \in \{-1,1\}^n$, because any $x \in \{-1,1\}^n$ can be obtained from e by successively changing the sign of some component i, which changes the inner product with b by $2b_i$, hence the parity is unchanged. Let $\mathscr{B}$ denote the set of all such vectors b,

$$\mathscr{B} := \{b \text{ integer } : e^T b = 1\}.$$

As a consequence, $b \in \mathscr{B}$ implies $|b^T x| \geqslant 1 \; \forall x \in \{-1,1\}^n$, and therefore

$$\langle bb^T, xx^T \rangle \geqslant 1.$$

Let us consider the (convex) set

$$HYP := \{X : \langle bb^T, X \rangle \geqslant 1 \; \forall b \in \mathscr{B}\},$$

which is the intersection of infinitely many halfspaces. Deza et al. [12] show that this set is in fact polyhedral, but it is currently open whether the separation problem $X \in HYP$ can be decided efficiently. The inequalities defining HYP are called *hypermetric inequalities*. Further information can be found e.g., in [3].

The following subclass of hypermetric inequalities is generated by $b \in \mathscr{B}$ with $b_i \in \{-1,0,1\}$ and only three nonzero entries, say b_i, b_j, b_k. Elements in this class are called *triangle inequalities*. The resulting polytope is sometimes called the *metric polytope MET*,

$$MET = \{X : x_{ij} + x_{ik} + x_{jk} \geqslant -1, x_{ij} - x_{ik} - x_{jk} \geqslant -1 \; \forall \text{ distinct } i,j,k\}.$$

These conditions follow from $b^T X b = x_{ii} + x_{jj} + x_{kk} + 2(x_{ij} + x_{ik} + x_{jk}) \geqslant 1$ in case $b_i = b_j = b_k = 1$ and the implicit assumption $\operatorname{diag}(X) = e$. The other inequalities follow by changing one of the signs in b.

Polynomial time separation for $X \in MET$ can be done trivially by enumerating all $4\binom{n}{3}$ constraints defining MET. This idea can be generalized to cliques of odd size $k > 3$.

Including all the constraints from MET in (18.14) results in the following SDP

$$\max\Big\{\frac{1}{4}\langle L, X\rangle : \operatorname{diag}(X) = e, \; X \in MET, \; X \succeq 0\Big\} \tag{18.30}$$

with $O(n^3)$ constraints, which is a computational challenge for current state-of-the-art software to solve SDP. Some computational experience with triangle inequalities and odd order clique constraints combined with (18.14) is given in [32].

Just as it was straightforward to transform 0/1 QP into MC, it is also possible to transform the triangle inequalities into the 0/1 setting. The resulting polyhedron was in fact studied independently under the name of *quadric Boolean polytope*, see e.g., [62, 71].

18.3.7 SDP, eigenvalues and the Hoffman-Wielandt inequality

We are now going to take a closer look at connections between eigenvalues of symmetric matrices, optimization over (sets of pairwise) orthogonal vectors and semidefinite programming. To illustrate these connections, we recall the following well known facts. Let $A \in \mathscr{S}_n$. Then we can express the largest eigenvalue $\lambda_{\max}(A)$ of A as

$$\lambda_{\max}(A) = \max_{x^T x = 1} x^T A x. \tag{18.31}$$

On the other hand, it is trivial to note that

$$\lambda_{\max}(A) = \min \lambda \text{ such that } \lambda I - A \succeq 0,$$

and this is a semidefinite program with dual

$$\lambda_{\max}(A) = \max \langle A, X \rangle \text{ such that } \mathrm{tr}(X) = 1,\ X \succeq 0. \tag{18.32}$$

Both problems have strictly feasible solutions, hence strong duality holds. The relations (18.31) and (18.32) are the simplest connections relating eigenvalues to optimization over orthogonal vectors on one hand and SDP on the other. It turns out that the following theorem provides a generalization of (18.31) with many applications in combinatorial optimization.

Theorem 18.4. *Let $A \in \mathscr{S}_n$, $B \in \mathscr{S}_k$ and $1 \leqslant k \leqslant n$. Let $\tilde{B} = \begin{pmatrix} B & 0 \\ 0 & 0 \end{pmatrix} \in \mathscr{S}_n$. Then*

$$\min_{X^T X = I_k} \langle X, AXB \rangle = \min_{\phi \text{ injection}} \sum_i \lambda_i(B) \lambda_{\phi(i)}(A) = \langle \lambda(\tilde{B}), \lambda(A) \rangle_-. \tag{18.33}$$

Remark 18.1. This theorem has a long history. John von Neumann [72] looks at the more general problem $\max Re(\mathrm{tr}(AUBV))$, where A and B are square complex matrices and the maximization is carried out over unitary U and V. $Re(x)$ denotes the real part of the complex number x. The above result follows as a special case. Hoffman and Wielandt [36] study the question of the "distance" of the eigenvalues of two normal matrices in terms of the matrix distance, and also prove the above result in disguised form for the case $k = n$. Berkowitz [7] and Marcus [55] investigate the problem $\min_{X^T X = I_k} E_m(B + X^T AX)$ where A and B are as in the theorem and $E_m(Y)$ is the m-th elementary symmetric function of degree m of the eigenvalues of Y, hence $E_1(Y) = \mathrm{tr}(Y)$. Therefore, their result proves the above theorem for the case that the smaller matrix is the identity.

Since we consider this a fundamental result, we include the following proof, which is inspired by [36] and which is in the spirit of combinatorial optimization. An argument based on first order optimality conditions in nonlinear optimization is given in [65].

Proof. First we observe that for $k < n$ we can extend X to an orthogonal matrix $Z = (X\ Y)$ through an appropriate choice of Y. Therefore

$$\langle Z, AZ\tilde{B}\rangle = \langle X, A(X\ Y)\begin{pmatrix} B \\ 0 \end{pmatrix}\rangle + \langle Y, AZ0\rangle = \langle X, AXB\rangle.$$

We also note for later use that $I = ZZ^T = XX^T + YY^T$, therefore

$$I - XX^T \succeq 0. \tag{18.34}$$

Hence we may consider $k = n$. A and B are symmetric, therefore they have an orthogonal diagonalization:

$$A = PD_A P^T,\ B = QD_B Q^T, \tag{18.35}$$

with $D_A = \text{Diag}(\lambda(A))$ and $D_B = \text{Diag}(\lambda(B))$. Let us assume that P and Q are chosen in such a way that the scalar product of the eigenvalues $\langle\lambda(A), \lambda(B)\rangle$ is minimal. This holds e.g., if the elements in $\lambda(A)$ are in nondecreasing order, and those of $\lambda(B)$ in nonincreasing order. Therefore we have

$$\min_{X^T X = I} \langle X, AXB\rangle \leqslant \langle PQ^T, A(PQ^T)B\rangle = \langle\lambda(A), \lambda(B)\rangle_-.$$

On the other hand, for any orthogonal X we have

$$\langle X, AXB\rangle = \text{tr} D_A (P^T XQ) D_B (P^T XQ)^T = \langle\lambda(A)\lambda(B)^T, (P^T XQ)\circ(P^T XQ)\rangle.$$

Now we observe that for the orthogonal matrix $Y = P^T XQ$ we have $Y \circ Y \in \Omega$. Therefore we get

$$\langle\lambda(A)\lambda(B)^T, Y\circ Y\rangle \geqslant \min_{Z\in\Omega} \sum_{ij} \lambda_i(A)\lambda_j(B) z_{ij} = \langle\lambda(A), \lambda(B)\rangle_-.$$

If $k < n$, it is a simple exercise to show that

$$\min_{\phi \text{ injection}} \sum_i \lambda_i(B)\lambda_{\phi(i)}(A) = \langle\lambda(\tilde{B}), \lambda(A)\rangle_-.$$

□

The semidefinite counterpart of this theorem, generalizing (18.32) has only recently been shown by Anstreicher and Wolkowicz [1] for the case $k = n$.

Theorem 18.5. *Let $A, B \in \mathcal{S}_n$. Then*

$$\min_{X^T X = I} \langle X, AXB\rangle = \max\{\text{tr}S + \text{tr}T :\ B\otimes A - I\otimes S - T\otimes I \succeq 0\}.$$

This result is obtained by taking the Lagrangian dual of the first term with respect to the constraints $X^T X = I$, $XX^T = I$. Hence there are two matrices S and T in the SDP. Moreover this is an SDP in the matrix space $\mathcal{S}_{n^2}$. The general case $k \leqslant n$ needs some slight modifications.

Theorem 18.6. *Let $A \in \mathscr{S}_n$, $B \in \mathscr{S}_k$ and $1 \leqslant k \leqslant n$. Then*

$$\min_{X^TX=I_k} \langle X, AXB \rangle = \max\{\mathrm{tr}S - \mathrm{tr}T : B \otimes A - S \otimes I_n + I_k \otimes T \succeq 0, \; T \succeq 0\}.$$

Remark 18.2. It is not hard to see that the case $k = n$, considered in [1] can be recovered from this result. We refer to [64] for further details.

Proof. We first recall that

$$\begin{aligned} z^* &:= \min_{\phi \text{ injection}} \sum_i \lambda_i(B)\lambda_{\phi(i)}(A) \\ &= \min\Big\{\sum_{ij} \lambda_i(B)\lambda_j(A) z_{ij} : Z = (z_{ij}) \; k \times n, \; Ze = e, \; Z^T e \leqslant e, \; Z \geqslant 0\Big\}. \end{aligned}$$

Linear programming duality shows that the last term is equal to

$$\max\Big\{\sum_i s_i - \sum_i t_i : \; s_i - t_j \leqslant \lambda_i(B)\lambda_j(A), \; t \geqslant 0\Big\}. \tag{18.36}$$

On the other hand, we also have

$$z^* = \min\{\langle X, AXB \rangle : X^T X = I_k, I_n - XX^T \succeq 0\}$$

after adding the redundant constraint $I - XX^T \succeq 0$, see (18.34). The factorization(18.35) suggests the transformation of variables $Y = P^T XQ$ and we get

$$z^* = \min\{\langle Y, D_A Y D_B \rangle : Y^T Y = I_k, I - YY^T \succeq 0\}.$$

Duality for semidefinite programs shows that the last term is equal to

$$\max\{\mathrm{tr}(S) - \mathrm{tr}(T) : D_B \otimes D_A - S \otimes I_n + I_k \otimes T \succeq 0, T \succeq 0\}.$$

Here we introduced the multiplier $S \in \mathscr{S}_k$ for the equation $Y^T Y = I_k$ and $T \in \mathscr{S}_n^+$ for $I - YY^T \succeq 0$. By restricting S and T to diagonal matrices we get

$$z^* \leqslant \max\Big\{\sum_i s_i - \sum_i t_i : \lambda_i(B)\lambda_j(A) - s_i + t_j \geqslant 0, t_i \geqslant 0\Big\}.$$

But this is again equal to (18.36), hence there is equality throughout. □

The Hoffman-Wielandt theorem was used in [23] to get eigenvalue based bounds for the *Quadratic Assignment Problem*

$$\min_{X \in \Pi} \mathrm{tr}\langle X, AXB \rangle.$$

Donath and Hoffman [13] use it to formulate the eigenvalue bound (18.29) for general k-partition.

Further reading: The previous sections have shown the rich structure of semidefinite optimization models applied to integer programming. The idea of using "matrix liftings" such as (18.1), has immediate generalizations. Since the extreme points of (18.1) have again 0/1 coordinates, one could apply another lifting based on this set. This iterated lifting raises interesting research questions. In [53] it is shown e.g., that n such liftings suffice to get to the integer optimum for a problem in n binary variables, see also [4, 69].

Optimization with polynomials has recently also turned out to be another fruitful area for the use of semidefinite optimization. The key observation here is that a polynomial $p(x)$ in n variables $x = (x_1, \dots, x_n)$ is certainly nonnegative for all $x \in \mathbb{R}^n$, if $p(x)$ can be written as a sum of squares of other polynomials in x. Such a sum of square representation can be found by solving a related SDP. Recent results on this can be found e.g., in [37, 44].

18.4 The theoretical power of SDP

Up to now we looked at semidefinite programs as a modeling tool for combinatorial optimization problems having some inherent "quadratic structure" in their integer formulations. The seminal work of Goemans and Williamson [20] from the early 1990's shows that SDP can also be used to generate provably good integer solutions. This approach now runs under the keyword "hyperplane rounding" and exploits the Gram representation of semidefinite matrices. It was applied quite successfully to various graph partition problems, the most prominent being Max-Cut. In this section we consider the hyperplane rounding idea applied to Max-Cut and graph coloring.

18.4.1 Hyperplane rounding for Max-Cut

Let us recall the cost function (18.13) of Max-Cut and its basic semidefinite relaxation (18.14). Given any matrix $X \succeq 0$ with $\text{diag}(X) = e$, Goemans and Williamson suggest to use it in the following way to generate a random bisection, given by $\xi \in \{-1, 1\}^n$.

Goemans-Williamson Hyperplane Rounding Algorithm

Input: $X \succeq 0$ with $\text{diag}(X) = e$, given by $X = V^T V$ with $n \times n$ matrix $V = (v_1, \dots, v_n)$.
Output: Bisection $\xi \in \{-1, 1\}^n$.

Select random vector $r \in \mathbb{R}^n$, uniformly distributed on the unit sphere.
$\xi_i = 1 \iff r^T v_i \geqslant 0$.

Goemans and Williamson observe the remarkable fact that the expectation value of the solution ξ has a simple form.

Theorem 18.7. *The expectation value of the solution ξ from the hyperplane rounding routine has value*

$$E(\sum_{ij\in E} a_{ij}\frac{1-\xi_i\xi_j}{2}) = \sum_{ij\in E} a_{ij}\frac{\arccos(v_i^T v_j)}{\pi}. \tag{18.37}$$

The proof of this result uses the following geometric fact. The probability that v_i and v_j are separated by the random vector r, drawn uniformly on the sphere, is proportional to the angle between the two vectors.

$$\text{Prob}(r \text{ separates } v_i \text{ and } v_j) = \frac{\arccos(v_i^T v_j)}{\pi}. \tag{18.38}$$

This together with the linearity of the expectation value gives the final result.

The idea now is to use the optimal solution X of the semidefinite relaxation (18.14) for hyperplane rounding. In order to estimate the quality of the resulting partition ξ, we need to relate the expectation value (18.37) somehow to the optimal solution value of (18.14). Goemans and Williamson observe the following lower bound on $\arccos(x)$.

Lemma 18.6. *For all $-1 \leqslant x \leqslant 1$ and $0.87856 < \alpha_{\text{GW}} < 0.87857$ it holds that*

$$\frac{\arccos(x)}{\pi} \geqslant \alpha_{\text{GW}}\frac{1}{2}(1-x).$$

They use it to show that a term by term lower estimate of the expectation value in theorem 18.7 in case $a_{ij} \geqslant 0$ leads to

$$\sum_{ij\in E} a_{ij}\frac{1}{\pi}\arccos x_{ij} \geqslant \alpha_{\text{GW}}\sum_{ij} a_{ij}\frac{1-x_{ij}}{2} = \alpha_{\text{GW}} z_{\text{sdp}}.$$

This is summarized as follows.

Theorem 18.8. *[20] Let $A \geqslant 0$. Then*

$$z_{\text{mc}} > 0.87856\, z_{\text{sdp}}.$$

Nesterov [60] proposes a different analysis. He uses the identity

$$\sum_{i<j} a_{ij}\frac{\arccos(x_{ij})}{\pi} = \frac{1}{2\pi}\sum_{ij} l_{ij}\arcsin(x_{ij}), \tag{18.39}$$

which follows easily from the definition of the Laplacian $L = (l_{ij})$ and the fact that

$$\arccos(x) + \arcsin(x) = \frac{\pi}{2}.$$

Let us set $Y = (y_{ij})$ with $y_{ij} = \arcsin(x_{ij})$. Then the right hand side in (18.39) becomes

$$\frac{1}{2\pi}\langle L,Y\rangle.$$

For comparing this to the optimal solution $\frac{1}{4}\langle L,X\rangle$ of (18.14), Nesterov shows that $Y \succeq X$. The proof of this result is based on the Schur-product theorem.

Theorem 18.9.

$$A,B \in \mathscr{S}^+ \Longrightarrow A \circ B \in \mathscr{S}^+.$$

Proof. The following elegant one-line proof of this result can be found in [6], see Lemma 4.3.1. We have the Gram representation $a_{ij} = a_i^T a_j$ and $b_{ij} = b_i^T b_j$ for appropriate choices of vectors a_i, b_i. Setting $C = (c_{ij}) = (a_{ij}b_{ij})$, we have

$$c_{ij} = a_i^T a_j b_i^T b_j = \langle a_i b_i^T, a_j b_j^T\rangle,$$

showing that C also has a Gram representation, and hence is semidefinite. □

Lemma 18.7. *Given* $X \succeq 0$ *with* $\mathrm{diag}(X) = e$ *set* $Y = (y_{ij})$ *with* $y_{ij} = \arcsin(x_{ij})$. *Then* $Y - X \succeq 0$.

Proof. Since $\arcsin(x) = x + \frac{1}{2\cdot 3}x^3 + \frac{1\cdot 3}{2\cdot 4\cdot 5}x^5 \ldots$ we have

$$Y - X = \frac{1}{6}X \circ X \circ X + \frac{3}{40}X \circ X \circ X \circ X \circ X \ldots$$

The Schur-product theorem shows that the right hand side is positive semidefinite. □

This gives the following more general estimate.

Theorem 18.10. *Suppose that the Laplacian* $L \succeq 0$. *Then*

$$z_{\mathrm{mc}} \geqslant \frac{2}{\pi} z_{sdp}.$$

Proof. If $L \succeq 0$, then $\langle L, Y - X\rangle \geqslant 0$, using the previous lemma. Therefore the expectation value

$$\frac{1}{2\pi}\langle L,Y\rangle \geqslant \frac{1}{2\pi}\langle L,X\rangle = \frac{2}{\pi} z_{sdp}.$$

□

Since $\frac{2}{\pi} \approx 0.636$, this result is weaker than the Goemans-Williamson analysis, but it is also more general, as $A \geqslant 0$ implies $L \succeq 0$, but L can be semidefinite even if A has negative entries.

Subsequently, the hyperplane rounding idea has been extended by Frieze and Jerrum [18] to the maximization version of bisection with $s = \frac{n}{2}$. Since the hyperplane rounding routine does not necessarily give a balanced bisection, the solution has to be modified to ensure $s = \frac{n}{2}$. Frieze and Jerrum [18] propose a 0.651-approximation algorithm. Later, Halperin and Zwick [24] improve this performance ratio to 0.7016 by strengthening the underlying SDP with triangle inequalities, and also by employing a more refined rounding heuristic.

It seems also quite natural to extend the hyperplane rounding idea to Max-k-Cut, starting from the SDP (18.27). We first note that a random partition into k blocks has two vertices i and j with probability $\frac{1}{k}$ in the same partition block. Therefore the expected weight of this partition is

$$\sum_{i<j} a_{ij}(1-\frac{1}{k}).$$

This immediately gives an approximation ratio of $1-\frac{1}{k}$ for Max-k-Cut. Frieze and Jerrum [18] provide better approximation ratios for all values of k. In particular, they have (not surprisingly) a_{GW} for $k=2$, 0.8327 for $k=3$ and 0.8503 for $k=4$. The last two values have been improved later on by De Klerk et al. [16] to 0.836 for $k=3$ and 0.8574 for $k=4$. The minimization versions of these problems are much harder.

18.4.2 Coloring

The hyperplane rounding idea underlying the theoretical estimate of theorem 18.8 can be applied to other types of graph problems. This approach is extended to graph coloring by Karger, Motwani and Sudan [39].

Determining $\chi(G)$ is well known to be NP-hard. Lund and Yannakakis [54] show that, even worse, there exists a constant $\varepsilon > 0$ for which no polynomial algorithm can color any graph with at most $n^{\varepsilon}\chi(G)$ colors, unless P=NP. In particular, coloring a three-colorable graph with at most four colors is NP-complete, see [42].

In view of these theoretical complexity results it is therefore a challenge to design polynomial algorithms to color three-colorable graphs with nontrivial upper bounds on the colors used. We first recall the following easy facts.

Lemma 18.8. *If $\chi(G)=3$ then $N(v)$ is bipartite for any $v \in V(G)$.*

Lemma 18.9. *Let G be a graph with maximum vertex degree Δ. Then G can be colored in polynomial time with $\Delta+1$ colors.*

Widgerson [73] observes that if $\Delta > \sqrt{n}$, then 2 colors suffice to color the neighbourhood of the vertex with largest degree, thereby legally coloring at least $\sqrt{n}$ vertices. If $\Delta \leqslant \sqrt{n}$, then the graph can be colored with $\sqrt{n}+1$ colors. Hence the following simple algorithm from Widgerson [73] colors any three-colorable graph with at most $3\sqrt{n}$ colors.

It turns out that the semidefinite program, underlying the ϑ function, can be used to improve Widgerson's algorithm. In fact, $\vartheta(G)$ is one of the ingredients for the currently strongest algorithms for three-colorable graphs.

Widgerson Algorithm for three-colorable graphs

Input: A graph G on n vertices with $\chi(G) = 3$.
Output: A coloring of G with at most $O(\sqrt{n})$ colors.

(1) While $\exists v \in V(G)$ with degree $\geqslant \sqrt{n}$
Color v with color 1 and $N(v)$ with (at most) two new colors.
Remove v and $N(v)$ from G and call the remaining graph G.
(2) Color G with at most $\Delta < \sqrt{n}$ colors.

Here is a first analysis. Suppose the optimal solution $V = (v_1, \ldots, v_n)$ of (18.26) has been computed. Therefore $\|v_i\| = 1$ and $v_i^T v_j = \lambda \leqslant -\frac{1}{2}$ for $ij \in E(G)$, because $\lambda \leqslant -\frac{1}{\chi(G)-1}$. We associate the vectors v_i to the vertices of G, and use them to define the following vertex partition of $V(G)$.

We first select t random hyperplanes through the origin (where t will be specified later). These hyperplanes partition $\mathbb{R}^n$ into 2^t regions and each v_i belongs to exactly one region. All vertices i with v_i in the same region R_j now are assigned to partition block j.

The main task now is to show that an appropriate choice of t will ensure that with high probability at least $\frac{n}{2}$ vertices in this partition are legally colored (= adjacent vertices are in different partition blocks).

First note that the probability that two vertices i and j are in the same region is equal to the probability that none of the t hyperplanes separates v_i and v_j, which is equal to

$$\big(1 - \frac{1}{\pi}\arccos(v_i^T v_j)\big)^t,$$

see (18.38). We have just seen that $ij \in E$ implies $v_i^T v_j = \lambda \leqslant -\frac{1}{2} = \cos(\frac{2\pi}{3})$. Therefore $\arccos(v_i^T v_j) \geqslant \frac{2\pi}{3}$, so

$$\big(1 - \frac{1}{\pi}\arccos(v_i^T v_j)\big)^t \leqslant \frac{1}{3^t}.$$

The expected value of edges with both endpoints in the same partition block (= monochromatic edges) is at most

$$|E|\frac{1}{3^t} \leqslant \frac{n\Delta}{2}3^{-t}.$$

Markov's inequality now tells us that with probability less than $\frac{1}{2}$ there are more than $n\Delta 3^{-t}$ monochromatic edges. After repeating this process several times, we have, with high probability, a coloring with 2^t colors, where at most $n\Delta 3^{-t}$ edges are monochromatic. Selecting $t = 2 + \log_3(\Delta)$ gives at most $\frac{n}{4}$ monochromatic edges. Therefore at most $\frac{n}{2}$ vertices are not colored legally. In other words, with high probability at least $\frac{n}{2}$ vertices are legally colored with at most $2^t \leqslant 8\Delta^{\log_3(2)}$ colors.

As $\log_3(2) \approx 0.631$ and Δ can be $O(n)$, this technique by itself does not improve Widgerson's result. In [39] it is suggested to run the while-loop of Widgerson's algorithm as long as $\Delta > n^{0.613}$. This uses up $O(n^{0.387})$ colors. Having $\Delta < O(n^{0.613})$, now $O((n^{0.613})^{\log_3(2)}) = O(n^{0.387})$ colors suffice for coloring the remaining graph. This gives, with high probability, a coloring using at most $O(n^{0.387})$ colors.

Karger, Motwani and Sudan provide another more refined analysis, which shows that in fact $\tilde{O}(n^{0.25})$ colors suffice. We use the $\tilde{O}$ notation to suppress polylogarithmic terms. We sketch this analysis, but follow the simplified presentation from [2]. The first observation is that if we can find a stable set of size $O(\frac{n}{s(n)})$ for some function $s(n)$, then we can use up one color for this stable set, and iterate at most $s(n)$ times to get a coloring using at most $\tilde{O}(s(n))$ colors. In [39] it is shown that a stable set of size $\tilde{O}(\frac{n}{\Delta^{1/3}})$ can be found by hyperplane rounding, see below. This is used as follows. Apply the while loop of Widgerson's algorithm as long as $\Delta \geqslant n^{0.75}$. This uses up $O(n^{0.25})$ colors and leaves a graph with $\Delta < n^{0.75}$. Now we apply the above stable set argument with $\Delta^{1/3} = O(n^{0.25})$ to find a coloring in the remaining graph using $\tilde{O}(n^{0.25})$ colors. In summary, we obtain a coloring using $\tilde{O}(n^{0.25})$ colors, once we can find stable sets of size $O(\frac{n}{\Delta^{1/3}})$.

We start with a solution of (18.26) of value $\lambda = -\frac{1}{2}$, which exists because $\chi(G) = 3$. We consider for a random vector r and $\varepsilon > 0$ to be specified later the set

$$V_r(\varepsilon) := \{i \in V : \langle r, v_i \rangle \geqslant \varepsilon\}.$$

We assume that the entries r_j of r are drawn independently from the standard normal distribution. This implies that for any unit vector v, $v^T r$ also has the standard normal distribution. Therefore

$$\text{Prob}(i \in V_r(\varepsilon)) = \int_{\varepsilon}^{\infty} \frac{1}{\sqrt{2\pi}} e^{-\frac{t^2}{2}} dt =: N(\varepsilon).$$

Let $I \subseteq V_r(\varepsilon)$ be the set of isolated vertices in $V_r(\varepsilon)$. Next note that

$$\text{Prob}(i \notin I | i \in V_r(\varepsilon)) = \text{Prob}(\exists j \in N(i) : \langle r, v_j \rangle \geqslant \varepsilon | \langle r, v_i \rangle \geqslant \varepsilon).$$

Having $j \in N(i)$ means that $\langle v_i, v_j \rangle = -\frac{1}{2}$. Therefore we can introduce $\tilde{v}_j \perp v_i$ such that

$$v_j = -\frac{1}{2} v_i + \frac{\sqrt{3}}{2} \tilde{v}_j.$$

Therefore $\tilde{v}_j = \frac{2}{\sqrt{3}} v_j + \frac{1}{\sqrt{3}} v_i$. If $\langle r, v_j \rangle \geqslant \varepsilon$ and $\langle r, v_i \rangle \geqslant \varepsilon$, then $\langle r, \tilde{v}_j \rangle \geqslant \sqrt{3}\varepsilon$. Therefore

$$\begin{aligned}
\text{Prob}(i \notin I \mid i \in V_r(\varepsilon)) &\leqslant \text{Prob}(\exists j \in N(i) : \langle r, \tilde{v}_j \rangle \geqslant \sqrt{3}\varepsilon) \\
&\leqslant \sum_{j \in N(i)} \text{Prob}(\langle r, \tilde{v}_j \rangle \geqslant \sqrt{3}\varepsilon) \\
&\leqslant \Delta N(\sqrt{3}\varepsilon).
\end{aligned}$$

Now $\varepsilon > 0$ is selected such that

$$N(\sqrt{3}\varepsilon) < \frac{1}{2\Delta}. \tag{18.40}$$

This implies that with probability less than $\frac{1}{2}$, that a vertex is not in I, thus $\mathrm{Prob}(i \in I) \geqslant \frac{1}{2}\mathrm{Prob}(i \in V_r(\varepsilon)) = \frac{1}{2}N(\varepsilon)$. Hence the expected cardinality of the stable set I is $\frac{n}{2}N(\varepsilon)$. To get the final result, one needs to see how the choice of ε in (18.40) relates to Δ and $N(\varepsilon)$. It can be shown that Δ is less than $\tilde{\Theta}(N(\varepsilon)^3)$, see [2], hence $\frac{n}{2}N(\varepsilon) = \tilde{\Omega}(\Delta^{-1/3}n)$.

In [2], a further refinement of the analysis is proposed, which together with a strengthened version of (18.26) gives an approximation of $\tilde{O}(n^{0.2111})$ colorings for three-colorable graphs, see also the recent dissertation [9].

18.5 Solving SDP in practice

We have just seen the great variety of modeling ideas leading to SDP. Moreover, some of these models even lead to relaxations where the approximation error can be determined a priori. It is therefore natural to ask for efficient methods to solve SDP. In this section we are going to describe the algorithmic machinery currently available for SDP.

18.5.1 Interior point algorithms

The most elegant way to solve SDP, and more generally linear optimization over closed convex cones, is based on Newton's method applied to a slightly modified version of the problem. The theoretical background for this approach goes back to the seminal work of Nesterov and Nemirovski from the late 1980's, see [61]. They showed that a family of convex optimization problems can be solved in polynomial time using self-concordant barrier functions.

Practical experience indicates that Newton's method in combination with the barrier idea works most efficiently in the primal-dual path-following setting, which will be briefly explained now. We recall the primal problem (18.2) and its dual (18.5),

$$\min\langle C, X\rangle \text{ such that } A(X) = b,\ X \succeq 0,$$

$$\max b^T y \text{ such that } Z = C - A^T(y) \succeq 0.$$

In the following we assume that both problems satisfy the Slater constraint qualification (18.6). In this case (X, y, Z) is optimal if and only if (18.7) holds. This is a system of $m + \binom{n+1}{2} + 1$ equations with the additional semidefiniteness constraints in $m + 2\binom{n+1}{2}$ variables. It is perhaps surprising that the additional semidefinitenes conditions in fact lead to an overdetermined system. Indeed note that $X \succeq 0, Z \succeq 0$ implies $X = UU^T, Z = VV^T$ for U and V of appropriate size. But then

$$0 = \langle Z, X \rangle = \langle VV^T, UU^T \rangle = \|V^T U\|^2$$

implies $V^T U = 0$ and therefore

$$ZX = VV^T UU^T = 0.$$

Thus the scalar equation

$$\langle Z, X \rangle = 0$$

together with $X \succeq 0, Z \succeq 0$ implies the matrix equation

$$ZX = 0.$$

Since ZX need not be symmetric, even if X and Z are symmetric, this equation has n^2 components.

Primal-dual path-following interior-point methods are based on solutions $X \succeq 0$, $Z \succeq 0$, of the following slightly modified optimality conditions for parameter $\mu > 0$.

$$A(X) - b = 0, \ \ C - Z - A^T(y) = 0, \ \ ZX - \mu I = 0. \tag{18.41}$$

Clearly, any solution of this system must satisfy $X \succ 0, Z \succ 0$, as $ZX = \mu I$ forces X and Z to be non-singular. It turns out that strict feasibility in fact characterizes unique solvability of (18.41).

Theorem 18.11. *(see for e.g., Theorem 10.2.1 in [74]) The following statements are equivalent:*

(i) (18.2) and (18.5) both have strictly feasible points.
(ii) (18.41) has a unique solution (X_μ, y_μ, Z_μ) for all $\mu > 0$.

The proof uses the following observation. Let $f : \mathscr{S}_n^{++} \mapsto \mathbb{R}$, $f(X) = \log\det X$. This function is strictly concave with $\nabla f = X^{-1}$,

$$f(X + h) = f(X) + \langle X^{-1}, h \rangle + o(\|h\|).$$

Consider the auxiliary problem, for fixed parameter $\mu > 0$.

$$(PB) \ \ \min \langle C, X \rangle - \mu f(X) \text{ such that } A(X) = b, \ X \succ 0.$$

This is a convex optimization problem with Lagrangian

$$L(X, y) = \langle C, X \rangle - \mu \log\det X + y^T(b - A(X)).$$

The necessary and sufficient optimality conditions therefore are

$$\begin{aligned} \nabla_X L &= C - \mu X^{-1} - A^T(y) = 0, \\ \nabla_y L &= b - A(X) = 0, \end{aligned}$$

together with the open set constraint $X \succ 0$. Setting $Z = \mu X^{-1}$, we recover (18.41). Problem (PB) is sometimes called the *primal barrier problem*. The term $-\mu \log\det X$ goes to infinity as $X \succ 0$ approaches the boundary of the cone of semidefinite matrices. The value of μ controls the influence of this term. The conditions (18.41) can also be derived by setting up a barrier problem for the dual SDP. Strict convexity of the cost function from the primal barrier problem shows that a minimizer, if it exists, must be unique. To show the existence of a minimizer requires an additional compactness argument. We refer to [74] for further details.

So strict feasibility of (18.2) and (18.5) insures unique solutions (X_μ, y_μ, Z_μ) for all $\mu > 0$. To show that these solutions form a smooth curve, parametrized by μ, we need to show that the system is differentiable with a non-singular Jacobian. While differentiability is obvious, the Jacobian of (18.41) is certainly not invertible, as it is not even square. The remedy is to replace the non-symmetric equation $ZX - \mu I = 0$ by an equivalent symmetric one. A general setup for this goes as follows. Let P be non-singular and define

$$H_P(M) := \frac{1}{2}(PMP^{-1} + (PMP^{-1})^T).$$

Replacing $ZX - \mu I = 0$ by $H_P(ZX) - \mu I = 0$ in (18.41) makes the Jacobian a square matrix, which can be shown to be non-singular for any invertible P and for all points (X, y, Z) "close" to solutions (X_μ, y_μ, Z_μ) of (18.41). Hence the solution set of (18.41) indeed defines a smooth curve which is often called *(primal-dual) central path*. Let us denote it by $CP := \{P_\mu = (X_\mu, y_\mu, Z_\mu) : \mu > 0\}$.

Primal-dual interior-point path-following methods use the Newton method to follow the primal-dual central path, maintaining $X \succ 0,\ Z \succ 0$ (interior points) for $\mu \to 0$. To follow the central path (at least approximately), we first need to come close to it. We do this iteratively as follows. Having the current interior iterate (X, y, Z) and a target value for μ, we try to reach a new interior point (X^+, y^+, Z^+) close to P_μ for given μ. Then the target value μ is reduced and we iterate. The main work is done in determining the search direction $(\Delta X, \Delta y, \Delta Z)$ which moves us from the current iterate to the new point. As mentioned before, there is a great variety to do so and here we show a very simple, popular and efficient variant. It was one of the first search directions proposed to solve SDP, see [34, 43, 57]. We linearize the equations in (18.41) and get

$$A(\Delta X) = b - A(X) := r_p, \tag{18.42}$$

$$\Delta Z + A^T(\Delta y) = C - A^T(y) - Z := r_d, \tag{18.43}$$

$$Z\Delta X + \Delta Z X = \mu I - ZX. \tag{18.44}$$

The vectors r_p and r_d express primal and dual infeasibility and would be 0 if feasible starting points were used. The last equation can be used to eliminate ΔX,

$$\Delta X = \mu Z^{-1} - X - Z^{-1}\Delta Z X,$$

the second one gives

$$\Delta Z = r_d - A^T(\Delta y).$$

Substitution into the first equation gives

$$A(Z^{-1}A^T(\Delta y)X) = v \text{ with } v = r_p - A(\mu Z^{-1} - X) + A(Z^{-1}r_d X).$$

The linear operator on the left hand side has the matrix representation

$$M\Delta y = v, \tag{18.45}$$

with $m_{ij} = \text{tr}(A_i Z^{-1} A_j X)$. This matrix can be shown to be positive definite (if $X \succ 0$, $Z \succ 0$ and the equations $A(X) = b$ are linearly independent). To determine the search direction we therefore need to solve the equation $M\Delta y = v$. Having Δy we get ΔZ and ΔX from backsubstitution. ΔX need not be symmetric, so the symmetric part of ΔX is taken.

This would give the new point $X + \Delta X, y + \Delta y, Z + \Delta Z$ except that the new point need not have definite matrices. This is repaired by a backtracking strategy, where starting with $t = 1$, t is reduced to a value $t^* > 0$ such that $X^+ = X + t^*\Delta X \succ 0$, $Z^+ = Z + t^*\Delta Z \succ 0$. This gives a new trial point (X^+, y^+, Z^+). The new target parameter for this point can be estimated to be

$$\mu^+ = \frac{1}{n}\text{tr}(X^+Z^+),$$

and a new iteration is started with μ^+ reduced by a multiplicative factor.

It was shown by Monteiro [57] that this rather simple and pragmatic approach in fact falls in the general class of search directions mentioned above.

To give some idea on the actual work involved, we consider the basic semidefinite relaxation for Max-Cut. Here $A(X) = \text{diag}(X)$ and $A^T(y) = \text{Diag}(y)$. It is not too hard to verify that in this case the matrix M in (18.45) has the simple form

$$M = X \circ Z^{-1}.$$

Hence the work in each iteration involves computing Z^{-1}, solving equation (18.45) and doing a few matrix multiplications to get ΔX. An efficient way to do the backtracking to stay inside the cone of semidefinite matrices consists in checking whether the Cholesky decomposition of $X + t\Delta X$ terminates successfully. This is a certificate that the matrix is semidefinite, hence an additional small reduction of t insures definiteness. To give some practical impression we tabulate computation times to solve this relaxation for some representative values of n. The iterations were stopped, once the relative error was below 10^{-7}, and it always took less than 20 iterations to reach this level of accuracy. The results clearly show that interior-point methods are indeed very efficient on smaller problems ($n \leqslant 1000$), but become prohibitive with respect to both time and space requirements, once n gets large.

We also tabulate timings to compute the ϑ-function in the computationally most expensive case of $m = \frac{1}{2}\binom{n}{2}$ equations, see Table 18.2. It is clear that once $n \approx 200$, the effort of interior point methods gets prohibitive.

Looking at the computation times in these two tables, it should be clear that interior-point methods become impractical, once m is substantially larger than say 5000, or once n is larger than about 1000. In the following sections we will consider algorithmic alternatives for larger problems.

n	time (secs.)
500	12
1000	75
1500	237
2000	586
2500	1109
3000	1900

Table 18.1 Interior-point computation times to solve (18.14) with relative accuracy 10^{-7}. Here $m = n$.

n	m	time (secs.)
100	2475	57
120	3570	161
140	4865	353
160	6360	757
180	8055	1520
200	9950	2817

Table 18.2 Interior-point computation times to solve (18.18) with relative accuracy 10^{-7}, $m = \frac{1}{2}\binom{n}{2}$.

18.5.2 Partial Lagrangian and the bundle method

We have just seen that interior-point methods are the method of choice to solve SDP and we also saw the limits of this approach both in terms of the dimension n of the matrix space and also in the number m of (primal) constraints. We also saw the need to be able to handle large-scale SDP to get good approximations of the underlying combinatorial optimization problem.

We are now going to describe a rather generic “work-around” for problems, where the matrix dimension is reasonable, but m can be arbitrary. Let us consider an SDP of the form

$$z = \max\{\langle C,X\rangle : A(X) = a,\ B(X) = b,\ X \succeq 0\}.$$

We have split the constraints into two sets. The motivation is that maintaining only the first set $A(X) = a$ would result in an SDP that is still manageable by interior point methods, but the inclusion of $B(X) = b$ makes the problem impractical for interior-point methods. For simplicity of exposition, we consider only equations. The presence of inequalities only leads to minor modifications (sign constraints on dual multipliers), which can be dealt with in the approach to be described.

Let us denote $\mathscr{X} := \{X : A(X) = a,\ X \succeq 0\}$. The idea now is to maintain $X \in \mathscr{X}$ explicitly and to put $B(X) = b$ into the cost function by taking the partial Lagrangian dual. Hence we get

$$z = \max\{\langle C,X\rangle : A(X) = a,\ B(X) = b,\ X \succeq 0\} = \max_{X\in\mathscr{X}} \min_{y} L(X,y),$$

where the partial Lagrangian L is given by $L(X,y) = \langle C,X\rangle + y^T(b - B(X))$. Applying the Minimax inequality we get under the usual strict feasibility conditions that

$$z = \min_{y} f(y) \leqslant f(y)\ \forall y,$$

where $f(y) = \max_{X\in\mathscr{X}} L(X,y)$. For evaluating f an SDP over $X \in \mathscr{X}$ has to be solved which we assume to be manageable. Suppose that for some y^*, we have $f(y^*) = L(X^*,y^*)$, so the maximum is attained at $X^* \in \mathscr{X}$. By setting

$$g^* = b - B(X^*)$$

we get for any y

$$f(y) \geqslant L(X^*,y) = f(y^*) + \langle g^*, y - y^*\rangle. \tag{18.46}$$

The inequality follows from the definition of f, the equality comes from substituting g^*. In the language of convex analysis, this inequality defines g^* to be a *subgradient of f* at y^*.

To compute z, we minimize f. This function is continuous and convex (pointwise maximum of linear functions), but it is not differentiable at points where the maximum is not unique. Hence we use some tools from non-smooth optimization to minimize f.

The *bundle method* will serve our purposes. It was introduced in the 1970's by Lemarechal, see [48, 49]. A comprehensive survey is also contained in [35]. We briefly explain its key features. The method iteratively approaches a minimizer of f. Let the current iterate be $\hat{y}$. Suppose we have evaluated f at $k \geqslant 1$ points $y_1,\dots,y_k$ with respective optimizers $X_1,\dots,X_k$ and subgradients $g_i = b - B(X_i)$. We also set $f_i = f(y_i)$. It is assumed that $\hat{y} \in \{y_1,\dots,y_k\}$. To get started we evaluate f at $\hat{y} = 0$ and set $y_1 = \hat{y}, k = 1$.

The subgradient inequality (18.46) implies that for all y

$$f(y) \geqslant \max_{i}\{f_i + g_i^T(y - y_i)\} =: \tilde{f}(y).$$

For simplification we set $h_i = f_i - g_i^T y_i$, $H = (h_1, \ldots, h_k)^T$ and $G = (g_1, \ldots, g_k)$. Then

$$\tilde{f}(y) = \max_{\lambda \in \Delta_k} \lambda^T (H + G^T y).$$

The bundle method now uses the minorant $\tilde{f}(y)$ as an approximation of the original f "close" to the current iterate $\hat{y}$. This is plausible because $\hat{y} \in \{y_1, \ldots, y_k\}$ implies $\tilde{f}(\hat{y}) = f(\hat{y})$. To insure that we stay close to $\hat{y}$, we add a regularization term and consider the following function

$$f_{bdl}(y) := \tilde{f}(y) + \frac{1}{2t} \|y - \hat{y}\|^2.$$

The bundle method minimizes this function over y to obtain a new iterate. The parameter $t > 0$ controls how close we stay to $\hat{y}$. Minimizing $f_{bdl}(y)$ is again a Min-Max problem which can be simplified as follows.

$$\min_y f_{bdl}(y) = \max_{\lambda \in \Delta} \min_y \lambda^T (H + G^T y) + \frac{1}{2t} \|y - \hat{y}\|^2.$$

The inner minimization is a strictly convex unconstrained quadratic optimization problem, hence we can replace the minimization by asking that the first order optimality conditions hold. Setting the derivative with respect to y equal to 0 gives

$$y = \hat{y} - tG\lambda.$$

After substitution, we get the equivalent problem

$$\max_{\lambda \in \Delta} \lambda^T (H + G^T \hat{y}) - \frac{t}{2} \|G\lambda\|^2.$$

This is a convex quadratic problem over the standard simplex Δ in $\mathbb{R}^k$ and can be solved by standard methods from convex optimization. The solution effort depends on k, which can be controlled by the user. Having the optimizer λ^* of this problem, we get the new estimate

$$y^{new} = \hat{y} - tG\lambda^*.$$

The bundle method now asks to evaluate the original function f at y^{new}. Some standard criteria are used to decide whether y^{new} becomes the new trial point, or whether we stay at $\hat{y}$. In any case, information from the new point is included as $y_{k+1} = y_{new}$ and the iteration is continued. The convergence theory of the bundle method is quite elaborate and can be found e.g., in [35].

Remark 18.3. Let us take a careful look at the above derivation of the bundle method. The specific form of the function $f(y)$ is in fact irrelevant, once the following property of f is satisfied. For any y^* we can determine a vector g^* such that

$$f(y) \geqslant f(y^*) + \langle g^*, y - y^* \rangle \;\; \forall y$$

holds. In words, we assume that f is convex and we are able to determine a subgradient of f at any point.

We close with an application of the bundle method applied to the semidefinite relaxation of Max-Cut which also includes the triangle inequalities, see (18.30). We maintain the equation $\operatorname{diag}(X) = e$ explicitly and dualize the triangle inequalities, which we formally denote by $B(X) \leqslant b$. Recall that there are $4\binom{n}{3}$ inequality constraints. Let us denote the violation of these contraints by

$$r := \min\{0, b - B(X)\}.$$

In Table 18.3 we consider the instance g3s of size $n = 300$ from [17]. This is a random graph with edge density of 10% and edge weights 1 and -1. The optimal value of the relaxation, as reported in [17] is 635.05. In the table we also provide information about the error r. We include the total violation of all constraints $\|r\|_1$, the maximal violation $\|r\|_\infty$ and the total number of violated constraints (last column). The results in this table clearly indicate that the bundle method is very efficient in getting close to the optimum quickly. The computation time for 100 bundle iterations was a few minutes only. The local convergence behaviour is obviously much weaker than in the case of interior-point methods. We refer to [17] for further details of this approach applied to (18.30).

iter	$f(y)$	$\|r\|_1$	$\|r\|_\infty$	contrs. viol.
1	679.3	152541.0	0.96	680822
10	660.4	21132.7	0.73	147094
20	648.1	1234.6	0.52	13605
30	642.2	193.7	0.32	2979
40	639.5	50.8	0.32	957
50	638.2	29.5	0.25	647
60	637.6	25.3	0.26	570
70	637.1	28.9	0.20	688
80	636.9	17.1	0.23	397
90	636.6	18.2	0.18	448
100	636.5	13.5	0.18	369

Table 18.3 The semidefinite relaxation of Max-Cut from (18.30) for a graph with $n = 300$. The vector r contains the violation of the triangle inequalities. The last column provides the number of violated constraints.

18.5.3 The spectral bundle method

The spectral bundle method introduced in [33] reduces the solution of SDP to the computation of the largest eigenvalue of a sequence of symmetric matrices.

The algorithmic machinery from numerical linear algebra provides methods to compute $\lambda_{\max}(C)$, which do not require to have C explicitly available, but only need a subroutine that evaluates the action of C. In other words, given x, we only need to be able to compute $y = Cx$.

Before describing the spectral bundle method in detail we first show that SDP with the constant trace property can equivalently be reformulated as an eigenvalue optimization problem.

The mapping A from (18.2) satisfies the *constant trace property* if the identity matrix is in the range of A^T, which means $\exists\eta$ such that $A^T(\eta) = I$. In this case any X such that $A(X) = b$, $X \succeq 0$ satisfies

$$tr(X) = \langle X, A^T(\eta)\rangle = \langle A(X), \eta\rangle = b^T\eta = a,$$

for some constant $a \geqslant 0$. The constant trace property therefore implies that feasible solutions of (18.2) have constant trace, equal to a. Excluding the case $a = 0$, which only has the zero matrix as feasible solution of SDP, we can assume without loss of generality that $a = 1$.

Let us consider SDP with the constant trace property. In this case we can add the redundant equation

$$\mathrm{tr}(X) = 1$$

to (18.2) and get for the dual, with multiplier λ for the new constraint:

$$\min\{b^T y + \lambda : \ A^T(y) + \lambda I - C = Z \succeq 0\}.$$

The optimality condition $ZX = 0$ together with $\mathrm{tr}(X) = 1$ implies that any optimal Z is singular. Hence, at the optimum we have

$$0 = \lambda_{\min}(Z) = \lambda_{\max}(-Z) = \lambda_{\max}(C - A^T(y)) - \lambda.$$

We conclude that the multiplier λ satisfies $\lambda = \lambda_{\max}(C - A^T(y))$. Substitution gives the following function $f(y) = b^T y + \lambda_{\max}(C - A^T(y))$. Solving the dual is therefore equivalent to the eigenvalue optimization problem

$$\min_y f(y).$$

The condition $Z \succeq 0$ is hidden in $\lambda_{\min}(Z) = 0$, which moves $\lambda_{\max}(C - A^T(y))$ into the cost function. It can easily be shown that f is convex. The cost function is smooth but not differentiable in case the largest eigenvalue has multiplicity larger than one. Suppose now that x^* is a unit-norm eigenvector to $\lambda_{\max}(C - A^T(y^*))$. Then, see (18.31),

$$\lambda_{\max}(C - A^T(y)) \geqslant \langle x^*, (C - A^T(y))x^*\rangle \ \forall y.$$

Let us define

$$g^* := b - A(x^* x^{*T}).$$

Then the above inequality shows that

$$f(y) \geqslant f(y^*) + \langle g^*, y - y^* \rangle \ \forall y.$$

Hence g^* is subgradient of f at y^*. Moreover, if the multiplicity of $\lambda_{\max}(C - A^T(y^*))$ is one, then x^* is unique up to multiplication by -1, hence g^* is unique as well, and $\nabla f(y^*) = g^*$. In view of all this, a first idea would be to use again the bundle method to minimize f. Indeed, Schramm and Zowe [67] apply it to compute $\vartheta(G)$. We will now see that we can in fact do better by exploiting the special form of the objective function. We recall from (18.32) that

$$\lambda_{\max}(A) = \max\{\langle A, W \rangle : \ \mathrm{tr}(W) = 1, \ W \succeq 0\}.$$

After substitution, we get the following min-max problem for the dual SDP.

$$\min_{y} \max_{\mathrm{tr}(W)=1, W \succeq 0} b^T y + \langle C - A^T(y), W \rangle \tag{18.47}$$

In the *spectral bundle method*, this problem is solved iteratively. We observe that evaluating f at $\hat{y}$ amounts to compute $\lambda_{\max}(C - A^T(\hat{y}))$ together with an eigenvector v. Having a current iterate $\hat{y}$, the following modifications are made in (18.47). First, the maximization is simplified by constraining W to be of the form

$$W = PVP^T$$

for given $n \times k$ matrix P such that $P^T P = I_k$. The idea is that P should contain "local" information of f around $\hat{y}$. In particular, we assume that the eigenvector v is contained in P. The new variable now is $V \in \mathscr{S}_k^+$. Since

$$\lambda_{\max}(C) = \max_{W \succeq 0, \ \mathrm{tr}(W)=1} \langle C, W \rangle \geqslant \max_{W \succeq 0, \ \mathrm{tr}(W)=1, \ W = PVP^T} \langle C, W \rangle = \lambda_{\max}(P^T C P),$$

we get the follwowing minorant

$$\tilde{f}(y) = \max_{V \succeq 0, \ \mathrm{tr}(V)=1} b^T y + \langle C - A^T(y), PVP^T \rangle \leqslant f(y)$$

of f. The inclusion of v in P insures that $f(\hat{y}) = \tilde{f}(\hat{y})$. To insure that the next iterate stays close to $\hat{y}$, we consider the following replacement of (18.47) for fixed parameter $t > 0$.

$$\min_{y} \max_{V \succeq 0, \ \mathrm{tr}(V)=1} b^T y + \langle C - A^T(y), PVP^T \rangle + \frac{1}{2t} \|y - \hat{y}\|^2.$$

Note the similarity to the standard bundle method from before. In the spectral bundle method, this min-max problem is solved to get the next trial point y. As before we exchange min and max and exploit the fact that the minimization with respect to y is again an unconstrained strictly convex quadratic problem. Therefore y is minimizer if and only if the partial derivative with respect to y is zero. This results in

$$y = \hat{y} - t(b - A(PVP^T)).$$

We now substitute this for y and get the quadratic SDP

$$\max_{V \succeq 0,\ \mathrm{tr}(V)=1} b^T \hat{y} + \langle C - A^T(\hat{y}), PVP^T \rangle - \frac{t}{2} \|b - A(PVP^T)\|^2.$$

This problem has just one scalar equation and can be solved by interior-point methods to get the optimal $V^* \in \mathscr{S}_k^+$. The new trial point $y^{new} = \hat{y} - t(b - A(PV^*P^T))$ is now used to compute the function value f, together with an eigenvector to $\lambda_{\max}$. We follow the usual bundle concept to decide whether or not y^{new} becomes the new trial point. In any case the matrix P is updated and a new iteration can be started. Helmberg and Rendl [33] explain in detail how the above quadratic SDP can be solved. Various update strategies for P are discussed and an elementary convergence analysis is given. Helmberg [28] describes implementation issues and presents computational results on a wide variety of SDP. Refinements of the spectral bundle method are given in [29] and [30].

Remark 18.4. The similarities of the spectral to the standard bundle method are quite obvious. In fact, constraining V to be a diagonal matrix (with diagonal entries λ_i) simplifies the above SDP to optimizing over $\lambda \in \Delta$, and we recover the standard bundle method in this case.

18.6 SDP and beyond

18.6.1 Copositive and completely positive matrices

In this section we will see that besides the cone of semidefinite matrices, there are several other cones in the space of symmetric matrices which have a close connection to integer programming. Let us define

$$\mathscr{C}^* := \{X \in \mathscr{S}_n : X = VV^T \text{ with } n \times k \text{ matrix } V \geqslant 0\} = \mathrm{conv}\{vv^T : v \in \mathbb{R}^n,\ v \geqslant 0\}.$$

Matrices in $\mathscr{C}^*$ are often called *completely positive*. The cone $\mathscr{C}^*$ has a dual, which we denote by $\mathscr{C}$ and which by definition is given as follows.

$$Y \in \mathscr{C} \iff \langle Y, X \rangle \geqslant 0\ \forall X \in \mathscr{C}^*.$$

This obviously holds if and only if

$$v^T Y v \geqslant 0\ \forall v \geqslant 0. \tag{18.48}$$

Matrices in this cone are usually called *copositive*. While $X \in \mathscr{S}^+$ has an efficient certificate, given e.g., through the Cholesky decomposition of X, it is NP-hard to decide whether $X \notin \mathscr{C}$, see [59].

We call problems of the form

$$\inf\{\langle C,X\rangle : A(X)=b,\ X\in\mathscr{C}\} \text{ and}$$

$$\inf\{\langle C,X\rangle : A(X)=b,\ X\in\mathscr{C}^*\}$$

copositive programs because either the problem or its dual involves optimization over copositive matrices.

18.6.2 Copositive relaxations

To see that copositive programs have some relevance in connection with integer programs we recall the following theorem from Motzkin and Strauss.

Theorem 18.12 ([58]). *Let A be the adjacency matrix of a graph. Then*

$$\frac{1}{\alpha(G)} = \min\{x^T(A+I)x : \ x\in\Delta\}.$$

Starting from this fact, it is not hard to show the following result, which was pointed out by De Klerk and Pasechnik, see [11].

Theorem 18.13. *Let A be the adjacency matrix of a graph. Then*

$$\alpha(G) = \max\{\langle J,X\rangle : \ \langle A+I,X\rangle = 1,\ X\in\mathscr{C}^*\} = \min\{\lambda : \ \lambda(A+I)-J\in\mathscr{C}\}.$$

Proof. Let S be a stable set of maximum cardinality $\alpha(G)$ with characteristic vector $\xi\in\{0,1\}^n$. Then $\frac{1}{\alpha}\xi\xi^T$ is feasible for the maximization problem and we get the first inequality in

$$\alpha \leqslant \sup\{\langle J,X\rangle : \ \langle A+I,X\rangle = 1,\ X\in\mathscr{C}^*\} \leqslant \inf\{\lambda : \ \lambda(A+I)-J\in\mathscr{C}\}.$$

Weak duality for conic programs implies the second inequality.

The Motzkin-Strauss theorem shows that

$$0 = \min\{x^T(A+I-\frac{1}{\alpha}ee^T)x : x\in\Delta\} = \min\{x^T(\alpha(A+I)-J)x : x\geqslant 0\}.$$

The second minimization being zero is the defining condition for $\alpha(A+I)-J$ to be in $\mathscr{C}$, see (18.48). Therefore the infimum above is at most α, but weak duality states that it is also at least α, hence there is equality throughout, and both the supremum and the infimum are attained (at $\frac{1}{\alpha}\xi\xi^T$ and $\lambda=\alpha$ respectively). □

DeKlerk and Pasechnik provide a proof for this result which is independent of the Motzkin-Strauss theorem. Let us put this result into perspective by recalling $\vartheta(G)$.

$$\vartheta(G) = \max\{\langle J,X\rangle : \ A_G(X)=0,\ \mathrm{tr}(X)=1,\ X\succeq 0\}.$$

Suppose now that we include the additional constraint $X\geqslant 0$, leading to an improved approximation $\vartheta^+(G)$ of $\alpha(G)$. This improvement was in fact suggested by Schrijver [68] and independently by [56]. The condition $A_G(X) = 0$ together with $X \geqslant 0$ can be simplified to $\langle A, X\rangle = 0$. In other words, the equations $A_G(X) = 0$ are added into just a scalar equation. We get

$$\alpha(G) \leqslant \vartheta^+(G) = \max\{\langle J, X\rangle : \langle A, X\rangle = 0,\ \mathrm{tr}(X) = 1,\ X \succeq 0,\ X \geqslant 0\}. \quad (18.49)$$

The above theorem therefore shows that replacing the cone $\{X : X \succeq 0, X \geqslant 0\}$ by $\mathscr{C}^*$ leaves no gap in (18.49).

This suggests to try a similar idea on the dual (18.25) of $\vartheta(G)$. Looking at the matrix tM with M from (18.21), it is clear that $M \in \mathscr{C}^*$, therefore we get the following improvement of $\vartheta(G)$ towards $\chi(G)$.

$$\chi(G) \geqslant \vartheta^C(G) = \min\{t : tI + A_{\bar{G}}(y) \in \mathscr{C},\ tI + A_{\bar{G}}(y) \succeq J\} \geqslant \vartheta(G).$$

It was recently shown in [15] that the improvement $\vartheta^C(G)$ is in fact equal to the fractional chromatic number $\chi_f(G)$. Another version to model the chromatic number was recently proposed by Gvozdenovic and Laurent, see [21, 22].

These results indicate that the modeling power of copositive programs is stronger than SDP. Burer [8] shows the following general result.

Theorem 18.14. *Let c and a_j be vectors from $\mathbb{R}^n$, $b \in \mathbb{R}^k$, $Q \in \mathscr{S}_n$ and $I \subseteq \{1,\dots,n\}$. The optimal values of the following two problems are equal.*

$$\min\{x^T Qx + c^T x : a_j^T x = b_j,\ x \geqslant 0,\ x_i \in \{0,1\}\ \forall i \in I\},$$

$$\min\{\mathrm{tr}(QX) + c^T x : a_j^T x = b_j, a_j^T X a_j = b_j^2,\ X_{ii} = x_i\ \forall i \in I,\ \begin{pmatrix} 1 & x^T \\ x & X \end{pmatrix} \in \mathscr{C}^*\}.$$

This result shows that it would be extremely interesting to have a better understanding of the cone of completely positive matrices. Outer approximations of $\mathscr{C}^*$, or equivalently, inner approximations of $\mathscr{C}$ would result in relaxations of the underlying optimization problem. First systematic attempts in this direction were proposed by Parrilo [63] and De Klerk and Pasechnik [11] who introduced hierarchies of relaxations based on sum of squares relaxations of polynomials. These relaxations are formulated as SDP of increasing dimension. A summary of this approach, which is far beyond the scope of this article, can be found in [46].

Inner approximations of $\mathscr{C}^*$ can be used as starting points for primal heuristics to combinatorial optimization problems. This is an area open for current research.

References

1. K.M. Anstreicher and H. Wolkowicz, *On Lagrangian relaxation of quadratic matrix constraints*, SIAM Journal on Matrix Analysis 22 (2000) 41–55.
2. S. Arora, E. Chlamtac, and M. Charikar, *New approximation guarantee for chromatic number*, Proceedings of the 38th STOC, Seattle, USA, 2006, pp. 215–224.

3. D. Avis and J. Umemoto, *Stronger linear programming relaxations for max-cut*, Mathematical Programming 97 (2003) 451–469.
4. E. Balas, S. Ceria, and G. Cornuéjols, *A lift-and-project cutting plane algorithm for mixed 0-1 programs*, Mathematical Programming 58 (1993) 295–324.
5. F. Barahona, M. Jünger, and G. Reinelt, *Experiments in quadratic 0-1 programming*, Mathematical Programming 44 (1989) 127–137.
6. A. Ben-Tal and A. Nemirovski, *Lectures on modern convex optimization*, MPS-SIAM Series on Optimization, 2001.
7. S. Berkowitz, *Extrema of elementary symmetric polynomials of the eigenvalues of the matrix* P^*KP+L, Linear Algebra Appl. 8 (1974) 273–280.
8. S. Burer, *On the copositive representation of binary and continuous nonconvex quadratic programs*, Mathematical Programming 120 (2009) 479–495.
9. E. Chlamtac, *Non-local analysis of sdp based approximation algorithms*, Ph.D. thesis, Princeton University, USA, 2009.
10. G.B. Dantzig, D.R. Fulkerson, and S.M. Johnson, *Solution of a large scale traveling salesman problem*, Journal of the Operations Research Society of America 2 (1954) 393–410.
11. E. de Klerk and D.V. Pasechnik, *Approximatin of the stability number of a graph via copositive programming*, SIAM Journal on Optimization 12 (2002) 875–892.
12. M. Deza, V.P. Grishukhin, and M. Laurent, *The hypermetric cone is polyhedral*, Combinatorica 13 (1993) 397–411.
13. W.E. Donath and A.J. Hoffman, *Lower bounds for the partitioning of graphs*, IBM Journal of Research and Developement 17 (1973) 420–425.
14. R.J. Duffin, *Infinite programs*, Ann. Math. Stud. 38 (1956) 157–170.
15. I. Dukanovic and F. Rendl, *Semidefinite programming relaxations for graph coloring and maximal clique problems*, Mathematical Programming 109 (2007) 345–365.
16. D.V. Pasechnik, E. de Klerk, and J.P. Warners, *On approximate graph colouring and MAX-k-CUT algorithms based on the* ϑ*-function*, Journal of Combinatorial Optimization 8 (2004) 267–294.
17. I. Fischer, G. Gruber, F. Rendl, and R. Sotirov, *Computational experience with a bundle method for semidefinite cutten plane relaxations of max-cut and equipartition*, Mathematical Programming 105 (2006) 451–469.
18. A. Frieze and M. Jerrum, *Improved approximation algorithms for MAX k-cut and MAX BISECTION*, Algorithmica 18 (1997) 67–81.
19. M.X. Goemans, *Semidefinite programming in combinatorial optimization*, Mathematical Programming 79 (1997) 143–162.
20. M.X. Goemans and D.P. Williamson, *Improved approximation algorithms for maximum cut and satisfiability problems using semidefinite programming*, Journal of the ACM 42 (1995) 1115–1145.
21. N. Gvozdenović and M. Laurent, *Computing semidefinite programming lower bounds for the (fractional) chromatic number via block-diagonalization*, SIAM Journal on Optimization 19 (2008) 592–615.
22. N. Gvozdenović and M. Laurent, *The operator* Ψ *for the chromatic number of a graph*, SIAM Journal on Optimization 19 (2008) 572–591.
23. S.W. Hadley, F. Rendl, and H. Wolkowicz, *A new lower bound via projection for the quadratic assignment problem*, Mathematics of Operations Research 17 (1992) 727–739.
24. E. Halperin and U. Zwick, *A unified framework for obtaining improved approximation algorithms for maximum graph bisection problems*, Lecture notes in Computer Science 2081, IPCO 2001, Springer Berlin, 2001, pp. 210–225.
25. P.L. Hammer, *Some network flow problems solved with pseudo-Boolean programming*, Operations Research 13 (1965) 388–399.
26. C. Helmberg, *Fixing variables in semidefinite relaxations*, SIAM J. Matrix Anal. Appl. 21 (2000) 952–969.
27. C. Helmberg, *Semidefinite programming*, European Journal of Operational Research 137 (2002) 461–482.

28. C. Helmberg, *Numerical validation of SBmethod*, Mathematical Programming 95 (2003) 381–406.
29. C. Helmberg, K.C. Kiwiel, and F. Rendl, *Incorporating inequality constraints in the spectral bundle method*, Integer Programming and combinatorial optimization (E.A. Boyd R.E. Bixby and R.Z. Rios-Mercado, eds.), Springer Lecture Notes in Computer Science 1412, 1998, pp. 423–435.
30. C. Helmberg and F. Oustry, *Bundle methods to minimize the maximum eigenvalue function*, Handbook of semidefinite programming: theory, algorithms and applications (R. Saigal H. Wolkowicz and L. Vandenberghe, eds.), Kluwer, 2000, pp. 307–337.
31. C. Helmberg, S. Poljak, F. Rendl, and H. Wolkowicz, *Combining semidefinite and polyhedral relaxations for integer programs*, Integer Programming and combinatorial optimization (E. Balas and J. Clausen, eds.), Springer Lecture Notes in Computer Science 920, 1995, pp. 124–134.
32. C. Helmberg and F. Rendl, *Solving quadratic (0,1)-problems by semidefinite programming and cutting planes*, Mathematical Programming 82 (1998) 291–315.
33. C. Helmberg and F. Rendl, *A spectral bundle method for semidefinite programming*, SIAM Journal on Optimization 10 (2000) 673–696.
34. C. Helmberg, F. Rendl, R. Vanderbei, and H. Wolkowicz, *An interior-point method for semidefinite programming*, SIAM Journal on Optimization 6 (1996) 342–361.
35. J.B. Hiriart-Urruty and C. Lemaréchal, *Convex analysis and minimization algorithms (vol. 1 and 2)*, Springer, 1993.
36. A.J. Hoffman and H.W. Wielandt, *The variation of the spectrum of a normal matrix*, Duke Math. Journal 20 (1953) 37–39.
37. D. Jibetean and M. Laurent, *Semidefinite approximations for global unconstrained polynomial optimization*, SIAM Journal on Optimization 16 (2005) 490–514.
38. D.S. Johnson, C.R. Aragon, L.A. McGeoch, and C. Schevon, *Optimization by simulated annealing: An experimental evaluation. I: Graph partitioning*, Operations Research 37 (1989) 865–892.
39. D. Karger, R. Motwani, and M. Sudan, *Approximate graph colouring by semidefinite programming*, Journal of the ACM 45 (1998) 246–265.
40. S.E. Karisch and F. Rendl, *Semidefinite programming and graph equipartition*, Fields Institute Communications 18 (1998) 77–95.
41. B.W. Kernighan and S. Lin, *An efficient heuristic procedure for partitioning graphs*, Bell System techn. Journal 49 (1970) 291–307.
42. S. Khanna, N. Linial, and S. Safra, *On the hardness of approximating the chromatic number*, Combinatorica 20 (2000) 393–415.
43. M. Kojima, S. Shindoh, and S. Hara, *Interior-point methods for the monotone semidefinite linear complementarity problem in symmetric matrices*, SIAM Journal on Optimization 7 (1997) 86–125.
44. J.B. Lasserre, *A sum of squares approximation of nonnegative polynomials*, SIAM Journal Optimization 16 (2006) 751–765.
45. M. Laurent, S. Poljak, and F. Rendl, *Connections between semidefinite relaxations of the max-cut and stable set problems*, Mathematical Programming 77 (1997) 225–246.
46. M. Laurent and F. Rendl, *Semidefinite programming and integer programming*, Discrete Optimization (K. Aardal, G.L. Nemhauser, and R. Weismantel, eds.), Elsevier, 2005, pp. 393–514.
47. E.L. Lawler, J.K. Lenstra, A.H.G. Rinnooy Kan, and D.B. Shmoys (eds.), *The traveling salesman problem, a guided tour of combinatorial optimization*, Wiley, Chicester, 1985.
48. C. Lemaréchal, *An extension of davidon methods to nondifferentiable problems*, Mathematical Programming Study 3 (1975) 95–109.
49. C. Lemaréchal, *Nonsmooth optimization and descent methods*, Tech. report, International Institute for Applied Systems Analysis, 1978.
50. A. Lisser and F. Rendl, *Graph partitioning using linear and semidefinite programming*, Mathematical Programming 95 (2002) 91–101.
51. L. Lovász, *On the shannon capacity of a graph*, IEEE Trans. Inform. Theory 25 (1979) 1–7.

52. L. Lovász, *Semidefinite programs and combinatorial optimization*, Recent advances in algorithms and combinatorics (B.A. Reed and C.L. Sales, eds.), CMS books in Mathematics, Springer, 2003, pp. 137–194.
53. L. Lovász and A. Schrijver, *Cones of matrices and set-functions and 0-1 optimization*, SIAM Journal on Optimization 1 (1991) 166–190.
54. C. Lund and M. Yannakakis, *On the hardness of approximating minimization problems*, Proceedings of the 25th ACM STOC, 1993, pp. 286–293.
55. M. Marcus, *Rearrangement and extremal results for Hermitian matrices*, Linear Algebra Appl. 11 (1975) 95–104.
56. R.J. McEliece, E.R. Rodemich, and H.C. Rumsey Jr., *The lovasz bound and some generalizations*, Journal of combinatorics and System Sciences 3 (1978) 134–152.
57. R.D.C. Monteiro, *Primal-dual path-following algorithms for semidefinite programming*, SIAM Journal on Optmization 7 (1997) 663–678.
58. T.S. Motzkin and E.G. Straus, *Maxima for graphs and a new proof of a theorem of turan*, Canadian Journal of Mathematics 17 (1965) 533–540.
59. K.G. Murty and S.N. Kabadi, *Some np-complete problems in quadratic and nonlinear programming*, Mathematical Programming 39 (1987) 117–129.
60. Y. Nesterov, *Quality of semidefinite relaxation for nonconvex quadratic optimization*, Tech. report, CORE, 1997.
61. Y. Nesterov and A.S. Nemirovski, *Interior point polynomial algorithms in convex programming*, SIAM Publications, SIAM, Philadelphia, USA, 1994.
62. M. Padberg, *The quadric Boolean polytope: some characteristics, facets and relatives*, Mathematical Programming 45 (1989) 139–172.
63. P. Parrilo, *Structured semidefinite programs and semialgebraic geometry methods in robustness and optimization*, Ph.D. thesis, California Institute of Technology, USA, 2000.
64. J. Povh and F. Rendl, *Approximating non-convex quadratic programs by semidefinite and copositive programming*, Proceedings of the 11th international conference on operational research (L. Neralic V. Boljuncic and K. Soric, eds.), Croation Operations Research Society, 2008, pp. 35–45.
65. F. Rendl and H. Wolkowicz, *Applications of parametric programming and eigenvalue maximization to the quadratic assignment problem*, Mathematical Programming 53 (1992) 63–78.
66. F. Rendl and H. Wolkowicz, *A projection technique for partitioning the nodes of a graph*, Annals of Operations Research 58 (1995) 155–179.
67. H. Schramm and J. Zowe, *A version of the bundle idea for minimizing a nonsmooth function: Conceptual idea, convergence analysis, numerical results*, SIAM Journal Optimization 2 (1992) 121–152.
68. A. Schrijver, *A comparison of the delsarte and lovasz bounds*, IEEE Transactions on Information Theory IT-25 (1979) 425–429.
69. H.D. Sherali and W.P. Adams, *A hierarchy of relaxations between the continuous and convex hull representations for zero-one programming problems*, SIAM Journal on Discrete Mathematics 3 (1990) 411–430.
70. H.D. Sherali and W.P. Adams, *A hierarchy of relaxations and convex hull characterizations for mixed-integer zero-one programming problems*, Discrete Applied Mathematics 52 (1994) 83–106.
71. C. De Simone, *The cut polytope and the Boolean quadric polytope*, Discrete Mathematics 79 (1990) 71–75.
72. J. von Neumann, *Some matrix inequalities and metrization of matrix space (1937)*, John von Neumann: Collected Works, Vol 4, MacMillan, 1962, pp. 205–219.
73. A. Widgerson, *Improving the performance guarantee for approximate graph colouring*, Journal of the ACM 30 (1983) 729– 735.
74. H. Wolkowicz, R. Saigal, and L. Vandenberghe (eds.), *Handbook of semidefinite programming*, Kluwer, 2000.

Chapter 19
The Group-Theoretic Approach in Mixed Integer Programming

Jean-Philippe, P. Richard and Santanu S. Dey

Abstract In this chapter, we provide an overview of the mathematical foundations and recent theoretical and computational advances in the study of the group-theoretic approach in mixed integer programming. We motivate the definition of group relaxation geometrically and present methods to optimize linear functions over this set. We then discuss fundamental results about the structure of group relaxations. We describe a variety of recent methods to derive valid inequalities for master group relaxations and review general proof techniques to show that candidate inequalities are strong (extreme) for these sets. We conclude by discussing the insights gained from computational studies aimed at gauging the strength of group-theoretic relaxations and cutting planes for mixed integer programs.

19.1 Introduction

In 1957, Dantzig [28] introduced the idea that numerous practical decision problems could be adequately modeled as linear optimization problems in which some or all the variables are restricted to be integer. Specifically, Dantzig highlighted the value of studying mixed integer programs (MIPs), i.e., optimization problems of the form:

Jean-Philippe P. Richard
Department of Industrial and Systems Engineering, University of Florida, Gainesville, USA
e-mail: richard@ise.ufl.edu

Santanu S. Dey
Center for Operations Research and Econometrics, Université Catholique de Louvain, Belgium
e-mail: Santanu.Dey@uclouvain.be

M. Jünger et al. (eds.), *50 Years of Integer Programming 1958-2008*,
DOI 10.1007/978-3-540-68279-0_19,

$$\min \sum_{j\in M} c_j x_j \tag{19.1}$$

$$(\texttt{MIP}) \quad s.t. \sum_{j\in M} A_j x_j = d \tag{19.2}$$

$$x_j \in \mathbb{Z}_+ \; \forall j \in I \tag{19.3}$$

$$x_j \in \mathbb{R}_+ \; \forall j \in C, \tag{19.4}$$

where $M = \{1,\ldots,m\}$, (I,C) is a partition of M with $I \neq \emptyset$, $d \in \mathbb{R}^{l\times 1}$, $c \in \mathbb{R}^{1\times m}$, $A \in \mathbb{R}^{l\times m}$ is a full row rank matrix and A_j is used to denote the jth column of A for $j \in M$. Vectors $x \in \mathbb{R}^m$ are considered to be feasible solutions for (MIP) if (*i*) they satisfy the linear constraints (19.2), (*ii*) the components of the vector x indexed by I are nonnegative integers, and (*iii*) the components of the vector x indexed by C are nonnegative real numbers. We refer to the set of all feasible solutions to (MIP) as F. A feasible solution x is said to be optimal for (MIP) if no other feasible solution in F yields a lower objective value for the linear objective function (19.1).

Although the formulation of practical problems as mixed integer programs is often simple, solving these problems to provable optimality can be computationally challenging. In fact, since the general mixed integer programming problem is NP-complete, see Schrijver [90] pages 245–247 for a proof, it is unlikely that an efficient algorithm will ever be designed to solve this problem. However since new applications of mixed integer programming are constantly being developed and are growing ever more complex, the quest for newer and faster algorithms has been an enduring area of research.

The first nontrivial algorithms for solving general IPs were proposed by Gomory [54] and Land and Doig [77] in 1958 and 1960 respectively; see Sections 12.4 and 12.5 for a historical perspective of these developments. These algorithms, although fundamentally different, both rely heavily on linear programming relaxations of (MIP). *Linear programming relaxations* are optimization problems obtained by removing the restrictions (19.3) that variables x_j for $j \in I$ are integer-valued. They are therefore linear programming problems, for which efficient solution algorithms can be used; see Vanderbei [96] for a textbook description of solution methodologies for linear programs. More generally, *relaxations* are optimization problems obtained by enlarging the feasible set F and/or decreasing the value of the objective function (19.1) over F. The concept of relaxation is important since it allows the substitution of mixed integer programs, which are typically difficult to solve, for optimization problems for which good solution algorithms are known. Clearly, if an optimal solution x^* to a relaxation of (MIP) belongs to F, then x^* is an optimal solution to (MIP). Land and Doig's and Gomory's algorithms differ in how they handle situations where x^* does not belong to F.

In Land and Doig's algorithm, the feasible region F of (MIP) is divided into subregions whenever the optimal relaxation solution x^* does not belong to F. The division is performed in such a way that the LP relaxations of the subregions do not contain x^*. For example, if $x_j^* \notin \mathbb{Z}$ for $j \in I$, one possible way to perform this division is to consider $F = F_1 \cup F_2$ where $F_1 = \{x \in F \mid x_j \leq \lfloor x_j^* \rfloor\}$ and $F_2 = \{x \in F \mid x_j \geq \lceil x_j^* \rceil\}$. It is easy to see that an optimal solution to (MIP) can be directly obtained from the optimal solutions of $\min\{cx \mid x \in F_1\}$ and $\min\{cx \mid x \in F_2\}$, problems that can be solved recursively. Over the years, different strategies have been developed following these lines. The resulting methods are grouped under the general vocable of branch-and-bound algorithms; see Achterberg et al. [2] for a recent description.

In Gomory's algorithm, an improved mixed integer programming formulation of (MIP) is created whenever x^* does not belong to F. This improved formulation is obtained through the addition of a new linear constraint that is satisfied by all solutions of F but not by the current linear programming solution x^*. We will refer to such an inequality as a *cut* in the remainder of this chapter. Gomory [54] showed that, if all variables are integer and bounded, optimal solutions of (MIP) can be obtained by sequentially adding a finite number of inequalities from a specific family of cuts. Over the years, different strategies have been proposed to generate cuts. The resulting methods are grouped under the general vocable of cutting-plane algorithms; see Marchand et al. [80] for a recent survey.

Fifty years after the publication of these seminal papers, the use of mixed integer programming as a modeling tool for practical optimization problems has become widespread. Successful commercial software have been developed that incorporate branch-and-bound and cutting-plane approaches. Further, new and stronger cutting planes have been developed, better branching strategies have been investigated, and stronger relaxations have been uncovered. One such relaxation that is applicable to all mixed integer programs is the *group relaxation*. This relaxation, introduced by Gomory [57], is particularly well-structured. As a result, it can be used as a substitute for LP relaxations in branch-and-bound procedures since there are simple algorithms to optimize linear functions over its feasible region and it can be used as a source of cuts in cutting plane approaches since its polyhedral structure is simple.

In this chapter, we provide an overview of the mathematical foundations and recent advances in the study of group relaxations of mixed integer programs. This overview is by no means exhaustive. However, we hope that it illustrates that, although the group approach is about as old as integer programming itself, its study is still swarming with open questions, computational possibilities and theoretical puzzles that might hold the key to future breakthroughs in the solution of mixed integer programs. The large growth of interest in this topic over the last years attests that this potential is currently being vigorously investigated.

In Section 19.2, we present a numerical example from which we provide an intuitive derivation of the corner relaxation that lays the foundation for the concepts that will be formalized in later sections. We then formally introduce Gomory's corner relaxation. In Section 19.3, we describe a method to optimize linear functions over Gomory's corner relaxation. We discuss conditions under which solving this relaxation is sufficient to solve the initial integer program. We also describe a hierarchy of extended group relaxations for integer programs and present algebraic results about their structure. In Section 19.4, we describe generalizations of corner relaxations that have been used for the generation of cutting planes in mixed integer programming. We discuss the properties and characteristics of the valid inequalities of these master corner relaxations. In particular, we present a hierarchy of valid inequalities of group relaxations based on their strength. In Section 19.5, we review recent approaches to derive candidate inequalities for master group relaxations and describe methods to prove that these candidate inequalities are strongest possible (extreme) among valid inequalities. We also present relationships that exist between the extreme inequalities of different master group relaxations. We conclude this section by providing a taxonomy of known extreme inequalities for master group problems. In Section 19.6, we discuss theoretical and computational studies that are aimed at evaluating the strength of group cuts. We conclude in Section 19.7 with a discussion of extensions of the group-theoretic approach, open questions related to group relaxations, and possible avenues of research.

19.2 The corner relaxation

In this section, we first review basic results about linear programming in Section 19.2.1. We then motivate, on a numerical example, the definition of Gomory's corner relaxation in Section 19.2.2. We conclude in Section 19.2.3 by formally defining Gomory's corner relaxation for both pure and mixed integer programs.

19.2.1 Linear programming relaxations

As LP relaxations are used in the cutting plane algorithm described in Section 19.1 and because optimal solutions of these relaxations are repeatedly obtained, we first review fundamental results in linear programming that we will use in the remainder of this chapter. Consider a linear programming problem of the form

$$(\mathtt{LP}) : \min\Big\{ \sum_{j\in M} c_j z_j \;\Big|\; \sum_{j\in M} A_j z_j = d,\ z \in \mathbb{R}^m_+ \Big\}$$

where $A = [A_1|A_2|\cdots|A_m] \in \mathbb{R}^{l\times m}$ is a matrix such that $rank(A) = l \le m$, $c \in \mathbb{R}^{1\times m}$ and $d \in \mathbb{R}^{l\times 1}$. The feasible region of $(\mathtt{LP})$ is a polyhedron. It can be verified that, when $(\mathtt{LP})$ has an optimal solution, it has an optimal solution that is a vertex of this polyhedron. To simplify the description of vertices, we introduce the following notation. Given a subset T of M, we refer to the submatrix of A obtained by selecting the columns whose indices belong to T as A_T and refer to the vector obtained from c by selecting those components whose indices belong to T as c_T. For any partition (B,N) of M, the linear constraints defining $(\mathtt{LP})$ can be rewritten as $A_B z_B + A_N z_N = d$. Further, if B is chosen in such a way that the matrix A_B is square and invertible, then B is said to form a basis of $(\mathtt{LP})$ and $(\mathtt{LP})$ can be reformulated as

$$\min\big\{c_B A_B^{-1} d + (c_N - c_B A_B^{-1} A_N) z_N \;\big|\; z_B + A_B^{-1} A_N z_N = A_B^{-1} d,\ z \in \mathbb{R}^m_+\big\}. \quad (19.5)$$

We refer to the above formulation of (LP) as the *simplex tableau* associated with basis B. We refer to the variables with indices in B as *basic* while we refer to the variables with indices in N as *nonbasic*. It is apparent from simplex tableau (19.5) that the solution $\tilde{z}_B = A_B^{-1}d$, $\tilde{z}_N = 0$ is feasible to (LP) whenever $\tilde{z}_B \geq 0$. Such a solution is referred to as a *basic feasible solution* of (LP). Further, if $\bar{c}_N = c_N - c_B A_B^{-1} A_N \geq 0$, it is clear that the corresponding basic feasible solution $(\tilde{z}_B, \tilde{z}_N)$ is optimal for (LP). A fundamental theorem of linear programming establishes that, if (LP) has an optimal solution, it has a basic feasible optimal solution with $\bar{c}_N \geq 0$; see Section 3 of Chvátal [22] for a more detailed discussion.

Theorem 19.1. *If* (LP) *has an optimal solution, then there exists a basis B for which $\bar{c}_N \geq 0$ and the associated basic solution is feasible for* (LP).

Various solution methodologies have been proposed to solve linear programs; see Vanderbei [96] for an exposition of existing methods. Among them, the simplex algorithm, introduced by Dantzig, is of particular interest to this chapter. This algorithm systematically examines bases of the linear programming problem until it discovers one that satisfies the conditions of Theorem 19.1. Although the simplex algorithm can be fooled into performing an exponential number of iterations, see Klee and Minty [75], it typically obtains optimal solutions of practical linear programs rapidly and is very well suited for re-optimization, which is of crucial importance in branch-and-cut techniques. When solving (MIP) using cutting planes, it is therefore natural to assume that a simplex tableau associated with an optimal basic solution of the LP relaxation of (MIP) will be known at each iteration of the algorithm. Therefore, tableau information can be used in the derivation of cuts. In the remainder of this section, we denote the rows of an optimal simplex tableau by

$$x_i + \sum_{j \in N} \bar{a}_{ij} x_j = \bar{d}_i, \quad \forall i \in B, \tag{19.6}$$

where B represents the set of basic variables, N represents the set of nonbasic variables, $\bar{a}_{\cdot j} = A_B^{-1} A_j$ and $\bar{d} = A_B^{-1} d$. We also assume that $\bar{d} \notin \mathbb{Z}^{l \times 1}$ since otherwise the basic feasible solution associated with simplex tableau (19.6) is optimal for (MIP).

19.2.2 Motivating example

To develop intuition on how the information contained in a simplex tableau can be used to generate a cut, we now consider the following simple example without continuous variables:

$$
\begin{aligned}
\min \quad & -2x_1 - x_2 & & \\
s.t. \quad & 1x_1 - 3x_2 \le & -3 & \\
& 35x_1 + 5x_2 \le & 258 & \\
& 5x_1 + 5x_2 \le & 63 & \\
& -10x_1 + 120x_2 \le & 771 & \\
& -215x_1 - 135x_2 \le & -1071 & \\
& x_1, x_2 \in \mathbb{Z}_+. & &
\end{aligned}
\tag{19.7}
$$

Although this problem does not directly fit the form of (MIP) since its linear constraints are not equalities, a nonnegative variable, called *slack*, can be added to each constraint to achieve this form. We denote by x_{i+2} the slack added to the ith constraint of (19.7) to make it an equality.

The polytope associated with the linear inequalities of (19.7) is the non-shaded polytope represented in Figure 19.1a. The feasible region F of the integer programming problem (19.7) consists of the integer points that lie within the polytope. Its convex hull conv(F), i.e., the smallest convex set containing all points of F, is the polytope that is shaded in Figure 19.1a. It can be verified that the solution $x^* = (6.5, 6.1)'$ highlighted with a star is optimal for the LP relaxation of (19.7) while the solution $\bar{x} = (6,6)'$ is optimal for (19.7). Since x^* is not optimal for (19.7), it is our goal to derive a valid inequality that improves the problem formulation and cuts x^* off.

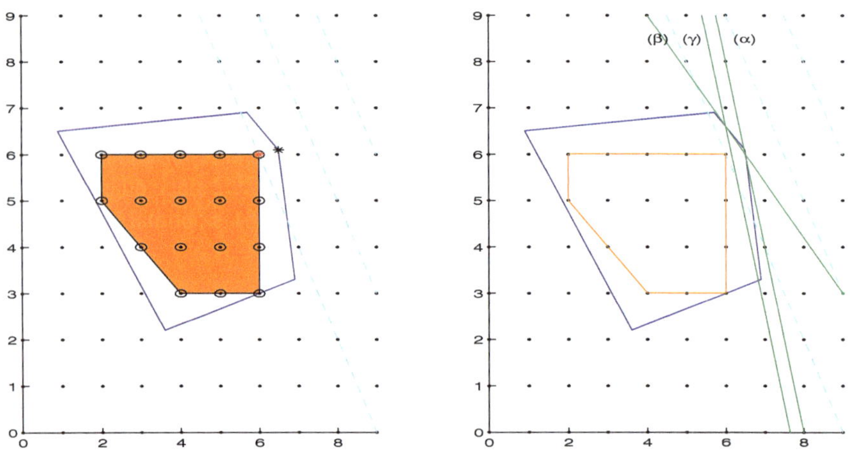

Fig. 19.1 LP relaxation, conv(S) and cuts for (19.7).

We know from Theorem 19.1 that there is a basic feasible solution of (19.7) representing x^*. The corresponding simplex tableau is:

$$
\begin{array}{lrrrl}
\min & -19.1 + & 1/30x_4 + & 5/30x_5 & \\
s.t. & x_1 + & 1/30x_4 - & 1/30x_5 & = 6.5 \\
 & x_2 - & 1/30x_4 + & 7/30x_5 & = 6.1 \\
 & x_3 - & 4/30x_4 + & 22/30x_5 & = 8.8 \\
 & x_6 + & 130/30x_4 - & 850/30x_5 & = 104 \\
 & x_7 + & 80/30x_4 + & 730/30x_5 & = 1150
\end{array}
\tag{19.8}
$$

where $B = \{1,2,3,6,7\}$ and $N = \{4,5\}$. Observe that the basic solution associated with this tableau is fractional. In particular the first three rows of (19.8) have integer infeasibilities that we will try to correct with the addition of a cutting plane.

From the first row of (19.8), we conclude that there is no solution of (19.7) in which the nonbasic variables x_4 and x_5 are both equal to zero. This is because, in this case, the basic variables x_1, x_2 and x_3 assume fractional values. This observation can be translated into the valid inequality $x_4 + x_5 \geq 1$ since the nonbasic variables x_4 and x_5 are known to be integer and nonnegative. Although this inequality is expressed in terms of slack variables x_4 and x_5, it can be transformed back to the space of original variables x_1, x_2 to yield the cut:

$$4x_1 + x_2 \leq 32. \tag{19.9}$$

The arguments underlying the above derivation apply to the generic tableau row (19.6) and yield the cut:

$$\sum_{j \in N} x_j \geq 1. \tag{19.10}$$

This cutting plane was proposed by Dantzig [29], and is commonly known as *Dantzig Cut* (DC). To evaluate the strength of the inequality (19.9), we represent it as inequality (α) in Figure 19.1b. In this figure, we observe that although (19.9) cuts off the current fractional solution, the portion of the LP relaxation that it removes is very small. Improved cuts can be obtained by observing that nonbasic variables with (*i*) zero or (*ii*) integer coefficients do not help in restoring the integrality of the row and therefore can be omitted from the summation in (19.10). These observations, due to Charnes and Cooper [19] and Ben-Israel and Charnes [14] respectively, yield the following general cut:

$$\sum_{j \in N | \bar{a}_{ij} \notin \mathbb{Z}} x_j \geq 1; \tag{19.11}$$

see Rubin and Graves [87] for other improvements. Inequality (19.11) is called *Modified Dantzig Cut* (MDC) in Bowman and Nemhauser [18]. It is at least as strong as DC. In our numerical example however, it reduces to (19.9) as all nonbasic variables in the first row of the simplex tableau (19.8) have fractional coefficients.

It is not surprising that DC and MDC typically remove only small portions of the LP relaxation of (MIP) since they are insensitive to the particular values of the variable coefficients in (19.6). Therefore, they associate very different geometries with the same inequality. We should therefore search for inequalities whose coefficients are more strongly influenced by the coefficients of the variables in the tableau rows.

A possible way to generate such a cut is to use the following three-step process implicitly used by Gomory [54] and explicitly proposed by Chvátal [21]; refer to Section 11.4.1 for a more detailed description and analysis. Starting from the first row of the simplex tableau (19.8), we round down the coefficients of all the variables. Since variables are nonnegative, this operation yields the valid inequality $x_1 - x_5 \leq 6.5$, which can be strengthened to $x_1 - x_5 \leq 6$ since the variables x_1 and x_5 are integer. Substituting slack x_5 with its definition, we obtain the cut:

$$6x_1 + 5x_2 \leq 69. \tag{19.12}$$

More generally, starting from a row of a simplex tableau that has a fractional right-hand-side $\bar{d}_i$, we obtain the inequality $x_i + \sum_{j\in N} \lfloor \bar{a}_{ij} \rfloor x_j \leq \lfloor \bar{d}_i \rfloor$ or equivalently, after subtracting this inequality from (19.6), the cut:

$$\sum_{j\in N} f_{ij} x_j \geq f_i \tag{19.13}$$

where $f_{ij} = \bar{a}_{ij} - \lfloor \bar{a}_{ij} \rfloor$ and $f_i = \bar{d}_i - \lfloor \bar{d}_i \rfloor$. This cutting plane is the corner stone of Gomory's finitely convergent algorithm for bounded pure integer programming; see Gomory [54]. It historically predates DCs and MDCs. It is known either as *Gomory's fractional cut* (GFC) or *Chvátal-Gomory cut*. We represent it as inequality (β) in Figure 19.1b where we observe that it is slightly stronger than (19.9).

Investigating the reason for the weakness of DC, we observe in (19.8) that, in order for the current basic solution to become integer, not only should one of the integer variables x_4 or x_5 be nonzero, it also should be sufficiently large. In particular, we see that when $x_5 = 0$, x_4 must take a value of at least 15 and when $x_4 = 0$, x_5 must take a value of at least 15. Since the variables x_4 and x_5 have coefficients with opposite signs, it is easy to verify that the inequality $\frac{1}{15}x_4 + \frac{1}{15}x_5 \geq 1$ is valid. Substituting for the slacks, we obtain

$$40x_1 + 10x_2 \leq 306. \tag{19.14}$$

We represent it as inequality (γ) in Figure 19.1b where it can be seen to provide a definite improvement over DC and GFC. The idea underlying the generation of this inequality can be extended to general tableau rows (19.6) as follows (we assume that all coefficients $\bar{a}_{ij}$ are fractional). Given any partition (N_1, N_2) of N, we rewrite (19.6) as

$$x_i + \sum_{j\in N_1} \lfloor \bar{a}_{ij} \rfloor x_j + \sum_{j\in N_2} \lceil \bar{a}_{ij} \rceil x_j - \lfloor \bar{d}_i \rfloor = f_i - \sum_{j\in N_1} f_{ij} \tilde{x}_j + \sum_{j\in N_2} (1 - f_{ij}) \tilde{x}_j \tag{19.15}$$

where $x_j = \tilde{x}_j$ for $j \in N$. Now consider the relaxation of (19.15) obtained by removing the requirement that $\tilde{x}_j = x_j$ and by weakening the condition that $\tilde{x}_j \in \mathbb{Z}_+$ to $\tilde{x}_j \geq 0$. As the left-hand-side of (19.15) is integral, the fractional part f_i of the right-hand-side must be "corrected" by the corresponding $\tilde{x}_j$ variables. The minimum value that variable $\tilde{x}_j$ can take when all other nonbasic variables are set to zero is given by

$$\tilde{x}_j \geq \frac{f_i}{f_{ij}} \quad \forall j \in N_1 \quad \text{and} \quad \tilde{x}_j \geq \frac{1-f_i}{1-f_{ij}} \quad \forall j \in N_2. \tag{19.16}$$

Since it can easily be verified that any feasible solution $(x, \tilde{x})$ of this relaxation can be expressed as a convex combination of feasible solutions of the form $(x, s_j e_j)$ where e_j is the jth unit vector in $\mathbb{R}^l$, we obtain that

$$\sum_{j \in N_1} \frac{f_{ij}}{f_i} \tilde{x}_j + \sum_{j \in N_2} \frac{1-f_{ij}}{1-f_i} \tilde{x}_j \geq 1$$

is a valid inequality. Since the choice of N_1 and N_2 is arbitrary, we select these sets optimally to obtain

$$\sum_{j \in N} \min\left\{\frac{f_{ij}}{f_i}, \frac{1-f_{ij}}{1-f_i}\right\} x_j \geq 1 \tag{19.17}$$

after the condition that $x_j = \tilde{x}_j$ is restored. This inequality is known as *Gomory Mixed Integer Cut* (GMIC) and was first obtained by Gomory [55].

At this point, it probably seems that there are myriads of other ways of generating cuts and so it is important to differentiate strong families of cuts from weak families. One possible criterion is to determine whether the cuts in the family under study are sufficiently strong to yield a finitely convergent cutting-plane algorithm. In this respect, it was proven in Gomory and Hoffman [58] that cutting-plane algorithms based on DCs are not always finitely convergent for pure integer programs. A variant based on MDCs however can be shown to be finitely convergent; see Bowman and Nemhauser [18]. GFCs can also be shown to yield a finitely convergent algorithm for bounded integer pograms; see Gomory [54] and Nourie and Venta [84] for a simplified proof. We note that although this algorithm is finitely convergent, the number of its steps can be arbitrarily large, even for two-variables problems; see Jeroslow and Kortanek [70]. As a result, finite convergence of cutting-plane algorithms is not a very strong criterion.

Another approach to evaluate the strength of these cutting planes is to analyze their commonalities and, in particular, to determine whether they are derived from a common relaxation of (19.6). In fact, we can see that in the derivation of all four inequalities (19.10), (19.11), (19.13) and (19.17), we used the fact that the basic variable x_i is integer, and that the nonbasic variables x_j for $j \in N$ are integer and nonnegative. However, we never used the fact that the basic variable x_i is nonnegative. In the case of DC and MDC, we used the fact that the basic variable x_i is integer to detect that not all the nonbasic variables can be zero. Second we used the fact that the nonbasic variables x_j are nonnegative and integer to impose that their sum must be greater or equal to 1. Similarly, to derive GFC, we used the fact that the basic variable is integer to detect that row i of the simplex tableau is problematic. We also used the fact that nonbasic variables are nonnegative to ensure that rounding yields a valid inequality. Finally, we used the fact that all variables are integer to round down the right-hand-side. In the derivation of GMIC, we used the fact that all variables were integer to conclude that the left-hand-side of (19.15) is integer. We also used the fact that nonbasic variables are nonnegative to obtain (19.16).

Therefore, to judge whether or not the inequalities we derived are strong, we could measure their strength with respect to the set of solutions to a single fractional tableau row in which the nonbasic variables x_j are assumed to be nonnegative and integer and in which the basic variable x_i is only assumed to be integer (and can take negative values). Formally, given a row i of (19.6) where $\bar{d} \notin \mathbb{Z}^l$, we define

$$C_i = \text{conv}\Big(\big\{(x_i, x_N) \in \mathbb{Z} \times \mathbb{Z}_+^{|N|} \mid x_i + \sum_{j \in N} \bar{a}_{ij} x_j = \bar{d}_i\big\} \setminus \big\{(0,0)\big\}\Big).$$

Note that the requirement that $(0,0)$ is removed from the set is used in the case where, although the tableau row i we consider has an integer right-hand-side, some other rows have fractional right-hand-sides. In this situation, we want to remove the point $(0,0)$ from C_i because we know from the other tableau rows that it does not yield an integer solution to (MIP). Note however that, if $\bar{d}_i$ is fractional, explicitly removing $(0,0)$ from the set is redundant as this point does not belong to the set in the first place. Since, in C_i, basic variable x_i can be adjusted independently and since the only requirement associated with it is that it is integer, we can also reformulate this set without using the variable x_i as

$$C_i' = \text{conv}\Big(\big\{x \in \mathbb{Z}_+^{|N|} \mid \sum_{j \in N} \bar{a}_{ij} x_j \equiv \bar{d}_i \pmod 1\big\} \setminus \big\{0\big\}\Big).$$

Here the notation "mod 1" signifies that $\sum_{j \in N} \bar{a}_{ij} x_j - \bar{d}_i \in \mathbb{Z}$. In Figure 19.2a, we shaded (in the space of x_1 and x_2 variables) the convex hull C_1' of all points that are obtained in this way when considering the first row of the simplex tableau (19.8) of our numerical example. The reason for using two-tone shading will be discussed later. From this figure, we observe that (19.14) is strong for the polyhedron C_1'. In Figures 19.2b and 19.2c, we present the two relaxations C_2' and C_5' in the space of x_1 and x_2 variables. We observe that not all the relaxations are as strong as each other and that some of these relaxations can be very weak. In particular, we observe that even if all the inequalities that can be generated from each of the relaxations C_i' for $i = 1,2,5$ were added to the LP formulation of (19.7), the linear programming relaxation solution would still not yield an optimal solution of the integer program. The question of determining how to generate stronger cuts therefore remains.

A natural idea is to investigate whether the information inferred from several rows of a tableau can be "combined" to obtain better cutting planes than those obtained by considering a single row. For our numerical example, the relaxation obtained by considering the convex hull of nonnegative integer solutions (x_4, x_5) for which the corresponding variables x_1 and x_2 in (19.6) are (possibly negative) integer is shaded in Figure 19.3a (in the space of x_1, x_2 variables). A similar picture is obtained in Figure 19.3b by considering the convex hull of nonnegative integer solutions (x_4, x_5) for which x_6 and x_7 are integer (possibly negative). These relaxations C'_{ij} can be stronger than the intersection of the relaxations C'_i and C'_j. Further, although C'_{12} does not coincide with the convex hull of integer solutions to (19.7), we note that if we were to add the defining inequalities of the relaxation of Figure 19.3a to the initial LP relaxation, an optimal solution to (19.7) would be exposed.

The above argument can be pushed further by considering more than two rows of the simplex tableau simultaneously. In this scheme, valid inequalities are obtained for the relaxation obtained by considering all basic variables to be integer (possibly negative) and nonbasic variables to be integer and nonnegative. This relaxation is known as *Gomory's corner relaxation* or *Gomory's group relaxation*. It is clearly as strong as those obtained by considering rows individually. Further, it has a very natural geometric interpretation. In fact, in the absence of degeneracy, i.e., when all components of $A_B^{-1}d$ are positive, it corresponds to the set of integer points located inside of the cone defined by all constraints that are active at the current basic feasible solution of the LP relaxation. Referring back to our example, the corner relaxation would be composed of integer solutions in the cone defined by constraints $35x_1 + 5x_2 \leq 258$ and $5x_1 + 5x_2 \leq 63$. Using this geometrical intuition, we can observe that the relaxation depicted in Figure 19.3a that was obtained by considering the first two rows of (19.8) is exactly the corner relaxation associated with basis $B = \{1,2,3,6,7\}$. Therefore in this case, all cuts valid for the corner relaxation can be obtained from considering the first two rows of the simplex tableau. In Figures 19.2 and 19.3 we use dark shading to highlight those points of the relaxations C'_i or C'_{ij} that do not belong the corner polyhedron.

19.2.3 Gomory's corner relaxation

As a generalization of the discussion given above, we now give a general definition of Gomory's corner relaxation. We assume that the linear programming relaxation of (`MIP`) is feasible. We consider first the case where $M = I$, i.e., all variables of the problem are integer. Given an optimal basis B of the LP relaxation of (`MIP`),

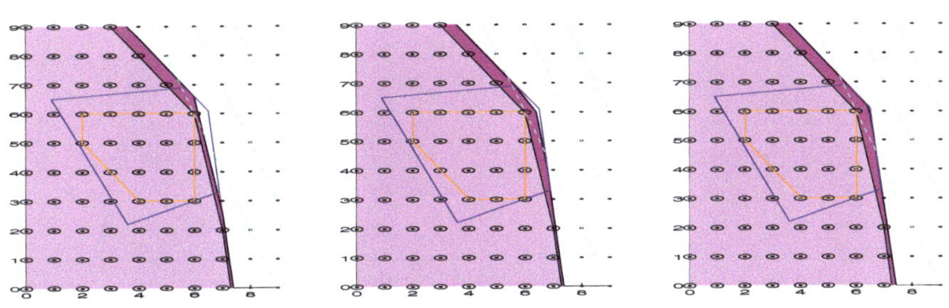

Fig. 19.2 One-row relaxations C'_1, C'_2 and C'_5.

we reformulate the feasible set F as

$$F = \{x \in \mathbb{Z}_+^m \mid x_B + A_B^{-1} A_N x_N = A_B^{-1} d\}.$$

Relaxing the nonnegativity condition on the basic variables x_B we obtain

$$\text{corner}_B(F) = \{x_N \in \mathbb{Z}_+^{|N|} \mid A_B^{-1} A_N x_N \equiv A_B^{-1} d \text{ (mod 1)}\}. \tag{19.18}$$

We refer to $\text{corner}_B(F)$ as the corner relaxation associated with basis B. We also define the corner polyhedron associated with basis B as the convex hull of solutions of $\text{corner}_B(F)$.

For mixed integer programs, we still want to consider the set of feasible mixed integer solutions in the cone spanned by the current basis B. Given this basis B, we reformulate the feasible set F as

$$F = \{(x_B, x_{NI}, x_{NC}) \in \mathbb{Z}^{|B|} \times \mathbb{Z}_+^{|NI|} \times \mathbb{R}_+^{|NC|} \mid x_B + A_B^{-1} A_N x_N = A_B^{-1} d\},$$

where $NI = N \cap I$ and $NC = N \cap C$. We then relax the lower bound on the basic variables. For basic variables that are continuous, relaxing the lower bound corresponds to relaxing the corresponding constraint. So we may assume that $B \cap C = \emptyset$ by discarding constraints with continuous basic variables if necessary. The corner relaxation associated with B then takes the form:

$$\text{corner}_B(F) = \{(x_{NI}, x_{NC}) \in \mathbb{Z}_+^{|NI|} \times \mathbb{R}_+^{|NC|} \mid A_B^{-1} A_N x_N \equiv A_B^{-1} d \text{ (mod 1)}\}. \tag{19.19}$$

Although we motivated the introduction of the corner relaxation by our desire to evaluate the quality of cutting planes, it follows trivially from its construction that it provides a relaxation of mixed integer programs. Therefore, it can be used both as a substitute to linear programming relaxations in branch-and-bound techniques and a source of cuts in cutting-plane algorithms. Next we describe both of these uses. In Section 19.3 we describe a family of methods to optimize over the corner relaxation. In Sections 19.4 and 19.5 we describe how to derive strong cutting planes for the corner polyhedron.

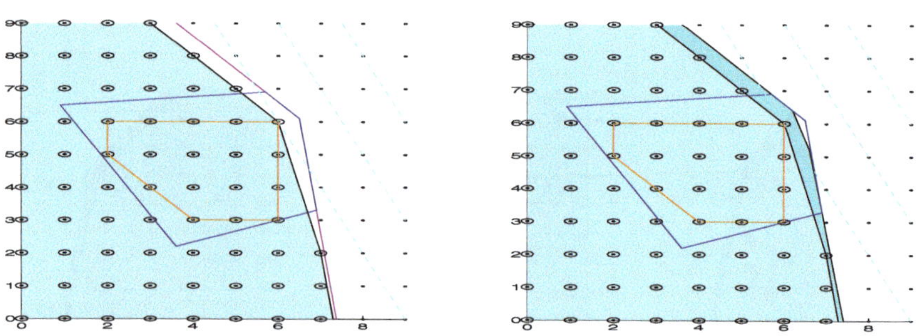

Fig. 19.3 Two-row relaxations C'_{12} and C'_{67}.

19.3 Group relaxations: optimal solutions and structure

In this section, we first focus on the problem of optimizing linear functions over Gomory's corner relaxation. Then we describe conditions under which an optimal solution of the corner relaxation solves (MIP) and discuss techniques that can be used to solve (MIP) when it does not. Finally, we present extended group relaxations which are tightenings of the corner relaxation and describe algebraic results about their structure.

19.3.1 Optimizing linear functions over the corner relaxation

The first methodology to optimize linear functions over the corner relaxation was introduced by Gomory [56] in 1965. Although the corner relaxation had not been formally introduced at the time, Gomory used a dynamic programming algorithm to investigate relations between continuous and integer optimal solutions to a linear program. This algorithm produces an optimal solution for the problem of maximizing a linear function over the corner relaxation. The search for more efficient solution algorithms experienced an explosion of activities in the late 1960s and early 1970s. Shapiro [91] proposed a shortest path algorithm variant of the dynamic programming algorithm of Gomory. Glover [52] and Hu [68] also developed algorithms. Chen and Zionts [20] investigated improvements to these algorithms and tested these variants computationally; see also Salkin and Morito [88]. We present one of these algorithms next.

We assume that the constraint matrix of the problem (MIP) is integral, i.e., $A \in \mathbb{Z}^{l\times m}$. We consider first a pure integer program, i.e., $C = \emptyset$. The problem we wish to solve is:

$$\min\{z^* + \bar{c}_N x_N \mid A_B^{-1}A_N x_N \equiv A_B^{-1}d \pmod 1), x_N \in \mathbb{Z}_+^{|N|}\} \tag{19.20}$$

where $z^* = c_B A_B^{-1}d$ and $\bar{c}_N = c_N - c_B A_B^{-1}A_N$.

In order for the algorithm to be simple to perform, we need the constraint coefficients of (19.20) to have only integer entries. This requirement can easily be achieved by multiplying all constraints by D, the determinant of the basis matrix A_B. However, a more efficient algorithm can be obtained if we consider an alternate representation of the group minimization problem using the Smith Normal Form of the basis matrix A_B; see Smith [93]. This representation makes use of unimodular matrices, i.e., matrices $U \in \mathbb{Z}^{k\times k}$ for which $\det(U) = \pm 1$.

Theorem 19.2 ([93]). *Let $F \in \mathbb{Z}^{l\times q}$ be such that $rank(F) = q \le l$. There exist unimodular matrices $U_1 \in \mathbb{Z}^{l\times l}$ and $U_2 \in \mathbb{Z}^{q\times q}$ for which $U_1 F U_2 = \binom{\Lambda}{0}$ where $\Lambda = diag(\lambda_1,\ldots,\lambda_q) \in \mathbb{Z}_+^{q\times q}$ is a diagonal matrix such that λ_i divides λ_{i+1} for $i = 1,\ldots,q-1$.*

In Theorem 19.2, the matrix Λ can be shown to be unique although the unimodular matrices U_1 and U_2 are not. Matrices U_1, Λ and U_2 can be computed using a variation of Gaussian elimination. In particular, they can be obtained as the product of permutation matrices and of matrices corresponding to integer elementary row and column operations. Further, matrices U_1, Λ and U_2 can be obtained in polynomial time; see Kannan and Bachem [73] for such an algorithm.

We now derive an equivalent form of (19.20). Consider an optimal basis B of the LP relaxation of (MIP). The group minimization problem associated with B is

$$\min\{z^* + \bar{c}_N x_N \mid A_B x_B + A_N x_N = d, (x_B, x_N) \in \mathbb{Z}^{|B|} \times \mathbb{Z}_+^{|N|}\}. \tag{19.21}$$

Using Theorem 19.2, we know there exist unimodular matrices U_1 and U_2 such that $U_1 A_B U_2 = \Lambda$. Since the matrices U_1 and U_2 are unimodular and A_B is a basis matrix, then $0 \neq |\det(A_B)| = |\det(U_1)||\det(\Lambda)||\det(U_2)| = \prod_{i=1}^{l} \lambda_i$, i.e., all the diagonal entries of Λ are nonzero. Problem (19.21) can be reformulated as:

$$\min\{z^* + \bar{c}_N x_N \mid \Lambda U_2^{-1} x_B + U_1 A_N x_N = U_1 d, (x_B, x_N) \in \mathbb{Z}^{|B|} \times \mathbb{Z}_+^{|N|}\} \tag{19.22}$$

after multiplying the defining constraint of (19.21) by U_1. It can be easily verified that U_2 being unimodular implies that U_2^{-1} is unimodular. Further, it can be shown that if $w = U_2^{-1} x_B$ then $w \in \mathbb{Z}^l$ if and only if $x_B \in \mathbb{Z}^l$. Therefore, formulation (19.22) ultimately reduces to

$$\min\{z^* + \bar{c}_N x_N \mid U_1 A_N x_N \equiv U_1 d \ (\mathrm{mod}\ \Lambda), x_N \in \mathbb{Z}_+^{|N|}\}. \tag{19.23}$$

In formulation (19.23), the "mod" notation is taken to mean that the two vectors $U_1 A_N x_N$ and $U_1 d$ differ by a vector z which is an integer multiple of Λ, i.e., $z = \Lambda x_B$ for some $x_B \in \mathbb{Z}^l$. As the matrix Λ is diagonal, it also signifies that the ith row of the formulation is considered modulo λ_i.

Example 19.1. For the numerical example of Section 19.2, it can be computed that, if

$$U_1 = \begin{pmatrix} 1 & 0 & 0 & 0 & 0 \\ -5 & 5 & -32 & 1 & 0 \\ 5 & 2 & 28 & 0 & 1 \\ 5 & -1 & 6 & 0 & 0 \\ 0 & -1 & 7 & 0 & 0 \end{pmatrix}, \ U_2 = \begin{pmatrix} 1 & 0 & 0 & -1 & 5 \\ 0 & 0 & 0 & 0 & 1 \\ 0 & 0 & 0 & 1 & -2 \\ 0 & 1 & 0 & 5 & -10 \\ 0 & 0 & 1 & -5 & 10 \end{pmatrix} \text{ and } \Lambda = \begin{pmatrix} 1 & 0 & 0 & 0 & 0 \\ 0 & 1 & 0 & 0 & 0 \\ 0 & 0 & 1 & 0 & 0 \\ 0 & 0 & 0 & 5 & 0 \\ 0 & 0 & 0 & 0 & 30 \end{pmatrix}$$

then $U_1 A_B U_2 = \Lambda$ where A_B is the basis matrix associated with $B = \{1,2,3,6,7\}$. Using the representation (19.23), the group minimization problem can therefore be restated as

$$
\begin{aligned}
\min \quad & -19.1 + 1/30x_4 + 1/6x_5 \\
s.t. \quad & 0x_4 + \quad 0x_5 \equiv \; -3 \pmod 1 \\
& 5x_4 - \; 32x_5 \equiv \; 60 \pmod 1 \\
& 2x_4 + \; 28x_5 \equiv 1194 \pmod 1 \\
& -1x_4 + \quad 6x_5 \equiv \; 105 \pmod 5 \\
& -1x_4 + \quad 7x_5 \equiv \; 183 \pmod{30}.
\end{aligned}
$$

In the above formulation, the first three constraints are trivially satisfied and can be omitted, yielding

$$
\begin{aligned}
\min \quad & -19.1 + 1/30x_4 + 1/6x_5 \\
s.t. \quad & 4x_4 + \quad 1x_5 \equiv 0 \pmod 5 \\
& 29x_4 + \quad 7x_5 \equiv 3 \pmod{30}
\end{aligned}
\tag{19.24}
$$

where the constraint coefficients and right-hand-sides have been reduced using the congruence relation.

We now focus on a methodology that reduces the solution of the group minimization problem (19.23) to that of a shortest path problem on an adequately defined network. A *network* (or directed graph) $\mathcal{N}$ consists of a finite set $\mathcal{V}$ (set of vertices) along with $\mathcal{A}$ (set of arcs) which is subset of directed pairs of vertices, i.e the set $\{(u,v) \mid u \in \mathcal{V}, v \in \mathcal{V}\}$. We define a dipath in $\mathcal{N}$ to be a sequence v_1, a_1, v_2, ..., v_{k-1}, a_k, v_k such that $a_i = (v_i, v_{i+1})$ for all $i = 1,\dots,k-1$. We refer to v_1 as the *origin* of the dipath while we refer to v_k as the *destination* of the dipath. In the preceding definition, the dipath can visit the same vertices several times. Given a cost vector c that associates a real value c_a to each arc $a \in \mathcal{A}$, we define the cost of dipath P to be $\sum_{i=1}^{k} c_{a_i}$. Given a cost vector $c \in \mathbb{R}^n$, and given origin and destination vertices s and t, the shortest dipath problem is the problem of finding a dipath in $\mathcal{N}$ from origin s to destination t that has smallest cost.

We now describe how the group minimization problem described in (19.23) can be solved as a shortest dipath problem. In the ensuing discussion, we say that a vector $a \in \mathbb{Z}^l$ is lexicographically positive if $a_{j^*} > 0$ where $j^* = \min\{i \in \{1,\dots,l\} \mid a_i \neq 0\}$. We also say that $a \in \mathbb{Z}^l$ is lexicographically smaller than $\bar{a} \in \mathbb{Z}^l$, and write $a \prec \bar{a}$ if $\bar{a} - a$ is lexicographically positive. Let $G = \{(a_1, a_2, \dots, a_l) \in \mathbb{Z}^l_+ \mid 0 \leq a_i \leq \lambda_i - 1, \forall i = 1,\dots,l\}$ and $D = \prod_{i=1}^{l} \lambda_i$. Clearly $|G| = D$. We sort the elements of G in lexicographically increasing order. We denote the sorted elements as g_k for $k = 0,\dots,D-1$. From now on, we refer to an element g of G and the vertex that it represents interchangeably. The network we create has one vertex for each element of G. We set the origin vertex $s = 0$ and the destination vertex $t = U_1 d$. For each column j of N and for each vertex $k = 0,\dots,D-1$, we create a directed arc between the vertex corresponding to element g_k and the vertex corresponding to element $(g_k + U_1 A_j) \pmod{\Lambda}$. The cost associated with this directed arc is set to $\bar{c}_j$. We refer to this network as a *group network*, a terminology first employed by Shapiro [91]. The above construction might create parallel arcs between vertices when there are two different columns $j \neq j' \in N$ for which $U_1A_{j'} = U_1A_j$. We claim that the dipaths from s to t correspond to feasible solutions of the group minimization problem (19.23) and that solutions to the group minimization problem can be extended into dipaths from s to t. To see this, consider any dipath P from s to t of cost c^*. For $j \in N$, let $\tilde{x}_j$ be the number of arcs of type U_1A_j that the dipath P contains. It is easy to verify that the solution $\tilde{x}_N = (\tilde{x}_j)_{j\in N}$ is feasible for the corner relaxation and that its objective function value equals c^*. Similarly, given any solution to the corner relaxation, $\tilde{x}_N$, we can create a dipath of cost c^* from vertex 0 to vertex U_1d by following a sequence of $\tilde{x}_j$ arcs of type U_1A_j for $j \in N$. It is clear however that the above dipath is not the unique dipath corresponding to $\tilde{x}_N$ and that re-orderings of its arcs could yield new dipaths from s to t of cost c^*. As a result of the above discussion, we conclude that it is possible to solve the group minimization problem (19.23) as a shortest dipath problem on the group network from vertex s to vertex t.

Example 19.2. In Figure 19.4 we represent (part of) the network that is created for solving (19.24) as a shortest path problem. As mentioned above, it is sufficient to consider the last two equations of (19.24). In the figure, the elements $g = (i, j)$ of G are represented to have x-coordinate i and y-coordinate j. For each vertex, there are two types of outgoing arcs, resulting in a graph with 300 arcs. To avoid overloading the picture with lines, the arcs $\{(0, j); (4, j-1 \pmod{30})\}$ and $\{(4, j); (0, j+7 \pmod{30})\}$ are not represented. The first type of arcs, represented with continuous lines, corresponds to column $(4,29)'$ of (19.24) and has cost $1/30$. The second type of arcs, represented with dashed lines, corresponds to column $(1,7)'$ of (19.24) and has cost $1/6$. In this network, we are looking for a shortest dipath from $(0,0)$ to $(0,3)$. These vertices are indicated with a double diamond.

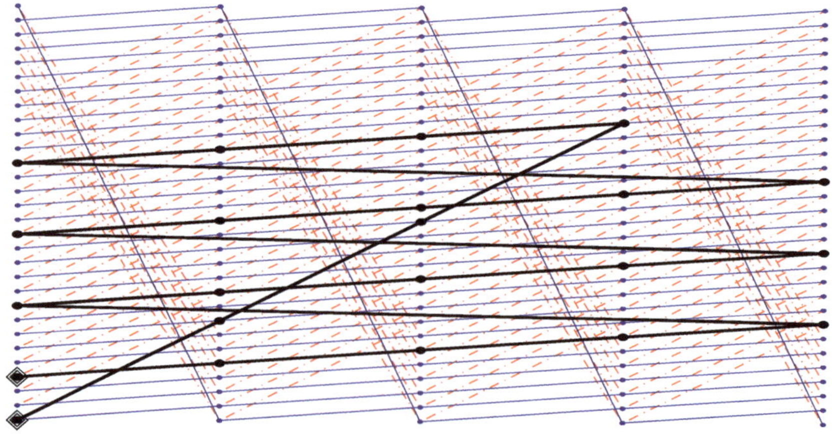

Fig. 19.4 Shortest path problem.

We now describe an algorithm that solves the group minimization problem in time $O(nD^2)$ where $n = |N|$. First, we assume for simplicity that the problem is not infeasible. We add an artificial arc (s,t) with very large cost so that the problem is feasible. Second, we assume that all components of the vector $\bar{c}_N$ are nonnegative for otherwise, the problem is unbounded from below. To see this, assume that $\bar{c}_j$ is negative for some $j \in N$. Since there is an arc of type j emanating from every vertex of the network, we can build an arbitrarily long dipath using only arcs of type j. Since the number of vertices in the network is finite, this implies the existence of a negative cost directed cycle, which proves the claim. Third, we assume that there is no more than one arc between any directed pair of vertices. This is because although parallel arcs might exist when $U_1A_j = U_1A_{j'}$ for $j \neq j' \in N$, there always exists an optimal solution in which only one of these arcs (one with minimum cost $\bar{c}_j$) is used. Finally, we remove all loops, i.e., arcs of the form (g_k, g_k) for some k, since they only add to the cost of dipaths.

We next describe the algorithm of Shapiro [91] as it achieves best computational results in the study of Chen and Zionts [20]. We define $\Gamma_{i+1}(g_k)$ as the cost of the shortest dipath from s to g_k uncovered after the $i+1^{\text{st}}$ iteration of the algorithm. This cost can be obtained using the recurrence:

$$\Gamma_{i+1}(g_k) = \min \begin{cases} \min_{j \in N} \left\{ \bar{c}_j + \Gamma_{i+1}(g_k - U_1A_j) \right\}, & \text{if } g_k \succ g_k - U_1A_j, \\ \min_{j \in N} \left\{ \bar{c}_j + \Gamma_i(g_k - U_1A_j) \right\}, & \text{if } g_k - U_1A_j \succ g_k, \\ \Gamma_i(g_k), \end{cases}$$

with $\Gamma_i(g_0) = 0$ and $\Gamma_0(g_k) = \infty$ for $i = 0, \ldots, D-1$ and $k = 1, \ldots, D-1$. It is proven in Shapiro [91] that $\Gamma_{D-1}(t)$ is the optimal value of the group minimization problem with right-hand-side t. The algorithm runs in time $O(nD^2)$, which can be improved upon. For example, using a Fibonacci heap implementation of Dijkstra's algorithm, see Ahuja et al. [3], shortest paths problems can be solved in time $O(|\mathscr{A}| + |\mathscr{V}| \log |\mathscr{V}|)$. Since in our case $|\mathscr{A}| = nD$, and $|\mathscr{V}| = D$, we obtain the following result.

Theorem 19.3. *The group minimization problem (19.23) associated with basis B can be solved in time $O(nD + D\log D)$ where n is the number of nonbasic variables and $D = \det(A_B)$.*

Theorem 19.3 does not take into consideration the time necessary to obtain the Smith Normal Form of A_B. The above algorithm is clearly not polynomial as D can be large even if the entries of A_B are small. In particular Hadamard [63] proved that for any complex $k \times k$ matrix A whose entries satisfy $|a_{ij}| \leq 1$, $|det(A)| \leq k^{k/2}$. Even for $(0,1)$ matrices, determinants can be large as $\frac{(k+1)^{(k+1)/2}}{2^k}$, an upper bound that is attained. Next, we discuss on an example the output of the algorithm.

Example 19.3. Solving the shortest path problem using the algorithm above, we obtain a shortest path from $(0,0)$ to $(0,3)$ that visits the following sequence of vertices $(0,0)$, $(1,7)$, $(2,14)$, $(3,21)$, $(2,20)$, $(1,19)$, $(0,18)$, $(4,17)$, $(3,16)$, $(2,15)$, $(1,14)$, $(0,13)$, $(4,12)$, $(3,11)$, $(2,10)$, $(1,9)$, $(0,8)$, $(4,7)$, $(3,6)$, $(2,5)$, $(1,4)$, $(0,3)$, where the first 3 arcs correspond to nonbasic column $(1,7)'$ and the last 18 arcs correspond to nonbasic column $(4,29)'$. This solution is represented with heavy continuous lines in Figure 19.4. We conclude that an optimal solution to the group minimization problem is $x_5 = 3$ and $x_4 = 18$. Using (19.8), we deduce that the value of variables $x_1 = 6$, $x_2 = 6$, $x_3 = 9$, $x_6 = 111$ and $x_7 = 1029$. This solution is feasible for the initial integer program (19.7) as all of its components are nonnegative. Therefore, it is optimal for (19.7).

When the group minimization problem to solve contains continuous variables, Wolsey [98] proposed to discretize its continuous variables so as to transform the problem into a purely integer group minimization problem. This method relies on the following result (which is valid for general mixed integer programs). Recall that we assumed that the constraint matrix of (MIP) has full row-rank.

Theorem 19.4 ([98]). *For* (MIP)*, define* $\mathscr{U} = \{B_i\}_{i\in\mathscr{B}}$ *to be the set of dual feasible bases involving only columns of* A_C*. Also define* $\bar{D} = l.c.m.\{|\det(A_{B_i})|\,|\,B_i \in \mathscr{U}\}$*. All optimal solutions to* (MIP) *can be obtained from the solution of the integer program*

$$\begin{aligned}
\min\ & \sum_{j\in I} c_j x_j + \sum_{j\in C} \frac{c_j}{\bar{D}} s_j \\
s.t.\ & \sum_{j\in I} \bar{D} A_j x_j + \sum_{j\in C} A_j s_j = \bar{D} d \\
& x_j \in \mathbb{Z}_+ \ \forall j \in I \\
& s_j \in \mathbb{Z}_+ \ \forall j \in C,
\end{aligned}$$

obtained by substituting $s_j = \bar{D} x_j$ *for* $j \in C$ *and constraining* s *to be a nonnegative integer.*

The magnitude of the multiplier $\bar{D}$ can be reduced when specifically dealing with a corner relaxation; see Lemma 3 in Wolsey [98].

19.3.2 Using corner relaxations to solve MIPs

In our numerical example, we observed that the solution to the group minimization problem is also an optimal solution to the initial integer program. Clearly, this is not always the case as the next example illustrates.

Example 19.4. Consider the variant of (19.7) where $d = (-3,258,63,771,-1071)'$ has been replaced with $d_2 = (8,193,28,756,-1097)'$. The right-hand-side of the simplex tableau (19.8) becomes $(5.5,0.1,2.8,799,99)'$ showing that the basis $B = \{1,2,3,6,7\}$ is still optimal for the LP relaxation of (19.7) and that the corresponding solution is fractional. Further since $U_1 d_2 = (8,785,113,15,3)' \equiv (0,0,0,0,3)')$ $(\text{mod } \Lambda)$ where $\Lambda = diag(1,1,1,5,30)$ we conclude that the shortest path problem to solve remains exactly that of Example 19.3. Therefore, an optimal solution to the group minimization problem is $x_5 = 3$ and $x_4 = 18$. Using (19.8), we deduce that the value of variables $x_1 = 5, x_2 = 0, x_3 = 3, x_6 = 806$ and $x_7 = -22$. The corresponding solution is not feasible as $x_7 < 0$.

Next we study sufficient conditions under which an optimal solution to the corner relaxation also solves the initial integer program. In view of Theorem 19.4, it is sufficient to consider the pure integer case; see Wolsey [98] for an explicit translation of the following results to the mixed integer case.

19.3.2.1 The asymptotic theorem

Consider an optimal solution $\bar{x}_N$ to the group minimization problem. Since the group minimization problem is a relaxation of (MIP), $\bar{x}_N$ will be optimal for (MIP) when the basic variables $\bar{x}_B$ corresponding to $\bar{x}_N$ are nonnegative i.e., $\bar{x}_B = A_B^{-1}d - A_B^{-1}A_N\bar{x}_N \geq 0$. Further if $(\bar{x}_B, \bar{x}_N)$ is optimal for (MIP), then $\bar{x}_N$ might remain optimal for the corner relaxation associated with B when d is changed, and $(A_B^{-1}d - A_B^{-1}A_N\bar{x}_N, \bar{x}_N)$ might be optimal for (MIP), as it is the case in Example 19.4. For this to be true for a particular right-hand-side d^*, we need *(i)* $A_B^{-1}d^* \geq 0$, i.e., the basis B remains optimal for the LP relaxation and so the optimal corner for d^* matches that for d, *(ii)* $U_1 d^* = U_1 d$, i.e., the destination of the shortest dipath remains unchanged, and *(iii)* $\bar{x}_B(d^*) \geq 0$ where $\bar{x}_B(\delta) = A_B^{-1}\delta - A_B^{-1}A_N\bar{x}_N$. The above conditions are very dependent on the knowledge of an optimal solution $\bar{x}_N$ to the group minimization. To obtain conditions that are easier to apply, we study the characteristics of optimal solutions to the group minimization problem.

Since all costs $\bar{c}_N$ are nonnegative, there always exists a shortest dipath in the group network that will not visit any vertex more than once. In fact, when a vertex is visited more than once, a directed cycle is created. Removing this cycle yields a new dipath whose cost is at least as small as before. Since the longest dipaths in the group network that do not repeat vertices have $D-1$ arcs, we obtain the following result.

Proposition 19.1 ([56], [91]). *If the group minimization problem has a feasible solution, it has an optimal solution* $\bar{x}_N$ *that satisfies* $\sum_{j \in N} \bar{x}_j \leq D-1$.

For optimal solutions $\bar{x}_N$ of the group minimization problem, upper bounds can also be obtained on individual variables x_j for $j \in N$. Remember that, for $j \in N$, there is a directed cycle composed only of arcs of type j. Denote by p_j the smallest length of a cycle using only arcs of type j in the group network. Clearly $p_j \leq D$. It follows that if p_j or more arcs of type j were to be used in a dipath from origin to destination, one could create a new dipath from origin to destination by deleting p_j arcs of type j.

Proposition 19.2 ([91]). *If the group minimization problem has a feasible solution, then there exists an optimal solution* $\bar{x}_N$ *that satisfies* $\bar{x}_j \leq p_j - 1$ *for all* $j \in N$.

Using these results, we see that when $d^* \in \mathscr{K}_B^l = \{\delta \in \mathbb{R}^l \mid A_B^{-1}\delta \geq (D-1)|A_B^{-1}A_N|e\}$ where e is the vector $(1,1,\ldots,1)'$ in $\mathbb{R}^l$ and when (MIP) is feasible for d^*, then d^* satisfies Condition *(i)*: $A_B^{-1}d^* \geq 0$. Further, let $\bar{x}_N$ be any optimal solution to the group minimization problem corresponding to basis B that satisfies $\sum_{j\in N}\bar{x}_j \leq D-1$. Then (*ii*) is not required and (*iii*) is satisfied as $\bar{x}_B(d^*) = A_B^{-1}d^* - A_B^{-1}A_N\bar{x}_N \geq (D-1)|A_B^{-1}A_N|e - A_B^{-1}A_N\bar{x}_N \geq 0$. It follows that all optimal solutions to the corner relaxation that satisfy $\sum_{j\in N}\bar{x}_j \leq D-1$ solve (MIP). Similarly when $d^* \in \mathscr{K}_B'' = \{\delta \in \mathbb{R}^l \mid A_B^{-1}\delta \geq |A_B^{-1}A_N|p\}$ where p is the vector $(p_1-1, p_2-1, \ldots, p_n-1)'$ and when (MIP) is feasible for d^*, then all optimal solutions to the corner relaxation that satisfy $\bar{x}_j \leq p_j - 1$ for all $j \in N$ solve (MIP).

Theorem 19.5 ([91]). *Let B be an optimal basis of the LP relaxation of* (MIP). *Assume that $d \in \mathscr{K}_B'$ or $d \in \mathscr{K}_B''$ and that* (MIP) *has a feasible solution. Then there is an optimal solution to* (MIP) *that minimizes cx over the corner relaxation associated with basis B.*

The above conditions indicate that an optimal solution to the group minimization problem from the corner associated with an optimal basis of the LP relaxation will solve the integer program provided that its right-hand-side d is sufficiently inside of the cone $\mathscr{K}_B = \{\delta \in \mathbb{R}^l \mid A_B^{-1}\delta \geq 0\}$. For a fixed cost vector c, $\mathscr{K}_B$ is the cone of right-hand-side vectors for which the dual feasible basis B is optimal for the LP relaxation. The difference between cone $\mathscr{K}_B$ and cones $\mathscr{K}_B'$ and $\mathscr{K}_B''$ describes the "fat" around $\mathscr{K}_B$ for which the result might not hold. A condition similar to that of Theorem 19.5 was given earlier by Gomory [56]. We present it next. Given the cone $\mathscr{K}_B$, we denote the set of points belonging to $\mathscr{K}_B$ that are at a Euclidean distance of d or more from the boundary of $\mathscr{K}_B$ by $\mathscr{K}_B(d)$. Clearly $\mathscr{K}_B = \mathscr{K}_B(0)$ and $\mathscr{K}_B(d) \subseteq \mathscr{K}_B$.

Theorem 19.6 ([56]). *Let B be an optimal basis of the LP relaxation of* (MIP). *Assume that $d \in K_B(l_{\max}(D-1))$ where $D = \det(A_B)$ and $l_{\max}$ is the Euclidian length of the longest nonbasic column of* (MIP). *Assume also that* (MIP) *has a feasible solution. Then there is an optimal solution to* (MIP) *that minimizes cx over the corner relaxation associated with basis B.*

The result of Theorem 19.6 is known under the name of *Asymptotic Theorem*. Its conditions are very conservative and the result holds in situations where the conditions do not. More generally, Balas [9] show that the sufficient condition of Theorem 19.6 (and in fact an even more general set of conditions) are so restrictive that they are never satisfied for 0-1 integer programs. Although it was believed for some time that corner relaxations would often be sufficient to solve IPs (see Hu [67]) the early large-scale study of group-theoretic approaches in IP conducted by Gorry et al. [63] concluded that "Computation has in fact revealed that solving the asymptotic problem does not solve the original problem for most real life IP problems of any size, say problems with more than 40 rows."

The derivation of the asymptotic theorem has an interesting consequence. Consider a basis B that is optimal for the LP relaxation of (MIP) for some right-hand-side d. Construct the group network associated with this basis, and for every one of its vertices g_k for which there exists a dipath from 0 to g_k, compute a shortest path from vertex 0 to vertex g_k that satisfies the conditions of Proposition 19.1. Denote the solutions of the corner relaxation corresponding to these dipaths as $\bar{x}_N^k$ for $k \in \{0,\ldots,D-1\}$. It follows from our derivation of Theorem 19.5 that for all right-hand-side vectors d^* in the cone $\mathscr{K}_B^l$ for which (MIP) has a feasible solution, one of the "corrections" $\bar{x}_N^k$ for $k \in \{0,\ldots,D-1\}$, say k^*, will be optimal for the corner relaxation and therefore the corresponding solution $(A_B^{-1}d^* - A_B^{-1}A_N\bar{x}_N^{k^*}, \bar{x}_N^{k^*})$ will be optimal for (MIP). Therefore, there is a finite set of corrections $\bar{x}_N^k$, $k \in \{0,\ldots,D-1\}$ from which we can build an optimal solution to each one of the feasible (MIP) (infinite in number) whose right-hand-sides d^* are sufficiently inside of the cone $\mathscr{K}_B$. This fact was observed by Gomory [56] and the idea of tabulating the values of $\bar{x}_N$ to solve (MIP) with different right-hand-sides was proposed by Gomory [57]. We record this result next.

Theorem 19.7. *Let B be an optimal basis of the LP relaxation of* (MIP) *for some right-hand-side d'. There exists a finite list of nonnegative integer vectors $\{x_N^k\}_{k\in Q_B}$ with the following property. For each $d \in \mathscr{K}_B^l$, the set*

$$Q_d = \{k \in Q_B \mid x_B^k = A_B^{-1}d - A_B^{-1}A_N x_N^k \geq 0 \text{ and integer } \}.$$

is such that

1. *$Q_d = \emptyset$ implying that* (MIP) *is infeasible, or*
2. *$Q_d \neq \emptyset$ implying that if $\bar{c}_N x_N^{k^*} = \min_{k\in Q_d}\{\bar{c}_N x_N^k\}$, the vector $(x_B^{k^*}, x_N^{k^*})$ is an optimal solution for* (MIP).

We will see in the following section that this result holds true even when the right-hand-sides d do not lie deep inside of the cone $\mathscr{K}_B$. To prove this result, first derived by Wolsey [100], it is not sufficient to tabulate shortest dipaths to every vertex in group networks. Instead, it is necessary to consider solutions to tighter relaxations.

19.3.2.2 When corner relaxations fail

Consider now an optimal solution of the corner relaxation of the problem obtained using the shortest path algorithm of Section 19.3.1 and assume that this solution is not feasible to (MIP). We describe next two families of approaches that can be used to obtain an optimal solution of (MIP).

Divide-and-conquer

Since the group minimization problem is a relaxation of (MIP), an optimal solution of (MIP) can be found among its feasible solutions. More precisely, given (MIP) and a corner relaxation based on basis B, one can obtain an optimal solution to (MIP) by enumerating all nonnegative integer vectors $\bar{x}_N$ of $\text{corner}_B(F)$ for which $\bar{x}_B = A_B^{-1}d - A_B^{-1}A_N\bar{x}_N$ is nonnegative and integer, and then choosing among them one that has lowest objective value. This exhaustive search might however be time-consuming and an implicit enumeration scheme would likely be more efficient in practice. In particular, one could think about using bounds obtained by solving the group minimization problem to trim down the search space.

One early idea along these lines was proposed by Shapiro [92]. When an optimal solution to the group minimization problem does not solve (MIP), its feasible region is divided into subsets, each one defined by imposing a different lower bound vector l_N on the nonbasic variables x_N. To obtain bounds on the best objective value achieved in each one of these subsets, the following optimization problem is solved

$$\tilde{z} = \min\{z^* + \bar{c}_N x_N \mid x_N \in \text{corner}_B(F), x_N \geq l_N\}. \tag{19.25}$$

Although this problem differs from the traditional group minimization problem, it is easy to see that its feasible solutions can still be considered to be dipaths from s to t in the group network. Further, this problem can be interpreted as that of finding a shortest dipath from origin s to destination t that uses at least l_j arcs of type j. Therefore, one simple way to solve this problem is to create a dipath from the origin by following l_j arcs of type j for all $j \in N$. This dipath leads to a new vertex that we call o'. It then remains to complete this dipath to reach destination vertex t. Therefore (19.25) can be interpreted as a shortest dipath problem from vertex o' to vertex t. Due to the symmetry of the group network, this problem reduces to finding a shortest dipath from vertex o to vertex $t - o'$. Solving this problem is simple as the dynamic programming algorithm we presented for solving the group minimization problem not only finds the shortest dipath from 0 to t ($\Gamma_{D-1}(t)$) but also from vertex 0 to every other vertex g ($\Gamma_{D-1}(g)$). Therefore, a solution to (19.25) can be easily found by re-using the previous dynamic programming table. A large-scale implementation of the above branch-and-bound algorithm, together with several improvements is described in Gorry et al. [63]. This algorithm also takes advantage of binary variables, when they are present in the problem. We refer to Nemhauser and Wolsey [83] pages 445–447 for a numerical example.

Feasible solutions to the corner relaxation can be seen as dipaths in the group network. Therefore, the branch-and-bound algorithm described above can be seen as a way to search the s-t dipaths in the group network for a shortest dipath that corresponds to a feasible solution to the integer program. In the case where the corner relaxation does not solve the integer program, this dipath is not a shortest dipath in the group network but is probably a "short" dipath in the group network. This suggests that one could find an optimal solution to the problem by developing algorithms that find the K shortest dipaths in the group network for K sufficiently large. This idea was first introduced by Wolsey [99] where a dynamic programming algorithm for finding the K best solutions to an integer program is described. This algorithm works directly on the solution of the corner relaxation rather than the associated group network as there can be many different dipaths in the group network that correspond to a single solution of the corner relaxation.

Relaxation improvement

When the group relaxation is not sufficiently strong to solve (MIP), we could search to produce a new, stronger, relaxation that does not contain the undesirable optimal solution. One way to achieve such stronger relaxation is to embed the problem into a finer group. The first algorithm based on this idea was proposed by Bell and Shapiro [13]. In this approach, the set of constraints of the problem is first duplicated. The first set is transferred to the objective with a penalty weight, following traditional Lagrangian relaxation principles. The feasible region associated with the second set is replaced with a group relaxation based on a group G. The resulting problem is therefore a relaxation of the initial integer program. The Lagrangian dual problem is solved using a combination of subgradient and primal-dual methods. If the solution to this relaxation is feasible to the initial integer program and satisfies optimality conditions, it is optimal for (MIP) and the algorithm is terminated. Otherwise a new group relaxation is produced using a group G' which is supergroup of G. This group G' is produced from optimal solutions of the primal-dual algorithm. The process is then iterated until an optimal solution to (MIP) is found. This procedure is proven to be finitely convergent for 0-1 problems in Bell and Shapiro [13]. We refer to Nemhauser and Wolsey [83] for a simplified version of the algorithm.

Another idea to strengthen the relaxation is to improve it directly where it is known to be weak. In particular, if the group minimization problem associated with basis B has an optimal solution $\bar{x}_N$ whose associated vector $\bar{x}_B$ is such that $\bar{x}_j < 0$ for $j \in B$, an improved relaxation of the integer program could be produced by requiring that the nonnegativity of basic variable x_j is not relaxed in the group relaxation. This idea, first proposed by Wolsey [97], leads to the concept of extended group relaxations. This concept has many ramifications. We devote the next section to its study.

19.3.3 Extended group relaxations

As mentioned above, a possible way to overcome situations where the corner relaxation is not sufficiently strong to solve (MIP) is to generate stronger relaxations by relaxing nonnegativity for only a subset of the basic variables. Wolsey [97] first proposed this idea in 1971. Again, assume that B is an optimal basis for the LP relaxation of (MIP) and let $(T,\bar{T})$ be a partition of the basic variables B. We are interested in solving extended group minimization problems of the form

$$\min\{z^* + \bar{c}_N x_N \mid A_B x_B + A_N x_N = d, x_T \in \mathbb{Z}^{|T|}, (x_{\bar{T}}, x_N) \in \mathbb{Z}_+^{|\bar{T}|+|N|}\} \quad (19.26)$$

in which the nonnegativity of basic variables in $\bar{T}$ is maintained. We refer to the feasible regions of extended group minimization problems as *extended group relaxations* and denote them by $EG_{B,T}(F)$, i.e.,

$$EG_{B,T}(F) = \{(x_T, x_{\bar{T}}, x_N) \in \mathbb{Z}^{|T|} \times \mathbb{Z}_+^{|\bar{T}|+|N|} \mid A_B x_B + A_N x_N = d\}.$$

It is clear that the strength of the extended group relaxation increases as the size of $|T|$ decreases. In particular, if an extended group relaxation is sufficiently strong to solve (MIP), then any of the extended group relaxations $EG_{B,T'}(F)$ with $T' \subseteq T$ will also solve (MIP). Further, since when $T = \emptyset$, the extended group relaxation corresponds to (MIP), there always exists an extended group relaxation that solves (MIP). It is therefore expected that extended group relaxations will become harder to solve as the size of $|T|$ decreases.

There are different ways of solving extended group relaxations. A first method, introduced by Wolsey [99], is an extension of the approach that was presented to solve group relaxations in Section 19.3.1. We again use the Smith Normal Form of a matrix, but this time, we compute the SNF of matrix A_T. Let U_1 and U_2 be unimodular matrices such that $U_1 A_T U_2 = \binom{\Lambda}{0}$. After multiplying the constraints of (19.26) throughout by matrix U_1, we obtain the following equivalent formulation of the problem:

$$\min \left\{ z^* + \bar{c}_N x_N \;\middle|\; \begin{array}{r} \begin{pmatrix} \Lambda \\ 0 \end{pmatrix} U_2^{-1} x_T + U_1 A_{\bar{T}} x_{\bar{T}} + U_1 A_N x_N = U_1 d \\ x_T \in \mathbb{Z}^{|T|}, (x_{\bar{T}}, x_N) \in \mathbb{Z}_+^{|\bar{T}|+|N|} \end{array} \right\} \tag{19.27}$$

which reduces to

$$\min \left\{ z^* + \bar{c}_N x_N \;\middle|\; \begin{array}{l} (U_1 A_{\bar{T}})^j x_{\bar{T}} + (U_1 A_N)^j x_N \equiv (U_1 d)^j \pmod{\lambda_j}, \text{for } j \in [T] \\ (U_1 A_{\bar{T}})^j x_{\bar{T}} + (U_1 A_N)^j x_N = (U_1 d)^j, \text{for } j \in [\bar{T}] \\ (x_{\bar{T}}, x_N) \in \mathbb{Z}_+^{|\bar{T}|+|N|} \end{array} \right\}, \tag{19.28}$$

where $[T] = \{1, \dots, |T|\}$, $[\bar{T}] = \{|T|+1, \dots, l\}$ and A^j is used to describe the jth row of matrix A. Wolsey [99] proposes to use a dynamic programming algorithm to solve (19.28). This algorithm is similar to that described in Section 19.3.1 and is therefore not repeated here. Extended group minimization problems can also be expressed as lattice programming problems, i.e., optimization problems of the form $\min\{cx \mid x \equiv d \pmod{\mathscr{L}}, x \in \mathbb{Z}_+^n\}$ where $c \in \mathbb{R}^n$, $\mathscr{L}$ is a sublattice of $\mathbb{Z}^n$ and $d \in \mathbb{Z}_+^n$. In particular, the extended group minimization problem (19.27) can be expressed as the lattice programming problem

$$\min\{\pi_T(c - c_B A_B^{-1} A) x_{\bar{T}} \mid x_{\bar{T}} \equiv \pi_T(u) \pmod{\mathscr{L}_T}, x_{\bar{T}} \in \mathbb{Z}^{|\bar{T}|}\}$$

where u is a feasible solution to (MIP), π_T is the operator that takes a vector v in $\mathbb{R}^m$ and maps it to the vector $\tilde{v}$ in $\mathbb{R}^{|\bar{T}|}$ obtained by deleting from v all components whose indices are in T and $\mathscr{L}_T = \pi_T(\mathscr{L})$ where $\mathscr{L}$ is the lattice $\{x \in \mathbb{Z}^n \mid Ax = 0\}$; see Thomas [95] for more detail. Lattice programming problems can be solved using Gröbner bases; see Sturmfels et al. [94].

Extended group relaxations can therefore be used to help solve integer programs when their corner relaxations do not. Further, by carefully selecting finite sets of points in these extended group relaxations, it is possible to generalize the result of Theorem 19.7 to right-hand-sides not deeply inside of the cones $\mathscr{K}_B$. In particular, Wolsey [100] first proved that there is a finite number of corrections $\bar{x}_N$ from which an optimal solution to each of the feasible integer programs

$$\texttt{IP}(d) : \min\{cx \mid Ax = d,\ x \in \mathbb{Z}^m_+\}$$

can be deduced. Formally,

Theorem 19.8 ([100]). *Assume* $\{u \in \mathbb{R}^l \mid u'A \geq c\} \neq \emptyset$. *Then there exists a finite list of nonnegative integer vectors* $\{x^q_N\}_{q \in Q_B}$ *for each dual feasible basis B with the following property: Let*

$$Q_d = \{q \in Q_B \mid x^q_B = A_B^{-1}d - A_B^{-1}A_N x^q_N \geq 0 \textit{ and integer }\}.$$

Either $Q_d = \emptyset$ *implying that* $\texttt{IP}(d)$ *is infeasible, or* $Q_d \neq \emptyset$ *implying that if* $\bar{c}_N x^{q*}_N = \min_{q \in Q_d}\{\bar{c}_N x^q_N\}$, *the vector* $(x^{q^*}_B, x^{q^*}_N)$ *is an optimal solution for* $\texttt{IP}(d)$.

Although the proof given in Wolsey [100] is constructive, it produces an overabundance of corrections and generates these corrections from all extended group relaxations. A more detailed characterization is obtained by Hoşten and Thomas [65] using algebraic arguments. In particular, they determine a minimal set of extended group relaxations that solves all feasible integer programs in $\texttt{IP}(d)$. We next present some of these results. The presentation is in line with that given in Thomas [95]. Recall that we are now interested in the family of problems $\texttt{IP}(d)$ in which the constraint matrix A and objective coefficients c are fixed but right-hand-sides d are allowed to take any value in $\mathbb{Z}^l$. We remove right-hand-sides d for which $\texttt{IP}(d)$ is infeasible by requiring that $d \in \mathscr{D} = \{\delta \in \mathbb{R}^l \mid \delta = Az \text{ for } z \in \mathbb{Z}^m_+\}$. We also assume that the vector c is generic, i.e., for all $d \in \mathscr{D}$, problem $\texttt{IP}(d)$ has a unique optimal solution. Further, instead of focusing solely on those extended group relaxations coming from an optimal basis of the LP relaxation of $\texttt{IP}(d)$, we consider extended relaxations from various sets of columns T that are not necessarily part of an optimal basis of the LP relaxation for d. As a result, we now refer to extended relaxations directly as $EG_T(F)$ instead of $EG_{B,T}(F)$. This extension allows some problems that otherwise would only be solved by trivial extended group relaxations to be solved by nontrivial extended group relaxations.

We define a cone complex Δ to be a collection of polyhedral cone called cells (or faces) for which *(i)* every face of a cell of Δ is a cell of Δ and *(ii)* the intersection of two cells of Δ is defined to be their common face. For nonempty cone complexes, the empty set is a cell of Δ since it is a face of every polyhedral cone.

Definition 19.1. The regular subdivision Δ_c of cone(A) is defined as the cone complex whose cells are all cones of the form cone(A_S) with $S \subseteq M$ for which there exists a vector $y \in \mathbb{R}^l$ for which $yA^j = c_j$ for $j \in S$ and $yA^j < c_j$ for $j \in M \backslash S$.

Although the elements of Δ_c are cones, we will refer to them simply through the index set S that they are defined from. The regular subdivision Δ_c is relevant to the study of extended relaxations as it defines the sets T for which extended group minimization problems are bounded.

Theorem 19.9 ([95]). *The extended group minimization defined on T for* $\texttt{IP}(d)$ *has a finite optimal solution if and only if $T \subseteq M$ is a face of Δ_c.*

As we mentioned above, if a relaxation $EG_T(F)$ solves $\texttt{IP}(d)$ then every extended group relaxation with $T' \subseteq T$ also solves the problem. Therefore, we are interested in those relaxations that solve the problem with $|T|$ as large as possible. This motivates the following definition.

Definition 19.2. A face T of the regular subdivision Δ_c is an associated set of $\texttt{IP}(d)$ if, for some $d \in \mathscr{D}$, the extended group minimization problem based on T solves $\texttt{IP}(d)$ but the extended group minimization problem based on T' does not solve $\texttt{IP}(d)$ for any face T' of Δ_c for which $T \subset T'$.

The fact that there is a finite number of associated sets does not imply Theorem 19.8, since it is conceivable that for each d solved by a particular associated set, a different correction vector $\bar{x}_T(d)$ would have to be created. However, only a finite number of corrections $\bar{x}_N$ are needed to obtain the optimal solution of all problems in $\texttt{IP}(d)$ for $d \in \mathscr{D}$ as we will see. First we denote by $\mathscr{O}_c$ the set of all optimal solutions of the problem $\texttt{IP}(d)$ that are obtained as d varies in $\mathscr{D}$. Since we assume that the family of problems is generic, there is a single solution for each $d \in \mathscr{D}$. Further, it can be proven that $\mathscr{O}_c$ is a down set, i.e., if $u \in \mathscr{O}_c$, $v \leq u$ and $v \in \mathbb{Z}_+$, then $v \in \mathscr{O}_c$. We are trying to show that the solutions of $\mathscr{O}_c$ can be explained by a finite number of carefully selected corrections. These corrections are underlying the following definition where we use the notation $S(u,T)$ to denote the set $\{u + \mathbb{N}e_i \mid i \in T\}$.

Definition 19.3. For $T \in \Delta_c$ and $u \in O_c$, (u,T) is called an admissible pair of O_c if

1. the support of u is contained in $\bar{T}$ and
2. $S(u,T) \subseteq O_c$.

Further, an admissible pair (u,T) is a standard pair of $\mathscr{O}_c$ if $S(u,T)$ is not properly contained in $S(v,T')$ for some other admissible pair (v,T').

Standard pairs help characterize what corrections and what extended group relaxations are needed to solve all problems $\texttt{IP}(d)$. For $d \in \mathbb{N}A$ and a standard pair (u,T), the system $A_T x = d - A_{\bar{T}} \pi_T(u)$ can be solved uniquely for x as T is a face of Δ_c. In other words, the vectors u can be thought of as one the correction vectors that can be used to obtain optimal solutions for $\texttt{IP}(d)$ as d varies in $\mathscr{D}$. Hoşten and Thomas [65] present an algorithm to compute standard pairs in a more general setting. Next, we describe a characterization of standard pairs using the concept of standard polytope. This notion arises from the reformulation of the extended group minimization problems as follows. Given a solution u in $\texttt{IP}(d)$ and assuming that $T = B$, we can reformulate the group minimization problem as

$$\begin{aligned}\min\{cx \mid x-u \in \mathscr{L}, x \in \mathbb{Z}^m_+\} &\iff \min\{cx \mid x-u=-Lz, x \geq 0, z \in \mathbb{Z}^{m-l}\}\\ &\iff \min\{(-cL)z \mid Lz \leq u, z \in \mathbb{Z}^{m-l}\}\end{aligned}$$

where $\mathscr{L}$ is the lattice $\{x \in \mathbb{Z}^m \mid Ax = 0\}$ and L is a matrix whose columns generate the lattice $\mathscr{L}$. Similarly, general extended group relaxations can be reformulated as

$$\min\{(-cL)z \mid L^T z \leq \pi_T(u), z \in \mathbb{Z}^{m-l}\}.$$

where L^T is the matrix obtained from L by deleting all of its rows indexed by T.

Definition 19.4. Given a face T of Δ_c and a point $u \in \mathbb{Z}^m_+$, we define the polytope $Q_u^{\bar{T}} = \{z \in \mathbb{R}^{m-l} \mid L^T z \leq \pi_T(u), (-cL)z \leq 0\}$ to be a standard polytope of $\mathtt{IP}(d)$ if $Q_u^{\bar{T}} \cap \mathbb{Z}^{m-l} = \{0\}$ and all relaxations of $Q_u^{\bar{T}}$ obtained by removing one inequality of $L^T z \leq \pi_T(u)$ contain one non-zero point in $\mathbb{Z}^m_+$.

The following result establishes the relations between standard pairs, standard polytopes and associated sets.

Theorem 19.10 ([65], [95]). *The admissible pair (u,T) is a standard pair of $\mathscr{O}_c$ if and only if the face T of Δ_c is associated to $\mathtt{IP}(d)$ if and only if the polytope $Q_u^{\bar{T}}$ is a standard polytope of $\mathtt{IP}(d)$.*

The set $\mathscr{O}_c$ is covered by the sets $S(u,T)$ corresponding to its standard pairs. It follows that the standard pair decomposition of $\mathscr{O}_c$ is unique since the standard pairs of $\mathscr{O}_c$ are obtained directly from standard polytopes. Further, since there is a finite number of standard polytopes of the form $Q_u^{\bar{T}}$ for all associated sets T, there is only a finite number of standard pairs. Associated sets induce a poset structure on the faces of the regular triangulation Δ_c with respect to set inclusion. Although the maximal elements of this poset are the maximal faces of Δ_c, its structure is complicated. However, the following important structural result holds.

Theorem 19.11 ([65], [95]). *If $T \in \Delta_c$ is an associated set of $\mathtt{IP}(d)$ and $|T| < l$, then there exists a face $T' \in \Delta_c$ that is also an associated set of $\mathtt{IP}(d)$ with the property that $T \subset T'$ and $|T' \backslash T| = 1$.*

The characterization of standard pairs opens various new lines of research. For example, we could search to determine families of integer programs $\mathtt{IP}(d)$ that are solved by the simplest extended group relaxations (those that correspond to maximal faces of Δ_c) for all $d \in \mathscr{D}$. Such a family of integer programs $\mathtt{IP}(d)$ is called a *Gomory family*. We refer to Hoşten and Thomas [66] for an analysis of these programs. In particular, the following result can be established.

Theorem 19.12 ([66], [95]). *The family of integer programs $\mathtt{IP}(d)$ is a Gomory family if and only if its standard polytopes are simplices.*

19.4 Master group relaxations: definitions and inequalities

Although it is simple to optimize linear functions over corner relaxations using shortest path algorithms, it is more difficult to study their polyhedral structure. In particular, although corner relaxations have a nice geometrical interpretation, they depend heavily on the actual instance of the mixed integer programming problem at hand. This suggests that results obtained from their polyhedral study would be very specific and therefore would have to be performed for every basis B and for every feasible set F, a dantesque task. Gomory [57] however pointed out that the polyhedral structure of $\text{conv}(\text{corner}_B(F))$ can be deduced from the study of more general master relaxations that are less dependent on the structure of the initial problem. We focus on these master relaxations and their structure in this section.

19.4.1 Groups

The master relaxations are based on the algebraic notion of a group. We therefore introduce first some basic definitions.

Definition 19.5. A group $(G, \oplus)$ is a set G with a binary operation $\oplus$ that satisfies the following four properties:

i. For all g_1, g_2 in G, $g_1 \oplus g_2$ belongs to G.
ii. For all g_1, g_2, g_3 in G, $(g_1 \oplus g_2) \oplus g_3 = g_1 \oplus (g_2 \oplus g_3)$.
iii. There exists $\bar{g} \in G$ (unique zero element) such that $g \oplus \bar{g} = \bar{g} \oplus g = g$ for all g in G.
iv. For all g_1 in G, there exists g_2 in G such that $g_1 \oplus g_2 = g_2 \oplus g_1 = \bar{g}$.

The definition of group does not require the binary operation to be commutative, i.e., $g_1 \oplus g_2$ is not necessarily equal to $g_2 \oplus g_1$. However, the groups that we will use in the remainder of this chapter satisfy this property. Given g in G and given $n \in \mathbb{Z}_+$, we define the element ng as $\underbrace{g \oplus g \oplus \cdots \oplus g}_{n \text{ times}}$. The following definition ensues.

Definition 19.6. The order of an element g of a finite group G is the smallest positive integer n such that $ng = \bar{g}$.

We define $(G', \oplus)$ to be a subgroup of $(G, \oplus)$ if $G' \subseteq G$ and $(G', \oplus)$ satisfies the definition of group. If G' is a subgroup of G, we use the notation $G' \preccurlyeq G$. Among all groups, we will focus on those that are subgroups of the group I^m that is defined next.

Definition 19.7. We refer to the group of real m-dimensional vectors with the addition modulo 1 componentwise as the infinite group I^m. We refer to m as the dimension of I^m.

Among the subgroups of I^m, the following are important both historically and practically.

Definition 19.8. Given a vector K of m positive integers, we refer to the group of real m-dimensional vectors whose components i are integer multiples of $\frac{1}{K_i}$ with the addition modulo 1 componentwise as the finite group C_K^m. We refer to m as the dimension of C_K^m. The subgroup $C_k^1 \preccurlyeq I^1$ is called the cyclic group of order k.

As we will often switch from coefficients of the simplex tableau (19.6) (coefficients in $\mathbb{R}^m$) to their representative in the corner relaxation $\text{corner}_B(F)$ (coefficients in I^m), we define $\mathscr{F} : \mathbb{R}^m \to I^m$ to be the function that maps a vector of real numbers to the vector of their fractional parts. Similarly we define $\mathscr{F}^{-1} : I^m \to \mathbb{R}^m$ to be the function that maps the element v of I^m to the element v' of $\mathbb{R}^m$ that is such that $\mathscr{F}(v') = v$ and $0 \leq v_i < 1$ for $i = 1, \ldots, m$.

We next give the definitions of group homomorphism and automorphism that play an important role in the study of relations between inequalities of various master group relaxations as we shall see in Section 19.5.

Definition 19.9. Given two groups $(G_1, \oplus)$ and $(G_2, \otimes)$, a function $\lambda : G_1 \to G_2$ is said to be a homomorphism if $\lambda(g_1 \oplus g_2) = \lambda(g_1) \otimes \lambda(g_2)$ for all $g_1, g_2 \in G_1$.

For I^1, it is easy to see that $\lambda(x) = nx \pmod 1$ where $n \in \mathbb{Z} \backslash \{0\}$ is a homomorphism of I^1 onto itself. This homomorphism can be easily extended to I^m by defining $\lambda(x) = (n_1 x_1, n_2 x_2, \ldots, n_m x_m)$ where $(n_1, \ldots, n_m) \neq 0$. We refer to $\lambda(x)$ as *multiplicative homomorphism*. An important characteristic of homomorphisms is their kernels.

Definition 19.10. We define the kernel of a homomorphism $\lambda : G_1 \to G_2$ as the set of elements of G_1 that are mapped to the zero of G_2, i.e., $\text{Kern}(\lambda) := \{g \in G_1 \mid \lambda(g) = 0\}$.

Among homomorphisms, those that are bijective are particularly interesting; see Section 19.5.

Definition 19.11. Given two groups $(G_1, \oplus)$ and $(G_2, \otimes)$, a function $\omega : G_1 \to G_2$ is said to be an isomorphism if ω is a homomorphism and if ω is also a bijection. We say that G_1 is isomorphic to G_2.

Given two groups $(G_1, \oplus)$ and $(G_2, \otimes)$, we define the group $G_1 \times G_2$ as the set $\{(u_1, u_2) \mid u_1 \in G_1, u_2 \in G_2\}$ together with the group operation between elements (u_1, u_2) and (v_1, v_2) taken as $(u_1 \oplus u_2, v_1 \otimes v_2)$. It is easily established that $C_{k_1}^1 \times C_{k_2}^1 \times \ldots \times C_{k_m}^1$ is isomorphic to the group C_K^m, where $K = (k_1, k_2, \ldots, k_m)$.

Isomorphisms establish links between groups that are stronger than those derived through homomorphisms. We will be particularly interested in isomorphisms of I^m onto itself.

Definition 19.12. An automorphism $\phi : G \to G$ is an isomorphism of a group to itself.

For I^2, it is easy to see that rotations (i.e., $\rho : I^2 \to I^2, \rho(x,y) = (1-y,x)$), reflections (i.e., $\varsigma : I^2 \to I^2$, $\varsigma(x,y) = (1-x,y)$) and their combinations form automorphisms. This observation naturally extends to I^m.

19.4.2 Master group relaxations of mixed integer programs

In this section, we further relax $\text{corner}_B(F)$ in such a way that results apply to a larger number of problems. We proceed in two steps. In the first step, we aggregate all nonbasic variables whose column coefficients fractional parts $\mathscr{F}(A_B^{-1}A_j)$ are identical. After this operation, all the variables in the relaxation have different coefficient vectors. In the second step, we obtain a subgroup G of I^m (G can be I^m itself) that contains all of the coefficient vectors $\mathscr{F}(A_B^{-1}A_j)$ for $j \in N$ and also $r := \mathscr{F}(A_B^{-1}d)$. We then relax the problem to a higher dimensional space by adding one variable for each element of the group G in the modular relation (19.18). We obtain

$$\text{mastercorner}(F) = \left\{ t \in \mathbb{Z}_+^{|G|} \;\middle|\; \sum_{g \in G} g t_g = r \right\} \tag{19.29}$$

which we refer to as *master corner relaxation* or *master group relaxation*. In (19.29) and wherever groups are used, the equality must be interpreted as modulo 1. Clearly, the set obtained is a relaxation of $\text{corner}_B(F)$. Further, the polyhedral structures of $\text{corner}_B(F)$ and $\text{mastercorner}(F)$ are intimately related; see Theorem 19.19 and Section 19.4.3. Observe also that, in the above definition of $\text{mastercorner}(F)$, we did not specify the nature of the group that must be used. In particular, the group can be chosen to be finite or infinite. We now give a formal definition of the master group relaxation based on G. We denote the 0 element of G as o.

Definition 19.13. Let G be a subgroup of I^m and let $r \in G \backslash \{o\}$. The master group problem $\text{MG}(G, \emptyset, r)$ is defined as the set of functions $t : G \to \mathbb{Z}$ such that
i. $\sum_{u \in G} u t(u) = r$,
ii. $t(u)$ is a non-negative integer for $u \in G$,
iii. t has a finite support, i.e., $t(u) > 0$ for only a finite subset of G.

In Definition 19.13, each feasible solution to the group relaxation is thought of as a function t on G since a variable is associated with every group element of G. The last condition of Definition 19.13 helps in ensuring that the expression $\sum_{u \in G} u t(u)$ is well-defined, even if the group used contains an infinite number of elements. When the group G used in the relaxation is finite, this last requirement is trivially satisfied. In Definition 19.13, we require that $r \neq o$ as we wish to generate cuts from tableaux whose basic solutions are fractional. It is however possible, as pointed out by Gomory [57] to define $\text{MG}(G, \emptyset, r)$ analogously for $r = o$. The only difference is that the function that is identically zero is removed from consideration. This definition is motivated by situations where we wish to generate cuts from a subset of tableaux rows whose basic variables are currently integer, but other tableaux rows (not considered in the group relaxation) have basic variables that are fractional. We will assume in the remainder of this chapter that $r \neq o$. We refer to Gomory [57] for a discussion of the case $r = o$.

We also mention that Gomory and Johnson [59] and Johnson [71] investigated variants of Definition 19.13 in which G is not chosen to be a subgroup of I^m but simply a subset of I^m. However, the results obtained under these less stringent conditions are less appealing and more difficult to apply.

Example 19.5. Consider the corner relaxation obtained for the simplex tableau (19.8). Introducing the notation $g_1 = (\frac{1}{30}, \frac{29}{30}, \frac{26}{30}, \frac{10}{30}, \frac{20}{30})'$, $g_2 = (\frac{29}{30}, \frac{7}{30}, \frac{22}{30}, \frac{20}{30}, \frac{10}{30})'$ and $r = (\frac{15}{30}, \frac{3}{30}, \frac{24}{30}, 0, 0)'$ and relabeling the variables x_4 as t_{g_1} and x_5 as t_{g_2}, we can rewrite (19.8) as

$$g_1 t_{g_1} + g_2 t_{g_2} \equiv r \pmod 1. \tag{19.30}$$

Master group relaxations of (19.30) are obtained by identifying a subgroup of I^5 that contains elements g_1 and g_2. One such group is $G = C^5_{(30,30,30,30,30)}$. After adding one variable for each element of the group G not already present in the modular expression (19.30), we obtain

$$\sum_{i=0}^{29}\sum_{j=0}^{29}\sum_{k=0}^{29}\sum_{l=0}^{29}\sum_{m=0}^{29} (\frac{i}{30}, \frac{j}{30}, \frac{k}{30}, \frac{l}{30}, \frac{m}{30})' t_{(\frac{i}{30}, \frac{j}{30}, \frac{k}{30}, \frac{l}{30}, \frac{m}{30})} \equiv r \pmod 1, \tag{19.31}$$

which is the defining inequality of the master relaxation based on G. Note that (19.31) defines an integer program that has 30^5 variables. Another master group relaxation could be obtained for (19.8) by using I^5 itself as group G to obtain

$$\sum_{g \in I^5} g t_g \equiv r \pmod 1. \tag{19.32}$$

The set of feasible solutions to (19.32) is clearly a relaxation of (19.30) since feasible solutions to (19.30) satisfy (19.32) after all variables t_g corresponding to elements of $I^5 \setminus \{g^1, g^2\}$ are padded with zeros.

The main advantage of master group relaxations over the more natural corner relaxation is that they are more general and therefore their study applies to a larger class of problems. As an example, valid inequalities for $\mathrm{MG}(I^m, \emptyset, r)$ apply to all problems whose right-hand-side fractional values equal r while valid inequalities for $\mathrm{corner}_B(F)$ apply only for those problems that have the same coefficient vectors fractional parts and right-hand-side fractional values as $\mathrm{corner}_B(F)$.

We now extend the definition of master group relaxation to the mixed integer case. We perform the following relaxation steps. First we aggregate all integer variables whose column coefficient fractional parts are identical, i.e., $\mathscr{F}(A_B^{-1}A_j) = \mathscr{F}(A_B^{-1}A_k)$ for j and $k \in NI$. We also define $r := \mathscr{F}(A_B^{-1}d)$. Second, we aggregate the continuous variables. Since all continuous variables are assumed to be nonnegative without explicit upper bounds, a column $A_B^{-1}A_j$ is essentially the same as any other column $A_B^{-1}A_k$ provided that $A_B^{-1}A_j = \alpha A_B^{-1}A_k$ for some $\alpha > 0$ and $j,k \in NC$. Therefore, for continuous variables, we aggregate all variables whose coefficients are proportional to each other. To streamline the representation of the set, it is customary to perform an additional normalization of the continuous variables that ensures that their coefficient vectors belong to the boundary S^m of the m-dimensional hypercube $[-1,1]^m$. Formally, we define S^m as the set of real m-dimensional vectors $w = (w_1, w_2, \ldots, w_m)$, such that $\max\{|w_i| \mid 1 \leqslant i \leqslant m\} = 1$. We then relax the problem to a higher-dimensional space as follows. First, we identify G, a subgroup of I^m that contains all the coefficient columns of integer variables. Second, we identify a subset S of S^m that contains all the (scaled) coefficient columns of the continuous variables in $\text{corner}_B(F)$. We obtain

$$\text{mastercorner}(F) = \Big\{(t,s) \in \mathbb{Z}_+^{|G|} \times \mathbb{R}_+^{|S|} \;\Big|\; \sum_{g\in G} g t_g + \mathscr{F}\big(\sum_{v\in S} v s_v\big) \equiv r \pmod 1\Big\}$$

which we refer to as *master corner relaxation* or *master group relaxation.*

Although the theory of Johnson [71] was established to handle cases where G is simply a subset of I^m, we restrict our presentation to those cases where G is a subgroup of I^m.

Definition 19.14. Let G be a subgroup of I^m, let S be a subset of S^m and let $r \in I^m \backslash \{0\}$. The mixed integer master group problem, $\text{MG}(G,S,r)$, is defined as the set of functions $t : G \to \mathbb{R}$ and $s : S \to \mathbb{R}$ that satisfy

i. $\sum_{u\in G} u t(u) + \mathscr{F}(\sum_{v\in S} v s(v)) = r$,
ii. $t(u)$ is a non-negative integer for $u \in G$, $s(v)$ is a non-negative real number for $v \in S$,
iii. t and s have finite supports. [1]

In Definition 19.13, we assumed that $r \in G$. This requirement was necessary to ensure that $\text{MG}(G,\emptyset,r)$ is not empty. In contrast, we do not require that $r \in G$ in Definition 19.14. However, when $r \notin G$, it is implicitly assumed that the set of columns of S is sufficiently rich to make the group problem feasible.

In Definition 19.13 and Definition 19.14 we introduced master group relaxations of integer and mixed integer programs. These relaxations can be "customized" through the choice of the subgroup G of I^m and the subset S of S^m. Remember that we have introduced these relaxations with the goal of obtaining cutting planes that are strong for mixed integer programs. However, we have not yet considered one of the main practical questions which is the ease with which valid inequalities can be described/derived for these relaxations.

---

[1] Note that in Definition 19.14 we used t and s to denote integer and continuous variables respectively. We chose this convention since it is classical and since it highlights the fact that the corner relaxation and the master corner relaxation have different sets of variables. We mention however that it is also common to choose x and y to represent integer and continuous variables when studying specific master group problems. In particular, we will use this later convention in Section 19.5.

19.4.3 A hierarchy of inequalities for master group problems

The beauty of the master relaxations of Gomory's corner polyhedron lies in the fact that, although the polyhedral structure of general integer programs can be extremely complicated, the structure of master group problems is surprisingly simple. We present these structural results next.

19.4.3.1 Valid inequalities: definition & representation

As is usual, we define a valid inequality to be an inequality that is satisfied by all solutions of the set considered. Among the valid inequalities of group relaxations, trivial inequalities such as $x_i \geq 0$ do not typically yield useful cutting planes (as they are already present in the problem formulation) and therefore, the focus is on those inequalities that are *nontrivial*. It is proven in Gomory [57] that cut coefficients must be nonnegative and right-hand-sides must be positive in all nontrivial inequalities of finite master group relaxations. Therefore, nontrivial inequalities can always be rescaled so that their right-hand-sides equal 1 and it suffices to specify the coefficient of every variable in a cut to completely describe it. Since in our case variables x_g exist for all $g \in G$, the coefficients of a valid inequality can be seen as forming a function ϕ from G to $\mathbb{R}_+$. This motivates the following definition of valid inequality for pure integer master group problems.

Definition 19.15. Let $r \neq o$. A function $\phi : G \to \mathbb{R}_+$ defines a valid inequality for $\mathrm{MG}(G,\emptyset,r)$ if $\sum_{u\in G} \phi(u)t(u) \geqslant 1$, $\forall t \in \mathrm{MG}(G,\emptyset,r)$, $\phi(o) = 0$ and $\phi(r) = 1$.

When the group used is I^m, the additional assumption that ϕ is continuous is sometimes posed in the definition of valid inequality. This is for example the case in Gomory and Johnson [61]. This additional requirement simplifies the derivation of some results and is motivated by the belief that, in practice, functions that are continuous yield cutting plane algorithms that are more numerically stable. However, since the continuity assumption is not strictly necessary in the definition of valid inequality, we will not pose it here.

For mixed integer group relaxations, the notion of valid inequality is defined similarly, except that additional coefficients must be specified for all continuous variables whose coefficients are in $S \subseteq S^m$.

Definition 19.16. Let $r \neq o$. A pair of functions, $\phi : G \to \mathbb{R}_+$ and $\mu : S \to \mathbb{R}_+$ defines a valid inequality for $\mathrm{MG}(G,S,r)$ if $\sum_{u\in G} \phi(u)t(u) + \sum_{v\in S} \mu(v)s(v) \geqslant 1$, $\forall (t,s) \in \mathrm{MG}(G,S,r)$, where $\phi(o) = 0$.

In the above definition, we do not require that $\phi(r) = 1$ as r might not belong to G. However, for instances of the mixed integer group problem where $r \in G$, we will assume implicitly that $\phi(r) = 1$ in valid inequalities.

Since valid inequalities for master group problems can be thought of as functions from G to $\mathbb{R}_+$ or (G,S) to $\mathbb{R}_+$, we will often refer to them as *valid functions*. Further, we will also often represent them as functions.

Example 19.6. In Section 19.2 we derived several valid inequalities for (19.8). In particular, we argued that MDC, GFC and GMIC were valid for single rows of the tableau with fractional right-hand-sides. This discussion also shows that the functions $\phi_r^{\text{MDC}}(u) = 1$ when $u \neq 0$ and $\phi_r^{\text{MDC}}(0) = 0$, $\phi_r^{\text{MDC}}(u) = u/r$ and $\phi_r^{\text{GMIC}}(u) = \min\{\frac{u}{r}, \frac{1-u}{1-r}\}$ are valid for $\text{MG}(C_K^1, \emptyset, r)$ for all K and $r \neq 0$. The same arguments also show that these functions are valid for $\text{MG}(I^1, \emptyset, r)$ for $r \neq 0$. In Figure 19.5a, we represent the function ϕ^{GMIC} for $\text{MG}(C_{11}^1, \emptyset, \frac{6}{11})$. To generate a valid function for a two-dimensional group problem, say $\text{MG}(I^2, \emptyset, r)$, one could use the following simple argument. First we add up the two defining constraints and then generate a GMIC from the combined row. Since the combined row defines a relaxation of the initial set and GMIC is valid for problems defined by a single constraint, we obtain the function $\phi_{(r_1,r_2)}^{\text{agg}}(u,v) = \phi_{r_1+r_2}^{\text{GMIC}}((u+v) \pmod 1)$ which is clearly valid for $\text{MG}(I^2, \emptyset, (r_1, r_2))$. In Figure 19.5b, we represent $\phi_{(r_1,r_2)}^{\text{agg}}$. The domain of definition for Figure 19.5a is finite while it is the entire I^2 for Figure 19.5b. It is also possible to represent the function μ describing the coefficients of continuous variables of a mixed integer master group cut in a similar manner.

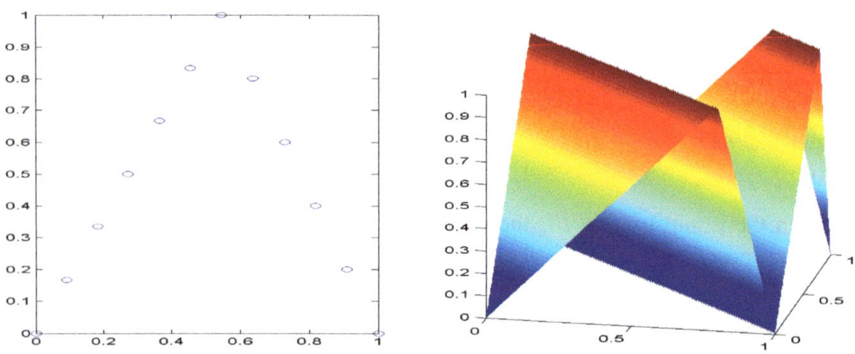

Fig. 19.5 Valid inequalities for group problems.

Valid inequalities for master group problems are important in that they translate into valid inequalities for mixed integer programs. However, valid inequalities might not even support the feasible region of master group relaxations and it is therefore important to characterize strong valid inequalities. We distinguish three levels of strength: *subadditive* inequalities, *minimal* inequalities and *extreme* inequalities. The following three subsections discuss this classification in more detail. We refer to Johnson [71] for a detailed discussion of this topic that contains the most general results about master group relaxations. We note however that most of the concepts presented in Johnson [71] had been introduced in various forms in the earlier work of Gomory [57] and Gomory and Johnson [59, 60].

19.4.3.2 Subadditive inequalities

Before we give the definition of a subadditive inequality, we first provide some motivation for this definition. We consider the case of a master group problem defined on a finite group G. In order for an inequality ϕ to be strong, it is reasonable to require that at least one of the feasible solutions of the master group problem satisfies it at equality, i.e.,

$$\sum_{u \in G} \phi(u)t(u) = 1 \tag{19.33}$$

for $t \in \mathrm{MG}(G,\emptyset,r)$. Clearly in order for the equality to hold, at least one of the variables must be positive. Assume without loss of generality that $t(w) \geq 1$. Now consider any group element $v \in G \backslash \{0\}$ and assume for simplicity that $w - v \neq v$. Clearly, we can create a new feasible solution for $\mathrm{MG}(G,\emptyset,r)$ as follows:

$$\tilde{t}(u) = \begin{cases} t(u), & \text{if } u \neq w, v, w-v, \\ t(u) - 1, & \text{if } u = w, \\ t(u) + 1, & \text{if } u = v, \\ t(u) + 1, & \text{if } u = w - v. \end{cases}$$

Since this solution belongs to $\mathrm{MG}(G,\emptyset,r)$, it must satisfy the valid inequality ϕ. It follows that

$$\sum_{u \in G} \phi(u)\tilde{t}(u) \geq 1. \tag{19.34}$$

Subtracting equality (19.33) from inequality (19.34), we obtain that $-\phi(w)+\phi(v)+\phi(w-v) \geq 0$. This line of ideas can be pushed further. In fact, one criterion for strength is to require that for each $w \in G$, there exists a solution t for which $t(w) \geq 1$ and $\sum_{u \in G} \phi(u)t(u) = 1$, since otherwise the coefficient of $t(w)$ in ϕ could be decreased. Therefore, it seems intuitive that strong inequalities for $\mathrm{MG}(G,\emptyset,r)$ should satisfy $\phi(v)+\phi(w-v) \geq \phi(w)$ for all $v, w \in G$. The previous intuitive ideas can be formalized; see Johnson [71] for a general description. Further, they can also be generalized so as to apply for situations where the underlying group is infinite and/or where continuous variables are present. The following definition then follows.

Definition 19.17. A subadditive valid inequality for $\mathrm{MG}(G,S,r)$ is a valid inequality for $\mathrm{MG}(G,S,r)$ that is such that

i. $\phi(u)+\phi(v) \geq \phi(u+v), \forall u, v \in G$,
ii. $\phi(u)+\sum_{v \in S} \mu(v)s(v) \geq \phi(w), \forall u, w \in G$ and S such that $u+\mathscr{F}(\sum_{v \in S} vs(v)) = w$,
iii. $\sum_{v \neq w} \mu(v)s(v) \geq \mu(w), \forall s$ such that $w = \sum_{v \neq w} vs(v)$ and $\sum_{v \neq w} vs(v) \in S$.

Theorem 19.13 ([59]). *Valid inequalities for master group problems are dominated by subadditive valid inequalities.*

Example 19.7. For pure integer master group problems, it suffices to show that $\phi(u)+\phi(v) \geq \phi(u+v)$ for all $u,v \in G$ to show that a function is subadditive. It is easily verified that all three valid functions ϕ_r^{MDC}, ϕ_r^{MDC} and ϕ_r^{GMIC} derived in Section 19.4.3.1 are subadditive. For example, for $u,v \in I^1$, $\phi_r^{\text{MDC}}(u)+\phi_r^{\text{MDC}}(v) = \frac{u}{r}+\frac{v}{r} \geq \frac{(u+v)(\text{mod } 1)}{r} = \phi_r^{\text{MDC}}(u+v)$.

19.4.3.3 Minimal inequalities

Reconsidering the notion of strength for inequalities of master group problems, a criterion stronger than subadditivity could be posed by requiring that the inequality considered is not coefficient-wise dominated by any other valid inequality of the master group problem. This naturally yields the following definition.

Definition 19.18. A valid inequality (ϕ,μ) for $\text{MG}(G,S,r)$ is said to be minimal for $\text{MG}(G,S,r)$ if there does not exist a valid inequality (ϕ^*,μ^*) for $\text{MG}(G,S,r)$ different from (ϕ,μ) such that $\phi^*(u) \leqslant \phi(u)$ $\forall u \in G$ and $\mu^*(v) \leqslant \mu(v)$ $\forall v \in S$.

The concept of minimal inequality is stronger than that of subadditive inequality as recorded in the following theorem.

Theorem 19.14 ([71]). *Minimal valid inequalities for* $\text{MG}(G,S,r)$ *are subadditive inequalities for* $\text{MG}(G,S,r)$.

The coefficients of minimal valid inequalities are well-characterized as we will describe in Theorem 19.15 and Theorem 19.16. Next, we give an intuitive derivation of these conditions. For any $u \in G$, we could consider the subadditive relations $\phi(u)+\phi(v) \geq \phi(u+v)$ for all $v \in G$. It is clear that for an inequality to be minimal, it should be the case that there exists v for which $\phi(u)+\phi(v) = \phi(u+v)$ for otherwise, a new inequality could be created that is still subadditive by reducing the coefficient $\phi(u)$. We therefore need to identify for each u, what v would make the subadditive inequality an equality. Assume for an instant that this v is exactly equal to $r-u$ for each u.

This would yield a minimal inequality as any inequality obtained from it by reducing one of the coefficients, say $\phi(u)$, by ε would be such that $\phi(u)+\phi(r-u) = 1-\varepsilon$. This is not possible as the solution $t(u)=1$, $t(r-u)=1$ and $t(v)=0$ for all other vs is feasible for $\text{MG}(G,\emptyset,r)$. Therefore $\phi(u)+\phi(r-u) \geq 1$. It can be proven that this condition is also necessary. The proof is by contradiction, assuming that $\phi(u)+\phi(r-u) = 1+\delta$ where $\delta > 0$. The contradiction is obtained by observing that the inequality

$$\rho(v) = \begin{cases} \frac{1}{1+\delta}\phi(v), & \text{if } v = u, \\ \phi(v), & \text{if } v \neq u, \end{cases}$$

is valid for $\text{MG}(G,\emptyset,r)$ and dominates ϕ. The statement is clear for all solutions that have $t(u)=0$ or $t(u) \geq \frac{1+\delta}{\phi(u)}$. When $1 \leq t(u) \leq \frac{1+\delta}{\phi(u)}$, we have

$$\sum_{v \in G} \rho(v)t(v) = \Big(\sum_{v \in G \setminus \{u\}} \phi(v)t(v) + \phi(u)(t(u)-1) \Big) + \phi(u) - \frac{\delta}{1+\delta}\phi(u)t(u)$$
$$\geq \phi(r-u) + \phi(u) - \frac{\delta}{1+\delta}\phi(u)t(u) \geq 1 + \delta - \delta = 1$$

where the first inequality holds because ϕ is subadditive. We now record this result in the following theorems. We start with the infinite group problem, as the notation for the finite case is more cumbersome than that used for the infinite case.

Theorem 19.15 ([71]). *Let (ϕ, μ) form an inequality for $\mathrm{MG}(I^m, S, r)$ where $r \neq o$ and $S \subseteq S^m$. The inequality (ϕ, μ) is minimal for $\mathrm{MG}(I^m, S, r)$ if and only if the three following conditions hold:*

i. $\phi(u) + \phi(v) \geq \phi(u+v)$, $\forall u, v \in I^m$
ii. $\mu(v) = \lim_{h \to 0} \frac{\phi(\mathscr{F}(hv))}{h}$, $\forall v \in S$
iii. $\phi(u) + \phi(r-u) = 1$, $\forall u \in I^m$.

Condition (ii) requires that coefficients of continuous variables be related to the slope of the function ϕ at the origin in the direction of their coefficients. It is possible to show that Conditions (i) and (ii) imply that the function (ϕ, μ) is subadditive; see Johnson [71].

We now describe the conditions that are required for inequalities to be minimal when the underlying group is a finite subgroup G of I^m. Although some conditions will be reminiscent of those presented in Theorem 19.15, the statement of the theorem is more technical.

Theorem 19.16 ([71]). *Let G be a finite subgroup of I^m and let W be a finite subset of S^m such that each element $w \in W$ has rational components. Let (ϕ, μ) form an inequality for $\mathrm{MG}(G, S, r)$ where $r \neq o$. Then (ϕ, μ) is a minimal valid inequality if and only if*

i. $\phi(u) + \phi(v) \geq \phi(u+v), \forall u, v \in G$
ii. $\sum_{w \in L} \mu(w)s(w) \geq \phi(v)$, *whenever* $v = \mathscr{F}(\sum_{w \in L} ws(w))$
iii. $\sum_{w \in L} \mu(w)s(w) \geq \mu(x)$, *whenever* $x = \sum_{w \in L} ws(w)$
iv. $\phi(v) + \sum_{w \in L} \mu(w)s(w) \geq 1$, *whenever* $r = \mathscr{F}(v + \sum_{w \in L} ws(w))$,

where in (ii), (iii) and (iv) the set L is a linearly independent subset of W, $s(w) \geq 0$, and we can assume that there is no $y \in W$, $y \notin L$, $y \neq \sum_{w \in L} ws(w)$ such that $y = \sum_{w \in L} ws'(w)$ for some $s'(w) \geq 0$, $w \in L$. In (ii), we can assume there is no $v' \in G$ and $s'(w)$, $0 \leq s'(w) \leq s(w)$ for $w \in L$, such that $v' = \mathscr{F}(\sum_{w \in L} ws'(w))$. Let

$$U_0 = \left\{ v \in G \,\middle|\, \begin{array}{c} \phi(v) + \sum_{w \in L} \mu(w)s(w) = 1 \\ \textit{for some } L \textit{ and } s(L) \textit{ with } r = v + \sum_{w \in L} ws(w) \end{array} \right\}$$

and let

$$W_0 = \left\{ \omega \in W \,\middle|\, \begin{array}{c} \phi(v) + \sum_{w \in L} \mu(w)s(w) = 1 \\ \textit{for some } v \in U_0 \textit{ and } L \textit{ with } \omega \in L \textit{ and } s(\omega) > 0 \end{array} \right\}.$$

For minimality, we need

v. *for every* $u \in G$, *either* $u \in U_0$ *or there exists* $v \in U_0$ *such that* $\phi(u)+\phi(v-u) = \phi(v)$

vi. *for every* $w \in W$, *either* $\mu(w) = 0$ *or* $w \in W_0$ *or there exists* $\omega_0 \in W_0$ *with* $\mu(\omega_0) = \sum_{\omega \in L} \mu(\omega)s(\omega)$ *where* L *is a linearly independent subset of* W, $w \in L$, $s(w) > 0$, *and* $\omega_0 = \sum_{\omega \in L} \omega s(\omega)$.

Although the expression of Theorem 19.16 is involved, its statement simplifies tremendously when considering pure integer group problems and when considering one-row master group relaxations. We present these two subcases next because of their relative simplicity and because of their historical significance. We consider first the pure integer case. In this case, $U_0 = \{r\}$ and $W_0 = \emptyset$.

Corollary 19.1. *Let* G *be a finite subgroup of* I^m *and let* $r \in G\backslash\{0\}$. *Let* ϕ *be a valid inequality for* $\mathrm{MG}(G,\emptyset,r)$. *Then* ϕ *is a minimal valid inequality if and only if*

i. $\phi(u)+\phi(v) \geq \phi(u+v), \forall u,v \in G$
ii. $\phi(u)+\phi(r-u) = \phi(r), \forall u \in G$.

For a finite group G, the function ϕ can also be recorded as a vector since its domain is finite. The result of Corollary 19.1 then states that the coefficients of minimal inequalities for finite group problems are completely determined by a set of linear constraints. In other words, minimal valid inequalities can be identified by selecting feasible solutions in the polytope

$$P^*(G,\emptyset,r) = \left\{ \phi \in \mathbb{R}^{|G|} \;\middle|\; \begin{array}{ll} \phi_u + \phi_v \geq \phi_{u+v}, & \forall u,v \in G \\ \phi_u + \phi_{r-u} = \phi_r, & \forall u \in G \\ \phi_r = 1, & \\ \phi_u \geq 0, & \forall u \in G \end{array} \right\}.$$

Example 19.8. Of the functions ϕ_r^{MDC}, ϕ_r^{MDC} and ϕ_r^{GMIC} defined in Section 19.4.3.1, we see that only ϕ_r^{GMIC} is minimal for $\mathrm{MG}(C_K^1,\emptyset,r)$. In fact, for any $u \in C_K^1\backslash\{o,r\}$, $\phi_r^{\mathrm{MDC}}(r-u) + \phi_r^{\mathrm{MDC}}(u) = 2 > 1 = \phi_r^{\mathrm{MDC}}(r)$. Further, when $0 < r < \frac{K-1}{K}$, then $\phi_r^{\mathrm{MDC}}(1-\frac{1}{K}) + \phi_r^{\mathrm{MDC}}(r+\frac{1}{K}) = \frac{r+1}{r} > 1 = \phi_r^{\mathrm{MDC}}(r)$. This proves that these two functions are coefficient-wise dominated by other inequalities. This is indeed the case as $\phi_r^{\mathrm{MDC}}(u) \geq \phi_r^{\mathrm{GMIC}}(u)$ for all $u \in C_K^1$ and $\phi_r^{\mathrm{MDC}}(u) \geq \phi_r^{\mathrm{GMIC}}(u)$ for all $u \in C_K^1$.

We now consider $\mathrm{MG}(G,\{-1,1\},r)$ where G is the finite cyclic group of order n. We only assume that $r \in I^1$ since continuous variables make the master group relaxation nonempty even if r does not belong to G. We introduce the notation r_- to denote the element of G immediately to the left of r while we define r_+ to denote the element of G immediately to the right of r. If $r \in G$, $r_+ = r_- = r$. Applying Theorem 19.16 with $S = \{-1,1\}$, we obtain the following characterization.

Corollary 19.2 ([59]). *Let* G *be a finite cyclic subgroup of* I^1 *and let* $r \in I^1\backslash\{0\}$. *Let* (ϕ,μ) *be a valid inequality for* $\mathrm{MG}(G,\{-1,1\},r)$. *Then* (ϕ,μ) *is a minimal valid inequality for* $\mathrm{MG}(G,\{-1,1\},r)$ *if and only if*

i. $\phi(u)+\phi(v) \geq \phi(u+v)$
ii. $\mu(+1)\frac{1}{n} \geq \phi(u_1)$ *where* $u_1 = \mathscr{F}(\frac{1}{n})$
iii. $\mu(-1)\frac{1}{n} \geq \phi(u_{n-1})$ *where* $u_{n-1} = \mathscr{F}(\frac{n-1}{n})$
iv. $\phi(r_-)+\mu(+1)|r-r_-| = 1$
v. $\phi(r_+)+\mu(-1)|r_+-r| = 1$
vi. $\phi(u)+\phi(r_- - u) = \phi(r_-)$ *or* $\phi(u)+\phi(r_+ - u) = \phi(r_+)$, *for* $u \in G$.

19.4.3.4 Extreme inequalities

Although minimal inequalities are not dominated by any other single valid inequality for the master group problem, they might still be obtained as a convex combination of other valid inequalities. This motivates the following definition.

Definition 19.19. A valid inequality (ϕ,μ) for $\text{MG}(G,S,r)$ is said to be extreme for $\text{MG}(G,S,r)$ if whenever $\phi = \frac{1}{2}\phi^1 + \frac{1}{2}\phi^2$ and $\mu = \frac{1}{2}\mu^1 + \frac{1}{2}\mu^2$ for some valid inequalities (ϕ^1,μ^1) and (ϕ^2,μ^2) of $\text{MG}(G,S,r)$ then $\phi = \phi^1 = \phi^2$ and $\mu = \mu^1 = \mu^2$.

The concept of extreme inequality is stronger than that of minimal inequality as shown next.

Theorem 19.17 ([71]). *Extreme valid inequalities for* $\text{MG}(G,S,r)$ *are minimal valid inequalities for* $\text{MG}(G,S,r)$.

Theorem 19.18 ([71]). *Extreme minimal valid inequalities for* $\text{MG}(G,S,r)$ *are extreme valid inequalities for* $\text{MG}(G,S,r)$.

For pure integer finite master group problems, the result of Corollary 19.1 and Theorem 19.18 yields a very elegant characterization of extreme valid inequalities. Observe from Corollary 19.1 that the coefficients of all minimal valid inequalities belong to the polytope $P^*(G,\emptyset,r)$. Therefore extreme valid inequalities correspond to extreme points of $P^*(G,\emptyset,r)$. This result is recorded formally in Theorem 19.20 and will be discussed in detail in Section 19.5.

Example 19.9. The function ϕ_r^{GMIC} defined in Section 19.4.3.1 is not only minimal for $\text{MG}(C_K^1,\emptyset,r)$, it is also extreme for this set. We refer to Section 19.5 for a proof of this claim.

The main utility of master group problems resides in the fact that, although the extreme inequalities present in a particular corner polyhedron are very dependent on the actual problem from which this corner polyhedron is extracted, they can all be obtained from the extreme inequalities of its master relaxations. Consider a corner relaxation and any of its master relaxation. Consider also a valid inequality for the master relaxation. We can construct an inequality for the initial corner relaxation by associating with each variable in the corner relaxation the coefficient it receives in the valid inequality of the master relaxation. The inequality so-created is valid for the initial corner relaxation. Further, this simple procedure is sufficient to produce all extreme inequalities of all corner relaxations as recorded in the following theorem.

Theorem 19.19 ([57]). *All extreme inequalities of* $\text{corner}_B(F)$ *can be obtained from the extreme inequalities of its master group relaxations.*

It is not true however that every extreme inequality of a master relaxation yields an extreme inequality of $\text{corner}_B(F)$; see Gomory [57] for a discussion.

19.5 Extreme inequalities

It follows from the discussion in Section 19.4.3 that extreme inequalities play a central role in the derivation of strong cuts for mixed integer programming. Although the definition of extreme inequality is common to both finite and infinite group problems, the tools that are used to prove that inequalities are extreme for finite and infinite group problems differ. We discuss in Section 19.5.1 and Section 19.5.2 extreme inequalities for finite and infinite master group problems respectively.

19.5.1 Extreme inequalities of finite master group problems

Gomory [57] provided a characterization of extreme inequalities for finite group problem that is presented in Theorem 19.20.

Theorem 19.20 ([57]). *Let* G *be a finite subgroup of* I^m *and let* $r \neq o$. *The extreme inequalities of* $\text{MG}(G,\emptyset,r)$ *are* $t_i \geq 0$ *for* $i \in G$ *and* $\sum_{i\in G}\phi_i t_i \geq 1$ *where* $\phi \in \mathbb{R}_+^{|G|}$ *are the extreme points of* $P^*(G,\emptyset,r)$, *where*

$$P^*(G,\emptyset,r) = \left\{ \phi \in \mathbb{R}^{|G|} \;\middle|\; \begin{array}{ll} \phi_u + \phi_v \geq \phi_{u+v}, & \forall u,v \in G \\ \phi_u + \phi_{r-u} = \phi_r, & \forall u \in G \\ \phi_r = 1, & \\ \phi_u \geq 0, & \forall u \in G \end{array} \right\}.$$

Proving that an inequality is extreme using Theorem 19.20 is conceptually simple as it reduces to proving that the vector of its coefficients corresponds to an extreme point of $P^*(G,\emptyset,r)$. However, it is often cumbersome in practice. We next discuss other options for deriving inequalities and proving they are extreme.

19.5.1.1 Results based on algebraic structure

Gomory [57] showed that the algebraic structure of groups can be used to explain a particular extreme inequality of one group problem as a transformed version of an extreme inequality for another group problem.

Theorem 19.21 ([57]). *Let $G^1, G^2 \preccurlyeq I^m$ where G^1 and G^2 are finite, and let $r \in G^1 \setminus \{o\}$. Let $\varsigma : G^1 \to G^2$ be a surjective homomorphism. Assume that $r \notin \mathrm{Kern}(\varsigma)$ and that $\phi : G^2 \to \mathbb{R}_+$ is an extreme inequality for $\mathrm{MG}(G^2, \emptyset, \varsigma(r))$. Then $\phi \circ \varsigma$ is extreme for $\mathrm{MG}(G^1, \emptyset, r)$.*

A closely related result is presented next.

Theorem 19.22 ([57]). *Let $\omega : G \to G$ be an automorphism. Then ϕ is an extreme inequality for $\mathrm{MG}(G, \emptyset, r)$ iff $\phi \circ \omega$ is extreme for $\mathrm{MG}(G, \emptyset, \omega^{-1}(r))$.*

19.5.1.2 Specific results for cyclic groups

We first show how extreme inequalities of $\mathrm{MG}(I^1, \emptyset, r)$ give rise to extreme inequalities of $\mathrm{MG}(C_n, \emptyset, r)$ for cyclic subgroups C_n of I^1.

Theorem 19.23 ([59, 60]). *If $\phi : I^1 \to \mathbb{R}_+$ is an extreme function for $\mathrm{MG}(I^1, \emptyset, r)$ that is continuous and piecewise linear, then for any cyclic $G \preccurlyeq I^1$ containing all the points at which ϕ is non-differentiable, the valid function $\phi|_G$ obtained by restricting ϕ to G is extreme for $\mathrm{MG}(G, \emptyset, r)$.*

We next provide an outline of the proof of Theorem 19.23. First, using the fact that ϕ is minimal for $\mathrm{MG}(I^1, \emptyset, r)$, it can be seen that $\phi|_G$ is minimal for $\mathrm{MG}(G, \emptyset, r)$. Then we assume by contradiction that $\phi|_G$ is not an extreme point of $P^*(G, \emptyset, r)$. Therefore there exist vectors $\hat{\phi}_1, \hat{\phi}_2 \in P^*(G, \emptyset, r)$ such that $\hat{\phi}_1 \neq \hat{\phi}_2$ and $\phi|_G = \frac{1}{2}\hat{\phi}_1 + \frac{1}{2}\hat{\phi}_2$. For any u in I^1, let u_1 and u_2 be the neighboring points to u in G. For $i = 1, 2$, we define the interpolation $\phi_i : I^1 \to \mathbb{R}_+$ of $\hat{\phi}_i : G \to \mathbb{R}_+$ as

$$\phi_i(u) = \begin{cases} \hat{\phi}_i(u), & \text{if } u \in G, \\ \frac{(\mathscr{F}^{-1}(u_2) - \mathscr{F}^{-1}(u))\hat{\phi}_i(u_1) + (\mathscr{F}^{-1}(u) - \mathscr{F}^{-1}(u_1))\hat{\phi}_i(u_2)}{\mathscr{F}^{-1}(u_2) - \mathscr{F}^{-1}(u_1)}, & \text{if } u \notin G. \end{cases} \tag{19.35}$$

It can be verified that ϕ_1 and ϕ_2 are valid subadditive inequalities for $\mathrm{MG}(I^1, \emptyset, r)$ using Proposition 19.3 which is discussed later. This is a contradiction to the fact that ϕ is extreme for $\mathrm{MG}(I^1, \emptyset, r)$, since our construction yields $\phi = \frac{1}{2}\phi_1 + \frac{1}{2}\phi_2$ with $\phi_1 \neq \phi_2$.

Given a subadditive valid inequality for $\mathrm{MG}(G, \emptyset, r)$ where G is a cyclic subgroup of I^1, the interpolation construction (19.35) generates a valid subadditive function for $\mathrm{MG}(I^1, \emptyset, r)$. However, the converse of Theorem 19.23 is not true, i.e., starting from an extreme inequality for a finite group problem and interpolating to I^1 does not always yield extreme inequalities; see Dey et al. [40] for an example. More generally, it is not well-understood how interpolation can be used to generate valid inequalities for multi-dimensional infinite group problems from valid inequalities of multi-dimensional finite group problems.

Tilting is another technique to generate extreme inequalities for group problems corresponding to cyclic subgroups. This technique is from Aráoz et al. [8]. We begin with a definition of a master knapsack polytope.

Definition 19.20. The feasible region of a master equality knapsack problem of size r, denoted as K_r is:

$$\left\{x \in \mathbb{Z}_+^r \;\middle|\; \sum_{i=1}^{r} i x_i = r\right\}. \tag{19.36}$$

Extreme inequalities of the master knapsack polytope satisfy "partial" subadditive conditions. The following result is from Aráoz [7].

Theorem 19.24 ([7]). *Extreme inequalities $\sum_{i=1}^r \rho_i x_i \geqslant \rho_r$ of the master knapsack polytope K_r are the extreme rays of the cone defined by*

i. $\rho_i + \rho_j \geqslant \rho_{i+j}\ \forall i,j,i+j \in \{1,\ldots,r\}$.
ii. $\rho_i + \rho_{r-i} = \rho_r\ \forall i \in \{1,\ldots,\lfloor \frac{r}{2} \rfloor\}$.

We next present the key steps in constructing valid inequalities for $\mathrm{MG}(C_n, \emptyset, \tilde{r})$, where $\tilde{r} = \frac{r}{n}$ $(r \in \{1,\ldots,n\})$ using tilting. First we pick any extreme inequality ρ of K_r (normalized so that $\rho_r = 1$) and construct the vector $\bar{\rho} \in \mathbb{R}^n_+$ as

$$\bar{\rho}_i = \begin{cases} \rho_i, & \text{if } i \leqslant r, \\ \frac{1-\frac{i}{n}}{1-\frac{r}{n}}, & \text{if } i \geqslant r. \end{cases} \tag{19.37}$$

Therefore $\bar{\rho}$ satisfies the following conditions:

i. $\bar{\rho}_i + \bar{\rho}_j \geqslant \bar{\rho}_{(i+j)(\mathrm{mod}\,(n))}$ if either all $i,j,i+j \in \{1,\ldots,r\}$ or if all $i,j,i+j \in \{r,\ldots,n\}$.
ii. $\bar{\rho}_i + \bar{\rho}_{(r-i)(\mathrm{mod}\,(n))} = \bar{\rho}_r\ \forall i \in \{1,\ldots,n\}$.

Hence treating $\bar{\rho}$ as a function over C_n, i.e., $\bar{\rho} : C_n \to \mathbb{R}_+$ satisfies complementarity and is "almost" subadditive. Second, we consider $\mathrm{MG}(C_n, \emptyset, \tilde{r})$ and the GMIC inequality ξ defined as

$$\xi_i = \begin{cases} \frac{i}{n}, & \text{if } i \leqslant r, \\ \frac{1-\frac{i}{n}}{1-\frac{r}{n}}, & \text{if } i \geqslant r. \end{cases} \tag{19.38}$$

Since GMIC is minimal, $\xi(u) + \xi(r-u) = 1$ and $\xi(u) + \xi(v) \geqslant \xi(u+v)\ \forall u, v \in C_n$. Finally, we attempt to construct a convex combination of ξ and $\bar{\rho}$ such that the resulting vector satisfies subadditivity and complementarity conditions.

Lemma 19.1 ([8]). *Set $\alpha \in \mathbb{R}$ as*

$$\alpha = \max \begin{cases} \frac{r}{n}(\rho_k - \rho_i - \rho_j), & \text{for } 1 \leqslant i,j,k \leqslant r, \\ \frac{r}{n(k-r)}((-\rho_i + \rho_j)(n-r) + (n-k)), & \text{for } 1 \leqslant i,j \leqslant r \text{ and } r \leqslant k < n, \\ \frac{r}{n}((\rho_k - \rho_i)\frac{n-r}{n-j} - 1), & \text{for } 1 \leqslant i,k \leqslant r \text{ and } r \leqslant j < n. \end{cases} \tag{19.39}$$

where $k \equiv (i+j)\ (\mathrm{mod}\, n)$. Let $\phi = \frac{\bar{\rho} + \alpha\xi}{1+\alpha}$. Then $\phi : C_n \to \mathbb{R}_+$ is subadditive and satisfies complementarity.

Note that if ρ is extreme for K_r, then it can always be assumed that $\rho_i \geqslant 0\ \forall i$ and $\rho_{i'} = 0$ for some $i' \in \{1,...,r\}$. This is because, given any vector $\rho \in \mathbb{R}^r$ that represents an extreme inequality of K_r, any vector $\rho' \in \mathbb{R}^r$ defined as $\rho'_i = \rho_i + \lambda i$ $\forall i \in \{1,...,r\}$ also represents an extreme inequality of K_r for any $\lambda \in \mathbb{R}$. Therefore, the function ϕ defined in Lemma 19.1 is a valid inequality for the finite one-row group problem. The following stronger result states that it is in fact extreme.

Theorem 19.25 ([8]). *Let ρ be a facet of K_r such that ρ is nonnegative, $\rho_r = 1$ and $\rho_{i'} = 0$ for some $i' \in \{1,...,r\}$. If α given by (19.39) is positive, then the function ϕ defined in Lemma 19.1 is extreme for* $\mathrm{MG}(C_n, \emptyset, r)$.

We refer to Aráoz et al. [8] for uses of Theorem 19.25 to derive extreme inequalities for $\mathrm{MG}(C_n, \emptyset, r)$.

19.5.2 Extreme inequalities for infinite group problems

For finite group problems, Theorem 19.20 gives a characterization of extreme inequalities. Since this characterization does not apply for infinite group problems, other techniques have been used to prove that candidate subadditive inequalities are extreme.

Often, continuity of the valid functions plays an important role in the proof that functions are extreme. We begin by formalizing the topology we use for the space I^m. All the results on continuity in this chapter are based on the topology induced by the following metric.

Definition 19.21. For $u, v \in I^m$, let $d(u,v) = \sqrt{\sum_{i=1}^m (d'(u_i,v_i))^2}$ where $d'(u_i,v_i) = \min\{|\mathscr{F}^{-1}(u_i) - \mathscr{F}^{-1}(v_i)|, 1 - |\mathscr{F}^{-1}(u_i) - \mathscr{F}^{-1}(v_i)|\}$.

We start with the presentation of a very common strategy that has been used to prove that functions are extreme. This strategy is presented in an algorithmic fashion in Table 19.1. Next we discuss some standard tools to handle the steps described in Table 19.1. We first consider the question of proving that a function is subadditive. Typically, this question is not simple to answer unless the function is piecewise linear and continuous.

Definition 19.22. Let $\phi : I^m \to \mathbb{R}_+$ be a piecewise linear and continuous function, i.e., I^m can be decomposed into finitely many polytopes with non-empty interiors $P_1, ..., P_k$, such that $\phi(u) = \alpha_t' u + \beta_t$, $\forall u \in P_t$, where $\alpha_t \in \mathbb{R}^m, \beta_t \in \mathbb{R}\ \forall t \in \{1,2,...k\}$. For a continuous and piecewise linear function ϕ, we say that a point l belongs to the boundary of ϕ, denoted $\mathbb{B}(\phi)$, if l belongs to the intersection of two polytopes P_i and P_j where $i \neq j$ and the gradient of ϕ in P_i is not equal to the gradient of ϕ in P_j.

Proposition 19.3 ([62], [38]). *Let ϕ be a continuous, piecewise linear and nonnegative function over I^m. The function ϕ is subadditive if and only if for all $l_1, l_2 \in \mathbb{B}(\phi)$*

$$\phi(l_1) + \phi(l_2) \geqslant \phi(l_1 + l_2) \tag{19.40}$$

$$\phi(l_1) + \phi(l_2 - l_1) \geqslant \phi(l_2). \tag{19.41}$$

0. Input: A function $\phi : I^m \to \mathbb{R}_+$ and $r \neq o$ such that $\phi(r) = 1$ and $\phi(o) = 0$.
i. Prove ϕ is subadditive, i.e., $\phi(u)+\phi(v) \geqslant 1\ \forall u, v \in I^m$. This shows that ϕ is a valid inequality for $\mathrm{MG}(I^m, \emptyset, r)$.
ii. Prove ϕ satisfies complementarity conditions, i.e., $\phi(u) + \phi(r-u) = 1\ \forall u \in I^m$. By Theorem 19.15, ϕ is a minimal valid inequality for $\mathrm{MG}(I^m, \emptyset, r)$.
iii. Assume by contradiction that ϕ is not extreme, i.e., $\phi = \frac{1}{2}\phi_1 + \frac{1}{2}\phi_2$ such that $\phi_1 \neq \phi_2$ and ϕ_1, ϕ_2 are valid inequalities for $\mathrm{MG}(I^m, \emptyset, r)$. It follows from Theorem 19.18 that ϕ_1 and ϕ_2 can be assumed to be minimal valid inequalities.
iv. Define the *equality set* of $E(\phi) = \{(u,v) \in I^m \times I^m \mid \phi(u)+\phi(v) = \phi(u+v)\}$. Since ϕ_1 and ϕ_2 are minimal functions, they are subadditive. Thus $E(\phi_1) \supseteq E(\phi)$ and $E(\phi_2) \supseteq E(\phi)$.
v. Obtain a contradiction by showing that if $E(\phi_1) \supseteq E(\phi)$, then $\phi_1 = \phi$.

Table 19.1 Proving a function $\phi : I^m \to \mathbb{R}_+$ is extreme for $\mathrm{MG}(I^m, \emptyset, r)$.

When $m = 1$, Proposition 19.3 implies that it is sufficient to consider the finite number of points at which the function ϕ is non-differentiable to prove that a continuous piecewise linear function is subadditive. We refer to Gomory and Johnson [62] and Dey and Richard [38] for more streamlined versions for Proposition 19.3 where it is assumed that the function ϕ satisfies the condition $\phi(u) + \phi(r-u) = 1$, and refer to Richard et al. [86] for a variant of Proposition 19.3 where ϕ is a general piecewise linear function that is not necessarily continuous.

Next we consider Step ii in Table 19.1. This step is typically not very difficult to verify and can be done with few calculations especially for piecewise linear functions. Steps iii and iv in Table 19.1 do not require computation. Next consider Step v in Table 19.1. For piecewise linear functions, a typical approach used is the following. Using the fact that $E(\phi_1) \supseteq E(\phi)$ it is verified that ϕ_1 is linear on polytopes $P_1, \ldots, P_t$, the same subsets of I^m on which ϕ is linear. Then the problem reduces to verifying that the gradients and constant terms in each of these subsets are equal for ϕ and ϕ_1. Results to prove that a function is linear in some polytope are usually referred to as *Interval Lemmas*. One such result is presented next.

Lemma 19.2 ([61]). *Let $U \equiv [u_1, u_2] \subset I^1$, $V \equiv [v_1, v_2] \subset I^1$ and $U+V \equiv [u_1 + v_1, u_2 + v_2]$ such that $u_1 \neq u_2$ and $v_1 \neq v_2$. If there exists a continuous real-valued function ϕ_i defined over U, V and $U+V$ such that $\phi_i(u) + \phi_i(v) = \phi_i(u+v)\ \forall u \in U, v \in V$, then ϕ_i must be a straight line with constant slope s over U, V and $U+V$.*

Observe that to use Lemma 19.2 for the function ϕ_1, we must know that ϕ_1 is continuous. We present next a sufficient condition for ϕ_1 to be continuous in some interval.

Lemma 19.3 ([40]). *Let $\phi : I^1 \to \mathbb{R}_+$ be a piecewise linear, subadditive and valid function such that $\phi(u) = cu\ \forall u \in U$, where $c > 0$, U is an interval $\{u \in I^1 | 0 \leqslant \mathscr{F}^{-1}(u) \leqslant w\}$ and w is a non-zero real number strictly less than 1. If $\phi = (1-\lambda)\phi^1 + \lambda\phi^2$, where $0 < \lambda < 1$ and ϕ^1 and ϕ^2 are subadditive valid functions for $\mathrm{MG}(I^1, \emptyset, r)$, then ϕ^1 and ϕ^2 are continuous at all points at which ϕ is continuous.*

Gomory and Johnson [62] and Dey and Richard [38] present different versions of Lemmas 19.2 and 19.3 for functions defined on I^1 and I^2. We now illustrate the framework presented in Table 19.1 on an example.

Example 19.10. Consider the GMIC introduced in Section 19.2. This inequality can be expressed as

$$\xi(u) = \begin{cases} \frac{u}{r}, & \text{if } u \le r, \\ \frac{1-u}{1-r}, & \text{if } u > r, \end{cases}$$

or equivalently as $\xi(u) = \min\{\frac{u}{r}, \frac{1-u}{1-r}\}$. We claim that ξ is extreme for $\mathrm{MG}(I^1,\emptyset,r)$. Following the scheme proposed in Table 19.1, we begin by showing that ξ is subadditive. Observe that $\mathbb{B}(\phi) = \{o,r\}$. Thus by Proposition 19.3, ξ is subadditive. Then we show that $\xi(u)+\xi(r-u) = 1$ for all $u \in I^1$. There are two cases. When $u \le r$, then $\xi(u)+\xi(r-u) = \frac{u}{r}+\frac{r-u}{r} = 1$. When $u > r$, then $\xi(u)+\xi(r-u) = \frac{1-u}{1-r}+\frac{1-(1+r-u)}{1-r} = 1$. Assume now by contradiction that ξ can be expressed as a convex combination of two other valid subadditive and minimal inequalities ξ_1 and ξ_2. By Lemma 19.3, we know that these inequalities are continuous. As ξ satisfies $\xi(u)+\xi(v) = \xi(u+v)$ for all $u,v \in [0,\frac{r}{2}]$, we conclude that $\xi_i(u)+\xi_i(v) = \xi_i(u+v)$ for all $u \in [0,\frac{r}{2}]$ and for $i = 1,2$. Using (Interval) Lemma 19.2 (we know the functions ξ_i are continuous), we conclude that ξ_1 and ξ_2 are linear over the interval $[0,r]$. However, we know that these functions are valid and therefore $\xi_i(0) = 0$ and $\xi_i(r) = 1$ for $i = 1,2$. Therefore, we obtain that $\xi_1(u) = \xi_2(u) = \xi(u)$ for all $u \in [0,r]$. Since ξ satisfies the additive relations $\xi(u)+\xi(v) = \xi(u+v)$ for all $u,v \in [\frac{1+r}{2},1]$, we conclude that $\xi_i(u)+\xi_i(v) = \xi_i(u+v)$ for all $u \in [\frac{1+r}{2},1]$ for all $i = 1,2$. It follows from (Interval) Lemma 19.2 that these functions are linear over $[r,1]$. However, we know that these functions are valid and therefore $\xi_i(r) = 1$ and $\xi_i(1) = 0$ for $i = 1,2$. Therefore, we obtain that $\xi_1(u) = \xi_2(u) = \xi(u)$ for all $u \in [r,1]$. This is the desired contradiction.

Using the methodology of Table 19.1, the following Two-Slope Theorem can be proven.

Theorem 19.26 ([59], [61]). *Continuous, piecewise linear, subadditive and minimal functions that have only two slopes are extreme for* $\mathrm{MG}(I^1,\emptyset,r)$.

In the next sections, we discuss methods for obtaining and proving extreme inequalities for $\mathrm{MG}(I^m,\emptyset,r)$.

19.5.2.1 Results based on algebraic structure

Similar to finite group problems, extreme inequalities for infinite group problems are also related through automorphisms and homomorphisms.

Theorem 19.27 ([71]). *Let* $\omega : I^m \to I^m$ *be an automorphism and let* ϕ *be an extreme inequality for* $\mathrm{MG}(I^m,\emptyset,r)$. *Then* $\phi \circ \omega$ *is extreme for* $\mathrm{MG}(I^m,\emptyset,\omega^{-1}(r))$.

It is shown in Johnson [71] that the only automorphisms for $\mathrm{MG}(I^m,\emptyset,r)$ are rotations, reflections and their combinations. It is also shown in [71] how automorphisms can be used for mixed integer group problems of the form $\mathrm{MG}(I^m,S^m,r)$. For homomorphisms, the following weaker result holds.

Proposition 19.4 ([71], [38]). *Let $\varsigma : G_1 \to G_2$ be a surjective map that is a homomorphism. The function ϕ is subadditive and minimal for $\mathrm{MG}(G_1,\emptyset,r)$ iff the function $\phi\circ\varsigma(x) = \phi(\varsigma(x))$ is subadditive and minimal for $\mathrm{MG}(G_2,\emptyset,v)$, for any v such that $\varsigma(v) = r$.*

For the multiplicative homomorphism described in Section 19.4.1 the following stronger result holds.

Theorem 19.28 ([61], [38]). *Let ς be a multiplicative homomorphism. Then ϕ is extreme for $\mathrm{MG}(I^m,\emptyset,r)$ iff $\phi\circ\varsigma$ is extreme for $\mathrm{MG}(I^m,\emptyset,v)$, where $\varsigma(v) = r$.*

19.5.2.2 Sequence of functions

It can be easily verified that the limit of a sequence of subadditive and minimal functions is itself subadditive and minimal. This result was first proven by Dash and Günlük [32, 33] for a specific family of functions and proven for general functions in Dey et al. [40].

Proposition 19.5 ([38]). *Let $\phi_i : I^m \to \mathbb{R}_+$ be valid, subadditive and minimal functions of $\mathrm{MG}(I^m,\emptyset,r)$ for $i = 1,2,\ldots$. If the sequence of functions $\{\phi_i\}_{i=1}^{\infty}$ converges to ϕ pointwise on I^m, then ϕ is a valid, subadditive and minimal function of $\mathrm{MG}(I^m,\emptyset,r)$.*

In general, the limit of a converging sequence of extreme functions is not necessarily extreme; see Dey et al. [40] for an example. However, it is possible to show that, if the breakpoints of the functions ϕ_i are "compatible" with the breakpoints of the function ϕ, then ϕ will be extreme.

Theorem 19.29 ([40]). *Let $\phi_i : I^1 \to \mathbb{R}_+$ be piecewise linear, continuous extreme functions of $\mathrm{MG}(I^m,\emptyset,r)$ for $i \geqslant 1$. Assume that the sequence of functions $\{\phi_i\}_{i=1}^{\infty}$ converges pointwise to ϕ and that*

i. ϕ is piecewise linear,
ii. The right-derivative of ϕ at zero, $\phi'_+(0)$, exists and satisfies $0 < \phi'_+(0) < \infty$,
iii. There exists a sequence of finite subgroups C_{k_i}, where $\lim_{i\to+\infty} k_i = +\infty$, that satisfy

a. $\phi_i(u) = \phi(u)\ \forall u \in C_{k_i}$, and
b. all the points at which the function ϕ_i is non-differentiable belong to C_{k_i}.

Then ϕ is an extreme function of $\mathrm{MG}(I^1,\emptyset,r)$.

19.5.2.3 Fill-in

Fill-in is a technique to obtain a valid subadditive inequality for $\mathrm{MG}(I^m, S^m, r)$ from a valid subadditive inequality for $\mathrm{MG}(G, S, r)$ where G is a subgroup of I^m and $S \subseteq S^m$. The fill-in procedure was first presented for one-row group problems in Gomory and Johnson [59, 60] and it can be used to explain all two-slope extreme inequalities for $\mathrm{MG}(I^1, S^1, r)$. The fill-in procedure is extended for the case of m-row group problems in Johnson [71]. We motivate the fill-in procedure on the following example.

Example 19.11. The key idea of the fill-in procedure is to express variables of a "larger size" group problem as aggregations of variables of a "smaller size" group problem in order to use the valid inequalities of the smaller size group problem as building blocks for valid inequalities of the larger size group problem.

Consider the problem $\mathrm{MG}(C_2 \times C_2, S, (1/3, 2/3)')$, where $S = \{e_1, -e_1, e_2, -e_2\}$, i.e.,[2]

$$\sum_{i=0}^{2}\sum_{j=0}^{2}\frac{1}{2}\begin{pmatrix} i \\ j \end{pmatrix} x_{i,j} + \begin{pmatrix} 1 \\ 0 \end{pmatrix} y_1 + \begin{pmatrix} -1 \\ 0 \end{pmatrix} y_2 + \begin{pmatrix} 0 \\ 1 \end{pmatrix} y_3 + \begin{pmatrix} 0 \\ -1 \end{pmatrix} y_4 \equiv \begin{pmatrix} \frac{1}{2} \\ \frac{2}{3} \end{pmatrix}. \quad (19.42)$$

Assume that

$$\sum_{i=0}^{2}\sum_{j=0}^{2} \alpha_i x_{i,j} + \sum_{k=1}^{4} \beta_k y_k \geqslant 1. \quad (19.43)$$

is a valid inequality for (19.42). Now suppose that we would like to construct a valid inequality for the "larger size" problem $\mathrm{MG}(C_6 \times C_6, \bar{S}, (1/3, 2/3)')$, where $\bar{S} = \{e_1, -e_1, e_2, -e_2, (1,1)'\} \supsetneq S$. Also note that $C_2 \times C_2$ is a subgroup of $C_6 \times C_6$.

We proceed in two steps. In the first step, we consider the problem $\mathrm{MG}(C_2 \times C_2, \bar{S}, (1/3, 2/3)')$, i.e., we introduce a continuous variable with coefficient $(1,1)'$ in (19.42). We can rewrite the extra column in this problem, $(1,1)'y_5$, as $(\mu e_1 + (\mu - 1)(-e_1) + \gamma e_2 + (\gamma - 1)(-e_2))y_5$. Thus $\mathrm{MG}(C_2 \times C_2, \bar{S}, (1/3, 2/3)')$ becomes

$$\sum_{i=0}^{2}\sum_{j=0}^{2}\frac{1}{2}\begin{pmatrix} i \\ j \end{pmatrix} x_{i,j} + \begin{pmatrix} 1 \\ 0 \end{pmatrix} (y_1 + \mu y_5) + \begin{pmatrix} -1 \\ 0 \end{pmatrix} (y_2 + (\mu - 1)y_5) +$$
$$\begin{pmatrix} 0 \\ 1 \end{pmatrix} (y_3 + \gamma y_5) + \begin{pmatrix} 0 \\ -1 \end{pmatrix} (y_4 + (\gamma - 1)y_5) \equiv \begin{pmatrix} \frac{1}{3} \\ \frac{2}{3} \end{pmatrix}.$$

Since (19.43) is a valid inequality for (19.42), we obtain the following valid cut for $\mathrm{MG}(C_2 \times C_2, \bar{S}, (1/3, 2/3)')$

[2] In the ensuing sections, integer variables in master group relaxations will be denoted by x while continuous variables will be denoted by y.

$$\sum_{i=0}^{2}\sum_{j=0}^{2} \alpha_i x_{i,j} + \sum_{k=1}^{4} \beta_k y_k + (\mu\beta_1 + (\mu-1)\beta_2 + \gamma\beta_3 + (\gamma-1)\beta_4)y_5 \geqslant 1, \quad (19.44)$$

if we assume $\mu \geqslant 0$, $\mu - 1 \geqslant 0$, $\gamma \geqslant 0$, and $\gamma - 1 \geqslant 0$. To obtain the best possible coefficient for y_5 we solve $\beta_5 = \min\{(\mu\beta_1 + (\mu-1)\beta_2 + \gamma\beta_3 + (\gamma-1)\beta_4) \mid \mu \geqslant 0, \mu - 1 \geqslant 0, \gamma \geqslant 0, \gamma - 1 \geqslant 0\}$. More generally, we can write

$$\beta_5 = \min\Big\{\sum_{j=1}^{4} y_j\beta_j \;\Big|\; e_1y_1 - e_1y_2 + e_2y_3 - e_2y_4 = (1,1)', y_j \geqslant 0\Big\}. \quad (19.45)$$

In the second step, we introduce the integer variable $\bar{x}$ corresponding to the column $(\frac{1}{6},\frac{1}{6})' \in C_6 \times C_6$ to the problem $\mathrm{MG}(C_2 \times C_2, \bar{S}, (1/3,2/3)')$. By rewriting the column, $(\frac{1}{6},\frac{1}{6})'\bar{x}$ as an aggregation of columns of the problem $\mathrm{MG}(C_2 \times C_2, \bar{S}, (1/3,2/3)')$ we can easily verify that the cutting plane,

$$\sum_{i=0}^{2}\sum_{j=0}^{2} \alpha_i x_{i,j} + \sum_{k=1}^{5} \beta_k y_k + \bar{\alpha}\bar{x} \geqslant 1. \quad (19.46)$$

is valid when $\bar{\alpha}$ is chosen as

$$\bar{\alpha} = \min\Bigg\{ \sum_{i=0}^{2}\sum_{j=0}^{2} \alpha_{i,j}x_{i,j} + \sum_{k=1}^{4} \beta_k y_k \;\Bigg|\; x_{i,j} \in \mathbb{Z}_+, y_k \geqslant 0, \\ \sum_{i=0}^{2}\sum_{j=0}^{2} \frac{1}{2}\begin{pmatrix} i \\ j \end{pmatrix} x_{i,j} + e_1y_1 - e_1y_2 + e_2y_3 - e_2y_4 \equiv \begin{pmatrix} \frac{1}{6} \\ \frac{1}{6} \end{pmatrix}\Bigg\}. \quad (19.47)$$

Adding the integer variables corresponding to the elements in the group $C_6 \times C_6$ one at a time by solving (19.47) (i.e., by changing the right-hand-side of (19.47) each time), we can build a valid inequality for $\mathrm{MG}(C_6 \times C_6, \bar{S}, (1/3,2/3)')$.

Now we generalize the ideas presented in Example 19.11. Let G be a subgroup of I^m and $S \subseteq S^m$. We assume that any $w \in \mathbb{R}^m$ can be written as a non-negative combination of elements in S. Let (ϕ, θ) be a valid subadditive function for $\mathrm{MG}(G,S,r)$. We can obtain a valid subadditive inequality (Φ, Θ) for $\mathrm{MG}(I^m, S^m, r)$ using the fill-in procedure presented in Table 19.2.

Observe that the optimization problem (19.48) solved in the first step of the fill-in procedure to obtain the function $\Theta : S^m \to \mathbb{R}_+$ is a generalization of (19.45). Note here that (19.48) is feasible because of the assumption that any $w \in \mathbb{R}^m$ can be written as a non-negative combination of elements in S. It is easily verified that $\tilde{\Theta} : \mathbb{R}^m \to \mathbb{R}_+$ constructed in Table 19.2 is subadditive over $\mathbb{R}^m$. Next, since $\phi : G \to \mathbb{R}_+$ is subadditive and $\tilde{\Theta} : \mathbb{R}^m \to \mathbb{R}_+$ is subadditive, a problem of the form $\min\{\sum_{v\in G}\phi(v)x(v) + \sum_{w\in L}\theta(w)y(w) \mid x(v) \in \mathbb{Z}_+, y(w) \geqslant 0, \sum_{v\in G} vx(v) + \mathscr{F}(\sum_{w\in L} wy(w)) = u\}$ will reduce to the form $\min\{\phi(v^*) + \tilde{\Theta}(w^*) \mid v^* \in G, w^* \in \mathbb{R}^m, v^* + \mathscr{F}(w^*) = u\}$. Therefore, (19.49), which is the optimization problem solved in the second step of the fill-in procedure to obtain the function $\Phi : I^m \to \mathbb{R}_+$, is a

i. Input: A subadditive valid function (ϕ, θ) for $\mathrm{MG}(G,S,r)$, where $S \subseteq S^m$ and G is a subgroup of I^m.
ii. Construct the function $\Theta : S^m \to \mathbb{R}_+$ as follows:

$$\Theta(w) = \min\Big\{ \sum_{v \in L} \theta(v) y(v) \;\Big|\; w = \sum_{v \in L} v y(v), y(v) \geqslant 0 \Big\}, \tag{19.48}$$

where L is a linearly independent subset of S.
iii. Construct the function $\Phi : I^m \to \mathbb{R}_+$ as follows:

- Let $\tilde{\Theta} : \mathbb{R}^m \to \mathbb{R}_+$ be the homogenous extension of Θ to $\mathbb{R}^m$, i.e., (here) $\tilde{\Theta}(w) = \lambda \Theta(w')$ where $w' \in S^m$, $\lambda \geqslant 0$, and $w = \lambda w'$.
- Compute $\Phi : I^m \to \mathbb{R}_+$ as follows:

$$\Phi(u) = \min\Big\{ \phi(v) + \tilde{\Theta}(w) \;\Big|\; v \in G, w \in S, v + \mathscr{F}(w) = u \Big\}. \tag{19.49}$$

iv. Output: A subadditive valid function (Φ, Θ) for $\mathrm{MG}(I^m, S^m, r)$.

Table 19.2 Fill-in procedure.

generalization of (19.47). We refer to Johnson [71] for a rigorous proof of the validity of the fill-in procedure presented in Table 19.2.

An important question is that of determining whether the fill-in procedure is strong. For one-dimensional group problems, fill-in reduces to building minimal two-slope inequalities and is therefore a strong procedure; see Theorem 19.26. Unfortunately, for group problems with multiple rows, the fill-in procedure only guarantees the subadditivity of the resulting function (Φ, Θ) for $\mathrm{MG}(I^m, S^m, r)$ even if (ϕ, θ) is an extreme inequality for $\mathrm{MG}(G,S,r)$. We next present sufficient conditions that guarantee the strength of functions obtained through fill-in.

Theorem 19.30 ([41], [42]). *Let (ϕ, π) be minimal for* $\mathrm{MG}(G, S^m, r)$ *where G is a finite subset of I^m and the homogenous extension of π to $\mathbb{R}^m$ is the gauge function of a polytope. Then the fill-in inequality (Φ, π) is extreme for* $\mathrm{MG}(I^m, S^m, r)$ *if and only if (ϕ, π) is extreme for* $\mathrm{MG}(G, S^m, r)$ *and (Φ, π) is minimal for* $\mathrm{MG}(I^m, S^m, r)$.

Theorem 19.30 implies that if we start with an extreme inequality for a group problem where all the columns of continuous variables are available, and then apply the fill-in procedure yielding a minimal inequality for the infinite group problem, then this minimal inequality is also extreme for the infinite group problem. Theorem 19.30 can be considered as the m-row version of the Two-Slope Theorem 19.26 for the one-row group problem. Since verifying minimality of a function is easier than proving extremality, Theorem 19.30 can be a useful tool to verify that a valid inequality is extreme for a group problem with multiple rows.

19.5.2.4 Continuous group problem

Most of the results discussed in the previous sections focused on integer variables. In this section, we consider the problem $\mathrm{MG}(\{o\},S^m,r)$. It is convenient to work with the problem $\mathrm{MG}(\{o\},\mathbb{R}^m,r)$ here.[3] The minimal inequalities for $\mathrm{MG}(\{o\},\mathbb{R}^m,r)$ can be characterized in terms of maximal lattice-free convex sets.

Definition 19.23 ([79]). A full-dimensional set $P \subseteq \mathbb{R}^m$ is a maximal lattice-free convex set if $\mathrm{interior}(P) \cap \mathbb{Z}^m = \emptyset$ and $\nexists$ a convex set $P' \subseteq \mathbb{R}^m$ such that $\mathrm{interior}(P') \cap \mathbb{Z}^m = \emptyset$, $P' \neq P$ and $P' \supseteq P$.

Theorem 19.31 ([79], [11]). *A full-dimensional set P is a maximal lattice-free convex set iff it is a lattice-free polyhedron with at least one integer point in the relative interior of each facet.*

The maximum number of inequalities required to define a m-dimensional maximal lattice-free convex set is 2^m; see Doignon [44], Bell [12], Scarf [89]. Suppose that $-r \in \mathrm{int}(P)$ and P is maximal lattice-free convex set. The set $P+r$ (obtained as the Minkowski sum of P and r) is a polyhedron containing the origin in its interior. Therefore, it may be written as $\{u \in \mathbb{R}^m \mid (a^i)'u \leqslant 1, i \in \{1,...,l\}\}$ where $l \leqslant 2^m$. Let π_P be the function $\pi_P(w) = \max_{i\in\{1,...l\}}\{(a^i)'w\}$.

Basu et al. [11] start with a very general definition of valid inequality for $\mathrm{MG}(\{o\},\mathbb{R}^m,r)$ of the form $\sum_{w\in\mathbb{R}^m}\pi(w)y(w) \geqslant \gamma$, where $\pi : \mathbb{R}^m \to \mathbb{R}$. Using this definition they show the following result.

Theorem 19.32 ([11]). *Every nontrivial valid linear inequality for* $\mathrm{MG}(\{o\},\mathbb{R}^m,r)$ *is dominated by a nontrivial minimal valid linear inequality for* $\mathrm{MG}(\{o\},\mathbb{R}^m,r)$. *Every nontrivial minimal valid linear inequality for* $\mathrm{MG}(\{o\},\mathbb{R}^m,r)$ *is equivalent to an inequality of the form* $\sum_{w\in\mathbb{R}^m}\pi(w)y(w) \geqslant 1$ *such that* $\pi(w) \geqslant 0\ \forall w$ *and the set* $P(\pi) := \{u \in \mathbb{R}^m \mid \pi(u+r) \leqslant 1\}$ *is a maximal lattice-free set with* $-r$ *in its interior. Moreover,* $\pi_{P(\pi)} = \pi$.

The statement of Theorem 19.32 remains the same if we replace $\mathrm{MG}(\{o\},\mathbb{R}^m,r)$ with $\mathrm{MG}(\{o\},\mathbb{Q}^m,r)$, except for redefining $P(\pi) = \mathrm{closure}(\{u \in \mathbb{Q}^m \mid \pi(u+r) \leqslant 1\})$. This result is proven in Borozan and Cornuéjols [17]. We refer to Johnson [71], Andersen et al. [6], Zambelli [101] for variants of the above result.

The following result from Cornuéjols and Margot [27] characterizes the subset of minimal inequalities that are extreme inequalities for $\mathrm{MG}(\{o\},\mathbb{Q}^2,r)$.

Theorem 19.33 ([27]). *Let* $r \in \mathbb{Q}^2$ *and* $\tilde{\pi} : \mathbb{Q}^2 \to \mathbb{R}_+$ *be a minimal valid inequality for* $\mathrm{MG}(\{o\},\mathbb{Q}^2,r)$. *Then* $\tilde{\pi} : \mathbb{Q}^2 \to \mathbb{R}_+$ *is extreme for* $\mathrm{MG}(\{o\},\mathbb{Q}^2,r)$ *iff*

- $P(\tilde{\pi})$ *is of the form* $\{(x_1,x_2) \mid c \leqslant a_1x_1 + a_2x_2 \leqslant c+1\}$ *where* $a_1,a_2,c \in \mathbb{Z}$ *and* $gcd(a_1,a_2) = 1$, *or*

[3] In the rest of the section, we present results where the columns w corresponding to continuous variables in the group problem are not assumed to be normalized so that $\max|w_i| = 1$. All the definitions of Section 19.4 are suitably modified. For example, we define $\mathrm{MG}(G,W,r)$ where $W \subseteq \mathbb{R}^m$ as the set of (x,y) such that: $x : G \to \mathbb{Z}_+$, $y : W \to \mathbb{R}_+$, $\sum_{u\in G} ux(u) + \sum_{w\in W} wy(w) = r$ and x and y have finite support.

- *$P(\tilde{\pi})$ is a maximal lattice-free triangle, or*
- *$P(\tilde{\pi})$ is a maximal lattice-free quadrilateral that satisfies the "ratio test", i.e., $\nexists h \in \mathbb{R}_+$ such that*

$$\frac{|b^i - a^i|}{|b^i - a^{i+1}|} = \begin{cases} h, & \text{if } i = 1,3, \\ \frac{1}{h}, & \text{if } i = 2,4, \end{cases} \tag{19.50}$$

where the points a^1, a^2, a^3, and a^4 represent the vertices of the lattice-free quadrilateral $P(\tilde{\pi})$ and b^1, b^2, b^3, and b^4 represent integer points in the relative interior of the sides a^1a^2, a^2a^3, a^3a^4 and a^4a^1 of the quadrilateral respectively.

Theorem 19.33 characterizes extreme inequalities for $\text{MG}(\{o\}, \mathbb{Q}^2, r)$. We can now apply the fill-in procedure discussed in Section 19.5.2.3 to obtain coefficients for integer variables. If the resulting fill-in inequalities are minimal, they must also be extreme by Theorem 19.30. We refer to Dey and Wolsey [41] for a characterization of the subset of extreme inequalities for $\text{MG}(I^2, S^2, r)$ that are obtained using fill-in.

19.5.2.5 Simple structures

In this section, we illustrate a technique to obtain valid inequalities for group problems by using valid inequalities for simple mixed integer sets. The prototypical procedure has three steps. It consists of (*i*) analyzing a simple mixed integer set Q (that typically has free integer variables and at least one non-negative continuous variable) and generating a valid inequality for this set, (*ii*) using aggregation of variables in the group problem/relaxation of the group problem to rewrite it in the form of Q and (*iii*) using the valid inequality for Q to obtain a valid inequality for the group problem.

This technique is similar to the fill-in procedure. To see the similarity, note that in Example 19.11 no property of a group is used in constructing a valid inequality of a larger size group problem from a valid inequality of a smaller size group problem. The main difference between this technique and the fill-in procedure is that in the fill-in process, we start from a valid inequality for a finite group problem, while here we start with a valid inequality for a simple mixed integer set Q.

To illustrate this procedure, we derive the two-step MIR inequalities of Dash and Günlük [33]. Consider the *two-step MIR set* defined as

$$Q^2 = \{(z_1, z_2, y) \in \mathbb{Z} \times \mathbb{Z}_+ \times \mathbb{R}_+ \mid z_1 + \alpha z_2 + y \geqslant b\}, \tag{19.51}$$

where $0 < \alpha < b - \lfloor b \rfloor$ and $\frac{1}{\alpha} \geqslant \left\lceil \frac{b - \lfloor b \rfloor}{\alpha} \right\rceil$. Assume for simplicity that $0 < b < 1$, i.e., $\lfloor b \rfloor = 0$.

The first step is to derive a valid inequality for (19.51). The following *two-step MIR inequality* is proven to be valid (in fact facet-defining) for Q^2 in Dash and Günlük [33],

$$\left\lceil \frac{b}{\alpha} \right\rceil z_1 + z_2 + \frac{y}{b - \alpha \lfloor \frac{b}{\alpha} \rfloor} \geqslant \left\lceil \frac{b}{\alpha} \right\rceil. \tag{19.52}$$

The name "two-step MIR inequality" is given to (19.52) because it can be derived using a sequence of two MIR inequalities as follows:

i. Treating the term $\alpha z_2 + y$ as a non-negative continuous variable, the MIR inequality for $z_1 + \alpha z_2 + y \geqslant b$, the defining constraint of Q^2, is

$$z_1 + \frac{\alpha z_2 + y}{b} \geqslant 1. \tag{19.53}$$

Multiplying the inequality $z_1 + \alpha z_2 + y \geqslant b$ by $\frac{1}{\alpha}$ and (19.53) by $\frac{b}{\alpha}$ the valid inequalities

$$\frac{1}{\alpha} z_1 + z_2 + \frac{y}{\alpha} \geqslant \frac{b}{\alpha} \tag{19.54}$$

$$\frac{b}{\alpha} z_1 + z_2 + \frac{y}{\alpha} \geqslant \frac{b}{\alpha} \tag{19.55}$$

can be obtained. By assumption, $\frac{1}{\alpha} \geqslant \lceil \frac{b}{\alpha} \rceil \geqslant \frac{b}{\alpha}$. Therefore there is a convex combination of (19.54) and (19.55) that is of the form

$$\left(\left\lceil \frac{b}{\alpha} \right\rceil z_1 + z_2 \right) + \frac{y}{\alpha} \geqslant \frac{b}{\alpha}. \tag{19.56}$$

ii. Now by applying the MIR inequality to (19.56), the two-step MIR inequality (19.52) is obtained.

Now (19.52) is used to obtain a valid inequality for the one-row mixed integer group problem. Consider the one row group problem (we consider the pure integer problem for simplicity)

$$x_0 + \sum_{i=1}^{n-1} \frac{i}{n} x_i = r, \tag{19.57}$$

where $x_0 \in \mathbb{Z}$ and $x_i \in \mathbb{Z}_+ \; \forall i = 1, \ldots, n-1$. Suppose $\alpha > 0$ is chosen such that $r > \alpha$ and $\frac{1}{\alpha} \geqslant \lceil \frac{r}{\alpha} \rceil = 2$. The following relaxation of (19.57) is constructed

$$\Big(x_0 + \sum_{i|i>nr} x_i\Big) + \alpha \Big(\sum_{i|(r-\alpha)n \leqslant i \leqslant nr} x_i \Big) + \Big(\sum_{i|i<n(r-\alpha)} \frac{i}{n} x_i + \sum_{i|\alpha n < i \leqslant nr} (\frac{i}{n} - \alpha) x_i \Big) \geqslant r. \tag{19.58}$$

The three terms in the left-hand-side of (19.58) satisfy the requirements on z_1, z_2 and y respectively. Therefore by applying inequality (19.52) to (19.58) and then substituting out x_0 using (19.57) a valid inequality for the group problem is obtained.

Using the arguments presented above, Dash and Günlük [33] prove the following general result for the one-row infinite group problem.

Theorem 19.34 ([33]). *Let* $0 < \alpha < r$ *and* $\frac{1}{\alpha} \geqslant \lceil \frac{r}{\alpha} \rceil > \frac{r}{\alpha}$. *Then the function* $g^{r,\alpha} : I^1 \to \mathbb{R}_+$ *defined as*

$$g^{r,\alpha}(u) = \begin{cases} \frac{u(1-\rho\tau)-k(u)(\alpha-\rho)}{\rho\tau(1-r)}, & \text{if } u - k(u)\alpha < \rho, \\ \frac{k(u)+1-\tau u}{\tau(1-r)}, & \text{if } u - k(u)\alpha \geqslant \rho, \end{cases} \tag{19.59}$$

where $\rho = r - \alpha \lfloor \frac{r}{\alpha} \rfloor$, $\tau = \lceil \frac{r}{\alpha} \rceil$, *and* $k(u) = \min\{\lceil \frac{u}{\alpha} \rceil, \tau\} - 1$ *represents a valid inequality for the one-row group problem with right-hand side r.*

Kianfar and Fathi [74] observe that the above procedure can be generalized. Instead of considering the two-step MIR set Q^2, they consider the following *n-step MIR* set

$$Q^n = \Big\{(z_1, z_2, ..., z_n, y) \in \mathbb{Z} \times \mathbb{Z}_+^{n-1} \times \mathbb{R}_+ \;\Big|\; \sum_{i=1}^{n} \alpha_i z_i + y \geqslant b\Big\}, \tag{19.60}$$

where $\alpha_i > 0 \ \forall i \in \{1, ..., n\}$, $\frac{b}{\alpha_1} < \lceil \frac{b}{\alpha_1} \rceil$, and $\frac{b^{i-1}}{\alpha_i} < \lceil \frac{b^{i-1}}{\alpha_i} \rceil \leqslant \frac{\alpha_{i-1}}{\alpha_i} \ \forall i \in \{2, ...n\}$ and where b^k is inductively defined as $b^k = b^{k-1} - \alpha_k \lfloor \frac{b^{k-1}}{\alpha_k} \rfloor$ with $b^0 = b$. First a *n-step MIR inequality* is proven to be facet-defining for Q^n, which is derived by the application of a sequence of n MIRs. We refer to Pochet and Wolsey [85] for similar inequalities for a set closely related to Q^n. Then using arguments similar to those used in the derivation of the two-step MIR inequality for the one row group problem, Kianfar and Fathi [74] present a real-valued function (in closed form) defined over I^1 that represents a valid inequality for the one-row group problem. They show that the function is minimal, continuous and has two-slopes. Using Two-Slope Theorem 19.26, the function can then be shown to be extreme.

19.5.2.6 Extreme inequalities for $\mathrm{MG}(I^m, \emptyset, r)$ using extreme inequalities for $\mathrm{MG}(I^1, \emptyset, r)$

In this section, we discuss some operations that can be used to create valid (and often extreme inequalities) for the group problem with multiple rows using valid inequalities for group problems with lesser number of rows. While the results are presented in the context of the infinite group problem, they can be suitably adapted for finite group problems.

If there exists a black-box algorithm to generate valid inequalities using one row of a simplex tableau, a natural way to construct valid inequalities using multiple rows of a simplex tableau is the following. First multiply each row of the tableau with a suitable number. Then add the weighted rows together to obtain one valid constraint. This constraint can then be used as an input for the black-box algorithm. This procedure is known in integer programming as *constraint aggegration* and has been shown to yield high quality cutting planes in practical implementations; see for example Marchand and Wolsey [81] and Andersen, Cornuéjols, and Li [4]. Interestingly, this procedure can be used to generate extreme inequalities for multiple-row group problems. We formally define this operation in the context of the group problem next.

Definition 19.24. Let $\phi : I^1 \to \mathbb{R}_+$ be a valid inequality for $\mathrm{MG}(I^1, \emptyset, r)$ and let $\lambda = (\lambda_1, ..., \lambda_m) \in \mathbb{Z}^m \backslash \{0\}$. We define the aggregation function $\phi^\lambda : I^m \to \mathbb{R}_+$ as $\phi^\lambda(x) = \phi(\sum_{i=1}^m \lambda_i x_i)$.

Clearly, if ϕ is a valid inequality for $\mathrm{MG}(I^1, \emptyset, r)$ then ϕ^λ is a valid inequality for $\mathrm{MG}(I^m, \emptyset, r')$ where $r' \in I^m$ satisfies $(\sum_{i=1}^m \lambda_i r_i') \pmod 1 = r$.

Theorem 19.35 ([36, 40, 38]). *If ϕ is an extreme inequality for $\mathrm{MG}(I^1, \emptyset, r)$, then ϕ^λ is an extreme inequality for $\mathrm{MG}(I^m, \emptyset, r')$ where $r' \in I^m$ satisfies $(\sum_{i=1}^m \lambda_i r_i') \pmod 1$ $= r$.*

We refer to Dey and Richard [38], Dey et al. [40] and Dey [36] for proofs of this result for different groups and different classes of functions ϕ. Similar to the aggregation scheme, a procedure called sequential-merge can be used to generate inequalities for multiple-row group problems using inequalities for group problems with lesser number of rows. The treatment here is from Dey and Richard [37, 39].

We begin by illustrating the key ideas of the sequential-merge procedure using two rows of a simplex tableau and using GMICs. We will then generalize the procedure.

i. Consider any two rows of a simplex tableau:

$$\sum_{i=1}^{n} a_i^1 x_i + \sum_{j=1}^{m} c_j^1 y_j = b^1 \tag{19.61}$$

$$\sum_{i=1}^{n} a_i^2 x_i + \sum_{j=1}^{m} c_j^2 y_j = b^2 \tag{19.62}$$

$$x_i \in \mathbb{Z}_+, y_j \in \mathbb{R}_+,$$

where b^1 and b^2 are fractional. As a first step, generate a MIR inequality from the second row:

$$\sum_{\{i|f_i^2 \leqslant r^2\}} \lfloor a_i^2 \rfloor x_i + \sum_{\{i|f_i^2 > r^2\}} \left(a_i^2 - \frac{1-f_i^2}{1-r^2} r^2 \right) x_i + \sum_{\{j|c_j^2 \leqslant 0\}} \frac{c_j^2}{1-r^2} y_j \leqslant \lfloor b^2 \rfloor, \tag{19.63}$$

where $r^k = b^k - \lfloor b^k \rfloor$, $f_i^k = a_i^k - \lfloor a_i^k \rfloor$ and $k \in \{1, 2\}$. The typical form of GMIC for the second row is essentially $\frac{(19.62)-(19.63)}{r^2}$. We will generalize this relationship between GMIC and MIR inequalities in Definition 19.25 to present the sequential-merge procedure for general group cuts and an arbitrary number of rows.

ii. Next add MIR inequality (19.63) to the first row of the simplex tableau (19.61) to obtain

$$\sum_{\{i|f_i^2<r^2\}} \left(a_i^1 + \lfloor a_i^2 \rfloor\right) x_i + \sum_{\{i|f_i^2 \geqslant r^2\}} \left(a_i^1 + a_i^2 - \frac{1-f_i^2}{1-r^2} r^2\right) x_i$$

$$+ \sum_{\{j|c_j^2 \leqslant 0\}} \left(c_j^1 + \frac{c_j^2}{1-r^2}\right) y_j + \sum_{\{j|c_j^2 \geqslant 0\}} c_j^1 y_j \leqslant b^1 + \lfloor b^2 \rfloor. \quad (19.64)$$

iii. Finally apply the MIR inequality to (19.64). The resulting MIR inequality is the sequential-merge inequality using two GMICs. We denote it as

$$\sum_i [\xi \Diamond \xi]^{-1}_{(r^1,r^2)} \begin{pmatrix} a_i^1 \\ a_i^2 \end{pmatrix} x_i + \sum_j [\xi \Diamond \xi]^{-1}_{(r^1,r^2)} \begin{pmatrix} c_j^1 \\ c_j^2 \end{pmatrix} y_j \leqslant [\xi \Diamond \xi]^{-1}_{(r^1,r^2)} \begin{pmatrix} b^1 \\ b^2 \end{pmatrix}, \quad (19.65)$$

where ξ is used to denote the GMIC. The notation in (19.65) will become clear at the end of the section.

iv. Similar to the MIR-GMIC relationship for one tableau row, the sequential-merge inequality (19.65) can be rewritten as $\frac{(19.61)+(19.62)-(19.65)}{r^1+r^2}$. The coefficients in the inequality $\frac{(19.61)+(19.62)-(19.65)}{r^1+r^2}$ depend now only on the fractional parts of the coefficients in the tableau.

The idea of the sequential-merge inequality can be generalized to other inequalities and to more rows as follows. First, derive a cut from a m-row group problem. Rewrite this inequality in the "MIR" form; see Definition 19.25. Second, add this inequality to the remaining row. Third, generate an inequality from this new one-row inequality constraint. Rewrite this inequality in a way that its coefficients depend only on the fractional part of the columns; see Definition 19.25.

We begin with a definition generalizing the relationship between MIR and GMIC.

Definition 19.25. Given a valid inequality ϕ for $\text{MG}(I^m,\emptyset,r)$, define the lifting-space representation $[\phi]_r : \mathbb{R}^m \to \mathbb{R}$ as $[\phi]_r(x) = \sum_{i=1}^m x_i - \sum_{i=1}^m r_i \phi(\mathscr{F}(x))$. Given a function $\psi : \mathbb{R}^m \to \mathbb{R}$ that satisfies $\psi(x+e_i) = \psi(x)+1$, where e_i is the ith unit vector of $\mathbb{R}^m$, define the group-space representation $[\psi]_r^{-1} : I^m \to \mathbb{R}$ as $[\psi]_r^{-1}(x) = \frac{\sum_{i=1}^m x_i - \psi(x)}{\sum_{i=1}^m \tilde{r}_i}$.

It is easily verified that applying MIR to a row of a simplex tableau yields $[\text{GMIC}]_r$ where r is the fractional part of the right-hand-side of the simplex tableau. It can also be verified that GMIC is $[\text{MIR}]_r^{-1}$. The following proposition establishes relationships between the lifting-space and group-space representations of a function.

Proposition 19.6 ([39]). *$[\phi]_r$ is superadditive iff ϕ is subadditive. Moreover, if ϕ is valid function for $\text{MG}(I^m,\emptyset,r)$, then given a simplex tableau $\sum_{i=1}^n a_i x_i = b$, $x \in \mathbb{Z}_+^n$ where $a_i \in \mathbb{R}^m$, $b \in \mathbb{Q}^m \setminus \mathbb{Z}^m$ and $r = \mathscr{F}(b)$, the inequality $\sum_{i=1}^n [\phi]_r(a_i) x_i \leqslant [\phi]_r(b)$ is valid.*

Now we present the general version of the sequential-merge procedure in Table 19.3.

i. Input: (1) The group problem $\mathrm{MG}(I^{m+1},\emptyset,(r^1,r^2)')$ where $r^1 \in I^1$, $r^2 \in I^m$ and $r^1 \neq o, r^2 \neq o$. (2) $h : I^m \to \mathbb{R}_+$, a valid subadditive inequality for $\mathrm{MG}(I^m,\emptyset,r^2)$. (3) $g : I^1 \to \mathbb{R}_+$ a valid subadditive inequality for $\mathrm{MG}(I^1,\emptyset,r^1)$ such that $[g]_{r^1} : \mathbb{R}^1 \to \mathbb{R}_+$ is a nondecreasing function.
ii. Construct the inequality $[h]_{r^2}$ based on Definition 19.25.
iii. Add the inequality $[h]_{r^2}$ to the first row of the group problem. We obtain the inequality,

$$\sum_{u=(u_1,u_2)\in(I^1\times I^m)} (u_1 + [h]_{r^2}(u_2))x(u) \leqslant r^1 + [h]_{r^2}(r^2). \tag{19.66}$$

iv. Apply the inequality $[g]_{r^1}$ to (19.66) as follows:

$$\sum_{u=(u_1,u_2)\in(I^1\times I^m)} [g]_{r^1}(u_1 + [h]_{r^2}(u_2))x(u) \leqslant [g]_{r^1}(r^1 + [h]_{r^2}(r^2)) \tag{19.67}$$

v. Apply the operation $[.]^{-1}_{r^1,r^2}$ to the coefficients of (19.67). The resulting inequality is the sequential-merge inequality denoted as

$$\sum_{u\in I^{m+1}} (g\Diamond h)(u)x(u) \geqslant 1. \tag{19.68}$$

Table 19.3 Sequential-merge procedure.

By Proposition 19.6, the function $[g]_{r^1}$ yields a valid inequality when applied to one row of a simplex tableau, i.e., a constraint with equality sign. However (19.67) is obtained by applying $[g]_{r^1}$ to (19.66) which is a constraint with less than or equal to sign. Therefore, the condition that $[g]_{r^1}$ is a non-decreasing function is required to prove the validity of (19.67) and (19.68).

The sequential-merge procedure can be used to obtain extreme inequalities. The following result describes sufficient conditions for sequential-merge inequalities to be extreme.

Theorem 19.36 ([39]). *Assume that g and h are continuous, piecewise linear valid functions for* $\mathrm{MG}(I^1,\emptyset,r^1)$ *and* $\mathrm{MG}(I^m,\emptyset,r^2)$ *respectively. Assume also that g and h are unique solutions of* $E(g)$ *and* $E(h)$ *respectively and that* $[g]_{r^1}$ *and* $[h]_{r^2}$ *are nondecreasing. Then* $g\Diamond h$ *is an extreme inequality for* $\mathrm{MG}(I^{m+1},\emptyset,(r^1,r^2)')$.

19.5.2.7 Approximate lifting

Richard et al. [86] and Miller et al. [82] propose an approximate lifting scheme to build inequalities for MIPs that can be proven to produce strong inequalities for master group relaxations. In particular, the inequalities that are produced can be shown to be extreme over a simpler (and smaller) set of functions. Among others, they consider CPL_n functions that are defined as follows.

Definition 19.26. Let $K \in \mathbb{R}_+$ and $r \in (0,K)$. Let $n \in \mathbb{Z}_+$, $z = (z_1, z_2, \cdots, z_n) \in \mathbb{R}^n_+$, and $\gamma = (\gamma_1, \gamma_2, \cdots, \gamma_n) \in \mathbb{R}^n_+$ be such that $\sum_{j=1}^n z_j = \frac{K-r}{2}$ and $\sum_{j=1}^n \gamma_j z_j = \frac{1}{2}$. A function $\psi : \mathbb{R}^1 \to \mathbb{R}_+$ is said to be a $\mathrm{CPL}_n(K;r;z;\gamma)$ function if, when u is restricted to $[0,K)$,

$$\psi(u) = \begin{cases} 0, & \text{if } u \in [0,r], \\ \Gamma_{i-1} + \gamma_i(u - r - Z_{i-1}), & \text{if } u \in (r+Z_{i-1}, r+Z_i), \\ \Gamma_i, & \text{if } u = r+Z_i, \\ 1-\Gamma_i, & \text{if } u = K - Z_i, \\ 1-\Gamma_{i-1} - \gamma_i(K - u - Z_{i-1}), & \text{if } u \in (K-Z_i, K-Z_{i-1}), \end{cases}$$

for $i = 1,\ldots,n$ where $Z_0 = 0$, $\Gamma_0 = 0$, $Z_i = \sum_{j=1}^{i} z_j$ and $\Gamma_i = \sum_{j=1}^{i} \gamma_j$ for $i = 1,\ldots,n$. For the sake of brevity, we call a $\mathrm{CPL}_n(K;r;z;\gamma)$ function a CPL_n function.

Superadditive CPL_n functions translate into valid inequalities for master group problems if the transformation $\phi(u) = \frac{u-\psi(Ku)}{\frac{r}{K}}$ is used. It can be shown in this case that ϕ is subadditive, $\phi(\frac{r}{K}) = 1$, $\phi(0) = 0$ and $\phi(u) \geq 0$ for all $u \in I^1$. Further, the parameters that describe superadditive CPL_n functions belong to a polyhedron that is reminiscent of $P^*(G,\emptyset,r)$.

Corollary 19.3 ([86]). *Let $z \in \mathbb{R}^n_+$ be such that $Z_n = \frac{K-r}{2}$. Parameter γ defines a superadditive CPL_n function if and only if γ belongs to the polyhedron*

$$\begin{aligned} P\Theta_n(z) = \{\gamma \in \mathbb{R}^{n-1}_+ \mid\ & \Gamma_i + \Gamma_j \leq \psi(2r + Z_i + Z_j), && 0 \leqslant i,j \leqslant n-1, \\ & \Gamma_i - \Gamma_j \leq \psi(r + K + Z_i - Z_j) - 1, && 0 \leqslant i,j \leqslant n-1, \\ & \Gamma_i + \Gamma_j \geq \psi(r + Z_i + Z_j), && 0 \leqslant i,j \leqslant n-1\}. \end{aligned}$$

The CPL_n functions and inequalities that correspond to the extreme points of $P\Theta_n(z)$ are extreme within their family. We refer to these functions as CPL_n-extreme functions and inequalities. It is clear that the resulting inequalities ϕ are not necessarily extreme among all valid inequalities of master group problems. However, we will show next, by considering the simplest case where $n = 2$, that many of them yield extreme inequalities for master group problems. This example is from Richard, Li and Miller [86]. Similar conclusions have been drawn for larger values of n; see Miller et al. [82].

Consider the case where $n = 2$. We can use Corollary 19.3 to derive the extreme points of $P\Theta_2(z_1)$ so as to derive the corresponding CPL_2-extreme inequalities. Clearly

$$P\Theta_2(z_1) = \{\gamma_1 \in \mathbb{R}_+ \mid \phi(r + 2z_1) \leq 2\gamma_1 \leq \phi(2r + 2z_1)\}.$$

Depending on the magnitude of z_1 with respect to the other parameters of the problems, we distinguish three cases. We only present the proof of the third case as the other two cases can be proven similarly. When $\frac{K-2r}{2} \leq z_1 \leq \frac{K-r}{2}$, we have that $2r + 2z_1 - K \in [0,r]$ and so $\phi(2r + 2z_1) = 1$. We now distinguish two subcases. First, when $z_1 > \frac{K-r}{3}$, then $r + 2z_1 \in [K - z_1, K]$ and $\phi(r + 2z_1) = 1 - \gamma_1 + \frac{\gamma_1}{z_1}(r + 3z_1 - K)$. Since $r + 2z_1 \leq K$, we have that $\frac{z_1}{K-r} \leq \frac{1}{2}$ and that $P\Theta_2(z_1) = [\frac{z_1}{K-r}, \frac{1}{2}]$. Second, when

$z_1 \leq \frac{K-r}{3}$, we have $r+2z_1 \in [r+z_1, K-z_1]$ and $\phi(r+2z_1) = \gamma_1 + \frac{1-2\gamma_1}{K-2z_1-r} z_1$. Since $r+2z_1 \leq K$, we have that $\frac{z_1}{K-r} \leq \frac{1}{2}$ and that $P\Theta_2(z_1) = [\frac{z_1}{K-r}, \frac{1}{2}]$.

As a result, we obtain the following theorem.

Theorem 19.37 ([86]). *The following are the only extreme points of $P\Theta_2(z_1)$:*

$$\gamma_1^1 = \Big\{ \frac{z_1}{K-r}, \quad z_1 \in [0, \tfrac{K-r}{2}],$$

$$\gamma_1^2 = \begin{cases} \frac{z_1+r}{K+r}, & z_1 \in [0, \frac{K-2r}{3}), \\ \frac{z_1}{K-2r}, & z_1 \in [\frac{K-2r}{3}, \frac{K-2r}{2}), \\ \frac{1}{2}, & z_1 \in [\frac{K-2r}{2}, \frac{K-r}{2}]. \end{cases}$$

It can be observed that $P\Theta_2(z_1)$ has exactly two extreme points for every admissible value of z_1. The CPL_2 inequalities generated by these points may however be dominated by other inequalities whose superadditive functions $\psi(a)$ do not belong to CPL_2. However, most of the CPL_2 inequalities are strong inequalities for the group problem as shown in the next theorem.

Theorem 19.38 ([86]). *Let $\phi(u)$ be the subadditive function induced by an extreme point given in Theorem 19.37. The vector π defined by $\pi_j = \phi(j)$ for $j = 1, ..., K-1$ satisfies the following properties:*

a) For the extreme point $\gamma_1^1 = \frac{z_1}{K-r}$, $\pi x \geqslant 1$ is a 2-slope extreme inequality for $\mathrm{MG}(C_K, \emptyset, \frac{r}{K})$.
b) For the extreme point $\gamma_1^2 = \frac{z_1+r}{K+r}$, $\pi x \geqslant 1$ is a 3-slope extreme inequality for $\mathrm{MG}(C_K, \emptyset, \frac{r}{K})$ when $K \geq 3r + 4z_1$.
c) For the extreme point $\gamma_1^2 = \frac{z_1}{K-2r}$, $\pi x \geqslant 1$ is a 2-slope extreme inequality for $\mathrm{MG}(C_K, \emptyset, \frac{r}{K})$ when $K = 2r + 2z_1$ and a 3-slope extreme inequality for $\mathrm{MG}(C_K, \emptyset, \frac{r}{K})$ when $r = 1$ and $K = 3r + 2z_1$.
d) For the extreme point $\gamma_1^2 = \frac{1}{2}$, $\pi x \geqslant 1$ is a 2-slope extreme inequality for $\mathrm{MG}(C_K, \emptyset, \frac{r}{K})$.

19.5.3 A compendium of known extreme inequalities for finite and infinite group problems

In this section, we present a taxonomy of the extreme inequalities that have been proposed for finite and infinite master group problems. Deriving an exhaustive taxonomy is not a trivial task as one may simply run a shooting experiment (see Section 19.6) with a group of large size and find a new inequality for that particular group that has not been described previously. So, in this section, we will focus on those inequalities that have been generalized into parameterized families that apply to various group sizes and/or right-hand-sides. We mention that extreme inequalities for some finite pure integer group problems are given in Gomory [57]. Extreme inequalities of one-dimensional finite mixed integer group problems up to order $n = 7$ are listed in Gomory and Johnson [60]. Similar results were also obtained in Evans [46]. Some families of filled-in inequalities are also presented in Johnson [71].

In Table 19.4, we present families of extreme inequalities. The table is divided in two parts. The first part presents those inequalities that are described from a specific constructive procedure that does not require the knowledge of another extreme inequality (`constituent functions`), while the second part describes those inequalities that are described through a generic constructive procedure, starting from another extreme inequality (`procedure`). In this table, `Name` describes the common name of the corresponding inequality, `Finite/Infinite` describes whether it applies for finite (F) and/or infinite (I) group problems, `Dimension` describes the number of rows in the group problem for which this inequality is defined, `Slopes/Gradients` describes the number of slopes or gradients the inequality has (for one-dimensional finite group problems, this characteristic is computed from the linear interpolation of the inequality), `Continuous/Discontinuous` reports whether the corresponding inequality is continuous (C) or discontinuous (D) (this applies only for infinite group problems), `Method` describes the method that was used to obtain this inequality (as classified in Section 19.5: analysis of the subadditive polytope $P^*(G,\emptyset,r)$ (AS), group relations (GR), approximate lifting (AL), simple sets (SS), simple procedures (SP), continuous problem (CP)), `Discovered` records the reference where the corresponding inequality was first described while `Proven` refers to the reference where the inequality was first proven to be extreme.

19.6 On the strength of group cuts and the group approach

In this section, we present different lines of results regarding the strength of the group-theoretic approach. In Section 19.6.1, we discuss the evaluation of the absolute strength of group cuts. In particular, we describe studies aimed at evaluating how much of the integrality gap can be closed using group cutting planes. In Section 19.6.2, we discuss the evaluation of the relative strength of different families of group cutting planes. In particular, we present simple measures to judge the strength of a group cutting plane, and we report the results of experiments aimed at determining if there are families of group cutting planes that are computationally more important than others. We make concluding remarks in Section 19.6.3.

19.6.1 Absolute strength of group relaxation

Since the group cutting planes used in practice are most often extreme inequalities of the convex hull of solutions to the master group relaxation (or the related *corner relaxation*), a study of the strength of the corner relaxation provides insight into

Name	Finite/ Infinite	Dimension	Slopes/ Gradients	Continuous/ Discontinuous	Method	Discovered	Proven
Constituent Functions							
GMIC	F	1	2	-	AS	[55]	[57]
GMIC	I	1	2	C	AS	[55]	[60]
GJ's 2-Slope	F/I	1	2	-/C	AS	[61]	[60]
2-Step MIR	F/I	1	2	-/C	SS	[33]	[60]
n-Step MIR	F/I	1	2	-/C	SS	[74]	[60]
Forward 3-Slope	F/I	1	3	-/C	AS	[61]	[61]
Backward 3-Slope	F	1	3	-	AS	[8]	[8]
Backward 3-Slope	I	1	3	C	GR	[40]	[40]
Improved GFC	I	1	1	D	CA	[78, 33]	[40]
2-Step MIR Limit	I	1	1	D	SS	[78, 33]	[40]
GJ's 2-Slope Limit	I	1	1	D	GR	[40]	[40]
3-Slope Limit	I	1	2	D	GR	[40]	[40]
3-Gradient	I	2	3	C	AS	[38]	[38]
Continuous Group Problem	F/I	2	2-4	C	CP	[17, 6]	[6, 27]
Procedures							
Homomorphism	F/I	1	-	-	-	[57]	[57]
λ-Homomorphism	I	n	-	-	-	[38]	[38]
Automorphism	F/I	n	-	-	-	[57]	[57]
Aggregation	I	n	-	-	-	[38]	[38]
Sequential-Merge	I	n	-	-	-	[39]	[39]
Interpolation	I	1	-	-	-	[60]	[40]*
Fill-in	I	n	-	-	-	[60, 71]	[42]*

* (Under some conditions)

Table 19.4 Taxonomy of extreme inequalities for group problems.

the strength of general group cutting planes. Fischetti and Monaci [49] conducted a set of computational experiments to gauge the strength of the group relaxation and of the related corner relaxation. We begin with the definition of a corner relaxation they used, as it differs slightly from that we presented before.

Definition 19.27 ([49]). For $A \in \mathbb{Q}^{m\times n}$, $c \in \mathbb{Q}^n$, and $b \in \mathbb{Q}^m$, let x^* be an optimal solution to the LP relaxation of the problem $\min\{c'x \mid Ax \leqslant b,\ x_j \in \mathbb{Z} \text{ for } j \in J\}$. Let K be the set of inequalities that are binding at x^*, i.e., $(a^k)'x^* = b^k\ \forall k \in K$. The FM-corner relaxation is defined as the problem $\min\{c'x \mid (a^k)'x \leqslant b^k \text{ for all } k \in K,\ x_j \in \mathbb{Z}, \text{ for all } j \in J\}$.

Observe that the definition of the FM-corner relaxation is independent of the basis of the optimal LP simplex tableau and depends only on the optimal solution of the LP relaxation. The traditional corner relaxation depends on the basis selected and therefore is not unique in the presence of primal degeneracy. The experiment conducted in [49] consists of the following steps:

i. Solve the LP relaxation of given MIP instance (considered to be a minimization problem).
ii. Obtain lower bounds on the objective function value of the MIP by the following five methods: (a) Add Gomory mixed integer cuts obtained from each

fractional row to the LP relaxation and resolve. (b) Add K-cuts (Cornuéjols [26] et al.) obtained from each fractional rows to the LP relaxation and resolve. (c) Add single-row interpolated group cuts (Fischetti and Saturni [50]) from each fractional rows to the LP relaxation and resolve. (d) Solve the FM-corner relaxation. (e) Solve the traditional corner relaxation.

iii. Using the lower bounds, compute the percentage of integrality gap closed in each case as:

$$\frac{\text{Lower bound} - \text{LP bound}}{\text{Optimal MIP solution} - \text{LP bound}}.$$

The test bed consists of instances from MIPLIB 2003 and 3.0 [15] that have known optimal solutions and instances from Atamtürk [1]. Fischetti and Monaci [49] make the following observations:

- On an average, approximately 79% and 34% of the integrality gap was closed by the FM-corner relaxation for the Atamtürk [1] and the MIPLIB instances respectively.
- Compiling data from Tables 1 and 2 in [49], we observe that the average gap closed by the FM-corner relaxation, among the MIPLIB instances tested in [49], is 25% for pure 0-1 IPs, and is 39% for problems having either continuous or general integer variables (or both). If only instances with general integer variables are considered, the FM-corner relaxation closes 56% of the gap. Therefore, the FM-corner relaxation seems to be significantly less strong for pure 0-1 IPs than for general MIPs. This corroborates the early observation made by Balas [9].
- When the LP relaxation is highly degenerate, the corner relaxation is weaker than the FM-corner relaxation. The value of the gap closed by the corner relaxation is 21% for pure 0-1 IPs, and is 25% for problems having either continuous or general integer variables. It is 37% for instances with general integer variables.
- Remarkably, the lower bound obtained by adding only GMICs to the LP relaxation is better than the lower bound obtained by the FM-corner relaxation, especially in the case of 0-1 IPs. This is not paradoxical, since this implies that once GMICs are added, the inequalities not binding at the current vertex of the LP relaxation become more important. Therefore in practice it can be expected that the integrality gap closed by the application of group cutting planes is larger than the integrality gap closed by the corner relaxation. We discuss this point in more detail in Section 19.6.2.2.

One of the most essential observations made by Fischetti and Monaci is that group cuts are most useful when some of the integer variables of the problems are not 0-1. Moreover, in cases where there is significant primal degeneracy, group cuts may not be very strong. However, given the fact that very few families of cutting planes are known for general mixed integer problems, the ease with which group cutting planes can be obtained, and the fact that group relaxations are relatively strong for general MIPs, there still seems to be an incentive to study new families of extreme inequalities for group relaxations.

19.6.2 Relative strength of different families of group cuts

In Section 19.5.3, we gave a list of known families of extreme inequalities for finite and infinite group problems. In this list, a number of extreme inequalities for the infinite group problems belonged to parameterized families and therefore provided a large number of valid inequalities for MIPs. Moreover, Gomory and Johnson [60] show that the number of extreme inequalities can increase exponentially as the size of the finite master group problem increases.

Example 19.12 ([60]). The number of two-slope extreme inequalities is at least $\frac{(2k)!}{k!k!}$ for the finite group problems $\mathrm{MG}(C_{20k}, \emptyset, \frac{1}{10})$ where $k \in \mathbb{Z}$ and $k \geqslant 1$. As k becomes large, this number approaches $\frac{2^{2k}}{\sqrt{\pi k}}$.

From a theoretical standpoint, this large number of inequalities suggests that it is probably very difficult to explain and document all the extreme inequalities of master group problems. From a practical standpoint, it also suggests that it is important to understand the relative strengths of different families of group cuts so as to focus only on those that are most useful computationally.

For a given polytope, various measures have been proposed to predict the relative importance of different extreme inequalities in a branch-and-bound framework; see Hunsaker [69] for details.

i. Best-case improvement in objective function value when the cut is added to a given relaxation: This measure was originally proposed by Goemans [53] for measuring the strength of facet-defining inequalities of the graphical Traveling Salesman Problem. Let E be the valid half-space defined by the inequality under study and let P be some relaxation of the problem. This measure is defined as

$$\sup_{c\in\mathbb{R}^n_+} \frac{\max\{c'x \mid x \in P\}}{\max\{c'x \mid x \in P \cap E\}}.$$

Larger values of this measure correspond to better cuts with respect to relaxation P.

ii. Shooting experiment: The shooting experiment was first proposed by Kuhn [76] for the Traveling Salesman Problem. To compute this measure, we select a point and shoot rays in random directions from this point. The first inequality hit is recorded. Inequalities receiving more shots are expected to be more important as they subtend a larger solid angle to the point from which the rays were shots. Observe however that the choice of point from which the rays are shot may yield different results and is of crucial importance when designing the experiment.

iii. Facet volume: As facets of polytopes are themselves polytopes in lower dimension, their volume can be computed as a measure of their importance. The larger the volume of a facet is, the more important it is considered.

iv. Chvátal-Gomory rank: Given a system of inequalities $(a^k)'x \leqslant b^k$, the Chvátal-Gomory procedure involves first taking a nonnegative combination to obtain the inequality, $\sum_i(\sum_k \lambda^k a^k_i)x_i \leqslant \sum_k \lambda^k b^k$. Then both sides are rounded down to obtain the valid inequality, $\sum_i \lfloor \sum_k \lambda^k a_i^k \rfloor x_i \leqslant \lfloor \sum_k \lambda^k b^k \rfloor$. Such inequalities are added for all nonnegative weights λ to obtain the first Chvátal-Gomory closure; see Section 11.8 for a more detailed exposition of this procedure. This process can be repeated to obtain the second closure and so forth. The Chvátal-Gomory rank of an inequality is the smallest integer k such that it is valid for the kth Chvátal-Gomory closure. Another related measure is the split rank; see Cornuéjols and Li [25] for details.

v. Empirical testing of cuts by measuring the size of the resulting branch-and-cut tree. This has the disadvantage of including factors other than cutting planes strength, such as branching strategies, ...

vi. Number of subadditive relations binding of the extreme inequality. This measure is specific to group problems and their relatives.

Hunsaker [69] computes each of the above measures for extreme inequalities of the master knapsack polytope: $\{\max c'x \mid \sum_{i=1}^{n} ix_i \leqslant n, x \in \mathbb{Z}_+^n\}$. The author observes that the size of the branch-and-bound tree is most strongly correlated to the results of the shooting experiment and the best-case improvement measure; see [69] for details of implementation. While the correlation between these measures may differ for different underlying problems, there are reasons to believe that the correlations obtained in these experiments are meaningful also for group relaxations since the master knapsack polytope is a facet of the finite master group problem; see Section 19.5.1.2 for a discussion on tilting and see Aráoz [8] for details.

In Section 19.6.2.1, we present the shooting method applied to the group cutting planes as a measure of their relative importance. We then present a related measure called the *merit index*. In Section 19.6.2.2, we outline some computational approaches that have been used to measure the relative importance of different group cutting planes.

19.6.2.1 Shooting experiment and the merit index

As discussed above, the shooting approach requires that all extreme inequalities of the polyhedron be known prior to shooting the rays. Since its extreme inequalities are well-characterized, there is a way of performing the shooting experiment for the finite master group problem without knowing all of its extreme inequalities explicitly.

Theorem 19.39 ([62]). *The extreme inequalities of* $\mathrm{MG}(G,\emptyset,r)$ *hit first by a random ray v (when shot from the origin) are the extreme inequalities π^* that correspond to optimal solutions of the following linear program:*

$$\begin{aligned} &\min\ v'\pi \\ &\text{s.t. } \pi(r) = 1 \\ &\quad \pi(g_i) + \pi(g_j) \geqslant \pi(g_i + g_j) \quad \forall g_i, g_j \in G \\ &\quad \pi(g_i) + \pi(r - g_i) = 1 \quad \forall g_i \in G \\ &\quad \pi(g_i) \geqslant 0 \quad \forall g_i \in G. \end{aligned} \tag{19.69}$$

Gomory et al. [62] conduct the shooting experiment as follows. First they generate a random vector and then solve (19.69) to obtain the corresponding extreme inequality. Since the point of shooting is the origin, which lies outside the group problem, there is a possibility that some large facets of the group problem are tilted in such a way that their projection on the unit sphere is small. However, it is empirically argued in Gomory et al. [62] that this is unlikely. The experiments in [62] are conducted for group problems where the order of underlying cyclic group is at most 30. The key observations are:

- Less than 10% of all the extreme inequalities are hit. Further, 50% of the hits are collected by a very small number of extreme inequalities.
- The extreme inequalities that receive the most hits are: GMIC, homomorphisms and automorphisms of the GMIC, and other two-slope inequalities. Given a cyclic group with non-trivial proper subgroups (i.e., a cyclic group whose order is not a prime), the most frequently occurring extreme inequalities are typically the inequalities obtained by "lifting up" inequalities for group problems corresponding to the subgroups. In other words, these inequalities are obtained using multiplicative homomorphisms; see Section 19.5.2.1. In a related result, Cornuéjols et al. [26] prove that for pure integer programming problems k-cuts which are homomorphisms of GMICs perform better variable-wise than GMIC with a probability of 50%.
- In addition to the structure corresponding to subgroups, there is significant persistence in the shape of extreme inequalities for group problems corresponding to different sizes. For example, the two extreme inequalities that are hit most often for $\text{MG}(C_{13}, \emptyset, 12/13)$ are similar in shape to the extreme inequalities that are hit most often for $\text{MG}(C_{19}, \emptyset, 18/19)$.

The most important observation of this study is that finite group problems have a very small list of extreme inequalities that play a dominant role in defining the convex hull of integer feasible points; see Gomory et al. [62] and Evans [46] for more detail.

The results of the shooting experiment are extended in Dash and Günlük [34]. In particular, the authors consider group problems with underlying cyclic groups of much larger order. For group problems where the order of the underlying cyclic group is less than 90, they conduct the same experiment as Gomory et al. [62]. For larger size problems (with underlying cyclic group of order up to 200) they conduct a simpler experiment, which they call *partial shooting*. In this set of experiments, given a random vector v, they first enumerate all the MIR based inequalities (such as GMIC, k-cuts, two-step MIRs [33]) to determine which inequality, $\bar{\pi}$, minimizes $v'\bar{\pi}$. Then they verify if $\bar{\pi}$ is a solution of (19.69). Thus this approach is faster than the original shooting experiment which makes it suitable for larger problems. The experiments in Dash and Günlük [34] with the much larger group problems reaffirm the results of Gomory et al. [62]. In particular, they observe that the extreme inequalities with the largest number of hits are typically k-cuts, GMICs and two-step MIRs. Dash and Günlük [34] also generalize the result of Cornuéjols et al. [26], proving that any two functions obtained by interpolation from extreme inequalities of the same finite group problem will outperform each other with equal probability, when compared on a row of a simplex tableau variable-wise.

For the infinite group problem, it is not possible to perform the shooting experiment. Gomory and Johnson [61] and Evans [46] however empirically observe that the number of subadditive relations that are binding for a group inequality has a strong correlation with the results of the shooting experiment. This is not surprising since the number of subadditive relations that are binding for a group inequality is directly related to the number of integer feasible points of the master group problem binding at the inequality; see Path Lemma in Gomory and Johnson [61]. Motivated by these observations, Gomory and Johnson [61] define the merit index to evaluate the importance of extreme inequalities of the infinite master group problem.

Definition 19.28 ([61]). Let $\pi : I^1 \to \mathbb{R}_+$ be a valid inequality for the infinite group problem $\mathrm{MG}(I^1, \emptyset, r)$. Let S_2 be the unit square in two dimensions. The merit index is twice the area of points $p = (u_1, u_2)$ in S_2 for which $\pi(u_1) + \pi(u_2) = \pi(u_1 + u_2)$ holds.

Since the area referred to in Definition 19.28 is a measure of the number of subadditive relations binding for a given inequality, the merit index is expected to strongly correlate to the results of the shooting experiment. Note that the definition of merit index can easily be generalized for a m-row group cut, by replacing S_2 with the unit cube in $2m$ dimensions.

The merit index for GMIC is $2r^2 - 2r + 1$, where r is the right-hand-side of the group problem. In many cases, for a given right-hand-side it is possible to create a homomorphism of GMIC (starting with a different right-hand-side) which has a larger merit index. Recently some discontinuous extreme inequalities for the one-row infinite group problem were found to have larger merit index than GMIC; see Dey et al. [40]. One interesting property of the merit index is that it is invariant under homomorphism; see [61].

19.6.2.2 Computational experiments

We now present results from computational studies aimed at evaluating the relative strength of different group cuts. The results are organized around three themes: *i*) comparison of the performance of specific one-row group cuts with the performance of GMICs, *ii*) comparison of the performance of general one-row group cuts with the performance of GMICs, and *iii*) performance of multi-row group cuts.

We begin this section by mentioning that although GMIC was proposed by Gomory [55] in 1960, it was not considered practical or computationally useful for a very long time. However, in 1996, this perception changed when Balas et al. [10] showed that by adding GMICs corresponding to all the rows with fractional variables, the performance of these cutting planes was significantly improved; see Cornuéjols [24] for an account of the dramatic turn-around on the perception of GMICs in the 1990's. Bixby and Rothberg [16] present a performance report on various general-purpose inequalities in CPLEX, showing that GMIC is one of the most useful cutting planes used; see Section 16.2.1.2 for a more detailed discussion.

Performance of specific cuts vs GMIC:

Cornuéjols et al. [26] considered inequalities that they call k-cuts (multiplicative homomorphism of GMICs). They tested these cuts on randomly generated 0-1 and bounded knapsack problems. Their approach is the following. First, they solve the LP relaxation. Second, they add k-cuts derived from all the rows of the simplex tableau. Third, they resolve and compute the amount of integrality gap closed. They observe that when k-cuts are applied for different values of k separately, GMIC (i.e., k-cut with $k = 1$) is not exceptionally better than other k-cuts. In fact, when all k-cuts for $2 \leqslant k \leqslant 10$ are simultaneously added, the gap closed is better than the gap closed with GMICs only. However, when the experiment is repeated on integer programs with multiple rows, k-cuts appear to be significantly less effective than GMICs. Finally, when k-cuts are applied to mixed integer knapsacks, the performance of k-cuts deteriorates as the value of k-increases. This is expected since increasing k leads to a weakening of the coefficients of continuous variables.

Letchford and Lodi [78] present some families of discontinuous group cuts that can be seen as a strict improvement of Gomory's fractional cuts; see also Dash and Günlük [32, 33], Richard et al. [86], Dey et al. [40] for related inequalities. Letchford and Lodi [78] test their class of inequalities on randomly generated multi-row 0-1 knapsack problems. They compare the cuts with GFCs which are not minimal for the one-row group problem. They observe that these cuts perform better than GFCs when used inside a branch-and-bound tree. However, it is believed that these inequalities may be more susceptible to numerical difficulties as the underlying function is discontinuous.

Dash and Günlük [33] analyze two-step MIR (Theorem 19.34) and present an algorithm for selecting optimal parameters of the two-step MIR inequalities for a given row of the simplex tableau. In their computational study, Dash et al. [31] compare the two-step MIR inequalities and the scaled MIR inequality with the Mixed Integer Rounding cuts/GMICs. The test set includes instances from MIPLIB 3.0 and 2003 [15], Atamtürk [1], Fischetti and Lodi [48], Fischetti et al. [47], and practical two-dimensional cutting stock problems arising in the steel industry. Their main conclusions are: (*i*) the two-step MIR inequalities perform very well on the Atamtürk [1] instances; on average, MIR inequalities, two-step MIR inequalities used along with MIR inequalities, and scaled MIR inequalities used along with MIR inequalities respectively close approximately 57%, 83% and 78% of the integrality gap on these instances, (*ii*) among the other instances, once MIR inequalities are added, no violated two-step MIR cuts can be found for 74% of the instances and (*iii*) if there exists a violated one-row group cut, then for most of these instances there exists a violated two-step MIR cut.

Performance of one-row group cuts vs GMIC:

Dash and Günlük [35] use the following procedure to determine the strength of general one-row group cutting planes with respect to GMIC. First, the LP relaxation is solved. Second, GMICs based on the rows of the optimal tableau are added. Third, the LP is resolved to obtain an optimal LP solution (v^*, x^*). Finally it is determined whether there exists a one-row group cut based on the original optimal tableau that cuts off (v^*, x^*). This step involves solving a separation LP for each row. The size of the LP depends on the fractional values in the simplex tableau. Dash and Günlük [35] present results regarding the size of the smallest possible LP for exact separation, techniques for relaxing the separation LP, methods to obtain lower bound on the separation problem, and heuristics for solving the separation problem. Based on the order of the cyclic group underlying the separation LP, heuristics and lower bounds are used to check if there exists a violated group cutting planes; see Section 5 of Dash and Günlük [35] for details. The test bed used are instances from MIPLIB 3.0 and 2003 [15], Fischetti and Lodi [48], and Fischetti et al. [47]. The key findings of this study are: (*i*) for 35% of the instances, GMICs are the only relevant one-row group cutting planes. Less than 20% of the pure integer instances fall in this category, while this behavior is found in more than 40% of the problems with continuous variables. Therefore, it appears that one crucial reason for the strength of GMICs is the strong coefficient of its continuous variables. (*ii*) whenever a violated one-row group cut exists, typically a violated two-step MIR inequality exists. Conversely, if no violated two-step MIR inequality is found, then in 75% of the cases, no other violated group inequality is found.

Since the underlying group in the master group relaxation can be very large in practice, Fischetti and Saturni [50] consider the problem of obtaining the most violated "interpolated subadditive function". However, constructing the separation problem with a smaller underlying cyclic subgroup becomes complicated. We refer the readers to [50] for details on how to determine the coefficients of the cut. The following procedure was used to determine the relative strength of different one-row group cutting planes. First, the LP relaxation is solved and an optimal solution x^* is obtained. Second, the tableau rows are stored. Third, from each row of the optimal simplex tableau, the most violated interpolated subadditive cut is generated and added to the LP relaxation, up to 200 cuts in one round. Cuts are added until no more cut can be separated. Fourth, the procedure is repeated for k-cuts and GMIC separately. The test bed in [50] is composed of instances from MIPLIB 3.0 and 2003 [15], Atamtürk [1], and randomly generated bounded and 0-1 single-constraint knapsack problems. The key observations of this study are: (*i*) although the interpolated subadditive cuts improve the quality of the LP relaxation, they rarely beat the performance of GMICs for MIPLIB instances. This confirms the theoretical findings in Dash and Günlük [34]. As the number of rows of the problem increases, the performance of the interpolated subadditive cuts decreases, (*ii*) for the instances of Atamtürk [1] the interpolated subadditive cutting planes perform better than GMICs. For single-constraint knapsack instances, very few interpolated subadditive cuts are able to improve GMIC bounds significantly, (*iii*) As the size of the underlying cyclic group used to generate the interpolated subadditive function is increased, the gap closed increases, although the improvement is not very significant.

Performance of multi-row group cuts vs GMIC:

All the studies presented above considered one-row group cutting planes. However, some valuable results can be found in Fischetti and Monaci [49] regarding multi-row group cuts. As mentioned earlier, traditional corner relaxation and FM-corner relaxations are both solved in [49] to determine their strength. A measure called GMI′ is computed for each instance. This measure represents the gap closed by considering a problem with all the original constraints defining the group relaxation (i.e., "the LP relaxation" of the group problem, in place of all the original constraints of the problem) and one round of GMIC cuts for each of the tableau rows. The difference between the parameter GMI′ and the bounds obtained by the solution of the group relaxation indicates the marginal benefit of exploiting group cutting planes (including multiple-row group cuts) other than GMICs. We present these parameters in Table 19.5 for the MIPLIB instances. These figures were obtained by compiling results in Tables 1 and 2 of [49]. For each of the categories, we see that the addition of general group cuts can potentially double or even triple the improvement in the gap closed by addition of GMICs only. This is significantly better than the performance obtained by the application of only one-row group cuts.

Recently, Espinoza [45] conducted experiments using cuts based on multiple-row group relaxations. These cuts are generated as follows. First the non-basic integer variables are relaxed to be continuous. Second, minimal group cuts of the problem $\text{MG}(\{o\}, \mathbb{Q}^m, r)$ are used to obtain cutting planes for the resulting relaxation of the simplex tableau. These cuts are derived using maximal lattice-free convex sets; see Section 19.5.2.4. Espinoza [45] uses two classes of lattice-free convex sets. The test bed used in this study is composed of a subset of MIPLIB 3.0 and 2003 instances (87 instances). Multiple-row group cuts are added after the CPLEX 10.2 default cutting planes are added at the root node. Gap closed at the root node and time to reach provable optimal solution in the branch-and-bound tree are compared with CPLEX defaults. The results of these experiments are encouraging. A geometric average speed-up over CPLEX 10.2 of 31% is observed for instances in which optimality was reached using the additional group cuts. CPLEX is faster by at least 5% on 10 instances. Interestingly, the performance of the group cuts, in terms of integrality gap closed, improves when these cuts are based on a larger number of rows of the simplex tableau. Moreover, if more rows are used to generate the cuts, a smaller number of cuts is required.

These experiments indicate that the use of multiple row group cutting planes may hold some potential. Observe however that very little is known about important non-GMIC extreme inequalities for multiple-row group problems. More extensive experiments must be conducted to determine if the improvements that can be obtained by using non-GMIC cutting planes can be harnessed.

Problem Type	Gap closed by GMI'	Gap closed by Group relaxation - Gap closed by GMI'
Pure $0-1$	10%	11%
Problems with continuous/ general integer variables	11%	14%
Problems with general integer variables	14%	22%

Table 19.5 Gap closed by GMI' and group relaxation; Averages obtained from Fischetti and Monaci [49].

19.6.3 Summary on strength of group cuts

Among the one-row group cutting planes, GMIC is clearly the most effective cut. Other group cutting planes that may be useful in practice are k-cuts and two-step MIR inequalities. However, it appears that the efficacy of these cutting planes reduces as the number of rows in the original problem increases and/or the number of continuous variables in the problem increases. The strength of GMIC may therefore be attributed to the following reasons: (i) GMIC has a very high merit index and therefore a large number of points *involving only the integer variables* of the group relaxation that are binding, (ii) GMIC gives the best possible coefficient for continuous variables among all one-row group cutting planes. Observe that the shooting experiment/merit index does not really consider the effect of the coefficients of the continuous variables. In fact, as pointed out in Section 19.6.2.1, the merit index is invariant under homomorphism, and therefore the k-cut and GMICs have the same merit index as GMICs while k-cuts have weaker coefficients of continuous variables than GMICs when $k > 1$. Therefore, the merit index can be intuitively thought of as a parameter to judge the strength of the integer coefficients.

While general one-row group cutting planes may not yield much improvement on the performance of GMIC, the experiments conducted in Fischetti and Monaci [49] and Espinoza [45] give hope that multi-row group cuts could be useful. However, this may heavily depend on whether a small subset of group cutting planes based on multiple rows are "important", since otherwise a very large library of cuts would be needed to provide any significant reduction in integrality gaps.

19.7 Conclusion and perspectives

The group-theoretic approach to the generation of cutting planes in mixed integer programming dates back to the early days of integer programming. Although the theoretical foundations of this approach were laid in the 1960s, its computational promise remained unfulfilled until the 1990s, when GMIC made a spectacular comeback in both theoretical research and computer implementations. Over the last decade, the group-theoretic approach has been the subject of an intense and renewed interest and the amount of literature on the subject has grown tremendously. However, many questions remain open.

On the theoretical front, many questions remain open. A number of techniques used to prove inequalities are extreme (especially for infinite group problems) are technical in nature. Better methods to search and prove that inequalities are extreme are needed. Given the large number of extreme inequalities, new methods, other than based on merit index or strength of continuous variables coefficients, are needed to predict the usefulness of extreme inequalities for group problems.

For group problems with single constraints, many families of extreme inequalities are known. Moreover, computational studies indicate that most of the important families are probably already known. However, constructive procedures to derive many of these inequalities are not known. In particular, simple constructive procedures now explain subfamilies of the two-slope inequalities, but none of the families of three-slope inequalities have constructive derivations.

For group problems with multiple constraints, very few families of extreme inequalities are known and many research questions are open. Recent computational studies suggest that stronger cutting-plane algorithms can be built by considering several rows. However, it is not clear which inequalities are most important for multi-row group problems. In particular, we are not aware of studies presenting the results of shooting experiments for multi-row group problems. The derivation of multi-row inequalities using lattice-free convex sets and the fill-in procedure yields a promising generic constructive procedure for extreme inequalities. The relation of this approach to disjunctive programming is an interesting research direction.

On the numerical front, the use of group cuts and especially those based on multi-row group problems open many research avenues. First, it is important to determine the contexts in which multi-row group cuts are most appropriate to apply. Second, it is also important to determine the best subsets of the tableau rows to generate cuts from and the optimal number of rows to use in these situations. Third, computational studies indicate that primal degeneracy can reduce the strength of group cuts. Methods for improving group cuts under such circumstances should be investigated. Fourth, numerical stability of group cuts is a computational problem that has been discussed much. Related to this question is that of generating low-density group cuts. Recently Cook et al. [23] have presented methods for numerically accurate implementations of GMIC. Whether such implementations for other group cuts are possible is another interesting question.

Finally, there is potential in considering extensions of the concept of group relaxations. For example, Dash et al. [30] analyzed a generalization of the master cyclic group polyhedron. There is also a possibility of improving group cutting planes by analyzing valid inequalities for extended group relaxations and by considering bounds on the variables. These directions of research, especially in the context of continuous group relaxations, are an active area of research; see Johnson [72], Dey and Wolsey [43], Basu et al. [11], Fukasawa and Günlük [51] and Andersen et al. [5]. Further analysis of these and related problems may yield stronger cutting planes for general mixed integer programs.

We believe the research related to group problems that is currently under way will not only answer these questions and many more, but also hold the key to new breakthrough and computational advances in MIP.

Acknowledgements The authors thank an anonymous referee for suggestions that helped broaden the scope of the exposition and improved its presentation.

References

1. A. Atamtürk, *On the facets of the mixed-integer knapsack polyhedron*, Mathematical Programming 98 (2003) 145–175.
2. T. Achterberg, T. Koch, and A. Martin, *Branching rules revisited*, Operation Research Letters 33 (2005) 42–54.
3. R.K. Ahuja, T.L. Magnanti, and J.B. Orlin, *Network flows: Theory, algorithms, and applications*, Prentice Hall, 1993.
4. K. Andersen, G. Cornuéjols, and Y. Li, *Reduce-and-split cuts: Improving the performance of mixed integer Gomory cuts*, Management Science 51 (2005) 1720–1732.
5. K. Andersen, Q. Louveaux, and R. Weismantel, *Geometric study of mixed-integer sets from 2 rows of 2 adjacent simplex bases*, Manuscript, 2009.
6. K. Andersen, Q. Louveaux, R. Weismantel, and L.A. Wolsey, *Cutting planes from two rows of a simplex tableau*, Proceedings 12th Conference on Integer and Combinatorial Optimization (M. Fischetti and D. P. Williamson, eds.), Springer-Verlag, 2007, pp. 30–42.
7. J. Aráoz, *Polyhedral neopolarities*, Ph.D. thesis, University of Waterloo, Department of Computer Sciences, Waterloo, Canada, 1974.
8. J. Aráoz, L. Evans, R.E. Gomory, and E.L. Johnson, *Cyclic groups and knapsack facets*, Mathematical Programming 96 (2003) 377–408.
9. E. Balas, *A note on the group-theoretic approach to integer programming and the 0-1 case*, Operations Research 21 (1973) 321–322.
10. E. Balas, S. Ceria, G. Cornuéjols, and N. Natraj, *Gomory cuts revisited*, Operations Research Letters 19 (1996) 1–9.
11. A. Basu, M. Conforti, G. Cornuéjols, and G. Zambelli, *Minimal inequalities for an infinite relaxation of integer programs.*, Manuscript, 2009.
12. D.E. Bell, *A theorem concerning the integer lattice*, Studies in Applied Mathematics 56 (1977) 187–188.
13. D.E. Bell and J.F. Shapiro, *A convergent duality theory for integer programming*, Operations Research 25 (1977) 419–434.
14. A. Ben-Israel and A. Charnes, *On some problems of diophantine programming*, Cahiers du Centre d'Études de Recherche Opérationelle 4 (1962) 215–280.
15. R.E. Bixby, E.A. Boyd, and R.R. Indovina, *MIPLIB: A test set of mixed integer programming problems*, SIAM News 25:2 (1992).
16. R.E. Bixby and E.E. Rothberg, *Progress in computational mixed integer programming - A look back from the other side of the tipping point*, Annals of Operations Research 149 (2007) 37–41.
17. V. Borozan and G. Cornuéjols, *Minimal inequalities for integer constraints*, http://integer.tepper.cmu.edu, 2007.
18. V.J. Bowman and G.L. Nemhauser, *A finiteness proof for modified Dantzig cuts in integer programming*, Naval Research Logistics Quarterly 17 (1970) 309–313.
19. A. Charnes and W.W. Cooper, *Management models and industrial applications of linear programming, ii*, Wiley, New York, 1961.
20. D.-S. Chen and S. Zionts, *Comparison of some algorithms for solving the group theoretic integer programming problem*, Operations Research 24 (1976) 1120–1128.

21. V. Chvátal, *Edmonds Polytopes and a Hierarchy of Combinatorial Problems*, Discrete Mathematics 4 (1973) 305–337.
22. V. Chvátal, *Linear programming*, W. H. Freeman and Company, New York, NY, 1983.
23. W. Cook, S. Dash, R. Fukasawa, and M. Goycoolea, *Numerically accurate Gomory mixed-integer cuts*, INFORMS Journal of Computing (To appear).
24. G. Cornuéjols, *Revival of the Gomory cuts in the 1990's*, Annals of Operations Research 149 (2007) 63–66.
25. G. Cornuéjols and Y. Li, *On the rank of mixed 0-1 polyhedra*, Mathematical Programming 91 (2002) 391–397.
26. G. Cornuéjols, Y. Li, and D. Vandenbussche, *K-cuts: a variation of Gomory mixed integer cuts from the LP tableau*, INFORMS Journal of Computing 15 (2003) 385–396.
27. G. Cornuéjols and F. Margot, *On the facets of mixed integer programs with two integer variables and two constraints*, Mathematical Programming 120 (2009) 429–456.
28. G.B. Dantzig, *Discrete-variable extremum problems*, Operations Research 5 (1957) 266–277.
29. G.B. Dantzig, *Note on solving linear programs in integers*, Naval Research Logistics Quarterly 6 (1959) 75–76.
30. S. Dash, R. Fukasawa, and O. Günlük, *On a generalization of the master cyclic group polyhedron*, Proceedings 12th Conference on Integer and Combinatorial Optimization (M. Fischetti and D. P. Williamson, eds.), Springer-Verlag, 2007, pp. 197–209.
31. S. Dash, M. Goycoolea, and O. Günlük, *Two step MIR inequalities for mixed integer programs.*, http://www.optimization-online.org/DBHTML/2006/07/1422.html, 2006.
32. S. Dash and O. Günlük, *Valid inequalities based on simple mixed-integer sets.*, Proceedings 10th Conference on Integer Programming and Combinatorial Optimization (D. Bienstock and G. Nemhauser, eds.), Springer-Verlag, 2004, pp. 33–45.
33. S. Dash and O. Günlük, *Valid inequalities based on simple mixed integer set*, Mathematical Programming 106 (2006) 29–53.
34. S. Dash and O. Günlük, *Valid inequalities based on the interpolation procedure*, Mathematical Programming 106 (2006) 111–136.
35. S. Dash and O. Günlük, *On the strength of Gomory mixed integer cuts as group cuts*, Mathematical Programming 115 (2008) 387–407.
36. S.S. Dey, *Strong cutting planes for unstructured mixed integer programs using multiple constraints*, Ph.D. thesis, Purdue University, West Lafayette, IN, USA, 2007.
37. S.S. Dey and J.-P.P. Richard, *Sequential-merge facets for two-dimensional group problems*, Proceedings 12th Conference on Integer and Combinatorial Optimization (M. Fischetti and D. P. Williamson, eds.), Springer-Verlag, 2007, pp. 30–42.
38. S.S. Dey and J.-P.P. Richard, *Facets of the two-dimensional infinite group problems*, Mathematics of Operations Research 33 (2008) 140–166.
39. S.S. Dey and J.-P.P. Richard, *Some relations between facets of low- and high-dimensional group problems*, Mathematical Programming (To appear).
40. S.S. Dey, J.-P.P. Richard, Y. Li, and L.A. Miller, *Extreme inequalities for infinite group problems.*, Mathematical Programming 121 (2010) 145–170.
41. S.S. Dey and L.A. Wolsey, *Lifting integer variables in minimal inequalities corresponding to lattice-free triangles*, Proceedings 13th Conference on Integer and Combinatorial Optimization (A. Lodi, A. Panconesi, and G. Rinaldi, eds.), Springer-Verlag, 2008, pp. 463–475.
42. S.S. Dey and L.A. Wolsey, *Two row mixed integer cuts via lifting*, Tech. Report CORE DP 30, Université catholique de Louvain, Louvain-la-Neuve, Belgium, 2008.
43. S.S. Dey and L.A. Wolsey, *Constrained infinite group relaxations of MIPs*, Tech. Report CORE DP 33, Université catholique de Louvain, Louvain-la-Neuve, Belgium, 2009.
44. J.P. Doignon, *Convexity in crystallographic lattices*, Journal of Geometry 3 (1973) 71–85.
45. D. Espinoza, *Computing with multiple-row Gomory cuts*, Proceedings 13th Conference on Integer Programming and Combinatorial Optimization (A. Lodi, A. Panconesi, and G. Rinaldi, eds.), Springer-Verlag, 2008, pp. 214–224.
46. L.A. Evans, *Cyclic group and knapsack facets with applications to cutting planes*, Ph.D. thesis, Georgia Institute of Technology, Atlanta, GA, USA, 2002.

47. M. Fischetti, F. Glover, and A. Lodi, *The feasibility pump*, Mathematical Programming 104 (2005) 91–104.
48. M. Fischetti and A. Lodi, *Local branching*, Mathematical Programming 98 (2003) 23–47.
49. M. Fischetti and M. Monaci, *How tight is the corner relaxation?*, Discrete Optimization 5 (2007) 262–269.
50. M. Fischetti and C. Saturni, *Mixed integer cuts from cyclic groups*, Mathematical Programming 109 (2007) 27–53.
51. R. Fukasawa and O. Günlük, *Strengthening lattice-free cuts using non-negativity*, http://www.optimization-online.org/DBHTML/2009/05/2296.html, 2009.
52. F. Glover, *Integer programming over a finite additive group*, SIAM Journal on Control 7 (1969) 213–231.
53. M.X. Goemans, *Worst-case comparison of valid inequalities for the TSP*, Mathematical Programming 69 (1995) 335–349.
54. R.E. Gomory, *Outline of an algorithm for integer solutions to linear programs*, Bulletin of the American Mathematical Society 64 (1958) 275–278.
55. R.E. Gomory, *An algorithm for the mixed integer problem*, Tech. Report RM-2597, RAND Corporation, 1960.
56. R.E. Gomory, *Some relation between integer and non-integer solutions of linear program*, Proceedings of National Academy of Science 53 (1965) 250–265.
57. R.E. Gomory, *Some polyhedra related to combinatorial problems*, Linear Algebra and its Application 2 (1969) 341–375.
58. R.E. Gomory and A.J. Hoffman, *On the convergence of an integer programming process*, Naval Research Logistics Quarterly 10 (1963) 121–124.
59. R.E. Gomory and E.L. Johnson, *Some continuous functions related to corner polyhedra, part I*, Mathematical Programming 3 (1972) 23–85.
60. R.E. Gomory and E.L. Johnson, *Some continuous functions related to corner polyhedra, part II*, Mathematical Programming 3 (1972) 359–389.
61. R.E. Gomory and E.L. Johnson, *T-space and cutting planes*, Mathematical Programming 96 (2003) 341–375.
62. R.E. Gomory, E.L. Johnson, and L. Evans, *Corner polyhedra and their connection with cutting planes*, Mathematical Programming 96 (2003) 321–339.
63. G.A. Gorry, W.D. Northup, and J.F. Shapiro, *Computational experience with a group theoretic integer programming algorithm*, Mathematical Programming 4 (1973) 171–192.
64. J. Hadamard, *Résolution d'une question relative aux déterminants*, Bulletin des Sciences Mathématiques 17 (1893) 30–31.
65. S. Hoşten and R.R. Thomas, *Standard pairs and group relaxations in integer programming*, Journal of Pure and Applied Algebra 139 (1999) 133–157.
66. S. Hoşten and R.R. Thomas, *Gomory integer programs*, Mathematical Programming 96 (2003) 271–292.
67. T.C. Hu, *Integer programming and network flows*, Addison-Wesley, Reading, MA, 1969.
68. T.C. Hu, *On the asymptotic integer algorithm*, Linear Algebra and its Applications 3 (1970) 279–294.
69. B. Hunsaker, *Measuring facets of polyhedra to predict usefulness in branch-and-cut algorithms*, Ph.D. thesis, Georgia Institute of Technology, Atlanta, USA, 2003.
70. R.G. Jeroslow and K.O. Kortanek, *On an algorithm of Gomory*, SIAM Journal on Applied Mathematics 21 (1971) 55–60.
71. E.L. Johnson, *On the group problem for mixed integer programming*, Mathematical Programming Study 2 (1974) 137–179.
72. E.L. Johnson, *Characterization of facets for multiple right-hand side choice linear programs*, Mathematical Programming Study 14 (1981) 112–142.
73. R. Kannan and R. Bachem, *Polynomial time algorithms for computing Smith and Hermite normal forms of an integer matrix*, SIAM Journal on Computation 8 (1979) 499–507.
74. K. Kianfar and Y. Fathi, *Generalized mixed integer rounding valid inequalities: Facets for infinite group polyhedra*, Mathematical Programming 120 (2009) 313–346.

75. V. Klee and G.J. Minty, *How good is the simplex algorithm?*, Inequalities (O. Shisha, ed.), vol. III, Academic Press, New York, 1972, pp. 159–175.
76. H.W. Kuhn, *On the origin of the Hungarian method*, History of Mathematical Programming: A collection of personal Reminiscences (J.K. Lenstra, K. Rinnooy, and A.H.G. Schrijver, eds.), Elsevier Science Publisher, 1991, pp. 77–81.
77. A.H. Land and A.G. Doig, *An automatic method for solving discrete programming problems*, Econometrica 28 (1960) 497– 520.
78. A.N. Letchford and A. Lodi, *Strengthening Chvátal-Gomory cuts and Gomory fractional cuts*, Operations Research Letters 30 (2002) 74–82.
79. L. Lovász, *Geometry of numbers and integer programming*, Mathematical Programming: Recent Developments and Applications (1989) 177–210.
80. H. Marchand, A. Martin, R. Weismantel, and L.A. Wolsey, *Cutting planes in integer and mixed integer programming*, Discrete Applied Mathematics 123 (2002) 397–446.
81. H. Marchand and L.A. Wolsey, *Aggregation and mixed integer rounding to solve MIPs*, Operations Research 49 (2001) 363–371.
82. L.A. Miller, Y. Li, and J.-P.P. Richard, *New facets for finite and infinite group problems from approximate lifting*, Naval Research Logistics 55 (2008) 172–191.
83. G.L. Nemhauser and L.A. Wolsey, *Integer and combinatorial optimization*, Wiley-Interscience, New York, NY, 1988.
84. F.J. Nourie and E.R. Venta, *An upper bound on the number of cuts needed in Gomory's method of integer forms*, Operations Research Letters 1 (1982) 129–133.
85. Y. Pochet and L.A. Wolsey, *Integer knapsacks and flow covers with divisible coefficients: polyhedra, optimization and separation*, Discrete Applied Mathematics 59 (1995) 57–74.
86. J.-P.P. Richard, Y. Li, and L.A. Miller, *Valid inequalities for MIPs and group polyhedra from approximate liftings*, Mathematical Programming 118 (2009) 253–277.
87. D.S. Rubin and R.L. Graves, *Strengthened Dantzig cuts for integer programming*, Operations Research 20 (1972) 178–182.
88. H.M. Salkin and S. Morito, *Integer programming by group theory: Some computational results*, Tech. report, Defense Technical Information Center OAI-PMH Repository [http://stinet.dtic.mil/oai/oai] (United States), 1975.
89. H.E. Scarf, *An observation on the structure of production sets with indivisibilities*, Proceedings of the National Academy of Sciences USA 74 (1977) 3637–3641.
90. A. Schrijver, *Theory of linear and integer programming*, John Wiley & Sons, Chichester, 1986.
91. J.F. Shapiro, *Dynamic programming algorithms for the integer programming problem - i: The integer programming problem viewed as a knapsack type problem*, Operations Research 16 (1968) 103–121.
92. J.F. Shapiro, *Group theoretic algorithms for the integer programming problem - ii: Extension to a general algorithm*, Operations Research 16 (1968) 928–947.
93. H.J.S. Smith, *On systems of indeterminate equations and congruences*, Philosophical Transactions of the Royal Society of London (A) 151 (1861) 293–326.
94. B. Sturmfels, R. Weismantel, and G. Ziegler, *Gröbner bases of lattices, corner polyhedra and integer programming*, Beiträge zur Algebra und Geometrie 36 (1995) 281–298.
95. R.R. Thomas, *The structure of group relaxations*, Handbooks in Operations Research and Management Science (R. Weismantel K. Aardal, G. Nemhauser, ed.), Elsevier, 2005, pp. 19–49.
96. R.J. Vanderbei, *Linear programming: Foundations and extensions*, Kluwer Academic Publishers, 2001.
97. L.A. Wolsey, *Extensions of the group theoretic approach in integer programming*, Management Science 18 (1971) 74–83.
98. L.A. Wolsey, *Group theoretic results in mixed integer programming*, Operations Research 19 (1971) 1691–1697.
99. L.A. Wolsey, *Generalized dynamic programming methods in integer programming*, Mathematical Programming 4 (1973) 222–232.

100. L.A. Wolsey, *The b-hull of an integer program*, Discrete Applied Mathematics 3 (1981) 193–201.
101. G. Zambelli, *On degenerate multi-row Gomory cuts*, Operations Research Letters 37 (2009) 21–22.

Part IV
DVD-Video / DVD-ROM

DOI 10.1007/978-3-540-68279-0,

The enclosed DVDs can be played on any DVD player without national restrictions. They contain more than 4 hours of video, including the talks of Gérard Cornuéjols, William Cook, and Laurence Wolsey (Part I), the introduction of the pioneer panel consisting of Egon Balas, Michel Balinski, Jack Edmonds, Ralph E. Gomory, Arthur M. Geoffrion, Richard M. Karp, and Susan Powell, who represented Ailsa H. Land, by George Nemhauser and William Pulleyblank and, as a special highlight, the panel discussion (Part II).

The DVD "Part I" can also be used as a DVD-ROM on a computer to view the slides of the talks and some selected photos taken during the conference.

MIX
Papier aus verantwortungsvollen Quellen
Paper from responsible sources
FSC® C105338

If you have any concerns about our products,
you can contact us on
ProductSafety@springernature.com

In case Publisher is established outside the EU,
the EU authorized representative is:
Springer Nature Customer Service Center GmbH
Europaplatz 3, 69115 Heidelberg, Germany

Printed by Libri Plureos GmbH
in Hamburg, Germany